Introduction to Digital Signal Processing

GW00482731

INTRODUCTION TO
Digital Signal
Processing

John G. Proakis *Northeastern University*

Dimitris G. Manolakis *Northeastern University*

Macmillan Publishing Company
NEW YORK

Collier Macmillan Publishers
LONDON

Macmillan Publishing Company
866 Third Avenue, New York, New York 10022

Collier Macmillan Canada, Inc.

Library of Congress Cataloging in Publication Data

Proakis, John G.
    Introduction to digital signal processing.

    Bibliography: p.
    Includes index.
    1. Signal processing–Digital techniques.
I. Manolakis, Dimitris   II. Title.
TK5102.5.P677   1988   621.38'043   87–26809
ISBN 0–02–396810–9 (Hardcover Edition)
ISBN 0–02–946253–3 (International Edition)

IE Printing: 1 2 3 4 5 6   Year: 9 0 1 2 3

ISBN 0-02-946253-3

Preface

This book has resulted from our teaching of undergraduate and graduate-level courses in digital signal processing over the past several years. In this book we present the fundamentals of discrete-time signals, systems, and modern digital processing algorithms and applications for students in electrical engineering, computer engineering, and computer science. As written, the book is suitable for either a one-semester or a two-semester undergraduate-level course in discrete systems and digital signal processing. It is also intended for use in a one-semester first-year graduate-level course in digital signal processing.

A balanced coverage is provided of both theory and practical applications. Emphasis is placed on software implementation of digital filters and digital signal processing algorithms. Appropriately designed programs in FORTRAN are provided to aid the student in implementing and exercising each algorithm immediately after exposition. A large number of well-designed problems and a number of computer experiments are also provided to help the student in mastering the subject matter. A solutions manual is available for the benefit of the instructor and can be obtained from the publisher.

It is assumed that the student in electrical and computer engineering has had undergraduate courses in circuit analysis, advanced calculus (preferably including ordinary differential equations), and linear systems for continuous-time signals, including an introduction to the Laplace transform. Although the Fourier series and Fourier transforms of periodic and aperiodic signals are described in Chapter 4, we expect that many students will have had this material in their first course in system analysis.

In Chapter 1 we describe the operations involved in the analog-to-digital conversion of an analog signal. The process of sampling a sinusoid is described in some detail and the problem of aliasing is explained. Signal quantization and digital-to-analog conversion are also described in general terms in Chapter 1, but the analysis is presented in subsequent chapters.

Chapter 2 is devoted entirely to the characterization and analysis of linear time-invariant (shift-invariant) discrete-time systems and discrete-time signals in the time domain. The convolution sum is derived and systems are categorized according to the duration of their impulse response as finite-duration impulse response (FIR) and infinite-duration impulse response (IIR). Linear time-invariant systems characterized by difference equations are presented and the solution of difference equations with initial conditions is obtained. The chapter concludes with a treatment of discrete-time correlation and its application to radar detection.

The z-transform is introduced in Chapter 3. Both the bilateral and the unilateral z-transform are presented, and methods for determining the inverse z-transform are described. The use of the z-transform in the analysis of linear time-invariant systems is illustrated, and the important properties of systems, such as causality and stability, are related to z-domain characteristics.

Chapter 4 treats the analysis of signals in the frequency domain. Fourier series and the Fourier transform are presented for both continuous-time and discrete-time signals. Also included in this chapter is a thorough treatment of sampling of a signal either in the time domain or in the frequency variable ω and the reconstruction of the signal from its samples. In this context we introduce the discrete Fourier transform (DFT) for finite-duration sequences.

Chapter 5 introduces the student to the frequency-domain characterization of discrete-time systems. The concept of a linear system acting as a filter is introduced and some simple FIR and IIR filter design methods are described. A number of important types of digital systems are described, including resonators, notch filters, comb filters, and oscillators. An important topic in this chapter is the use of the DFT in linear filtering of signals, which includes the description of the overlap-add and overlap-save methods for linear filtering of iong data sequences.

Several topics are treated in Chapter 6, including bandpass signals and systems, amplitude modulation, sampling of bandpass signals, signal reconstruction from the samples of a signal, inverse filters, deconvolution, and system identification. In this chapter the student is introduced to several practical applications of digital signal processing. For example, the discussion on amplitude modulation with discrete-time signals leads to a description of a digital implementation of a QAM (quadrature amplitude modulation) modem. Deconvolution and system identification are motivated with examples in seismic signal processing and adaptive equalization.

Chapter 7 treats the realization of IIR and FIR systems. This treatment includes direct-form, cascade, parallel, lattice, and lattice-ladder realizations. The chapter concludes with a thorough treatment of state-space analysis and structures for discrete-time systems.

Techniques for design of digital FIR and IIR filters are presented in Chapter 8. The design techniques include both direct design methods in discrete time and techniques that involve the conversion of analog filters into digital filters by various transformations. In this chapter we also include a discussion of sampling-rate conversion techniques.

Chapter 9 treats the computation of the DFT. Radix-2 and radix-4 fast Fourier transform (FFT) algorithms are derived and the applications of the FFT algorithms to the computation of convolution and correlation are also described. The Goertzel algorithm and the chirp-z transform are introduced as two methods by which the DFT can be computed via linear filtering.

Finite-word-length effects in digital filtering and in the computation of the DFT are the major topics treated in Chapter 10. The discussion emphasizes implementations based on fixed-point arithmetic and includes quantization errors in analog-to-digital conversion, parameter quantization in digital filters, limit cycles, the problem of overflows in addition, and round-off errors in multiplication. A brief review of probabilistic and statistical concepts is presented as an appendix to this chapter.

Power spectrum estimation is the topic treated in Chapter 11. Our coverage of this subject includes a description of nonparametric and model-based (parametric) methods for power spectrum estimation.

The final chapter treats adaptive filters. This topic has received considerable attention by researchers in digital signal processing over the past 15 years, and many computationally efficient algorithms for adaptive filtering have been developed during this period. We describe two basic algorithms, the LMS algorithm, which is based on a gradient optimization with a single adjustable parameter, and recursive least-squares algorithms, which include both direct-form FIR and lattice realizations.

At Northeastern University, we have used the first five chapters and a part of Chapter 6 for a one-semester course in discrete systems and digital signal processing. For this purpose we have introduced some simple filter design methods in Chapter 5. It is also possible to include the radix-2 decimation-in-time FFT algorithm in conjunction with the discussion in Chapter 5 on the use of the DFT in linear filtering.

A one-semester senior-level course for students who have had prior exposure to discrete systems may use the material in Chapters 1 through 4 for a quick review and then proceed to cover Chapters 5 through 10. In such a course the students may be introduced to the use of software packages that are now readily available for digital filter design.

In a first-year graduate-level course in digital signal processing, the early chapters provide the student with a good review of discrete-time systems. The instructor may move quickly through most of this material and then cover Chapters 5 through 10, followed by either Chapter 11 or Chapter 12. In such a course we highly recommend the use of available software packages for digital filter design and power spectrum estimation.

The authors are indebted to a number of graduate students and faculty colleagues who provided valuable suggestions and assistance in the preparation of the manuscript. We wish to thank Mr. A. El-Jaroudi for preparing the solutions manual, Mr. A. L. Kok for assisting in the preparation of the figures, and Professors J. Deller, V. Ingle, C. Keller, L. Merakos, P. Monticciolo, and M. Schetzen for their comments and suggestions on various drafts of the manuscript. We especially wish to express our appreciation to Professors C. L. Nikias, H. J. Trussell, and S. G. Wilson, who reviewed the manuscript and provided many important suggestions that have been incorporated. Finally, we wish to thank Ms. Gloria Proakis for typing the entire manuscript.

John G. Proakis
Dimitris G. Manolakis

Contents

CHAPTER 1

Introduction

Digital signal processing is an area of science and engineering that has developed rapidly over the past 20 years. This rapid development is a result of the significant advances in digital computer technology and integrated-circuit fabrication. The digital computers and associated digital hardware of two decades ago were relatively large and expensive and, as a consequence, their use was limited to general-purpose non-real-time (off-line) scientific computations and business applications. The rapid developments in integrated-circuit technology, starting with medium-scale integration (MSI) and progressing to large-scale integration (LSI), and now, very-large-scale integration (VLSI) of electronic circuits has spurred the development of powerful, smaller, faster, and cheaper digital computers and special-purpose digital hardware. These inexpensive and relatively fast digital circuits have made it possible to construct highly sophisticated digital systems that are capable of performing complex digital signal processing functions and tasks, which are usually too difficult and/or too expensive to be performed by analog circuitry or analog signal processing systems. Hence many of the signal processing tasks that were conventionally performed by analog means are realized today by less expensive and often more reliable digital hardware.

We do not wish to imply that digital signal processing is the proper solution for all signal processing problems. Indeed, for many signals with extremely wide bandwidths real-time processing is a requirement. For such signals, analog or, perhaps, optical signal processing is the only possible solution. However, where digital circuits are available and have sufficiently high speed to perform the signal processing, they are usually preferable.

Not only do digital circuits yield cheaper and more reliable systems for signal processing, but they have other advantages as well. In particular, digital processing hardware allows us to have programmable operations. Through software, one can more easily modify the signal processing functions to be performed by the hardware. Thus digital hardware and associated software provide a greater degree of flexibility in system design. Also, there is often a higher order of precision achievable with digital hardware and software compared with analog circuits and analog signal processing systems. For all these reasons, there has been an explosive growth in digital signal processing theory and applications over the past decade.

In this book our objective is to present an introduction of the basic analysis tools and techniques for digital processing of signals. We begin, in this chapter, by introducing some of the necessary terminology and describing the important operations associated with the process of converting an analog signal to digital form that is suitable for digital processing. As we shall see, digital processing of analog signals has some drawbacks. First, and foremost, conversion of an analog signal to digital form, which is accomplished by sampling the signal and quantizing the samples, results in a distortion that prevents

1

us from reconstructing the original analog signal from the quantized samples. Control of the amount of this distortion is achieved by proper choice of the sampling rate and the precision in the quantization process. Second, there are finite precision effects that must be considered in the digital processing of the quantized samples. These important issues are considered in some detail in this book. The emphasis, however, is on the analysis and design of digital signal processing systems and computational techniques.

1.1 Signals, Systems, and Signal Processing

A *signal* is defined as any physical quantity that varies with time, space, or any other independent variable or variables. Mathematically, we describe a signal as a function of one or more independent variables. For example, the functions

$$s_1(t) = 5t \qquad\qquad (1.1.1)$$
$$s_2(t) = 20t^2$$

describe two signals, one that varies linearly with the independent variable t (time) and a second that varies quadratically with t. As another example, consider the function

$$s(x, y) = 3x + 2xy + 10y^2 \qquad\qquad (1.1.2)$$

This function describes a signal of two independent variables x and y which might represent the two spatial coordinates in a plane.

The signals described by (1.1.1) and (1.1.2) belong to a class of signals that are precisely defined by specifying the functional dependence on the independent variable. However, there are cases where such a functional relationship is unknown or too highly complicated to be of any practical use.

For example, a speech signal (see Fig. 1.1) cannot be described functionally by expressions such as (1.1.1). In general, a segment of speech may be represented to a high accuracy as a sum of several sinusoids of different amplitudes and frequencies, that is, as

$$\sum_{i=1}^{N} A_i(t) \sin\left[2\pi F_i(t)t + \theta_i(t)\right] \qquad\qquad (1.1.3)$$

where $\{A_i(t)\}$, $\{F_i(t)\}$, and $\{\theta_i(t)\}$ are the sets of (possibly time-varying) amplitudes, frequencies, and phases, respectively, of the sinusoids. In fact, one way to interpret the information content or message conveyed by any short time segment of the speech signal

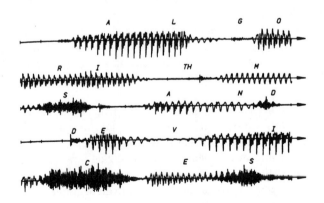

is to measure the amplitudes, frequencies, and phases contained in the short time segment of the signal.

Another example of a natural signal is an electrocardiogram (ECG). Such a signal provides a doctor with information about the operation of the patient's heart. Similarly, an electroencephalogram (EEG) signal provides information about the activity of the brain.

Speech, electrocardiogram, and electroencephalogram signals are examples of information-bearing signals that evolve as functions of a single independent variable, namely, time. An example of a signal that is a function of two independent variables is an image signal. The independent variables in this case are the spatial coordinates. These are but a few examples of the countless numbers of natural signals that are encountered in practice.

Associated with natural signals is the means by which such signals are generated. For example, speech signals are generated by forcing air through the vocal cords. Images are obtained by exposing a photographic film to a scene or an object. Thus signal generation is usually associated with a *system* that responds to a stimulus or force. In a speech signal, the system consists of the vocal cords and the vocal tract, also called the vocal cavity. The stimulus in combination with the system is called a *signal source*. Thus we have speech sources, images sources, and various other types of signal sources.

A *system* may also be defined as a physical device that performs an operation on a signal. For example, a filter that is used to reduce the noise and interference corrupting a desired information-bearing signal is called a system. In this case the filter performs some operation(s) on the signal which has the effect of reducing (filtering) the noise and interference from the desired information-bearing signal.

When we pass a signal through a system, as in filtering, for example, we say that we have processed the signal. In this case the processing of the signal involves filtering the noise and interference from the desired signal. In general, the system is characterized by the type of operation that it performs on the signal. For example, if the operation is linear, the system is called linear. If the operation on the signal is nonlinear, the system is said to be nonlinear, and so forth. Such operations are usually referred to as *signal processing*.

For our purposes, we find it convenient to broaden the definition of a system to include not only physical devices, but also software realizations of operations on a signal. In digital processing of signals on a digital computer, the operations performed on a signal consist of a number of mathematical operations as specified by a software program. In this case, the program represents an implementation of the system in *software*. Thus we have a system that is realized on a digital computer by means of a sequence of mathematical operations; that is, we have a digital signal processing system realized in software. For example, a digital computer can be programmed to perform digital filtering. Alternatively, the digital processing on the signal may be performed by digital *hardware* (logic circuits) that is configured to perform the desired specified operations. In such a realization, we have a physical device that performs the specified operations. More broadly, a digital system can be realized by a combination of digital hardware and software, each of which performs its own set of specified operations.

This book deals with the processing of signals by digital means, either in software or in hardware. Since many of the signals that are encountered in practice are analog signals, we will also consider the problem of converting an analog signal into a digital signal for processing. Thus we will be dealing primarily with digital systems. The operations performed by such a system usually can be specified mathematically. The method or the set of rules for implementing the system by a program that performs the corresponding mathematical operations is called an *algorithm*. Usually, there are many ways or algo-

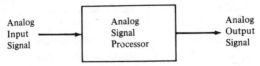

FIGURE 1.2 Analog signal processing.

rithms by which a system can be implemented either in software or in hardware to perform the desired operations and computations. In practice, we have an interest in devising algorithms that are computationally efficient, fast, and easily implemented. Thus a major topic in our study of digital signal processing is the discussion of efficient algorithms for performing such operations as filtering, correlation, and spectral analysis.

1.1.1 Basic Elements of a Digital Signal Processing System

Most of the signals encountered in science and engineering are analog in nature. That is, the signals are functions of a continuous variable, such as time or space and usually take on values in a continuous range. Such signals may be processed directly by appropriate analog systems such as filters or frequency analyzers or frequency multipliers for the purpose of changing their characteristics or extracting some desired information. In such a case we say that the signal has been processed directly in its analog form, as illustrated in Fig. 1.2. Both the input signal and the output signal are in analog form.

Digital signal processing provides an alternative method for processing the analog signal, as illustrated in Fig. 1.3. In order to perform the processing digitally, there is a need for an interface between the analog signal and the digital processor. This interface is called an *analog-to-digital (A/D) converter*. The output of the A/D converter is a digital signal that is appropriate as an input to the digital processor.

The digital signal processor may be a large programmable digital computer or a small microprocessor that is programmed to perform the desired operations on the input signal. It may also be a hardwired digital processor that is configured to perform a specified set of operations on the input signal. Programmable machines provide the flexibility to change the signal processing operations through a change in the software, whereas hardwired machines are difficult to reconfigure. Consequently, programmable signal processors are very common in practice. On the other hand, when the signal processing operations are well defined, as in some applications, a hardwired implementation of the operations can be optimized so that it results in a cheaper signal processor and, usually, one that runs at a faster speed than its programmable counterpart. In applications where the digital output from the digital signal processor is to be given to the user in analog form, as in speech communications, for example, we must provide another interface from the digital to the analog domain. Such an interface is called a *digital-to-analog (D/A) converter*. Thus the signal is provided to the user in analog form, as illustrated in the block diagram of Fig. 1.3. There are other practical applications, however, in-

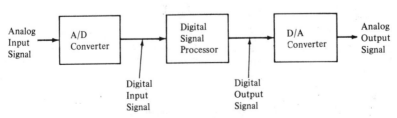

FIGURE 1.3 Block diagram of a digital signal processing system.

volving signal analysis applications where the desired information is conveyed in digital form and no D/A converter is required. For example, in the digital processing of radar signals, the information extracted from the radar signal, such as the position of the aircraft and its speed, may simply be printed on paper. There is no need for a D/A converter in this case.

1.1.2 Advantages of Digital Over Analog Signal Processing

There are many reasons why digital signal processing of an analog signal may be preferable to processing the signal directly in the analog domain, as mentioned briefly earlier. First, a digital programmable system allows flexibility in reconfiguring the digital signal processing operations simply by changing the program. Reconfiguration of an analog system usually implies a redesign of the hardware, testing, and verification that it operates properly.

Accuracy considerations also play an important role in determining the form of the signal processor. Digital signal processing provides better control of accuracy requirements. Tolerances in analog circuit components make it extremely difficult for the system designer to control the accuracy of an analog signal processing system. On the other hand, a digital system provides much better control of accuracy requirements. Such requirements, in turn, result in specifying the accuracy requirements in the A/D converter and the digital signal processor, in terms of word length, floating-point versus fixed-point arithmetic, and similar factors.

Digital signals are easily stored on magnetic media (tape or disk) without deterioration or loss of signal fidelity beyond that introduced in the A/D conversion. As a consequence, the signals become transportable and can be processed off-line in a remote laboratory. The digital signal processing method also allows for the implementation of more sophisticated signal processing algorithms. It is usually very difficult to perform precise mathematical operations on signals in analog form. However, these operations can be routinely implemented on a digital computer by means of software.

In some cases a digital implementation of the signal processing system is cheaper than its analog counterpart. The lower cost may be due to the fact that the digital hardware is cheaper, or perhaps it is a result of the flexibility for modifications provided by the digital implementation.

As a consequence of the advantages cited above, digital signal processing has been applied in practical systems covering a broad range of disciplines. We cite, for example, the application of digital signal processing techniques in speech processing and signal transmission on telephone channels, image processing and transmission, in seismology and geophysics, in oil exploration, in the detection of nuclear explosions, in the processing of signals received from outer space, and in a vast variety of other applications. Some of these applications are cited in subsequent chapters.

As already indicated, however, digital implementation has its limitations. One practical limitation is the speed of operation of A/D converters and digital signal processors. We shall observe below that signals having extremely wide bandwidths require fast-sampling-rate A/D converters and fast digital signal processors. Hence there are analog signals with large bandwidths for which a digital processing approach is beyond the state of the art of digital hardware.

1.2 Classification of Signals

The methods we are going to use in processing a signal or in analyzing the response of a system to a signal depend heavily on the characteristic attributes of the specific signal.

There are techniques that apply only to specific families of signals. Consequently, any investigation in signal processing should start with a classification of signals involved in the specific application.

1.2.1 Multichannel and Multidimensional Signals

As explained in Section 1.1, a signal is described by a function of one or more independent variables. The value of the function (i.e., the dependent variable) can be a real-valued scalar quantity, a complex-valued quantity, or perhaps a vector. For example, the signal

$$s_1(t) = A \sin 3\pi t$$

is a real-valued signal. However, the signal

$$s_2(t) = Ae^{j3\pi t} = A \cos 3\pi t + jA \sin 3\pi t$$

is complex valued.

In some applications, signals are generated by multiple sources or multiple sensors. Such signals, in turn, can be represented in vector form. For example, consider an array of eight sensors placed on the human head as shown in Fig. 1.4, to monitor brain activity. If $s_k(t)$ denotes the electrical signal from the kth sensor of the array, then the set of eight signals as a function of time can be represented by a vector $s(t)$, defined as

$$s(t) = \begin{bmatrix} s_1(t) \\ s_2(t) \\ \vdots \\ s_8(t) \end{bmatrix}$$

We refer to such a vector of signals as a *multichannel signal*. Specifically, we call $s(t)$ an eight-channel signal.

In general, each element of the vector $s(t)$ may be represented as real or complex valued. In some signal processing applications it is convenient to adopt a representation of a signal as a complex-valued function of an independent variable. Specifically, suppose that we have L sinusoidal signals of the same frequency F but which differ in amplitude and phase, for example,

$$\begin{aligned} s_k(t) &= A_k(t) \cos \left[2\pi F t + \theta_k(t) \right] \\ &= \text{real part of } [A_k(t)e^{j\theta_k(t)}e^{j2\pi Ft}] \\ &= \text{real part of } [u_k(t)e^{j2\pi Ft}] \end{aligned}$$

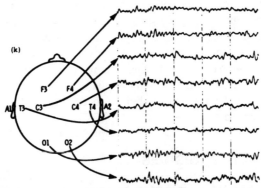

FIGURE 1.4 Example of an eight-channel EEG signal. [From paper by Gevins et al. (1975). Reprinted with permission from the IEEE.]

where

$$u_k(t) = A_k(t)e^{k\theta_k(t)}$$

We may represent the signal vector $s(t)$ as

$$s(t) = \text{real part of } [U(t)e^{j2\pi Ft}]$$

where the complex-valued signal vector $U(t)$ is defined as

$$U(t) = \begin{bmatrix} u_1(t) \\ u_2(t) \\ \vdots \\ u_L(t) \end{bmatrix}$$

Thus the real signal vector $s(t)$ may be represented by the equivalent complex-valued signal vector $U(t)$. It should be emphasized, however, that observable signals such as $s_k(t)$, $k = 1, 2, \ldots, L$, are real valued. The complex representation given above is used for mathematical convenience.

Let us now turn our attention to the independent variable(s). If the signal is a function of a single independent variable, the signal is called a *one-dimensional* signal. On the other hand, a signal is called *M-dimensional* if its value is a function of M independent variables. The following example provides further clarification of these definitions.

EXAMPLE 1.2.1

The output voltage of a sinusoidal generator at the frequency F is a single-channel, one-dimensional signal described functionally as

$$s(t) = A \sin 2\pi Ft$$

where A is the signal amplitude.

The output of an electrocardiograph having three leads (three sensors), as shown in Fig. 1.5, is an example of a three-channel ECG signal. The signals may be represented as a vector having three components.

The picture shown in Fig. 1.6 is an example of a two-dimensional signal, since the intensity or brightness $I(x, y)$ at each point is a function of two independent variables. On the other hand, a black-and-white television picture may be represented as $I(x, y, t)$ since the brightness is a function of time. Hence the TV picture may be treated as a three-dimensional signal. In contrast, a color TV picture may be described by three intensity functions of the form $I_r(x, y, t)$, $I_g(x, y, t)$ and $I_b(x, y, t)$, corresponding to the brightness of the three principal colors (red, green, blue) as functions of time. Hence the color TV picture is a three-channel, three-dimensional signal, which can be represented by the vector

$$I(x, y, t) = \begin{bmatrix} I_r(x, y, t) \\ I_g(x, y, t) \\ I_b(x, y, t) \end{bmatrix}$$

In this book we deal mainly with single-channel, one-dimensional real- or complex-valued signals and we will refer to them simply as signals. In mathematical terms these signals are described by a function of a single independent variable. Although the independent variable need not be time, it is common practice to use t as the independent

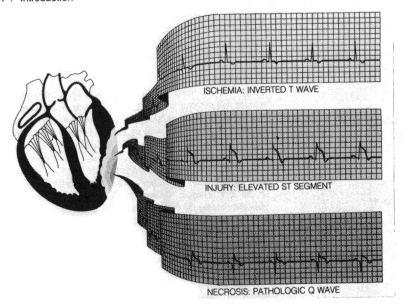

FIGURE 1.5 Example of a three-channel ECG signal. (Reprinted with permission from Reading EKG's Correctly. Copyright © 1984 Springhouse Corporation. All rights reserved.)

variable. In many cases the signal processing operations and algorithms developed in this text for one-dimensional, single-channel signals can be extended to multichannel and multidimensional signals.

1.2.2 Continuous-Time Versus Discrete-Time Signals

Signals may be further classified into four different categories depending on the characteristics of the time (independent) variable and the values they take.

Continuous-time signals or *analog signals* are defined for every value of time and they take on values in the continuous interval (a, b), where a may be $-\infty$ and b may

FIGURE 1.6 Example of a two-dimensional signal.

be ∞. Mathematically, these signals can be described by functions of a continuous variable. The speech waveform in Fig. 1.1 and the signals $x_1(t) = \cos \pi t$, $x_2(t) = e^{-|t|}$, $-\infty < t < \infty$ are examples of analog signals.

Discrete-time signals are defined only at discrete values of time. These time instants need not be equidistant, but in practice they are usually taken equally spaced for computational convenience and mathematical tractability. The signal $x(t_n) = e^{-|t_n|}$, $n = 0$, ± 1, ± 2, . . . provides an example of a discrete-time signal. If we use the index n of the discrete-time instants as the independent variable, the signal value becomes a function of an integer variable (i.e., a sequence of numbers). Thus a discrete-time signal can be represented mathematically by a sequence of real or complex numbers. To emphasize the discrete-time nature of a signal, we shall denote such a signal as $x(n)$ instead of $x(t)$. If the time instants t_n are equally spaced (i.e., $t_n = nT$), the notation $x(nT)$ is also used. For example, the sequence

$$x(n) = \begin{cases} 0.8^n & \text{if } n \geq 0 \\ 0 & \text{otherwise} \end{cases} \tag{1.2.1}$$

is a discrete-time signal, which is represented graphically as in Fig. 1.7.

In applications, discrete-time signals may arise in two ways:

1. By selecting values of an analog signal at discrete-time instants. This process is called *sampling* and is discussed in more detail in Section 1.4. All measuring instruments that take measurements at a regular interval of time provide discrete-time signals. For example, the signal $x(n)$ in Fig. 1.7 can be obtained by sampling the analog signal $x(t) = 0.8^t$, $t \geq 0$ and $x(t) = 0$, $t < 0$ once every second.
2. By accumulating a variable over a period of time. For example, counting the number of cars using a given street every hour, and recording the value of gold every day, both result in discrete-time signals. Figure 1.8 shows a graph of the Wölfer sunspot numbers. Each sample of this discrete-time signal provides the number of sunspots observed during an interval of 1 year.

1.2.3 Continuous-Valued Versus Discrete-Valued Signals

The values of a continuous-time or discrete-time signal may be continuous or discrete. If a signal takes on all possible values on a finite or an infinite range, it is said to be continuous-valued signal. Alternatively, if the signal takes on values from a finite set of possible values it is said to be a discrete-valued signal. Usually, these values are equidistant and hence can be expressed as an integer multiple of the distance between two successive values.

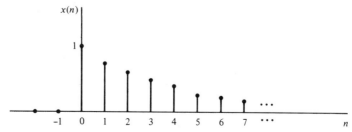

FIGURE 1.7 Graphical representation of the discrete time signal $x(n) = 0.8^n$ for $n > 0$ and $x(n) = 0$ for $n < 0$.

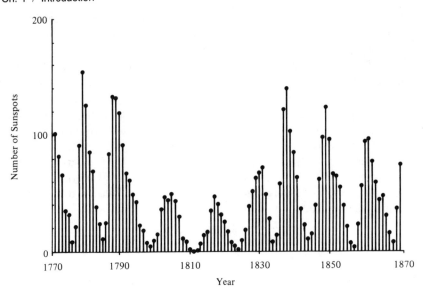

FIGURE 1.8 Wölfer annual sunspot numbers (1770–1869).

A discrete-time signal having a set of discrete values is called a *digital signal*. Figure 1.9 shows a digital signal that takes on one of four possible values.

In order for a signal to be processed digitally, it must be discrete in time and its values must be discrete (i.e., it must be a digital signal). If the signal to be processed is in analog form, it is first converted to a digital signal by sampling the signal at discrete instants in time, to obtain a discrete-time signal, and by *quantizing* its values to a set of discrete values, as described later in the chapter. The process of converting a continuous-valued signal into a discrete-valued signal, called *quantization*, is basically an approximation process. It may be accomplished simply by rounding or truncation. For example, if the allowable signal values in the digital signal are integers, say 0 through 15, the continuous-value signal will be quantized into these integer values. Thus the signal value 8.58 will be approximated by the value 8 if the quantization process is performed by truncation or by 9 if the quantization process is performed by rounding to the nearest integer. A better explanation of the analog-to-digital conversion process is given later in the chapter.

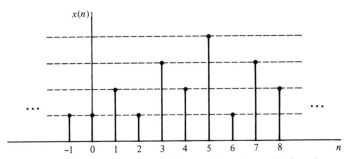

FIGURE 1.9 Digital signal with four different amplitude values.

1.3 The Concept of Frequency in Continuous-Time and Discrete-Time Signals

The concept of frequency is familiar to students in engineering and the sciences. This concept is basic in, for example, the design of a radio receiver, a high-fidelity system, or a special filter for color photography. From physics we know that frequency is closely related to a specific type of periodic motion called harmonic oscillation, which is described by sinusoidal functions. The concept of frequency is directly related to the concept of time. Actually, it has the dimension of inverse time. Thus we expect that the nature of time (continuous or discrete) should affect accordingly the nature of the frequency.

1.3.1 Continuous-Time Sinusoidal Signals

A simple harmonic oscillation is mathematically described by the following continuous-time sinusoidal signal:

$$x_a(t) = A \cos (\Omega t + \theta) \qquad -\infty < t < \infty \tag{1.3.1}$$

shown in Fig. 1.10. The subscript a used with $x(t)$ denotes an analog signal. This signal is completely characterized by three parameters: A is the *amplitude* of the sinusoid, Ω is the *frequency* in radians per second (rad/s), and θ is the *phase* in radians. Instead of Ω, we often use the frequency F in cycles per second or hertz (Hz), where

$$\Omega = 2\pi F \tag{1.3.2}$$

In terms of F, (1.3.1) can be written as

$$x_a(t) = A \cos (2\pi F t + \theta) \qquad -\infty < t < \infty \tag{1.3.3}$$

We will use both forms, (1.3.1) and (1.3.3), in representing sinusoidal signals.

The analog sinusoidal signal in (1.3.3) is characterized by the following properties:

A1. For every fixed value of the frequency F, $x_a(t)$ is periodic. Indeed, it can easily be shown, using elementary trigonometry, that

$$x_a(t + T_p) = x_a(t)$$

where $T_p = 1/F$ is the fundamental period of the sinusoidal signal.

A2. Continuous-time sinusoidal signals with distinct (different) frequencies are themselves distinct.

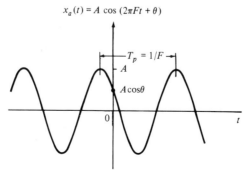

$$x_a(t) = A \cos (2\pi F t + \theta)$$

FIGURE 1.10 Example of an analog sinusoidal signal.

A3. Increasing the frequency F results in an increase in the rate of oscillation of the signal, in the sense that more periods are included in a given time interval.

We observe that for $F = 0$, the value $T_p = \infty$ is consistent with the fundamental relation $F = 1/T_p$. Due to continuity of the time variable t, we can increase the frequency F, without limit, with a corresponding increase in the rate of oscillation.

The relationships we have described for sinusoidal signals carry over to the class of complex exponential signals

$$x_a(t) = Ae^{j(\Omega t + \theta)} \tag{1.3.4}$$

This can easily be seen by expressing these signals in terms of sinusoids using the Euler identity

$$e^{\pm j\phi} = \cos\phi \pm j\sin\phi \tag{1.3.5}$$

By definition, frequency is an inherently positive physical quantity. This is obvious if we interpret frequency as the number of cycles per unit time in a periodic signal. However, in many cases, only for mathematical convenience, we need to introduce negative frequencies. To see this we recall that the sinusoidal signal (1.3.1) may be expressed as

$$x_a(t) = A\cos(\Omega t + \theta) = \frac{A}{2}e^{j(\Omega t + \theta)} + \frac{A}{2}e^{-j(\Omega t + \theta)} \tag{1.3.6}$$

which follows from (1.3.5). Note that a sinusoidal signal can be obtained by adding two equal-amplitude complex-conjugate exponential signals, sometimes called phasors. This is illustrated in Fig. 1.11. As time progresses the phasors rotate in opposite directions with angular frequencies $\pm\Omega$ radians per second. Since a *positive frequency* corresponds to counter-clockwise uniform angular motion, a *negative frequency* simply corresponds to clockwise angular motion.

For mathematical convenience, we use negative frequencies throughout this book. Hence the frequency range for analog sinusoids is $-\infty < F < \infty$.

1.3.2 Discrete-Time Sinusoidal Signals

A discrete-time sinusoidal signal may be expressed as

$$x(n) = A\cos(\omega n + \theta) \qquad -\infty < n < \infty \tag{1.3.7}$$

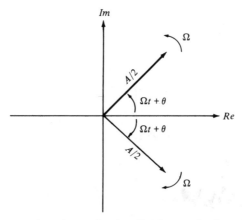

FIGURE 1.11 Representation of a cosine function by a pair of complex-conjugate exponentials (phasors).

where n is an integer variable, called the *sample number*, A is the *amplitude* of the sinusoid, ω is the *frequency* in radians per sample, and θ is the *phase* in radians.

If instead of ω we use the frequency variable f defined by

$$\omega \equiv 2\pi f \tag{1.3.8}$$

the relation (1.3.7) becomes

$$x(n) = A\cos(2\pi fn + \theta) \qquad -\infty < n < \infty \tag{1.3.9}$$

The frequency f has dimensions of cycles per sample. In Section 1.4, where we consider the sampling of analog sinusoids, we relate the frequency variable f of a discrete-time sinusoid to the frequency F in cycles per second for the analog sinusoid. For the moment we consider the discrete-time sinusoid in (1.3.7) independently of the continuous-time sinusoid given in (1.3.1). Figure 1.12 shows a sinusoid with frequency $\omega = \pi/6$ radians per sample ($f = \frac{1}{12}$ cycles per sample) and phase $\theta = \pi/3$.

In contrast to continuous-time sinusoids, the discrete-time sinusoids are characterized by the following properties:

B1. *A discrete-time sinusoid is periodic only if its frequency f is a rational number.*

By definition, a discrete-time signal $x(n)$ is periodic with period N ($N > 0$) if and only if

$$x(n + N) = x(n) \qquad \text{for all } n \tag{1.3.10}$$

The smallest value of N for which (1.3.10) is true is called the *fundamental period*.

The proof of the periodicity property is simple. For a sinusoid with frequency f_0 to be periodic, we should have

$$\cos[2\pi f_0(N + n) + \theta] = \cos(2\pi f_0 n + \theta)$$

This relation is true if and only if there exist an integer k such that

$$2\pi f_0 N = 2k\pi$$

or, equivalently,

$$f_0 = \frac{k}{N} \tag{1.3.11}$$

According to (1.3.11), a discrete-time sinusoidal signal is periodic only if its frequency f_0 can be expressed as the ratio of two integers (i.e., f_0 is rational).

To determine the fundamental period N of a periodic sinusoid, we express its frequency

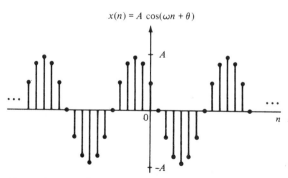

$$x(n) = A\cos(\omega n + \theta)$$

FIGURE 1.12 Example of a discrete-time sinusoidal signal ($\omega = \pi/6$ and $\theta = \pi/3$).

f_0 as in (1.3.11) and cancel common factors so that k and N are prime numbers. Then the fundamental period of the sinusoid is equal to N. Observe that a small change in frequency can result in a large change in the period. For example, note that $f_1 = 31/60$ implies that $N_1 = 60$, whereas $f_2 = 30/60$ results in $N_2 = 2$.

B2. *Discrete-time sinusoids whose frequencies are separated by an integer multiple of 2π are identical.*

To prove this assertion, let us consider the sinusoid $\cos(\omega_0 n + \theta)$. It easily follows that

$$\cos[(\omega_0 + 2\pi)n + \theta] = \cos(\omega_0 n + 2\pi n + \theta) = \cos(\omega_0 n + \theta) \quad (1.3.12)$$

As a result, all sinusoidal sequences

$$x_k(n) = A\cos(\omega_k n + \theta) \qquad k = 0, 1, 2, \ldots \qquad (1.3.13)$$

where

$$\omega_k = \omega_0 + 2k\pi \qquad -\pi \le \omega_0 \le \pi$$

are *indistinguishable* (i.e., *identical*). On the other hand, the sequences of any two sinusoids with frequencies in the range $-\pi \le \omega \le \pi$ or $-\frac{1}{2} \le f \le \frac{1}{2}$ are distinct. Consequently, discrete-time sinusoidal signals with frequencies $|\omega| \le \pi$ or $|f| \le \frac{1}{2}$ are unique. Any sequence resulting from a sinusoid with a frequency $|\omega| > \pi$, or $|f| > \frac{1}{2}$, is identical to a sequence obtained from a sinusoidal signal with frequency $|\omega| < \pi$. Because of this similarity, we call the sinusoid having the frequency $|\omega| > \pi$ an alias of a corresponding sinusoid with frequency $|\omega| < \pi$. Thus we regard frequencies in the range $-\pi \le \omega \le \pi$, or $-\frac{1}{2} < f < \frac{1}{2}$ as unique and all frequencies $|\omega| > \pi$, or $|f| > \frac{1}{2}$, as aliases. The reader should notice the difference between discrete-time sinusoids and continuous-time sinusoids, where the latter result in distinct signals for Ω or F in the entire range $-\infty < \Omega < \infty$ or $-\infty < F < \infty$.

B3. *The highest rate of oscillation in a discrete-time sinusoid is attained when $\omega = \pi$ (or $\omega = -\pi$) or, equivalently, $f = \frac{1}{2}$ (or $f = -\frac{1}{2}$).*

To illustrate this property, let us investigate the characteristics of the sinusoidal signal sequence

$$x(n) = \cos\omega_0 n$$

when the frequency varies from 0 to π. To simplify the argument, we take values of $\omega_0 = 0, \pi/8, \pi/4, \pi/2, \pi$ corresponding to $f = 0, \frac{1}{16}, \frac{1}{8}, \frac{1}{4}, \frac{1}{2}$, which result in periodic sequences having periods $N = \infty, 16, 8, 4, 2$, as depicted in Fig. 1.13. We note that the period of the sinusoid decreases as the frequency increases. In fact, we can see that the rate of oscillation increases as the frequency increases.

To see what happens for $\pi \le \omega_0 \le 2\pi$, we consider the sinusoids with frequencies $\omega_1 = \omega_0$ and $\omega_2 = 2\pi - \omega_0$. Note that as ω_1 varies from π to 2π, ω_2 varies from π to 0. It can be easily seen that

$$x_1(n) = A\cos\omega_1 n = A\cos\omega_0 n$$
$$x_2(n) = A\cos\omega_2 n = A\cos(2\pi - \omega_0)n \qquad (1.3.14)$$
$$= A\cos(-\omega_0 n) = x_1(n)$$

Hence ω_2 is an alias of ω_1. If we had used a sine function instead of a cosine function, the result is basically the same, except for a 180° phase difference between the sinusoids $x_1(n)$ and $x_2(n)$. In any case, as we increase the relative frequency ω_0 of a discrete-time sinusoid from π to 2π, its rate of oscillation decreases. For $\omega_0 = 2\pi$ the result is a

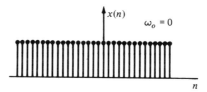

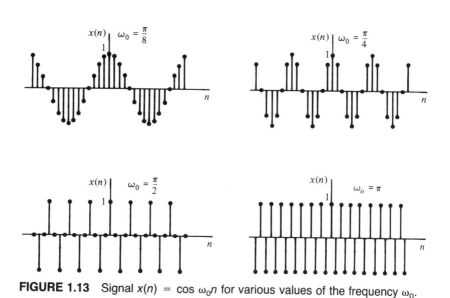

FIGURE 1.13   Signal $x(n) = \cos \omega_0 n$ for various values of the frequency ω_0.

constant signal as in the case for $\omega_0 = 0$. Obviously, for $\omega_0 = \pi$ (or $f = \frac{1}{2}$) we have the highest rate of oscillation.

As for the case of continuous-time signals, negative frequencies can be introduced as well for discrete-time signals. For this purpose we use the identity

$$x(n) = A \cos (\omega n + \theta) = \frac{A}{2} e^{j(\omega n + \theta)} + \frac{A}{2} e^{-j(\omega n + \theta)} \qquad (1.3.15)$$

Since discrete-time sinusoidal signals with frequencies that are separated by an integer multiple of 2π are identical, it follows that the frequencies in any interval $\omega_1 \leq \omega \leq \omega_1 + 2\pi$ constitute *all* the existing discrete-time sinusoids or complex exponentials. Hence the frequency range for discrete-time sinusoids is finite with duration 2π. Usually, we choose the range $0 \leq \omega \leq 2\pi$ or $-\pi \leq \omega \leq \pi$ ($0 \leq f \leq 1$, $-\frac{1}{2} \leq f \leq \frac{1}{2}$), which we call the *fundamental range*.

1.3.3 Harmonically Related Complex Exponentials

Sinusoidal signals and complex exponentials play a major role in the analysis of signals and systems. In some cases we deal with sets of *harmonically related* complex exponentials (or sinusoids). These are sets of periodic complex exponentials with fundamental frequencies that are multiples of a single positive frequency. Although we confine our discussion to complex exponentials clearly the same properties hold for sinusoidal signals. We consider harmonically related complex exponentials in both continuous time and discrete time.

Continuous-Time Exponentials. The basic signals for continuous-time, harmonically related exponentials are

$$s_k(t) = e^{jk\Omega_0 t} = e^{j2\pi kF_0 t} \qquad k = 0, \pm 1, \pm 2, \ldots \qquad (1.3.16)$$

We note that for each value of k, $s_k(t)$ is periodic with fundamental period $1/(kF_0) = T_p/k$ or fundamental frequency kF_0. Since a signal that is periodic with period T_p/k is also periodic with period $k(T_p/k) = T_p$ for any positive integer k, we see that all of the $s_k(t)$ have a common period of T_p. Furthermore, according to Section 1.3.1, F_0 is allowed to take any value and all members of the set are distinct, in the sense that if $k_1 \neq k_2$, then $s_{k1}(t) \neq s_{k2}(t)$.

From the basic signals in (1.3.16) we can construct a linear combination of harmonically related complex exponentials of the form

$$x_a(t) = \sum_{k=-\infty}^{\infty} c_k s_k(t) = \sum_{k=-\infty}^{\infty} c_k e^{jk\Omega_0 t} \qquad (1.3.17)$$

where c_k, $k = 0, \pm 1, \pm 2, \ldots$ are arbitrary complex constants. The signal $x_a(t)$ is periodic with fundamental period $T_p = 1/F_0$, and its representation in terms of (1.3.17) is called the *Fourier series* expansion for $x_a(t)$. The complex-valued constants are the Fourier series coefficients and the signal $s_k(t)$ is called the kth harmonic of $x_a(t)$.

Discrete-time Exponentials. Since a discrete-time complex exponential is periodic if its relative frequency is a rational number, we choose $f_0 = 1/N$ and we define the sets of harmonically related complex exponentials by

$$s_k(n) = e^{j2\pi kf_0 n} \qquad k = 0, \pm 1, \pm 2, \ldots \qquad (1.3.18)$$

In contrast to the continuous-time case, we note that

$$s_{k+N}(n) = e^{j2\pi n(k+N)/N} = e^{j2\pi n}s_k(n) = s_k(n)$$

This means that consistent with (1.3.10), there are only N distinct periodic complex exponentials in the set described by (1.3.18). Furthermore, all members of the set have a common period of N samples. Clearly, we can choose any consecutive N complex exponentials, say from $k = n_0$ to $k = n_0 + N - 1$ to form a harmonically related set with fundamental frequency $f_0 = 1/N$. Most often, for convenience, we choose the set that corresponds to $n_0 = 0$, that is, the set

$$s_k(n) = e^{j2\pi kn/N} \qquad k = 0, 1, 2, \ldots, N - 1 \qquad (1.3.19)$$

EXAMPLE 1.3.1 ──

Stored in the memory of a digital signal processor is one cycle of the sinusoidal signal

$$x(n) = \sin(2\pi n/N + \theta)$$

where $\theta = 2\pi q/N$, where q and N are integers.

(a) Determine how this table of values can be used to obtain values of harmonically related sinusoids having the same phase.
(b) Determine how this table can be used to obtain sinusoids of the same frequency but different phase.

Solution: (a) Let $x_k(n)$ denote the sinusoidal signal sequence

$$x_k(n) = \sin\left(\frac{2\pi nk}{N} + \theta\right)$$

This is a sinusoid with frequency $f_k = k/N$, which is harmonically related to $x(n)$. But $x_k(n)$ may be expressed as

$$x_k(n) = \sin\left[\frac{2\pi(kn)}{N} + \theta\right]$$
$$= x(kn)$$

Thus we observe that $x_k(0) = x(0)$, $x_k(1) = x(k)$, $x_k(2) = x(2k)$, and so on. Hence the sinusoidal sequence $x_k(n)$ can be obtained from the table of values of $x(n)$ by taking every kth value of $x(n)$, beginning with $x(0)$. In this manner we can generate the values of all harmonically related sinusoids with frequencies $f_k = k/N$ for $k/N \leq \frac{1}{2}$.

(b) We can control the phase θ of the sinusoid with frequency $f_k = k/N$ by taking the first value of the sequence from memory location $q = \theta N/2\pi$, where q is an integer. Thus the initial phase θ controls the starting location in the table and we wrap around the table each time the index (kn) exceeds N.

As in the case of continuous-time signals, it is obvious that the linear combination

$$x(n) = \sum_{k=0}^{N-1} c_k s_k(n) = \sum_{k=0}^{N-1} c_k e^{j2\pi kn/N} \qquad (1.3.20)$$

results in a periodic signal with fundamental period N. As we shall see later, this is the Fourier series representation for a periodic discrete-time sequence with Fourier coefficients $\{c_k\}$. The sequence $s_k(n)$ is called the kth harmonic of $x(n)$.

1.4 Analog-to-Digital and Digital-to-Analog Conversion

Most signals of practical interest, such as speech, biological signals, seismic signals, radar signals, sonar signals, and various communications signals such as audio and video signals, are analog. To process analog signals by digital means, it is first necessary to convert them into digital form, that is, to convert them to a sequence of numbers having finite precision. This procedure is called *analog-to-digital (A/D) conversion*, and the corresponding devices are called *A/D converters* (ADCs).

Conceptually, we view A/D conversion as a two-step process. This process is illustrated in Fig. 1.14.

1. *Sampling*. This is the conversion of a continuous-time signal into a discrete-time signal obtained by taking "samples" of the continuous-time signal at discrete-time instants. Thus, if $x_a(t)$ is the input to the sampler, the output is $x_a(nT) \equiv x(n)$, where T is called the *sampling interval*.

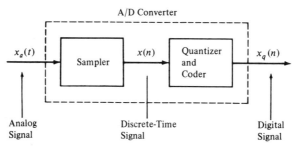

FIGURE 1.14   Basic parts of an analog-to-digital (A/D) converter.

2. *Quantization and coding.* This is the conversion of a discrete-time continuous-valued signal into a discrete-time, discrete-valued (digital) signal. The value of each signal sample is represented by a value selected from a finite set of possible values. In the coding process each discrete value is represented by a b-bit number, which we may denote as $x_q(n)$. The difference between the unquantized sample $x(n)$ and the quantized output $x_q(n)$ is called the quantization error.

Although we model the A/D converter as a sampler followed by a quantizer and coder, in practice the A/D conversion is performed by a single device that takes $x_a(t)$ and produces $x_q(n)$.

In many cases of practical interest (e.g., speech processing) it is desirable to convert the processed digital signals into analog form. (Obviously, we cannot listen to the sequence of samples representing a speech signal or see the numbers corresponding to a TV signal.) The process of converting a digital signal into an analog signal is known as *digital-to-analog (D/A) conversion.* All D/A converters "connect the dots" in a digital signal by performing some kind of interpolation, whose accuracy depends on the quality of the D/A conversion process. Figure 1.15 illustrates a simple form of D/A conversion, called a zero-order hold or a staircase approximation. Other approximations are possible, such as linearly connecting a pair of successive samples (linear interpolation), fitting a quadratic through three successive samples (quadratic interpolation), and so on. Is there an optimum (ideal) interpolator? For signals having a *limited frequency content* (finite bandwidth), the sampling theorem introduced below specifies the optimum form of interpolation.

Sampling and quantization are discussed in below. In particular, we demonstrate that sampling does not result in a loss of information, nor does it introduce distortion in the signal if the signal bandwidth is finite. In principle, the analog signal can be reconstructed from the samples, provided that the sampling rate is sufficiently high to avoid the problem commonly called *aliasing.* On the other hand, quantization is a noninvertible or irreversible process which results in signal distortion. We shall show that the amount of distortion depends on the accuracy, as measured by the number of bits, in the A/D conversion process. The factors affecting the choice of the desired accuracy of the A/D converter are cost and sampling rate. In general, the cost increases with an increase in accuracy and/or sampling rate.

1.4.1 Sampling of Analog Signals

There are many ways to sample an analog signal. We will limit our discussion to *periodic* or *uniform sampling*, which is the type of sampling used most often in practice. This is

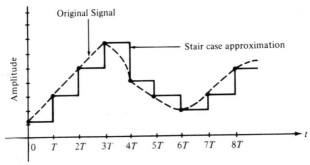

FIGURE 1.15   Zero-order-hold digital-to-analog (D/A) conversion.

described by the relation

$$x(n) = x_a(nT) \qquad -\infty < n < \infty \qquad (1.4.1)$$

where $x(n)$ is the discrete-time signal obtained by "taking samples" of the analog signal $x_a(t)$ every T seconds. This procedure is illustrated in Fig. 1.16. The time interval T between successive samples is called the *sampling period* or *sample interval* and its reciprocal $1/T = F_s$ is called the *sampling rate* (samples per second) or the *sampling frequency* (hertz).

Periodic sampling establishes a relationship between the time variables t and n of continuous-time and discrete-time signals, respectively. Indeed, these variables are linearly related through the sampling period T or, equivalently, through the sampling rate $F_s = 1/T$, as

$$t = nT = \frac{n}{F_s} \qquad (1.4.2)$$

As a consequence of (1.4.2), there exists a relationship between the frequency variable F (or Ω) for analog signals and the frequency variable f (or ω) for discrete-time signals. To establish this relationship, we consider an analog sinusoidal signal of the form

$$x_a(t) = A \cos (2\pi F t + \theta) \qquad (1.4.3)$$

which, when sampled periodically at a rate $F_s = 1/T$ samples per second, yields

$$x_a(nT) \equiv x(n) = A \cos (2\pi F nT + \theta) \qquad (1.4.4)$$

$$= A \cos \left(\frac{2\pi n F}{F_s} + \theta \right)$$

If we compare (1.4.4) with (1.3.9), we note that the frequency variables F and f are linearly related as

$$f = \frac{F}{F_s} \qquad (1.4.5)$$

or, equivalently, as

$$\omega = \Omega T \qquad (1.4.6)$$

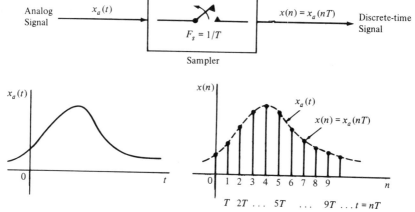

FIGURE 1.16 Periodic sampling of an analog signal.

The relation in (1.4.5) justifies the name *relative* or *normalized frequency*, which is sometimes used to describe the frequency variable f. As (1.4.5) implies, we can use f to determine the frequency F in hertz only if the sampling frequency F_s is known.

We recall from Section 1.3.1 that the range of the frequency variable F or Ω for continuous-time sinusoids are

$$-\infty < F < \infty \qquad (1.4.7)$$
$$-\infty < \Omega < \infty$$

However, the situation is different for discrete-time sinusoids. From Section 1.3.2 we recall that

$$-\tfrac{1}{2} < f < \tfrac{1}{2} \qquad (1.4.8)$$
$$-\pi < \omega < \pi$$

By substituting from (1.4.5) and (1.4.6) into (1.4.8), we find that the frequency of the continuous-time sinusoid when sampled at a rate $F_s(1/T)$ must fall in the range

$$-\frac{F_s}{2} \le F \le \frac{F_s}{2} \qquad (1.4.9)$$
$$-\frac{1}{2T} \le F \le \frac{1}{2T}$$

or, equivalently,

$$-\pi F_s \le \Omega \le \pi F_s \qquad (1.4.10)$$
$$-\frac{\pi}{T} \le \Omega \le \frac{\pi}{T}$$

These relations are summarized in Table 1.1.

From the relations above we observe that the fundamental difference between continuous-time and discrete-time signals is in their range of values of the frequency variables F and f, or Ω and ω. Periodic sampling of a continuous-time signal implies a mapping of an infinite frequency range for the variable F (or Ω) into a finite frequency range for the variable f (or ω). Since the highest frequency in a discrete-time signal is $\omega = \pi$ or $f = \tfrac{1}{2}$, it follows that, with a sampling rate F_s, the corresponding highest values of F and Ω are

$$F_{max} = \frac{F_s}{2} = \frac{1}{2T} \qquad (1.4.11)$$

$$\Omega_{max} = \pi F_s = \frac{\pi}{T}$$

TABLE 1.1 Relations Between Frequency and Relative Frequency

Analog Signals		Discrete-Time Signals	
$\Omega = 2\pi F$ $\dfrac{\text{radians}}{\text{sec}}$ hertz		$\omega = 2\pi f$ $\dfrac{\text{radians}}{\text{sample}}$ $\dfrac{\text{cycles}}{\text{sample}}$	
$-\infty < \Omega < \infty$	$\omega = \Omega T$	$-\pi \le \Omega \le \pi$	
$-\infty < F < \infty$	$f = F/F_s \longrightarrow$	$-\tfrac{1}{2} \le f \le \tfrac{1}{2}$	
$-\pi/T \le \Omega \le \pi/T$	$\Omega = \omega/T$		
$-F_s/2 \le F \le F_s/2$	$\longleftarrow$ $F = f \cdot F_s$		

Therefore, sampling introduces an ambiguity, since the highest frequency in a continuous-time signal that can be uniquely distinguished when such a signal is sampled at a rate $F_s = 1/T$ is $F_{max} = F_s/2$ or $\Omega_{max} = \pi F_s$.

The full implications of these frequency relations can be appreciated by considering the two analog sinusoidal signals

$$x_1(t) = \cos 2\pi(10)t$$
$$x_2(t) = \cos 2\pi(50)t \tag{1.4.12}$$

which are sampled at a rate $F_s = 40$ Hz. The corresponding discrete-time signals or sequences are

$$x_1(n) = \cos 2\pi \left(\frac{10}{40}\right) n = \cos \frac{\pi}{2} n$$

$$x_2(n) = \cos 2\pi \left(\frac{50}{40}\right) n = \cos \frac{5\pi}{2} n \tag{1.4.13}$$

However, $\cos 5\pi n/2 = \cos(2\pi n + \pi n/2) = \cos \pi n/2$. Hence $x_2(n) = x_1(n)$. Thus the sinusoidal signals are identical and, consequently, indistinguishable. If we are given the sampled values generated by $\cos(\pi/2)n$, there is an ambiguity as to whether these sampled values correspond to $x_1(t)$ or $x_2(t)$. Since $x_2(t)$ yields exactly the same values as $x_1(t)$ when the two are sampled at $F_s = 40$ samples per second, we say that the frequency $F_2 = 50$ Hz is an *alias* of the frequency $F_1 = 10$ Hz at the sampling rate of 40 samples per second.

It is important to note that F_2 is not the only alias of F_1. In fact, at the sampling rate of 40 samples per second, the frequency $F_3 = 90$ Hz is also an alias of F_1, so is the frequency $F_4 = 130$ Hz, and so on. All of the sinusoids $\cos 2\pi(F_1 + 40k)t$, $k = 1, 2, 3, 4, \ldots$, sampled at 40 samples per second yield identical values. Consequently, they are all aliases of $F_1 = 10$ Hz.

In general, the sampling of a continuous-time sinusoidal signal

$$x_a(t) = A \cos(2\pi F_0 t + \theta) \tag{1.4.14}$$

with a sampling rate $F_s = 1/T$ results in a discrete-time signal

$$x(n) = A \cos(2\pi f_0 n + \theta) \tag{1.4.15}$$

where $f_0 = F_0/F_s$ is the relative frequency of the sinusoid. If we assume that $-F_s/2 \le F_0 \le F_s/2$, the frequency f_0 of $x(n)$ is in the range $-\frac{1}{2} \le f_0 \le \frac{1}{2}$, which is the frequency range for a discrete-time signals. In this case, the relationship between F_0 and f_0 is one-to-one, and hence it is possible to identify (or reconstruct) the analog signal $x_a(t)$ from the samples $x(n)$.

On the other hand, if the sinusoids

$$x_a(t) = A \cos(2\pi F_k t + \theta) \tag{1.4.16}$$

where

$$F_k = F_0 + kF_s \qquad k = \pm 1, \pm 2, \ldots \tag{1.4.17}$$

are sampled at a rate F_s, it is clear that the frequency F_k is outside the fundamental frequency range $-F_s/2 \le F \le F_s/2$. Consequently, the sampled signal is

$$x(n) \equiv x_a(nT) = A \cos\left(2\pi \frac{F_0 + kF_s}{F_s} n + \theta\right)$$

$$= A \cos(2\pi n F_0/F_s + \theta + 2\pi kn)$$

$$= A \cos(2\pi f_0 n + \theta)$$

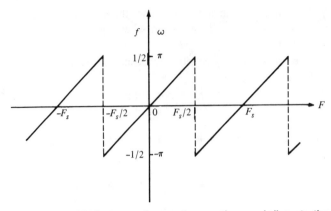

FIGURE 1.17 Relationship between the continuous-time and discrete-time frequency variables in the case of periodic sampling.

which is identical to the discrete-time signal in (1.4.15) obtained by sampling (1.4.14). Thus an infinite number of continuous-time sinusoids is represented after sampling by the *same* discrete-time signal (i.e., by the same set of samples). Consequently, if we are given the sequence $x(n)$ there is an ambiguity as to which continuous-time signal $x_a(t)$ these values represent. Equivalently, we can say that the frequencies $F_k = F_0 + kF_s$, $-\infty < k < \infty$ are indistinguishable from the frequency F_0 after sampling and hence they are aliases of F_0. The relationship between the frequency variables of the continuous-time and discrete-time signals is illustrated in Fig. 1.17.

An example of aliasing is illustrated in Fig. 1.18, where two sinusoids with frequencies $F_0 = \frac{1}{8}$ Hz and $F_1 = -\frac{7}{8}$ Hz yield identical samples when a sampling rate of $F_s = 1$ Hz is used. From (1.4.17) it easily follows that for $k = -1$, $F_0 = F_1 + F_s = (-\frac{7}{8} + 1)$ Hz $= \frac{1}{8}$ Hz.

Since $F_s/2$, which corresponds to $\omega = \pi$, is the highest frequency that can be represented uniquely with a sampling rate F_s, it is a simple matter to determine the mapping of any (alias) frequency above $F_s/2$ ($\omega = \pi$) into the equivalent frequency below $F_s/2$. We may use $F_s/2$ or $\omega = \pi$ as the pivotal point and reflect or "fold" the alias frequency to the range $0 \leq \omega \leq \pi$. Since the point of reflection is $F_s/2$ ($\omega = \pi$), the frequency $F_s/2$ ($\omega = \pi$) is called the *folding frequency*.

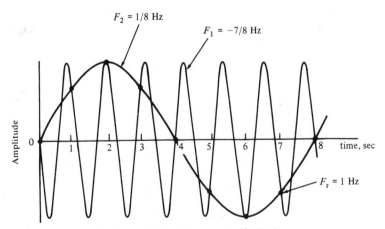

FIGURE 1.18 Illustration of aliasing.

EXAMPLE 1.4.1

Consider the analog signal

$$x_a(t) = 3 \cos 100\pi t$$

(a) Determine the minimum required sampling rate to avoid aliasing.
(b) Suppose that the signal is sampled at the rate $F_s = 200$ Hz. What is the discrete-time signal obtained after sampling?
(c) Suppose that the signal is sampled at the rate $F_s = 75$ Hz. What is the discrete-time signal obtained after sampling?
(d) What is the frequency $F < F_s/2$ of a sinusoid that yields samples identical to those obtained in part (c)?

Solution: **(a)** The frequency of the analog signal is $F = 50$ Hz. Hence the minimum sampling rate required to avoid aliasing is $F_s = 100$ Hz.
(b) If the signal is sampled at $F_s = 200$ Hz, the discrete-time signal is

$$x(n) = 3 \cos \frac{100\pi}{200} n = 3 \cos \frac{\pi}{2} n$$

(c) If the signal is sampled at $F_s = 75$ Hz, the discrete-time signal is

$$x(n) = 3 \cos \frac{100\pi}{75} n = 3 \cos \frac{4\pi}{3} n$$

$$= 3 \cos \left(2\pi - \frac{2\pi}{3} \right) n$$

$$= 3 \cos \frac{2\pi}{3} n$$

(d) For the sampling rate of $F_s = 75$ Hz, we have

$$F = fF_s = 75f$$

The frequency of the sinusoid in part (c) is $f = \frac{1}{3}$. Hence

$$F = 25 \text{ Hz}$$

Clearly, the sinusoidal signal

$$y_a(t) = 3 \cos 2\pi F t$$
$$= 3 \cos 50\pi t$$

sampled at $F_s = 75$ samples/s yields identical samples. Hence $F = 50$ Hz is an alias of $F = 25$ Hz for the sampling rate $F_s = 75$ Hz.

1.4.2 The Sampling Theorem

Given any analog signal, how should we select the sampling period T or, equivalently, the sampling rate F_s? To answer this question, we must have some information about the characteristics of the signal to be sampled. In particular, we must have some general information concerning the *frequency content* of the signal. Such information is generally available to us. For example, we know generally that the major frequency components of a speech signal fall below 3000 Hz. On the other hand, television signals, in general, contain important frequency components up to 5 MHz. The information content of such signals is contained in the amplitudes, frequencies, and phases of the various frequency components. On the other hand, detailed knowledge of the characteristics of such signals is not available to us prior to obtaining the signals. In fact, the purpose of processing the signals is usually to extract this detailed information. However, if we know the maximum frequency content of the general class of signals (e.g., the class of speech signals, the class of video signals, etc.), we can specify the sampling rate necessary to convert the analog signals to digital signals.

Let us suppose that any analog signal can be represented as a sum of sinusoids of different amplitudes, frequencies, and phases, that is,

$$x_a(t) = \sum_{i=1}^{N} A_i \cos\left(2\pi F_i t + \theta_i\right) \qquad (1.4.18)$$

where N denotes the number of frequency components. All signals such as speech and video lend themselves to such a representation over any short time segment. The amplitudes, frequencies, and phases usually change slowly with time from one time segment to another. However, suppose that the frequencies do not exceed some known frequency, say F_{max}. For example, $F_{max} = 3000$ Hz for the class of speech signals and $F_{max} = 5$ MHz for television signals. Since the maximum frequency may vary slightly from different realizations among signals of any given class (e.g., it may vary slightly from speaker to speaker), we may wish to ensure that F_{max} does not exceed some predetermined value by passing the analog signal through a filter that severely attenuates frequency components above F_{max}. Thus we are certain that no signal in the class contains frequency components (having significant amplitude or power) above F_{max}. Such filtering is commonly used in practice prior to sampling.

From our knowledge of F_{max}, we can select the appropriate sampling rate. We know that the highest frequency in an analog signal that can be unambiguously reconstructed when the signal is sampled at a rate $F_s = 1/T$ is $F_s/2$. Any frequency above $F_s/2$ or below $-F_s/2$ results in samples that are identical with a corresponding frequency in the range $-F_s/2 \le F \le F_s/2$. To avoid the ambiguities resulting from aliasing, we must select the sampling rate to be sufficiently high. That is, we must select $F_s/2$ to be greater than F_{max}. Thus to avoid the problem of aliasing, F_s is selected so that

$$F_s > 2F_{max} \qquad (1.4.19)$$

where F_{max} is the largest frequency component in the analog signal. With the sampling rate selected in this manner, any frequency component, say $|F_i| < F_{max}$, in the analog signal is mapped into a discrete-time sinusoid with a frequency

$$-\frac{1}{2} \le f_i = \frac{F_i}{F_s} \le \frac{1}{2} \qquad (1.4.20)$$

or, equivalently,

$$-\pi \le \omega_i = 2\pi f_i \le \pi \qquad (1.4.21)$$

Since, $|f| = \frac{1}{2}$ or $|\omega| = \pi$ is the highest (unique) frequency in a discrete-time signal, the choice of sampling rate according to (1.4.19) avoids the problem of aliasing. In other words, the condition $F_s > 2F_{max}$ ensures that all the sinusoidal components in the analog signal are mapped into corresponding discrete-time frequency components with frequencies in the fundamental interval. Thus all the frequency components of the analog signal are represented in sampled form without ambiguity, and hence the analog signal can be reconstructed without distortion from the sample values using an "appropriate" interpolation (digital-to-analog conversion) method. The "appropriate" or ideal interpolation formula is specified in the following theorem, which is called the *sampling theorem*. Its proof is deferred to Section 4.5.

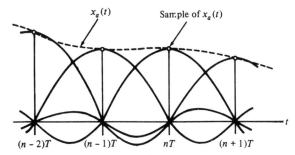

FIGURE 1.19   Ideal D/A conversion (interpolation).

SAMPLING THEOREM: *If the highest frequency contained in an analog signal $x_a(t)$ is $F_{max} = B$ and the signal is sampled at a rate $F_s > 2F_{max} \equiv 2B$, then $x_a(t)$ can be exactly recovered from its sample values using the interpolation function*

$$g(t) = \frac{\sin 2\pi Bt}{2\pi Bt} \tag{1.4.22}$$

Thus $x_a(t)$ may be expressed as

$$x_a(t) = \sum_{n=-\infty}^{\infty} x_a\left(\frac{n}{F_s}\right) g\left(t - \frac{n}{F_s}\right) \tag{1.4.23}$$

where $x_a(n/F_s) = x_a(nT) \equiv x(n)$ are the samples of $x_a(t)$.

When the sampling of $x_a(t)$ is done at the minimum sampling rate $F_s = 2B$, the reconstruction formula in (1.4.23) becomes

$$x_a(t) = \sum_{n=-\infty}^{\infty} x_a\left(\frac{n}{2B}\right) \frac{\sin 2\pi B(t - n/2B)}{2\pi B(t - n/2B)} \tag{1.4.24}$$

The sampling rate $F_N = 2B = 2F_{max}$ is called the *Nyquist rate*. Figure 1.19 illustrates the ideal D/A conversion process using the interpolation function in (1.4.22).

As can be observed from either (1.4.23) or (1.4.24), the reconstruction of $x_a(t)$ from the sequence $x(n)$ is a complicated process, involving a weighted sum of the interpolation function $g(t)$ and its time-shifted versions $g(t - nT)$ for $-\infty < n < \infty$, where the weighting factors are the samples $x(n)$. Because of the complexity and the infinite number of samples required in (1.4.23) or (1.4.24), these reconstruction formulas are primarily of theoretical interest. Practical interpolation methods are treated in Section 6.2.

EXAMPLE 1.4.2 _____

Consider the analog signal

$$x_a(t) = 3 \cos 50\pi t + 10 \sin 300\pi t - \cos 100\pi t$$

What is the Nyquist rate for this signal?

Solution: The frequencies present in the signal above are

$$F_1 = 25 \text{ Hz} \qquad F_2 = 150 \text{ Hz} \qquad F_3 = 50 \text{ Hz}$$

Thus $F_{max} = 150$ Hz and according to (1.4.2),

$$F_s > 2F_{max} = 300 \text{ Hz}$$

The Nyquist rate is $F_N = 2F_{max}$. Hence

$$F_N = 300 \text{ Hz}$$

Discussion: It should be observed that the signal component $10 \sin 300\pi t$, sampled at the Nyquist rate $F_N = 300$, results in the samples $10 \sin \pi n$, which are identically zero. In other words, we are sampling the analog sinusoid at its zero-crossing points, and hence we miss this signal component completely. This situation would not occur if the sinusoid is offset in phase by some amount θ. In such a case we have $10 \sin (300\pi t + \theta)$ sampled at the Nyquist rate $F_N = 300$ samples per second, which yields the samples

$$
\begin{aligned}
10 \sin (\pi n + \theta) &= 10(\sin \pi n \cos \theta + \cos \pi n \sin \theta) \\
&= 10 \sin \theta \cos \pi n \\
&= (-1)^n 10 \sin \theta
\end{aligned}
$$

Thus if $\theta \neq 0$ or π, the samples of the sinusoid taken at the Nyquist rate are not all zero. However, we still cannot obtain the correct amplitude from the samples because the phase θ is unknown. A simple remedy that avoids this potentially troublesome situation is to sample the analog signal at a rate higher than the Nyquist rate.

EXAMPLE 1.4.3 ———————————————————————————

Consider the analog signal

$$x_a(t) = 3 \cos 2000\pi t + 5 \sin 6000\pi t + 10 \cos 12{,}000\pi t$$

(a) What is the Nyquist rate for this signal?
(b) Assume now that we sample this signal using a sampling rate $F_s = 5000$ samples/s. What is the discrete-time signal obtained after sampling?
(c) What is the analog signal $y_a(t)$ we can reconstruct from the samples if we use ideal interpolation?

Solution: (a) The frequencies existing in the analog signal are

$$F_1 = 1 \text{ kHz} \qquad F_2 = 3 \text{ kHz} \qquad F_3 = 6 \text{ kHz}$$

Thus $F_{max} = 6$ kHz, and according to the sampling theorem,

$$F_s > 2F_{max} = 12 \text{ kHz}$$

The Nyquist rate is

$$F_N = 12 \text{ kHz}$$

(b) Since we have chosen $F_s = 5$ kHz, the folding frequency is

$$\frac{F_s}{2} = 2.5 \text{ kHz}$$

and this is the maximum frequency that can be represented uniquely by the sampled signal. By making use of (1.4.5) we obtain

$$x(n) = x_a(nT) = x_a\left(\frac{n}{F_s}\right)$$

$$
\begin{aligned}
&= 3 \cos 2\pi(\tfrac{1}{5})n + 5 \sin 2\pi(\tfrac{3}{5})n + 10 \cos 2\pi(\tfrac{6}{5})n \\
&= 3 \cos 2\pi(\tfrac{1}{5})n + 5 \sin 2\pi(1 - \tfrac{2}{5})n + 10 \cos 2\pi(1 + \tfrac{1}{5})n \\
&= 3 \cos 2\pi(\tfrac{1}{5})n + 5 \sin 2\pi(-\tfrac{2}{5})n + 10 \cos 2\pi(\tfrac{1}{5})n
\end{aligned}
$$

Finally, we obtain

$$x(n) = 13 \cos 2\pi(\tfrac{1}{5})n - 5 \sin 2\pi(\tfrac{2}{5})n$$

The same result can be obtained using Fig. 1.17. Indeed, since $F_s = 5$ kHz, the folding frequency is $F_s/2 = 2.5$ kHz. This is the maximum frequency that can be represented uniquely by the sampled signal. From (1.4.17) we have $F_0 = F_k - kF_s$. Thus F_0 can be obtained by substracting from F_k an integer multiple of F_s such that $-F_s/2 \le F_0 \le F_s/2$. The frequency F_1 is less than $F_s/2$ and thus it is not affected by aliasing. However, the other two frequencies are above the folding frequency and they will be changed by the aliasing effect. Indeed,

$$F_2' = F_2 - F_s = -2 \text{ kHz}$$
$$F_3' = F_3 - F_s = 1 \text{ kHz}$$

From (1.4.5) it follows that $f_1 = \tfrac{1}{5}$, $f_2 = -\tfrac{2}{5}$, and $f_3 = \tfrac{1}{5}$, which are in agreement with the result above.

(c) Since only the frequency components at 1 kHz and 2 kHz are present in the sampled signal, the analog signal we can recover is

$$y_a(t) = 13 \cos 2000\pi t - 5 \sin 4000\pi t$$

which is obviously different from the original signal $x_a(t)$. This distortion of the original analog signal was caused by the aliasing effect, due to the low sampling rate used.

Although aliasing is a pitfall to be avoided, there are two useful practical applications which are based on the exploitation of the aliasing effect. These applications are the stroboscope and the sampling oscilloscope. Both instruments are designed to operate as aliasing devices in order to represent high frequencies as low ones.

To elaborate, let us consider a signal with high-frequency components confined to a given frequency band $B_1 < F < B_2$, where $B_2 - B_1 \equiv B$ is defined as the bandwidth of the signal. We assume that $B \ll B_1 < B_2$. This condition means that the frequency components in the signal are much larger than the bandwidth B of the signal. Such signals are usually called passband or narrowband signals. Now, if this signal is sampled at a rate $F_s \ge 2B$, but $F_s \ll B_1$, then all the frequency components contained in the signal will be aliases of frequencies in the range $0 < F < F_s/2$. Consequently, if we observe the frequency content of the signal in the fundamental range $0 < F < F_s/2$, we know precisely the frequency content of the analog signal since we know the frequency band $B_1 < F < B_2$ under consideration. Consequently, if the signal is a narrowband (passband) signal, we can reconstruct the original signal from the samples, provided that the signal is sampled at a rate $F_s > 2B$, when B is the bandwidth. This statement constitutes another form of the sampling theorem, which we call the *passband form*, in order to distinguish it from the previous form of the sampling theorem, which applies in general to all types of signals. The latter is sometimes called the *baseband form*. The *passband form* of the sampling theorem is described in detail in Section 6.1.5.

1.4.3 Quantization and Coding

As we have seen, a digital signal is a sequence of numbers (samples) in which each number is represented by a finite number of digits (finite precision).

The process of converting a discrete-time signal into a digital signal by expressing each sample value as a finite (instead of an infinite) number of digits, is called *quanti-*

zation. The error introduced in representing the continuous-valued signal by a finite set of discrete value levels is called *quantization error* or *quantization noise*.

We denote the quantizer operation on the samples $x(n)$ as $Q[x(n)]$ and let $x_q(n)$ denote the sequence of quantized samples at the output of the quantizer. Hence

$$x_q(n) = Q[x(n)]$$

Then the quantization error is a sequence $e_q(n)$ defined as the difference between the quantized value and the actual sample value. Thus

$$e_q(n) = x_q(n) - x(n) \tag{1.4.25}$$

We will illustrate the quantization process with an example. Let us consider the discrete-time signal

$$x(n) = \begin{cases} 0.9^n & n \geq 0 \\ 0 & n < 0 \end{cases}$$

which was obtained by sampling the analog exponential signal $x_a(t) = 0.9^t$, $t \geq 0$ with a sampling frequency $F_s = 1$ Hz (see Fig. 1.20(a). Observation of Table 1.2, which shows the values of the first 10 samples of $x(n)$, reveals that the description of the sample value $x(n)$ requires n significant digits. It is obvious that this signal cannot be processed by using a calculator or a digital computer since only the first few samples can be stored and manipulated. For example, most calculators process numbers with only eight significant digits.

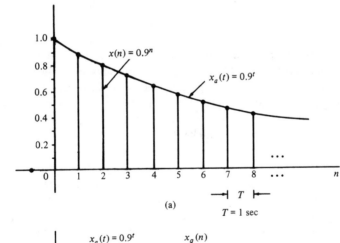

(a)

$T = 1$ sec

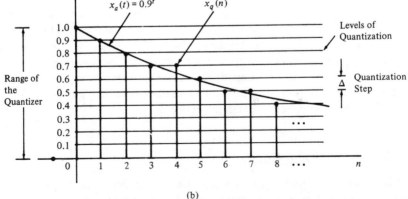

(b)

FIGURE 1.20 Illustration of quantization.

TABLE 1.2 Numerical Illustration of Quantization with One Significant Digit Using Truncation or Rounding

n	$x(n)$ Discrete-Time Signal	$x_q(n)$ (Truncation)	$x_q(n)$ (rounding)	$e_q(n) = x_q(n) - x(n)$ (Rounding)
0	1	1.0	1.0	0.0
1	0.9	0.9	0.9	0.0
2	0.81	0.8	0.8	−0.01
3	0.729	0.7	0.7	−0.029
4	0.6561	0.6	0.7	0.0439
5	0.59049	0.5	0.6	0.00951
6	0.531441	0.5	0.5	−0.031441
7	0.4782969	0.4	0.5	0.0217031
8	0.43046721	0.4	0.4	−0.03046721
9	0.387420489	0.3	0.4	0.012579511

However, let us assume that we want to use only one significant digit. To eliminate the excess digits we can either simply discard them (*truncation*) or discard them by rounding the resulting number (*rounding*). The resulting quantized signals $x_q(n)$ are shown in Table 1.2. We will discuss only quantization by rounding, although it is just as easy to treat truncation. The rounding process is graphically illustrated in Fig. 1.20b. The allowed values in the digital signal are called the *quantization levels*, whereas the distance Δ between two successive quantization levels is called *quantization step size* or *resolution*. The rounding quantizer assigns each sample of $x(n)$ to the nearest quantization level. In contrast, a quantizer that performs truncation would have assigned each sample of $x(n)$ to the quantization level below it. The quantization error $e_q(n)$ in rounding is limited to the range of $-\Delta/2$ to $\Delta/2$, that is,

$$-\frac{\Delta}{2} \le e_q(n) \le \frac{\Delta}{2} \tag{1.4.26}$$

In other words, the instantaneous quantization error cannot exceed half of the quantization step (see Table 1.2).

If $x_{\min}$ and $x_{\max}$ represent the minimum and maximum value of $x(n)$ and m is the number of quantization levels, then

$$\Delta = \frac{x_{\max} - x_{\min}}{m - 1} \tag{1.4.27}$$

We define the *dynamic range* of the signal as $x_{\max} - x_{\min}$. In our example we have $x_{\max} = 1$, $x_{\min} = 0$, and $m = 11$, which leads to $\Delta = 0.1$. Note that if the dynamic range is fixed, increasing the number of quantization levels m results in a decrease of the quantization step size. Thus the quantization error decreases and the accuracy of the quantizer increases. In practice we can reduce the quantization error to an insignificant amount by choosing a sufficient number of quantization levels.

Theoretically, quantization of analog signals always results in a loss of information. This is a result of the ambiguity introduced by quantization. Indeed, quantization is an irreversible or noninvertible process (i.e., a many-to-one mapping) since all samples in a distance $\Delta/2$ about a certain quantization level are assigned the same value. This ambiguity makes the exact quantitative analysis of quantization extremely difficult. This subject is discussed further in Chapter 10, where we use statistical analysis.

Finally, the *coding* process in an A/D converter assigns a unique binary number to each quantization level. If we have m levels we need at least m different binary numbers. With a word length of b bits we can create 2^b different binary numbers. Hence we have $2^b \geq m$, or quivalently, $b \geq \log_2 m$. Thus the number of bits required in the coder is the smallest integer greater than or equal to $\log_2 m$. In our example it can easily be seen that we need a coder with $b = 4$ bits. Commercially available A/D converters may be bought with finite precision of $b = 16$ or less. Generally, the higher the sampling speed and the finer the quantization, the more expensive the device becomes.

1.4.4 Digital-to-Analog Conversion

To convert a digital signal into an analog signal we may use a digital-to-analog (D/A) converter. A D/A converter is characterized by the number of bits it uses and the sampling period that is specified by the user. As stated previously, the task of a D/A converter is to interpolate between samples.

The sampling theorem specifies the optimum interpolation for a bandlimited signal. However, this type of interpolation is too complicated and, hence impractical, as indicated previously. From a practical viewpoint, the simplest D/A converter is the zero-order hold shown in Fig. 1.15, which simply holds constant the value of one sample until the next one. Additional improvement can be obtained by using linear interpolation as shown in Fig. 1.21 to connect successive samples with straight-line segments. The zero-order hold and linear interpolator are analyzed in Section 6.2. Better interpolation can be achieved by using more sophisticated higher-order interpolation techniques.

In general, suboptimum interpolation techniques result in passing frequencies above the folding frequency. Such frequency components are undesirable and are usually removed by passing the output of the interpolator through a proper analog filter, which is called a *postfilter* or *smoothing filter*. Thus D/A conversion usually involves a suboptimum interpolator followed by a postfilter.

1.4.5 Analysis of Digital Signals and Systems Versus Discrete-Time Signals and Systems

We have seen that a digital signal is defined as a function of an integer independent variable and its values are taken from a finite set of possible values. The usefulness of such signals is a consequence of the possibilities offered by digital computers. Computers operate on numbers, which are represented by a string of 0's and 1's. The length of this string (*word length*) is fixed and finite and usually is 8, 12, 16 or 32 bits. The finite word length effects in computations cause complications in the analysis of digital signal processing systems. To avoid these complications, we neglect the quantized nature of

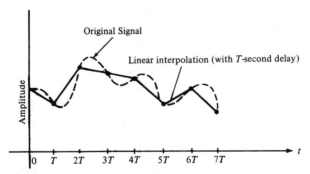

FIGURE 1.21 Linear point connector (with T-second delay).

digital signals and systems in much of our analysis and consider them as discrete-time signals and systems.

In Chapter 10 we investigate the consequences of using a finite word length. This is an important topic, because many digital signal processing problems are solved with small computers or microprocessors that employ fixed-point arithmetic. Consequently, one must look carefully at the problem of finite-precision arithmetic and take account of its effects in the design of software and hardware that performs the desired signal processing tasks.

1.5 Summary and References

In this introductory chapter we have attempted to provide the motivation for digital signal processing as an alternative to analog signal processing. We presented the basic elements of a digital signal processing system and defined the operations needed to convert an analog signal into a digital signal for processing. Of particular importance is the sampling theorem, which was introduced by Nyquist (1928) and later popularized in the classic paper by Shannon (1949). The sampling theorem as described in Section 1.4.2 is derived in Section 4.5. Sinusoidal signals were introduced primarily for the purpose of illustrating the aliasing phenomenon and for the subsequent development of the sampling theorem.

Quantization effects that are inherent in the A/D conversion of a signal were also introduced in this chapter. Signal quantization is best treated in statistical terms, as described in Chapter 10.

Finally, the topic of signal reconstruction, or D/A conversion, was described briefly. Signal reconstruction based on staircase or linear interpolation methods is treated in Section 6.2.

The reader who is interested in delving into a number of practical applications of digital signal processing (e.g., speech processing, image processing; radar signal processing, sonar signal processing, and geophysical signal processing) will find the book edited by Oppenheim (1978) an excellent reference.

PROBLEMS

1.1 Classify the following signals according to whether they are (1) one- or multi-dimensional; (2) single or multichannel, (3) continuous time or discrete time, and (4) analog or digital (in amplitude). Give a brief explanation.
  (a) Closing prices of utility stocks on the New York Exchange.
  (b) A color movie.
  (c) Position of the steering wheel of a car in motion relative to car's reference frame.
  (d) Position of the steering wheel of a car in motion relative to ground reference frame.
  (e) Weight and height measurements of a child taken every month.

1.2 Determine which of the following sinusoids are periodic and compute their fundamental period.

(a) $\cos 0.01 \pi n$ **(b)** $\cos\left(\pi \dfrac{30n}{105}\right)$ **(c)** $\cos 3\pi n$ **(d)** $\sin 3n$

(e) $\sin\left(\pi \dfrac{62n}{10}\right)$

1.3 Determine whether or not each of the following signals is periodic. In case a signal is periodic, specify its fundamental period.

(a) $x_a(t) = 3 \cos (5t + \pi/6)$

(b) $x(n) = 3 \cos (5n + \pi/6)$

(c) $x(n) = 2 \exp [j(n/6 - \pi)]$

(d) $x(n) = \cos (n/8) \cos (\pi n/8)$

(e) $x(n) = \cos (\pi n/2) - \sin (\pi n/8) + 3 \cos (\pi n/4 + \pi/3)$

1.4 (a) Show that the fundamental period N_p of the signals

$$s_k(n) = e^{j2\pi kn/N} \qquad k = 0, 1, 2, \ldots$$

is given by $N_p = N/\text{GCD}(k, N)$, where GCD is the greatest common divisor of k and N.

(b) What is the fundamental period of this set for $N = 7$?

(c) What is it for $N = 16$?

1.5 Consider the following analog sinusoidal signal:

$$x_a(t) = 3 \sin (100\pi t)$$

(a) Sketch the signal $x_a(t)$ for $0 \le t \le 30$ ms.

(b) The signal $x_a(t)$ is sampled with a sampling rate $F_s = 300$ samples/s. Determine the frequency of the discrete-time signal $x(n) = x_a(nT)$, $T = 1/F_s$, and show that it is periodic.

(c) Compute the sample values in one period of $x(n)$. Sketch $x(n)$ on the same diagram with $x_a(t)$. What is the period of the discrete-time signal in milliseconds?

(d) Can you find a sampling rate F_s such that the signal $x(n)$ reaches its peak value of 3? What is the minimum F_s suitable for this task?

1.6 A continuous-time sinusoid $x_a(t)$ with fundamental period $T_p = 1/F_0$ is sampled at a rate $F_s = 1/T$ to produce a discrete-time sinusoid $x(n) = x_a(nT)$.

(a) Show that $x(n)$ is periodic if $T/T_p = k/N$ (i.e., T/T_p is a rational number).

(b) If $x(n)$ is periodic, what is its fundamental period T_p in seconds?

(c) Explain the statement: $x(n)$ is periodic if its fundamental period T_p, in seconds, is equal to an integer number of periods of $x_a(t)$.

1.7 An analog signal contains frequencies up to 10 kHz.

(a) What range of sampling frequencies will allow exact reconstruction of this signal from its samples?

(b) Suppose that we sample this signal with a sampling frequency $F_s = 8$ kHz. Examine what will happen to the frequency $F_1 = 5$ kHz.

(c) Repeat part (b) for a frequency $F_2 = 9$ kHz.

1.8 An analog electrocardiogram (ECG) signal contains useful frequencies up to 100 Hz.

(a) What is the Nyquist rate for this signal?

(b) Suppose that we sample this signal at a rate of 250 samples/s. What is the highest frequency that can be represented uniquely at this sampling rate?

1.9 An analog signal $x_a(t) = \sin (480\pi t) + 3 \sin (720\pi t)$ is sampled 600 times per second.

(a) Determine the Nyquist sampling rate for $x_a(t)$.

(b) Determine the folding frequency.

(c) What are the frequencies, in radians, in the resulting discrete signal $x(n)$?

(d) If $x(n)$ is passed through an ideal D/A converter, what is the reconstructed signal $y_a(t)$?

1.10 A digital communication link carries binary-coded words representing samples of an input signal

$$x_a(t) = 3 \cos 600\pi t + 2 \cos 1800\pi t$$

The link is operated at 10,000 bits/s and each input sample is quantized into 1024 different voltage levels.
(a) What is the sampling frequency and the folding frequency?
(b) What is the Nyquist rate for the signal $x_a(t)$?
(c) What are the frequencies in the resulting discrete-time signal $x(n)$?
(d) What is the resolution Δ?

1.11 Consider the simple signal processing system shown in Fig. P1.11. The sampling periods of the A/D and D/A converters are $T = 5$ ms and $T' = 1$ ms, respectively. Determine the output $y_a(t)$ of the system, if the input is

$$x_a(t) = 3 \cos 100\pi t + 2 \sin 250\pi t \qquad (t \text{ in seconds})$$

The postfilter removes any frequency component above $F_s/2$.

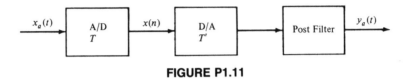

FIGURE P1.11

1.12 **(a)** Derive the expression for the discrete-time signal $x(n)$ in Example 1.4.2 using the periodicity properties of sinusoidal functions.
(b) What is the analog signal we can obtain from $x(n)$ if in the reconstruction process we assume that $F_s = 10$ kHz?

1.13 The discrete-time signal $x(n) = 6.35 \cos (\pi/10)n$ is quantized with a resolution (a) $\Delta = 0.1$ or (b) $\Delta = 0.02$. How many bits are required in the A/D converter in each case?

1.14 How many bits are required for the storage of a seismic signal if the sampling rate is $F_s = 20$ samples/s and we use an 8-bit A/D converter? What is the maximum frequency that can be present in the resulting digital seismic signal?

COMPUTER EXPERIMENTS

1.15 *Sampling of sinusoidal signals: aliasing*   Consider the following continuous-time sinusoidal signal

$$x_a(t) = \sin 2\pi F_0 t \qquad -\infty < t < \infty$$

Since $x_a(t)$ is described mathematically, its sampled version can be described by values every T seconds using the built-in mathematical functions. The sampled signal is described by the formula

$$x(n) = x_a(nT) = \sin 2\pi \frac{F_0}{F_s} n \qquad -\infty < n < \infty$$

where $F_s = 1/T$ is the sampling frequency.
(a) Plot the signal $x(n)$, $0 \le n \le 99$ for $F_s = 5$ kHz and $F_0 = 0.5, 2, 3,$ and 4.5 kHz. Explain the similarities and differences between the various plots.

(b) Suppose that $F_0 = 2$ kHz and $F_s = 50$ kHz.
   (1) Plot the signal $x(n)$. What is the frequency f_0 of the signal $x(n)$?
   (2) Plot the signal $y(n)$ created by taking the even-numbered samples of $x(n)$. Is this a sinusoidal signal? Why? If yes, what is its frequency?

1.16 *Quantization error in A/D conversion of a sinuoidal signal* Let $x_q(n)$ be the signal obtained by quantizing the signal $x(n) = \sin 2\pi f_0 n$. The quantization error power P_q is defined by

$$P_q = \frac{1}{N} \sum_{n=0}^{N-1} e^2(n) = \frac{1}{N} \sum_{n=0}^{N-1} [x_q(n) - x(n)]^2$$

The "quality" of the quantized signal can be measured by the signal-to-quantization noise ratio (SQNR) defined by

$$\text{SQNR} = 10 \log_{10} \frac{P_x}{P_q}$$

where P_x is the power of the unquantized signal $x(n)$.
   (a) For $f_0 = 1/50$ and $N = 200$, write a program to quantize the signal $x(n)$, using truncation, to 64, 128, and 256 quantization levels. In each case plot the signals $x(n)$, $x_q(n)$, and $e(n)$ and compute the corresponding SQNR.
   (b) Repeat part (a) by using rounding instead of truncation.
   (c) Comment on the results obtained in parts (a) and (b).

CHAPTER 2

Discrete-Time Signals and Systems

In Chapter 1 we introduced the reader to a number of important types of signals and described the sampling process by which an analog signal is converted to a discrete-time signal. In addition, we presented in some detail the characteristics of discrete-time sinusoidal signals.

The sinusoid is an important elementary signal that serves as a basic building block in more complex signals. However, there are other elementary signals that are important in our treatment of signal processing. These discrete-time signals are introduced in this chapter and are used as basis functions or building blocks to describe more complex signals.

The major emphasis in this chapter is the characterization of discrete-time systems in general and the class of linear time-invariant (LTI) systems in particular. A number of important time-domain properties of LTI systems are defined and developed, and an important formula is derived, called the convolution formula, which allows us to determine the output of an LTI system to any given arbitrary input signal. In addition to the convolution formula, difference equations are introduced as an alternative method for describing the input–output relationship of an LTI system, and recursive and nonrecursive realizations of LTI systems are treated.

Our motivation for the emphasis on the study of LTI systems is twofold. First, there is a large collection of mathematical techniques that can be applied to the analysis of LTI systems. Second, many practical systems are either LTI systems or can be approximated by LTI systems.

Because of its importance in digital signal processing applications and its close resemblance to the convolution formula, we also introduce the correlation between two signals. The autocorrelation and crosscorrelation of signals are defined and their properties are presented. The computation of these correlation sequences is described and applied to detection and estimation of signals corrupted by noise.

2.1 Discrete-Time Signals

As we discussed in Chapter 1, a discrete-time signal $x(n)$ is a function of an independent variable that is an integer. It is graphically represented as in Fig. 2.1. It is important to note that a discrete-time signal is *not defined* at instants between two successive samples. Also, it is incorrect to think that $x(n)$ is equal to zero if n is not an integer. Simply, the signal $x(n)$ is not defined for noninteger values of n.

In the sequel we will assume that a discrete-time signal is defined for every integer value n for $-\infty < n < \infty$. By tradition, we will refer to $x(n)$ as the ''nth sample'' of the signal even if the signal $x(n)$ may be inherently discrete time (i.e., it was not obtained by sampling an analog signal). If, indeed, $x(n)$ was obtained from sampling an analog signal $x_a(t)$, then $x(n) \equiv x_a(nT)$, where T is the sampling period (i.e., the time between successive samples).

Besides the graphical representation of a discrete-time signal or sequence as illustrated in Fig. 2.1, there are some alternative representations that are often more convenient to use. These are:

1. Functional representation, such as

$$x(n) = \begin{cases} 1 & \text{for } n = 1, 3 \\ 4 & \text{for } n = 2 \\ 0 & \text{elsewhere} \end{cases} \tag{2.1.1}$$

2. Tabular representation, such as

n	$\cdots$	-2	-1	0	1	2	3	4	5	$\cdots$
$x(n)$	$\cdots$	0	0	0	1	4	1	0	0	$\cdots$

3. Sequence representation

An infinite-duration signal or sequence with the time origin ($n = 0$) indicated by the symbol $\uparrow$ is represented as

$$x(n) = \left\{ \ldots\ 0, 0, \underset{\uparrow}{1}, 4, 1, 0, 0, \ldots \right\} \tag{2.1.2}$$

A sequence $x(n)$, which is zero for $n < 0$, may be represented as

$$x(n) = \left\{ \underset{\uparrow}{0}, 1, 4, 1, 0, 0, \ldots \right\} \tag{2.1.3}$$

The time origin for a sequence $x(n)$, which is zero for $n < 0$, is understood to be the first (leftmost) point in the sequence.

A finite duration sequence may be represented as

$$x(n) = \left\{ 3, -1, -2, \underset{\uparrow}{5}, 0, 4, -1 \right\} \tag{2.1.4}$$

whereas a finite-duration sequence that satisfies the condition $x(n) = 0$ for $n < 0$ may be represented as

$$x(n) = \left\{ \underset{\uparrow}{0}, 1, 4, 1 \right\} \tag{2.1.5}$$

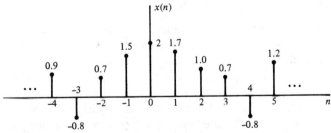

FIGURE 2.1 Graphical representation of a discrete-time signal.

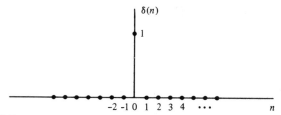

FIGURE 2.2 Graphical representation of the unit sample signal.

The signal in (2.1.4) consists of seven samples or points (in time), so it is called or identified as a seven-point sequence. Similarly, the sequence given by (2.1.5) is a four-point sequence.

2.1.1 Some Elementary Discrete-Time Signals

In our study of discrete-time signals and systems there are a number of basic signals that appear often and play an important role. These signals are defined below.

1. The *unit sample sequence* is denoted as $\delta(n)$ and is defined as

$$\delta(n) \equiv \begin{cases} 1 & \text{for } n = 0 \\ 0 & \text{for } n \neq 0 \end{cases} \tag{2.1.6}$$

In words, the unit sample sequence is a signal that is zero everywhere, except at $n = 0$ where its value is unity. This signal is sometimes referred to as a *unit impulse*. In contrast to the analog signal $\delta(t)$, which is also called a unit impulse and is defined to be infinite at $t = 0$, zero everywhere else, and has unit area, the unit sample sequence is much less mathematically complicated. The graphical representation of $\delta(n)$ is shown in Fig. 2.2.

2. The *unit step signal* is denoted as $u(n)$ and is defined as

$$u(n) \equiv \begin{cases} 1 & \text{for } n \geq 0 \\ 0 & \text{for } n < 0 \end{cases} \tag{2.1.7}$$

Figure 2.3 illustrates the unit step signal.

3. The *unit ramp signal* is denoted as $u_r(n)$ and is defined as

$$u_r(n) \equiv \begin{cases} n & \text{for } n \geq 0 \\ 0 & \text{for } n < 0 \end{cases} \tag{2.1.8}$$

This signal is illustrated in Fig. 2.4.

4. The *exponential signal* is a sequence of the form

$$x(n) = a^n \qquad \text{for all } n \tag{2.1.9}$$

If the parameter a is real, then $x(n)$ is a real signal. Figure 2.5 illustrates $x(n)$ for various values of the parameter a.

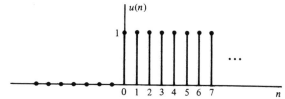

FIGURE 2.3 Graphical representation of the unit step signal.

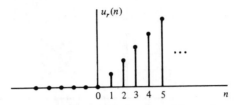

FIGURE 2.4 Graphical representation of the unit ramp signal.

When the parameter a is complex valued, it can be expressed as

$$a \equiv re^{j\theta}$$

where r and θ are now the parameters. Hence we may express $x(n)$ as

$$x(n) = r^n e^{j\theta n} \qquad (2.1.10)$$
$$= r^n(\cos \theta n + j \sin \theta n)$$

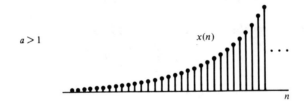

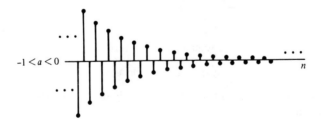

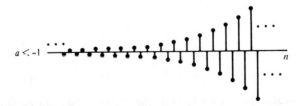

FIGURE 2.5 Graphical representation of exponential signals.

Since $x(n)$ is now complex valued, it can be represented graphically by plotting the real part

$$x_R(n) \equiv r^n \cos \theta n \qquad (2.1.11)$$

as a function of n, and separately plotting the imaginary part

$$x_I(n) \equiv r^n \sin \theta n \qquad (2.1.12)$$

as a function of n. Figure 2.6 illustrates the graphs of $x_R(n)$ and $x_I(n)$ for $r = 0.9$ and

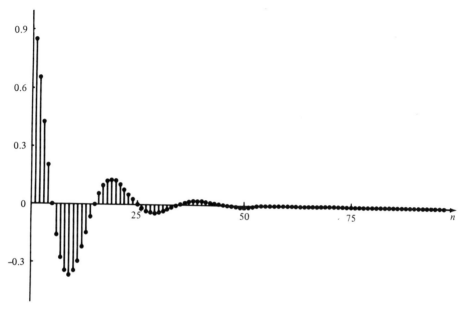

(a) Graph of $x_R(n) = (0.9)^n \cos \dfrac{\pi n}{10}$

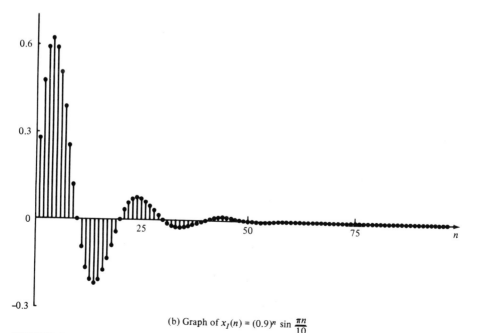

(b) Graph of $x_I(n) = (0.9)^n \sin \dfrac{\pi n}{10}$

FIGURE 2.6 Graph of the real and imaginary components of a complex-valued exponential signal.

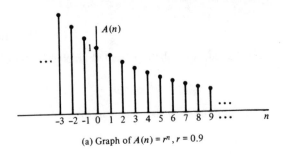

(a) Graph of $A(n) = r^n$, $r = 0.9$

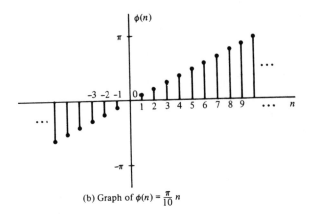

(b) Graph of $\phi(n) = \frac{\pi}{10} n$

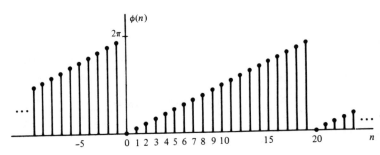

(c) Graph of $\phi(n) = \frac{\pi}{10} n$, modulo 2π plotted in the range $(0, 2\pi)$.

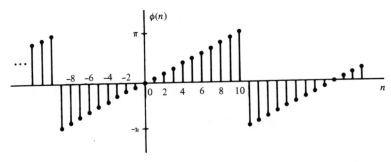

(d) Graph of $\phi(n) = \frac{\pi}{10} n$, modulo 2π plotted in the range $(-\pi, \pi)$

FIGURE 2.7 Graph of amplitude and phase function of a complex-valued exponential signal: (a) graph of $A(n) = r^n$, $r = 0.9$; (b) graph of $\phi(n) = (\pi/10)n$; (c) graph of $\phi(n) = (\pi/10)n$, modulo 2π plotted in the range $[0, 2\pi)$; (d) graph of $\phi(n) = (\pi/10)n$, modulo 2π plotted in the range $(-\pi, \pi]$.

$\theta = \pi/10$. We observe that the signals $x_R(n)$ and $x_I(n)$ are a damped (decaying exponential) cosine function and a damped sine function. The angle variable θ is simply the frequency of the sinusoid, previously denoted by the (normalized) frequency variable ω. Clearly, if $r = 1$, the damping disappears and $x_R(n)$, $x_I(n)$, and $x(n)$ have a fixed amplitude, which is unity.

Alternatively, the signal $x(n)$ given by (2.1.10) may be graphically represented by the amplitude function

$$|x(n)| = A(n) \equiv r^n \qquad (2.1.13)$$

and the phase function

$$\angle x(n) = \phi(n) \equiv \theta n \qquad (2.1.14)$$

Figure 2.7 illustrates $A(n)$ and $\phi(n)$ for $r = 0.9$ and $\theta = \pi/10$. We observe that the phase function is linear with n. However, the phase is defined only over the interval $-\pi < \theta \leq \pi$ or, equivalently, over the interval $0 \leq \theta < 2\pi$. Consequently, it is conventional to plot $\phi(n)$ over the finite interval $-\pi < \theta \leq \pi$ or $0 \leq \theta < 2\pi$. In other words, we subtract multiplies of 2π from $\phi(n)$ before plotting. In one case, $\phi(n)$ is constrained to the range $-\pi < \theta \leq \pi$ and in the other case $\phi(n)$ is constrained to the range $0 \leq \theta < 2\pi$. The subtraction of multiples of 2π from $\phi(n)$ is equivalent to interpreting the function $\phi(n)$ as $\phi(n)$, modulo 2π. The two possible graphs for $\phi(n)$, modulo 2π, are shown in Fig. 2.7c and d. Note that the two graphs are equivalent, since one can be obtained from the other by a shift of the axis.

2.1.2 Classification of Discrete-Time Signals

The mathematical methods that are employed in the analysis of discrete-time signals and systems depend on the characteristics of the signals. In this section we classify discrete-time signals according to a number of different characteristics.

Energy Signals and Power Signals. The energy E of a signal $x(n)$ is defined as

$$E \equiv \sum_{n=-\infty}^{\infty} |x(n)|^2 \qquad (2.1.15)$$

We have used the magnitude-squared values of $x(n)$, so that our definition applies to complex-valued signals as well as real-valued signals. The energy of a signal may be finite or infinite. If E is finite (i.e., $0 < E < \infty$), then $x(n)$ is called an *energy signal*. Sometimes we add a subscript x to E and write E_x to emphasize that E_x is the energy of the signal $x(n)$.

EXAMPLE 2.1.1. _____

Determine the energy of the sequence

$$x(n) = \begin{cases} (\tfrac{1}{2})^n & n \geq 0 \\ 3^n & n < 0 \end{cases}$$

Solution: From the definition in (2.1.15) we have

$$E = \sum_{n=-\infty}^{\infty} |x(n)|^2$$

$$= \sum_{n=0}^{\infty} (\tfrac{1}{2})^{2n} + \sum_{n=-\infty}^{-1} 3^{2n}$$

$$= \frac{1}{1 - \tfrac{1}{4}} + \sum_{n=1}^{\infty} (\tfrac{1}{3})^{2n}$$

$$= \tfrac{4}{3} + \tfrac{9}{8} - 1 = \tfrac{35}{24}$$

Since E is finite, this is an energy signal.

EXAMPLE 2.1.2 _____

Determine the energy of the unit step sequence $u(n)$.

Solution: Beginning with the definition given in (2.1.15), we have

$$E = \sum_{n=-\infty}^{\infty} |x(n)|^2 = \sum_{n=-\infty}^{\infty} u^2(n) = \sum_{n=0}^{\infty} 1$$

which is infinite. Consequently, the unit step sequence is not an energy signal.

EXAMPLE 2.1.3 _____

Determine the energy of the constant-amplitude complex-valued exponential sequence

$$x(n) = Ae^{j\omega_0 n}$$

where A and ω_0 are constants.

Solution: From the definition in (2.1.15), we have

$$E = \sum_{n=-\infty}^{\infty} |Ae^{j\omega_0 n}|^2 = \sum_{n=-\infty}^{\infty} A^2$$

which is also infinite. Therefore, the complex-valued constant-amplitude exponential signal is not a finite-energy signal.

Many signals that possess infinite energy, however, have a finite average power. The average power of a discrete-time signal $x(n)$ is defined as

$$P = \lim_{N\to\infty} \frac{1}{2N+1} \sum_{n=-N}^{N} |x(n)|^2 \qquad (2.1.16)$$

If we define the signal energy of $x(n)$ over the finite interval $-N \le n \le N$ as

$$E_N \equiv \sum_{n=-N}^{N} |x(n)|^2 \qquad (2.1.17)$$

then we may express the signal energy E as

$$E \equiv \lim_{N\to\infty} E_N \qquad (2.1.18)$$

and the average power of the signal $x(n)$ as

$$P \equiv \lim_{N \to \infty} \frac{1}{2N + 1} E_N \qquad (2.1.19)$$

Clearly, if E is finite, $P = 0$. On the other hand, if E is infinite, the average power P may be either finite or infinite. If P is finite (and nonzero), the signal is called a *power signal*. The following examples illustrate these two cases.

EXAMPLE 2.1.4 _____

In Example 2.1.2 we showed that the unit step sequence has infinite energy. The average power of the signal is

$$P = \lim_{N \to \infty} \frac{1}{2N + 1} \sum_{n=0}^{N} u^2(n)$$

$$= \lim_{N \to \infty} \frac{N + 1}{2N + 1} = \lim_{N \to \infty} \frac{1 + 1/N}{2 + 1/N} = \frac{1}{2}$$

Consequently, the unit step sequence is a power signal.

_____

EXAMPLE 2.1.5 _____

In Example 2.1.3 we found that the constant-amplitude complex-valued exponential sequence

$$x(n) = Ae^{j\omega_0 n}$$

has infinite energy. The average power of this signal is

$$P = \lim_{N \to \infty} \frac{1}{2N + 1} \sum_{n=-N}^{N} |Ae^{j\omega_0 n}|^2$$

$$= \lim_{N \to \infty} \frac{1}{2N + 1} \sum_{n=-N}^{N} A^2$$

$$= \lim_{N \to \infty} \frac{1}{2N + 1} (2N + 1)A^2 = A^2$$

Therefore, the exponential sequence is also a power signal.

_____

EXAMPLE 2.1.6 _____

Determine the average power and the energy of the unit ramp sequence

$$x(n) = \begin{cases} n & n \geq 0 \\ 0 & n < 0 \end{cases}$$

Solution: Over the interval $-N \leq n \leq N$, this signal has energy

$$E_N = \sum_{n=0}^{N} n^2$$

$$= \frac{N(N + 1)(2N + 1)}{6}$$

Clearly, $E = \infty$. The average power is

$$P = \lim_{N \to \infty} \frac{1}{2N + 1} E_N$$

$$= \lim_{N \to \infty} \frac{N(N + 1)}{6}$$

which is also infinite. Consequently, the unit ramp has infinite energy and infinite average power. Hence it is neither an energy signal nor a power signal.

Table 2.1 summarizes the results of the previous six examples.

TABLE 2.1 Examples of Energy and Power Signals

Signal	E	P	Type
$\delta(n)$	1	0	Energy
$u(n)$	∞	$\frac{1}{2}$	Power
$u_r(n)$	∞	∞	Undefined
$Ae^{j\omega_0 n}$	∞	A^2	Power

Periodic Signals and Aperiodic Signals. As defined on Section 1.3, a signal $x(n)$ is periodic with period N ($N > 0$) if and only if

$$x(n + N) = x(n) \qquad \text{for all } n \tag{2.1.20}$$

The smallest value of N for which (2.1.20) holds is called the (fundamental) period. If there is no value of N that satisfies (2.1.20), the signal is called *nonperiodic* or *aperiodic*.
    We have already observed that the sinusoidal signal of the form

$$x(n) = A \sin 2\pi f_0 n \tag{2.1.21}$$

is periodic when f_0 is a rational number, that is, if f_0 can be expressed as

$$f_0 = \frac{k}{N} \tag{2.1.22}$$

where k and N are integers.
    The energy of a periodic signal $x(n)$ over a single period, say, over the interval $0 \leq n \leq N - 1$, is finite if $x(n)$ takes on finite values over the period. However, the energy of the periodic signal for $-\infty \leq n \leq \infty$ is infinite. On the other hand, the average power of the periodic signal is finite and it is equal to the average power over a single period. Thus if $x(n)$ is a periodic signal with fundamental period N and takes on finite values, its power is given by

$$P = \frac{1}{N} \sum_{n=0}^{N-1} |x(n)|^2 \tag{2.1.23}$$

Consequently, periodic signals are power signals.

Symmetric (Even) and Antisymmetric (Odd) Signals. A real-valued signal $x(n)$ is called symmetric (even) if

$$x(-n) = x(n) \tag{2.1.24}$$

On the other hand, a signal $x(n)$ is called antisymmetric (odd) if

$$x(-n) = -x(n) \tag{2.1.25}$$

We note that if $x(n)$ is odd, then $x(0) = 0$. Examples of signals with even and odd symmetry are illustrated in Fig. 2.8.

We wish to illustrate that any arbitrary signal can be expressed as the sum of two signal components, one of which is even and the other odd. The even signal component is formed by adding $x(n)$ to $x(-n)$ and dividing by 2, that is,

$$x_e(n) = \tfrac{1}{2}[x(n) + x(-n)] \tag{2.1.26}$$

Clearly, $x_e(n)$ satisfies the symmetry condition (2.1.24). Similarly, we form an odd signal component $x_o(n)$ according to the relation

$$x_o(n) = \tfrac{1}{2}[x(n) - x(-n)] \tag{2.1.27}$$

Again, it is clear that $x_o(n)$ satisfies (2.1.25); hence it is indeed odd. Now, if we add the two signal components, defined by (2.1.26) and (2.1.27), we obtain $x(n)$, that is,

$$x(n) = x_e(n) + x_o(n) \tag{2.1.28}$$

Thus any arbitrary signal can be expressed as in (2.1.28).

2.1.3 Simple Manipulations of Discrete-Time Signals

In this section we consider some simple modifications or manipulations involving the independent variable and the signal amplitude (dependent variable).

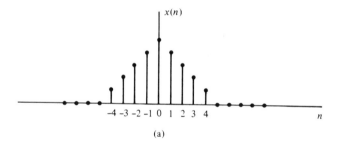

(a)

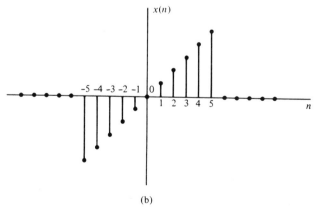

(b)

FIGURE 2.8   Example of even (a) and odd (b) signals.

Transformation of the Independent Variable (Time). A signal $x(n)$ may be shifted in time by replacing the independent variable n by $n - k$, where k is an integer. If k is a positive integer, the time shift results in a delay of the signal by k units of time. If k is a negative integer, the time shift results in an advance of the signal by $|k|$ units in time.

EXAMPLE 2.1.7 _____

A signal $x(n)$ is graphically illustrated in Fig. 2.9a. Show a graphical representation of the signals $x(n - 3)$ and $x(n + 2)$.

Solution: The signal $x(n - 3)$ is obtained by delaying $x(n)$ by three units in time. The result is illustrated in Fig. 2.9b. On the other hand, the signal $x(n + 2)$ is obtained by advancing $x(n)$ by two units in time. The result is illustrated in Fig. 2.9c. Note that delay corresponds to shifting a signal to the right, whereas advance implies shifting the signal to the left on the time axis.

_____

If the signal $x(n)$ is stored on magnetic tape or on a disk or, perhaps, in the memory of a computer, it is a relatively simple operation to modify the base by introducing a delay or an advance. On the other hand, if the signal is not stored but is being generated by some physical phenomenon in real time, it is not possible to advance the signal in time, since such an operation involves signal samples that have not yet been generated.

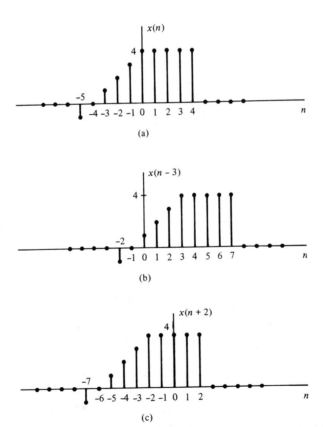

FIGURE 2.9 Graphical representation of a signal, and its delayed and advanced versions.

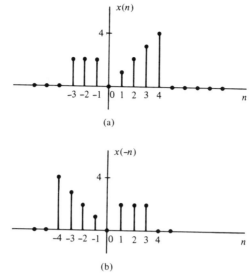

(a)

(b)

FIGURE 2.10 Graphical illustration of the folding operation.

Whereas it is always possible to insert a delay into signal samples that have already been generated, it is physically impossible to view the future signal samples. Consequently, in real-time signal processing applications, the operation of advancing the time base of the signal is physically unrealizable.

Another useful modification of the time base is to replace the independent variable n by $-n$. The result of this operation is a *folding* or a *reflection* of the signal about the time origin $n = 0$.

EXAMPLE 2.1.8 _____

Show the graphical representation of the signal $x(-n)$, where $x(n)$ is the signal illustrated in Fig. 2.10a.

Solution: The new signal $y(n) = x(-n)$ is shown in Fig. 2.10b. Note that $y(0) = x(0)$, $y(1) = x(-1)$, $y(2) = x(-2)$, and so on. Also, $y(-1) = x(1)$, $y(-2) = x(2)$, and so on. Therefore, $y(n)$ is simply $x(n)$ reflected or folded about the time origin $n = 0$.

EXAMPLE 2.1.9 _____

Show the graphical representation of the signal $y(n) = x(-n + 2)$, where $x(n)$ is the signal illustrated in Fig. 2.10a.

Solution: Figure 2.11a illustrates the folded signal $x(-n)$. The signal $y(n) = x(-n + 2)$ is simply $x(-n)$ delayed by two units in time. The resulting signal is illustrated in Fig. 2.11b. A simple way to verify that the result in Fig. 2.11b is correct is to compute samples, such as $y(0) = x(2)$, $y(1) = x(1)$, $y(2) = x(0)$, $y(-1) = x(3)$, and so on.

It is important to note that the operations of folding and time delaying (or advancing) a signal are not commutative. If we denote the time-delay operation by TD and the

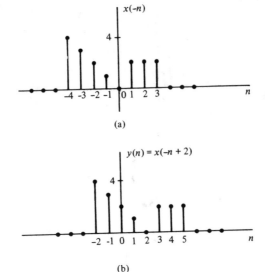

(a)

(b)

FIGURE 2.11 Graphical illustration of a combined folding and delaying operation.

folding operation by FD, we can write

$$TD_k[x(n)] = x(n - k) \qquad k > 0 \qquad (2.1.29)$$
$$FD[x(n)] = x(-n)$$

Now

$$TD_k\{FD[x(n)]\} = TD_k[x(-n)] \qquad = x(-n + k) \qquad (2.1.30)$$

whereas

$$FD\{TD_k[x(n)]\} = FD[x(n - k)] = x(-n - k) \qquad (2.1.31)$$

Note that because the signs of n and k in $x(n - k)$ and $x(-n + k)$ are different, the result is a shift of the signals $x(n)$ and $x(-n)$ to the right by k samples, corresponding to a time delay.

A third modification of the independent variable involves replacing n by μn, where μ is an integer. We refer to this time-base modification as *time scaling* or *down sampling*.

EXAMPLE 2.1.10

Show the graphical representation of the signal $y(n) = x(2n)$, where $x(n)$ is the signal illustrated in Fig. 2.12a.

Solution: We note that the signal $y(n)$ is obtained from $x(n)$ by taking every other sample from $x(n)$, starting with $x(0)$. Thus $y(0) = x(0)$, $y(1) = x(2)$, $y(2) = x(4)$, . . . and $y(-1) = x(-2)$, $y(-2) = x(-4)$, and so on. In other words, we have skipped the odd-numbered samples in $x(n)$ and retained the even-numbered samples. The resulting signal is illustrated in Fig. 2.12b.

If the signal $x(n)$ was originally obtained by sampling an analog signal $x_a(t)$, then $x(n) = x_a(nT)$, where T is the sampling interval. Now, $y(n) = x(2n) = x_a(2Tn)$. Hence the time-scaling operation described in Example 2.1.10 is equivalent to changing the

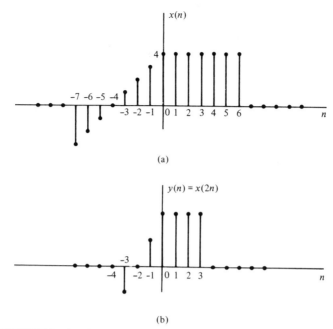

(a)

(b)

FIGURE 2.12 Graphical illustration of down-sampling operation.

sampling rate from $1/T$ to $1/2T$, that is, to decreasing the rate by a factor of 2. This is a *down-sampling* operation.

Addition, Multiplication, and Scaling of Sequences. Let us define the *addition, multiplication,* and *scaling* of discrete-time signals.

Amplitude scaling of a signal by a constant A is accomplished by multiplying the value of every signal sample by A. Consequently, we obtain

$$y(n) = Ax(n) \qquad -\infty < n < \infty$$

The *sum* of two signals $x_1(n)$ and $x_2(n)$ is a signal $y(n)$, whose value at any instant is equal to the sum of the values of these two signals at that instant, that is,

$$y(n) = x_1(n) + x_2(n) \qquad -\infty < n < \infty$$

The *product* of two signals is similarly defined on a sample-to-sample basis as

$$y(n) = x_1(n)x_2(n) \qquad -\infty < n < \infty$$

2.2 Discrete-Time Systems

In many applications of digital signal processing we wish to design a device or an algorithm that performs some prescribed operation on a discrete-time signal. Such a device or algorithm is called a discrete-time system. More specifically, a *discrete-time system* is a device or algorithm that operates on a discrete-time signal, called the *input* or *excitation*, according to some well-defined rule, to produce another discrete-time signal called the *output* or *response* of the system. In general, we view a system as an operation or a set of operations performed on an input signal $x(n)$ to produce the output signal $y(n)$. We say that the input signal $x(n)$ is *transformed* by the system into a signal $y(n)$, and

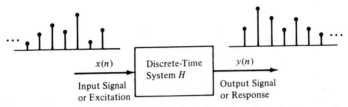

FIGURE 2.13 Block diagram representation of a discrete-time system.

express the general relationship between $x(n)$ and $y(n)$ as

$$y(n) \equiv H[x(n)] \tag{2.2.1}$$

where the symbol H denotes the transformation (also called an operator), or processing performed by the system on $x(n)$ to produce $y(n)$. The mathematical relationship in (2.2.1) is depicted graphically in Fig. 2.13.

There are various ways to describe the characteristics of the system and the operation it performs on $x(n)$ to produce $y(n)$. In this chapter we shall be concerned with the time-domain characterization of systems. We shall begin with an input–output description of the system. The input–output description focuses on the terminal behavior of the system and ignores the detailed internal construction or realization of the system. Later, in Section 7.5, we introduce the state-space description of a system. In this description we develop mathematical equations that not only describe the input–output behavior of the system but specify its internal behavior and structure.

2.2.1 Input–Output Description of Systems

The input–output description of a discrete-time system consists of a mathematical expression or a rule, which explicitly defines the relation between the input and output signals (*input–output relationship*). The exact internal structure of the system is either unknown or it is ignored. Thus the only way to interact with the system is by using its input and output terminals (i.e., the system is assumed to be a ''black box'' to the user). To reflect this philosophy we use the graphical representation depicted in Fig. 2.13, and the general input–output relationship in (2.2.1) or, alternatively, the notation

$$x(n) \xrightarrow{\quad H \quad} y(n) \tag{2.2.2}$$

which simply means that $y(n)$ is the response of the system H to the excitation $x(n)$. The following examples illustrate several different systems.

EXAMPLE 2.2.1 ———————————————————————————————

Determine the response of the following sytems to the input signal

$$x(n) = \begin{cases} |n| & -3 \le n \le 3 \\ 0 & \text{otherwise} \end{cases}$$

(a) $y(n) = x(n)$
(b) $y(n) = x(n - 1)$
(c) $y(n) = x(n + 1)$
(d) $y(n) = \frac{1}{3}[x(n + 1) + x(n) + x(n - 1)]$
(e) $y(n) = \max\{x(n + 1), x(n), x(n - 1)\}$
(f) $y(n) = \sum_{k=-\infty}^{n} x(k) = x(n) + x(n - 1) + x(n - 2) + \cdots$ $\qquad$ (2.2.3)

Solution: First, we determine explicitly the sample values of the input signal

$$x(n) = \left\{ \ldots, 0, 3, 2, 1, 0, 1, 2, 3, 0, \ldots \right\}$$
$$\uparrow$$

Next, we determine the output of each system using its input–output relationship.

(a) In this case the output is exactly the same as the input signal. Such a system is known as the *identity* system.

(b) This system simply delays the input by one sample. Thus its output is given by

$$x(n) = \left\{ \ldots, 0, 3, 2, 1, 0, 1, 2, 3, 0, \ldots \right\}$$
$$\uparrow$$

(c) In this case the system "advances" the input one sample into the future. For example, the value of the output at time $n = 0$ is $y(0) = x(1)$. The response of this system to the given input is

$$x(n) = \left\{ \ldots, 0, 3, 2, 1, 0, 1, 2, 3, 0, \ldots \right\}$$
$$\uparrow$$

(d) The output of this system at any time is the mean value of the present, the immediate past, and the immediate future samples. For example, the output at time $n = 0$ is

$$y(0) = \tfrac{1}{3}[x(-1) + x(0) + x(1)] = \tfrac{1}{3}[1 + 0 + 1] = \tfrac{2}{3}$$

Repeating this computation for every value of n, we obtain the output signal

$$y(n) = \left\{ \ldots, 0, 1, \tfrac{5}{3}, 2, 1, \tfrac{2}{3}, 1, 2, \tfrac{5}{3}, 1, 0, \ldots \right\}$$
$$\uparrow$$

(e) This system selects as its output at time n the maximum value of the three input samples $x(n - 1)$, $x(n)$, and $x(n + 1)$. Thus the response of this system to the input signal $x(n)$ is

$$y(n) = \left\{ 0, 3, 3, 3, 2, 1, 2, 3, 3, 3, 0, \ldots \right\}$$
$$\uparrow$$

(f) This system is basically an *accumulator* that computes the running sum of all the past input values up to present time. The response of this system to the given input is

$$y(n) = \left\{ \ldots, 0, 3, 5, 6, 6, 7, 9, 12, 0, \ldots \right\}$$
$$\uparrow$$

We observe that for several of the systems considered in Example 2.2.1 the output at time $n = n_0$ depends not only on the value of the input at $n = n_0$ [i.e., $x(n_0)$], but also on the values of the input applied to the system before and after $n = n_0$. Consider, for instance, the accumulator in the example. We see that the output at time $n = n_0$ depends not only on the input at time $n = n_0$, but also on $x(n)$ at times $n = n_0 - 1, n_0 - 2$, and so on. By a simple algebraic manipulation the input–output relation of the accumulator may be written as

$$y(n) = \sum_{k=-\infty}^{n} x(k) = \sum_{k=-\infty}^{n-1} x(k) + x(n)$$

$$= y(n-1) + x(n)$$

(2.2.4)

which justifies the term *accumulator*. Indeed, the system computes the current value of the output by adding (accumulating) the current value of the input to the previous output value.

There are some interesting conclusions that can be drawn by taking a close look into this apparently simple system. Suppose that we are given the input signal $x(n)$ for $n \geq n_0$, and we wish to determine the output $y(n)$ of this system for $n \geq n_0$. For $n = n_0, n_0 + 1, \ldots,$ (2.2.4) gives

$$y(n_0) = y(n_0 - 1) + x(n_0)$$
$$y(n_0 + 1) = y(n_0) + x(n_0 + 1)$$

and so on. Note that we have a problem in computing $y(n_0)$, since it depends on $y(n_0 - 1)$. However,

$$y(n_0 - 1) = \sum_{k=-\infty}^{n_0-1} x(k)$$

that is, $y(n_0 - 1)$ "summarizes" the effect on the system from all the inputs which had been applied to the system before time n_0. Thus the response of the system for $n \geq n_0$ to the input $x(n)$ that is applied at time n_0 is the combined result of this input and all inputs that had been applied previously to the system. Consequently, $y(n)$, $n \geq n_0$ is not uniquely determined by the input $x(n)$ for $n \geq n_0$.

The additional information required to determine $y(n)$ for $n \geq n_0$ is the *initial condition* $y(n_0 - 1)$. This value summarizes the effect of all previous inputs to the system. Thus the initial condition $y(n_0 - 1)$ together with the input sequence $x(n)$ for $n \geq n_0$ uniquely determine the output sequence $y(n)$ for $n \geq n_0$.

If the accumulator had no excitation prior to n_0, the initial condition is $y(n_0 - 1) = 0$. In such a case we say that the system is *initially relaxed*. Since $y(n_0 - 1) = 0$, the output sequence $y(n)$ depends only on the input sequence $x(n)$ for $n \geq n_0$.

It is customary to assume that every system is relaxed at $n = -\infty$. In this case, if an input $x(n)$ is applied at $n = -\infty$, the corresponding output $y(n)$ is *solely* and *uniquely* determined by the given input.

EXAMPLE 2.2.2 _____

The accumulator described by (2.2.3) is excited by the sequence $x(n) = nu(n)$. Determine its output under the condition that:

(a) It is initially relaxed [i.e., $y(-1) = 0$].
(b) Initially, $y(-1) = 1$.

Solution: The output of the system is defined as

$$y(n) = \sum_{k=-\infty}^{n} x(k)$$

$$= \sum_{k=-\infty}^{-1} x(k) + \sum_{k=0}^{n} x(k)$$

$$= y(-1) + \sum_{k=0}^{n} x(k)$$

But

$$\sum_{k=0}^{n} x(k) = \frac{n(n+1)}{2}$$

(a) If the system is initially relaxed, $y(-1) = 0$ and hence

$$y(n) = \frac{n(n+1)}{2} \qquad n \geq 0$$

(b) On the other hand, if the initial condition is $y(-1) = 1$, then

$$y(n) = 1 + \frac{n(n+1)}{2} = \frac{n^2 + n + 2}{2} \qquad n \geq 0$$

2.2.2 Block Diagram Representation of Discrete-Time Systems

It is useful at this point to introduce a block diagram representation of discrete-time systems. For this purpose we need to define some basic building blocks that can be interconnected to form complex systems.

An Adder. Figure 2.14 illustrates a system (adder) that performs the addition of two signal sequences to form another (the sum) sequence, which we denote as $y(n)$. Note that it is not necessary to store either one of the sequences in order to perform the addition. In other words, the addition operation is *memoryless*.

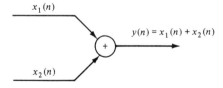

FIGURE 2.14 Graphical representation of an adder.

A Constant Multiplier. This operation is depicted by Fig. 2.15, and simply represents applying a scale factor on the input $x(n)$. Note that this operation is also memoryless.

$$\xrightarrow{\;x(n)\;} a \xrightarrow{\;y(n) = ax(n)\;}$$

FIGURE 2.15 Graphical representation of a constant multiplier.

A Signal Multiplier. Figure 2.16 illustrates the multiplication of two signal sequences to form another (the product) sequence, denoted in the figure as $y(n)$. As in the previous two cases, we may view the multiplication operation as memoryless.

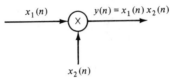

$$x_1(n) \qquad \qquad y(n) = x_1(n) x_2(n)$$

$$x_2(n)$$

FIGURE 2.16   Graphical representation of a signal multiplier.

A Unit Delay Element. The unit delay is a special system that simply delays the signal passing through it by one sample. Figure 2.17 illustrates such a system. If the input signal is $x(n)$, the output is $x(n - 1)$. In fact, the sample $x(n - 1)$ is stored in memory at time $n - 1$ and it is recalled from memory at time n to form

$$y(n) = x(n - 1)$$

Thus this basic building block requires memory. The use of the symbol z^{-1} to denote the unit of delay will become apparent when we discuss the z-transform in Chapter 3.

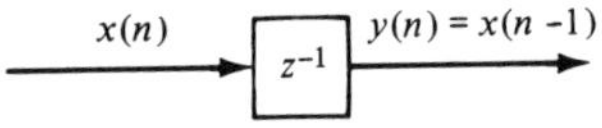

$$x(n) \qquad \boxed{z^{-1}} \qquad y(n) = x(n-1)$$

FIGURE 2.17   Graphical representation of the unit delay element.

A Unit Advance Element. In contrast to the unit delay, a unit advance moves the input $x(n)$ ahead by one sample in time to yield $x(n + 1)$. Figure 2.18 illustrates this operation, with the operator z being used to denote the unit advance. We observe that any such advance is physically impossible in real time, since, in fact, it involves looking into the future of the signal. On the other hand, if we store the signal in the memory of the computer, we can recall any sample at any time. In such a non-real-time application, it is possible to advance the signal $x(n)$ in time.

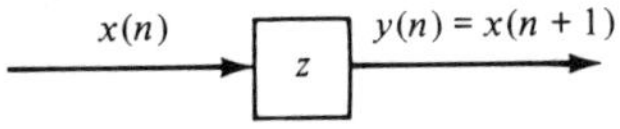

$$x(n) \qquad \boxed{z} \qquad y(n) = x(n + 1)$$

FIGURE 2.18   Graphical representation of the unit advance element.

EXAMPLE 2.2.3 ────────────────────────────────────

Using basic building blocks introduced above, sketch the block diagram representation of the discrete-time system described by the input–output relation.

$$y(n) = \tfrac{1}{2}x(n) + \tfrac{1}{2}x(n - 1) \qquad\qquad (2.2.5)$$

where $x(n)$ is the input and $y(n)$ is the output of the system.

Solution: According to (2.2.5), the output $y(n)$ is obtained by multiplying the input $x(n)$ by 0.5, multiplying the previous input $x(n - 1)$ by 0.5, and adding the two products. Figure 2.19a illustrates this block diagram realization of the system. A simple rearrangement of (2.2.5), namely,

$$y(n) = \tfrac{1}{2}[x(n) + x(n - 1)] \qquad\qquad (2.2.6)$$

Black Box

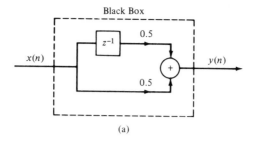

(a)

Black Box

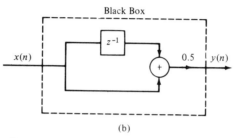

(b)

FIGURE 2.19 Block diagram realizations of the system $y(n) = 0.5x(n) + 0.5x(n - 1)$.

leads to the block diagram realization shown in Fig. 2.19b. Note that if we treat "the system" from the "viewpoint" of an input–output or an external description, we are not concerned about how the system is realized. On the other hand, if we adopt an internal description of the system, we know exactly how the system building blocks are configured. In terms of such a realization, we can see that a system is *relaxed* at time $n = n_0$ if the outputs of all the *delays* existing in the system are zero at $n = n_0$ (i.e., all memory is *filled* with zeros).

EXAMPLE 2.2.4

Sketch the block diagram representation of the discrete-time system described by the input–output equation

$$y(n) = \tfrac{1}{2}y(n - 1) + 5x(n) + 2x(n - 2) \qquad (2.2.7)$$

Solution: This system can be realized by using three multipliers and two adders. There are also three unit delays, two of which are associated with the input $x(n)$ and one delay associated with the output $y(n)$. Consequently, we must provide for three storage registers. The block diagram representation is illustrated in Fig. 2.20.

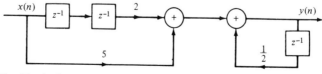

FIGURE 2.20 Block diagram representation of the system $y(n) = \tfrac{1}{2}y(n - 1) + 5x(n) + 2x(n - 2)$.

2.2.3 Classification of Discrete-Time Systems

In the analysis as well as in the design of systems, it is desirable to classify the systems according to the general properties that they satisfy. In fact, the mathematical techniques that we will develop in this and in subsequent chapters for analyzing and designing discrete-time systems depend heavily on the general characteristics of the systems that are being considered. For this reason it is necessary for us to develop a number of properties or categories that can be used to describe the general characteristics of systems.

Static Versus Dynamic Systems. A discrete-time system is called *static* or memoryless if its output at any instant n depends at most on the input sample at the same time, but not on past or future samples of the input. In any other case, the system is said to be *dynamic* or to have memory. If the output of a system at time n is completely determined by the input samples in the interval from $n - N$ to n $(N \geq 0)$, the system is said to have *memory* of duration N. If $N = 0$, the system is static. If $0 < N < \infty$, the system is said to have *finite memory*, whereas if $N = \infty$, the system is said to have *infinite memory*.

The systems described by the following input–output equations

$$y(n) = ax(n) \tag{2.2.8}$$
$$y(n) = nx(n) + bx^3(n) \tag{2.2.9}$$

are both static or memoryless. Note that there is no need to have stored any of the past inputs or outputs in order to compute the present output. On the other hand, the systems described by the following input–output relations

$$y(n) = x(n) + 3x(n - 1) \tag{2.2.10}$$

$$y(n) = \sum_{k=0}^{N} x(n - k) \tag{2.2.11}$$

$$y(n) = \sum_{k=0}^{\infty} x(n - k) \tag{2.2.12}$$

are dynamic systems or systems with memory. The systems described by (2.2.10) and (2.2.11) have finite memory, whereas the system described by (2.2.12) has infinite memory.

We observe that static or memoryless systems are described in general by input–output equations of the form

$$y(n) = F[x(n), n]$$

and they do not include delay elements (memory).

Time-Invariant Versus Time-Variant Systems. We may subdivide the general class of systems into the two broad categories, time-invariant systems and time-variant systems. A system is called time-invariant if its input–output characteristics do not change with time. To elaborate, suppose that we have a system H in a relaxed state which, when excited by an input signal $x(n)$, produces an output signal $y(n)$. Thus we write

$$y(n) = H[x(n)]$$

Now suppose that the same input signal is delayed by k units of time to yield $x(n - k)$, and again applied to the same system. If the characteristics of the system do not change with time, the output of the relaxed system will be $y(n - k)$. That is, the output will be the same as the response to $x(n)$, except that it will be delayed by the same k units in time that the input was delayed. This leads us to define a time-invariant or shift-invariant system as follows.

DEFINITION: *A relaxed system H is* time invariant *or* shift invariant *if and only if*

$$x(n) \xrightarrow{\ \ H\ \ } y(n)$$

implies that

$$x(n - k) \xrightarrow{\ \ H\ \ } y(n - k) \tag{2.2.13}$$

for every input signal x(n) and every time shift k.

To determine whether any given system is time invariant or not, we need to perform the test specified by the definition given above. Basically, we excite the system with an arbitrary input sequence $x(n)$, which will produce an output denoted as $y(n)$. Next we delay the input sequence by same amount k and recompute the output. In general, we may write the output as

$$y(n, k) = H[x(n - k)]$$

Now if this output $y(n, k) = y(n - k)$, for all possible values of k, the system is time invariant. On the other hand, if the output $y(n, k) \neq y(n - k)$, even for one value of k, the system is time variant.

EXAMPLE 2.2.5 _____

Determine if the systems shown in Fig. 2.21 are time invariant or time variant.

Solution: **(a)** This system is described by the input–output equations

$$y(n) = H[x(n)] = x(n) - x(n - 1) \tag{2.2.14}$$

Now if the input is delayed by k units in time and applied to the system, it is clear from the block diagram that the output will be

$$y(n, k) = x(n - k) - x(n - k - 1) \tag{2.2.15}$$

On the other hand, from (2.2.14) we note that if we delay $y(n)$ by k units in time, we obtain

$$y(n - k) = x(n - k) - x(n - k - 1) \tag{2.2.16}$$

Since the right-hand sides of (2.2.15) and (2.2.16) are identical, it follows that $y(n, k) = y(n - k)$. Therefore, the system is time invariant.
(b) The input–output equation for this system is

$$y(n) = H[x(n)] = nx(n) \tag{2.2.17}$$

The response of this system to $x(n - k)$ is

$$y(n, k) = nx(n - k) \tag{2.2.18}$$

Now if we delay $y(n)$ in (2.2.17) by k units in time, we obtain

$$y(n - k) = (n - k)x(n - k) \tag{2.2.19}$$
$$= nx(n - k) - kx(n - k)$$

This system is time variant, since $y(n, k) \neq y(n - k)$.

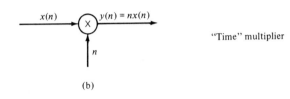

(a)

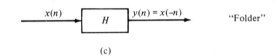

(b)

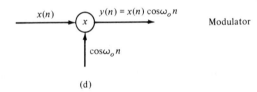

(c)

$x(n)$ $\times$ $y(n) = x(n) \cos\omega_0 n$ Modulator

$\cos\omega_0 n$

(d)

FIGURE 2.21 Examples of a time-invariant (a) and some time-variant systems (b)–(d).

(c) This system is described by the input–output relation

$$y(n) = H[x(n)] = x(-n) \tag{2.2.20}$$

The response of this system to $x(n - k)$ is (see Example 2.1.9)

$$y(n, k) = H[x(n - k)] = x(-n - k) \tag{2.2.21}$$

Now, if we delay the output $y(n)$, as given by (2.2.20), by k units in time, the result will be

$$y(n - k) = x(-n + k) \tag{2.2.22}$$

Since $y(n, k) \neq y(n - k)$, the system is time variant.

(d) The input–output equation for this system is

$$y(n) = x(n) \cos \omega_0 n \tag{2.2.23}$$

The response of this system to $x(n - k)$ is

$$y(n, k) = x(n - k) \cos \omega_0 n \tag{2.2.24}$$

If the expression in (2.2.23) is delayed by k units and the result is compared to (2.2.24), it is evident that the system is time variant.

Linear Versus Nonlinear Systems. The general class of systems may also be subdivided into linear systems and nonlinear systems. A linear system is one that satisfies the *superposition principle*. Simply stated, the principle of superposition requires that the response of the system to a weighted sum of signals is equal to the corresponding weighted sum of the responses (outputs) of the system to each of the individual input signals.

To elaborate, suppose that we have two input signals $x_1(n)$ and $x_2(n)$. Suppose that the system under consideration produces the response

$$y_1(n) = H[x_1(n)] \qquad (2.2.25)$$
$$y_2(n) = H[x_2(n)] \qquad (2.2.26)$$

to the two signals applied separately. Next, suppose that we form a new signal $x_3(n)$ by forming a linear combination of $x_1(n)$ and $x_2(n)$, that is,

$$x_3(n) = a_1 x_1(n) + a_2 x_2(n) \qquad (2.2.27)$$

where a_1 and a_2 are arbitrary, possibly complex constants. The response of the system to $x_3(n)$ is denoted as

$$
\begin{aligned}
y_3(n) &= H[x_3(n)] \\
&= H[a_1 x_1(n) + a_2 x_2(n)]
\end{aligned} \qquad (2.2.28)
$$

The principle of superposition requires that

$$
\begin{aligned}
y_3(n) &= a_1 y_1(n) + a_2 y_2(n) \\
&= a_1 H[x_1(n)] + a_2 H[x_2(n)]
\end{aligned} \qquad (2.2.29)
$$

That is, the response of the system to the weighted linear combination of input signals (2.2.29) is equal to the corresponding weighted combination of output signals given by (2.2.25) and (2.2.26). The superposition principle is pictorially illustrated in Fig. 2.22.

The development above leads us to following definition of linearity.

DEFINITION: *A relaxed system H is linear if and only if*

$$H[a_1 x_1(n) + a_2 x_2(n)] = a_1 H[x_1(n)] + a_2 H[x_2(n)] \qquad (2.2.30)$$

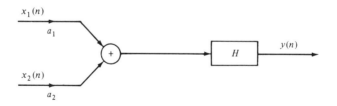

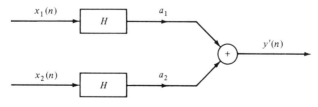

If $y'(n) = y(n)$, then H is linear

FIGURE 2.22 Graphical representation of the superposition principle. H is linear if and only if $y(n) = y'(n)$.

for any arbitrary input sequences $x_1(n)$ and $x_2(n)$, and any arbitrary constants a_1 and a_2.

The superposition principle embodied in the relation (2.2.30) can be separated into two parts. First, suppose that $a_2 = 0$. Then (2.2.30) reduces to

$$H[a_1 x_1(n)] = a_1 H[x_1(n)] = a_1 y_1(n) \qquad (2.2.31)$$

where

$$y_1(n) = H[x_1(n)]$$

The relation (2.2.31) demonstrates the *multiplicative* or *scaling property* of a linear system. That is, if the response of the system to the input $x_1(n)$ is $y_1(n)$, the response to $a_1 x_1(n)$ is simply $a_1 y_1(n)$. Thus any scaling of the input results in an identical scaling of the corresponding output.

Second, suppose that $a_1 = a_2 = 1$ in (2.2.30). Then

$$\begin{aligned} H[x_1(n) + x_2(n)] &= H[x_1(n)] + H[x_1(n)] \qquad (2.2.32) \\ &= y_1(n) + y_2(n) \end{aligned}$$

This relation demonstrates the *additivity property* of a linear system. The additivity and multiplicative properties constitute the superposition principle as it applies to linear systems.

The linearity condition embodied in (2.2.30) can be extended to any arbitrary weighted linear combination of signals based on induction. Indeed, suppose that

$$x(n) = \sum_{k=1}^{M-1} a_k x_k(n) \xrightarrow{\quad H \quad} y(n) = \sum_{k=1}^{M-1} a_k y_k(n)$$

where

$$y_k(n) = H[x_k(n)] \qquad k = 1, 2, \ldots, M-1$$

is true for the $M - 1$ signals. Then for the signal

$$\sum_{k=1}^{M} a_k x_k(n) = x(n) + a_M x_M(n)$$

the output of the system is

$$\begin{aligned} H\left[\sum_{k=1}^{M} a_k x_k(n) \right] &= H[x(n) + a_M x_M(n)] \\ &= H[x(n)] + a_M H[x_M(n)] \\ &= \sum_{k=1}^{M-1} a_k y_k(n) + a_M y_M(n) \\ &= \sum_{k=1}^{M} a_k y_k(n) \end{aligned}$$

Therefore, the relation holds in general for M signals, where $M = 2, 3, \ldots$.

In general, we observe from (2.2.31) that if $a_1 = 0$, then $y(n) = 0$. In other words, a relaxed, linear system with zero input produces a zero output. If a system produces a nonzero output with a zero input, the system may be either nonrelaxed or nonlinear. If a relaxed system does not satisfy the superposition principle as given by the definition above, it is called *nonlinear*.

EXAMPLE 2.2.6

Determine if the systems described by the following input–output equations are linear or nonlinear.

(a) $y(n) = nx(n)$
(b) $y(n) = x(n^2)$
(c) $y(n) = x^2(n)$
(d) $y(n) = Ax(n) + B$
(e) $y(n) = e^{x(n)}$

Solution: (a) For two input sequences $x_1(n)$ and $x_2(n)$, the corresponding outputs are

$$\begin{aligned} y_1(n) &= nx_1(n) \\ y_2(n) &= nx_2(n) \end{aligned} \qquad (2.2.33)$$

A linear combination of the two input sequences results in the output

$$\begin{aligned} y_3(n) &= H[a_1x_1(n) + a_2x_2(n)] = n[a_1x_1(n) + a_2x_2(n)] \\ &= a_1nx_1(n) + a_2nx_2(n) \end{aligned} \qquad (2.2.34)$$

On the other hand, a linear combination of the two outputs in (2.2.33) results in the output

$$a_1y_1(n) + a_2y_2(n) = a_1nx_1(n) + a_2nx_2(n) \qquad (2.2.35)$$

Since the right-hand sides of (2.2.34) and (2.2.35) are identical, the system is linear.
(b) As in part (a), we find the response of the system to two separate input signals $x_1(n)$ and $x_2(n)$. The result is

$$\begin{aligned} y_1(n) &= x_1(n^2) \\ y_2(n) &= x_2(n^2) \end{aligned} \qquad (2.2.36)$$

The output of the system to a linear combination of $x_1(n)$ and $x_2(n)$ is

$$y_3(n) = H[a_1x_1(n) + a_2x_2(n)] = a_1x_1(n^2) + a_2x_2(n^2) \qquad (2.2.37)$$

Finally, a linear combination of the two outputs in (2.2.36) yields

$$a_1y_1(n) + a_2y_2(n) = a_1x_1(n^2) + a_2x_2(n^2) \qquad (2.2.38)$$

By comparing (2.2.37) with (2.2.38), we conclude that the system is linear.
(c) The output of the system is the square of the input. (Electronic devices that have such an input–output characteristic and are called square-law devices.) From our previous discussion it is clear that such a system is memoryless. We now illustrate that this system is nonlinear.

The responses of the system to two separate input signals are

$$\begin{aligned} y_1(n) &= x_1^2(n) \\ y_2(n) &= x_2^2(n) \end{aligned} \qquad (2.2.39)$$

The response of the system to a linear combination of these two input signals is

$$\begin{aligned} y_3(n) &= H[a_1x_1(n) + a_2x_2(n)] \\ &= [a_1x_1(n) + a_2x_2(n)]^2 \\ &= a_1^2x_1^2(n) + 2a_1a_2x_1(n)x_2(n) + a_2^2x_2^2(n) \end{aligned} \qquad (2.2.40)$$

On the other hand, if the system is linear, it would produce a linear combination of the two outputs in (2.2.39), namely,

$$a_1y_1(n) + a_2y_2(n) = a_1x_1^2(n) + a_2x_2^2(n) \qquad (2.2.41)$$

Since the actual output of the system, as given by (2.2.40), is not equal to (2.2.41), the system is nonlinear.

(d) Assuming that the system is excited by $x_1(n)$ and $x_2(n)$ separately, we obtain the corresponding outputs

$$y_1(n) = Ax_1(n) + B \qquad\qquad (2.2.42)$$
$$y_2(n) = Ax_2(n) + B$$

A linear combination of $x_1(n)$ and $x_2(n)$ produces the output

$$\begin{aligned} y_3(n) &= H[a_1x_1(n) + a_2x_2(n)] \\ &= A[a_1x_1(n) + a_2x_2(n)] + B \qquad (2.2.43) \\ &= Aa_1x_1(n) + a_2Ax_2(n) + B \end{aligned}$$

On the other hand, if the system were linear, its output to the linear combination of $x_1(n)$ and $x_2(n)$ would be a linear combination of $y_1(n)$ and $y_2(n)$, that is,

$$a_1y_1(n) + a_2y_2(n) = a_1Ax_1(n) + a_1B + a_2Ax_2(n) + a_2B \qquad (2.2.44)$$

Clearly, (2.2.43) and (2.2.44) are different and hence the system fails to satisfy the linearity test.

The reason that this system fails to satisfy the linearity test is not that the system is nonlinear (in fact, the system is described by a linear equation) but it is the presence of the constant B. Consequently, the output depends on both the input excitation and on the parameter $B \neq 0$. Hence, for $B \neq 0$, the system is not relaxed. If we set $B = 0$, the system is now relaxed and the linearity test is satisfied.

(e) Note that the system described by the input–output equation

$$y(n) = e^{x(n)}$$

is relaxed. If $x(n) = 0$, we find that $y(n) = 1$. This is an indication that the system is nonlinear. This, in fact, is the conclusion reached when the linearity test, as described above, is applied.

Causal Versus Noncausal Systems. We begin with the definition of causal discrete-time systems.

DEFINITION: *A system is said to be* causal *if the output of the system at any time n [i.e., $y(n)$] depends only on present and past inputs [i.e., $x(n)$, $x(n - 1)$, $x(n - 2)$, . . .], but does not depend on future inputs [i.e., $x(n + 1)$, $x(n + 2)$, . . .]. In mathematical terms, the output of a causal system satisfies an equation of the form*

$$y(n) = F[x(n), x(n - 1), x(n - 2), . . .] \qquad (2.2.45)$$

where $F[\cdot]$ is some arbitrary function.

If a system does not satisfy this definition, it is called *noncausal*. Such a system has an output that depends not only on present and past inputs but also on future inputs.

It is apparent that in real-time signal processing applications we cannot observe future values of the signal, and hence a noncausal system is physically unrealizable (i.e., it cannot be implemented). On the other hand, if the signal is recorded so that the processing is done off-line (non-real time), it is possible to implement a noncausal system, since all values of the signal are available at the time of processing. This is often the case in the processing of geophysical signals and images.

EXAMPLE 2.2.7 _____

Determine if the systems described by the following input–output equations are causal or noncausal.

(a) $y(n) = x(n) - x(n - 1)$
(b) $y(n) = \sum_{k=-\infty}^{n} x(k)$
(c) $y(n) = ax(n)$
(d) $y(n) = x(n) + 3x(n + 4)$
(e) $y(n) = x(n^2)$
(f) $y(n) = x(2n)$
(g) $y(n) = x(-n)$

Solution: The systems described in parts (a), (b), and (c) are clearly causal, since the output depends only on the present and past inputs. On the other hand, the systems in parts (d), (e), and (f) are clearly noncausal, since the output depends on future values of the input. The system in (g) is also noncausal, as we note by selecting, for example, $n = -1$, which yields $y(-1) = x(1)$. Thus the output at $n = -1$ depends on the input at $n = 1$, which is two units of time into the future.

_____

Stable Versus Unstable Systems. Stability is an important property that must be considered in any practical application of a system. Unstable systems usually exhibit erratic and extreme behavior and cause overflow in any practical implementation. Below, we define mathematically what we mean by a stable system, and later, in Section 2.3.6, we explore the implications of this definition for linear, time-invariant systems.

DEFINITION: *An arbitrary relaxed system is said to be bounded input–bounded output (BIBO) stable if and only if every bounded input produces a bounded output.*

The conditions that the input sequence $x(n)$ and the output sequence $y(n)$ are bounded is simply translated mathematically to mean that there exist some finite numbers, say M_x and M_y, such that

$$|x(n)| \leq M_x < \infty, \qquad |y(n)| \leq M_y < \infty$$

for all n. If for some bounded input sequence $x(n)$, the output is unbounded (infinite), the system is classified as unstable.

2.2.4 Interconnection of Discrete-Time Systems

Discrete-time systems may be interconnected to form larger systems. There are two basic ways in which systems can be interconnected: in cascade (series) or in parallel. These interconnections are illustrated in Fig. 2.23. Note that the two interconnected systems are different.

In the cascade interconnection the output of the first system is

$$y_1(n) = H_1[x(n)] \tag{2.2.46}$$

and the output of the second system is

$$y(n) = H_2[y_1(n)] \\ = H_2\{H_1[x(n)]\} \tag{2.2.47}$$

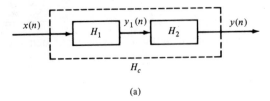

(a)

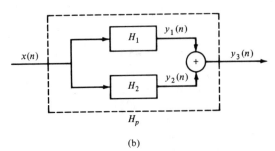

(b)

FIGURE 2.23 Cascade (a) and parallel (b) interconnections of systems.

We observe that systems H_1 and H_2 can be combined or consolidated into a single overall system

$$H_c \equiv H_2 H_1 \qquad (2.2.48)$$

Consequently, we may express the output of the combined system as

$$y(n) = H_c[x(n)]$$

In general, the order in which the operations H_1 and H_2 are performed is important. That is,

$$H_2 H_1 \neq H_1 H_2$$

for arbitrary systems. However, if the systems H_1 and H_2 are linear and time invariant, then (a) H_c is time invariant and (b) $H_2 H_1 = H_1 H_2$, that is, the order in which the systems process the signal is not important. $H_2 H_1$ and $H_1 H_2$ yield identical output sequences.

The proof of (a) is given below. The proof of (b) is given in Section 2.3.4. To prove time invariance, suppose that H_1 and H_2 are time invariant; then

$$x(n - k) \xrightarrow{\;\;H_1\;\;} y_1(n - k)$$

and

$$y_1(n - k) \xrightarrow{\;\;H_2\;\;} y(n - k)$$

Thus

$$x(n - k) \xrightarrow{\;H_c = H_2 H_1\;} y(n - k)$$

and therefore, H_c is time invariant.

In the parallel interconnection, the output of the system H_1 is $y_1(n)$ and the output of the system H_2 is $y_2(n)$. Hence the output of the parallel interconnection is

$$\begin{aligned}
y_3(n) &= y_1(n) + y_2(n) \\
&= H_1[x(n)] + H_2[x(n)] \\
&= (H_1 + H_2)[x(n)] \\
&= H_p[x(n)]
\end{aligned}$$

where $H_p = H_1 + H_2$.

In general, we may use parallel and cascade interconnection of systems to construct larger, more complex systems. Conversely, we may take a larger system and break it down into smaller subsystems for purposes of analysis and implementation. We shall use these notions later, in the design and implementation of digital filters.

2.3 Analysis of Discrete-Time Linear Time-Invariant Systems

In Section 2.2 we classified systems in accordance with a number of characteristic properties or categories: namely, linearity, causality, stability, and time invariance. Having done so, we now turn our attention to the analysis of the important class of linear time-invariant (LTI) systems. In particular, we shall demonstrate that such systems are simply characterized in the time domain by their response to a unit sample sequence. We shall also demonstrate that any arbitrary input signal can be decomposed and represented as a weighted sum of unit sample sequences. As a consequence of the linearity and time-invariance properties of the system, it follows that the response of the system to any arbitrary input signal can be expressed in terms of the unit sample response of the system. The general form for the expression that relates the unit sample response of the system and the arbitrary input signal to the output signal, which is called the convolution sum or the convolution formula, will also be derived. Thus we are able to determine the output of any linear time-invariant system to any arbitrary input signal.

2.3.1 Techniques for the Analysis of Linear Systems

There are two basic methods for analyzing the behavior or response of a linear system to a given input signal. One method is based on the direct solution of the input–output equation for the system, which, in general, has the form

$$y(n) = F[y(n - 1), y(n - 2), \ldots, y(n - N), x(n), x(n - 1), \ldots, x(n - M)]$$

where $F[\cdot]$ denotes some function of the quantities in brackets. Specifically, for an LTI system, we shall see later that the general form of the input–output relationship is

$$y(n) = -\sum_{k=1}^{N} a_k y(n - k) + \sum_{k=0}^{M} b_k x(n - k) \qquad (2.3.1)$$

where $\{a_k\}$ and $\{b_k\}$ are constant parameters that specify the system and are independent of $x(n)$ and $y(n)$. The input–output relationship in (2.3.1) is called a difference equation and represents one way to characterize the behavior of a discrete-time LTI system. The solution of (2.3.1) is the subject of Section 2.4.

The second method for analyzing the behavior of a linear system to a given input signal is first to decompose or resolve the input signal into a sum of elementary signals. The elementary signals are selected so that the response of the system to each signal component is easily determined. Then, using the linearity property of the system, the responses of the system to the elementary signals are added to obtain the total response of the system to the given input signal. This second method is the one described in this section.

To elaborate, let us suppose that the input signal $x(n)$ is resolved into a weighted sum of elementary signal components $\{x_k(n)\}$ so that

$$x(n) = \sum_{k} c_k x_k(n) \qquad (2.3.2)$$

where the $\{c_k\}$ are the set of amplitudes (weighting coefficients) in the decomposition of the signal $x(n)$. Now suppose that the response of the system to the elementary signal component $x_k(n)$ is $y_k(n)$. Thus assuming that the system is relaxed,

$$y_k(n) \equiv H[x_k(n)] \qquad (2.3.3)$$

and the response to $c_k x_k(n)$ is $c_k y_k(n)$, as a consequence of the scaling property of the linear system.

Finally, the total response to the input $x(n)$ is

$$
\begin{aligned}
y(n) = H[x(n)] &= H\left[\sum_k c_k x_k(n)\right] \\
&= \sum_k c_k H[x_k(n)] \qquad (2.3.4) \\
&= \sum_k c_k y_k(n)
\end{aligned}
$$

In (2.3.4) we used the additivity property of the linear system.

Although the choice of the elementary signals appears to be arbitrary, to a large extent, our selection is heavily dependent on the class of input signals that we wish to consider. If we place no restriction on the characteristics of the input signals, its resolution into a weighted sum of unit sample (impulse) sequences proves to be mathematically convenient and completely general. On the other hand, if we restrict our attention to a subclass of input signals, there may be another set of elementary signals that is more convenient mathematically in the determination of the output. For example, if the input signal $x(n)$ is periodic with period N, we have already observed in Section 1.3.5 that a mathematically convenient set of elementary signals is the set of exponentials

$$x_k(n) = e^{j\omega_k n} \qquad k = 0, 1, \ldots, N - 1 \qquad (2.3.5)$$

where the frequenceis $\{\omega_k\}$ are harmonically related, that is,

$$\omega_k = \left(\frac{2\pi}{N}\right) k \qquad k = 0, 1, \ldots, N - 1 \qquad (2.3.6)$$

The frequency $2\pi/N$ is called the fundamental, and all higher-frequency components are multiples of the fundamental frequency component. This subclass of input signals is considered in more detail later.

With the resolution of the input signal into a weighted sum of unit sample sequences, we must first determine the response of the system to a unit sample sequence and then use the scaling and multiplicative properties of the linear system to determine the formula for the output given any arbitrary input. This development is described in detail below.

2.3.2 Resolution of a Discrete-Time Signal into Impulses

Suppose we have an arbitrary signal $x(n)$ that we wish to resolve into a sum of unit sample sequences. To bring forth the notation established in the preceding section, we select the elementary signals $x_k(n)$ to be

$$x_k(n) = \delta(n - k) \qquad (2.3.7)$$

where k represents the delay of the unit sample sequence. To handle an arbitrary signal $x(n)$ that may have nonzero values over an infinite duration, the set of unit impulses must also be infinite, to encompass the infinite number of delays.

Now suppose that we multiply the two sequences $x(n)$ and $\delta(n - k)$. Since $\delta(n - k)$ is zero everywhere except at $n = k$, where its value is unity, the result of this multiplication is another sequence which is zero everywhere except at $n = k$, where its value is $x(k)$, as illustrated in Fig. 2.24. Thus

$$x(n)\delta(n - k) = x(k)\delta(n - k) \qquad (2.3.8)$$

is a sequence which is zero everywhere except at $n = k$, where its value is $x(k)$. If we were to repeat the multiplication of $x(n)$ with $\delta(n - m)$, where m is another delay ($m \neq k$), the result will be a sequence that is zero everywhere except at $n = m$, where its value is $x(m)$. Hence

$$x(n)\delta(n - m) = x(m)\delta(n - m) \qquad (2.3.9)$$

as illustrated in Fig. 2.24d and e. In other words, each multiplication of the signal $x(n)$ by a unit impulse at some delay k, [i.e., $\delta(n - k)$], in essence picks out the single value

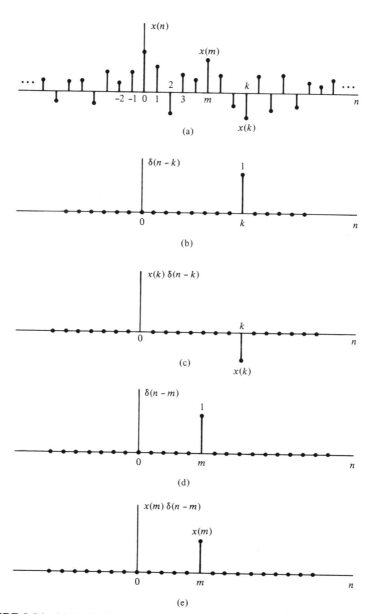

FIGURE 2.24 Multiplication of a signal $x(n)$ with a shifted unit sample sequence.

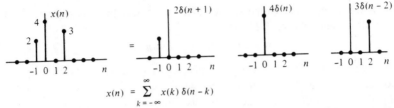

$$x(n) = \sum_{k=-\infty}^{\infty} x(k)\,\delta(n-k)$$

FIGURE 2.25 Graphical illustration of the decomposition of a signal into a superposition of scaled and shifted impulses.

$x(k)$ of the signal $x(n)$ at the delay where the unit impulse is nonzero. Consequently, if we repeat this multiplication over all possible delays, $-\infty < k < \infty$, and sum all the product sequences, the result will be a sequence that is equal to the sequence $x(n)$, that is,

$$x(n) = \sum_{k=-\infty}^{\infty} x(k)\delta(n-k) \qquad (2.3.10)$$

We emphasize that the right-hand side of (2.3.10) is the summation of an infinite number of unit sample sequences where the unit sample sequence $\delta(n-k)$ has an amplitude value $x(k)$. Thus the right-hand side of (2.3.10) gives the resolution or decomposition of any arbitrary signal $x(n)$ into a weighted (scaled) sum of shifted unit sample sequences.

EXAMPLE 2.3.1 ————————————————————————————

Consider the special case of a finite-duration sequence given as

$$x(n) = \left\{ 2, 4, 0, 3 \atop \uparrow \right\}$$

Resolve the sequence $x(n)$ into a sum of weighted impulse sequences.

Solution: Since the sequence $x(n)$ is nonzero for the time instants $n = -1, 0, 2$, we need three impulses at delays $k = -1, 0, 2$. Following (2.3.10) we find that

$$x(n) = 2\delta(n+1) + 4\delta(n) + 3\delta(n-2)$$

Figure 2.25 illustrates the decomposition of $x(n)$ into weighted (scaled) unit sample sequences.

2.3.3 Response of LTI Systems to Arbitrary Inputs: The Convolution Sum

Having resolved an arbitrary input signal $x(n)$ into a weighted sum of impulses, we are now ready to determine the response of any relaxed linear system to any input signal. First, we define the response $y(n, k)$ of the system to the input unit sample sequence at $n = k$ by the special symbol $h(n, k)$, $-\infty < k < \infty$. That is,

$$y(n, k) \equiv h(n, k) = H[\delta(n-k)] \qquad (2.3.11)$$

In (2.3.11) we note that n is the time index and k is a parameter showing the location of the input impulse. If the impulse at the input is scaled by an amount $c_k \equiv x(k)$, the response of the system is the correspondingly scaled output, that is,

$$c_k y(n, k) = x(k)h(n, k) \tag{2.3.12}$$

Finally, if the input is the arbitrary signal $x(n)$ which is expressed as a sum of weighted impulses, that is,

$$x(n) = \sum_{k=-\infty}^{\infty} x(k)\delta(n - k) \tag{2.3.13}$$

then the response of the system to $x(n)$ is the corresponding sum of weighted outputs, that is,

$$y(n) = H[x(n)] = H\left[\sum_{k=-\infty}^{\infty} x(k)\delta(n - k)\right]$$

$$= \sum_{k=-\infty}^{\infty} x(k)H[\delta(n - k)]$$

$$= \sum_{k=-\infty}^{\infty} x(k)h(n, k) \tag{2.3.14}$$

Clearly, (2.3.14) follows from the superposition property of the linear system, and is known as the *superposition summation*.

We note that (2.3.14) is an expression for the response of a linear system to any arbitrary input sequence $x(n)$. This expression is a function of both $x(n)$ and the responses $h(n, k)$ of the system to the unit impulses $\delta(n - k)$ for $-\infty < k < \infty$. In deriving (2.3.14) we used the linearity property of the system but not its time-invariance property. Thus the expression in (2.3.14) applies to any relaxed linear (time-variant) system.

If, in addition, the system is time invariant, the formula in (2.3.14) simplifies considerably. In fact, if the response of the LTI system to the unit sample sequence $\delta(n)$ is denoted as $h(n)$, that is,

$$h(n) \equiv H[\delta(n)] \tag{2.3.15}$$

then by the time-invariance property, the response of the system to the delayed unit sample sequence $\delta(n - k)$ is

$$h(n - k) = H[\delta(n - k)] \tag{2.3.16}$$

Consequently, the formula in (2.3.14) reduces to

$$y(n) = \sum_{k=-\infty}^{\infty} x(k)h(n - k) \tag{2.3.17}$$

Now we observe that the relaxed LTI system is completely characterized by a single function $h(n)$, namely, its response to the unit sample sequence $\delta(n)$. In contrast, a time-variant linear system requires an infinite number of unit sample response functions $h(n, k)$, one for each possible delay, to characterize its output, in general.

The formula in (2.3.17) that gives the response $y(n)$ of the LTI system as a function of the input signal $x(n)$ and the unit sample (impulse) response $h(n)$ is called a *convolution sum*. We say that the input $x(n)$ is convolved with the impulse response $h(n)$ to yield the output $y(n)$. We shall now explain the procedure for computing the response $y(n)$, both mathematically and graphically, given the input $x(n)$ and the impulse response $h(n)$ of the system.

Suppose that we wish to compute the output of the system at some time instant, say

$n = n_0$. According to (2.3.17), the response at $n = n_0$ is given as

$$y(n_0) = \sum_{k=-\infty}^{\infty} x(k)h(n_0 - k) \tag{2.3.18}$$

Our first observation is that the index in the summation is k, and hence both the input signal $x(k)$ and the impulse response $h(n_0 - k)$ are functions of k. Second, we observe that the sequences $x(k)$ and $h(n_0 - k)$ are multiplied together to form a product sequence. The output $y(n_0)$ is simply the sum over all values of the product sequence. The sequence $h(n_0 - k)$ is obtained from $h(k)$ by, first, folding $h(k)$ about $k = 0$ (the time origin), which results in the sequence $h(-k)$. Then the folded sequence is shifted by n_0 to yield $h(n_0 - k)$. In summary, then, the process of computing the convolution between $x(k)$ and $h(k)$ involves the following four steps.

1. *Folding.* Fold $h(k)$ about $k = 0$ to obtain $h(-k)$.
2. *Shifting.* Shift $h(-k)$ by n_0 to the right (left) if n_0 is positive (negative), to obtain $h(n_0 - k)$.
3. *Multiplication.* Multiply $x(k)$ by $h(n_0 - k)$ to obtain the product sequence $v_{n_0}(k) \equiv x(k)h(n_0 - k)$.
4. *Summation.* Sum all the values of the product sequence $v_{n_0}(k)$ to obtain the value of the output at time $n = n_0$.

We note that this procedure results in the response of the system at a single time instant, say $n = n_0$. In general, we are interested in evaluating the response of the system over all time instants $-\infty < n < \infty$. Consequently, the steps 2 through 4 outlined above must be repeated, in general, for all possible time shifts $-\infty < n < \infty$.

In order to gain a better understanding of the procedure for evaluating the convolution sum we shall demonstrate the process graphically. The graphs will aid us in explaining the four steps involved in the computation of the convolution sum.

EXAMPLE 2.3.2 ———————————————————————————————

The impulse response of a linear time-invariant system is

$$h(n) = \begin{cases} 1 & n = -1 \\ 2 & n = 0 \\ 1 & n = 1 \\ -1 & n = 2 \\ 0 & \text{otherwise} \end{cases} \tag{2.3.19}$$

Determine the response of the system to the input signal

$$x(n) = \begin{cases} 1 & n = 0 \\ 2 & n = 1 \\ 3 & n = 2 \\ 1 & n = 3 \\ 0 & \text{otherwise} \end{cases} \tag{2.3.20}$$

Solution: We shall compute the convolution according to the formula (2.3.17), but we shall use graphs of the sequences to aid us in the computation. In Fig. 2.26a we illustrate the input signal sequence $x(k)$ and the impulse response $h(k)$ of the system, using k as the time index in order to be consistent with (2.3.17).

The first step in the computation of the convolution sum is to fold $h(k)$. The folded

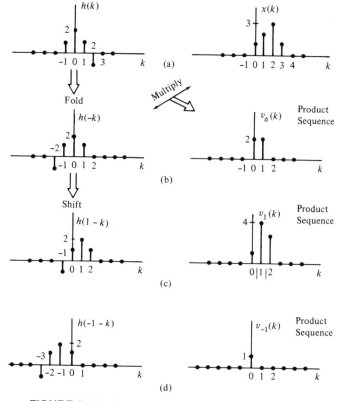

FIGURE 2.26   Graphical computation of convolution.

sequence $h(-k)$ is illustrated in Fig. 2.26b. Now we can compute the output at $n = 0$, according to (2.3.17), which is

$$y(0) = \sum_{k=-\infty}^{\infty} x(k)h(-k) \tag{2.3.21}$$

Since the shift $n = 0$, we use $h(-k)$ directly without shifting it. The product sequence

$$v_0(k) \equiv x(k)h(-k) \tag{2.3.22}$$

is also shown in Fig. 2.26b. Finally, the sum of all the terms in the product sequence yields

$$y(0) = \sum_{h=-\infty}^{\infty} v_0(k) = 4$$

We continue the computation by evaluating the response of the system at $n = 1$. According to (2.3.17),

$$y(1) = \sum_{h=-\infty}^{\infty} x(k)h(1 - k) \tag{2.3.23}$$

The sequence $h(1 - k)$ is simply the folded sequence $h(-k)$ shifted to the right by one unit in time. This sequence is illustrated in Fig. 2.26c. The product sequence

$$v_1(k) = x(k)h(1 - k) \tag{2.3.24}$$

is also illustrated in Fig. 2.26c. Finally, the sum of all the values in the product sequence yields

$$y(1) = \sum_{k=-\infty}^{\infty} v_1(k) = 8$$

In a similar manner, we obtain $y(2)$ by shifting $h(-k)$ two units to the right, forming the product sequence $v_2(k) = x(k)h(2 - k)$ and then summing all the terms in the product sequence. Thus we obtain $y(2) = 8$. By shifting $h(-k)$ farther to the right, multiplying the corresponding sequence, and summing over all the values of the resulting product sequences, we obtain $y(3) = 3$, $y(4) = -2$, $y(5) = -1$. For $n > 5$, we find that $y(n) = 0$ because the product sequences contain all zeros. Thus we have obtained the response $y(n)$ for $n > 0$.

Next we wish to evaluate $y(n)$ for $n < 0$. We begin with $n = -1$. Then

$$y(-1) = \sum_{k=-\infty}^{\infty} x(k)h(-1 - k) \qquad (2.3.25)$$

Now the sequence $h(-1 - k)$ is simply the folded sequence $h(-k)$ shifted one time unit to the left. The resulting sequence is illustrated in Fig. 2.26d. The corresponding product sequence is also shown in Fig. 2.26d. Finally, summing over the values of the product sequence, we obtain

$$y(-1) = 1$$

From observation of the graphs of Fig. 2.26, it is clear that any further shifts of $h(-1 - k)$ to the left always results in an all-zero product sequence, and hence

$$y(n) = 0 \qquad \text{for } n \le -2$$

Now we have the entire response of the system for $-\infty < n < \infty$, which we summarize below as

$$y(n) = \left\{ \ldots, 0, 0, 1, 4, 8, 8, 3, -2, -1, 0, 0, \ldots \right\} \qquad (2.3.26)$$

In Example 2.3.2 we illustrated the computation of the convolution sum, using graphs of the sequences to aid us in visualizing the steps involved in the computation procedure.

Before working out another example, we wish to show that the convolution operation is symmetric in the sense that it is irrelevant which of the two sequences is folded and shifted. Indeed, if we begin with (2.3.17) and make a change in the variable of the summation, from k to m, by defining a new index $m = n - k$, then $k = n - m$, and, hence (2.3.17) becomes

$$y(n) = \sum_{m=-\infty}^{\infty} x(n - m)h(m) \qquad (2.3.27)$$

Since m is a dummy index, we may simply replace m by k so that

$$y(n) = \sum_{k=-\infty}^{\infty} x(n - k)h(k) \qquad (2.3.28)$$

The expression in (2.3.28) involves leaving the impulse response $h(k)$ unaltered, while the input sequence is folded and shifted. Although the output $y(n)$ in (2.3.28) is identical to (2.3.17), the product sequences in the two forms of the convolution formula are not identical. In fact, if we define the two product sequences as

$$v_n(k) = x(k)h(n - k)$$
$$w_n(k) = x(n - k)h(k)$$

it can be easily shown that

$$v_n(k) = w_n(n - k)$$

and therefore,

$$y(n) = \sum_{k=-\infty}^{\infty} v_n(k) = \sum_{k=-\infty}^{\infty} w_n(n - k)$$

since both sequences contain the same sample values in a different arrangement.

EXAMPLE 2.3.3 ———————————————————————————————————

Determine the output $y(n)$ of a relaxed linear time-invariant system with impulse response

$$h(n) = a^n u(n) \qquad |a| < 1$$

when the input is a unit step sequence, that is,

$$x(n) = u(n)$$

Solution: In this case both $h(n)$ and $x(n)$ are infinite-duration sequences. We use the form of the convolution formula given by (2.3.28) in which $x(k)$ is folded. The sequences $h(k)$, $x(k)$, and $x(-k)$ are shown in Fig. 2.27. The product sequence $v_0(k)$, $v_1(k)$, and $v_2(k)$ corresponding to $x(-k)h(k)$, $x(1 - k)h(k)$, and $x(2 - k)h(k)$ are illustrated in Fig. 2.27c, d, and e, respectively. Thus we obtain the outputs

$$y(0) = 1$$
$$y(1) = 1 + a$$
$$y(2) = 1 + a + a^2$$

Clearly, for $n > 0$, the output is

$$y(n) = 1 + a + a^2 + \cdots + a^n \qquad\qquad (2.3.29)$$
$$= \frac{1 - a^{n+1}}{1 - a}$$

On the other hand, for $n < 0$, the product sequences consist of all zeros. Hence

$$y(n) = 0 \qquad n < 0$$

A graph of the output $y(n)$ is illustrated in Fig. 2.27f for the case $0 < a < 1$. Note the exponential rise in the output as a function of n. Since $|a| < 1$, the final value of the output as n approaches infinity is

$$y(\infty) = \lim_{n \to \infty} y(n) = \frac{1}{1 - a} \qquad\qquad (2.3.30)$$

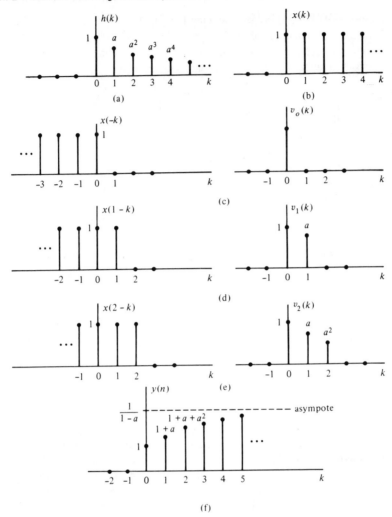

FIGURE 2.27 Graphical computation of convolution in Example 2.3.4.

To summarize, the convolution formula provides us with a means for computing the response of a relaxed linear time-invariant system to any arbitrary input signal $x(n)$. It takes one of two equivalent forms, either (2.3.17) or (2.3.28), where $x(n)$ is the input signal to the system, $h(n)$ is the impulse response of the system, and $y(n)$ is the *output* of the system to the input signal $x(n)$. The evaluation of the convolution formula involves four operations: namely, *folding* either the impulse response as specified by (2.3.17) or the input sequence as specified by (2.3.28) to yield either $h(-k)$ or $x(-k)$, respectively, *shifting* the folded sequence by n units in time to yield either $h(n-k)$ or $x(n-k)$, *multiplying* the two sequences to yield the product sequence, either $x(k)h(n-k)$ or $x(n-k)h(k)$, and finally *summing* all the values in the product sequence to yield the output $y(n)$ of the system at time n. The folding operation is done only once. However, the other three operations are repeated for all possible shifts $-\infty < n < \infty$ in order to obtain $y(n)$ for $-\infty < n < \infty$.

A FORTRAN subroutine for computing the convolution of two sequences is given in Fig. 2.28.

```
       SUBROUTINE CONVL (X, N, H, M, Y, LY)
C
C      SUBROUTINE CONVL COMPUTES THE CONVOLUTION
C      BETWEEN THE SEQUENCES X AND H
C      PARAMETERS
C         X    : ARRAY CONTAINING SEQ X
C         N    : THE LENGTH OF SEQ X
C         H    : ARRAY CONTAINING SEQ H
C         M    : THE LENGTH OF SEQ H
C         Y    : ARRAY CONTAINING THE CONVOLUTION OF X AND H
C         LY   : THE LENGTH OF SEQ Y
C
       DIMENSION X(1),H(1),Y(1)
       LY=N+M-1
       DO 1 K=1,LY
   1   Y(K)=0.
       DO 2 I=1,N
       DO 2 J=1.,M
       K=I+J-1
   2   Y(K)=Y(K)+X(I)*H(J)
       RETURN
       END
```

FIGURE 2.28   Subroutine for convolution.

2.3.4 Properties of Convolution and the Interconnection of LTI Systems

In this section we investigate some important properties of convolution and interpret these properties in terms of interconnecting linear time-invariant systems. We should stress that these properties hold for every input signal.

It is convenient to simplify the notation by using an asterisk to denote the convolution operation. Thus

$$y(n) = x(n) * h(n) \equiv \sum_{k=-\infty}^{\infty} x(k)h(n - k) \tag{2.3.31}$$

In this notation the sequence following the asterisk [i.e., the impulse response $h(n)$] is folded and shifted. The input to the system is $x(n)$. On the other hand, we also showed that

$$y(n) = h(n) * x(n) \equiv \sum_{k=-\infty}^{\infty} h(k)x(n - k) \tag{2.3.32}$$

In this form of the convolution formula, it is the input signal that is folded. Alternatively, we may interpret this form of the convolution formula as resulting from an interchange of the roles of $x(n)$ and $h(n)$. In other words, we may regard $x(n)$ as the impulse response of the system and $h(n)$ as the excitation or input signal. Figure 2.29 illustrates this interpretation.

We may view convolution more abstractly as a mathematical operation between two signal sequences, say $x(n)$ and $h(n)$, that satisfies a number of properties. The property embodied in (2.3.31) and (2.3.32) is called the commutative law.

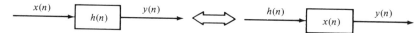

FIGURE 2.29   Interpretation of the commutative property of convolution.

Commutative Law

$$x(n) * h(n) = h(n) * x(n) \qquad (2.3.33)$$

Viewed mathematically, the convolution operation also satisfies the associative law, which may be stated as follows.

Associative Law

$$[x(n) * h_1(n)] * h_2(n) = x(n) * [h_1(n) * h_2(n)] \qquad (2.3.34)$$

From a physical point of view, we may interpret $x(n)$ as the input signal to a linear time-invariant system with impulse response $h_1(n)$. The output of this system, denoted as $y_1(n)$, becomes the input to a second linear time-invariant system with impulse response $h_2(n)$. Then the output is

$$\begin{aligned} y(n) &= y_1(n) * h_2(n) \\ &= [x(n) * h_1(n)] * h_2(n) \end{aligned}$$

which is precisely the left-hand side of (2.3.34). Thus the left-hand side of (2.3.34) corresponds to having two linear time-invariant systems in cascade. Now the right-hand side of (2.3.34) indicates that the input $x(n)$ is applied to an equivalent system having an impulse response, say $h(n)$, which is equal to the convolution of the two impulse responses. That is,

$$h(n) = h_1(n) * h_2(n)$$

and

$$y(n) = x(n) * h(n)$$

Furthermore, since the convolution operation satisfies the commutative property, one can interchange the order of the two systems with responses $h_1(n)$ and $h_2(n)$ without altering the overall input–output relationship. Figure 2.30 graphically illustrates the associative property.

The generalization of the associative law to more than two systems in cascade follows easily from the discussion given above. Thus if we have L linear time-invariant systems in cascade with impulse responses $h_1(n), h_2(n), \ldots, h_L(n)$, there is an equivalent linear time-invariant system having an impulse response that is equal to the $(L - 1)$ – fold convolution of the impulse responses. That is,

$$h(n) = h_1(n) * h_2(n) * \cdots * h_L(n) \qquad (2.3.35)$$

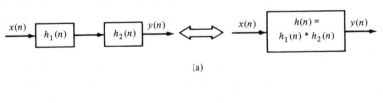

(a)

(b)

FIGURE 2.30 Implications of the associative (a) and the associative and commutative (b) properties of convolution.

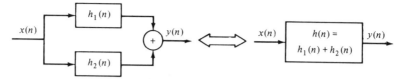

FIGURE 2.31 Interpretation of the distributive property of convolution: two LTI systems connected in parallel can be replaced by a single system with $h(n) = h_1(n) + h_2(n)$.

The commutative law implies that the order in which the convolutions are performed is immaterial. Conversely, any linear time-invariant system can be decomposed into a cascade interconnection of subsystems. A method for accomplishing the decomposition will be described later.

A third property that is satisfied by the convolution operation is the distributive law, which may be stated as follows.

Distributive Law

$$x(n) * [h_1(n) + h_2(n)] = x(n) * h_1(n) + x(n) * h_2(n) \qquad (2.3.36)$$

Interpreted physically, this law implies that if we have two linear time-invariant systems with impulse responses $h_1(n)$ and $h_2(n)$ excited by the same input signal $x(n)$, the sum of the two responses is identical to the response of an overall system with impulse response

$$h(n) = h_1(n) + h_2(n)$$

Thus the overall system is viewed as a parallel combination of the two linear time-invariant systems as illustrated in Fig. 2.31.

The generalization of (2.3.36) to more than two linear time-invariant systems in parallel follows easily by mathematical induction. Thus the interconnection of L linear time-invariant systems in parallel with impulse responses $h_1(n)$, $h_2(n)$, . . . , $h_L(n)$ and excited by the same input $x(n)$ is equivalent to one overall system with impulse response

$$h(n) = \sum_{j=1}^{L} h_j(n) \qquad (2.3.37)$$

Conversely, any linear time-invariant system can be decomposed into a parallel interconnection of subsystems.

EXAMPLE 2.3.4 _____

Determine the impulse response for the cascade of two linear time-invariant systems having impulse responses

$$h_1(n) = (\tfrac{1}{2})^n u(n)$$

and

$$h_2(n) = (\tfrac{1}{4})^n u(n)$$

Solution: To determine the overall impulse response of the two systems in cascade, we simply convolve $h_1(n)$ with $h_2(n)$. Hence

$$h(n) = \sum_{k=-\infty}^{\infty} h_1(k)h_2(n - k)$$

where $h_2(n)$ is folded and shifted. We define the product sequence

$$v_n(k) = h_1(k)h_2(n - k)$$
$$= (\tfrac{1}{2})^k(\tfrac{1}{4})^{n-k}$$

which is nonzero for $k \geq 0$ and $n - k \geq 0$ or $n \geq k \geq 0$. On the other hand, for $n < 0$, we have $v_n(k) = 0$ for all k, and hence

$$h(n) = 0 \qquad n < 0$$

For $n \geq k \geq 0$, the sum of the values of the product sequence $v_n(k)$ over all k yields

$$h(n) = \sum_{k=0}^{n} (\tfrac{1}{2})^k(\tfrac{1}{4})^{n-k}$$

$$= (\tfrac{1}{4})^n \sum_{k=0}^{n} 2^k$$

$$= (\tfrac{1}{4})^n(2^{n+1} - 1)$$

$$= (\tfrac{1}{2})^n[2 - (\tfrac{1}{2})^n] \qquad n \geq 0$$

2.3.5 Causal Linear Time-Invariant Systems

In Section 2.2.3 we defined a causal system as one whose output at time n depends only on present and past inputs but does not depend on future inputs. In other words, the output of the system at some time instant n, say $n = n_0$, depends only on values of $x(n)$ for $n \leq n_0$.

In the case of a linear, time-invariant system, causality can be translated to a condition on the impulse response. To determine this relationship, let us consider a linear time-invariant system having an output at time $n = n_0$ given by the convolution formula

$$y(n_0) = \sum_{k=-\infty}^{\infty} h(k)x(n_0 - k)$$

Suppose that we subdivide the sum into two sets of terms, one set involving present and past values of the input [i.e., $x(n)$ for $n \leq n_0$] and one set involving future values of the input [i.e., $x(n)$, $n > n_0$]. Thus we obtain

$$y(n_0) = \sum_{k=0}^{\infty} h(k)x(n_0 - k) + \sum_{k=-\infty}^{-1} h(k)x(n_0 - k)$$
$$= [h(0)x(n_0) + h(1)x(n_0 - 1) + h(2)x(n_0 - 2) + \cdots]$$
$$+ [h(-1)x(n_0 + 1) + h(-2)x(n_0 + 2) + \cdots]$$

We observe that the terms in the first sum involve $x(n_0)$, $x(n_0 - 1)$, . . . , which are the present and past values of the input signal. On the other hand, the terms in the second sum involve the input signal components $x(n_0 + 1)$, $x(n_0 + 2)$, ... Now if the output at time $n = n_0$ is to depend only on the present and past inputs, then, clearly, the impulse response of the system must satisfy the condition

$$h(n) = 0 \qquad n < 0 \qquad\qquad (2.3.38)$$

Since $h(n)$ is the response of the relaxed linear time-invariant system to a unit impulse applied at $n = 0$, it follows that $h(n) = 0$ for $n < 0$ is both a necessary and a sufficient condition for causality. Hence a relaxed LTI system is causal if and only if its impulse response is zero for negative values of n.

Since for a causal system, $h(n) = 0$ for $n < 0$, the limits on the summation of the convolution formula may be modified to reflect this restriction. Thus we have the two equivalent forms

$$y(n) = \sum_{k=0}^{\infty} h(k)x(n - k) \tag{2.3.39}$$

$$= \sum_{k=-\infty}^{n} x(k)h(n - k) \tag{2.3.40}$$

As indicated previously, causality is required in any real-time signal processing application, since at any given time n we have no access to future values of the input signal. Only the present and past values of the input signal are available in computing the present output.

It is sometimes convenient to call a sequence that is zero for $n < 0$, a *causal sequence*, and one that is nonzero for $n < 0$ and $n > 0$, a *noncausal sequence*. This terminology means that such a sequence could be the unit sample response of a causal or a noncausal system, respectively.

If the input to a causal linear time-invariant system is a causal sequence [i.e., if $x(n) = 0$ for $n < 0$], the limits on the convolution formula are further restricted. In this case the two equivalent forms of the convolution formula become

$$y(n) = \sum_{k=0}^{n} h(k)x(n - k) \tag{2.3.41}$$

$$= \sum_{k=0}^{n} x(k)h(n - k) \tag{2.3.42}$$

We observe that in this case, the limits on the summations for the two alternative forms are identical, and the upper limit is growing with time. Clearly, the response of a causal system to a causal input sequence is causal, since $y(n) = 0$ for $n < 0$.

EXAMPLE 2.3.5 _____

Determine the step response of the linear time-invariant system with impulse response

$$h(n) = a^n u(n)$$

Solution: Since the input signal is a unit step, which is a causal signal, and the system is also causal, we can use one of the special forms of the convolution formula, either (2.3.41) or (2.3.42). Since $x(n) = 1$ for $n \geq 0$, (2.3.41) is simpler to use. Because of the simplicity of this problem, one can skip the steps involved with sketching the folded and shifted sequences. Instead, we use direct substitution of the signals sequences in (2.3.41) and obtain

$$y(n) = \sum_{k=0}^{n} a^k$$

$$= \frac{1 - a^{n+1}}{1 - a}$$

and $y(n) = 0$ for $n < 0$. We note that this result is identical to that obtained in Example 2.3.3. In this simple case, however, we were able to compute the convolution algebraically without resorting to the detailed procedure outlined previously.

2.3.6 Stability of Linear Time-Invariant Systems

As indicated previously, stability is an important property that must be considered in any practical implementation of a system. We defined an arbitrary relaxed system as BIBO stable if and only if its output sequence $y(n)$ is bounded for every bounded input $x(n)$.

If $x(n)$ is bounded, there exists a constant M_x such that

$$|x(n)| \leq M_x < \infty$$

Similarly, if the output is bounded, there exists a constant M_y such that

$$|y(n)| < M_y < \infty$$

for all n.

Now, given such a bounded input sequence $x(n)$ to a linear time-invariant system, let us investigate the implications of the definition of stability on the characteristics of the system. Toward this end, we work again with the convolution formula

$$y(n) = \sum_{k=-\infty}^{\infty} h(k)x(n-k)$$

If we take the absolute value of both sides of this equation, we obtain

$$|y(n)| = \left| \sum_{k=-\infty}^{\infty} h(k)x(n-k) \right|$$

Now, the absolute value of the sum of terms is always less than or equal to the sum of the absolute values of the terms. Hence

$$|y(n)| \leq \sum_{k=-\infty}^{\infty} |h(k)| \, |x(n-k)|$$

If the input is bounded, there exists a finite number M_x such that $|x(n)| \leq M_x$. By substituting this upper bound for $x(n)$ in the equation above, we obtain

$$|y(n)| \leq M_x \sum_{k=-\infty}^{\infty} |h(k)|$$

From this expression we observe that the output will be bounded if the impulse response of the system satisfies the condition

$$\sum_{k=-\infty}^{\infty} |h(k)| < \infty \qquad (2.3.43)$$

That is, *a linear time-invariant system is stable if its impulse response is absolutely summable*. This condition is not only sufficient but it is also necessary to ensure the stability of the system.

The condition in (2.3.43) implies that the impulse response $h(n)$ goes to zero as n approaches infinity. As a consequence, the output of the system will go to zero as n approaches infinity if the input is set to zero beyond $n > n_0$. To prove this, suppose that $|x(n)| < M_x$ for $n < n_0$ and $x(n) = 0$ for $n \geq n_0$. Then, at $n = n_0 + N$, the system output is

$$y(n_0 + N) = \sum_{k=-\infty}^{N-1} h(k)x(n_0 + N - k) + \sum_{k=N}^{\infty} h(k)x(n_0 + N - k)$$

But the first sum is zero since $x(n) = 0$ for $n \geq n_0$. For the remaining part, we take the absolute value of the output, which is

$$\left| y(n_0 + N) \right| = \left| \sum_{k=N}^{\infty} h(k)x(n_0 + N - k) \right| \leq \sum_{k=N}^{\infty} \left| h(k) \right| \left| x(n_0 + N - k) \right|$$

$$\leq M_x \sum_{k=N}^{\infty} \left| h(k) \right|$$

Now, as N approaches infinity,

$$\lim_{N \to \infty} \sum_{k=N}^{\infty} \left| h(n) \right| = 0$$

and hence

$$\lim_{N \to \infty} \left| y(n_0 + N) \right| = 0$$

This result implies that any excitation at the input to the system, which is of a finite duration, will produce an output that will be "transient" in nature; that is, its amplitude will decay with time and die out eventually, when the system is stable.

EXAMPLE 2.3.6

Determine the range of values of the parameter a for which the linear time-invariant system with impulse response

$$h(n) = a^n u(n)$$

is stable.

Solution: First, we note that the system is causal. Consequently, the lower index on the summation in (2.3.43) begins with $k = 0$. Hence

$$\sum_{k=0}^{\infty} \left| a^k \right| = \sum_{k=0}^{\infty} \left| a \right|^k = 1 + \left| a \right| + \left| a \right|^2 + \cdots$$

Clearly, this geometric series converges to

$$\sum_{k=0}^{\infty} \left| a \right|^k = \frac{1}{1 - \left| a \right|}$$

provided that $\left| a \right| < 1$. Otherwise, it diverges. Therefore, the system is stable if $\left| a \right| < 1$. Otherwise, it is unstable. In effect, $h(n)$ must decay exponentially toward zero as n approaches infinity for the system to be stable.

EXAMPLE 2.3.7

Determine the range of values of a and b for which the linear time-invariant system with impulse response

$$h(n) = \begin{cases} a^n & n \geq 0 \\ b^n & n < 0 \end{cases}$$

is stable.

Solution: This system is noncasual. The condition on stability given by (2.3.43) yields

$$\sum_{n=-\infty}^{\infty} |h(n)| = \sum_{n=0}^{\infty} |a|^n + \sum_{n=-\infty}^{-1} |b|^n$$

From Example 2.3.6 we have already determined that the first sum converges for $|a| < 1$. The second sum can be manipulated as follows:

$$\sum_{n=-\infty}^{-1} |b|^n = \sum_{n=1}^{\infty} \frac{1}{|b|^n} = \frac{1}{|b|}\left(1 + \frac{1}{|b|} + \frac{1}{|b|^2} + \cdots\right)$$

$$= \beta(1 + \beta + \beta^2 + \cdots) = \frac{\beta}{1 - \beta}$$

where $\beta = 1/|b|$ must be less than unity for the geometric series to converge. Consequently, the system is stable if both $|a| < 1$ and $|b| > 1$ are satisfied.

2.3.7 Systems with Finite-Duration and Infinite-Duration Impulse Response

Up to this point we have characterized a linear time-invariant system in terms of its impulse response $h(n)$. It is also convenient, however, to subdivide the class of linear time-invariant systems into two types, those that have a finite-duration impulse response (FIR) and those that have an infinite-duration impulse response (IIR). Thus an FIR system has an impulse response that is zero outside of some finite time interval. Without loss of generality, we focus our attention on causal FIR systems, so that

$$h(n) = 0 \qquad n < 0 \text{ and } n \geq M$$

The convolution formula for such a system reduces to

$$y(n) = \sum_{k=0}^{M-1} h(k)x(n - k)$$

A useful interpretation of this expression is obtained by observing that the output at any time n is simply a weighted linear combination of the input signal samples $x(n)$, $x(n - 1), \ldots, x(n - M + 1)$. In other words, the system simply weights, by the values of the impulse response $h(k)$, $k = 0, 1, \ldots, M - 1$, the most recent M signal samples and sums the M products. In effect, the system acts as a *window* that views only the most recent M input signal samples in forming the output. It neglects or simply "forgets" all prior input samples [i.e., $x(n - M), x(n - M - 1), \ldots$]. Thus we say that an FIR system has a finite memory of length-M samples.

In contrast, an IIR linear time-invariant system has an infinite-duration impulse response. Its output, based on the convolution formula, is

$$y(n) = \sum_{k=0}^{\infty} h(k)x(n - k)$$

where causality has been assumed, although this assumption is not necessary. Now, the system output is a weighted [by the impulse response $h(k)$] linear combination of the input signal samples $x(n), x(n - 1), x(n - 2), \ldots$. Since this weighted sum involves the present and all the past input samples, we say that the system has an infinite memory.

We investigate the characteristics of FIR and IIR systems in more detail in subsequent chapters.

2.3.8 The Step Response of a Linear Time-Invariant System

We have observed the important role that the impulse response of a linear time-invariant system plays in determining its output. We have also seen that any arbitrary input signal can be represented as a sum of weighted impulse sequences, and as a consequence, we derived the convolution formula for determining the output of any linear time-invariant system to any arbitrary input, given the impulse response of the system.

Although the impulse response plays a key role in the analysis and synthesis of linear time-invariant systems, it is sometimes of interest to deal with the step response of the system. The step response can be obtained by exciting the input of the system by a unit step sequence, that is,

$$x(n) = u(n)$$

Now, for this input the convolution formula gives the output

$$y(n) \equiv s(n) = \sum_{k=-\infty}^{\infty} h(k)u(n - k)$$

or

$$s(n) = \sum_{k=-\infty}^{n} h(k) \tag{2.3.44}$$

Thus (2.3.44) relates the impulse response to the step response of the system.

We may wish to express the impulse response explicitly in terms of the step response. This can be accomplished as follows. From (2.3.44) we may write

$$s(n) = h(n) + \sum_{k=-\infty}^{n-1} h(k)$$

$$= h(n) + s(n - 1)$$

Hence

$$h(n) = s(n) - s(n - 1) \tag{2.3.45}$$

Now, using (2.3.45), we can substitute the step response in place of $h(n)$ in the various expressions we had derived previously which involved $h(n)$. For example, the convolution formula becomes

$$y(n) = \sum_{k=-\infty}^{\infty} x(k)[s(n - k) - s(n - k - 1)]$$

$$= \sum_{k=-\infty}^{\infty} x(k)s(n - k) - \sum_{k=-\infty}^{\infty} x(k)s(n - 1 - k)$$

Given the step response $s(n)$ of the system and any arbitrary input signal $x(n)$, we define the convolution of $x(n)$ with $s(n)$ as

$$y_s(n) = \sum_{k=-\infty}^{\infty} x(k)s(n - k) \tag{2.3.46}$$

Then the response $y(n)$ of the system to the input signal $x(n)$ may be expressed as

$$y(n) = y_s(n) - y_s(n - 1) \tag{2.3.47}$$

EXAMPLE 2.3.8 ───

The step response of a linear time-invariant system is

$$s(n) = \frac{1 - a^{n+1}}{1 - a} u(n) \qquad |a| < 1 \qquad (2.3.48)$$

Determine the response of the system to the input signal

$$x(n) = \left\{ 1, 2, 3, 1 \atop \uparrow \right\}$$

Solution: First, we shall evaluate the convolution of $x(n)$ with $s(n)$ as indicated in (2.3.46). Then we shall apply (2.3.47) to determine $y(n)$. We have

$$y_s(n) = \sum_{k=0}^{3} x(k)s(n - k)$$
$$= s(n) + 2s(n - 1) + 3s(n - 2) + s(n - 3)$$

The output of the system is

$$y(n) = y_s(n) - y_s(n - 1)$$
$$= s(n) + s(n - 1) + s(n - 2) - 2s(n - 3) - s(n - 4)$$

Hence, by substituting for $s(n)$, we obtain

$$y(0) = 1 \qquad y(1) = 2 + a \qquad y(2) = 3 + 2a + a^2$$

and, in general,

$$y(n) = a^{n-3} + 3a^{n-2} + 2a^{n-1} + a^n \qquad n > 3$$

───

It is apparent from the discussion above that any linear time-invariant system can be completely characterized by its step response, and its response to any arbitrary signal can be expressed in terms of a convolution of the input with the step response. However, no obvious advantages result from a step response characterization of the system compared with the impulse response. In view of this fact, the emphasis will be on the impulse response for characterizing the behavior of linear time-invariant systems.

2.4 Discrete-Time Systems Described by Difference Equations

Up to this point we have treated linear and time-invariant systems that are characterized by their unit sample response $h(n)$. In turn, $h(n)$ allows us to determine the output $y(n)$ of the system for any given input sequence $x(n)$ by means of the convolution summation,

$$y(n) = \sum_{k=-\infty}^{\infty} h(k)x(n - k) \qquad (2.4.1)$$

In general, then, we have shown that any *relaxed*, linear time-invariant system is characterized by the input–output relationship in (2.4.1). Moreover, the convolution summation formula in (2.4.1) suggests a means for the realization of the system. In the case of FIR systems such a realization involves additions, multiplications, and a finite number of memory locations. Consequently, an FIR system is readily implemented directly, as implied by the convolution summation.

If the system is IIR, however, its practical implementation as implied by convolution is clearly impossible, since it requires an infinite number of memory locations, multiplications, and additions. A question that naturally arises, then, is whether or not it is possible to realize IIR systems other than in the form suggested by the convolution summation. Fortunately, the answer is yes, there is a practical and computationally efficient means for implementing a family of IIR systems, as will be demonstrated in this section. Within the general class of IIR systems, this family of discrete-time systems are more conveniently described by difference equations. This family or subclass of IIR systems is very useful in a variety of practical applications, including the implementation of digital filters, and the modeling of physical phenomena and physical systems.

2.4.1 Recursive and Nonrecursive Discrete-Time Systems

As indicated above, the convolution summation formula expresses the output of the linear time-invariant system explicitly in terms of the input signal only. However, this need not be the case, as is shown below. There are many systems where it is either necessary or desirable to express the output of the system not only in terms of the present and past values of the input, but also in terms of already available past output values. The following problem illustrates this point.

Suppose that we wish to compute the *cumulative average* of a signal $x(n)$ in the interval $0 \le k \le n$, defined as

$$y(n) = \frac{1}{n+1} \sum_{k=0}^{n} x(k) \qquad n = 0, 1, \ldots \qquad (2.4.2)$$

As implied by (2.4.2), the computation of $y(n)$ requires the storage of all the input samples $x(k)$ for $0 \le k \le n$. Since n is increasing, we appararently need a memory that is growing linearly with time.

Our intuition suggests, however, that $y(n)$ can be computed more efficiently by utilizing the previous output value $y(n-1)$. Indeed, by a simple algebraic rearrangement of (2.4.2), we obtain

$$(n+1)y(n) = \sum_{k=0}^{n-1} x(k) + x(n)$$

$$= ny(n-1) + x(n)$$

and hence

$$y(n) = \frac{n}{n+1} y(n-1) + \frac{1}{n+1} x(n) \qquad (2.4.3)$$

Now, the cumulative average $y(n)$ can be computed by multiplying the previous output value $y(n-1)$ by $n/(n+1)$, multiplying the present input $x(n)$ by $1/(n+1)$, and adding the two products. Thus the computation of $y(n)$ by means of (2.4.3) requires two multiplications, one addition, and one memory location, as illustrated in Fig. 2.32.

DEFINITION: *A system whose output $y(n)$ at time n depends on any number of past output values $y(n-1)$, $y(n-2)$, . . . is called a* recursive *system.*

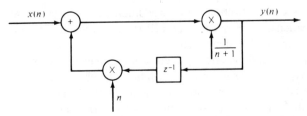

FIGURE 2.32 Realization of a recursive cumulative averaging system.

To determine the computation of the recursive system in (2.4.3) in more detail, suppose that we begin the process with $n = 0$ and proceed forward in time. Thus, according to (2.4.3), we obtain

$$
\begin{aligned}
y(0) &= x(0) \\
y(1) &= \tfrac{1}{2}y(0) + \tfrac{1}{2}x(1) \\
y(2) &= \tfrac{2}{3}y(1) + \tfrac{1}{3}x(2)
\end{aligned}
$$

and so on. If one grows fatigued with this computation and wishes to pass the problem to someone else at some time, say $n = n_0$, the only information with which one needs to provide his or her successor is the past value $y(n_0 - 1)$ and the new input samples $x(n)$, $x(n + 1)$, and so on. Thus the successor begins with

$$
y(n_0) = \frac{n_0}{n_0 + 1} y(n_0 - 1) + \frac{1}{n_0 + 1} x(n_0)
$$

and proceeds forward in time until some time, say $n = n_1$, when he or she becomes fatigued and passes the computational burden to someone else with the information on the value $y(n_1 - 1)$, and so on.

The point we wish to make in this discussion is that if one wishes to compute the response (in this case, the cumulative average) of the system (2.4.3) to an input signal $x(n)$ applied at $n = n_0$, we need the value $y(n_0 - 1)$ and the input samples $x(n)$ for $n \ge n_0$. The term $y(n_0 - 1)$ is called the *initial condition* for the system in (2.4.3) and contains all the essential information needed to determine the response of the system for $n \ge n_0$ to the input signal $x(n)$, independent of what has occurred in the past.

The following example illustrates the use of a (nonlinear) recursive system to compute the square root of a number.

EXAMPLE 2.4.1 Square-Root Algorithm _____

Many computers and calculators compute the square root of a positive number A, using the iterative algorithm

$$
s_n = \frac{1}{2}\left(s_{n-1} + \frac{A}{s_{n-1}}\right) \qquad n = 0, 1, \ldots
$$

where s_{-1} is an initial guess (estimate) of $\sqrt{A}$. As the iteration converges we have $s_n \approx s_{n-1}$. Then it easily follows that $s_n \approx \sqrt{A}$.

Consider now the recursive system

$$
y(n) = \frac{1}{2}\left[y(n - 1) + \frac{x(n)}{y(n - 1)}\right] \tag{2.4.4}
$$

which is realized as in Fig. 2.33. If we excite this system with a step of amplitude A [i.e., $x(n) = Au(n)$] and use as an initial condition $y(-1)$ an estimate of $\sqrt{A}$, the response

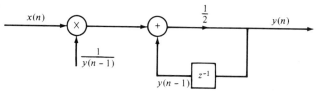

FIGURE 2.33 Realization of the square-root system.

$y(n)$ of the system will tend toward $\sqrt{A}$ as n increases. Note that in contrast to the system (2.4.3), we do not need to specify exactly the initial condition. A rough estimate is sufficient for the proper performance of the system. For example, if we let $A = 2$ and $y(-1) = 1$, we obtain $y(0) = \frac{3}{2}$, $y(1) = 1.4166667$, $y(2) = 1.4142157$. Similarly, for $y(-1) = 1.5$, we have $y(0) = 1.416667$, $y(1) = 1.4142157$. Compare these values with the $\sqrt{2}$, which is approximately 1.4142136.

We have now introduced two simple recursive systems, where the output $y(n)$ depends on the previous output value $y(n - 1)$ and the current input $x(n)$. Both systems are causal. In general, we can formulate more complex causal recursive systems in which the output $y(n)$ is a function of several past output values and present and past inputs. The system should have a finite number of delays or, equivalently, should require a finite number of storage locations in order to be practically implemented. Thus the output of a causal and practically realizable recursive system can be expressed in general as

$$y(n) = F[y(n - 1), y(n - 2), \ldots, y(n - N), x(n), x(n - 1) \ldots, x(n - M)]$$
(2.4.5)

where $F[\cdot]$ denotes some function of its arguments. This is a recursive equation specifying a procedure for computing the system output in terms of previous values of the output and present and past inputs.

In contrast, if $y(n)$ depends only on the present and past inputs, then

$$y(n) = F[x(n), x(n - 1), \ldots, x(n - M)]$$
(2.4.6)

Such a system is called *nonrecursive*. We hasten to add that the causal FIR systems described in Section 2.3.7 in terms of the convolution sum formula have the form of (2.4.6). Indeed, the convolution summation for a causal FIR system is

$$
\begin{aligned}
y(n) &= \sum_{k=0}^{M} h(k)x(n - k) \\
&= h(0)x(n) + h(1)x(n - 1) + \cdots + h(M)x(n - M) \\
&= F[x(n), x(n - 1), \ldots, x(n - M)]
\end{aligned}
$$

where the function $F[\cdot]$ is simply a linear weighted sum of present and past inputs and the impulse response values $h(n)$, $0 \leq n \leq M$, constitute the weighting coefficients. Consequently, the causal linear time-invariant FIR systems described by the convolution formula in Section 2.3.7 are nonrecursive. The basic differences between nonrecursive and recursive systems are illustrated in Fig. 2.34. A simple inspection of this figure reveals that the fundamental difference between these two systems is the feedback loop in the recursive system, which feeds back the output of the system into the input. This feedback loop contains a delay element. The presence of this delay is crucial for the realizability of the system, since the absence of this delay would force the system to compute $y(n)$ in terms of $y(n)$, which is not possible for discrete-time systems.

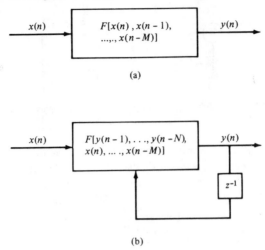

FIGURE 2.34 Basic form for a causal and realizable (a) nonrecursive and (b) recursive system.

The presence of the feedback loop or, equivalently, the recursive nature of (2.4.5) creates another important difference between recursive and nonrecursive systems. For example, suppose that we wish to compute the output $y(n_0)$ of a system when it is excited by an input applied at time $n = 0$. If the system is recursive, in order to compute $y(n_0)$, we need to compute all the previous values $y(0), y(1), \ldots, y(n_0 - 1)$ first. In contrast, if the system is nonrecursive, we can compute the output $y(n_0)$ immediately without having $y(n_0 - 1), y(n_0 - 2), \ldots$ In conclusion, the output of a recursive system should be computed in order [i.e., $y(0), y(1), y(2), \ldots$], whereas for a nonrecursive system, the output can be computed in any order [i.e., $y(200), y(15), y(3), y(300)$, etc.]. This feature is desirable in some practical applications.

2.4.2 Linear Time-Invariant Systems Characterized by Constant-Coefficient Difference Equations

In Section 2.3 we treated linear time-invariant systems and characterized them in terms of their impulse response. In this subsection we focus our attention on a family of linear time-invariant systems described by an input–output relation called a difference equation with constant coeffficients. Systems described by constant-coefficient linear difference equations are a subclass of the recursive and nonrecursive systems introduced in the preceding subsection. To bring out the important ideas we begin by treating a simple recursive system described by a first-order difference equation.

Suppose that we have a recursive system with an input–output equation

$$y(n) = ay(n - 1) + x(n) \qquad (2.4.7)$$

where a is a constant. Figure 2.35 shows a block diagram realization of the system. In comparing this system with the cumulative averaging system described by the input–output equation (2.4.3), we observe that the system in (2.4.7) has a constant coefficient (independent of time), whereas the system described in (2.4.3) has time-variant coefficients. As we show below, (2.4.7) is an input–output equation for a linear time-invariant system, whereas (2.4.3) describes a linear time-variant system.

Now, suppose that we apply an input signal $x(n)$ to the system for $n \geq 0$. We make no assumptions about the input signal for $n < 0$, but we do assume the existence of the

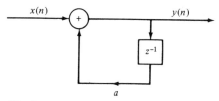

FIGURE 2.35   Block diagram realization of a simple recursive system.

initial condition $y(-1)$. Since (2.4.7) describes the system output implicitly, we must solve this equation to obtain an explicit expression for the system output. Suppose that we compute successive values of $y(n)$ for $n \geq 0$, beginning with $y(0)$. Thus

$$y(0) = ay(-1) + x(0)$$
$$y(1) = ay(0) + x(1) = a^2y(-1) + ax(0) + x(1)$$
$$y(2) = ay(1) + x(2) = a^3y(-1) + a^2x(0) + ax(1) + x(2)$$
$$\vdots$$

$$y(n) = ay(n - 1) + x(n)$$
$$\quad = a^{n+1}y(-1) + a^nx(0) + a^{n-1}x(1) + \cdots + ax(n - 1) + x(n)$$

or, more compactly,

$$y(n) = a^{n+1}y(-1) + \sum_{k=0}^{n} a^k x(n - k) \qquad n \geq 0 \qquad (2.4.8)$$

The response $y(n)$ of the system as given by the right-hand side of (2.4.8) consists of two parts. The first part, which contains the term $y(-1)$, is a result of the initial condition $y(-1)$ of the system. The second part is the response of the system to the input signal $x(n)$.

If the system is initially relaxed at time $n = 0$, then its memory (i.e., the output of the delay) should be zero. Hence $y(-1) = 0$. Thus a recursive system is relaxed if it starts with zero initial conditions. Because the memory of the system describes, in some sense, its "state," we say that the system is at zero state and its corresponding output is called the *zero-state response* or *forced response*, and is denoted by $y_{zs}(n)$. Obviously, the zero-state response or forced response of the system (2.4.7) is given by

$$y_{zs}(n) = \sum_{k=0}^{n} a^k x(n - k) \qquad n \geq 0 \qquad (2.4.9)$$

It is interesting to note that (2.4.9) is a convolution summation involving the input signal convolved with the impulse response

$$h(n) = a^n u(n) \qquad (2.4.10)$$

We also observe that the system described by the first-order difference equation in (2.4.7) is causal. As a result, the lower limit on the convolution summation in (2.4.9) is $k = 0$. Furthermore, the condition $y(-1) = 0$ implies that the input signal can be assumed causal and hence the upper limit on the convolution summation in (2.4.9) is n, since $x(n - k) = 0$ for $k > n$. In effect, we have obtained the result that the relaxed recursive system described by the first-order difference equation in (2.4.7) is a linear time-invariant IIR system with impulse response given by (2.4.10).

Now, suppose that the system described by (2.4.7) is initially nonrelaxed [i.e., $y(-1) \neq 0$] and the input $x(n) = 0$ for all n. Then the output of the system with zero input is called the *zero-input response* or *natural response* and is denoted by $y_{zi}(n)$. From (2.4.7), with $x(n) = 0$ for $-\infty < n < \infty$, we obtain

$$y_{zi}(n) = a^{n+1}y(-1) \qquad n \geq 0 \qquad (2.4.11)$$

We observe that a recursive system with nonzero initial condition is nonrelaxed in the sense that it can produce an output without being excited. Note that the zero-input response is due to the memory of the system.

To summarize, the zero-input response is obtained by setting the input signal to zero, and hence it is independent of the input. It depends only on the nature of the system and the initial condition. Thus the zero-input response is a characteristic of the system itself, and hence it is also known as the *natural* or *free response* of the system. On the other hand, the zero-state response depends on the nature of the system and the input signal. Since this output is a response forced upon it by the input signal, it is usually called the *forced response* of the system.

The system described by the first-order difference equation in (2.4.7) is the simplest possible recursive system in the general class of recursive systems described by linear constant-coefficient difference equations. The general form for such an equation is

$$y(n) = -\sum_{k=1}^{N} a_k y(n - k) + \sum_{k=0}^{M} b_k x(n - k) \qquad (2.4.12)$$

or, equivalently,

$$\sum_{k=0}^{N} a_k y(n - k) = \sum_{k=0}^{M} b_k x(n - k) \qquad a_0 \equiv 1 \qquad (2.4.13)$$

The integer N is called the *order* of the difference equation or the order of the system. The negative sign on the right-hand side of (2.4.12) is introduced as a matter of convenience to allow us to express the difference equation in (2.4.13) without any negative signs.

Equation (2.4.12) expresses the output of the system at time n directly as a weighted sum of past outputs $y(n - 1)$, $y(n - 2)$, . . . $y(n - N)$ as well as past and present input signals samples. We observe that in order to determine $y(n)$ for $n \geq 0$, we need the input $x(n)$ for all $n \geq 0$ and the initial conditions $y(-1)$, $y(-2)$, . . . , $y(-N)$. In other words, the initial conditions summarize all that we need to know about the past history of the response of the system in order to compute the present and future outputs. The general solution of the N-order constant-coefficient difference equation is considered in the following subsection.

At this point we restate the properties of linearity, time invariance, and stability in the context of recursive systems described by linear constant-coefficient difference equations. As we have observed, a recursive system may be relaxed or nonrelaxed, depending on the initial conditions. Hence the definitions of these properties must take into account the presence of the initial conditions.

We begin with the definition of linearity.

DEFINITION: *A system is linear if it satisfies the following three requirements:*

1. The total response is equal to the sum of the zero-input and zero-state responses [i.e., $y(n) = y_{zi}(n) + y_{zs}(n)$].
2. The principle of superposition applies to the zero-state response (*zero-state linear*).
3. The principle of superposition applies to the zero-input response (*zero-input linear*).

A system that does not satisfy *all three* separate requirements is by definition nonlinear. Obviously, for a relaxed system, $y_{zi}(n) = 0$, and thus requirement 2, which is the definition of linearity given in Section 2.2.4, is sufficient.

We illustrate the application of these requirements by a simple example.

EXAMPLE 2.4.2

Determine if the recursive system defined by the difference equation

$$y(n) = ay(n - 1) + x(n)$$

is linear.

Solution: By combining (2.4.9) and (2.4.11), we obtain (2.4.8), which may be expressed as

$$y(n) = y_{zi}(n) + y_{zs}(n)$$

Thus the first requirement for linearity is satisfied.

To check for the second requirement, let us assume that $x(n) = c_1 x_1(n) + c_2 x_2(n)$. Then (2.4.9) gives

$$y_{zs}(n) = \sum_{k=0}^{n} a^k [c_1 x_1(n - k) + c_2 x_2(n - k)]$$

$$= c_1 \sum_{k=0}^{n} a^k x_1(n - k) + c_2 \sum_{k=0}^{n} a^k x_2(n - k)$$

$$= c_1 y_{zs}^{(1)}(n) + c_2 y_{zs}^{(2)}(n)$$

Hence $y_{zs}(n)$ satisfies the principle of superposition, and thus the system is zero-state linear.

Now let us assume that $y(-1) = c_1 y_1(-1) + c_2 y_2(-1)$. From (2.4.11) we obtain

$$y_{zi}(n) = a^{n+1} [c_1 y_1(-1) + c_2 y_2(-1)]$$
$$= c_1 a^{n+1} y_1(-1) + c_2 a^{n+1} y_2(-1)$$
$$= c_1 y_{zi}^{(1)}(n) + c_2 y_{zi}^{(2)}(n)$$

Hence the system is zero-input linear.

Since the system satisfies all three conditions for linearity, it is linear.

Although it is somewhat tedious, the procedure used in Example 2.4.2 to demonstrate linearity for the system described by the first-order difference equation carries over directly to the general recursive systems described by the constant-coefficient difference equation given in (2.4.13). Hence a recursive system described by the linear difference equation in (2.4.13) also satisfies all three conditions in the definition of linearity, and hence it is linear.

The next question that arises is whether or not the causal linear system described by the linear constant-coefficient difference equation in (2.4.13) is time invariant. This is a fairly easy task, when dealing with systems described by explicit input–output mathematical relationships. Clearly, the system described by (2.4.13) is time invariant because the coefficients a_k and b_k are constants. On the other hand, if one or more of these coefficients depends on time, the system is time variant, since its properties change as a function of time. Thus we conclude that *the recursive system described by a linear constant-coefficient difference equation is linear and time invariant.*

The final issue is the stability of the recursive system described by the linear, constant-coefficient difference equation in (2.4.13). In Section 2.3.6 we introduced the concept of bounded input–bounded output (BIBO) stability for relaxed systems. For nonrelaxed systems that may be nonlinear, BIBO stability should be viewed with some care. However, in the case of a linear time-invariant recursive system described by the linear constant-coefficient difference equation in (2.4.13), it suffices to state that such a system is BIBO stable if and only if for every bounded input and every bounded initial condition, the total system response is bounded.

EXAMPLE 2.4.3

Determine if the linear time-invariant recursive system described by the difference equation given in (2.4.7) is stable.

Solution: Let us assume that the input signal $x(n)$ is bounded in amplitude, that is, $|x(n)| \leq M_x < \infty$ for all $n \geq 0$. From (2.4.8) we have

$$|y(n)| \leq |a^{n+1}y(-1)| + \left| \sum_{k=0}^{n} a^k x(n-k) \right| \qquad n \geq 0$$

$$\leq |a|^{n+1} |y(-1)| + M_x \sum_{k=0}^{n} |a|^k \qquad n \geq 0$$

$$\leq |a|^{n+1} |y(-1)| + M_x \frac{1 - |a|^{n+1}}{1 - |a|} = M_y \qquad n \geq 0$$

If n is finite, the bound M_y is finite and the output is bounded independently of the value of a. However, as $n \rightarrow \infty$, the bound M_y remains finite only if $|a| < 1$, because $|a|^n \rightarrow 0$ as $n \rightarrow \infty$. Then $M_y = 1/(1 - |a|)$.

Thus the system is stable only if $|a| < 1$.

For the simple first-order system in Example 2.4.3, we were able to express the condition for BIBO stability in terms of the system parameter a, namely $|a| < 1$. We should stress, however, that this task becomes more difficult for higher-order systems. Fortunately, other simple and more efficient techniques exist for investigating the stability of recursive systems, as we shall see in subsequent chapters.

2.4.3 Solution of Linear Constant-Coefficient Difference Equations

Given a linear constant-coefficient difference equation as the input–output relationship describing a linear time-invariant system, our objective in this subsection is to determine an explicit expression for the output $y(n)$. The method that is developed is termed the *direct method*. An alternative method based on the z-transform is described in Chapter 3. For reasons that will become apparent later, the z-transform approach is called the *indirect method*.

The Homoegeneous Solution and the Zero-Input Response. We begin the problem of solving the linear constant-coefficient difference equation given by (2.4.13) by assuming that the input $x(n) = 0$. Thus we will first obtain the solution to the *homogeneous difference equation*

$$\sum_{k=0}^{N} a_k y(n-k) = 0 \qquad (2.4.14)$$

from which we can obtain the zero-input response.

The procedure for solving a linear constant-coefficient difference equation directly is very similar to the procedure for solving a linear constant-coefficient differential equation. Basically, we assume that the solution is in the form of an exponential, that is,

$$y_h(n) = \lambda^n \qquad (2.4.15)$$

when the subscript h on $y(n)$ is used to denote the solution to the homogeneous difference equation. If we substitute this assumed solution in (2.4.14), we obtain the polynomial equation

$$\sum_{k=0}^{N} a_k \lambda^{n-k} = 0$$

or

$$\lambda^{n-N}(\lambda^N + a_1 \lambda^{N-1} + a_2 \lambda^{N-2} + \cdots + a_{N-1}\lambda + a_N) = 0 \qquad (2.4.16)$$

The polynomial in parentheses is called the *characteristic polynomial* of the system. In general, it has N roots, which we will denote as $\lambda_1, \lambda_2, \ldots, \lambda_N$. The roots may be real or complex valued. If the coefficients $a_1, a_2, \ldots, a_N$ are real, as is usually the case in practice, complex-valued roots will occur in complex-conjugate pairs. Some of the N roots may be identical, in which case we have multiple-order roots.

For the moment, let us assume that the roots are distinct; that is, there are no multiple-order roots. Then the most general solution to the homogeneous difference equation in (2.4.14) is

$$y_h(n) = C_1\lambda_1^n + C_2\lambda_2^n + \cdots + C_N\lambda_N^n \qquad (2.4.17)$$

where $C_1, C_2, \ldots, C_N$ are weighting coefficients.

These coefficients are determined from the initial conditions specified for the system. Since the input $x(n) = 0$, (2.4.17) may be used to obtain the *zero-input response* of the system. The following examples illustrate the procedure.

EXAMPLE 2.4.4

Determine the zero-input response of the system described by the first-order difference equation

$$y(n) + a_1 y(n-1) = 0 \qquad (2.4.18)$$

Solution: The assumed solution is

$$y_h(n) = \lambda^n$$

When we substitute this solution in (2.4.18), we obtain

$$\lambda^n + a_1 \lambda^{n-1} = 0$$
$$\lambda^{n-1}(\lambda + a_1) = 0$$
$$\lambda = -a_1$$

Therefore, the solution to the homogeneous difference equation is

$$y_k(n) = C\lambda^n = C(-a_1)^n \qquad (2.4.19)$$

To determine the zero-input response of the system, we must evaluate C by using an initial condition. Now, from (2.4.18) we have

$$y(0) = -a_1 y(-1)$$

On the other hand, from (2.4.19) we have

$$y_h(0) = C$$

and hence the zero-input response of the system is

$$y_{zi}(n) = (-a_1)^{n+1}y(-1) \qquad n \geq 0 \qquad (2.4.20)$$

With $a = -a_1$, this result is consistent with (2.4.11) for the first-order system, which was obtained earlier by iteration of the difference equation.

EXAMPLE 2.4.5 ───

Determine the zero-input response of the system described by the second-order difference equation

$$y(n) - 3y(n - 1) - 4y(n - 2) = 0 \qquad (2.4.21)$$

Solution: First we determine the solution to the homogeneous equation. We assume the solution to be the exponential

$$y_h(n) = \lambda^n$$

Upon substitution of this solution into (2.4.21), we obtain the characteristic equation

$$\lambda^n - 3\lambda^{n-1} - 4\lambda^{n-2} = 0$$
$$\lambda^{n-2}(\lambda^2 - 3\lambda - 4) = 0$$

Therefore, the roots are $\lambda = -1, 4$, and the general form of the solution to the homogeneous equation is

$$y_h(n) = C_1\lambda_1^n + C_2\lambda_2^n \qquad (2.4.22)$$
$$= C_1(-1)^n + C_2(4)^n$$

The zero-input response of the system can be obtained from the homogenous solution by evaluating the constants in (2.4.22), given the initial conditions $y(-1)$ and $y(-2)$. From the difference equation in (2.4.21) we have

$$y(0) = 3y(-1) + 4y(-2)$$
$$y(1) = 3y(0) + 4y(-1)$$
$$\quad = 3[3y(-1) + 4y(-2)] + 4y(-1)$$
$$\quad = 13y(-1) + 12y(-2)$$

On the other hand, from (2.4.22) we obtain

$$y(0) = C_1 + C_2$$
$$y(1) = -C_1 + 4C_2$$

By equating these two sets of relations, we have

$$C_1 + C_2 = 3y(-1) + 4y(-2)$$
$$-C_1 + 4C_2 = 13y(-1) + 12y(-2)$$

The solution of these two equations is

$$C_1 = -\tfrac{1}{5}y(-1) + \tfrac{4}{5}y(-2)$$
$$C_2 = \tfrac{16}{5}y(-1) + \tfrac{16}{5}y(-2)$$

Therefore, the zero input response of the system is

$$y_{zi}(n) = [-\tfrac{1}{5}y(-1) + \tfrac{4}{5}y(-2)](-1)^n \qquad (2.4.23)$$
$$+ [\tfrac{16}{5}y(-1) + \tfrac{16}{5}y(-2)](4)^n \qquad n \geq 0$$

For example, if $y(-2) = 0$ and $y(-1) = 5$, then $C_1 = -1$, $C_2 = 16$, and hence

$$y_{zi}(n) = (-1)^{n+1} + (4)^{n+2} \qquad n \geq 0$$

The examples above illustrate the method for obtaining the homogeneous solution and the zero-input response of the system when the characteristic equation contains distinct roots. On the other hand, if the characteristic equation contains multiple roots, the form of the solution given in (2.4.17) must be modified. For example, if λ_1 is a root of multiplicity m, then (2.4.17) becomes

$$y_h(n) = C_1\lambda_1^n + C_2 n\lambda_1^n + C_3 n^2\lambda_1^n + \cdots + C_m n^{m-1}\lambda_1^n \qquad (2.4.24)$$
$$+ C_{m+1}\lambda_{m+1}^n + \cdots + C_N\lambda_n$$

The Particular Solution of the Difference Equation. Having determined the solution of the Nth-order difference equation with $x(n) = 0$, let us now determine the solution when $x(n)$ is nonzero. Basically, our approach is the same as that used to solve the homogeneous difference equation, namely, to assume a form of the solution. Thus we obtain the *particular solution* to the difference equation. The following examples demonstrate the procedure.

EXAMPLE 2.4.6

Determine the particular solution of the first-order difference equation

$$y(n) + a_1 y(n-1) = x(n) \qquad |a_1| < 1 \qquad (2.4.25)$$

when the input $x(n)$ is a unit step sequence, that is,

$$x(n) = u(n)$$

Solution: Since the input sequence $x(n)$ is a constant for $n \geq 0$, the form of the solution that we assume is also a constant. Hence the assumed solution of the difference equation to the forcing function $x(n)$, called the *particular solution* of the difference equation, is

$$y_p(n) = Ku(n)$$

where K is a scale factor which is to be determined so that it satisfies (2.4.25), and the subscript p on $y(n)$ denotes that this is the particular solution. Upon substitution of this assumed solution into (2.4.25), we obtain

$$Ku(n) + a_1 Ku(n-1) = u(n)$$

To determine K, we must evaluate this equation for any $n \geq 1$, where none of the terms vanish (i.e., in the steady state). Thus

$$K + a_1 K = 1$$
$$K = \frac{1}{1 + a_1}$$

Therefore, the particular solution to the difference equation is

$$y_p(n) = \frac{1}{1 + a_1} u(n) \qquad (2.4.26)$$

In the example above, the input $x(n)$, $n \geq 0$, was a constant and the form assumed for the particular solution is also a constant. If $x(n)$ is an exponential, we would assume that the particular solution is also an exponential. If $x(n)$ were a sinusoid, then $y_p(n)$ would also be a sinusoid. Thus our assumed form for the particular solution takes the basic form of the signal $x(n)$. Table 2.2 provides the general form of the particular solution for several types of excitation.

TABLE 2.2   General Form of the Particular Solution for Several Types of Input Signals

Input Signal, $x(n)$	Particular Solution, $y_p(n)$
A (constant)	K
AM^n	KM^n
An^M	$K_0 n^M + K_1 n^{M-1} + \cdots + K_M$
$A^n n^M$	$A^n(K_0 n^M + K_1 n^{M-1} + \cdots + K_M)$
$\left\{ \begin{matrix} A \cos \omega_0 n \\ A \sin \omega_0 n \end{matrix} \right\}$	$K_1 \cos \omega_0 n + K_2 \sin \omega_0 n$

EXAMPLE 2.4.7

Determine the particular solution of the difference equation

$$y(n) = \tfrac{5}{6} y(n - 1) - \tfrac{1}{6} y(n - 2) + x(n) \qquad (2.4.27)$$

when the forcing function $x(n) = 2^n$, $n \geq 0$ and zero elsewhere.

Solution: The form of the particular solution is

$$y_p(n) = K2^n \qquad n \geq 0$$

Upon substitution of $y_p(n)$ into the difference equation (2.4.27), we obtain

$$K2^n u(n) = \tfrac{5}{6} K2^{n-1} u(n - 1) - \tfrac{1}{6} K2^{n-2} u(n - 2) + 2^n u(n)$$

To determine the value of K, we can evaluate this equation for any $n \geq 2$, where none of the terms vanish. Thus we obtain

$$4K = \tfrac{5}{6}(2K) - \tfrac{1}{6}K + 4$$

and hence $K = \tfrac{8}{5}$. Therefore, the particular solution is

$$y_p(n) = \tfrac{8}{5} 2^n \qquad n \geq 0$$

We have now demonstrated how to determine the two components of the solution to a difference equation with constant coefficients. These two components are the homogeneous solution and the particular solution. From these two components, we construct the total solution from which we can obtain the zero-state response.

The Total Solution and the Zero-State Response of the System. The linearity property of the linear constant-coefficient difference equation allows us to add the homogeneous solution and the particular solution in order to obtain the *total solution*. Thus

$$y(n) = y_h(n) + y_p(n)$$

The resultant sum $y(n)$ contains the constant parameters $\{C_i\}$ embodied in the homogeneous solution component $y_h(n)$. These constants can be determined to satisfy the initial conditions. The following example illustrates the procedure.

EXAMPLE 2.4.8

Determine the total solution $y(n)$, $n \geq 0$, to the difference equation

$$y(n) + a_1 y(n - 1) = x(n) \tag{2.4.28}$$

when $x(n)$ is a unit step sequence [i.e., $x(n) = u(n)$] and $y(-1)$ is the initial condition.

Solution: From (2.4.19) of Example 2.4.4, the homogeneous solution is

$$y_h(n) = C(-a_1)^n$$

and from (2.4.26) of Example 2.4.6, the particular solution is

$$y_p(n) = \frac{1}{1 + a_1}$$

Consequently, the total solution is

$$y(n) = C(-a_1)^n + \frac{1}{1 + a_1} \qquad n \geq 0 \tag{2.4.29}$$

where the constant C is determined to satisfy the initial condition $y(-1)$.

In particular, suppose that we wish to obtain the zero-state response of the system described by the first-order difference equation in (2.4.28). Then we set $y(-1) = 0$. To evaluate C, we evaluate (2.4.28) at $n = 0$. Thus we obtain

$$y(0) + a_1 y(-1) = 1$$
$$y(0) = 1$$

On the other hand, (2.4.29) evaluated at $n = 0$ yields

$$y(0) = C + \frac{1}{1 + a_1}$$

Consequently,

$$C + \frac{1}{1 + a_1} = 1$$

$$C = \frac{a_1}{1 + a_1}$$

Substitution for C into (2.4.29) yields the zero-state response of the system

$$y_{zs}(n) = \frac{1 - (-a_1)^{n+1}}{1 + a_1} \qquad n \geq 0$$

If we evaluate the parameter C in (2.4.29) under the condition that $y(-1) \neq 0$, the total solution will include the zero-input response as well as the zero-state response of the system. In this case (2.4.28) yields

$$y(0) + a_1 y(-1) = 1$$
$$y(0) = -a_1 y(-1) + 1$$

On the other hand, (2.4.29) yields

$$y(0) = C + \frac{1}{1 + a_1}$$

By equating these two relations, we obtain

$$C + \frac{1}{1 + a_1} = -a_1 y(-1) + 1$$

$$C = -a_1 y(-1) + \frac{a_1}{1 + a_1}$$

Finally, if we substitute this value of C into (2.4.29), we obtain

$$y(n) = (-a_1)^{n+1} y(-1) + \frac{1 - (-a_1)^{n+1}}{1 + a_1} \qquad n \geq 0 \tag{2.4.30}$$

$$= y_{zi}(n) + y_{zs}(n)$$

We observe that the system response as given by (2.4.30) is consistent with the response $y(n)$ given in (2.4.8) for the first-order system (with $a = -a_1$), which was obtained by solving the difference equation iteratively. Furthermore, we note that the value of the constant C depends both on the initial condition $y(-1)$ and on the excitation function. Consequently, the value of C influences both the zero-input response and the zero-state response. On the other hand, if we wish to obtain the zero-state response only, we simply solve for C under the condition that $y(-1) = 0$, as demonstrated in Example 2.4.8.

We further observe that the particular solution to the difference equation can be obtained from the zero-state response of the system. Indeed, if $|a_1| < 1$, which is the condition for stability of the system, as will be shown in Section 2.4.4, the limiting value of $y_{zs}(n)$ as n approaches infinity, is the particular solution, that is,

$$y_p(n) = \lim_{n \to \infty} y_{zs}(n) = \frac{1}{1 + a_1}$$

Since this component of the system response does not go to zero as n approaches infinity, it is usually called the *steady-state response* of the system. This response persists as long as the input persists. The component that dies out as n approaches infinity is called the *transient response* of the system.

EXAMPLE 2.4.9

Determine the response $y(n)$, $n \geq 0$, of the system described by the second-order difference equation

$$y(n) - 3y(n - 1) - 4y(n - 2) = x(n) + 2x(n - 1) \tag{2.4.31}$$

when the input sequence is

$$x(n) = 4^n u(n)$$

Solution: We have already determined the solution to the homogeneous difference equation for this system in Example 2.4.5. From (2.4.22) we have

$$y_h(n) = C_1(-1)^n + C_2(4)^n \tag{2.4.32}$$

The particular solution to (2.4.31) is assumed to be an exponential sequence of the same form as $x(n)$. Normally, we could assume a solution of the form

$$y_p(n) = K(4)^n u(n)$$

However, we observe that $y_p(n)$ is already contained in the homogeneous solution, so that this particular solution is redundant. Instead, we select the particular solution to be linearly independent of the terms contained in the homogeneous solution. In fact, we treat this situation in the same manner as we have already treated multiple roots in the characteristic equation. Thus we assume that

$$y_p(n) = Kn(4)^n u(n) \tag{2.4.33}$$

Upon substitution of (2.4.33) into (2.4.31), we obtain

$$Kn(4)^n u(n) - 3K(n - 1)(4)^{n-1} u(n - 1) - 4K(n - 2)(4)^{n-2} u(n - 2)$$
$$= (4)^n u(n) + 2(4)^{n-1} u(n - 1)$$

To determine K, we evaluate this equation for any $n \geq 2$, where none of the unit step terms vanish. To simplify the arithmetic, we select $n = 2$, from which we obtain $K = \frac{6}{5}$. Therefore,

$$y_p(n) = \tfrac{6}{5} n(4)^n u(n) \tag{2.4.34}$$

The total solution to the difference equation is obtained by adding (2.4.32) to (2.4.34). Thus

$$y(n) = C_1(-1)^n + C_2(4)^n + \tfrac{6}{5} n(4)^n \qquad n \geq 0 \tag{2.4.35}$$

where the constants C_1 and C_2 are determined to satisfy the initial conditions. To accomplish this, we return to (2.4.31), from which we obtain

$$y(0) = 3y(-1) + 4y(-2) + 1$$
$$y(1) = 3y(0) + 4y(-1) + 6$$
$$= 13y(-1) + 12y(-2) + 9$$

On the other hand, (2.4.35) evaluated at $n = 0$ and $n = 1$ yields

$$y(0) = C_1 + C_2$$
$$y(1) = -C_1 + 4C_2 + \tfrac{24}{5}$$

We can now equate these two sets of relations to obtain C_1 and C_2. In so doing, we will have the response due to initial conditions $y(-1)$ and $y(-2)$ (the zero-input response) and the zero-state or forced response.

Since we have already solved for the zero-input response in Example 2.4.5, we can simplify the computations above by setting $y(-1) = y(-2) = 0$. Then we have

$$C_1 + C_2 = 1$$
$$-C_1 + 4C_2 + \tfrac{24}{5} = 9$$

Hence $C_1 = -\frac{1}{25}$ and $C_2 = \frac{26}{25}$. Finally, we have the zero-state response to the forcing function $x(n) = (4)^n u(n)$ in the form

$$y_{zs}(n) = -\tfrac{1}{25}(-1)^n + \tfrac{26}{25}(4)^n + \tfrac{6}{5} n(4)^n \qquad n \geq 0 \tag{2.4.36}$$

The total response of the system, which includes the response to arbitrary initial conditions, is the sum of (2.4.23) and (2.4.36).

2.4.4 The Impulse Response of a Linear Time-Invariant Recursive System

The impulse response of a relaxed linear time-invariant system was previously defined as the response of the system to a unit sample excitation [i.e., $x(n) = \delta(n)$]. In the case of a recursive system, $h(n)$ is simply equal to the zero-state response of the system when the input $x(n) = \delta(n)$.

For example, in the simple first-order recursive system given in (2.4.7) the zero-state response, given in (2.4.8), is

$$y_{zs}(n) = \sum_{k=0}^{n} a^k x(n - k) \qquad (2.4.37)$$

With $x(n) = \delta(n)$ is substituted into (2.4.37), we obtain

$$y_{zs}(n) = \sum_{k=0}^{n} a^k \delta(n - k)$$

$$= a^n \qquad n \geq 0$$

Hence the impulse response of the first-order recursive system described by (2.4.7) is

$$h(n) = a^n u(n) \qquad (2.4.38)$$

as indicated in Section 2.4.2.

In the general case of an arbitrary, linear time-invariant recursive system, the zero-state response expressed in terms of the convolution summation is

$$y_{zs}(n) = \sum_{k=0}^{n} h(k)x(n - k) \qquad n \geq 0 \qquad (2.4.39)$$

When the input is an impulse [i.e., $x(n) = \delta(n)$], (2.4.39) reduces to

$$y_{zs}(n) = h(n) \qquad (2.4.40)$$

Now, let us consider the problem of determining the impulse response $h(n)$ given a linear constant-coefficient difference equation description of the system. In terms of our discussion in the preceding subsection, we have established the fact that the total response of the system to any excitation function consists of the sum of two solutions of the difference equation: the solution to the homogeneous equation plus the particular solution to the excitation function. In the case where the excitation is an impulse, the particular solution is zero, since $x(n) = 0$ for $n > 0$, that is,

$$y_p(n) = 0$$

Consequently, the response of the system to an impulse consists only of the solution to the homogeneous equation, with the $\{C_k\}$ parameters evaluated to satisfy the initial conditions dictated by the impulse. The following example illustrates the procedure for obtaining $h(n)$ given the difference equation for the system.

EXAMPLE 2.4.10 ————————————————————————————

Determine the impulse response $h(n)$ for the system described by the second-order difference equation

$$y(n) - 3y(n - 1) - 4y(n - 2) = x(n) + 2x(n - 1) \qquad (2.4.41)$$

Solution: We have already determined in Example 2.4.5 that the solution to the homogeneous difference equation for this system is

$$y_h(n) = C_1(-1)^n + C_2(4)^n \qquad n \geq 0 \qquad (2.4.42)$$

Since the particular solution is zero when $x(n) = \delta(n)$, the impulse response of the system is simply given by (2.4.42), where C_1 and C_2 must be evaluated to satisfy (2.4.41).

For $n = 0$ and $n = 1$, (2.4.41) yields

$$y(0) = 1$$
$$y(1) = 3y(0) + 2 = 5$$

where we have imposed the conditions $y(-1) = y(-2) = 0$, since the system must be relaxed. On the other hand, (2.4.42) evaluated at $n = 0$ and $n = 1$ yields

$$y(0) = C_1 + C_2$$
$$y(1) = -C_1 + 4C_2$$

By solving these two sets of equations for C_1 and C_2, we obtain

$$C_1 = -\tfrac{1}{5} \qquad C_2 = \tfrac{6}{5}$$

Therefore, the impulse response of the system is

$$h(n) = [-\tfrac{1}{5}(-1)^n + \tfrac{6}{5}(4)^n]u(n)$$

We make the observation that both the simple first-order recursive system and the second-order recursive system have impulse responses that are infinite in duration. In other words, both of these recursive systems are IIR systems. In fact, due to the recursive nature of the system, any recursive system described by a linear constant-coefficient difference equation is an IIR system. The converse is not true, however. That is, not every linear time-invariant IIR system can be described by a linear constant-coefficient difference equation. In other words, recursive systems described by linear constant-coefficient difference equations are a subclass of linear time-invariant IIR systems.

The extension of the approach demonstrated above for determining the impulse response of the first- and second-order systems generalizes in a straightforward manner. When the system is described by an Nth-order linear difference equation of the type given in (2.4.13), the solution of the homogeneous equation is

$$y_h(n) = \sum_{k=1}^{N} C_k \lambda_k^n$$

when the roots $\{\lambda_k\}$ of the characteristic polynomial are distinct. Hence the impulse response of the system is identical in form, that is,

$$h(n) = \sum_{k=1}^{N} C_k \lambda_k^n \qquad (2.4.43)$$

where the parameters $\{C_k\}$ are determined by setting the initial conditions $y(-1) = \cdots = y(-N) = 0$.

The form of $h(n)$ given above allows us easily to relate the stability of a system described by an Nth-order difference equation to the values of the roots of the characteristic polynomial. Indeed, since BIBO stability requires that the impulse response be absolutely summable, then, for a causal system, we have

$$\sum_{n=0}^{\infty} |h(n)| = \sum_{n=0}^{\infty} \left| \sum_{k=1}^{N} C_k \lambda_k^n \right|$$

$$\leq \sum_{k=1}^{N} |C_k| \sum_{n=0}^{\infty} |\lambda_k|^n$$

Now if $|\lambda_k| < 1$ for all k, then

$$\sum_{n=0}^{\infty} |\lambda_k|^n < \infty$$

and hence

$$\sum_{n=0}^{\infty} |h(n)| < \infty$$

On the other hand, if one or more of the $|\lambda_k| \geq 1$, $h(n)$ is no longer absolutely summable, and consequently, the system is unstable. Therefore, a necessary and sufficient condition for the stability of a causal IIR system described by a linear constant-coefficient difference equation is that all roots of the characteristic polynomial be less than unity in magnitude. The reader may verify that this condition carries over to the case where the system has roots of multiplicity m.

2.4.5 Structures for the Realization of Linear Time-Invariant Systems

In this subsection we describe structures for the realization of systems described by linear constant-coefficient difference equations. Additional structures for these systems are introduced in Chapter 7.

As a beginning, let us consider the first-order system

$$y(n) = -a_1 y(n-1) + b_0 x(n) + b_1 x(n-1) \qquad (2.4.44)$$

which is realized as in Fig. 2.36a. This realization uses separate delays (memory) for both the input and output signal samples and it is called a *direct form I structure*. Note that this system can be viewed as two linear time-invariant systems in cascade. The first is a nonrecursive, system described by the equation

$$v(n) = b_0 x(n) + b_1 x(n-1) \qquad (2.4.45)$$

whereas the second is a recursive system described by

$$y(n) = -a_1 y(n-1) + v(n) \qquad (2.4.46)$$

As we have seen in Section 2.3.4, however, if we interchange the order of linear time-invariant systems in cascade, the overall system response remains the same. Thus if we interchange the order of the recursive and nonrecursive systems, we obtain an alternative structure for the realization of the system described by (2.4.44). The resulting system is shown in Fig. 2.36b. From this figure we obtain the two difference equations

$$w(n) = -a_1 w(n-1) + x(n) \qquad (2.4.47)$$
$$y(n) = b_0 w(n) + b_1 w(n-1) \qquad (2.4.48)$$

which provide an alternative algorithm for computing the output of the system described by the single difference equation given in (2.4.44). In other words, the two difference equations (2.4.47) and (2.4.48) are equivalent to the single difference equation (2.4.44).

A close observation of Fig. 2.36b reveals that the two delay elements contain the same input $w(n)$ and hence the same output $w(n-1)$. Consequently, these two elements can be merged into one delay, as shown in Fig. 2.36c. In contrast to the direct form I structure, this new realization requires only one delay for the auxiliary quantity $w(n)$, and hence it is more efficient in terms of memory requirements. It is called the *direct form II structure* and it is used extensively in practical applications.

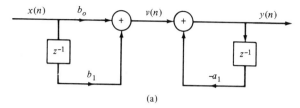

(a)

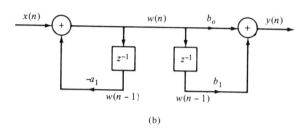

(b)

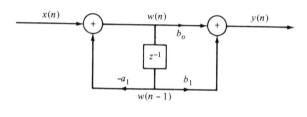

(c)

FIGURE 2.36 Steps in converting from the direct form I realization in (a) to the direct form II realization in (c).

These structures can readily be generalized for the general linear time-invariant recursive system described by the difference equation

$$y(n) = -\sum_{k=1}^{N} a_k y(n - k) + \sum_{k=0}^{M} b_k x(n - k) \qquad (2.4.49)$$

Figure 2.37 illustrates the direct form I structure for this system. This structure requires $M + N$ delays and $N + M + 1$ multiplications. It can be viewed as the cascade of a nonrecursive system

$$v(n) = \sum_{k=0}^{M} b_k x(n - k) \qquad (2.4.50)$$

and a recursive system

$$y(n) = -\sum_{k=1}^{N} a_k y(n - k) + v(n) \qquad (2.4.51)$$

By reversing the order of these two systems as was previously done for the first-order system, we obtain the direct form II structure shown in Fig. 2.38 for $N > M$. This structure is the cascade of a recursive system

$$w(n) = -\sum_{k=1}^{N} a_k w(n - k) + x(n) \qquad (2.4.52)$$

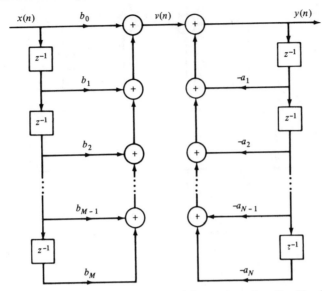

FIGURE 2.37   Direct form I structure of the system described by (2.4.49).

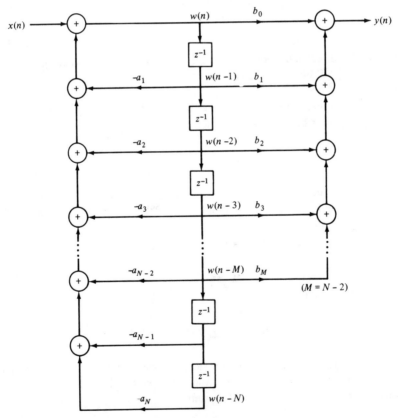

FIGURE 2.38   Direct form II structure for the system described by (2.4.49).

followed by a nonrecursive system

$$y(n) = \sum_{k=0}^{M} b_k w(n - k) \tag{2.4.53}$$

We observe that if $N \geq M$, this structure requires a number of delays equal to the order N of the system. However, if $M > N$, the required memory is specified by M. Figure 2.38 can easily by modified to handle this case. Thus the direct form II structure requires $M + N + 1$ multiplications and max $\{M, N\}$ delays. Because it requires the minimum number of delays for the realization of the system described by (2.4.49), it is sometimes called a *canonic form*.

A special case of (2.4.49) occurs if we set the system parameters $a_k = 0$, $k = 1$, $\ldots$, N. Then the input–output relationship for the system reduces to

$$y(n) = \sum_{k=0}^{M} b_k x(n - k) \tag{2.4.54}$$

which is a nonrecursive linear time-invariant system. This system views only the most recent $M + 1$ input signal samples and weights each sample by the appropriate coefficient b_k from the set $\{b_k\}$, prior to addition. In other words, the system output is basically a *weighted moving average* of the input signal. For this reason it is sometimes called a *moving average (MA) system*. Such a system is an FIR system with an impulse response $h(k)$ equal to the coefficients b_k, that is,

$$h(k) = \begin{cases} b_k & 0 \leq k \leq M \\ 0 & \text{otherwise} \end{cases} \tag{2.4.55}$$

If we return to (2.4.49) and set $M = 0$, the general linear time-invariant system reduces to a "purely recursive" system described by the difference equation

$$y(n) = -\sum_{k=1}^{N} a_k y(n - k) + b_0 x(n) \tag{2.4.56}$$

In this case the system output is a weighted linear combination of N past outputs and the present input.

Linear time-invariant systems described by a second-order difference equation are an important subclass of the more general systems described by (2.4.49) or (2.4.53) or (2.4.56). The reason for their importance will be explained later when we discuss quantization effects. Suffice it to say at this point that second-order systems are usually used as basic building blocks for realizing higher-order systems.

The most general second-order system is described by the difference equation

$$y(n) = -a_1 y(n - 1) - a_2 y(n - 2) + b_0 x(n) + b_1 x(n - 1) + b_2 x(n - 2) \tag{2.4.57}$$

which is obtained from (2.4.49) by setting $N = 2$ and $M = 2$. The direct form II structure for realizing this system is shown in Fig. 2.39a. If we set $a_1 = a_2 = 0$, then (2.4.57) reduces to

$$y(n) = b_0 x(n) + b_1 x(n - 1) + b_2 x(n - 2) \tag{2.4.58}$$

which is a special case of the FIR system described by (2.4.54). The structure for realizing this system is shown in Fig. 2.39b. Finally, if we set $b_1 = b_2 = 0$ in (2.4.57), we obtain the purely recursive second-order system described by the difference equation

$$y(n) = -a_1 y(n - 1) - a_2 y(n - 2) + b_0 x(n) \tag{2.4.59}$$

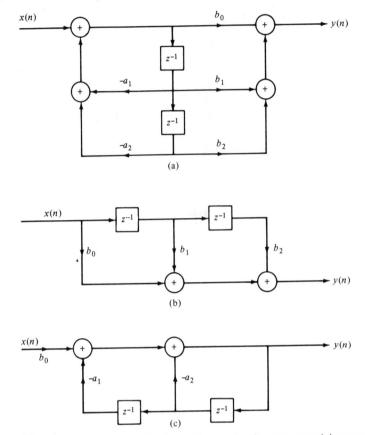

FIGURE 2.39 Structures for the realization of second-order systems: (a) general second-order system; (b) FIR system; (c) "purely recursive system"

which is a special case of (2.4.56). The structure for realizing this system is shown in Fig. 2.39c.

2.4.6 Recursive and Nonrecursive Realizations of FIR Systems

We have already made the distinction between FIR and IIR systems, based on whether the impulse response $h(n)$ of the system has a finite duration or an infinite duration. We have also made the distinction between recursive and nonrecursive systems. Basically, a causal recursive system is described by an input–output equation of the form

$$y(n) = F[y(n-1), \ldots, y(n-N), x(n), \ldots, x(n-M)] \qquad (2.4.60)$$

and for a linear time-invariant system specifically, by the difference equation

$$y(n) = -\sum_{k=1}^{N} a_k y(n-k) + \sum_{k=0}^{M} b_k x(n-k) \qquad (2.4.61)$$

On the other hand, causal nonrecursive systems do not depend on past values of the output and hence are described by an input–output equation of the form

$$y(n) = F[x(n), x(n-1), \ldots, x(n-M)] \qquad (2.4.62)$$

and for linear time-invariant systems specifically, by the difference equation in (2.4.61) with $a_k = 0$ for $k = 1, 2, \ldots, N$.

In the case of FIR systems, we have already observed that it is always possible to realize such systems nonrecursively. In fact, with $a_k = 0, k = 1, 2, \ldots, N$, in (2.4.61), we have a system with an input–output equation

$$y(n) = \sum_{k=0}^{M} b_k x(n - k) \tag{2.4.63}$$

This system is nonrecursive and FIR. As indicated in (2.4.55), the impulse response of the system is simply equal to the coefficients $\{b_k\}$. Hence every FIR system can be realized nonrecursively. On the other hand, any FIR system can also be realized recursively. Although the general proof of this statement is given later, we shall give a simple example now which will illustrate the point.

Suppose that we have an FIR system of the form

$$y(n) = \frac{1}{M + 1} \sum_{k=0}^{M} x(n - k) \tag{2.4.64}$$

for computing the *moving average* of a signal $x(n)$. Clearly, this system is FIR with impulse response

$$h(n) = \frac{1}{M + 1} \qquad 0 \le n \le M$$

Figure 2.40 illustrates the structure of the nonrecursive realization of the system. Now, suppose that we express (2.4.64) as

$$y(n) = \frac{1}{M + 1} \sum_{k=0}^{M} x(n - 1 - k) + \frac{1}{M + 1} [x(n) - x(n - 1 - M)] \tag{2.4.65}$$

$$= y(n - 1) + \frac{1}{M + 1} [x(n) - x(n - 1 - M)]$$

Now, (2.4.65) represents a recursive realization of the FIR system. The structure of this recursive realization of the moving average system is illustrated in Fig. 2.41.

In summary, we may think of the terms FIR and IIR as general characteristics that distinguish a type of linear time-invariant system, and of the terms *recursive* and *non-recursive* as descriptions of the structures for realizing or implementing the system.

2.4.7 Software Implementation of Discrete-Time Systems

The main objective of this brief treatment of the software implementation of FIR and IIR systems is to equip the reader with the necessary tools to simulate the operation of such systems on a digital computer and to observe their response to various types of input signals.

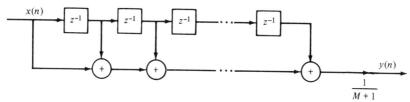

FIGURE 2.40   Nonrecursive realization of an FIR moving average system.

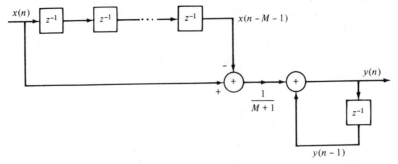

FIGURE 2.41 Recursive realization of an FIR moving average system.

In Sections 2.4.5 and 2.4.6 we described different realizations of FIR and IIR systems. In this section we discuss briefly the implementation of these two types of linear time-invariant systems in software. In particular, we consider the direct-form realization of these systems.

Let us consider the direct-form realization of the FIR system described by the equation

$$y(n) = \sum_{k=0}^{M} h(k)x(n - k) \qquad (2.4.66)$$

where the $\{h(k)\}$ are the $M + 1$ system coefficients. At time $n = n_0$, the output consists of the sum of $M + 1$ terms, that is,

$$y(n_0) = h(0)x(n_0) + h(1)x(n_0 - 1) + \cdots + h(M)x(n - M) \qquad (2.4.67)$$

The data required for the computation are $x(n)$, $n = n_0, n_0 - 1, \ldots, n_0 - M$, which are stored in a buffer and the set of $M + 1$ coefficients of the FIR system, which are also stored in memory. The entire computation may be programmed in FORTRAN with a simple DO loop, as follows:

```
        S = 0.0
        DO 10, K=1, MP1
          S = S + H(K)*XM(K)                    (2.4.68)
        10 CONTINUE
        YOUT = S
```

where MP1 $= M + 1$.

Once we have computed $y(n_0)$, we are ready to accept the new input sample $x(n_0 + 1)$ and begin computation of $y(n_0 + 1)$. If the input buffer stores the most recent $M + 1$ samples of $x(n)$, prior to the arrival of $x(n_0 + 1)$, we must discard the oldest sample [i.e., $x(n_0 - M)$] to make room for the new sample $x(n_0 + 1)$. A simple way to accomplish this is to shift the contents of the data buffer one memory location down and then to store the new sample $x(n_0 + 1)$ in the first location, which becomes available after the shifting operation. Then we repeat the computation performed with the DO loop in (2.4.68) to obtain $y(n_0 + 1)$, and so on. The shifting operation can be accomplished by another DO loop as follows:

```
        DO 20 K=1, M
          XM(MP2-K) = X(MP1-K)                  (2.4.69)
        20 CONTINUE
```

where MP2 $= M + 2$.

The shifting operation of the input data vector may be inefficient and can be eliminated by the use of a "circular buffer," in which the new input data sample is stored in the location of the oldest data sample. Thus $x(n_0 + 1)$ is stored in $x(n_0 - M)$, $x(n_0 + 2)$ is stored in $x(n_0 + 1 - M)$, and so on. A moving pointer moves in a circular fashion, with the newest data sample in order to indicate the starting point in the computation of the output $y(n)$ according to (2.4.66). If this approach is adopted, the DO loop in (2.4.68) must be modified.

Figure 2.42 gives the listing of a FORTRAN subroutine FIRDF that implements the FIR system in (2.4.66) based on the DO loops in (2.4.68) and (2.4.69). The program can be initialized by calling the subroutine once with INIT = 0. This sets XM(K) = 0, K = 1, 2, . . . , M + 1. Thereafter, each call of the subroutine processes another input point. For example, a FORTRAN program that uses the subroutine FIRDF on a computer-generated signal is given in Fig. 2.43.

Now, let us consider the software implementation of the IIR system described by the difference equation

$$y(n) = -\sum_{k=1}^{N} a_k y(n - k) + \sum_{k=0}^{M} b_k x(n - k) \qquad (2.4.70)$$

We will program (2.4.70) for the direct form I structure. Figure 2.44 illustrates the information needed to compute (2.4.70) and indicates the general approach. We note that (2.4.70) can be subdivided into the computations of the variables $v(n)$ and $y(n)$,

```
C
C      FIR DIGITAL FILTER
C
C      INPUT  : H(K),K=1,2,...,M+1  ,XIN
C      OUTPUT : YOUT
C
       SUBROUTINE FIRDF (XIN,YOUT,M,H,XM,INIT)
       DIMENSION H(1),XM(1)
       IF (INIT.NE.0) GOTO 6
C
C      INITIALIZATION
C
       MP1=M+1
       MP2=M+2
       DO 5 K=1,MP1
         XM(K)=0.0
    5 CONTINUE
       RETURN
    6 CONTINUE
C
C      SHIFT INPUT DATA
C
       DO 10 K=1,M
         XM(MP2-K)=XM(MP1-K)
   10 CONTINUE
       XM(1)=XIN
C
C      COMPUTE THE CURRENT OUTPUT SAMPLE
C
       S=0.0
       DO 20 K=1,MP1
         S=S+H(K)*XM(K)
   20 CONTINUE
       YOUT=S
       RETURN
       END
```

FIGURE 2.42 Subroutine for the realization of a FIR system in software.

```
C        PROGRAM FIR FILTERING
C        DECLARATION OF REQUIRED ARRAYS
         DIMENSION X(100),XM(3),H(3),Y(100)
C        LOAD THE SYSTEM PARAMETERS
         M=3
         H(1)=0.25
         H(2)=0.50
         H(3)=0.25
C        CREATE INPUT SIGNAL
         PI=4.0*ATAN(1.0)
         DO 50 N=1,100
            NM1=N-1
            X(N)=COS(NM1*PI/20.0)+COS(PI*NM1)
      50 CONTINUE
C        INITIALIZATION
         CALL FIRDF(XIN,YOUT,M,H,XM,0)
C        FILTER OPERATION
         DO 100 N=1,100
            XIN=X(N)
            CALL FIRDF(XIN,YOUT,M,H,XM,1)
            Y(N)=YOUT
     100 CONTINUE
C        PRINT OR PLOT INPUT/OUTPUT SIGNALS
         END
```

FIGURE 2.43 A FORTRAN program that uses the subroutine FIRDF to implement a system with $h(0) = \frac{1}{4}$, $h(1) = \frac{1}{2}$, and $h(2) = \frac{1}{4}$.

where

$$v(n) = \sum_{k=0}^{M} b_k x(n - k) \qquad (2.4.71)$$

and

$$y(n) = - \sum_{k=1}^{N} a_k y(n - k) + v(n) \qquad (2.4.72)$$

The nonrecursive system in (2.4.71) corresponds to the left part of the block diagram in Fig. 2.44 and can easily be implemented as the FIR system described above. Although

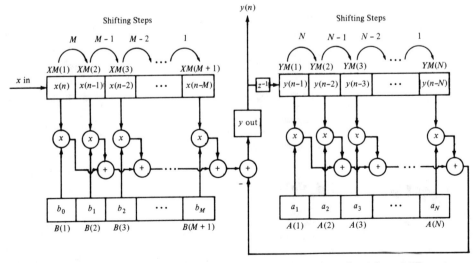

FIGURE 2.44 Block diagram illustrating the computations in a direct form I IIR system.

```
C
C        DIRECT FORM 1   IIR DIGITAL FILTER
C
C        INPUT   : B(1),B(2),...,B(M+1),A(1),A(2),...A(N),XIN
C        OUTPUT : YOUT
C        NOTE THAT   : B(K+1) = b   , K=0,1,...M
C
         SUBROUTINE IIRDF1 (XIN,YOUT,M,N,B,A,XM,YM,INIT)
         DIMENSION A(1),B(1),XM(1),YM(1)
         IF (INIT.NE.0) GOTO 6
C        INITIALIZATION
         NM1=N-1
         MP1=M+1
         MP2=M+2
         DO 1 K=1,MP1
           XM(K)=0.0
       1 CONTINUE
         DO 5 K=1,N
           YM(K)=0.0
       5 CONTINUE
         RETURN
       6 CONTINUE
C        SHIFT INPUT DATA
         DO 10 K=1,M
           XM(MP2-K)=XM(MP1-K)
      10 CONTINUE
         XM(1)=XIN
C        COMPUTE THE CURRENT OUTPUT SAMPLE
         S=0.0
         DO 20 K=1,MP1
           S=S+B(K)*XM(K)
      20 CONTINUE
         DO 30 K=1,N
           S=S+A(K)*YM(K)
      30 CONTINUE
         YOUT=S
C        SHIFT OUTPUT DATA
         DO 40 K=1,NM1
           YM(N+1-K)=YM(N-K)
      40 CONTINUE
         YM(1)=YOUT
         RETURN
         END
```

FIGURE 2.45   Subroutine IIRDF1 for the realization of a direct form I IIR system in software.

the system described by (2.4.72) is purely recursive, from a computational viewpoint, the summation in (2.4.72) can be computed in the same manner as the values of the output and a buffer for the system parameter $\{a_k\}$.

A FORTRAN subroutine IIRDF1 which implements (2.4.71) and (2.4.72) is given in Fig. 2.45. The system is initialized by calling the subroutine once with INIT $= 0$.

2.5 Correlation of Discrete-Time Signals

A mathematical operation that closely resembles convolution is correlation. Just as in the case of convolution, two signal sequences are involved in correlation. In contrast to convolution, however, our objective in computing the correlation between the two signals is to measure the degree to which the two signals are similar and thus to extract some information that depends to a large extent on the application. Correlation of signals is often encountered in radar, sonar, digital communications, and other areas in science and engineering.

To be specific, let us suppose that we have two signal sequences $x(n)$ and $y(n)$ that we wish to compare. In radar and active sonar applications, $x(n)$ may represent the (sampled version) of the transmitted signal and $y(n)$ may represent the sampled version of the received signal at the output of the analog-to-digital (A/D) converter. If a target is present in the space being searched by the radar or sonar, the received signal $y(n)$ consists of a delayed version of the transmitted signal, reflected from the target, and corrupted by additive noise. Figure 2.46 depicts the radar signal reception problem.

We may represent the received signal sequence as

$$y(n) = \alpha x(n - D) + w(n) \qquad (2.5.1)$$

where α is some attenuation factor representing the signal loss involved in the round-trip transmission of the signal $x(n)$, D is the round-trip delay, which is assumed to be an integer multiple of the sampling interval, and $w(n)$ represents the additive noise that is picked up by the antenna and any noise generated by the electronic components and amplifiers contained in the front end of the receiver. On the other hand, if there is no target in the space searched by the radar and sonar, the received signal $y(n)$ consists of noise alone.

Having the two signal sequences, $x(n)$, which is called the reference signal or transmitted signal, and $y(n)$, the received signal, the problem in radar and sonar detection is to compare $y(n)$ and $x(n)$ in order to determine if a target is present and, if so, to determine the time delay D from which one can compute the distance to the target. In practice, the signal $x(n - D)$ is heavily corrupted by the additive noise to the point where a visual inspection of $y(n)$ will not reveal the presence or absence of the desired signal reflected from the target. Correlation provides us with a means for extracting this important information from $y(n)$.

Digital communications is another area where correlation is often used. In digital communications the information to be transmitted from one point to another is usually converted to binary from, that is, a sequence of zeros and ones, which are then transmitted to the intended receiver. To transmit a 0 we may transmit the signal sequence $x_0(n)$ for

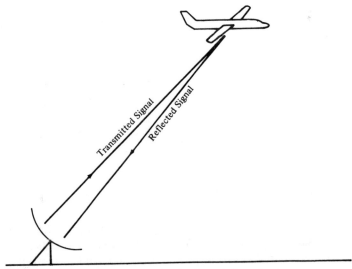

FIGURE 2.46 Radar target detection.

$0 \le n \le L - 1$, and to transmit a 1 we may transmit the signal sequence $x_1(n)$ for $0 \le n \le L - 1$, where L is some integer that denotes the number of samples in each of the two sequences. Very often, $x_1(n)$ is selected to be the negative of $x_0(n)$. The signal received by the intended receiver may be represented as

$$y(n) = x_i(n) + w(n) \qquad i = 0, 1 \quad 0 \le n \le L - 1 \qquad (2.5.2)$$

where now the uncertainty is whether $x_0(n)$ or $x_1(n)$ is the signal component in $y(n)$ and $w(n)$ represents the additive noise and other interference that is inherent in any communication system. Again, such noise has its origins in the electronic components contained in the front end of the receiver. In any case, the receiver knows the possible transmitted sequences $x_0(n)$ and $x_1(n)$ and is faced with the task of comparing the received signal $y(n)$ with both $x_0(n)$ and $x_1(n)$ to determine which of the two signals better matches $y(n)$. This comparison process is performed by means of the correlation operation described below.

2.5.1 Crosscorrelation and Autocorrelation Sequences

Suppose that we have two signal sequences $x(n)$ and $y(n)$ each of which has finite energy. The *crosscorrelation* of $x(n)$ and $y(n)$ is a sequence $r_{xy}(l)$, which is defined as

$$r_{xy}(l) = \sum_{n=-\infty}^{\infty} x(n)y(n - l) \qquad l = 0, \pm 1, \pm 2, \ldots \qquad (2.5.3)$$

or, equivalently, as

$$r_{xy}(l) = \sum_{n=-\infty}^{\infty} x(n + l)y(n) \qquad l = 0, \pm 1, \pm 2, \ldots \qquad (2.5.4)$$

The index l is the (time) shift (or *lag*) parameter and the subscripts xy on the crosscorrelation sequence $r_{xy}(l)$ indicate the sequences being correlated. The order of the subscripts, with x preceding y, indicates the direction in which one sequence is shifted, relative to the other. To elaborate, in (2.5.3), the sequence $x(n)$ is left unshifted and $y(n)$ is shifted by l units in time, to right for l positive and to the left for l negative. Equivalently, in (2.5.4), the sequence $y(n)$ is left unshifted and $x(n)$ is shifted by l units in time, to the left for l positive and to the right for l negative. But shifting $x(n)$ to the left by l units relative to $y(n)$ is equivalent to shifting $y(n)$ to the right by l units relative to $x(n)$. Hence the computations (2.5.3) and (2.5.4) yield identical crosscorrelation sequences.

If we reverse the roles of $x(n)$ and $y(n)$ in (2.5.3) and (2.5.4) and hence reverse the order of the indices xy, we obtain the crosscorrelation sequence

$$r_{yx}(l) = \sum_{n=-\infty}^{\infty} y(n)x(n - l) \qquad (2.5.5)$$

or, equivalently,

$$r_{yx}(l) = \sum_{n=-\infty}^{\infty} y(n + l)x(n) \qquad (2.5.6)$$

By comparing (2.5.3) with (2.5.6) or (2.5.4) with (2.5.5), we conclude that

$$r_{xy}(l) = r_{yx}(-l) \qquad (2.5.7)$$

Therefore, $r_{yx}(l)$ is simply the folded version of $r_{xy}(l)$, where the folding is done with respect to $l = 0$. Hence $r_{yx}(l)$ provides exactly the same information as $r_{xy}(l)$ about the similarity of $x(n)$ to $y(n)$.

EXAMPLE 2.5.1

Determine the crosscorrelation sequence $r_{xy}(l)$ of the sequences

$$x(n) = \left\{ \ldots, 0, 0, 2, -1, 3, 7, \underset{\uparrow}{1}, 2, -3, 0, 0, \ldots \right\}$$

$$y(n) = \left\{ \ldots, 0, 0, 1, -1, 2, -2, 4, \underset{\uparrow}{1}, -2, 5, 0, 0, \ldots \right\}$$

Solution: Let us use the definition in (2.5.3) to compute $r_{xy}(l)$. For $l = 0$ we have

$$r_{xy}(0) = \sum_{n=-\infty}^{\infty} x(n)y(n)$$

The product sequence $v_0(n) = x(n)y(n)$ is

$$v_0(n) = \left\{ \ldots, 0, 0, 2, 1, 6, -14, \underset{\uparrow}{4}, 2, 6, 0, 0, \ldots \right\}$$

and hence the sum over all values of n is

$$r_{xy}(0) = 7$$

For $l > 0$, we simply shift $y(n)$ to right relative to $x(n)$ by l units, compute the product sequence $v_l(n) = x(n)y(n - l)$, and finally, we sum over all values of the product sequence. Thus we obtain

$$\begin{aligned} r_{xy}(1) &= 13 & r_{xy}(2) &= -18 & r_{xy}(3) &= 16 & r_{xy}(4) &= -7 \\ r_{xy}(5) &= 5 & r_{xy}(6) &= -3 & r_{xy}(l) &= 0 & l &\geq 7 \end{aligned}$$

For $l < 0$, we shift $y(n)$ to the left relative to $x(n)$ by l units, compute the product sequence $v_l(n) = x(n)y(n - l)$, and sum over all values of the product sequence. Thus we obtain the values of the crosscorrelation sequence

$$\begin{aligned} r_{xy}(-1) &= 0 & r_{xy}(-2) &= 33 & r_{xy}(-3) &= -14 & r_{xy}(-4) &= 36 \\ r_{xy}(-5) &= 19 & r_{xy}(-6) &= -9 & r_{xy}(-7) &= 10 & r_{xy}(l) &= 0, & l &\leq -8 \end{aligned}$$

Therefore, the crosscorrelation sequence of $x(n)$ and $y(n)$ is

$$r_{xy}(l) = \left\{ 10, -9, 19, 36, -14, 33, 0, \underset{\uparrow}{7}, 13, -18, 16, -7, 5, -3 \right\}$$

The similarities between the computation of the crosscorrelation of two sequences and the convolution of two sequences is apparent. In the computation of convolution, one of the sequences is folded, then shifted, then multiplied by the other sequence to form the product sequence for that shift, and finally, the values of the product sequence are summed. Except for the folding operation, the computation of the crosscorrelation sequence involves the same operations: shifting one of the sequences, multiplication of the two sequences, and summing over all values of the product sequence. Consequently, if we have a computer program that performs convolution, we may use it to perform crosscorrelation by providing as inputs to the program the sequence $x(n)$ and the folded sequence $y(-n)$. Then the convolution of $x(n)$ with $y(-n)$ yields the crosscorrelation $r_{xy}(l)$, that is,

$$r_{xy}(l) = x(n) * y(-n) \tag{2.5.8}$$

In the special case where $y(n) = x(n)$, we have the *autocorrelation* of $x(n)$, which is defined as the sequence

$$r_{xx}(l) = \sum_{n=-\infty}^{\infty} x(n)x(n-l) \tag{2.5.9}$$

or, equivalently, as

$$r_{xx}(l) = \sum_{n=-\infty}^{\infty} x(n+l)x(n) \tag{2.5.10}$$

EXAMPLE 2.5.2

Compute the crosscorrelation sequence $r_{yx}(l)$ of the sequences $x(n)$ and $y(n)$, where

$$x(n) = \left\{ 3, 3, 3, -1, -1, -1 \atop \uparrow \right\}$$

and

$$y(n) = x(n-1) + w(n) \qquad n = 0, 1, \ldots$$

The sequence $w(n)$ represents an additive noise sequence with values

$$w(n) = \left\{ 0.75, -1.4, -1.1, 0.4, -0.42, -0.7 \atop \uparrow \right\}$$

The numbers for $w(n)$ were obtained from a uniform random number generator with output values in the range $(-1.5, 1.5)$. The signal $x(n-1)$ represents the desired signal delayed by one unit in time and $w(n)$ represents an additive noise disturbance.

Solution: The crosscorrelation sequence $r_{yx}(l)$ may be expressed as

$$r_{yx}(l) = \sum_{n=-\infty}^{\infty} y(n)x(n-l)$$

$$= \sum_{n=-\infty}^{\infty} [x(n-1) + w(n)]x(n-l)$$

$$= \sum_{n=-\infty}^{\infty} x(n-1)x(n-l) + \sum_{n=-\infty}^{\infty} w(n)x(n-l)$$

$$= r_{xx}(l-1) + r_{wx}(l)$$

Thus the crosscorrelation of the noisy sequence $y(n)$ with $x(n)$ may be viewed as the sum of the autocorrelation sequence $r_{xx}(l-1)$ and the crosscorrelation sequence $r_{wx}(l)$. It is clear that $r_{xx}(l-1)$ has a peak at $l = 1$, which corresponds to the value of the delay in the signal sequence $x(n)$.

The values of the autocorrelation $r_{xx}(l)$ are

$$r_{xx}(l) = \left\{ -3, -6, -9, 4, 17, 30, 17, 4, -9, -6, -3 \right\}$$
$$\uparrow$$

while the crosscorrelation sequence between $x(n)$ and $w(n)$ is

$$r_{wx}(l) = \left\{ -0.75, 0.65, 1.75, 5.15, -0.03, -3.73, -7.58, -5.06, \right.$$
$$\uparrow$$
$$\left. -4.56, -3.36, -2.1 \right\}$$

To determine the crosscorrelation sequence $r_{yx}(l)$ we add $r_{xx}(l - 1)$ to $r_{wx}(l)$. Thus we obtain

$$r_{yx}(l) = \left\{ -0.75, -2.35, -4.25, -3.85, 3.97, 13.27, 22.42, 11.94, \right.$$
$$\uparrow$$
$$\left. -0.56, -12.36, -8.1, 3 \right\}$$

As we observe, $r_{yx}(l)$ has a peak at $l = 1$. From this result we conclude that the signal $x(n)$ is contained in $y(n)$ at a delay of $l = 1$. Figure 2.47 illustrates the sequences $w(n)$ and $y(n)$ and the crosscorrelation $r_{yx}(l)$.

$$x(n) = \{3, 3, 3, -1, -1, -1\}$$

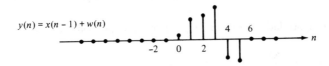

$w(n) = \{0.75, -1.4, -1.1, -0.4, -0.42, -0.7\}$

$\Delta = 3$

$y(n) = x(n - 1) + w(n)$

$r_{yx}(l)$

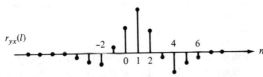

FIGURE 2.47   Crosscorrelation of $x(n)$ with $y(n) = x(n - 1) + w(n)$ in Example 2.5.2.

$x(n) = \{3, 3, 3, -1, -1, -1\}$

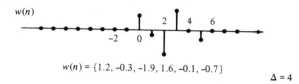

$w(n) = \{1.2, -0.3, -1.9, 1.6, -0.1, -0.7\}$

$\Delta = 4$

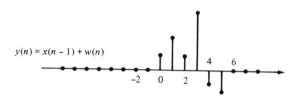

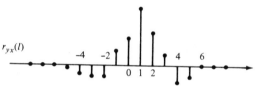

FIGURE 2.48 Crosscorrelation of $x(n)$ with $y(n) = x(n - 1) + w(n)$ in Example 2.5.2.

The computations above may be repeated with different noise sequences. For example, the sequence

$$w(n) = \{1.2, -0.3, -1.9, 1.6, -0.1, -0.7\}$$

obtained from a uniform random number generator with output in the range $(-2, 2)$, resulted in the crosscorrelation sequence $r_{yx}(l)$ illustrated in Fig. 2.48.

In dealing with finite-duration sequences, it is customary to express the autocorrelation and crosscorrelation in terms of the finite limits on the summation. In particular, if $x(n)$ and $y(n)$ are causal sequences of length N [i.e., $x(n) = y(n) = 0$ for $n < 0$ and $n \geq N$], the crosscorrelation and autocorrelation sequences may be expressed as

$$r_{xy}(l) = \sum_{n=0}^{N-|l|-1} x(n)y(n - l) \tag{2.5.11}$$

and

$$r_{xx}(l) = \sum_{n=0}^{N-|l|-1} x(n)x(n - l) \tag{2.5.12}$$

These expressions are equivalent to the definitions for $r_{xy}(l)$ and $r_{xx}(l)$ given above since $x(n) = y(n) = 0$ for $n < 0$ and $n \geq N$.

2.5.2 Properties of the Autocorrelation and Crosscorrelation Sequences

The autocorrelation and crosscorrelation sequences have a number of important properties that are presented below. To develop these properties, let us assume that we have two sequences $x(n)$ and $y(n)$ with finite energy from which we form the linear combination

$$ax(n) + by(n - l)$$

where a and b are arbitrary constants and l is some time shift. The energy in this signal is

$$\sum_{n=-\infty}^{\infty} [ax(n) + by(n - l)]^2 = a^2 \sum_{n=-\infty}^{\infty} x^2(n) + b^2 \sum_{n=-\infty}^{\infty} y^2(n - l)$$

$$+ 2ab \sum_{n=-\infty}^{\infty} x(n)y(n - l) \qquad (2.5.13)$$

$$= a^2 r_{xx}(0) + b^2 r_{yy}(0) + 2ab r_{xy}(l)$$

First, we note that $r_{xx}(0) = E_x$ and $r_{yy}(0) = E_y$, the energies of $x(n)$ and $y(n)$, respectively. It is obvious that

$$a^2 r_{xx}(0) + b^2 r_{yy}(0) + 2ab r_{xy}(l) \geq 0 \qquad (2.5.14)$$

Now, assuming that $b \neq 0$, we may divide (2.5.14) by b^2 to obtain

$$r_{xx}(0) \left(\frac{a}{b}\right)^2 + 2x_{xy}(l) \left(\frac{a}{b}\right) + r_{yy}(0) \geq 0$$

We view this equation as a quadratic with coefficients $r_{xx}(0)$, $2r_{xy}(l)$, and $r_{yy}(0)$. Since the quadratic is nonnegative, it follows that the discriminant of this quadratic must be nonpositive, that is,

$$4[r_{xy}^2(l) - r_{xx}(0)r_{yy}(0)] \leq 0$$

Therefore, the crosscorrelation sequence satisfies the condition that

$$|r_{xy}(l)| \leq \sqrt{r_{xx}(0)r_{yy}(0)} = \sqrt{E_x E_y} \qquad (2.5.15)$$

In the special case where $y(n) = x(n)$, (2.5.15) reduces to

$$|r_{xx}(l)| \leq r_{xx}(0) = E_x \qquad (2.5.16)$$

This means that the autocorrelation sequence of a signal attains its maximum value at zero lag. This result is consistent with the notation that a signal matches perfectly with itself at zero shift. As the lag increases we expect the autocorrelation sequence $r_{xx}(l)$ to decrease. That is, $r_{xx}(l)$ should tend toward zero as l approaches infinity.

In the case of the crosscorrelation sequence, the upper bound on its values is given in (2.5.15). We note that if

$$y(n) = \pm cx(n - n_0) \qquad (2.5.17)$$

where c is an arbitrary scale factor and n_0 is some time shift, the crosscorrelation $r_{xy}(l)$ becomes

$$r_{xy}(l) = \pm c r_{xx}(l - n_0) \qquad (2.5.18)$$

and the autocorrelation of $y(n)$ at $n = 0$ is simply

$$r_{yy}(0) = c^2 r_{xx}(0) \qquad (2.5.19)$$

By substituting (2.5.18) and (2.5.19) in (2.5.15), the inequality reduces to

$$|r_{xy}(l)| = |\pm c r_{xx}(l - n_0)| \leq c r_{xx}(0)$$

In this case the range of values of $r_{xy}(l)$ is bounded in the range

$$-cr_{xx}(0) \leq r_{xy}(l) \leq cr_{xx}(0) \tag{2.5.20}$$

In practice it is often desirable to normalize the autocorrelation and crosscorrelation sequences to the range from -1 to 1. In the case of the autocorrelation sequence, we may simply divide by $r_{xx}(0)$. Thus the normalized autocorrelation sequence is defined as

$$\rho_{xx}(l) = \frac{r_{xx}(l)}{r_{xx}(0)} \tag{2.5.21}$$

Similarly, we define the normalized crosscorrelation sequence

$$\rho_{xy}(l) = \frac{r_{xy}(l)}{\sqrt{r_{xx}(0)r_{yy}(0)}} \tag{2.5.22}$$

Now $|\rho_{xx}(l)| \leq 1$ and $|\rho_{xy}(l)| \leq 1$, and hence these sequence are independent of signal scaling.

Finally, as we have already demonstrated, the crosscorrelation sequence satisfies the property

$$r_{xy}(l) = r_{yx}(-l)$$

With $y(n) = x(n)$, this relation results in the following important property for the auto-correlation sequence:

$$r_{xx}(l) = r_{xx}(-l) \tag{2.5.23}$$

Hence the autocorrelation function is an even function. Consequently, it suffices to compute $r_{xx}(l)$ for $l \geq 0$.

EXAMPLE 2.5.3 _____

Compute the autocorrelation of the signal

$$x(n) = a^n u(n) \qquad -1 < a < 1$$

Solution: Since $x(n)$ is an infinite-duration signal, its autocorrelation also has infinite duration. We distinguish two cases.

If $l \geq 0$, from Fig. 2.49 we observe that

$$r_{xx}(l) = \sum_{n=l}^{\infty} x(n)x(n-l) = \sum_{n=l}^{\infty} a^n a^{n-l} = a^{-l} \sum_{n=l}^{\infty} (a^2)^n$$

Since $|a| < 1$, the infinite series *converges* and we obtain

$$r_{xx}(l) = \frac{1}{1-a^2} a^l \qquad l \geq 0$$

For $l < 0$ we have

$$r_{xx}(l) = \sum_{n=0}^{\infty} x(n)x(n-l) = a^{-l} \sum_{n=0}^{\infty} (a^2)^n = \frac{1}{1-a^2} a^{-l} \qquad l < 0$$

But when l is negative, $a^{-l} = a^{|l|}$. Thus the two relations for $r_{xx}(l)$ can be combined into the following expression:

$$r_{xx}(l) = \frac{1}{1-a^2} a^{|l|} \qquad -\infty < l < \infty \tag{2.5.24}$$

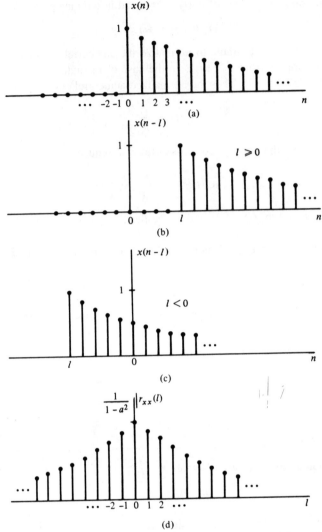

FIGURE 2.49   Computation of the autocorrelation of the signal $x(n) = a^n$, $-1 < a < 1$.

The sequence $r_{xx}(l)$ is shown in Fig. 2.49(d). We observe that

$$r_{xx}(-l) = r_{xx}(l)$$

and

$$r_{xx}(0) = \frac{1}{1 - a^2}$$

Therefore, the normalized autocorrelation sequence is

$$\rho_{xx}(l) = \frac{r_{xx}(l)}{r_{xx}(0)} = a^{|l|} \qquad -\infty < l < \infty \qquad (2.5.25)$$

2.5.3 Correlation of Periodic Sequences

In Section 2.5.1 we defined the crosscorrelation and autocorrelation sequences of energy signals. In this section we consider the correlation sequences of power signals and, in particular, periodic signals.

Let $x(n)$ and $y(n)$ be two power signals. Their crosscorrelation sequence is defined as

$$r_{xy}(l) = \lim_{M \to \infty} \frac{1}{2M + 1} \sum_{n=-M}^{M} x(n)y(n - l) \tag{2.5.26}$$

If $x(n) = y(n)$, we have the definition of the autocorrelation sequence of a power signal as

$$r_{xx}(l) = \lim_{M \to \infty} \frac{1}{2M + 1} \sum_{n=-M}^{M} x(n)x(n - l) \tag{2.5.27}$$

In particular, if $x(n)$ and $y(n)$ are two periodic sequences, each with period N, the averages indicated in (2.5.26) and (2.5.27) over the infinite interval are identical to the averages over a single period, so that (2.5.26) and (2.5.27) reduce to

$$r_{xy}(l) = \frac{1}{N} \sum_{n=0}^{N-1} x(n)y(n - l) \tag{2.5.28}$$

and

$$r_{xx}(l) = \frac{1}{N} \sum_{n=0}^{N-1} x(n)x(n - l) \tag{2.5.29}$$

It is clear that $r_{xy}(l)$ and $r_{xx}(l)$ are periodic correlation sequences with period N. The factor $1/N$ may be viewed as a normalization scale factor.

In some practical applications, correlation is used to identify periodicities in an observed physical signal which may be corrupted by random interference. For example, consider a signal sequence $y(n)$ of the form

$$y(n) = x(n) + w(n) \tag{2.5.30}$$

where $x(n)$ is a periodic sequence of some unknown period N and $w(n)$ represents an additive random interference. Suppose that we observe M samples of $y(n)$, say $0 \le n \le M - 1$, where $M \gg N$. For all practical purposes, we may assume that $y(n) = 0$ for $n < 0$ and $n \ge M$. Now the autocorrelation sequence of $y(n)$, using the normalization factor of $1/M$, is

$$r_{yy}(l) = \frac{1}{M} \sum_{n=0}^{M-1} y(n)y(n - l) \tag{2.5.31}$$

If we substitute for $y(n)$ from (2.5.30) into (2.5.31) we obtain

$$r_{yy}(l) = \frac{1}{M} \sum_{n=0}^{M-1} [x(n) + w(n)][x(n - l) + w(n - l)]$$

$$= \frac{1}{M} \sum_{n=0}^{M-1} x(n)x(n - l) + \frac{1}{M} \sum_{n=0}^{M-1} [x(n)w(n - l) + w(n)x(n - l)]$$

$$+ \frac{1}{M} \sum_{n=0}^{M-1} w(n)w(n - l) \tag{2.5.32}$$

$$= r_{xx}(l) + r_{xw}(l) + r_{wx}(l) + r_{ww}(l)$$

The first factor on the right-hand side of (2.5.32) is the autocorrelation sequence of $x(n)$. Since $x(n)$ is periodic, its autocorrelation sequence will exhibit the same periodicity, thus containing relatively large peaks at $l = 0, N, 2N$, and so on. However, as the shift l approaches M, the peaks will be reduced in amplitude due to the fact that we have a finite data record of M samples so that many of the products $x(n)x(n - l)$ will be zero. Consequently, we should avoid computing $r_{yy}(l)$ for large lags, say, $l > M/2$.

The crosscorrelations $r_{xw}(l)$ and $r_{wx}(l)$ between the signal $x(n)$ and the additive random interference are expected to be relatively small as a result of the expectation that $x(n)$ and $w(n)$ will be totally unrelated. Finally, the last term on the right-hand side of (2.5.32) is the autocorrelation sequence of the random sequence $w(n)$. This correlation sequence will certainly contain a peak at $l = 0$, but because of its random characteristics, $r_{ww}(l)$ is expected to decay rapidly toward zero. Consequently, only $r_{xx}(l)$ is expected to have

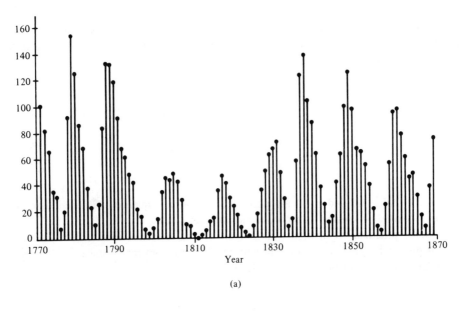

(a)

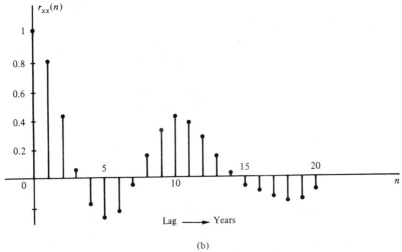

(b)

FIGURE 2.50 Identification of periodicity in the Wölfer sunspot numbers: (a) annual Wölfer sunspot numbers; (b) autocorrelation sequence.

TABLE 2.3 Yearly Wölfer Sunspot Numbers

1770	101	1795	21	1820	16	1845	40
1771	82	1796	16	1821	7	1846	62
1772	66	1797	6	1822	4	1847	98
1773	35	1798	4	1823	2	1848	124
1774	31	1799	7	1824	8	1849	96
1775	7	1800	14	1825	17	1850	66
1776	20	1801	34	1826	36	1851	64
1777	92	1802	45	1827	50	1852	54
1778	154	1803	43	1828	62	1853	39
1779	125	1804	48	1829	67	1854	21
1780	85	1805	42	1830	71	1855	7
1781	68	1806	28	1831	48	1856	4
1782	38	1807	10	1832	28	1857	23
1783	23	1808	8	1833	8	1858	55
1784	10	1809	2	1834	13	1859	94
1785	24	1810	0	1835	57	1860	96
1786	83	1811	1	1836	122	1861	77
1787	132	1812	5	1837	138	1862	59
1788	131	1813	12	1838	103	1863	44
1789	118	1814	14	1839	86	1864	47
1790	90	1815	35	1840	63	1865	30
1791	67	1816	46	1841	37	1866	16
1792	60	1817	41	1842	24	1867	7
1793	47	1818	30	1843	11	1868	37
1794	41	1819	24	1844	15	1869	74

large peaks for $l > 0$. This behavior allows us to detect the presence of the periodic signal $x(n)$ buried in the interference $w(n)$ and to identify its period.

An example that illustrates the use of autocorrelation to identify a hidden periodicity in an observed physical signal is shown in Fig. 2.50. This figure illustrates the autocorrelation (normalized) sequence for the Wölfer sunspot numbers for $0 \leq l \leq 20$, where any value of l corresponds to one year. These numbers are given in Table 2.3 for the 100-year period 1770–1869. There is clear evidence in this figure that a periodic trend exists, with a period of 10 to 11 years.

EXAMPLE 2.5.4

Suppose that a signal sequence $x(n) = \sin(\pi/5)\, n$, for $0 \leq n \leq 99$ is corrupted by an additive noise sequence $w(n)$, where the values of the additive noise are selected independently from sample to sample from a uniform distribution over the range $(-\Delta/2)$, and Δ is a parameter of the distribution. The observed sequence is $y(n) = x(n) + w(n)$. Determine the autocorrelation sequence $r_{yy}(n)$ and thus determine the period of the signal $x(n)$.

Solution: The assumption is that the signal sequence $x(n)$ has some unknown period that we are attempting to determine from the noise-corrupted observations $\{y(n)\}$. Although $x(n)$ is periodic with period 10, we have only a finite-duration sequence of length $M = 100$ [i.e., 10 periods of $x(n)$]. The noise power level P_N in the sequence $w(n)$ is determined by the parameter Δ. We simply state that $P_w = \Delta^2/12$. The signal power level is $P_x =$

$\frac{1}{2}$. Therefore, the signal-to-noise ratio (SNR) is defined as

$$\frac{P_x}{P_w} = \frac{\frac{1}{2}}{\Delta^2/12} = \frac{6}{\Delta^2}$$

Usually, the SNR is expressed on a logarithmic scale in decibels (dB) as $10 \log_{10}(P_x/P_w)$.

Figure 2.51 illustrates a sample of a noise sequence $w(n)$ and the observed sequence $y(n) = x(n) + w(n)$ when the SNR $= 1$ dB. The autocorrelation sequence $r_{yy}(l)$ is illustrated in Fig. 2.51c. We observe that the periodic signal $x(n)$ embedded in $y(n)$ results in a periodic autocorrelation function $r_{xx}(l)$ with period $N = 10$. The effect of the additive noise is to add to the peak value at $l = 0$, but for $l \neq 0$, the correlation sequence $r_{ww}(l) \approx 0$ as a result of the fact that values of $w(n)$ were generated independently. Such noise is usually called *white noise*. The presence of this noise explains

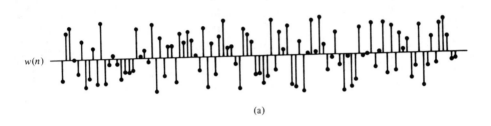

$w(n)$

(a)

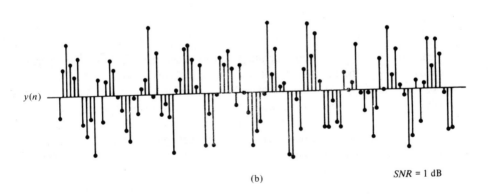

$y(n)$

(b)   SNR = 1 dB

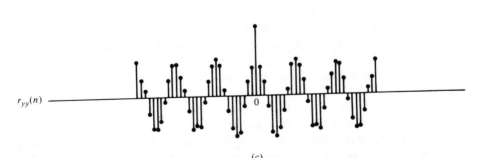

$r_{yy}(n)$

(c)

FIGURE 2.51 Use of autocorrelation to detect the presence of a periodic signal corrupted by noise.

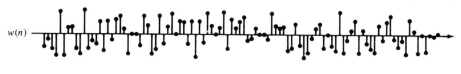

(a)

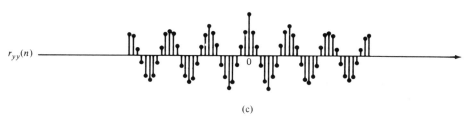

(b)

$SNR = 5$ dB

(c)

FIGURE 2.52 Use of autocorrelation to detect the presence of a periodic signal corrupted by noise.

the reason for the large peak at $l = 0$. The smaller, nearly equal peaks at $l = \pm 10$, ± 20, . . . are due the periodic characteristics of $x(n)$.

Figure 2.52 illustrates the noise sequence $w(n)$, the noise-corrupted signal $y(n)$, and the autocorrelation sequence $r_{yy}(l)$ for the same signal embedded is a smaller noise level. In this case, the SNR $= 5$ dB. Even with this relatively small noise level, the periodicity of the signal is not easily determined from observation of $y(n)$. However, it is clearly evident from observation of the autocorrelation sequence $r_{yy}(n)$.

2.5.4 Computation of Correlation Sequences

As indicated on Section 2.5.1, the procedure for computing the crosscorrelation sequence between $x(n)$ and $y(n)$ involves shifting one of the sequences, say $x(n)$, to obtain $x(n - l)$, multiplying the shifted sequence by $y(n)$ to obtain the product sequence $y(n)x(n - l)$ and then summing all the values of the product sequence to obtain $r_{yx}(l)$.

This procedure is repeated for different values of the lag l. Except for the folding operation that is involved in convolution, the basic operations given above for computing the correlation sequence are identical to those in convolution.

The procedure for computing the convolution is directly applicable to computing the correlation of two sequences. Specifically, if we fold $y(n)$ to obtain $y(-n)$, then the convolution of $x(n)$ with $y(-n)$ is identical to the crosscorrelation of $x(n)$ with $y(n)$. That is,

$$r_{xy}(l) = x(n) * y(-n)|_{n=l} \tag{2.5.33}$$

As a consequence, the computational procedure described for convolution can be applied directly to the computation of the correlation sequence.

We will now describe an algorithm that can be easily programmed to compute the crosscorrelation sequence of two finite-duration signals $x(n)$, $0 \le n \le N - 1$, and $y(n)$, $0 \le n \le M - 1$.

The algorithm computes $r_{xy}(l)$ for positive lags. According to the relation $r_{xy}(-l) = r_{yx}(l)$, the values of $r_{xy}(l)$ for negative lags can be obtained by using the same algorithm for positive lags and interchanging the roles of $x(n)$ and $y(n)$. We observe that if $M \le N$, $r_{xy}(l)$ can be computed by the relations

$$r_{xy}(l) = \begin{cases} \displaystyle\sum_{n=l}^{M-1+l} x(n)y(n - l) & 0 \le l \le N - M \\[2em] \displaystyle\sum_{n=l}^{N-1} x(n)y(n - l) & N - M < l \le N - 1 \end{cases} \tag{2.5.34}$$

On the other hand, if $M > N$, the formula for the crosscorrelation becomes

$$r_{xy}(l) = \sum_{n=l}^{N-1} x(n)y(n - l) \qquad 0 \le l \le N - 1 \tag{2.5.35}$$

The formulas in (2.5.34) and (2.5.35) can be combined and computed by means of the following simple algorithm illustrated in the flowchart in Fig. 2.53. A subroutine for this algorithm is given in Fig. 2.54. By interchanging the roles of $x(n)$ and $y(n)$ and recomputing the crosscorrelation sequence, we obtain the values of $r_{xy}(l)$ corresponding to negative shifts l.

If we wish to compute the autocorrelation sequence $r_{xx}(l)$, we set $y(n) = x(n)$ and $M = N$. The computation of $r_{xx}(l)$ can be done by means of the same algorithm for positive shifts only and deleting the IF statement in the program listing. A subroutine for the resulting computation of $r_{xx}(l)$ is given in Fig. 2.55.

2.6 Summary and References

The major theme of this chapter is the characterization of discrete-time signals and systems in the time domain. Of particular importance is the class of linear time-invariant (LTI) systems which are widely used in the design and implementation of digital signal processing systems. We characterized LTI systems by their unit sample response $h(n)$ and derived the convolution summation, which is a formula for determining the response $y(n)$ of the system characterized by $h(n)$ to any given input sequence $x(n)$.

The class of LTI systems characterized by linear difference equations with constant coefficients is by far the most important of the LTI systems in the theory and application

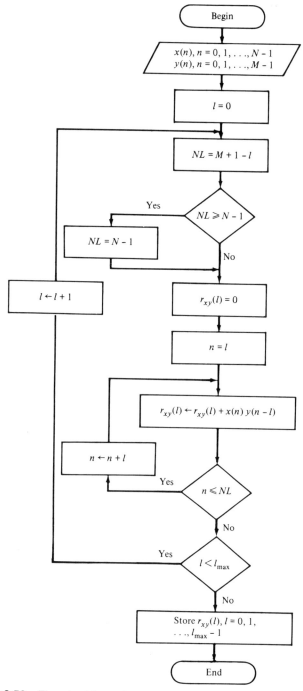

FIGURE 2.53 Flowchart for software implementation of crosscorrelation.

of digital signal processing. The general solution of a linear difference equation with constant coefficients was derived in this chapter and shown to consist of two components: the solution of the homogeneous equation which represents the natural response of the system when the input is zero, and the particular solution, which represents the response of the system to the input signal. From the difference equation, we also demonstrated how to derive the unit sample response of the LTI system.

```
          SUBROUTINE CROS (X,N,Y,M,R,LMAX)
C
C     SUBROUTINE CROS COMPUTES THE CROSSCORRELATION
C     SEQUENCE BETWEEN X AND Y
C     PARAMETERS
C       X   : ARRAY CONTAINING SEQ X
C       N   : THE LENGTH OF SEQ X
C       Y   : ARRAY CONTAINING SEQ Y
C       M   : THE LENGTH OF SEQ Y
C       R   : ARRAY CONTAINING THE CORRELATION
C     LMAX  :THE LENGTH OF THE CORRELATION
C
      DIMENSION X(1),Y(1),R(1)
      DO 10 L=1,LMAX
      NL=M+1-L
      IF (NL.GE.N-1) NL=N-1
      R(L)=0.0
      DO 10 K=L,NL
      R(L)=R(L)+X(K)*Y(K-L)
   10 CONTINUE
      RETURN
      END
```

FIGURE 2.54 Subroutine for computing the crosscorrelation of two sequences.

Linear time-invariant systems were generally subdivided into FIR (finite-duration impulse response) and IIR (infinite-duration impulse response) depending on whether $h(n)$ has finite duration or infinite duration, respectively. The realizations of such systems were briefly described and their implementation in terms of software was also discussed. Furthermore, in the realization of FIR systems, we made the distinction between recursive and nonrecursive realizations. On the other hand, we observe that IIR systems can be implemented recursively, only.

There are a number of good texts on discrete-time signals and systems. We mention as examples the books by McGillem and Cooper (1984), Oppenheim and Willsky (1983), and Siebert (1986). Linear constant-coefficient difference equations are treated in depth in the books by Hildebrand (1952) and Levy and Lessman (1961).

The last topic in this chapter, on correlation of discrete-time signals, plays an important role in digital signal processing, especially in applications dealing with digital communications, radar detection and estimation, sonar, and geophysics. In our treatment of correlation sequences, we avoided the use of statistical concepts. Correlation was simply defined as a mathematical operation between two sequences, which produces another

```
C
C     COMPUTATION OF AUTOCORELATION FUNCTION
C
      SUBROUTINE AUTOC (X,N,R,L)
      DIMENSION X(1),R(1)
      DO 1 K=1,L
        R(K)=0.0
        NK=N-K+1
        DO 1 ND=1,NK
          NDK=ND+K-1
          R(K)=R(K)+X(ND)*X(ND)
    1 CONTINUE
      RETURN
      END
```

FIGURE 2.55 Subroutine for computing the autocorrelation of a sequence.

sequence, called either the *crosscorrelation sequence* when the two sequences are different, or the *autocorrelation sequence* when the two sequences are identical.

In practical applications in which correlation is used, one (or both) of the sequences is (are) contaminated by noise and, perhaps, other forms of interference. In such a case, the noisy sequence is called a *random sequence* and it is characterized in statistical terms. The corresponding correlation sequence becomes a function of the statistical characteristics of the noise and any other interference. In effect, the correlation computed from the noisy sequence represents an estimate of the noise-free correlation.

The statistical characterization of sequences and their correlation is treated in Chapter 10 (Appendix 10A). Supplementary reading on probabilistic and statistical concepts dealing with correlation can be found in the excellent books by Davenport (1970), Helstrom (1984), Papoulis (1984), and Peebles (1987).

APPENDIX 2A: RANDOM NUMBER GENERATORS

In some of the examples given in Section 2.5, random numbers were generated to simulate the effect of noise on signals and to illustrate how the method of correlation can be used to detect the presence of a signal buried in noise. In the case of periodic signals, the correlation technique also allowed us to estimate the period of the signal.

Random number generators are often used in practice to simulate the effect of noiselike signals and other random phenomena that are encountered in the physical world. Such noise is present in electronic devices and systems and usually limits our ability to communicate over large distances and to detect relatively weak signals. By generating such noise on a computer, we are able to study its effects, through simulation of communication systems, radar detection systems, and the like and to assess the performance of such systems in the presence of noise.

Most computer software libraries include a uniform random number generator. Such a random number generator generates a number between zero and 1 with equal probability. We call the output of the random number generator a random variable. If A denotes such a random variable, its range is the interval $0 \le A \le 1$.

We know that the numerical output of digital computer has limited precision, and as a consequence it is impossible to represent the continuum of numbers in the interval $0 \le A \le 1$. However, we may assume that our computer represents each output by a large number of bits in either fixed point or floating point. Consequently, for all practical purposes, the number of outputs in the interval $0 \le A \le 1$ is sufficiently large, so that we are justified in assuming that any value in the interval is a possible output from the generator.

The uniform probability density function for the random variable A, denoted as $p(A)$, is illustrated in Fig. 2A.1a. We note that the average value or mean value of A, denoted as m_A, is $m_A = \frac{1}{2}$. The integral of the probability density function, which represents the area under $p(A)$, is called the probability distribution function of the random variable A and is defined as

$$F(A) = \int_{-\infty}^{A} p(x)\, dx \qquad (2A.1)$$

For any random variable, this area must always be unity, which is the maximum value that can be achieved by a distribution function. Hence

$$F(1) = \int_{-\infty}^{1} p(x)\, dx = 1 \qquad (2A.2)$$

and the range of $F(a)$ is $0 \le F(A) \le 1$ for $0 \le A \le 1$.

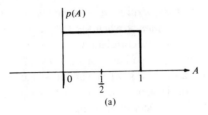

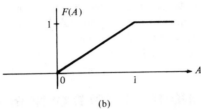

FIGURE 2A.1 Probability density function $p(A)$ and the probability distribution function $F(A)$ of a uniformly distributed random variable A.

If we wish to generate uniformly distributed noise in an interval $(b, b + 1)$ it can simply be accomplished by using the output A of the random number generator and shifting it by an amount b. Thus a new random variable B can be defined as

$$B = A + b \qquad (2A.3)$$

which now has a mean value $m_B = b + \frac{1}{2}$. For example, if $b = -\frac{1}{2}$, the random variable B is uniformly distributed in the interval $(-\frac{1}{2}, \frac{1}{2})$, as shown in Fig. 2A.2a. Its probability distribution function $F(B)$ is shown in Fig. 2A.2b.

A uniformly distributed random variable in the range $(0, 1)$ can be used to generate random variables with other probability distribution functions. For example, suppose that we wish to generate a random variable C with probability distribution function $F(C)$, as

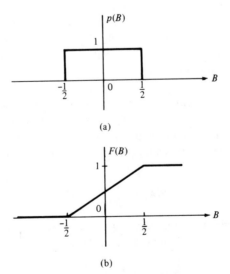

FIGURE 2A.2 Probability density function and the probability distribution function of a zero-mean uniformly distributed random variable.

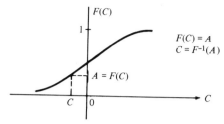

FIGURE 2A.3 Inverse mapping from the uniformly distributed random variable A to the new random variable C.

illustrated in Fig. 2A.3. Since the range of $F(C)$ is the interval $(0, 1)$, we begin by generating a uniformly distributed random variable A in the range $(0, 1)$. If we set

$$F(C) = A \qquad (2A.4)$$

then

$$C = F^{-1}(A) \qquad (2A.5)$$

Thus we solve (2A.4) for C and the solution in (2A.5) provides the value of C for which $F(C) = A$. By this means we obtain a new random variable C with probability distribution $F(C)$. This inverse mapping from A to C is illustrated in Fig. 2A.3.

EXAMPLE 2A.1

Generate a random variable C that has the linear probability density function shown in Fig. 2A.4a, that is,

$$p(C) = \begin{cases} \dfrac{C}{2} & 0 \le C \le 2 \\ 0 & \text{otherwise} \end{cases}$$

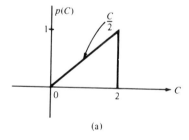

(a)

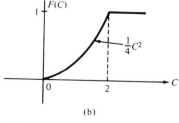

(b)

FIGURE 2A.4 Linear probability density function and the corresponding probability distribution function.

Solution: This random variable has a probability distribution function

$$F(C) = \begin{cases} 0 & C < 0 \\ \frac{1}{4}C^2 & 0 \le C \le 2 \\ 1 & C > 2 \end{cases}$$

which is illustrated in Fig. 2A.4b. We generate a uniformly distributed random variable
A and set $F(C) = A$. Hence

$$F(C) = \tfrac{1}{4}C^2 = A$$

Upon solving for C, we obtain

$$C = 2\sqrt{A}$$

Thus we generate a random variable C with probability function $F(C)$, as shown in Fig.
2A.4b.

In Example 2A.1 the inverse mapping $C = F^{-1}(A)$ was simple. In some cases it is
not. This problem arises in trying to generate random numbers that have a normal
distribution function.

Noise encountered in physical systems is often characterized by the normal or Gaussian
probability distribution, which is illustrated in Fig. 2A.5. The probability density function
is given by

$$p(C) = \frac{1}{\sqrt{2\pi}\,\sigma} e^{-C^2/2\sigma^2} \qquad -\infty < C < \infty \qquad (2A.6)$$

where σ^2 is the variance of C, which is a measure of the spread of the probability density
function $p(C)$. The probability distribution function $F(C)$ is the area under $p(C)$ over
the range $(-\infty, C)$. Thus

$$F(C) = \int_{-\infty}^{C} p(x)\, dx \qquad (2A.7)$$

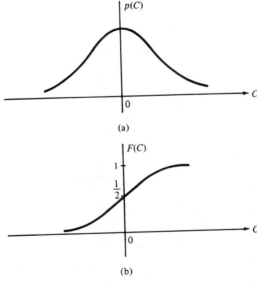

(a)

(b)

FIGURE 2A.5   Gaussian probability density function and the corresponding probability
distribution function.

Unfortunately, the integral in (2A.7) cannot be expressed in terms of simple functions. Consequently, the inverse mapping is difficult to achieve.

A way has been found to cirmumvent this problem. From probability theory it is known that a (Rayleigh distributed) random variable R, with probability distribution function

$$F(R) = \begin{cases} 0 & R < 0 \\ 1 - e^{-R^2/2\sigma^2} & R \geq 0 \end{cases} \qquad (2A.8)$$

is related to a pair of Gaussian random variables C and D, through the transformation

$$C = R \cos \Theta \qquad (2A.9)$$
$$D = R \sin \Theta \qquad (2A.10)$$

where Θ is a uniformly distributed variable in the interval $(0, 2\pi)$. The parameter σ^2 is the variance of C and D. Since (2A.8) is easily inverted, we have

$$F(R) = 1 - e^{-R^2/2\sigma^2} = A$$

and hence

$$R = \sqrt{2\sigma^2 \ln [1/(1 - A)]} \qquad (2A.11)$$

where A is a uniformly distributed random variable in the interval $(0, 1)$. Now if we generate a second uniformly distributed random variable B and define

$$\Theta = 2\pi B \qquad (2A.12)$$

then from (2A.9) and (2A.10), we obtain two statistically independent Gaussian distributed random variables C and D.

The method described above is often used in practice to generate Gaussian distributed random variables. As shown in Fig. 2A.5, these random variables have a mean value of zero and a variance σ^2. If a nonzero mean Gaussian random variable is desired, then C and D can be translated by the addition of the mean value.

A subroutine that implements the method above for generating Gaussian distributed random variables is given in Fig. 2A.6.

```
C       SUBROUTINE GAUSS CONVERTS A UNIFORM RANDOM
C       SEQUENCE XIN IN [0,1] TO A GAUSSIAN RANDOM
C       SEQUENCE WITH G(0,SIGMA**2)
C       PARAMETERS   :
C            XIN     :UNIFORM IN [0,1] RANDOM NUMBER
C              B     :UNIFORM IN [0,1] RANDOM NUMBER
C            SIGMA   :STANDARD DEVIAION OF THE GAUSSIAN
C            YOUT    :OUTPUT FROM THE GENERATOR
C
        SUBROUTINE GAUSS (XIN,B,SIGMA,YOUT)
        PI=4.0*ATAN(1.0)
        B=2.0*PI*B
        R=SQRT(2.0*(SIGMA**2)*ALOG(1.0/(1.0-XIN)))
        YOUT=R*COS(B)
        RETURN
        END

C       NOTE: TO USE THE ABOVE SUBROUTINE FOR A
C             GAUSSIAN RANDOM NUMBER GENERATOR
C             YOU MUST PROVIDE AS INPUT TWO UNIFORM RANDOM NUMBERS
C             XIN AND B
C             XIN AND B MUST BE STATISTICALLY INDEPENDENT
```

FIGURE 2A.6 Subroutine for generating Gaussian random variables.

PROBLEMS

2.1 A discrete-time signal $x(n)$ is defined as

$$x(n) = \begin{cases} 1 + \dfrac{n}{3} & -3 \le n \le -1 \\ 1 & 0 \le n \le 3 \\ 0 & \text{elsewhere} \end{cases}$$

(a) Determine its values and sketch the signal $x(n)$.
(b) Sketch the signals that result if we:
  (1) First fold $x(n)$ and then delay the resulting signal by four samples.
  (2) First delay $x(n)$ by four samples and then fold the resulting signal.
(c) Sketch the signal $x(-n + 4)$.
(d) Compare the results in parts (b) and (c) and derive a rule for obtaining the signal $x(-n + k)$ from $x(n)$.
(e) Can you express the signal $x(n)$ in terms of signals $\delta(n)$ and $u(n)$?

2.2 A discrete-time signal $x(n)$ is shown in Fig. P2.2. Sketch the label carefully each of the following signals.
(a) $x(n - 2)$ (b) $x(4 - n)$ (c) $x(n + 2)$ (d) $x(n)u(2 - n)$
(e) $x(n - 1)\delta(n - 3)$ (f) $x(n^2)$ (g) even part of $x(n)$
(h) odd part of $x(n)$

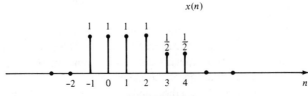

$x(n)$

FIGURE P2.2

2.3 Show that
(a) $\delta(n) = u(n) - u(n - 1)$
(b) $u(n) = \sum_{k=-\infty}^{n} \delta(k) = \sum_{k=0}^{\infty} \delta(n - k)$

2.4 Show that any signal can be decomposed into an even and an odd component. Is the decomposition unique? Illustrate your arguments using the signal

$$x(n) = \left\{ 2, 3, 4, \underset{\uparrow}{5}, 6 \right\}$$

2.5 Show that the energy (power) of a real-valued energy (power) signal is equal to the sum of the energies (powers) of its even and odd components.

2.6 Consider the system

$$y(n) = H[x(n)] = x(n^2)$$

(a) Determine if the system is time invariant.
(b) To clarify the result in part (a) assume that the signal

$$x(n) = \begin{cases} 1 & 0 \le n \le 3 \\ 0 & \text{elsewhere} \end{cases}$$

is applied into the system.

(1) Sketch the signal $x(n)$.

(2) Determine and sketch the signal $y(n) = H[x(n)]$.

(3) Sketch the signal $y_2'(n) = y(n - 2)$.

(4) Determine and sketch the signal $x_2(n) = x(n - 2)$.

(5) Determine and sketch the signal $y_2(n) = H[x_2(n)]$.

(6) Compare the signals $y_2(n)$ and $y(n - 2)$. What is your conclusion?

(c) Repeat part (b) for the system

$$y(n) = x(n) - x(n - 1)$$

Can you use this result to make any statement about the time invariance of this system? Why?

(d) Repeat parts (b) and (c) for the system

$$y(n) = H[x(n)] = nx(n)$$

2.7 A discrete-time system can be

(1) Static or dynamic

(2) Linear or nonlinear

(3) Time invariant or time varying

(4) Causal or noncausal

(5) Stable or unstable

Examine the following systems with respect to the properties above.

(a) $y(n) = \cos [x(n)]$

(b) $y(n) = \sum_{k=-\infty}^{n+1} x(k)$

(c) $y(n) = x(n) \cos (\omega_0 n)$

(d) $y(n) = x(-n + 2)$

(e) $y(n) = \text{Trun}[x(n)]$, where $\text{Trun}[x(n)]$ denotes the integer part of $x(n)$, obtained by truncation

(f) $y(n) = \text{Round}[x(n)]$, where $\text{Round}[x(n)]$ denotes the integer part of $x(n)$ obtained by rounding

Remark: The systems in parts (e) and (f) are quantizers by truncation and rounding, respectively.

(g) $y(n) = |x(n)|$

(h) $y(n) = x(n)u(n)$

(i) $y(n) = x(n) + nx(n + 1)$

(j) $y(n) = x(2n)$

(k) $y(n) = \begin{cases} x(n) & \text{if } x(n) \geq 0 \\ 0 & \text{if } x(n) < 0 \end{cases}$

(l) $y(n) = x(-n)$

(m) $y(n) = \text{sign}[x(n)]$

(n) The ideal sampling system with input $x_a(t)$ and output $x(n) = x_a(nT)$, $-\infty < n < \infty$

2.8 Two discrete-time systems H_1 and H_2 are connected in cascade to form a new system H as shown in Fig. P2.8. Prove or disprove the following statements.

(a) If H_1 and H_2 are linear, then H is linear (i.e., the cascade connection of two linear systems is linear).

(b) If H_1 and H_2 are time invariant, then H is time invariant.

(c) If H_1 and H_2 are causal, then H is causal.

(d) If H_1 and H_2 are linear and time invariant, the same holds for H.

(e) If H_1 and H_2 are linear and time invariant, then interchanging their order does not change the system H.

(f) As in part (e) except that H_1, H_2 are now time varying. (*Hint:* Use an example.)

(g) If H_1 and H_2 are nonlinear, then H is nonlinear.

(h) If H_1 and H_2 are stable, then H is stable.

(i) Show by an example that the inverse of parts (c) and (h) do not hold in general.

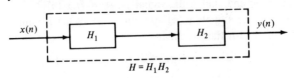

$$H = H_1 H_2$$

FIGURE P2.8

2.9 Let H be an LTI, relaxed, and BIBO stable system with input $x(n)$ and output $y(n)$. Show that:

(a) If $x(n)$ is periodic with period N [i.e., $x(n) = x(n + N)$ for all $n \geq 0$], the output $y(n)$ tends to a periodic signal with the same period.

(b) If $x(n)$ is bounded and tends to a constant, the output will also tend to a constant.

(c) If $x(n)$ is an energy signal, the output $y(n)$ will also be an energy signal.

2.10 Show that the necessary and sufficient condition for a relaxed LTI system to be BIBO stable is

$$\sum_{n=-\infty}^{\infty} |h(n)| \leq M_h < \infty$$

for some constant M_n.

2.11 Show that:

(a) A relaxed linear system is causal if and only if for any input $x(n)$ such that

$$x(n) = 0 \text{ for } n < n_0 \Rightarrow y(n) = 0 \qquad \text{for } n < n_0$$

(b) A relaxed LTI system is causal if and only if

$$h(n) = 0 \qquad \text{for } n < 0$$

2.12 (a) Show that for any real or complex constant a, and any finite integer numbers M and N, we have

$$\sum_{n=M}^{N} a^n = \begin{cases} \dfrac{a^M - a^{N+1}}{1 - a} & \text{if } a \neq 1 \\[2mm] N - M + 1 & \text{if } a = 1 \end{cases}$$

(b) Show that if $|a| < 1$, then

$$\sum_{n=0}^{\infty} a^n = \frac{1}{1 - a}$$

2.13 Compute the convolution $y(n) = x(n) * h(n)$ of the following signals and check the correctness of the results by using the test

$$\Sigma_x \, \Sigma_h = \Sigma_y$$

where $\Sigma_x \equiv \Sigma x(n)$.

(a) $x(n) = \{1, 2, 4\}$, $h(n) = \{1, 1, 1, 1, 1\}$

(b) $x(n) = \{1, 2, -1\}$, $h(n) = x(n)$

(c) $x(n) = \{0, 1, -2, 3, -4\}$, $h(n) = \{\frac{1}{2}, \frac{1}{2}, 1, \frac{1}{2}\}$

(d) $x(n) = \{1, 2, 3, 4, 5\}$, $h(n) = \{1\}$

(e) $x(n) = \left\{ 1, -2, 3 \atop \uparrow \right\}$, $h(n) = \left\{ 0, 0, 1, 1, 1, 1 \atop \uparrow \right\}$

(f) $x(n) = \left\{ 0, 0, 1, 1, 1, 1 \atop \uparrow \right\}$, $h(n) = \left\{ 1, -2, 3 \atop \uparrow \right\}$

(g) $x(n) = \left\{ 0, 1, 4, -3 \atop \uparrow \right\}$, $h(n) = \left\{ 1, 0, -1, -1 \atop \uparrow \right\}$

(h) $x(n) = \left\{ 1, 1, 2 \atop \uparrow \right\}$, $h(n) = u(n)$

(i) $x(n) = \left\{ 1, 1, 0, 1, 1 \atop \uparrow \right\}$, $h(n) = \left\{ 1, -2, -3, 4 \atop \uparrow \right\}$

(j) $x(n) = \left\{ 1, 2, 0, 2, 1 \atop \uparrow \right\}$ $h(n) = x(n)$

(k) $x(n) = (\tfrac{1}{2})^n u(n)$, $h(n) = (\tfrac{1}{4})^n u(n)$

2.14 Compute and plot the convolutions $x(n) * h(n)$ and $h(n) * x(n)$ for the pairs of signals shown in Fig. P2.14.

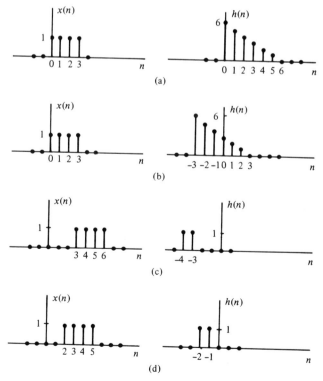

FIGURE P2.14

2.15 Determine and sketch the convolution $y(n)$ of the signals

$$x(n) = \begin{cases} \frac{1}{3}n & 0 \leq n \leq 6 \\ 0 & \text{elsewhere} \end{cases}$$

$$h(n) = \begin{cases} 1 & -2 \leq n \leq 2 \\ 0 & \text{elsewhere} \end{cases}$$

(a) Numerically
(b) Analytically

2.16 Compute the convolution $y(n)$ of the signals

$$x(n) = \begin{cases} \alpha^n & -3 \leq n \leq 5 \\ 0 & \text{elsewhere} \end{cases}$$

$$h(n) = \begin{cases} 1 & 0 \leq n \leq 4 \\ 0 & \text{elsewhere} \end{cases}$$

2.17 Consider the following three operations.
(a) Multiply the integer numbers: 131 and 122.
(b) Compute the convolution of signals: $\{1, 3, 1\} * \{1, 2, 2\}$.
(c) Multiply the polynomials: $1 + 3z + z^2$ and $1 + 2z + 2z^2$.
(d) Repeat part (a) for the numbers 1.31 and 12.2.
(e) Comment on your results.

2.18 Compute the convolution $y(n) = x(n) * h(n)$ of the following pairs of signals.
(a) $x(n) = a^n u(n)$, $h(n) = b^n u(n)$ when $a \neq b$ and when $a = b$

(b) $x(n) = \begin{cases} 1 & n = -2, 0, 1 \\ 2 & n = -1 \\ 0 & \text{elsewhere} \end{cases}$

$h(n) = \delta(n) - \delta(n - 1) + \delta(n - 4) + \delta(n - 5)$
(c) $x(n) = u(n + 1) - u(n - 4) - \delta(n - 5)$
$h(n) = [u(n + 2) - u(n - 3)] \cdot (3 - |n|)$
(d) $x(n) = u(n) - u(n - 5)$
$h(n) = u(n - 2) - u(n - 8) + u(n - 11) - u(n - 17)$

2.19 Let $x(n)$ be the input signal to a discrete-time filter with impulse response $h_i(n)$ and let $y_i(n)$ be the corresponding output.
(a) Compute and sketch $x(n)$ and $y_i(n)$ in the following cases, using the same scale in all figures.

$$x(n) = \{1, 4, 2, 3, 5, 3, 3, 4, 5, 7, 6, 9\}$$
$$h_1(n) = \{1, 1\}$$
$$h_2(n) = \{1, 2, 1\}$$
$$h_3(n) = \{\tfrac{1}{2}, \tfrac{1}{2}\}$$
$$h_4(n) = \{\tfrac{1}{4}, \tfrac{1}{2}, \tfrac{1}{4}\}$$
$$h_5(n) = \{\tfrac{1}{4}, -\tfrac{1}{2}, \tfrac{1}{4}\}$$

Sketch $x(n)$, $y_1(n)$, $y_2(n)$ on one graph and $x(n)$, $y_3(n)$, $y_4(n)$, $y_5(n)$ on another graph
(b) What is the difference between $y_1(n)$ and $y_2(n)$, and between $y_3(n)$ and $y_4(n)$?
(c) Comment on the smoothness of $y_2(n)$ and $y_4(n)$. Which factors affect the smoothness?

(d) Compare $y_4(n)$ with $y_5(n)$. What is the difference? Can you explain it?

(e) Let $h_6(n) = \{\frac{1}{2}, -\frac{1}{2}\}$. Compute $y_6(n)$. Sketch $x(n)$, $y_2(n)$, and $y_6(n)$ on the same figure and comment on the results.

2.20 Let $x(n)$, $N_1 \le n \le N_2$ and $h(n)$, $M_1 \le n \le M_2$ be two finite-duration signals.

(a) Determine the range $L_1 \le n \le L_2$ of their convolution, in terms of N_1, N_2, M_1 and M_2.

(b) Determine the limits of the cases of partial overlap from the left, full overlap, and partial overlap from the right. For convenience, assume that $h(n)$ has shorter duration than $x(n)$.

(c) Illustrate the validity of your results by computing the convolution of the signals

$$x(n) = \begin{cases} 1 & -2 \le n \le 4 \\ 0 & \text{elsewhere} \end{cases}$$

$$h(n) = \begin{cases} 2 & -1 \le n \le 2 \\ 0 & \text{elsewhere} \end{cases}$$

2.21 Determine the impulse response and the unit step response of the systems described by the difference equation

(a) $y(n) = 0.6y(n-1) - 0.08y(n-2) + x(n)$

(b) $y(n) = 0.7y(n-1) - 0.1y(n-2) + 2x(n) - x(n-2)$

2.22 Consider a system with impulse response

$$h(n) = \begin{cases} (\frac{1}{2})^n & 0 \le n \le 4 \\ 0 & \text{elsewhere} \end{cases}$$

Determine the input $x(n)$ for $0 \le n \le 8$ that will generate the output sequence

$$y(n) = \left\{ \underset{\uparrow}{1}, 2, 2.5, 3, 3, 3, 2, 1, 0, \ldots \right\}$$

2.23 Consider the interconnection of LTI systems as shown in Fig. P2.23.

(a) Express the overall impulse response in terms of $h_1(n)$, $h_2(n)$, $h_3(n)$, and $h_4(n)$.

(b) Determine $h(n)$ when

$$h_1(n) = \{\tfrac{1}{2}, \tfrac{1}{4}, \tfrac{1}{2}\}$$
$$h_2(n) = h_3(n) = (n+1)u(n)$$
$$h_4(n) = \delta(n-2)$$

(c) Determine the response of the system in part (b) if

$$x(n) = \delta(n+2) + 3\delta(n-1) - 4\delta(n-3)$$

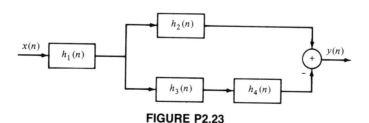

FIGURE P2.23

2.24 Consider the system in Fig. P2.24 with $h(n) = a^n u(n)$, $-1 < a < 1$. Determine the response $y(n)$ of the system to the excitation

$$x(n) = u(n + 5) - u(n - 10)$$

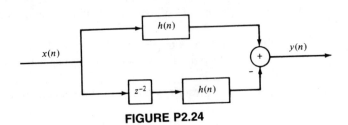

FIGURE P2.24

2.25 Compute and sketch the step response of the system

$$y(n) = \frac{1}{M} \sum_{k=0}^{M-1} x(n - k)$$

2.26 Determine the range of values of the parameter a for which the linear time-invariant system with impulse response

$$h(n) = \begin{cases} a^n & n \geq 0, \ n \text{ even} \\ 0 & \text{otherwise} \end{cases}$$

is stable.

2.27 Determine the response of the system with impulse response

$$h(n) = a^n u(n)$$

to the input signal

$$x(n) = u(n) - u(n - 10)$$

(*Hint:* The solution can be obtained easily and quickly by applying the linearity and time-invariance properties to the result in Example 2.3.5.)

2.28 Determine the response of the (relaxed) system characterized by the impulse response

$$h(n) = (\tfrac{1}{2})^n u(n)$$

to the input signal

$$x(n) = \begin{cases} 1 & 0 \leq n < 10 \\ 0 & \text{otherwise} \end{cases}$$

2.29 Determine the response of the (relaxed) system characterized by the impulse response

$$h(n) = (\tfrac{1}{2})^n u(n)$$

to the input signals
(a) $x(n) = 2^n u(n)$
(b) $x(n) = u(-n)$

2.30 Three systems with impulse responses $h_1(n) = \delta(n) - \delta(n - 1)$, $h_2(n) = h(n)$, and $h_3(n) = u(n)$, are connected in cascade.

(a) What is the impulse response, $h_c(n)$, of the overall system?

(b) Does the order of the interconnection affect the overall system?

2.31 (a) Prove and explain graphically the difference between the relations
$x(n)\delta(n - n_0) = x(n_0)$ and $x(n) * \delta(n - n_0) = x(n - n_0)$

(b) Show that a discrete-time system, which is described by a convolution summation, is LTI and relaxed,

(c) What is the impulse response of the system described by $y(n) = x(n - n_0)$?

2.32 Two signals $s(n)$ and $v(n)$ are related through the following difference equations

$$s(n) + a_1 s(n - 1) + \cdots + a_N s(n - N) = b_0 v(n)$$

Design the block diagram realization of:

(a) The system that generates $s(n)$ when excited by $v(n)$.

(b) The system that generates $v(n)$ when excited by $s(n)$.

(c) What is the impulse response of the cascade interconnection of systems in parts (a) and (b)?

2.33 Compute the zero-state response of the system described by the difference equation

$$y(n) + \tfrac{1}{2}y(n - 1) = x(n) + 2x(n - 2)$$

to the input

$$x(n) = \left\{ 1, 2, 3, 4, 2, 1 \atop \uparrow \right\}$$

by solving the difference equation recursively.

2.34 Determine the direct form II realization for each of the following LTI systems.

(a) $2y(n) + y(n - 1) - 4y(n - 3) = x(n) + 3x(n - 5)$

(b) $y(n) = x(n) - x(n - 1) + 2x(n - 2) - 3x(n - 4)$

2.35 Consider the discrete-time system shown in Fig. P2.35.

(a) Compute the 10 first samples of its impulse response.

(b) Find the input–output relation.

(c) Apply the input $x(n) = \left\{ 1, 1, 1, \ldots \atop \uparrow \right\}$ and compute the first 10 samples of

the output.

(d) Compute the first 10 samples of the output for the input given in part (c) by using convolution.

(e) Is the system causal? Is it stable?

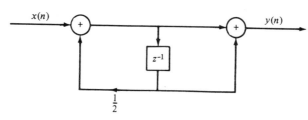

FIGURE P2.35

2.36 Consider the system described by the difference equation

$$y(n) = ay(n - 1) + bx(n)$$

(a) Determine b in terms of a so that

$$\sum_{n=-\infty}^{\infty} h(n) = 1$$

(b) Compute the zero-state step response $s(n)$ of the system and choose b so that $s(\infty) = 1$.

(c) Compare the values of b obtained in parts (a) and (b). What did you notice?

2.37 A discrete-time system is realized by the structure shown in Fig. P2.37.

(a) Compute the impulse response.

(b) Determine a realization for its inverse system, that is, the system which produces $x(n)$ as an output when $y(n)$ is used as an input.

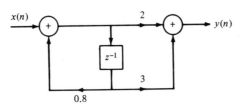

FIGURE P2.37

2.38 Consider the discrete-time system shown in Fig. P2.38.

(a) Compute the first six values of the impulse response of the system.

(b) Compute the first six values of the zero-state step response of the system.

(c) Determine an analytical expression for the impulse response of the system.

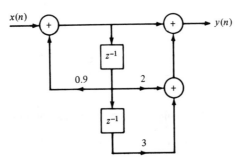

FIGURE P2.38

2.39 Determine and sketch the impulse response of the following systems for $n = 0$, 1, . . ., 9.

(a) Fig. P2.39(a).

(b) Fig. P2.39(b).

(c) Fig. P2.39(c).

(d) Classify the systems above as FIR or IIR.

(e) Find an explicit expression for the impulse response of the system in part (c).

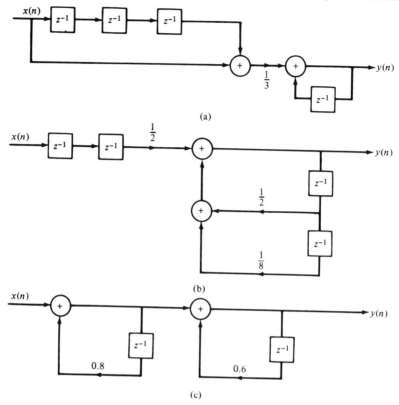

FIGURE P2.39

2.40 Consider the systems shown in Fig. P2.40.
   (a) Determine and sketch their impulse responses $h_1(n)$, $h_2(n)$, and $h_3(n)$.
   (b) Is it possible to choose the coefficients of these systems in such a way that

$$h_1(n) = h_2(n) = h_3(n)$$

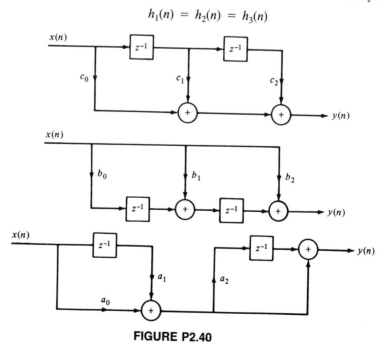

FIGURE P2.40

2.41 Consider the system shown in Fig. P2.41.

(a) Determine its impulse response $h(n)$.

(b) Show that $h(n)$ is equal to the convolution of the following signals.

$$h_1(n) = \delta(n) + \delta(n-1)$$
$$h_2(n) = (\tfrac{1}{2})^n u(n)$$

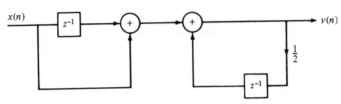

FIGURE P2.41

2.42 Compute the sketch the convolution $y_i(n)$ and correlation $r_i(n)$ sequences for the following pair of signals and comment on the results obtained.

(a) $x_1(n) = \left\{ \begin{array}{c} 1, 2, 4 \\ \uparrow \end{array} \right\}$. $h_1(n) = \left\{ \begin{array}{c} 1, 1, 1, 1, 1 \\ \uparrow \end{array} \right\}$.

(b) $x_2(n) = \left\{ \begin{array}{c} 0, 1, -2, 3, -4 \\ \uparrow \end{array} \right\}$, $h_2(n) = \left\{ \begin{array}{c} \tfrac{1}{2}, 1, 2, 1, \tfrac{1}{2} \\ \uparrow \end{array} \right\}$

(c) $x_3(n) = \left\{ \begin{array}{c} 1, 2, 3, 4 \\ \uparrow \end{array} \right\}$, $h_3(n) = \left\{ \begin{array}{c} 4, 3, 2, 1 \\ \uparrow \end{array} \right\}$

(d) $x_4(n) = \left\{ \begin{array}{c} 1, 2, 3, 4 \\ \uparrow \end{array} \right\}$, $h_4(n) = \left\{ \begin{array}{c} 1, 2, 3, 4 \\ \uparrow \end{array} \right\}$

2.43 The zero-state response of a causal LTI system to the input $x(n) =$

$\left\{ \begin{array}{c} 1, 3, 3, 1 \\ \uparrow \end{array} \right\}$ is $y(n) = \left\{ \begin{array}{c} 1, 4, 6, 4, 1 \\ \uparrow \end{array} \right\}$.

Determine its impulse response.

2.44 Prove by direct substitution the equivalence of equations (2.4.52) and (2.4.53), which describe the direct form II structure, to the relation (2.4.49), which describes the direct form I structure.

2.45 Determine the response $y(n)$, $n \geq 0$ of the system described by the second-order difference equation

$$y(n) - 4y(n-1) + 4y(n-2) = x(n) - x(n-1)$$

when the input is

$$x(n) = (-1)^n u(n)$$

and the initial conditions are $y(-1) = y(-2) = 0$.

2.46 Determine the impulse response $h(n)$ for the system described by the second-order difference equation

$$y(n) - 4y(n - 1) + 4y(n - 2) = x(n) - x(n - 1)$$

2.47 Show that any discrete-time signal $x(n)$ can be expressed as

$$x(n) = \sum_{k=-\infty}^{\infty} [x(k) - x(k - 1)]u(n - k)$$

where $u(n - k)$ is a unit step delayed by k units in time, that is,

$$u(n - k) = \begin{cases} 1 & n \geq k \\ 0 & \text{otherwise} \end{cases}$$

2.48 Show that the output of an LTI system can be expressed in terms of its unit step response $s(n)$ as follows.

$$y(n) = \sum_{k=-\infty}^{\infty} [s(k) - s(k - 1)]x(n - k) = \sum_{k=-\infty}^{\infty} [x(k) - x(k - 1)]s(n - k)$$

2.49 Compute the correlation sequences $r_{xx}(l)$ and $r_{xy}(l)$ for the following signal sequences.

$$x(n) = \begin{cases} 1 & n_0 - N \leq n \leq n_0 + N \\ 0 & \text{otherwise} \end{cases}$$

$$y(n) = \begin{cases} 1 & -N \leq n \leq N \\ 0 & \text{otherwise} \end{cases}$$

2.50 Determine the autocorrelation sequences of the following signals.

(a) $x(n) = \left\{ 1, 2, 1, 1 \atop \uparrow \right\}$

(b) $y(n) = \{1, 1, 2, 1\}$
    $\quad\;\; \uparrow$

What is your conclusion?

2.51 What is the normalized autocorrelation sequence of the signal $x(n)$ given by

$$x(n) = \begin{cases} 1 & -N \leq n \leq N \\ 0 & \text{otherwise} \end{cases}$$

COMPUTER EXPERIMENTS

2.52 *Time-delay estimation in radar.* Let $x_a(t)$ be the transmitted signal and $y_a(t)$ be the received signal in a radar system, where

$$y_a(t) = ax_a(t - t_d) + v_a(t)$$

and $v_a(t)$ is additive random noise. The signals $x_a(t)$ and $y_a(t)$ are sampled in the receiver, according to the sampling theorem, and are processed digitally to determine the time delay and hence the distance of the object. The resulting discrete-time signals are

$$\begin{aligned} x(n) &= x_a(nT) \\ y(n) &= y_a(nT) = ax_a(nT - DT) + v_a(nT) \\ &\triangleq ax(n - D) + v(n) \end{aligned}$$

(a) Explain how we can measure the delay D by computing the crosscorrelation $r_{xy}(l)$.

(b) Let $x(n)$ be the 13-point *Barker sequence*

$$x(n) = \{+1, +1, +1, +1, +1, -1, -1, +1, +1, -1, +1, -1, +1\}$$

and $v(n)$ be a Gaussian random sequence with zero mean and variance $\sigma^2 = 0.01$. Write a program that generates the sequence $y(n)$, $0 \le n \le 199$ for $a = 0.9$ and $D = 20$. Plot the signals $x(n)$, $y(n)$, $0 \le n \le 199$.

(c) Compute and plot the crosscorrelation $r_{xy}(l)$, $0 \le l \le 59$. Use the plot to estimate the value of the delay D.

(d) Repeat parts (b) and (c) for $\sigma^2 = 0.1$ and $\sigma^2 = 1$.

(e) Repeat parts (b) and (c) for the signal sequence

$$x(n) = \{-1, -1, -1, +1, +1, +1, +1, -1, +1, -1,$$
$$+1, +1, -1, -1, +1\}$$

which is obtained from the four-stage feedback shift register shown in Fig. P2.52. Note that $x(n)$ is just one period of the periodic sequence obtained from the feedback shift register.

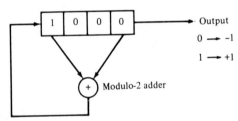

FIGURE P2.52 Linear feedback shift register.

(f) Repeat parts (b) and (c) for a sequence of period $N = 2^7 - 1$, which is obtained from a seven-stage feedback shift register. Table 2.4 gives the stages connected to the modulo-2 adder for (maximal-length) shift-register sequences of length $N = 2^m - 1$.

2.53 *Implementation of LTI systems.* Consider the recursive discrete-time system described by the difference equation

$$y(n) = -a_1 y(n-1) - a_2 y(n-2) + b_0 x(n)$$

where $a_1 = -0.8$, $a_2 = 0.64$, and $b_0 = 0.866$.

(a) Write a program to compute and plot the impulse response $h(n)$ of the system for $0 \le n \le 49$.

(b) Write a program to compute and plot the zero-state step response $s(n)$ of the system for $0 \le n \le 100$.

(c) Define an FIR system with impulse response $h_{FIR}(n)$ given by

$$h_{FIR}(n) = \begin{cases} h(n) & 0 \le n \le 19 \\ 0 & \text{elsewhere} \end{cases}$$

where $h(n)$ is the impulse response computed in part (a). Write a program to compute and plot its step response.

(d) Compare the results obtained in parts (b) and (c) and explain their similarities and differences.

TABLE 2.4   Shift-Register Connections
for Generating Maximal-Length
Sequences

m	Stages Connected to Modulo-2 Adder
1	1
2	1, 2
3	1, 3
4	1, 4
5	1, 4
6	1, 6
7	1, 7
8	1, 5, 6, 7
9	1, 6
10	1, 8
11	1, 10
12	1, 7, 9, 12
13	1, 10, 11, 13
14	1, 5, 9, 14
15	1, 15
16	1, 5, 14, 16
17	1, 15

2.54 Use the subroutines introduced in this chapter to write a computer program that computes the overall impulse response $h(n)$ of the system shown in Fig. P2.54 for $0 \le n \le 99$. The systems H_1, H_2, H_3, and H_4 are specified by

$$H_1: h_1(n) = \left\{ \underset{\uparrow}{1}, \tfrac{1}{2}, \tfrac{1}{4}, \tfrac{1}{8}, \tfrac{1}{16}, \tfrac{1}{32} \right\}$$

$$H_2: h_2(n) = \left\{ \underset{\uparrow}{1}, 1, 1, 1, 1 \right\}$$

$$H_3: y_3(n) = \tfrac{1}{4}x(n) + \tfrac{1}{2}x(n-1) + \tfrac{1}{4}x(n-2)$$
$$H_4: y(n) = 0.9y(n-1) - 0.81y(n-2) + v(n) + v(n-1)$$

Plot $h(n)$ for $0 \le n \le 99$.

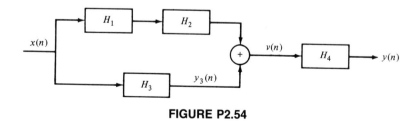

FIGURE P2.54

3

The z-Transform and Its Application to the Analysis of LTI Systems

Transform techniques are an important tool in the analysis of signals and linear time-invariant (LTI) systems. In this chapter we introduce the z-transform, develop its properties, and demonstrate its importance in the analysis and characterization of linear time-invariant systems.

The z-transform plays the same role in the analysis of discrete-time signals and LTI systems as the Laplace transform does in the analysis of continuous-time signals and LTI systems. For example, we shall see that in the z-domain (complex z-plane) the convolution of two time-domain signals is equivalent to multiplication of their corresponding z-transforms. This property greatly simplifies the analysis of the response of an LTI system to various signals. In addition, the z-transform provides us with a means of characterizing an LTI system, and its response to various signals, by its pole–zero locations.

We begin this chapter by defining the z-transform. Its important properties are presented in Section 3.2. In Section 3.3 the transform is used to characterize signals in terms of their pole–zero patterns. Section 3.4 describes methods for inverting the z-transform of a signal so as to obtain the time-domain representation of the signal. The one-sided z-transform is treated in Section 3.5 and used to solve linear difference equations with nonzero initial conditions. The chapter concludes with a discussion on the use of the z-transform in the analysis of LTI systems.

3.1 The z-Transform

In this section we introduce the z-transform of a discrete-time signal, investigate its convergence properties, and conclude with a brief discussion of the inverse z-transform.

3.1.1 The Direct z-Transform

The z-transform of a discrete-time signal $x(n)$ is defined as the power series

$$X(z) \equiv \sum_{n=-\infty}^{\infty} x(n)z^{-n} \tag{3.1.1}$$

where z is a complex variable. The relation (3.1.1) is sometimes called the *direct z-transform* because it transforms the time-domain signal $x(n)$ into its complex-plane representation $X(z)$. The inverse procedure [i.e., obtaining $x(n)$ from $X(z)$] is called the *inverse z-transform* and is examined briefly in Section 3.1.2 and in more detail in Section 3.4.

For convenience, the z-transform of a signal $x(n)$ will be denoted by

$$X(z) \equiv Z\{x(n)\} \tag{3.1.2}$$

whereas the relationship between $x(n)$ and $X(z)$ will be indicated by

$$x(n) \xleftarrow{\quad z \quad} X(z) \tag{3.1.3}$$

Since the z-transform is an infinite power series, it exists only for those values of z for which this series converges. The *region of convergence* (ROC) of $X(z)$ is the set of all values of z for which $X(z)$ attains a finite value. Thus any time we cite a z-transform we should also indicate its ROC.

We illustrate these concepts by some simple examples.

EXAMPLE 3.1.1

Determine the z-transforms of the following *finite-duration* signals.

(a) $x_1(n) = \{1, 2, 5, 7, 0, 1\}$

(b) $x_2(n) = \{1, 2, 5, 7, 0, 1\}$
$\qquad\qquad\qquad \uparrow$

(c) $x_3(n) = \{0, 0, 1, 2, 5, 7, 0, 1\}$

(d) $x_4(n) = \{2, 4, 5, 7, 0, 1\}$
$\qquad\qquad\qquad \uparrow$

(e) $x_5(n) = \delta(n)$

(f) $x_6(n) = \delta(n - k),\ k > 0$

(g) $x_7(n) = \delta(n + k),\ k > 0$

Solution: From definition (3.1.1), we have

(a) $X_1(z) = 1 + 2z^{-1} + 5z^{-2} + 7z^{-3} + z^{-5}$, ROC: entire z-plane except $z = 0$

(b) $X_2(z) = z^2 + 2z + 5 + 7z^{-1} + z^{-3}$, ROC: entire z-plane except $z = 0$ and $z = \infty$

(c) $X_3(z) = z^{-2} + 2z^{-3} + 5z^{-4} + 7z^{-5} + z^{-7}$, ROC: entire z-plane except $z = 0$

(d) $X_4(z) = 2z^2 + 4z + 5 + 7z^{-1} + z^{-3}$, ROC: entire z-plane except $z = 0$ and $z = \infty$

(e) $X_5(z) = 1$ [i.e., $\delta(n) \xleftarrow{\quad z \quad} 1$], ROC: entire z-plane

(f) $X_6(z) = z^{-k}$ [i.e., $\delta(n - k) \xleftarrow{\quad z \quad} z^{-k}$], $k > 0$, ROC: entire z-plane except $z = 0$

(g) $X_7(z) = z^k$ [i.e., $\delta(n + k) \xleftarrow{\quad z \quad} z^k$], $k > 0$, ROC: entire z-plane except $z = \infty$

From this example it is easily seen that the ROC of a *finite-duration signal* is the entire z-plane, except possibly the points $z = 0$ and/or $z = \infty$. These points are excluded, because z^k $(k > 0)$ becomes unbounded for $z = \infty$ and z^{-k} $(k > 0)$ becomes unbounded for $z = 0$.

From a mathematical point of view the z-transform is simply an alternative representation of a signal. This is nicely illustrated in Example 3.1.1, where we see that the coefficient of z^{-n}, in a given transform, is the value of the signal at time n. In other words, the exponent of z contains the time information we need to identify the samples of the signal.

In many cases we can express the sum of the finite or infinite series for the z-transform in a closed-form expression. In such cases the z-transform offers a compact alternative representation of the signal.

EXAMPLE 3.1.2 ──

Determine the z-transform of the signal

$$x(n) = (\tfrac{1}{2})^n u(n)$$

Solution: The signal $x(n)$ consists of an infinite number of nonzero samples

$$x(n) = \{1, (\tfrac{1}{2}), (\tfrac{1}{2})^2, (\tfrac{1}{2})^3, \ldots, (\tfrac{1}{2})^n, \ldots\}$$

The z-transform of $x(n)$ is the infinite power series

$$X(z) = 1 + \tfrac{1}{2} z^{-1} + (\tfrac{1}{2})^2 z^{-2} + (\tfrac{1}{2})^n z^{-n} + \cdots$$

$$= \sum_{n=0}^{\infty} (\tfrac{1}{2})^n z^{-n} = \sum_{n=0}^{\infty} (\tfrac{1}{2} z^{-1})^n$$

This is an infinite geometric series. We recall that

$$1 + A + A^2 + A^3 + \cdots = \frac{1}{1 - A} \qquad \text{if } |A| < 1$$

Consequently, for $|\tfrac{1}{2} z^{-1}| < 1$, or equivalently, for $|z| > \tfrac{1}{2}$, $X(z)$ converges to

$$X(z) = \frac{1}{1 - \tfrac{1}{2}z^{-1}} \qquad \text{ROC: } |z| > \tfrac{1}{2}$$

We see that in this case, the z-transform provides a compact alternative representation of the signal $x(n)$.

──

Let us express the complex variable z in polar form as

$$z = re^{j\theta} \tag{3.1.4}$$

where $r = |z|$ and $\theta = \angle z$. Then $X(z)$ may be expressed as

$$X(z)|_{z=re^{j\theta}} = \sum_{n=-\infty}^{\infty} x(n) r^{-n} e^{-j\theta n}$$

In the ROC of $X(z)$, $|X(z)| < \infty$. But

$$|X(z)| = \left| \sum_{n=-\infty}^{\infty} x(n) r^{-n} e^{-j\theta n} \right| \tag{3.1.5}$$

$$\leq \sum_{n=-\infty}^{\infty} |x(n) r^{-n} e^{-j\theta n}| = \sum_{n=-\infty}^{\infty} |x(n) r^{-n}|$$

Hence $|X(z)|$ is finite if the sequence $x(n) r^{-n}$ is absolutely summable.

The problem of finding the ROC for $X(z)$ is equivalent to determining the range of values of r for which the sequence $x(n) r^{-n}$ is absolutely summable. To elaborate, let us express (3.1.5) as

$$|X(z)| \leq \sum_{n=-\infty}^{-1} |x(n)r^{-n}| + \sum_{n=0}^{\infty} \left| \frac{x(n)}{r^n} \right|$$

$$\leq \sum_{n=1}^{\infty} |x(-n)r^n| + \sum_{n=0}^{\infty} \left| \frac{x(n)}{r^n} \right|$$

(3.1.6)

If $X(z)$ converges in some region of the complex plane, both summations in (3.1.6) must be finite in that region. If the first sum in (3.1.6) converges, there must exist values of r small enough such that the product sequence $x(-n)r^n$, $1 \leq n < \infty$, is absolutely summable. Therefore, the ROC for the first sum consists of all points in a circle of some radius r_1, where $r_1 < \infty$, as illustrated in Fig. 3.1a. On the other hand, if the second sum in (3.1.6) converges, there must exist values of r large enough such that the product sequence $x(n)/r^n$, $0 \leq n < \infty$, is absolutely summable. Hence the ROC for the second sum in (3.1.6) consists of all points outside a circle of radius $r > r_2$, as illustrated in Fig. 3.1b.

Since the convergence of $X(z)$ requires that both sums in (3.1.6) be finite, it follows

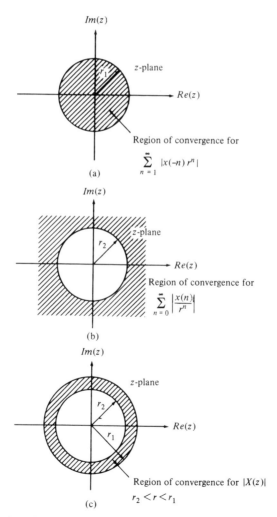

FIGURE 3.1 Region of convergence for $X(z)$ and its corresponding causal and anti-causal components.

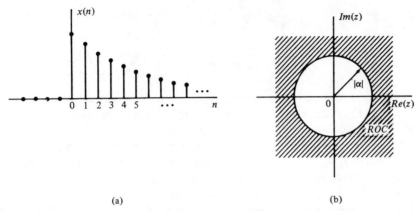

(a)   (b)

FIGURE 3.2   Causal exponential signal $x(n) = \alpha^n u(n)$ (a), and the ROC of its z-transform (b).

that the ROC of $X(z)$ is generally specified as the annular region in the z-plane, $r_2 < r < r_1$, which is the common region where both sums are finite. This region is illustrated in Fig. 3.1c. On the other hand, if $r_2 > r_1$, there is no common region of convergence for the two sums and hence $X(z)$ does not exist.

The following examples illustrate these important concepts.

EXAMPLE 3.1.3 _____

Determine the z-transform of the signal

$$x(n) = \alpha^n u(n) = \begin{cases} \alpha^n & n \geq 0 \\ 0 & n < 0 \end{cases}$$

Solution: From the definition (3.1.1) we have

$$X(z) = \sum_{n=0}^{\infty} \alpha^n z^{-n} = \sum_{n=0}^{\infty} (\alpha z^{-1})^n$$

If $|\alpha z^{-1}| < 1$ or equivalently $|z| > |\alpha|$, this power series converges to $1/(1 - \alpha z^{-1})$. Thus we have the z-transform pair

$$x(n) = \alpha^n u(n) \xleftarrow{\ z\ } X(z) = \frac{1}{1 - \alpha z^{-1}} \qquad \text{ROC: } |z| > |\alpha| \qquad (3.1.7)$$

The ROC is the exterior of a circle having radius $|\alpha|$. Figure 3.2 shows a graph of the signal $x(n)$ and its corresponding ROC. Note that, in general, α need not be real.

If we set $\alpha = 1$ in (3.1.7), we obtain the z-transform of the unit step signal

$$x(n) = u(n) \xleftarrow{\ z\ } X(z) = \frac{1}{1 - z^{-1}} \qquad \text{ROC: } |z| > 1 \qquad (3.1.8)$$

_____

EXAMPLE 3.1.4 _____

Determine the z-transform of the signal

$$x(n) = -\alpha^n u(-n - 1) = \begin{cases} 0 & n \geq 0 \\ -\alpha^n & n \leq -1 \end{cases}$$

Solution: From the definition (3.1.1) we have

$$X(z) = \sum_{n=-\infty}^{-1} (-\alpha^n)z^{-n} = -\sum_{l=1}^{\infty} (\alpha^{-1}z)^l$$

where $l = -n$. Using the formula

$$A + A^2 + A^3 + \cdots = A(1 + A + A^2 + \cdots) = \frac{A}{1-A}$$

when $|A| < 1$ gives

$$X(z) = -\frac{\alpha^{-1}z}{1 - \alpha^{-1}z} = \frac{1}{1 - \alpha z^{-1}}$$

provided that $|\alpha^{-1}z| < 1$ or, equivalently, $|z| < |\alpha|$. Thus

$$x(n) = -\alpha^n u(-n-1) \xleftarrow{\quad z \quad} X(z) = \frac{1}{1 - \alpha z^{-1}} \qquad \text{ROC: } |z| < |\alpha| \qquad (3.1.9)$$

The ROC is now the interior of a circle having radius $|\alpha|$. This is shown in Fig. 3.3.

Examples 3.1.3 and 3.1.4 illustrate two very important issues. The first concerns the uniqueness of the z-transform. From (3.1.7) and (3.1.9) we see that the causal signal $\alpha^n u(n)$ and the anticausal signal $-\alpha^n u(-n-1)$ have identical closed-form expressions for the z-transform, that is,

$$Z\{\alpha^n u(n)\} = Z\{-\alpha^n u(-n-1)\} = \frac{1}{1 - \alpha z^{-1}}$$

This implies that a closed-form expression for the z-transform does not uniquely specify the signal in the time domain. The ambiguity can be resolved only if in addition to the closed-form expression, the ROC is specified. In summary, *a discrete-time signal $x(n)$ is uniquely determined by its z-transform $X(z)$ and the region of convergence of $X(z)$*. In this text the term "z-transform" will be used to refer to both the closed-form expression and the corresponding ROC. Example 3.1.3 also illustrates the point that *the ROC of a causal signal is the exterior of a circle of some radius r_2 while the ROC of an anticausal signal is the interior of a circle of some radius r_1.* The following example considers a sequence that is nonzero for $-\infty < n < \infty$.

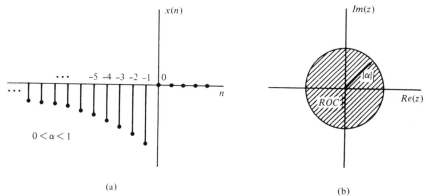

(a)   (b)

FIGURE 3.3 Anticausal signal $x(n) = -\alpha^n u(-n-1)$ (a), and the ROC of its z-transform (b).

EXAMPLE 3.1.5 _____

Determine the z-transform of the signal

$$x(n) = \alpha^n u(n) + b^n u(-n - 1)$$

Solution: From definition (3.1.1) we have

$$X(z) = \sum_{n=0}^{\infty} \alpha^n z^{-n} + \sum_{n=-\infty}^{-1} b^n z^{-n} = \sum_{n=0}^{\infty} (\alpha z^{-1})^n + \sum_{l=1}^{\infty} (b^{-1} z)^l$$

The first power series converges if $|\alpha z^{-1}| < 1$ or $|z| > |\alpha|$. The second power series converges if $|b^{-1} z| < 1$ or $|z| < |b|$.

In determining the convergence of $X(z)$, we consider two different cases.

Case 1 $|b| < |\alpha|$: In this case the two ROC above do not overlap, as shown in Fig. 3.4(a). Consequently, we cannot find values of z for which both power series converge simultaneously. Clearly, in this case, $X(z)$ does not exist.

Case 2 $|b| > |\alpha|$: In this case there is a ring in the z-plane where both power series converge simultaneously, as shown in Fig. 3.4(b). Then we obtain

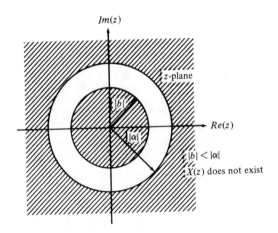

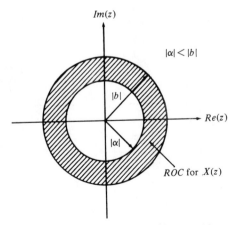

FIGURE 3.4   ROC for z-transform in Example 3.1.5.

$$X(z) = \frac{1}{1 - \alpha z^{-1}} - \frac{1}{1 - b z^{-1}}$$

$$= \frac{b - \alpha}{\alpha + b - z - \alpha b z^{-1}}$$

(3.1.10)

The ROC of $X(z)$ is $|\alpha| < |z| < |b|$.

This example shows that *if there is a ROC for an infinite duration two-sided signal, it is a ring (annular region) in the z-plane.* From Examples 3.1.1, 3.1.3, 3.1.4, and 3.1.5, we see that the ROC of a signal depends both on its duration (finite or infinite) and whether it is causal, anticausal, or two-sided. These facts are summarized in Table 3.1.

TABLE 3.1 Characteristic Families of Signals with Their Corresponding ROC

Signal	ROC

Finite-Duration Signals

Causal

Entire z-plane except $z = 0$

Anticausal

Entire z-plane except $z = \infty$

Two-Sided

Entire z-plane except $z = 0$ and $z = \infty$

Infinite-Duration Signals

Causal

$|z| > r_2$

Anticausal

$|z| < r_1$

Two-Sided

$r_2 < |z| < r_1$

Finally, we note that the z-transform defined by (3.1.1) is sometimes referred to as the *two-sided* or *bilateral z-transform*, to distinguish it from the *one-sided* or *unilateral z-transform* given by

$$X^+(z) = \sum_{n=0}^{\infty} x(n)z^{-n} \tag{3.1.11}$$

The one-sided z-transform is examined in Section 3.5. In this text we use the expression z-transform exclusively to mean the two-sided z-transform defined by (3.1.1). The term "two-sided" will be used only in cases where we want to resolve any ambiguities. Clearly, if $x(n)$ is causal [i.e., $x(n) = 0$ for $n < 0$], the one-sided and two-sided z-transforms are equivalent. In any other case, they are different.

3.1.2 The Inverse z-Transform

Often, we have the z-transform $X(z)$ of a signal and we must determine the signal sequence. The procedure for transforming from the z-domain to the time domain is called the *inverse z-transform*. An inversion formula for obtaining $x(n)$ from $X(z)$ can be derived by using the *Cauchy integral theorem*, which is an important theorem in the theory of complex variables.

To begin, we have the z-transform defined by (3.1.1) as

$$X(z) = \sum_{k=-\infty}^{\infty} x(k)z^{-k} \tag{3.1.12}$$

Suppose that we multiply both sides of (3.1.2) by z^{n-1} and integrate both sides over a closed contour within the ROC of $X(z)$ which encloses the origin. Such a contour is illustrated in Fig. 3.5. Thus we have

$$\oint_C X(z)z^{n-1}\, dz = \oint_C \sum_{k=-\infty}^{\infty} x(k)z^{n-1-k}\, dz \tag{3.1.13}$$

where C denotes the closed contour in the ROC of $X(z)$, taken in a counterclockwise direction. Since the series converges on this contour, we may interchange the order of integration and summation on the right-hand side of (3.1.13). Thus (3.1.13) becomes

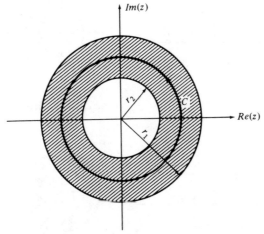

FIGURE 3.5   Contour C for integral in (3.1.13).

$$\oint_C X(z)z^{n-1}\, dz = \sum_{k=-\infty}^{\infty} x(k) \oint_C z^{n-1-k}\, dz \qquad (3.1.14)$$

Now we may invoke the Cauchy integral theorem, which states that

$$\frac{1}{2\pi j} \oint_C z^{n-1-k}\, dz = \begin{cases} 1 & k = n \\ 0 & k \ne n \end{cases} \qquad (3.1.15)$$

where C is any contour that encircles the origin. By applying (3.1.15), the right-hand side of (3.1.14) reduces to $2\pi jx(n)$ and hence the desired inversion formula

$$x(n) = \frac{1}{2\pi j} \oint_C X(z)z^{n-1}\, dz \qquad (3.1.16)$$

Although the contour integral in (3.1.16) provides the desired inversion formula for determining the sequence $x(n)$ from the z-transform, we shall not use (3.1.16) directly in our evaluation of inverse z-transforms. In our treatment we deal with signals and systems in the z-domain which have rational z-transforms (i.e., z-transforms that are a ratio of two polynomials). For such z-transforms we develop a simpler method for inversion.

3.2 Properties of the z-Transform

The z-transform is a very powerful tool for the study of discrete-time signals and systems. The power of this transform is a consequence of some very important properties that the transform possesses. In this section we examine some of these properties.

In the treatment below it should be remembered that when we combine several z-transforms, the ROC of the overall transform is, at least, the intersection of the ROC of the individual transforms. This will become more apparent later, when we discuss specific examples.

Linearity. If

$$x_1(n) \xleftarrow{\quad z \quad} X_1(z)$$

and

$$x_2(n) \xleftarrow{\quad z \quad} X_2(z)$$

then

$$x(n) = a_1 x_1(n) + a_2 x_2(n) \xleftarrow{\quad z \quad} X(z) = a_1 X_1(z) + a_2 X_2(z) \qquad (3.2.1)$$

for any constants a_1 and a_2. The proof of this property follows immediately from the definition of linearity and is left as an exercise for the reader.

The linearity property can easily be generalized for an arbitrary number of signals. Basically, it implies that the z-transform of a linear combination of signals is the same linear combination of their z-transforms. Thus the linearity property helps us to find the z-transform of a signal by expressing the signal as a sum of elementary signals for which the z-transform is already known.

EXAMPLE 3.2.1 _____

Determine the z-transform of the signal

$$x(n) = 3\delta(n + 1) + 2\delta(n) + 6\delta(n - 3) - \delta(n - 4)$$

Solution: From the linearity property, we have

$$X(z) = 3Z\{\delta(n + 1)\} + 2Z\{\delta(n)\} + 6Z\{\delta(n - 3)\} - Z\{\delta(n - 4)\}$$

Using the transform pairs (e), (f), and (g) in Example 3.1.1, we obtain

$$X(z) = 3z + 2 + 6z^{-3} - z^{-4}$$

The ROC is the entire z-plane except $z = 0$ and $z = \infty$.

The same result can be obtained by using the definition of the z-transform and the fact that

$$x(n) = \left\{ 3, 2, 0, 0, 6, -1 \atop \uparrow \right\}$$

_____

EXAMPLE 3.2.2 _____

Determine the z-transform of the signal

$$x(n) = [3(2^n) - 4(3^n)]u(n)$$

Solution: If we define the signals

$$x_1(n) = 2^n u(n)$$

and

$$x_2(n) = 3^n u(n),$$

then $x(n)$ can be written as

$$x(n) = 3x_1(n) - 4x_2(n)$$

According to (3.2.1), its z-transform is

$$X(z) = 3X_1(z) - 4X_2(z)$$

From (3.1.7) we recall that

$$\alpha^n u(n) \xleftarrow{\ z\ } \frac{1}{1 - \alpha z^{-1}} \qquad \text{ROC: } |z| > |\alpha| \qquad\qquad (3.2.2)$$

By setting $\alpha = 2$ and $\alpha = 3$ in (5.2.2), we obtain

$$x_1(n) = 2^n u(n) \xleftarrow{\ z\ } X_1(z) = \frac{1}{1 - 2z^{-1}} \qquad \text{ROC: } |z| > 2$$

$$x_2(n) = 3^n u(n) \xleftarrow{\ z\ } X_2(z) = \frac{1}{1 - 3z^{-1}} \qquad \text{ROC: } |z| > 3$$

The intersection of the ROC of $X_1(z)$ and $X_2(z)$ is $|z| > 3$. Thus the overall transform $X(z)$ is

$$X(z) = \frac{3}{1 - 2z^{-1}} - \frac{4}{1 - 3z^{-1}} \qquad \text{ROC: } |z| > 3$$

EXAMPLE 3.2.3

Determine the z-transform of the signals

(a) $x(n) = (\cos \omega_0 n)u(n)$
(b) $x(n) = (\sin \omega_0 n)u(n)$

Solution: (a) By using Euler's identity, the signal $x(n)$ can be expressed as

$$x(n) = (\cos \omega_0 n)u(n) = \tfrac{1}{2}e^{j\omega_0 n}u(n) + \tfrac{1}{2}e^{-j\omega_0 n}u(n)$$

Thus (3.2.1) implies that

$$X(z) = \tfrac{1}{2}Z\{e^{j\omega_0 n}u(n)\} + \tfrac{1}{2}Z\{e^{-j\omega_0 n}u(n)\}$$

If we set $\alpha = e^{\pm j\omega_0} (|\alpha| = |e^{\pm j\omega_0}| = 1)$ in (3.2.2), we obtain

$$e^{j\omega_0 n}u(n) \xleftarrow{\quad z \quad} \frac{1}{1 - e^{j\omega_0}z^{-1}} \qquad \text{ROC: } |z| > 1$$

and

$$e^{-j\omega_0 n}u(n) \xleftarrow{\quad z \quad} \frac{1}{1 - e^{-j\omega_0}z^{-1}} \qquad \text{ROC: } |z| > 1$$

Thus

$$X(z) = \frac{1}{2}\frac{1}{1 - e^{j\omega_0}z^{-1}} + \frac{1}{2}\frac{1}{1 - e^{-j\omega_0}z^{-1}} \qquad \text{ROC: } |z| > 1$$

After some simple algebraic manipulations we obtain the desired result, namely,

$$(\cos \omega_0 n)u(n) \xleftarrow{\quad z \quad} \frac{1 - z^{-1}\cos \omega_0}{1 - 2z^{-1}\cos \omega_0 + z^{-2}} \qquad \text{ROC: } |z| > 1 \qquad (3.2.3)$$

(b) From Euler's identity,

$$x(n) = (\sin \omega_0 n)u(n) = \frac{1}{2j}[e^{j\omega_0 n}u(n) - e^{-j\omega_0 n}u(n)]$$

Thus

$$X(z) = \frac{1}{2j}\left(\frac{1}{1 - e^{j\omega_0}z^{-1}} - \frac{1}{1 - e^{-j\omega_0}z^{-1}}\right) \qquad \text{ROC: } |z| > 1$$

and finally,

$$(\sin \omega_0 n)u(n) \xleftarrow{\quad z \quad} \frac{z^{-1}\sin \omega_0}{1 - 2z^{-1}\cos \omega_0 + z^{-2}} \qquad \text{ROC: } |z| > 1 \qquad (3.2.4)$$

Time Shifting. If

$$x(n) \xleftarrow{\quad z \quad} X(z)$$

then

$$x(n - k) \xleftarrow{\quad z \quad} z^{-k}X(z) \tag{3.2.5}$$

The ROC of $z^{-k}X(z)$ is the same as that of $X(z)$ except for $z = 0$ if $k > 0$ and $z = \infty$ if $k < 0$. The proof of this property follows immediately from the definition of the z-transform given in (3.1.1)

The properties of linearity and time shifting are the key features that make the z-transform extremely useful for the analysis of discrete-time LTI systems.

EXAMPLE 3.2.4

By applying the time-shifting property, determine the z-transform of the signals $x_2(n)$ and $x_3(n)$ in Example 3.1.1 from the z-transform of $x_1(n)$.

Solution: It can easily be seen that

$$x_2(n) = x_1(n + 2)$$

and

$$x_3(n) = x_1(n - 2)$$

Thus from (3.2.5) we obtain

$$X_2(z) = z^2X_1(z) = z^2 + 2z + 5 + 7z^{-1} + z^{-3}$$

and

$$X_3(z) = z^{-2}X_1(z) = z^{-2} + 2z^{-3} + 5z^{-4} + 7z^{-5} + z^{-7}$$

Note that because of the multiplication by z^2, the ROC of $X_2(z)$ does not include the point $z = \infty$, even if it is contained in the ROC of $X_1(z)$.

Example 3.2.4 provides additional insight in understanding the meaning of the shifting property. Indeed, if we recall that the coefficient of z^{-n} is the sample value at time n, it is immediately seen that delaying a signal by k ($k > 0$) samples [i.e., $x(n) \rightarrow x(n - k)$] corresponds to multiplying all terms of the z-transform by z^{-k}. The coefficient of z^{-n} becomes the coefficient of $z^{-(n+k)}$.

EXAMPLE 3.2.5

Determine the transform of the signal

$$x(n) = \begin{cases} 1 & 0 \le n \le N - 1 \\ 0 & \text{elsewhere} \end{cases} \tag{3.2.6}$$

Solution: We can determine the z-transform of this signal by using the definition (3.1.1). Indeed,

$$X(z) = \sum_{n=0}^{N-1} 1 \cdot z^{-n} = 1 + z^{-1} + \cdots + z^{-(N-1)} = \begin{cases} N & \text{if } z = 1 \\ \dfrac{1 - z^{-N}}{1 - z^{-1}} & \text{if } z \neq 1 \end{cases}$$

$$(3.2.7)$$

Since $x(n)$ has finite duration, its ROC is the entire z-plane, except $z = 0$.

Let us also derive this transform by using the linearity and time shifting properties. Note that $x(n)$ can be expressed in terms of two unit step signals

$$x(n) = u(n) - u(n - N)$$

By using (3.2.1) and (3.2.5) we have

$$X(z) = Z\{u(n)\} - Z\{u(n - N)\} = (1 - z^{-N})Z\{u(n)\} \qquad (3.2.8)$$

However, from (3.1.8) we have

$$Z\{u(n)\} = \frac{1}{1 - z^{-1}} \qquad \text{ROC: } |z| > 1$$

which, when combined with (3.2.8), leads to (3.2.7).

Example 3.2.5 helps to clarify a very important issue regarding the ROC of the combination of several z-transforms. If the linear combination of several signals has finite duration, the ROC of its z-transform is exclusively dictated by the finite-duration nature of this signal, not by the ROC of the individual transforms.

Scaling in the z-Domain. If

$$x(n) \xleftarrow{\quad z \quad} X(z) \qquad \text{ROC: } r_1 < |z| < r_2$$

then

$$a^n x(n) \xleftarrow{\quad z \quad} X(a^{-1}z) \qquad \text{ROC: } |a|r_1 < |z| < |a|r_2 \qquad (3.2.9)$$

for any constant a, real or complex.

PROOF. From the definition (3.1.1)

$$Z\{a^n x(n)\} = \sum_{n=-\infty}^{\infty} a^n x(n)z^{-n} = \sum_{n=-\infty}^{\infty} x(n)(a^{-1}z)^{-n}$$
$$= X(a^{-1}z)$$

Since the ROC of $X(z)$ is $r_1 < |z| < r_2$, the ROC of $X(a^{-1}z)$ is

$$r_1 < |a^{-1}z| < r_2$$

or

$$|a|r_1 < |z| < |a|r_2$$

EXAMPLE 3.2.6 _____

Determine the z-transforms of the signals

(a) $x(n) = a^n(\cos \omega_0 n)u(n)$

(b) $x(n) = a^n(\sin \omega_0 n)u(n)$

Solution: **(a)** From (3.2.3) and (3.2.9) we easily obtain

$$a^n(\cos \omega_0 n)u(n) \xrightarrow{\quad z \quad} \frac{1 - az^{-1} \cos \omega_0}{1 - 2az^{-1} \cos \omega_0 + a^2 z^{-2}} \qquad |z| > |a| \qquad (3.2.10)$$

(b) Similarly, (3.2.4) and (3.2.9) yield

$$a^n(\sin \omega_0 n)u(n) \xrightarrow{\quad z \quad} \frac{az^{-1} \sin \omega_0}{1 - 2az^{-1} \cos \omega_0 + a^2 z^{-2}} \qquad |z| > |a| \qquad (3.2.11)$$

Time Reversal. If

$$x(n) \xrightarrow{\quad z \quad} X(z) \qquad \text{ROC: } r_1 < |z| < r_2$$

then

$$x(-n) \xrightarrow{\quad z \quad} X(z^{-1}) \qquad \text{ROC: } \frac{1}{r_2} < |z| < \frac{1}{r_1} \qquad (3.2.12)$$

PROOF. From the definition (3.1.1), we have

$$Z\{x(-n)\} = \sum_{n=-\infty}^{\infty} x(-n)z^{-n} = \sum_{l=-\infty}^{\infty} x(l)(z^{-1})^{-l} = X(z^{-1})$$

where the change of variable $l = -n$ is made. The ROC of $X(z^{-1})$ is

$$r_1 < |z^{-1}| < r_2 \quad \text{or equivalently} \quad \frac{1}{r_2} < |z| < \frac{1}{r_1}$$

Note that the ROC for $x(n)$ is the inverse of that for $x(-n)$. This means that if z_0 belongs to the ROC of $x(n)$, then $1/z_0$ is in the ROC for $x(-n)$.

An intuitive proof of (3.2.12) is the following. When we fold a signal, the coefficient of z^{-n} becomes the coefficient of z^n. Thus, folding a signal is equivalent to replacing z by z^{-1} in the z-transform formula. In other words, reflection in the time domain corresponds to inversion in the z-domain.

EXAMPLE 3.2.7 _____

Determine the z-transform of the signal

$$x(n) = u(-n)$$

Solution: It is known from (3.1.8) that

$$u(n) \xrightarrow{\quad z \quad} \frac{1}{1 - z^{-1}} \qquad \text{ROC: } |z| > 1$$

By using (3.2.12), we easily obtain

$$u(-n) \xrightarrow{\quad z \quad} \frac{1}{1 - z} \qquad \text{ROC: } |z| < 1 \qquad (3.2.13)$$

Differentiation in the z-Domain. If

$$x(n) \xrightarrow{\quad z \quad} X(z)$$

then

$$nx(n) \xleftarrow{\quad z \quad} -z \frac{dX(z)}{dz} \qquad (3.2.14)$$

PROOF. By differentiating both sides of (3.1.1), we have

$$\frac{dX(z)}{dz} = \sum_{n=-\infty}^{\infty} x(n)(-n)z^{-n-1} = -z^{-1} \sum_{n=-\infty}^{\infty} [nx(n)]z^{-n}$$

$$= -z^{-1}Z\{nx(n)\}$$

Note that both transforms have the same ROC.

EXAMPLE 3.2.8 _____

Determine the z-transform of the signal

$$x(n) = na^n u(n)$$

Solution: The signal $x(n)$ can be expressed as $nx_1(n)$, where $x_1(n) = a^n u(n)$. From (3.2.2) we have that

$$x_1(n) = a^n u(n) \xleftarrow{\quad z \quad} X_1(z) = \frac{1}{1 - az^{-1}} \qquad \text{ROC: } |z| > |a|$$

Thus, by using (3.2.14), we obtain

$$na^n u(n) \xleftarrow{\quad z \quad} X(z) = -z \frac{dX_1(z)}{dz} = \frac{az^{-1}}{(1 - az^{-1})^2} \qquad \text{ROC: } |z| > |a| \quad (3.2.15)$$

If we set $a = 1$ in (3.2.15), we find the z-transform of the unit ramp signal

$$nu(n) \xleftarrow{\quad z \quad} \frac{z^{-1}}{(1 - z^{-1})^2} \qquad \text{ROC: } |z| > 1 \qquad (3.2.16)$$

Convolution Property. If

$$x_1(n) \xleftarrow{\quad z \quad} X_1(z)$$

$$x_2(n) \xleftarrow{\quad z \quad} X_2(z)$$

then

$$x(n) = x_1(n) * x_2(n) \xleftarrow{\quad z \quad} X(z) = X_1(z)X_2(z) \qquad (3.2.17)$$

The ROC of $X(z)$ is, at least, the intersection of that for $X_1(z)$ and $X_2(z)$.
PROOF. The convolution of $x_1(n)$ and $x_2(n)$ is defined as

$$x(n) = \sum_{k=-\infty}^{\infty} x_1(k)x_2(n - k)$$

The z-transform of $x(n)$ is

$$X(z) = \sum_{n=-\infty}^{\infty} x(n)z^{-n} = \sum_{n=-\infty}^{\infty} \left[\sum_{k=-\infty}^{\infty} x_1(k)x_2(n - k) \right] z^{-n}$$

Upon interchanging the order of the summations and applying the time-shifting property in (3.2.5), we obtain

$$X(z) = \sum_{k=-\infty}^{\infty} x_1(k) \left[\sum_{n=-\infty}^{\infty} x_2(n-k)z^{-n} \right]$$

$$= X_2(z) \sum_{k=-\infty}^{\infty} x_1(k)z^{-k} = X_2(z)X_1(z)$$

EXAMPLE 3.2.9

Compute the convolution $x(n)$ of the signals

$$x_1(n) = \{1, -2, 1\}$$

$$x_2(n) = \begin{cases} 1 & 0 \leq n \leq 5 \\ 0 & \text{elsewhere} \end{cases}$$

Solution: From (3.1.1), we have

$$X_1(z) = 1 - 2z^{-1} + z^{-2}$$
$$X_2(z) = 1 + z^{-1} + z^{-2} + z^{-3} + z^{-4} + z^{-5}$$

According to (3.2.17), we carry out the multiplication of $X_1(z)$ and $X_2(z)$. Thus

$$X(z) = X_1(z)X_2(z) = 1 - z^{-1} - z^{-6} + z^{-7}$$

Hence

$$x(n) = \left\{ \underset{\uparrow}{1}, -1, 0, 0, 0, 0, -1, 1 \right\}$$

The same result can also be obtained by noting that

$$X_1(z) = (1 - z^{-1})^2$$

$$X_2(z) = \frac{1 - z^{-6}}{1 - z^{-1}}$$

Then

$$X(z) = (1 - z^{-1})(1 - z^{-6}) = 1 - z^{-1} - z^{-6} + z^{-7}$$

The reader is encouraged to obtain the same result by explicitly using the convolution summation formula (time-domain approach).

The convolution property is one of the most powerful properties of the z-transform, because it converts the convolution of two signals (time domain) to multiplication of their transforms. To compute the convolution of two signals, using the z-transform requires the following steps:

Step 1. Compute the z-transforms of the signals to be convolved.

$$X_1(z) = Z\{x_1(n)\}$$

$$\text{(time domain} \longrightarrow \text{z-domain)}$$

$$X_2(z) = Z\{x_2(n)\}$$

Step 2. Multiply the two z-transforms.

$$X(z) = X_1(z)X_2(z) \qquad \text{(z-domain)}$$

Step 3. Find the inverse z-transform of $x(z)$.

$$x(n) = Z^{-1}\{X(z)\} \qquad \text{(z-domain} \longrightarrow \text{time domain)}$$

This procedure is, in many cases, computationally easier than the direct evaluation of the convolution summation.

Correlation Property. If

$$x_1(n) \xleftarrow{\quad z \quad} X_1(z)$$

$$x_2(n) \xleftarrow{\quad z \quad} X_2(z)$$

then

$$r_{x_1 x_2}(l) = \sum_{n=-\infty}^{\infty} x_1(n)x_2(n - l) \xleftarrow{\quad z \quad} R_{x_1 x_2}(z) = X_1(z)X_2(z^{-1}) \qquad (3.2.18)$$

PROOF. We recall that

$$r_{x_1 x_2}(l) = x_1(l) * x_2(-l)$$

Using the convolution and time-reversal properties, we easily obtain

$$R_{x_1 x_2}(z) = Z\{x_1(l)\}Z\{x_2(-l)\} = X_1(z)X_2(z^{-1})$$

The ROC of $R_{x_1 x_2}(z)$ is at least the intersection of that for $X_1(z)$ and $X_2(z^{-1})$.

As in the case of convolution, the crosscorrelation of two signals is more easily done via polynomial multiplication according to (3.2.18) and inverse transforming the result.

EXAMPLE 3.2.10 _____

Determine the autocorrelation sequence of the signal

$$x(n) = a^n u(n) \qquad -1 < a < 1$$

Solution: Since the autocorrelation sequence of a signal is its correlation with itself, (3.2.18) gives

$$R_{xx}(z) = Z\{r_{xx}(l)\} = X(z)X(z^{-1})$$

From (3.2.2) we have

$$X(z) = \frac{1}{1 - az^{-1}} \qquad \text{ROC: } |z| > |a| \qquad \text{(causal signal)}$$

and by using (3.2.15), we obtain

$$X(z^{-1}) = \frac{1}{1 - az} \qquad \text{ROC: } |z| < \frac{1}{|a|} \qquad \text{(anticausal signal)}$$

Thus

$$R_{xx}(z) = \frac{1}{1 - az^{-1}} \frac{1}{1 - az} = \frac{1}{1 - a(z + z^{-1}) + a^2} \qquad \text{ROC: } |a| < |z| < \frac{1}{|a|}$$

Since the ROC of $R_{xx}(z)$ is a ring, $r_{xx}(l)$ is a two-sided signal, even if $x(n)$ is causal.

To obtain $r_{xx}(l)$, we observe that the z-transform of the sequence in Example 3.1.5 with $b = 1/a$ is simply $(1 - a^2)R_{xx}(z)$. Hence it follows that

$$r_{xx}(l) = \frac{1}{1 - a^2} a^{|l|} \qquad -\infty < l < \infty$$

The reader is encouraged to compare this approach with the time-domain solution of the same problem given in Section 2.5.

Multiplication of Two Sequences.   If

$$x_1(n) \xleftarrow{\quad z \quad} X_1(z)$$

$$x_2(n) \xleftarrow{\quad z \quad} X_2(z)$$

then

$$x_3(n) = x_1(n)x_2(n) \xleftarrow{\quad z \quad} X_3(z) = \frac{1}{2\pi j} \oint_C X_1(v)X_2\left(\frac{z}{v}\right) v^{-1}\, dv \quad (3.2.19)$$

where C is a closed contour that encloses the origin and lies within the common region of convergence of $X_1(v)$ and $X_2(1/v)$.

PROOF.   The z-transform of $x_3(n)$ is

$$X_3(z) = \sum_{n=-\infty}^{\infty} x_3(n)z^{-n} = \sum_{n=-\infty}^{\infty} x_1(n)x_2(n)z^{-n}$$

Let us substitute the inverse transform

$$x_1(n) = \frac{1}{2\pi j} \oint_C X_1(v)v^{n-1}\, dv$$

for $x_1(n)$ in the z-transform $X_3(z)$ and interchange the order of summation and integration. Thus we obtain

$$X_3(z) = \frac{1}{2\pi j} \oint_C X_1(v) \left[\sum_{n=-\infty}^{\infty} x_2(n) \left(\frac{z}{v}\right)^{-n} \right] v^{-1}\, dv$$

The sum in the brackets is simply the transform $X_2(z)$ evaluated at z/v. Therefore,

$$X_3(z) = \frac{1}{2\pi j} \oint_C X_1(v)X_2\left(\frac{z}{v}\right) v^{-1}\, dv$$

which is the desired result.

To obtain the ROC of $X_3(z)$ we note that if $X_1(v)$ converges for $r_{1l} < |v| < r_{1u}$ and $X_2(z)$ converges for $r_{2l} < |z| < r_{2u}$, then the ROC of $X_2(z/v)$ is

$$r_{2l} < \left| \frac{z}{v} \right| < r_{2u}$$

Hence the ROC for $X_3(z)$ is at least

$$r_{1l}r_{2l} < |z| < r_{1u}r_{2u} \qquad (3.2.20)$$

Although this property will not be used immediately, it will prove useful later, especially in our treatment of filter design based on the window technique, where we multiply

TABLE 3.2 Properties of the z-Transform

Property	Time Domain	z-Domain	ROC
Notation	$x(n)$ $x_1(n)$ $x_2(n)$	$X(z)$ $X_1(z)$ $X_2(z)$	$r_2 < \lvert z \rvert < r_1$ ROC_1 ROC_2
Linearity	$a_1 x_1(n) + a_2 x_2(n)$	$a_1 X_1(z) + a_2 X_2(z)$	At least the intersection of ROC_1 and ROC_2
Time shifting	$x(n-k)$	$z^{-k}X(z)$	That of $X(z)$, except $z = 0$ if $k > 0$ and $z = \infty$ if $k < 0$
Scaling in the z-domain	$a^n x(n)$	$X(a^{-1}z)$	$\lvert a \rvert r_2 < \lvert z \rvert < \lvert a \rvert r_1$
Time reversal	$x(-n)$	$X(z^{-1})$	$\dfrac{1}{r_1} < \lvert z \rvert < \dfrac{1}{r_2}$
Differentiation in the z-domain	$n x(n)$	$-z \dfrac{dX(z)}{dz}$	$r_2 < \lvert z \rvert < r_1$
Convolution	$x_1(n) * x_2(n)$	$X_1(z)X_2(z)$	At least, the intersection of ROC_1 and ROC_2
Correlation	$r_{x_1 x_2}(l) = x_1(l) * x_2(-l)$	$R_{x_1 x_2}(z) = X_1(z)X_2(z^{-1})$	At least, the intersection of ROC of $X_1(z)$ and $X_2(z^{-1})$
Multiplication	$x_1(n)x_2(n)$	$\dfrac{1}{2\pi j}\oint_C X_1(v)X_2\left(\dfrac{z}{v}\right) v^{-1}\, dv$	At least $r_{1l}r_{2l} < \lvert z \rvert < r_{1u}r_{2u}$
Initial value theorem	If $x(n)$ causal	$x(0) = \lim\limits_{z \to \infty} X(z)$	

the impulse response of an IIR system by a finite-duration "window" which serves to truncate the impulse response of the IIR system.

The Initial Value Theorem. If $x(n)$ is *causal* [i.e., $x(n) = 0$ for $n < 0$], then

$$x(0) = \lim_{z \to \infty} X(z) \tag{3.2.21}$$

PROOF. Since $x(n)$ is causal, (3.1.1) gives

$$X(z) = \sum_{n=0}^{\infty} x(n)z^{-n} = x(0) + x(1)z^{-1} + x(2)z^{-2} + \cdots$$

Obviously, as $z \to \infty$, $z^{-n} \to 0$ since $n > 0$ and (3.2.21) follows.

All the properties of the z-transform we have presented in this section are summarized in Table 3.2 for easy reference. They are listed in the same order as they have been introduced in the text.

We have now derived most of the z-transforms that are encountered in many practical applications. These z-transform pairs are summarized in Table 3.3 for easy reference. A simple inspection of this table shows that all these z-transforms are *rational functions* (i.e., ratios of polynomials in z^{-1}). As will soon become apparent, rational z-transforms are encountered not only as the z-transforms of various important signals but also in the characterization of discrete-time linear time-invariant systems described by constant-coefficient difference equations.

TABLE 3.3 Some Common z-Transform Pairs

	Signal $x(n)$	z-Transform $X(z)$	ROC
1	$\delta(n)$	1	All z
2	$u(n)$	$\dfrac{1}{1 - z^{-1}}$	$\|z\| > 1$
3	$nu(n)$	$\dfrac{z^{-1}}{(1 - z^{-1})^2}$	$\|z\| > 1$
4	$a^n u(n)$	$\dfrac{1}{1 - az^{-1}}$	$\|z\| > \|a\|$
5	$na^n u(n)$	$\dfrac{az^{-1}}{(1 - az^{-1})^2}$	$\|z\| > \|a\|$
6	$-a^n u(-n - 1)$	$\dfrac{1}{1 - az^{-1}}$	$\|z\| < \|a\|$
7	$-na^n u(-n - 1)$	$\dfrac{az^{-1}}{(1 - az^{-1})^2}$	$\|z\| < \|a\|$
8	$(\cos \omega_0 n)u(n)$	$\dfrac{1 - z^{-1} \cos \omega_0}{1 - 2z^{-1} \cos \omega_0 + z^{-2}}$	$\|z\| > 1$
9	$(\sin \omega_0 n)u(n)$	$\dfrac{z^{-1} \sin \omega_0}{1 - 2z^{-1} \cos \omega_0 + z^{-2}}$	$\|z\| > 1$
10	$(a^n \cos \omega_0 n)u(n)$	$\dfrac{1 - az^{-1} \cos \omega_0}{1 - 2az^{-1} \cos \omega_0 + a^2 z^{-2}}$	$\|z\| > \|a\|$
11	$(a^n \sin \omega_0 n)u(n)$	$\dfrac{az^{-1} \sin \omega_0}{1 - 2az^{-1} \cos \omega_0 + a^2 z^{-2}}$	$\|z\| > \|a\|$

3.3 Rational z-Transforms

As indicated in Section 3.2, an important family of z-transforms are those for which $X(z)$ is a rational function, that is, a ratio of two polynomials in z^{-1} (or z). In this section we discuss some very important issues regarding the class of rational z-transforms.

3.3.1 Poles and Zeros

The *zeros* of a z-transform $X(z)$ are the values of z for which $X(z) = 0$. The *poles* of a z-transform are the values of z for which $X(z) = \infty$. If $X(z)$ is a rational function, then

$$X(z) = \frac{N(z)}{D(z)} = \frac{b_0 + b_1 z^{-1} + \cdots + b_M z^{-M}}{a_0 + a_1 z^{-1} + \cdots + a_N z^{-N}} = \frac{\sum_{k=0}^{M} b_k z^{-k}}{\sum_{k=0}^{N} a_k z^{-k}} \tag{3.3.1}$$

If $a_0 \neq 0$ and $b_0 \neq 0$, we can avoid the negative powers of z by factoring out the terms $b_0 z^{-M}$ and $a_0 z^{-N}$ as follows:

$$X(z) = \frac{N(z)}{D(z)} = \frac{b_0 z^{-M}}{a_0 z^{-N}} \frac{z^M + (b_1/b_0)z^{M-1} + \cdots + b_M/b_0}{z^N + (a_1/a_0)z^{N-1} + \cdots + a_N/a_0}$$

Since $N(z)$ and $D(z)$ are polynomials in z, they can be expressed in factored form as

$$X(z) = \frac{N(z)}{D(z)} = \frac{b_0}{a_0} z^{-M+N} \frac{(z - z_1)(z - z_2) \cdots (z - z_M)}{(z - p_1)(z - p_2) \cdots (z - p_N)}$$

$$X(z) = G z^{N-M} \frac{\prod_{k=1}^{M} (z - z_k)}{\prod_{k=1}^{N} (z - p_k)} \tag{3.3.2}$$

where $G \equiv b_0/a_0$. Thus $X(z)$ has M finite zeros at $z = z_1, z_2, \ldots, z_M$ (the roots of the numerator polynomial), N finite poles at $z = p_1, p_2, \ldots, p_N$ (the roots of the denominator polynomial), and $|N - M|$ zeros (if $N > M$) or poles (if $N < M$) at the origin $z = 0$. Poles or zeros may also occur at $z = \infty$. A zero exists at $z = \infty$ if $X(\infty) = 0$ and a pole exists at $z = \infty$ if $X(\infty) = \infty$. If we count the poles and zeros at zero and infinity, we find that $X(z)$ has exactly the same number of poles as zeros.

We can represent $X(z)$ graphically by a *pole–zero plot* (or *pattern*) in the complex plane, which shows the location of poles by crosses ($\times$) and the location of zeros by circles ($\bigcirc$). The multiplicity of multiple-order poles or zeros is indicated by a number close to the corresponding cross or circle. Obviously, by definition, the ROC of a z-transform should not contain any poles.

EXAMPLE 3.3.1 _____

Determine the pole–zero plot for the signal

$$x(n) = a^n u(n) \qquad a > 0$$

Solution: From Table 3.3 we find that

$$X(z) = \frac{1}{1 - az^{-1}} = \frac{z}{z - a} \qquad \text{ROC: } |z| > a$$

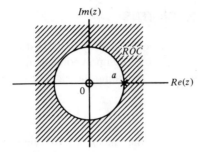

FIGURE 3.6 Pole–zero plot for the causal exponential signal $x(n) = a^n u(n)$.

Thus $X(z)$ has one zero at $z_1 = 0$ and one pole at $p_1 = a$. The pole–zero plot is shown in Fig. 3.6. Note that the pole $p_1 = a$ is not included in the ROC since the z-transform does not converge at a pole.

___

EXAMPLE 3.3.2 _____

Determine the pole–zero plot for the signal

$$x(n) = \begin{cases} a^n & 0 \le n \le M - 1 \\ 0 & \text{elsewhere} \end{cases}$$

where $a > 0$.

Solution: From the definition (3.1.1) we obtain

$$X(z) = \sum_{n=0}^{M-1} (az^{-1})^n = \frac{1 - (az^{-1})^M}{1 - az^{-1}} = \frac{z^M - a^M}{z^{M-1}(z - a)}$$

Since $a > 0$, the equation $z^M = a^M$ has M roots at

$$z_k = a e^{j2\pi k/M} \qquad k = 0, 1, \ldots, M - 1$$

The zero $z_0 = a$ cancels the pole at $z = a$. Thus

$$X(z) = \frac{(z - z_1)(z - z_2) \cdots (z - z_{M-1})}{z^{M-1}}$$

which has $M - 1$ zeros and $M - 1$ poles, located as shown in Fig. 3.7 for $M = 8$.

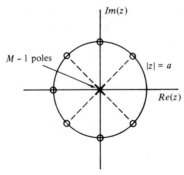

FIGURE 3.7 Pole–zero pattern for the finite-duration signal $x(n) = a^n$, $0 \le n \le M - 1$ $(a > 0)$, for $M = 8$.

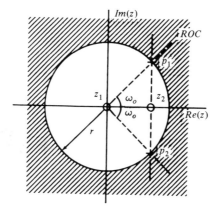

FIGURE 3.8 Pole–zero pattern for Example 3.3.3.

Note that the ROC is the entire z-plane except $z = 0$ because of the $M - 1$ poles located at the origin.

Clearly, if we are given a pole–zero plot, we can determine $X(z)$, by using (3.3.2), to within a scaling factor G. This is illustrated in the following example.

EXAMPLE 3.3.3

Determine the z-transform and the signal that corresponds to the pole–zero plot of Fig. 3.8.

Solution: There are two zeros ($M = 2$) at $z_1 = 0$, $z_2 = r \cos \omega_0$ and two poles ($N = 2$) at $p_1 = re^{j\omega_0}$, $p_2 = re^{-j\omega_0}$. By substitution of these relations into (3.3.2), we obtain

$$X(z) = G \frac{(z - z_1)(z - z_2)}{(z - p_1)(z - p_2)} = G \frac{z(z - r \cos \omega_0)}{(z - re^{j\omega_0})(z - re^{-j\omega_0})} \qquad \text{ROC: } |z| > r$$

After some simple algebraic manipulations, we obtain

$$X(z) = G \frac{1 - rz^{-1} \cos \omega_0}{1 - 2rz^{-1} \cos \omega_0 + r^2 z^{-2}} \qquad \text{ROC: } |z| > r$$

From Table 3.3 we find that

$$x(n) = G(r^n \cos \omega_0 n)u(n)$$

From Example 3.3.3 we see that the product $(z - p_1)(z - p_2)$ results in a polynomial with real coefficients, when p_1 and p_2 are complex conjugates. In general, if a polynomial has real coefficients, its roots are either real or occur in complex-conjugate pairs.

As we have seen, the z-transform $X(z)$ is a complex function of the complex variable $z = \text{Re}(z) + j\,\text{Im}(z)$. Obviously, $|X(z)|$, the magnitude of $X(z)$, is a real and positive function of z. Since z represents a point in the complex plane, $|X(z)|$ is a two-dimensional function and describes a "surface." This is illustrated in Fig. 3.9 for the z-transform

$$X(z) = \frac{z^{-1} - z^{-2}}{1 + 1.2732z^{-1} + 0.81z^{-2}} \tag{3.3.3}$$

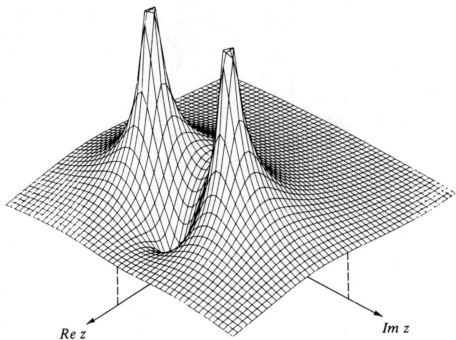

$Re\ z$ $Im\ z$

FIGURE 3.9   Graph of $|X(z)|$ for the z-transform in (3.3.3). [Reproduced with permission from *Introduction to Systems Analysis*, by T. H. Glisson, © 1985 by McGraw-Hill Book Company.]

which has one zero $z_1 = 1$ and two poles at $p_1, p_2 = 0.9e^{\pm j135°}$. Note the high peaks near the singularities (poles) and the deep valley close to the zero.

3.3.2 Pole Location and Time-Domain Behavior for Causal Signals

In this subsection we consider the relation between the z-plane location of a pole pair and the form (shape) of the corresponding signal in the time domain. The discussion is based generally on the collection of z-transforms pairs given in Table 3.3 and the results in the preceding subsection. We deal exclusively with real, causal signals. In particular, we will see that the characteristic behavior of causal signals will depend on whether the poles of the transform are contained in the region $|z| < 1$, or in the region $|z| > 1$, or on the circle $|z| = 1$. Since the circle $|z| = 1$ has a radius of 1, it is called the *unit circle*.

If a real signal has a z-transform with one pole, this pole has to be real. The only such signal is the real exponential

$$x(n) = a^n u(n) \xrightarrow{\quad z \quad} X(z) = \frac{1}{1 - az^{-1}} \qquad \text{ROC: } |z| > |a|$$

having one zero at $z_1 = 0$ and one pole at $p_1 = a$ on the real axis. Figure 3.10 illustrates the behavior of the signal with respect to the location of the pole relative to the unit circle. The signal is decaying if the pole is inside the unit circle, fixed if the pole is on the unit circle, and growing if the pole is outside the unit circle. In addition, a negative pole results in a signal that alternates in sign. Obviously, causal signals with poles outside the unit circle become unbounded, cause overflow in digital systems, and in general, should be avoided.

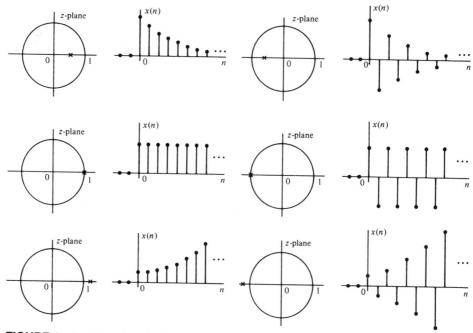

FIGURE 3.10 Time-domain behavior of a single-real pole causal signal as a function of the location of the pole with respect to the unit circle.

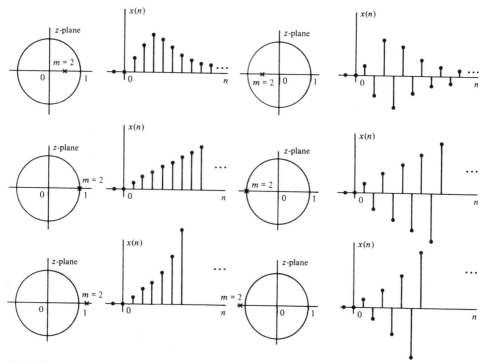

FIGURE 3.11 Time-domain behavior of causal signals corresponding to a double ($m = 2$) real pole, as a function of the pole location.

173

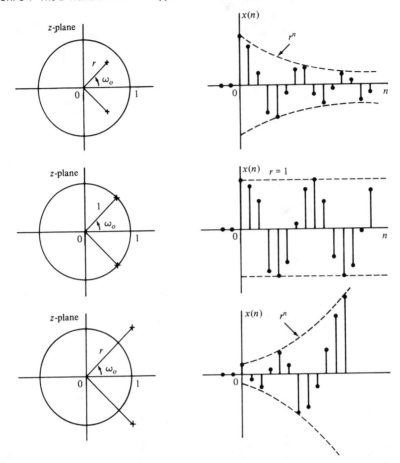

FIGURE 3.12   A pair of complex-conjugate poles corresponds to causal signals with oscillatory behavior.

A causal real signal with a double real pole has the form

$$x(n) = na^n u(n)$$

(see Table 3.3) and its behavior is illustrated in Fig. 3.11. Note that in contrast to the single-pole signal, a double real pole on the unit circle results in an unbounded signal.

Figure 3.12 illustrates the case of a pair of complex-conjugate poles. According to Table 3.3, this configuration of poles results in an exponentially weighted sinusoidal signal. The distance r of the poles from the origin determines the envelope of the sinusoidal signal and their angle with the real positive axis, its relative frequency. Note that the amplitude of the signal is growing if $|p| > 1$, constant if $|p| = 1$ (sinusoidal signals), and decaying if $|p| < 1$.

Finally, Fig. 3.13 shows the behavior of a causal signal with a double pair of poles on the unit circle. This reinforces the corresponding results in Fig. 3.11 and illustrates that multiple poles on the unit circle should be treated with great care.

To summarize, causal real signals with simple real poles or simple complex-conjugate pairs of poles which are inside or on the unit circle are always bounded in amplitude. Furthermore, a signal with a pole (or a complex-conjugate pair of poles) near the origin decays more rapidly than one associated with a pole near (but inside) the unit circle.

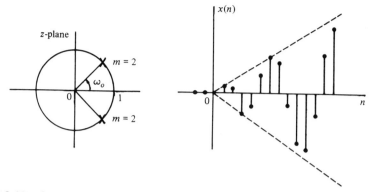

FIGURE 3.13 Causal signal corresponding to a double pair of complex-conjugate poles on the unit circle.

Thus the time behavior of a signal strongly depends on the location of its poles relative to the unit circle. Zeros also affect the behavior of a signal but not as strongly as poles. For example, in the case of sinusoidal signals the presence and location of zeros affects only their phase.

3.3.3 The System Function of a Linear Time-Invariant System

In Chapter 2 we demonstrated that the output of a (relaxed) linear, time-invariant system to an input sequence $x(n)$ may be obtained by computing the convolution of $x(n)$ with the unit sample response of the system. The convolution property derived in Section 3.2 allows us to express this relationship in the z-domain as

$$Y(z) = H(z)X(z) \tag{3.3.4}$$

where $Y(z)$ is the z-transform of the output sequence $y(n)$, $X(z)$ is the z-transform of the input sequence $x(n)$, and $H(z)$ is the z-transform of the unit sample response $h(n)$.

If we know $h(n)$ and $x(n)$, we can determine their corresponding z-transforms $H(z)$ and $X(z)$, multiply them to obtain $Y(z)$, and thus we can determine $y(n)$ by evaluating the inverse z-transform of $Y(z)$. Alternatively, if we know $x(n)$ and we observe the output $y(n)$ of the system, we can determine the unit sample response by first solving for $H(z)$ from the relation

$$H(z) = \frac{Y(z)}{X(z)} \tag{3.3.5}$$

and then evaluating the inverse z-transform of $H(z)$.

Since

$$H(z) = \sum_{n=-\infty}^{\infty} h(n)z^{-n} \tag{3.3.6}$$

it is clear that $H(z)$ represents the z-domain characterization of a system, whereas $h(n)$ is the corresponding time-domain characterization of the system. In other words, $H(z)$ and $h(n)$ are equivalent descriptions of a system in the two domains. The transform $H(z)$ is called the *system function*.

The relation in (3.3.5) is particularly useful in obtaining $H(z)$ when the system is described by a linear constant-coefficient difference equation of the form

$$y(n) = -\sum_{k=1}^{N} a_k y(n-k) + \sum_{k=0}^{M} b_k x(n-k) \tag{3.3.7}$$

In this case the system function can be determined directly from (3.3.7) by computing the z-transform of both sides of (3.3.7). Thus, by applying the time-shifting property, we obtain

$$Y(z) = -\sum_{k=1}^{N} a_k Y(z) z^{-k} + \sum_{k=0}^{M} b_k X(z) z^{-k}$$

$$Y(z) \left(1 + \sum_{k=1}^{N} a_k z^{-k} \right) = X(z) \left(\sum_{k=0}^{M} b_k z^{-k} \right)$$

$$\frac{Y(z)}{X(z)} = \frac{\displaystyle\sum_{k=0}^{M} b_k z^{-k}}{1 + \displaystyle\sum_{k=1}^{N} a_k z^{-k}}$$

or, equivalently,

$$H(z) = \frac{\displaystyle\sum_{k=0}^{M} b_k z^{-k}}{1 + \displaystyle\sum_{k=1}^{N} a_k z^{-k}} \tag{3.3.8}$$

Therefore, a linear time-invariant system described by a constant-coefficient difference equation has a rational system function.

This is the general form for the system function of a system described by a linear constant-coefficient difference equation. From this general form we obtain two important special forms. First, if $a_k = 0$ for $1 \le k \le N$, (3.3.8) reduces to

$$H(z) = \sum_{k=0}^{M} b_k z^{-k} \tag{3.3.9}$$

$$= \frac{1}{z^M} \sum_{k=0}^{M} b_k z^{M-k}$$

In this case, $H(z)$ contains M zeros, whose values are determined by the system parameters $\{b_k\}$, and an Mth-order pole at the origin $z = 0$. Since the system contains only trivial poles (at $z = 0$) and M nontrivial zeros, it is called an *all-zero system*. Clearly, such a system has a finite-duration impulse response (FIR) and it is called an FIR system or a moving average (MA) system.

On the other hand, if $b_k = 0$ for $1 \le k \le M$, the system function reduces to

$$H(z) = \frac{b_0}{1 + \displaystyle\sum_{k=1}^{N} a_k z^{-k}} \tag{3.3.10}$$

$$= \frac{b_0 z^N}{\displaystyle\sum_{k=0}^{N} a_k z^{N-k}} \qquad a_0 \equiv 1$$

In this case $H(z)$ consists of N poles, whose values are determined by the system parameters $\{a_k\}$ and an Nth order zero at the origin $z = 0$. We usually do not make reference to these trivial zeros. Consequently, the system function in (3.3.10) contains only non-trivial poles and the corresponding system is called an *all-pole system*. Due to the presence of poles, the impulse response of such a system is infinite in duration, and hence it is an IIR system.

The general form of the system function given by (3.3.8) contains both poles and zeros, and hence the corresponding system is called a *pole–zero system*, with N poles and M zeros. Poles and/or zeros at $z = 0$ and $z = \infty$ are implied but are not counted explicitly. Due to the presence of poles, a pole–zero system is an IIR system.

The following example illustrates the procedure for determining the system function and the unit sample response from the difference equation.

EXAMPLE 3.3.4

Determine the system function and the unit sample response of the system described by the difference equation

$$y(n) = \tfrac{1}{2} y(n - 1) + 2x(n)$$

Solution: By computing the z-transform of the difference equation, we obtain

$$Y(z) = \tfrac{1}{2} z^{-1} Y(z) + 2X(z)$$

Hence the system function is

$$\frac{Y(z)}{X(z)} \equiv H(z) = \frac{2}{1 - \tfrac{1}{2} z^{-1}}$$

This system has a pole at $z = \tfrac{1}{2}$ and a zero at the origin. Using Table 3.3 we obtain the inverse transform

$$h(n) = 2(\tfrac{1}{2})^n u(n)$$

This is the unit sample response of the system.

We have now demonstrated that rational z-transforms are encountered in the characterization of linear time-invariant systems and commonly used signals. In section 3.4 we describe several methods for determining the inverse z-transform of rational functions.

3.3.4 Interconnection of Linear Time-Invariant Systems

In Chapter 2 we indicated that we can interconnect small systems to form larger systems. There are basically two types of interconnections: the cascade and the parallel interconnections.

Two systems with system functions $H_1(z)$ and $H_2(z)$, when connected in cascade, form an overall system having the system function $H_3(z) = H_1(z)H_2(z)$. This follows from the fact that the impulse response $h_3(n)$ of the overall system is the convolution of $h_1(n)$ and $h_2(n)$. On the other hand, if the two systems are in parallel, their overall system function is $H_3(z) = H_1(z) + H_2(z)$. In the time domain, the corresponding relationship for the impulse response is $h_3(n) = h_1(n) + h_2(n)$.

From these two basic types of interconnections it is possible to construct complex systems. Conversely, it is possible to decompose a large system into smaller subsystems involving a cascade or a parallel interconnection of basic building blocks.

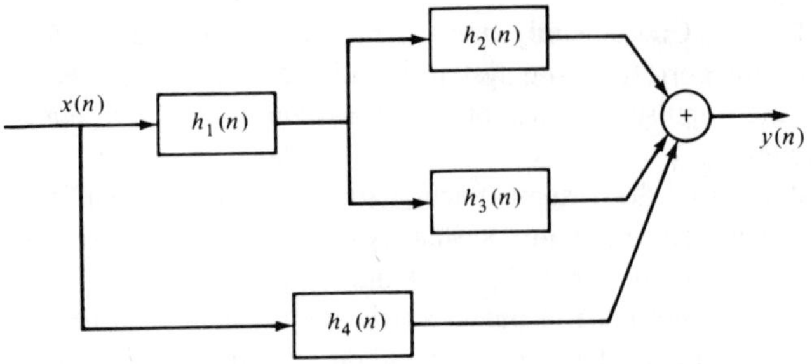

FIGURE 3.14   Interconnection of LTI systems.

EXAMPLE 3.3.5 _____

Determine the system function of the overall system shown in Fig. 3.14.

Solution: The parallel connection of $H_2(z)$ and $H_3(z)$ can be represented as a $H_2(z)$ + $H_3(z)$. In turn, this combination is in cascade with $H_1(z)$, which results in the equivalent system $H_1(z)[H_2(z) + H_3(z)]$. Finally, this combination is in parallel with $H_4(z)$. Consequently, the overall system has a system function $H_t(z)$ given as

$$H_t(z) = H_4(z) + H_1(z)[H_2(z) + H_3(z)]$$

3.4 Inversion of the z-Transform

As we saw in Section 3.1.2, the inverse z-transform is formally given by

$$x(n) = \frac{1}{2\pi j} \oint_C X(z)z^{n-1}\, dz \qquad (3.4.1)$$

where the integral is a contour integral over a closed path C that encircles the origin and lies within the region of convergence of $X(z)$. For simplicity, C can be taken as a circle in the ROC of $X(z)$ in the z-plane.

There are three methods that are often used for the evaluation of the inverse z-transform in practice:

1. Direct evaluation of (3.4.1), using the residue theorem.
2. Expansion into a series of terms, in the variables z and z^{-1}.
3. Partial-fraction expansion and table look-up.

We will limit our attention to methods 2 and 3, since they are more than sufficient for most practical applications. In general, method 1 is more complicated and requires knowledge of complex-variable theory.

3.4.1 Expansion of the z-Transform Into a Series

The basic idea in this method is the following: Given a z-transform $X(z)$ with its corresponding ROC, we can expand $X(z)$ into a power series of the form

$$X(z) = \sum_{n=-\infty}^{\infty} c_n z^{-n} \qquad (3.4.2)$$

which converges in the given ROC. Then, by the uniqueness of the z-transform, $x(n) = c_n$ for all n. When $X(z)$ is rational, the expansion may be performed by long division.

To illustrate this technique, we will invert some z-transforms involving the same expression for $X(z)$, but different ROC. This will also serve to emphasize again the importance of the ROC in dealing with z-transforms.

EXAMPLE 3.4.1 _____

Determine the inverse z-transform of

$$X(z) = \frac{1}{1 - 1.5z^{-1} + 0.5z^{-2}}$$

when

(a) ROC: $|z| > 1$
(b) ROC: $|z| < 0.5$

Solution: (a) Since the ROC is the exterior of a circle, we expect $x(n)$ to be a causal signal. Thus we seek a power series expansion in negative powers of z. By dividing the numerator of $X(z)$ by its denominator, we obtain the power series

$$X(z) = \frac{1}{1 - \frac{3}{2}z^{-1} + \frac{1}{2}z^{-2}} = 1 + \frac{3}{2}z^{-1} + \frac{7}{4}z^{-2} + \frac{15}{8}z^{-3} + \frac{31}{16}z^{-4} + \cdots$$

By comparing this relation with (3.1.1), we conclude that

$$x(n) = \left\{ 1, \tfrac{3}{2}, \tfrac{7}{4}, \tfrac{15}{8}, \tfrac{31}{16}, \ldots \right\}$$
$$\uparrow$$

Note that in each step of the long-division process, we eliminate the lowest-power term of z^{-1}.

(b) In this case the ROC is the interior of a circle. Consequently, the signal $x(n)$ is anticausal. To obtain a power series expansion in positive powers of z, we perform the long division in the following way:

$$
\begin{array}{r}
2z^2 + 6z^3 + 14z^4 + 30z^5 + 62z^6 + \cdots \\
\tfrac{1}{2}z^{-2} - \tfrac{3}{2}z^{-1} + 1 \overline{)1} \\
\underline{1 - 3z + 2z^2} \\
3z - 2z^2 \\
\underline{3z - 9z^2 + 6z^3} \\
7z^2 - 6z^3 \\
\underline{7z^2 - 21z^3 + 14z^4} \\
15z^3 - 14z^4 \\
\underline{15z^3 - 45z^4 + 30z^5} \\
31z^4 - 30z^5
\end{array}
$$

Thus

$$X(z) = \frac{1}{1 - \frac{3}{2}z^{-1} + \frac{1}{2}z^{-2}} = 2z^2 + 6z^3 + 14z^4 + 30z^5 + 62z^6 + \cdots$$

In this case $x(n) = 0$ for $n \geq 0$. By comparing this result to (3.1.1), we conclude that

$$x(n) = \left\{ \cdots 62, 30, 14, 6, 2, 0, 0 \atop \uparrow \right\}$$

We observe that in each step of the long-division process, the lowest-power term of z is eliminated. We emphasize that in the case of anticausal signals we simply carry out the long division by writing down the two polynomials in "reverse" order (i.e., starting with the most negative term on the left).

From the example above we note that, in general, the method of long division will not provide answers for $x(n)$ when n is large because the long-division becomes tedious. Although, the method provides a direct evaluation of $x(n)$, a closed-form solution is not possible, except if the resulting pattern is simple enough to infer the general term $x(n)$. Hence this method is used only if one wished to determine the values of the first few samples of the signal.

3.4.2 Partial-Fraction Expansion Method

In the table look-up method, we attempt to express the function $X(z)$ as a linear combination

$$X(z) = \alpha_1 X_1(z) + \alpha_2 X_2(z) + \cdots + \alpha_K X_K(z) \qquad (3.4.3)$$

where $X_1(z), \ldots, X_K(z)$ are expressions with inverse transforms $x_1(n), \ldots, x_K(n)$ available in a table of z-transform pairs. If such a decomposition is possible, then $x(n)$, the inverse z-transform of $X(z)$, can easily be found using the linearity property as

$$x(n) = \alpha_1 x_1(n) + \alpha_2 x_2(n) + \cdots + \alpha_K x_K(n) \qquad (3.4.4)$$

This approach is particularly useful if $X(z)$ is a rational function, as in (3.3.1). Without loss of generality, we assume that $a_0 = 1$, so that (3.3.1) can be expressed as

$$X(z) = \frac{N(z)}{D(z)} = \frac{b_0 + b_1 z^{-1} + \cdots + b_M z^{-M}}{1 + a_1 z^{-1} + \cdots + a_N z^{-N}} \qquad (3.4.5)$$

Note that if $a_0 \neq 1$, we can obtain (3.4.5) from (3.3.1) by dividing both numerator and denominator by a_0.

A rational function of the form (3.4.5) is called *proper* if $a_N \neq 0$ and $M < N$. From (3.3.2) it follows that this is equivalent to saying that the number of finite zeros is less than the number of finite poles.

An improper rational function ($M \geq N$) can always be written as the sum of a polynomial and a proper rational function. This procedure is illustrated by the following example.

EXAMPLE 3.4.2 ────────────────────────────────────

Express the improper rational transform

$$X(z) = \frac{1 + 3z^{-1} + \frac{11}{6} z^{-2} + \frac{1}{3} z^{-3}}{1 + \frac{5}{6} z^{-1} + \frac{1}{6} z^{-2}}$$

in terms of a polynomial and a proper function.

Solution: First, we note that we should reduce the numerator so that the terms z^{-2} and z^{-3} are eliminated. Thus we should carry out the long division with these two polynomials written in *reverse* order. We stop the division when the order of the remainder becomes z^{-1}. Then we obtain

$$X(z) = 1 + 2z^{-1} + \frac{\frac{1}{6}z^{-1}}{1 + \frac{5}{6}z^{-1} + \frac{1}{6}z^{-2}}$$

In general, any improper rational function ($M \geq N$) can be expressed as

$$X(z) = \frac{N(z)}{D(z)} = c_0 + c_1 z^{-1} + \cdots + c_{M-N} z^{-(M-N)} + \frac{N_1(z)}{D(z)} \qquad (3.4.6)$$

The inverse z-transform of the polynomial can easily be found by inspection. We focus our attention on the inversion of proper rational transforms, since any improper function can be transformed to a proper one by using (3.4.6). We carry out the development in two steps. First, we perform a partial-fraction expansion of the proper rational function and then we invert each of the terms.

Let $X(z)$ be a proper rational function, that is,

$$X(z) = \frac{N(z)}{D(z)} = \frac{b_0 + b_1 z^{-1} + \cdots + b_M z^{-M}}{1 + a_1 z^{-1} + \cdots + a_N z^{-N}} \qquad (3.4.7)$$

where

$$a_N \neq 0 \qquad \text{and} \qquad M < N$$

To simplify our discussion we eliminate negative powers of z by multiplying both the numerator and denominator of (3.4.7) by z^N. This results in

$$X(z) = \frac{b_0 z^N + b_1 z^{N-1} + \cdots + b_M z^{N-M}}{z^N + a_1 z^{N-1} + \cdots + a_N} \qquad (3.4.8)$$

which contains only positive powers of z. Since $N > M$, the function

$$\frac{X(z)}{z} = \frac{b_0 z^{N-1} + b_1 z^{N-2} + \cdots + b_M z^{N-M-1}}{z^N + a_1 z^{N-1} + \cdots + a_N} \qquad (3.4.9)$$

is also always proper.

Our task in performing a partial-fraction expansion is to express (3.4.9) or, equivalently, (3.4.7) as a sum of simple fractions. For this purpose we first factor the denominator polynomial in (3.4.9) into factors that contain the poles $p_1, p_2, \ldots, p_N$ of $X(z)$. We distinguish two cases.

Distinct Poles. Suppose that the poles $p_1, p_2, \ldots, p_N$ are all different (distinct). Then we seek an expansion of the form

$$\frac{X(z)}{z} = \frac{A_1}{z - p_1} + \frac{A_2}{z - p_2} + \cdots + \frac{A_N}{z - p_N} \qquad (3.4.10)$$

The problem is to determine the coefficients $A_1, A_2, \ldots, A_N$. There are two ways to solve this problem. We illustrate them with a simple example.

EXAMPLE 3.4.3 _____

Determine the partial-fraction expansion of the proper function

$$X(z) = \frac{1}{1 - 1.5z^{-1} + 0.5z^{-2}} \tag{3.4.11}$$

Solution: First we eliminate the negative powers, by multiplying both numerator and denominator by z^2. Thus

$$X(z) = \frac{z^2}{z^2 - 1.5z + 0.5}$$

The poles of $X(z)$ are $p_1 = 1$ and $p_2 = 0.5$. Consequently, the expansion of the form (3.4.10) is

$$\frac{X(z)}{z} = \frac{z}{(z - 1)(z - 0.5)} = \frac{A_1}{z - 1} + \frac{A_2}{z - 0.5} \tag{3.4.12}$$

A very simple method to determine A_1 and A_2 is to multiply the equation by the denominator term $(z - 1)(z - 0.5)$. Thus we obtain

$$z = (z - 0.5)A_1 + (z - 1)A_2 \tag{3.4.13}$$

Now if we set $z = p_1 = 1$ in (3.4.13), we eliminate the term involving A_2. Hence

$$1 = (1 - 0.5)A_1$$

Thus we obtain the result $A_1 = 2$. Next we return to (3.4.13) and set $z = p_2 = 0.5$. Thus we eliminate the term involving A_1, so we have

$$0.5 = (0.5 - 1)A_2$$

and hence $A_2 = -1$. Therefore, the result of the partial-fraction expansion is

$$\frac{X(z)}{z} = \frac{2}{z - 1} - \frac{1}{z - 0.5} \tag{3.4.14}$$

_____

The example given above suggests that we can determine the coefficients $A_1, A_2, \ldots,$ A_N by multiplying both sides of (3.4.10) by each of the terms $(z - p_k)$, $k = 1, 2, \ldots,$ N and evaluating the resulting expressions at the corresponding pole positions, $p_1, p_2,$ $\ldots, p_N$. Thus we have, in general,

$$\frac{(z - p_k)X(z)}{z} = \frac{(z - p_k)A_1}{z - p_1} + \cdots + A_k + \cdots + \frac{(z - p_k)A_N}{z - p_N} \tag{3.4.15}$$

Consequently, with $z = p_k$, (3.4.15) yields the kth coefficient as

$$A_k = \left. \frac{(z - p_k)X(z)}{z} \right|_{z = p_k} \qquad k = 1, 2, \ldots, N \tag{3.4.16}$$

EXAMPLE 3.4.4 _____

Determine the partial-fraction expansion of

$$X(z) = \frac{1 + z^{-1}}{1 - z^{-1} + 0.5z^{-2}} \tag{3.4.17}$$

Solution: To eliminate negative powers of z in (3.4.17), we multiply both numerator and denominator by z^2. Thus

$$\frac{X(z)}{z} = \frac{z + 1}{z^2 - z + 0.5}$$

The poles of $X(z)$ are complex conjugates:

$$p_1 = \tfrac{1}{2} + j\tfrac{1}{2}$$

and

$$p_2 = \tfrac{1}{2} - j\tfrac{1}{2}$$

Since $p_1 \neq p_2$, we will seek an expansion of the form (3.4.10). Thus

$$\frac{X(z)}{z} = \frac{z + 1}{(z - p_1)(z - p_2)} = \frac{A_1}{z - p_1} + \frac{A_2}{z - p_2}$$

To obtain A_1 and A_2, we use the formula (3.4.16). Thus we obtain

$$A_1 = \frac{(z - p_1)X(z)}{z}\bigg|_{z=p_1} = \frac{z + 1}{z - p_2}\bigg|_{z=p_1} = \frac{\tfrac{1}{2} + j\tfrac{1}{2} + 1}{\tfrac{1}{2} + j\tfrac{1}{2} - \tfrac{1}{2} + j\tfrac{1}{2}} = \tfrac{1}{2} - j\tfrac{3}{2}$$

$$A_2 = \frac{(z - p_2)X(z)}{z}\bigg|_{z=p_2} = \frac{z + 1}{z - p_1}\bigg|_{z=p_2} = \frac{\tfrac{1}{2} - j\tfrac{1}{2} + 1}{\tfrac{1}{2} - j\tfrac{1}{2} - \tfrac{1}{2} - j\tfrac{1}{2}} = \tfrac{1}{2} + j\tfrac{3}{2}$$

The expansion (3.4.10) and the formula (3.4.16) hold for both real and complex poles. The only constraint is that all poles be distinct. We also note that $A_2 = A_1^*$. It can be easily seen that this is a consequence of the fact that $p_2 = p_1^*$. In other words, *complex-conjugate poles result in complex-conjugate coefficients in the partial-fraction expansion*. This simple result will prove very useful later in our discussion.

Multiple-Order Poles. If $X(z)$ has a pole of multiplicity l, that is, it contains in its denominator the factor $(z - p_k)^l$, then the expansion (3.4.10) is no longer true. In this case a different expansion is needed. First, we investigate the case of a double pole (i.e., $l = 2$).

EXAMPLE 3.4.5 _____

Determine the partial-fraction expansion of

$$X(z) = \frac{1}{(1 + z^{-1})(1 - z^{-1})^2} \tag{3.4.18}$$

Solution: First, we express (3.4.18) in terms of positive powers of z, in the form

$$\frac{X(z)}{z} = \frac{z^2}{(z + 1)(z - 1)^2}$$

$X(z)$ has a simple pole at $p_1 = -1$ and a double pole $p_2 = p_3 = 1$. In such a case the appropriate partial-fraction expansion is

$$\frac{X(z)}{z} = \frac{z^2}{(z + 1)(z - 1)^2} = \frac{A_1}{z + 1} + \frac{A_2}{z - 1} + \frac{A_3}{(z - 1)^2} \tag{3.4.19}$$

The problem is to determine the coefficients A_1, A_2, and A_3.

We proceed as in the case of distinct poles. To determine A_1, we multiply both sides of (3.4.19) by $(z + 1)$ and evaluate the result at $z = -1$. Thus (3.4.19) becomes

$$\frac{(z + 1)X(z)}{z} = A_1 + \frac{z + 1}{z - 1}A_2 + \frac{z + 1}{(z - 1)^2}A_3$$

which, when evaluated at $z = -1$, yields

$$A_1 = \frac{(z + 1)X(z)}{z}\bigg|_{z=-1} = \frac{1}{4}$$

Next, if we multiply both sides of (3.4.19) by $(z - 1)^2$, we obtain

$$\frac{(z - 1)^2 X(z)}{z} = \frac{(z - 1)^2}{z + 1}A_1 + (z - 1)A_2 + A_3 \qquad (3.4.20)$$

Now, if we evaluate (3.4.20) at $z = 1$, we obtain A_3. Thus

$$A_3 = \frac{(z - 1)^2 X(z)}{z}\bigg|_{z=1} = \frac{1}{2}$$

The remaining coefficient A_2 can be obtained by differentiating both sides of (3.4.20) with respect to z and evaluating the result at $z = 1$. Note that it is not necessary formally to carry out the differentiation of the right-hand side of (3.4.20), since all terms except A_2 vanish when we set $z = 1$. Thus

$$A_2 = \frac{d}{dz}\left[\frac{(z - 1)^2 X(z)}{z}\right]_{z=1} = \frac{3}{4} \qquad (3.4.21)$$

The generalization of the procedure in the example above to the case of an lth-order pole $(z - p_k)^l$ straightforward. The partial-fraction expansion must contain the terms

$$\frac{A_{1k}}{z - p_k} + \frac{A_{2k}}{(z - p_k)^2} + \cdots + \frac{A_{lk}}{(z - p_k)^l}$$

The coefficients $\{A_{ik}\}$ can be evaluated through differentiation as illustrated in the example above for $l = 2$.

Now that we have performed the partial-fraction expansion, we are ready to take the final step in the inversion of $X(z)$. First, let us consider the case in which $X(z)$ contains distinct poles. From the partial-fraction expansion (3.4.10), it easily follows that

$$X(z) = A_1 \frac{1}{1 - p_1 z^{-1}} + A_2 \frac{1}{1 - p_2 z^{-1}} + \cdots + A_N \frac{1}{1 - p_N z^{-1}} \qquad (3.4.22)$$

The inverse z-transform, $x(n) = Z^{-1}\{X(z)\}$, can be obtained by inverting each term in (3.4.22) and taking the corresponding linear combination. From Table 3.3 it follows that these terms can be inverted using the formula

$$Z^{-1}\left\{\frac{1}{1 - p_k z^{-1}}\right\} = \begin{cases} (p_k)^n u(n) & \text{if ROC: } |z| > |p_k| \\ & \text{(causal signals)} \\ -(p_k)^n u(-n - 1) & \text{if ROC: } |z| < |p_k| \\ & \text{(anticausal signals)} \end{cases} \qquad (3.4.23)$$

If the signal $x(n)$ is causal, the ROC is $|z| > p_{max}$ where $p_{max} = \max\{|p_1|, |p_2|, \ldots, |p_N|\}$. In this case all terms in (3.4.22) result in causal signal components and the signal

$x(n)$ is given by

$$x(n) = (A_1 p_1^n + A_2 p_2^n + \cdots + A_N p_N^n)u(n) \tag{3.4.24}$$

If all poles are real, (3.4.24) is the desired expression for the signal $x(n)$. Thus a causal signal, having a z-transform that contains real and distinct poles, is a linear combination of real exponential signals.

Suppose now that all poles are distinct but some of them are complex. In this case some of the terms in (3.4.22) result in complex exponential components. However, if the signal $x(n)$ is real, we should be able to reduce these terms into real ones. If $x(n)$ is real, the polynomials appearing in $X(z)$ have real coefficients. In this case, as we have seen in Section 3.3, if p_j is a pole, its complex conjugate p_j^* is also a pole. As was demonstrated in Example 3.4.4, the corresponding coefficeints in the partial-fraction expansion are also complex conjugates. Thus the contribution of two complex-conjugate poles is of the form

$$x_k(n) = [A_k(p_k)^n + A_k^*(p_k^*)^n]u(n) \tag{3.4.25}$$

These two terms can be combined to form a real signal component. First, we express A_j and p_j in polar form (i.e., amplitude and phase) as

$$A_k = |A_k|e^{j\alpha_k} \tag{3.4.26}$$
$$p_k = r_k e^{j\beta_k} \tag{3.4.27}$$

where α_k and β_k are the phase components of A_k and p_k. Substitution of these relations into (3.4.25) gives

$$x_k(n) = |A_k|r_k^n[e^{j(\beta_k n + \alpha_k)} + e^{-j(\beta_k n + \alpha_k)}]u(n)$$

or, equivalently,

$$x_k(n) = 2|A_k|r_k^n \cos(\beta_k n + \alpha_k)u(n) \tag{3.4.28}$$

Thus we conclude that

$$Z^{-1}\left(\frac{A_k}{1 - p_k z^{-1}} + \frac{A_k^*}{1 - p_k^* z^{-1}}\right) = 2|A_k|r_k^n \cos(\beta_k n + \alpha_k)u(n) \tag{3.4.29}$$

if the ROC is $|z| > |p_k| = r_k$.

From (3.4.29) we observe that each pair of complex-conjugate poles in the z-domain results in a causal sinusoidal signal component with an exponential envelope. The distance r_k of the pole from the origin determines the exponential weighting (growing if $r_k > 1$, decaying if $r_k < 1$, constant if $r_k = 1$). The angle of the poles with the positive real axis provides the frequency of the sinusoidal signal. The zeros, or equivalently the numerator of the rational transform, affect only indirectly the amplitude and the phase of $x_k(n)$ through A_k.

In the case of *multiple* poles, either real or complex, the inverse transform of terms of the form $A/(z - p_k)^n$ is required. In the case of a double pole the following transform pair (see Table 3.3) is quite useful:

$$Z^{-1}\left\{\frac{pz^{-1}}{(1 - pz^{-1})^2}\right\} = np^n u(n) \tag{3.4.30}$$

provided that the ROC is $|z| > |p|$. The generalization to the case of poles with higher multiplicity is straightforward.

EXAMPLE 3.4.6

Determine the inverse z-transform of

$$X(z) = \frac{1}{1 - 1.5z^{-1} + 0.5z^{-2}}$$

if

(a) ROC: $|z| > 1$
(b) ROC: $|z| < 0.5$
(c) ROC: $0.5 < |z| < 1$

Solution: This is the same problem that we treated in Example 3.4.1. The partial-fraction expansion for $X(z)$ was determined in Example 3.4.3. The partial-fraction expansion of $X(z)$ yields

$$X(z) = \frac{2}{1 - z^{-1}} - \frac{1}{1 - 0.5z^{-1}} \qquad (3.4.31)$$

To invert $X(z)$ we should apply (3.4.23) for $p_1 = 1$ and $p_2 = 0.5$. However, this requires the specification of the corresponding ROC.

(a) In case when the ROC is $|z| > 1$, the signal $x(n)$ is causal and both terms in (3.4.31) are causal terms. According to (3.4.23), we obtain

$$x(n) = 2(1)^n u(n) - (0.5)^n u(n) = (2 - 0.5^n)u(n) \qquad (3.4.32)$$

which agrees with the result in Example 3.4.1(a).
(b) When the ROC is $|z| < 0.5$, the signal $x(n)$ is anticausal. Thus both terms in (3.4.31) result in anticausal components. From (3.4.23) we obtain

$$x(n) = [-2 + (0.5)^n]u(-n - 1) \qquad (3.4.33)$$

(c) In this case the ROC $0.5 < |z| < 1$ is a ring, which implies that the signal $x(n)$ is two-sided. Thus one of the terms corresponds to a causal signal and the other to an anticausal signal. Obviously, the given ROC is the overlapping of the regions $|z|$ 0.5 and $|z| < 1$. Hence the pole $p_2 = 0.5$ provides the causal part and the pole $p_1 = 1$ the anticausal. Thus

$$x(n) = -2(1)^n u(-n - 1) - (0.5)^n u(n) \qquad (3.4.34)$$

EXAMPLE 3.4.7

Determine the causal signal $x(n)$ whose z-transform is given by

$$X(z) = \frac{1 + z^{-1}}{1 - z^{-1} + 0.5z^{-2}}$$

Solution: In Example 3.4.4 we have obtained the partial-fraction expansion as

$$X(z) = \frac{A_1}{1 - p_1 z^{-1}} + \frac{A_2}{1 - p_2 z^{-1}}$$

where

$$A_1 = A_2^* = \tfrac{1}{2} - j\tfrac{3}{2}$$

and

$$p_1 = p_2^* = \tfrac{1}{2} + j\tfrac{1}{2}$$

Since we have a pair of complex-conjugate poles, we should use (3.4.29). The polar forms of A_1 and p_1 are

$$A_1 = \frac{\sqrt{10}}{2} e^{-j71.565}$$

$$p_1 = \frac{1}{\sqrt{2}} e^{j\pi/4}$$

Hence

$$x(n) = \sqrt{10} \left(\frac{1}{\sqrt{2}}\right)^n \cos\left(\frac{\pi n}{4} - 71.565°\right) u(n)$$

EXAMPLE 3.4.8 _____

Determine the causal signal $x(n)$ having the z-transform

$$X(z) = \frac{1}{(1 + z^{-1})(1 - z^{-2})}$$

Solution: From Example 3.4.5 we have

$$X(z) = \frac{1}{4}\frac{1}{1 + z^{-1}} + \frac{3}{4}\frac{1}{1 - z^{-1}} + \frac{1}{2}\frac{z^{-1}}{(1 - z^{-1})^2}$$

By applying the inverse transform relations in (3.4.23) and (3.4.30), we obtain

$$x(n) = \frac{1}{4}(-1)^n u(n) + \frac{3}{4}u(n) + \frac{1}{2}nu(n) = \left[\frac{1}{4}(-1)^n + \frac{3}{4} + \frac{n}{2}\right]u(n)$$

3.4.3 Decomposition of Rational z-Transforms

At this point it is appropriate to discuss some additional issues concerning the decomposition of rational z-transforms, which will prove very useful in the implementation of discrete-time systems.

Suppose that we have a rational z-transform $X(z)$ expressed as

$$X(z) = \frac{\displaystyle\sum_{k=0}^{M} b_k z^{-k}}{1 + \displaystyle\sum_{k=1}^{N} a_k z^{-k}} = b_0 \frac{\displaystyle\prod_{k=1}^{M}(1 - z_k z^{-1})}{\displaystyle\prod_{k=1}^{N}(1 - p_k z^{-1})} \tag{3.4.35}$$

where, for simplicity, we have assumed that $a_0 \equiv 1$. If $M \geq N$ [i.e., $X(z)$ is improper], we convert $X(z)$ to a sum of a polynomial and a proper function

$$X(z) = \sum_{k=0}^{M-N} c_k z^{-k} + X_{pr}(z) \tag{3.4.36}$$

If the poles of $X_{pr}(z)$ are distinct, it can be expanded in partial fractions as

$$X_{pr}(z) = A_1 \frac{1}{1 - p_1 z^{-1}} + A_2 \frac{1}{1 - p_2 z^{-1}} + \cdots + A_N \frac{1}{1 - p_N z^{-1}} \qquad (3.4.37)$$

As we have already observed, there may be some complex-conjugate pairs of poles in (3.4.37). Since we usually deal with real signals, we should avoid complex coefficients in our decomposition. This can be achieved by grouping and combining terms containing complex-conjugate poles, in the following way:

$$\frac{A}{1 - pz^{-1}} + \frac{A^*}{1 - p^* z^{-1}} = \frac{A - Ap^* z^{-1} + A^* - A^* pz^{-1}}{1 - pz^{-1} - p^* z^{-1} + pp^* z^{-2}} \qquad (3.4.38)$$

$$= \frac{b_0 + b_1 z^{-1}}{1 + a_1 z^{-1} + a_2 z^{-2}}$$

where

$$\begin{aligned} b_0 &= 2 \, \text{Re} \, (A) & a_1 &= -2 \, \text{Re} \, (p) \\ b_1 &= 2 \, \text{Re} \, (Ap^*) & a_2 &= |p|^2 \end{aligned} \qquad (3.4.39)$$

are the desired coefficients. Obviously, any rational transform of the form (3.4.38) with coefficients given by (3.4.39), which is the case when $a_1^2 - 4a_2 < 0$, can be inverted using (3.4.29). By combining (3.4.36) (3.4.37), and (3.4.38) we obtain a partial-fraction expansion for the z-transform with *distinct* poles that contains real coefficients. The general result is

$$X(z) = \sum_{k=0}^{M-N} c_k z^{-k} + \sum_{k=1}^{K_1} \frac{b_k}{1 + a_k z^{-1}} + \sum_{k=1}^{K_2} \frac{b_{0k} + b_{1k} z^{-1}}{1 + a_{1k} z^{-1} + a_{2k} z^{-2}} \qquad (3.4.40)$$

where $K_1 + 2K_2 = N$. Obviously, if $M = N$, the first term is just a constant, and when $M < N$, this term vanishes. When there are also multiple poles, some additional higher-order terms should be included in (3.4.40).

An alternative from is obtained by expressing $X(z)$ as a product of simple terms as in (3.4.35). However, the complex-conjugate poles and zeros should be combined to avoid complex coefficients in the decomposition. Such combinations result in second-order rational terms of the following form:

$$\frac{(1 - z_k z^{-1})(1 - z^*_k z^{-1})}{(1 - p_k z^{-1})(1 - p^*_k z^{-1})} = \frac{1 + b_{1k} z^{-1} + b_{2k} z^{-2}}{1 + a_{1k} z^{-1} + a_{2k} z^{-2}} \qquad (3.4.41)$$

where

$$\begin{aligned} b_{1k} &= -2 \, \text{Re} \, (z_k) & a_{1k} &= -2 \, \text{Re} \, (p_k) \\ b_{2k} &= |z_k|^2 & a_{2k} &= |p_k|^2 \end{aligned} \qquad (3.4.42)$$

Assuming for simplicity that $M = N$, we see that $X(z)$ can be decomposed in the following way:

$$X(z) = b_0 \prod_{k=1}^{K_1} \frac{1 + b_k z^{-1}}{1 + a_k z^{-1}} \prod_{k=1}^{K_2} \frac{1 + b_{1k} z^{-1} + b_{2k} z^{-2}}{1 + a_{1k} z^{-1} + a_{2k} z^{-2}} \qquad (3.4.43)$$

where $N = K_1 + 2K_2$. We will return to these important forms in Chapters 7 and 8.

3.5 The One-Sided z-Transform

The two-sided z-transform requires that the corresponding signals be specified for the entire time range $-\infty < n < \infty$. This requirement prevents its use from a very useful family of practical problems, namely the evaluation of the output of nonrelaxed systems. As we recall, these systems are described by difference equations with nonzero initial conditions. Since the input is applied at a finite time, say n_0, both input and output signals are specified for $n \geq n_0$, but by no means are zero for $n < n_0$. Thus the two-sided z-transform cannot be used. In this section we develop the one-sided z-transform which can be used to solve difference equations with initial conditions.

3.5.1 Definition and Properties

The *one-sided* or *unilateral* z-transform of a signal $x(n)$ is defined by

$$X^+(z) \equiv \sum_{n=0}^{\infty} x(n)z^{-n} \tag{3.5.1}$$

We will also use the notations $Z^+\{x(n)\}$ and

$$x(n) \xleftarrow{\quad z^+ \quad} X^+(z)$$

The one-sided z-transform differs from the two-sided transform in the lower limit of the summation, which is always zero whether or not the signal $x(n)$ is zero for $n < 0$ (i.e., causal). Due to this choice of lower limit, the one-sided z-transform has the following characteristics:

1. It does not contain information about the signal $x(n)$ for negative values of time (i.e., for $n < 0$).
2. It is *unique* only for causal signals, because only these signals are zero for $n < 0$.
3. The one-sided z-transform $X^+(z)$ of $x(n)$ is identical to the two-sided z-transform of the signal $x(n)u(n)$. Since $x(n)u(n)$ is causal, the ROC of its transform, and hence the ROC of $X^+(z)$, is always the exterior of a circle. Thus when we deal with one-sided z-transforms, it is not necessary to refer to their ROC.

EXAMPLE 3.5.1 _____

Determine the one-sided z-transform of the signal in Example 3.1.1.

Solution: From the definition (3.5.1), we obtain

$$x_1(n) = \left\{1, 2, 5, 7, 0, 1 \atop \uparrow\right\} \xleftarrow{\quad z^+ \quad} X_1^+(z) = 1 + 2z^{-1} + 5z^{-2} + 7z^{-3} + z^{-5}$$

$$x_2(n) = \left\{1, 2, 5, 7, 0, 1 \atop \uparrow\right\} \xleftarrow{\quad z^+ \quad} X_2^+(z) = 5 + 7z^{-1} + z^{-3}$$

$$x_3(n) = \left\{0, 0, 1, 2, 5, 7, 0, 1 \atop \uparrow\right\} \xleftarrow{\quad z^+ \quad} X_3^+(z) = z^{-2} + 2z^{-3} + 5z^{-4} + 7z^{-5} + z^{-7}$$

$$x_4(n) = \left\{2, 4, 5, 7, 0, 1 \atop \uparrow\right\} \xleftarrow{\quad z^+ \quad} X_4^+(z) = 5 + 7z^{-1} + z^{-3}$$

$$x_5(n) = \delta(n) \xleftarrow{\quad z^+ \quad} X_5^+(z) = 1$$

$$x_6(n) = \delta(n - k) \qquad k > 0 \xleftarrow{\;z^+\;} X_6^+(z) = z^{-k}$$

$$x_7(n) = \delta(n + k) \qquad k > 0 \xleftarrow{\;z^+\;} X_7^+(z) = 0$$

Note that for a noncausal signal, the one-sided z-transform is not unique. Indeed, $X_2^+(z) = X_4^+(z)$ but $x_2(n) \neq x_4(n)$. Also for anticausal signals, $X^+(z)$ is always zero.

Almost all properties we have studied for the two-sided z-transform carry over to the one-sided z-transform with the exception of the *shifting* property.

Shifting Property. CASE 1: TIME DELAY. If

$$x(n) \xleftarrow{\;z^+\;} X^+(z)$$

then

$$x(n - k) \xleftarrow{\;z^+\;} z^{-k}[X^+(z) + \sum_{n=1}^{k} x(-n)z^n] \qquad k > 0 \qquad (3.5.2)$$

In case $x(n)$ is causal, then

$$x(n - k) \xleftarrow{\;z^+\;} z^{-k}X^+(z) \qquad (3.5.3)$$

PROOF. From the defintion (3.5.1) we have

$$Z^+\{x(n - k)\} = z^{-k}\left[\sum_{l=-k}^{-1} x(l)z^{-l} + \sum_{l=0}^{\infty} x(l)z^{-l}\right]$$

$$= z^{-k}\left[\sum_{l=-1}^{-k} x(l)z^{-l} + X^+(z)\right]$$

By changing the index from l to $n = -l$, the result in (3.5.2) is easily obtained.

EXAMPLE 3.5.2 ───

Determine the one-sided z-transform of the signals

(a) $x(n) = a^n u(n)$
(b) $x_1(n) = x(n - 2)$, $x(n) = a^n$

Solution: (a) From (3.5.1) we easily obtain

$$X^+(z) = \frac{1}{1 - az^{-1}}$$

(b) We will apply the shifting property for $k = 2$. Indeed, we have

$$Z^+\{x(n - 2)\} = z^{-2}[X^+(z) + x(-1)z + x(-2)z^2]$$
$$= z^{-2}X^+(z) + x(-1)z^{-1} + x(-2)$$

Thus since $x(-1) = a^{-1}$, $x(-2) = a^{-2}$, we obtain

$$X^+_1(z) = \frac{z^{-2}}{1 - az^{-1}} + a^{-1}z^{-1} + a^{-2}$$

The meaning of the shifting property can be intuitively explained, if we write (3.5.2) as follows:

$$Z^+\{x(n - k)\} = [x(-k) + x(-k + 1)z^{-1} + \cdots + x(-1)z^{-k+1}] \quad (3.5.4)$$
$$+ z^{-k}X^+(z) \quad k > 0$$

To obtain $x(n - k)$ ($k > 0$) from $x(n)$, we should shift $x(n)$ by k samples to the right. Then k "new" samples, $x(-k)$, $x(-k + 1)$, . . . , $x(-1)$, enter the positive time axis with $x(-k)$ located at time zero. The first term in (3.5.4) stands for the z-transform of these samples. The "old" samples of $x(n - k)$ are the same as those of $x(n)$ simply shifted by k samples to the right. Their z-transform is obviously $z^{-k}X^+(z)$, which is the second term in (3.5.4).

CASE 2: TIME ADVANCE. If

$$x(n) \xleftarrow{\quad z^+ \quad} X^+(z)$$

then

$$x(n + k) \xleftarrow{\quad z^+ \quad} z^k \left[X^+(z) - \sum_{n=0}^{k-1} x(n)z^{-n} \right] \quad k > 0 \quad (3.5.5)$$

PROOF. From (3.5.1) we have

$$Z^+\{x(n + k)\} = \sum_{n=0}^{\infty} x(n + k)z^{-n} = z^k \sum_{l=k}^{\infty} x(l)z^{-l}$$

where we have changed the index of summation from n to $l = n + k$. Now, from (3.5.1) we obtain

$$X^+(z) = \sum_{l=0}^{\infty} x(l)z^{-l} = \sum_{l=0}^{k-1} x(l)z^{-l} + \sum_{l=k}^{\infty} x(l)z^{-l}$$

By combining the last two relations, we easily obtain (3.5.5).

EXAMPLE 3.5.3

With $x(n)$, as given in Example 3.5.2, determine the one-sided z-transform of the signal

$$x_2(n) = x(n + 2)$$

Solution: We will apply the shifting theorem for $k = 2$. From (3.5.5), with $k = 2$, we obtain

$$Z^+\{x(n + 2)\} = z^2 X^+(z) - x(0)z^2 - x(1)z$$

But $x(0) = 1$, $x(1) = a$, and $X^+(z) = 1/(1 - az^{-1})$. Thus

$$Z^+\{x(n + 2)\} = \frac{z^2}{1 - az^{-1}} - z^2 - az$$

The case of a time advance can be explained intuitively as follows. To obtain $x(n + k)$, $k > 0$, we should shift $x(n)$ by k samples to the left. As a result, the samples $x(0)$, $x(1)$, . . . , $x(k - 1)$ "leave" the positive time axis. Thus we first remove their contribution to the $X^+(z)$, and then we multiply what remains by z^k to compensate for the shifting of the signal by k samples.

The importance of the shifting property lies in its application to the solution of difference equations with constant coefficients and nonzero initial conditions. This makes the one-sided z-transform a very useful tool for the analysis of recursive linear time-invariant discrete-time systems.

An important theorem useful in the analysis of signals and systems is the final value theorem.

FINAL VALUE THEOREM: *If*

$$x(n) \xleftarrow{\quad z^+ \quad} X^+(z)$$

then

$$\lim_{n \to \infty} x(n) = \lim_{z \to 1} (z - 1)X^+(z) \qquad (3.5.6)$$

The limit in (3.5.6) exists if the ROC of $(z - 1)X^+(z)$ includes the unit circle.

The proof of this theorem is left as an exercise for the reader.

This theorem is useful when we are interested in the asymptotic behavior of a signal $x(n)$ and we know its z-transform but not the signal itself. In such cases, especially if it is complicated to invert $X^+(z)$, we can use the final value theorem to determine the limit of $x(n)$ as n goes to infinity.

EXAMPLE 3.5.4

The impulse response of a relaxed linear time-invariant system is $h(n) = \alpha^n u(n)$, $|\alpha| < 1$. Determine the value of the step response of the system as $n \to \infty$.

Solution: The step response of the system is

$$y(n) = h(n) * x(n)$$

where

$$x(n) = u(n)$$

Obviously, if we excite a causal system with a causal input the output will be causal. Since $h(n)$, $x(n)$, $y(n)$ are causal signals, the one-sided and two-sided z-transforms are identical. From the convolution property (3.2.17) we know that the z-transforms of $h(n)$ and $x(n)$ must be multiplied to yield the z-transform of the output. Thus

$$Y(z) = \frac{1}{1 - \alpha z^{-1}} \frac{1}{1 - z^{-1}} = \frac{z^2}{(z - 1)(z - \alpha)} \qquad \text{ROC: } |z| > |\alpha|$$

Now

$$(z - 1)Y(z) = \frac{z^2}{z - \alpha} \qquad \text{ROC: } |z| > |\alpha|$$

Since $|\alpha| < 1$ the ROC of $(z - 1)Y(z)$ includes the unit circle. Consequently, we can apply (3.5.6) and obtain

$$\lim_{n \to \infty} y(n) = \lim_{z \to 1} \frac{z^2}{z - \alpha} = \frac{1}{1 - \alpha}$$

3.5.2 Solution of Diffference Equations

The one-sided z-transform is a very efficient tool for the solution of difference equations with nonzero initial conditions. It achieves that by reducing the difference equation relating the two time-domain signals to an equivalent algebraic equation relating their one-sided z-transforms. This equation can be solved easily to obtain the transform of the desired signal. The signal in the time domain is obtained by inverting the resulting z-transform. We will illustrate this approach with two examples.

EXAMPLE 3.5.5 _____

The well-known Fibonacci sequence of integer numbers is obtained by computing each term as the sum of the two previous ones. The first few terms of the sequence are

$$1, 1, 2, 3, 5, 8, \ldots$$

Determine a closed-form expression for the nth term of the Fibonacci sequence.

Solution: Let $y(n)$ be the nth term of the Fibonacci sequence. Clearly, $y(n)$ satisfies the difference equation

$$y(n) = y(n-1) + y(n-2) \tag{3.5.7}$$

with initial conditions

$$y(0) = y(-1) + y(-2) = 1 \tag{3.5.8a}$$
$$y(1) = y(0) + y(-1) = 1 \tag{3.5.8b}$$

From (3.5.8b) we have $y(-1) = 0$. Then (3.5.8a) gives $y(-2) = 1$. Thus we have to determine $y(n)$, $n \geq 0$ which satisfies (3.5.7), with initial conditions $y(-1) = 0$ and $y(-2) = 1$.

By taking the one-sided z-transform of (3.5.7) and using the shifting property (3.5.2), we obtain

$$Y^+(z) = [z^{-1}Y^+(z) + y(-1)] + [z^{-2}Y^+(z) + y(-2) + y(-1)z^{-1}]$$

or

$$Y^+(z) = \frac{1}{1 - z^{-1} - z^{-2}} = \frac{z^2}{z^2 - z - 1} \tag{3.5.9}$$

where we have used the fact that $y(-1) = 0$ and $y(-2) = 1$.

We can invert $Y^+(z)$ by the partial-fraction expansion method. The poles of $Y^+(z)$ are

$$p_1 = \frac{1 + \sqrt{5}}{2} \qquad p_2 = \frac{1 - \sqrt{5}}{2}$$

and the corresponding coefficients are $A_1 = p_1/\sqrt{5}$ and $A_2 = -p_2/\sqrt{5}$. Therefore,

$$y(n) = \left[\frac{1 + \sqrt{5}}{2\sqrt{5}} \left(\frac{1 + \sqrt{5}}{2} \right)^n - \frac{1 - \sqrt{5}}{2\sqrt{5}} \left(\frac{1 - \sqrt{5}}{2} \right)^n \right] u(n)$$

or, equivalently,

$$y(n) = \frac{1}{\sqrt{5}} \left(\frac{1}{2} \right)^{n+1} [(1 + \sqrt{5})^{n+1} - (1 - \sqrt{5})^{n+1}] u(n) \tag{3.5.10}$$

EXAMPLE 3.5.6 _____

Determine the step response of the system

$$y(n) = \alpha y(n - 1) + x(n) \qquad -1 < \alpha < 1 \qquad (3.5.11)$$

when the initial condition is $y(-1) = 1$.

Solution: By taking the one-sided z-transform of both sides of (3.5.11), we obtain

$$Y^+(z) = \alpha[z^{-1}Y^+(z) + y(-1)] + X^+(z)$$

Upon substitution for $y(-1)$ and $X^+(z)$ and solving for $Y^+(z)$, we obtain the result

$$Y^+(z) = \frac{\alpha}{1 - \alpha z^{-1}} + \frac{1}{(1 - \alpha z^{-1})(1 - z^{-1})} \qquad (3.5.12)$$

By performing a partial-fraction expansion and inverse transforming the result, we have

$$y(n) = \alpha^{n+1}u(n) + \frac{1 - \alpha^{n+1}}{1 - \alpha} u(n) \qquad (3.5.13)$$

$$= \frac{1}{1 - \alpha} (1 - \alpha^{n+2})u(n)$$

3.6 Analysis of Linear Time-Invariant Systems in the z-Domain

In Section 3.3.3 we introduced the system function of a linear time-invariant system and related it to the unit sample response and to the difference equation description of systems. In this section we describe the use of the system function in the determination of the response of the system to some excitation signal. Furthermore, we extend this method of analysis to nonrelaxed systems. Our attention is focused on the important class of pole–zero systems represented by linear constant-coefficient difference equations with arbitrary initial conditions.

We also consider the topic of stability of linear time-invariant systems and describe a test for determining the stability of a system based on the coefficients of the denominator polynomial in the system function. Finally, we provide a detailed analysis of second-order systems, which form the basic building block for the realization of higher-order systems.

3.6.1 Response of Relaxed Pole–Zero Systems

Let us consider a pole–zero system described by the general linear constant-coefficient difference equation in (3.3.7) and the corresponding system function in (3.3.8). We represent $H(z)$ as a ratio of two polynomials $B(z)/A(z)$, where $B(z)$ is the numerator polynomial that contains the zeros of $H(z)$ and $A(z)$ is the denominator polynomial that determines the poles of $H(z)$. Furthermore, let us assume that the input signal $x(n)$ has a rational z-transform $X(z)$ of the form

$$X(z) = \frac{N(z)}{Q(z)} \qquad (3.6.1)$$

This assumption is not overly restrictive, since most signals of practical interest have rational z-transforms, as indicated previously.

If the system is initially relaxed, that is, the initial conditions for the difference equation are zero, $y(-1) = y(-2) = \cdots = y(-N) = 0$, the z-transform of the output of the system has the form

$$Y(z) = H(z)X(z) = \frac{B(z)N(z)}{A(z)Q(z)} \tag{3.6.2}$$

Now suppose that the system contains simple poles $p_1, p_2, \ldots, p_N$ and the z-transform of the input signal contains poles $q_1, q_2, \ldots, q_L$, where $p_k \neq q_m$ for all $k = 1, 2, \ldots, N$ and $m = 1, 2, \ldots, L$. In addition, we assume that the zeros of the numerator polynomials $B(z)$ and $N(z)$ do not coincide with the poles $\{p_k\}$ and $\{q_k\}$, so that there is no pole–zero cancellation. Then a partial-fraction expansion of $Y(z)$ yields

$$Y(z) = \sum_{k=1}^{N} \frac{A_k}{1 - p_k z^{-1}} + \sum_{k=1}^{L} \frac{Q_k}{1 - q_k z^{-1}} \tag{3.6.3}$$

The inverse transform of $Y(z)$ yields the output signal from the system in the form

$$y(n) = \sum_{k=1}^{N} A_k(p_k)^n u(n) + \sum_{k=1}^{L} Q_k(q_k)^n u(n) \tag{3.6.4}$$

We observe that the output sequence $y(n)$ can be subdivided into two parts. The first part is a function of the poles $\{p_k\}$ of the system and is called the *natural response* of the system. The influence of the input signal on this part of the response is through the scale factors $\{A_k\}$. The second part of the response is a function of the poles $\{q_k\}$ of the input signal and is called the *forced response* of the system. The influence of the system on this response is exerted through the scale factors $\{Q_k\}$.

We should emphasize that the scale factors $\{A_k\}$ and $\{Q_k\}$ are functions of both sets of poles $\{p_k\}$ and $\{q_k\}$. For example, if $X(z) = 0$ so that the input is zero, then $Y(z) = 0$, and consequently, the output is zero. Clearly, then, the natural response of the system is zero. This implies that the natural response of the system is different from the zero-input response.

When $X(z)$ and $H(z)$ have one or more poles in common or when $X(z)$ and/or $H(z)$ contain multiple-order poles, then $Y(z)$ will have multiple-order poles. Consequently, the partial-fraction expansion of $Y(z)$ will contain factors of the form $1/(1 - p_l z^{-1})^k$, $k = 1, 2, \ldots, m$, where m is the pole order. The inversion of these factors will produce terms of the form $n^{k-1} p_l^n$ in the output $y(n)$ of the system, as indicated in Section 3.4.2.

3.6.2 Response of Pole–Zero Systems with Nonzero Initial Conditions

Suppose that the signal $x(n)$ is applied to the pole–zero system at $n = 0$. Thus the signal $x(n)$ is assumed to be causal. The effects of all previous input signals to the system are reflected in the initial conditions $y(-1), y(-2), \ldots, y(-N)$. Since the input $x(n)$ is causal and since we are interested in determining the output $y(n)$ for $n \geq 0$, we may use the one-sided z-transform, which allows us to deal with the initial conditions. Thus the one-sided z-transform of (3.3.7) becomes

$$Y^+(z) = -\sum_{k=1}^{N} a_k z^{-k} \left[Y^+(z) + \sum_{n=1}^{k} y(-n)z^n \right] + \sum_{k=0}^{M} b_k z^{-k} X^+(z) \tag{3.6.5}$$

Since $x(n)$ is causal, we may set $X^+(z) = X(z)$. In any case (3.6.5) may be expressed as

$$Y^+(z) = \frac{\sum_{k=0}^{M} b_k z^{-k}}{1 + \sum_{k=1}^{N} a_k z^{-k}} X(z) - \frac{\sum_{k=1}^{N} a_k z^{-k} \sum_{n=1}^{k} y(-n)z^n}{1 + \sum_{k=1}^{N} a_k z^{-k}}$$

$$= H(z)X(z) + \frac{N_0(z)}{A(z)} \tag{3.6.6}$$

where

$$N_0(z) = - \sum_{k=1}^{N} a_k z^{-k} \sum_{n=1}^{k} y(-n)z^n \tag{3.6.7}$$

From (3.6.6) it is apparent that the output of the system with nonzero initial conditions can be subdivided into two parts. The first is the zero-state response of the system, defined in the z-domain as

$$Y_{zs}(z) = H(z)X(z) \tag{3.6.8}$$

The second component corresponds to the output resulting from the nonzero initial conditions. This output is the zero-input response of the system, which is defined in the z-domain as

$$Y_{zi}^+(z) = \frac{N_0(z)}{A(z)} \tag{3.6.9}$$

Hence the total response is the sum of these two output components, which can be expressed in the time domain by determining the inverse z-transforms of $Y_{zs}(z)$ and $Y_{zi}(z)$ separately, and adding the results. Thus

$$y(n) = y_{zs}(n) + y_{zi}(n) \tag{3.6.10}$$

Since the denominator of $Y_{zi}^+(z)$, is $A(z)$, its poles are $p_1, p_2, \ldots, p_N$. Consequently, the zero-input response has the form

$$y_{zi}(n) = \sum_{k=1}^{N} D_k(p_k)^n u(n) \tag{3.6.11}$$

This may be added to (3.6.4) and the terms involving the poles $\{p_k\}$ may be combined to yield the total repsonse in the form

$$y(n) = \sum_{k=1}^{N} A_k'(p_k)^n u(n) + \sum_{k=1}^{L} Q_k(q_k)^n u(n) \tag{3.6.12}$$

where, by definition,

$$A_k' = A_k + D_k \tag{3.6.13}$$

The development above indicates clearly that the effect of the initial conditions is to alter the natural response of the system through modification of the scale factors $\{A_k\}$. There are no new poles introduced by the nonzero initial conditions. Furthermore, there is no effect on the forced response of the system. These important points are reinforced in the following example.

EXAMPLE 3.6.1 _____

Determine the unit step response of the system described by the difference equation

$$y(n) = 0.9y(n-1) - 0.81y(n-2) + x(n)$$

under the following initial conditions:

(a) $y(-1) = y(-2) = 0$
(b) $y(-1) = y(-2) = 1$

Solution: The system function is

$$H(z) = \frac{1}{1 - 0.9z^{-1} + 0.81z^{-2}}$$

This system has two complex-conjugate poles at

$$p_1 = 0.9e^{j\pi/3} \qquad p_2 = 0.9e^{-j\pi/3}$$

The z-transform of the unit step sequence is

$$X(z) = \frac{1}{1 - z^{-1}}$$

Therefore,

$$Y_{zs}(z) = \frac{1}{(1 - 0.9e^{j\pi/3}z^{-1})(1 - 0.9e^{-j\pi/3}z^{-1})(1 - z^{-1})}$$

$$= \frac{0.542 - j0.049}{1 - 0.9e^{j\pi/3}z^{-1}} + \frac{0.542 + j0.049}{1 - 0.9e^{-j\pi/3}z^{-1}} + \frac{1.099}{1 - z^{-1}}$$

and hence the zero-state response is

$$y_{zs}(n) = \left[1.099 + 1.088(0.9)^n \cos\left(\frac{\pi}{3}n - 5.2°\right)\right]u(n)$$

(a) Since the initial conditions are zero in this case, we conclude that $y(n) = y_{zs}(n)$.
(b) For the initial conditions $y(-1) = y(-2) = 1$, the additional component in the z-transform is

$$Y_{zi}(z) = \frac{N_0(z)}{A(z)} = \frac{0.09 - 0.81z^{-1}}{1 - 0.9z^{-1} + 0.81z^{-2}}$$

$$= \frac{0.026 + j0.4936}{1 - 0.9e^{j\pi/3}z^{-1}} + \frac{0.026 - j0.4936}{1 - 0.9e^{-j\pi/3}z^{-1}}$$

Consequently, the zero-input response is

$$y_{zi}(n) = 0.988(0.9)^n \cos\left(\frac{\pi}{3}n + 87°\right)u(n)$$

In this case the total response has the z-transform

$$Y(z) = Y_{zs}(z) + Y_{zi}(z)$$

$$= \frac{1.099}{1 - z^{-1}} + \frac{0.568 + j0.445}{1 - 0.9e^{j\pi/3}z^{-1}} + \frac{0.568 - j0.445}{1 - 0.9e^{-j\pi/3}z^{-1}}$$

The inverse transform yields the total response in the form

$$y(n) = 1.099u(n) + 1.44(0.9)^n \cos\left(\frac{\pi}{3}n + 38°\right)u(n)$$

3.6.3 Transient and Steady-State Responses

As we have seen from our previous discussion, the response of a system to a given input can be separated into two components, the natural response and the forced response. The natural response of a causal system has the form

$$y_{nr}(n) = \sum_{k=1}^{N} A_k(p_k)^n u(n) \tag{3.6.14}$$

where $\{p_k\}$, $k = 1, 2, \ldots, N$ are the poles of the system and $\{A_k\}$ are scale factors that depend on the characteristics of the input sequence and on the initial conditions.

If $|p_k| < 1$ for all k, then, $y_{nr}(n)$ decays to zero as n approaches infinity. In such a case we refer to the natural response of the system as the *transient response*. The rate at which $y_{nr}(n)$ decays toward zero depends on the magnitude of the pole positions. If all the poles have small magnitudes, the decay is very rapid. On the other hand, if one or more poles are located near the unit circle, the corresponding terms in $y_{nr}(n)$ will decay slowly toward zero and the transient will persist for a relatively long time.

The forced response of the system has the form

$$y_{fr}(n) = \sum_{k=1}^{L} Q_k(q_k)^n u(n) \tag{3.6.15}$$

where $\{q_k\}$, $k = 1, 2, \ldots, L$ are the poles in the forcing function and $\{Q_k\}$ are scale factors that depend on the characteristics of the system and on the input sequence. If all the poles of the input signal fall inside the unit circle, $y_{fr}(n)$ will decay toward zero as n approaches infinity, just as in the case of the natural response. This should not be surprising since the input signal is also a transient signal. On the other hand, when the causal input signal is a sinusoid, the poles fall on the unit circle and consequently, the forced response is also a sinusoid that persists for all $n \geq 0$. In this case, the forced response is called the *steady-state response* of the system. Thus for the system to sustain a steady-state output for $n \geq 0$, the input signal must persist for all $n \geq 0$.

The following example illustrates the presence of the steady-state response.

EXAMPLE 3.6.2 ———

Determine the transient and steady-state responses of the system characterized by the difference equation

$$y(n) = 0.5y(n - 1) + x(n)$$

when the input signal is $x(n) = 10 \cos(\pi n/4)u(n)$. The system is initially at rest (i.e., it is relaxed).

Solution: The system function for this system is

$$H(z) = \frac{1}{1 - 0.5z^{-1}}$$

and therefore the system has a pole at $z = 0.5$. The z-transform of the input signal is (from Table 3.3)

$$X(z) = \frac{10(1 - (1/\sqrt{2})z^{-1})}{1 - \sqrt{2}\,z^{-1} + z^{-2}}$$

Consequently,

$$Y(z) = H(z)X(z)$$

$$= \frac{10(1 - 1/\sqrt{2})z^{-1})}{(1 - 0.5z^{-1})(1 - e^{j\pi/4}z^{-1})(1 - e^{-j\pi/4}z^{-1})}$$

$$= \frac{6.3}{1 - 0.5z^{-1}} + \frac{6.78e^{-j28.7°}}{1 - e^{j\pi/4}z^{-1}} + \frac{6.78e^{j28.7°}}{1 - e^{-j\pi/4}z^{-1}}$$

The natural or transient response is

$$y_{nr}(n) = 6.3(0.5)^n u(n)$$

and the forced or steady-state response is

$$y_{fr}(n) = [6.78e^{-j28.7°}(e^{j\pi n/4}) + 6.78e^{j28.7°}e^{-j\pi n/4}]u(n)$$

$$= 13.56 \cos \left(\frac{\pi}{4}n - 28.7° \right) u(n)$$

Thus we see that the steady-state response persists for all $n \geq 0$, just as the input signal persists for all $n \geq 0$.

3.6.4 Causality and Stability

As defined previously, a causal linear time-invariant system is one whose unit sample response $h(n)$ satisfies the condition

$$h(n) = 0 \qquad n < 0$$

We have also shown that the ROC of the z-transform of a causal sequence is the exterior of a circle. Consequently, *a linear time-invariant system is causal if and only if the ROC of the system function is the exterior of a circle of radius $r < \infty$.*

The stability of a linear time-invariant system can also be expressed in terms of the characteristics of the system function. As we recall from our previous discussion, a necessary and sufficient condition for a linear time-invariant system to be BIBO stable is

$$\sum_{n=-\infty}^{\infty} |h(n)| < \infty$$

In turn, this condition implies that $H(z)$ must contain the unit circle within its ROC. Indeed, since

$$H(z) = \sum_{n=-\infty}^{\infty} h(n)z^{-n}$$

it follows that

$$|H(z)| \leq \sum_{n=-\infty}^{\infty} |h(n)z^{-n}| = \sum_{n=-\infty}^{\infty} |h(n)| \, |z^{-n}|$$

When evaluated on the unit circle (i.e., $|z| = 1$),

$$|H(z)| \leq \sum_{n=-\infty}^{\infty} |h(n)|$$

Hence, if the system is BIBO stable, the unit circle is contained in the ROC of $H(z)$. The converse is also true. Therefore, *a linear time-invariant system is BIBO stable if and only if the ROC of the system function includes the unit circle.*

We should stress, however, that the conditions for causality and stability are different and that one does not imply the other. For example, a causal system may be stable or unstable, just as a noncausal system may be stable or unstable. Similarly, an unstable system may be either causal or noncausal, just as a stable system may be causal or noncausal.

For a causal system, however, the condition on stability can be narrowed to some extent. Indeed, a causal system is characterized by a system function $H(z)$ having as a ROC the exterior of some circle of radius r. For a stable system, the ROC must include the unit circle. Consequently, a causal and stable system must have a system function that converges for $|z| > r < 1$. Since the ROC cannot contain any poles of $H(z)$, it follows that *a causal linear time-invariant system is BIBO stable if and only if all the poles of $H(z)$ are inside the unit circle.*

EXAMPLE 3.6.3 _____

A linear time-invariant system is characterized by the system function

$$H(z) = \frac{3 - 4z^{-1}}{1 - 3.5z^{-1} + 1.5z^{-2}}$$

$$= \frac{1}{1 - \frac{1}{2}z^{-1}} + \frac{2}{1 - 3z^{-1}}$$

Specify the ROC of $H(z)$ and determine $h(n)$ for the following conditions:

(a) The system is stable.
(b) The system is causal.
(c) The system is purely anticausal.

Solution: The system has poles at $z = \frac{1}{2}$ and $z = 3$.

(a) Since the system is stable, its ROC must include the unit circle and hence it is $\frac{1}{2} < |z| < 3$. Consequently, $h(n)$ is noncausal and is given as

$$h(n) = (\tfrac{1}{2})^n u(n) - 2(3)^n u(-n-1)$$

(b) Since the system is causal, its ROC is $|z| > 3$. In this case

$$h(n) = (\tfrac{1}{2})^n u(n) + 2(3)^n u(n)$$

This system is unstable.

(c) If the system is anticausal, its ROC is $|z| < 0.5$. Hence

$$h(n) = -[(\tfrac{1}{2})^n + 2(3)^n]u(-n-1)$$

In this case the system is unstable.

_____

3.6.5 Pole–Zero Cancellations

When a z-transform has a pole that is at the same location as a zero, the pole is canceled by the zero and, consequently, the term containing that pole in the inverse z-transform vanishes. Such pole–zero cancellations are very important in the analysis of pole–zero systems.

Pole–zero cancellations may occur either in the system function itself or in the product of the system function with the z-transform of the input signal. In the first case we say that the order of the system is reduced by one. In the latter case we say that the pole of the system is suppressed by the zero in the input signal, or vice versa. Thus, by properly selecting the position of the zeros of the input signal, it is possible to suppress one or more system modes (pole factors) in the response of the system. Similarly, by proper selection of the zeros of the system function, it is possible to suppress one or more modes of the input signal from the response of the system.

When the zero is located very near the pole but not exactly at the same location, the term in the response has a very small amplitude. For example, nonexact pole–zero cancellations may occur in practice as a result of finite numerical precision used in representing the coefficients of the system. Consequently, one should not attempt to stabilize an inherently unstable system by placing a zero in the input signal at the location of the pole.

EXAMPLE 3.6.4 _____

Determine the unit sample response of the system characterized by the difference equation

$$y(n) = 2.5y(n - 1) - y(n - 2) + x(n) - 5x(n - 1) + 6x(n - 2)$$

Solution: The system function is

$$H(z) = \frac{1 - 5z^{-1} + 6z^{-2}}{1 - 2.5z^{-1} + z^{-2}}$$

$$= \frac{1 - 5z^{-1} + 6z^{-2}}{(1 - \frac{1}{2}z^{-1})(1 - 2z^{-1})}$$

This system has poles at $p_1 = 2$ and $p_1 = \frac{1}{2}$. Consequently, at first glance it appears that the unit sample response is

$$Y(z) = H(z)X(z) = \frac{1 - 5z^{-1} + 6z^{-2}}{(1 - \frac{1}{2}z^{-1})(1 - 2z^{-1})}$$

$$= z\left(\frac{A}{z - \frac{1}{2}} + \frac{B}{z - 2}\right)$$

By evaluating the constants at $z = \frac{1}{2}$ and $z = 2$, we find that

$$A = \frac{5}{2} \qquad B = 0$$

The fact that $B = 0$ indicates that there exists a zero at $z = 2$ which cancels the pole at $z = 2$. In fact, the zeros occur at $z = 2$ and $z = 3$. Consequently, $H(z)$ reduces to

$$H(z) = \frac{1 - 3z^{-1}}{1 - \frac{1}{2}z^{-1}} = \frac{z - 3}{z - \frac{1}{2}}$$

$$= 1 - \frac{2.5z^{-1}}{1 - \frac{1}{2}z^{-1}}$$

and therefore

$$h(n) = \delta(n) - 2.5(\tfrac{1}{2})^{n-1}u(n - 1)$$

The reduced-order system obtained by canceling the common pole and zero is characterized by the difference equation

$$y(n) = \tfrac{1}{2}y(n - 1) + x(n) - 3x(n - 1)$$

Although the original system is also BIBO stable due to the pole–zero cancellation, in a practical implementation of this second-order system we may encounter an instability due to imperfect cancellation of the pole and the zero.

EXAMPLE 3.6.5

Determine the response of the system

$$y(n) = \tfrac{5}{6}y(n-1) - \tfrac{1}{6}y(n-2) + x(n)$$

to the input signal $x(n) = \delta(n) - \tfrac{1}{3}\delta(n-1)$.

Solution: The system function is

$$H(z) = \frac{1}{1 - \tfrac{5}{6}z^{-1} + \tfrac{1}{6}z^{-2}}$$

$$= \frac{1}{(1 - \tfrac{1}{2}z^{-1})(1 - \tfrac{1}{3}z^{-1})}$$

This system has two poles, one at $z = \tfrac{1}{2}$ and the other at $z = \tfrac{1}{3}$. The z-transform of the input signal is

$$X(z) = 1 - \tfrac{1}{3}z^{-1}$$

In this case the input signal contains a zero at $z = \tfrac{1}{3}$ which cancels the pole at $z = \tfrac{1}{3}$. Consequently,

$$Y(z) = H(z)X(z)$$

$$Y(z) = \frac{1}{1 - \tfrac{1}{2}z^{-1}}$$

and hence the response of the system is

$$y(n) = (\tfrac{1}{2})^n u(n)$$

Clearly, the mode $(\tfrac{1}{3})^n$ is suppressed from the output as a result of the pole–zero cancellation.

3.6.6 Multiple-Order Poles and Stability

As we have observed, a necessary and sufficient condition for a causal linear time-invariant system to be BIBO stable is that all its poles lie inside the unit circle. The input signal is bounded if its z-transform contains poles $\{q_k\}$, $k = 1, 2, \ldots, L$, which satisfy the condition $|q_k| \leq 1$ for all k. We note that the forced response of the system, given in (3.6.15), is also bounded, even when the input signal contains one or more distinct poles on the unit circle.

In view of the fact that a bounded input signal may have poles on the unit circle, it might appear that a stable system may also have poles on the unit circle. This is not the case, however, since such a system will produce an unbounded response when excited by an input signal that also has a pole at the same position on the unit circle. The following example illustrates this point.

EXAMPLE 3.6.6

Determine the step response of the causal system described by the difference equation

$$y(n) = y(n-1) + x(n)$$

Solution: The system function for the system is

$$H(z) = \frac{1}{1 - z^{-1}}$$

We note that the system contains a pole on the unit circle at $z = 1$. The z-transform of the input signal $x(n) = u(n)$ is

$$X(z) = \frac{1}{1 - z^{-1}}$$

which also contains a pole at $z = 1$. Hence the output signal has the transform

$$Y(z) = H(z)X(z)$$

$$= \frac{1}{(1 - z^{-1})^2}$$

which contains a double pole at $z = 1$.
The inverse z-transform of $Y(z)$ is

$$y(n) = (n + 1)u(n)$$

which is a ramp sequence. Thus $y(n)$ is unbounded, even when the input is bounded. Consequently, the system is unstable.

Example 3.6.6 demonstrates clearly that BIBO stability requires that the system poles be strictly inside the unit circle. If the system poles are all inside the unit circle and the excitation sequence $x(n)$ contains one or more poles that coincide with the poles of the system, the output $Y(z)$ will contain multiple-order poles. As indicated previously, such multiple-order poles will result in an output sequence that contains terms of the form

$$A_k n^b (p_k)^n u(n)$$

where $0 \le b \le m - 1$ and m is the order of the pole. If $|p_k| < 1$, these terms decay to zero as n approaches infinity because the exponential factor $(p_k)^n$ dominates over the term n^b. Consequently, no bounded input signal can produce an unbounded output signal if the system poles are all inside the unit circle.

Finally, we should state that the only useful systems which contain poles on the unit circle are the digital oscillators discussed in Chapter 5. We call such systems *marginally stable*.

3.6.7 The Schür–Cohn Stability Test

We have stated previously that the stability of a system is determined by the position of the poles. The poles of the system are the roots of the denominator polynomial of $H(z)$, namely,

$$A(z) = 1 + a_1 z^{-1} + a_2 z^{-2} + \cdots + a_N z^{-N} \qquad (3.6.16)$$

When the system is causal all the roots of $A(z)$ must lie inside the unit circle for the system to be stable.

There are several computational procedures that aid us in determining if any of the roots of $A(z)$ lie outside the unit circle. These procedures are called *stability criteria*. Below we describe the Schür–Cohn test procedure for the stability of a system characterized by the system function $H(z) = B(z)/A(z)$.

Before we describe the Schür–Cohn test we need to establish some useful notation. We denote a polynomial of degree m by

$$A_m(z) = \sum_{k=0}^{m} \alpha_m(k)z^{-k} \qquad \alpha_m(0) = 1 \qquad (3.6.17)$$

The *reciprocal* or *reverse polynomial* $B_m(z)$ of degree m is defined as

$$B_m(z) = z^{-m}A_m(z^{-1}) \qquad (3.6.18)$$

$$= \sum_{k=0}^{m} \alpha_m(m - k)z^{-k}$$

We observe that the coefficients of $B_m(z)$ are the same as those of $A_m(z)$, but in reverse order.

In the Schür–Cohn stability test, to determine if the polynomial $A(z)$ has all its roots inside the unit circle, we compute a set of coefficients, called *reflection coefficients*, $K_1, K_2, \ldots, K_N$ from the polynomials $A_m(z)$. First, we set

$$A_N(z) = A(z)$$

and

$$(3.6.19)$$

$$K_N = \alpha_N(N)$$

Then we compute the lower-degree polynomials $A_m(z)$, $m = N, N - 1, N - 2, \ldots, 1$, according to the recursive equation

$$A_{m-1}(z) = \frac{A_m(z) - K_m B_m(z)}{1 - K_m^2} \qquad (3.6.20)$$

where the coefficients K_m are defined as

$$K_m = \alpha_m(m) \qquad (3.6.21)$$

The Schür–Cohn stability test states that *the polynomial $A(z)$ given by (3.6.16) has all its roots inside the unit circle if and only if the coefficients K_m satisfy the condition* $|K_m| < 1$ *for all* $m = 1, 2, \ldots, N$.

We shall not provide a proof of the Schür–Cohn test. Instead, we illustrate the computational procedure by means of the following example.

EXAMPLE 3.6.7 _____

Determine if the system having the system function

$$H(z) = \frac{1}{1 - \frac{7}{4}z^{-1} - \frac{1}{2}z^{-2}}$$

is stable.

Solution: We begin with $A_2(z)$, which is defined as

$$A_2(z) = 1 - \frac{7}{4}z^{-1} - \frac{1}{2}z^{-2}$$

Hence

$$K_2 = -\frac{1}{2}$$

Now

$$B_2(z) = -\frac{1}{2} - \frac{7}{4}z^{-1} + z^{-2}$$

and

$$A_1(z) = \frac{A_2(z) - K_2 B_2(z)}{1 - K_2^2}$$

$$= 1 - \tfrac{7}{2}z^{-1}$$

Therefore,

$$K_1 = -\tfrac{7}{2}$$

Since $|K_1| > 1$ it follows that the system is unstable. This fact is easily established in this example, since the denominator is easily factored to yield the two poles at $p_1 = -2$ and $p_2 = \tfrac{1}{4}$. However, for higher-degree polynomials, the Schür–Cohn test provides a simpler test for stability than direct factoring of $H(z)$.

The Schür–Cohn stability test can be easily programmed in a digital computer and it is very efficient in terms of arithmetic operations. Specifically, it requires only N^2 multiplications to determine the coefficients $\{K_m\}$, $m = 1, 2, \ldots, N$. The recursive equation in (3.6.20) can be expressed in terms of the polynomial coefficients by expanding the polynomials in both sides of (3.6.20) and equating the coefficients corresponding to equal powers. Indeed, it is easily established that (3.6.20) is equivalent to the following algorithm: Set

$$\alpha_N(k) = \alpha_k \qquad k = 1, 2, \ldots, N \tag{3.6.22}$$

$$K_N = \alpha_N(N) \tag{3.6.23}$$

Then, for $m = N, N - 1, \ldots, 1$, compute

$$K_m = \alpha_m(m) \qquad \alpha_{m-1}(0) = 1$$

and

$$\alpha_{m-1}(k) = \frac{\alpha_m(k) - K_m b_m(k)}{1 - K_m^2} \qquad k = 1, 2, \ldots, m - 1 \tag{3.6.24}$$

where

$$b_m(k) = \alpha_m(m - k) \qquad k = 0, 1, \ldots, m \tag{3.6.25}$$

A FORTRAN subroutine that implements the algorithm above for the Schür–Cohn stability test is given in Fig. 3.15. Note that the algorithm breaks down if one of the $\{K_m\}$ coefficients becomes exactly equal to ± 1. The parameter IER in the subroutine provides the information on the occurrence of this event.

The recursive algorithm given above for the computation of the coefficients $\{K_m\}$ finds application in signal processing problems and, especially, in speech signal processing.

3.6.8 Second-Order Systems

In this section we provide a detailed analysis of a system having two poles. As we shall see in Chapter 7, two-pole systems form the basic building blocks for the realization of higher-order systems.

Let us consider a causal two-pole system described by the second-order difference equation

$$y(n) = -a_1 y(n - 1) - a_2 y(n - 2) + b_0 x(n) \tag{3.6.26}$$

```
C
C      SCHUR - COHN  STABILITY TEST
C
C      INPUT  : A(1),A(2),...A(IP)
C      OUTPUT : RK(1),RK(2),...RK(IP)
C
       SUBROUTINE SCHUR (A,B,RK,IP,IER)
       DIMENSION A(1),B(1),RK(1)
       IER=0
       IP2=IP+2
       DO 10 M=1,IP
         RK(M)=A(M)
    10 CONTINUE
       DO 30 L=2,IP
         M=IP2-1
         DO 20 J=1,M
           B(J)=RK(M-J)
    20    CONTINUE
         IF (ABS(RK(M)).GE.1.0) IER=1
         HELP=1.0-RK(M)*RK(M)
         HELP=1.0/HELP
         M1=M-1
         DO 30 J=1,M1
           RK(J)=RK(J)-RK(M)*B(J)
           RK(J)=HELP*RK(J)
    30 CONTINUE
       RETURN
       END
```

FIGURE 3.15   FORTRAN subroutine for Schür–Cohn test.

The system function is

$$H(z) = \frac{Y(z)}{X(z)} = \frac{b_0}{1 + a_1 z^{-1} + a_2 z^{-1}}$$

(3.6.27)

$$= \frac{b_0 z^2}{z^2 + a_1 z + a_2}$$

This system has two zeros at the origin and poles at

$$p_1, p_2 = -\frac{a_1}{2} \pm \sqrt{\frac{a_1^2 - 4a_2}{2}}$$

(3.6.28)

The system is BIBO stable if the poles lie inside the unit circle, that is, if $|p_1| < 1$ and $|p_2| < 1$. These conditions can be related to the values of the coefficients a_1 and a_2. In particular, the roots of a quadratic equation satisfy the relations

$$a_1 = -(p_1 + p_2)$$ (3.6.29)
$$a_2 = p_1 p_2$$ (3.6.30)

From (3.6.29) and (3.6.30) we easily obtain the conditions that a_1 and a_2 must satisfy for stability. First, a_2 must satisfy the condition

$$|a_2| = |p_1 p_2| = |p_1| \, |p_2| < 1$$ (3.6.31)

The condition for a_1 may be expressed as

$$|a_1| < 1 + a_2$$ (3.6.32)

The conditions in (3.6.31) and (3.6.32) can also be derived from the Schür–Cohn stability test. From the recursive equations in (3.6.22) through (3.6.25), we find that

$$K_1 = \frac{a_1}{1 + a_2} \tag{3.6.33}$$

and

$$K_2 = a_2 \tag{3.6.34}$$

The system is stable if and only if $|K_1| < 1$ and $|K_2| < 1$. Consequently,

$$-1 < a_2 < 1$$

or equivalently $|a_2| < 1$, which agrees with (3.6.31). Also,

$$-1 < \frac{a_1}{1 + a_2} < 1$$

or, equivalently,

$$a_1 < 1 + a_2$$
$$a_1 > -1 - a_2$$

which are in agreement with (3.6.32). Therefore, a two-pole system is stable if and only if the coefficients a_1 and a_2 satisfy the conditions in (3.6.31) and (3.6.32).

The stability conditions given in (3.6.31) and (3.6.32) define a region in the coefficient plane (a_1, a_2) which is in the form of a triangle, as shown in Fig. 3.16. The system is stable if and only if the point (a_1, a_2) lies inside the triangle, which we call the *stability triangle*.

The characteristics of the two-pole system depend on the location of the poles or, equivalently, on the location of the point (a_1, a_2) in the stability triangle. The poles of the system may be real or complex conjugate, depending on the value of the discriminant $\Delta = a_1^2 - 4a_2$. The parabola $a_2 = a_1^2/4$ splits the stability traingle into two regions, as illustrated in Fig. 3.16. The region below the parabola $(a_1^2 > 4a_2)$ corresponds to real

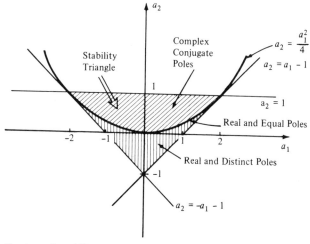

FIGURE 3.16 Region of stability (stability triangle) in the (a_1, a_2) coefficient plane for a second-order system.

and distinct poles. The points on the parabola ($a_1^2 = 4a_2$) result in real and equal (double) poles. Finally, the points above the parabola correspond to complex-conjugate poles.

Additional insight into the behavior of the system can be obtained from the unit sample responses for these three cases.

Real and Distinct Poles ($a_1^2 > 4a_2$). Since p_1, p_2 are real and $p_1 \neq p_2$, the system function may be expressed in the form

$$H(z) = \frac{A_1}{1 - p_1 z^{-1}} + \frac{A_2}{1 - p_2 z^{-1}} \tag{3.6.35}$$

where

$$A_1 = \frac{b_0 p_1}{p_1 - p_2} \qquad A_2 = \frac{-b_0 p_2}{p_1 - p_2} \tag{3.6.36}$$

Consequently, the unit sample response is

$$h(n) = \frac{b_0}{p_1 - p_2} (p_1^{n+1} - p_2^{n+1}) u(n) \tag{3.6.37}$$

Therefore, the unit sample response is the difference of two decaying exponential sequences. Figure 3.17 illustrates a typical graph for $h(n)$ when the poles are distinct.

Real and Equal Poles ($a_1^2 = 4a_2$). In this case $p_1 = p_2 = p = -a_1/2$. The system function is

$$H(z) = \frac{b_0}{(1 - pz^{-1})^2} \tag{3.6.38}$$

and hence the unit sample response of the system is

$$h(n) = b_0(n + 1)p^n u(n) \tag{3.6.39}$$

We observe that $h(n)$ is the product of a ramp sequence and a real decaying exponential sequence. The graph of $h(n)$ is shown in Fig. 3.18.

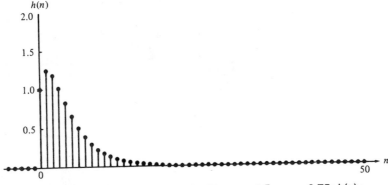

FIGURE 3.17 Plot of $h(n)$ given by (3.6.37) with $p_1 = 0.5$, $p_2 = 0.75$; $h(n) = [1/(p_1 - p_2)](p_1^{n+1} - p_2^{n+1})u(n)$.

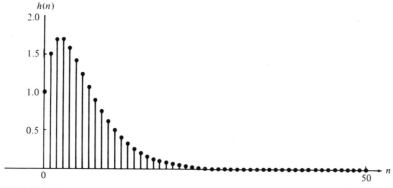

FIGURE 3.18 Plot of $h(n)$ given by (3.6.39) with $p = \frac{3}{4}$; $h(n) = (n + 1)p^n u(n)$.

Complex-Conjugate Poles $(a_1^2 < 4a_2)$. Since the poles are complex conjugate, the system function can be factored and expressed as

$$H(z) = \frac{A}{1 - pz^{-1}} + \frac{A^*}{1 - p^*z^{-1}}$$

$$= \frac{A}{1 - re^{j\omega_0}z^{-1}} + \frac{A^*}{1 - re^{-j\omega_0}z^{-1}} \qquad (3.6.40)$$

where $p = re^{j\omega}$ and $0 < \omega_0 < \pi$. Note that when the poles are complex conjugates, the parameters a_1 and a_2 are related to r and ω_0 according to

$$a_1 = -2r \cos \omega_0 \qquad (3.6.41)$$
$$a_2 = r^2$$

The constant A in the partial-fraction expansion of $H(z)$ is easily shown to be

$$A = \frac{b_0 p}{p - p^*} = \frac{b_0 re^{j\omega_0}}{r(e^{j\omega_0} - e^{-j\omega_0})}$$

$$= \frac{b_0 e^{j\omega_0}}{j2 \sin \omega_0} \qquad (3.6.42)$$

Consequently, the unit sample response of a system with complex-conjugate poles is

$$h(n) = \frac{b_0 r^n}{\sin \omega_0} \frac{e^{j(n+1)\omega_0} - e^{-j(n+1)\omega_0}}{2j} u(n)$$

$$= \frac{b_0 r^n}{\sin \omega_0} \sin (n + 1)\omega_0 u(n) \qquad (3.6.43)$$

In this case $h(n)$ has an oscillatory behavior with an exponentially decaying envelope when $r < 1$. The angle ω_0 of the poles determines the frequency of oscillation and the distance r of the poles from the origin determines the rate of decay. When r is close to unity, the decay is slow. When r is close to the origin, the decay is fast. A typical graph of $h(n)$ is illustrated in Fig. 3.19.

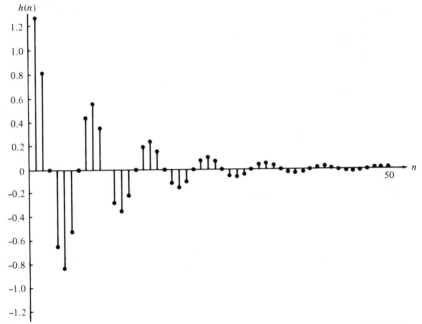

FIGURE 3.19 Plot of $h(n)$ given by (3.6.43) with $b_0 = 1$, $\omega_0 = \pi/4$, $r = 0.9$; $h(n) = [b_0 r^n/(\sin \omega_0)] \sin [(n + 1)\omega_0] u(n)$.

3.7 Summary and References

The z-transform plays the same role in discrete-time signals and systems as the Laplace transform does in continuous-time signals and systems. In this chapter we derived the important properties of the z-transform, which are extremely useful in the analysis of discrete-time systems. Of particular importance is the convolution property, which transforms the convolution of two sequences into a product of their z-transforms.

In the context of LTI systems, the convolution property results in the product of the z-transform $X(z)$ of the input signal with the system function $H(z)$, where the latter is the z-transform of the unit sample response of the system. This relationship allows us to determine the output of an LTI system to an input with transform $X(z)$ by computing the product $Y(z) = H(z)X(z)$ and then determining the inverse z-transform of $Y(z)$ to obtain the output sequence $y(n)$.

We observed that many signals of practical interest have rational z-transforms. Moreover, LTI systems characterized by constant-coefficient linear difference equations also possess rational system functions. Consequently, in determining the inverse of z-transform, we naturally emphasized the inversion of rational transforms. For such transforms, the partial-fraction expansion method is relatively easy to apply, in conjunction with the ROC, to determine the corresponding sequence in the time domain. The one-sided z-transform was introduced and used to solve for the response of causal systems excited by causal input signals with nonzero initial conditions.

Finally, we considered the characterization of LTI systems in the z-transform domain. In particular, we related the pole–zero locations of a system to its time-domain characteristics and restated the requirements for stability and causality of LTI systems in terms of the pole locations. We demonstrated that a causal system has a system function $H(z)$ with a ROC $|z| > r_1$, where $0 < r_1 \leq \infty$. In a stable and causal system, the poles of $H(z)$ lie inside the unit circle. On the other hand, if the system is noncausal, the condition for stability requires that the unit circle be contained in the ROC of $H(z)$. Hence a noncausal stable LTI system will have a system function with poles both inside and outside the unit circle with an annular ROC that includes the unit circle. The Schür–Cohn test for the stability of a causal LTI system was described and a software program was presented for computing the reflection coefficients $\{K_i\}$ from which the stability of the system is determined.

An excellent comprehensive treatment of the z-transform and its application to the analysis of LTI systems is given in the text by Jury (1964). The Schür–Cohn test for stability is treated in several texts. Our presentation was given in the context of reflection coefficients which are used in linear predictive coding of speech signals. The text by Markel and Gray (1976) is a good reference for the Schür–Cohn test and its application to speech signal processing.

PROBLEMS

3.1 Determine the z-transform of the following signals.

(a) $x(n) = \left\{ 3, 0, 0, 0, 0, 6, 1, -4 \right\}$
$\qquad\qquad\qquad\quad \uparrow$

(b) $x(n) = \begin{cases} (\frac{1}{2})^n & n \geq 5 \\ 0 & n \leq 4 \end{cases}$

3.2 Compute the z-transforms of the following signals and sketch the corresponding pole–zero patterns.
(a) $x(n) = (1 + n)u(n)$
(b) $x(n) = (a^n + a^{-n})u(n)$, a real
(c) $x(n) = (-1)^n 2^{-n} u(n)$
(d) $x(n) = (na^n \sin \omega_0 n)u(n)$
(e) $x(n) = (na^n \cos \omega_0 n)u(n)$
(f) $x(n) = Ar^n \cos (\omega_0 n + \phi)u(n)$, $0 < r < 1$
(g) $x(n) = \frac{1}{2}(n^2 + n)(\frac{1}{3})^{n-1}u(n - 1)$
(h) $x(n) = (\frac{1}{2})^n[u(n) - u(n - 10)]$

3.3 Compute the z-transforms and sketch the ROC of the following signals.
(a) $x_1(n) = \begin{cases} (\frac{1}{3})^n & n \geq 0 \\ (\frac{1}{2})^{-n} & n < 0 \end{cases}$

(b) $x_2(n) = \begin{cases} (\frac{1}{3})^n - 2^n & n \geq 0 \\ 0 & n < 0 \end{cases}$

(c) $x_3(n) = x_1(n + 4)$
(d) $x_4(n) = x_1(-n)$

3.4 Determine the z-transform of the following signals.
(a) $x(n) = n(-1)^n u(n)$
(b) $x(n) = n^2 u(n)$
(c) $x(n) = -na^n u(-n - 1)$

(d) $x(n) = (-1)^n \left(\cos \dfrac{\pi}{3} n \right) u(n)$

(e) $x(n) = (-1)^n u(n)$
(f) $x(n) = \{1, 0, -1, 0, 1, -1, \ldots\}$

3.5 (a) Determine the z-transform of the signal

$$x(n) = \alpha^{|n|} \qquad |\alpha| < 1$$

(b) What is the z-transform of the constant signal $x(n) = 1$, $-\infty < n < \infty$?

3.6 Express the z-transform of

$$y(n) = \sum_{k=-\infty}^{n} x(k)$$

in terms of $X(z)$. [*Hint:* Find the difference $y(n) - y(n-1)$.]

3.7 Compute the convolution of the following signals by means of the z-transform.

$$x_1(n) = \begin{cases} (\frac{1}{3})^n & u \geq 0 \\ (\frac{1}{2})^{-n} & n < 0 \end{cases}$$

$$x_2(n) = (\frac{1}{2})^n u(n)$$

3.8 Use the convolution property to:
(a) Express the z-transform of

$$y(n) = \sum_{k=-\infty}^{n} x(k)$$

in terms of $X(z)$.
(b) Determine the z-transform of $x(n) = (n + 1)u(n)$. [*Hint:* Show first that $x(n) = u(n) * u(n)$.]

3.9 The z-transform $X(z)$ of a real signal $x(n)$ includes a pair of complex-conjugate zeros and a pair of complex-conjugate poles. What happens to these pairs if we multiply $x(n)$ by $e^{j\omega_0 n}$? (*Hint:* Use the scaling theorem in the z-domain.)

3.10 Apply the final value theorem to determine $x(\infty)$ for the signal

$$x(n) = \begin{cases} 1 & \text{if } n \text{ is even} \\ 0 & \text{otherwise} \end{cases}$$

3.11 Using long-division, determine the inverse z-transform of

$$X(z) = \frac{1 + 2z^{-1}}{1 - 2z^{-1} + z^{-2}}$$

if **(a)** $x(n)$ is causal and **(b)** $x(n)$ is anticausal.

3.12 Determine the causal signal $x(n)$ having the z-transform

$$X(z) = \frac{1}{(1 - 2z^{-1})(1 - z^{-1})^2}$$

3.13 Let $x(n)$ be a sequence with z-transform $X(z)$. Determine, in terms of $X(z)$, the z-transforms of the following signals.

(a) $x_1(n) = \begin{cases} x\left(\dfrac{n}{2}\right) & \text{if } n \text{ even} \\ 0 & \text{if } n \text{ odd} \end{cases}$

(b) $x_2(n) = x(2n)$

3.14 Determine the causal signal $x(n)$ if its z-transform $X(z)$ is given by:

(a) $X(z) = \dfrac{1 + 3z^{-1}}{1 + 3z^{-1} + 2z^{-2}}$

(b) $X(z) = \dfrac{1}{1 - z^{-1} + \frac{1}{2}z^{-2}}$

(c) $X(z) = \dfrac{z^{-6} + z^{-7}}{1 - z^{-1}}$

(d) $X(z) = \dfrac{1 + 2z^{-2}}{1 + z^{-2}}$

(e) $X(z) = \dfrac{1}{4}\dfrac{1 + 6z^{-1} + z^{-2}}{(1 - 2z^{-1} + 2z^{-2})(1 - 0.5z^{-1})}$

(f) $X(z) = \dfrac{2 - 1.5z^{-1}}{1 - 1.5z^{-1} + 0.5z^{-2}}$

(g) $X(z) = \dfrac{1 + 2z^{-1} + z^{-2}}{1 + 4z^{-1} + 4z^{-2}}$

(h) $X(z)$ is specified by a pole–zero pattern in Fig. P3.14. The constant $G = \frac{1}{4}$.

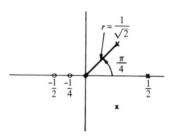

FIGURE P3.14

(i) $X(z) = \dfrac{1 - \frac{1}{4}z^{-1}}{1 + \frac{1}{2}z^{-1}}$

3.15 Determine all possible signals $x(n)$ associated with the z-transform

$$X(z) = \dfrac{5z^{-1}}{(1 - 2z^{-1})(3 - z^{-1})}$$

3.16 Determine the convolution of the following pairs of signals by means of the z-transform.

(a) $x_1(n) = (\frac{1}{4})^n u(n - 1), \quad x_2(n) = [1 + (\frac{1}{2})^n]u(n)$
(b) $x_1(n) = u(n), \quad x_2(n) = \delta(n) + (\frac{1}{2})^n u(n)$
(c) $x_1(n) = (\frac{1}{2})^n u(n), \quad x_2(n) = \cos \pi n\, u(n)$
(d) $x_1(n) = nu(n), \quad x_2(n) = 2^n u(n - 1)$

3.17 Prove the final value theorem for the one-sided z-transform.

3.18 If $X(z)$ is the z-transform of $x(n)$, show that:

(a) $Z\{x^*(n)\} = X^*(z^*)$
(b) $\operatorname{Re}[x(n)] = \frac{1}{2}[X(z) + X^*(z^*)]$
(c) $\operatorname{Im}[x(n)] = \dfrac{1}{2j}[X(z) - X^*(z^*)]$

(d) If

$$x_k(n) = \begin{cases} x\left(\dfrac{n}{k}\right) & \text{if } n/k \text{ integer} \\ 0 & \text{otherwise} \end{cases}$$

then

$$X_k(z) = X(z^k)$$

(e) $Z\{e^{j\omega_0 n}x(n)\} = X(ze^{-j\omega_0})$

3.19 By first differentiating $X(z)$ and then using appropriate properties of the z-transform, determine $x(n)$ for the following transforms.
   (a) $X(z) = \log(1 - 2z)$, $|z| < \frac{1}{2}$
   (b) $X(z) = \log(1 - \frac{1}{2}z^{-1})$, $|z| > \frac{1}{2}$

3.20 (a) Draw the pole–zero pattern for the signal

$$x_1(n) = (r^n \sin \omega_0 n)u(n) \qquad 0 < r < 1$$

   (b) Compute the z-transform, $X_2(z)$, which corresponds to the pole–zero pattern in part (a).
   (c) Compare $X_1(z)$ with $X_2(z)$. Are they indentical? If not, indicate a method to derive $X_1(z)$ from the pole–zero pattern.

3.21 Show that the roots of a polynomial with real coefficients are real or form complex-conjugate pairs. The inverse is not true, in general.

3.22 Prove the convolution and correlation properties of the z-transform using only its definition.

3.23 Determine the signal $x(n)$ with z-transform

$$X(z) = e^z + e^{1/z} \qquad |z| \neq 0$$

3.24 Determine, in closed form, the causal signals $x(n)$ whose z-transforms are given by:

   (a) $X(z) = \dfrac{1}{1 + 1.5z^{-1} - 0.5z^{-2}}$

   (b) $X(z) = \dfrac{1}{1 - 0.5z^{-1} + 0.6z^{-2}}$

   Check partially your results by computing $x(0)$, $x(1)$, $x(2)$, and $x(\infty)$ by an alternative method.

3.25 Determine all possible signals that can have the following z-transforms.

   (a) $X(z) = \dfrac{1}{1 - 1.5z^{-1} + 0.5z^{-2}}$

   (b) $X(z) = \dfrac{1}{1 - \frac{1}{2}z^{-1} + \frac{1}{4}z^{-2}}$

3.26 Determine the signal $x(n)$ with z-transform

$$X(z) = \dfrac{3}{1 - \frac{10}{3}z^{-1} + z^{-2}}$$

if $X(z)$ converges on the unit circle.

3.27 Let $x(n)$, $0 \le n \le N - 1$ be a finite-duration sequence, which is also real-valued and even. Show that the zeros of the polynomial $X(z)$ occur in mirror-image pairs about the unit circle; that is, if $z = re^{j\theta}$ is a zero of $X(z)$, then $z = (1/r)e^{j\theta}$ is also a zero.

3.28 Compute the convolution of the following pair of signals in the time domain and by using the one-sided z-transform.

(a) $x_1(n) = \left\{ 1, 1, 1, 1, 1 \atop \uparrow \right\}$

$\qquad x_2(n) = \left\{ 1, 1, 1 \atop \uparrow \right\}$

(b) $x_1(n) = (\frac{1}{2})^n u(n)$, $\quad x_2(n) = (\frac{1}{3})^n u(n)$

(c) $x_1(n) = \left\{ 1, 2, 3, 4 \atop \uparrow \right\}$

$\qquad x_2(n) = \left\{ 4, 3, 2, 1 \atop \uparrow \right\}$

(d) $x_1(n) = \left\{ 1, 1, 1, 1, 1 \atop \uparrow \right\}$

$\qquad x_2(n) = \left\{ 1, 1, 1 \atop \uparrow \right\}$

Did you obtain the same results by both methods? Explain.

3.29 Determine the one-sided z-transform of the constant signal $x(n) = 1$, $-\infty < n < \infty$.

3.30 Prove that the Fibonacci sequence can be thought of as the impulse response of the system described by the difference equation $y(n) = y(n - 1) + y(n - 2) + x(n)$. Then determine $h(n)$ using z-transform techniques.

3.31 Use the one-sided z-transform to determine $y(n)$, $n \ge 0$ in the following cases.
(a) $y(n) + \frac{1}{2}y(n - 1) - \frac{1}{4}y(n - 2) = 0$; $\quad y(-1) = y(-2) = 1$
(b) $y(n) - 1.5y(n - 1) + 0.5y(n - 2) = 0$; $\quad y(-1) = 1, y(-2) = 0$
(c) $y(n) = \frac{1}{2}y(n - 1) + x(n)$
$\qquad x(n) = (\frac{1}{3})^n u(n)$, $\quad y(-1) = 1$
(d) $\quad y(n) = \frac{1}{4}y(n - 2) + x(n)$
$\qquad x(n) = u(n)$
$\qquad y(-1) = 0$; $\quad y(-2) = 1$

3.32 Show that the following systems are equivalent.
(a) $y(n) = 0.2y(n - 1) + x(n) - 0.3x(n - 1) + 0.02x(n - 2)$
(b) $y(n) = x(n) - 0.1x(n - 1)$

3.33 Compute the unit step response of the system with impulse response

$$h(n) = \begin{cases} 3^n & n < 0 \\ (\frac{2}{5})^n & n \geq 0 \end{cases}$$

3.34 Compute the zero-state response for the following pairs of systems and input signals.

(a) $h(n) = (\frac{1}{3})^n u(n)$, $x(n) = (\frac{1}{2})^n \left(\cos \frac{\pi}{3} n \right) u(n)$

(b) $h(n) = (\frac{1}{2})^n u(n)$, $x(n) = (\frac{1}{3})^n u(n) + (\frac{1}{2})^{-n} u(-n-1)$

(c) $y(n) = -0.1y(n-1) + 0.2y(n-2) + x(n) + x(n-1)$
$x(n) = (\frac{1}{3})^n u(n)$

(d) $y(n) = \frac{1}{2}x(n) - \frac{1}{2}x(n-1)$
$x(n) = 10 \left(\cos \frac{\pi}{2} n \right) u(n)$

(e) $y(n) = -y(n-2) + 10x(n)$
$x(n) = 10 \left(\cos \frac{\pi}{2} n \right) u(n)$

(f) $h(n) = (\frac{2}{5})^n u(n)$, $x(n) = u(n) - u(n-7)$

(g) $h(n) = (\frac{1}{2})^n u(n)$, $x(n) = (-1)^n$, $-\infty < n < \infty$

(h) $h(n) = (\frac{1}{2})^n u(n)$, $x(n) = (n+1)(\frac{1}{4})^n u(n)$

3.35 Consider the system

$$H(z) = \frac{1 - 2z^{-1} + 2z^{-2} - z^{-3}}{(1 - z^{-1})(1 - 0.5z^{-1})(1 - 0.2z^{-1})} \qquad \text{ROC: } 0.5 < |z| < 1$$

(a) Sketch the pole–zero pattern. Is the system stable?

(b) Determine the impulse response of the system.

3.36 Compute the response of the system

$$y(n) = 0.7y(n-1) - 0.12y(n-2) + x(n-1) + x(n-2)$$

to the input $x(n) = nu(n)$. Is the system stable?

3.37 Determine the impulse response, $h(n)$, of the system shown in Fig. P3.37 if $h_1(n) = (\frac{1}{3})^n u(n)$, $h_2(n) = (\frac{1}{2})^n u(n)$, and $h_3(n) = (\frac{1}{5})^n u(n)$.

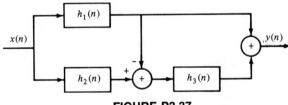

FIGURE P3.37

3.38 Determine the impulse response and the step response of the following causal systems. Plot the pole–zero patterns and determine which of the systems are stable.

(a) $y(n) = \frac{3}{4}y(n-1) - \frac{1}{8}y(n-2) + x(n)$

(b) $y(n) = y(n-1) - 0.5y(n-2) + x(n) + x(n-1)$

(c) $H(z) = \dfrac{z^{-1}(1 + z^{-1})}{(1 - z^{-1})^3}$

(d) $y(n) = 0.6y(n-1) - 0.08y(n-2) + x(n)$

(e) $y(n) = 0.7y(n-1) - 0.1y(n-2) + 2x(n) - x(n-2)$

3.39 We want to design a causal discrete-time LTI system with the property that if the input is

$$x(n) = (\tfrac{1}{2})^n u(n) - \tfrac{1}{4}(\tfrac{1}{2})^{n-1} u(n-1)$$

then the output is

$$y(n) = (\tfrac{1}{3})^n u(n)$$

(a) Determine the impulse response $h(n)$ and the system function $H(z)$ of a system that satisfies the foregoing conditions.
(b) Find the difference equation that characterizes this system.
(c) Determine a realization of the system that requires the minimum possible amount of memory.
(d) Determine if the system is stable.

3.40 Determine the stability region for the causal system

$$H(z) = \frac{1}{1 + a_1 z^{-1} + a_2 z^{-2}}$$

by computing its poles and restricting them to be inside the unit circle.

3.41 Consider the interconnection of systems shown in Fig. P3.41, where $h(n) = a^n u(n)$, $-1 < a < 1$.

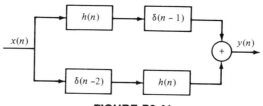

FIGURE P3.41

(a) Determine the impulse response of the overall system and determine if it is causal and stable.
(b) Draw a realization of the system using the minimum number of adders, multipliers, and delay elements.

3.42 Consider the system

$$H(z) = \frac{z^{-1} + \tfrac{1}{2}z^{-2}}{1 - \tfrac{3}{5}z^{-1} + \tfrac{2}{25}z^{-2}}$$

Determine:
(a) The impulse response
(b) The zero-state step response
(c) The step response if $y(-1) = 1$ and $y(-2) = 2$

3.43 Determine the system function, impulse response, and zero-state step response of the system shown in Fig P3.43.

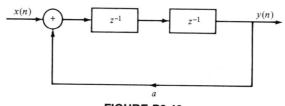

FIGURE P3.43

3.44 Consider the causal system

$$y(n) = -a_1 y(n-1) + b_0 x(n) + b_1 x(n-1)$$

Determine:
(a) The impulse response
(b) The zero-state step response
(c) The step response if $y(-1) = A \neq 0$
(d) The response to the input

$$x(n) = \cos \omega_0 n \qquad 0 \le n < \infty$$

3.45 Determine the zero-state response of the system

$$y(n) = \tfrac{1}{2} y(n-1) + 4x(n) + 3x(n-1)$$

to the input

$$x(n) = e^{j\omega_0 n} u(n)$$

What is the steady-state response of the system?

3.46 Consider the causal system defined by the pole–zero pattern shown in Fig. P3.46.

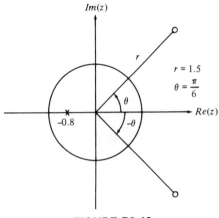

FIGURE P3.46

(a) Determine the system function and the impulse response of the system given that $H(z)|_{z=1} = 1$.
(b) Is the system stable?
(c) Sketch a possible implementation of the system and determine the corresponding difference equations.

3.47 An FIR LTI system has an impulse response $h(n)$, which is real valued, even, and has finite duration $2N + 1$. Show that if $z_1 = re^{j\omega_0}$ is a zero of the system, then $z_1 = (1/r)e^{j\omega_0}$ is also a zero.

CHAPTER 4

Frequency Analysis of Signals

The z-transform is one of several mathematical tools that is useful in the analysis and design of LTI systems. Another set of useful mathematical tools developed in this chapter is the Fourier series and Fourier transform. These signal representations basically involve the decomposition of the signals in terms of sinusoidal (or complex exponential) components. With such a decomposition, a signal is said to be represented in the *frequency domain*.

As we shall demonstrate, most signals of practical interest can be decomposed into a sum of sinusoidal signal components. For the class of periodic signals, such a decomposition is called a *Fourier series*. For the class of finite energy signals, the decomposition is called the *Fourier transform*. These decompositions are extremely important in the analysis of LTI systems, as will be shown in Chapter 5, because the response of an LTI system to a sinusoidal input signal is a sinusoid of the same frequency but of different amplitude and phase. Furthermore, the linearity property of the LTI system implies that a linear sum of sinusoidal components at the input produces a similar linear sum of sinusoidal components at the output, which differ only in the amplitudes and phases from the input sinusoids. This characteristic behavior of LTI systems renders the sinusoidal decompositions of signals very important. Although many other decompositions of signals are possible, only the class of sinusoidal (or complex exponential) signals possess this desirable property in passing through an LTI system.

We begin our study of frequency analysis of signals with the representation of continuous-time periodic and aperiodic signals by means of the Fourier series and the Fourier transform, respectively. This is followed by a parallel treatment of discrete-time periodic and aperiodic signals. The properties of the Fourier transform are described in detail and a number of time-frequency dualities are presented. We conclude the chapter with a discussion of sampling of signals in the time and frequency domain. This discussion leads us naturally to the treatment of the Discrete Fourier Transform (DFT) for finite-duration sequences. The DFT plays a very important role not only in the frequency analysis of digital signals but also in other applications, as we shall observe in subsequent chapters.

4.1 Frequency Analysis of Analog Signals

It is well known that a prism can be used to break up white light (sunlight) into the colors of the rainbow (see Fig. 4.1a). In a paper submitted in 1672 to the Royal Society, Isaac Newton used the term *spectrum* to describe the *continuous* bands of colors produced by this apparatus. To understand this phenomenon, Newton placed another prism upside-down with respect to the first and showed that the colors blended back into white light,

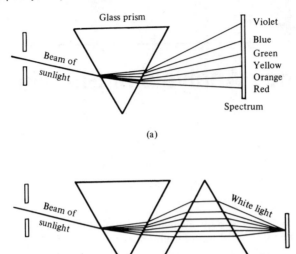

(a)

(b)

FIGURE 4.1   (a) Analysis and (b) synthesis of the white light (sunlight) using glass prisms.

as in Fig. 4.1b. By inserting a slit between the two prisms and blocking one or more colors from hitting the second prism, he showed that the remixed light is no longer white. Hence the light passing through the first prism is simply analyzed into its component colors without any other change. However, only if we mix again all of these colors do we obtain the original white light.

Later, Joseph Fraunhofer (1787–1826), in making measurements of light emitted by the sun and stars, discovered that the spectrum of the observed light consists of distinct color lines. A few years later (mid-1800s) Gustav Kirchhoff and Robert Bunsen found that each chemical element, when heated to incandescence, radiates its own distinct color of light. As a consequence, each chemical element can be identified by its own *line spectrum*.

From physics we know that each color corresponds to a specific frequency of the visible spectrum. Hence the analysis of light into colors is actually a form of *frequency analysis*.

Frequency analysis of a signal involves the resolution of the signal into its frequency (sinusoidal) components. Instead of light, our signal waveforms are basically functions of time. The role of the prism is played by the Fourier analysis tools that we develop below: the Fourier series and the Fourier transform. The recombination of the sinusoidal components to reconstruct the original signal is basically a Fourier synthesis problem. The problem of signal analysis is basically the same for the case of a signal waveform and for the case of the light from heated chemical compositions. Just as in the case of chemical compositions, different signal waveforms have different spectra. Thus the spectrum of a signal provides an "identity" or a signature for the signal in the sense that no other signal has the same spectrum. As we will see below, this attribute is related to the mathematical viewpoint of frequency-domain techniques.

If we decompose a waveform into sinusoidal components, in much the same way that a prism separates white light into different colors, the sum of these sinusoidal components results in the original waveform. On the other hand, if any of these components is missing, the result is a different signal.

In our treatment of frequency analysis we will develop the proper mathematical tools ("prisms") for the decomposition of signals ("light") into sinusoidal frequency components (colors). Furthermore, the tools ("inverse prisms") for synthesis of a given signal from its frequency components will also be developed.

The basic motivation for developing the frequency analysis tools is to provide a mathematical and pictorial representation for the frequency components that are contained in any given signal. As in physics, the term *spectrum* is used when referring to the frequency content of a signal. The process of obtaining the spectrum of a given signal using the basic mathematical tools described in this chapter is known as *frequency* or *spectral analysis*. In contrast, the process of determining the spectrum of a signal in practice, based on actual measurements of the signal, is called *spectrum estimation*. This distinction is very important. In a practical problem the signal to be analyzed does not lend itself to an exact mathematical description. The signal is usually some information-bearing signal from which we are attempting to extract the relevant information. If the information that we wish to extract can be obtained either directly or indirectly from the spectral content of the signal, we can perform *spectrum estimation* on the information-bearing signal and thus obtain an estimate of the signal spectrum. In fact, we may view spectral estimation as a type of spectral analysis performed on signals obtained from physical sources (e.g., speech, EEG, ECG, etc.). The instruments or software programs used to obtain spectral estimates of such signals are known as *spectrum analyzers*.

Below, we will deal with spectral analysis. However, in Chapter 11 we shall treat the subject of power spectrum estimation.

4.1.1 Frequency Analysis for Continuous-Time Periodic Signals

In this section we present the frequency analysis tools for continuous-time periodic signals. Examples of periodic signals encountered in practice are square waves, rectangular waves, triangular waves, and of course, sinusoids and complex exponentials.

The basic mathematical representation of periodic signals is the Fourier series, which is a linear weighted sum of harmonically related sinusoids or complex exponentials. Jean Baptiste Joseph Fourier (1768–1830), a French mathematician, used such trigonometric series expansions in describing the phenomenon of heat conduction and temperature distribution through bodies. Although his work was motivated by the problem of heat conduction, the mathematical techniques that he developed during the early part of the nineteenth century now find application in a variety of problems encompassing many different fields, including optics, vibrations in mechanical systems, system theory, and electromagnetics.

From Chapter 1 we recall that a linear combination of harmonically related complex exponentials of the form

$$x(t) = \sum_{k=-\infty}^{\infty} c_k e^{j2\pi k F_0 t} \tag{4.1.1}$$

is a periodic signal with fundamental period $T_p = 1/F_0$. Hence we may think of the exponential signals

$$\{e^{j2\pi k F_0 t}, \quad k = 0, \pm 1, \pm 2, \ldots\}$$

as the basic "building blocks" from which we can construct periodic signals of various types by properly choosing the fundamental frequency and the coefficients $\{c_k\}$. F_0 determines the fundamental period of $x(t)$ and the coefficients $\{c_k\}$ specify the shape of the waveform.

Suppose that we are given a periodic signal $x(t)$ with period T_p. We may represent the periodic signal by the series (4.1.1), called a *Fourier series*, where the fundamental frequency F_0 is selected to be the reciprocal of the given period T_p. To determine the expression for the coefficients $\{c_k\}$, we first multiply both sides of (4.1.1) by the complex exponential

$$e^{-j2\pi F_0 lt}$$

where l is an integer and then we integrate both sides of the resulting equation over a single period, say from 0 to T_p, or more generally, from t_0 to $t_0 + T_p$, where t_0 is an arbitrary but mathematically convenient starting value. Thus we obtain

$$\int_{t_0}^{t_0+T_p} x(t)e^{-j2\pi lF_0 t}\, dt = \int_{t_0}^{t_0+T_p} e^{-j2\pi lF_0 t}\left(\sum_{k=-\infty}^{\infty} c_k e^{+j2\pi kF_0 t}\right) dt \qquad (4.1.2)$$

To evaluate the integral on the right-hand side of (4.1.2), we interchange the order of the summation and integration and combine the two exponentials. Hence

$$\sum_{k=-\infty}^{\infty} c_k \int_{t_0}^{t_0+T_p} e^{j2\pi F_0(k-l)t}\, dt = \sum_{k=-\infty}^{\infty} c_k \left[\frac{e^{j2\pi F_0(k-l)t}}{j2\pi F_0(k-l)}\right]_{t_0}^{t_0+T_p} \qquad (4.1.3)$$

The right-hand side of (4.1.3) evaluated at the lower and upper limits, t_0 and $t_0 + T_p$, respectively, yields zero. On the other hand, if $k = l$, we have

$$\int_{t_0}^{t_0+T_p} dt = t \Big]_{t_0}^{t_0+T_p} = T_p$$

Consequently, (4.1.2) reduces to

$$\int_{t_0}^{t_0+T_p} x(t)e^{-j2\pi lF_0 t}\, dt = c_l T_p$$

and therefore the expression for the Fourier coefficients in terms of the given periodic signal becomes

$$c_l = \frac{1}{T_p}\int_{t_0}^{t_0+T_p} x(t)e^{-j2\pi lF_0 t}\, dt$$

Since t_0 is arbitrary, this integral can be evaluated over any interval of length T_p, that is, over any interval equal to the period of the signal $x(t)$. Consequently, the integral for the Fourier series coefficients will be written as

$$c_l = \frac{1}{T_p}\int_{T_p} x(t)e^{-j2\pi lF_0 t}\, dt \qquad (4.1.4)$$

An important issue that arises in the representation of the periodic signal $x(t)$ by the Fourier series is whether or not the series converges to $x(t)$ for every value of t. It is known that if $x(t)$ is continuous within a period, the Fourier series does indeed converge to $x(t)$. It is also known that if $x(t)$ has a discontinuity at some t, the Fourier series converges to the midpoint (average value) of the discontinuity.

Of particular importance is the class of periodic signals that are square integrable over a period, that is,

$$\int_{T_p} |x(t)|^2\, dt < \infty \qquad (4.1.5)$$

If this condition is satisfied, the Fourier coefficients $\{c_k\}$ are finite.

An alternative set of conditions that ensure the existence of the Fourier series for $x(t)$ is the set of *Dirichlet conditions*. These are:

1. The signal $x(t)$ has a finite number of discontinuities in any period.
2. The signal $x(t)$ contains a finite number of maxima and minima during any period.
3. The signal $x(t)$ is absolutely integrable in any period, that is,

$$\int_{T_p} |x(t)| \, dt < \infty \qquad (4.1.6)$$

All periodic signals of practical interest satisfy these conditions.

In summary, if $x(t)$ is periodic and satisfies the Dirichlet conditions, it can be represented in a Fourier series as in (4.1.1), where the coefficients are specified by (4.1.4). These relations are summarized below.

Synthesis equation

$$x(t) = \sum_{k=-\infty}^{\infty} c_k e^{j2\pi kF_0 t} \qquad (4.1.7)$$

Analysis equation

$$c_k = \frac{1}{T_p} \int_{T_p} x(t) e^{-j2\pi kF_0 t} \, dt \qquad (4.1.8)$$

In general, the Fourier coefficients c_k are complex valued. Moreover, it is easily shown that if the periodic signal is real, c_k and c_{-k} are complex conjugates. As a result, if

$$c_k = |c_k| e^{j\theta_k}$$

then

$$c_{-k} = |c_k|^{-j\theta_k}$$

Consequently, the Fourier series may also be represented in the form

$$x(t) = c_0 + 2 \sum_{k=1}^{\infty} |c_k| \cos (2\pi kF_0 t + \theta_k) \qquad (4.1.9)$$

where c_o is real valued when $x(t)$ is real.

Finally, we should indicate that yet another form for the Fourier series can be obtained by expanding the cosine function in (4.1.9) as

$$\cos (2\pi kF_0 t + \theta_k) = \cos 2\pi kF_0 t \cos \theta_k - \sin 2\pi kF_0 t \sin \theta_k$$

Consequently, we may rewrite (4.1.9) in the form

$$x(t) = a_0 + \sum_{k=1}^{\infty} (a_k \cos 2\pi kF_0 t - b_k \sin 2\pi kF_0 t) \qquad (4.1.10)$$

where

$$a_0 = c_0$$
$$a_k = 2|c_k| \cos \theta_k$$
$$b_k = 2|c_k| \sin \theta_k$$

The expressions in (4.1.7), (4.1.9), and (4.1.10) constitute three equivalent forms for the Fourier series representation of a periodic signal.

4.1.2 Power Density Spectrum of Periodic Signals

A periodic signal has infinite energy and a finite average power, which is given as

$$P_x = \frac{1}{T_p} \int_{T_p} |x(t)|^2 \, dt \tag{4.1.11}$$

If we take the complex conjugate of (4.1.7) and substitute for $x^*(t)$ in (4.1.11), we obtain

$$P_x = \frac{1}{T_p} \int_{T_p} x(t) \sum_{k=-\infty}^{\infty} c_k^* e^{-j2\pi k F_0 t} \, dt$$

$$= \sum_{k=-\infty}^{\infty} c_k^* \left[\frac{1}{T_p} \int_{T_p} x(t) e^{-j2\pi k F_0 t} \, dt \right] \tag{4.1.12}$$

$$= \sum_{k=-\infty}^{\infty} |c_k|^2$$

Therefore, we have established the relation

$$P_x = \frac{1}{T_p} \int_{T_p} |x(t)|^2 \, dt = \sum_{k=-\infty}^{\infty} |c_k|^2 \tag{4.1.13}$$

which is called *Parseval's relation* for power signals.

To illustrate the physical meaning of (4.1.13), suppose that $x(t)$ consists of a single complex exponential

$$x(t) = c_k e^{j2\pi k F_0 t}$$

In this case, all the Fourier series coefficients except c_k are zero. Consequently, the average power in the signal is

$$P_x = |c_k|^2$$

It is obvious that $|c_k|^2$ represents the power in the kth harmonic component of the signal. Hence the total average power in the periodic signal is simply the sum of the average powers in all the harmonics.

If we plot the $|c_k|^2$ as a function of the frequencies kF_0, $k = 0, \pm 1, \pm 2, \ldots$, the diagram that we obtain shows how the power of the periodic signal is distributed among the various frequency components. This diagram, which is illustrated in Fig. 4.2, is called the *power density spectrum* of the periodic signal $x(t)$. Since the power in a periodic signal exists only at discrete values of frequencies (i.e., $F = 0, \pm F_0, \pm 2F_0, \ldots$), the signal is said to have a *line spectrum*. The spacing between two consecutive spectral lines is equal to the reciprocal of the fundamental period T_p, whereas the shape of the spectrum (i.e., the power distribution of the signal), depends on the time-domain characteristics of the signal.

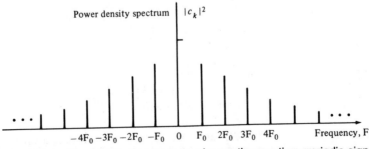

FIGURE 4.2 Power density spectrum of a continuous-time periodic signal.

As indicated in the preceding section, the Fourier series coefficients $\{c_k\}$ are complex valued, that is, they may be represented as

$$c_k = |c_k| e^{j\theta_k}$$

where

$$\theta_k = \angle c_k$$

Instead of plotting the power spectrum, we may plot the magnitude voltage spectrum $\{|c_k|\}$ and the phase spectrum $\{\theta_k\}$ as a function of frequency. Clearly, the power spectral density in the periodic signal is simply the square of the magnitude spectrum. The phase information is totally destroyed (or does not appear) in the power spectral density.

If the periodic signal is real valued, the Fourier series coefficients $\{c_k\}$ satisfy the condition

$$c_{-k} = c_k^*$$

Consequently, $|c_k|^2 = |c_k^*|^2$. Hence the power spectrum is a symmetric function of frequency. This condition also means that the magnitude spectrum is symmetric (even function) about the origin and the phase spectrum is an odd function. As a consequence of the symmetry, it is sufficient to specify the spectrum of a real periodic signal for positive frequencies only. Furthermore, the total average power may be expressed as

$$P_x = c_0^2 + 2 \sum_{k=1}^{\infty} |c_k|^2 \tag{4.1.14}$$

$$= a_0^2 + \frac{1}{2} \sum_{k=1}^{\infty} (a_k^2 + b_k^2) \tag{4.1.15}$$

which follows directly from the relationships given in Section 4.1.1 among $\{a_k\}$, $\{b_k\}$, and $\{c_k\}$ coefficients in the Fourier series expressions.

EXAMPLE 4.1.1

Determine the Fourier series and the power density spectrum of the rectangular pulse train signal illustrated in Fig. 4.3.

Solution: The signal is periodic with fundamental period T_p and, clearly, satisfies the Dirichlet conditions. Consequently, we may represent the signal in the Fourier series given by (4.1.7) with the Fourier coefficients specified by (4.1.8).

Since $x(t)$ is an even signal [i.e., $x(t) = x(-t)$], it is convenient to select the integration interval from $-T_p/2$ to $T_p/2$. Thus (4.1.8) evaluated for $k = 0$ yields

$$c_0 = \frac{1}{T_p} \int_{-T_p/2}^{T_p/2} x(t)\, dt = \frac{1}{T_p} \int_{-\tau/2}^{\tau/2} A\, dt = \frac{A\tau}{T_p} \tag{4.1.16}$$

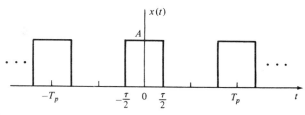

FIGURE 4.3   Continuous-time periodic train of rectangular pulses.

The term c_0 represents the average value (dc, component) of the signal $x(t)$. For $k \neq 0$ we have

$$c_k = \frac{1}{T_p} \int_{-\tau/2}^{\tau/2} A e^{-j2\pi kF_0 t} \, dt = \frac{A}{T_p} \frac{e^{-j2\pi F_0 kt}}{-j2\pi kF_0} \Bigg]_{-\tau/2}^{\tau/2}$$

$$= \frac{A}{\pi F_0 kT_p} \frac{e^{j\pi kF_0 \tau} - e^{-j\pi kF_0 \tau}}{j2} \qquad (4.1.17)$$

$$= \frac{A\tau}{T_p} \frac{\sin \pi kF_0 \tau}{\pi kF_0 \tau} \qquad k = \pm 1, \pm 2, \ldots$$

It is interesting to note that the right-hand side of (4.1.17) has the form $(\sin \phi)/\phi$, where $\phi = \pi kF_0 \tau$. In this case ϕ takes on discrete values since F_0 and τ are fixed and the index k varies. However, if we plot $(\sin \phi)/\phi$ with ϕ as a continuous parameter over the range $-\infty < \phi < \infty$, we obtain the graph shown in Fig. 4.4. We observe that this function decays to zero as $\phi \to \pm\infty$, has a maximum value of unity at $\phi = 0$, and is zero at multiples of π (i.e., at $\phi = m\pi$, $m = \pm 1, \pm 2, \ldots$). It is clear that the Fourier coefficients given by (4.1.17) are the sample values of the $(\sin \phi)/\phi$ function for $\phi = \pi kF_0 \tau$ and scaled in amplitude by $A\tau/T_p$.

Since the periodic function $x(t)$ is even, the Fourier coefficients c_k are real. Consequently, the phase spectrum is either zero, when c_k is positive, or π when c_k is negative. Instead of plotting the magnitude and phase spectra separately, we may simply plot $\{c_k\}$ on a single graph, showing both the positive and negative values c_k on the graph. This is commonly done in practice when the Fourier coefficients $\{c_k\}$ are real.

Figure 4.5 illustrates the Fourier coefficients of the rectangular pulse train when T_p is fixed and the pulse width τ is allowed to vary. In this case $T_p = 0.25$ second, so that $F_0 = 1/T_p = 4H_z$ and $\tau = 0.05T_p$, $\tau = 0.1T_p$, and $\tau = 0.2T_p$. We observe that the effect of decreasing τ while keeping T_p fixed is to spread out the signal power over the frequency range. The spacing between adjacent spectral lines is $F_0 = 4$ Hz, independent of the value of the pulse width τ.

On the other hand, it is also instructive to fix τ and vary the period T_p when $T_p > \tau$. Figure 4.6 illustrates this condition when $T_p = 5\tau$, $T_p = 10\tau$, and $T_p = 20\tau$. In this case, the spacing between adjacent spectral lines decreases as T_p increases. In the limit as $T_p \to \infty$, the Fourier coefficients c_k approach zero due to the factor of T_p in the denominator of (4.1.17). This behavior is consistent with the fact that as $T_p \to \infty$ and τ remains fixed, the resulting signal is no longer a power signal. Instead, it becomes an energy signal and its average power is zero. The spectra of finite energy signals are described in the next section.

We also note that if $k \neq 0$ and $\sin (\pi kF_0 \tau) = 0$, then $c_k = 0$. The harmonics with zero power occur at frequencies kF_0 such that $\pi(kF_0)\tau = m\pi$, $m = \pm 1, \pm 2, \ldots$, or at $kF_0 = m/\tau$. For example, if $F_0 = 4$ Hz and $\tau = 0.2T_p$, it follows that the spectral

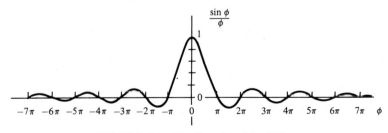

FIGURE 4.4 The function $(\sin \phi)/\phi$.

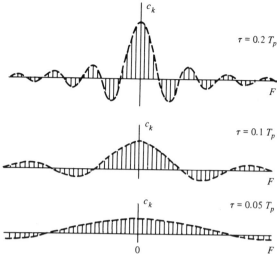

FIGURE 4.5   Fourier coefficients of the rectangular pulse train when T_p is fixed and the pulse width τ varies.

components at ± 20 Hz, ± 40 Hz, . . . have zero power. These frequencies correspond to the Fourier coefficients c_k, $k = \pm 5, \pm 10, \pm 15, \ldots$. On the other hand, if $\tau = 0.1T_p$, the spectral components with zero power are $k = \pm 10, \pm 20, \pm 30, \ldots$.

The power density spectrum for the rectangular pulse train is

$$|c_k|^2 = \begin{cases} \left(\dfrac{A\tau}{T_p}\right)^2 & k = 0 \\[3mm] \left(\dfrac{A\tau}{T_p}\right)^2 \left(\dfrac{\sin \pi kF_0\tau}{\pi kF_0\tau}\right)^2 & k = \pm 1, \pm 2, \ldots \end{cases} \tag{4.1.18}$$

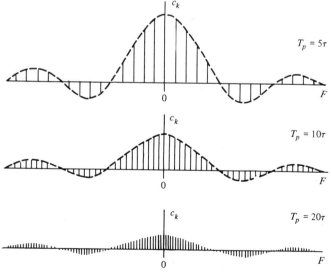

FIGURE 4.6   Fourier coefficient of a rectangular pulse train with fixed pulse width τ and varying period T_p.

4.1.3 The Fourier Transform for Continuous-Time Aperiodic Signals

In Section 4.1.1 we developed the Fourier series to represent a periodic signal as a linear combination of harmonically related complex exponentials. As a consequence of the periodicity we saw that these signals possess line spectra with equidistant lines. The line spacing is equal to the fundamental frequency, which in turn is the inverse of the fundamental period of the signal. We may view the fundamental period as providing the number of lines per unit of frequency (line density), as illustrated in Fig. 4.6.

With this interpretation in mind, it is apparent that if we allow the period to increase without limit, the line spacing tends toward zero. In the limit, when the period becomes infinite, the signal becomes aperiodic and its spectrum becomes continuous. This argument suggests that the spectrum of an aperiodic signal will be the envelope of the line spectrum in the corresponding periodic signal obtained by repeating the aperiodic signal with some period T_p.

Let us consider an aperiodic signal $x(t)$ with finite duration as shown in Fig. 4.7a. From this aperiodic signal, we can create a periodic signal $x_p(t)$ with period T_p, as shown in Fig. 4.7b. Clearly, $x_p(t) = x(t)$ in the limit as $T_p \to \infty$, that is,

$$x(t) = \lim_{T_p \to \infty} x_p(t) \tag{4.1.19}$$

This interpretation implies that we should be able to obtain the spectrum of $x(t)$ from the spectrum of $x_p(t)$ simply by taking the limit as $T_p \to \infty$.

We begin with the Fourier series representation of $x_p(t)$,

$$x_p(t) = \sum_{k=-\infty}^{\infty} c_k e^{j2\pi k F_0 t} \qquad F_0 = \frac{1}{T_p} \tag{4.1.20}$$

where

$$c_k = \frac{1}{T_p} \int_{-T_p/2}^{T_p/2} x_p(t) e^{-j2\pi k F_0 t} \, dt \tag{4.1.21}$$

Since $x_p(t) = x(t)$ for $-T_p/2 \le t \le T_p/2$, (4.1.21) may be expressed as

$$c_k = \frac{1}{T_p} \int_{-T_p/2}^{T_p/2} x(t) e^{-j2\pi k F_0 t} \, dt \tag{4.1.22}$$

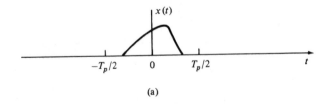

(a)

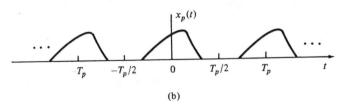

(b)

FIGURE 4.7   (a) Aperiodic signal $x(t)$ and (b) periodic signal $x_p(t)$ constructed by repeating $x(t)$ with a period T_p.

It is also true that $x(t) = 0$ for $|t| > T_p/2$. Consequently, the limits on the integral in (4.1.22) may be replaced by $-\infty$ and ∞. Hence

$$c_k = \frac{1}{T_p} \int_{-\infty}^{\infty} x(t)e^{-j2\pi kF_0t}\, dt \tag{4.1.23}$$

Let us now define a function $X(F)$, called the *Fourier transform* of $x(t)$, as

$$X(F) = \int_{-\infty}^{\infty} x(t)e^{-j2\pi Ft}\, dt \tag{4.1.24}$$

$X(F)$ is a function of the continuous variable F. It does not depend on T_p or F_0. However, if we compare (4.1.23) and (4.1.24), it is clear that the Fourier coefficients c_k can be expressed in terms of $X(F)$ as

$$c_k = \frac{1}{T_p} X(kF_0)$$

or equivalently,

$$T_p c_k = X(kF_0) = X\left(\frac{k}{T_p}\right) \tag{4.1.25}$$

Thus the Fourier coefficients are samples of $X(F)$ taken at multiples of F_0 and scaled by F_0 (multiplied by $1/T_p$). Substitution for c_k from (4.1.25) into (4.1.20) yields

$$x_p(t) = \frac{1}{T_p} \sum_{k=-\infty}^{\infty} X\left(\frac{k}{T_p}\right) e^{j2\pi kF_0t} \tag{4.1.26}$$

We wish to take the limit of (4.1.26) as T_p approaches infinity. First, we define $\Delta F = 1/T_p$. With this substitution, (4.1.26) becomes

$$x_p(t) = \sum_{k=-\infty}^{\infty} X(k\,\Delta F)e^{j2\pi k\,\Delta Ft}\,\Delta F \tag{4.1.27}$$

It is clear that in the limit as T_p approaches infinity, $x_p(t)$ reduces to $x(t)$. Also, ΔF becomes the differential dF and $k\,\Delta F$ becomes the continuous frequency variable F. In turn, the summation in (4.1.27) becomes an integral over the frequency variable F. Thus

$$\lim_{T_p \to \infty} x_p(t) = x(t) = \lim_{\Delta F \to 0} \sum_{k=-\infty}^{\infty} X(k\,\Delta F)e^{-j2\pi k\,\Delta Ft}\,\Delta F$$

$$x(t) = \int_{-\infty}^{\infty} X(F)e^{j2\pi Ft}\, dF \tag{4.1.28}$$

This integral relationship yields $x(t)$ when $X(F)$ is known, and it is called the *inverse Fourier transform*.

This concludes our heuristic derivation of the Fourier transform pair given by (4.1.24) and (4.1.28) for an aperiodic signal $x(t)$. Although the derivation is not mathematically rigorous, it has led to the desired Fourier transform relationships with relatively simple intuitive arguments. In summary, the frequency analysis of continuous-time aperiodic signals involves the following Fourier transform pair.

Synthesis equation, inverse transform

$$x(t) = \int_{-\infty}^{\infty} X(F)e^{j2\pi Ft}\, dF \tag{4.1.29}$$

Analysis equation, direct transform

$$X(F) = \int_{-\infty}^{\infty} x(t)e^{-j2\pi Ft} \, dt \tag{4.1.30}$$

It is apparent that the essential difference between the Fourier series and the Fourier transform is that the spectrum in the latter is continuous and hence the synthesis of an aperiodic signal from its spectrum is accomplished by means of integration instead of summation.

Finally, we wish to indicate that the Fourier transform pair in (4.1.29) and (4.1.30) can be expressed in terms of the radian frequency variable $\Omega = 2\pi F$. Since $dF = d\Omega/2\pi$, (4.1.29) and (4.1.30) become

$$x(t) = \frac{1}{2\pi} \int_{-\infty}^{\infty} X(\Omega)e^{j\Omega t} \, d\Omega \tag{4.1.31}$$

$$X(\Omega) = \int_{-\infty}^{\infty} x(t)e^{-j\Omega t} \, dt \tag{4.1.32}$$

The Fourier transform exists if the signal $x(t)$ has a finite energy, that is, if

$$\int_{-\infty}^{\infty} |x(t)|^2 \, dt < \infty \tag{4.1.33}$$

An alternative set of conditions that are sufficient for the existence of the Fourier transform is the *Dirichlet conditions*, which may be expressed as:

1. The signal $x(t)$ has a finite number of finite discontinuities.
2. The signal $x(t)$ has a finite number of maxima and minima.
3. The signal $x(t)$ is absolutely integrable, that is,

$$\int_{-\infty}^{\infty} |x(t)| \, dt < \infty \tag{4.1.34}$$

The third condition follows easily from the definition of the Fourier transform, given in (4.1.30). Indeed,

$$|X(F)| = \left| \int_{-\infty}^{\infty} x(t)e^{-j2\pi Ft} \, dt \right| \le \int_{-\infty}^{\infty} |x(t)| \, dt$$

Hence $|X(F)| < \infty$ if (4.1.34) is satisfied.

If a signal $x(t)$ is absolutely integrable, it will also have finite energy. That is,

$$E_x = \int_{-\infty}^{\infty} |x(t)|^2 \, dt \le \left[\int_{-\infty}^{\infty} |x(t)| \, dt \right]^2 \tag{4.1.35}$$

Consequently, if

$$\int_{-\infty}^{\infty} |x(t)| \, dt < \infty$$

then

$$\int_{-\infty}^{\infty} |x(t)|^2 \, dt < \infty$$

However, the converse is not true. That is, a signal may have finite energy but may not be absolutely integrable. For example, the signal

$$x(t) = \frac{\sin 2\pi t}{\pi t} \tag{4.1.36}$$

is square integrable but is not absolutely integrable. This signal has the Fourier transform

$$X(F) = \begin{cases} 1 & |F| \leq 1 \\ 0 & |F| > 1 \end{cases} \tag{4.1.37}$$

Since this signal violates (4.1.34), it is apparent that the Dirichlet conditions are sufficient but not necessary for the existence of the Fourier transform. In any case, nearly all finite energy signals have a Fourier transform, so that we need not worry about the pathological signals, which are seldom encountered in practice.

4.1.4 Energy Density Spectrum of Aperiodic Signals

Let $x(t)$ be any finite energy signal with Fourier transform $X(F)$. Its energy is

$$E_x = \int_{-\infty}^{\infty} |x(t)|^2 \, dt$$

which, in turn, may be expressed in terms of $X(F)$ as follows:

$$\begin{aligned} E_x &= \int_{-\infty}^{\infty} x(t)x^*(t) \, dt \\ &= \int_{-\infty}^{\infty} x(t) \, dt \left[\int_{-\infty}^{\infty} X^*(F)e^{-j2\pi Ft} \, dF \right] \\ &= \int_{-\infty}^{\infty} X^*(F) \, dF \left[\int_{-\infty}^{\infty} x(t)e^{-j2\pi Ft} \, dt \right] \\ &= \int_{-\infty}^{\infty} |X(F)|^2 \, dF \end{aligned}$$

Therefore, we conclude that

$$E_x = \int_{-\infty}^{\infty} |x(t)|^2 \, dt = \int_{-\infty}^{\infty} |X(F)|^2 \, dF \tag{4.1.38}$$

This is *Parseval's relation* for aperiodic, finite energy signals and expresses the principle of conservation of energy in the time and frequency domains.

The spectrum $X(F)$ of a signal is in general complex valued. Consequently, it is usually expressed in polar forms as

$$X(F) = |X(F)|e^{j\Theta(F)}$$

where $|X(F)|$ is the magnitude spectrum and $\Theta(F)$ is the phase spectrum,

$$\Theta(F) = \angle X(F)$$

On the other hand, the quantity

$$S_{xx}(F) = |X(F)|^2 \tag{4.1.39}$$

which is the integrand in (4.1.38), represents the distribution of energy in the signal as a function of frequency. Hence $S_{xx}(F)$ is called the *energy density spectrum* of $x(t)$. The integral of $S_{xx}(F)$ over all frequencies gives the total energy in the signal. Viewed in another way, the energy in the signal $x(t)$ over a band of frequencies $F_1 \leq F \leq F_1 + \Delta F$ is

$$\int_{F_1}^{F_1 + \Delta F} S_{xx}(F) \, dF$$

From (4.1.39) we observe that $S_{xx}(F)$ does not contain any phase information [i.e.,

$S_{xx}(F)$ is purely real and nonnegative]. Since the phase spectrum of $x(t)$ is not contained in $S_{xx}(F)$, it is impossible to reconstruct the signal given $S_{xx}(F)$.

Finally, as in the case of Fourier series, it is easily shown that if the signal $x(t)$ is real, then

$$|X(-F)| = |X(F)| \tag{4.1.40}$$
$$\angle X(-F) = -\angle X(F) \tag{4.1.41}$$

By combining (4.1.40) and (4.1.39), we obtain

$$S_{xx}(-F) = S_{xx}(F) \tag{4.1.42}$$

In other words, the energy density spectrum of a real signal has even symmetry.

EXAMPLE 4.1.2 ————————————————————————————————

Determine the Fourier transform and the energy density spectrum of a rectangular pulse signal defined as

$$x(t) = \begin{cases} A & |t| \le \tau/2 \\ 0 & |t| > \tau/2 \end{cases} \tag{4.1.43}$$

and illustrated in Fig. 4.8(a).

Solution: Clearly, this signal is aperiodic and satisfies the Dirichlet conditions. Hence its Fourier transform exists. By applying (4.1.29), we find that

$$X(F) = \int_{-\tau/2}^{\tau/2} A e^{-j2\pi F t}\, dt = A\tau \frac{\sin \pi F \tau}{\pi F \tau} \tag{4.1.44}$$

We observe that $X(F)$ is real and hence it can be depicted graphically using only one diagram, as shown in Fig. 4.8(b). Obviously, $X(F)$ has the shape of the $(\sin \phi)/\phi$ function shown in Fig. 4.4. Hence the spectrum of the rectangular pulse is the envelope of the line spectrum (Fourier coefficients) of the periodic signal obtained by periodically repeating the pulse with period T_p as in Fig. 4.3. In other words, the Fourier coefficients c_k in the corresponding periodic signal $x_p(t)$ are simply samples of $X(F)$ at frequencies $kF_0 = k/T_p$. Specifically,

$$c_k = \frac{1}{T_p} X(kF_0) = \frac{1}{T_p} X\left(\frac{k}{T_p}\right) \tag{4.1.45}$$

From (4.1.44) we note that the zero crossings of $X(F)$ occur at multiples of $1/\tau$. Furthermore, the width of the main lobe, which contains most of the signal energy, is equal to $1/\tau$. As the pulse duration τ decreases (increases), the main lobe becomes broader (narrower) and more energy is moved to the higher (lower) frequencies, as illustrated in Fig. 4.9. Thus as the signal pulse is expanded (compressed) in time, its transform is compressed (expanded) in frequency. This behavior between the time function and its spectrum is a type of uncertainty principle that appears in different forms in various branches of science and engineering.

Finally, the energy density spectrum of the rectangular pulse is

$$S_{xx}(F) = (A\tau)^2 \left(\frac{\sin \pi F \tau}{\pi F \tau}\right)^2 \tag{4.1.46}$$

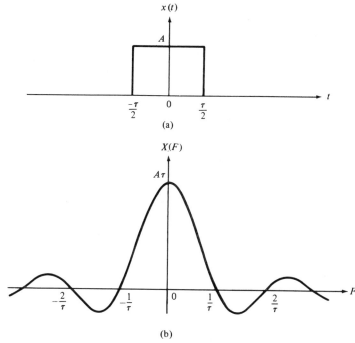

FIGURE 4.8 (a) Rectangular pulse and (b) its Fourier transform.

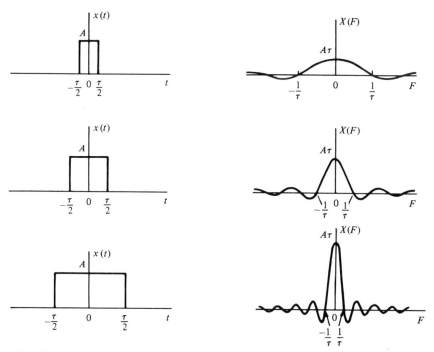

FIGURE 4.9 Fourier transform of a rectangular pulse for various width values.

4.2 Frequency Analysis of Discrete-Time Signals

In Section 4.1 we developed the Fourier series representation for continuous-time periodic (power) signals and the Fourier transform for finite energy aperiodic signals. In this section we repeat the development for the class of discrete-time signals.

As we have observed from the discussion of Section 4.1, the Fourier series representation of a continuous-time periodic signal may consist of an infinite number of frequency components, where the frequency spacing between two successive harmonically related frequencies is $1/T_p$, and where T_p is the fundamental period. Since the frequency range for continuous-time signals extends from $-\infty$ to ∞, it is possible to have signals that contain an infinite number of frequency components. In contrast, the frequency range for discrete-time signals is unique over the interval $(-\pi, \pi)$ or $(0, 2\pi)$. A discrete-time signal of fundamental period N may consist of frequency components separated by $2\pi/N$ radians or $f = 1/N$ cycles. Consequently, the Fourier series representation of the discrete-time periodic signal will contain at most N frequency components. This is the basic difference between the Fourier series representations for continuous-time and discrete-time periodic signals.

4.2.1 The Fourier Series for Discrete-Time Periodic Signals

Suppose that we are given a periodic sequence $x(n)$ with period N, that is, $x(n) = x(n + N)$ for all n. The Fourier series representation for $x(n)$ consists of N harmonically related exponential functions

$$e^{j2\pi kn/N} \quad k = 0, 1, \ldots, N - 1$$

and is expressed as

$$x(n) = \sum_{k=0}^{N-1} c_k e^{j2\pi kn/N} \tag{4.2.1}$$

where the $\{c_k\}$ are the coefficients in the series representation.

To derive the expression for the Fourier coefficients, we use the following formula:

$$\sum_{n=0}^{N-1} e^{j2\pi kn/N} = \begin{cases} N & k = 0, \pm N, \pm 2N, \ldots \\ 0 & \text{otherwise} \end{cases} \tag{4.2.2}$$

Note the similarity of (4.2.2) with the continuous-time counterpart in (4.1.3). The proof of (4.2.2) follows immediately from the application of the geometric summation formula

$$\sum_{n=0}^{N-1} a^n = \begin{cases} N & a = 1 \\ \dfrac{1 - a^N}{1 - a} & a \neq 1 \end{cases} \tag{4.2.3}$$

The expression for the Fourier coefficients c_k can be obtained by multiplying both sides of (4.2.1) by the exponential $e^{-j2\pi ln/N}$ and summing the product from $n = 0$ to $n = N - 1$. Thus

$$\sum_{n=0}^{N-1} x(n) e^{-j2\pi ln/N} = \sum_{n=0}^{N-1} \sum_{k=0}^{N-1} c_k e^{j2\pi (k-l)n/N} \tag{4.2.4}$$

If we perform the summation over n first, in the right-hand side of (4.2.4), we obtain

$$\sum_{n=0}^{N-1} e^{j2\pi(k-l)n/N} = \begin{cases} N & k - l = 0, \pm N, \pm 2N, \ldots \\ 0 & \text{otherwise} \end{cases} \tag{4.2.5}$$

where we have made use of (4.2.2). Therefore, the right-hand side of (4.2.4) reduces to Nc_l and hence

$$c_l = \frac{1}{N} \sum_{n=0}^{N-1} x(n)e^{-j2\pi ln/N} \qquad l = 0, 1, \ldots, N - 1 \tag{4.2.6}$$

Thus we have the desired expression for the Fourier coefficients in terms of the signal $x(n)$.

The relationships (4.2.1) and (4.2.6) for the frequency analysis of discrete-time signals are summarized below.

Synthesis equation

$$x(n) = \sum_{k=0}^{N-1} c_k e^{j2\pi kn/N} \tag{4.2.7}$$

Analysis equation

$$c_k = \frac{1}{N} \sum_{n=0}^{N-1} x(n)e^{-j2\pi kn/N} \tag{4.2.8}$$

Equation (4.2.7) is often called the discrete-time Fourier series (DTFS). The Fourier coefficients $\{c_k\}$, $k = 0, 1, \ldots, N - 1$ provide the description of $x(n)$ in the frequency domain, in the sense that c_k represents the amplitude and phase associated with the frequency components

$$s_k(n) = e^{j2\pi kn/N} = e^{j\omega_k n}$$

where $\omega_k = 2\pi k/N$.

We recall from Section 1.3.3 that the functions $s_k(n)$ are periodic with period N. Hence $s_k(n) = s_k(n + N)$. In view of this periodicity, it follows that the Fourier coefficients c_k, when viewed beyond the range $k = 0, 1, \ldots, N - 1$, also satisfy a periodicity condition. Indeed, from (4.2.8), which holds for every value of k, we have

$$c_{k+N} = \frac{1}{N} \sum_{n=0}^{N-1} x(n)e^{-j2\pi(k+N)n/N} = \frac{1}{N} \sum_{n=0}^{N-1} x(n)e^{-j2\pi kn/N} = c_k \tag{4.2.9}$$

Therefore, the Fourier series coefficients $\{c_k\}$ form a periodic sequence when extended outside of the range $k = 0, 1, \ldots, N - 1$. Hence

$$c_{k+N} = c_k$$

that is, $\{c_k\}$ is a periodic sequence with fundamental period N. *Thus the spectrum of a signal $x(n)$, which is periodic with period N, is a periodic sequence with period N.* Consequently, any N consecutive samples of the signal or its spectrum provide a complete description of the signal in the time or frequency domains.

Although the Fourier coefficients form a periodic sequence, we will focus our attention on the single period with range $k = 0, 1, \ldots, N - 1$. This is convenient, since in the frequency domain this amounts to covering the fundamental range $0 \leq \omega_k = 2\pi k/N < 2\pi$, for $0 \leq k < N - 1$. In contrast, the frequency range $-\pi < \omega_k = 2\pi k/N \leq \pi$ corresponds to $-N/2 < k \leq N/2$, which creates an inconvenience when N is odd.

Clearly, if we use a sampling frequency F_s, the range $0 \le k \le N - 1$ corresponds to the frequency range $0 \le F < F_s$.

EXAMPLE 4.2.1 —————————————————————————

Determine the spectrum of the signal

$$x(n) = \cos \omega_0 n$$

when **(a)** $\omega_0 = \sqrt{2}\pi$ and **(b)** $\omega_0 = \pi/3$.

Solution: **(a)** For $\omega_0 = \sqrt{2}\pi$, we have $f_0 = 1/\sqrt{2}$. Since f_0 is not a rational number, the signal is not periodic. Consequently, this signal cannot be expanded in a Fourier series. Nevertheless, the signal does possess a spectrum. Its spectral content consists of the single frequency component at $\omega = \omega_0 = \sqrt{2}\pi$.
(b) In this case $f_0 = \frac{1}{6}$ and hence $x(n)$ is periodic with fundamental period $N = 6$. From (4.2.8) we have

$$c_k = \tfrac{1}{6} \sum_{n=0}^{5} x(n)e^{-j2\pi kn/6} \qquad k = 0, 1, \ldots, 5$$

However, $x(n)$ can be expressed as

$$x(n) = \cos \frac{2\pi n}{6} = \tfrac{1}{2}e^{j2\pi n/6} + \tfrac{1}{2}e^{-j2\pi n/6}$$

which is already in the form of the exponential Fourier series in (4.2.7). In comparing the two exponential terms in $x(n)$ with (4.2.7), it is apparent that $c_1 = \frac{1}{2}$. The second exponential in $x(n)$ corresponds to the term $k = -1$ in (4.2.7). However, this term can also be written as

$$e^{-j2\pi n/6} = e^{j2\pi(5-6)n/6} = e^{j2\pi(5n)/6}$$

which means that $c_{-1} = c_5$. But this is consistent with (4.2.9), and our previous observation that the Fourier series coefficients form a periodic sequence of period N. Consequently, we conclude that

$$c_0 = c_2 = c_3 = c_4 = 0$$
$$c_1 = \tfrac{1}{2} \qquad c_5 = \tfrac{1}{2}$$

Figure 4.10 illustrates the spectral content of the signal $x(n)$.

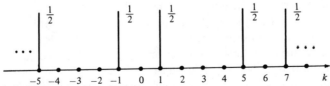

FIGURE 4.10 Spectrum of the periodic signal in Example 4.2.1.

4.2.2 Power Density Spectrum of Periodic Signals

The average power of a discrete-time periodic signal with period N was defined in (2.1.23) as

$$P_x = \frac{1}{N} \sum_{n=0}^{N-1} |x(n)|^2 \tag{4.2.10}$$

We shall now derive an expression for P_x in terms of the Fourier coefficient $\{c_k\}$.
   If we use the relation (4.2.7) in (4.2.10), we have

$$P_x = \frac{1}{N} \sum_{n=0}^{N-1} x(n)x^*(n)$$

$$= \frac{1}{N} \sum_{n=0}^{N-1} x(n) \left(\sum_{k=0}^{N-1} c_k^* e^{-j2\pi kn/N} \right)$$

Now, we can interchange the order of the two summations and make use of (4.2.8). Thus we obtain

$$P_x = \sum_{k=0}^{N-1} c_k^* \left[\frac{1}{N} \sum_{n=0}^{N-1} x(n)e^{-j2\pi kn/N} \right] \tag{4.2.11}$$

$$= \sum_{k=0}^{N-1} |c_k|^2 = \frac{1}{N} \sum_{n=0}^{N-1} |x(n)|^2$$

which is the desired expression for the average power in the periodic signal. In other words, the average power in the signal is the sum of the powers of the individual frequency components. We view (4.2.11) as a Parseval's relation for discrete-time periodic signals. The sequence $|c_k|^2$ for $k = 0, 1, \ldots, N - 1$ is the distribution of power as a function of frequency and is called the *power density spectrum* of the periodic signal.
   If we are interested in the energy of the sequence $x(n)$ over a single period, (4.2.11) implies that

$$E_N = \sum_{n=0}^{N-1} |x(n)|^2 = N \sum_{k=0}^{N-1} |c_k|^2 \tag{4.2.12}$$

which is consistent with our previous results for continuous-time periodic signals.
   If the signal $x(n)$ is real [i.e., $x^*(n) = x(n)$], then, proceeding as in Section 4.2.1, we can easily show that

$$c_k^* = c_{-k} \tag{4.2.13}$$

or equivalently,

$$|c_{-k}| = |c_k| \qquad \text{(even symmetry)} \tag{4.2.14}$$

$$-\angle c_{-k} = \angle c_k \qquad \text{(odd symmetry)} \tag{4.2.15}$$

These symmetry properties for the magnitude and phase spectra of a periodic signal, in conjunction with the periodicity property, have very important implications on the frequency range of discrete-time signals.
   Indeed, by combining (4.2.9) with (4.2.4) and (4.2.15), we obtain

$$|c_k| = |c_{N-k}| \tag{4.2.16}$$

and

$$\angle c_k = -\angle c_{N-k} \tag{4.2.17}$$

More specifically, we have

$$
\begin{array}{lll}
|c_0| = |c_N| & \angle c_0 = -\angle c_N = 0 & \\
|c_1| = |c_{N-1}| & \angle c_1 = -\angle c_{N-1} & \\
|c_{N/2}| = |c_{N/2}| & \angle c_{N/2} = 0 & \text{if } N \text{ is even} \\
|c_{(N-1)/2}| = |c_{(N+1)/2}| & \angle c_{(N-1)/2} = -\angle c_{(N+1)/2} & \text{if } N \text{ is odd}
\end{array}
\tag{4.2.18}
$$

Thus, for a real signal, the spectrum c_k, $k = 0, 1, \ldots, N/2$ for N even or $k = 0$, $1, \ldots, (N-1)/2$ for N odd, completely specifies the signal in the frequency domain. Clearly, this is consistent with the fact that the highest relative frequency that can be represented by a discrete-time signal is equal to π. Indeed, if $0 \le \omega_k = 2\pi k/N \le \pi$, then $0 \le k \le N/2$.

By making use of these symmetry properties of the Fourier series coefficients of a real signal, the Fourier series in (4.2.7) can also be expressed in the alternative forms

$$
x(n) = c_0 + 2 \sum_{k=1}^{L} |c_k| \cos\left(\frac{2\pi}{N} kn + \theta_k\right)
\tag{4.2.19}
$$

$$
= a_0 + \sum_{k=1}^{L} \left(a_k \cos\frac{2\pi}{N} kn - b_k \sin\frac{2\pi}{N} kn\right)
\tag{4.2.20}
$$

where $a_0 = c_0$, $a_k = 2|c_k| \cos\theta_k$, $b_k = 2|c_k| \sin\theta_k$, and $L = N/2$ if N is even and $L = (N-1)/2$ if N is odd.

Finally, we note that as in the case of continuous-time signals, the power density spectrum $|c_k|^2$ does not contain any phase information. Furthermore, the spectrum is discrete and periodic with a fundamental period equal to that of the signal itself.

EXAMPLE 4.2.2

Determine the Fourier series coefficients and the power density spectrum of the periodic signal shown in Fig. 4.11.

Solution: By applying the analysis equation (4.2.8) to the signal shown in Fig. 4.11, we obtain

$$
c_k = \frac{1}{N} \sum_{n=0}^{N-1} x(n)e^{-j2\pi kn/N} = \frac{1}{N} \sum_{n=0}^{L-1} Ae^{-j2\pi kn/N} \qquad k = 0, 1. \ldots, N - 1
$$

which is a geometric summation. Now we can use (4.2.3) to simplify the summation above. Thus we obtain

$$
c_k = \frac{A}{N} \sum_{n=0}^{L-1} (e^{-j2\pi k/N})^n =
\begin{cases}
\dfrac{AL}{N} & k = 0 \\[2ex]
\dfrac{A}{N} \dfrac{1 - e^{-j2\pi kL/N}}{1 - e^{-j2\pi k/N}} & k = 1, 2, \ldots, N - 1
\end{cases}
$$

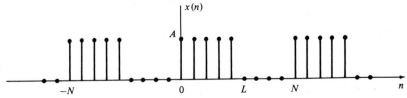

FIGURE 4.11 Discrete-time periodic square-wave signal.

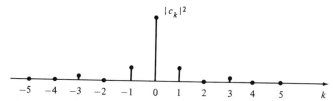

FIGURE 4.12 Plot of the power density spectrum given by (4.2.22).

The last expression can be simplified further if we note that

$$\frac{1 - e^{-j2\pi kL/N}}{1 - e^{-j2\pi k/N}} = \frac{e^{-j\pi kL/N}}{e^{-j\pi k/N}} \frac{e^{j\pi kL/N} - e^{-j\pi kL/N}}{e^{j\pi k/N} - e^{-j\pi k/N}}$$

$$= e^{-j\pi k(L-1)/N} \frac{\sin (\pi kL/N)}{\sin (\pi k/N)}$$

Therefore,

$$c_k = \begin{cases} \dfrac{AL}{N} & K = 0, \pm N, \pm 2N, \ldots \\[2ex] \dfrac{A}{N} e^{-j\pi k(L-1)/N} \dfrac{\sin (\pi kL/N)}{\sin (\pi k/N)} & \text{otherwise} \end{cases} \tag{4.2.21}$$

The power density spectrum of this periodic signal is

$$|c_k|^2 = \begin{cases} \left(\dfrac{AL}{N}\right)^2 & k = 0, +N, \pm 2N, \ldots \\[2ex] \left(\dfrac{A}{N}\right)^2 \left(\dfrac{\sin \pi kL/N}{\sin \pi k/N}\right)^2 & \text{otherwise} \end{cases} \tag{4.2.22}$$

Figure 4.12 illustrates the plot of $|c_k|^2$ for $L = 5$, $N = 10$, and $A = 1$.

4.2.3 Measurement of Harmonic Distortion in Digital Sinusoidal Generators

Consider an ideal sinusoidal generator that produces the signal

$$x(n) = \cos 2\pi f_0 n \qquad -\infty < n < \infty$$

If the relative frequency $f_0 = k_0/N$, where k_0 and N are relatively prime numbers, the signal $x(n)$ is periodic with fundamental period N. Since $x(n)$ can be expressed as

$$x(n) = \cos \frac{2\pi k_0 n}{N} = \tfrac{1}{2}(e^{j2\pi k_0 n/N} + e^{-j2\pi k_0 n/N})$$

$$= \tfrac{1}{2}[e^{j2\pi k_0 n/N} + e^{j2\pi(N-k_0)/N}] \tag{4.2.23}$$

it is clear that the spectrum of an ideal sinusoid with frequency $f_0 = k_0/N$ consists of two lines at frequencies $2\pi k_0/N$ and $2\pi(N - k_0)/N$ radians, as illustrated in Fig. 4.13. The Fourier coefficients of the signal are

$$c_k = \begin{cases} \tfrac{1}{2} & k = k_0, N - k_0 \\ 0 & 0 \le k \le N - 1, k \ne k_0, N - k_0 \end{cases} \tag{4.2.24}$$

In practice, the approximations made in computing the samples of a sinusoid of relative frequency f_0 result in a certain amount of power falling into other frequencies. This

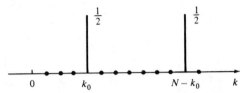

FIGURE 4.13 Spectrum of ideal sinusoid with frequency $f = k_0/N$.

spurious power results in distortion, which is referred to as *harmonic distortion*. Harmonic distortion is usually measured in terms of the *total harmonic distortion* (THD), which is defined as the ratio

$$\text{THD} = \frac{\text{spurious harmonic power}}{\text{total power}} \qquad (4.2.25)$$

The total power P_x can be evaluated in either the time or the frequency domain using Parseval's relation (4.2.11). Thus

$$P_x = \frac{1}{N} \sum_{k=0}^{N-1} |x(n)|^2 = \sum_{k=0}^{N-1} |c_k|^2 \qquad (4.2.26)$$

The amount of power in the desired frequency $f_0 = k_0/N$ is

$$|c_{k_0}|^2 + |c_{N-k_0}|^2 = 2|c_{k_0}|^2$$

since $x(n)$ is a real-valued signal. Consequently, the spurious harmonic power is

$$P_s = P_x - 2|c_{k_0}|^2$$

Therefore, the THD is given by

$$\text{THD} = \frac{P_x - 2|c_{k_0}|^2}{P_x} = 1 - 2\frac{|c_{k_0}|^2}{P_x} \qquad (4.2.27)$$

where

$$c_{k_0} = \frac{1}{N} \sum_{n=0}^{N-1} x(n)e^{-j2\pi k_0 n/N} \qquad (4.2.28)$$

Table 4.1 provides the total harmonic distortion resulting from the generation of $\cos \phi$ by means of the Taylor series approximation.

$$\cos \phi = 1 - \frac{\phi^2}{2!} + \frac{\phi^4}{4!} - \frac{\phi^6}{6!} + \cdots \qquad (4.2.29)$$

Note the reduction of THD as we add more terms in the approximation formula.

TABLE 4.1 Total Harmonic Distortion

Number of Terms	Frequency, f_0		
	$\frac{1}{32}$	$\frac{1}{28}$	$\frac{1}{256}$
2	0.52477	0.52650	0 52702
3	0.36097	0.36950	0.37105
4	0.35306	0.34814	0.34780
5	0.07586	0.09563	0.09912
6	0.00389	0.00572	0.00609
7	0.00006	0.00009	0.00010
8	0.00000	0.00000	0.00000

4.2.4 Frequency Analysis of Discrete-Time Aperiodic Signals

Just as in the case of continuous-time aperiodic energy signals, the frequency analysis of discrete-time aperiodic finite-energy signals involves a Fourier transform of the time-domain signal. Consequently, the development in this section parallels to a large extent that given in Section 4.1.3.

The Fourier transform of a finite-energy discrete-time signal $x(n)$ is defined as

$$X(\omega) = \sum_{n=-\infty}^{\infty} x(n)e^{-j\omega n} \tag{4.2.30}$$

Physically, $X(\omega)$ represents the frequency content of the signal $x(n)$. In other words, $X(\omega)$ is a decomposition of $x(n)$ into its frequency components.

We observe two basic differences between the Fourier transform of a discrete-time finite-energy signal and the Fourier transform of a finite-energy analog signal. First, for continuous-time signals, the Fourier transform, and hence the spectrum of the signal, have a frequency range of $(-\infty, \infty)$. In contrast, the frequency range for a discrete-time signal is unique over the range $(-\pi, \pi)$ or, equivalently, $(0, 2\pi)$. This property is reflected in the Fourier transform of the signal. Indeed, $X(\omega)$ is periodic with period 2π, that is,

$$
\begin{aligned}
X(\omega + 2\pi k) &= \sum_{n=-\infty}^{\infty} x(n)e^{-j(\omega + 2\pi k)n} \\
&= \sum_{n=-\infty}^{\infty} x(n)e^{-j\omega n}e^{-j2\pi kn} \\
&= \sum_{n=-\infty}^{\infty} x(n)e^{-j\omega n} = X(\omega)
\end{aligned}
\tag{4.2.31}
$$

Hence $X(\omega)$ is periodic with period 2π. But this property is just a consequence of the fact that the frequency range for any discrete-time signal is limited to $(-\pi, \pi)$ or $(0, 2\pi)$, and any frequency outside this interval is equivalent to a frequency within the interval.

The second basic difference is also a consequence of the discrete-time nature of the signal. Since the signal is discrete in time, the Fourier transform of the signal involves a summation of terms instead of an integral, as in the case of continuous-time signals.

Since $X(\omega)$ is a periodic function of the frequency variable ω, it has a Fourier series expansion, provided that the conditions for the existence of the Fourier series, described previously, are satisfied. In fact, from the definition of the Fourier transform $X(\omega)$ of the sequence $x(n)$, given by (4.2.30), we observe that $X(\omega)$ has the form of a Fourier series. The Fourier coefficients in this series expansion are the values of the sequence $x(n)$.

To demonstrate this point, let us evaluate the sequence $x(n)$ from $X(\omega)$. First, we multiply both sides (4.2.30) by $e^{j\omega m}$ and integrate over the interval $(-\pi, \pi)$. Thus we have

$$\int_{-\pi}^{\pi} X(\omega)e^{j\omega m}\, d\omega = \int_{-\pi}^{\pi} \left[\sum_{n=-\infty}^{\infty} x(n)e^{-j\omega n} \right] e^{j\omega m}\, d\omega \tag{4.2.32}$$

The integral on the right-hand side of (4.2.32) can be evaluated by first interchanging the order of summation and integration. Thus we obtain

$$\sum_{n=-\infty}^{\infty} x(n) \int_{-\pi}^{\pi} e^{j\omega(m-n)}\, d\omega = \begin{cases} 2\pi x(m) & m = n \\ 0 & m \neq n \end{cases} \tag{4.2.33}$$

By combining (4.2.32) and (4.2.33), we obtain the desired result that

$$x(n) = \frac{1}{2\pi} \int_{-\pi}^{\pi} X(\omega)e^{j\omega n} \, d\omega \tag{4.2.34}$$

If we compare the integral in (4.2.34) with (4.1.8), we note that this is just the expression for the Fourier series coefficient for a function that is periodic with period 2π. The only difference between (4.1.8) and (4.2.34) is the sign on the exponent in the integrand, which is a consequence of our definition of the Fourier transform as given by (4.2.30). Therefore, the Fourier transform of the sequence $x(n)$ defined by (4.2.30) has the form of a Fourier series expansion.

In summary, the *Fourier transform pair for discrete-time signals* is as follows.

Synthesis equation, inverse transform

$$x(n) = \frac{1}{2\pi} \int_{2\pi} X(\omega)e^{j\omega n} \, d\omega \tag{4.2.35}$$

Analysis equation, direct transform

$$X(\omega) = \sum_{n=-\infty}^{\infty} x(n)e^{-j\omega n} \tag{4.2.36}$$

Conditions for the existence of the Fourier transform may be established by returning to the expression (4.2.30). Clearly, $X(\omega)$ exists if the series in (4.2.30) converges. The series converges if and only if $x(n)$ is absolutely summable, that is, if

$$\sum_{n=-\infty}^{\infty} |x(n)| < \infty \tag{4.2.37}$$

Now it is always true that

$$\left[\sum_{n=-\infty}^{\infty} |x(n)| \right]^2 \geq \sum_{n=-\infty}^{\infty} |x(n)|^2 = E_x$$

Consequently, if $x(n)$ is absolutely summable, it is a finite-energy signal.

4.2.5 Energy Density Spectrum of Aperiodic Signals

Recall that the energy of a discrete-time signal $x(n)$ was defined as

$$E_x = \sum_{n=-\infty}^{\infty} |x(n)|^2 \tag{4.2.38}$$

Let us now express the energy E_x in terms of the spectral characteristic $X(\omega)$. First we have

$$E_x = \sum_{n=-\infty}^{\infty} x(n)x^*(n) = \sum_{n=-\infty}^{\infty} x(n) \left[\frac{1}{2\pi} \int_{-\pi}^{\pi} X^*(\omega)e^{-j\omega n} \, d\omega \right]$$

If we interchange the order of integration and summation in the equation above, we obtain

$$E_x = \frac{1}{2\pi} \int_{-\pi}^{\pi} X^*(\omega) \left[\sum_{n=-\infty}^{\infty} x(n)e^{-j\omega n} \right] d\omega$$

$$= \frac{1}{2\pi} \int_{-\pi}^{\pi} |X(\omega)|^2 \, d\omega$$

Therefore, the energy relation between $x(n)$ and $X(\omega)$ is

$$E_x = \sum_{n=-\infty}^{\infty} |x(n)|^2 = \frac{1}{2\pi} \int_{-\pi}^{\pi} |X(\omega)|^2 \, d\omega \qquad (4.2.39)$$

This is Parseval's relation for discrete-time aperiodic signals with finite energy.

The spectrum $X(\omega)$ is, in general, a complex-valued function of frequency. It may be expressed as

$$X(\omega) = |X(\omega)|e^{j\Theta(\omega)} \qquad (4.2.40)$$

where

$$\Theta(\omega) = \angle X(\omega)$$

is the phase spectrum and $|X(\omega)|$ is the magnitude spectrum.

As in the case of continuous-time signals, the quantity

$$S_{xx}(\omega) = |X(\omega)|^2 \qquad (4.2.41)$$

represents the distribution of energy as a function of frequency and it is called the *energy density spectrum* of $x(n)$. Clearly, $S_{xx}(\omega)$ does not contain any phase information.

Suppose now that the signal $x(n)$ is real. Then it easily follows that

$$X^*(\omega) = X(-\omega) \qquad (4.2.42)$$

or equivalently,

$$|X(-\omega)| = |X(\omega)| \qquad \text{(even symmetry)} \qquad (4.2.43)$$

and

$$\angle X(-\omega) = -\angle X(\omega) \qquad \text{(odd symmetry)} \qquad (4.2.44)$$

From (4.2.41) it also follows that

$$S_{xx}(-\omega) = S_{xx}(\omega) \qquad \text{(even symmetry)} \qquad (4.2.45)$$

From these symmetry properties we conclude that the frequency range of discrete-time signals can be limited further to the range $0 \le \omega \le \pi$ (i.e., one-half of the period). Indeed, if we know $X(\omega)$ in the range $0 \le \omega \le \pi$, we can determine it for the range $-\pi \le \omega < 0$ using the symmetry properties given above. As we have already observed, similar results hold for discrete-time periodic signals. Therefore, the frequency-domain description of a real, discrete-time signal is completely specified by its spectrum in the frequency range $0 \le \omega \le \pi$.

Usually, we will work with the fundamental interval $0 \le \omega \le \pi$ or $0 \le F \le F_s/2$, in terms of hertz. We will draw more than half a period only when this is required by the specific application.

EXAMPLE 4.2.3 _____

Determine and sketch the energy density spectrum $S_{xx}(\omega)$ of the signal

$$x(n) = a^n u(n) \qquad -1 < a < 1$$

for $a = 0.5$ and $a = -0.5$.

Solution: Since $|a| < 1$, the sequence $x(n)$ is absolutely summable, as can be verified by applying the geometric summation formula, that is,

$$\sum_{n=-\infty}^{\infty} |x(n)| = \sum_{n=0}^{\infty} |a|^n = \frac{1}{1 - |a|} < \infty$$

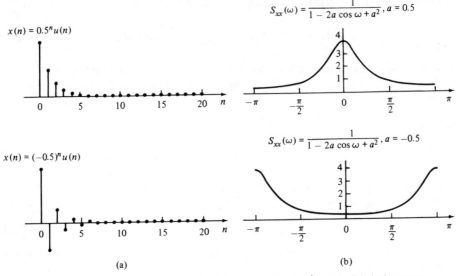

$$S_{xx}(\omega) = \frac{1}{1 - 2a\cos\omega + a^2}, a = 0.5$$

$x(n) = 0.5^n u(n)$

$$S_{xx}(\omega) = \frac{1}{1 - 2a\cos\omega + a^2}, a = -0.5$$

$x(n) = (-0.5)^n u(n)$

(a)   (b)

FIGURE 4.14 (a) Sequence $x(n) = (\frac{1}{2})^n u(n)$ and $x(n) = (-\frac{1}{2})^n u(n)$; (b) their energy density spectra.

Hence the Fourier transform of $x(n)$ exists and is obtained by applying (4.2.36). Thus

$$X(\omega) = \sum_{n=0}^{\infty} a^n e^{-j\omega n} = \sum_{n=0}^{\infty} (ae^{-j\omega})^n$$

Since $|ae^{-j\omega}| = |a| < 1$, use of the geometric summation formula again yields

$$X(\omega) = \frac{1}{1 - ae^{-j\omega}}$$

The energy density spectrum is given by

$$S_{xx}(\omega) = |X(\omega)|^2 = X(\omega)X^*(\omega) = \frac{1}{(1 - ae^{-j\omega})(1 - ae^{j\omega})}$$

or, equivalently, as

$$S_{xx}(\omega) = \frac{1}{1 - 2a\cos\omega + a^2}$$

Note that $S_{xx}(-\omega) = S_{xx}(\omega)$ in accordance with (4.2.45).

Figure 4.14 shows the signal $x(n)$ and its corresponding spectrum for $a = 0.5$ and $a = -0.5$. Note that for $a = -0.5$ the signal has more rapid variations and as a result its spectrum has stronger high frequencies.

EXAMPLE 4.2.4 _____

Determine the signal $x(n)$ corresponding to the spectrum

$$X(\omega) = \begin{cases} 1 & |\omega| \le \omega_c \\ 0 & \text{otherwise} \end{cases}$$

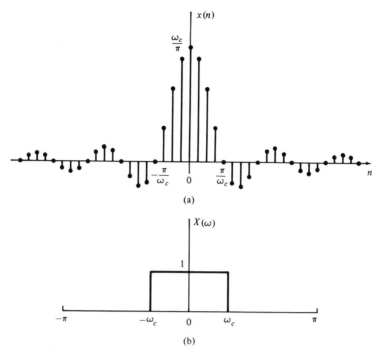

FIGURE 4.15   Fourier transform pair in Example 4.2.4.

Solution: From (4.2.35) we have

$$x(n) = \frac{1}{2\pi} \int_{-\pi}^{\pi} X(\omega) e^{j\omega n} \, d\omega$$

$$= \frac{1}{2\pi} \int_{-\omega_c}^{\omega_c} e^{j\omega n} \, d\omega$$

$$= \frac{\omega_c}{\pi} \frac{\sin \omega_c n}{\omega_c n} \qquad n \neq 0$$

For $n = 0$ we obtain

$$x(n) = \frac{1}{2\pi} \int_{-\omega_c}^{\omega_c} d\omega = \frac{\omega_c}{\pi}$$

Therefore,

$$x(n) = \begin{cases} \dfrac{\omega_c}{\pi} & n = 0 \\[2mm] \dfrac{\omega_c}{\pi} \dfrac{\sin \omega_c n}{\omega_c n} & n \neq 0 \end{cases}$$

This transform pair is illustrated in Fig. 4.15.

We make the observation that the sequence $x(n)$ has finite energy, $E_x = \omega_c/\pi$, but it is not absolutely summable.

EXAMPLE 4.2.5 ——————————————————————————————————

Determine the Fourier transform and the energy density spectrum of the sequence

$$x(n) = \begin{cases} A & 0 \le n \le L - 1 \\ 0 & \text{otherwise} \end{cases} \qquad (4.2.46)$$

which is illustrated in Fig. 4.16.

Solution: Before computing the Fourier transform, we observe that

$$\sum_{n=-\infty}^{\infty} |x(n)| = \sum_{n=0}^{L-1} |A| = L|A| < \infty$$

Hence $x(n)$ is absolutely summable and its Fourier transform exists. Furthermore, we note that $x(n)$ is a finite-energy signal with $E_x = |A|^2 L$.
  The Fourier transform of this signal is

$$
\begin{aligned}
X(\omega) &= \sum_{n=0}^{L-1} A e^{-j\omega n} \\
&= A \frac{1 - e^{-j\omega L}}{1 - e^{-j\omega}} \qquad (4.2.47) \\
&= A e^{-j(\omega/2)(L-1)} \frac{\sin(\omega L/2)}{\sin(\omega/2)}
\end{aligned}
$$

For $\omega = 0$ the transform in (4.2.17) yields $X(0) = AL$, which is easily established by setting $\omega = 0$ in the defining equation for $X(\omega)$ or by using l'Hospital's rule in (4.2.47) to resolve the indeterminate form when $\omega = 0$.
  The magnitude and phase specta of $x(n)$ are

$$|X(\omega)| = \begin{cases} |A|L & \omega = 0 \\ |A| \left| \dfrac{\sin(\omega L/2)}{\sin(\omega/2)} \right| & \text{otherwise} \end{cases} \qquad (4.2.48)$$

and

$$\angle X(\omega) = \angle A - \frac{\omega}{2}(L-1) + \angle \frac{\sin(\omega L/2)}{\sin(\omega/2)} \qquad (4.2.49)$$

where we should remember that the phase of a real quantity is zero if the quantity is positive and π if it is negative.
  The spectra $|X(\omega)|$ and $\angle X(\omega)$ are shown in Fig. 4.17 for the case $A = 1$ and $L = 5$. The energy density spectrum is simply the square of the expression given in (4.2.48).

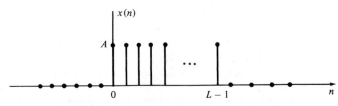

FIGURE 4.16   Discrete-time rectangular pulse.

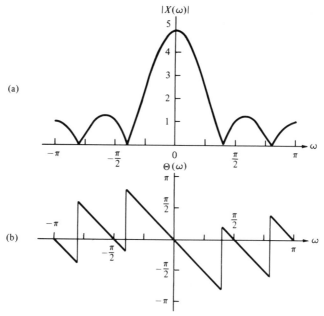

FIGURE 4.17 Mangitude and phase of Fourier transform of the discrete-time rectangular pulse in Fig. 4.16.

There is an interesting relationship that exists between the Fourier transform of the constant amplitude pulse in Example 4.2.5 and the periodic rectangular wave considered in Example 4.2.2. If we evaluate the Fourier transform as given in (4.2.47) at a set of equally spaced (harmonically related) frequencies

$$\omega_k = \frac{2\pi}{N} k \qquad k = 0, 1, \ldots, N - 1$$

we obtain

$$X\left(\frac{2\pi}{N} k\right) = Ae^{-j(\pi/N)k(L-1)} \frac{\sin\left[(\pi/N)kL\right]}{\sin\left[(\pi/N)k\right]} \qquad (4.2.50)$$

If we compare this result with the expression for the Fourier series coefficients given in (4.2.22) for the periodic rectangular wave, we find that

$$X\left(\frac{2\pi}{N} k\right) = Nc_k \qquad k = 0, 1, \ldots, N - 1 \qquad (4.2.51)$$

To elaborate, we have established that the Fourier transform of the rectangular pulse, which is identical with a single period of the periodic rectangular pulse train, evaluated at the frequencies $\omega = 2\pi k/N$, $k = 0, 1, \ldots, N - 1$, which are identical to the harmonically related frequency components used in the Fourier series representation of the periodic signal, is simply a multiple of the Fourier coefficients $\{c_k\}$ at the corresponding frequencies.

The relationship given in (4.2.51) for the Fourier transform of the rectangular pulse evaluated at $\omega = 2\pi k/N$, $k = 0, 1, \ldots, N - 1$, and the Fourier coefficients of the corresponding periodic signal is not only true for these two signals but, in fact, holds in general. This relationship is developed further in Section 4.5.

4.2.6 Relationship of the Fourier Transform to the z-Transform

The z-transform of a sequence $x(n)$ is defined as

$$X(z) = \sum_{n=-\infty}^{\infty} x(n)z^{-n} \qquad \text{ROC: } r_2 < |z| < r_1 \qquad (4.2.52)$$

where $r_2 < |z| < r_1$ is the region of convergence of $X(z)$. Let us express the complex variable z in polar form as

$$z = re^{j\omega} \qquad (4.2.53)$$

where $r = |z|$ and $\omega = \angle z$. Then within the region of convergence of $X(z)$, we may substitute $z = re^{j\omega}$ into (4.2.52). This yields

$$X(z)\big|_{z=re^{j\omega}} = \sum_{n=-\infty}^{\infty} x(n)r^{-n}e^{-j\omega n} \qquad (4.2.54)$$

From the relationship in (4.2.54) we note that $X(z)$ can be interpreted as the Fourier transform of the signal sequence $x(n)r^{-n}$. The weighting factor r^{-n} is growing with n if $r < 1$ and decaying if $r > 1$. Alternatively, if $X(z)$ converges for $|z| = 1$, then

$$X(z)\big|_{z=e^{j\omega}} \equiv X(\omega) = \sum_{n=-\infty}^{\infty} x(n)e^{-j\omega n} \qquad (4.2.55)$$

Therefore, the Fourier transform may be viewed as the z-transform of the sequence, evaluated on the unit circle. If $X(z)$ does not converge in the region $|z| = 1$ [i.e., if the unit circle is not contained in the region of convergence of $X(z)$], the Fourier transform $X(\omega)$ does not exist. Conversely, if $X(\omega)$ exists, then $X(z)$ converges on the unit circle.

EXAMPLE 4.2.6 _____

Determine the z-transform and the Fourier transform of the sequence

$$x(n) = (\tfrac{1}{2})^n u(n)$$

Solution: The z-transform for the sequence is

$$X(z) = \sum_{n=0}^{\infty} (\tfrac{1}{2})^n z^{-n} = \frac{1}{1 - \tfrac{1}{2}z^{-1}} \qquad |z| > \tfrac{1}{2}$$

Since $X(z)$ converges on the unit circle, $X(\omega)$ may be obtained by setting $z = e^{j\omega}$. Thus we obtain the Fourier transform

$$X(\omega) = \frac{1}{1 - \tfrac{1}{2}e^{-j\omega}}$$

4.2.7 The Fourier Transform of Signals with Poles on the Unit Circle

As was shown in the preceding section, the Fourier transform of a discrete-time-signal can be determined by evaluating its z-transform on the unit circle, provided that the unit circle lies within the region of convergence of $X(z)$. Otherwise, the Fourier transform does not exist.

There are some very useful signals that have poles on the unit circle. These signals occur very often in practical applications, and thus it would be very fruitful to investigate their Fourier transforms.

EXAMPLE 4.2.7

Determine the Fourier transform of the following signals.

(b) $x_1(n) = u(n)$
(b) $x_2(n) = (-1)^n u(n)$
(c) $x_3(n) = (\cos \omega_0 n)u(n)$

by evaluating their z-transforms on the unit circle.

Solution: **(a)** From Table 3.3 we find that

$$X_1(z) = \frac{1}{1 - z^{-1}} = \frac{z}{z - 1} \qquad \text{ROC: } |z| > 1$$

$X_1(z)$ has a pole, $p_1 = 1$, on the unit circle.
   If we evaluate $X_1(z)$ on the unit circle, we obtain

$$X_1(\omega) = \frac{e^{j\omega/2}}{2j \sin(\omega/2)} = \frac{1}{2 \sin(\omega/2)} e^{j(\omega - \pi)/2}$$

Thus

$$|X_1(\omega)| = \frac{1}{2|\sin(\omega/2)|}$$

$$\angle X_1(\omega) = \begin{cases} \dfrac{\omega}{2} - \dfrac{\pi}{2} & \text{if } \sin \dfrac{\omega}{2} \ge 0 \\[2mm] \dfrac{\omega}{2} + \dfrac{\pi}{2} & \text{if } \sin \dfrac{\omega}{2} < 0 \end{cases}$$

The presence of a pole at $z = 1$ (i.e., at $\omega = 0$) creates a problem only when we want to compute $|X_1(\omega)|$ at $\omega = 0$, because $|X_1(0)| = \infty$. For any other value of ω, $X_1(\omega)$ is finite (i.e., well behaved). Although, at first glance one might expect the signal to have zero-frequency components at all frequencies except at $\omega = 0$, this is not the case. This happens because the signal $x_1(n)$ is not a constant for all $-\infty < n < \infty$. Instead, it is *turned on* at $n = 0$. This abrupt jump creates all frequency components existing in the range $0 < \omega \le \pi$. Generally, all signals which start at a finite time will have non-zero-frequency components everywhere in the frequency axis from zero up to the folding frequency.

(b) From Table 3.3 we find that the z-transform of $a^n u(n)$ with $a = -1$ reduces to

$$X_2(z) = \frac{1}{1 + z^{-1}} = \frac{z}{z + 1} \qquad \text{ROC: } |z| > 1$$

which has a pole at $z = -1 = e^{j\pi}$. The Fourier transform is

$$X_2(\omega) = \frac{e^{j\omega/2}}{2 \cos(\omega/2)}$$

Hence the magnitude is

$$|X_2(\omega)| = \frac{1}{2|\cos(\omega/2)|}$$

and the phase is

$$
\angle X_2(\omega) = \begin{cases} \dfrac{\omega}{2} & \text{if } \cos\dfrac{\omega}{2} \geq 0 \\[3mm] \dfrac{\omega}{2} + \pi & \text{if } \cos\dfrac{\omega}{2} < 0 \end{cases}
$$

Note that due to the presence of the pole at $a = -1$ (i.e., at frequency $\omega = \pi$), the magnitude of the Fourier transform becomes infinite. Now the frequency component at which $|X(\omega)| = \infty$ is at $\omega = \pi$ instead of $\omega = 0$. We observe that $(-1)^n u(n) = (\cos \pi n)$, which is the fastest-possible oscillating signal in discrete time.

(c) From the discussion above, it follows that $X_3(\omega)$ will be infinite at the frequency component at $\omega = \omega_0$. Indeed, from Table 3.3, we find that

$$
x_3(n) = (\cos \omega_0 n)u(n) \xleftarrow{\ z\ } X_3(z) = \frac{1 - z^{-1} \cos \omega_0}{1 - 2z^{-1} \cos \omega_0 + z^{-2}} \qquad \text{ROC: } |z| > 1
$$

The Fourier transform is

$$
X_3(\omega) = \frac{1 - e^{-j\omega} \cos \omega_0}{(1 - e^{-j(\omega - \omega_0)})(1 - e^{j(\omega + \omega_0)})}
$$

The magnitude of $X_3(\omega)$ is given by

$$
|X_3(\omega)| = \frac{|1 - e^{-j\omega} \cos \omega_0|}{|1 - e^{-j(\omega - \omega_0)}|\,|1 - e^{-j(\omega + \omega_0)}|}
$$

Now if $\omega = -\omega_0$ or $\omega = \omega_0$, $|X(\omega)|$ becomes infinite. For all other frequencies, the Fourier transform is well behaved.

We conclude from the example above that the Fourier transform of signals with poles on the unit circle does not exist at frequencies equal to the angle that these poles form with the positive real axis. We also note that for all these signals the use of Fourier series is not possible, simply because they are aperiodic.

4.3 Properties of the Fourier Transform for Discrete-Time Signals

The Fourier transform for aperiodic finite-energy discrete-time signals described in the preceding section possesses a number of properties that are very useful in reducing the complexity of frequency analysis problems in many practical applications. In this section we develop the important properties of the Fourier transform. Similar properties hold for the Fourier transform of aperiodic finite-energy continuous-time signals.

For convenience, we adopt the notation

$$
X(\omega) \equiv F\{x(n)\} = \sum_{n=-\infty}^{\infty} x(n)e^{-j\omega n} \tag{4.3.1}
$$

for the direct transform (analysis equation) and

$$
x(n) \equiv F^{-1}\{X(\omega)\} = \frac{1}{2\pi} \int_{2\pi} X(\omega)e^{j\omega n}\,d\omega \tag{4.3.2}
$$

for the inverse transform (synthesis equation). We also refer to $x(n)$ and $X(\omega)$ as a *Fourier transform pair* and denote this relationship with the notation

$$x(n) \xleftarrow{\quad F \quad} X(\omega) \tag{4.3.3}$$

Recall that $X(\omega)$ is periodic with period 2π. Consequently, any interval of length 2π is sufficient for the specification of the spectrum. Usually, we plot the spectrum in the fundamental interval $[-\pi, \pi]$. We emphasize that all the spectral information contained in the fundamental interval is necessary for the complete description or characterization of the signal. For this reason the range of integration in (4.3.2) is always 2π, independent of the specific characteristics of the signal within the fundamental interval.

Linearity. If

$$x_1(n) \xleftarrow{\quad F \quad} X_1(\omega)$$

and

$$x_2(n) \xleftarrow{\quad F \quad} X_2(\omega)$$

then

$$a_1 x_1(n) + a_2 x_2(n) \xleftarrow{\quad F \quad} a_1 X_1(\omega) + a_2 X_2(\omega) \tag{4.3.4}$$

Simply stated, the Fourier transformation, viewed as an operation on a signal $x(n)$, is a linear transformation. Thus the Fourier transform of a linear combination of two or more signals is equal to the same linear combination of the Fourier transforms of the individual signals. This property is easily proved by using (4.3.1). The linearity property makes the Fourier transform suitable for the study of linear systems.

EXAMPLE 4.3.1 _____

Determine the Fourier transform of the signal

$$x(n) = a^{|n|} \qquad -1 < a < 1 \tag{4.3.5}$$

Solution: First, we observe that $x(n)$ can be expressed as

$$x(n) = x_1(n) + x_2(n)$$

where

$$x_1(n) = \begin{cases} a^n & n \geq 0 \\ 0 & n < 0 \end{cases}$$

and

$$x_2(n) = \begin{cases} a^{-n} & n < 0 \\ 0 & n \geq 0 \end{cases}$$

Beginning with the definition of the Fourier transform in (4.3.1), we have

$$X_1(\omega) = \sum_{n=-\infty}^{\infty} X_1(n) e^{-j\omega n} = \sum_{n=0}^{\infty} a^n e^{-j\omega n} = \sum_{n=0}^{\infty} (ae^{-j\omega})^n$$

The summation is a geometric series that converges to

$$X_1(\omega) = \frac{1}{1 - ae^{-j\omega}}$$

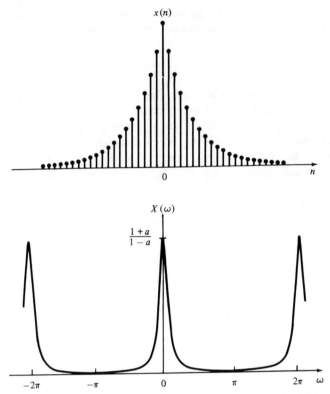

FIGURE 4.18 Sequence $x(n)$ and its Fourier transform in Example 4.3.1 with $a = 0.8$.

provided that

$$|ae^{-j\omega}| = |a| \cdot |e^{-j\omega}| = |a| < 1$$

which is a condition that is satisfied in this problem. Similarly, the Fourier transform of $x_2(n)$ is

$$X_2(\omega) = \sum_{n=-\infty}^{\infty} x_2(n)e^{-j\omega n} = \sum_{n=-\infty}^{-1} a^{-n}e^{-j\omega n}$$

$$= \sum_{n=-\infty}^{-1} (ae^{j\omega})^{-n} = \sum_{k=1}^{\infty} (ae^{j\omega})^k$$

$$= \frac{ae^{j\omega}}{1 - ae^{j\omega}}$$

By combining these two transforms, we obtain the Fourier transform of $x(n)$ in the form

$$X(\omega) = X_1(\omega) + X_2(\omega)$$

$$= \frac{1 - a^2}{1 - 2a \cos \omega + a^2} \tag{4.3.6}$$

Figure 4.18 illustrates $x(n)$ and $X(\omega)$ for the case in which $a = 0.8$.

Symmetry Properties. When a signal contains some symmetry properties in the time domain, it is possible to infer its frequency-domain characteristics. A discussion of various symmetry properties and the implications of these properties in the frequency domain is given below.

Suppose that both the signal $x(n)$ and its transform $X(\omega)$ are complex valued. Then they can be expressed in rectangular form as

$$x(n) = x_R(n) + jx_I(n) \tag{4.3.7}$$
$$X(\omega) = X_R(\omega) + jX_I(\omega) \tag{4.3.8}$$

By substituting (4.3.7) and $e^{-j\omega} = \cos \omega - j \sin \omega$ into (4.3.1) and separating the real and imaginary parts, we obtain

$$X_R(\omega) = \sum_{n=-\infty}^{\infty} [x_R(n) \cos \omega n + x_I(n) \sin \omega n] \tag{4.3.9}$$

$$X_I(\omega) = -\sum_{n=-\infty}^{\infty} [x_R(n) \sin \omega n - x_I(n) \cos \omega n] \tag{4.3.10}$$

In a similar manner, by substituting (4.3.8) and $e^{j\omega} = \cos \omega + j \sin \omega$ into (4.3.2), we obtain

$$x_R(n) = \frac{1}{2\pi} \int_{2\pi} [X_R(\omega) \cos \omega n - X_I(\omega) \sin \omega n] \, d\omega \tag{4.3.11}$$

$$x_I(n) = \frac{1}{2\pi} \int_{2\pi} [X_R(\omega) \sin \omega n + X_I(\omega) \cos \omega n] \, d\omega \tag{4.3.12}$$

Now, let us investigate some special cases.

REAL SIGNALS. If $x(n)$ is real, then $x_R(n) = x(n)$ and $x_I(n) = 0$. Hence (4.3.9) and (4.3.10) reduce to

$$X_R(\omega) = \sum_{n=-\infty}^{\infty} x(n) \cos \omega n \tag{4.3.13}$$

and

$$X_I(\omega) = -\sum_{n=-\infty}^{\infty} x(n) \sin \omega n \tag{4.3.14}$$

Since $\cos(-\omega)n = \cos \omega n$ and $\sin(-\omega)n = -\sin \omega n$, it follows from (4.3.13) and (4.3.14) that

$$X_R(-\omega) = X_R(\omega) \qquad \text{(even)} \tag{4.3.15}$$
$$X_I(-\omega) = -X_I(\omega) \qquad \text{(odd)} \tag{4.3.16}$$

If we combine (4.3.15) and (4.3.16) into a single equation, we have

$$X^*(\omega) = X(-\omega) \tag{4.3.17}$$

In this case we say that the spectrum of a real signal has *Hermitian symmetry*.

With the aid of Fig. 4.19, we observe that the magnitude and phase spectra for real signals are

$$|X(\omega)| = \sqrt{X_R^2(\omega) + X_I^2(\omega)} \tag{4.3.18}$$

$$\angle X|\omega| = \tan^{-1} \frac{X_I(\omega)}{X_R(\omega)} \tag{4.3.19}$$

As a consequence of (4.3.15) and (4.3.16), the magnitude and phase spectra also possess the symmetry properties

$$|X(\omega)| = |X(-\omega)| \qquad \text{(even)} \tag{4.3.20}$$
$$\angle X(-\omega) = -\angle X(\omega) \qquad \text{(odd)} \tag{4.3.21}$$

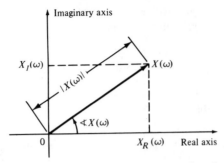

FIGURE 4.19 Magnitude and phase functions.

In the case of the inverse transform of a real-valued signal, i.e., $x(n) = x_R(n)$, (4.3.11) implies that

$$x(n) = \frac{1}{2\pi} \int_{2\pi} [X_R(\omega) \cos \omega n - X_I(\omega) \sin \omega n] \, d\omega \qquad (4.3.22)$$

Since both products $X_R(\omega) \cos \omega n$ and $X_I(\omega) \sin \omega n$ are even functions of ω, we have

$$x(n) = \frac{1}{\pi} \int_0^\pi [X_R(\omega) \cos \omega n - X_I(\omega) \sin \omega n] \, d\omega \qquad (4.3.23)$$

REAL AND EVEN SIGNALS. If $x(n)$ is real and even [i.e., $x(-n) = x(n)$], then $x(n) \cos \omega n$ is even and $x(n) \sin \omega n$ is odd. Hence, from (4.3.13), (4.3.14), and (4.3.23) we obtain

$$X_R(\omega) = x(0) + 2 \sum_{n=1}^\infty x(n) \cos \omega n \qquad \text{(even)} \qquad (4.3.24)$$

$$X_I(\omega) = 0 \qquad (4.3.25)$$

$$x(n) = \frac{1}{\pi} \int_0^\pi X_R(\omega) \cos \omega n \, d\omega \qquad (4.3.26)$$

Thus real and even signals possess real-valued spectra, which, in addition, are even functions of the frequency variable ω.

REAL AND ODD SIGNALS. If $x(n)$ is real and odd [i.e., $x(-n) = -x(n)$], then $x(n) \cos \omega n$ is odd and $x(n) \sin \omega n$ is even. Consequently, (4.3.13) and (4.3.23) imply that

$$X_R(\omega) = 0 \qquad (4.3.27)$$

$$X_I(\omega) = -2 \sum_{n=1}^\infty x(n) \sin \omega n \qquad \text{(odd)} \qquad (4.3.28)$$

$$x(n) = -\frac{1}{\pi} \int_0^\pi X_I(\omega) \sin \omega n \, d\omega \qquad (4.3.29)$$

Thus real-valued odd signals possess purely imaginary-valued spectral characteristics, which, in addition, are odd functions of the frequency variable ω.

PURELY IMAGINARY SIGNALS. In this case $x_R(n) = 0$ and $x(n) = jx_I(n)$. Thus (4.3.9), (4.3.10), and (4.3.12) reduce to

$$X_R(\omega) = \sum_{n=-\infty}^{\infty} x_I(n) \sin \omega n \qquad \text{(odd)} \tag{4.3.30}$$

$$X_I(\omega) = \sum_{n=-\infty}^{\infty} X_I(n) \cos \omega n \qquad \text{(even)} \tag{4.3.31}$$

$$x_I(n) = \frac{1}{\pi} \int_0^{\pi} [X_R(\omega) \sin \omega n + X_I(\omega) \cos \omega n] \, d\omega \tag{4.3.32}$$

If $x_I(n)$ is odd [i.e., $x_I(-n) = -x_I(n)$], then

$$X_R(\omega) = 2 \sum_{n=1}^{\infty} x_I(n) \sin \omega n \qquad \text{(odd)} \tag{4.3.33}$$

$$X_I(\omega) = 0 \tag{4.3.34}$$

$$x_I(n) = \frac{1}{\pi} \int_0^{\pi} X_R(\omega) \sin \omega n \, d\omega \tag{4.3.35}$$

Similarly, if $x_I(n)$ is even [i.e., $x_I(-n) = x_I(n)$], we have

$$X_R(\omega) = 0 \tag{4.3.36}$$

$$X_I(\omega) = x_I(0) + 2 \sum_{n=1}^{\infty} x_I(n) \cos \omega n \qquad \text{(even)} \tag{4.3.37}$$

$$x_I(n) = \frac{1}{\pi} \int_0^{\pi} X_I(\omega) \cos \omega n \, d\omega \tag{4.3.38}$$

These important symmetry properties of the Fourier transform are summarized in Fig. 4.20 and are often used to simplify Fourier transform calculations in practice.

EXAMPLE 4.3.2 _____

Determine and sketch $X_R(\omega)$, $X_I(\omega)$, $|X(\omega)|$, and $\angle X(\omega)$ for the Fourier transform

$$X(\omega) = \frac{1}{1 - ae^{-j\omega}} \qquad -1 < a < 1 \tag{4.3.39}$$

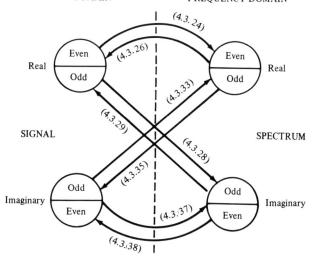

FIGURE 4.20 Summary of symmetry properties for the Fourier transform.

Solution: By multiplying both the numerator and denominator of (4.3.39) by the complex conjugate of the denominator, we obtain

$$X(\omega) = \frac{1 - ae^{j\omega}}{(1 - ae^{-j\omega})(1 - ae^{j\omega})} = \frac{1 - a\cos\omega - ja\sin\omega}{1 - 2a\cos\omega + a^2}$$

This expression can be subdivided into real and imaginary parts. Thus we obtain

$$X_R(\omega) = \frac{1 - a\cos\omega}{1 - 2a\cos\omega + a^2}.$$

$$X_I(\omega) = -\frac{a\sin\omega}{1 - 2a\cos\omega + a^2}$$

Substitution of the last two equations into (4.3.18) and (4.3.19) yields the magnitude and phase spectra as

$$|X(\omega)| = \frac{1}{\sqrt{1 - 2a\cos\omega + a^2}} \qquad (4.3.40)$$

and

$$\angle X(\omega) = -\tan^{-1}\frac{a\sin\omega}{1 - a\cos\omega} \qquad (4.3.41)$$

Figures 4.21 and 4.22 show the graphical representation of these spectra for $a = 0.8$. The reader can easily verify that all symmetry properties for the spectra of real signals apply to this case, as expected.

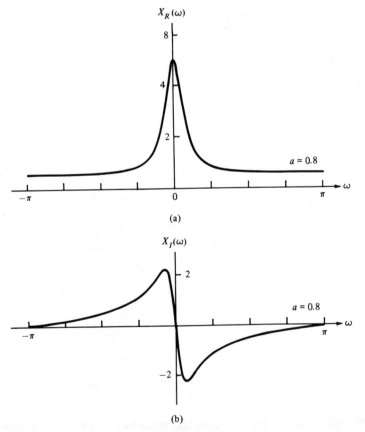

(a)

(b)

FIGURE 4.21 Graph of $X_R(\omega)$ and $X_I(\omega)$ for the transform in Example 4.3.2.

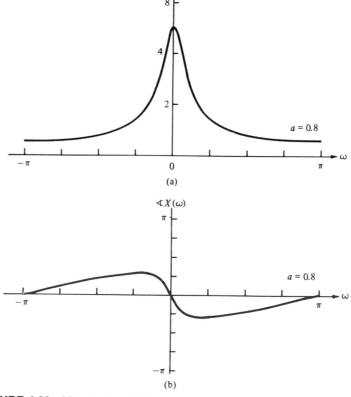

FIGURE 4.22 Magnitude and phase spectra of the transform in Example 4.3.2.

EXAMPLE 4.3.3

Determine the Fourier transform of the signal

$$x(n) = \begin{cases} A & -M \le n \le M \\ 0 & \text{elsewhere} \end{cases} \tag{4.3.42}$$

Solution: Clearly, $x(-n) = x(n)$. Thus $x(n)$ is a real and even signal. From (4.3.24) we obtain

$$X(\omega) = X_R(\omega) = A\left(1 + 2\sum_{n=1}^{M} \cos \omega n\right)$$

If we use the identity given in Problem 4.11, we obtain the simpler form

$$X(\omega) = A\,\frac{\sin (M + \frac{1}{2})\omega}{\sin (\omega/2)}$$

Since $X(\omega)$ is real, the magnitude and phase spectra are given by

$$|X(\omega)| = \left|A\,\frac{\sin (M + \frac{1}{2})\omega}{\sin (\omega/2)}\right| \tag{4.3.43}$$

and

$$\angle X(\omega) = \begin{cases} 0 & \text{if } X(\omega) > 0 \\ \pi & \text{if } X(\omega) < 0 \end{cases} \tag{4.3.44}$$

Figure 4.23 shows the graphs for $X(\omega)$.

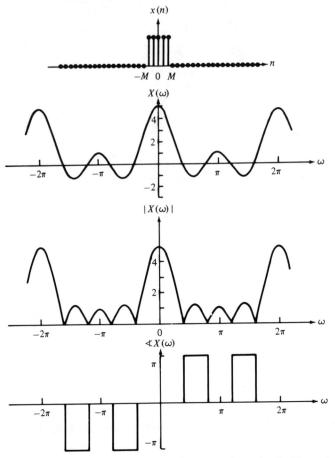

FIGURE 4.23 Spectral characteristics of rectangular pulse in Example 4.3.3.

Time Shifting. If

$$x(n) \xleftarrow{\quad F \quad} X(\omega)$$

then

$$x(n - k) \xleftarrow{\quad F \quad} e^{-j\omega k}X(\omega) \qquad (4.3.45)$$

The proof of this property follows immediately from the equivalent z-transform property given in (3.2.5). By evaluating the z-transform on the unit circle, we obtain

$$F\{x(n - k)\} = X(\omega)e^{-j\omega k}$$

By using (4.2.40), (4.3.45) may be expressed as

$$F\{x(n - k)\} = |X(\omega)|e^{j[\angle X(\omega) - \omega k]} \qquad (4.3.46)$$

This relation means that if a signal is shifted in the time domain by k samples, its magnitude spectrum remains unchanged. However, the phase spectrum is changed by an amount $-\omega k$. This result can easily be explained if we recall that the frequency content of a signal depends only on its shape. From a mathematical point of view, we can say that shifting by k in the time domain is equivalent to multiplying the spectrum by $e^{-j\omega k}$ in the frequency domain.

Time Reversal. If

$$x(n) \xleftarrow{\quad F \quad} X(\omega)$$

then

$$x(-n) \xleftarrow{\quad F \quad} X(-\omega) \tag{4.3.47}$$

This property can be established from the equivalent property for the z-transform given in (3.2.12). If we substitute $z = e^{j\omega}$, we obtain

$$F\{x(-n)\} = \sum_{l=-\infty}^{\infty} x(l)e^{j\omega l} = X(-\omega)$$

If $x(n)$ is real, then from (4.3.20) and (4.3.21) we obtain

$$F\{x(-n)\} = X(-\omega) = |X(-\omega)|e^{j\angle X(-\omega)}$$
$$= |X(\omega)|e^{-j\angle X(\omega)}$$

This means that if a signal is folded about the origin in time, its magnitude spectrum remains unchanged and the phase spectrum undergoes a change in sign (phase reversal).

Convolution Theorem. If

$$x_1(n) \xleftarrow{\quad F \quad} X_1(\omega)$$

and

$$x_2(n) \xleftarrow{\quad F \quad} X_2(\omega)$$

then

$$x(n) = x_1(n) * x_2(n) \xleftarrow{\quad F \quad} X(\omega) = X_1(\omega)X_2(\omega) \tag{4.3.48}$$

To prove this property we may use the equivalent z-transform relation $X(z) = X_1(z)X_2(z)$. Then, with the substitution $z = e^{j\omega}$, we obtain (4.3.48).

The convolution theorem is one of the most powerful tools in linear systems analysis. It means that if we convolve two signals in the time domain, this is equivalent to multiplying their spectra in the frequency domain. In later chapters we will see that the convolution theorem provides an important computational tool for many digital signal processing applications.

EXAMPLE 4.3.4 _____

By use of (4.3.48), determine the convolution of the sequences

$$x_1(n) = x_2(n) = \left\{ 1, \underset{\uparrow}{1}, 1 \right\}$$

Solution: By using (4.3.24), we obtain

$$X_1(\omega) = X_2(\omega) = 1 + 2 \cos \omega$$

Then

$$X(\omega) = X_1(\omega)X_2(\omega) = (1 + 2 \cos \omega)^2$$
$$= 3 + 4 \cos \omega + 2 \cos 2\omega$$
$$= 3 + 2(e^{j\omega} + e^{-j\omega}) + (e^{j2\omega} + e^{-j2\omega})$$

Hence the convolution of $x_1(n)$ with $x_2(n)$ is

$$x(n) = \left\{ 1\ 2\ \underset{\uparrow}{3}\ 2\ 1 \right\}$$

Figure 4.24 illustrates the foregoing relationships.

The Correlation Theorem. If

$$x_1(n) \xleftarrow{\quad F \quad} X_1(\omega)$$

and

$$x_2(n) \xleftarrow{\quad F \quad} X_2(\omega)$$

then

$$r_{x_1 x_2}(l) \xleftarrow{\quad F \quad} S_{x_1 x_2}(\omega) = X_1(\omega)X_2(-\omega) \qquad (4.3.49)$$

To prove this property, we may again use the equivalent relationship for the z-transform, given by (3.2.18) and let $z = e^{j\omega}$. The result in (4.3.49) follows immediately. The function $S_{x_1 x_2}(\omega)$ is called the *cross-energy density spectrum* of the signals $x_1(n)$ and $x_2(n)$.

The Wiener–Khintchine Theorem. Let $x(n)$ be a real signal. Then

$$r_{xx}(l) \xleftarrow{\quad F \quad} S_{xx}(\omega) \qquad (4.3.50)$$

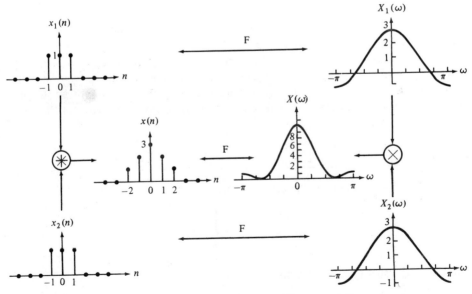

FIGURE 4.24 Graphical representation of the convolution property.

That is, the energy spectral density of an energy signal is the Fourier transform of its autocorrelation sequence. This is a special case of (4.3.49).

This is a very important result. It means that the autocorrelation sequence of a signal and its energy spectral density contain the same information about the signal. Since, neither of these contains any phase information, it is impossible to reconstruct the signal uniquely from the autocorrelation function or the energy density spectrum.

EXAMPLE 4.3.5 _____

Determine the energy density spectrum of the signal

$$x(n) = a^n u(n) \qquad -1 < a < 1$$

Solution: From Example 2.5.2 we found that the autocorrelation function for this signal is

$$r_{xx}(l) = \frac{1}{1 - a^2} a^{|l|} \qquad \infty < l < \infty$$

By using the result in (4.3.6) for the Fourier transform of $a^{|l|}$, derived in Example 4.3.1, we have

$$F\{r_{xx}(l)\} = \frac{1}{1 - a^2} F\{a^{|l|}\} = \frac{1}{1 - 2a \cos \omega + a^2}$$

Thus, according to the Wiener–Kintchine theorem,

$$S_{xx}(\omega) = \frac{1}{1 - 2a \cos \omega + a^2}$$

Frequency Shifting. If

$$x(n) \xleftarrow{\quad F \quad} X(\omega)$$

then

$$e^{j\omega_0 n} x(n) \xleftarrow{\quad F \quad} X(\omega - \omega_0) \tag{4.3.51}$$

This property is easily proved by direct substitution into the analysis equation (4.3.1). According to this property, multiplication of a sequence $x(n)$ by $e^{j\omega_0 n}$ is equivalent to a frequency translation of the spectrum $X(\omega)$ by ω_0. This frequency translation is illustrated in Fig. 4.25. Since the spectrum $X(\omega)$ is periodic, the shift ω_0 applies to the spectrum of the signal in every period.

The Modulation Theorem. If

$$x(n) \xleftarrow{\quad F \quad} X(\omega)$$

then

$$x(n) \cos \omega_0 n \xleftarrow{\quad F \quad} \tfrac{1}{2}X(\omega + \omega_0) + \tfrac{1}{2}X(\omega - \omega_0) \tag{4.3.52}$$

To prove the modulation theorem, we first express the signal $\cos \omega_0 n$ as

$$\cos \omega_0 n = \tfrac{1}{2}(e^{j\omega_0 n} + e^{-j\omega_0 n})$$

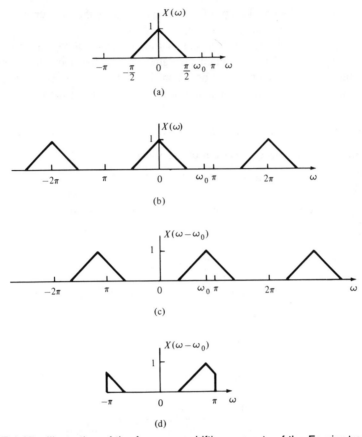

FIGURE 4.25   Illustration of the frequency-shifting property of the Fourier transform.

Upon multiplying $x(n)$ by these two exponentials and using the frequency-shifting property described in the preceding section, we obtain the desired result in (4.3.52).

Although the property given in (4.3.51) may also be viewed as (complex) modulation, in practice we prefer to use (4.3.52) because the signal $x(n) \cos \omega_0 n$ is real. Clearly, in this case the symmetry properties (4.3.15) and (4.3.16) are preserved.

The modulation theorem is illustrated in Fig. 4.26, which contains a plot of the spectra of the signals $x(n)$, $y_1(n) = x(n) \cos 0.5\pi n$ and $y_2(n) = x(n) \cos \pi n$.

Parseval's Theorem. If

$$x_1(n) \xleftarrow{\quad F \quad} X_1(\omega)$$

and

$$x_2(n) \xleftarrow{\quad F \quad} X_2(\omega)$$

then

$$\sum_{n=-\infty}^{\infty} x_1(n)x_2^*(n) = \frac{1}{2\pi} \int_{-\pi}^{\pi} X_1(\omega)X_2^*(\omega)\, d\omega \qquad (4.3.53)$$

To prove this theorem, we use (4.3.1) to eliminate $X_1(\omega)$ on the right-hand side of (4.3.53). Thus we have

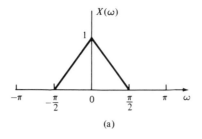

(a)

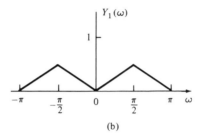

(b)

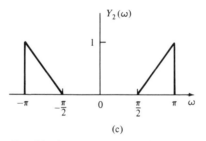

(c)

FIGURE 4.26 Graphical representation of the modulation theorem.

$$\frac{1}{2\pi} \int_{2\pi} \left[\sum_{n=-\infty}^{\infty} x_1(n)e^{-j\omega n} \right] X_2^*(\omega) \, d\omega$$

$$= \sum_{n=-\infty}^{\infty} x_1(n) \frac{1}{2\pi} \int_{2\pi} X_2^*(\omega)e^{-j\omega n} \, d\omega = \sum_{n=-\infty}^{\infty} x_1(n)x_2^*(n)$$

In the special case where $x_2(n) = x_1(n) = x(n)$, Parseval's relation (4.3.53) reduces to

$$\sum_{n=-\infty}^{\infty} |x(n)|^2 = \frac{1}{2\pi} \int_{2\pi} |X(\omega)|^2 \, d\omega \qquad (4.3.54)$$

We observe that the left-hand side of (4.3.54) is simply the energy E_x of the signal $x(n)$. It is also equal to the autocorrelation of $x(n)$, $r_{xx}(l)$, evaluated at $l = 0$. The integrand in the right-hand side of (4.3.54) is equal to the energy density spectrum, so the integral over the interval $-\pi \le \omega \le \pi$ yields the total signal energy. Therefore, we conclude that

$$E_x = r_{xx}(0) = \sum_{n=-\infty}^{\infty} |x(n)|^2 = \frac{1}{2\pi} \int_{2\pi} |X(\omega)|^2 \, d\omega = \frac{1}{2\pi} \int_{-\pi}^{\pi} S_{xx}(\omega) \, d\omega \qquad (4.3.55)$$

Multiplication of Two Sequences. If

$$x_1(n) \xleftarrow{\quad F \quad} X_1(\omega)$$

and

$$x_2(n) \xleftarrow{\quad F \quad} X_2(\omega)$$

then

$$x_3(n) \equiv x_1(n)x_2(n) \xleftarrow{\quad F \quad} X_3(\omega) = \frac{1}{2\pi} \int_{-\pi}^{\pi} X_1(\lambda)X_2(\omega - \lambda) \, d\lambda \quad (4.3.56)$$

The integral on the right-hand side of (4.3.56) represents the convolution of the Fourier transforms $X_1(\omega)$ and $X_2(\omega)$. This relation is the dual of the time-domain convolution. In other words, the multiplication of two time-domain sequences is equivalent to the convolution of their Fourier transforms. On the other hand, the convolution of two time-domain sequences is equivalent to the multiplication of their Fourier transforms.

The proof of (4.3.56) follows from the z-transform relation in (3.2.19). If $X_1(v)$ and $X_2(1/v)$ converge on the unit circle defined by $v = e^{j\lambda}$, $-\pi < \lambda < \pi$, we may select the unit circle as the contour for the integration specified by (3.2.19). By substituting $v = e^{j\lambda}$ and $z = e^{j\omega}$ in (3.2.19) and changing the variable of intergration from v to λ, we obtain the result in (4.3.56), where

$$X_1(\lambda) \equiv X_1(v)\big|_{v = e^{j\lambda}}$$

and

$$X_2(\omega - \lambda) \equiv X_2\left(\frac{z}{v}\right)\bigg|_{v = e^{j\lambda}, \ z = e^{j\omega}}$$

The Fourier transform pair in (4.3.56) will prove useful in our treatment of FIR filter design based on the window technique.

TABLE 4.2   Properties of the Fourier Transform for Discrete-Time Signals

Property	Time Domain	Frequency Domain
Notation	$x(n)$	$X(\omega)$
	$x_1(n)$	$X_1(\omega)$
	$x_2(n)$	$X_2(\omega)$
Linearity	$a_1x_1(n) + a_2x_2(n)$	$a_1X_1(\omega) + a_2X_2(\omega)$
Time shifting	$x(n - k)$	$e^{-j\omega k}X(\omega)$
Time reversal	$x(-n)$	$X(-\omega)$
Convolution	$x_1(n) * x_2(n)$	$X_1(\omega)X_2(\omega)$
Correlation	$r_{x_1 x_2}(l) = x_1(l) * x_2(-l)$	$R_{x_1 x_2}(\omega) = X_1(\omega)X_2(-\omega)$
		$= X_1(\omega)X_2^*(\omega)$
		[if $x_2(n)$ is real]
Wiener–Khintchine theorem	$r_{xx}(l)$	$S_{xx}(\omega)$
Frequency shifting	$e^{j\omega_0 n}x(n)$	$X(\omega - \omega_0)$
Modulation	$x(n) \cos \omega_0 n$	$\frac{1}{2}X(\omega + \omega_0) + \frac{1}{2}X(\omega - \omega_0)$
Multiplication	$x_1(n)x_2(n)$	$\dfrac{1}{2\pi} \displaystyle\int_{-\pi}^{\pi} X_1(\lambda)X_2(\omega - \lambda) \, d\lambda$
Differentiation in the frequency domain	$nx(n)$	$j\dfrac{dX(\omega)}{d\omega}$
Conjugation	$x^*(n)$	$X^*(-\omega)$
Parseval's theorem	$\displaystyle\sum_{n=-\infty}^{\infty} x_1(n)x_2^*(n) = \dfrac{1}{2\pi} \displaystyle\int_{-\pi}^{\pi} X_1(\omega)X_2^*(\omega) \, d\omega$	

Differentiation in the Frequency Domain. If

$$x(n) \xleftarrow{\quad F \quad} X(\omega)$$

then

$$nx(n) \xleftarrow{\quad F \quad} j \frac{dX(\omega)}{d\omega} \tag{4.3.57}$$

To prove this property, we use the definition of the Fourier transform in (4.3.1) and differentiate the series term by term with respect to ω. Thus we obtain

$$\frac{dX(\omega)}{d\omega} = \frac{d}{d\omega} \left[\sum_{n=-\infty}^{\infty} x(n) e^{-j\omega n} \right]$$

$$= \sum_{n=-\infty}^{\infty} x(n) \frac{d}{d\omega} e^{-j\omega n}$$

$$= -j \sum_{n=-\infty}^{\infty} nx(n) e^{-j\omega n}$$

Now we multiply both sides of the equation by j to obtain the desired result in (4.3.57).

TABLE 4.3   Some Useful Fourier Transform Pairs for Discrete-Time Aperiodic Signals

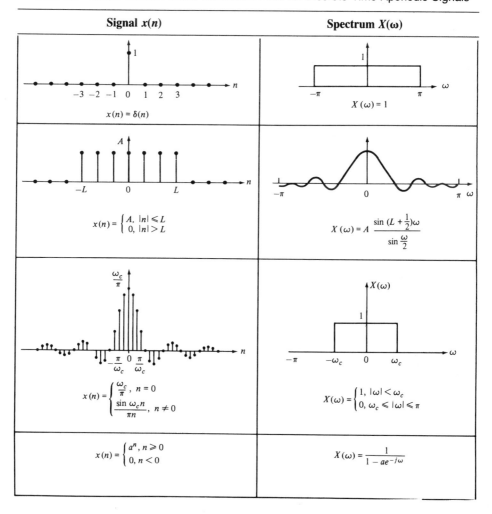

Signal $x(n)$	Spectrum $X(\omega)$
$x(n) = \delta(n)$	$X(\omega) = 1$
$x(n) = \begin{cases} A, & \lvert n \rvert \leqslant L \\ 0, & \lvert n \rvert > L \end{cases}$	$X(\omega) = A \dfrac{\sin\left(L + \frac{1}{2}\right)\omega}{\sin \frac{\omega}{2}}$
$x(n) = \begin{cases} \dfrac{\omega_c}{\pi}, & n = 0 \\ \dfrac{\sin \omega_c n}{\pi n}, & n \neq 0 \end{cases}$	$X(\omega) = \begin{cases} 1, & \lvert \omega \rvert < \omega_c \\ 0, & \omega_c \leqslant \lvert \omega \rvert \leqslant \pi \end{cases}$
$x(n) = \begin{cases} a^n, & n \geqslant 0 \\ 0, & n < 0 \end{cases}$	$X(\omega) = \dfrac{1}{1 - ae^{-j\omega}}$

The properties derived in this section are summarized in Table 4.2, which serves as a convenient reference. Table 4.3 illustrates some useful Fourier transform pairs that will be encountered in later chapters.

4.4 Frequency-Domain Characterization of Signals and Time-Frequency Dualities

The frequency analysis tools that we have developed in this chapter find widespread use in many practical problems. Natural signals, such as seismic, biological, and electromagnetic signals, can be characterized by their frequency content. Usually, the frequency content of the signal provides us with the important information that we are interested in extracting from the signal. For example, in the case of biological signals, frequency analysis is often used for diagnostic purposes (e.g., to diagnose an abnormality in a physical system, such as illness in a person). In seismic signals, frequency analysis may serve as an aid in exploring the structure of the earth and in searching for oil. In the case of electromagnetic signals, such as radar, frequency analysis provides important information on the speed of an airplane.

In this section we describe the frequency-domain characteristics of commonly encountered physical signals. In addition, we discuss some important dualities between the time-domain and frequency-domain characteristics of signals.

4.4.1 Physical and Mathematical Dualities

In the previous sections of the chapter we have introduced several methods for the frequency analysis of signals. The presentation of several methods was necessary to accommodate the different types of signals. To summarize, the following frequency analysis tools have been introduced:

1. The Fourier series for continuous-time periodic signals.
2. The Fourier transform for continuous-time aperiodic signals.
3. The Fourier series for discrete-time periodic signals.
4. The Fourier transform for discrete-time aperiodic signals.

Figure 4.27 summarizes the analysis and synthesis formulas for these types of signals.

As we have already indicated several times, there are two time-domain characteristics that determine the type of signal spectrum we obtain: whether the time variable is continuous or discrete, and whether the signal is periodic or aperiodic. Let us briefly summarize the results of the previous sections.

CONTINUOUS-TIME SIGNALS HAVE APERIODIC SPECTRA. A close inspection of the Fourier series and Fourier transform analysis formulas for continuous-time signals does not reveal any kind of periodicity in the spectral domain. This lack of periodicity is a consequence of the fact that the complex exponential $\exp(j2\pi Ft)$ is a function of the continuous variable t, and hence it is not periodic in F. Thus the frequency range of continuous-time signals extends from $F = 0$ to $F = \infty$.

DISCRETE-TIME SIGNALS HAVE PERIODIC SPECTRA. Indeed, both the Fourier series and the Fourier transform for discrete-time signals are periodic with period $\omega = 2\pi$. As a result of this periodicity, the frequency range of discrete-time signals is finite and extends from $\omega = 0$ to $\omega = \pi$ radians, where $\omega = \pi$ corresponds to the highest possible rate of oscillation.

PERIODIC SIGNALS HAVE DISCRETE SPECTRA. As we have observed, periodic signals are described by means of Fourier series. The Fourier series coefficients provide the "lines" that constitute the discrete spectrum. The line spacing ΔF or Δf is equal to the inverse of the period T_p or N, respectively, in the time domain. That is, $\Delta F = 1/T_p$ for continuous-time periodic signals and $\Delta f = 1/N$ for discrete-time signals.

APERIODIC FINITE ENERGY SIGNALS HAVE CONTINUOUS SPECTRA. This property is a direct consequence of the fact that both $X(F)$ and $X(\omega)$ are functions of $\exp(j2\pi Ft)$ and $\exp(j\omega n)$, respectively, which are continuous functions of the variables F and ω. The continuity in frequency is necessary in order to break the harmony and thus create aperiodic signals.

In summary, we can conclude that *periodicity with "period"* α *in one domain automatically implies discretization with "spacing"* $1/\alpha$ *in the other domain, and vice versa.*

If we keep in mind that "period" in the frequency domain means the frequency range and "spacing" in the time domain is the sampling period T and in the frequency domain the line spacing ΔF, then $\alpha = T_p$ implies that $1/\alpha = 1/T_p = \Delta F$, $\alpha = N$ implies that $\Delta f = 1/N$, and $\alpha = F_s$ implies that $T = 1/F_s$.

These time-frequency dualities are apparent from observation of Fig. 4.27. We stress, however, that the illustrations used in this figure do not correspond to any actual transform pairs. Thus any comparison among them should be avoided.

A careful inspection of Fig. 4.27 also reveals some mathematical symmetries and dualities among the several frequency analysis relationships. In particular, we observe that there are dualities between the following analysis and synthesis equations:

1. The analysis and synthesis equations of the continuous-time Fourier transform.
2. The analysis and synthesis equations of the discrete-time Fourier series.
3. The analysis equation of the continuous-time Fourier series and the synthesis equation of the discrete-time Fourier transform.
4. The analysis equation of the discrete-time Fourier transform and the synthesis equation of the continuous-time Fourier series.

Note that all dual relations differ only in the sign of the exponent of the corresponding complex exponential. It is interesting to note that this change in sign can be thought of either as a folding of the signal or a folding of the spectrum, since

$$e^{-j2\pi Ft} = e^{j2\pi(-F)t} = e^{j2\pi F(-t)}$$

If we turn our attention now to the spectral density of signals, we recall that we have used the term *energy density spectrum* for characterizing finite-energy aperiodic signals and the term *power density spectrum* for periodic signals. This terminology is consistent with the fact that periodic signals are power signals and aperiodic signals with finite energy are energy signals.

4.4.2 Frequency-Domain Classification of Signals: The Concept of Bandwidth

Just as we have classified signals according to their time-domain characteristics, it is also desirable to classify signals according to their frequency-domain characteristics. It is common practice to classify signals in rather broad terms according to their frequency content.

In particular, if a power signal (or energy signal) has its power density spectrum (or its energy density spectrum) concentrated about zero frequency, such a signal is called a *low-frequency signal*. Figure 4.28a illustrates the spectral characteristics of such a signal. On the other hand, if the signal power density spectrum (or the energy density spectrum) is concentrated at high frequencies, the signal is called a *high-frequency signal*.

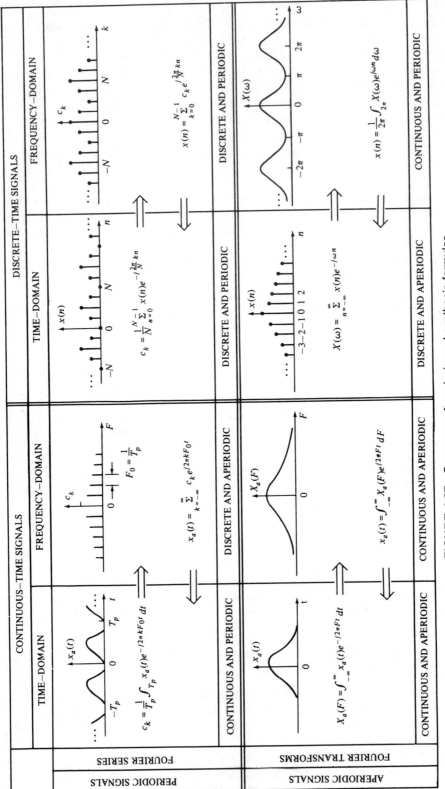

FIGURE 4.27 Summary of analysis and synthesis formulas.

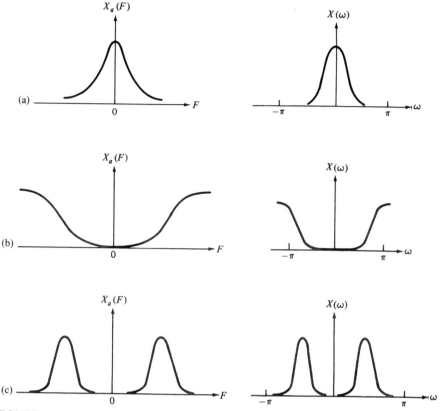

FIGURE 4.28 (a) Low-frequency, (b) high-frequency, and (c) medium-frequency signals.

Such a signal spectrum is illustrated in Fig. 4.28b. A signal having a power density spectrum (or an energy density spectrum) concentrated somewhere in the broad frequency range between low frequencies and high frequencies is called a *medium-frequency signal* or a *bandpass signal*. Figure 4.28(c) illustrates such a signal spectrum.

In addition to this relatively broad frequency-domain classification of signals, it is often desirable to express quantitatively the range of frequencies over which the power or energy density spectrum is concentrated. This quantitative measure is called the *bandwidth* of a signal. For example, suppose that a continuous-time signal has 95% of its power (or energy) density spectrum concentrated in the frequency range $F_1 \le F \le F_2$. Then the 95% bandwidth of the signal is $F_2 - F_1$. In a similar manner, we may define the 75% or 90% or 99% bandwidth of the signal.

In the case of a bandpass signal, the term *narrowband* is used to describe the signal if its bandwidth $F_2 - F_1$ is much smaller (say, by a factor of 10 or more) than the median frequency $(F_2 + F_1)/2$. Otherwise, the signal is called *wideband*.

We shall say that a signal is *bandlimited* if its spectrum is zero outside the frequency range $|F| \ge B$. For example, a continuous-time finite-energy signal $x(t)$ is bandlimited if its Fourier transform $X(F) = 0$ for $|F| > B$. A discrete-time finite-energy signal $x(n)$ is said to be *periodically bandlimited* if

$$|X(\omega)| = 0 \qquad \text{for } \omega_0 < |\omega| < \pi$$

Similarly, a periodic continuous-time signal $x_p(t)$ is periodically bandlimited if its Fourier coefficients $c_k = 0$ for $|k| > M$, where M is some positive integer. A periodic discrete-

time signal with fundamental period N is periodically bandlimited if the Fourier coeffi-
cients $c_k = 0$ for $k_0 < |k| < N$. Figure 4.29 illustrates the four types of bandlimited
signals.

By exploiting the duality between the frequency domain and the time domain, we can
provide similar means for characterizing signals in the time domain. In particular, a
signal $x(t)$ will be called *time-limited* if

$$x(t) = 0 \qquad |t| > \tau$$

If the signal is periodic with period T_p, it will be called *periodically time-limited* if

$$x_p(t) = 0 \qquad \tau < |t| < T_p/2$$

If we have a discrete-time signal $x(n)$ of finite duration, that is,

$$x(n) = 0 \qquad |n| > N$$

it is also called time-limited. When the signal is periodic with fundamental period N, it
is said to be periodically time-limited if

$$x(n) = 0 \qquad n_0 < |n| < N$$

We state, without proof, that a basic characteristic of any signal is that it cannot be
time-limited and bandlimited simultaneously. Furthermore, a reciprocal relationship exists between the time duration and the frequency duration of a signal. To elaborate, if
we have a short-duration rectangular pulse in the time domain, its spectrum has a width
that is inversely proportional to the duration of the time-domain pulse. The narrower the
pulse becomes in the time domain, the larger the bandwidth of the signal becomes.
Consequently, the product of the time duration and the bandwidth of a signal cannot be
made arbitrarily small. A short-duration signal has a large bandwidth and a small bandwidth signal has a long duration. Thus, for any signal, the time–bandwidth product is
fixed and cannot be made arbitrarily small.

4.4.3 The Frequency Ranges of Some Natural Signals

The frequency analysis tools that we have developed in this chapter are usually applied
to a variety of signals that are encountered in practice (e.g., seismic, biological, and

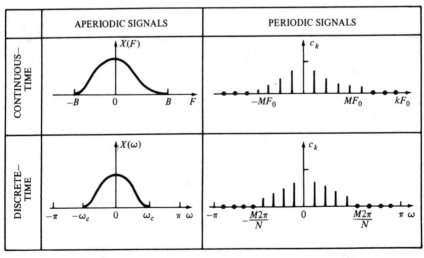

FIGURE 4.29 Some examples of bandlimited signals.

electromagnetic signals). In general, the frequency analysis is performed for the purpose of extracting information from the observed signal. For example, in the case of biological signals, such as an ECG signal, the analysis tools are used to extract information that is relevant for diagnostic purposes. In the case of seismic signals, we may be interested in detecting the presence of a nuclear explosion or in determining the characteristics and location of an earthquake. An electromagnetic signal, such as a radar signal reflected from an airplane, contains information on the position of the plane and its radial velocity. These parameters can be estimated from observation of the received radar signal.

In processing any signal for the purpose of measuring parameters or extracting other types of information, one must know approximately the range of frequencies contained by the signal. For reference, Tables 4.4, 4.5, and 4.6 give approximate limits in the frequency domain for biological, seismic, and electromagnetic signals.

TABLE 4.4 Frequency Ranges of Some Biological Signals

Type of Signal	Frequency Range (Hz)
Electroretinogram[a]	0–20
Electronystagmogram[b]	0–20
Pneumogram[c]	0–40
Electrocardiogram (ECG)	0–100
Electroencephalogram (EEG)	0–100
Electromyogram[d]	10–200
Sphygmomanogram[e]	0–200
Speech	100–4000

[a]A graphic recording of retina characteristics.
[b]A graphic recording of involuntary movement of the eyes.
[c]A graphic recording of respiratory activity.
[d]A graphic recording of muscular action, such as muscular contraction.
[e]A recording of blood pressure.

TABLE 4.5 Frequency Ranges of Some Seismic Signals

Type of Signal	Frequency Range (Hz)
Wind noise	100–1000
Seismic exploration signals	10–100
Earthquake and nuclear explosion signals	0.01–10
Seismic noise	0.1–1

TABLE 4.6 Frequency Ranges of Electromagnetic Signals

Type of Signal	Wavelength (m)	Frequency Range (Hz)
Radio broadcast	10^4-10^2	$3 \times 10^4 - 3 \times 10^6$
Shortwave radio signals	10^2-10^{-2}	$3 \times 10^6 - 3 \times 10^{10}$
Radar, satellite communications, space communications, common-carrier microwave	$1-10^{-2}$	$3 \times 10^8 - 3 \times 10^{10}$
Infared	$10^{-3}-10^{-6}$	$3 \times 10^{11} - 3 \times 10^{14}$
Visible light	$3.9 \times 10^{-7} - 8.1 \times 10^{-7}$	$3.7 \times 10^{14} - 7.7 \times 10^{14}$
Ultraviolet	$10^{-7}-10^{-8}$	$3 \times 10^{15} - 3 \times 10^{16}$
Gamma rays and x-rays	$10^{-9}-10^{-10}$	$3 \times 10^{17} - 3 \times 10^{18}$

4.5 Sampling of Signals in the Time and Frequency Domains

As explained in Chapter 1, sampling is a basic prerequisite for digital processing of continuous-time signals. Usually, the original continuous-time signal is converted into a sequence of numbers by periodic sampling. After the sequence is processed, the resulting sequence may be converted back to a continuous-time signal, as, for example, in speech signal processing. In other cases, where the objective of the signal processing is to extract signal parameters or other relevant information from the signal, there is no need to reconvert the signal to analog form. Radar and sonar signal processing are examples of the latter.

Sampling may also be performed on the spectrum of a signal. For example, if the signal is an aperiodic finite-energy signal, its spectrum is continuous, and hence its computation is possible, in practice, only at a set of discrete frequencies. Since we are observing the spectrum at discrete frequencies, we refer to such a computation as *frequency-domain sampling*. Such frequency-domain sampling occurs in spectrum analysis and spectrum estimation.

Analysis of periodic sampling of continuous-time signals or continuous spectra using the frequency analysis tools introduced in this chapter is the subject of this section.

4.5.1 Time-Domain Sampling and Reconstruction of Analog Signals

To process a continuous-time signal by digital signal processing techniques, it is necessary to convert the signal into a sequence of numbers. As was discussed in Section 1.4, this is usually done by sampling the analog signal, say $x_a(t)$, periodically every T seconds to produce a discrete-time signal $x(n)$ given by

$$x(n) = x_a(nT) \qquad -\infty < n < \infty \qquad (4.5.1)$$

The samples must then be quantized to a discrete set of amplitude levels, and the resulting digital signal is passed to the digital processor. Figure 4.30 illustrates a typical configuration of a system for processing an analog signal digitally. In the following discussion we neglect the quantization errors that are inherent in the A/D conversion process.

The relationship (4.5.1) describes the sampling process in the time domain. As discussed in Chapter 1, the sampling frequency $F_s = 1/T$ must be selected large enough such that the sampling process will not result in any loss of spectral information (no aliasing). Indeed, if the spectrum of the analog signal can be recovered from the spectrum of the discrete-time signal, there is no loss of information. Consequently, we investigate the sampling process by finding the relationship between the spectra of signals $x_a(t)$ and $x(n)$.

If $x_a(t)$ is an aperiodic signal with finite energy, its (voltage) spectrum is given by the Fourier transform relation

$$X_a(F) = \int_{-\infty}^{\infty} x_a(t)e^{-j2\pi Ft}\, dt \qquad (4.5.2)$$

FIGURE 4.30   Configuration of system for digital processing of an analog signal.

whereas the signal $x_a(t)$ can be recovered from its spectrum by the inverse Fourier transform

$$x_a(t) = \int_{-\infty}^{\infty} X_a(F)e^{j2\pi Ft} \, dF \tag{4.5.3}$$

Note that utilization of all frequency components in the infinite frequency range $-\infty < F < \infty$ is necessary to recover the signal $x_a(t)$ if it is not bandlimited.

The spectrum of a discrete-time signal $x(n)$ obtained by sampling $x_a(t)$ is given by the Fourier transform relation

$$X(\omega) = \sum_{n=-\infty}^{\infty} x(n)e^{-j\omega n}$$

or $\hspace{10cm}$ (4.5.4)

$$X(f) = \sum_{n=-\infty}^{\infty} x(n)e^{-j2\pi fn}$$

The sequence $x(n)$ can be recovered from its spectrum $X(\omega)$ or $X(f)$ by the inverse transform

$$\begin{aligned} x(n) &= \frac{1}{2\pi} \int_{-\pi}^{\pi} X(\omega)e^{j\omega n} \, d\omega \\ &= \int_{-1/2}^{1/2} X(f)e^{j2\pi fn} \, df \end{aligned} \tag{4.5.5}$$

In order to determine the relationship between the spectra of the discrete-time signal and the analog signal, we note that periodic sampling imposes a relationship between the independent variables t and n in the signals $x_a(t)$ and $x(n)$, respectively. That is,

$$t = nT = \frac{n}{F_s} \tag{4.5.6}$$

This relationship in the time domain implies a corresponding relationship between the frequency variables F and f in $X_a(F)$ and $X(f)$, respectively.

Indeed, substitution of (4.5.6) into (4.5.3) yields

$$x(n) \equiv x_a(nT) = \int_{-\infty}^{\infty} X_a(F)e^{j2\pi nF/F_s} \, dF \tag{4.5.7}$$

If we compare (4.5.5) with (4.5.7), we conclude that

$$\int_{-1/2}^{1/2} X(f)e^{j2\pi fn} \, df = \int_{-\infty}^{\infty} X_a(F)e^{j2\pi nF/F_s} \, dF \tag{4.5.8}$$

From the development in Chapter 1 we know that periodic sampling imposes a relationship between the frequency variables F and f of the corresponding analog and discrete-time signals, respectively. That is,

$$f = \frac{F}{F_s} \tag{4.5.9}$$

With the aid of (4.5.9), we may make a simple change in variable (4.5.8), and thus we obtain the result

$$\frac{1}{F_s} \int_{-F_s/2}^{F_s/2} X\left(\frac{F}{F_s}\right) e^{j2\pi nF/F_s} \, dF = \int_{-\infty}^{\infty} X_a(F)e^{j2\pi nF/F_s} \, dF \tag{4.5.10}$$

Now we turn our attention to the integral on the right-hand side of (4.5.10). The integration range of this integral can be divided into an infinite number of intervals of width F_s. Thus the integral over the infinite range can be expressed as a sum of integrals, that is,

$$\int_{-\infty}^{\infty} X_a(F)e^{j2\pi nF/F_s}\, dF = \sum_{k=-\infty}^{\infty} \int_{(k-1/2)F_s}^{(k+1/2)F_s} X_a(F)e^{j2\pi nF/F_s}\, dF \qquad (4.5.11)$$

We observe that $X_a(F)$ in the frequency interval $(k - \frac{1}{2})F_s$ to $(k + \frac{1}{2})F_s$ is identical to $X_a(F - kF_s)$ in the interval $-F_s/2$ to $F_s/2$. Consequently,

$$\sum_{k=-\infty}^{\infty} \int_{(k-1/2)F_s}^{(k+1/2)F_s} X_a(F)e^{j2\pi nF/F_s}\, dF = \sum_{k=-\infty}^{\infty} \int_{-F_s/2}^{F_s/2} X_a(F - kF_s)e^{j2\pi nF/F_s}\, dF$$

$$= \int_{-F_s/2}^{F_s/2} \left[\sum_{k=-\infty}^{\infty} X_a(F - kF_s) \right] e^{j2\pi nF/F_s}\, dF$$

$$(4.5.12)$$

where we have used the periodicity of the exponential, namely,

$$e^{j2\pi n(F + kF_s)/F_s} = e^{j2\pi nF/F_s}$$

Upon comparison of (4.5.12), (4.5.11), and (4.5.10), we conclude that

$$X\left(\frac{F}{F_s}\right) = F_s \sum_{k=-\infty}^{\infty} X_a(F - kF_s) \qquad (4.5.13)$$

or, equivalently,

$$X(f) = F_s \sum_{k=-\infty}^{\infty} X_a[(f - k)F_s] \qquad (4.5.14)$$

This is the desired relationship between the spectrum $X(F/F_s)$ or $X(f)$ of the discrete-time signal and the spectrum $X_a(F)$ of the analog signal. The right-hand side of (4.5.13) or (4.5.14) consists of a periodic repetition of the scaled spectrum $F_s X_a(F)$ with period F_s. This periodicity is necessary because the spectrum $X(f)$ or $X(F/F_s)$ of the discrete-time signal is periodic with period $f_p = 1$ or $F_p = F_s$.

For example, suppose that the spectrum of a band-limited analog signal is as shown in Fig. 4.31a. The spectrum is zero for $|F| \geq B$. Now, if the sampling frequency F_s is selected to be greater than $2B$, the spectrum $X(F/F_s)$ of the discrete-time signal will appear as shown in Fig. 4.31b. Thus, if the sampling frequency F_s is selected such that $F_s \geq 2B$, where $2B$ is the Nyquist rate, then

$$X\left(\frac{F}{F_s}\right) = F_s X_a(F) \qquad |F| \leq F_s/2 \qquad (4.5.15)$$

In this case there is no aliasing, and hence the spectrum of the discrete-time signal is identical (within the scale factor F_s) to the spectrum of the analog signal, within the fundamental frequency range $|F| \leq F_s/2$ or $|f| \leq \frac{1}{2}$.

On the other hand, if the sampling frequency F_s is selected such that $F_s < 2B$, the periodic continuation of $X_a(F)$ results in spectral overlap, as illustrated in Fig. 4.31c and d. Thus the spectrum $X(F/F_s)$ of the discrete-time signal contains aliased frequency components of the analog signal spectrum $X_a(F)$. The end result is that the aliasing which occurs prevents us from recovering the original signal $x_a(t)$ from the samples.

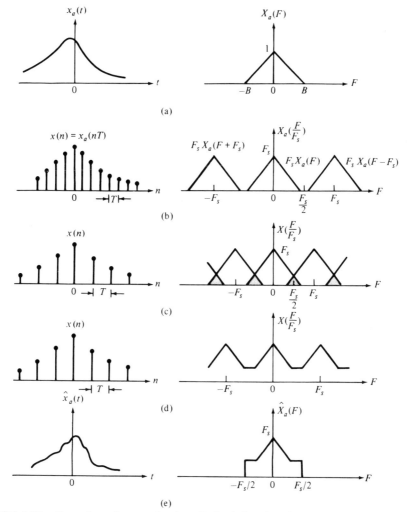

FIGURE 4.31 Sampling of an analog bandlimited signal and aliasing of spectral components.

Given the discrete-time signal $x(n)$ with the spectrum $X(F/F_s)$, as illustrated in Fig. 4.31(b), with no aliasing, it is now possible to reconstruct the original analog signal from the samples $x(n)$. Since in the absence of aliasing

$$X_a(F) = \begin{cases} \dfrac{1}{F_s} X\left(\dfrac{F}{F_s}\right) & |F| \le F_s/2 \\[2mm] 0, & |F| > F_s/2 \end{cases}$$

(4.5.16)

and by the Fourier transform relationship (4.5.4),

$$X\left(\frac{F}{F_s}\right) = \sum_{n=-\infty}^{\infty} x(n)e^{-j2\pi Fn/F_s}$$

(4.5.17)

the inverse Fourier transform of $X_a(F)$ is

$$x_a(t) = \int_{-F_s/2}^{F_s/2} X_a(F)e^{j2\pi Ft}\, dF$$

(4.5.18)

To be specific, we assume that $F_s = 2B$. With the substitution of (4.5.16) into (4.5.18), we have

$$x_a(t) = \frac{1}{F_s} \int_{-F_s/2}^{F_s/2} \left[\sum_{n=-\infty}^{\infty} x(n)e^{-j2\pi Fn/F_s} \right] e^{j2\pi Ft} \, dF$$

$$= \frac{1}{F_s} \sum_{n=-\infty}^{\infty} x(n) \int_{-F_s/2}^{F_s/2} e^{j2\pi F(t-n/F_s)} \, dF$$

$$= \sum_{n=-\infty}^{\infty} x_a(nT) \frac{\sin(\pi/T)(t-nT)}{(\pi/T)(t-nT)} \tag{4.5.19}$$

where $x(n) = x_a(nT)$ and where $T = 1/F_s = 1/2B$ is the sampling interval. This is the reconstruction formula give by (1.4.24) in our discussion of the sampling theorem.

The reconstruction formula in (4.5.19) involves the function

$$g(t) = \frac{\sin(\pi/T)t}{(\pi/T)t} \tag{4.5.20}$$

appropriately shifted by nT, $n = 0, \pm 1, \pm 2, \ldots$, and multiplied or weighted by the corresponding samples $x_a(nT)$ of the signal. We call (4.5.19) an interpolation formula for reconstructing $x_a(t)$ from its samples, and $g(t)$, given in (4.5.20), is the interpolation function. We note that at $t = kT$, the interpolation function $g(t - nT)$ is zero except at $k = n$. Consequently, $x_a(t)$ evaluated at $t = kT$ is simply the sample $x_a(kT)$. At all other times the weighted sum of the time shifted versions of the interpolation function combine to yield exactly $x_a(t)$. This combination is illustrated in Fig. 4.32.

The formula in (4.5.19) for reconstructing the analog signal $x_a(t)$ from its samples is called the *ideal interpolation formula*. It forms the basis for the *sampling theorem*, which may be stated as follows.

SAMPLING THEOREM: *A bandlimited continuous-time signal, with highest frequency (bandwidth) B hertz, can be uniquely recovered from its samples provided that the sampling rate $F_s \geq 2B$ samples per second.*

According to the sampling theorem and the reconstruction formula in (4.5.19), the recovery of $x_a(t)$ from its samples $x(n)$ requires an infinite number of samples. However, in practice, we deal with a finite number of samples of the signal and finite-duration signals. As a consequence we are concerned only with reconstructing a finite-duration signal from a finite number of samples.

When aliasing occurs due to too low a sampling rate, this effect can be described by a multiple folding of the frequency axis of the frequency variable F for the analog signal.

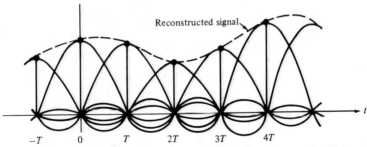

FIGURE 4.32 Reconstruction of a continuous-time signal using ideal interpolation.

Figure 4.33a shows the spectrum $X_a(F)$ of an analog signal. According to (4.5.13), sampling of the signal with a sampling frequency F_s results in a periodic repetition of $X_a(F)$ with period F_s. If $F_s < 2B$, the shifted replicas of $X_a(F)$ overlap. The overlap that occurs within the fundamental frequency range $-F_s/2 \leq F \leq F_s/2$ is illustrated in Fig. 4.33b. The corresponding spectrum of the discrete-time signal within the fundamental frequency range is obtained by adding all the shifted portions within the range $|f| \leq \frac{1}{2}$, to yield the spectrum shown in Fig. 4.33c.

A careful inspection of Fig. 4.33a and b reveals that the aliased spectrum in Fig. 4.33c can be obtained by folding the original spectrum like an accordian with pleats at every odd multiple of $F_s/2$. Consequently, the frequency $F_s/2$ is called the *folding frequency*, as indicated in Chapter 1. Clearly, then, periodic sampling automatically forces a folding of the frequency axis of an analog signal at odd multiples of $F_s/2$ and results in the relationship $F = fF_s$ between the frequencies for continuous-time signals and discrete-time signals. Due to the folding of the frequency axis, the relationship $F = fF_s$ is not truly linear, but piecewise linear, to accommodate for the aliasing effect. This relationship is illustrated in Fig. 4.34.

If the analog signal is bandlimited to $B \leq F_s/2$, the relationship between f and F is linear and one-to-one. In other words, there is no aliasing. In practice, prefiltering is usually employed prior to sampling to ensure that frequency components of the signal above $F \geq B$ are sufficiently attenuated so that, if aliased, they cause negligible distortion on the desired signal.

The following examples serve to illustrate the problem of aliasing of frequency components.

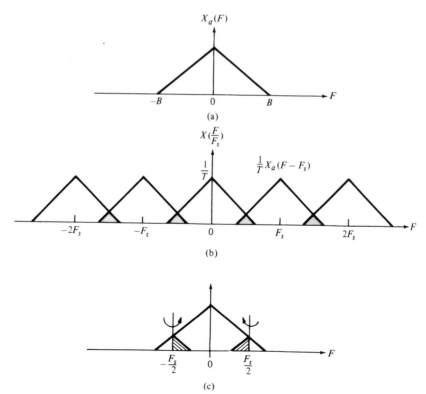

FIGURE 4.33   Illustration of aliasing around the folding frequency.

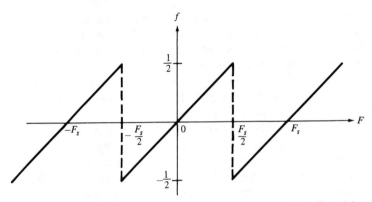

FIGURE 4.34 Relationship between frequency variables F and f.

EXAMPLE 4.5.1 Aliasing in Sinusoidal Signals _____

The continuous-time signal

$$x_a(t) = \cos 2\pi F_0 t = \tfrac{1}{2} e^{j2\pi F_0 t} + \tfrac{1}{2} e^{-j2\pi F_0 t}$$

has a discrete spectrum with spectral lines at $F = \pm F_0$, as shown in Fig. 4.35a. The process of sampling this signal with a sampling frequency F_s introduces replicas of the spectrum about multiples of F_s. This is illustrated in Fig. 4.35b for $F_s/2 < F_0 < F_s$.

To reconstruct the continuous-time signal, we should select the frequency components inside the fundamental frequency range $|F| \le F_s/2$. The resulting spectrum is shown in Fig. 4.35c. The reconstructed signal is

$$x_a(t) = \cos 2\pi (F_s - F_0) t$$

Now, if F_s is selected such that $F_s < F_0 < 3F_s/2$, the spectrum of the sampled signal is shown in Fig. 4.35d. The reconstructed signal, shown in Fig. 4.35e, is

$$x_a(t) = \cos 2\pi (F_0 - F_s) t$$

In both cases, aliasing has occurred, so that the frequency of the reconstructed signal is an aliased version of the frequency of the original signal.

EXAMPLE 4.5.2 Sampling a Nonbandlimited Signal _____

Consider the continuous-time signal

$$x_a(t) = e^{-A|t|} \qquad A > 0$$

whose spectrum is given by

$$X_a(F) = \frac{2A}{A^2 + (2\pi F)^2}$$

Determine the spectrum of the sampled signal $x(n) \equiv x_a(nT)$.

Solution: If we sample $x_a(t)$ with a sampling frequency $F_s = 1/T$, we have

$$x(n) = x_a(nT) = e^{-AT|n|} = (e^{-AT})^{|n|} \qquad -\infty < n < \infty$$

The spectrum of $x(n)$ can be found easily if we use a direct computation of the Fourier transform. We find that

$$X\left(\frac{F}{F_s}\right) = \frac{1 - e^{-2AT}}{1 - 2e^{-AT} \cos 2\pi FT + e^{-2AT}} \qquad T = \frac{1}{F_s}$$

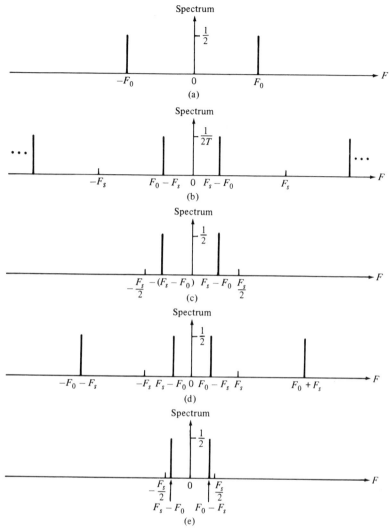

FIGURE 4.35   Aliasing of sinusoidal signals.

Clearly, since $\cos 2\pi FT = \cos 2\pi(F/F_s)$ is periodic with period F_s, so is $X(F/F_s)$.

Since $X_a(F)$ is not bandlimited, aliasing cannot be avoided. The spectrum of the reconstructed signal $\hat{x}_a(t)$ is

$$
\hat{X}_a(F) = \begin{cases} TX\left(\dfrac{F}{F_s}\right) & |F| \le \dfrac{F_s}{2} \\ 0 & |F| > \dfrac{F_s}{2} \end{cases}
$$

Figure 4.36a shows the original signal $x_a(t)$ and its spectrum $X_a(F)$ for $A = 1$. The sampled signal $x(n)$ and its spectrum $X(F/F_s)$ are shown in Fig. 4.36b for $F_s = 1$ Hz. The aliasing distortion is clearly noticeable in the frequency domain. The reconstructed signal $\hat{x}_a(t)$ is shown in Fig. 4.36c. The distortion due to aliasing can be reduced significantly by increasing the sampling rate. For example, Fig. 4.36d illustrates the reconstructed signal corresponding to a sampling rate $F_s = 20$ Hz. It is interesting to note that in every case $x_a(nT) = \hat{x}_a(nT)$, but $x_a(t) \ne \hat{x}_a(t)$ at other values of time.

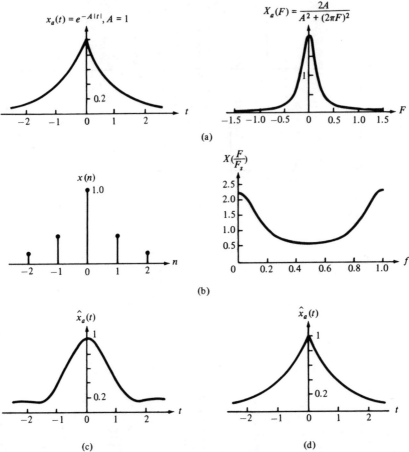

FIGURE 4.36 (a) Analog signal $x_a(t)$ and its spectrum $X_a(F)$; (b) $x(n) = x_a(nT)$ and the spectrum of $x(n)$ for $A = 1$ and $F_s = 1$; (c) reconstructed signal $\hat{x}_a(t)$ for $F_s = 1$; (d) reconstructed signal $\hat{x}_a(t)$ for $F_s = 20$.

Up to this point in our discussion of sampling we have assumed that all the signals under consideration were lowpass signals. Clearly, if we sample a bandpass signal at twice its highest frequency, the original signal can be perfectly reconstructed. However, as we will see in Section 6.1.5, it is possible to sample a bandpass signal at a minimum rate of twice its bandwidth without any loss of information.

Finally, we mention that as an interpolation formula, the sampling theorem was known in mathematics a long time ago. However, it was Shannon (1949) who introduced it in the communication theory literature in terms of frequency-domain concepts.

4.5.2 Frequency-Domain Sampling and Reconstruction of Analog Signals

From our previous discussion we recall that continuous-time finite-energy signals have continuous spectra. In this section we consider the sampling of such signals periodically and the reconstruction of the signals from samples of their spectra.

Consider an analog signal $x_a(t)$ with a continuous spectrum $X_a(F)$, which is the Fourier transform of $x_a(t)$, given by (4.5.2). Now, suppose that we obtain samples of $X_a(F)$ every

δF hertz, as shown in Fig. 4.37. Is it possible to recover $X_a(F)$ or, equivalently, $x_a(t)$ from its samples $\{X_a(k\ \delta F),\ -\infty < k < \infty\}$?

This problem is mathematically the dual to that of sampling a continuous-time signal in the time domain. Indeed, if we begin with (4.5.2) and sample the spectrum every δF hertz, we obtain

$$X_a(k\ \delta F) = \int_{-\infty}^{\infty} x_a(t)e^{-j2\pi k\delta Ft}\, dt \tag{4.5.21}$$

It is convenient to define the reciprocal of δF as

$$T_s = \frac{1}{\delta F}$$

Then (4.5.21) can be expressed as

$$X_a(k\ \delta F) = \int_{-\infty}^{\infty} x_a(t)e^{-j2\pi kt/T_s}\, dt \tag{4.5.22}$$

which is analogous to (4.5.7) for time-domain sampling. Proceeding as in the case of time-domain sampling, we subdivide the integration range of the integral in (4.5.22) into an infinite number of intervals of width T_s and change the variable of integration so as to translate each integral into the fundamental range $-T_s/2 \leq t \leq T_s/2$. These steps lead to the result

$$X_a(k\ \delta F) = \int_{-T_s/2}^{T_s/2} \left[\sum_{n=-\infty}^{\infty} x_a(t - nT_s) \right] e^{-j2\pi kt/T_s}\, dt \tag{4.5.23}$$

which is the dual of (4.5.12).

The signal

$$x_p(t) \equiv \sum_{n=-\infty}^{\infty} x_a(t - nT_s) \tag{4.5.24}$$

is clearly periodic with fundamental period $T_s = 1/\delta F$. Hence it can be expanded into a Fourier series

$$x_p(t) = \sum_{k=-\infty}^{\infty} c_k e^{j2\pi k\delta Ft} \tag{4.5.25}$$

where

$$c_k = \frac{1}{T_s} \int_{-T_s/2}^{T_s/2} x_p(t)e^{-j2\pi k\delta Ft}\, dt \tag{4.5.26}$$

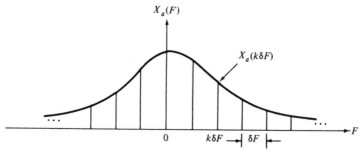

FIGURE 4.37 Evaluating the Fourier transform at a set of equally spaced discrete frequencies.

Comparison of (4.5.26) with (4.5.23) gives the result

$$c_k = \frac{1}{T_s} X_a(k \, \delta F)$$

$$= \delta F X_a(k \, \delta F) \qquad -\infty < k < \infty$$

(4.5.27)

We conclude from (4.5.27) that the samples of the spectrum $X_a(F)$ correspond to (within the scale factor δF) the Fourier coefficients of a periodic signal $x_p(t)$ with period $T_s = 1/\delta F$. The periodic signal $x_p(t)$ consists of the sum of periodically extended versions of the analog signal $x_a(t)$ as given by (4.5.24) and illustrated in Fig. 4.38.

It is apparent from Fig. 4.38 that recovery of the signal $x_a(t)$ from its periodic extension $x_p(t)$ is possible if $x_a(t)$ is time-limited to $|t| \leq \tau$ [i.e., $x_a(t) = 0$ for $|t| > \tau$, where $\tau < T_s/2$]. If $\tau > T_s/2$, exact reconstruction of $x_a(t)$ is not possible due to "time-domain aliasing."

If the analog signal $x_a(t)$ is time-limited to τ seconds and T_s is selected such that $T_s > 2\tau$, there is no aliasing and the signal spectrum $X_a(F)$ can be perfectly reconstructed from the samples $X_a(k \, \delta F)$ by using the interpolation formula

$$X_a(F) = \sum_{k=-\infty}^{\infty} X_a(k \, \delta F) \frac{\sin \left[(\pi/\delta F)(F - k \, \delta F) \right]}{(\pi/\delta F)(F - k \, \delta F)}$$

(4.5.28)

which is the dual of (4.5.19).

The development given above is important primarily for conceptual purposes. Although frequency sampling of continuous-time signals is encountered in practical frequency analysis applications, in many applications of this type, the analog signal is converted to a discrete-time signal and the frequency sampling is performed on the spectrum of the discrete-time signal. Frequency sampling of discrete-time signal spectra is considered below.

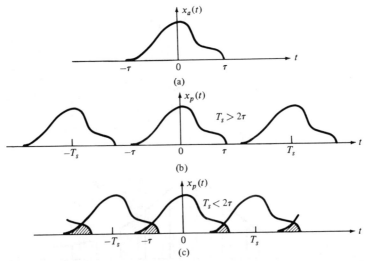

FIGURE 4.38 Periodic signal resulting from sampling a finite energy, time-limited signal in the frequency domain. Aliasing in the time domain is illustrated in (c).

4.5.3 Frequency-Domain Sampling and Reconstruction of Discrete-Time Signals

We recall that aperiodic finite-energy signals have continuous spectra. Let us consider such an aperiodic discrete-time signal $x(n)$ with Fourier transform

$$X(\omega) = \sum_{n=-\infty}^{\infty} x(n)e^{-j\omega n} \qquad (4.5.29)$$

Suppose that we sample $X(\omega)$ periodically in frequency at a spacing of $\delta\omega$ radians between successive samples. Since $X(\omega)$ is periodic with period 2π, only samples in the fundamental frequency range are necessary. For convenience, we take N equidistant samples in the interval $0 \leq \omega < 2\pi$ with spacing $\delta\omega = 2\pi/N$, as shown in Fig. 4.39. First, we consider the selection of N, the number of samples in the frequency domain.

If we evaluate (4.5.29) at $\omega = 2\pi k/N$, we obtain

$$X\left(\frac{2\pi}{N}k\right) = \sum_{n=-\infty}^{\infty} x(n)e^{-j2\pi kn/N} \qquad k = 0, 1, \ldots, N-1 \qquad (4.5.30)$$

The summation in (4.5.29) can be subdivided into an infinite number of summations, where each sum contains N terms. Thus

$$X\left(\frac{2\pi}{N}k\right) = \cdots + \sum_{n=-N}^{-1} x(n)e^{-j2\pi kn/N} + \sum_{n=0}^{N-1} x(n)e^{-j2\pi kn/N} + \sum_{n=N}^{2N-1} x(n)e^{-j2\pi kn/N}$$

$$= \sum_{l=-\infty}^{\infty} \sum_{n=lN}^{lN+N-1} x(n)e^{-j2\pi kn/N}$$

If we change the index in the inner summation from n to $n - lN$ and interchange the order of the summation, we obtain the result

$$X\left(\frac{2\pi}{N}k\right) = \sum_{n=0}^{N-1}\left[\sum_{l=-\infty}^{\infty} x(n - lN)\right] e^{-j2\pi kn/N} \qquad (4.5.31)$$

for $k = 0, 1, 2, \ldots, N-1$.

The signal

$$x_p(n) = \sum_{l=-\infty}^{\infty} x(n - lN) \qquad (4.5.32)$$

obtained by the periodic repetition of $x(n)$ every N samples, is clearly periodic with fundamental period N. Consequently, it can be expanded in a Fourier series as

$$x_p(n) = \sum_{k=0}^{N-1} c_k e^{j2\pi kn/N} \qquad n = 0, 1, \ldots, N-1 \qquad (4.5.33)$$

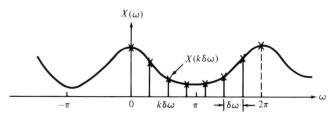

FIGURE 4.39   Frequency-domain sampling of the Fourier transform.

with Fourier coefficients

$$c_k = \frac{1}{N} \sum_{n=0}^{N-1} x_p(n) e^{-j2\pi kn/N} \qquad k = 0, 1, \ldots, N - 1 \qquad (4.5.34)$$

Upon comparing (4.5.34) with (4.5.31), we conclude that

$$c_k = \frac{1}{N} X\left(\frac{2\pi}{N} k\right) \qquad k = 0, 1, \ldots, N - 1 \qquad (4.5.35)$$

Therefore,

$$x_p(n) = \frac{1}{N} \sum_{k=0}^{N-1} X\left(\frac{2\pi}{N} k\right) e^{j2\pi kn/N} \qquad n = 0, 1, \ldots, N - 1 \qquad (4.5.36)$$

The relationship in (4.5.36) provides the reconstruction of the periodic signal $x_p(n)$ from the samples of the spectrum $X(\omega)$. However, it does not imply that we can recover $X(\omega)$ or $x(n)$ from the samples. To accomplish this, we need to consider the relationship between $x_p(n)$ and $x(n)$.

Since $x_p(n)$ is the periodic extension of $x(n)$ as given by (4.5.32), it is clear that $x(n)$ can be recovered from $x_p(n)$ if there is no aliasing in the time domain, that is, if $x(n)$ is time-limited to less than the period N of $x_p(n)$. This situation is illustrated in Fig. 4.40, where without loss of generality, we consider a finite-duration sequence $x(n)$, which is nonzero in the interval $0 \le n \le L - 1$. We observe that when $N \ge L$,

$$x(n) = x_p(n) \qquad 0 \le n \le N - 1$$

so that $x(n)$ can be recovered from $x_p(n)$ without ambiguity. On the other hand, if $N < L$, it is not possible to recover $x(n)$ from its periodic extension due to *time-domain aliasing*. Thus we conclude that the spectrum of an aperiodic discrete-time signal with finite duration L can be exactly recovered from its samples at frequencies $\omega_k = 2\pi k/N$ if $N \ge L$. The procedure is to compute $x_p(n)$, $n = 0, 1, \ldots, N - 1$ from (4.5.36); then

$$x(n) = \begin{cases} x_p(n) & 0 \le n \le N - 1 \\ 0 & \text{elsewhere} \end{cases} \qquad (4.5.37)$$

and finally, $X(\omega)$ can be computed from (4.5.29).

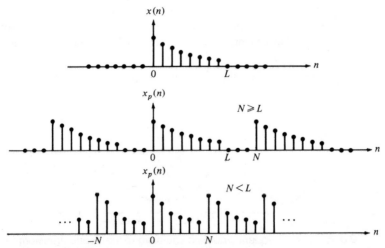

FIGURE 4.40   Aperiodic sequence $x(n)$ of length L and its periodic extension for $N \ge L$ (no aliasing) and $N < L$ (aliasing).

As in the case of continuous-time signals it is possible to express the spectrum $X(\omega)$ directly in terms of its samples $X(2\pi k/N)$, $k = 0, 1, \ldots , N - 1$. To derive such an interpolation formula from $X(\omega)$, we assume that $N \geq L$ and begin with (4.5.36). Since $x(n) = x_p(n)$ for $0 \leq n \leq N - 1$,

$$x(n) = \frac{1}{N} \sum_{k=0}^{N-1} X\left(\frac{2\pi}{N} k\right) e^{j2\pi kn/N} \qquad 0 \leq n \leq N - 1 \qquad (4.5.38)$$

If we use (4.5.29) and substitute for $x(n)$, we obtain

$$
\begin{aligned}
X(\omega) &= \sum_{n=0}^{N-1} \left[\frac{1}{N} \sum_{k=0}^{N-1} X\left(\frac{2\pi}{N} k\right) e^{j2\pi kn/N} \right] e^{-j\omega n} \\
&= \sum_{k=0}^{N-1} X\left(\frac{2\pi}{N} k\right) \left[\frac{1}{N} \sum_{n=0}^{N-1} e^{-j(\omega - 2\pi k/N)n} \right]
\end{aligned}
\qquad (4.5.39)
$$

The inner summation term in the brackets of (4.5.39) represents the basic interpolation function shifted by $2\pi k/N$ in frequency. Indeed, if we define

$$
\begin{aligned}
P(\omega) &= \frac{1}{N} \sum_{n=0}^{N-1} e^{-j\omega n} = \frac{1}{N} \frac{1 - e^{-j\omega N}}{1 - e^{-j\omega}} \\
&= \frac{\sin (\omega N/2)}{N \sin (\omega/2)} e^{-j\omega(N-1)/2}
\end{aligned}
\qquad (4.5.40)
$$

then (4.5.39) can be expressed as

$$X(\omega) = \sum_{k=0}^{N-1} X\left(\frac{2\pi}{N} k\right) P\left(\omega - \frac{2\pi}{N} k\right) \qquad (4.5.41)$$

The interpolation function $P(\omega)$ is not the familiar $(\sin \theta)/\theta$ but, instead, it is a periodic counterpart of it, due to the periodic nature of $X(\omega)$. The phase shift in (4.5.40) reflects the fact that the signal $x(n)$ is a causal, finite-duration sequence of length N. The function $\sin (\omega N/2)/N \sin (\omega/2)$ is plotted in Fig. 4.41 for $N = 5$. We observe that the function $P(\omega)$ has the property

$$P\left(\frac{2\pi}{N} k\right) = \begin{cases} 1 & k = 0 \\ 0 & k = 1, 2, \ldots , N - 1 \end{cases} \qquad (4.5.42)$$

Consequently, the interpolation formula in (4.5.41) gives exactly the sample values $X(2\pi k/N)$ for $\omega = 2\pi k/N$. At all other frequencies, the formula provides a properly weighted linear combination of the original spectral samples.

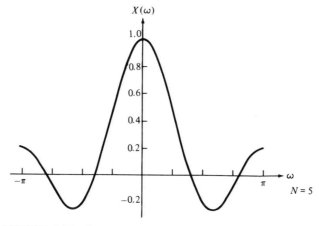

FIGURE 4.41   Plot of the function $[\sin (\omega N/2)]/[N \sin (\omega/2)]$.

The following example illustrates the frequency-domain sampling of a discrete-time signal and the time-domain aliasing that results.

EXAMPLE 4.5.3 ────────────────────────────────

Consider the signal

$$x(n) = a^n u(n) \qquad 0 < a < 1$$

The spectrum of this signal is sampled at frequencies $\omega_k = 2\pi k/N$, $k = 0, 1, \ldots$, $N - 1$. Determine the reconstructed spectra for $a = 0.8$ when $N = 5$ and $N = 50$.

Solution: The Fourier transform of the sequence $x(n)$ is

$$X(\omega) = \sum_{n=0}^{\infty} a^n e^{-j\omega n} = \frac{1}{1 - ae^{-j\omega}}$$

When evaluated for a $a = 0.8$ at the N frequencies, we obtain

$$X\left(\frac{2\pi}{N}k\right) = \frac{1}{1 - 0.8e^{-j2\pi k/N}}$$

The periodic sequence $x_p(n)$ corresponding to the frequency samples $X(2\pi k/N)$, $k = 0$, $1, \ldots, N - 1$ is obtained from (4.5.36). Thus

$$x(n) \equiv x_p(n) = \frac{1}{N} \sum_{k=0}^{N-1} X\left(\frac{2\pi k}{N}\right) e^{j2\pi kn/N} \qquad n = 0, 1, \ldots, N - 1$$

The results of this computation are illustrated in Fig. 4.42 for $N = 5$ and $N = 50$. For comparison, the original sequence $x(n)$ and its spectrum are also shown. The effect of aliasing is clearly evident for $N = 5$. However, for $N = 50$, the aliased components are very small and, consequently, $\hat{x}(n) \approx x(n)$, for $n = 0, 1, \ldots, N - 1$.

4.5.4 The Discrete Fourier Transform for Finite-Duration Sequences

The development in the preceding section was concerned with the frequency-domain sampling of an aperiodic finite-energy sequence $x(n)$. In general, the equally spaced frequency samples $X(2\pi k/N)$, $k = 0, 1, \ldots, N - 1$, do not represent the original sequence $x(n)$ uniquely when $x(n)$ has infinite duration. Instead, the frequency samples $X(2\pi k/N)$, $k = 0, 1, \ldots, N - 1$, correspond to a periodic sequence $x_p(n)$ of period N, where $x_p(n)$ is an aliased version of $x(n)$, as indicated by the relation in (4.5.32), that is,

$$x_p(n) = \sum_{l=-\infty}^{\infty} x(n - lN) \qquad (4.5.43)$$

When the sequence $x(n)$ has a finite duration of length $L \leq N$, then $x_p(n)$ is simply a periodic repetition of $x(n)$, where $x_p(n)$ over a single period is given as

$$x_p(n) = \begin{cases} x(n) & 0 \leq n \leq L - 1 \\ 0 & L \leq n \leq N - 1 \end{cases} \qquad (4.5.44)$$

Consequently, the frequency samples $X(2\pi k/N)$, $k = 0, 1, \ldots, N - 1$ uniquely represent the finite-duration sequence $x(n)$. Since $x(n) \equiv x_p(n)$ over a single period (padded by $N - L$ zeros), the original finite-duration sequence $x(n)$ can be obtained from the frequency samples $\{X(2\pi k/N\}$ by means of the formula (4.5.36).

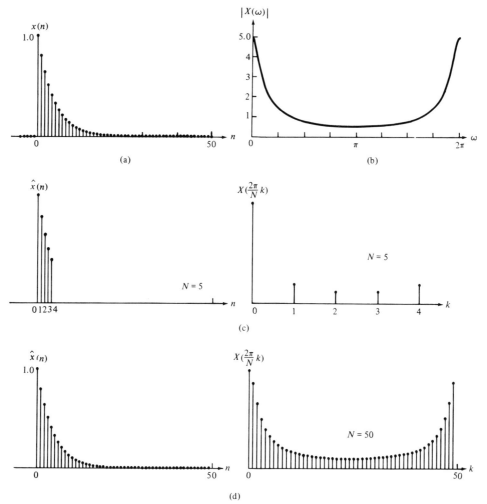

FIGURE 4.42 (a) Plot of sequence $x(n) = (0.8)^n u(n)$; (b) its Fourier transform (magnitude only); (c) effect of aliasing with $N = 5$; (d) reduced effect of aliasing with $N = 50$.

In summary, a finite-duration sequence $x(n)$ of length L [i.e., $x(n) = 0$ for $n < 0$ and $n \geq L$] has a Fourier transform

$$X(\omega) = \sum_{n=0}^{L-1} x(n)e^{-j\omega n} \qquad 0 \leq \omega \leq 2\pi \qquad (4.5.45)$$

where the upper and lower indices in the summation reflect the fact that $x(n) = 0$ outside the range $0 \leq n \leq L - 1$. When we sample $X(\omega)$ at equally spaced frequencies $\omega_k = 2\pi k/N$, $k = 0, 1, 2, \ldots, N - 1$, where $N \geq L$, the resultant samples are

$$X(k) \equiv X\left(\frac{2\pi k}{N}\right) = \sum_{n=0}^{L-1} x(n)e^{-j2\pi kn/N}$$

$$X(k) = \sum_{n=0}^{N-1} x(n)e^{-j2\pi kn/N} \qquad k = 0, 1, 2, \ldots, N - 1 \qquad (4.5.46)$$

where for convenience, the upper index in the sum has been increased from $L - 1$ to $N - 1$ since $x(n) = 0$ for $n \geq L$.

The relation in (4.5.46) is a formula for transforming a sequence $\{x(n)\}$ of length $L \leq N$ into a sequence of frequency samples $\{X(k)\}$ of length N. Since the frequency samples are obtained by evaluating the Fourier transform $X(\omega)$ at a set of N (equally spaced) discrete frequencies, the relation in (4.5.46) is called the *discrete Fourier transform* (DFT) of $x(n)$. In turn, the relation given by (4.5.36), which allows us to recover the sequence $x(n)$ from the frequency samples, that is,

$$x(n) = \frac{1}{N} \sum_{k=0}^{N-1} X(k) e^{j2\pi kn/N} \qquad n = 0, 1, \ldots, N - 1 \qquad (4.5.47)$$

is called the *inverse DFT* (IDFT). Clearly, when $x(n)$ has length $L < N$, the N-point IDFT will yield $x(n) = 0$ for $L \leq n \leq N - 1$. To summarize, the formulas for the DFT and IDFT are

DFT

$$X(k) = \sum_{n=0}^{N-1} x(n) e^{-j2\pi kn/N} \qquad k = 0, 1, 2, \ldots, N - 1 \qquad (4.5.46)$$

IDFT

$$x(n) = \frac{1}{N} \sum_{k=0}^{N-1} X(k) e^{j2\pi kn/N} \qquad n = 0, 1, 2, \ldots, N - 1 \qquad (4.5.47)$$

EXAMPLE 4.5.4 _____

A finite-duration sequence of length L is given as

$$x(n) = \begin{cases} 1 & 0 \leq n \leq L - 1 \\ 0 & \text{otherwise} \end{cases}$$

Determine the N-point DFT of this sequence for $N \geq L$.

Solution: The Fourier transform of this sequence is

$$X(\omega) = \sum_{n=0}^{L-1} x(n) e^{-j\omega n}$$

$$= \sum_{n=0}^{L-1} e^{-j\omega n} = \frac{1 - e^{-j\omega L}}{1 - e^{-j\omega}} = \frac{\sin(\omega L/2)}{\sin(\omega/2)} e^{-j\omega(L-1)/2}$$

The magnitude and phase of $X(\omega)$ are illustrated in Fig. 4.43 for $L = 10$. The N-point DFT of $x(n)$ is simply $X(\omega)$ evaluated at the set of N equally spaced frequencies $\omega_k = 2\pi k/N, k = 0, 1, \ldots, N - 1$. Hence

$$X(k) = \frac{1 - e^{-j2\pi kL/N}}{1 - e^{-j2\pi k/N}} \qquad k = 0, 1, \ldots, N - 1$$

$$= \frac{\sin(\pi kL/N)}{\sin(\pi k/N)} e^{-j\pi k(L-1)/N}$$

If N is selected such that $N = L$, then the DFT becomes

$$X(k) = \begin{cases} L & k = 0 \\ 0 & k = 1, 2, \ldots, L - 1 \end{cases}$$

Thus there is only one nonzero value in the DFT. This is apparent from observation of $X(\omega)$, since $X(\omega) = 0$ at the frequencies $\omega_k = 2\pi k/L, k \neq 0$. The reader should verify that $x(n)$ can be recovered from $X(k)$ by performing an L-point IDFT.

Although the L-point DFT is sufficient to represent uniquely the sequence $x(n)$ in the frequency domain, it is apparent that it does not provide sufficient detail to yield a good picture of the spectral characteristics of $x(n)$. If we wish to have better picture, we must evaluate (interpolate) $X(\omega)$ at closer spaced frequencies, say $\omega_k = 2\pi k/N$, where $N > L$. In effect, we may view this computation as expanding the size of the sequence from L points to N points by appending $N - L$ zeros to the sequence $x(n)$. This is sometimes called padding the sequence $x(n)$ with $N - L$ zeros. Then the N-point DFT provides finer interpolation than the L-point DFT.

Figure 4.44 provides a plot of the N-point DFT, in magnitude and phase, for $L = 10$, $N = 50$, and $N = 100$. Now the spectral characteristics of the sequence are more clearly evident, as one will conclude by comparing these spectra with the continuous spectrum $X(\omega)$.

It is instructive to view the DFT and IDFT as linear transformations on sequences $\{x_N(n)\}$ and $\{X(k)\}$, respectively. Let us define an N-point vector $\mathbf{x}_N(n)$ of the signal sequence $x(n)$, $n = 0, 1, \ldots, N - 1$, an N-point vector $\mathbf{X}_N(k)$ of frequency samples, and an $N \times N$ matrix $\mathbf{W}_N$ as

$$\mathbf{x}_N = \begin{bmatrix} x(0) \\ x(1) \\ \vdots \\ x(N-1) \end{bmatrix} \qquad \mathbf{X}_N = \begin{bmatrix} X(0) \\ X(1) \\ \vdots \\ X(N-1) \end{bmatrix}$$

$$\mathbf{W}_N = \begin{bmatrix} 1 & 1 & 1 & \cdots & 1 \\ 1 & W_N & W_N^2 & \cdots & W_N^{N-1} \\ & W_N^2 & W_N^4 & \cdots & W_N^{2(N-1)} \\ \vdots & \vdots & \vdots & \cdots & \vdots \\ 1 & W_N^{N-1} & W_N^{2(N-1)} & \cdots & W_N^{(N-1)(N-1)} \end{bmatrix} \tag{4.5.48}$$

where, by definition,

$$W_N = e^{-j2\pi/N}$$

which is an Nth root of unity.

With these definitions, the N-point DFT may be expressed in matrix form as

$$\mathbf{X}_N = \mathbf{W}_N \mathbf{x}_N \tag{4.5.49}$$

where W_N is the matrix of the linear transformation. We observe that W_N is a symmetric matrix. If we assume that the inverse of W_N exists, then (4.5.49) can be inverted by premultiplying both sides by W^{-1}. Thus we obtain

$$\mathbf{x}_N = \mathbf{W}_N^{-1} \mathbf{X}_N \tag{4.5.50}$$

But this is just an expression for the IDFT.

In fact, the IDFT as given by (4.5.47) may be expressed in matrix form as

$$\mathbf{x}_N = \frac{1}{N} \mathbf{W}_N^* \mathbf{X}_N \tag{4.5.51}$$

where $\mathbf{W}_N^*$ denotes the complex conjugate of the matrix $\mathbf{W}_N$. Comparison of (4.5.51) with (4.5.50) leads us to conclude that

$$\mathbf{W}_N^{-1} = \frac{1}{N} \mathbf{W}_N^* \tag{4.5.52}$$

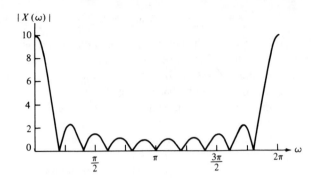

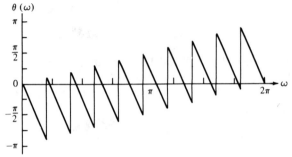

FIGURE 4.43 Magnitude and phase characteristics of the Fourier transform for signal in Example 4.5.4.

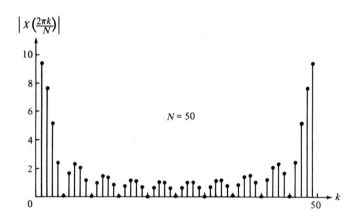

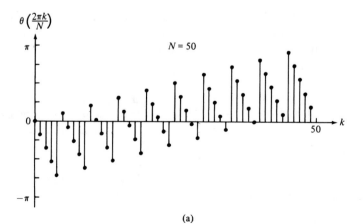

(a)

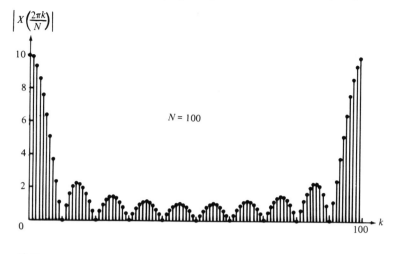

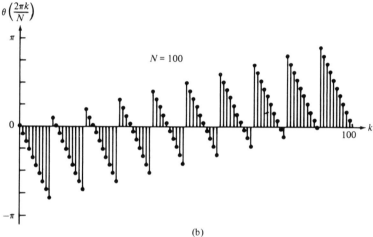

(b)

FIGURE 4.44 Magnitude and phase of an N-point DFT in Example 4.5.4: (a) $L = 10$, $N = 50$; (b) $L = 10$, $N = 100$.

which, in turn, implies that

$$\mathbf{W}_N \mathbf{W}_N^* = N\mathbf{I}_N \tag{4.5.53}$$

where $\mathbf{I}_N$ is an $N \times N$ identity matrix. Therefore, the matrix $\mathbf{W}_N$ in the transformation is an orthogonal matrix. Furthermore, its inverse exists and is given as $\mathbf{W}_N^*/N$. Of course, the existence of the inverse of $\mathbf{W}_N$ was established previously from our derivation of the IDFT.

The DFT and IDFT play a very important role in many digital signal processing applications, such as frequency analysis (spectrum analysis) of signals, power spectrum estimation, and linear filtering. The importance of the DFT and IDFT in such practical applications is due to a large extent on the existence of computationally efficient algorithms, known collectively as Fast Fourier Transform (FFT) algorithms, for computing the DFT and IDFT. This class of algorithms is described in Chapter 9.

4.6 Summary and References

The Fourier series and the Fourier transform are the mathematical tools for analyzing the characteristics of signals in the frequency domain. The Fourier series is appropriate for representing a periodic signal as a weighted sum of harmonically related sinusoidal components, where the weighting coefficients represent the strengths of each of the harmonics and the magnitude squared of each weighting coefficient represents the power of the corresponding harmonic. As we have indicated, the Fourier series is one of many possible orthogonal series expansions for a periodic signal. Its importance stems from the characteristic behavior of LTI systems, as we shall see in Chapter 5.

The Fourier transform is appropriate for representing the spectral characteristics of aperiodic signals with finite energy. It is closely related to the z-transform introduced in Chapter 3. We observed that the Fourier transform of a signal exists if its z-transform converges on the unit circle and can be obtained by evaluating the z-transform on the unit circle. Clearly, there are signals for which the Fourier transform does not exist, but their z-transform does exist.

The important properties of the Fourier transform were also presented in this chapter. We observed that these properties closely parallel those for the z-transform.

The last major topic of this chapter was concerned with the sampling of signals in the time and frequency domains. We derived the sampling theorem, which was introduced in Chapter 1, and demonstrated the aliasing effects due to undersampling. We observed that sampling the spectrum of a signal is the dual of time-domain sampling, in the sense that time-domain aliasing occurs when the frequency samples are far apart. Furthermore, a reconstruction formula was derived for recovering the spectrum of the signal from the samples of the spectrum, which is akin to the reconsruction formula obtained for time-domain sampling. Similar results were obtained in the case of discrete-time signals. When the signal sequence has a finite duration L, we demonstrated that N samples of the Fourier transform where $N > L$, uniquely represent the signal. The N equally spaced samples of the Fourier transform are called the discrete Fourier transform (DFT) of the L-point sequence and the inverse relationship for obtaining the sequence from its DFT is called the inverse DFT (IDFT). The DFT is one of the most important transforms in digital signal processing. It is widely used for linear filtering applications, as we shall see in the following chapter and for performing spectrum analysis of signals as described in Chapter 11. The efficient computation of the DFT on a digital computer is treated in Chapter 9.

There are many excellent texts on Fourier series and Fourier transforms. For reference, we include the texts by Bracewell (1978), Davis (1963), Dym and McKean (1972), and Papoulis (1962).

PROBLEMS

4.1 Consider the full-wave rectified sinusoid in Fig. P4.1.

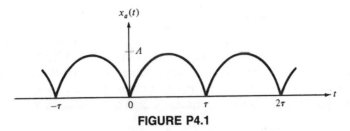

FIGURE P4.1

(a) Determine its spectrum $X_a(F)$.
(b) Compute the power of the signal.
(c) Plot the power spectral density.
(d) Check the validity of Parseval's relation for this signal.

4.2 Compute and sketch the magnitude and phase spectra for the following signals $(a > 0)$.

(a) $x_a(t) = \begin{cases} Ae^{-at} & t \geq 0 \\ 0 & t < 0 \end{cases}$

(b) $x_a(t) = Ae^{-a|t|}$

4.3 Consider the signal

$$x(t) = \begin{cases} 1 - |t|/\tau & |t| \leq \tau \\ 0 & \text{elsewhere} \end{cases}$$

(a) Determine and sketch its magnitude and phase spectra, $|X_a(F)|$ and $\angle X_a(F)$, respectively.
(b) Create a periodic signal $x_p(t)$ with fundamental period $T_p \geq 2\tau$, so that $x(t) = x_p(t)$ for $|t| < T_p/2$. What are the Fourier coefficients c_k for the signal $x_p(t)$?
(c) Using the results in parts (a) and (b), show that $c_k = (1/T_p)X_a(k/T_p)$.

4.4 Consider the following periodic signal:

$$x(n) = \left\{ \ldots, 1, 0, 1, 2, 3, 2, 1, 0, 1, \ldots \right\}$$
$$\uparrow$$

(a) Sketch the signal $x(n)$ and its magnitude and phase spectra.
(b) Using the results in part (a), verify Parseval's relation by computing the power in the time and frequency domains.

4.5 Consider the signal

$$x(n) = 2 + 2 \cos \frac{\pi n}{4} + \cos \frac{\pi n}{2} + \frac{1}{2} \cos \frac{3\pi n}{4}$$

(a) Determine and sketch its power density spectrum.
(b) Evaluate the power of the signal.

4.6 Determine and sketch the magnitude and phase spectra of the following periodic signals.

(a) $x(n) = 4 \sin \dfrac{\pi(n - 2)}{3}$

(b) $x(n) = \cos \dfrac{2\pi}{3} n + \sin \dfrac{2\pi}{5} n$

(c) $x(n) = \cos \dfrac{2\pi}{3} n \sin \dfrac{2\pi}{5} n$

(d) $x(n) = \left\{ \ldots, -2, -1, 0, 1, 2, -2, -1, 0, 1, 2, \ldots \right\}$
$$\uparrow$$

(e) $x(n) = \left\{ \ldots, -1, 2, 1, 2, -1, 0, -1, 2, 1, 2, \ldots \right\}$
$$\uparrow$$

(f) $x(n) = \left\{ \ldots, 0, 0, 1, 1, 0, 0, 0, 1, 1, 0, 0, \ldots \right\}$
$\uparrow$

(g) $x(n) = 1, \quad -\infty < n < \infty$
(h) $x(n) = (-1)^n, \quad -\infty < n < \infty$

4.7 Determine the periodic signals $x(n)$, with fundamental period $N = 8$, if their Fourier coefficients are given by:

(a) $c_k = \cos \dfrac{k\pi}{4} + \sin \dfrac{3k\pi}{4}$

(b) $c_k = \begin{cases} \sin \dfrac{k\pi}{3} & 0 \le k \le 6 \\ 0 & k = 7 \end{cases}$

(c) $\{c_k\} = \left\{ \ldots, 0, \frac{1}{4}, \frac{1}{2}, 1, 2, 1, \frac{1}{2}, \frac{1}{4}, 0 \ldots \right\}$
$\uparrow$

4.8 Compute the Fourier transform of the following signals.
    (a) $x(n) = u(n) - u(n - 6)$
    (b) $x(n) = 2^n u(-n)$
    (c) $x(n) = (\frac{1}{4})^n u(n + 4)$
    (d) $x(n) = (\alpha^n \sin \omega_0 n) u(n), \quad |\alpha| < 1$
    (e) $x(n) = |\alpha|^n \sin \omega_0 n, \quad |\alpha| < 1$
    (f) $x(n) = \begin{cases} 2 - (\frac{1}{2})n & |n| \le 4 \\ 0 & \text{elsewhere} \end{cases}$

    (g) $x(n) = \left\{ -2, -1, 0, 1, 2 \right\}$
                    $\uparrow$

    (h) $x(n) = \begin{cases} A(2M + 1 - |n|) & |n| \le M \\ 0 & |n| > M \end{cases}$

    Sketch the magnitude and phase spectra for parts (a), (f), and (g).

4.9 Determine the signals having the following Fourier transforms.
    (a) $X(\omega) = \begin{cases} 0 & 0 \le |\omega| \le \omega_0 \\ 1 & \omega_0 < |\omega| \le \pi \end{cases}$
    (b) $X(\omega) = \cos^2 \omega$
    (c) $X(\omega) = \begin{cases} 1 & \omega_0 - \delta\omega/2 \le |\omega| \le \omega_0 + \delta\omega/2 \\ 0 & \text{elsewhere} \end{cases}$
    (d) The signal shown in Fig. P4.9.

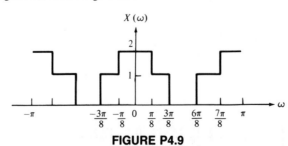

FIGURE P4.9

4.10 Consider the signal

$$x(n) = \left\{ 1, 0, -1, 2, 3 \atop \uparrow \right\}$$

with Fourier transform $X(\omega) = X_R(\omega) + j(X_I(\omega))$. Determine and sketch the signal $y(n)$ with Fourier transform

$$Y(\omega) = X_I(\omega) + X_R(\omega)e^{j2\omega}$$

4.11 In Example 4.3.3, the Fourier transform of the signal

$$x(n) = \begin{cases} 1 & -M \le n \le M \\ 0 & \text{otherwise} \end{cases}$$

was shown to be

$$X(\omega) = 1 + 2 \sum_{n=1}^{M} \cos \omega n$$

Show that the Fourier transform of

$$x_1(n) = \begin{cases} 1 & 0 \le n \le M \\ 0 & \text{otherwise} \end{cases}$$

and

$$x_2(n) = \begin{cases} 1 & -M \le n \le -1 \\ 0 & \text{otherwise} \end{cases}$$

are, respectively,

$$X_1(\omega) = \frac{1 - e^{-j\omega(M+1)}}{1 - e^{-j\omega}}$$

$$X_2(\omega) = \frac{e^{j\omega} - e^{j\omega(M+1)}}{1 - e^{j\omega}}$$

Thus prove that

$$X(\omega) = X_1(\omega) + X_2(\omega)$$
$$= \frac{\sin (M + \frac{1}{2})\omega}{\sin (\omega/2)}$$

and therefore,

$$1 + 2 \sum_{n=1}^{M} \cos \omega n = \frac{\sin (M + \frac{1}{2})\omega}{\sin (\omega/2)}$$

4.12 Consider the signal

$$x(n) = \left\{ -1, 2, -3, 2, -1 \atop \uparrow \right\}$$

with Fourier transform $X(\omega)$. Computer the following quantities, without explicitly computing $X(\omega)$:

(a) $X(0)$ (b) $\angle X(\omega)$ (c) $\int_{-\pi}^{\pi} X(\omega) \, d\omega$ (d) $X(\pi)$
(e) $\int_{-\pi}^{\pi} |X(\omega)|^2 \, d\omega$

4.13 The center of gravity of a signal $x(n)$ is defined as

$$c = \frac{\sum_{n=-\infty}^{\infty} nx(n)}{\sum_{n=-\infty}^{\infty} x(n)}$$

and provides a measure of the "time delay" of the signal.
(a) Express c in terms of $X(\omega)$.
(b) Compute c for the signal $x(n)$ whose Fourier transform is shown in Fig. P4.13.

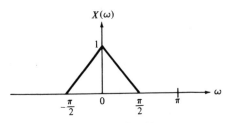

$X(\omega)$

FIGURE P4.13

4.14 Compute the N-point DFTs of the signals
(a) $x(n) = \delta(n)$
(b) $x(n) = \delta(n - n_0),\ 0 < n_0 < N$
(c) $x(n) = a^n,\ 0 \le n \le N - 1$
(d) $x(n) = \begin{cases} 1 & 0 \le n \le 7 \\ 0 & 8 \le n \le 15 \end{cases}$
(e) $x(n) = e^{j(2\pi/N)k_0 n},\ 0 \le n \le N - 1$
(f) $x(n) = \cos \dfrac{2\pi}{N} k_0 n,\ 0 \le n \le N - 1$
(g) $x(n) = \sin \dfrac{2n}{N} k_0 n,\ 0 \le n \le N - 1$

4.15 Consider the finite-duration signal

$$x(n) = \{1, 2, 3, 1\}$$

(a) Compute its four-point DFT by solving explicitly the 4-by-4 system of linear equations defined by the inverse DFT formula.
(b) Check the answer in part (a) by computing the four-point DFT, using its definition.

4.16 (a) Compute the Fourier transform $X(\omega)$ of the signal

$$x(n) = \left\{ 1, 2, 3, 2, 1, 0 \atop \uparrow \right\}$$

(b) Compute the 6-point DFT $V(k)$ of the signal

$$v(n) = \{3, 2, 1, 0, 1, 2\}$$

(c) Is there any relation between $X(\omega)$ and $V(k)$? Explain.

4.17 Let $x(n)$ be an arbitrary signal, not necessarily real-valued, with Fourier transform $X(\omega)$. Express the Fourier transforms of the following signals in terms of $X(\omega)$.

(a) $x^*(n)$

(b) $x^*(-n)$

(c) $y(n) = x(n) - x(n-1)$

(d) $y(n) = \displaystyle\sum_{k=-\infty}^{n} x(k)$

(e) $y(n) = x(2n)$

(f) $y(n) = \begin{cases} x(n/2) & n \text{ even} \\ 0 & n \text{ odd} \end{cases}$

4.18 Determine and sketch the Fourier transforms $X_1(\omega)$, $X_2(\omega)$, and $X_3(\omega)$ of the following signals.

(a) $x_1(n) = \left\{ 1, 1, 1, 1, 1 \atop \uparrow \right\}$

(b) $x_2(n) = \left\{ 1, 0, 1, 0, 1, 0, 1, 0, 1 \atop \qquad\quad\uparrow \right\}$

(c) $x_3(n) = \left\{ 1, 0, 0, 1, 0, 0, 1, 0, 0, 1, 0, 0, 1 \atop \qquad\qquad\quad\uparrow \right\}$

(d) Is there any relation between $X_1(\omega)$, $X_2(\omega)$, and $X_3(\omega)$? What is its physical meaning?

(e) Show that if

$$x_k(n) = \begin{cases} x\left(\dfrac{n}{k}\right) & \text{if } n/k \text{ integer} \\ 0 & \text{otherwise} \end{cases}$$

then

$$X_k(\omega) = X(k\omega)$$

4.19 Let $x(n)$ be a signal with Fourier transform as shown in Fig. P4.19. Determine and sketch the Fourier transforms of the following signals.

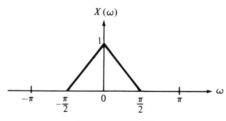

$X(\omega)$

FIGURE P4.19

(a) $x_1(n) = \left(\cos \dfrac{\pi}{4} n \right) x(n)$

(b) $x_2(n) = \left(\sin \dfrac{\pi}{2} n \right) x(n)$

(c) $x_3(n) = \left(\cos \dfrac{\pi}{2} n \right) x(n)$

(d) $x_4(n) = (\cos \pi n) x(n)$

COMPUTER EXPERIMENTS

4.20 *Frequency-domain sampling* Consider the following discrete-time signal

$$x(n) = \begin{cases} a^{|n|} & |n| \leq L \\ 0 & |n| > L \end{cases}$$

where $a = 0.95$ and $L = 10$

(a) Compute and plot the signal $x(n)$.

(b) Show that

$$X(\omega) = \sum_{n=-\infty}^{\infty} x(n)e^{-j\omega n} = x(0) + 2 \sum_{n=1}^{L} x(n) \cos \omega n$$

Plot $X(\omega)$ by computing it at $\omega = \pi k/100$, $k = 0, 1, \ldots, 100$.

(c) Compute

$$c_K = \frac{1}{N} X\left(\frac{2\pi}{N} K\right) \qquad K = 0, 1, \ldots, N - 1$$

for $N = 30$.

(d) Determine and plot the signal

$$\tilde{x}(n) = \sum_{k=0}^{N-1} c_k e^{j(2\pi/N)kn}$$

What is the relation between the signals $x(n)$ and $\tilde{x}(n)$? Explain.

(e) Compute and plot the signal $\tilde{x}_1(n) = \sum_{l=-\infty}^{\infty} x(n - lN)$, $-L \leq n \leq L$ for $N = 30$. Compare the signals $\tilde{x}(n)$ and $\tilde{x}_1(n)$.

(f) Repeat parts (c) to (e) for $N = 15$.

4.21 *Frequency-domain sampling* The signal $x(n) = a^{|n|}$, $-1 < a < 1$ has a Fourier transform

$$X(\omega) = \frac{1 - a^2}{1 - 2a \cos \omega + a^2}$$

(a) Plot $X(\omega)$ for $0 \leq \omega \leq 2\pi$, $a = 0.8$.

Reconstruct and plot $X(\omega)$ from its samples $X(2\pi k/N)$, $0 \leq k \leq N - 1$ for:

(b) $N = 20$

(c) $N = 100$

(d) Compare the spectra obtained in parts (b) and (c) with the original spectrum $X(\omega)$ and explain the differences.

(e) Illustrate the time-domain aliasing when $N = 20$.

4.22 *Measurement of the total harmonic distortion in quantized sinusoids* Let $x(n)$ be a periodic sinusoidal signal with frequency $f_0 = k/N$, that is,

$$x(n) = \sin 2\pi f_0 n$$

(a) Write a computer program that quantizes the signal $x(n)$ into b bits or equivalently into $L = 2^b$ levels by using rounding. The resulting signal is denoted by $x_q(n)$.

(b) For $f_0 = 1/50$ compute the THD of the quantized signals $x_q(n)$ obtained by using $b = 4, 6, 8$, and 16 bits.

(c) Repeat part (b) for $f_0 = 1/100$.

(d) Comment on the results obtained in parts (b) and (c).

4.23 *Time-domain sampling* Consider the continuous-time signal

$$x_a(t) = \begin{cases} e^{-j2\pi F_0 t} & t \geq 0 \\ 0 & t < 0 \end{cases}$$

(a) Compute analytically the spectrum $X_a(F)$ of $x_a(t)$.

(b) Compute analytically the spectrum of the signal $x(n) = x_a(nT)$, $T = 1/F_s$.

(c) Plot the magnitude spectrum $|X_a(F)|$ for $F_0 = 10$ Hz.

(d) Plot the magnitude spectrum $|X(F)|$ for $F_s = 10, 20, 40$, and 100 Hz.

(e) Explain the results obtained in part (d) in terms of the aliasing effect.

CHAPTER 5

Analysis and Design of Discrete-Time Systems in the Frequency Domain

In Chapter 4 we were concerned with the representation of signals in the frequency domain. Having developed a frequency-domain representation for signals, we are now prepared to discuss the characterization of discrete-time systems in the frequency domain. This development allows us to view the operation and functional characteristics of linear time-invariant systems in the frequency domain, and leads to frequency-domain input–output relations for this class of systems. We illustrate that a linear time-invariant system is characterized in the frequency domain by its frequency response function, which is simply the Fourier transform of its impulse response. This characterization leads naturally to the viewpoint that a linear time-invariant system acts as a filter on the various frequency components in the input. Then some design techniques for simple frequency-selective filters are described.

The use of the Discrete Fourier Transform (DFT) in linear filtering is considered in Section 5.4. In this section, another form of convolution, called circular convolution, is also described and related to the multiplication of two DFTs.

The z-transform introduced in Chapter 3 plays a very important role in the analysis and design of discrete-time systems and in digital signal processing in general. In this chapter the pole–zero characterization of linear time-invariant systems is used to design a variety of different types of filters and systems, including lowpass, highpass, and bandpass filters, notch filters, comb filters, all-pass filters, digital resonators, and digital sinusoidal generators.

5.1 Frequency-Domain Characteristics of Linear Time-Invariant Systems

In this section we develop the characterization of linear time-invariant systems in the frequency domain. The basic excitation signals in this development are the complex exponentials and sinusoidal functions. The characteristics of the system are described by a function of the frequency variable ω called the frequency response, which is related to the system function $H(z)$ and the impulse response $h(n)$ of the system.

The frequency response function completely characterizes a linear time-invariant system in the frequency domain and allows us to determine the steady-state response of the system to any arbitrary weighted linear combination of sinusoids or complex exponentials. Since periodic sequences, in particular, lend themselves to a Fourier series decomposition as a weighted sum of harmonically related complex exponentials, it becomes a simple matter to determine the response of a linear time-invariant system to this class of signals.

300

5.1.1 Response to Complex Exponential and Sinusoidal Signals

In Chapter 2 it was demonstrated that the response of any relaxed linear time-invariant system to an arbitrary input signal $x(n)$ is given by the convolution sum formula

$$y(n) = \sum_{k=-\infty}^{\infty} h(k)x(n-k) \tag{5.1.1}$$

In this input–output relationship, the system is characterized in the time domain by its unit sample response $\{h(n), -\infty < n < \infty\}$.

To develop a freqeuncey-domain characterization of the system, let us excite the system with the complex exponential

$$x(n) = Ae^{j\omega n} \qquad -\infty < n < \infty \tag{5.1.2}$$

where A is the amplitude and ω is any arbitrary frequency confined to the frequency interval $[-\pi, \pi]$. By substituting (5.1.2) into (5.1.1), we obtain the response

$$y(n) = \sum_{k=-\infty}^{\infty} h(k)[Ae^{j\omega(n-k)}] \tag{5.1.3}$$

$$= A\left[\sum_{k=-\infty}^{\infty} h(k)e^{-j\omega k}\right] e^{j\omega n}$$

We observe that the term in brackets in (5.1.3) is a function of the frequency variable ω. In fact, this term is the Fourier transform of the unit sample response $h(k)$ of the system. Hence we denote this function as

$$H(\omega) = \sum_{k=-\infty}^{\infty} h(k)e^{-j\omega k} \tag{5.1.4}$$

Clearly, the function $H(\omega)$ exists if the system is BIBO stable, that is, if

$$\sum_{n=-\infty}^{\infty} |h(n)| < \infty$$

We may also view $H(\omega)$ as the z-transform $H(z)$ of the unit sample response $h(n)$, evaluated on the unit circle.

With the definition in (5.1.4), the response of the system to the complex exponential given in (5.1.2) is

$$y(n) = AH(\omega)e^{j\omega n} \tag{5.1.5}$$

We note that the response is also in the form of a complex exponential, of the same frequency as the input, but altered by the multiplicative factor $H(\omega)$.

As a result of this characteristic behavior, the exponential signal in (5.1.2) is called an *eigenfunction* of the system. In other words, an eigenfunction of a system is an input signal that produces an output that differs from the input by a constant multiplicative factor. The multiplicative factor is called an *eigenvalue* of the system. In this case a complex exponential signal of the form (5.1.2) is an eigenfunction of a linear time-invariant system, and $H(\omega)$ evaluated at the frequency of the input signal is the corresponding eigenvalue.

EXAMPLE 5.1.1

Determine the output sequence of the system with impulse response

$$h(n) = (\tfrac{1}{2})^n u(n) \tag{5.1.6}$$

when the input is the complex exponential sequence

$$x(n) = Ae^{j\pi n/2} \qquad -\infty < n < \infty$$

Solution: First we evaluate the Fourier transform of the impulse response $h(n)$, and then we use (5.1.5) to determine $y(n)$. From Example 4.2.3 we recall that

$$H(\omega) = \sum_{n=-\infty}^{\infty} h(n)e^{-j\omega n} = \frac{1}{1 - \frac{1}{2}e^{-j\omega}} \qquad (5.1.7)$$

At $\omega = \pi/2$, (5.1.7) yields

$$H\left(\frac{\pi}{2}\right) = \frac{1}{1 + j\frac{1}{2}} = \frac{2}{\sqrt{5}} e^{-j26.6°}$$

and therefore the output is

$$y(n) = A\left(\frac{2}{\sqrt{5}} e^{-j26.6°}\right) e^{j\pi n/2}$$

$$y(n) = \frac{2}{\sqrt{5}} Ae^{j(\pi n/2 - 26.6°)} \qquad -\infty < n < \infty \qquad (5.1.8)$$

This example clearly illustrates that the only effect of the system on the input signal is to scale the amplitude by $2/\sqrt{5}$ and shift the phase by $-26.6°$. Thus the output is also a complex exponential of frequency $\pi/2$, amplitude $2A/\sqrt{5}$, and phase $-26.6°$.

If we alter the frequency of the input signal, the effect of the system on the input also changes and hence the output changes. In particular, if the input sequence is a complex exponential of frequency π, that is,

$$x(n) = Ae^{j\pi n} \qquad -\infty < n < \infty \qquad (5.1.9)$$

then, at $\omega = \pi$,

$$H(\pi) = \frac{1}{1 - \frac{1}{2}e^{-j\pi}} = \frac{1}{\frac{3}{2}} = \frac{2}{3}$$

and the output of the system is

$$y(n) = \frac{2}{3}Ae^{j\pi n} \qquad -\infty < n < \infty \qquad (5.1.10)$$

We note that $H(\pi)$ is purely real [i.e., the phase associated with $H(\omega)$ is zero at $\omega = \pi$]. Hence the input is scaled in amplitude by the factor $H(\pi) = \frac{2}{3}$, but the phase shift is zero.

In general, $H(\omega)$ is a complex-valued function of the frequency variable ω. Hence it may be expressed in polar form as

$$H(\omega) = |H(\omega)| e^{j\Theta(\omega)} \qquad (5.1.11)$$

where $|H(\omega)|$ is the magnitude of $H(\omega)$ and

$$\Theta(\omega) = \angle H(\omega)$$

which is the phase shift imparted on the input signal by the system at the frequency ω.

An important property of $H(\omega)$ is that this function is periodic with period 2π. Indeed, from the definition of $H(\omega)$ in (5.1.4), we have, for any integer m,

$$H(\omega + 2\pi m) = H(\omega)$$

In fact, we may view (5.1.4) as the exponential Fourier series expansion for $H(\omega)$, with $h(k)$ as the Fourier series coefficients. Consequently, the unit impulse $h(k)$ is related to $H(\omega)$ through the integral expression

$$h(k) = \frac{1}{2\pi} \int_{-\pi}^{\pi} H(\omega) e^{j\omega k} \, d\omega \tag{5.1.12}$$

which was developed in Chapter 4.

For a linear time-invariant system with a real-valued impulse response, the magnitude and phase functions possess symmetry properties which are given below. From the definition of $H(\omega)$, we have

$$
\begin{aligned}
H(\omega) &= \sum_{k=-\infty}^{\infty} h(k) e^{-j\omega k} \\
&= \sum_{k=-\infty}^{\infty} h(k) \cos \omega k - j \sum_{k=-\infty}^{\infty} h(k) \sin \omega k \\
&= H_R(\omega) + jH_I(\omega) \\
&= \sqrt{H_R^2(\omega) + H_I^2(\omega)} \; e^{j \, \tan^{-1}[H_I(\omega)/H_R(\omega)]}
\end{aligned} \tag{5.1.13}
$$

where $H_R(\omega)$ and $H_I(\omega)$ denote the real and imaginary components of $H(\omega)$, defined as

$$H_R(\omega) = \sum_{k=-\infty}^{\infty} h(k) \cos \omega k \tag{5.1.14}$$

$$H_I(\omega) = - \sum_{k=-\infty}^{\infty} h(k) \sin \omega k$$

It is clear from (5.1.12) that the magnitude and phase of $H(\omega)$, expressed in terms of $H_R(\omega)$ and $H_I(\omega)$, are

$$|H(\omega)| = \sqrt{H_R^2(\omega) + H_I^2(\omega)} \tag{5.1.15}$$

$$\Theta(\omega) = \tan^{-1} \frac{H_I(\omega)}{H_R(\omega)}$$

We note that $H_R(\omega) = H_R(-\omega)$ and $H_I(\omega) = -H_I(-\omega)$, so that $H_R(\omega)$ is an even function of ω and $H_I(\omega)$ is an odd function of ω. As a consequence, it follows that $|H(\omega)|$ is an even function of ω and $\Theta(\omega)$ is an odd function of ω. Hence if we know $|H(\omega)|$ and $\Theta(\omega)$ for $0 \le \omega \le \pi$, we also know these functions for $-\pi \le \omega \le 0$.

EXAMPLE 5.1.2

Determine the magnitude and phase of $H(\omega)$ for the three-point moving average (MA) system

$$y(n) = \tfrac{1}{3}[x(n+1) + x(n) + x(n-1)]$$

and plot these two functions for $0 \le \omega \le \pi$.

Solution: Since

$$h(n) = \left\{ \tfrac{1}{3}, \underset{\uparrow}{\tfrac{1}{3}}, \tfrac{1}{3} \right\}$$

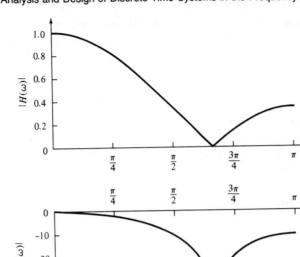

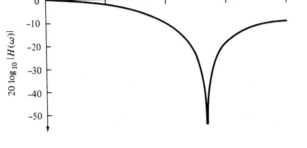

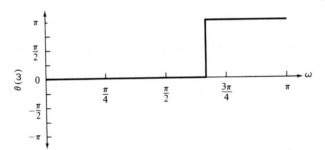

FIGURE 5.1 Magnitude and phase responses for the MA system in Example 5.1.2.

it follows that

$$H(\omega) = \tfrac{1}{3}(e^{j\omega} + 1 + e^{-j\omega}) = \tfrac{1}{3}(1 + 2\cos\omega)$$

Hence

$$|H(\omega)| = \tfrac{1}{3}|1 + 2\cos\omega| \tag{5.1.16}$$

$$\Theta(\omega) = \begin{cases} 0 & 0 \le \omega \le 2\pi/3 \\ \pi & 2\pi/3 \le \omega < \pi \end{cases}$$

Figure 5.1 illustrates the graphs of the magnitude and phase of $H(\omega)$. As indicated previously, $|H(\omega)|$ is an even function and $\Theta(\omega)$ is an odd function of frequency. It is apparent from the frequency response characteristic $H(\omega)$ that this moving average filter smooths the input data, as we would expect from the input–output equation.

The symmetry properties satisfied by the magnitude and phase functions of $H(\omega)$ and the fact that a sinusoid can be expressed as a sum or difference of two complex-conjugate exponential functions imply that the response of a linear time-invariant system to a sinusoid is similar in form to the response when the input is a complex exponential. Indeed, if the input is

$$x_1(n) = Ae^{j\omega n}$$

the output is

$$y_1(n) = A|H(\omega)|e^{j\Theta(\omega)}e^{j\omega n}$$

On the other hand, if the input is

$$x_2(n) = Ae^{-j\omega n}$$

the response of the system is

$$\begin{aligned} y_2(n) &= A|H(-\omega)|e^{j\Theta(-\omega)}e^{-j\omega n} \\ &= A|H(\omega)|e^{-j\Theta(\omega)}e^{-j\omega n} \end{aligned}$$

where, in the last expression, we have made use of the symmetry properties $|H(\omega)| = |H(-\omega)|$ and $\Theta(\omega) = -\Theta(-\omega)$. Now, by applying the superposition property of the linear time-invariant system, we find that the response of the system to the input

$$x(n) = \tfrac{1}{2}[x_1(n) + x_2(n)] = A \cos \omega n$$

is

$$\begin{aligned} y(n) &= \tfrac{1}{2}[y_1(n) + y_2(n)] \\ y(n) &= A|H(\omega)| \cos [\omega n + \Theta(\omega)] \end{aligned} \tag{5.1.17}$$

Similarly, if the input is

$$x(n) = \frac{1}{j2} [x_1(n) - x_2(n)] = A \sin \omega n$$

the response of the system is

$$\begin{aligned} y(n) &= \frac{1}{j2} [y_1(n) - y_2(n)] \\ y(n) &= A|H(\omega)| \sin [\omega n + \Theta(\omega)] \end{aligned} \tag{5.1.18}$$

It is apparent from the discussion above that $H(\omega)$ or, equivalently, $|H(\omega)|$ and $\Theta(\omega)$ completely characterize the effect of the system on a sinusoidal input signal of any arbitrary frequency. Indeed, we note that $|H(\omega)|$ determines the amplification ($|H(\omega)| > 1$) or attenuation ($|H(\omega)| < 1$) imparted by the system on the input sinusoid. The phase $\Theta(\omega)$ determines the amount of phase shift imparted by the system on the input sinusoid. Consequently, by knowing $H(\omega)$, we are able to determine the response of the system to any sinusoidal input signal. Since $H(\omega)$ specifies the response of the system in the frequency domain, it is called the *frequency response* of the system. Correspondingly, $|H(\omega)|$ is called *magnitude response* and $\Theta(\omega)$ is called the *phase response* of the system.

If the input to the system consists of more than one sinusoid, the superposition property of the linear system can be used to determine the response. The following examples illustrate the use of the superposition property.

EXAMPLE 5.1.3 _____

Determine the response of the system in Example 5.1.1 to the input signal

$$x(n) = 10 - 5 \sin \frac{\pi}{2} n + 20 \cos \pi n \qquad -\infty < n < \infty$$

Solution: The frequency response of the system is given in (5.1.7) as

$$H(\omega) = \frac{1}{1 - \frac{1}{2}e^{-j\omega}}$$

The first term in the input signal is a fixed signal component, corresponding to $\omega = 0$. Thus

$$H(0) = \frac{1}{1 - \frac{1}{2}} = 2$$

The second term in $x(n)$ has a frequency $\pi/2$. At this frequency the frequency response of the system is

$$H\left(\frac{\pi}{2}\right) = \frac{2}{\sqrt{5}}e^{-j26.6°}$$

Finally, the third term in $x(n)$ has a frequency $\omega = \pi$. At this frequency

$$H(\pi) = \tfrac{2}{3}$$

Hence the response of the system to $x(n)$ is

$$y(n) = 20 - \frac{10}{\sqrt{5}}\sin\left(\frac{\pi}{2}n - 26.6°\right) + \frac{40}{3}\cos \pi n \qquad -\infty < n < \infty$$

EXAMPLE 5.1.4 _____

A linear time-invariant system is described by the following difference equation:

$$y(n) = ay(n - 1) + bx(n) \qquad 0 < a < 1$$

(a) Determine the magnitude and phase of the frequency response $H(\omega)$ of the system.
(b) Choose the parameter b so that the maximum value of $|H(\omega)|$ is unity, and sketch $|H(\omega)|$ and $\angle H(\omega)$ for $a = 0.9$.
(c) Determine the output of the system to the input signal

$$x(n) = 5 + 12 \sin \frac{\pi}{2}n - 20 \cos\left(\pi n + \frac{\pi}{4}\right)$$

Solution: The impulse response of the system is

$$h(n) = ba^n u(n)$$

Since $|a| < 1$, the system is BIBO stable and hence $H(\omega)$ exists.

(a) The frequency response is

$$H(\omega) = \sum_{n=-\infty}^{\infty} h(n)e^{-j\omega n}$$

$$= \frac{b}{1 - ae^{-j\omega}}$$

Since

$$1 - ae^{-j\omega} = (1 - a \cos \omega) + ja \sin \omega$$

it follows that

$$|1 - ae^{-j\omega}| = \sqrt{(1 - a \cos \omega)^2 + (a \sin \omega)^2}$$
$$= \sqrt{1 + a^2 - 2a \cos \omega}$$

and

$$\angle (1 - ae^{-j\omega}) = \tan^{-1} \frac{a \sin \omega}{1 - a \cos \omega}$$

Therefore,

$$|H(\omega)| = \frac{|b|}{\sqrt{1 + a^2 - 2a \cos \omega}}$$

$$\angle H(\omega) = \Theta(\omega) = \angle b - \tan^{-1} \frac{a \sin \omega}{1 - a \cos \omega}$$

(b) Since the parameter a is positive, the denominator of $|H(\omega)|$ attains a minimum at $\omega = 0$. Therefore, $|H(\omega)|$ attains its maximum value at $\omega = 0$. At this frequency we have

$$|H(0)| = \frac{|b|}{1 - a} = 1$$

which implies that $b = \pm(1 - a)$. We choose $b = 1 - a$, so that

$$|H(\omega)| = \frac{1 - a}{\sqrt{1 + a^2 - 2a \cos \omega}}$$

and

$$\Theta(\omega) = -\tan^{-1} \frac{a \sin \omega}{1 - a \cos \omega}$$

The frequency response plots for $|H(\omega)|$ and $\Theta(\omega)$ are illustrated in Fig. 5.2. We observe that this system attenuates high frequency signals.

(c) The input signal consists of components of frequencies $\omega = 0$, $\pi/2$, and π. For $\omega = 0$, $|H(0)| = 1$ and $\Theta(0) = 0$. For $\omega = \pi/2$,

$$\left| H\left(\frac{\pi}{2}\right) \right| = \frac{1 - a}{\sqrt{1 + a^2}} = \frac{0.1}{\sqrt{1.81}} = 0.074$$

$$\Theta\left(\frac{\pi}{2}\right) = -\tan^{-1} a = -42°$$

For $\omega = \pi$,

$$|H(\pi)| = \frac{1 - a}{1 + a} = \frac{0.1}{1.9} = 0.053$$

$$\Theta(\pi) = 0$$

Therefore, the output of the systems is

$$y(n) = 5|H(0)| + 12 \left| H\left(\frac{\pi}{2}\right) \right| \sin\left[\frac{\pi}{2} n + \Theta\left(\frac{\pi}{2}\right)\right]$$

$$- 20 |H(\pi)| \cos\left[\pi n + \frac{\pi}{4} + \Theta(\pi)\right]$$

$$= 5 + 0.888 \sin\left(\frac{\pi}{2} n - 42°\right) - 1.06 \cos\left(\pi n + \frac{\pi}{4}\right) \qquad -\infty < n < \infty$$

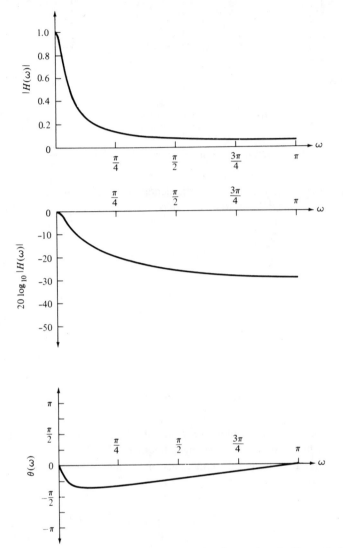

FIGURE 5.2 Magnitude and phase responses for the system in Example 5.1.4, with $a = 0.9$.

In the most general case, if the input to the system consists of an arbitrary linear combination of sinusoids of the form

$$x(n) = \sum_{i=1}^{L} A_i \cos (\omega_i n + \phi_i) \qquad -\infty < n < \infty$$

where $\{A_i\}$ and $\{\phi_i\}$ are the amplitudes and phases of the corresponding sinusoidal components, the response of the system is simply

$$y(n) = \sum_{i=1}^{L} A_i |H(\omega_i)| \cos [\omega_i n + \phi_i + \Theta(\omega_i)] \qquad (5.1.19)$$

where $|H(\omega_i)|$ and $\Theta(\omega_i)$ are the magnitude and phase, respectively, imparted by the system to the individual frequency components of the input signal.

It is clear that depending on the frequency response $H(\omega)$ of the system, input sinusoids of different frequencies will be affected differently by the system. For example, some sinusoids may be completely suppressed by the system if $H(\omega) = 0$ at the frequencies of these sinusoids. Other sinusoids may receive no attenuation (or perhaps, some amplification) by the system. In effect, we may view the linear time-invariant system functioning as a filter to sinusoids of different frequencies, passing some of the frequency components to the output and suppressing or preventing other frequency components from reaching the output. In fact, as discussed in Chapter 8, the basic digital filter design problem involves the determination of the parameters of a linear time-invariant system to achieve a desired frequency response $H(\omega)$.

5.1.2 Steady-State and Transient Response to Sinusoidal Input Signals

In the discussion of the preceding section we determined the response of a linear time-invariant system to exponential and sinusoidal input signals applied to the system at $n = -\infty$. We usually call such signals eternal exponentials or eternal sinusoids, because they were applied at $n = -\infty$. In such a case, the response that we observe at the output of the system is the steady-state response. There is no transient response in this case.

On the other hand, if the exponential or sinusoidal signal is applied at some finite time instant, say at $n = 0$, the response of the system consists of two terms, the transient response and the steady-state response. To demonstrate this behavior, let us consider, as an example, the system described by the first-order difference equation

$$y(n) = ay(n - 1) + x(n) \tag{5.1.20}$$

This system was considered in Section 2.4.2. Its response to any input $x(n)$ applied at $n = 0$ is given by (2.4.8) as

$$y(n) = a^{n+1}y(-1) + \sum_{k=0}^{n} a^k x(n - k) \qquad n \geq 0 \tag{5.1.21}$$

where $y(-1)$ is the initial condition.

Now, let us assume that the input to the system is the complex exponential

$$x(n) = Ae^{j\omega n} \qquad n \geq 0 \tag{5.1.22}$$

applied at $n = 0$. When we substitute (5.1.22) into (5.1.21), we obtain

$$
\begin{aligned}
y(n) &= a^{n+1}y(-1) + A \sum_{k=0}^{n} a^k e^{j\omega(n-k)} \\
&= a^{n+1}y(-1) + A \left[\sum_{k=0}^{n} (ae^{-j\omega})^k \right] e^{j\omega n} \\
&= a^{n+1}y(-1) + A \frac{1 - a^{n+1}e^{-j\omega(n+1)}}{1 - ae^{-j\omega}} e^{j\omega n} \qquad n \geq 0 \\
&= a^{n+1}y(-1) - \frac{Aa^{n+1}e^{-j\omega(n+1)}}{1 - ae^{-j\omega}} e^{j\omega n} + \frac{A}{1 - ae^{-j\omega}} e^{j\omega n} \qquad n \geq 0
\end{aligned}
\tag{5.1.23}
$$

We recall that the system in (5.1.20) is BIBO stable if $|a| < 1$. In this case the two terms involving a^{n+1} in (5.1.23) decay toward zero as n approaches infinity. Consequently, we are left with the steady-state response

$$
\begin{aligned}
y_{ss}(n) &= \lim_{n \to \infty} y(n) = \frac{A}{1 - ae^{-j\omega}} e^{j\omega n} \\
&= AH(\omega)e^{j\omega n}
\end{aligned}
\tag{5.1.24}
$$

The first two terms in (5.1.23) constitute the transient response of the system, that is,

$$y_{tr}(n) = a^{n+1}y(-1) - \frac{Aa^{n+1}e^{-j\omega(n+1)}}{1 - ae^{-j\omega}}e^{j\omega n} \qquad n \geq 0 \qquad (5.1.25)$$

which decay toward zero as n approaches infinity. The first term in the transient response is the zero-input response of the system, while the second term is the transient produced by the exponential input signal.

In general, all linear time-invariant BIBO systems behave in a similar fashion when excited by a complex exponential or a sinusoid at $n = 0$ or at some other finite time instant. That is, the transient response decays toward zero as $n \to \infty$, leaving only the steady-state response that we determined in the preceding section. In many practical applications, the transient response of the system is unimportant, and hence it is usually ignored in dealing with the response of the system to sinusoidal inputs.

5.1.3 Steady-State Response to Periodic Input Signals

Suppose that the input to a stable linear time-invariant system is a periodic signal $x(n)$ with fundamental period N. Since such a signal exists from $-\infty < n < \infty$, the total response of the system at any time instant n is simply equal to the steady-state response.

To determine the response $y(n)$ of the system, we make use of the Fourier series representation of the periodic signal, which is

$$x(n) = \sum_{k=0}^{N-1} c_k e^{j2\pi kn/N} \qquad k = 0, 1, \ldots, N - 1 \qquad (5.1.26)$$

where the $\{c_k\}$ are the Fourier series coefficients. Now the response of the system to the complex exponential signal

$$x_k(n) = c_k e^{j2\pi kn/N} \qquad k = 0, 1, \ldots, N - 1$$

is

$$y_k(n) = c_k H\left(\frac{2\pi}{N}k\right)e^{j2\pi kn/N} \qquad k = 0, 1, \ldots, N - 1 \qquad (5.1.27)$$

where

$$H\left(\frac{2\pi k}{N}\right) = H(\omega)|_{\omega = 2\pi k/N} \qquad k = 0, 1, \ldots, N - 1$$

By using the superposition principle for linear systems, we obtain the response of the system to the periodic signal $x(n)$ in (5.1.26) as

$$y(n) = \sum_{k=0}^{N-1} c_k H\left(\frac{2\pi k}{N}\right)e^{j2\pi kn/N} \qquad -\infty < n < \infty \qquad (5.1.28)$$

This result implies that the response of the system to the periodic input signal $x(n)$ is also periodic with the same period N. The Fourier series coefficients for $y(n)$ are

$$d_k \equiv c_k H\left(\frac{2\pi k}{N}\right) \qquad k = 0, 1, \ldots, N - 1 \qquad (5.1.29)$$

Hence the linear system may change the shape of the periodic input signal by scaling the amplitude and shifting the phase of the Fourier series components, but it does not affect the period of the periodic input signal.

5.1.4 Relationships Between the System Function and the Frequency Response Function

In Section 5.1.1 we indicated that the system function $H(z)$ evaluated on the unit circle yields the frequency response $H(\omega)$ of the system. Thus

$$H(\omega) = H(z)\big|_{z=e^{j\omega}}$$

$$= \sum_{n=-\infty}^{\infty} h(n)z^{-n}\Big|_{z=e^{j\omega}} \tag{5.1.30}$$

$$= \sum_{n=-\infty}^{\infty} h(n)e^{-j\omega n}$$

It is often desirable to express the magnitude squared $|H(\omega)|^2$ in terms of $H(z)$. First, we note that

$$|H(\omega)|^2 = H(\omega)H^*(\omega)$$
$$= H(\omega)H(-\omega)$$

where the relationship $H^*(\omega) = H(-\omega)$ holds when the unit sample response of the system is real. Under this condition,

$$|H(\omega)|^2 = H(\omega)H(-\omega) = H(z)H(z^{-1})\big|_{z=e^{j\omega}} \tag{5.1.31}$$

We observe that the function $H(z^{-1})$ has zeros and poles that are the reciprocal of the zeros and poles of $H(z)$. Consequently, the product $H(z)H(z^{-1})$ contains poles and zeros that occur in reciprocal pairs. In other words, if z_i is a root of either the numerator or denominator polynomials of $H(z)H(z^{-1})$, then $1/z_i$ is also a root. Thus, if z_i is a root inside the unit circle, then $1/z_i$ is outside the unit circle. Furthermore, we note that if $H(z)$ has all its poles inside the unit circle, so that $H(\omega)$ exists, then $H(z^{-1})$ has all its poles outside the unit circle and its ROC includes the unit circle. To elaborate, if z_m is the largest (in magnitude) pole of $H(z)$, then the ROC of $H(z)$ is $|z| > |z_m|$. The mapping $z \to z^{-1}$ inverts all points in the z-plane so that the smallest (in magnitude) pole of $H(z^{-1})$ is $1/z_m$ and the ROC of $H(z^{-1})$ is $|z| < 1/|z_m|$. Since $|z_m| < 1$, $H(z^{-1})$ also converges on the unit circle.

When $H(z)$ is the system function of a causal system, another interpretation of $H(z^{-1})$ is that it is the system function of a purely anticausal system. In other words, the unit sample response corresponding to the system with system function $H(z^{-1})$ is zero for $n \geq 0$.

Now, if we express $H(z)$ as a ratio of two polynomials in z, that is,

$$H(z) = \frac{B(z)}{A(z)} = \frac{\displaystyle\sum_{k=0}^{M} b_k z^{-k}}{1 + \displaystyle\sum_{k=1}^{N} a_k z^{-k}} \tag{5.1.32}$$

then

$$H(z)H(z^{-1}) = \frac{B(z)B(z^{-1})}{A(z)A(z^{-1})} \tag{5.1.33}$$

According to the correlation theorem for the z-transform (see Table 3.2), the function $H(z)H(z^{-1})$ is the z-transform of the autocorrelation of the unit sample response. Then it follows from the Wiener–Kintchine theorem that $|H(\omega)|^2$ is the Fourier transform of the autocorrelation sequence of $h(n)$.

Similarly, the transforms $D(z) = B(z)B(z^{-1})$ and $C(z) = A(z)A(z^{-1})$ are the z-transforms of the autocorrelation sequences $\{c_l\}$ and $\{d_l\}$, where

$$c_l = \sum_{k=0}^{N-|l|} a_k a_{k+l} \qquad -N \le l \le N \qquad (5.1.34)$$

$$d_l = \sum_{k=0}^{M-|l|} b_k b_{k+l} \qquad -M \le l \le M \qquad (5.1.35)$$

Since the system parameters $\{a_k\}$ and $\{b_k\}$ are real valued, it follows that $c_l = c_{-l}$ and $d_l = d_{-l}$. By using this symmetry property, $|H(\omega)|^2$ may be expressed as

$$|H(\omega)|^2 = \frac{d_0 + 2\sum_{k=1}^{M} d_k \cos k\omega}{c_0 + 2\sum_{k=1}^{N} c_k \cos k\omega} \qquad (5.1.36)$$

Finally, we note that $\cos k\omega$ can be expressed as a polynomial function of $\cos \omega$. That is,

$$\cos k\omega = \sum_{m=0}^{k} \beta_m (\cos \omega)^m \qquad (5.1.37)$$

where $\{\beta_m\}$ are the coefficients in the expansion. Consequently, the numerator and denominator of $|H(\omega)|^2$ may be viewed as polynomial functions of $\cos \omega$. The following example illustrates the foregoing relationships.

EXAMPLE 5.1.5

Determine $|H(\omega)|^2$ for the system

$$y(n) = -0.1y(n-1) + 0.2y(n-2) + x(n) + x(n-1)$$

Solution: The system function is

$$H(z) = \frac{1 + z^{-1}}{1 + 0.1z^{-1} - 0.2z^{-2}}$$

and its ROC is $|z| > 0.5$. Hence $H(\omega)$ exists. Now

$$H(z)H(z^{-1}) = \frac{1 + z^{-1}}{1 + 0.1z^{-1} - 0.2z^{-2}} \cdot \frac{1 + z}{1 + 0.1z - 0.2z^2}$$

$$= \frac{2 + z + z^{-1}}{1.05 + 0.08(z + z^{-1}) - 0.2(z^{-2} + z^{-2})}$$

By evaluating $H(z)H(z^{-1})$ on the unit circle, we obtain

$$|H(\omega)|^2 = \frac{2 + 2\cos \omega}{1.05 + 0.16\cos \omega - 0.4\cos 2\omega}$$

However, $\cos 2\omega = 2\cos^2 \omega - 1$. Consequently, $|H(\omega)|^2$ may be expressed as

$$|H(\omega)|^2 = \frac{2(1 + \cos \omega)}{1.45 + 0.16\cos \omega - 0.8\cos^2 \omega}$$

5.2 Frequency-Domain Analysis of Linear Time-Invariant Systems

The frequency-domain approach developed in the preceding section for determining the steady-state response of a stable linear time-invariant system to a periodic input signal can be generalized to solve the problem of computing the zero-state response to aperiodic finite-energy signals. The mathematical tool required for this analysis is the Fourier transform for discrete-time signals, introduced in Section 4.2.

5.2.1 Input–Output Relations in the Frequency Domain

In this section we develop the frequency-domain input–output relations for a linear time-invariant system excited by a finite-energy signal $x(n)$. Our starting point is the convolution sum formula

$$y(n) = \sum_{k=-\infty}^{\infty} h(k)x(n-k) \tag{5.2.1}$$

which applies to a relaxed linear time-invariant system. By applying the convolution theorem to (5.2.1), we obtain

$$Y(\omega) = H(\omega)X(\omega) \tag{5.2.2}$$

This is the desired input–output relation in the frequency domain. It means that the spectrum of the signal at the output of the system is equal to the spectrum of the signal at the input multiplied by the frequency response of the system.

The expression in (5.2.2) can be written in polar form as

$$\begin{aligned} Y(\omega) &= |H(\omega)|e^{j\Theta_h(\omega)} |X(\omega)|e^{j\Theta_x(\omega)} \\ &= |H(\omega)| |X(\omega)|e^{j[\Theta_x(\omega) + \Theta_h(\omega)]} \end{aligned} \tag{5.2.3}$$

Consequently, the magnitude and phase of $Y(\omega)$ can be expressed as

$$|Y(\omega)| = |H(\omega)| |X(\omega)| \tag{5.2.4}$$

and

$$\angle Y(\omega) = \angle X(\omega) + \angle H(\omega) \tag{5.2.5}$$

or, equivalently,

$$\Theta_y(\omega) = \Theta_x(\omega) + \Theta_h(\omega)$$

By its very nature, a finite-energy aperiodic signal contains a continuum of frequency components. The linear time-invariant system, through its frequency response function, attenuates some frequency components of the input signal and amplifies other frequency components. Observation of the graph of $|H(\omega)|$ shows which frequency components are amplified and which are attenuated. On the other hand, the angle of $H(\omega)$ determines the phase shift imparted in the continuum of frequency components of the input signal as a function of frequency.

We also observe that *the output of a linear time-invariant system cannot contain frequency components that are not contained in the input signal.* In other words, the system cannot create new frequency components. As we shall see in Chapter 6, it takes either a linear time-variant system or a nonlinear system to create frequency components that are not necessarily contained in the input signal.

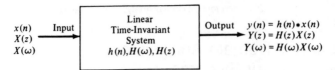

FIGURE 5.3 Time- and frequency-domain input–output relationships in initially relaxed systems.

Figure 5.3 illustrates schematically the time-domain and frequency-domain approaches to the analysis of initially relaxed BIBO-stable LTI systems. We observe that in time-domain analysis we deal with the convolution of the input signal with the impulse response of the system to obtain the response of the system in the time domain. On the other hand, in frequency-domain analysis, we deal with the signal spectrum $X(\omega)$ and the frequency response $H(\omega)$ of the system, which are related through multiplication, to yield the spectrum of the signal at the output of the system. Alternatively, we may use the z-transform equivalents $H(z)$ and $X(z)$, from which we can determine $Y(z)$ and $y(n)$.

Now, let us return to the basic relation in (5.2.2). Suppose that we have $Y(\omega)$ and it is desired to express the output of the system in the time domain. This can be accomplished by determining the inverse Fourier transform

$$y(n) = \frac{1}{2\pi} \int_{-\pi}^{\pi} Y(\omega) e^{j\omega n} \, d\omega \tag{5.2.6}$$

If we return to the basic input–output relation in (5.2.3) and we compute the squared magnitude of both sides, we obtain

$$|Y(\omega)|^2 = |H(\omega)|^2 |X(\omega)|^2 \tag{5.2.7}$$
$$S_{yy}(\omega) = |H(\omega)|^2 S_{xx}(\omega)$$

where $S_{yy}(\omega)$ and $S_{xx}(\omega)$ are the energy density spectra of $y(n)$ and $x(n)$, respectively. This leads to Parseval's relation for the energy of the system output, namely

$$E_y = \frac{1}{2\pi} \int_{-\pi}^{\pi} |Y(\omega)|^2 \, d\omega$$
$$= \frac{1}{2\pi} \int_{-\pi}^{\pi} |H(\omega)|^2 S_{xx}(\omega) \, d\omega \tag{5.2.8}$$

EXAMPLE 5.2.1 _____

A linear time-invariant system is characterized by its impulse response

$$h(n) = (\tfrac{1}{2})^n u(n)$$

Determine the spectrum and the energy density spectrum of the output signal when the system is excited by the signal

$$x(n) = (\tfrac{1}{4})^n u(n)$$

Solution: From Example 5.1.1 we found that the frequency response of this system is

$$H(\omega) = \frac{1}{1 - \tfrac{1}{2} e^{-j\omega}}$$

On the other hand, the input sequence $x(n)$ has a Fourier transform

$$X(\omega) = \frac{1}{1 - \tfrac{1}{4} e^{-j\omega}}$$

The spectrum of the signal at the output of the system is

$$Y(\omega) = H(\omega)X(\omega)$$

$$= \frac{1}{(1 - \frac{1}{2}e^{-j\omega})(1 - \frac{1}{4}e^{-j\omega})}$$

The corresponding energy density spectrum is

$$S_{yy}(\omega) = |Y(\omega)|^2$$

$$= |H(\omega)|^2|X(\omega)|^2$$

$$= \frac{1}{(\frac{5}{4} - \cos\omega)(\frac{17}{16} - \frac{1}{2}\cos\omega)}$$

5.2.2 Frequency Response of an Interconnection of Systems

In Chapter 2 we characterized an interconnection of linear time-invariant systems in the time domain. In particular, if we have L systems connected in parallel, the impulse response $h(n)$ of the resultant system is given by

$$h(n) = \sum_{k=1}^{L} h_k(n) \tag{5.2.9}$$

where $h_k(n)$, $k = 1, \ldots, L$ are the impulse responses of the individual systems. By using the linearity property of the Fourier transform, we find that the frequency response of the overall system is

$$H(\omega) = \sum_{k=1}^{L} H_k(\omega) \tag{5.2.10}$$

where $H_k(\omega)$ is the frequency response corresponding to the impulse response $h_k(n)$. This result is consistent with the equivalent z-transform relationship.

If the L linear time-invariant systems are connected in cascade, the impulse response of the overall system is

$$h(n) = h_1(n) * h_2(n) \cdots * h_L(n) \tag{5.2.11}$$

By using the convolution theorem of the Fourier transform, it is easy to show that

$$H(\omega) = H_1(\omega)H_2(\omega) \cdots H_L(\omega) \tag{5.2.12}$$

We observe that the parallel interconnection of systems involves additivity in both the time and frequency domain. However, the cascade connection involves convolution of the impulse responses in the time domain and multiplication of the frequency responses $H_k(\omega)$, $k = 1, 2, \ldots, L$ in the frequency domain.

Figure 5.4 illustrates the parallel and cascade interconnections for two systems.

5.2.3 Computation of the Frequency Response Function

The frequency response function $H(\omega)$ of a linear time-invariant system may be computed from the Fourier transform relationship

$$H(\omega) = \sum_{n=-\infty}^{\infty} h(n)e^{-j\omega n} \tag{5.2.13}$$

when the impulse response of the system is available.

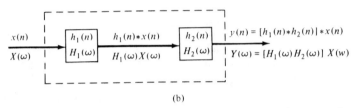

(a)

(b)

FIGURE 5.4   LTI systems in (a) parallel and (b) cascade connections.

If the system is characterized by a linear constant-coefficient difference equation of the form

$$y(n) = -\sum_{k=1}^{N} a_k y(n - k) + \sum_{k=0}^{M} b_k x(n - k) \qquad (5.2.14)$$

the frequency response may be simply obtained by evaluating $H(z)$ on the unit circle. Thus from (3.3.8) we obtain

$$H(\omega) = \frac{\displaystyle\sum_{k=0}^{M} b_k e^{-j\omega k}}{1 + \displaystyle\sum_{k=1}^{N} a_k e^{-j\omega k}} \qquad (5.2.15)$$

Equation (5.2.15) clearly illustrates that the frequency response $H(\omega)$ of the system characterized by a linear constant-coefficient difference equation depends only on the coefficients $\{a_k\}$ and $\{b_k\}$. Special cases are easily obtained from (5.2.15). For example, if the system is an FIR system, $a_k = 0$, $k = 1, 2, \ldots, N$, and hence (5.2.15) reduces to

$$H(\omega) = \sum_{k=0}^{M} b_k e^{-j\omega k} \qquad (5.2.16)$$

This is consistent with our previous statement that an FIR system has an impulse response

$$h(n) = \begin{cases} b_n & n = 0, 1, \ldots, M \\ 0 & \text{otherwise} \end{cases} \qquad (5.2.17)$$

If the system is purely recursive, $b_k = 0$, $k = 1, 2, \ldots, M$, and hence

$$H(\omega) = \frac{b_0}{1 + \displaystyle\sum_{k=1}^{N} a_k e^{-j\omega k}} \qquad (5.2.18)$$

EXAMPLE 5.2.2 Hanning Filter _____

Determine and sketch the magnitude and phase responses for the FIR system characterized by the (moving average) difference equation

$$y(n) = \tfrac{1}{4}x(n) + \tfrac{1}{2}x(n-1) + \tfrac{1}{4}x(n-2)$$

Solution: By applying (5.2.16), we obtain

$$H(\omega) = \tfrac{1}{4} + \tfrac{1}{2}e^{-j\omega} + \tfrac{1}{4}e^{-j2\omega}$$

This expression can be simplified somewhat by noting that

$$\begin{aligned} H(\omega) &= \tfrac{1}{4}(1 + 2e^{-j\omega} + e^{-j2\omega}) \\ &= \tfrac{1}{4}e^{-j\omega}(e^{j\omega} + 2 + e^{-j\omega}) \\ &= \tfrac{1}{2}(1 + \cos \omega)e^{-j\omega} \end{aligned}$$

Consequently,

$$\begin{aligned} |H(\omega)| &= \tfrac{1}{2}(1 + \cos \omega) \\ \Theta(\omega) &= -\omega \end{aligned} \tag{5.2.19}$$

Figure 5.5 shows plots of these functions.

We observe that the Hanning filter has a lowpass frequency response characteristic. Its magnitude response is unity at $\omega = 0$ (dc) and rolls off or decays to zero at $\omega = \pi$. Its phase response is a linear function of frequency. This simple filter is used for data smoothing in many applications.

_____

An alternative method for evaluating the frequency response $H(\omega)$ given by (5.2.15) is based on a geometric approach. To describe this method, we begin with the system function $H(z)$, corresponding to (5.2.15), which is expressed in terms of poles and zeros as

$$H(z) = Gz^{-M+N}\frac{(z - z_1)(z - z_2) \cdots (z - z_M)}{(z - p_1)(z - p_2) \cdots (z - p_N)} \tag{5.2.20}$$

The Fourier transform corresponding to $H(z)$ is

$$H(\omega) = Ge^{j\omega(N-M)}\frac{(e^{j\omega} - z_1)(e^{j\omega} - z_2) \cdots (e^{j\omega} - z_M)}{(e^{j\omega} - p_1)(e^{j\omega} - p_2) \cdots (e^{j\omega} - p_N)} \tag{5.2.21}$$

Now we express each of the factors in (5.2.21) in polar form, as

$$e^{j\omega} - z_k = V_k(\omega)e^{j\Theta_k(\omega)} \tag{5.2.22}$$

and

$$e^{j\omega} - p_k = U_k(\omega)e^{j\Phi_k(\omega)} \tag{5.2.23}$$

where

$$V_k(\omega) \equiv |e^{j\omega} - z_k| \qquad \Theta_k(\omega) \equiv \angle (e^{j\omega} - z_k) \tag{5.2.24}$$

and

$$U_k(\omega) \equiv |e^{j\omega} - p_k| \qquad \Phi_k(\omega) = \angle (e^{j\omega} - p_k) \tag{5.2.25}$$

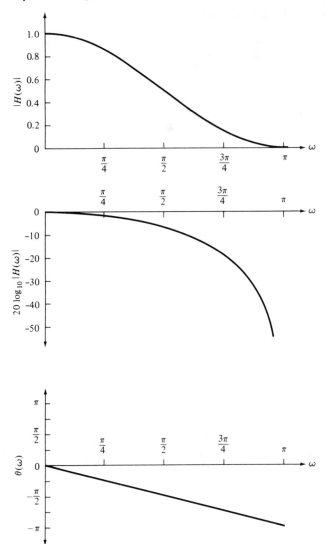

FIGURE 5.5   Magnitude and phase response of the Hanning filter in Example 5.2.2.

The magnitude of $H(\omega)$ is equal to the product of magnitudes of all terms in (5.2.21). Thus, using (5.2.21) through (5.2.23), we obtain

$$|H(\omega)| = |G| \frac{V_1(\omega) \cdots V_M(\omega)}{U_1(\omega)U_2(\omega) \cdots U_N(\omega)} \tag{5.2.26}$$

since the magnitude of $e^{j\omega(N-M)}$ is 1.

The phase of $H(\omega)$ is the sum of the phases of the numerator factors, minus the phases of the denominator factors. Thus, by combining (5.2.21) through (5.2.24), we have

$$H(\omega) = \angle\, G + \omega(N - M) + \Theta_1(\omega) + \Theta_2(\omega) + \cdots + \Theta_M(\omega) \\ - [\Phi_1(\omega) + \Phi_2(\omega) + \cdots + \Phi_N(\omega)] \tag{5.2.27}$$

The phase of the gain term G is zero or π, depending on whether G is positive or negative.

Clearly, if we know the zeros and the poles of the system function $H(z)$, we can evaluate the frequency response from (5.2.26) and (5.2.27). Figure 5.6a and b contain

```
C     MAGNITUDE AND PHASE RESPONSE OF A DIGITAL FILTER WITH COEFFICIENTS A(I), B(I)
C
C     INPUT PARAMETERS
C       A        ARRAY CONTAINING THE DENOMINATOR COEF
C       B        ARRAY CONTAINING THE NUMERATOR COEF
C       N        DIMENSION OF A
C       M        DIMENSION OF B
C       G        GAIN FACTOR
C       L        NUMBER OF POINTS IN WHERE THE FUNCTION IS EVALUATED
C     OUTPUT PARAMETERS
C     HMAG(K),  K=1,2,...,L :MAGNITUDE RESPONSE
C     HPHA(K),  K=1,2,...,L :PHASE RESPONSE
C
      SUBROUTINE FRESP    (N,A,M,B,G,HMAG,HPHA,L)
      DIMENSION A(1),B(1),HMAG(1),HPHA(1)
      PI=4.0*ATAN(1.0)
C
C     COMPUTE MAGNITUDE AND PHASE RESPONSE
C
      DO 400 I=1,L
        OMEGA=(I-1)*PI/FLOAT(L-1)
C       COMPUTE REAL AND COMPLEX PARTS OF NUMERATOR
        TEMP1=B(1)
        TEMP2=0.0
        DO 100 J=2,M+1
          JM1=J-1
          TEMP1=TEMP1+B(J)*COS(JM1*OMEGA)
          TEMP2=TEMP2+B(J)*SIN(JM1*OMEGA)
  100   CONTINUE
        HMAG(I)=SQRT(TEMP1*TEMP1+TEMP2*TEMP2)
        IF (TEMP1.NE.0.0) HPHA(I)=ATAN(TEMP2/TEMP1)
        IF (TEMP1.EQ.0.0) HPHA(I)=PI/2.
C
C       ADJUST QUADRANTS
C
        IF ((TEMP2.LT.0.0).AND.(TEMP1.LE.0.0)) HPHA(I)=HPHA(I)-PI
        IF ((TEMP2.GT.0.0).AND.(TEMP1.LT.0.0)) HPHA(I)=HPHA(I)+PI
        IF (N.EQ.0) GOTO 900
C
C       COMPUTE REAL AND COMPLEX PARTS OF DENOMINATOR
C
        TEMP1=1.0
        TEMP2=0.0
        DO 200 J=1,N
          TEMP1=TEMP1+A(J)*COS(J*OMEGA)
          TEMP2=TEMP2+A(J)*SIN(J*OMEGA)
  200   CONTINUE
        IF ((TEMP1.EQ.0.0).AND.(TEMP2.EQ.0.0)) GOTO 500
        HMAG(I)=HMAG(I)/SQRT(TEMP1*TEMP1+TEMP2*TEMP2)
        IF (TEMP1.NE.0.) HPHA(I)=HPHA(I)-ATAN(TEMP2/TEMP1)
        IF (TEMP1.EQ.0.0) HPHA(I)=HPHA(I)-PI/2
C
C       ADJUST QUADRANTS
C
  900   IF ((TEMP2.LT.0.0).AND.(TEMP1.LE.0.)) HPHA(I)=HPHA(I)-PI
        IF ((TEMP2.GT.0.0).AND.(TEMP1.LT.0.)) HPHA(I)=HPHA(I)+PI
        IF (HPHA(I).GT.PI) HPHA(I)=HPHA(I)-2.*PI
        IF (HPHA(I).LT.-PI) HPHA(I)=HPHA(I)+2.*PI
  400 CONTINUE
  500 RETURN
      END
```

FIGURE 5.6 Programs for computing the magnitude and phase response of an IIR system from (a) filter coefficients and (b) pole–zero pattern.

```
      COMPUTATION OF THE MAGNITUDE AND PHASE OF THE
      FOURIER TRANSFORM FROM THE POLE - ZERO PATTERN


      R(K) , TH(K) ,K=1,2,...N :MAGNITUDE AND PHASE OF ZEROS
      P(K)   PH(K) ,K=1,2,...   :MAGNITUDE AND PHASE OF POLES
      G : GAIN FACTOR
      XMAG(J) , J=1,2,...,L :MAGNITUDE OF THE FOURIER TRANSFORM
      XPHA(J) , J=1,2,...,L :PHASE OF THE FOURIER TRANSFORM

      SUBROUTINE FTPLZ (M,R,TH,N,P,PH,G,XMAG,XPHA,L)
      DIMENSION R(1),TH(1),P(1),PH(1),XMAG(1),XPHA(1)
      PI=4.0*ATAN(1.0)
      DO 50 J=1,M
        T1=R(J)
        T2=PI*TH(J)/180.0
        R(J)=T1*COS(T2)
        TH(J)=T1*SIN(T2)
50    CONTINUE
      DO 55 J=1,N
        T1=P(J)
        T2=PI*PH(J)/180.0
        P(J)=T1*COS(T2)
        PH(J)=T1*SIN(T2)
55    CONTINUE

      COMPUTE MAGNITUDE AND PHASE RESPONSE

      DO 400 I=1,L
        OMEGA=(I-1)*PI/FLOAT(L-1)
        COMPUTE THE MAG AND PHASE OF DISTANCE VECTOR FROM UNIT CIRCLE
        TEMP1 ACCUMULATES THE MAG AND TEMP2 ACCUMULATES THE PHASE
        TEMP1=1.0
        TEMP2=0.0
        DO 100 J=1,M
          TEMP=COS(OMEGA)-R(J)
          TEMPJ=SIN(OMEGA)-TH(J)
          TEMP1=TEMP1*SQRT(TEMP*TEMP+TEMPJ*TEMPJ)
          IF (TEMP.EQ.0.) ANGLE=PI/2*TEMPJ/ABS(TEMPJ)
          IF (TEMP.LT.0.) ANGLE=ATAN(TEMPJ/TEMP)+PI
          IF (TEMP.GT.0.) ANGLE=ATAN(TEMPJ/TEMP)
          TEMP2=TEMP2+ANGLE
100     CONTINUE
        DO 150 J=1,N
          TEMP=COS(OMEGA)-P(J)
          TEMPJ=SIN(OMEGA)-PH(J)
          TEMP1=TEMP1/SQRT(TEMP*TEMP+TEMPJ*TEMPJ)
          IF (TEMP.EQ.0.) ANGLE=PI/2*TEMPJ/ABS(TEMPJ)
          IF (TEMP.LT.0.) ANGLE=ATAN(TEMPJ/TEMP)+PI
          IF (TEMP.GT.0.) ANGLE=ATAN(TEMPJ/TEMP)
          TEMP2=TEMP2-ANGLE
150     CONTINUE

      INCLUDE GAIN FACTOR

      TEMP1=TEMP1*ABS(G)
        IF (G.LT.0.) TEMP2=TEMP2-PI
        XMAG(I)=TEMP1
        XPHA(I)=TEMP2

      ADJUST PHASE BETWEEN -PI AND PI

      IF (XPHA(I).GT.PI) XPHA(I)=XPHA(I)-PI*2.
        IF (XPHA(I).GT.PI) GOTO 200
      IF (XPHA(I).LT.-PI) XPHA(I)=XPHA(I)+2.*PI
        IF (XPHA(I).LT.-PI) GOTO 300
      CONTINUE
        RETURN
        END
```

FIGURE 5.6   (continued)

FORTRAN subroutines for calculating $|H(\omega)|$ and $\angle H(\omega)$ from the coefficients $\{a_i\}$, $\{b_i\}$ and the pole–zero locations.

There is a geometric interpretation of the quantities appearing in (5.2.26) and (5.2.27). Let us consider a pole p_k and a zero z_k located at points A and B of the z-plane, as shown in Fig. 5.7a. Assume that we wish to compute $H(\omega)$ at a specific value of frequency ω. The given value of ω determines the angle of $e^{j\omega}$ with the positive real axis. The tip of the vector $e^{j\omega}$ specifies a point L on the unit circle. The evaluation of the Fourier transform for the given value of ω is equivalent to evaluating the z-transform at the point L of the complex plane. Let us draw the vectors $\mathbf{AL}$ and $\mathbf{BL}$ from the pole and zero locations to the point L, at which we wish to compute the Fourier transform. From Fig. 5.7a it follows that

$$\mathbf{CL} = \mathbf{CA} + \mathbf{AL}$$

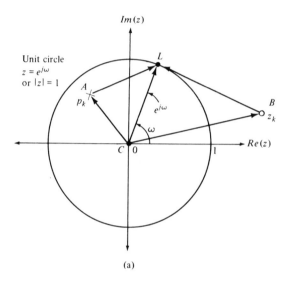

(a)

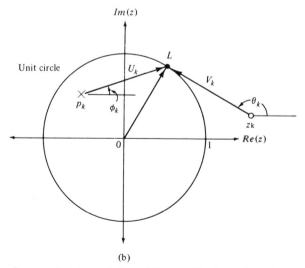

(b)

FIGURE 5.7 Geometric interpretation of the contribution of a pole and a zero to the Fourier transform (1) magnitude: the factor V_k/U_k, (2) phase: the factor $\theta_k - \phi_k$.

and

$$CL = CB + BL$$

However, $CL = e^{j\omega}$, $CA = p_k$ and $CB = z_k$. Thus

$$AL = e^{j\omega} - p_k \tag{5.2.28}$$

and

$$BL = e^{j\omega} - z_k \tag{5.2.29}$$

By combining these relations with (5.2.22) and (5.2.23), we obtain

$$AL = e^{j\omega} - p_k = U_k(\omega)e^{j\Phi_k(\omega)} \tag{5.2.30}$$
$$BL = e^{j\omega} - z_k = V_k(\omega)e^{j\Theta_k(\omega)} \tag{5.2.31}$$

Thus $U_k(\omega)$ is the length of **AL**, that is, the distance of the pole p_k from the point L corresponding to $e^{j\omega}$, whereas $V_k(\omega)$ is the distance of the zero z_k from the same point L. The phases $\Phi_k(\omega)$ and $\Theta_k(\omega)$ are the angles of the vectors **AL** and **BL** with the positive real axis, respectively. These geometric interpretations are shown in Fig. 5.7b.

Geometric interpretations are very useful in understanding how the location of poles and zeros affects the magnitude and phase of the Fourier transform. Suppose that a zero, say z_k, and a pole, say p_k, are on the unit circle as shown in Fig. 5.8. We note that at $\omega = \angle z_k$, $V_k(\omega)$ and consequently $|H(\omega)|$ become zero. Similarly, at $\omega = \angle p_k$ the length $U_k(\omega)$ becomes zero and hence $|H(\omega)|$ becomes infinite. Clearly, the evaluation of phase in these cases has no meaning.

From this discussion we can easily see that the presence of a zero close to the unit circle will cause the magnitude of the frequency response at frequencies that correspond to points of the unit circle close to that point to be small. In contrast, the presence of a pole close to the unit circle will cause the magnitude of the frequency response to be large at frequencies close to that point. Thus poles have the opposite effect of zeros. Also, placing a zero close to a pole cancels the effect of the pole, and vice versa. This can be also seen from (5.2.21), since if $z_k = p_k$, the terms $e^{j\omega} - z_k$ and $e^{j\omega} - p_k$ will cancel. Obviously, the presence of both poles and zeros in a transform results in a greater

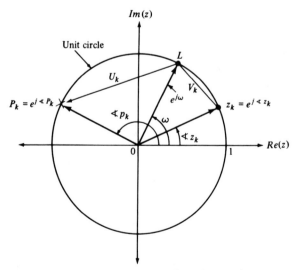

FIGURE 5.8   A zero on the unit circle causes $|H(\omega)| = 0$ at $\omega = \angle z_k$. In contrast, a pole on the unit circle results in $|H(\omega)| = \infty$ at $\omega = \angle p_k$.

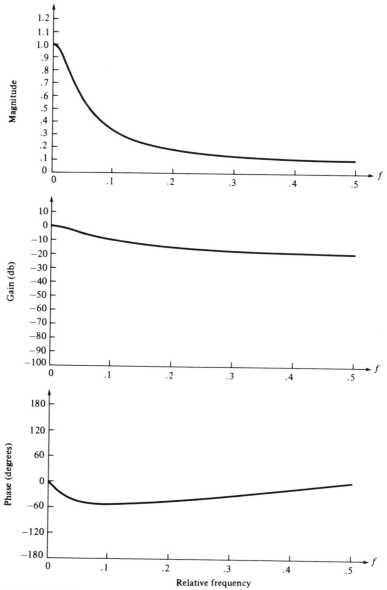

FIGURE 5.9 Magnitude and phase of system with $H(z) = 1/(1 - 0.8z^{-1})$.

variety of shapes for $|H(\omega)|$ and $\angle\, H(\omega)$. This observation is very important in the design of digital filters. We conclude our discussion with the following example illustrating these concepts, as well as the use of the FORTRAN subroutine shown in Fig. 5.6.

EXAMPLE 5.2.3 _____

Evaluate the frequency response of the system described by the system function

$$H(z) = \frac{1}{1 - 0.8z^{-1}} = \frac{z}{z - 0.8}$$

Solution: Clearly, $H(z)$ has a zero at $z = 0$ and a pole at $p = 0.8$. Hence the system is described by the parameters

$$G = 1 \quad M = 1 \quad N = 1 \quad z_1 = 0 \quad p_1 = 0.8$$

The subroutine FTPLZ is called with the following parameters:

$$G = 1 \quad M = 1 \quad R(1) = 0.0 \quad \text{TH}(1) = 0.0$$
$$P(1) = 0.8 \quad \text{PH}(1) = 0.0 \quad L = 101$$

The output of the computation is $|H(\omega)|$ and

$$\angle H(\omega_k) \quad \text{for } \omega_k = \pi \frac{k}{L-1} \quad k = 0, 1, \ldots, L-1$$

Note that if $V_k(\omega)$ or $U_k(\omega)$ are zero, we set $\Theta_k(\omega)$ or $\Phi_k(\omega)$ as zero, although these values are not used by the subroutine in the actual computation. $|H(\omega)|$ and $\angle H(\omega)$ are plotted in Fig. 5.9.

Finally, we note that if we want the computation of $|H(\omega)|$ in decibels, (5.2.26) should be replaced by

$$|H(\omega)|_{\text{dB}} = 20 \log_{10} |G| + 20 \sum_{k=1}^{M} \log_{10} V_k(\omega) - 20 \sum_{k=1}^{N} \log_{10} U_k(\omega) \quad (5.2.32)$$

5.3 Linear Time-Invariant Systems as Frequency-Selective Filters

The term *filter* is commonly used to describe a device that discriminates, according to some attribute of the objects applied at its input, what passes through it. For example, an air filter allows air to pass through it but prevents dust particles that are present in the air from passing through. An oil filter performs a similar function, with the exception that oil is the substance allowed to pass through the filter, while particles of dirt are collected at the input to the filter and prevented from passing through. In photography, an ultraviolet filter is often used to prevent ultraviolet light, which is present in sunlight, and which is not a part of visible light, from passing through and affecting the chemicals on the film.

As we have observed in the preceding two sections, a linear time-invariant system also performs a type of discrimination or filtering among the various frequency components at its input. The nature of this filtering action is determined by the frequency response characteristics $H(\omega)$, which in turn depends on the choice of the system parameters (e.g., the coefficients $\{a_k\}$ and $\{b_k\}$ in the difference equation characterization of the system). Thus, by proper selection of the coefficients, we can design frequency-selective filters that pass signals with frequency components in some bands while they attenuate signals that contain frequency components in other frequency bands.

In general, a linear time-invariant system modifies the input signal spectrum $X(\omega)$ according to its frequency response $H(\omega)$ to yield an output signal with spectrum $Y(\omega) = H(\omega)X(\omega)$. In a sense, $H(\omega)$ acts as a *weighting function* or a *spectral shaping function* to the different frequency components in the input signal. When viewed in this context, any linear time-invariant system can be considered to be a frequency-shaping filter, even though it may not necessarily completely block any or all frequency components. Consequently, the terms "linear time-invariant system" and "filter" are synonymous and are often used interchangeably.

In this book we use the term *filter* to describe a linear time-invariant system which is used to perform spectral shaping or frequency-selective filtering. Filtering is used in digital signal processing in a variety of ways: for example, to remove undesirable noise from desired signals; for spectral shaping such as equalization of communication channels; for signal detection in radar, sonar, and communications; for performing spectral analysis of signals; and so on.

5.3.1 Ideal Frequency-Selective Filters and Distortionless Transmission

In many practical applications, we have the problem of separating signals that have nonoverlapping spectra, with the added constraint that the desired signals be undistorted by the filters that perform the signal separation. This problem usually arises in communications, where multiple signals are frequency-division multiplexed (stacked together in nonoverlapping frequency bands) and transmitted over a common channel (coaxial cable, fiber optic cable, or satellite channel). At the receiving end of the communication system the signals must be separated by frequency-selective filters and delivered, or transmitted, to their final destination. The frequency-selective filters must be designed to introduce negligible distortion in the signals that pass through them.

Let us consider a signal $x(n)$ with frequency content in a band of frequencies $\omega_1 < \omega < \omega_2$. Hence $X(\omega)$ is a bandlimited signal, that is,

$$X(\omega) = 0 \qquad \omega \geq \omega_2, \quad \omega \leq \omega_1$$

Suppose that this signal is passed through a filter with frequency response

$$H(\omega) = \begin{cases} Ce^{-j\omega k} & \omega_1 < \omega < \omega_2 \\ 0 & \text{otherwise} \end{cases} \qquad (5.3.1)$$

where C and k are positive constants. The signal at the output of the filter has the spectrum

$$\begin{aligned} Y(\omega) &= X(\omega)H(\omega) \\ &= CX(\omega)e^{-j\omega k} \qquad \omega_1 < \omega < \omega_2 \end{aligned} \qquad (5.3.2)$$

Recall that the time-shifting property of the Fourier transform is

$$x(n - k) \longleftrightarrow X(\omega)e^{-j\omega k}$$

If we compare this relation with (5.3.2), it is apparent that the output of the filter is

$$y(n) = Cx(n - k) \qquad (5.3.3)$$

Consequently, the filter output is simply a delayed and amplitude-scaled version of the input signal. A pure delay is usually tolerable and is not considered a distortion of the signal. Neither is amplitude scaling. Thus the filter characterized by the transfer function (5.3.1) is called an *ideal* (bandpass) *filter*. Its amplitude is a constant, that is,

$$|H(\omega)| = C \qquad \omega_1 < \omega < \omega_2$$

and its phase is a linear function of frequency, that is,

$$\Theta(\omega) = -\omega k$$

These frequency response characteristics are illustrated in Fig. 5.10 for $C = 1$, $k = 4$, $\omega_1 = \pi/4$, and $\omega_2 = 3\pi/4$.

In general, any deviation of the frequency response characteristics of a linear filter from the ideal results in signal distortion. If the filter has a frequency-variable magnitude

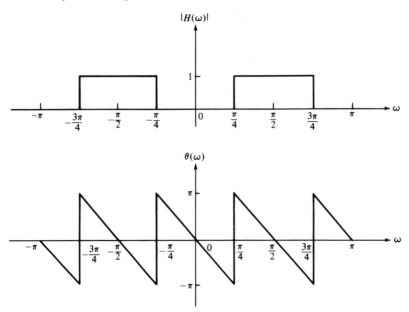

FIGURE 5.10   Magnitude and phase characteristics of an ideal bandpass filter.

response characteristic within the frequency band occupied by the signal, the filter introduces *amplitude distortion*. If the phase characteristic is not linear within the desired frequency band, the signal undergoes *phase distortion*.

The derivative of the phase with respect to frequency has the units of delay. Hence we may define the signal delay as a function of frequency as

$$\tau(\omega) = -\frac{d\Theta(\omega)}{d\omega} \qquad (5.3.4)$$

and observe that for a linear-phase filter, the delay is a constant, independent of frequency. Consequently, a filter that causes phase distortion has a variable-frequency delay and one that has linear phase has a constant delay within the desired frequency range. If the delay is not a constant within the desired frequency range, we say that the filter introduces *delay distortion*. Thus delay distortion is synonymous with phase distortion.

Just as in the case of signals, filters are classified according to their frequency response characteristics. In particular, an *ideal lowpass filter* has the frequency response characteristic

$$H(\omega) = \begin{cases} Ce^{-j\omega k} & |\omega| \leq \omega_c \\ 0 & \text{otherwise} \end{cases} \qquad (5.3.5)$$

where ω_c is called the cutoff frequency. Similarly, an *ideal highpass filter* is defined as one having the frequency response

$$H(\omega) = \begin{cases} Ce^{-j\omega k} & |\omega| \geq \omega_c \\ 0 & \text{otherwise} \end{cases} \qquad (5.3.6)$$

We have already defined an ideal bandpass filter. Finally, we define an *ideal bandelimination* or *bandstop filter* as one having the frequency response characteristic

$$H(\omega) = \begin{cases} Ce^{-j\omega k} & |\omega| \leq \omega_1 \text{ and } |\omega| \geq \omega_2 \\ 0 & \omega_1 < |\omega| < \omega_2 \end{cases} \qquad (5.3.7)$$

These ideal frequency response characteristics are illustrated in Fig. 5.11.

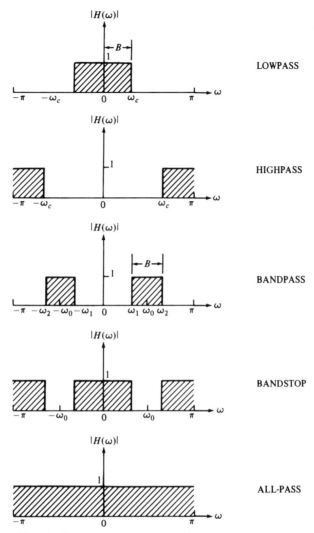

FIGURE 5.11 Magnitude responses for some ideal frequency-selective discrete-time filters.

5.3.2 Causality and Its Implications

Is it possible to realize an ideal filter in practice? Let us attempt to answer this question by examining the impulse response $h(n)$ of an ideal lowpass filter with frequency response characteristic

$$H(\omega) = \begin{cases} 1 & |\omega| \leq \omega_c \\ 0 & \omega_c < \omega \leq \pi \end{cases} \tag{5.3.8}$$

The impulse response of this filter is

$$h(n) = \begin{cases} \dfrac{\omega_c}{\pi} & n = 0 \\[4mm] \dfrac{\omega_c}{\pi} \dfrac{\sin \omega_c n}{\omega_c n} & n \neq 0 \end{cases} \tag{5.3.9}$$

A plot of $h(n)$ for $\omega_c = \pi/4$ is illustrated in Fig. 5.12.

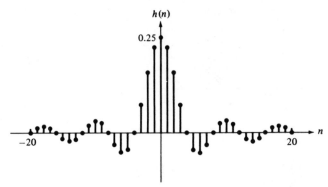

FIGURE 5.12 Unit sample response of an ideal lowpass filter.

It is clear that the ideal lowpass filter is noncausal and hence it cannot be realized in practice. In addition, $h(n)$ is not absolutely summable. Consequently, the ideal lowpass filter is unstable.

We also observe that the width of the main lobe of $h(n)$ is inversely proportional to the bandwidth ω_c of the filter. As the filter bandwidth increases the impulse response becomes narrower, and vice versa. For $\omega_c = \pi$, the filter becomes all-pass and its impulse response becomes a unit sample sequence.

If the impulse response is delayed by n_0 samples, the ideal lowpass filter has a linear phase, that is,

$$h(n - n_0) \xleftarrow{\quad F \quad} H(\omega)e^{-j\omega n_0}$$

However, no finite value of delay can result in a causal filter. In other words, the ideal lowpass filter is physically unrealizable.

One possible solution is to introduce a large delay n_0 in $h(n)$ and arbitrarily to set $h(n) = 0$ for $n < n_0$. However, the resulting system no longer has an ideal frequency response characteristic.

Although our discussion above has been limited to the realization of a lowpass filter, our conclusions hold, in general, for all the other ideal filter characteristics. In brief, none of the ideal filter characteristics illustrated in Fig. 5.11 is causal; hence all are physically unrealizable.

A question that arises naturally at this point is the following: What are the necessary and sufficient conditions that a frequency response characteristic $H(\omega)$ must satisfy in order for the resulting filter to be causal? The answer to this question is given by the Paley–Wiener theorem, which may be stated as follows:

PALEY–WIENER THEOREM: *If $h(n)$ has finite energy and $h(n) = 0$ for $n < 0$, then [for a reference, see Wiener and Paley (1934)]*

$$\int_{-\pi}^{\pi} |\ln|H(\omega)||\, d\omega < \infty \qquad (5.3.10)$$

Conversely, if $|H(\omega)|$ is square integrable and if the integral in (5.3.10) is finite, then we can associate with $|H(\omega)|$ a phase response $\Theta(\omega)$, so that the resulting filter with frequency response

$$H(\omega) = |H(\omega)|e^{j\Theta(\omega)}$$

is causal.

One important conclusion that we draw from the Paley–Wiener theorem is that the magnitude function $|H(\omega)|$ may be zero at some frequencies, but it cannot be zero over any finite band of frequencies, since the integral will become infinite. Consequently, any ideal filter will be noncausal.

Apparently, causality imposes some tight constraints on a linear time-invariant system. In addition to the Paley–Wiener condition, causality also implies a strong relationship between $H_R(\omega)$ and $H_I(\omega)$, the real and imaginary components of the frequency response $H(\omega)$. To illustrate this dependence, we decompose $h(n)$ into an even and an odd sequence, that is,

$$h(n) = h_e(n) + h_o(n) \tag{5.3.11}$$

where

$$h_e(n) = \tfrac{1}{2}[h(n) + h(-n)] \tag{5.3.12}$$

and

$$h_o(n) = \tfrac{1}{2}[h(n) - h(-n)] \tag{5.3.13}$$

Now, if $h(n)$ is causal, it is possible to recover $h(n)$ from its even part $h_e(n)$ for $0 \le n \le \infty$ or from its odd component $h_o(n)$ for $1 \le n \le \infty$.

Indeed, if we define a sequence

$$u_+(n) = \begin{cases} 2 & n \ge 0 \\ 1 & n = 0 \\ 0 & n < 0 \end{cases} \tag{5.3.14}$$

then it easily follows that

$$h(n) = h_e(n)u_+(n) \qquad n \ge 0 \tag{5.3.15}$$

Consequently, $h(n)$ is recovered entirely from its even part. On the other hand, we have

$$h(n) = h_o(n)u_+(n) + h(0)\,\delta(n) \qquad n \ge 1 \tag{5.3.16}$$

Since $h_o(n) = 0$ for $n = 0$, we cannot recover $h(0)$ from $h_o(n)$ and hence we also must know $h(0)$. In any case, it is apparent that $h_o(n) = h_e(n)$ for $n \ge 1$, so there is a strong relationship between $h_o(n)$ and $h_e(n)$.

If $h(n)$ is absolutely summable (i.e., BIBO stable), the frequency response $H(\omega)$ exists, and

$$H(\omega) = H_R(\omega) + jH_I(\omega) \tag{5.3.17}$$

In addition, if $h(n)$ is real valued and causal, the symmetry properties of the Fourier transform imply that

$$h_e(n) \xleftarrow{\quad F \quad} H_R(\omega)$$
$$h_o(n) \xleftarrow{\quad F \quad} H_I(\omega) \tag{5.3.18}$$

Since $h(n)$ is completely specified by $h_e(n)$, it follows that $H(\omega)$ is completely determined if we know $H_R(\omega)$. Alternatively, $H(\omega)$ is completely determined from $H_I(\omega)$ and $h(0)$. In short, $H_R(\omega)$ and $H_I(\omega)$ are interdependent and cannot be specified independently if the system is causal. Equivalently, the magnitude and phase responses of a causal filter are interdependent and hence cannot be specified independently.

Now that we know the restrictions that causality imposes on the frequency response characteristic and the fact that ideal filters are not achievable in practice, we can return to the class of linear time-invariant systems specified by the difference equation

$$y(n) = -\sum_{k=1}^{N} a_k y(n-k) + \sum_{k=0}^{M} b_k x(n-k) \qquad (5.3.19)$$

which are causal and physically realizable. As we have demonstrated, such systems have a frequency response

$$H(\omega) = \frac{\displaystyle\sum_{k=0}^{M} b_k e^{-j\omega k}}{1 + \displaystyle\sum_{k=1}^{N} a_k e^{-j\omega k}} \qquad (5.3.20)$$

The basic digital filter design problem is to approximate any of the ideal frequency response characteristics with a system of the form (5.3.19), which has the frequency response (5.3.20), by properly selecting the coefficients $\{a_k\}$ and $\{b_k\}$. The next two subsections treat this approximation problem.

5.3.3 Nonideal Frequency-Selective Filters

As we observed from our discussion of the preceding section, ideal filters are noncausal and hence physically unrealizable for real-time signal processing applications. Causality implies that the frequency response characteristic $H(\omega)$ of the filter cannot be zero, except at a finite set of points in frequency. In addition, $H(\omega)$ cannot have an infinitely sharp cutoff from passband to stopband, that is, $H(\omega)$ cannot drop from unity to zero abruptly.

Although the frequency response characteristics possessed by ideal filters may be desirable, they are not absolutely necessary in most practical applications. If we relax these conditions, it is possible to realize causal filters that approximate the ideal filters as closely as we desire. In particular, it is not necessary to insist that the magnitude $|H(\omega)|$ be constant in the entire passband of the filter. A small amount of ripple in the passband, as illustrated in Fig. 5.13, is usually tolerable. Similarly, it is not necessary for the filter response $|H(\omega)|$ to be zero in the stopband. A small, nonzero value or a small amount of ripple in the stopband is also tolerable.

The transition of the frequency response from passband to stopband defines the *transition band* or *transition region* of the filter, as illustrated in Fig. 5.13. The band-edge

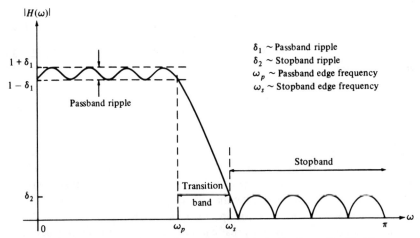

FIGURE 5.13 Magnitude characteristics of physically realizable filters.

frequency ω_p defines the edge of the passband, while the frequency ω_s denotes the beginning of the stopband. Thus the width of the transition band is $\omega_s - \omega_p$. The width of the passband is usually called the *bandwidth* of the filter. For example, if the filter is lowpass with a passband edge frequency ω_p, its bandwidth is ω_p.

If there is ripple in the passband of the filter, its value is denoted as δ_1, and the magnitude $|H(\omega)|$ varies between the limits $1 \pm \delta_1$. The ripple in the stopband of the filter is denoted as δ_2.

To accommodate a large dynamic range in the graph of the frequency response of any filter, it is common practice to use a logarithmic scale for the magnitude $|H(\omega)|$. In particular, we compute and plot $20 \log_{10}|H(\omega)|$, which has the units of decibels (abbreviated dB), as a function of ω. Consequently, the ripple in the passband is $20 \log_{10} \delta_1$ decibels.

In any filter design problem we may specify (1) the maximum tolerable passband ripple, (2) the maximum tolerable stopband ripple, (3) the passband edge frequency ω_p, and (4) the stopband edge frequency ω_s. Based on these specifications, we may select the parameters $\{a_k\}$ and $\{b_k\}$ in the frequency response characteristic, given by (5.3.20), which best approximates the desired specification. The degree to which $H(\omega)$ approximates the specifications depends in part on the criterion used in the selection of the filter coefficients $\{a_k\}$ and $\{b_k\}$ as well as the numbers (M, N) of coefficients.

In the following section we present a method for designing linear-phase FIR filters. Additional design techniques for such filters are described in Chapter 8.

5.3.4 Design of Linear-Phase FIR Filters

One of the simplest types of filters that we can design is an FIR filter with linear phase. As we shall see later, only FIR filters can be designed to have linear phase. IIR filters cannot have linear phase. In many practical applications such as in digital communications, where significant phase (or delay) distortion cannot be tolerated, linear-phase filters are widely used.

An FIR filter of length M has a frequency response

$$H(\omega) = \sum_{k=0}^{M-1} b_k e^{-j\omega k} \tag{5.3.21}$$

where the filter coefficients $\{b_k\}$ are also the values of the unit sample response of the filter, that is,

$$h(n) = \begin{cases} b_n & 0 \le n \le M - 1 \\ 0 & \text{otherwise} \end{cases} \tag{5.3.22}$$

To conform to well-established convention in the technical literature, the length of the FIR filter is chosen as M (instead of $M + 1$) in this as well as in subsequent sections dealing with filter design. The linear-phase condition is obtained by imposing symmetry conditions on the unit sample response of the filter.

In particular, we consider two different symmetry conditions for $h(n)$. Let us begin with the symmetry condition

$$h(n) = h(M - 1 - n) \tag{5.3.23}$$

To demonstrate that the filter satisfying this symmetry condition has linear phase, we consider the cases where M is odd and M is even separately. For example, if $M = 5$, the symmetry condtion is $h(0) = h(4)$, $h(1) = h(3)$, and $h(2)$ does not have a matching term. Thus the filter impulse response is symmetric about the point $h(2)$. The corresponding frequency response is

$$
\begin{aligned}
H(\omega) &= h(0) + h(1)e^{-j\omega} + h(2)e^{-j2\omega} + h(3)e^{-3\omega} + h(4)e^{-4\omega} \\
&= e^{-j2\omega}[h(2) + h(0)e^{j2\omega} + h(4)e^{-2\omega} + h(1)e^{j\omega} + h(3)e^{-j\omega}]
\end{aligned}
$$

Since $h(0) = h(4)$ and $h(1) = h(3)$, $H(\omega)$ can be expressed as

$$
H(\omega) = e^{-j2\omega}[h(2) + 2h(0) \cos 2\omega + 2h(1) \cos \omega] \tag{5.3.24}
$$

In (5.3.24), the term in brackets is real for all values of ω and hence we denote it as

$$
H_r(\omega) = h(2) + 2h(0) \cos 2\omega + 2h(1) \cos \omega \tag{5.3.25}
$$

We should be careful not to confuse $H_r(\omega)$ as defined in (5.3.25) with $H_R(\omega)$, previously defined in (5.3.17) as the real part of $H(\omega)$. Clearly, the magnitude of the frequency response $H(\omega)$ is

$$
|H(\omega)| = |H_r(\omega)| \tag{5.3.26}
$$

The phase characteristic of the filter is

$$
\Theta(\omega) = \begin{cases} -2\omega & \text{if } H_r(\omega) > 0 \\ -2\omega + \pi & \text{if } H_r(\omega) < 0 \end{cases} \tag{5.3.27}
$$

We observe that the phase characteristic $\Theta(\omega)$ is a linear function of ω provided that $H_r(\omega)$ is positive (or negative). When $H_r(\omega)$ changes sign from positive to negative (or vice versa), the phase undergoes an abrupt change of π radians. If these phase changes occur outside the passband of the filter (i.e., in the stopband), we really do not care, since the desired signal passing through the filter has no frequency content in the stopband.

Now, if M is even, say $M = 4$, for example, the symmetry condition in (5.3.23) implies that $h(0) = h(3)$ and $h(1) = h(2)$. In this case each value of the unit sample response has a matching term. The corresponding frequency response is

$$
\begin{aligned}
H(\omega) &= h(0) + h(1)e^{-j\omega} + h(2)e^{-j2\omega} + h(3)e^{-j3\omega} \\
&= e^{-j3\omega/2}[h(0)e^{j3\omega/2} + h(3)e^{-j3\omega/2} + h(1)e^{j\omega/2} + h(2)e^{-j\omega/2}]
\end{aligned}
$$

With the aid of the symmetry relations, this expression simplifies to

$$
H(\omega) = e^{-j3\omega/2}[2h(0) \cos (3\omega/2) + 2h(1) \cos (\omega/2)] \tag{5.3.28}
$$

As in the case where M is odd, we observe that the term in the brackets is real, so that $H(\omega)$ can be expressed as

$$
H(\omega) = H_r(\omega)e^{-j3\omega/2}
$$

where $|H(\omega)| = |H_r(\omega)|$ and

$$
\Theta(\omega) = \begin{cases} -\dfrac{3\omega}{2} & \text{if } H_r(\omega) > 0 \\ -\dfrac{3\omega}{2} + \pi & \text{if } H_r(\omega) < 0 \end{cases}
$$

Again, the phase of the filter is linear, with phase jumps of π radians at frequencies where $H_r(\omega)$ changes sign from positive to negative and vice versa.

From these two special cases, we can extrapolate to the general case of a filter with arbitrary length M. In general, the frequency response of an FIR filter having a unit sample response $h(n)$ that satisfies the symmetry condition in (5.3.23) may be expressed as

$$H(\omega) = H_r(\omega)e^{-j\omega(M-1)/2} \tag{5.3.29}$$

where

$$H_r(\omega) = h\left(\frac{M-1}{2}\right) + 2\sum_{n=0}^{(M-3)/2} h(n)\cos\omega\left(\frac{M-1}{2} - n\right) \qquad M \text{ odd} \tag{5.3.30}$$

$$H_r(\omega) = 2\sum_{n=0}^{(M/2)-1} h(n)\cos\omega\left(\frac{M-1}{2} - n\right) \qquad M \text{ even} \tag{5.3.31}$$

The phase characteristic of the filter for both M odd and M even is

$$\Theta(\omega) = \begin{cases} -\omega\left(\dfrac{M-1}{2}\right) & \text{if } H_r(\omega) > 0 \\[2mm] -\omega\left(\dfrac{M-1}{2}\right) + \pi & \text{if } H_r(\omega) < 0 \end{cases} \tag{5.3.32}$$

The second symmetry condition that yields a linear-phase FIR filter is

$$h(n) = -h(M - 1 - n) \tag{5.3.33}$$

In this case we call the unit sample response *antisymmetric*. When M is odd, the center point of the antisymmetric $h(n)$ is $n = (M - 1)/2$. The condition (5.3.33) implies that

$$h\left(\frac{M-1}{2}\right) = 0$$

For example, if $M = 5$, we have $h(0) = -h(4)$, $h(1) = -h(3)$, and $h(2) = 0$. However, if M is even, each term in $h(n)$ has a matching term of opposite sign.

It is straightforward to show that the frequency response of an FIR filter with an antisymmetric unit sample response may be expressed as

$$H(\omega) = H_r(\omega)e^{j[-\omega(M-1)/2 + \pi/2]} \tag{5.3.34}$$

where

$$H_r(\omega) = 2\sum_{n=0}^{(M-3)/2} h(n)\sin\omega\left(\frac{M-1}{2} - n\right) \qquad M \text{ odd} \tag{5.3.35}$$

$$H_r(\omega) = 2\sum_{n=0}^{(M/2)-1} h(n)\sin\omega\left(\frac{M-1}{2} - n\right) \qquad M \text{ even} \tag{5.3.36}$$

The phase characteristic of the filter for both M odd and M even is

$$\Theta(\omega) = \begin{cases} \dfrac{\pi}{2} - \omega\left(\dfrac{M-1}{2}\right) & \text{if } H_r(\omega) > 0 \\[2mm] \dfrac{3\pi}{2} - \omega\left(\dfrac{M-1}{2}\right) & \text{if } H_r(\omega) < 0 \end{cases} \tag{5.3.37}$$

These general frequency response formulas can be used to design linear phase FIR filters with symmetric and antisymmetric unit sample responses. We note that the number of filter coefficients that specify the frequency response is $(M + 1)/2$ when M is odd or $M/2$ when M is even in (5.3.30) and (5.3.31), respectively. On the other hand, if the unit sample response is antisymmetric,

$$h\left(\frac{M-1}{2}\right) = 0$$

so that there are $(M - 1)/2$ filter coefficients when M is odd and $M/2$ coefficients when M is even to be specified.

The choice of a symmetric or antisymmetric unit sample response depends on the application. As we shall see later, a symmetric unit sample response is suitable for some applications, while an antisymmetric unit sample response is more suitable for other applications. For example, if $h(n) = -h(M - 1 - n)$ and M is odd, (5.3.35) implies that $H_r(0) = 0$ and $H_r(\pi) = 0$. Consequently, (5.3.35) is not suitable as either a lowpass filter or a highpass filter. Similarly, the antisymmetric unit sample response with M even also results in $H_r(0) = 0$, as can be easily verified from (5.3.36). Consequently, we would not use the antisymmetric condition in the design of a lowpass linear-phase FIR filter. On the other hand, the symmetry condition $h(n) = h(M - 1 - n)$ yields a linear phase FIR filter with a nonzero response at $\omega = 0$, if desired, that is,

$$H_r(0) = h\left(\frac{M - 1}{2}\right) + 2 \sum_{n=0}^{(M-3)/2} h(n) \qquad M \text{ odd} \qquad (5.3.38)$$

$$H_r(0) = 2 \sum_{n=0}^{(M/2)-1} h(n) \qquad M \text{ even} \qquad (5.3.39)$$

Each of the equations in (5.3.30), (5.3.31), (5.3.35), and (5.3.36) constitutes a set of linear equations for determining the coefficients of an FIR filter. Consequently, if we specify the frequency response at either $(M + 1)/2$ or $(M - 1)/2$ or $M/2$ points in ω, we can solve the corresponding set of linear equations for the coefficients $\{h(n)\}$. Although the values of ω can be chosen arbitrarily, it is usually desirable to select equally spaced points in frequency, in the range $0 \le \omega \le \pi$. Thus if we select the frequencies as

$$\omega_k = \frac{2\pi k}{M} \qquad \begin{array}{ll} k = 0, 1, \ldots, \dfrac{M - 1}{2} & M \text{ odd} \\[3mm] k = 0, 1, \ldots, \dfrac{M}{2} - 1 & M \text{ even} \end{array} \qquad (5.3.40)$$

and define

$$a_{kn} = 2 \cos \omega_k \left(\frac{M - 1}{2} - n\right) \qquad (5.3.41)$$

$$a_{kn} = 1 \qquad n = \frac{M - 1}{2}, \quad \text{all } k$$

then the linear equations in (5.3.30) and (5.3.31) for the symmetric FIR filter become

$$\sum_{n=0}^{(M-1)/2} a_{kn} h(n) = H_r(\omega_k) \qquad k = 0, 1, \ldots, \frac{M - 1}{2}, \quad M \text{ odd} \qquad (5.3.42)$$

$$\sum_{n=0}^{(M/2)-1} a_{kn} h(n) = H_r(\omega_k) \qquad k = 0, 1, \ldots, \frac{M}{2} - 1, \quad M \text{ even} \qquad (5.3.43)$$

In the case of an antisymmetric unit sample response we require specifications of the frequency response at $(M - 1)/2$ points when M is odd and at $M/2$ points when M is even. Since (5.3.35) and (5.3.36) imply that $H_r(0) = 0$, independent of our choice of $\{h(n)\}$, it is obvious that the frequency $\omega = 0$ cannot be used in the specification. For M odd there is no problem, since we can specify $H_r(\omega)$ at $(M - 1)/2$ equally spaced points in frequency. These can be chosen as $\omega_k = 2\pi k/M$ for $k = 1, 2, \ldots, (M - 1)/2$. When M is even, we require $M/2$ frequencies, so that if we cannot use $\omega = 0$, we may use $\omega = \pi$ in the specification. Thus we may define the frequencies

$\{\omega_k\}$ as

$$\omega_k = \frac{2\pi k}{M} \qquad k = 1, 2, \ldots, \frac{M-1}{2}, \quad M \text{ odd}$$

$$k = 1, 2, \ldots, \frac{M}{2}, \quad M \text{ even}$$

$$(5.3.44)$$

An alternative choice of equally spaced frequencies that completely avoids the zeros at $\omega = 0$ (and $\omega = \pi$) is the set

$$\omega_k = \frac{2\pi(k + \frac{1}{2})}{M} \qquad k = 0, 1, 2, \ldots, \frac{M-1}{2}, \quad M \text{ odd}$$

$$k = 0, 1, 2, \ldots, \frac{M}{2} - 1, \quad M \text{ even}$$

$$(5.3.45)$$

This set of frequencies is simply offset by π/M from the set given in (5.3.40). In any case we may define

$$b_{kn} = 2 \sin \omega_k \left(\frac{M-1}{2} - n \right)$$

Then the linear equations in (5.3.35) and (5.3.36) for the antisymmetric FIR filter become

$$\sum_{n=0}^{(M-3)/2} b_{kn}h(n) = H_r(\omega_k) \qquad k = 1, 2, \ldots, (M-1)/2, \quad M \text{ odd} \qquad (5.3.46)$$

$$\sum_{n=0}^{(M/2)-1} b_{kn}h(n) = H_r(\omega_k) \qquad k = 1, 2, \ldots, \frac{M}{2}, \quad M \text{ even} \qquad (5.3.47)$$

The alternative set of frequencies given by (5.3.45) may also be used in (5.3.42) and (5.3.43), instead of the frequencies given by (5.3.40).

EXAMPLE 5.3.1

Determine the unit sample response $\{h(n)\}$ of a linear phase FIR filter of length $M = 4$ for which the frequency response at $\omega = 0$ and $\omega = \pi/2$ is specified as

$$H_r(0) = 1 \qquad H_r\left(\frac{\pi}{2}\right) = \frac{1}{2}$$

Solution: This is a very simple filter design problem involving two unknown filter parameters $h(0)$ and $h(1)$. The set of linear equations are

$$a_{00}h(0) + a_{01}h(1) = H_r(0) = 1$$

$$a_{10}h(0) + a_{11}h(1) = H_r\left(\frac{\pi}{2}\right) = \frac{1}{2}$$

where

$$a_{00} = 2 \qquad a_{01} = 2 \qquad a_{10} = -\sqrt{2} \qquad a_{11} = \sqrt{2}$$

In matrix form these equations are

$$\mathbf{Ah} = \mathbf{H}_r$$

where

$$\mathbf{A} = \begin{bmatrix} 2 & 2 \\ -\sqrt{2} & \sqrt{2} \end{bmatrix} \qquad \mathbf{h} = \begin{bmatrix} h(0) \\ h(1) \end{bmatrix} \qquad \mathbf{H}_r = \begin{bmatrix} 1 \\ \frac{1}{2} \end{bmatrix}$$

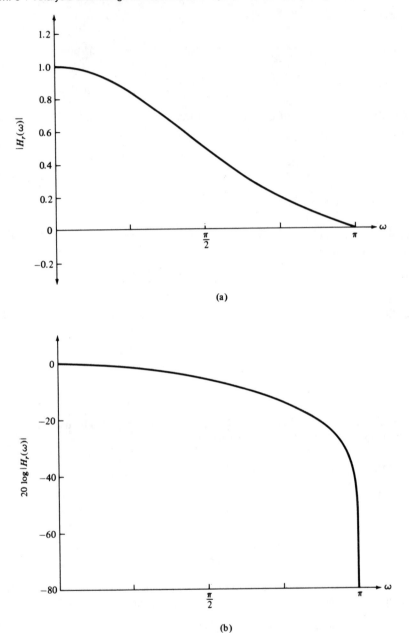

FIGURE 5.14   Magnitude response characteristics of $M = 4$, linear-phase FIR filter in Example 5.3.1.

The solution is

$$h(0) = \frac{1}{4\sqrt{2}} (\sqrt{2} - 1) = 0.0732232 = h(3)$$

$$h(1) = \frac{1}{4\sqrt{2}} (\sqrt{2} + 1) = 0.4267766 = h(2)$$

The frequency response of this filter is

$$H(\omega) = H_r(\omega)e^{-3\omega/2}$$

where

$$H_r(\omega) = \frac{\sqrt{2}}{4} \left[(\sqrt{2} - 1) \cos \frac{2\omega}{2} + (\sqrt{2} + 1) \cos \frac{\omega}{2} \right]$$

$|H_r(\omega)|$ and $20 \log H_r(\omega)$ are plotted in Fig. 5.14. We observe that the filter characteristic is indeed lowpass and that $|H_r(\omega)|$ decays monotonically in the frequency band $0 \le \omega \le \pi$. The filter has a 3-dB bandwidth of $\pi/2$ radians.

The preceding example involves a relatively small filter. In Example 5.3.2 we consider the design of a much longer filter. Now $h(n)$ cannot be found without a digital computer to provide the solution of the linear equations.

EXAMPLE 5.3.2 _____

Determine the coefficients $\{h(n)\}$ of a linear-phase FIR filter of length $M = 15$ which has a symmetric unit sample response and a frequency response that satisfies the condition

$$H_r \left(\frac{2\pi k}{15} \right) = \begin{cases} 1 & k = 0, 1, 2, 3 \\ 0 & k = 4, 5, 6, 7 \end{cases}$$

Solution: With the aid of a digital computer we can solve the set of linear equations given by (5.3.42). The filter coefficients obtained from the solution are

$$\begin{aligned}
h(0) &= h(14) = 0.04981588 \\
h(1) &= h(13) = 0.04120224 \\
h(2) &= h(12) = 0.06666674 \\
h(3) &= h(11) = -0.03648787 \\
h(4) &= h(10) = -0.1078689 \\
h(5) &= h(9) \ = 0.03407801 \\
h(6) &= h(8) \ = 0.3188924 \\
h(7) &= \qquad\ \ = 0.4666666
\end{aligned}$$

The frequency response of the resulting filter is illustrated in Fig. 5.15. We observe that the filter has an overshoot at the edge of the passband, just prior to the transition region. It also has relatively large sidelobes in the stopband. The largest sidelobe is -15 dB.

The example above serves to illustrate a problem in the design of linear-phase FIR filters in which the specifications change abruptly from the passband, where $H_r(\omega_k) = 1$, to the stopband, where $H_r(\omega)$ is specified to be zero at the discrete frequencies $\{\omega_k\}$. Instead of having an abrupt change, if we specify an intermediate value for $H_r(\omega)$ in the transition region, the resulting frequency response has significantly smaller sidelobes in the stopband. The following example illustrates this point.

EXAMPLE 5.3.3 _____

Repeat the filter design problem in Example 5.3.2 with the frequency response specifications

$$H_r \left(\frac{2\pi k}{15} \right) = \begin{cases} 1 & k = 0, 1, 2, 3 \\ 0.4 & k = 4 \\ 0 & k = 5, 6, 7 \end{cases}$$

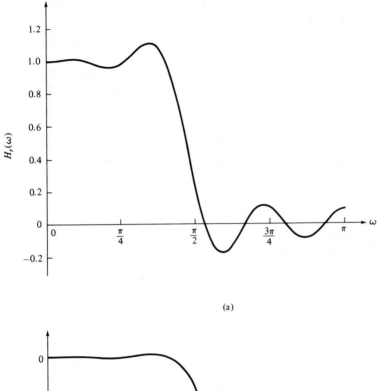

FIGURE 5.15 Frequency response of M = 15 linear-phase FIR filter in Example 5.3.2.

Solution: In this case, the filter coefficients obtained by solving the linear equations in (5.3.42) are

$$
\begin{aligned}
h(0) &= h(14) = -0.01412893 \\
h(1) &= h(13) = -0.001945309 \\
h(2) &= h(12) = 0.04000004 \\
h(3) &= h(11) = 0.01223454 \\
h(4) &= h(10) = -0.09138802 \\
h(5) &= h(9) \;\;= -0.01808986 \\
h(6) &= h(8) \;\;= 0.3133176 \\
h(7) &\qquad\quad\; = 0.52
\end{aligned}
$$

The frequency response of the resulting filter is shown in Fig. 5.16. We note that the sidelobes are now significantly lower than in the previous design. The largest sidelobe is now −41 dB.

This example clearly illustrates the advantages in allowing for a small transition in the design of filters. The only disadvantage is that the width of the transition region is increased. Specifically, the transition region is now about twice as wide as in Example 5.3.2. However, the benefits usually outweigh the one disadvantage.

Further reduction in the stopband sidelobes can be obtained by allowing for one or

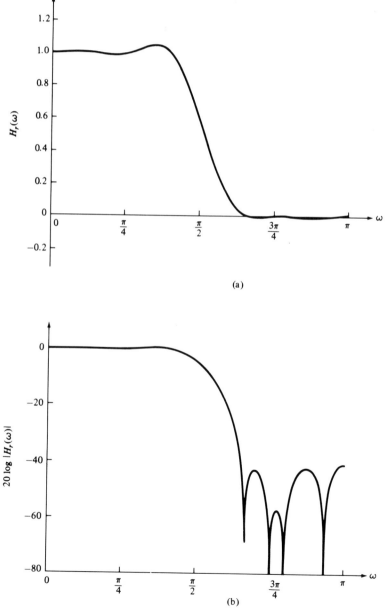

(a)

(b)

FIGURE 5.16   Frequency response of M = 15 linear-phase FIR filter in Example 5.3.3.

more additional frequency specifications in the transition band. In effect, the transition band is widened further to achieve additional reduction in the stopband sidelobes.

The question that we have not addressed is concerned with the values of the specifications in the transition band. This problem has been considered in the technical literature by Rabiner, et al. (1970). In this paper the optimum values of the specifications in the transition region are tabulated for a large variety of filter lengths. The values given are optimum in the sense that they result in minimizing the largest sidelobe in the stopband. These tables are reprinted in Appendix 5A. We conclude this section with an example of a design of a bandpass filter.

EXAMPLE 5.3.4 ⎯⎯⎯⎯⎯⎯⎯⎯⎯⎯⎯⎯⎯⎯⎯⎯⎯⎯⎯⎯⎯⎯⎯⎯⎯⎯⎯⎯⎯⎯⎯⎯

Determine the coefficients $\{h(n)\}$ of a bandpass linear-phase FIR filter of length $M = 16$ which has a symmetric unit sample response and a frequency response that satisfies the condition

$$H_r\left(\frac{2\pi k}{16}\right) = \begin{cases} 0 & k = 0, 1, 7 \\ 0.456 & k = 2, 6 \\ 1 & k = 3, 4, 5 \end{cases}$$

Solution: The values of the frequency response in the transition regions were selected from the tabulated results in the reference cited above. By solving the set of linear equations given in (5.3.43), we obtain the following values for the unit sample response:

$$\begin{aligned}
h(0) &= h(15) = 0.01051756 \\
h(1) &= h(14) = 0.02774795 \\
h(2) &= h(13) = 0.04067173 \\
h(3) &= h(12) = -0.02057317 \\
h(4) &= h(11) = 0.04840164 \\
h(5) &= h(10) = -0.155752 \\
h(6) &= h(9) = -0.266221 \\
h(7) &= h(8) = 0.3362424
\end{aligned}$$

The frequency response or the bandpass filter is illustrated in Fig. 5.17. The largest sidelobe in the stopbands is down to -34 dB.

⎯⎯⎯

To summarize, we have introduced a method for designing linear phase FIR filters based on specification of the desired frequency response at a set of equally spaced frequencies, either $\omega_k = 2\pi k/M$ or $\omega_k = \pi(2k + 1)/M$. The choice of the set of frequencies is left to the filter designer. The selection of the set of frequencies may be made on the basis of which set provides a closer match to the desired critical frequencies that determine the passband(s) and stopband(s) of the filter (i.e., ω_p and ω_s for the lowpass filter). For example, if a filter as length $M = 16$ is desired, the frequency specifications

$$H_r\left(\frac{2\pi k}{16}\right) = \begin{cases} 1 & k = 0, 1, 2, 3 \\ 0.389 & k = 4 \\ 0 & k = 5, 6, 7 \end{cases} \tag{5.3.48}$$

result in a filter having the frequency response illustrated in Fig. 5.18. On the other hand, the frequency specifications

$$H_r\left(\frac{\pi(2k + 1)}{16}\right) = \begin{cases} 1 & k = 0, 1, 2, 3 \\ 0.363 & k = 4 \\ 0 & k = 5, 6, 7 \end{cases} \tag{5.3.49}$$

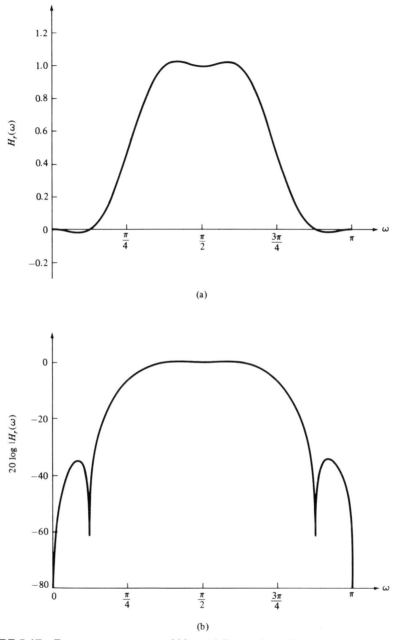

FIGURE 5.17   Frequency response of M = 16 linear-phase FIR filter in Example 5.3.4.

result in a filter having the frequency response shown in Fig. 5.19. The only basic difference between these two response characteristics is that the one specified by (5.3.49) has a wider bandwidth than the one specified by (5.3.48), by about $\pi/16$. This result is due to the frequency specifications. Consequently, our choice of either (5.3.48) or (5.3.49) would depend on a choice of the desired critical frequencies ω_p and ω_s.

The selection of a set of equally spaced frequencies is desirable for the purpose of simplicity and because there exists an alternative FIR filter structure, called a frequency sampling realization and described in Chapter 7, which lends itself to the frequency

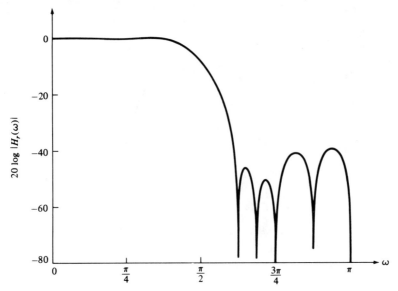

FIGURE 5.18 Frequency response of $M = 16$ linear-phase FIR filter with specifications given by (5.3.48).

specifications above. On the other hand, the filter designer may select any arbitrary set of frequencies in the range $0 \leq \omega \leq \pi$, not necessarily uniformly spaced, that best fits the desired frequency response and initial frequencies.

5.3.5 A Simple Lowpass-to-Highpass Filter Transformation

Suppose that we have designed a prototype lowpass filter with impulse response $h_{lp}(n)$. By using the frequency translation property of the Fourier transform, it is possible to convert the prototype filter to either a bandpass or a highpass filter. Frequency transfor-

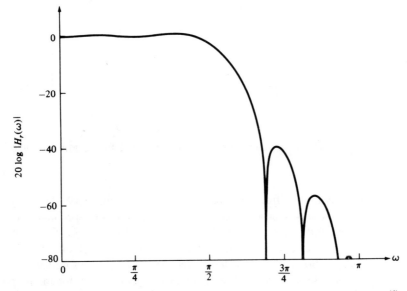

FIGURE 5.19 Frequency response of $M = 16$ linear-phase FIR filter with specifications given by (5.3.49).

mations for converting a prototype lowpass filter into a filter of another type are described in detail in Section 8.3. In this section we present a simple-frequency transformation for converting a lowpass filter into a highpass filter, and vice versa.

If $h_{lp}(n)$ denotes the impulse response of a lowpass filter with frequency response $H_{lp}(\omega)$, a highpass filter can be obtained by translating $H_{lp}(\omega)$ by π radians (i.e., replacing ω by $\omega - \pi$. Thus

$$H_{hp}(\omega) = H_{lp}(\omega - \pi) \qquad (5.3.50)$$

where $H_{hp}(\omega)$ is the frequency response of the highpass filter. Since a frequency translation of π radians is equivalent to multiplication of the impulse response $h_{lp}(n)$ by $e^{j\pi n}$, the impulse response of the highpass filter is

$$\begin{aligned} h_{hp}(n) &= (e^{j\pi})^n h_{lp}(n) \\ &= (-1)^n h_{lp}(n) \end{aligned} \qquad (5.3.51)$$

Therefore, the impulse response of the highpass filter is simply obtained from the impulse response of the lowpass filter by changing the signs of the odd-numbered samples in $h_{lp}(n)$. Conversely,

$$h_{lp}(n) = (-1)^n h_{hp}(n) \qquad (5.3.52)$$

If the lowpass filter is described by the difference equation

$$y(n) = -\sum_{k=1}^{N} a_k y(n - k) + \sum_{k=0}^{M} b_k x(n - k) \qquad (5.3.53)$$

its frequency response is

$$H_{lp}(\omega) = \frac{\displaystyle\sum_{k=0}^{M} b_k e^{-j\omega k}}{1 + \displaystyle\sum_{k=1}^{N} a_k e^{-j\omega k}} \qquad (5.3.54)$$

Now, if we replace ω by $\omega - \pi$, in (5.3.54), then

$$H_{hp}(\omega) = \frac{\displaystyle\sum_{k=0}^{M} (-1)^k b_k e^{-j\omega k}}{1 + \displaystyle\sum_{k=1}^{N} (-1)^k a_k e^{-j\omega k}} \qquad (5.3.55)$$

which corresponds to the difference equation

$$y(n) = -\sum_{k=1}^{N} (-1)^k a_k y(n - k) + \sum_{k=0}^{M} (-1)^k b_k x(n - k) \qquad (5.3.56)$$

EXAMPLE 5.3.5 _____

Convert the lowpass filter described by the difference equation

$$y(n) = 0.9y(n - 1) + 0.1x(n)$$

into a highpass filter.

Solution: The difference equation for the highpass filter, according to (5.3.56), is

$$y(n) = -0.9y(n - 1) + 0.1x(n)$$

and its frequency response is

$$H_{hp}(\omega) = \frac{0.1}{1 + 0.9e^{-j\omega}}$$

The reader may verify that $H_{hp}(\omega)$ is indeed highpass.

5.4 Linear Filtering Methods Based on the DFT

In Section 4.5.4 we defined the discrete Fourier transform (DFT) as the sampled version of the Fourier transform $X(\omega)$ for a finite-duration sequence $x(n)$. The sampling is performed at N equally spaced frequencies $\omega_k = 2\pi k/N$, $k = 0, 1, \ldots, N - 1$, yielding the result

$$X(k) \equiv X(\omega)|_{\omega_k = 2\pi k/N} \qquad k = 0, 1, \ldots, N - 1 \qquad (5.4.1)$$

We demonstrated that the set of N complex numbers $\{X(k)\}$ provides a frequency-domain representation that uniquely characterizes a finite-duration sequence $x(n)$ of length $L \le N$. In other words, if the DFT of $x(n)$ is

$$X(k) = \sum_{n=0}^{N-1} x(n)e^{-j2\pi kn/N} \qquad k = 0, 1, \ldots, N - 1 \qquad (5.4.2)$$

then the sequence $\{x(n)\}$ can be recovered via the inverse DFT (IDFT)

$$x(n) = \frac{1}{N}\sum_{k=0}^{N-1} X(k)e^{j2\pi kn/N} \qquad n = 0, 1, \ldots, N - 1 \qquad (5.4.3)$$

as demonstrated in Section 4.5.4.

Since the DFT provides a discrete frequency representation of a finite-duration sequence in the frequency domain, it is interesting to explore its use as a computational tool for linear system analysis and, especially, for linear filtering. We have already established that when a system with frequency response $H(\omega)$ is excited with an input signal that has a spectrum $X(\omega)$, the output of the system has the spectrum

$$Y(\omega) = X(\omega)H(\omega) \qquad (5.4.4)$$

The output sequence $y(n)$ is determined from its spectrum via the inverse Fourier transform

$$y(n) = \frac{1}{2\pi}\int_{-\pi}^{\pi} Y(\omega)e^{j\omega n}\, d\omega \qquad (5.4.5)$$

The expressions in (5.4.4) and (5.4.5) provide an alternative means to convolution for determining the output $y(n)$ of a system to a given input sequence $x(n)$. In this approach the input signal $x(n)$ is Fourier transformed to yield $X(\omega)$ and the system is represented by the frequency response $H(\omega)$. Since convolution of $x(n)$ with $h(n)$ is equivalent to multiplication of $X(\omega)$ with $H(\omega)$, we obtain $Y(\omega) = X(\omega)H(\omega)$. Finally, the output sequence $y(n)$ is obtained from its spectrum by computing the inverse Fourier transform given by (5.4.5).

The problem with the frequency-domain approach as outlined above is the frequency-domain functions $X(\omega)$, $H(\omega)$, and $Y(\omega)$ are functions of the continuous variable ω. As a consequence, the computations cannot be done on a digital computer, since the computer can only store and perform computations on quantities at discrete frequencies or in discrete time. In other words, a computer is suitable for performing convolution of two sequences, but it is not suitable for performing the computations in (5.4.4) and (5.4.5).

On the other hand, the DFT does lend itself to computation on a digital computer. In the discussion that follows, we describe how the DFT can be used to perform linear filtering in the frequency domain. In particular, we present a computational procedure that serves as an alternative to time-domain convolution. In fact, the frequency-domain approach based on the DFT is computationally more efficient than time-domain convolution due to the existence of efficient algorithms for computing the DFT. These algorithms, which are described in Chapter 9, are collectively called the fast Fourier transform (FFT) algorithms.

5.4.1 Multiplication of Two DFTs and Circular Convolution

Suppose that we have two finite-duration sequences of length N, $x_1(n)$ and $x_2(n)$. Their respective N-point DFTs are

$$X_1(k) = \sum_{n=0}^{N-1} x_1(n)e^{-j2\pi nk/N} \qquad k = 0, 1, \ldots, N-1 \qquad (5.4.6)$$

$$X_2(k) = \sum_{n=0}^{N-1} x_2(n)e^{-j2\pi nk/N} \qquad k = 0, 1, \ldots, N-1 \qquad (5.4.7)$$

If we multiply the two DFTs together, the result is a DFT, say $X_3(k)$, of a sequence $x_3(n)$ of length N. Let us determine the relationship between $x_3(n)$ and the sequences $x_1(n)$ and $x_2(n)$.

We have

$$X_3(k) = X_1(k)X_2(k) \qquad k = 0, 1, \ldots, N-1 \qquad (5.4.8)$$

The IDFT of $\{X_3(k)\}$ is

$$
\begin{aligned}
x_3(m) &= \frac{1}{N}\sum_{k=0}^{N-1} X_3(k)e^{j2\pi km/N} \\
&= \frac{1}{N}\sum_{k=0}^{N-1} X_1(k)X_2(k)e^{j2\pi km/N}
\end{aligned}
\qquad (5.4.9)
$$

Suppose that we substitute for $X_1(k)$ and $X_2(k)$ in (5.4.9), using the DFTs given in (5.4.6) and (5.4.7). Thus we obtain

$$
\begin{aligned}
x_3(m) &= \frac{1}{N}\sum_{k=0}^{N-1}\left[\sum_{n=0}^{N-1} x_1(n)e^{-j2\pi kn/N}\right]\left[\sum_{l=0}^{N-1} x_2(l)e^{-j2\pi kl/N}\right]e^{j2\pi km/N} \\
&= \frac{1}{N}\sum_{n=0}^{N-1} x_1(n)\sum_{l=0}^{N-1} x_2(l)\left[\sum_{k=0}^{N-1} e^{j2\pi k(m-n-l)/N}\right]
\end{aligned}
\qquad (5.4.10)
$$

The inner sum in the brackets in (5.4.10) has the form

$$\sum_{k=0}^{N-1} a^k = \begin{cases} N & a = 1 \\ \dfrac{1-a^N}{1-a} & a \neq 1 \end{cases} \qquad (5.4.11)$$

where a is defined as

$$a = e^{j2\pi(m-n-l)/N}$$

We observe that $a = 1$ when $m - n - l$ is multiple of N. On the other hand, $a^N = 1$ for any value of $a \neq 0$. Consequently, (5.4.11) reduces to

$$\sum_{k=0}^{N-1} a^k = \begin{cases} N & l = m - n + pN = (m - n)(\text{mod } N), \quad p \text{ an integer} \\ 0 & \text{otherwise} \end{cases} \quad (5.4.12)$$

If we substitute the result in (5.4.12) into (5.4.10), we obtain the desired expression for $x_3(m)$ in the form

$$x_3(m) = \sum_{n=0}^{N-1} x_1(n)x_2(m - n, (\text{mod } N)) \quad m = 0, 1, \ldots, N - 1 \quad (5.4.13)$$

The expression in (5.4.13) has the form of a convolution sum. However, it is not the ordinary linear convolution that was introduced in Chapter 2, which relates the output sequence $y(n)$ of a linear system to the input sequence $x(n)$ and the impulse response $h(n)$. Instead, the convolution sum in (5.4.13) involves the index $(m - n)(\text{mod } N)$ and is called *circular convolution*. Thus we conclude that multiplication of the DFTs of two sequences is equivalent to the circular convolution of the two sequences in the time domain.

The following example llustrates the operations involved in circular convolution.

EXAMPLE 5.4.1 _____

Perform the circular convolution of the following two sequences:

$$x_1(n) = \left\{ \begin{array}{c} 2,1,2,1 \\ \uparrow \end{array} \right\}$$

$$x_2(n) = \left\{ \begin{array}{c} 1, 2, 3, 4 \\ \uparrow \end{array} \right\}$$

Solution: Each sequence consists of four nonzero points. For purposes of illustrating the operations involved in circular convolution it is desirable to graph each sequence as points on a circle. Thus the sequences $x_1(n)$ and $x_2(n)$ are graphed as illustrated in Fig. 5.20a. We note that the sequences are graphed in a counterclockwise direction on a circle. This establishes the reference direction in rotating one of the sequences relative to the other.

Now, $x_3(m)$ is obtained by circularly convolving $x_1(n)$ with $x_2(n)$ as specified by (5.4.13). Beginning with $m = 0$ we have

$$x_3(0) = \sum_{n=0}^{3} x_1(n)x_2(-n, (\text{mod } 4))$$

But $x_2(-n, (\text{mod } 4))$ is simply the sequence $x_2(n)$ folded and graphed on a circle as illustrated in Fig. 5.20b. In other words, the folded sequence is simply $x_2(n)$ graphed in a clockwise direction. Another way to view this folding operation is to note that

$$x_2(0, (\text{mod } 4)) = x_2(0)$$
$$x_2(-1, (\text{mod } 4)) = x_2(3)$$
$$x_2(-2, (\text{mod } 4)) = x_2(2)$$
$$x_2(-3, (\text{mod } 4)) = x_2(1)$$

which is consistent with the graph in Fig. 5.20b.

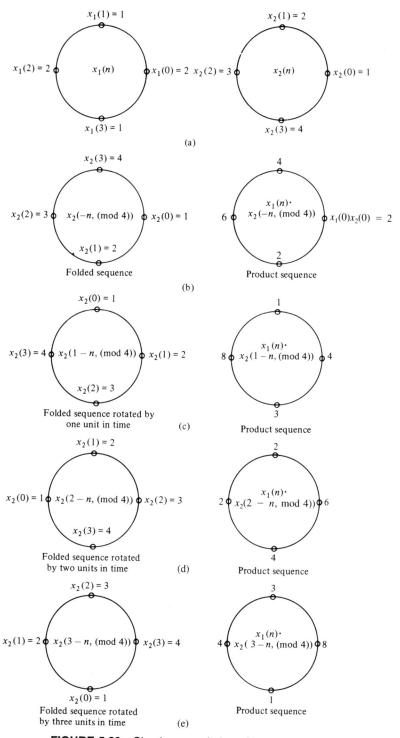

FIGURE 5.20 Circular convolution of two sequences.

The product sequence is obtained by multiplying $x_1(n)$ with $x_2(-n, \text{(mod 4)})$, point by point. This sequence is also illustrated in Fig. 5.20b. Finally, we sum the values in the product sequence to obtain

$$x_3(0) = 14$$

For $m = 1$ we have

$$x_3(1) = \sum_{n=0}^{3} x_1(n)x_2(1 - n, \text{(mod 4)})$$

It is easily verified that $x_2(1 - n, \text{(mod 4)})$ is simply the sequence $x_2(-n, \text{(mod 4)})$ rotated counterclockwise by one unit in time as illustrated in Fig. 5.20c. This rotated sequence multiplies $x_1(n)$ to yield the product sequence, also illustrated in Fig. 5.20c. Finally, we sum the values in the product sequence to obtain $x_3(1)$. Thus

$$x_3(1) = 16$$

For $m = 2$ we have

$$x_3(2) = \sum_{n=0}^{3} x_1(n)x_2(2 - n, \text{(mod 4)})$$

Now $x_2(2 - n, \text{(mod 4)})$ is the folded sequence in Fig. 5.20b rotated two units of time in the counterclockwise direction. The resultant sequence is illustrated in Fig. 5.20d along with the product sequence $x_1(n)x_2(2 - n, \text{(mod 4)})$. By summing the four terms in the product sequence, we obtain

$$x_3(2) = 14$$

For $m = 3$ we have

$$x_3(3) = \sum_{n=0}^{3} x_1(n)x_2(3 - n, \text{(mod 4)})$$

The folded sequence $x_2(-n, \text{(mod 4)})$ is now rotated by three units in time to yield $x_2(3 - n, \text{(mod 4)})$ and the resultant sequence is multiplied by $x_1(n)$ to yield the product sequence as illustrated in Fig. 5.20e. The sum of the values in the product sequence is

$$x_3(3) = 16$$

We observe that if the computation above is continued beyond $m = 3$, we simply repeat the sequence of four values obtained above. Therefore, the circular convolution of the two sequences $x_1(n)$ and $x_2(n)$ yields the sequence

$$x_3(n) = \left\{ \begin{array}{l} 14,\ 16,\ 14,\ 16 \\ \uparrow \end{array} \right\}$$

From the example above we observe that circular convolution involves basically the same four steps as the ordinary *linear convolution* introduced in Chapter 2: *folding* one sequence, *shifting* the folded sequence, *multiplying* the two sequences to obtain a product sequence, and finally, *summing* the values of the product sequence. The basic difference between these two types of convolution is that, in circular convolution, the folding and shifting (rotating) operations are performed in a circular fashion by computing the index of one of the sequences modulo N. In linear convolution, there is no modulo N operation.

The reader may easily show from our previous development that either one of the two sequences may be folded and rotated without changing the result of the circular convolution. Thus

$$x_3(m) = \sum_{n=0}^{N-1} x_2(n)x_1(m - n, (\text{mod } N)) \qquad m = 0, 1, \ldots, N - 1 \quad (5.4.14)$$

The following example serves to illustrate the computation of $x_3(n)$ by means of the DFT and IDFT.

EXAMPLE 5.4.2

By means of the DFT and IDFT, determine the sequence $x_3(n)$ corresponding to the circular convolution of the sequences $x_1(n)$ and $x_2(n)$ given in Example 5.4.1.

Solution: First we compute the DFTs of $x_1(n)$ and $x_2(n)$. The four-point DFT of $x_1(n)$ is

$$X_1(k) = \sum_{n=0}^{3} x_1(n)e^{-j2\pi nk/4} \qquad k = 0, 1, 2, 3$$

$$= 2 + e^{-j\pi k/2} + 2e^{-j\pi k} + e^{-j3\pi k/2}$$

Thus

$$X_1(0) = 6 \qquad X_1(1) = 0 \qquad X_1(2) = 2 \qquad X_1(3) = 0$$

The DFT of $x_2(n)$ is

$$X_2(k) = \sum_{n=0}^{3} x_2(n)e^{-j2\pi nk/4} \qquad k = 0, 1, 2, 3$$

$$= 1 + 2e^{-j\pi k/2} + 3e^{-j\pi k} + 4e^{-j3\pi k/2}$$

Thus

$$X_2(0) = 10 \qquad X_2(1) = -2 + j2 \qquad X_2(2) = -2 \qquad X_2(3) = -2 - j2$$

When we multiply the two DFTs, we obtain the product

$$X_3(k) = X_1(k)X_2(k)$$

or, equivalently,

$$X_3(0) = 60 \qquad X_3(1) = 0 \qquad X_3(2) = -4 \qquad X_3(3) = 0$$

Now, the IDFT of $X_3(k)$ is

$$x_3(n) = \tfrac{1}{4} \sum_{k=0}^{3} X_3(k)e^{j2\pi nk/4} \qquad n = 0, 1, 2, 3$$

$$= \tfrac{1}{4}(60 - 4 e^{j\pi n})$$

Thus

$$x_3(0) = 14 \qquad x_3(1) = 16 \qquad x_3(2) = 14 \qquad x_3(3) = 16$$

which is the result obtained in Example 5.4.1 from circular convolution.

5.4.2 Use of the DFT in Linear Filtering

In the preceding section it was demonstrated that the product of two DFTs is equivalent to the circular convolution of the corresponding time-domain sequences. Unfortunately, circular convolution is of no use to us if our objective is to determine the output of a linear filter to a given input sequence. In this case we seek a frequency-domain methodology that is equivalent to linear convolution.

Suppose that we have a finite-duration sequence $x(n)$ of length L which excites an FIR filter of length M. Without loss of generality, let

$$
\begin{aligned}
x(n) &= 0 \qquad n < 0 \text{ and } n \geq L \\
h(n) &= 0 \qquad n < 0 \text{ and } n \geq M
\end{aligned}
$$

where $h(n)$ is the impulse response of the FIR filter.

The output sequence $y(n)$ of the FIR filter may be expressed in the time domain as the convolution of $x(n)$ and $h(n)$, that is

$$
y(n) = \sum_{k=0}^{M-1} h(k)x(n-k) \tag{5.4.15}
$$

Since $h(n)$ and $x(n)$ are finite-duration sequences, their convolution is also finite in duration. In fact, the duration of $y(n)$ is $L + M - 1$.

The frequency-domain equivalent to (5.4.15) is

$$
Y(\omega) = X(\omega)H(\omega) \tag{5.4.16}
$$

If the sequence $y(n)$ is to be represented uniquely in the frequency domain by samples of its spectrum $Y(\omega)$ at a set of discrete frequencies, the number of distinct samples must equal or exceed $L + M - 1$. Therefore, a DFT of size $N \geq L + M - 1$ is required to represent $\{y(n)\}$ in the frequency domain.

Now if

$$
\begin{aligned}
Y(k) &\equiv Y(\omega)\big|_{\omega = 2\pi k/N} & k = 0, 1, \ldots, N-1 \\
&= X(\omega)H(\omega)\big|_{\omega = 2\pi k/N} & k = 0, 1, \ldots, N-1
\end{aligned}
$$

then

$$
Y(k) = X(k)H(k) \qquad k = 0, 1, \ldots, N-1 \tag{5.4.17}
$$

where $\{X(k)\}$ and $\{H(k)\}$ are the N-point DFTs of the corresponding sequences $x(n)$ and $h(n)$, respectively. Since the sequences $x(n)$ and $h(n)$ have a duration less than N, we simply pad these sequences with zeros to increase their length to N. This increase in the size of the sequences does not alter their spectra $X(\omega)$ and $H(\omega)$, which are continuous spectra, since the sequences are aperiodic. However, by sampling their spectra at N equally spaced points in frequency (computing the N-point DFTs) we have increased the number of samples that represent these sequences in the frequency domain beyond the minimum number (L or M, respectively).

Since the $N = L + M - 1$-point DFT of the output sequence $y(n)$ is sufficient to represent $y(n)$ in the frequency domain, it follows that the multiplication of the N-point DFTs $X(k)$ and $H(k)$, according to (5.4.17), followed by the computation of the N-point IDFT must yield the sequence $\{y(n)\}$. In turn, this implies that the N-point circular convolution of $x(n)$ with $h(n)$ must be equivalent to the linear convolution of $x(n)$ with $h(n)$. In other words, by increasing the length of the sequences $x(n)$ and $h(n)$ to N points (by appending zeros), and then circularly convolving the resulting sequences, we obtain the same result as would be obtained with linear convolution. Thus with zero padding, the DFT can be used to perform linear filtering.

The following example illustrates the methodology in the use of the DFT in linear filtering.

EXAMPLE 5.4.3

By means of the DFT and IDFT, determine the response of the FIR filter with impulse response

$$h(n) = \left\{ \underset{\uparrow}{1}, 2, 3 \right\}$$

to the input sequence

$$x(n) = \left\{ \underset{\uparrow}{1}, 2, 2, 1 \right\}$$

Solution: The input sequence has length $L = 4$ and the impulse response has length $M = 3$. Linear convolution of these two sequences produces a sequence of length $N = 6$. Consequently, the size of the DFTs must be at least six.

For simplicity we compute eight-point DFTs. We should also mention that the efficient computation of the DFT via the fast Fourier transform (FFT) algorithm is usually performed for a length N that is a power of 2. Hence the eight-point DFT of $x(n)$ is

$$X(k) = \sum_{n=0}^{7} x(n)e^{-j2\pi kn/8}$$

$$= 1 + 2e^{-j\pi k/4} + 2e^{-j\pi k/2} + e^{-j\pi k/4} \qquad k = 0, 1, \ldots, 7$$

This computation yields

$$X(0) = 6 \qquad X(1) = \frac{2 + \sqrt{2}}{2} - j\left(\frac{4 + 3\sqrt{2}}{2}\right)$$

$$X(2) = -1 - j \qquad X(3) = \frac{2 - \sqrt{2}}{2} + j\left(\frac{4 - 3\sqrt{2}}{2}\right)$$

$$X(4) = 0 \qquad X(5) = \frac{2 - \sqrt{2}}{2} - j\frac{4 - 3\sqrt{2}}{2}$$

$$X(6) = -1 + j \qquad X(7) = \frac{2 + \sqrt{2}}{2} + j\left(\frac{4 + 3\sqrt{2}}{2}\right)$$

The eight-point DFT of $h(n)$ is

$$H(k) = \sum_{n=0}^{7} h(n)e^{-j2\pi kn/8}$$

$$= 1 + 2e^{-j\pi k/4} + 3e^{-j\pi k/2}$$

Hence

$$H(0) = 6 \qquad H(1) = 1 + \sqrt{2} - j(3 + \sqrt{2}) \qquad H(2) = -2 - j2$$
$$H(3) = 1 - \sqrt{2} + j(3 - \sqrt{2}) \qquad H(4) = 2$$
$$H(5) = 1 - \sqrt{2} - j(3 - \sqrt{2}) \qquad H(6) = -2 + j2$$
$$H(7) = 1 + \sqrt{2} + j(3 + \sqrt{2})$$

The product of these two DFTs yields $Y(k)$, which is

$Y(0) = 36$ $Y(1) = -14.07 - j17.48$ $Y(2) = j4$ $Y(3) = 0.07 + j0.515$
$Y(4) = 0$ $Y(5) = 0.07 - j0.515$ $Y(6) = -j4$ $Y(7) = -14.07 + j17.48$

Finally, the eight-point IDFT is

$$y(n) = \tfrac{1}{8} \sum_{k=0}^{7} Y(k)e^{j2\pi kn/8} \qquad n = 0, 1, \ldots, 7$$

This computation yields the result

$$y(n) = \left\{ 1, 4, 9, 11, 8, 3, 0, 0 \atop \uparrow \right\}$$

We observe that the first six values of $y(n)$ constitute the set of desired output values. The last two values are zero because we used an eight-point DFT and IDFT, when, in fact, the minimum number of points required is six.

Although the multiplication of two DFTs corresponds to circular convolution in the time domain, we have observed that padding the sequences $x(n)$ and $h(n)$ with a sufficient number of zeros forces the circular convolution to yield the same output sequence as linear convolution. In the case of the FIR filtering problem in Example 5.4.3, it is a simple matter to demonstrate that the six-point circular convolution of the sequences

$$h(n) = \left\{ 1, 2, 3, 0, 0, 0 \atop \uparrow \right\} \tag{5.4.18}$$

$$x(n) = \left\{ 1, 2, 2, 1, 0, 0 \atop \uparrow \right\} \tag{5.4.19}$$

results in the output sequence

$$y(n) = \left\{ 1, 4, 9, 11, 8, 3 \atop \uparrow \right\} \tag{5.4.20}$$

which is the same sequence obtained from linear convolution.

It is important for us to understand the aliasing that results in the time domain when the size of the DFTs is smaller than $L + M - 1$. The following example focuses on the aliasing problem.

EXAMPLE 5.4.4

Determine the sequence $y(n)$ that results from the use of four point DFTs in Example 5.4.3.

Solution: The four-point DFT of $h(n)$ is

$$H(k) = \sum_{n=0}^{3} h(n)e^{-j2\pi kn/4}$$

$$H(k) = 1 + 2e^{-j\pi k/2} + 3e^{-jk\pi} \qquad k = 0, 1, 2, 3$$

Hence

$$H(0) = 6 \qquad H(1) = -2 - j2 \qquad H(2) = 2 \qquad H(3) = -2 + j2$$

The four-point DFT of $x(n)$ is

$$X(k) = 1 + 2e^{-j\pi k/2} + 2e^{-j\pi k} + 3e^{-j3\pi k/2} \qquad k = 0, 1, 2, 3$$

Hence

$$X(0) = 6 \qquad X(1) = -1 - j \qquad X(2) = 0 \qquad X(3) = -1 + j$$

The product of these two four-point DFTs is

$$\hat{Y}(0) = 36 \qquad \hat{Y}(1) = j4 \qquad \hat{Y}(2) = 0 \qquad \hat{Y}(3) = -j4$$

The four-point IDFT yields

$$\hat{y}(n) = \tfrac{1}{4} \sum_{k=0}^{3} \hat{Y}(k)e^{j2\pi kn/4} \qquad n = 0, 1, 2, 3$$

$$= \tfrac{1}{4}(36 + j4e^{j\pi n/2} - j4e^{j3\pi n/2})$$

Therefore,

$$\hat{y}(n) = \left\{ \underset{\uparrow}{9}, 7, 9, 11 \right\}$$

The reader may verify that the four-point circular convolution of $h(n)$ with $x(n)$ yields the same sequence $\hat{y}(n)$.

If we compare the result $\hat{y}(n)$ obtained from four-point DFTs with the sequence $y(n)$ obtained from the use of eight-point (or six-point) DFTs the time-domain aliasing effects derived in Section 4.5.3 are clearly evident. In particular, $y(4)$ is aliased into $y(0)$ to yield

$$\hat{y}(0) = y(0) + y(4) = 9$$

Similarly, $y(5)$ is aliased into $y(1)$ to yield

$$\hat{y}(1) = y(1) + y(5) = 7$$

All other aliasing has no effects since $y(n) = 0$ for $n \geq 6$. Consequently, we have

$$\hat{y}(2) = y(2) = 9$$
$$\hat{y}(3) = y(3) = 11$$

Therefore, only the first two points of $\hat{y}(n)$ are corrupted by the effect of aliasing [i.e., $\hat{y}(0) \neq y(0)$ and $\hat{y}(1) \neq y(1)$]. This observation has important ramifications in the discussion of the following section, in which we treat the filtering of long sequences.

5.4.3 Filtering of Long Data Sequences

In practical applications involving linear filtering of signals, the input sequence $x(n)$ is often a very long sequence. This is especially true in some real-time signal processing applications concerned with signal monitoring and analysis.

Since linear filtering performed via the DFT involves operations on a block of data, which by necessity must be limited in size due to limited memory of a digital computer, a long input signal sequence must be segmented to fixed-size blocks prior to processing. Since the filtering is linear, successive blocks may be processed one at a time via the DFT and the output blocks are fitted together to form the overall output signal sequence.

Below, we describe two methods for linear FIR filtering a long sequence on a block-by-block basis using the DFT. The input sequence is segmented into blocks and each block is processed via the DFT and IDFT to produce a block of output data. The output blocks are fitted together to form an overall output sequence which is identical to the sequence obtained if the long block had been processed via time-domain convolution.

The two methods to be described are called the *overlap-save method* and the *overlap-add method*. For both methods we assume that the FIR filter has duration M. The input data sequence is segmented into blocks of L points, where, by assumption, $L >> M$ without loss of generality.

Overlap-Save Method. In this method the size of the input data blocks is $N = L + M - 1$ and the size of the DFTs and IDFT are of length N. Each data block consists of the last $M - 1$ data points of the previous data block followed by L new data points to form a data sequence of length $N = L + M - 1$. An N-point DFT is computed for each data block. The impulse response of the FIR filter is increased in length by appending $L - 1$ zeros and an N-point DFT of the sequence is computed once and stored. The multiplication of the two N-point DFTs $\{H(k)\}$ and $\{X_m(k)\}$ for the mth block of data yields

$$\hat{Y}_m(k) = H(k)X_m(k) \qquad k = 0, 1, \ldots, N - 1 \qquad (5.4.21)$$

Then the N-point IDFT yields the result

$$\hat{Y}_m(n) = \{\hat{y}_m(0)\hat{y}_m(1) \cdots \hat{y}_m(M - 1)\hat{y}_m(M) \cdots \hat{y}_m(N - 1)\} \qquad (5.4.22)$$

Since the data record is of length N, the first $M - 1$ points of $y_m(n)$ are corrupted by aliasing and must be discarded. The last L points of $y_m(n)$ are exactly the same as the result from linear convolution and, as a consequence,

$$\hat{y}_m(n) = y_m(n) \qquad n = M, M + 1, \ldots, N - 1 \qquad (5.4.23)$$

To avoid loss of data due to aliasing, the last $M - 1$ points of each data record are saved and these points become the first $M - 1$ data points of the subsequent record, as indicated above. To begin the processing, the first $M - 1$ points of the first record are set to zero. Thus the blocks of data sequences are

$$x_1(n) = \{\underbrace{0, 0, \ldots, 0,}_{M - 1 \text{ points}} x(0), x(1), \ldots, x(L - 1)\} \qquad (5.4.24)$$

$$x_2(n) = \{\underbrace{x(L - M + 1), \ldots, x(L - 1),}_{\substack{M - 1 \text{ data points} \\ \text{from } x_1(n)}} \underbrace{x(L), \ldots, x(2L - 1)}_{L \text{ new data points}}\} \qquad (5.4.25)$$

$$x_3(n) = \{\underbrace{x(2L - M + 1), \ldots, x(2L - 1),}_{\substack{M - 1 \text{ data points} \\ \text{from } x_2(n)}} \underbrace{x(2L), \ldots, x(3L - 1)}_{L \text{ new data points}}\} \qquad (5.4.26)$$

and so forth. The resulting data sequences from the IDFT are given by (5.4.22), where the first $M - 1$ points are discarded due to aliasing and the remaining L points constitute the desired result from linear convolution. This segmentation of the input data and the fitting of the output data blocks together to form the output sequence are graphically illustrated in Fig. 5.21.

Overlap-Add Method. In this method the size of the input data block is L points and the size of the DFTs and IDFT is $N = L + M - 1$. To each data block we append $M - 1$ zeros and compute the N-point DFT. Thus the data blocks may be represented as

$$x_1(n) = \{x(0), x(1), \ldots, x(L - 1), \underbrace{0, 0, \ldots, 0}_{M - 1 \text{ zeros}}\} \tag{5.4.27}$$

$$x_2(n) = \{x(L), x(L + 1), \ldots, x(2L - 1), \underbrace{0, 0, \ldots, 0}_{M - 1 \text{ zeros}}\} \tag{5.4.28}$$

$$x_3(n) = \{x(2L), \ldots, x(3L - 1), \underbrace{0, 0, \ldots, 0}_{M - 1 \text{ zeros}}\} \tag{5.4.29}$$

and so forth. The two N-point DFTs are multiplied together to form

$$Y_m(k) = H(k)X_m(k) \qquad k = 0, 1, \ldots, N - 1 \tag{5.4.30}$$

The IDFT yields data blocks of length N that are free of aliasing since the size of the DFTs and IDFT is $N = L + M - 1$ and the sequences are increased to N-points by appending zeros to each block.

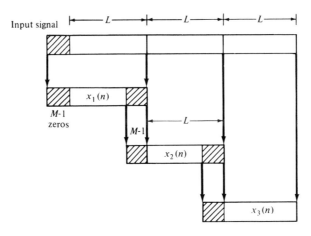

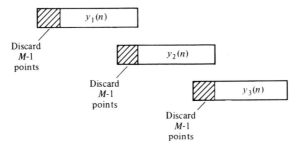

FIGURE 5.21 Linear FIR filtering by the overlap-save method.

Since each data block is terminated with $M - 1$ zeros, the last $M - 1$ points from each output block must be overlapped and added to the first $M - 1$ points of the succeeding block. Hence this method is called the overlap-add method. This overlapping and adding yields the output sequence

$$y(n) = \{y_1(0), y_1(1), \ldots, y_1(L - 1), y_1(L) + y_2(0), y_1(L + 1) +$$
$$y_2(1), \ldots, y_1(N - 1) + y_2(M - 1), y_2(M), \ldots\} \quad (5.4.31)$$

The segmentation of the input data into blocks and the fitting of the output data blocks to form the output sequence are graphically illustrated in Fig. 5.22.

At this point it may appear to the reader that the use of the DFT in linear FIR filtering is not only an indirect method of computing the output of an FIR filter, but it may also be more expensive computationally since the input data must be converted to the frequency domain via the DFT, multiplied by the DFT of the FIR filter, and finally, converted back to the time domain via the IDFT. On the contrary, however, by using the fast Fourier transform algorithm, as will be shown later, the DFTs and IDFT require fewer computations to compute the output sequence than the direct realization of the FIR filter in the time domain. This computational efficiency is the basic advantage of using the DFT to compute the output of an FIR filter.

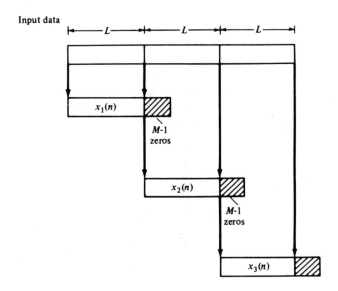

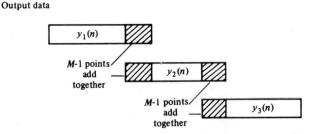

FIGURE 5.22   Linear FIR filtering by the overlap-add method.

5.5 Design of Digital Filters by Placement of Poles and Zeros in the z-Plane

In this section we treat the design of some simple digital filters by placement of poles and zeros in the z-plane. We have already described how the location of poles and zeros affects the frequency response characteristics of the system. In particular, in Section 5.2.3 we presented a graphical method for computing the frequency response characteristics from the pole–zero plot. This same approach can be used to design a number of simple but important digital filters with desirable frequency response characteristics.

The basic principle underlying the pole–zero placement method is to locate poles near points of the unit circle corresponding to frequencies to be emphasized and to place zeros near the frequencies to be deemphasized. Furthermore, the following constraints must be imposed:

1. All poles should be placed inside the unit circle in order for the filter to be stable. However, zeros can be placed anywhere in the z-plane.
2. All complex zeros and poles must occur in complex-conjugate pairs in order for the filter coefficients to be real.

From our previous discussion we recall that for a given pole–zero pattern, the system function $H(z)$ may be expressed as

$$H(z) = \frac{\sum\limits_{k=0}^{M} b_k z^{-k}}{1 + \sum\limits_{k=1}^{N} a_k z^{-k}}$$

$$= G \frac{\prod\limits_{k=1}^{M} (1 - z_k z^{-1})}{\prod\limits_{k=1}^{N} (1 - p_k z^{-1})} \tag{5.5.1}$$

where G is a gain constant that is selected to normalize the frequency response at some specified frequency, that is, G is selected such that

$$|H(\omega_0)| = 1 \tag{5.5.2}$$

where ω_0 is a frequency in the passband of the filter. Usually, N is selected to equal or exceed M, so that the filter has more nontrivial poles than zeros.

Below we illustrate the method of pole–zero placement in the design of some simple lowpass, highpass, and bandpass filters, digital resonators, and comb filters. The design procedure is facilitated when carried out interactively on a digital computer with a graphics terminal.

5.5.1 Lowpass, Highpass, and Bandpass Filters

In the design of lowpass digital filters the poles should be placed near the unit circle at points corresponding to low frequencies (near $\omega = 0$) and zeros should be placed near or on the unit circle at points corresponding to high frequencies (near $\omega = \pi$). The opposite holds true for highpass filters.

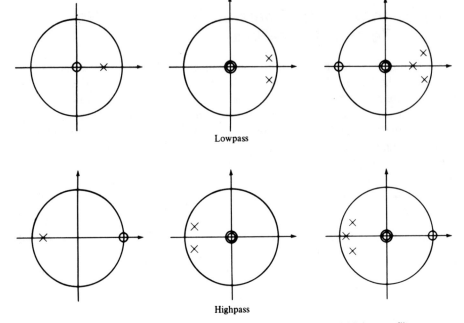

Lowpass

Highpass

FIGURE 5.23   Pole–zero patterns for several lowpass and highpass filters.

Figure 5.23 illustrates the pole–zero placement of three lowpass and three highpass filters. The magnitude and phase responses for the single-pole filter with system function

$$H_1(z) = \frac{1 - a}{1 - az^{-1}} \qquad (5.5.3)$$

are illustrated in Fig. 5.24 for $a = 0.9$. The gain G was selected as $1 - a$, so that the filter has unity gain at $\omega = 0$. The gain of this filter at high frequencies is relatively small.

The addition of a zero at $z = -1$ further attenuates the response of the filter at high frequencies. This leads to a filter with a system function

$$H_2(z) = \frac{1 - a}{2} \frac{1 + z^{-1}}{1 - az^{-1}} \qquad (5.5.4)$$

and a frequency response characterstic that is also illustrated in Fig. 5.24. In this case the magnitude of $H_2(z)$ goes to zero at $\omega = \pi$.

Similarly, we can obtain simple highpass filters by reflecting (folding) the pole–zero locations of the lowpass filters about the imaginary axis in the z-plane. Thus we obtain the system function

$$H_3(z) = \frac{1 - a}{2} \frac{1 - z^{-1}}{1 + az^{-1}} \qquad (5.5.5)$$

which has the frequency response characteristics illustrated in Fig. 5.25 for $a = 0.9$.

EXAMPLE 5.5.1 _____

A two-pole lowpass filter has the system function

$$H(z) = \frac{G}{(1 - pz^{-1})^2}$$

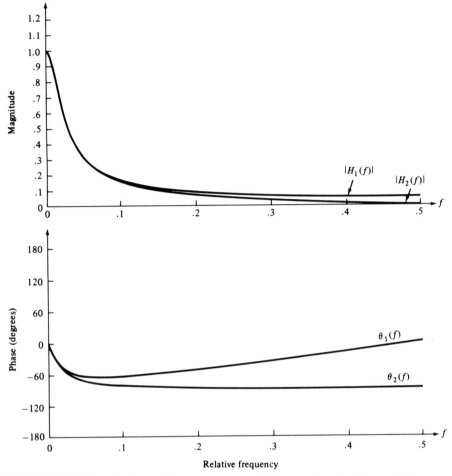

FIGURE 5.24   Magnitude and phase response of (1) a single-pole filter and (2) a one-pole, one-zero filter; $H_1(z) = (1 - a)/(1 - az^{-1})$, $H_2(z) = [(1 - a)/2] [(1 + z^{-1})/ (1 - az^{-1})]$.

Determine the values of G and p such that the frequency response $H(\omega)$ satisfies the conditions

$$H(0) = 1$$

and

$$\left| H\left(\frac{\pi}{4}\right) \right|^2 = \frac{1}{2}$$

Solution: At $\omega = 0$ we have

$$H(0) = \frac{G}{(1 - p)^2} = 1$$

Hence

$$G = (1 - p)^2$$

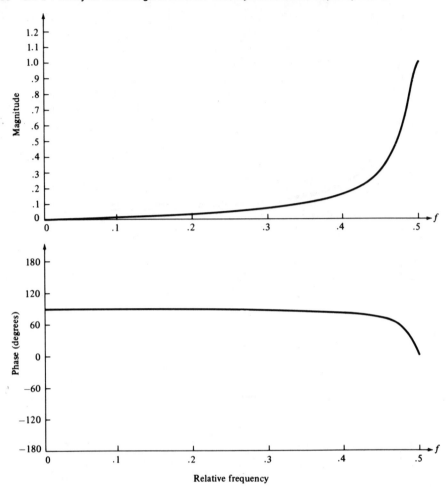

FIGURE 5.25 Magnitude and phase response of a simple highpass filter; $H(z) = [(1 - a)/2] [(1 - z^{-1})/(1 + az^{-1})]$.

At $\omega = \pi/4$,

$$H\left(\frac{\pi}{4}\right) = \frac{(1 - p)^2}{(1 - pe^{-j\pi/4})^2}$$

$$= \frac{(1 - p)^2}{(1 - p \cos(\pi/4) + jp \sin(\pi/4))^2}$$

$$= \frac{(1 - p)^2}{(1 - p/\sqrt{2} + jp/\sqrt{2})^2}$$

Hence

$$\frac{(1 - p)^4}{(1 - p/\sqrt{2})^2 + p^2/2} = \frac{1}{2}$$

or, equivalently,

$$2(1 - p)^4 = 1 + p^2 - \sqrt{2}p$$

The value of $p = 0.23$ satisfies this equation. Consequently, the system function for the desired filter is

$$H(z) = \frac{0.458}{(1 - 0.32z^{-1})^2}$$

The same principles can be applied for the design of bandpass filters. Basically, the bandpass filter should contain one or more pairs of complex-conjugate poles near the unit circle, in the vicinity of the frequency band that constitutes the passband of the filter. The following example serves to illustrate the basic ideas.

EXAMPLE 5.5.2 _____

Design a two-pole bandpass filter that has the center of its passband at $\omega = \pi/2$, zero in its frequency response characteristic at $\omega = 0$ and $\omega = \pi$, and its magnitude response is $1/\sqrt{2}$ at $\omega = 4\pi/9$.

Solution: Clearly, the filter must have poles at

$$p_{1,2} = re^{\pm j\pi/2}$$

and zeros at $z = 1$ and $z = \pi$. Consequently, the system function is

$$H(z) = G \frac{(z - 1)(z + 1)}{(z - jr)(z + jr)}$$

$$= G \frac{z^2 - 1}{z^2 + r^2}$$

The gain factor is determined by evaluating the frequency response $H(\omega)$ of the filter at $\omega = \pi/2$. Thus we have

$$H\left(\frac{\pi}{2}\right) = G \frac{2}{1 - r^2} = 1$$

$$G = \frac{1 - r^2}{2}$$

The value of r is determined by evaluating $H(\omega)$ at $\omega = 4\pi/9$. Thus we have

$$\left| H\left(\frac{4\pi}{9}\right) \right|^2 = \frac{(1 - r^2)^2}{4} \frac{2 - 2\cos(8\pi/9)}{1 + r^4 + 2r^2 \cos(8\pi/9)} = \frac{1}{2}$$

or, equivalently,

$$1.94(1 - r^2)^2 = 1 - 1.88r^2 + r^4$$

The value of $r^2 = 0.7$ satisfies this equation. Therefore, the system function for the desired filter is

$$H(z) = 0.15 \frac{1 - z^{-2}}{1 + 0.7z^{-2}}$$

Its frequency response is illustrated in Fig. 5.26.

It should be emphasized that the main purpose of the foregoing methodology for designing simple digital filters by pole–zero placement is to provide insight into the effect

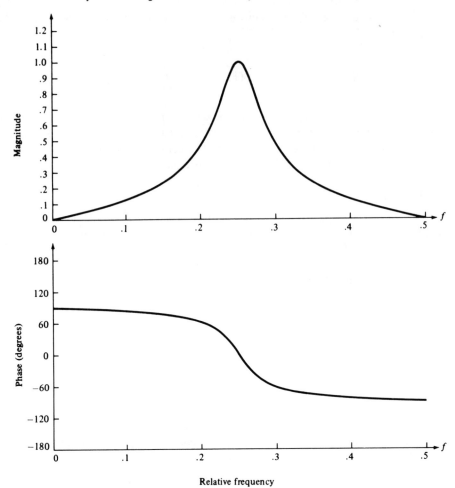

FIGURE 5.26 Magnitude and phase response of a simple bandpass filter in Example 5.5.2; $H(z) = 0.15[(1 - z^{-2})/(1 + 0.7z^{-2})]$.

that poles and zeros have on the frequency response characteristic of systems. The methodology is not intended to provide a good method for designing digital filters with well-specified passband and stopband characteristics. Systematic methods for the design of sophisticated digital filters for practical applications are discussed in Chapter 8.

5.5.2 Digital Resonators

A *digital resonator* is a special two-pole bandpass filter with the pair of complex-conjugate poles located near the unit circle as shown in Fig. 5.27a. The magnitude of the frequency response of the filter is shown in Fig. 5.27b. The name resonator refers to the fact that the filter has a large magnitude response (i.e., it resonates) in the vicinity of the pole location. The angular position of the pole determines the resonant frequency of the filter. Digital resonators are useful in many applications, including simple bandpass filtering and speech generation.

In the design of a digital resonator with a resonant peak at or near $\omega = \omega_0$, we select the complex-conjugate poles at

$$p_{1,2} = re^{\pm j\omega} \qquad 0 < r < 1 \qquad (5.5.6)$$

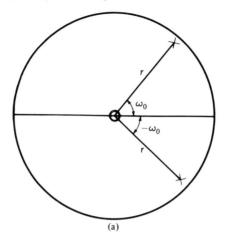

(a)

(b)

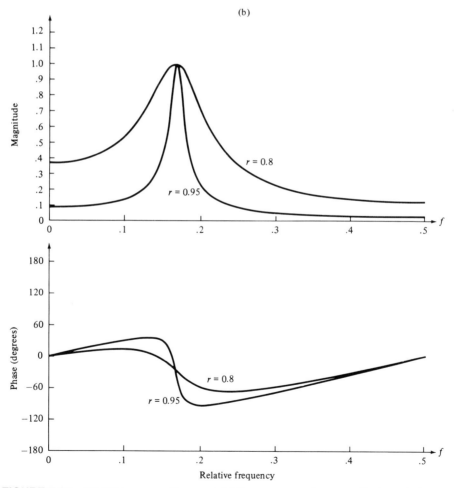

FIGURE 5.27 (a) Pole–zero pattern and (b) the corresponding magnitude and phase response of a digital resonator with (1) $r = 0.8$, (2) $r = 0.95$.

In addition, we may select up to two zeros. Although there are many possible choices, two cases are of special interest. One choice is to locate the zeros at the origin. The other choice is to locate a zero at $z = 1$ and a zero at $z = -1$. This choice completely eliminates the response of the filter at frequencies $\omega = 0$ and $\omega = \pi$, and it is useful in many practical applications.

The system function of the digital resonator with zeros at the origin is

$$H(z) = \frac{G}{(1 - re^{j\omega_0}z^{-1})(1 - re^{-j\omega_0}z^{-1})} \tag{5.5.7}$$

$$H(z) = \frac{G}{1 - (2r\cos\omega_0)z^{-1} + r^2 z^{-2}} \tag{5.5.8}$$

Since $|H(\omega)|$ has its peak at or near $\omega = \omega_0$, we select the gain G so that $|H(\omega_0)| = 1$. From (5.5.7) we obtain

$$H(\omega_0) = \frac{G}{(1 - re^{j\omega_0}e^{-j\omega_0})(1 - re^{-j\omega_0}e^{-j\omega_0})} \tag{5.5.9}$$

$$= \frac{G}{(1 - r)(1 - re^{-j2\omega_0})}$$

and hence

$$|H(\omega_0)| = \frac{G}{(1 - r)\sqrt{1 + r^2 - 2r\cos 2\omega_0}} = 1$$

Thus the desired normalization factor is

$$G = (1 - r)\sqrt{1 + r^2 - 2r\cos 2\omega_0} \tag{5.5.10}$$

The frequency response of the resonator in (5.5.7) may be expressed as

$$|H(\omega)| = \frac{G}{U_1(\omega)U_2(\omega)} \tag{5.5.11}$$

$$\theta(\omega) = 2\omega - \Phi_1(\omega) - \Phi_2(\omega)$$

where $U_1(\omega)$ and $U_2(\omega)$ are the magnitudes of the vectors from p_1 and p_2 to the point ω in the unit circle and $\Phi_1(\omega)$ and $\Phi_2(\omega)$ are the corresponding angles of these two vectors. The magnitudes $U_1(\omega)$ and $U_2(\omega)$ may be expressed as

$$U_1(\omega) = \sqrt{1 + r^2 - 2r\cos(\omega_0 - \omega)} \tag{5.5.12}$$

$$U_2(\omega) = \sqrt{1 + r^2 - 2r\cos(\omega_0 + \omega)}$$

For any value of r, $U_1(\omega)$ takes its minimum value $(1 - r)$ at $\omega = \omega_0$. The product $U_1(\omega)U_2(\omega)$ reaches a minimum value at the frequency

$$\omega_r = \cos^{-1}\left(\frac{1 + r^2}{2r}\cos\omega_0\right) \tag{5.5.13}$$

which defines precisely the resonant frequency of the filter. We observe that when r is very close to unity, $\omega_r \approx \omega_0$, which is the angular position of the pole. We also observe that as r approaches unity the resonance peak becomes sharper because $U_1(\omega)$ changes more rapidly in relative size in the vicinity of ω_0. A quantitative measure of the sharpness of the resonance is provided by the 3-dB bandwidth $\Delta\omega$ of the filter. For values of r close to unity,

$$\Delta\omega \approx 2(1 - r) \tag{5.5.14}$$

Figure 5.27b illustrates the magnitude and phase of digital resonators with $\omega_0 = \pi/3$, $r = 0.8$ and $\omega_0 = \pi/3$, $r = 0.95$. We note that the phase response undergoes its greatest rate of change near the resonant frequency.

If the zeros of the digital resonator are placed at $z = 1$ and $z = -1$, the resonator has the system function

$$
\begin{aligned}
H(z) &= G \frac{(1 - z^{-1})(1 + z^{-1})}{(1 - re^{j\omega_0}z^{-1})(1 - re^{-j\omega_0}z^{-1})} \\
&= G \frac{1 - z^{-2}}{1 - (2r \cos \omega_0)z^{-1} + r^2 z^{-2}}
\end{aligned}
\tag{5.5.15}
$$

and a frequency response characteristic

$$
H(\omega) = G \frac{1 - e^{-j2\omega}}{[1 - re^{j(\omega_0 - \omega)}][1 - re^{-j(\omega_0 + \omega)}]}
\tag{5.5.16}
$$

We observe that the zeros at $z = \pm 1$ affect both the magnitude and phase response of the resonator. For example, the magnitude response is

$$
|H(\omega)| = G \frac{N(\omega)}{U_1(\omega)U_2(\omega)}
\tag{5.5.17}
$$

where $N(\omega)$ is defined as

$$
N(\omega) = \sqrt{2(1 - \cos 2\omega)}
$$

Due to the presence of the zero factor, the resonant frequency is altered from that given by the expression in (5.5.13). The bandwidth of the filter is also altered. Although exact values for these two parameters are rather tedious to derive, we may easily compute the frequency response in (5.5.16) and compare the result with the previous case in which the zeros are located at the origin.

Figure 5.28 illustrates the magnitude and phase characteristics for $\omega = \pi/3$, $r = 0.8$ and $\omega = \pi/3$, $r = 0.95$. We observe that this filter has a slightly smaller bandwidth than the resonator, which has zeros at the origin. In addition, there appears to be a very small shift in the resonant frequency due to the presence of the zeros.

5.5.3 Notch Filters

A notch filter is a filter that contains one or more deep notches or, ideally, perfect nulls in its frequency response characteristics. Figure 5.29 illustrates the frequency response characteristic of a notch filter with nulls at frequencies ω_0 and ω_1. Notch filters are useful in many applications where specific frequency components must be eliminated. For example, instrumentation and recording systems require that the power-line frequency of 60 Hz and harmonics be eliminated.

To create a null in the frequency response of a filter at a frequency ω_0, we simply introduce a pair of complex-conjugate zeros on the unit circle at an angle ω_0. That is,

$$
z_{1,2} = e^{\pm j\omega_0}
$$

Thus the system function for an FIR notch filter is simply

$$
\begin{aligned}
H(z) &= G(1 - e^{j\omega_0}z^{-1})(1 - e^{-j\omega_0}z^{-1}) \\
&= G(1 - 2 \cos \omega_0 z^{-1} + z^{-2})
\end{aligned}
\tag{5.5.18}
$$

As an illustration, Fig. 5.30 shows the magnitude response for a notch filter having a null at $\omega = \pi/4$.

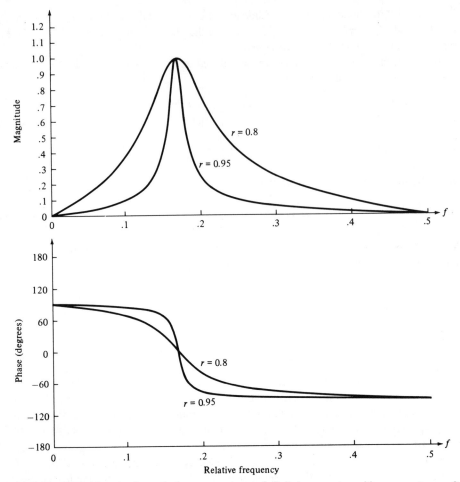

FIGURE 5.28 Magnitude and phase response of digital resonator with zeros at $\omega = 0$ and $\omega = \pi$ and (1) $r = 0.8$ and (2) $r = 0.95$.

The problem with the FIR notch filter is that the notch has a relatively large bandwidth, which means that other frequency components around the desired null are severely attenuated. To reduce the bandwidth of the null, we may resort to a more sophisticated, longer FIR filter designed according to a criterion described in Chapter 8. Alternatively, we may attempt to improve on the frequency response characteristics, in an ad hoc manner, by introducing poles in the system function.

Suppose that we place a pair of complex-conjugate poles at

$$p_{1,2} = re^{\pm j\omega_0}$$

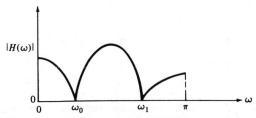

FIGURE 5.29 Frequency response characteristic of a notch filter.

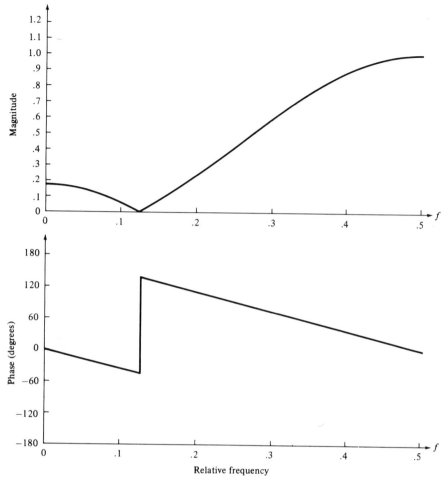

FIGURE 5.30 Frequency response characteristics of a notch filter with a notch at $\omega = \pi/4$ or $f = 1/8$; $H(z) = G[1 - 2\cos\omega_0 z^{-1} + z^{-2}]$.

The effect of the poles is to introduce a resonance in the vicinity of the null and thus to reduce the bandwidth of the notch. The system function for the resulting filter is

$$H(z) = G \frac{1 - 2\cos\omega_0 z^{-1} + z^{-2}}{1 - 2r\cos\omega_0 z^{-1} + r^2 z^{-2}} \qquad (5.5.19)$$

The magnitude response $|H(\omega)|$ of the filter in (5.5.19) is plotted in Fig. 5.31 for $\omega_0 = \pi/4$, $r = 0.85$ and $\omega_0 = \pi/4$, $r = 0.95$. When compared with the frequency response of the FIR filter in Fig. 5.30, we note that the effect of the poles is to reduce the bandwidth of the notch.

In addition to reducing the bandwidth of the notch, the introduction of a pole in the vicinity of the null may result in a small ripple in the passband of the filter, due to the resonance created by the pole. The effect of the ripple can be reduced by introducing additional poles and/or zeros in the system function of the notch filter. The major problem with this approach is that it is basically an ad hoc, trial-and-error method.

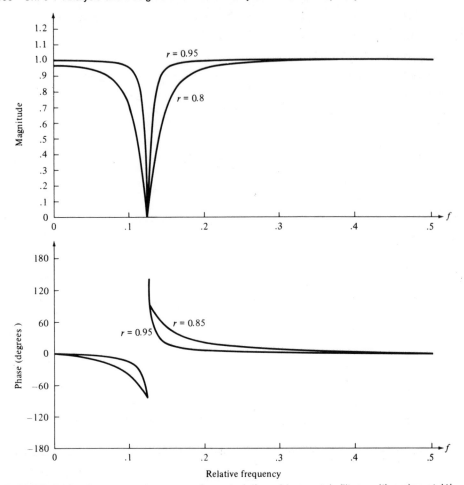

FIGURE 5.31 Frequency response characteristics of two notch filters with poles at (1) $r = 0.85$ and (2) $r = 0.95$; $H(z) = G[(1 - 2 \cos \omega_0 z^{-1} + z^{-2})/(1 - 2r \cos \omega_0 z^{-1} + r^2 z^{-2})]$.

5.5.4 Comb Filters

In its simplest form, a comb filter may be viewed as a notch filter in which the nulls occur periodically across the frequency band, hence the analogy to an ordinary comb that has periodically spaced teeth. Comb filters find applications in a wide range of practical systems, such as in the rejection of power-line harmonics, in the separation of solar and lunar components from ionospheric measurements of electron concentration, and in the suppression of clutter from fixed objects in moving-target-indicator (MTI) radars.

To illustrate a simple form of a comb filter, consider a moving average (FIR) filter that is described by the difference equation

$$y(n) = \frac{1}{M + 1} \sum_{k=0}^{M} x(n - k) \qquad (5.5.20)$$

The system function of this FIR filter is

$$H(z) = \frac{1}{M+1} \sum_{k=0}^{M} z^{-k}$$

$$= \frac{1}{M+1} \frac{[1 - z^{-(M+1)}]}{(1 - z^{-1})}$$

(5.5.21)

and its frequency response is

$$H(\omega) = \frac{e^{-j\omega M/2}}{M+1} \frac{\sin \omega \left(\dfrac{M+1}{2} \right)}{\sin (\omega/2)}$$

(5.5.22)

From (5.5.21) we observe that the filter has zeros on the unit circle at

$$z = e^{j2\pi k/(M+1)} \qquad k = 1, 2, 3, \ldots, M$$

(5.5.23)

Note that the pole at $z = 1$ is actually canceled by the zero at $z = 1$, so that in effect the FIR filter does not contain poles outside $z = 0$.

A plot of the magnitude characteristic of (5.5.22) clearly illustrates the existence of the periodically spaced zeros in frequency at $\omega_k = 2\pi k/(M+1)$ for $k = 1, 2, \ldots$, M. Figure 5.32 shows $|H(\omega)|$ for $M = 10$.

In more general terms, we can create a comb filter by taking an FIR filter with system function

$$H(z) = \sum_{k=0}^{M} h(k) z^{-k}$$

(5.5.24)

and replacing z by z^L, where L is a positive integer. Thus the new FIR filter has a system function

$$H_L(z) = \sum_{k=0}^{M} h(k) z^{-kL}$$

(5.5.25)

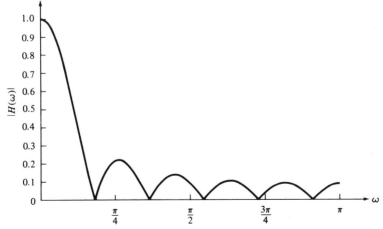

FIGURE 5.32 Magnitude response characteristic for the comb filter given by (5.5.22) with $M = 10$.

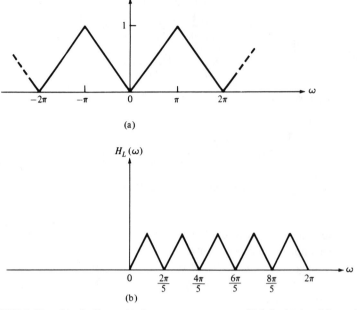

(a)

(b)

FIGURE 5.33 Comb filter with frequency response $H_L(\omega)$ obtained from $H(\omega)$.

If the frequency response of the original FIR filter is $H(\omega)$, the frequency response of the FIR in (5.5.25) is

$$H_L(\omega) = \sum_{k=0}^{M} h(k)e^{-jkL\omega}$$
$$= H(L\omega)$$

(5.5.26)

Consequently, the frequency response characteristic $H_L(\omega)$ is simply an L-order repetition of $H(\omega)$ in the range $0 \le \omega \le 2\pi$. Figure 5.33 illustrates the relationship between $H_L(\omega)$ and $H(\omega)$ for $L = 5$.

Now, suppose that the original FIR filter with system function $H(z)$ has a spectral null (i.e., a zero), at some frequency ω_0. Then the filter with system function $H_L(z)$ has periodically spaced nulls at $\omega_k = \omega_0 + 2\pi k/L$, $k = 0, 1, 2, \ldots, L - 1$. As an illustration, Fig. 5.34 shows an FIR comb filter with $M = 3$ and $L = 3$. This FIR filter may be viewed as an FIR filter of length 10, but only four of the 10 filter coefficients are nonzero.

Let us now return to the moving average filter with system function given by (5.5.21). Suppose that we replace z by z^L. Then the resulting comb filter has the system function

$$H_L(z) = \frac{1}{M+1} \frac{1 - z^{-L(M+1)}}{1 - z^{-L}}$$

(5.5.27)

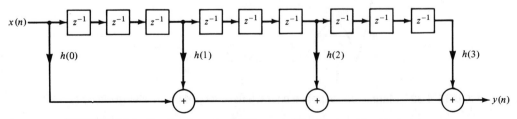

FIGURE 5.34 Realization of an FIR comb filter having $M = 3$ and $L = 3$.

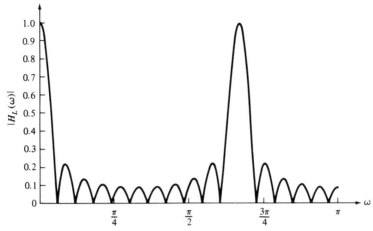

FIGURE 5.35   Magnitude response characteristic for a comb filter given by (5.5.28), with $L = 3$ and $M = 10$.

and a frequency response

$$H_L(\omega) = \frac{1}{M + 1} \frac{\sin \omega L(M + 1)/2}{\sin \omega L/2} e^{-j\omega LM/2} \tag{5.5.28}$$

This filter has zeros on the unit circle at

$$z_k = e^{j2\pi k/L(M+1)} \tag{5.5.29}$$

for all integer values of k except $k = 0, L, 2L, \ldots, ML$. Figure 5.35 illustrates $|H_L(\omega)|$ for $L = 3$ and $M = 10$.

The comb filter described by (5.5.27) finds application in the separation of solar and lunar spectral components in ionospheric measurements of electron concentration as described in the paper by Bernhardt et al. (1976). The solar period is $T_s = 24$ hours and results in a solar component of one cycle per day and its harmonics. The lunar period is $T_L = 24.84$ hours and provides spectral lines at 0.96618 cycle per day and its harmonics. Figure 5.36a shows a plot of the power density spectrum of the unfiltered ionospheric measurements of the electron concentration. Note that the weak lunar spectral components are almost hidden by the strong solar spectral components.

The two sets of spectral components can be separated by the use of comb filters. If we wish to obtain the solar components we may use a comb filter with a narrow passband at multiples of one cycle per day. This can be achieved by selecting L such that $F_s/L = 1$ cycle per day, where F_s is the corresponding sampling frequency. The result is a filter that has peaks in its frequency response at multiples of one cycle per day. By selecting $M = 58$, the filter will have nulls at multiples of $(F_s/L)/(M + 1) = 1/59$ cycle per day. These nulls are very close to the lunar components and result in good rejection. Figure 5.36b illustrates the power spectral density of the output of the comb filter that isolates the solar components. A comb filter that rejects the solar components and passes the lunar components can be designed in a similar manner. Figure 5.36c illustrates the power spectral density at the output of such a lunar filter.

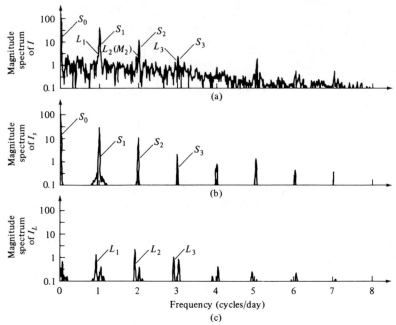

FIGURE 5.36   (a) Spectrum of unfiltered electron content data; (b) spectrum of output of solar filter; (c) spectrum of output of lunar filter. [From paper by Bernhardt et al. (1976). Reprinted with permission of the American Geophysical Union.]

5.5.5 All-Pass Filters

An all-pass filter is defined as a system that has a constant magnitude response for all frequencies, that is,

$$|H(\omega)| = 1 \qquad 0 \le \omega \le \pi \tag{5.5.30}$$

The simplest example of an all-pass filter is a pure delay system with system function

$$H(z) = z^{-k} \tag{5.5.31}$$

This system passes all signals without modification except for a delay of k samples. This is a trivial all-pass system that has a linear phase response characteristic.

A more interesting all-pass filter is one that is described by the system function

$$H(z) = \frac{a_N + a_{N-1}z^{-1} + \cdots + a_1 z^{-N+1} + z^{-N}}{1 + a_1 z^{-1} + \cdots + a_N z^{-N}} \tag{5.5.32}$$

$$= \frac{\displaystyle\sum_{k=0}^{N} a_k z^{-N+k}}{\displaystyle\sum_{k=0}^{N} a_k z^{-k}} \qquad a_0 = 1$$

where all the filter coefficients $\{a_k\}$ are real. If we define the polynomial $A(z)$ as

$$A(z) = \sum_{k=0}^{N} a_k z^{-k} \qquad a_0 = 1 \tag{5.5.33}$$

then (5.5.32) can be expressed as

$$H(z) = z^{-N} \frac{A(z^{-1})}{A(z)} \tag{5.5.34}$$

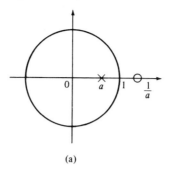

(a)

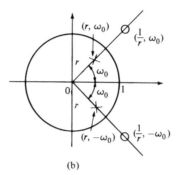

(b)

FIGURE 5.37   Pole–zero patterns of (a) a first-order and (b) a second-order all-pass filter.

Since

$$|H(\omega)|^2 = H(z)H(z^{-1})|_{z=e^{j\omega}} = 1$$

the system given by (5.5.34) is all-pass. Furthermore, if z_0 is a pole of $H(z)$, then $1/z_0$ is a zero of $H(z)$ (i.e., the poles and zeros are reciprocals of one another). Figure 5.37 illustrates typical pole–zero patterns for a single-pole, single-zero filter and a two-pole, two-zero filter. A plot of the phase characteristics of these filters is shown in Fig. 5.38 for $a = 0.6$ and $r = 0.9$, $\omega_0 = \pi/4$.

All-pass filters find application as phase equalizers. When placed in cascade with a system that has an undesired phase response, a phase equalizer is designed to compensate for the poor phase characteristics of the system and thus to produce an overall linear-phase response.

5.5.6 Digital Sinusoidal Oscillators

A *digital sinusoidal oscillator* may be viewed as a limiting form of a two-pole resonator for which the complex-conjugate poles lie on the unit circle. From our previous discussion of second-order systems we recall that a system with system function

$$H(z) = \frac{b_0}{1 + a_1 z^{-1} + a_2 z^{-2}} \tag{5.5.35}$$

and parameters

$$a_1 = -2r \cos \omega_0 \quad \text{and} \quad a_2 = r^2 \tag{5.5.36}$$

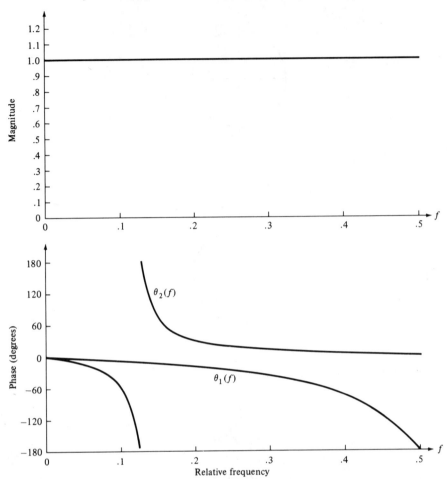

FIGURE 5.38   Frequency response characteristics of an all-pass filter with system functions (1) $H(z) = (0.6 + z^{-1})/(1 + 0.6z^{-1})$, (2) $H(z) = (r^2 - 2r \cos \omega_0 z^{-1} + z^{-2})/(1 - 2r \cos \omega_0 z^{-1} + r^2 z^{-2})$, $r = 0.9$, $\omega_0 = \pi/4$.

has complex-conjugate poles at $p = re^{\pm j\omega_0}$, and a unit sample response

$$h(n) = \frac{b_0 r^n}{\sin \omega_0} \sin (n + 1)\omega_0 \, u(n) \tag{5.5.37}$$

If the poles are placed on the unit circle ($r = 1$) and b_0 is set to $A \sin \omega_0$, then

$$h(n) = A \sin (n + 1)\omega_0 u(n) \tag{5.5.38}$$

Thus the impulse response of the second-order system with complex-conjugate poles on the unit circle is a sinusoid and the system is called a *digital sinusoidal oscillator* or a *digital sinusoidal generator*. A digital sinusoidal generator is a basic component of a digital frequency synthesizer.

The block diagram representation of the system function given by (5.5.35) is illustrated in Fig. 5.39. The corresponding difference equation for this system is

$$y(n) = -a_1 y(n - 1) - y(n - 2) + b_0 \, \delta(n) \tag{5.5.39}$$

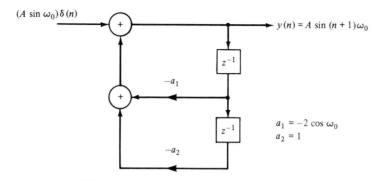

FIGURE 5.39   Digital sinusoidal generator.

where the parameters are $a_1 = -2 \cos \omega_0$ and $b_0 = A \sin \omega_0$, and the initial conditions are $y(-1) = y(-2) = 0$. By iterating the difference equation in (5.5.39) we obtain

$$y(0) = A \sin \omega_0$$
$$y(1) = 2 \cos \omega_0 y(0) = 2A \sin \omega_0 \cos \omega_0 = A \sin 2\omega_0$$
$$y(2) = 2 \cos \omega_0 y(1) - y(0)$$
$$= 2A \cos \omega_0 \sin 2\omega_0 - A \sin \omega_0$$
$$= A(4 \cos^2 \omega_0 - 1) \sin \omega_0$$
$$= 3A \sin \omega_0 - 4 \sin^3 \omega_0 = A \sin 3\omega_0$$

and so forth. We note that the application of the impulse at $n = 0$ serves the purpose of beginning the sinusoidal oscillation. Thereafter, the oscillation is self-sustaining because the system has no damping (i.e., $r = 1$).

It is interesting to note that the sinusoidal oscillation obtained from the system in (5.5.39) can also be obtained by setting the input to zero and the initial conditions to $y(-1) = 0$, $y(-2) = -A \sin \omega_0$. Thus the zero-input response to the second-order system described by the homogeneous difference equation

$$y(n) = -a_1 y(n-1) - y(n-2) \qquad (5.5.40)$$

with initial conditions $y(-1) = 0$ and $y(-2) = -A \sin \omega_0$ is exactly the same as the response of (5.5.39) to an impulse excitation. In fact, the difference equation in (5.5.40) can be obtained directly from the trigonometric identity

$$\sin \alpha + \sin \beta = 2 \sin \frac{\alpha + \beta}{2} \cos \frac{\alpha - \beta}{2} \qquad (5.5.41)$$

where, by definition, $\alpha = (n + 1)\omega_0$, $\beta = (n - 1)\omega_0$, and $y(n) = \sin (n + 1)\omega_0$.

In some practical applications involving modulation of two sinusoidal carrier signals in phase quadrature, there is a need to generate the sinusoids $A \sin \omega_0 n$ and $A \cos \omega_0 n$. These signals can be generated from the so-called *coupled-form oscillator*, which can be obtained from the trigonometric formulas

$$\cos (\alpha + \beta) = \cos \alpha \cos \beta - \sin \alpha \sin \beta$$
$$\sin (\alpha + \beta) = \sin \alpha \cos \beta + \cos \alpha \sin \beta$$

where, by definition, $\alpha = n\omega_0$, $\beta = \omega_0$, and

$$y_c(n) = \cos n\omega_0 u(n) \qquad (5.5.42)$$
$$y_s(n) = \sin n\omega_0 u(n) \qquad (5.5.43)$$

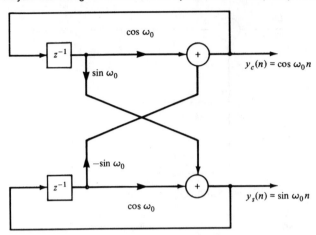

FIGURE 5.40   Realization of the coupled-form oscillator.

Thus we obtain the two coupled difference equations

$$y_c(n) = (\cos \omega_0)y_c(n - 1) - (\sin \omega_0)y_s(n - 1) \qquad (5.5.44)$$
$$y_s(n) = (\sin \omega_0)y_c(n - 1) + (\cos \omega_0)y_s(n - 1) \qquad (5.5.45)$$

which can also be expressed in matrix form as

$$\begin{bmatrix} y_c(n) \\ y_s(n) \end{bmatrix} = \begin{bmatrix} \cos \omega_0 & -\sin \omega_0 \\ \sin \omega_0 & \cos \omega_0 \end{bmatrix} \begin{bmatrix} y_c(n - 1) \\ y_s(n - 1) \end{bmatrix} \qquad (5.5.46)$$

The structure for the realization of the coupled-form oscillator is illustrated in Fig. 5.40. We note that this is a two-output system which is not driven by any input, but which requires the initial conditions $y_c(-1) = A \cos \omega_0$ and $y_s(-1) = -A \sin \omega_0$ in order to begin its self-sustaining oscillations.

Finally, it is interesting to note that (5.5.46) corresponds to vector rotation in the two-dimensional coordinate system with coordinates $y_c(n)$ and $y_s(n)$. As a consequence, the coupled-form oscillator can also be implemented by use of the so-called CORDIC algorithm [see the book by Kung et al. (1985)].

5.6 Summary and References

In this chapter we have focused on the frequency-domain characteristics of LTI systems. We showed that an LTI system is characterized in the frequency domain by its frequency response function $H(\omega)$, which is the Fourier transform of the impulse response of the system. We also observed that the frequency response function determines the effect of the system on any input signal. In fact, by transforming the input signal into the frequency domain, we observed that it is a simple matter to determine the effect of the system on the signal and to determine the system output. When viewed in the frequency domain, an LTI system performs spectral shaping or spectral filtering on the input signal.

The design of filters is one of the important topics treated in this chapter. A frequency sampling method was described for designing linear phase FIR filters. Such filters may be implemented directly or via the DFT. In the context of using the DFT for linear filtering, we demonstrated that the multiplication of two DFTs is equivalent to circular convolution of the corresponding sequences in the time domain. To avoid the aliasing effects of circular convolution, zero padding of the sequences was used, so that the circular convolution of the padded sequences was equivalent to linear convolution. Thus the DFT and IDFT can be used to perform linear convolution or, equivalently, linear filtering with an FIR filter. Due to the similarity of correlation with convolution, the DFT and IDFT may also be used to perform crosscorrelation of two sequences via the frequency domain.

The frequency sampling method for designing linear-phase FIR filters was first described in papers by Gold and Jordan (1969) and by Rabiner et al. (1970). The use of the DFT and IDFT for efficiently computing the convolution or correlation of two sequences was proposed in papers by Helms (1967) and by Stockham (1966).

The design of some simple IIR filters was also considered in this chapter, from the viewpoint of pole–zero placement. By means of this method, we were able to design simple digital resonators, notch filters, comb filters, all-pass filters, and digital sinusoidal generators.

The design of more complex IIR filters is treated in detail in Chapter 8, where several references are also given. Digital sinusoidal generators find use in frequency synthesis applications. A comprehensive treatment of frequency synthesis techniques is given in the text edited by Gorski-Popiel (1975).

APPENDIX 5A. TABLES OF TRANSITION COEFFICIENTS FOR THE DESIGN OF LINEAR-PHASE FIR FILTERS

In Section 5.3.4 we described a design method for linear-phase FIR filters that involved the specification of $H_r(\omega)$ at a set of equally spaced frequencies $\omega_k = 2\pi(k + \alpha)/M$, where $\alpha = 0$ or $\alpha = \frac{1}{2}$, $k = 0, 1, \ldots , (M - 1)/2$ for M odd and $k = 0, 1, 2, \ldots ,$ $(M/2) - 1$ for M even, where M is the length of the filter. Within the passband of the filter, we select $H_r(\omega_k) = 1$, and in the stopband, $H_r(\omega_k) = 0$. For frequencies in the transition band, the values of $H(\omega_k)$ are optimized to minimize the maximum sidelobe in the stopband. This is called a *minimax optimization criterion*.

The optimization of the values of $H(\omega)$ in the transition band has been performed by Rabiner et al. (1970) and tables of transition values have been provided in the published paper. A selected number of the tables for lowpass FIR filters are included in this appendix.

Four tables are given below. Table 5.1 lists the transition coefficients for the case $\alpha = 0$ and one coefficient in the transition band for both M odd and M even. Table 5.2 lists the transition coefficients for the case $\alpha = 0$, and two coefficients in the transition band for M odd and M even. Table 5.3 lists the transition coefficients for the case $\alpha = \frac{1}{2}$, M even and one coefficient in the transition band. Finally, Table 5.4 lists the transition coefficients for the case $\alpha = \frac{1}{2}$, M even, and two coefficients in the transition band. The tables also include the level of the maximum sidelobe and a bandwidth parameter, denoted as BW.

To use the tables, we begin with a set of specifications, including (1) the bandwidth of the filter, which may be defined as $(2\pi/M)$ (BW + α), where BW is the number of consecutive frequencies at which $H(\omega_k) = 1$, (2) the width of the transition region, which is roughly $2\pi/M$ times the number of transition coefficients, and (3) the maximum tolerable sidelobe in the stopband. The length of the filter may be selected from the tables to satisfy the specifications.

TABLE 5.1 Transition Coefficients for $\alpha = 0$

	M Odd			M Even	
BW	Minimax	T_1	BW	Minimax	T_1
	$M = 15$			$M = 16$	
1	− 42.30932283	0.43378296	1	− 39.75363827	0.42631836
2	− 41.26299286	0.41793823	2	− 37.61346340	0.40397949
3	− 41.25333786	0.41047636	3	− 36.57721567	0.39454346
4	− 41.94907713	0.40405884	4	− 35.87249756	0.38916626
5	− 44.37124538	0.39268189	5	− 35.31695461	0.38840332
6	− 56.01416588	0.35766525	6	− 35.51951933	0.40155639
	$M = 33$			$M = 32$	
1	− 43.03163004	0.42994995	1	− 42.24728918	0.42856445
2	− 42.42527962	0.41042481	2	− 41.29370594	0.40773926
3	− 42.40898275	0.40141601	3	− 41.03810358	0.39662476
4	− 42.45948601	0.39641724	4	− 40.93496323	0.38925171
6	− 42.52403450	0.39161377	6	− 40.85183477	0.37897949
8	− 42.44085121	0.39039917	8	− 40.75032616	0.36990356
10	− 42.11079407	0.39192505	10	− 40.54562140	0.35928955
12	− 41.92705250	0.39420166	12	− 39.93450451	0.34487915
14	− 44.69430351	0.38552246	14	− 38.91993237	0.34407349
15	− 56.18293285	0.35360718			
	$M = 65$			$M = 64$	
1	− 43.16935968	0.42919312	1	− 42.96059322	0.42882080
2	− 42.61945581	0.40903320	2	− 42.30815172	0.40830689
3	− 42.70906305	0.39920654	3	− 42.32423735	0.39807129
4	− 42.86997318	0.39335937	4	− 42.43565893	0.39177246
5	− 43.01999664	0.38950806	5	− 42.55461407	0.38742065
6	− 43.14578819	0.38679809	6	− 42.66526604	0.38416748
10	− 43.44808340	0.38129272	10	− 43.01104736	0.37609863
14	− 43.54684496	0.37946167	14	− 43.28309965	0.37089233
18	− 43.48173618	0.37955322	18	− 43.56508827	0.36605225
22	− 43.19538212	0.38162842	22	− 43.96245098	0.35977783
26	− 42.44725609	0.38746948	26	− 44.60516977	0.34813232
30	− 44.76228619	0.38417358	30	− 43.81448936	0.29973144
31	− 59.21673775	0.35282745			
	$M = 125$			$M = 128$	
1	− 43.20501566	0.42899170	1	− 43.15302420	0.42889404
2	− 42.66971111	0.40867310	2	− 42.59092569	0.40847778
3	− 42.77438974	0.39868774	3	− 42.67634487	0.39838257
4	− 42.95051050	0.39268189	4	− 42.84038544	0.39226685
6	− 43.25854683	0.38579101	5	− 42.99805641	0.38812256
8	− 43.47917461	0.38195801	7	− 43.25537014	0.38281250
10	− 43.63750410	0.37954102	10	− 43.52547789	0.3782638
18	− 43.95589399	0.37518311	18	− 43.93180990	0.37251587
26	− 44.05913115	0.37384033	26	− 44.18097305	0.36941528
34	− 44.05672455	0.37371826	34	− 44.40153408	0.36686401
42	− 43.94708776	0.37470093	42	− 44.67161417	0.36394653
50	− 43.58473492	0.37797851	50	− 45.17186594	0.35902100
58	− 42.14925432	0.39086304	58	− 46.92415667	0.34273681
59	− 42.60623264	0.39063110	62	− 49.46298973	0.28751221
60	− 44.78062010	0.38383713			
61	− 56.22547865	0.35263062			

Source: Rabiner et al. (1970); © 1970 IEEE; reprinted with permission.

TABLE 5.2 Transition Coefficients for $\alpha = 0$

	M Odd				*M* Even		
BW	**Minimax**	T_1	T_2	**BW**	**Minimax**	T_1	T_2
	M = 15				*M* = 16		
1	−70.60540585	0.09500122	0.58995418	1	−65.27693653	0.10703125	0.60559357
2	−69.26168156	0.10319824	0.59357118	2	−62.85937929	0.12384644	0.62201631
3	−69.91973495	0.10083618	0.58594327	3	−62.96594906	0.12827148	0.62855407
4	−75.51172256	0.08407953	0.55715312	4	−66.03942485	0.12130127	0.61952704
5	−103.46078300	0.05180206	0.49917424	5	−71.73997498	0.11066284	0.60979204
	M = 33				*M* = 32		
1	−70.60967541	0.09497070	0.58985167	1	−67.37020397	0.09610596	0.59045212
2	−68.16726971	0.10585937	0.59743846	2	−63.93104696	0.11263428	0.60560235
3	−67.13149548	0.10937500	0.59911696	3	−62.49787903	0.11931763	0.61192546
5	−66.53917217	0.10965576	0.59674101	5	−61.28204536	0.12541504	0.61824023
7	−67.23387909	0.10902100	0.59417456	7	−60.82049131	0.12907715	0.62307031
9	−67.85412312	0.10502930	0.58771575	9	−59.74928167	0.12068481	0.60685586
11	−69.08597469	0.10219727	0.58216391	11	−62.48683357	0.13004150	0.62821502
13	−75.86953640	0.08137207	0.54712777	13	−70.64571857	0.11017914	0.60670943
14	−104.04059029	0.05029373	0.49149549				
	M = 65				*M* = 64		
1	−70.66014957	0.09472656	0.58945943	1	−70.26372528	0.09376831	0.58789222
2	−68.89622307	0.10404663	059.476127	2	−67.20729542	0.10411987	0.59421778
3	−67.90234470	0.10720215	0.59577449	3	−65.80684280	0.10850220	0.59666158
4	−67.24003792	0.10726929	0.59415763	4	−64.95227051	0.11038818	0.59730067
5	−66.86065960	0.10689087	0.59253047	5	−64.42742348	0.11113281	0.59698496
9	−66.27561188	0.10548706	0.58845983	9	−63.41714096	0.10936890	0.59088884
13	−65.96417046	0.10466309	0.58660485	13	−62.72142410	0.10828857	0.58738641
17	−66.16404629	0.10649414	0.58862042	17	−62.37051868	0.11031494	0.58968142
21	−66.76456833	0.10701904	0.58894575	21	−62.04848146	0.11254273	0.59249461
25	−68.13407993	0.10327148	0.58320831	25	−61.88074064	0.11994629	0.60564501
29	−75.98313046	0.08069458	0.54500379	29	−70.05681992	0.10717773	0.59842159
30	−104.92083740	0.04978485	0.48965181				
	M = 125				*M* = 128		
1	−70.68010235	0.09464722	0.58933268	1	−70.58992958	0.09445190	0.58900996
2	−68.94157696	0.10390015	0.59450024	2	−68.62421608	0.10349731	0.59379058
3	−68.19352627	0.10682373	0.59508549	3	−67.66701698	0.10701294	0.59506081
5	−67.34261131	0.10668945	0.59187505	4	−66.95196629	0.10685425	0.59298926
7	−67.09767151	0.10587158	0.59821869	6	−66.32718945	0.10596924	0.58953845
9	−67.05801296	0.10523682	0.58738706	9	−66.01315498	0.10471191	0.58593906
17	−67.17504501	0.10372925	0.58358265	17	−65.89422417	0.10288086	0.58097354
25	−67.22918987	0.10316772	0.58224835	25	−65.92644215	0.10182495	0.57812308
33	−67.11609936	0.10303955	0.58198956	33	−65.95577812	0.10096436	0.57576437
41	−66.71271324	0.10313721	0.58245499	41	−65.97698021	0.10094604	0.57451694
49	−66.62364197	0.10561523	0.58629534	49	−65.67919827	0.09865112	0.56927420
57	−69.28378487	0.10061646	0.57812192	57	−64.61514568	0.09845581	0.56604486
58	−70.35782337	0.09663696	0.57121235	61	−71.76589394	0.10496826	0.59452277
59	−75.94700718	0.08054886	0.54451285				
60	−104.09012318	0.04991760	0.48963264				

Source: Rabiner et al. (1970); © 1970 IEEE; reprinted with permission.

TABLE 5.3 Transition Coefficients for $\alpha = \frac{1}{2}$

BW	Minimax	T_1
	$M = 16$	
1	−51.60668707	0.26674805
2	−47.48000240	0.32149048
3	−45.19746828	0.34810181
4	−44.32862616	0.36308594
5	−45.68347692	0.36661987
6	−56.63700199	0.34327393
	$M = 32$	
1	−52.64991188	0.26073609
2	−49.39390278	0.30878296
3	−47.72596645	0.32984619
4	−46.68811989	0.34217529
6	−45.33436489	0.35704956
8	−44.30730963	0.36750488
10	−43.11168003	0.37810669
12	−42.97900438	0.38465576
14	−56.32780266	0.35030518
	$M = 64$	
1	−52.90375662	0.25923462
2	−49.74046421	0.30603638
3	−48.38088989	0.32510986
4	−47.47863007	0.33595581
5	−46.88655186	0.34287720
6	−46.46230555	0.34774170
10	−45.46141434	0.35859375
14	−44.85988188	0.36470337
18	−44.34302616	0.36983643
22	−43.69835377	0.37586059
26	−42.45641375	0.38624268
30	−56.25024033	0.35200195
	$M = 128$	
1	−52.96778202	0.25885620
2	−49.82771969	0.30534668
3	−48.51341629	0.32404785
4	−47.67455149	0.33443604
5	−47.11462021	0.34100952
7	−46.43420267	0.34880371
10	−45.88529110	0.35493774
18	−45.21660566	0.36182251
26	−44.87959814	0.36521607
34	−44.61497784	0.36784058
42	−44.32706451	0.37066040
50	−43.87646437	0.37500000
58	−42.30969715	0.38807373
62	−56.23294735	0.35241699

Source: Rabiner et al. (1970); © 1970 IEEE; reprinted with permission.

TABLE 5.4 Transition Coefficients for $\alpha = \frac{1}{2}$

BW	Minimax	T_1	T_2
	$M = 16$		
1	-77.26126766	0.05309448	0.41784180
2	-73.81026745	0.07175293	0.49369211
3	-73.02352142	0.07862549	0.51966134
4	-77.95156193	0.07042847	0.51158076
5	-105.23953247	0.04587402	0.46967784
	$M = 32$		
1	-80.49464130	0.04725342	0.40357383
2	-73.92513466	0.07094727	0.49129255
3	-72.40863037	0.08012695	0.52153983
5	-70.95047379	0.08935547	0.54805908
7	-70.22383976	0.09403687	0.56031410
9	-69.94402790	0.09628906	0.56637987
11	-70.82423878	0.09323731	0.56226952
13	-104.85642624	0.04882812	0.48479068
	$M = 64$		
1	-80.80974960	0.04658203	0.40168723
2	-75.11772251	0.06759644	0.48390015
3	-72.66662025	0.07886963	0.51850058
4	-71.85610867	0.08393555	0.53379876
5	-71.34401417	0.08721924	0.54311474
9	-70.32861614	0.09371948	0.56020256
13	-69.34809303	0.09761963	0.56903714
17	-68.06440258	0.10051880	0.57543691
21	-67.99149132	0.10289307	0.58007699
25	-69.32065105	0.10068359	0.57729656
29	-105.72862339	0.04923706	0.48767025
	$M = 128$		
1	-80.89347839	0.04639893	0.40117195
2	-77.22580583	0.06295776	0.47399521
3	-73.43786240	0.07648926	0.51361278
4	-71.93675232	0.08345947	0.53266251
6	-71.10850430	0.08880615	0.54769675
9	-70.53600121	0.09255371	0.55752959
17	-69.95890045	0.09628906	0.56676912
25	-69.29977322	0.09834595	0.57137301
33	-68.75139713	0.10077515	0.57594641
41	-67.89687920	0.10183716	0.57863142
49	-66.76120186	0.10264282	0.58123560
57	-69.21525850	0.10157471	0.57946395
61	-104.57432938	0.04970703	0.48900685

Source: Rabiner et al. (1970); © 1970 IEEE; reprinted with permission.

As an illustration, the filter design in Example 5.2.2 for which $M = 15$ and

$$H_r\left(\frac{2\pi k}{M}\right) = \begin{cases} 1 & k = 0, 1, 2, 3 \\ 0.4 & k = 4 \\ 0 & k = 5, 6, 7 \end{cases}$$

corresponds to $\alpha = 0$, BW $= 4$, since $H_r(\omega_k) = 1$ at the four consecutive frequencies $\omega_k = 2\pi k/15$, $k = 0, 1, 2, 3$, and the transition coefficient is 0.4 at the frequency $\omega_k = 8\pi/15$. The value given in Table 5.1 for $M = 15$ and BW $= 4$ is $T_1 = 0.40405884$, which we truncated to $T_1 = 0.4$ in Example 5.2.2. The maximum sidelobe is at -41.9 dB, according to Table 5.1.

PROBLEMS

5.1 Consider an LTI system with impulse response $h(n) = (\frac{1}{3})^{|n|}$
  (a) Determine and sketch the magnitude and phase response $|H(\omega)|$ and $\angle H(\omega)$, respectively.
  (b) Determine and sketch the magnitude and phase spectra for the input and output signals for the following inputs:

  (1) $x(n) = \cos\dfrac{3\pi n}{8}$, $-\infty < n < \infty$

  (2) $x(n) = \left\{\ldots, 1, 0, 0, 1, 1, 1, 0, 1, 1, 1, 0, 1, \ldots\right\}$
  $\phantom{(2) x(n) = \{\ldots, 1, 0, 0, 1,}\uparrow$

5.2 Determine and sketch the magnitude and phase response of the following systems:
  (a) $y(n) = \frac{1}{2}[x(n) + x(n-1)]$
  (b) $y(n) = \frac{1}{2}[x(n) - x(n-1)]$
  (c) $y(n) = \frac{1}{2}[x(n+1) - x(n-1)]$
  (d) $y(n) = \frac{1}{2}[x(n+1) + x(n-1)]$
  (e) $y(n) = \frac{1}{2}[x(n) + x(n-2)]$
  (f) $y(n) = \frac{1}{2}[x(n) - x(n-2)]$
  (g) $y(n) = \frac{1}{3}[x(n) + x(n-1) + x(n-2)]$
  (h) $y(n) = x(n) - x(n-8)$
  (i) $y(n) = 2x(n-1) - x(n-2)$
  (j) $y(n) = \frac{1}{4}[x(n) + x(n-1) + x(n-2) + x(n-3)]$
  (k) $y(n) = \frac{1}{8}[x(n) + 3x(n-1) + 3x(n-2) + x(n-3)]$
  (l) $y(n) = x(n-4)$
  (m) $y(n) = x(n+4)$
  (n) $y(n) = \frac{1}{4}[x(n) - 2x(n-1) + x(n-2)]$

5.3 Consider the digital filter shown in Fig. P5.3.

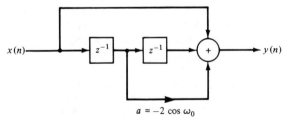

$a = -2\cos\omega_0$

FIGURE P5.3

  (a) Determine the input–output relation and the impulse response $h(n)$.
  (b) Determine and sketch the magnitude $|H(\omega)|$ and the phase response $\angle H(\omega)$ of the filter and find which frequencies are completely blocked by the filter.

(c) When $\omega_0 = \pi/2$, determine the output $y(n)$ to the input

$$x(n) = 3 \cos\left(\frac{\pi}{3}n + 30°\right) \qquad -\infty < n < \infty$$

5.4 Consider the FIR filter

$$y(n) = x(n) - x(n - 4)$$

(a) Compute and sketch its magnitude and phase response.
(b) Compute its response to the input

$$x(n) = \cos\frac{\pi}{2}n + \cos\frac{\pi}{4}n \qquad -\infty < n < \infty$$

(c) Explain the results obtained in part (b) in terms of the answer given in part (a).

5.5 Determine the steady-state response of the system

$$y(n) = \tfrac{1}{2}[x(n) - x(n - 2)]$$

to the input signal

$$x(n) = 5 + 3 \cos\left(\frac{\pi}{2}n + 60°\right) + 4 \sin(\pi n + 45°) \qquad -\infty < n < \infty$$

5.6 From our discussions it is apparent that an LTI system cannot produce frequencies at its output that are different from those applied in its input. Thus if a system creates "new" frequencies, it must be nonlinear and/or time varying. Verify these arguments by determining the outputs of the following system to the input signal

$$x(n) = A \cos \omega n$$

(a) $y(n) = x(2n)$
(b) $y(n) = x^2(n)$
(c) $y(n) = (\cos \pi n)x(n)$

5.7 Consider an LTI system with impulse response

$$h(n) = \left[\left(\frac{1}{4}\right)^n \cos\left(\frac{\pi}{4}n\right)\right]u(n)$$

(a) Determine its system function $H(z)$.
(b) Is it possible to implement this system using a finite number of adders, multipliers, and unit delays? If yes, how?
(c) Provide a rough sketch of $|H(\omega)|$ using the pole–zero plot.
(d) Determine the response of the system to the input

$$x(n) = (\tfrac{1}{4})^n u(n)$$

5.8 An FIR filter is described by the difference equation

$$y(n) = x(n) - x(n - 10)$$

(a) Compute and sketch its magnitude and phase response.
(b) Determine its response to the inputs

$$(1) \ x(n) = \cos\frac{\pi}{10}n + 3 \sin\left(\frac{\pi}{3}n + \frac{\pi}{10}\right) \qquad -\infty < n < \infty$$

$$(2) \ x(n) = 5 + 6 \cos\left(\frac{2\pi}{5}n + \frac{\pi}{2}\right) \qquad -\infty < n < \infty$$

5.9 The frequency response of an ideal bandpass filter is given by

$$H(\omega) = \begin{cases} 0 & |\omega| \leq \dfrac{\pi}{8} \\ 1 & \dfrac{\pi}{8} < |\omega| < \dfrac{3\pi}{8} \\ 0 & \dfrac{3\pi}{8} \leq |\omega| \leq \pi \end{cases}$$

(a) Determine its impulse response
(b) Show that this impulse response can be expressed as the product of $\cos(n\pi/4)$ and the impulse response of a lowpass filter.

5.10 Consider the system described by the difference equation

$$y(n) = \tfrac{1}{2}y(n-1) + x(n) + \tfrac{1}{2}x(n-1)$$

(a) Determine its impulse response.
(b) Determine its frequency response:
(1) From the impulse response
(2) From the difference equation
(c) Determine its response to the input

$$x(n) = \cos\left(\frac{\pi}{2}n + \frac{\pi}{4}\right) \qquad -\infty < n < \infty$$

5.11 Sketch roughly the magnitude $|X(\omega)|$ of the Fourier transforms corresponding to the pole–zero patterns given in Fig. P5.11.

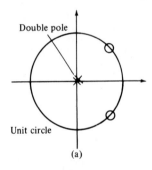

(a)

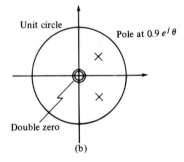

(b)

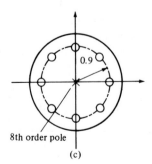

(c)

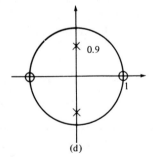

(d)

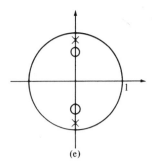

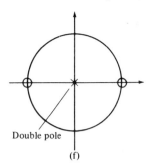

Double pole

(e)   (f)

FIGURE P5.11

5.12 Design an FIR filter that completely blocks the frequency $\omega_0 = \pi/4$ and then compute its output if the input is

$$x(n) = \left(\sin \frac{\pi}{4} n \right) u(n)$$

for $n = 0, 1, 2, 3, 4$. Does the filter fulfill your expectations? Explain.

5.13 A digital filter is characterized by the following properties:
    (1) It is high pass and has one pole and one zero.
    (2) The pole is at a distance $r = 0.9$ from the origin of the z-plane.
    (3) Constant signals do not pass through the system.
    (a) Plot the pole–zero pattern of the filter and determine it system function $H(z)$.
    (b) Compute the magnitude response $|H(\omega)|$ and the phase response $\angle H(\omega)$ of the filter.
    (c) Normalize the frequency response $H(\omega)$ so that $|H(\pi)| = 1$.
    (d) Determine the input–output relation (difference equation) of the filter in the time domain.
    (e) Compute the output of the system if the input is

$$x(n) = 2 \cos \left(\frac{\pi}{6} n + 45° \right) \qquad -\infty < n < \infty$$

(You can use either algebraic or geometrical arguments.)

5.14 A causal first-order digital filter is described by the system function

$$H(z) = G \frac{1 + bz^{-1}}{1 + az^{-1}}$$

    (a) Sketch the direct form I and direct form II realizations of this filter and find the corresponding difference equations.
    (b) For $a = 0.5$ and $b = -0.6$, sketch the pole–zero pattern. Is the system stable? Why?
    (c) For $a = -0.5$ and $b = 0.5$, determine G, so that the maximum value of $|H(\omega)|$ is equal to 1.
    (d) Sketch the magnitude response $|H(\omega)|$ and the phase response $\angle H(\omega)$ of the filter obtained in part (c).
    (e) In a specific application it is known that $a = 0.8$. Does the resulting filter amplify high frequencies or low frequencies in the input? Choose the value of b so as to improve the characteristics of this filter (i.e., make it a better lowpass or a better highpass filter).

5.15 Show geometrically that the system specified by the pole–zero plot in Fig. P5.15 is all-pass (i.e., $|H(\omega)|$ = constant).

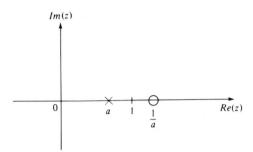

FIGURE P5.15

5.16 Determine the coefficients of a linear-phase FIR filter

$$y(n) = b_0 x(n) + b_1 x(n - 1) + b_2 x(n - 2)$$

such that:
(a) It rejects completely any frequency component at $\omega_0 = 2\pi/3$.
(b) Its frequency response is normalized so that $H(0) = 1$.
(c) Compute and sketch the magnitude and phase response of the filter to check if it satisfies the requirements.

5.17 Determine the frequency response $H(\omega)$ of the following moving average filters.

(a) $y(n) = \dfrac{1}{2M + 1} \displaystyle\sum_{k=-M}^{M} x(n - k)$

(b) $y(n) = \dfrac{1}{4M} x(n + M) + \dfrac{1}{2M} \displaystyle\sum_{k=-M+1}^{M-1} x(n - k) + \dfrac{1}{4M} x(n - M)$

Which filter provides better smoothing? Why?

5.18 The convolution $x(t)$ of two continuous-time signals $x_1(t)$ and $x_2(t)$, from which at least one is nonperiodic, is defined by

$$x(t) \triangleq x_1(t) * x_2(t) \triangleq \int_{-\infty}^{\infty} x_1(\lambda) x_2(t - \lambda) \, d\lambda$$

(a) Show that $X(F) = X_1(F)X_2(F)$, where $X_1(F)$ and $X_2(F)$ are the spectra of $x_1(t)$ and $x_2(t)$, respectively.
(b) Compute $x(t)$ if $x_1(t) = x_2(t) = \begin{cases} 1 & |t| < \tau/2 \\ 0 & \text{elsewhere} \end{cases}$
(c) Determine the spectrum of $x(t)$ using the results in part (a).

5.19 Compute the magnitude and phase response of a filter with system function

$$H(z) = 1 + z^{-1} + z^{-2} + \cdots + z^{-8}$$

If the sampling frequency is $F_s = 1$ kHz, determine the frequency of the analog sinusoids that cannot pass through the filter.

5.20 A second-order system has a double pole at $p_{1,2} = 0.5$ and two zeros at

$$z_{1,2} = e^{\pm j3\pi/4}$$

Using geometric arguments, choose the gain G of the filter so that $|H(0)| = 1$.

5.21 Compute the 3-dB bandwidth of the filters ($0 < a < 1$)

$$H_1(z) = \frac{1 - a}{1 - az^{-1}}$$

$$H_2(z) = \frac{1 - a}{2} \frac{1 + z^{-1}}{1 - az^{-1}}$$

Which is a better lowpass filter?

5.22 Design a digital oscillator with adjustable phase, that is, a digital filter which produces the signal

$$y(n) = \cos(\omega_0 n + \theta) u(n)$$

5.23 This problem provides another derivation of the structure for the coupled-form oscillator by considering the system

$$y(n) = ay(n - 1) + x(n)$$

for $a = e^{j\omega_0}$.

Let $x(n)$ be real. Then $y(n)$ is complex. Thus

$$y(n) = y_R(n) + jy_I(n)$$

(a) Determine the equations describing a system with one input $x(n)$ and the two outputs $y_R(n)$ and $y_I(n)$.

(b) Determine a block diagram realization

(c) Show that if $x(n) = \delta(n)$, then

$$y_R(n) = (\cos \omega_0 n)u(n)$$
$$y_I(n) = (\sin \omega_0 n)u(n)$$

(d) Compute $y_R(n)$, $y_I(n)$, $n = 0, 1, \ldots, 9$ for $\omega_0 = \pi/6$. Compare these with the true values of the sine and cosine.

5.24 Consider a filter with system function

$$H(z) = \frac{(1 - e^{j\omega_0}z^{-1})(1 - e^{-j\omega_0}z^{-1})}{(1 - re^{j\omega_0}z^{-1})(1 - re^{-j\omega_0}z^{-1})}$$

(a) Sketch the pole–zero pattern.

(b) Using geometric arguments, show that for $r \simeq 1$, the system is a notch filter and provide a rough sketch of its magnitude response if $\omega_0 = 60°$.

(c) For $\omega_0 = 60°$, choose G so that the maximum value of $|H(\omega)|$ is 1.

(d) Draw a direct form II realization of the system

(e) Determine the approximate 3-dB bandwidth of the system.

5.25 Design an FIR digital filter that will reject a very strong 60-Hz sinusoidal interference contaminating a 200-Hz useful sinusoidal signal. Determine the gain of the filter so that the useful signal does not change amplitude. The filter works at a sampling frequency $F_s = 500$ samples/s. Compute the output of the filter if the input is a 60-Hz sinusoid or a 200-Hz sinusoid with unit amplitude. How does the performance of the filter compare with your requirements?

5.26 Determine the gain G for the digital resonator described by (5.5.16) so that $|H(\omega_0)| = 1$.

5.27 Demonstrate that the difference equation given in (5.5.40) can be obtained by applying the trigonometric identity

$$\cos \alpha + \cos \beta = 2 \cos \frac{\alpha + \beta}{2} \cos \frac{\alpha - \beta}{2}$$

where $\alpha = (n + 1)\omega_0$, $\beta = (n - 1)\omega_0$, and $y(n) = \cos \omega_0 n$. Thus show that the sinusoidal signal $y(n) = A \cos \omega_0 n$ can be generated from (5.5.40) by use of the initial conditions $y(-1) = A \cos \omega_0$ and $y(-2) = A \cos 2\omega_0$.

5.28 Use the trigonometric identity in (5.5.41) with $\alpha = n\omega_0$ and $\beta = (n - 2)\omega_0$ to derive the difference equation for generating the sinusoidal signal $y(n) = A \sin n\omega_0$. Determine the corresponding initial conditions.

5.29 Using the z-transform pairs 8 and 9 in Table 3.3, determine the difference equations for the digital oscillators that have impulse responses $h(n) = A \cos n\omega_0 \, u(n)$ and $h(n) = A \sin n\omega_0 \, u(n)$, respectively.

5.30 Determine the structure for the coupled-form oscillator by combining the structure for the digital oscillators obtained in Problem 5.29.

5.31 Convert the highpass filter with system function

$$H(z) = \frac{1 - z^{-1}}{1 - az^{-1}} \qquad a < 1$$

into a notch filter that rejects the frequency $\omega_0 = \pi/4$ and its harmonics.
(a) Determine the difference equation.
(b) Sketch the pole–zero pattern.
(c) Sketch the magnitude response for both filters.

5.32 Choose L and M for a lunar filter that must have narrow passbands at $(k \pm \Delta F)$ cycles/day, where $k = 1, 2, 3, \ldots$ and $\Delta F = 0.067726$.

5.33 (a) Show that the system corresponding to the pole–zero pattern of Fig. 5.37 is all-pass.
(b) What is the number of delays and multipliers required for the efficient implementation of a second-order all-pass system?

5.34 A digital notch filter is required to remove an undesirable 60-Hz hum associated with a power supply in an ECG recording application. The sampling frequency used is $F_s = 500$ samples/s. (a) Design a second-order FIR notch filter and (b) a second-order pole–zero notch filter for this purpose. In both cases choose the gain G so that $|H(\omega)| = 1$ for $\omega = 0$.

5.35 Determine the coefficients $\{h(n)\}$ of a highpass linear phase FIR filter of length $M = 4$ which has an antisymmetric unit sample response and a frequency response that satisfies the condition

$$H_r\left(\frac{\pi}{4}\right) = \frac{1}{2} \qquad H_r\left(\frac{3\pi}{4}\right) = 1$$

5.36 In an attempt to design a four-pole bandpass digital filter with desired magnitude response

$$|H_d(\omega)| = \begin{cases} 1 & \frac{\pi}{6} \le \omega \le \frac{\pi}{2} \\ 0 & \text{elsewhere} \end{cases}$$

we select the four poles at

$$p_{1,2} = 0.8e^{\pm j2\pi/9}$$
$$p_{3,4} = 0.8e^{\pm j4\pi/9}$$

and four zeros at

$$z_1 = 1 \qquad z_2 = -1 \qquad z_{3,4} = e^{\pm 3\pi/4}$$

(a) Determine the value of the gain so that

$$\left| H\left(\frac{5\pi}{12}\right) \right| = 1$$

(b) Determine the system function $H(z)$.
(c) Determine the magnitude of the frequency response $H(\omega)$ for $0 \le \omega \le \pi$ and compare it with the desired response $|H_d(\omega)|$.

5.37 Let $x_1(t)$ and $x_2(t)$ be two periodic signals with the same period T_p. It can easily be seen that their convolution does not converge. In such cases it is useful to consider the so-called *periodic convolution* defined by

$$y(t) \triangleq \int_{T_p} x_1(\lambda)x_2(t - \lambda) \, d\lambda \triangleq x_1(t) \circledast x_2(t)$$

where we *integrate always exactly over one period.*
(a) Show that $y(t)$ is also periodic with period T_p.
(b) Compute $y(t)$ if $x_1(t) = x_2(t) = x(t)$, where $x(t)$ is shown in Fig. P5.37.

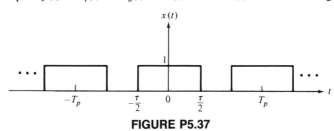

FIGURE P5.37

(c) If a_k, b_k, and c_k are the Fourier coefficients for $x_1(t)$, $x_2(t)$, and $y(t)$, respectively, show that

$$c_k = T_p a_k b_k$$

5.38 The ideal analog differentiator is described by

$$y_a(t) = \frac{dx_a(t)}{dt}$$

where $x_a(t)$ is the input and $y_a(t)$ the output signal.
(a) Determine its frequency response by exciting the system with the input $x_a(t) = e^{j2\pi Ft}$.
(b) Sketch the magnitude and phase response of an ideal analog differentiator bandlimited to B hertz.
(c) The ideal digital differentiator is defined as

$$H(\omega) = j\omega \qquad |\omega| \le \pi$$

Justify this definition by comparing the frequency response $|H(\omega)|$, $\angle H(\omega)$ with that in part (b).

(d) By computing the frequency response $H(\omega)$, show that the discrete-time system

$$y(n) = x(n) - x(n - 1)$$

is a good approximation to a differentiator at low frequencies.

(e) Compute the response of the system to the input

$$x(n) = A \cos (\omega_0 n + \theta)$$

5.39 A discrete-time system with input $x(n)$ and output $y(n)$ is described in the frequency domain by the relation

$$Y(\omega) = e^{-j2\pi\omega}X(\omega) + \frac{dX(\omega)}{d\omega}$$

(a) Compute the response of the system to the input $x(n) = \delta(n)$.

(b) Check if the system is LTI and stable.

5.40 Consider an ideal lowpass filter with impulse response $h(n)$ and frequency response

$$H(\omega) = \begin{cases} 1 & |\omega| \leq \omega_c \\ 0 & \omega_c < |\omega| < \pi \end{cases}$$

What is the frequency response of the filter defined by

$$g(n) = \begin{cases} h\left(\dfrac{n}{2}\right) & n \text{ even} \\ 0 & n \text{ odd} \end{cases}$$

5.41 Consider the system shown in Fig. P5.41. Determine its impulse response and its frequency response if the system $H(\omega)$ is:

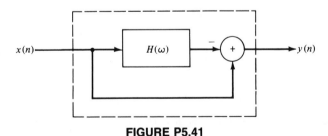

FIGURE P5.41

(a) Lowpass with cutoff frequency ω_c.

(b) Highpass with cutoff frequency ω_c.

5.42 Frequency inverters have been used for many years for speech scrambling. Indeed, a voice signal $x(n)$ becomes unintelligible if we invert its spectrum as shown in Fig. P5.42.

(a) Determine how frequency inversion can be performed in the time domain.

(b) Design an unscrambler.

(*Hint:* The required operations are very simple and can easily be done in real time.)

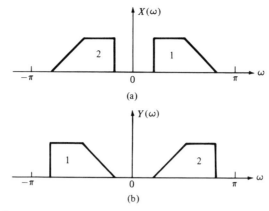

FIGURE P5.42   (a) Original spectrum; (b) frequency-inverted spectrum.

5.43 A lowpass filter is described by the difference equation

$$y(n) = 0.9y(n - 1) + 0.1x(n)$$

(a) By performing a frequency translation of $\pi/2$, transform the filter into a bandpass filter.

(b) What is the impulse response of the bandpass filter?

(c) What is the major problem with the frequency translation method for transforming a prototype lowpass filter into a bandpass filter?

5.44 Consider a system with a real-valued impulse response $h(n)$ and frequency response

$$H(\omega) = |H(\omega)|e^{j\theta(\omega)}$$

The quantity

$$D = \sum_{n=-\infty}^{\infty} n^2 h^2(n)$$

provides a measure of the "effective duration" of $h(n)$.

(a) Express D in terms of $H(\omega)$.

(b) Show that D is minimized for $\theta(\omega) = 0$.

5.45 The impulse response $h(n)$ of a causal system is real-valued and the real part of its frequency response $H_R(\omega)$ is given by

$$H_R(\omega) = 1 + a \cos 2\omega$$

where a is real-valued.

(a) Determine $h(n)$ and $H(\omega)$.

(b) Show that $h(n)$ can be completely recovered from $H_I(\omega)$ and $h(0)$.

5.46 Consider the lowpass filter

$$y(n) = ay(n - 1) + bx(n) \qquad 0 < a < 1$$

(a) Determine b so that $|H(0)| = 1$.

(b) Determine the 3-dB bandwidth ω_3 for the normalized filter in part (a).

(c) How does the choice of the parameter a affect ω_3?

(d) Repeat parts (a) through (c) for the highpass filter obtained by choosing $-1 < a < 0$.

5.47 Sketch the magnitude and phase response of the multipath channel

$$y(n) = x(n) + \alpha x(n - M) \qquad \alpha > 0$$

for $\alpha \ll 1$.

5.48 Determine and sketch the magnitude and phase response of the systems shown in Fig. P5.48(a) through (c).

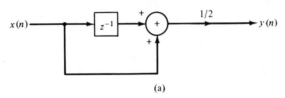

(a)

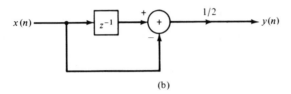

(b)

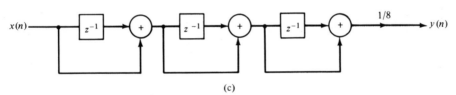

(c)

FIGURE P5.48

5.49 The signals $x(n)$ and $h(n)$ are nonzero in the ranges $0 \le n \le 7$ and $0 \le n \le 19$, respectively. The 20-point DFTs of each of the sequences are multiplied and the inverse DFT is computed. Let $s(n)$ denote the inverse DFT. Specify which points in $s(n)$ correspond to the output of an LTI system with input $x(n)$ and impulse response $h(n)$.

5.50 Determine and sketch the impulse response and the magnitude and phase responses of the FIR filter shown in Fig. P5.50 for $b = 1$ and $b = -1$.

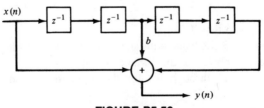

FIGURE P5.50

5.51 The causal system

$$H(z) = \frac{1}{1 + \sum\limits_{k=1}^{N} a_k z^{-k}}$$

is known to be unstable.

We modify this system by changing its impulse response $h(n)$ to

$$h'(n) = \lambda^n h(n) u(n)$$

(a) Show that by properly choosing λ we can obtain a new stable system.

(b) What is the difference equation describing the new system?

COMPUTER EXPERIMENTS

5.52 Given a signal $x(n)$, we can create echoes and reverberations by delaying and scaling the signal as follows

$$y(n) = \sum\limits_{k=0}^{\infty} g_k x(n - kD)$$

where D is positive integer and $g_k > g_{k-1} > 0$.

(a) Explain why the comb filter

$$H(z) = \frac{1}{1 - az^{-D}}$$

can be used as a reverberator (i.e., as a device to produce artificial reverberations).

(*Hint:* Determine and sketch its impulse response.)

(b) The all-pass comb filter

$$H(z) = \frac{z^{-D} - a}{1 - az^{-D}}$$

is used in practice to build digital reverberators by cascading three to five such filters and properly choosing the parameters a and D. Compute and plot the impulse response of two such reverberators each obtained by cascading three sections with the following parameters.

	Unit 1			Unit 2	
Section	D	a	Section	D	a
1	50	0.7	1	50	0.7
2	40	0.665	2	17	0.77
3	32	0.63175	3	6	0.847

(c) The difference between echo and reverberation is that with pure echo there are clear repetitions of the signal, but with reverberations, there are not. How is this reflected in the shape of the impulse response of the reverberator? Which unit in part (b) is a better reverberator?

(d) If the delays D_1, D_2, D_3 in a certain unit are prime numbers, the impulse response of the unit is more "dense." Explain why.

(e) Plot the phase response of units 1 and 2 and comment on them.

(f) Plot $h(n)$ for D_1, D_2, and D_3 being nonprime. What do you notice?

More details about this application can be found in a paper by J. A. Moorer, "Signal Processing Aspects of Computer Music: A Survey," *Proc, IEEE*, vol. 65, No. 8, Aug. 1977, pp. 1108–1137.

5.53 By trial-and-error design a third-order lowpass filter with cutoff frequency at $\omega_c = \pi/9$ radians/sample interval. Start your search with

(a) $z_1 = z_2 = z_3 = 0$, $p_1 = r$, $p_{2,3} = re^{\pm j\omega_c}$, $r = 0.8$

(b) $r = 0.9$, $z_1 = z_2 = z_3 = -1$

5.54 A speech signal with bandwidth $B = 10$ kHz is sampled at $F_2 = 20$ kHz. Suppose that the signal is corrupted by four sinusoids with frequencies

$$F_1 = 10{,}000 \text{ Hz} \qquad F_3 = 7778 \text{ Hz}$$
$$F_2 = 8889 \text{ Hz} \qquad F_4 = 6667 \text{ Hz}$$

(a) Design a FIR filter that eliminates these frequency components.

(b) Choose the gain of the filter so that $|H(0)| = 1$ and then plot the log magnitude response and the phase response of the filter.

(c) Does the filter fulfill your objectives? Do you recommend the use of this filter in a practical application?

5.55 Compute and sketch the frequency response of a digital resonator with $\omega = \pi/6$ and $r = 0.6, 0.9, 0.99$. In each case compute the bandwidth and the resonance frequency from the graph and check if they are in agreement with the theoretical results.

5.56 Consider the discrete-time system

$$y(n) = ay(n - 1) + (1 - a)x(n) \qquad n \geq 0$$

where $a = 0.9$ and $y(-1) = 0$.

(a) Compute and sketch the output $y_i(n)$ of the system to the input signals

$$x_i(n) = \sin 2\pi f_i n \qquad 0 \leq n \leq 100$$

where $f_1 = \frac{1}{4}$, $f_2 = \frac{1}{5}$, $f_3 = \frac{1}{10}$, $f_4 = \frac{1}{2}$.

(b) Compute and sketch the magnitude and phase response of the system and use these results to explain the response of the system to the signals given in part (a).

Modulation, Inverse Systems, and Deconvolution

As indicated in Chapter 1, digital signal processing has found widespread use in a number of fields, including communications, geophysics, and in diagnostic medicine. The applications of digital signal processing techniques in these fields is extremely large, and any attempt to cover even a few is an ambitious task.

In this chapter we develop some basic mathematical relationships and formulas that are useful in the application of digital signal processing to communications and geophysics. In particular, we treat the topic of modulation and demodulation of signals, in both continuous time and discrete time. Thus we develop the mathematical tools needed to describe the basic building blocks in the implementation of a modem for transmitting data at a high rate over telephone channels. Our treatment of amplitude modulation includes the representation of bandpass signals and systems and leads us to the development of the sampling theorem for bandpass signals.

In addition to our discussion of modulation and demodulation, we also treat the major topics of deconvolution, inverse systems, and system identification. These three basic topics find many practical applications in a variety of fields, most notably in geophysics (e.g., oil exploration) and in communications (e.g., channel equalization and echo cancellation).

The other major topic treated in this chapter is signal reconstruction based on suboptimum albeit practical interpolation methods. In particular, we describe the zero-order hold (staircase interpolation), the first-order hold (linear interpolation), and linear interpolation with delay.

6.1 Bandpass Signals, Systems, and Amplitude Modulation

In Section 5.3 we defined a bandpass filter or a bandpass system as a system that passes frequency components of a signal in some narrow frequency band above zero frequency. By direct analogy, we define a bandpass signal as a signal with frequency content concentrated in a narrow band of frequencies above zero frequency. Bandpass signals and systems arise frequently in practice, most notably in communications where information-bearing signals are transmitted over channels such as satellites, telephone channels, radio channels, and fiber optic cable.

In this section we describe the characteristics of bandpass signals and systems and apply our results to the important problem of transmitting an information-bearing signal by amplitude modulation over a bandpass channel. In digital communication systems, the amplitude-modulated signal is often synthesized digitally, converted to analog form for transmission over the communication channel and then converted back to digital form at the receiving end for digital processing. This processing, called *demodulation*, involves the extraction of the information-bearing signal from the amplitude-modulated signal.

The modulator–demodulator pair in the transmitter and the receiver is usually called a *modem*. It is often implemented digitally in practice. Such a digital implementation for amplitude-modulated signals is described in this section. The final topic of this section is a treatment of sampling of bandpass signals.

6.1.1 Representation of Bandpass Signals and Systems

An analog signal $x(t)$ with frequency content concentrated in a narrow band of frequencies around some frequency F_c is called a *bandpass signal* and is represented in general as

$$x(t) = A(t) \cos [2\pi F_c t + \phi(t)] \tag{6.1.1}$$

where $A(t)$ is called the *amplitude* or *envelope* of the signal and $\phi(t)$ is called the *phase* of the signal. The frequency F_c may be selected to be any frequency within the frequency band occupied by the signal. Usually, it is the frequency at the center of the frequency band and, in amplitude modulation, it is called the *carrier frequency*, as we shall observe below.

By expanding the cosine function in (6.1.1), we obtain a second representation of the bandpass signal, namely,

$$\begin{aligned} x(t) &= A(t) \cos \phi(t) \cos 2\pi F_c t - A(t) \sin \phi(t) \sin 2\pi F_c t \\ &= u_c(t) \cos 2\pi F_c t - u_s(t) \sin 2\pi F_c t \end{aligned} \tag{6.1.2}$$

where, by definition,

$$\begin{aligned} u_c(t) &= A(t) \cos \phi(t) \\ u_s(t) &= A(t) \sin \phi(t) \end{aligned} \tag{6.1.3}$$

Since the sinusoids $\cos 2\pi F_c t$ and $\sin 2\pi F_c t$ differ by $90°$ and are usually represented as two rotating phasors, as shown in Fig. 6.1, we say that they are in phase quadrature (i.e., they are perpendicular). In turn, the signal components $u_c(t)$ and $u_s(t)$ that multiply $\cos 2\pi F_c t$ and $\sin 2\pi F_c t$, respectively, are called the *in-phase* and *quadrature components* or, simply, the quadrature components of the signal $x(t)$.

Finally, a third representation for the bandpass signal $x(t)$ is obtained by defining the *complex envelope*

$$u(t) = u_c(t) + ju_s(t) \tag{6.1.4}$$

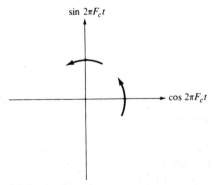

FIGURE 6.1 Sine and cosine signals in phase quadrature.

so that

$$x(t) = \text{Re } [u(t)e^{j2\pi F_c t}] \tag{6.1.5}$$

where Re $[q]$ denotes the real part of the quantity q.

In the frequency domain, the signal $x(t)$ is represented by its Fourier transform. That is,

$$X(F) = \int_{-\infty}^{\infty} x(t)e^{-j2\pi Ft} \, dt \tag{6.1.6}$$

If we substitute from (6.1.5) into (6.1.6) for $x(t)$ and use the identity

$$\text{Re } [q] = \tfrac{1}{2}(q + q^*) \tag{6.1.7}$$

we obtain

$$\begin{aligned}X(F) &= \tfrac{1}{2} \int_{-\infty}^{\infty} \left[u(t)e^{j2\pi F_c t} + u^*(t)e^{-j2\pi F_c t} \right] e^{-j2\pi Ft} \, dt \\[2mm]&= \tfrac{1}{2} \int_{-\infty}^{\infty} u(t)e^{-j2\pi(F - F_c)t} \, dt + \tfrac{1}{2} \int_{-\infty}^{\infty} u^*(t)e^{-j2\pi(F + F_c)t} \, dt\end{aligned} \tag{6.1.8}$$

Now, if $U(F)$ denotes the Fourier transform of the complex signal $u(t)$, then (6.1.8) may be expressed as

$$X(F) = \tfrac{1}{2}[U(F - F_c) + U^*(-F - F_c)] \tag{6.1.9}$$

It is apparent from (6.1.9) that the spectrum of the bandpass signal $x(t)$ can be obtained from the spectrum of the complex signal $u(t)$ by a frequency translation. To be more precise, suppose that the spectrum of the signal $u(t)$ is as shown in Fig. 6.2a. Then the spectrum of $X(F)$ for positive frequencies is simply $U(F)$ translated in frequency to the right by F_c and scaled in amplitude by $\tfrac{1}{2}$. The spectrum of $X(F)$ for negative frequencies is obtained by first folding $U(F)$ about $F = 0$ to obtain $U(-F)$, conjugating $U(-F)$ to obtain $U^*(-F)$, translating $U^*(-F)$ in frequency to the left by F_c and scaling the result by $\tfrac{1}{2}$. The folding and conjugation of $U(F)$ for the negative-frequency component of the spectrum result in a magnitude spectrum $|X(F)|$ that is even and a phase spectrum $\angle X(F)$ that is odd. These symmetry properties must hold since the signal $x(t)$ is real valued. However, they do not apply to the spectrum of the complex signal $u(t)$.

The development above implies that *any bandpass signal $x(t)$ can be represented by an equivalent lowpass signal $u(t)$*. In general, the equivalent lowpass signal $u(t)$ is complex valued, whereas the bandpass signal $x(t)$ is real. The latter can be obtained from the former through the time-domain relation in (6.1.5) or through the frequency-domain relation in (6.1.9).

The representation of a bandpass signal in terms of an equivalent lowpass signal conveniently carries over to analog bandpass filters. To be specific, suppose that we have a bandpass filter with a frequency response characteristic $H(F)$, as illustrated in Fig. 6.3a. Clearly, we can define an equivalent lowpass filter with frequency response $C(F)$, as illustrated in Fig. 6.3b, such that $H(F)$ is related to $C(F)$ through a frequency translation. Specifically,

$$H(F) = C(F - F_c) + C^*(-F - F_c) \tag{6.1.10}$$

This expression resembles (6.1.9) except for the scale factor of $\tfrac{1}{2}$ that appears in (6.1.9). We deliberately defined $C(F)$ without this scale factor for a reason that will become apparent below.

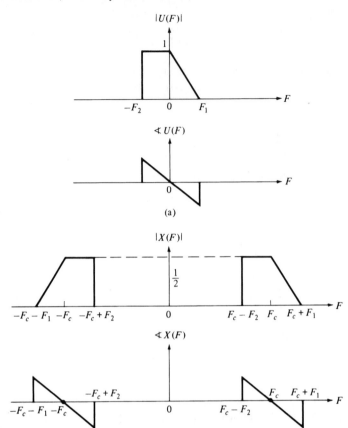

FIGURE 6.2 (a) Spectrum of the lowpass signal and (b) the corresponding spectrum for the bandpass signal.

Now, the inverse transform of $H(F)$ yields the impulse response $h(t)$ of the bandpass filter. If $c(t)$ denotes the inverse Fourier transform of $C(F)$, then

$$c(t)e^{j2\pi F_c t} \xleftarrow{\quad F \quad} C(F - F_c)$$

and

$$c^*(t)e^{-j2\pi F_c t} \xleftarrow{\quad F \quad} C^*(-F - F_c)$$

Consequently, the time-domain relation corresponding to (6.1.10) has the form

$$\begin{aligned} h(t) &= c(t)e^{j2\pi F_c t} + c^*(t)e^{-j2\pi F_c t} \\ &= 2\,\text{Re}\,[c(t)e^{j2\pi F_c t}] \end{aligned} \tag{6.1.11}$$

Thus the bandpass filter with impulse response $h(t)$ may be represented by an equivalent lowpass filter with impulse response $c(t)$, or $2c(t)$ if we wish to include the scale factor.

Let us now consider the response of a bandpass system with frequency response $H(F)$ to a bandpass input signal $x(t)$ having the spectrum $X(F)$. In terms of frequency-domain quantities, the spectrum of the filter output $Y(F)$ is simply the product of $H(F)$ with $X(F)$, that is,

$$Y(F) = H(F)X(F) \tag{6.1.12}$$

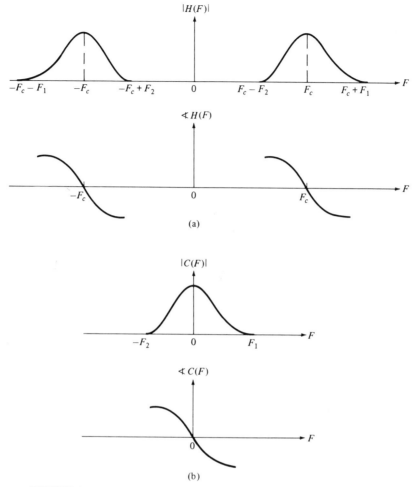

FIGURE 6.3   (a) Bandpass filter and (b) its equivalent lowpass filter.

The corresponding time-domain relation for the output $y(t)$ is given by the *convolution integral*

$$y(t) = \int_{-\infty}^{\infty} h(\tau)x(t - \tau) \, d\tau \qquad (6.1.13)$$

We observe that (6.1.12) and (6.1.13), which apply to continuous-time signals and systems, are akin to their discrete-time counterparts given by (5.2.2) and (5.2.1), respectively. In other words, the response of a linear time-invariant system with impulse response $h(t)$ to an input signal $x(t)$ is determined by convolving $h(t)$ with $x(t)$, where the convolution takes the form of an integral, because $h(t)$ and $x(t)$ are continuous-time functions. By computing the Fourier transform of (6.1.13), it is easy to show that (6.1.12) follows. Consequently, convolution in the time domain is equivalent to multiplying the corresponding spectra in the frequency domain, just as in the case of discrete-time signals and systems.

Now, if we use (6.1.9) and (6.1.10) to substitute for $X(F)$ and $H(F)$ in (6.1.12), we obtain

$$\begin{aligned} Y(F) &= [C(F - F_c) + C^*(-F - F_c)] [\tfrac{1}{2}U(F - F_c) + \tfrac{1}{2}U^*(-F - F_c)] \\ &= \tfrac{1}{2}[C(F - F_c) U(F - F_c) + C^*(-F - F_c)U^*(-F - F_c)] \end{aligned} \qquad (6.1.14)$$

where the cross terms $C(F - F_c)U^*(-F - F_c)$ and $C^*(-F - F_c)U(F - F_c)$ vanish because the spectra do not overlap. By defining an equivalent lowpass spectrum $V(F)$ as

$$V(F) = C(F)U(F) \qquad (6.1.15)$$

it follows that $Y(F)$ can be expressed as

$$Y(F) = \tfrac{1}{2}[V(F - F_c) + V^*(-F - F_c)] \qquad (6.1.16)$$

which is the desired result. In turn, the inverse Fourier transform of $V(F)$ is simply the convolution of the equivalent lowpass input signal $u(t)$ with the equivalent lowpass impulse response $c(t)$, that is,

$$v(t) = \int_{-\infty}^{\infty} c(\tau)u(t - \tau)\, d\tau \qquad (6.1.17)$$

where $v(t)$ is related to the real bandpass signal input $y(t)$ according to the formula

$$y(t) = \text{Re}\,[v(t)e^{j2\pi F_c t}] \qquad (6.1.18)$$

We note that the spectrum of the output signal given in (6.1.16) resembles the form of the spectrum in the input signal given by (6.1.9). Both expressions have the scale factor $\tfrac{1}{2}$, which is made possible as a result of our definition in (6.1.10). The results above indicate that it is possible to model bandpass signals and systems by their equivalent lowpass counterparts. Analysis and synthesis of such signals can be performed entirely at low frequencies and the desired bandpass signals and systems can easily be obtained from their equivalent lowpass counterparts by a simple frequency translation. Figure 6.4 summarizes these relationships.

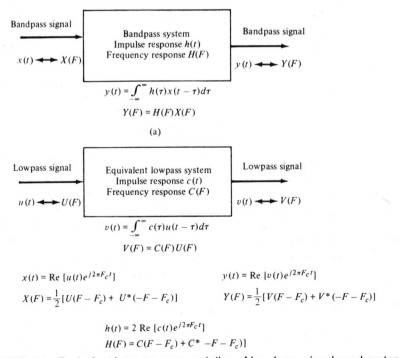

FIGURE 6.4   Equivalent lowpass representation of bandpass signals and systems.

6.1.2 Amplitude-Modulated Signals

In both analog and digital communications, information-bearing signals are usually translated in frequency and transmitted from the source to the destination via some physical
medium called the communication channel. The channel may be a pair of wires, or
coaxial cable, or a fiber optic cable, or a radio-frequency channel. The frequency translation is usually required to either match the signal spectrum to the frequency passband
of the channel and/or to allow many users simultaneously to use the available channel
bandwidth without interfering with one another.

Figure 6.5 illustrates the simultaneous transmission of 12 information-bearing signals,
as is usually done in communication of voice signals over telephone channels, for example. In this case, each signal occupies a bandwidth of approximately 3 kHz, in the
audio-frequency band, which defines the range of its frequency content. Each of the 12
signals is translated by a different amount (i.e., by a different frequency) and transmitted
over the same telephone channel (e.g., a coaxial cable). This technique of stacking a
number of signals in frequency so that their spectra do not overlap (i.e., they do not
interfere with one another) is called *frequency-division multiplexing* (FDM). In telephone
communications, each signal occupies a *voice-band channel,* and the 12 channels together
constitute a *group channel.* This represents the first level of multiplexing in a hierarchy
of wider-bandwidth FDM channels.

At the receiving end of the communication system (in the telephone system, the
receiving end is usually called a *central office* to which the receiving subscriber is
connected) the signals are separated by appropriately designed filters, translated in fre-

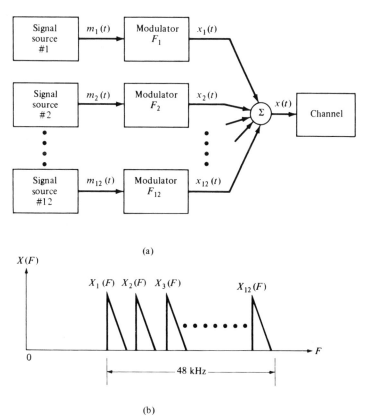

FIGURE 6.5   Frequency-division multiplexing of different information-bearing signals.

quency down to the audio frequency band (demodulated) and delivered to the user via a pair of wires from the central office to the subscriber telephone.

Frequency translation is easily accomplished by multiplying the information-bearing signal, denoted here by $m(t)$, with a sinusoid cos $2\pi F_c t$, generated by an oscillator, where F_c is the frequency of the sinusoid. The result of this multiplication is the signal

$$x(t) = m(t) \cos 2\pi F_c t \tag{6.1.19}$$

The sinusoid is usually called the *carrier signal* and $m(t)$ is called the *modulating signal*. Thus, through multiplication, the information-bearing signal $m(t)$ is impressed on the amplitude of the carrier and the resulting signal is called a *double-sideband suppressed carrier amplitude-modulated signal*. The device that performs the modulation, in this case the multiplication of the two signals, is called an amplitude modulator.

To illustrate that amplitude modulation simply involves a frequency translation of the spectrum of the signal, we express the sinusoid in (6.1.19) as a sum of two exponentials. Thus

$$x(t) = \tfrac{1}{2}m(t)e^{j2\pi F_c t} + \tfrac{1}{2}m(t)e^{-j2\pi F_c t} \tag{6.1.20}$$

The Fourier transform of $x(t)$ is

$$
\begin{aligned}
X(F) &= \int_{-\infty}^{\infty} x(t)e^{-j2\pi F t} \, dt \\
&= \tfrac{1}{2} \int_{-\infty}^{\infty} [m(t)e^{j2\pi F_c t} + m(t)e^{-j2\pi F_c t}]e^{-j2\pi F t} \, dt \\
&= \tfrac{1}{2}[M(F - F_c) + M(F + F_c)]
\end{aligned}
\tag{6.1.21}
$$

where $M(F)$ is the spectrum of $m(t)$.

Figure 6.6 illustrates the spectrum $M(F)$ of the lowpass signal and the spectrum $X(F)$ of the modulated signal. As we observe from Fig. 6.6, one of the characteristics of the modulated signal is that its spectrum for $F > 0$ covers the range $F_c - F_1 \le F \le F_c + F_1$. In other words, its bandwidth is $2F_1$, whereas the lowpass signal for $F \ge 0$ has frequency content in the band $0 \le F \le F_1$. Hence the frequency translation results in a signal occupying twice the bandwidth of the lowpass signal.

We know that the signal $m(t)$ is real and hence its magnitude spectrum $|M(F)|$ is symmetric about $F = 0$, and its phase spectrum is an odd function. Consequently, if we know $M(F)$ for $0 \le F \le F_1$, we can construct $M(F)$ for $-F_1 \le F \le 0$, and vice versa. In other words, it is sufficient to know $M(F)$ for $0 \le F \le F_1$ or $-F_1 \le F \le 0$.

In the case of the bandpass (modulated) signal, the symmetry property above implies that the frequency components in the range $F_c \le F \le F_c + F_1$ (or $F_c - F_1 \le F \le F_c$) are sufficient to characterize the signal completely. These frequency components constitute the *upper sideband* of the signal $x(t)$, while the frequency components in the range $F_c - F_1 \le F \le F_c$ constitute the *lower sideband*. The amplitude-modulated signal $x(t)$ is called a *double-sideband signal*, as indicated above, and its frequency content is given by (6.1.21). Since the carrier only serves the purpose of translating the signal $m(t)$ to a different frequency range, as indicated by (6.1.21), we say that *the carrier is suppressed*.

The main point of the discussion above is that the transmission of a single sideband suffices to characterize the modulated signal completely. Thus we can save 50% of the channel bandwidth. Whether we transmit the upper or lower sideband is immaterial. The removal of one of the sidebands results in a *single-sideband signal* having the spectrum illustrated in Fig. 6.7.

The method commonly used to generate a single-sideband signal is illustrated by the block diagram in Fig. 6.8. The modulating signal $m(t)$ is multiplied by the sinusoidal

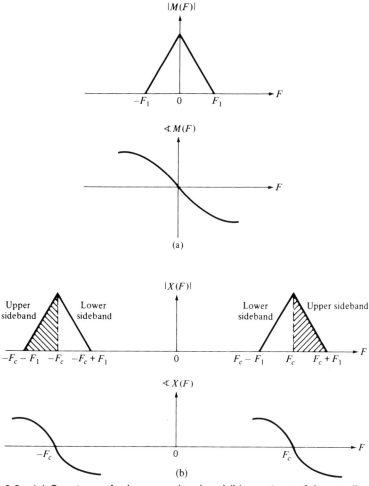

FIGURE 6.6   (a) Spectrum of a lowpass signal and (b) spectrum of the amplitude-modulated bandpass signal.

carrier $\cos 2\pi F_c t$, generated by an oscillator. In addition, the signal $m(t)$ is passed through a filter with system function

$$H(F) = \begin{cases} e^{-j\pi/2} & F > 0 \\ e^{j\pi/2} & F < 0 \end{cases} \qquad (6.1.22)$$

and the output $m(t)$ of the filter is multiplied by the quadrature carrier $\sin 2\pi F_c t$. If the two product signals are added, the result is a single-sideband signal $x_L(t)$ containing the lower sideband of $x(t)$. On the other hand, if the two product signals are subtracted, we obtain a single-sideband signal $x_U(t)$ containing the upper sideband.

Ideally, the filter with transfer function $H(F)$ is an all-pass filter which introduces a constant $90°$ phase shift in all of the components of the input signal, as indicated by (6.1.22). From a practical point of view, the magnitude $|H(F)|$ need only be constant within the frequency band of the signal $m(t)$ (i.e., for $F \leq F_1$). Beyond F_1 its magnitude characteristic can be designed to decay toward zero. This filter is called a *Hilbert transformer*. Its impulse response corresponding to $H(F)$ in (6.1.22) is

$$h(t) = \frac{1}{\pi t} \qquad (6.1.23)$$

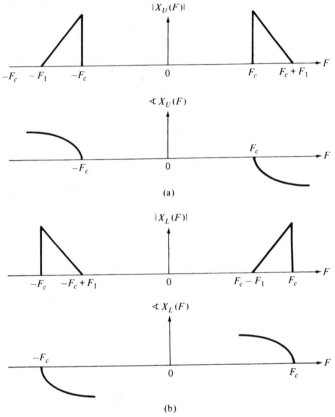

(a)

(b)

FIGURE 6.7 (a) Spectrum of the upper sideband signal and (b) the spectrum of the lower sideband signal.

When excited by an input $m(t)$, the output of the filter is the convolution of $h(t)$ with $m(t)$, which can be expressed as

$$\hat{m}(t) = \frac{1}{\pi} \int_{-\infty}^{\infty} \frac{m(\tau)}{t - \tau} \, d\tau \qquad (6.1.24)$$

We call $\hat{m}(t)$ the *Hilbert transform* of the signal $m(t)$. The time-domain and frequency-domain characteristics of the Hilbert transformer are illustrated in Fig. 6.9.

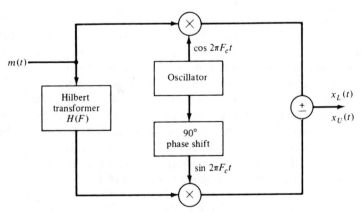

FIGURE 6.8 Method for generating a single-sideband signal.

We observe that the ideal Hilbert transformer is noncausal and since

$$\int_{-\infty}^{\infty} |h(t)| \, dt = \infty$$

the filter is unstable. However, it can be approximated in practice by a causal and stable filter, as we will show later, in the design of digital Hilbert transformers.

To prove that the method described above produces a single-sideband signal, let us consider the signal

$$
\begin{aligned}
x_U(t) &= m(t) \cos 2\pi F_c t - \hat{m}(t) \sin 2\pi F_c t \\
&= \mathrm{Re}\,\{[m(t) + j\hat{m}(t)]e^{j2\pi F_c t}\} \\
&= \mathrm{Re}\,[u(t)e^{j2\pi F_c t}]
\end{aligned}
\tag{6.1.25}
$$

where, by definition,

$$u(t) = m(t) + j\hat{m}(t) \tag{6.1.26}$$

The complex-valued signal $u(t)$ has a spectrum

$$
\begin{aligned}
U(F) &= M(F) + j\hat{M}(F) \\
&= M(F) + jH(F)M(F) \\
&= M(F)[1 + jH(F)]
\end{aligned}
\tag{6.1.27}
$$

Now, for $F > 0$, $H(F) = -j$, hence

$$U(F) = 2M(F) \qquad F > 0$$

For $F < 0$, $H(F) = j$, hence

$$U(F) = 0 \qquad F < 0$$

Consequently, $U(F)$ has the lowpass signal spectrum equal to $2M(F)$ for positive frequencies and zero for negative frequencies, as illustrated in Fig. 6.10a. Finally, (6.1.2) is akin to (6.1.5), and therefore the spectrum of $x_U(t)$ is simply that given by (6.1.9). That is,

$$X_U(F) = \tfrac{1}{2}[U(F - F_c) + U^*(-F - F_c)] \tag{6.1.28}$$

which can also be expressed as

$$
X_U(F) = \begin{cases}
M(F - F_c) & F_c \le F \le F_c + F_1 \\
M^*(-F - F_c) & -F_c - F_1 \le F \le -F_c \\
0 & \text{otherwise}
\end{cases}
\tag{6.1.29}
$$

The spectrum $X_U(F)$ is illustrated in Fig. 6.10b.

The interested reader may verify that by adding the two product signals instead of subtracting them, we obtain a single-sideband signal that consists of the lower sideband.

The reverse operation of modulation is called *demodulation* and takes place at the receiving end of the communication system. In effect, demodulation involves the frequency translation of a signal from bandpass to lowpass and recovery of $m(t)$. Such a translation may be accomplished by multiplication of the bandpass signal with the carrier.

Specifically, suppose that we have the double-sideband signal in (6.1.19). If we multiply this signal by $\cos 2\pi F_c t$, we obtain

$$
\begin{aligned}
x(t) \cos 2\pi F_c t &= m(t) \cos^2 2\pi F_c t \\
&= \tfrac{1}{2}m(t)(1 + \cos 4\pi F_c t) \\
&= \tfrac{1}{2}m(t) + \tfrac{1}{2}m(t) \cos 4\pi F_c t \\
&= \tfrac{1}{2}m(t) + \tfrac{1}{4}m(t)(e^{j4\pi F_c t} + e^{-j4\pi F_c t})
\end{aligned}
\tag{6.1.30}
$$

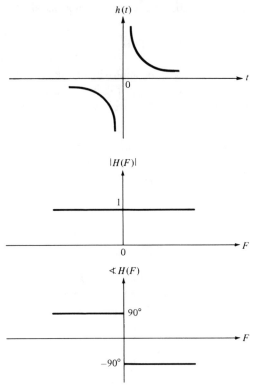

FIGURE 6.9   Time-domain and frequency-domain characteristics of Hilbert transform filter.

We observe that the resulting signal consists of a component proportional to $m(t)$ and an amplitude-modulated component centered at twice the carrier frequency $2F_c$. We call this second component the *double-frequency term*. The spectrum of the signal in (6.1.30) is illustrated in Fig. 6.11.

It is apparent that if we wish to recover the information-bearing signal $m(t)$, at the receiving end, and to reject the double-frequency component, we may pass the signal in (6.1.30) through a lowpass filter that effectively attenuates frequency components above F_1. This may be readily accomplished in practice to yield the desired demodulated signal $m(t)$.

It should be emphasized that the recovery of $m(t)$ from the received signal was made possible because we demodulated with the carrier $\cos 2\pi F_c t$. If the demodulation is performed with a phase-offset carrier, say $\cos (2\pi F_c t + \phi)$, the result of the demodulation process prior to lowpass filtering is

$$x(t) \cos (2\pi F_c t + \phi) = \tfrac{1}{2}m(t) \cos \phi + \tfrac{1}{2}m(t) \cos (4\pi F_c t + \phi) \quad (6.1.31)$$

and after lowpass filtering it is

$$\tfrac{1}{2}m(t) \cos \phi$$

We observe that the phase offset reduces the amplitude of the demodulated signal, and in the extreme case where $\phi = 90°$, the demodulated signal vanishes completely.

Phase offsets between the transmitted and the received signals occur primarily because of two reasons. One is that the oscillators at the transmitter and the receiver drift with time, and hence their sinusoidal outputs are not phase locked. A second reason is that the propagation delay that the signal encounters in transmission is not known precisely,

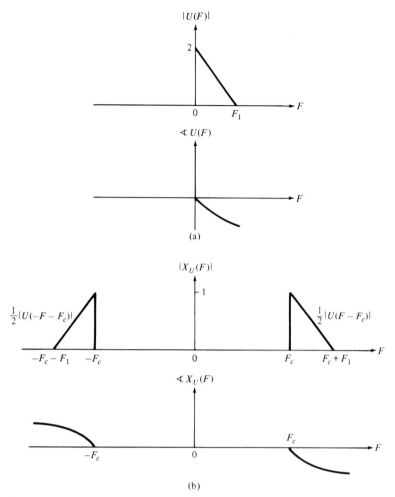

FIGURE 6.10   (a) Lowpass and (b) bandpass single-sideband signal spectra.

and since transmission delay is equivalent to a phase offset, there is uncertainty in the carrier phase of the received signal.

The discussion above clearly points to the need at the receiver for a carrier that is frequency locked and phase locked to the frequency and phase of the received signal.

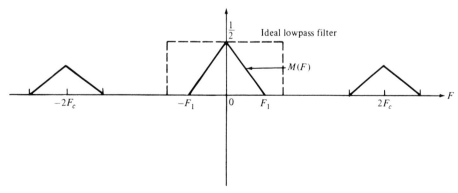

FIGURE 6.11   Spectrum of the demodulated signal.

Electronic devices called *phase-locked loops* are often used at the receiver to generate a local carrier that is locked in phase and frequency to that of the received signal.

To describe briefly the operation of the phase-locked loop, let us consider the basic block diagram of such a loop that is illustrated in Fig. 6.12. We assume that the received signal has some energy at the carrier F_c that can be extracted by a narrowband filter tuned to F_c. If the received signal contains no energy at F_c, there are various ways that involve nonlinear operations on the received signal which can be used to generate a carrier signal component. In any case, let us assume that such a carrier component has been generated and has the form

$$y(t) = \sin (2\pi F_c t + \phi)$$

at the output of the bandpass filter. The output of the voltage-controlled oscillator (VCO) is assumed to be

$$v(t) = \cos (2\pi F_c t + \hat{\phi})$$

where $\hat{\phi}$ denotes an estimate of ϕ. These two signals are multiplied to yield the product signal $e(t)$, which is

$$
\begin{aligned}
e(t) = y(t)v(t) &= \sin (2\pi F_c t + \phi) \cos (2\pi F_c t + \hat{\phi}) \\
&= \tfrac{1}{2} \sin (\phi - \hat{\phi}) \\
&\quad + \tfrac{1}{2} \sin (4\pi F_c t + \phi + \hat{\phi})
\end{aligned}
\tag{6.1.32}
$$

The term $\sin (\phi - \hat{\phi})$ is a zero-frequency (dc) component while the second term in the product signal is at twice the carrier. Hence a lowpass filter will pass only the $\sin (\phi - \hat{\phi})$ signal. In fact, if $\phi = \hat{\phi}$, the excitation to the lowpass filter is zero and, consequently, the output of the lowpass filter will be zero. In such a case the VCO input is maintained at the nominal value which results in the output phase $\phi = \hat{\phi}$. On the other hand, if $\phi - \hat{\phi} = \delta$, where δ is some phase offset relative to ϕ, the excitation to the filter is $\sin \delta$. This excitation will result in a filter response that is used to drive the VCO off its nominal excitation so as to correct for the phase offset. Thus the VCO regulates the output phase depending on the value of the phase offset δ.

A single-sideband signal can be demodulated in much the same way as a double-sideband amplitude-modulated signal. In particular, if we had the single-sideband signal

$$x_U(t) = m(t) \cos 2\pi F_c t - \hat{m}(t) \sin 2\pi F_c t$$

and multiplied it by the carrier $\cos (2\pi F_c t + \phi)$, the result is

$$
\begin{aligned}
x_U(t) \cos (2\pi F_c t + \phi) &= \tfrac{1}{2}m(t) \cos \phi + \tfrac{1}{2}m(t) \cos (4\pi F_c t + \phi) \\
&\quad + \tfrac{1}{2}\hat{m}(t) \sin \phi + \tfrac{1}{2}\hat{m}(t) \sin (4\pi F_c t + \phi)
\end{aligned}
$$

The product signal contains lowpass signal components and signal components at twice the carrier. With the aid of an ideal lowpass filter, we can exclude the double-frequency

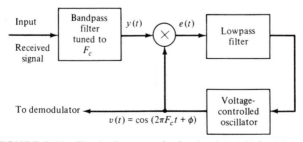

FIGURE 6.12 Block diagram of a basic phase-locked loop.

components and thus we have

$$\tfrac{1}{2}[m(t) \cos \phi + \hat{m}(t) \sin \phi]$$

It is observed that if we employ a phase-locked carrier, with phase $\phi = 0$, we recover $m(t)$. On the other hand, if $\phi \neq 0$, there is interference from $\hat{m}(t)$, which is undesirable. This problem again serves to emphasize the importance of having a phase-locked carrier in order to perform the demodulation of the signal.

Our discussion above was concerned with modulation and demodulation of analog signals. With this as a background, we now treat the equivalent problem in discrete time.

6.1.3 Amplitude-Modulated Discrete-Time Signals

In digital communications, a digital information sequence is usually transmitted via amplitude modulation from the source to the destination. At the source end, it is often convenient to generate the modulated signal in digital form and then to convert it to analog form by means of a D/A converter for transmission over the analog channel to the receiver. At the receiving end of the communication system, the received analog signal may be digitized by passage through an A/D converter and demodulated digitally (in discrete time). Below we describe the modulation and demodulation techniques for discrete-time signals, and in the following section we apply these techniques to a digital implementation of a modem (modulator–demodulator).

A discrete-time sequence $m(k)$ with a lowpass spectrum

$$M(\omega) = \sum_{k=-\infty}^{\infty} m(k)e^{-j\omega k} \qquad (6.1.33)$$

may be translated in frequency and converted to a bandpass signal by multiplication with the carrier $\cos \omega_c k$, which may be generated by a digital sinusoidal oscillator described in Section 5.5.6. The result is the signal

$$x(k) = m(k) \cos \omega_c k \qquad (6.1.34)$$

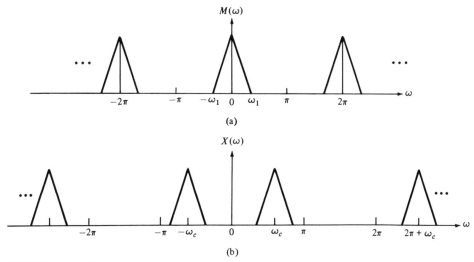

FIGURE 6.13 (a) Spectrum of the lowpass signal and (b) the spectrum of the bandpass signal.

which has the spectrum

$$X(\omega) = \tfrac{1}{2}[M(\omega - \omega_c) + M(\omega + \omega_c)] \qquad -\pi \le \omega \le \pi \qquad (6.1.35)$$

as can be easily demonstrated by substituting (6.1.34) into (6.1.33). Figure 6.13 illustrates the spectrum of the lowpass signal $m(k)$ and the resulting double-sideband spectrum obtained via amplitude modulation. We recall that $X(\omega)$ is periodic with period 2π.

As in the case of a real analog signal, one of the two sidebands of $X(\omega)$ is redundant and may be removed to yield a single-sideband signal. This may be accomplished by means of the method previously illustrated in Fig. 6.8, where, now, we employ a *digital Hilbert transformer*, having (ideally) the frequency response

$$H(\omega) = \begin{cases} -j & 0 \le \omega \le \pi \\ j & -\pi \le \omega < 0 \end{cases} \qquad (6.1.36)$$

The corresponding impulse response of the ideal Hilbert transformer is

$$\begin{aligned} h(n) &= \frac{1}{2\pi} \int_{-\pi}^{\pi} H(\omega)e^{j\omega n}\, d\omega \\ &= \frac{1}{2\pi} \int_{-\pi}^{0} je^{j\omega n}\, d\omega - \frac{1}{2\pi} \int_{0}^{\pi} je^{j\omega n}\, d\omega \qquad (6.1.37) \\ &= \begin{cases} 0 & n = 0 \\ \dfrac{2}{\pi} \dfrac{\sin^2 (\pi n/2)}{n} & n \ne 0 \end{cases} \end{aligned}$$

Figure 6.14 illustrates the frequency response characteristic and the impulse response of the ideal digital Hilbert transformer. We observe that $h(n)$ is physically unrealizable and unstable, since

$$\sum_{n=-\infty}^{\infty} |h(n)| = \infty$$

However, it is possible to approximate the discrete Hilbert transform by an FIR filter, as will be shown in Chapter 8. For example, one may simply truncate the impulse response in (6.1.37) and insert a delay so that $h(n)$ is causal. Alternatively, one may use the design method described in Section 5.3.4 for linear-phase FIR filters, where the bandwidth of the Hilbert transform can be selected to match the spectral characteristics of the signal sequence $\{m(k)\}$.

A single-sideband amplitude modulated signal is represented as

$$x_U(k) = m(k) \cos \omega_c k - \hat{m}(k) \sin \omega_c k \qquad (6.1.38)$$

when the signal consists of the upper sideband, and as

$$x_L(k) = m(k) \cos \omega_c k + \hat{m}(k) \sin \omega_c k \qquad (6.1.39)$$

when the signal consists of the lower sideband. These signals are conveniently written in the form

$$\begin{aligned} x_U(k) &= \text{Re } [u(k)e^{j\omega_c k}] \\ x_L(k) &= \text{Re } [u^*(k)e^{j\omega_c k}] \end{aligned} \qquad (6.1.40)$$

where, by definition,

$$u(k) = m(k) + j\hat{m}(k) \qquad (6.1.41)$$

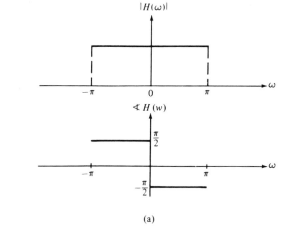

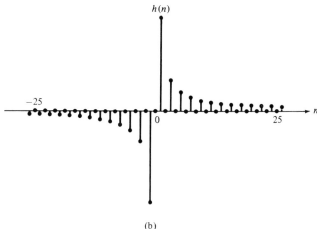

(b)

FIGURE 6.14 Frequency response (a) and the impulse response (b) of an ideal discrete-time Hilbert transform filter.

Then it follows from our previous discussion that the spectrum of $x_U(k)$ is

$$X_U(\omega) = \tfrac{1}{2}[U(\omega - \omega_c) + U^*(-\omega - \omega_c)] \qquad (6.1.42)$$

Now, from the linearity of the Fourier transform, we have

$$
\begin{aligned}
U(\omega) &= M(\omega) + j\hat{M}(\omega) \\
&= M(\omega) + jH(\omega)M(\omega) \qquad (6.1.43) \\
&= M(\omega)[1 + jH(\omega)]
\end{aligned}
$$

For $0 \le \omega \le \pi$, $H(\omega) = -j$ and hence (6.1.42) yields $U(\omega) = 2M(\omega)$. On the other hand, for $-\pi \le \omega < 0$, $H(\omega) = j$ and hence $U(\omega) = 0$. Therefore,

$$
U(\omega) = \begin{cases} 2M(\omega) & 0 \le \omega \le \pi \\ 0 & -\pi \le \omega < 0 \end{cases}
$$

Consequently, the spectrum of the single-sideband signal is

$$
X_U(\omega) = \begin{cases} M(\omega - \omega_c) & \omega_c \le \omega \le \omega_c + \omega_1 \\ M^*(-\omega - \omega_c) & -\omega_c - \omega_1 \le \omega \le -\omega_c \\ 0 & \text{otherwise} \end{cases} \qquad (6.1.44)
$$

It is clear from Fig. 6.13 that the carrier frequency must be selected such that $\omega_c >$ ω_1. Furthermore, it is also apparent that the bandwidth of the signal $m(k)$ must be less than $\pi/2$ (i.e., $\omega_1 < \pi/2$) to avoid aliasing effects. These conditions lead to the restriction of ω_c to the range

$$\omega_1 < \omega_c < \pi - \omega_1 \qquad (6.1.45)$$

Most physical communication channels are analog. If the digital single-sideband signal synthesized as described above is to be transmitted over an analog channel, it must be converted to an analog signal by passing it through a D/A converter. Additional frequency translation may be introduced in the analog domain to shift the spectrum of the signal to another frequency band. At the receiver, the analog signal may be converted back to a digital signal by passing it through an A/D converter.

In the demodulation process that takes place at the receiving end of the communication system, the digital amplitude-modulated signal is converted to lowpass by multiplication with the carrier $\cos \omega_c k$. As in the case of analog signals, the multiplication generates a double-frequency component in addition to the desired lowpass component. The double-frequency component can be eliminated by a lowpass filter that passes only the desired information-bearing signal $m(k)$.

In the synthesis of the single-sideband amplitude-modulated signal, the Hilbert-transformed signal sequence $\hat{m}(k)$ is used to cancel one of the signal sidebands and thus to reduce the frequency bandwidth required to transmit the signal over the channel by a factor of 2. An alternative modulation technique that achieves the same bandwidth efficiency as in single-sideband is *double-sideband quadrature amplitude modulation* (QAM) in which the two quadrature carriers $\cos \omega_c k$ and $\sin \omega_c k$ are amplitude modulated by two separate information-bearing signals as illustrated in Fig. 6.15. Thus the synthesized digital signal sequence may be expressed as

$$x(k) = m_1(k) \cos \omega_c k + m_2(k) \sin \omega_c k \qquad (6.1.46)$$

where $m_1(k)$ and $m_2(k)$ are the two separate information-bearing signals. The QAM signal sequence $x(k)$ now has the spectrum

$$X(\omega) = \tfrac{1}{2}[M_1(\omega - \omega_c) + M_1(\omega + \omega_c)] - j\tfrac{1}{2}[M_2(\omega - \omega_c) - M_2(\omega + \omega_c)] \qquad (6.1.47)$$

Although the spectrum components of the information-bearing signals overlap, the quadrature phase relationship in the carrier components $\cos \omega_c k$ and $\sin \omega_c k$ allows us to separate the two signals at the receiving end of the communication system.

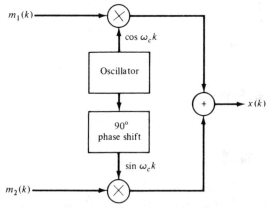

FIGURE 6.15 Generation of a QAM signal.

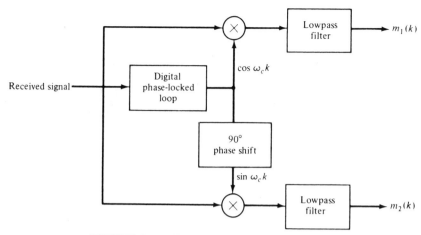

FIGURE 6.16 Demodulation of a QAM signal.

The demodulation is performed as illustrated in Fig. 6.16. By using a digital phase-locked loop to obtain the carrier component $\cos \omega_c k$ and to generate $\sin \omega_c k$, the received sequence is multiplied by the two quadrature carriers. This multiplication process or mixing process results in the two signal sequences

$$x(k) \cos \omega_c k = \tfrac{1}{2} m_1(k) + \tfrac{1}{2} m_1(k) \cos 2\omega_c k + m_2(k) \cos \omega_c k \sin \omega_c k$$
$$= \tfrac{1}{2} m_1(t) + \tfrac{1}{2} m_1(k) \cos 2\omega_c k + \tfrac{1}{2} m_2(k) \sin 2\omega_c k \tag{6.1.48}$$
$$x(k) \sin \omega_c k = \tfrac{1}{2} m_2(k) - \tfrac{1}{2} m_2(k) \cos 2\omega_c k + \tfrac{1}{2} m_1(k) \sin 2\omega_c k \tag{6.1.49}$$

The information-bearing signal components $m_1(k)$ and $m_2(k)$ can be recovered by passing each of the sequences given by (6.1.48) and (6.1.49) through a lowpass filter that rejects the double-frequency terms centered at $2\omega_c$.

The digital phase-locked loop used in the demodulation of the digital QAM signal and the single-sideband signal is similar in form to the analog phase-locked loop described above. The basic building blocks illustrated in Fig. 6.12 for the analog phase-locked loop can be implemented in discrete time on a digital machine.

6.1.4 Digital Implementation of a QAM MODEM

Modems are used in digital communication systems to transmit data over a communication channel from one point to another. Typical channels are satellite channels, telephone channels, radio channels, and fiber optic cable.

Most point-to-point communication systems are designed to transmit and receive data in both directions. Consequently, there is a modulator and a demodulator or modem required at each end of the system. In this section we briefly describe a digital implementation of the basic building blocks in a modem that is suitable for synchronous communication over a channel. By "synchronous communication" we mean that data symbols (bits or groups of bits) are transmitted periodically at a specified fixed rate controlled by a clock. To be specific, let us assume that the channel is a telephone line that has a nominal bandwidth of approximately 2400 Hz covering the frequency range between 600 and 3000 Hz. This channel may be viewed as a bandpass channel with center frequency $F_c = 1800$ Hz. The equivalent lowpass channel may be viewed as having a nominal bandwidth of 1200 Hz and centered at zero frequency as illustrated in Fig. 6.17. Consequently, the equivalent lowpass signal that is transmitted over the channel must have a bandwidth of approximately 1200 Hz. Correspondingly, the Nyquist rate

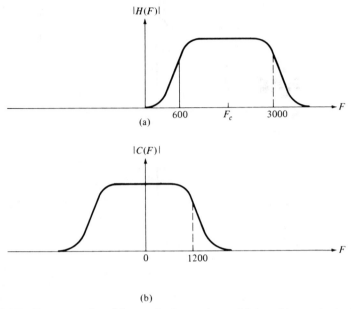

FIGURE 6.17 Frequency band for a telephone channel (a) and its equivalent lowpass characteristic (b).

for sampling such a signal is 2400 samples per second, if the channel is an ideal lowpass channel with bandwidth 1200 Hz.

The digital symbols may be multivalued so that the bit rate over the channel may be a multiple of the symbol rate, which is 2400. For example, with QAM, utilizing four possible amplitude levels per symbol on each of the two quadrature carriers, we can achieve two bits per symbol for each quadrature carrier signal or, equivalently, a bit rate of 9600 bits per second over the telephone channel. We shall use this bit rate in our modem design. Of course, we are assuming that the channel noise corrupting the transmitted signal is sufficiently small to allow us to detect and discriminate among the four possible amplitude levels.

There is no unique method for implementing the digital modem. Our implementation will use basic building blocks that have been described previously; FIR filters, multipliers, A/D and D/A converters, and the like. The block diagrams for the modulator and demodulator are illustrated in Figs. 6.18 and 6.19, respectively. To begin, we assume that the data to be transmitted is a synchronous sequence of binary digits (zeros and ones) at the fixed rate of 9600 bits per second. At the modulator, blocks of four bits at a time are split into a pair of two-bit symbols. Each two-bit symbol is mapped into an amplitude value according to the mapping

$$
\begin{aligned}
00 &\longrightarrow 3 \\
01 &\longrightarrow 1 \\
11 &\longrightarrow -1 \\
10 &\longrightarrow -3
\end{aligned}
\qquad (6.1.50)
$$

The grouping of bits and the mapping into amplitudes is performed in the functional block labeled "signal point selection."

Note that the amplitude values are selected to be equidistant. Thus we generate a pair of amplitudes, say $(A_c(n), A_s(n))$ once every $T = 1/2400$ second, where the subscripts indicate the transmission of these amplitudes on the cosine and sine carriers, respectively.

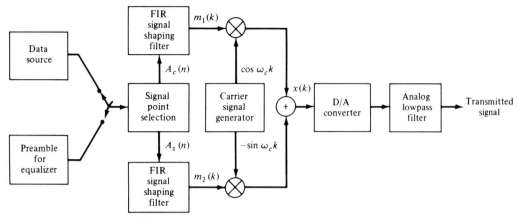

FIGURE 6.18   Block diagram of QAM modulator.

Each pair of amplitudes may be viewed as the coordinates of a point in the complex plane. When viewed in this manner, we obtain the signal point constellation consisting of 16 points, as illustrated in Fig. 6.20. Each signal point corresponds to a block of four information bits.

The sequence of signal amplitudes $\{A_c(n), A_s(n)\}$ is used to excite a pair of lowpass filters having a nominal bandwidth of 1200 Hz. To avoid distortion effects, linear-phase FIR filters are employed. To avoid spectrum aliasing effects, these filters are clocked at a multiple of the symbol rate of 2400 symbols. If $T = 1/2400$, the sampling interval for the FIR filters is $T_s = T/N$, where N is an integer greater than unity. We shall select $N = 6$ for our design. Then since the symbol rate is six times slower, the input sequence to the FIR filters consists of the pair of signal amplitudes $(A_c(n), A_s(n))$ followed by five pairs of zeros.

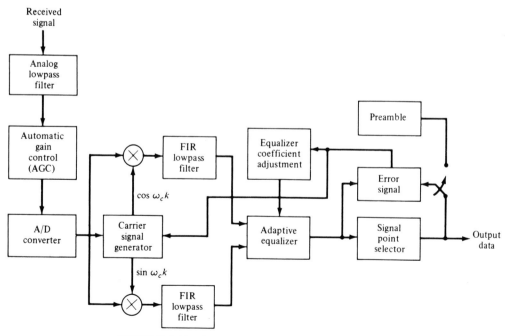

FIGURE 6.19   Block diagram of QAM demodulator.

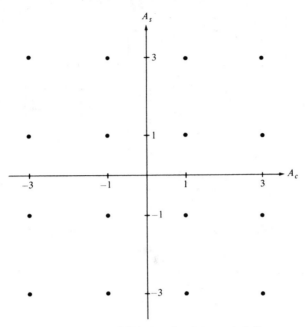

FIGURE 6.20 QAM signal-point constellation.

The linear phase FIR filters are designed to have the "raised-cosine" impulse response

$$
h(n) = \frac{\sin \frac{\pi}{N}\left(k - \frac{M-1}{2}\right) \cos \frac{2\pi\beta}{N}\left(k - \frac{M-1}{2}\right)}{\frac{\pi}{N}\left(k - \frac{M-1}{2}\right) \; 1 - \frac{4\beta^2}{N^2}\left(k - \frac{M-1}{2}\right)}
\tag{6.1.51}
$$

where M is the length of the filter and β is a parameter in the range $0 \le \beta \le 1$ that controls the bandwidth of the filter. In our design we pick $\beta = 0.2$, which yields a small bandwidth, and $N = 6$, as indicated above. The filter length is selected to be $M = 85$. The impulse response and the frequency response (magnitude only) are illustrated in Fig. 6.21. It should be observed that $h(nN) = 0$, $n \ne 0$, when $(M - 1)/2$ is selected to be a multiple N. This is a desirable property that ensures no intersymbol interference among successive symbols at the sampling instants. The output sequences from the two FIR filters are denoted as $\{m_1(k)\}$ and $\{m_2(k)\}$, where $\{m_1(k)\}$ is the sequence obtained with the input amplitude sequence $\{A_c(n)\}$ and $\{m_2(k)\}$ is the output sequence obtained with the input amplitude sequence $\{A_s(n)\}$.

The sequences $\{m_1(k)\}$ and $\{m_2(k)\}$ modulate the cosine and sine carriers, respectively. Since the desired carrier frequency is $F_c = 1800$ Hz and the sampling rate is six times the symbol rate, the desired cosine carrier is

$$
\begin{aligned}
c(k) &= \cos 2\pi F_c t|_{t=kT_s} \\
&= \cos 2\pi(1800)kT_s \tag{6.1.52} \\
&= \cos \frac{\pi}{4} k
\end{aligned}
$$

and similarly, the sine carrier is

$$
s(k) = \sin \frac{\pi}{4} k \tag{6.1.53}
$$

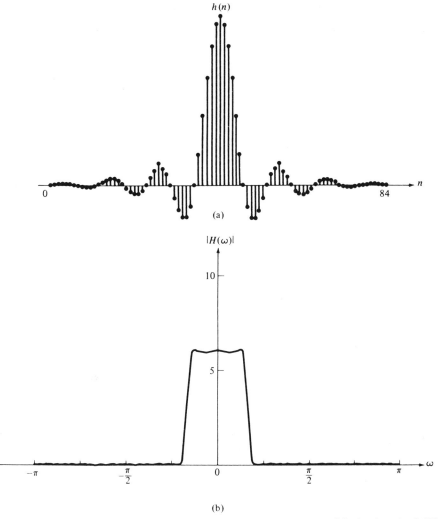

FIGURE 6.21   (a) Impulse response and (b) frequency response of "raised-cosine" FIR filters.

In view of the extremely simple nature of these carrier signals, they are easily stored in a read-only memory (ROM) and addressed appropriately for use in the modulation process.

The synthesized digital signal sequence at the modulator is

$$x(k) = m_1(k) \cos \frac{\pi}{4} k - m_2(k) \sin \frac{\pi}{4} k \qquad (6.1.54)$$

This sequence is passed through a D/A converter followed by an analog lowpass filter that is used to attenuate signal frequency components severely above 3000 Hz. This analog signal is then transmitted over the telephone channel.

At the demodulator side, as shown in Fig. 6.19, the analog signal is first filtered by an analog filter to remove high-frequency noise components. This filter is followed by an automatic gain control (AGC) device that is used to ensure that the signal levels do not saturate the A/D converter. Thus the signal values will fall within the dynamic range of the A/D converter and no overflow will occur. In addition, the AGC allows us to use fixed thresholds in the detection of the amplitude levels, since it compensates for gain variations in the channel. The AGC has a long time constant and hence does not respond to the rapid amplitude variations of the data signal. The sequence at the output of the A/D converter is demodulated by multiplication with the locally generated carrier signal generator, which has embedded in it a digital phase-locked loop, and the resulting products are lowpass filtered by a pair of FIR filters, identical in form to the filters used at the modulator. This filtering removes the double-frequency components.

To compensate for signal distortion effects introduced by the channel in the transmission of the signal, an additional filter is usually employed, which is called a channel equalizer. Since the channel characteristics are unknown a priori, the equalizer is an adaptive FIR filter with adjustable coefficients. Initially, the adaptation is performed with the aid of a preamble that is transmitted over the channel for the purpose of training the equalizer. An error signal is formed, which is the difference between the desired output symbol (preamble) and the actual output of the equalizer. The coefficients of the equalizer are then adjusted to minimize some function (usually the mean square value) of this error. Once the equalizer is trained on the preamble to make reliable estimates of the transmitted symbols, we continue to adjust its coefficients when data are transmitted, by assuming that the data decisions are correct and then forming an error signal as in the case where the preamble was used. Although the equalizer is shown to operate on two input sequences simultaneously, it may be split into two separate adaptive FIR equalizers.

A simple digital phase-locked loop that can be used in the implementation of the modem is illustrated in Fig. 6.22. The lowpass loop filter has three parameters a_1, a_2, and a_3 that can be adjusted to provide the desired response characteristic. The sequence at the output of the loop filter is proportional to the phase error between the output phase of the VCO and the signal carrier phase. A phase detector or phase comparator is used to compare the phase of the input carrier and the phase of the signal from the VCO. The phase detector generates the signal at the input to the loop filter.

All the signal processing at the demodulator prior to the equalizer is performed at the rate of $N = 6$ times the symbol rate. On the other hand, the equalizer output is not required to be computed at that rate. In fact, the equalizer output rate should be one pair of signal samples at the symbol rate. The desired pair of samples corresponds to $A_c(n)$ and $A_s(n)$. However, if we do not know in advance the exact timing for sampling the output of the equalizer, we may have to compute more than one output for each pair of

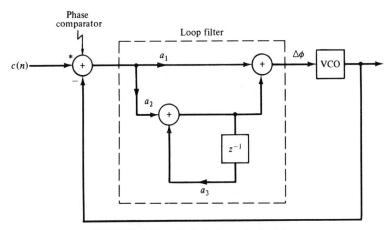

FIGURE 6.22 Digital phase-locked loop.

input symbols. For example, we may choose to compute three samples per symbol and select the largest output of the three. In such a case the equalizer input may be taken as every other sample.

Finally, the output of the equalizer at the proper sampling instant is compared with the possible signals points in the signal constellation and a decision is made in favor of the signal point closest to the received point.

Although the description above of the modem implementation did not go into the details of the design, we hope that the reader has obtained an appreciation of the simplicity of this digital implementation and sufficient understanding of basic concepts involved in the modulation and demodulation processes.

6.1.5 Sampling of Analog Bandpass Signals

We have already demonstrated in Section 4.5.1 that any signal with highest frequency B can be uniquely represented by samples taken at the minimum rate (Nyquist rate) of $2B$ samples per second. However, if the signal is a bandpass signal with frequency components in the band $B_1 \leq F \leq B_2$, as shown in Fig. 6.23, a blind application of the sampling theorem would have us sampling the signal at a rate of $2B_2$ samples per second.

If that were the case and B_2 was an extremely high frequency, it would certainly be advantageous to perform a frequency shift of the bandpass signal by an amount

$$F_c = \frac{B_1 + B_2}{2} \tag{6.1.55}$$

and sampling the equivalent lowpass signal. As we have learned from our discussion of modulation such a frequency shift can be achieved by multiplying the bandpass signal as given in (6.1.2) by the quadrature carriers $\cos 2\pi F_c t$ and $\sin 2\pi F_c t$ and lowpass filtering the products to eliminate the signal components at $2F_c$. Clearly, the multiplication and the subsequent filtering are performed in the analog domain and the outputs of the filters are then sampled. The resulting equivalent lowpass signal has a bandwidth $B/2$, where $B = B_2 - B_1$; hence it can be represented uniquely by samples taken at the rate of B samples per second for each of the quadrature components. Thus the sampling can be performed on each of the lowpass filter outputs at the rate of B samples per second, as indicated in Fig. 6.24. Therefore, the resulting rate is $2B$ samples per second.

In view of the fact that frequency conversion to lowpass allows us to reduce the sampling rate to $2B$ samples per second, it should be possible to sample the bandpass signal at a comparable rate. In fact, it is.

Suppose that the upper frequency $F_c + B/2$ is a multiple of the bandwidth B (i.e., $F_c + B/2 = kB$), where k is a positive integer. If we sample $x(t)$ at the rate $2B = 1/T$

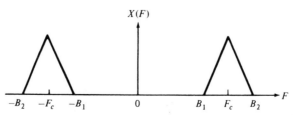

FIGURE 6.23   Bandpass signal with frequency components in the range $B_1 \leq F \leq B_2$.

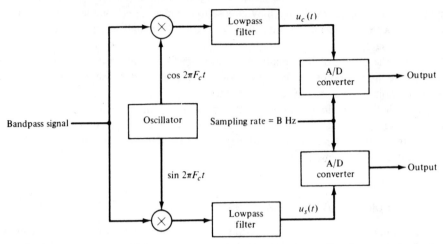

FIGURE 6.24 Sampling of a bandpass signal by first converting to an equivalent lowpass signal.

samples per second, we have

$$x(nT) = u_c(nT) \cos 2\pi F_c nT - u_s(nT) \sin 2\pi F_c nT \qquad (6.1.56)$$

$$= u_c(nT) \cos \frac{\pi n(2k-1)}{2} - u_s(nT) \sin \frac{\pi n(2k-1)}{2}$$

where the last step is obtained by substituting $F_c = kB - B/2$ and $T = 1/2B$.
For n even, say $n = 2m$, (6.1.56) reduces to

$$x(2mT) \equiv x(mT_1) = u_c(mT_1) \cos \pi m(2k-1) = (-1)^m u_c(mT_1) \qquad (6.1.57)$$

where $T_1 = 2T = 1/B$. For n odd, say $n = 2m - 1$, (6.1.56) reduces to

$$x(2mT - T) \equiv x\left(mT_1 - \frac{T_1}{2}\right) = u_s\left(mT_1 - \frac{T_1}{2}\right)(-1)^{m+k+1} \qquad (6.1.58)$$

Therefore, the even-numbered samples of $x(t)$, which occur at the rate of B samples per second, produce samples of the lowpass signal component $u_c(t)$, while the odd-numbered samples of $x(t)$, which also occur at the rate of B samples per second, produce samples of the lowpass signal component $u_s(t)$.

Now, the samples $\{u_c(mT_1)\}$ and the samples $\{u_s(mT_1 - T_1/2)\}$ can be used to reconstruct the equivalent lowpass signals. Thus, according to the sampling theorem for lowpass signals with $T_1 = 1/B$,

$$u_c(t) = \sum_{m-\infty}^{\infty} u_c(mT_1) \frac{\sin (\pi/T_1) (t - mT_1)}{(\pi/T_1)(t - mT_1)} \qquad (6.1.59)$$

$$u_s(t) = \sum_{m=-\infty}^{\infty} u_s\left(mT_1 - \frac{T_1}{2}\right) \frac{\sin (\pi/T_1) (t - mT_1 + T_1/2)}{(\pi T_1)(t - mT_1 + T_1/2)} \qquad (6.1.60)$$

Furthermore, the relations in (6.1.57) and (6.1.58) allow us to express $u_c(t)$ and $u_s(t)$ directly in terms of samples of $x(t)$. Now, since $x(t)$ is expressed as

$$x(t) = u_c(t) \cos 2\pi F_c t - u_s(t) \sin 2\pi F_c t \qquad (6.1.61)$$

substitution from (6.1.60), (6.1.59), (6.1.58), and (6.1.57) into (6.1.61) yields

$$
\begin{aligned}
x(t) = \sum_{m=-\infty}^{\infty} \Bigg\{ & (-1)^m x(2mT) \frac{\sin{(\pi/2T)(t - 2mT)}}{(\pi/2T)(t - 2mT)} \cos 2\pi F_c t \\
& + (-1)^{m+k} x((2m - 1)T) \frac{\sin{(\pi/2T)(t - 2mT + T)}}{(\pi/2T)(t - 2mT + T)} \Bigg\}
\end{aligned}
\tag{6.1.62}
$$

But $(-1)^m \cos 2\pi F_c t = \cos 2\pi F_c(t - 2mT)$ and $(-1)^{m+k} \sin 2\pi F_c t = \cos 2\pi F_c(t - 2mT + T)$. With these substitutions, (6.1.62) reduces to

$$
x(t) = \sum_{m=-\infty}^{\infty} x(mT) \frac{\sin{(\pi/2T)(t - mT)}}{(\pi/2T)(t - mT)} \cos 2\pi F_c(t - mT)
\tag{6.1.63}
$$

where $T = 1/2B$. This is the desired reconstruction formula for the bandpass signal $x(t)$ with samples taken at the rate of 2B samples per second, for the special case in which the upper band frequency $F_c + B/2$ is a multiple of the signal bandwidth B.

In the general case where only the condition $F_c \geq B/2$ is assumed to hold, let us define the integer part of the ratio $F_c + B/2$ to B as

$$
r = \left\lfloor \frac{F_c + B/2}{B} \right\rfloor
\tag{6.1.64}
$$

While holding the upper cutoff frequency $F_c + B/2$ constant, we increase the bandwidth from B to B' such that

$$
\frac{F_c + B/2}{B'} = r
\tag{6.1.65}
$$

Furthermore, it is convenient to define a new center frequency for the increased bandwidth signal as

$$
F_c' = F_c + \frac{B}{2} - \frac{B'}{2}
\tag{6.1.66}
$$

Clearly, the increased signal bandwidth B' includes the original signal spectrum of bandwidth B.

Now the upper cutoff frequency $F_c + B/2$ is a multiple of B'. Consequently, the signal reconstruction formula in (6.1.63) holds with F_c replaced by F_c' and T replaced by T', where $T' = 1/2B'$, that is,

$$
x(t) = \sum_{n=-\infty}^{\infty} x(nT') \frac{\sin{(\pi/2T')(t - mT')}}{(\pi/2T')(t - mT')} \cos 2\pi F_c'(t - mT')
\tag{6.1.67}
$$

This proves that $x(t)$ can be represented by samples taken at the uniform rate $1/T' = 2Br'/r$, where r' is the ratio

$$
r' = \frac{F_c + B/2}{B} = \frac{F_c}{B} + \frac{1}{2}
\tag{6.1.68}
$$

and $r = \lfloor r' \rfloor$.

We observe that when the upper cutoff frequency $F_c + B/2$ is not an integer multiple of the bandwidth B, the sampling rate for the bandpass signal must be increased by the factor r'/r. However, note that as F_c/B increases, the ratio r'/r tends toward unity. Consequently, the percent increase in sampling rate tends to zero.

The derivation given above also illustrates the fact that the lowpass signal components

$u_c(t)$ and $u_s(t)$ may be expressed in terms of samples of the bandpass signal. Indeed, from (6.1.57), (6.1.58), (6.1.59), and (6.1.60), we obtain the result

$$u_c(t) = \sum_{n=-\infty}^{\infty} (-1)^n x(2nT') \frac{\sin(\pi/2T')(t - 2nT')}{(\pi/2T')(t - 2nT')} \tag{6.1.69}$$

and

$$u_s(t) = \sum_{n=-\infty}^{\infty} (-1)^{n+r} x(2nT' - T') \frac{\sin(\pi/2T')(t - 2nT' + T')}{(\pi/2T')(t - 2nT' + T')} \tag{6.1.70}$$

where $r = \lfloor r' \rfloor$.

In conclusion, we have demonstrated that a bandpass signal can be represented uniquely by samples taken at a rate

$$2B \le F_s < 4B$$

where B is the bandwidth of the signal. The lower limit applies when the upper frequency $F_c + B/2$ is a multiple of B. The upper limit on F_s is obtained under worst-case conditions when $r = 1$ and $r' \approx 2$.

6.2 Signal Reconstruction Techniques

In Section 4.5.1 we demonstrated that a bandlimited lowpass analog signal, which has been sampled at the Nyquist rate (or faster), can be reconstructed from its samples without distortion. The ideal reconstruction formula or ideal interpolation formula derived in Section 4.5.1 is

$$x(t) = \sum_{n=-\infty}^{\infty} x(nT) \frac{\sin(\pi/T)(t - nT)}{(\pi/T)(t - nT)} \tag{6.2.1}$$

where the sampling interval $T = 1/F_s = 1/2B$, F_s is the sampling frequency and B is the bandwidth of the analog signal.

We have viewed the reconstruction of the signal $x(t)$ from its samples as an interpolation problem and have described the function

$$g(t) = \frac{\sin(\pi t/T)}{\pi t/T} \tag{6.2.2}$$

as the ideal interpolation function. The interpolation formula for $x(t)$, given by (6.2.1), is basically a linear superposition of time-shifted versions of $g(t)$, with each $g(t - nT)$ weighted by the corresponding signal sample $x(nT)$.

Alternatively, we may view the reconstruction of the signal from its samples as a linear filtering process in which a discrete-time sequence of short pulses (ideally impulses) with amplitudes equal to the signal samples excites an analog filter, as illustrated in Fig. 6.25. The analog filter corresponding to the ideal interpolator has a frequency response

$$H(F) = \begin{cases} T & |F| \le \dfrac{1}{2T} = \dfrac{F_s}{2} \\[2mm] 0 & |F| > \dfrac{1}{2T} \end{cases} \tag{6.2.3}$$

$H(F)$ is simply the Fourier transform of the interpolation function $g(t)$. In other words, $H(F)$ is the frequency response of an analog reconstruction filter whose impulse response is $h(t) = g(t)$. As shown in Fig. 6.26, the ideal reconstruction filter is an ideal lowpass

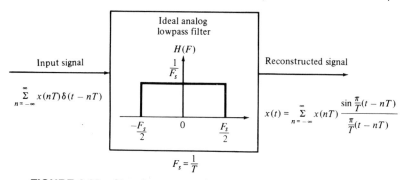

FIGURE 6.25   Signal reconstruction viewed as a filtering process.

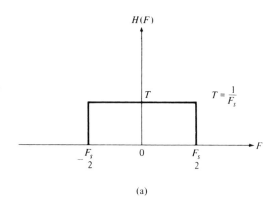

(a)

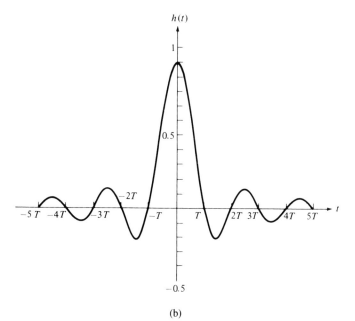

(b)

FIGURE 6.26   Frequency response (a) and the impulse response (b) of an ideal lowpass filter.

filter and its impulse response extends for all time. Hence the filter is noncausal and physically nonrealizable.

In this section we present some practical, albeit nonideal interpolation techniques and interpret them as linear filters. Although many sophisticated polynomial interpolation techniques can be devised and analyzed, our discussion is limited to constant and linear interpolation. Quadratic and higher polynomial interpolation is often used in numerical analysis, but is less likely to be used in digital signal processing.

6.2.1 Zero-Order Hold

A zero-order hold approximates the analog signal by a series of rectangular pulses whose height is equal to the corresponding value of the signal pulse. Figure 6.27a illustrates the approximation of the analog signal $x(t)$ by a zero-order hold. As shown, the approximation, denoted as $\hat{x}(t)$, is basically a staircase function which takes the signal sample and holds it for T seconds. When the next sample arrives, it jumps to the next value and holds it for T seconds, and so on.

When viewed as a linear filter, as shown in Fig. 4.27b, the zero-order hold has an impulse response

$$h(t) = \begin{cases} 1 & 0 \le t \le T \\ 0 & \text{otherwise} \end{cases} \qquad (6.2.4)$$

This is illustrated in Fig. 6.27c. The corresponding frequency response is

$$H(F) = \int_{-\infty}^{\infty} h(t)e^{-j\pi Ft}\, dt$$

$$= \int_{0}^{T} e^{-j2\pi Ft}\, dt \qquad (6.2.5)$$

$$= T\left(\frac{\sin \pi FT}{\pi FT}\right) e^{-j\pi FT}$$

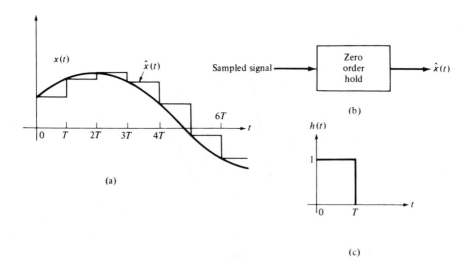

FIGURE 6.27 (a) Approximation of an analog signal by a zero-order hold; (b) linear filtering interpretation; (c) impulse response of the zero-order hold filter.

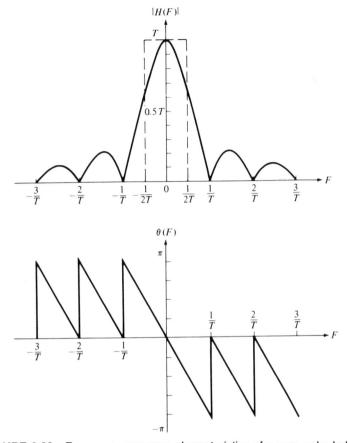

FIGURE 6.28 Frequency response characteristics of a zero-order hold.

The magnitude and phase of $H(F)$ are plotted in Figs. 6.28. For comparison, the frequency response of the ideal interpolator is superimposed on the magnitude characteristics.

It is apparent that the zero-order hold does not possess a sharp cutoff frequency response characteristic. This is due to a large extent on the sharp transitions of its impulse response $h(t)$. As a consequence, the zero-order hold passes undesirable aliased frequency components (frequencies above $F_s/2$) to its output. To remedy this problem, it is common practice to filter $\hat{x}(t)$ by passing it through a lowpass filter which highly attenuates frequency components above $F_s/2$. In effect, the lowpass filter following the zero-order hold smooths the signal $\hat{x}(t)$ and removes the sharp discontinuities.

6.2.2 First-Order Hold

A first-order hold approximates $x(t)$ by straight-line segments which have a slope that is determined by the current sample $x(nT)$ and the previous sample $x(nT - T)$. An illustration of this signal reconstruction techniques is given in Fig. 6.29.

The mathematical relationship between the input samples and the output waveform is

$$\hat{x}(t) = x(nT) + \frac{x(nT) - x(nT - T)}{T}(t - nT) \qquad nT \le t < (n + 1)T \qquad (6.2.6)$$

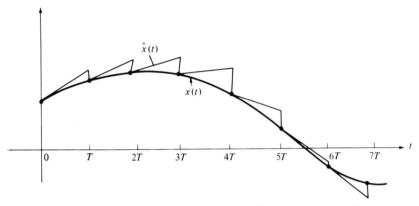

FIGURE 6.29 Signal reconstruction with a first-order hold.

When viewed as a linear filter, the impulse response of the first-order hold is

$$h(t) = \begin{cases} 1 + \dfrac{t}{T} & 0 \le t \le T \\ 1 - \dfrac{t}{T} & T \le t < 2T \\ 0 & \text{otherwise} \end{cases} \tag{6.2.7}$$

This impulse response is depicted in Fig. 6.30a. The Fourier transform of $h(t)$ yields the frequency response, which can be expressed in the form

$$H(F) = T(1 + 4\pi F^2 T^2)^{1/2} \left(\frac{\sin \pi FT}{\pi FT} \right)^2 e^{j\Theta(F)} \tag{6.2.8}$$

where the phase $\Theta(F)$ is

$$\Theta(F) = -\pi FT + \tan^{-1} 2\pi FT \tag{6.2.9}$$

These frequency response characteristics are graphically illustrated in Fig. 6.30b and c.

Since this reconstruction technique also suffers from distortion due to passage of frequency components above $F_s/2$, as can be observed from Fig. 6.30b, it is followed by a lowpass filter that significantly attenuates frequencies above the folding frequency $F_s/2$.

The peaks in $H(F)$ within the band $|F| \le F_s/2$ may be undesirable in some applications. In such a case it is possible to modify the impulse response by reducing the slope by some factor $\beta < 1$. This results in the impulse response $h(t)$ illustrated in Fig. 6.31a. The corresponding frequency response is given by

$$H(F) = T \left[1 - \beta + \beta(1 + j2\pi FT) \frac{\sin \pi FT}{\pi FT} e^{-j\pi FT} \right] \frac{\sin \pi FT}{\pi FT} \tag{6.2.10}$$

The magnitude $|H(F)|$ is illustrated in Fig. 6.31b for $\beta = 0.5$, $\beta = 0.3$, and $\beta = 0.1$. We note that the peak in $H(F)$ is relatively small for $\beta = 0.3$ and does not exist when $\beta = 0.1$. Thus this modified first-order hold exhibits better frequency response charac- teristics in the frequency range $|F| \le F_s/2$.

6.2.3 Linear Interpolation with Delay

The first-order hold performs signal reconstruction by computing the slope of the straight line based on the current sample $x(nT)$ and the past sample $x(nT - T)$ of the signal. In

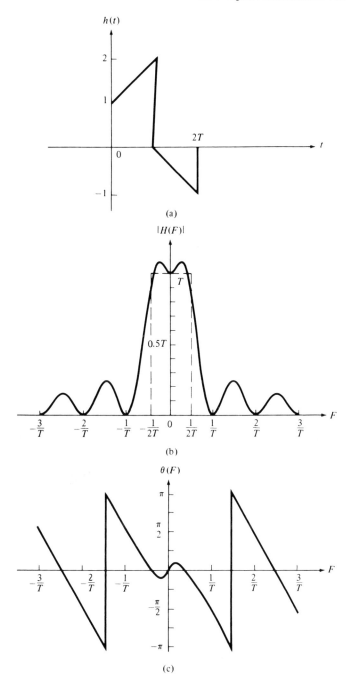

FIGURE 6.30 Impulse response (a) and frequency response characteristics (b) and (c) for a first-order hold.

effect, this technique *linearly extrapolates* or attempts to *linearly predict* the next sample of the signal based on the samples $x(nT)$ and $x(nT - T)$. As a consequence, the estimated signal waveform $\hat{x}(t)$ contains jumps at the sample points.

The jumps in $\hat{x}(t)$ can be avoided by providing a one-sample delay in the reconstruction process. Then successive sample points can be connected by straight-line segments. Thus

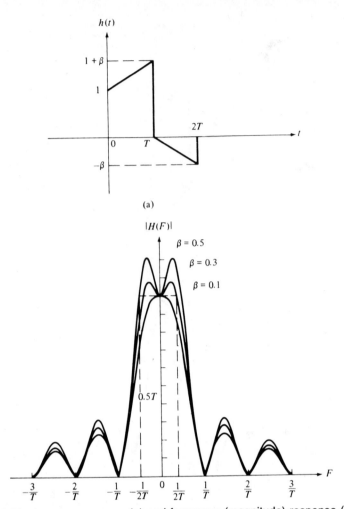

(a)

FIGURE 6.31   Impulse response (a) and frequency (magnitude) response (b) for a modified first-order hold.

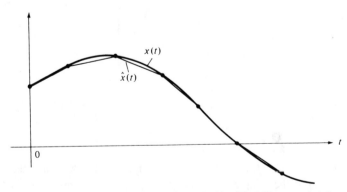

FIGURE 6.32   Linear interpolation of $x(t)$ with a T-second delay.

the resulting interpolated signal $\hat{x}(t)$ can be expressed as

$$\hat{x}(t) = x(nT - T) + \frac{x(nT) - x(nT - T)}{T}(t - nT) \qquad nT \leq t < (n + 1)T \quad (6.2.11)$$

We observe that at $t = nT$, $\hat{x}(nT) = x(nT - T)$ and at $t = nT + T$, $\hat{x}(nT + T) = x(nT)$. Therefore, $x(t)$ has an inherent delay of T seconds in interpolating the actual signal $x(t)$. Figure 6.32 illustrates this linear interpolation technique.

Viewed as a linear filter, the linear interpolator with a T-second delay has an impulse response

$$h(t) = \begin{cases} t/T & 0 \leq t < T \\ 2 - t/T & T \leq t < 2T \\ 0 & \text{otherwise} \end{cases} \qquad (6.2.12)$$

The corresponding frequency response is

$$H(F) = \int_0^T \frac{t}{T} e^{-j2\pi Ft} \, dt + \int_T^{2T} \left(2 - \frac{t}{T}\right) e^{-j2\pi Ft} \, dt$$

$$= T \left(\frac{\sin \pi FT}{\pi FT}\right)^2 e^{-j2\pi Ft} \qquad (6.2.13)$$

The impulse response and frequency response characteristics of this interpolation filter are illustrated in Fig. 6.33. We observe that the magnitude characteristic falls off rapidly and contains small sidelobes beyond the sampling frequency F_s. Furthermore, its phase characteristic is linear due to the delay T. By following this interpolator with a lowpass filter that has a sharp cutoff beyond the frequency $F_s/2$, the high-frequency components in $\hat{x}(t)$ can be further reduced.

This concludes our discussion of signal reconstruction based on simple interpolation techniques. The techniques that we have described are easily incorporated into the design of practical D/A converters for reconstructing analog signals from digital signals. We shall consider interpolation again in Chapter 8 in the context of changing the sampling rate in a digital signal processing system.

6.3 Inverse Systems, Deconvolution, and System Identification

As we have seen, a linear time-invariant system takes an input signal $x(n)$ and produces an output signal $y(n)$, which is the convolution of $x(n)$ with the unit sample response $h(n)$ of the system. In many practical applications we are given an input signal $x(n) = x_d(n) + x_i(n)$, where $x_d(n)$ represents a desired signal sequence and $x_i(n)$ represents some undesired interference or noise component, and we are asked to design a system that will suppress the undesired interference component. In such a case, the objective is to design a system that filters out the additive interference and noise while preserving the characteristics of the desired input signal $x_d(n)$. This is basically the filter design problem that is treated in detail in Chapter 8.

There is another class of practical problems in which we are given an output signal from a system whose characteristics are unknown and we are asked to determine the input signal. For example, in the transmission of digital information at high data rates over telephone channels, it is well known that the channel distorts the signal and causes intersymbol interference among the data symbols. The intersymbol interference may cause errors when we attempt to recover the data. If the data are transmitted over the

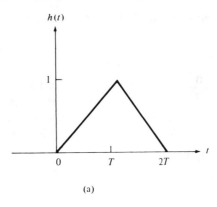

(a)

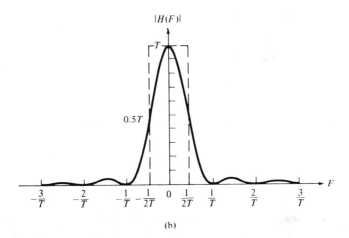

(b)

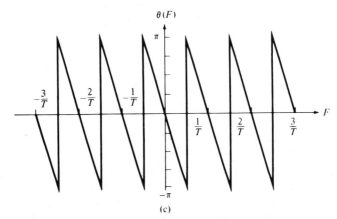

(c)

FIGURE 6.33 Impulse response (a) and frequency response characteristics (b) and (c) for the linear interpolator with delay.

direct dial network, the channel characteristics that cause the distortion vary considerably from one telephone call to another. Therefore, it is appropriate to assume that the channel, which is well modeled as a linear system, is unknown to the receiver that must recover the digital information. In such a case the problem is to design a corrective system which, when cascaded with the original system, produces an output that, in some sense, corrects for the distortion caused by the channel and thus yields a replica of the desired transmitted signal. In digital communications such a corrective system is called an *equalizer*. In the general context of linear systems theory, however, we call the corrective system an *inverse system*, because the corrective system has a frequency response which is basically the reciprocal of the frequency response of the system that caused the distortion. Furthermore, since the distortive system yields an output $y(n)$ that is the convolution of the input $x(n)$ with the impulse response $h(n)$, the inverse system operation that takes $y(n)$ and produces $x(n)$ is called *deconvolution*.

If the distortive system is unknown, it is often necessary, when possible, to excite the system with a known signal, observe the output, compare it with the input, and in some manner, determine the characteristics of the system. For example, in the digital communication problem described above, the measurement of the channel frequency response can be accomplished by transmitting a set of equal amplitude sinusoids at different frequencies with a specified set of phases, within the frequency band of the channel. The channel will attenuate and phase shift each of the sinusoids. By comparing the received signal with the transmitted signal, the receiver obtains a measurement of the channel frequency response which can be used to design the inverse system. The process of determining the characteristics of the unknown system, either $h(n)$ or $H(\omega)$, by a set of measurements performed on the system is called *system identification*.

The term "deconvolution" is often used in seismic signal processing and, more generally, in geophysics to describe the operation of separating the input signal from the characteristics of the system which are being measured. The deconvolution operation is actually intended to identify the characteristics of the system, in this case, the earth, and may also be viewed as a system identification problem. The "inverse system," in this case, has a frequency response that is the reciprocal of the input signal spectrum that has been used to excite the system.

In this section we describe the methods for system identification and deconvolution. As examples of these methods, we cite applications in the fields of communications and geophysics.

6.3.1 Invertibility of Linear Time-Invariant Systems

A system is said to be *invertible* if there is a one-to-one correspondence between its input and output signals. This definition implies that if we know the output sequence $y(n)$, $-\infty < n < \infty$ of an invertible system H, we can uniquely determine its input $x(n)$, $-\infty < n < \infty$. The *inverse system* with input $y(n)$ and output $x(n)$ is denoted by H^{-1}. Clearly, the cascade connection of a system and its inverse is equivalent to the identity system, since

$$w(n) = H^{-1}[y(n)] = H^{-1}\{H[x(n)]\} = x(n) \qquad (6.3.1)$$

as illustrated in Fig. 6.34. For example, the systems defined by the input–output relations $y(n) = ax(n)$ and $y(n) = x(n - 5)$ are invertible, whereas the input–output relations $y(n) = x^2(n)$ and $y(n) = 0$ represent noninvertible systems.

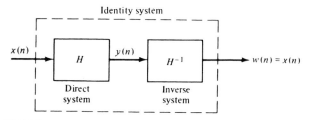

FIGURE 6.34   System H in cascade with its inverse H^{-1}.

As indicated above, inverse systems are important in many practical applications, including geophysics and digital communications. Let us begin by considering the problem of determining the inverse of a given system. We limit our discussion to the class of linear time-invariant discrete-time systems.

Now, suppose that the linear time-invariant system H has an impulse response $h(n)$ and let $h_I(n)$ denote the impulse response of the inverse system H^{-1}. Then (6.3.1) is equivalent to the convolution equation

$$w(n) = h_I(n) * h(n) * x(n) = x(n) \tag{6.3.2}$$

But (6.3.2) implies that

$$h(n) * h_I(n) = \delta(n) \tag{6.3.3}$$

The convolution equation in (6.3.3) can be used to solve for $h_I(n)$ for a given $h(n)$. However, the solution of (6.3.3) in the time domain is usually difficult. A simpler approach is to transform (6.3.3) into the z-domain and solve for H^{-1}. Thus in the z-transform domain, (6.3.3) becomes

$$H(z)H_I(z) = 1$$

and therefore the system function for the inverse system is

$$H_I(z) = \frac{1}{H(z)} \tag{6.3.4}$$

If $H(z)$ has a rational system function

$$H(z) = \frac{B(z)}{A(z)} \tag{6.3.5}$$

then

$$H_I(z) = \frac{A(z)}{B(z)} \tag{6.3.6}$$

Thus the zeros of $H(z)$ become the poles of the inverse system, and vice versa. Furthermore, if $H(z)$ is an FIR system, then $H_I(z)$ is an all-pole system, and if $H(z)$ is an all-pole system, then $H_I(z)$ is an FIR system.

EXAMPLE 6.3.1

Determine the inverse of the system with impulse response

$$h(n) = (\tfrac{1}{2})^n u(n)$$

Solution: The system function corresponding to $h(n)$ is

$$H(z) = \frac{1}{1 - \tfrac{1}{2}z^{-1}} \qquad \text{ROC: } |z| > \tfrac{1}{2}$$

This system is both causal and stable. Since $H(z)$ is an all-pole system, its inverse is FIR and is given by the system function

$$H_I(z) = 1 - \tfrac{1}{2}z^{-1}$$

Hence its impulse response is

$$h_I(n) = \delta(n) - \tfrac{1}{2}\delta(n - 1)$$

EXAMPLE 6.3.2 _____

Determine the inverse of the system with impulse response

$$h(n) = \delta(n) - \tfrac{1}{2}\delta(n - 1)$$

Solution: This is an FIR system and its system function is

$$H(z) = 1 - \tfrac{1}{2}z^{-1} \qquad \text{ROC: } |z| > 0$$

The inverse system has the system function

$$H_I(z) = \frac{1}{H(z)} = \frac{1}{1 - \tfrac{1}{2}z^{-1}} = \frac{z}{z - \tfrac{1}{2}}$$

Thus $H_I(z)$ has a zero at the origin and a pole at $z = \tfrac{1}{2}$. In this case there are two possible regions of convergence and hence two possible inverse systems, as illustrated in Fig. 6.35. If we take the ROC of $H_I(z)$ as $|z| > \tfrac{1}{2}$, the inverse transform yields

$$h_I(n) = (\tfrac{1}{2})^n u(n)$$

which is the impulse response of a causal and stable system. On the other hand, if the ROC is assumed to be $|z| < \tfrac{1}{2}$, the inverse system has an impulse response

$$h_I(n) = -(\tfrac{1}{2})^n u(-n - 1)$$

In this case the inverse system is anticausal and unstable.

_____

We observe that (6.3.3) cannot be solved uniquely by using (6.3.3) unless we specify the region of convergence for the system function of the inverse system.

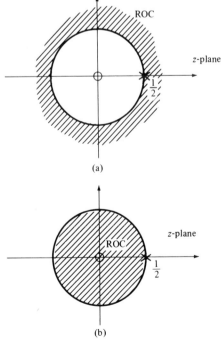

(a)

(b)

FIGURE 6.35 Two possible regions of convergence for $H(z) = z/(z - \tfrac{1}{2})$.

In some practical applications the impulse response $h(n)$ does not possess a z-transform that can be expressed in closed form. As an alternative we may solve (6.3.3) directly using a digital computer. Since (6.3.3) does not possess a unique solution, in general, we assume that the system and its inverse are causal. Then (6.3.3) simplifies to the equation

$$\sum_{k=0}^{n} h(k)h_I(n-k) = \delta(n) \tag{6.3.7}$$

By assumption, $h_I(n) = 0$ for $n < 0$. For $n = 0$ we obtain

$$h_I(0) = 1/h(0) \tag{6.3.8}$$

The values of $h_I(n)$ for $n \geq 1$ can be obtained recursively from the equation

$$h_I(n) = \sum_{k=1}^{n} \frac{h(n)h_I(n-k)}{h(0)} \qquad n \geq 1 \tag{6.3.9}$$

This recursive relation can easily be programmed on a digital computer.

There are two problems associated with (6.3.9). First, the method does not work if $h(0) = 0$. However, this problem can easily be remedied by introducing an appropriate delay in the right-hand-side of (6.3.7), that is, by replacing $\delta(n)$ by $\delta(n-m)$, where $m = 1$ if $h(0) = 0$ and $h(1) \neq 0$, and so on. Second, the recursion in (6.3.9) gives rise to round-off errors which grow with n and, as a result, the numerical accuracy of $h(n)$ deteriorates for large n.

EXAMPLE 6.3.3 _____

Determine the causal inverse of the FIR system with impulse response

$$h(n) = \delta(n) - \alpha \, \delta(n-1)$$

Solution: Since $h(0) = 1$, $h(1) = -\alpha$, and $h(n) = 0$ for $n \geq \alpha$, we have

$$h_I(0) = 1/h(0) = 1$$

and

$$h_I(n) = \alpha h_I(n-1) \qquad n \geq 1$$

Consequently,

$$h_I(1) = \alpha, \quad h_I(2) = \alpha^2, \quad \ldots, \quad h_I(n) = \alpha^n$$

which corresponds to a causal IIR system as expected.

_____

6.3.2 Minimum-Phase, Maximum-Phase, and Mixed-Phase Systems

The invertibility of a linear time-invariant system is intimately related to the characteristics of the phase spectral function of the system. To illustrate this point, let us consider two FIR systems, characterized by the system functions

$$H_1(z) = 1 + \tfrac{1}{2}z^{-1} = z^{-1}(z + \tfrac{1}{2}) \tag{6.3.10}$$
$$H_2(z) = \tfrac{1}{2} + z^{-1} = z^{-1}(\tfrac{1}{2}z + 1) \tag{6.3.11}$$

The system in (6.3.10) has a zero at $z = \tfrac{1}{2}$ and an impulse response $h(0) = 1$, $h(1) = 1/2$. The system in (6.3.11) has a zero at $z = -2$ and an impulse response $h(0) =$

$1/2$, $h(1) = 1$, which is the reverse of the system in (6.3.10). This is due to the reciprocal relationship between the zeros of $H_1(z)$ and $H_2(z)$.

In the frequency domain, the two systems are characterized by their frequency response functions, which can be expressed as

$$|H_1(\omega)| = |H_2(\omega)| = \sqrt{\tfrac{5}{4} + \cos \omega} \tag{6.3.12}$$

and

$$\Theta_1(\omega) = -\omega + \tan^{-1} \frac{\sin \omega}{\tfrac{1}{2} + \cos \omega} \tag{6.3.13}$$

$$\Theta_2(\omega) = -\omega + \tan^{-1} \frac{\sin \omega}{2 + \cos \omega} \tag{6.3.14}$$

The magnitude characteristics for the two systems are identical because the zeros of $H_1(z)$ and $H_2(z)$ are reciprocals.

The graphs of $\Theta_1(\omega)$ and $\Theta_2(\omega)$ are illustrated in Fig. 6.36. We observe that the phase characteristic $\Theta_1(\omega)$ for the first system begins at zero phase at the frequency $\omega = 0$ and terminates at zero phase at the frequency $\omega = \pi$. Hence the net phase change, $\Theta_1(\pi) - \Theta_1(0)$ is zero. On the other hand, the phase characteristic for the system with the zero outside the unit circle undergoes a net phase change $\Theta_2(\pi) - \Theta_2(0) = \pi$ radians. As a consequence of these different phase characteristics, we call the first system a *minimum-phase system* and the second system is called a *maximum-phase system*.

These definitions are easily extended to an FIR system of arbitrary length. To be specific, an FIR system of length $M + 1$ has M zeros. Its frequency response can be expressed as

$$H(\omega) = b_0(1 - z_1 e^{-j\omega})(1 - z_2 e^{-j\omega}) \cdots (1 - z_M e^{-j\omega}) \tag{6.3.15}$$

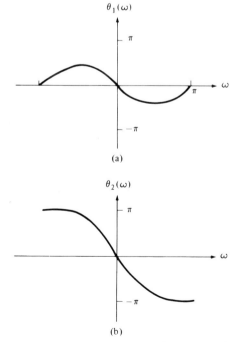

$\theta_1(\omega)$

(a)

$\theta_2(\omega)$

(b)

FIGURE 6.36 Phase response characteristics for the systems in (6.3.10) and (6.3.11).

where $\{z_i\}$ denote the zeros and b_0 is an arbitrary constant. When all the zeros are inside the unit circle, each term in the product of (6.3.15), corresponding to a real-valued zero, will undergo a net phase change of zero between $\omega = 0$ and $\omega = \pi$. Also, each pair of complex-conjugate factors in $H(\omega)$ will undergo a net phase change of zero. Therefore,

$$\angle\, H(\pi) - \angle\, H(0) = 0 \qquad (6.3.16)$$

and hence the system is called a minimum-phase system. On the other hand, when all the zeros are outside the unit circle, a real-valued zero will contribute a net phase change of π radians as the frequency varies from $\omega = 0$ to $\omega = \pi$, and each pair of complex-conjugate zeros will contribute a net phase change of 2π radians. Therefore,

$$\angle\, H(\pi) - \angle\, H(0) = M\pi \qquad (6.3.17)$$

which is the largest possible phase change for an FIR system with M zeros. Hence the system is called maximum phase. It follows from the discussion above that

$$\angle\, H_{\max}(\pi) \geq \angle\, H_{\min}(\pi) \qquad (6.3.18)$$

If the FIR system with M zeros has some of its zeros inside the unit circle and the remaining outside the unit circle, it is called a *mixed-phase system* or a *non-minimum-phase system*.

Since the derivative of the phase characteristic of the system is a measure of the time delay that signal frequency components undergo in passing through the system, a minimum phase characteristic implies a minimum delay function, while a maximum-phase characteristic implies that the delay characteristic is also maximum.

Another characteristic that distinguishes a minimum-phase FIR system from a maximum-phase FIR system is the *partial energy* in the impulse response, which is defined as

$$E(n) = \sum_{k=0}^{n} h^2(k) \qquad (6.3.19)$$

A minimum-phase system has most of its energy near $n = 0$, while a maximum phase system peaks at $n = M$. Consequently,

$$E_{\min}(n) \geq E_{\max}(n) \qquad n = 0, 1, \ldots, M \qquad (6.3.20)$$

Now suppose that we have an FIR system with real coefficients. Then the magnitude square value of its frequency response is

$$|H(\omega)|^2 = H(z)H(z^{-1})\big|_{z = e^{j\omega}} \qquad (6.3.21)$$

This relationship implies that if we replace a zero z_k of the system by its inverse $1/z_k$, the magnitude characteristic of the system does not change. Thus if we reflect a zero z_k that is inside the unit circle into a zero $1/z_k$ outside the unit circle, the magnitude characteristic of the frequency response is invariant to such a change.

It is apparent from the discussion above that if $|H(\omega)|^2$ is the magnitude square frequency response of an FIR system having M zeros, there are 2^M possible configurations for the M zeros, some of which are inside the unit circle and the remaining are outside the unit circle. Clearly, one configuration has all the zeros inside the unit circle, which corresponds to the minimum-phase system. A second configuration has all the zeros outside the unit circle, which corresponds to the maximum-phase system. The remaining $2^M - 2$ configurations correspond to mixed-phase systems. However, not all $2^M - 2$ mixed-phase configurations necessarily correspond to FIR systems with real-valued coefficients. Specifically, any pair of complex-conjugate zeros result in only two possible configurations, whereas a pair of real-valued zeros yield four possible configurations.

EXAMPLE 6.3.4

Determine the zeros for the following FIR systems and indicate whether the system is minimum phase, maximum phase, or mixed phase.

$$H_1(z) = 6 + z^{-1} - z^{-2}$$
$$H_2(z) = 1 - z^{-1} - 6z^{-2}$$
$$H_3(z) = 1 - \tfrac{5}{2}z^{-1} - \tfrac{3}{2}z^{-2}$$
$$H_4(z) = 1 + \tfrac{5}{3}z^{-1} - \tfrac{2}{3}z^{-2}$$

Solution: By factoring the system functions we find the zeros for the four systems are

$$H_1(z) \longrightarrow z_{1,2} = -\tfrac{1}{2}, \tfrac{1}{3} \longrightarrow \text{minimum phase}$$
$$H_2(z) \longrightarrow z_{1,2} = -2, 3 \longrightarrow \text{maximum phase}$$
$$H_3(z) \longrightarrow z_{1,2} = -\tfrac{1}{2}, 3 \longrightarrow \text{mixed phase}$$
$$H_4(z) \longrightarrow z_{1,2} = -2, \tfrac{1}{3} \longrightarrow \text{mixed phase}$$

Since the zeros of the four systems are reciprocals of one another, it follows that all four systems have identical magnitude frequency response characteristics but different phase characteristics.

The minimum-phase characteristic property of FIR systems carries over to IIR systems with rational system functions. Specifically, a *stable* IIR system with system function

$$H(z) = \frac{B(z)}{A(z)} \tag{6.3.22}$$

is minimum phase if all the zeros of $H(z)$ are inside the unit circle [i.e., all the roots of $B(z)$ are inside the unit circle]. If some of the zeros are outside the unit circle, the system is called mixed phase, and when all the zeros are outside the unit circle, the system is called maximum phase.

The discussion above brings us to an important point that should be emphasized. That is, a *stable* pole–zero system that is minimum phase has a stable inverse which is also minimum phase. The inverse system has the system function

$$H^{-1}(z) = \frac{A(z)}{B(z)} \tag{6.3.23}$$

Hence the minimum-phase property of $H(z)$ ensures the stability of the inverse system $H^{-1}(z)$ and the stability of $H(z)$ implies the minimum-phase property of $H^{-1}(z)$. Mixed-phase systems and maximum-phase systems result in unstable inverse systems.

6.3.3 System Identification and Deconvolution

Suppose that we excite an unknown linear time-invariant system with an input sequence $x(n)$ and we observe the output sequence $y(n)$. From the output sequence we wish to determine the impulse response of the unknown system. This is a problem in *system identification*, which can be solved by *deconvolution*. Thus we have

$$y(n) = h(n) * x(n)$$
$$= \sum_{k=-\infty}^{\infty} h(k)x(n-k) \tag{6.3.24}$$

An analytical solution of the deconvolution problem can be obtained by working with the z-transform of (6.3.24). In the z-transform domain we have

$$Y(z) = H(z)X(z)$$

and hence

$$H(z) = \frac{Y(z)}{X(z)} \tag{6.3.25}$$

$X(z)$ and $Y(z)$ are the z-transforms of the available input signal $x(n)$ and the observed output signal $y(n)$, respectively. This approach is attractive when there are closed-form expressions for $X(z)$ and $Y(z)$.

EXAMPLE 6.3.5

A causal system produces the output sequence

$$y(n) = \begin{cases} 1 & n = 0 \\ \frac{7}{10} & n = 1 \\ 0 & \text{otherwise} \end{cases}$$

when excited by the input sequence

$$x(n) = \begin{cases} 1 & n = 0 \\ -\frac{7}{10} & n = 1 \\ \frac{1}{10} & n = 2 \\ 0 & \text{otherwise} \end{cases}$$

Determine its impulse response and its input–output equation.

Solution: The system function is easily determined by taking the z-transforms of $x(n)$ and $y(n)$. Thus we have

$$H(z) = \frac{Y(z)}{X(z)} = \frac{1 + \frac{7}{10}z^{-1}}{1 - \frac{7}{10}z^{-1} + \frac{1}{10}z^{-2}}$$

$$= \frac{1 + \frac{7}{10}z^{-1}}{(1 - \frac{1}{2}z^{-1})(1 - \frac{1}{5}z^{-1})}$$

Since the system is causal, its ROC is $|z| > \frac{1}{2}$. The system is also stable since its poles lie inside the unit circle.

The input–output difference equation for the system is

$$y(n) = \frac{7}{10}y(n-1) - \frac{1}{10}y(n-2) + x(n) + \frac{7}{10}x(n-1)$$

Its impulse response is determined by performing a partial-fraction expansion of $H(z)$ and inverse transforming the result. This computation yields

$$h(n) = [4(\tfrac{1}{2})^n - 3(\tfrac{1}{5})^n]u(n)$$

We observe that (6.3.25) determines the unknown system uniquely if it is known that the system is causal. This assumption is certainly justified in most cases of practical interest.

In practical applications where no simple closed-form expressions exist for $X(z)$ and $Y(z)$, we can deal directly with the time-domain expression given by (6.3.24). If we assume that the system and the input signal are causal, then

$$y(n) = \sum_{k=0}^{n} h(k)x(n - k) \qquad n \geq 0 \qquad (6.3.26)$$

For $n = 0$ we obtain

$$y(0) = h(0)x(0)$$

and, therefore,

$$h(0) = y(0)/x(0) \qquad (6.3.27)$$

For $n \geq 1$, (6.3.26) can be rearranged to yield $h(n)$ recursively. That is,

$$y(n) = h(n)x(0) + \sum_{k=0}^{n-1} h(k)x(n - k)$$

and hence

$$h(n) = \frac{y(n) - \sum_{k=0}^{n-1} h(k)x(n - k)}{x(0)} \qquad n \geq 1 \qquad (6.3.28)$$

The recursive relation in (6.3.28) requires, of course, that $x(0) \neq 0$.

EXAMPLE 6.3.6

Determine the impulse response of a causal system given the following input and output sequences:

$$x(n) = \left\{ \begin{array}{c} 2, 1, 2 \\ \uparrow \end{array} \right\}$$

$$y(n) = \left\{ \begin{array}{c} 4, 6, 8, 5, 2 \\ \uparrow \end{array} \right\}$$

Solution: By applying (6.3.27) and (6.3.28), we obtain the following results:

$$h(0) = y(0)/x(0) = 2$$

$$h(1) = \frac{y(1) - h(0)x(1)}{x(0)} = 2$$

$$h(2) = \frac{y(2) - h(0)x(2) - h(1)x(1)}{x(0)} = 1$$

$$h(3) = \frac{y(3) - h(0)x(3) - h(1)x(2) - h(2)x(1)}{x(0)} = 0$$

$$h(4) = \frac{y(4) - h(0)x(4) - h(1)x(3) - h(2)x(2) - h(3)x(1)}{x(0)} = 0$$

Clearly, $h(n) = 0$ for $n \geq 3$. Therefore, the desired system has the impulse response

$$h(n) = \left\{ \begin{array}{c} 2, 2, 1 \\ \uparrow \end{array} \right\}$$

It is apparent from the development above that a causal system can be uniquely identified from observation of the input and output signal sequences.

As indicated previously, a related problem is one in which a signal is observed at the output of a system and the problem is to determine the input signal. When the impulse response of the system is known, we have the deconvolution problem that was treated in Section 6.3.1 using an inverse system. Alternatively, we may use the recursive equation in (6.3.28) and interchange the roles of $h(n)$ and $x(n)$. On the other hand, when the impulse response of the system is unknown, as in the case of digital communications over a distortive channel, the system can first be identified by observing its output to a known input signal (each as a sequence of sinusoids of different frequencies) and then using the "measured" system in performing the deconvolution.

In the following section we describe a system identification method that is based on correlation techniques and yields the frequency response characteristic of the unknown system. This development attempts to avoid statistical concepts and the use of random process theory. Hence we employ the definitions of correlation functions based on time averages, introduced in Section 2.5.1. The reader who is familiar with statistical concepts should have no difficulty in interpreting the results in terms of statistical notions. Two additional methods for system identification based on the least-squares method are described in Section 6.3.5.

6.3.4 System Identification Based on Correlation Techniques

In the preceding section we described a system identification method based on knowing both the input sequence $x(n)$ to the system and observing the output sequence $y(n)$. A recursive formula was derived for obtaining $h(n)$ given $x(n)$ and $y(n)$.

In this section we present an alternative method in which the system parameters are obtained by correlation methods. As indicated previously, the frequency response of a system may simply be measured by exciting the system with sinusoids of different frequencies that cover the frequency bandwidth of the system and measuring the attenuation and phase shift imparted on the sinusoids by the unknown system. This approach requires that we decide in advance how many sinusoids we should use in the measurement, what frequency separation we should select, and how to resolve the various sinusoids at the output of the system in order to perform the necessary magnitude and phase measurements.

These issues can be avoided by selecting as an input signal $x(n)$ one that has a continuous frequency content instead of discrete frequency sinusoids and covers the bandwidth of the system. Then the system identification can be performed by an alternative method based on correlation techniques. To explain these techniques, we first develop some basic input–output relations for correlation sequences and their spectra for linear time-invariant discrete systems. The following development is based on the assumption that all signals are finite energy signals, although a parallel development can also be carried out for power signals.

Suppose that a signal $x(n)$ with known autocorrelation sequence $r_{xx}(n)$ is applied to a stable linear time-invariant discrete system with impulse response $h(n)$. The system output is

$$y(n) = \sum_{k=-\infty}^{\infty} h(k)x(n-k)$$

The crosscorrelation sequence $r_{yx}(l)$ is

$$r_{yx}(l) = \sum_{n=-\infty}^{\infty} y(n)x(n-l)$$

$$= \sum_{n=-\infty}^{\infty} \left[\sum_{k=-\infty}^{\infty} h(k)x(n-k) \right] x(n-l)$$

$$= \sum_{k=-\infty}^{\infty} h(k) \left[\sum_{n=-\infty}^{\infty} x(n-k)x(n-l) \right] \qquad (6.3.29)$$

But the inner sum in (6.3.29) is the autocorrelation sequence $r_{xx}(l-k)$. Consequently, (6.3.29) becomes

$$r_{yx}(l) = \sum_{k=-\infty}^{\infty} h(k)r_{xx}(l-k) \qquad (6.3.30)$$

Since (6.3.30) is a convolution of $h(k)$ with $r_{xx}(l)$, the z-transform of this equation is

$$S_{yx}(z) = H(z)S_{xx}(z) \qquad (6.3.31)$$

If we evaluate (6.3.31) on the unit circle (i.e., let $z = e^{j\omega}$), we have Fourier transform of (6.3.30) in the form

$$S_{yx}(\omega) = H(\omega)S_{xx}(\omega) \qquad (6.3.32)$$

where the quantity $S_{xx}(\omega)$ is the energy density spectrum of the input signal, defined in Table 4.2 as

$$S_{xx}(\omega) = |X(\omega)|^2 \qquad (6.3.33)$$

and $S_{yx}(\omega)$ is *cross-energy density* spectrum of the signals $x(n)$ and $y(n)$. Thus we see that either (6.3.31) or (6.3.32) provides a method for estimating the system function or the frequency response of an unknown system $H(\omega)$ if we know $S_{yx}(z)$ and $S_{xx}(z)$ or the energy spectral density $S_{xx}(\omega)$ of the input and the cross-energy density spectrum $S_{yx}(\omega)$. That is, $H(z) = S_{yx}(z)/S_{xx}(z)$ or, equivalently,

$$H(\omega) = \frac{S_{yx}(\omega)}{S_{xx}(\omega)} \qquad (6.3.34)$$

EXAMPLE 6.3.7 _____

An input sequence $x(n)$ with autocorrelation

$$r_{xx}(l) = \begin{cases} 1 & l = 0 \\ \frac{1}{2} & l = \pm 1 \\ 0 & \text{otherwise} \end{cases}$$

is used to excite a causal system with unknown system function $H(z)$. The crosscorrelation between the system input and output is measured as

$$r_{yx}(l) = \begin{cases} \frac{1}{2} & l = -1 \\ 1 & l = 0 \\ -1 & l = 2 \\ -\frac{1}{2} & l = 3 \\ 0 & \text{otherwise} \end{cases}$$

Determine $H(z)$ and $H(\omega)$.

Solution: The z-transforms of $r_{xx}(l)$ and $r_{yx}(l)$ are

$$S_{xx}(z) = \tfrac{1}{2}z + 1 + \tfrac{1}{2}z^{-1}$$
$$S_{yx}(z) = \tfrac{1}{2}z + 1 - z^{-2} - \tfrac{1}{2}z^{-3}$$

$H(z)$ is the ratio of $S_{yx}(z)$ to $S_{xx}(z)$. Thus

$$H(z) = \frac{\tfrac{1}{2}z + 1 - z^{-2} - \tfrac{1}{2}z^{-3}}{\tfrac{1}{2}z + 1 + \tfrac{1}{2}z^{-1}}$$

By dividing the denominator into the numerator, we obtain the system in the form

$$H(z) = 1 - z^{-2}$$

and the corresponding frequency response is

$$
\begin{aligned}
H(\omega) &= 1 - e^{-j2\omega}\\
&= e^{-j\omega}(e^{j\omega} - e^{-j\omega})\\
&= (2 \sin \omega)e^{-j(\omega - \pi/2)}
\end{aligned}
$$

If the input signal $x(n)$ is designed to have a relatively flat (white) spectrum over the bandwidth of the system, then

$$H(\omega) = CS_{yx}(\omega) \tag{6.3.35}$$

where $C = 1/S_{xx}(\omega)$ is a constant, independent of frequency over the frequency band of $H(\omega)$. Thus $H(\omega)$ is proportional to the Fourier transform of the crosscorrelation sequence between $x(n)$ and $y(n)$. Figure 6.37 illustrates, by means of a block diagram, the procedure for determining $H(\omega)$.

In practice the computation of the crosscorrelation sequence is performed only over a finite set of lags, with input and output sequences, $x(n)$ and $y(n)$, respectively, that are finite in duration. Consequently, the crosscorrelation sequence is an estimate of the true crosscorrelation and hence the procedure above yields an estimate $H(\omega)$ of the actual system frequency response. The input sequence $x(n)$ is usually a spectrally flat (white noise) sequence generated either by a noise source or pseudorandomly on a digital computer (see Appendix 2A).

In the event that the input signal $x(n)$ is not available, but only the output $y(n)$ of the system is available for system identification, we may compute the autocorrelation sequence $\{r_{yy}(l)\}$ of the output and its Fourier transform. This yields the result

$$S_{yy}(\omega) = |H(\omega)|^2 S_{xx}(\omega) \tag{6.3.36}$$

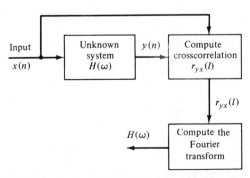

FIGURE 6.37   System identification by the crosscorrelation technique.

Now, if the input signal energy spectrum $S_{xx}(\omega)$ is flat in the frequency band of $H(\omega)$, then

$$S_{yy}(\omega) = C|H(\omega)|^2 \qquad (6.3.37)$$

where C is an arbitrary constant. In this case we obtain the magnitude of $H(\omega)$ as

$$|H(\omega)| = \sqrt{\frac{S_{yy}(\omega)}{C}} \qquad (6.3.38)$$

However, there is no phase information in $S_{yy}(\omega)$.

The time-domain equivalent of (6.3.37), with $C = 1$ for mathematical convenience, is

$$r_{yy}(l) = h(l) * h(-l) \qquad (6.3.39)$$

where $h(l)$ is the impulse response of the unknown system. The z-transform of (6.3.39) yields

$$S_{yy}(z) = H(z)H(z^{-1}) \qquad (6.3.40)$$

Although the autocorrelation technique does not provide the phase spectrum of the system, if the system is known to be minimum phase, $H(z)$ can be determined uniquely. The following example illustrates this point.

EXAMPLE 6.3.8 _____

An unknown system is excited by a white noise sequence and from the response $y(n)$ we have obtained the autocorrelation sequence

$$r_{yy}(l) = \begin{cases} 38 & l = 0 \\ 5 & l = \pm 1 \\ -6 & l = \pm 2 \\ 0 & \text{otherwise} \end{cases}$$

Determine the system function corresponding to the minimum-phase system.

Solution: Since the input sequence is white, $S_{xx}(z) = 1$ and $S_{yy}(z)$ is given by (6.3.40). The z-transform of the autocorrelation $r_{yy}(l)$ is

$$\begin{aligned} S_{yy}(z) &= -6z^2 + 5z + 38 + 5z^{-1} - 6z^{-2} \\ &= H(z)H(z^{-1}) \end{aligned}$$

We can solve for the four roots of $S_{yy}(z)$, which are of the form $z_1, z_2, 1/z_1, 1/z_2$, where z_1 and z_2 are inside the unit circle. Then $H(z)$ has the form

$$\begin{aligned} H(z) &= b_0(1 - z_1 z^{-1})(1 - z_2 z^{-1}) \\ &= b_0 + b_1 z^{-1} + b_2 z^{-2} \end{aligned}$$

It is easily verified that

$$\begin{aligned} b_0^2 + b_1^2 + b_2^2 &= 38 \\ b_0 b_1 + b_1 b_2 &= 5 \\ b_0 b_2 &= -6 \end{aligned}$$

The minimum-phase solution of these equations is $b_0 = 6$, $b_1 = 1$, $b_2 = -1$. Hence

$$\begin{aligned} H(z) &= 6 + z^{-1} - z^{-2} \\ &= 6z^{-2} (z + \tfrac{1}{2})(z - \tfrac{1}{3}) \end{aligned}$$

which is the desired minimum-phase system.

On the other hand, the maximum-phase solution is $b_0 = -1$, $b_1 = 1$, $b_2 = 6$, which corresponds to a system with zeros at $z = -2$ and $z = 3$. Two mixed-phase solutions are also possible.

6.3.5 System Identification Based on the Least-Squares Method

The method of least squares is an alternative technique for performing system identification. In this method we postulate a model for the system and determine the parameters of the model that minimize, in the least-squares sense, the error between the actual system response and the response of the model.

We shall describe two techniques. One is based on an FIR model for the system and the second is based on an all-pole IIR model for the system. Although a pole–zero model for an IIR system is not treated in this section, this model is treated in Section 8.4 in the context of digital filter design.

Let us begin with the FIR model for an unknown system $H(z)$. The model has a system function

$$\hat{H}(z) = \sum_{k=0}^{M} b_k z^{-k} \qquad (6.3.41)$$

Both the unknown system and the model are excited by the same input sequence $x(n)$ as shown in Fig. 6.38. Let $y(n)$ be the observed output of the unknown system and let $\hat{y}(n)$ be the output of the model. Hence

$$\hat{y}(n) = \sum_{k=0}^{M} b_k x(n - k) \qquad (6.3.42)$$

The difference between the output $y(n)$ of the unknown system and the output $\hat{y}(n)$ of the model is the error $e(n)$. Thus $e(n)$ is defined as

$$e(n) = y(n) - \hat{y}(n) \qquad (6.3.43)$$

$$e(n) = y(n) - \sum_{k=0}^{M} b_k x(n - k) \qquad n = 0, 1, \ldots$$

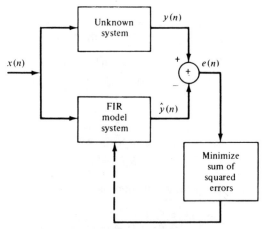

FIGURE 6.38 System identification based on FIR model and least-squares criterion.

In the method of least squares, the parameters $\{b_k\}$ of the model are selected to minimize the sum of the squared-error sequence $e(n)$, that is,

$$\mathcal{E} = \sum_{n=0}^{\infty} e^2(n) \tag{6.3.44}$$

$$= \sum_{n=0}^{\infty} \left[y(n) - \sum_{k=0}^{M} b_k x(n-k) \right]^2$$

By differentiating with respect to each of the model coefficients $\{b_k\}$, we obtain the set of linear equations

$$\sum_{k=0}^{M} b_k r_{xx}(l-k) = r_{yx}(l) \qquad l = 0, 1, \ldots, M \tag{6.3.45}$$

where, by definition,

$$r_{xx}(k,l) = \sum_{n=0}^{\infty} x(n-k)x(n-l) = r_{xx}(k-l) \tag{6.3.46}$$

$$r_{yx}(l) = \sum_{n=0}^{\infty} y(n)x(n-l) \tag{6.3.47}$$

We observe that $r_{xx}(l)$ is the autocorrelation of the excitation sequence $x(n)$ and $r_{yx}(l)$ is the crosscorrelation of the input $x(n)$ with the output $y(n)$ of the unknown system. In practice, the upper limit on the summations in (6.3.46) and (6.3.47) is, of course, finite, but large compared to M.

It is interesting to note that the set of linear equations in (6.3.45), obtained with the least-squares optimization method, resembles the form of the linear equations in (6.3.30), which were obtained directly by crosscorrelating the input $x(n)$ to the unknown system with the output $y(n)$. In fact, the equations in (6.3.30) reduce to the equations in (6.3.45) if we impose the restriction that the unknown system is a causal FIR system of length $M + 1$, and the input is causal. Therefore, the crosscorrelation method for system identification, described in the preceding section, is also an optimum method in the sense that it results in a least-squares error solution.

EXAMPLE 6.3.9

In a radio communication system, the signal transmitted over the channel propagates over two different paths as shown in Fig. 6.39. The direct path has an unknown gain (attenuation) factor b_0 and the reflected path is characterized by a gain factor $b_1 < b_0$,

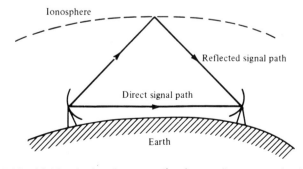

FIGURE 6.39 Multipath signal propagation in a radio communication system.

and a relative time delay t_d. Thus the received signal can be expressed (in continuous time) as

$$y(t) = b_0 x(t) + b_1 x(t - t_d)$$

The reflected path (multipath component) constitutes an interference component that can be eliminated at the receiver by use of an inverse system through which $y(t)$ is processed. To implement the inverse system, we first identify the channel system function using the method of least squares. For this purpose the input signal $x(t)$ is assumed known at the receiver.

Solution: We assume that the received signal is sampled sufficiently fast to avoid aliasing and so that the time delay t_d is an integer multiple of the sampling interval. Thus the signal $y(t)$ can be expressed in sampled form as

$$y(n) = b_0 x(n) + b_1 x(n - D)$$

where D is an integer delay. The crosscorrelation of $y(n)$ with $x(n)$ results in the equation

$$r_{yx}(l) = b_0 r_{xx}(l) + b_1 r_{xx}(l - D)$$

where $r_{yx}(l)$ and $r_{xx}(l)$ are given by (6.3.47) and (6.3.46), respectively. The crosscorrelation $r_{yx}(l)$ will exhibit two peaks, one at $l = 0$, corresponding to the direct path and one at $l = D$ corresponding to the reflected path. Thus D can be determined from the crosscorrelation sequence $r_{yx}(l)$.

Once D is determined, the parameters b_0 and b_1 can be determined from the following two equations:

$$b_0 r_{xx}(0) + b_1 r_{xx}(D) = r_{yx}(0)$$
$$b_0 r_{xx}(D) + b_1 r_{xx}(0) = r_{yx}(D)$$

Thus the system function of the unknown channel is

$$H(z) = b_0 + b_1 z^{-D}$$

and the desired inverse system is

$$\hat{H}_I(z) = \frac{1}{H(z)} = \frac{1}{b_0 (1 + (b_1/b_0)z^{-D})}$$

Since $b_0 > b_1$, the inverse system is stable. In this case the channel is FIR and its inverse (the channel equalizer) $\hat{H}_I(z)$ is IIR.

Let us now consider the system identification of an all-pole system with system function

$$H(z) = \frac{b_0}{1 + \sum_{k=1}^{N} a_k z^{-k}} \tag{6.3.48}$$

We should view (6.3.48) as the model for the actual unknown system. The actual system may differ to some degree from the model, however. The problem is to identify the parameters b_0 and $\{a_k\}$ of this model. Suppose that we place an FIR system in cascade with the unknown system, which has the system function $1/H(z)$. If the unknown system is indeed all-pole, as in (6.3.48), the cascade of the two systems results in an identity system, as illustrated in Fig. 6.40.

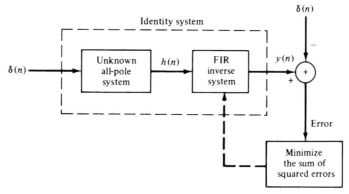

FIGURE 6.40 Identification of an all-pole system by use of a least-squares FIR inverse system.

Now, suppose that the unknown system is excited by an impulse $\delta(n)$. Its output is the sequence $\{h(n)\}$, its impulse response. In the cascade configuration of Fig. 6.40, $h(n)$ is the input to the FIR inverse system and $y(n)$ is its output. Ideally, the desired output is

$$y_d(n) = \begin{cases} 1 & n = 0 \\ 0 & n > 0 \end{cases} \tag{6.3.49}$$

The actual output of the FIR system is

$$y(n) = \frac{1}{b_0}\left[h(n) + \sum_{k=1}^{N} a_k h(n - k) \right] \tag{6.3.50}$$

The condition that $y(0) = y_d(0) = 1$ is satisfied by selecting $b_0 = h(0)$. For $n > 0$, $y(n)$ represents the error between the desired output $y_d(n) = 0$ from the ideal inverse system and the actual FIR inverse system. The parameters $\{a_k\}$ will be selected to minimize the sum of squares of the error sequence, that is,

$$\mathscr{E} = \sum_{n=1}^{\infty} y^2(n) \tag{6.3.51}$$

$$= \sum_{n=1}^{\infty}\left[h(n) + \sum_{k=1}^{N} a_k h(n - k) \right]^2 \bigg/ h^2(0)$$

By differentiating with respect to the parameters $\{a_k\}$, it is easily established that we obtain the set of linear equations of the form

$$\sum_{k=1}^{N} a_k r_{hh}(k, l) = -r_{hh}(l, 0) \qquad l = 1, 2, \ldots , N \tag{6.3.52}$$

where, by definition,

$$r_{hh}(k, l) = \sum_{n=1}^{\infty} h(n - k)h(n - l) \tag{6.3.53}$$

$$= \sum_{n=0}^{\infty} h(n)h(n + k - l) = r_{hh}(k - l)$$

The solution of (6.3.52) yields the desired parameters for the inverse system $1/H(z)$. Thus we have solved the identification problem for the unknown system based on an all-pole model for the system.

In practice, the correlation sequence in (6.3.53) is estimated from a finite number of observations of $h(n)$. Consequently, the upper limit on the sum in (6.3.53) is finite. In such a case we solve the equations in (6.3.52) based on an estimate of $r_{hh}(k, l)$ computed over a finite data record.

Although the least-squares FIR inverse method outlined above was described in the context of identifying a system based on an all-pole model, the same method can be used to design approximate inverses for either an FIR or a pole–zero system. This problem is considered in the following section.

6.3.6 FIR Least-Squares Inverse Filters

In the preceding section we described the use of the least-squares error criterion in the identification of FIR and IIR systems. In this section we use a similar approach to determine a least-squares FIR inverse filter to a known system.

Recall that the inverse to a linear time-invariant system with impulse response $h(n)$ and system function $H(z)$ is defined as the system whose impulse response $h_I(n)$ and system function $H_I(z)$, satisfy the respective equations.

$$h(n) * h_I(n) = \delta(n) \tag{6.3.54}$$
$$H(z)H_I(z) = 1 \tag{6.3.55}$$

In general, $H_I(z)$ will be IIR, unless $H(z)$ is an all-pole system, in which case $H_I(z)$ will be FIR.

In many practical applications, it is desirable to restrict the inverse filter to be FIR. Obviously, one simple method is to truncate $h_I(n)$. In so doing, we incur a total squared approximation error equal to

$$\mathcal{E}_t = \sum_{n=M+1}^{\infty} h_I^2(n) \tag{6.3.56}$$

where $M+1$ is the length of the truncated filter and $\mathcal{E}_t$ represents the energy in the tail of the impulse response $h_I(n)$.

Alternatively, we may use the least-squares error criterion to optimize the $M + 1$ coefficients of the FIR filter. First, let $d(n)$ denote the *desired output sequence* of the FIR filter of length $M + 1$ and let $h(n)$ be the input sequence. Then, if $y(n)$ is the output sequence of the filter, as illustrated in Fig. 6.41, the error sequence between the desired output and the actual output is

$$e(n) = d(n) - \sum_{k=0}^{M} b_k h(n - k) \tag{6.3.57}$$

where the $\{b_k\}$ are the FIR filter coefficients.

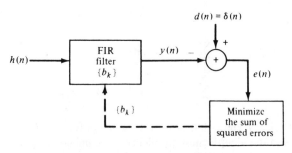

FIGURE 6.41 Least-squares FIR inverse filter.

The sum of squares of the error sequence is

$$\mathcal{E} = \sum_{n=0}^{\infty} \left[d(n) - \sum_{k=0}^{M} b_k h(n-k) \right]^2 \tag{6.3.58}$$

When $\mathcal{E}$ is minimized with respect to the filter coefficients, we obtain the set of linear equations

$$\sum_{k=0}^{M} b_k r_{hh}(k-l) = r_{dh}(l) \qquad l = 0, 1, \ldots, M \tag{6.3.59}$$

where $r_{hh}(l)$ is the autocorrelation of $h(n)$, defined as

$$r_{hh}(l) = \sum_{n=0}^{\infty} h(n)h(n-l) \tag{6.3.60}$$

and $r_{dh}(n)$ is the crosscorrelation between the desired output $d(n)$ and the input sequence $h(n)$, defined as

$$r_{dh}(l) = \sum_{n=0}^{\infty} d(n)h(n-l) \tag{6.3.61}$$

In view of the similarity of this problem to the least-squares FIR system identification problem considered in the preceding section, it is not surprising that the linear equations in (6.3.59) are identical in form to the linear equations in (6.3.45). The optimum, in the least-squares sense, FIR filter that satisfies the linear equations in (6.3.59) is called the *Wiener filter*, after the famous mathematician Norbert Wiener, who introduced optimum least-squares filtering in engineering [see book by Wiener (1949)].

If the optimum least-squares FIR filter is to be an approximate inverse filter, the desired response is

$$d(n) = \delta(n) \tag{6.3.62}$$

The crosscorrelation between $d(n)$ and $h(n)$ reduces to

$$r_{dh}(l) = \begin{cases} h(0) & l = 0 \\ 0 & \text{otherwise} \end{cases} \tag{6.3.63}$$

Therefore, the coefficients of the least-squares FIR filter are obtained from the solution of the linear equations in (6.3.59), which can be expressed in matrix form as

$$\begin{bmatrix} r_{hh}(0) & r_{hh}(1) & r_{hh}(2) & \cdots & r_{hh}(M) \\ r_{hh}(1) & r_{hh}(0) & r_{hh}(1) & \cdots & r_{hh}(M-1) \\ \vdots & \vdots & \vdots & & \vdots \\ r_{hh}(M) & r_{hh}(M-1) & \cdots & \cdots & r_{hh}(0) \end{bmatrix} \begin{bmatrix} b_0 \\ b_1 \\ \vdots \\ b_M \end{bmatrix} = \begin{bmatrix} h(0) \\ 0 \\ \vdots \\ 0 \end{bmatrix} \tag{6.3.64}$$

or, equivalently, as

$$\mathbf{R}_{hh}\mathbf{b} = \mathbf{c} \tag{6.3.65}$$

where $\mathbf{b}$ is the vector of filter coefficients and $\mathbf{c} = [h(0), 0, \ldots, 0]'$.

We observe that the matrix $\mathbf{R}_{hh}$ is not only symmetric but it also has the special property that all the elements along any diagonal are equal. Such a matrix is called a Toeplitz matrix and lends itself to efficient inversion by means of an algorithm due to

Levinson (1947) and Durbin (1959), which requires a number of computations proportional to M^2 instead of the usual M^3. The Levinson–Durbin algorithm is described in
Appendix 6A. In the following example, we derive the least-squares FIR inverse to a
simple FIR filter.

EXAMPLE 6.3.10

Determine the least-squares FIR inverse filter of length 2 to the system with impulse
response

$$h(n) = \begin{cases} 1 & n = 0 \\ -\alpha & n = 1 \\ 0 & \text{otherwise} \end{cases}$$

where $|\alpha| < 1$. Compare the least-squares solution with the approximate inverse obtained
by truncating $h_I(n)$.

Solution: Since the system has a system function $H(z) = 1 - \alpha z^{-1}$, the exact inverse
is IIR and is given by

$$H_I(z) = \frac{1}{1 - \alpha z^{-1}}$$

or, equivalently,

$$h_I(n) = \alpha^n u(n)$$

If this is truncated after n terms, the residual energy in the tail is

$$\mathscr{E}_t = \sum_{k=n}^{\infty} \alpha^{2k}$$

$$\mathscr{E}_t = \alpha^{2n}(1 + \alpha^2 + \alpha^4 + \cdots)$$

$$\mathscr{E}_t = \frac{\alpha^{2n}}{1 - \alpha^2}$$

From (6.3.64) the least-squares FIR filter of length 2 satisfies the equations

$$\begin{bmatrix} 1 + \alpha^2 & -\alpha \\ -\alpha & 1 + \alpha^2 \end{bmatrix} \begin{bmatrix} b_0 \\ b_1 \end{bmatrix} = \begin{bmatrix} 1 \\ 0 \end{bmatrix}$$

which have the solution

$$b_0 = \frac{1 + \alpha^2}{1 + \alpha^2 + \alpha^4}$$

$$b_1 = \frac{\alpha}{1 + \alpha^2 + \alpha^4}$$

For purposes of comparison, the truncated inverse filter of length 2 has the coefficients
$b_0 = 1, b_1 = \alpha$.

The minimum value of the least-squares error obtained with the optimum FIR filter
is

$$\mathscr{E}_{min} = \sum_{n=0}^{\infty} \left[d(n) - \sum_{k=0}^{M} b_k h(n - k) \right] d(n)$$

$$= \sum_{n=0}^{\infty} d^2(n) - \sum_{k=0}^{M} b_k r_{dh}(k)$$

(6.3.66)

In the case where the FIR filter is the least-squares inverse filter, $d(n) = \delta(n)$ and $r_{dh}(n) = h(0)\,\delta(n)$. Therefore,

$$\mathcal{E}_{min} = 1 - h(0)b_0$$

It is interesting to note that in the example given above

$$\mathcal{E}_{min} = \frac{\alpha^4}{1 + \alpha^2 + \alpha^4}$$

which compares with

$$\mathcal{E}_t = \frac{\alpha^4}{1 - \alpha^2}$$

for the truncated approximate inverse. Clearly, $\mathcal{E}_t > \mathcal{E}_{min}$, so that the least-squares FIR inverse filter is superior.

In the example given above, the impulse response $h(n)$ of the system was minimum phase. In such a case we selected the desired response to be $d(0) = 1$ and $d(n) = 0$, $n \geq 1$. On the other hand, if the system is nonminimum phase, a delay should be inserted in the desired response in order to obtain a good filter design. The value of the appropriate delay depends on the characteristics of $h(n)$. In any case we can compute the least-squares error filter for different delays and select the filter that produces the smallest error. The following example illustrates the effect of the delay.

EXAMPLE 6.3.11 _____

Determine the least-squares FIR inverse of length 2 to the system with impulse response

$$h(n) = \begin{cases} -\alpha & n = 0 \\ 1 & n = 1 \\ 0 & \text{otherwise} \end{cases}$$

where $|\alpha| < 1$.

Solution: This is a maximum-phase system. If we select $d(n) = [1 \quad 0]$ we obtain the same solution as in Example 6.3.10, with a minimum least-squares error

$$\mathcal{E}_{min} = 1 - h(0)b_0$$

$$= 1 + \alpha \frac{1 + \alpha^2}{1 + \alpha^2 + \alpha^4}$$

If $0 < \alpha < 1$, $\mathcal{E}_{min} > 1$, which represents a poor inverse filter. If $-1 < \alpha < 0$, $\mathcal{E}_{min} < 1$. In particular, for $\alpha = \frac{1}{2}$, we obtain $\mathcal{E}_{min} = 1.57$. For $\alpha = -\frac{1}{2}$, $\mathcal{E}_{min} = 0.81$, which is still a very large value for the squared error.

Now suppose that the desired response is specified as $d(n) = \delta(n - 1)$. Then the set of equations for the filter coefficients, obtained from (6.3.59), are the solution to the equations

$$\begin{bmatrix} 1 + \alpha^2 & -\alpha \\ -\alpha & 1 + \alpha^2 \end{bmatrix} \begin{bmatrix} b_0 \\ b_1 \end{bmatrix} = \begin{bmatrix} h(1) \\ h(0) \end{bmatrix} = \begin{bmatrix} 1 \\ -\alpha \end{bmatrix}$$

The solution of these equations is

$$b_0 = \frac{1}{1 + \alpha^2 + \alpha^4}$$

$$b_1 = \frac{-\alpha^3}{1 + \alpha^2 + \alpha^4}$$

The least-squares error, given by (6.3.66), is

$$\mathcal{E}_{min} = 1 - b_0 r_{dh}(0) - b_1 r_{dh}(1)$$

$$= 1 - b_0 h(1) - b_1 h(0)$$

$$\mathcal{E}_{min} = 1 - \frac{1}{1 + \alpha^2 + \alpha^4} + \frac{\alpha^4}{1 + \alpha^2 + \alpha^4}$$

$$\mathcal{E}_{min} = 1 - \frac{1 - \alpha^4}{1 + \alpha^2 + \alpha^4}$$

In particular, suppose that $\alpha = \pm\frac{1}{2}$. Then $\mathcal{E}_{min} = 0.29$. Consequently, the desired response $d(n) = \delta(n - 1)$ results in a significantly better inverse filter. Further improvement is possible by increasing the length of the inverse filter.

In general, when the desired response is specified to contain a delay D, then the crosscorrelation $r_{dh}(l)$, defined in (6.3.61), becomes

$$r_{dh}(l) = h(D - l) \qquad l = 0, 1, \ldots, M$$

The set of linear equations for the coefficients of the least-squares FIR inverse filter given by (6.3.59) reduce to

$$\sum_{k=0}^{M} b_k r_{hh}(k - l) = h(D - l) \qquad l = 0, 1, \ldots, M \tag{6.3.67}$$

Then the expression for the corresponding least-squares error, given in general by (6.3.66), becomes

$$\mathcal{E}_{min} = 1 - \sum_{k=0}^{M} b_k h(D - k) \tag{6.3.68}$$

Least-squares FIR inverse filters are often used in many practical applications for deconvolution, including communications and seismic signal processing. The following section describes its application to the deconvolution of seismograms.

6.3.7 Deconvolution Based on Least-Squares Prediction

Least-squares methods have found widespread use in the field of geophysics and in particular in seismic signal processing for oil exploration, which involves the deconvolution of seismograms. Below we describe deconvolution by means of the least-squares method to remove the reverberation and signal excitation effects in seismograms.

In seismic exploration of the earth's subsurface, an acoustic source is used to emit a signal waveform that can be represented in sampled form as a discrete-time sequence $x(n)$. As this short pulse of acoustic energy propagates through the earth, some of the energy is reflected back to the surface. The reflections result from discontinuities in the composition of the earth subsurface. These observations have led to mathematical models of the earth subsurface as a layered medium, where each layer has a different composition and hence different propagation characteristics. Reflections of the signal occur at the boundaries of the different layers.

Figure 6.42 illustrates a very simple model to account for signal reflections. With such a model, the signal $y(n)$ received and recorded at a seismometer can be expressed as

$$y(n) = \sum_{i=1}^{L} k_i x(n - D_i) \tag{6.3.69}$$

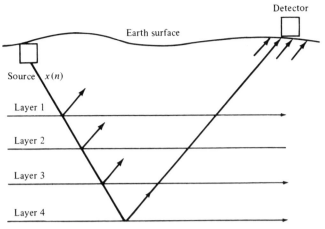

FIGURE 6.42   Signal propagation in a layered medium.

where the $\{k_i\}$ are the reflection coefficients at the interfaces between the various layers of the earth and $\{D_i\}$ represent the corresponding propagation delays. The number L of reflection coefficients is usually large.

The received signal $y(n)$ is viewed as the convolution of the excitation $x(n)$ with the sequence

$$\epsilon(n) = \sum_{k=1}^{L} k_i\, \delta(n - D_i) \tag{6.3.70}$$

where $\epsilon(n)$ represents the characteristics of the medium. The geophysicist is interested in recovering the sequence $\epsilon(n)$ from $y(n)$ and thus in extracting the information contained in $\epsilon(n)$ concerning the composition of the earth. Consequently, this is a problem in deconvolution or inverse filtering to remove the effects of the excitation $x(n)$.

A more complex signal propagation model is obtained when there is a layer of water between the signal source and the earth layers. It is well known that large amounts of oil and gas lie in the rock formations below the bottom of the sea. In seismic exploration to identify the location of these deposits, a short signal pulse is imparted into the water by a seismic source. The bottom of the sea reflects a portion of the pulse energy back to the surface while the remaining energy propagates into the rock formations and is reflected back to the surface from the various layers in the rock formations of the earth. The pulse energy that remains in the water is reflected from the surface back to the bottom. Some of this energy from the second bounce is reflected back to the surface while the remaining energy propagates through into the rock formation, and so on. The received signal $y(n)$ is recorded at the surface by a seismic detector. A model for this rather complex signal propagation is illustrated in Fig. 6.43.

The signal energy that propagates within the water is called a reverberation signal. It is relatively strong and tends to mask the reflections from the rock formations in the earth. In general, we can represent the received signal sequence at the detector as

$$y(n) = x(n) * c(n) * \epsilon(n) \tag{6.3.71}$$

where $\epsilon(n)$ represents the response of the layered earth to a unit impulse, $c(n)$ is an impulse train representing the reflectivity properties of the water, and $x(n)$ is the input signal pulse. The signal

$$p(n) = x(n) * c(n) \tag{6.3.72}$$

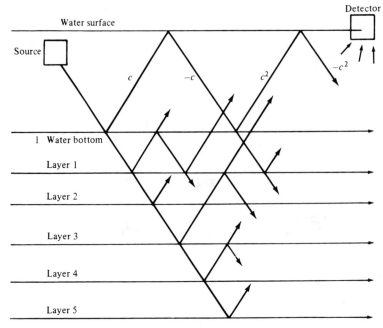

FIGURE 6.43   Reverberation of signal in water.

represents the undesirable reverberation component that we wish to eliminate through deconvolution.

The reverberation signal corresponding to a signal path may be expressed as

$$p_1(n) = x(n) - c_1 x(n - D) + c_1^2 x(n - 2D) - c_1^3 x(n - 3D) + \cdots$$

and in the z-transform domain as

$$P_1(z) = (1 - c_1 z^{-D} + c_1^2 z^{-2D} - c_1^3 z^{-3D} + \cdots)X(z) \qquad (6.3.73)$$

$$= \frac{1}{1 - c_1 z^{-D}} X(z)$$

where D is the delay in propagation from the surface to the bottom of the sea. Therefore, the system function for the propagation of the reverberating signal in the water is

$$C(z) = \frac{1}{1 - c_1 z^{-D}} \qquad (6.3.74)$$

Since $c_1 < 1$, on physical grounds, the system is stable. Its inverse is FIR and minimum phase. Below we describe the method for deconvolving $y(n)$ with respect to the excitation $x(n)$ and the reverberating medium $c(n)$ to obtain $\epsilon(n)$. The deconvolution to remove the effects of both $x(n)$ and $c(n)$ is performed simultaneously in one step. Thus the inverse filter is the combined inverse to $p(n) = x(n) * c(n)$.

To derive the inverse system for deconvolution, it is necessary for us to adopt a statistical approach. We assume that the desired sequence $\epsilon(n)$ is a sequence of uncorrelated reflections. Consequently, $\epsilon(n)$ resembles white noise with an autocorrelation sequence

$$r_{\epsilon\epsilon}(l) = \begin{cases} E_\epsilon & l = 0 \\ 0 & l \neq 0 \end{cases} \qquad (6.3.75)$$

where E_ϵ is some arbitrary constant. On the other hand, the sequence $p(n) = x(n) * c(n)$ is highly correlated. In other words, successive samples of $p(n)$ do not change much. We take advantage of this correlation or slow variation in $p(n)$ by estimating $p(n)$ from past samples and then subtracting the estimate from $y(n)$.

To proceed, suppose that we form an estimate of $p(n)$ by forming a weighted linear combination of the past M samples of $y(n)$, that is, the samples, $y(n - 1)$, $y(n - 2)$, $\ldots$, $y(n - M)$. Since we are forming a weighted linear combination of the past M samples to estimate the future sample $y(n)$, we call this estimate a predicted value, and the process of forming the estimate is called linear prediction. Let $\hat{y}(n)$ be the predicted value. It is expressed as

$$\hat{y}(n) = \sum_{k=1}^{M} b_k y(n - k) \tag{6.3.76}$$

where the $\{b_k\}$ are called the *prediction coefficients*. In fact, the equation in (6.3.76) basically describes the operation of an FIR filter of length M with coefficients $\{b_k\}$. The filter coefficients are selected to minimize the sum of the squared-error sequence $e(n)$, that is,

$$\mathcal{E} = \sum_{n=0}^{\infty} e^2(n) \tag{6.3.77}$$

$$\mathcal{E} = \sum_{n=0}^{\infty} \left[y(n) - \sum_{k=1}^{M} b_k y(n - k) \right]^2$$

The minimization of $\mathcal{E}$ with respect to the coefficients $\{b_k\}$ leads to the set of linear equations of the form

$$\sum_{k=1}^{M} b_k r_{yy}(k - l) = r_{yy}(l) \qquad l = 1, 2, \ldots, M \tag{6.3.78}$$

where $r_{yy}(l)$ is the autocorrelation of the sequence $y(n)$, defined as

$$r_{yy}(l) = \sum_{n=0}^{\infty} y(n)y(n - l) \tag{6.3.79}$$

We express the equations given by (6.3.78) in the matrix form

$$\begin{bmatrix} r_{yy}(0) & r_{yy}(1) & \cdots & r_{yy}(M - 1) \\ r_{yy}(1) & r_{yy}(0) & \cdots & r_{yy}(M - 2) \\ \vdots & & & \\ r_{yy}(M - 1) & r_{yy}(M - 2) & \cdots & r_{yy}(0) \end{bmatrix} \begin{bmatrix} b_1 \\ b_2 \\ \vdots \\ b_M \end{bmatrix} = \begin{bmatrix} r_{yy}(1) \\ r_{yy}(2) \\ \vdots \\ r_{yy}(M) \end{bmatrix} \tag{6.3.80}$$

or, equivalently, as

$$\mathbf{R}_{yy}\mathbf{b} = \mathbf{r}_{yy} \tag{6.3.81}$$

These equations are often called the *normal equations* or the *Yule–Walker equations* [for a reference, see the paper by Makhoul (1975)].

We observe again that the matrix $\mathbf{R}_{yy}$ is a Toeplitz matrix and hence it lends itself to the efficient inversion algorithm due to Levinson and Durbin (see Appendix 6A) which requires a number of computations proportional to M^2 instead of the usual M^3.

Now the autocorrelation sequence $r_{yy}(l)$ can be expressed in the spectral domain as

$$S_{yy}(\omega) = S_{pp}(\omega)S_{\epsilon\epsilon}(\omega) \tag{6.3.82}$$

where $S_{yy}(\omega)$ is the Fourier transform of $r_{yy}(l)$, $S_{pp}(\omega)$ is the Fourier transform of the autocorrelation $r_{pp}(l)$, and $S_{\epsilon\epsilon}(\omega)$ is the Fourier transform of the autocorrelation $r_{\epsilon\epsilon}(l)$. Alternatively, in the correlation domain, the relationship is

$$r_{yy}(l) = r_{pp}(l) * r_{\epsilon\epsilon}(l) \tag{6.3.83}$$

Since the correlation sequence $\{r_{\epsilon\epsilon}(l)\}$ is an impulse (a unit sample) it follows that

$$r_{yy}(l) = E_\epsilon r_{pp}(l) \tag{6.3.84}$$

Therefore, except for the proportionality factor E_ϵ, the autocorrelation sequence for $y(n)$ is identical to the autocorrelation of the sequence $p(n)$. This implies that the coefficients of the FIR linear prediction filter depend only on the correlation properties of the sequence $p(n)$. There is no dependence on the sequence $\epsilon(n)$.

In view of the foregoing correlation properties of the signals, the FIR filter predicts the sequence $p(n)$ from past samples of $y(n)$, as indicated in (6.3.76). The predicted value $y(n)$ is basically an estimate of $p(n)$, which is subtracted from the observed value $y(n)$ to yield an estimate of the desired sequence $\epsilon(n)$, as illustrated in Fig. 6.44.

The overall system in Fig. 6.44 may be viewed as an FIR filter with coefficients

$$g_k = \begin{cases} 1 & k = 0 \\ -b_k & k = 1, 2, \ldots, M \end{cases} \tag{8.3.85}$$

It remains to be shown that this FIR filter with system function

$$G(z) = \sum_{k=0}^{M} g_k z^{-k} \tag{8.3.86}$$

is the least-squares inverse filter to the sequence $p(n)$, which represents the combined reverberation and signal excitation as defined by (6.3.72). Toward this end, we write the linear equation in (6.3.81) in the form

$$\mathbf{r}_{pp} - \mathbf{R}_{pp}\mathbf{b} = O \tag{6.3.87}$$

where the autocorrelation matrix $\mathbf{R}_{pp}$ and vector $\mathbf{r}_{pp}$ are related to $\mathbf{R}_{yy}$ and $\mathbf{r}_{yy}$ by (6.5.84). Equivalently, (6.3.87) may be written as a set of M linear equations of the form

$$\sum_{k=0}^{M} g_k r_{pp}(k - l) = 0 \qquad l = 1, 2, \ldots, M \tag{6.3.88}$$

where the $\{g_k\}$ are defined by (6.5.85).

Let us augment the set of linear equations in (6.3.88) with the equation

$$\sum_{k=0}^{M} g_k r_{pp}(k) = \beta \tag{6.3.89}$$

where β is a scale factor. Then the combination of (6.3.89) with (6.3.88) results in a

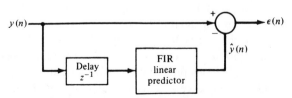

FIGURE 6.44   Deconvolution by linear prediction.

set of $M + 1$ linear equations of the form

$$
\begin{bmatrix}
r_{pp}(0) & r_{pp}(1) & \cdots\cdots & r_{pp}(M) \\
r_{pp}(1) & r_{pp}(0) & r_{pp}(1) & \cdot\cdot & r_{pp}(M-1) \\
\vdots & & & \\
r_{pp}(M) & r_{pp}(M-1) & \cdots\cdots & r_{pp}(0)
\end{bmatrix}
\begin{bmatrix}
g_0 \\ g_1 \\ \vdots \\ g_M
\end{bmatrix}
=
\begin{bmatrix}
\beta \\ 0 \\ \vdots \\ 0
\end{bmatrix}
\tag{6.3.90}
$$

Clearly, this set of equations is identical in form to the set of equation for the least-squares FIR inverse filter given by (6.3.64). Therefore, the coefficients obtained by solving the set of $M + 1$ equations given by (6.3.90) or, equivalently, by solving the reduced set of M equations given by (6.3.87) are the coefficients of the least-squares FIR inverse filter.

The least-squares FIR inverse filter of length $M + 1$, with coefficients $\{g_k\}$, is called the *prediction error filter*. The least-squares method described in this section for deconvolution based on linear prediction is called *predictive deconvolution*.

6.4 Summary and References

Modulation is a basic operation in communications where we are concerned with the transmission of information by many users over the same physical channel. Among the three basic modulation methods (amplitude, phase, and frequency), amplitude modulation is by far the most widely used in practice. In this chapter we presented the mathematical characterization of bandpass signals and systems, and emphasized amplitude modulation, as well as demodulation, of both analog and discrete-time signals. In our discussion of the implementation of a QAM modem, we indicated that it is often advantageous to synthesize the QAM signal digitally and then to pass the digital signal through a D/A converter. Our treatment of bandpass signals also included the development of the sampling theorem for bandpass signals.

The second topic treated in this chapter was the use of some simple reconstruction techniques for converting the digital signal to analog form. The zero-order hold (ZOH) is by far the most widely used interpolation method. This is usually followed by a post-filter for smoothing the output of the ZOH and thus reducing the power from aliased frequency components at the filter output. We also observed that linear interpolation techniques are better at rejecting frequency components above $F_s/2$.

Inverse filtering, deconvolution, and system identification are important operations that are frequently encountered in digital signal processing. In Section 6.3 we presented several methods for performing inverse filtering and deconvolution. The least-squares methods described in Sections 6.3.5, 6.3.6, and 6.3.7 for performing system identification, inverse filtering, and deconvolution are by far the most important practical methods. These techniques are particularly robust in the presence of additive noise perturbations superimposed on the signal.

There are a number of fine texts that provide an in-depth coverage of modulation and demodulation of signals. For reference we mention the books by Black (1953), Carlson (1975), Schwartz (1980), Stremler (1982), and Taub and Schilling (1986). The derivation of the bandpass sampling theorem presented in Section 6.1.5 is due to Brown (1980).

A vast amount of technical literature exists on the topics of inverse filtering, deconvolution, and system identification. In the context of communications, system identification as performed for channel equalization, the reader may refer to the book by Proakis (1983). Deconvolution techniques are widely applied in seismic signal processing. For reference, we suggest the papers by Wood and Treitel (1975), Peacock and Treitel (1969), Robinson and Treitel (1978), and the book by Robinson (1962). The Levinson–Durbin algorithm for efficiently solving for the prediction coefficients was first proposed by Levinson (1947) and later modified by Durbin (1959).

APPENDIX 6A. THE LEVINSON–DURBIN ALGORITHM

In this appendix we describe an efficient algorithm for solving a set of linear equations of the form

$$\mathbf{R}_N \mathbf{a}_N = \mathbf{c}_N \tag{6A.1}$$

where $\mathbf{a}_N$ is an $N \times 1$ vector of coefficients expressed as $\mathbf{a}_N' = [a_N(1) \quad a_N(2) \quad \cdots \quad a_N(N)]$, $\mathbf{c}_N$ is an $N \times 1$ vector with elements $c(k)$, $k = 1, 2, \ldots, N$ and $\mathbf{R}_N$ is an $N \times N$ Toeplitz correlation matrix of the form

$$
\begin{bmatrix}
r(0) & r(1) & r(2) & \cdots & r(N-1) \\
r(1) & r(0) & r(1) & \cdots & r(N-2) \\
r(2) & r(1) & r(0) & \cdots & r(N-3) \\
\vdots & \vdots & \vdots & & \vdots \\
r(N-1) & r(N-2) & r(N-3) & & r(0)
\end{bmatrix}
\tag{6A.2}
$$

$\mathbf{R}_N$ and $\mathbf{c}_N$ are given and $\mathbf{a}_N$ is to be determined.

The solution that we seek is

$$\mathbf{a}_N = \mathbf{R}_N^{-1} \mathbf{c}_N \tag{6A.3}$$

The inversion of $\mathbf{R}_N$ by standard techniques, such as Gauss elimination, has a computational complexity (number of multiplications) proportional to N^3. However, $\mathbf{R}_N$ is Toeplitz and, consequently, the Levinson–Durbin algorithm presented below for solving (6A.1) has a complexity proportional to N^2.

It is desirable that the reader gain some physical insight into the solution presented below. For this reason we formulate the solution in terms of a least-squares estimation problem, which arises frequently in practice as we have observed from our study of Section 6.3.

Let us assume that we are given a sequence of data $x(n)$, $-\infty < n < \infty$, from which we form a linear estimate of the form

$$\hat{y}(n) = -\sum_{k=1}^{N} a_N(k)x(n-k) \qquad -\infty < n < \infty \tag{6A.4}$$

The negative sign in the definition of $\hat{y}(n)$ given by (6A.4) is used to conform with current practice in the technical literature. Note that the estimate $\hat{y}(n)$ may be viewed as the output of an FIR filter of length N excited by the sequence $x(n)$. Suppose that the sequence $\hat{y}(n)$ is an estimate of another sequence, denoted as $d(n)$, $-\infty < n < \infty$. The difference

$$e(n) = d(n) - \hat{y}(n) \qquad -\infty < n < \infty$$

represents the error between the desired sequence and the estimated sequence.

If we adopt the least-squares criterion, we select the coefficients $\{a_N(k)\}$ to minimize the quantity

$$\mathscr{E}_N = \sum_{n=-\infty}^{\infty} e^2(n) \tag{6A.5}$$

$$= \sum_{n=-\infty}^{\infty} \left[d(n) + \sum_{k=1}^{N} a_N(k)x(n-k) \right]^2$$

where the subscript on $\mathscr{E}_N$ denotes the length of the FIR filter. The minimization of $\mathscr{E}_N$ with respect to the coefficients $\{a_N(k)\}$ yields the set of linear equations having the form

$$\sum_{k=1}^{N} a_N(k)r_{xx}(k-l) = -r_{dx}(l) \qquad l = 1, 2, \ldots, N \tag{6A.6}$$

where $r_{xx}(k)$ is the autocorrelation sequence of $x(n)$ and $r_{dx}(l)$ is the crosscorrelation of the sequences $x(n)$ and $d(n)$. These equations have the form of the matrix equation given in (6A.1).

A special case of (6A.6) is obtained when the desired signal is $d(n) = x(n)$. In this case the estimate $\hat{y}(n)$ is called the linear predicted value of $x(n)$ and the linear equations in (6A.6) become

$$\sum_{k=1}^{N} a_N(k)r_{xx}(k-l) = -r_{xx}(l) \qquad l = 1, 2, \ldots, N \tag{6A.7}$$

These equations are called the *normal equations* or *the Yule–Walker equations*, as indicated in Section 6.3.

We shall first solve (6A.7) and then (6A.6). The key to the method of solution is to proceed recursively, beginning with a predictor of order 1 (one coefficient) and increasing the order recursively, using the lower-order solutions to obtain the solution to the next higher order. Thus the solution to the first-order predictor obtained from (6A.7) is

$$a_1(1) = -\frac{r_{xx}(1)}{r_{xx}(0)} \tag{6A.8}$$

The resulting least-squares error is

$$\mathscr{E}_1 = \sum_{n=-\infty}^{\infty} [x(n) + a_1(1)x(n)]^2$$

$$= \sum_{n=-\infty}^{\infty} x^2(n) + 2a_1(1) \sum_{n=-\infty}^{\infty} x(n)x(n-1) + a_1^2(1) \sum_{n=-\infty}^{\infty} x^2(n-1) \tag{6A.9}$$

$$\mathscr{E}_1 = r_{xx}(0) + 2a_1(1)r_{xx}(1) + a_1^2(1)r_{xx}(0)$$

If we substitute the solution (6A.8) for $r_{xx}(1)$ into (6A.9), we obtain

$$\mathscr{E}_1 = r_{xx}(0)(1 - a_1^2(1)) \tag{6A.10}$$

The next step is to solve for the coefficients $\{a_2(1), a_2(2)\}$ of the second-order predictor and express the solution in terms of $a_1(1)$. The two equations obtained from (6A.7) are

$$a_2(1)r_{xx}(0) + a_2(2)r_{xx}(1) = -r_{xx}(1)$$
$$a_2(1)r_{xx}(1) + a_2(2)r_{xx}(0) = -r_{xx}(2) \tag{6A.11}$$

By using (6A.8) to eliminate $r_{xx}(1)$ from (6A.11), we obtain the solution

$$a_2(2) = -\frac{r_{xx}(2) + a_1(1)r_{xx}(1)}{r_{xx}(0) - a_1^2(1)r_{xx}(0)}$$

$$= -\frac{r_{xx}(2) + a_1(1)r_{xx}(1)}{\mathscr{E}_1} \tag{6A.12}$$

where $\mathscr{E}_1$ is defined in (6A.10). Furthermore, if we substitute $r_{xx}(1) = -a_1(1)r_{xx}(0)$ into the first equation in (6A.11) and divide the result by $r_{xx}(0)$, we obtain the desired equation for $a_2(1)$ as

$$a_2(1) = a_1(1) + a_2(2)a_1(1) \tag{6A.13}$$

Thus we have $a_2(1)$ in terms of $a_1(1)$ (the lower-order predictor) and $a_2(2)$.

In general, we may express the solution for the coefficients of an mth-order predictor in terms of the coefficients of an $(m-1)$st-order predictor. Thus we express the coefficients vector $\mathbf{a}_m$ as the sum of two vectors, namely

$$\mathbf{a}_m = \begin{bmatrix} a_m(1) \\ a_m(2) \\ \vdots \\ a_m(m) \end{bmatrix} = \begin{bmatrix} \mathbf{a}_{m-1} \\ \cdots \\ 0 \end{bmatrix} + \begin{bmatrix} \mathbf{d}_{m-1} \\ \cdots \\ K_m \end{bmatrix} \tag{6A.14}$$

where the vector $\mathbf{d}_{m-1}$ and the scalar K_m are to be determined. Now we partition the $m \times m$ autocorrelation matrix $\mathbf{R}_m$ as

$$\mathbf{R}_m = \begin{bmatrix} \mathbf{R}_{m-1} & \mathbf{r}_{m-1}^b \\ \cdots & \cdots \\ \mathbf{r}_{m-1}^{bt} & r_{xx}(0) \end{bmatrix} \tag{6A.15}$$

where $\mathbf{r}_{m-1}^{bt} = [r_{xx}(m-1) \quad r_{xx}(m-2) \quad \cdots \quad r_{xx}(1)] = (\mathbf{r}_{m-1}^b)^t$. The superscript b on $\mathbf{r}_{m-1}$ denotes the vector $\mathbf{r}_{m-1}^t = [r_{xx}(1) \quad r_{xx}(2) \quad \cdots \quad r_{xx}(m-1)]$ with elements taken in reverse order.

The solution to the equation $\mathbf{R}_m\mathbf{a}_m = -\mathbf{r}_m$ may be expressed as

$$\begin{bmatrix} \mathbf{R}_{m-1} & \mathbf{r}_{m-1}^b \\ \cdots & \cdots \\ \mathbf{r}_{m-1}^{bt} & r_{xx}(0) \end{bmatrix} \left\{ \begin{bmatrix} \mathbf{a}_{m-1} \\ \cdots \\ 0 \end{bmatrix} + \begin{bmatrix} \mathbf{d}_{m-1} \\ \cdots \\ K_m \end{bmatrix} \right\} = -\begin{bmatrix} \mathbf{r}_{m-1} \\ \cdots \\ r_{xx}(m) \end{bmatrix} \tag{6A.16}$$

From (6A.16) we obtain two equations, namely,

$$\mathbf{R}_{m-1}\mathbf{a}_{m-1} + \mathbf{R}_{m-1}\mathbf{d}_{m-1} + K_m\mathbf{r}_{m-1}^b = -\mathbf{r}_{m-1} \tag{6A.17}$$

$$\mathbf{r}_{m-1}^{bt}\mathbf{a}_{m-1} + \mathbf{r}_{m-1}^{bt}\mathbf{d}_{m-1} + K_m r_{xx}(0) = -r_{xx}(m) \tag{6A.18}$$

Since $\mathbf{R}_{m-1}\mathbf{a}_{m-1} = -\mathbf{r}_{m-1}$, (6A.17) yields the solution

$$\mathbf{d}_{m-1} = -K_m\mathbf{R}_{m-1}^{-1}\mathbf{r}_{m-1}^b \tag{6A.19}$$

But $\mathbf{r}_{m-1}^b$ is just $\mathbf{r}_{m-1}$ with elements taken in reverse order. Therefore, the solution in (6A.19) is simply $\mathbf{a}_{m-1}$ in reverse order, that is,

$$\mathbf{d}_{m-1} = K_m\mathbf{a}_{m-1}^b = K_m \begin{bmatrix} a_{m-1}(m-1) \\ a_{m-1}(m-2) \\ \vdots \\ a_{m-1}(1) \end{bmatrix} \tag{6A.20}$$

The equation (6A.18) is a scalar equation, which can be used to solve for K_m. If we use (6A.20) to eliminate $\mathbf{d}_{m-1}$, we obtain

$$K_m[r_{xx}(0) + \mathbf{r}^{bt}_{m-1}\mathbf{a}_{m-1}] = -r_{xx}(m) - \mathbf{r}^{bt}_{m-1}\mathbf{a}_{m-1}$$

and hence

$$K_m = -\frac{r_{xx}(m) + \mathbf{r}^{bt}_{m-1}\mathbf{a}_{m-1}}{r_{xx}(0) + \mathbf{r}^{bt}_{m-1}\mathbf{a}_{m-1}} = -\frac{r_{xx}(m) + \mathbf{r}^{bt}_{m-1}\mathbf{a}_{m-1}}{\mathscr{E}_{m-1}} \tag{6A.21}$$

By returning to (6A.14) and substituting the solution in (6A.20), we obtain the desired recursion for the coefficient as

$$a_m(m) = K_m = -\frac{r_{xx}(m) + \mathbf{r}^{bt}_{m-1}\mathbf{a}_{m-1}}{\mathscr{E}_{m-1}}$$

$$\begin{aligned}a_m(k) &= a_{m-1}(k) + K_m a_{m-1}(m-k) \\ &= a_{m-1}(k) + a_m(m)a_{m-1}(m-k)\end{aligned} \qquad \begin{aligned}k &= 1, 2, \ldots, m-1 \\ m &= 1, 2, \ldots, N\end{aligned} \tag{6A.22}$$

The least-squares error $\mathscr{E}_m$ can also be computed recursively. For the mth-order predictor, the error is

$$\begin{aligned}\mathscr{E}_m &= r_{xx}(0) + 2\sum_{k=1}^{m} a_m(k)r_{xx}(k) + \sum_{k=1}^{m} a_m(k)\sum_{j=1}^{m} a_m(j)r_{xx}(j-k) \\ &= r_{xx}(0) + \sum_{k=1}^{m} a_m(k)r_{xx}(k) + \sum_{k=1}^{m} a_m(k)\left[\sum_{j=1}^{m} a_m(j)r_{xx}(j-k) + r_{xx}(k)\right]\end{aligned} \tag{6A.23}$$

But the third term on the right-hand side of (6A.23) is zero for the optimum choice of the coefficients. Hence

$$\mathscr{E}_m = r_{xx}(0) + \sum_{k=1}^{m} a_m(k)r_{xx}(k) \tag{6A.24}$$

Now we can use (6A.22) to substitute for $a_m(k)$ in (6A.24). Thus we obtain

$$\begin{aligned}\mathscr{E}_m &= r_{xx}(0) + \sum_{k=1}^{m-1} a_{m-1}(k)r_{xx}(k) + a_m(m)\left[r_{xx}(m) + \sum_{k=1}^{m-1} a_{m-1}(m-k)r_{xx}(k)\right] \\ &= \mathscr{E}_{m-1} + a_m(m)[r_{xx}(m) + \mathbf{r}^{bt}_{m-1}\mathbf{a}_{m-1}] \\ &= \mathscr{E}_{m-1}(1 - a_m^2(m)) = \mathscr{E}_{m-1}(1 - K_m^2)\end{aligned} \tag{6A.25}$$

where the last step follows from the expression for $a_m(m)$ in (6A.22).

The solution for the coefficients given in (6A.22) applies to the equations given in (6A.7), where the right-hand side has elements of the autocorrelation sequence. In the more general set of linear equations

$$\mathbf{R}_N\mathbf{b}_N = \mathbf{c}_N \tag{6A.26}$$

the solution may be expressed in terms of (6A.22). We leave it as an exercise to the reader to show that the solution to (6A.26) may be obtained recursively as

$$b_m(m) = \frac{c(m) - \mathbf{r}^{bt}_{m-1}\mathbf{b}_{m-1}}{\mathscr{E}_{m-1}}$$

$$b_m(k) = b_{m-1}(k) - b_m(m)a_{m-1}(m-k) \qquad \begin{aligned}k &= 1, 2, \ldots, m-1 \\ m &= 1, 2, \ldots, N\end{aligned} \tag{6A.27}$$

where $b_1(1) = c(1)/r_{xx}(0)$ and $a_m(k)$ is the solution given in (6A.22). Therefore, two sets of recursions are required to solve (6A.26), the set given in (6A.22) and the recursion in (6A.27).

PROBLEMS

6.1 Consider the system

$$y(n) = x(n) - 0.95x(n - 6)$$

(a) Sketch its pole–zero pattern.
(b) Sketch its magnitude response using the pole–zero plot.
(c) Determine the system function of its causal inverse system.
(d) Sketch the magnitude response of the inverse system using the pole–zero plot.

6.2 Let H_1 be the causal system described by the difference equation

$$y(n) = \tfrac{7}{12}y(n - 1) - \tfrac{1}{12}y(n - 2) + x(n - 1) - \tfrac{1}{2}x(n - 2)$$

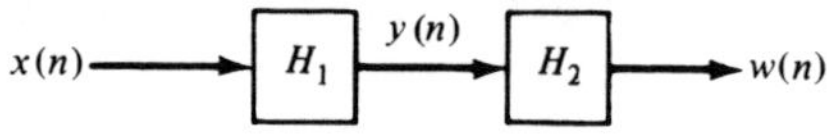

$$x(n) \longrightarrow \boxed{H_1} \xrightarrow{y(n)} \boxed{H_2} \longrightarrow w(n)$$

FIGURE P6.2

(a) Determine the system H_2 in Fig. P6.2 so that $w(n) = x(n)$. Is the inverse system H_2 causal?
(b) Determine the system H_2 in Fig. P6.2 so that $w(n) = x(n - 1)$. Is the system H_2 causal? Explain.
(c) Determine the difference equations for the system H_2 in parts (a) and (b).

6.3 Determine the impulse response and the difference equation for all possible systems specified by the system functions

(a) $H(z) = \dfrac{z^{-1}}{1 - z^{-1} - z^{-2}}$

(b) $H(z) = \dfrac{1}{1 - e^{-4a}z^{-4}}, \quad 0 < a < 1$

6.4 Determine the impulse response of a causal LTI system which produces the response

$$y(n) = \left\{ \begin{array}{c} 1, -1, 3, -1, 6 \\ \uparrow \end{array} \right\}$$

when excited by the input signal

$$x(n) = \left\{ \begin{array}{c} 1, 1, 2 \\ \uparrow \end{array} \right\}$$

6.5 The system

$$y(n) = \tfrac{1}{2}y(n - 1) + x(n)$$

is excited with the input

$$x(n) = (\tfrac{1}{4})^n u(n)$$

Determine the sequences $r_{xx}(l)$, $r_{hh}(l)$, $r_{xy}(l)$, and $r_{yy}(l)$.

6.6 Determine a causal and stable system with magnitude response

$$|H(\omega)|^2 = \frac{2 + 2 \cos \omega}{2 \cos^2 \omega + 3 \cos \omega + 1.25}$$

6.7 Consider the signal

$$x(n) = a^n u(n) \qquad -1 < a < 1$$

(a) Show that the optimum first-order one-step predictor, defined by

$$e(n) = x(n) - \hat{x}(n) = x(n) - a_1 x(n - 1)$$

is specified by

$$a_1 = a$$

and the minimum total squared error is given by

$$E_1 = 1 - a^2$$

(b) Determine the second-order predictor

$$e(n) = x(n) - a_1 x(n - 1) - a_2 x(n - 2)$$

and the corresponding minimum total squared error.

6.8 Consider the system

$$H(z) = \frac{1 + bz^{-1}}{1 + a_1 z^{-1} + a_2 z - 2}$$

and let $r_{hh}(l)$ be the autocorrelation sequence of $h(n)$.
(a) Determine $r_{hh}(0)$, $r_{hh}(1)$, $r_{hh}(2)$, and $r_{hh}(3)$ in terms of b, a_1, and a_2.
(b) Can this process be generalized to an arbitrary pole–zero system?

6.9 Determine the least-squares FIR inverse of length 3 to the system with impulse response

$$h(n) = \begin{cases} 2 & n = 0 \\ 1 & n = 1 \\ 0 & \text{otherwise} \end{cases}$$

Also, determine the minimum squared error $\mathscr{E}_{\min}$.

6.10 Determine the least-squares FIR inverse filter of length 3 for the system with impulse response $h(n)$ given in Example 6.5.11, when $\alpha = \frac{1}{2}$ and the desired response is specified as $d(n) = \delta(n - 2)$. Also compute the minimum least-squares error.

6.11 If $h(0) = h(1) = \cdots h(m - 1) = 0$ and $h(m) \neq 0$, determine the recursive equation for the impulse response of the causal inverse system that allows for a delay of m time units in the definition of the inverse, that is, if $H(z)$ is the causal system, the causal inverse system has the system function

$$H_I(z) = \frac{z^{-m}}{H(z)}$$

when $h(0) = h(1) = \cdots = h(m - 1) = 0$.

6.12 By choosing $n = 3$, illustrate that (6.3.7) corresponds to a lower-triangular set of linear equation and explain the reason that the solution is not possible if $h(0) = 0$.

6.13 Determine if the following FIR systems are minimum phase.

(a) $h(n) = \left\{ \underset{\uparrow}{10},\ 9,\ -7,\ -8,\ 0,\ 5,\ 3 \right\}$

(b) $h(n) = \left\{ \underset{\uparrow}{5},\ 4,\ -3,\ -4,\ 0,\ 2,\ 1 \right\}$

6.14 Can you determine the coefficients of the all-pole system

$$H(z) = \frac{1}{1 + \sum\limits_{k=1}^{N} a_k z^{-k}}$$

if you know its order N and the values $h(0)$, $h(1)$, . . . , $h(L-1)$ of its impulse response? How? What happens if you do not know N?

6.15 Consider a system with impulse response

$$h(n) = b_0\, \delta(n) + b_1\, \delta(n - D) + b_2\, \delta(n - 2D)$$

(a) Explain why the system generates echoes spaced D samples apart.
(b) Determine the magnitude and phase response of the system.
(c) Show that for $|b_0 + b_2| << |b_1|$, the locations of maxima and minima of $|H(\omega)|^2$ are at

$$\omega = \pm \frac{k}{D}\, \pi \qquad k = 0, 1, 2, \ldots$$

(d) Plot $|H(\omega)|$ and $< H(\omega)$ for $b_0 = 0.1$, $b_1 = 1$, and $b_2 = 0.05$ and discuss the results.

6.16 Consider the pole–zero system

$$H(z) = \frac{B(z)}{A(z)} = \frac{i + bz^{-1}}{1 + az^{-1}} = \sum_{n=0}^{\infty} h(n) z^{-n}$$

(a) Determine $h(0)$, $h(1)$, $h(2)$, and $h(3)$ in terms of a and b.
(b) Let $r_{hh}(l)$ be the autocorrelation sequence of $h(n)$. Determine $r_{hh}(0)$, $r_{hh}(1)$, $r_{hh}(2)$, and $r_{hh}(3)$ in terms of a and b.

6.17 Consider the sampling of the bandpass signal whose spectrum is illustrated in Fig. P6.17. Determine the minimum sampling rate F_s to avoid aliasing.

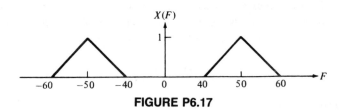

FIGURE P6.17

6.18 Consider the sampling of the bandpass signal whose spectrum is illustrated in Fig. P6.18. Determine the minimum sampling rate F_s to avoid aliasing.

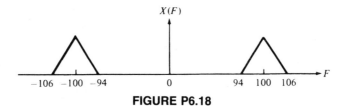

FIGURE P6.18

6.19 The input $x(n)$ to the discrete-time system

$$y(n) = ay(n - 1) + x(n) \qquad 0 < a < 1$$

is restricted to be periodic with fundamental period N. Determine in terms of the impulse response $h(n)$, an FIR system, which yields the same steady-state response as the system above, when both systems are excited by the given $x(n)$.

6.20 A linear time-invariant system has an input sequence $x(n)$ and an output sequence $y(n)$. The user has access only to the system output $y(n)$. In addition, the following information is available.

(1) The input signal is periodic with a given fundamental period N and has a flat spectral envelope, that is,

$$x(n) = \sum_{k=0}^{N-1} c_k^x e^{j(2\pi/N)kn} \qquad \text{all } n$$

where $c_k^x = 1$ for all k.

(2) The system $H(z)$ is all-pole, that is,

$$H(z) = \frac{1}{1 + \displaystyle\sum_{k=1}^{P} a_k z^{-k}}$$

but the order p and the coefficients $(a_k, 1 \leq k \leq p)$ are unknown. Is it possible to determine the order p and the numerical values of the coefficients $\{a_k, 1 \leq k \leq p\}$ by taking measurements on the output $y(n)$? If yes, explain how. Is this possible for every value of p?

6.21 Repeat Problem 6.20 for a system with system function

$$H(z) = \frac{\displaystyle\sum_{k=0}^{M} b_k z^{-k}}{1 + \displaystyle\sum_{k=1}^{P} a_k z^{-k}}$$

6.22 Prove that if the inverse of a linear time-invariant system exists, it is also linear and time invariant.

COMPUTER EXPERIMENTS

6.23 *Frequency analysis of amplitude-modulated discrete-time signal* The discrete-time signal

$$x(n) = \cos 2\pi f_1 n + \cos 2\pi f_2 n$$

where $f_1 = \frac{1}{18}$ and $f_2 = \frac{5}{128}$, modulates the amplitude of the carrier

$$x_c(n) = \cos 2\pi f_c n$$

where $f_c = \frac{50}{128}$. The resulting amplitude-modulated signal is

$$x_{am}(n) = x(n) \cos 2\pi f_c n$$

(a) Sketch the signals $x(n)$, $x_c(n)$, and $x_{am}(n)$, $0 \leq n \leq 255$.
(b) Compute and sketch the 128-point DFT of the signal $x_{am}(n)$, $0 \leq n \leq 127$.
(c) Compute and sketch the 128-point DFT of the signal $x_{am}(n)$, $0 \leq n \leq 99$.
(d) Compute and sketch the 256-point DFT of the signal $x_{am}(n)$, $0 \leq n \leq 179$.
(e) Explain the results obtained in parts (b) through (d), by deriving the spectrum of the amplitude-modulated signal and comparing it with the experimental results.

6.24 *FIR system modeling* Consider an "unknown" FIR system with impulse response $h(n)$, $0 \leq n \leq 11$ given by

$$
\begin{aligned}
h(0) &= h(11) = 0.309828 \times 10^{-1} \\
h(1) &= h(10) = 0.416901 \times 10^{-1} \\
h(2) &= h(9) \ \ = -0.577081 \times 10^{-1} \\
h(3) &= h(8) \ \ = -0.852502 \times 10^{-1} \\
h(4) &= h(7) \ \ = 0.147157 \times 10^{0} \\
h(5) &= h(6) \ \ = 0.449188 \times 10^{0}
\end{aligned}
$$

A potential user has access to the input and output of the system but does not have any information about its impulse response other than that it is FIR. In an effort to determine the impulse response of the system, the user excites it with a zero mean, uniformly distributed in the range $[-0.5, 0.5]$ random sequence $x(n)$ and records the signal $x(n)$ and the corresponding output $y(n)$ for $0 \leq n \leq 199$.
(a) By using the available information that the unknown system is FIR, the user employs the method of least-squares to obtain an FIR model $h(n)$, $0 \leq n \leq M - 1$. Set up the system of linear equations, specifying the parameters $h(0)$, $h(1)$, . . . , $h(M - 1)$. Specify formulas we should use to determine the necessary autocorrelation and crosscorrelation values.
(b) Since the order of the system is unknown, the user decides to try models of different orders and check the corresponding total squared error. Clearly, this error will be zero (or very close to it if the order of the model becomes equal to the order of the system). Compute the FIR models $h_M(n)$, $0 \leq n \leq M - 1$ for $M = 8, 9, 10, 11, 12, 13, 14$ as well as the corresponding total squared errors E_M, $M = 8, 9, . . . , 14$. What do you observe?
(c) Determine and plot the frequency response of the system and the models for $M = 11, 12, 13$. Comment on the results.
(d) Suppose now that the output of the system is corrupted by additive noise, so instead of the signal $y(n)$, $0 \leq n \leq 199$, we have available the signal

$$v(n) = y(n) + 0.01w(n)$$

where $w(n)$ is a Gaussian random sequence with zero mean and variance $\sigma^2 = 1$.

Repeat part (b) by using $v(n)$ instead of $y(n)$ and comment on the results. The quality of the model can be also determined by the quantity

$$Q = \frac{\sum\limits_{n=0}^{\infty} [h(n) - \hat{h}(n)]^2}{\sum\limits_{n=0}^{\infty} h^2(n)}$$

7

CHAPTER

Realization of Discrete-Time Systems

The focus of this chapter is the realization of linear time-invariant discrete-time systems in either software or hardware. As we noted in Chapter 2, there are various configurations or structures for the realization of any FIR and IIR discrete-time system. In Chapter 2 we described the simplest of these structures, namely, the direct-form realizations. However, there are other more practical realizations that offer some distinct advantages, especially when quantization effects are taken into consideration.

Of particular importance are the cascade, parallel, and lattice structures, which exhibit robustness in finite-word-length implementations. Also described in this chapter is the frequency-sampling realization for an FIR system, which often has the advantage of being computationally efficient when compared with alternative FIR realizations.

A major part of this chapter is concerned with the state-space formulation for linear time-invariant systems. An analysis of systems characterized by the state-variable form is presented and several state-space structures are described.

7.1 Structures for the Realization of Discrete-Time Systems

Let us consider the important class of linear time-invariant discrete-time systems characterized by the general linear constant-coefficient difference equation

$$
y(n) = -\sum_{k=1}^{N} a_k y(n-k) + \sum_{k=0}^{M} b_k x(n-k)
\qquad (7.1.1)
$$

As we have shown by means of the z-transform, such a class of linear time-invariant discrete-time systems are also characterized by the rational system function

$$
H(z) = \frac{\displaystyle\sum_{k=0}^{M} b_k z^{-k}}{1 + \displaystyle\sum_{k=1}^{N} a_k z^{-k}}
\qquad (7.1.2)
$$

which is a ratio of two polynomials in z^{-1}. From the latter characterization, we obtain the zeros and poles of the system function, which depend on the choice of the system parameters $\{b_k\}$ and $\{a_k\}$ and which determine the frequency response characteristics of the system.

Our focus in this chapter is on the various methods of implementing (7.1.1) or (7.1.2) in either hardware or in software on a programmable digital computer. We shall show

467

that (7.1.1) or (7.1.2) can be implemented in a variety of different ways which depend on the form in which these two characterizations are arranged.

In general, we may view (7.1.1) as a computational procedure (an algorithm) for determining the output sequence $y(n)$ of the system from the input sequence $x(n)$. However, the computations in (7.1.1) can be arranged in various ways, into equivalent sets of difference equations. Each set of equations defines a computational procedure or an algorithm for implementing the system. From each set of equations we can construct a block diagram consisting of an interconnection of delay elements, multipliers, and adders. In Section 2.4.5 we referred to such a block diagram as a *realization* of the system or, equivalently, as a *structure* for realizing the system.

If the system is to be implemented in software, the block diagram or, equivalently, the set of equations that are obtained by rearranging (7.1.1) can be converted into a program that runs on a digital computer. Alternatively, the structure in block diagram form implies a hardware configuration for implementing the system.

Perhaps, the one issue that may not be clear to the reader at this point is why we are considering any rearrangements of (7.1.1) or (7.1.2). Why should't one simply implement (7.1.1) or (7.1.2) directly without any rearrangement? If either (7.1.1) or (7.1.2) is rearranged in some manner, what are the benefits to be derived in the corresponding implementation?

These are important questions for which answers are given in this chapter as well as in Chapter 10. At this point in our development, we simply state that the major factors that influence our choice of a specific realization are computational complexity, memory requirements, and finite-word-length effects in the computations.

Computational complexity refers to the number of arithmetic operations (multiplications, divisions, and additions) required to compute an output value $y(n)$ for the system. In the past, these were the only items used to measure computational complexity. However, with recent developments in the design and fabrication of rather sophisticated programmable digital signal processing chips, other factors, such as the number of times a fetch from memory is performed or the number of times a comparison between two numbers is performed per output sample, have become important in assessing the computational complexity of a given realization of a system.

Memory requirements refers to the number of memory locations required to store the system parameters, past inputs, past outputs, and any intermediate computed values.

Finite-word-length effects or finite-precision effects refer to the quantization effects that are inherent in any digital implementation of the system, either in hardware or in software. The parameters of the system must necessarily be represented with finite precision. The computations that are performed in the process of computing an output from the system must be rounded-off or truncated to fit within the limited precision constraints of the computer or the hardware used in the implementation. Whether the computations are performed in fixed-point or floating-point arithmetic is another consideration. All these problems are usually called finite-word-length effects and are extremely important in influencing our choice of a system realization. We shall see that different structures of a system, which are equivalent for infinite precision, exhibit different behavior when finite-precision arithmetic is used in the implementation. Therefore, it is very important in practice to select a realization that is not very sensitive to finite-word-length effects.

Although the three factors listed above are the major ones in influencing our choice of the realization of a system of the type described by either (7.1.1) or (7.1.2), other factors, such as whether the structure or the realization lends itself to parallel processing, or whether the computations can be pipelined, may play a role in our selection of the specific implementation. These additional factors are usually important in the realization of more complex digital signal processing algorithms.

In our discussion of alternative realizations, we will concentrate on the three major factors outlined above. Occasionally, we will include some additional factors that may be important in some implementations. In particular, the emphasis on this chapter is on computational complexity and memory requirements. The topic of finite-word-length effects is discussed in Chapter 10.

7.2 Realizations for FIR Systems

In general, an FIR system is described by the difference equation

$$y(n) = \sum_{k=0}^{M-1} b_k x(n - k) \tag{7.2.1}$$

or, equivalently, by the system function

$$H(z) = \sum_{k=0}^{M-1} b_k z^{-k} \tag{7.2.2}$$

Furthermore, the unit sample response of the FIR system is identical to the coefficients $\{b_k\}$, that is,

$$h(n) = \begin{cases} b_n & 0 \leq n \leq M - 1 \\ 0 & \text{otherwise} \end{cases} \tag{7.2.3}$$

We shall present several methods for implementing an FIR system, beginning with the simplest structure, called the direct form. A second structure is the cascade-form realization. The third structure that we shall describe is the frequency-sampling realization, which is related to the FIR filter design method described in Section 5.3.4. Finally, we present a lattice realization of an FIR system. In this discussion we follow the convention used often in the technical literature, which is to use $\{h(n)\}$ for the parameters of an FIR system.

In addition to the four realizations indicated above, an FIR system may be realized by means of the DFT, as described in Section 5.4. From one point of view, the DFT may be considered as a computational procedure rather than a structure for an FIR system. However, when the computational procedure is implemented in hardware, there is a corresponding structure for the FIR system. In practice, hardware implementations of the DFT are based on the use of the fast Fourier transform (FFT) algorithms described in Chapter 9. Consequently, we shall defer further discussion on FIR filtering via the DFT until we have described the FFT algorithms in Chapter 9.

7.2.1 Direct-Form Realization

The direct-form realization follows immediately from the nonrecursive difference equation given by (7.2.1) or, equivalently, by the convolution summation

$$y(n) = \sum_{k=0}^{M-1} h(k) x(n - k) \tag{7.2.4}$$

The structure is illustrated in Fig. 7.1.

We observe that this structure requires $M - 1$ memory locations for the purpose of storing the $M - 1$ previous inputs, and has a complexity of M multiplications and $M - 1$ additions per output point. Since the output consists of a weighted linear combination of $M - 1$ past values of the input and the weighted current value of the input,

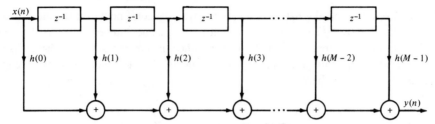

FIGURE 7.1 Direct-form realization of FIR system.

the structure in Fig. 7.1, resembles a tapped delay line or a transversal system. Consequently, the direct-form realization is often called a transversal or tapped-delay-line filter.

When the FIR system has linear phase, the unit sample response of the system satisfies
either the symmetry or asymmetry condition

$$h(n) = \pm h(M - 1 - n) \tag{7.2.5}$$

For such a system the number of multiplications is reduced from M to $M/2$ for M even
and to $(M - 1)/2$ for M odd. For example, the structure that takes advantage of this
symmetry is illustrated in Fig. 7.2 for the case in which M is odd.

7.2.2 Cascade-Form Realizations

The cascade realization follows naturally from the system function given by (7.2.2). It
is a simple matter to factor $H(z)$ into second-order FIR systems so that

$$H(z) = G \prod_{k=1}^{K} H_k(z) \tag{7.2.6}$$

where

$$H_k(z) = 1 + b_{k1}z^{-1} + b_{k2}z^{-2} \qquad k = 1, 2, \ldots, K \tag{7.2.7}$$

and K is the integer part of $(M + 1)/2$. The gain parameter G may be equally distributed
among the K filter sections, such that $G = G_1 G_2 \cdots G_K$. The zeros of $H(z)$ are grouped
in pairs to produce the second-order FIR systems of the form (7.2.7). It is always
desirable to form pairs of complex-conjugate roots so that the coefficients $\{b_{ki}\}$ in (7.2.7)
are real valued. On the other hand, real-valued roots may be paired in any arbitrary

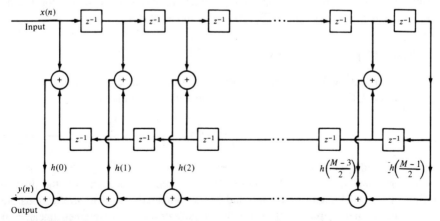

FIGURE 7.2 Direct-form realization of linear-phase FIR system (M odd).

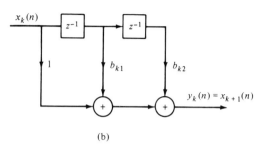

(a)

(b)

FIGURE 7.3 Cascade realization of an FIR system.

manner. The cascade-form realization along with the basic second-order section are shown in Fig. 7.3.

In the case of linear-phase FIR filters, we recall that the symmetry in $h(n)$ implies that the zeros of $H(z)$ also exhibit a form of symmetry. In particular, if z_k and z_k^* are a pair of complex-conjugate zeros then $1/z_k$ and $1/z_k^*$ are also a pair of complex-conjugate zeros. Consequently, we gain some simplification by forming fourth-order sections of the FIR system as follows:

$$H_k(z) = (1 - z_k z^{-1})(1 - z_k^* z^{-1})(1 - z^{-1}/z_k)(1 - z^{-1}/z_k^*) \qquad (7.2.8)$$
$$= 1 + c_{k1} z^{-1} + c_{k2} z^{-2} + c_{k1} z^{-3} + z^{-4}$$

where the coefficients $\{c_{k1}\}$ and $\{c_{k2}\}$ are functions of z_k. Thus by combining the two pairs of poles, to form a fourth-order filter section, we have reduced the number of multiplications from four to two (i.e., by a factor of 50%). Figure 7.4 illustrates the basic fourth-order FIR filter structure.

7.2.3 Frequency-Sampling Realization

The frequency-sampling realization is an alternative structure for an FIR filter in which the parameters that characterize the filter are the values of the desired frequency response

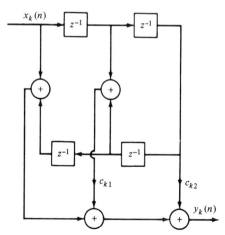

FIGURE 7.4 Fourth-order section in a cascade realization of an FIR system.

instead of the impulse response $h(n)$. To derive the frequency-sampling structure, we recall that in Section 5.3.4 we specified the desired frequency response at a set of equally spaced frequencies, namely

$$\omega_k = \frac{2\pi}{M} (k + \alpha) \qquad k = 0, 1, \ldots, \frac{M-1}{2}, \quad M \text{ odd}$$

$$k = 0, 1, \ldots, \frac{M}{2} - 1, \quad M \text{ even}$$

$$\alpha = 0 \text{ or } \tfrac{1}{2}$$

and solved for the unit sample response $h(n)$ from these equally spaced frequency specifications. Thus we may write the frequency response as

$$H(\omega) = \sum_{n=0}^{M-1} h(n)e^{-j\omega n}$$

and the values of $H(\omega)$ at frequencies $\omega_k = (2\pi/M)(k + \alpha)$ are simply

$$H(k + \alpha) = H\left(\frac{2\pi}{M}(k + \alpha)\right) \tag{7.2.9}$$

$$= \sum_{n=0}^{M-1} h(n)e^{-j2\pi(k+\alpha)n/M} \qquad k = 0, 1, \ldots, M - 1$$

The set of values $\{H(k + \alpha)\}$ are called the frequency samples of $H(\omega)$. In the case where $\alpha = 0$, $\{H(k)\}$ corresponds to the M-point DFT of $\{h(n)\}$.

It is a simple matter to invert (7.2.9) and express $h(n)$ in terms of the frequency samples. The result is

$$h(n) = \frac{1}{M} \sum_{k=0}^{M-1} H(k + \alpha)e^{j2\pi(k+\alpha)n/M} \qquad n = 0, 1, \ldots, M - 1 \tag{7.2.10}$$

When $\alpha = 0$, (7.2.10) is simply the IDFT of $\{H(k)\}$. Now if we use (7.2.10) to substitute for $h(n)$ in the z-transform $H(z)$, we have

$$H(z) = \sum_{n=0}^{M-1} h(n)z^{-n} \tag{7.2.11}$$

$$= \sum_{n=0}^{M-1} \left[\frac{1}{M} \sum_{k=0}^{M-1} H(k + \alpha)e^{j2\pi(k+\alpha)n/M} \right] z^{-n}$$

By interchanging the order of the two summations in (7.2.11) and performing the summation over the index n we obtain

$$H(z) = \sum_{k=0}^{M-1} H(k + \alpha) \left[\frac{1}{M} \sum_{n=0}^{M-1} (e^{j2\pi(k+\alpha)/M} z^{-1})^n \right] \tag{7.2.12}$$

$$= \frac{1 - z^{-M}e^{j2\pi\alpha}}{M} \sum_{k=0}^{M-1} \frac{H(k + \alpha)}{1 - e^{j2\pi(k+\alpha)/M} z^{-1}}$$

Thus the system function $H(z)$ is characterized by the set of frequency samples $\{H(k + \alpha)\}$ instead of $\{h(n)\}$.

We view this FIR filter realization as a cascade of two filters [i.e., $H(z) = H_1(z)H_2(z)$]. One is an all-zero filter, or a comb filter, with system function

$$H_1(z) = \frac{1}{M} (1 - z^{-M}e^{j2\pi\alpha}) \tag{7.2.13}$$

Its zeros are located at equally spaced points on the unit circle at

$$z_k = e^{j2\pi(k+\alpha)/M} \qquad k = 0, 1, \ldots, M - 1$$

The second filter with system function

$$H_2(z) = \sum_{k=0}^{M-1} \frac{H(k + \alpha)}{1 - e^{j2\pi(k+\alpha)/M}z^{-1}} \qquad (7.2.14)$$

consists of a parallel bank of single-pole filters with resonant frequencies

$$p_k = e^{j2\pi(k+\alpha)/M} \qquad k = 0, 1, \ldots, M - 1$$

Note that the pole locations are identical to the zero locations and both occur at $\omega_k = 2\pi(k + \alpha)/M$, which are the frequencies at which the desired frequency response is specified. The gains of the parallel bank of resonant filters are simply the complex-valued parameters $\{H(k + \alpha)\}$. This cascade realization is illustrated in Fig. 7.5.

When the desired frequency response characteristic of the FIR filter is narrowband, most of the gain parameters $\{H(k + \alpha)\}$ will be zero. Consequently, the corresponding resonant filters can be eliminated and only the filters with nonzero gains need to be retained. The net result is a filter that requires a fewer number of computations (multi-

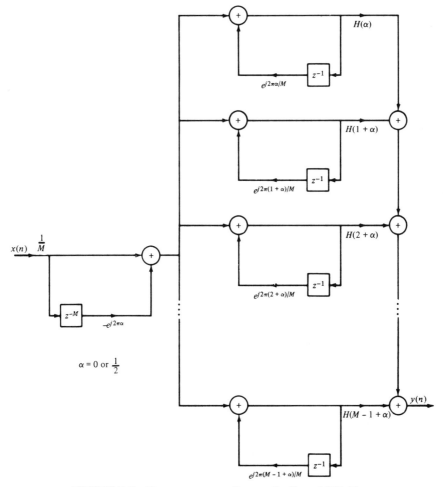

FIGURE 7.5   Frequency-sampling realization of FIR filter.

plications and additions) than the corresponding direct-form realization. Thus we have obtained a more efficient realization.

The frequency-sampling filter structure can be simplified further by exploiting the symmetry in $H(k + \alpha)$, namely, $H(k) = H^*(M - k)$ for $\alpha = 0$ and

$$H(k + \tfrac{1}{2}) = H^*(M - k - \tfrac{1}{2}) \qquad \text{for } \alpha = \tfrac{1}{2}$$

These relations are easily deduced from (7.2.9). As a result of this symmetry, a pair of single-pole filters can be combined to form a single two-pole filter with real-valued parameters. Thus for $\alpha = 0$ the system function $H_2(z)$ reduces to

$$H_2(z) = \frac{H(0)}{1 - z^{-1}} + \sum_{k=1}^{(M-1)/2} \frac{A(k) + B(k)z^{-1}}{1 - 2\cos(2\pi k/M)z^{-1} + z^{-2}} \qquad M \text{ odd}$$

$$H_2(z) = \frac{H(0)}{1 - z^{-1}} + \frac{H(M/2)}{1 + z^{-1}} + \sum_{k=1}^{(M/2)-1} \frac{A(k) + B(k)z^{-1}}{1 - 2\cos(2\pi k/M)z^{-1}) + z^{-2}} \qquad M \text{ even}$$

$$(7.2.15)$$

where, by definition,

$$A(k) = H(k) + H(M - k) \tag{7.2.16}$$
$$B(k) = H(k)e^{-j2\pi k/M} + H(M - k)e^{j2\pi k/M}$$

Similar expressions can be obtained for $\alpha = \tfrac{1}{2}$.

EXAMPLE 7.2.1

Sketch the block diagram for the direct-form realization and the frequency-sampling realization of the $M = 32$, $\alpha = 0$, linear-phase FIR filter which has frequency samples

$$H\left(\frac{2\pi k}{32}\right) = \begin{cases} 1, & k = 0, 1, 2 \\ \dfrac{1}{2}, & k = 3 \\ 0, & k = 4, 5, \ldots, 15 \end{cases}$$

Compare the computational complexity of these two structures.

Solution: Since the filter is symmetric, we exploit this symmetry and thus reduce the number of multiplications per output point by a factor of 2, from 32 to 16 in the direct-form realization. The number of additions per output point is 31. The block diagram of the direct realization is illustrated in Fig. 7.6.

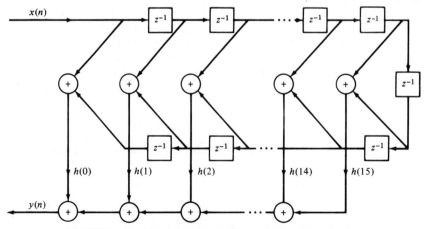

FIGURE 7.6 Direct-form realization of $M = 32$ FIR filter.

We use the form in (7.2.13) and (7.2.15) for the frequency-sampling realization and drop all terms that have zero-gain coefficients $\{H(k)\}$. The nonzero coefficients are $H(k)$, and the corresponding pairs $H(M - k)$, for $k = 0, 1, 2, 3$. The block diagram of the resulting realization is shown in Fig. 7.7. Since $H(0) = 1$, the single-pole filter requires no multiplication. The three double-pole filter sections require three multiplications each for a total of nine multiplications. The total number of additions is 13. Therefore, the frequency-sampling realization of this FIR filter is computationally more efficient than the direct-form realization.

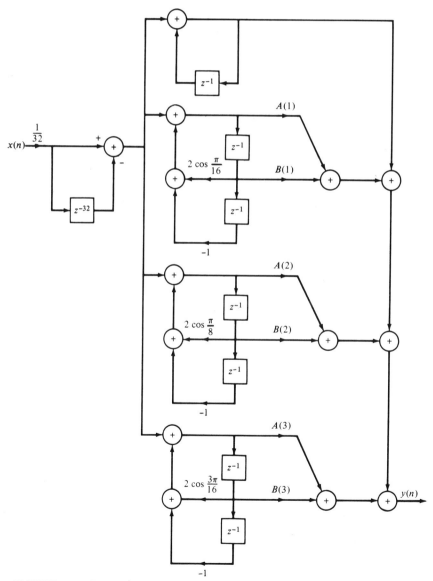

FIGURE 7.7 Frequency-sampling realization for the FIR filter in Example 7.2.1.

7.2.4 Lattice Realization

In this section we introduce another FIR filter structure, called the lattice filter or the lattice realization. Lattice filters are extensively used in digital speech processing and in the implementation of adaptive filters.

Let us begin the development by considering a sequence of FIR filters with system functions

$$H_m(z) = A_m(z) \qquad m = 0, 1, 2, \ldots, M - 1 \qquad (7.2.17)$$

where, by definition, $A_m(z)$ is the polynomial

$$A_m(z) = 1 + \sum_{k=1}^{m} \alpha_m(k)z^{-k} \qquad m \geq 1 \qquad (7.2.18)$$

and $A_0(z) = 1$. The unit sample response of the mth filter is $h_m(0) = 1$ and $h_m(k) = \alpha_m(k)$, $k = 1, 2, \ldots, m$. The subscript m on the polynomial $A_m(z)$ denotes the degree of the polynomial.

It is desirable to view the FIR filters as linear predictors. In this context, the input data sequence $x(n - 1), x(n - 2), \ldots, x(n - m)$ is used to predict the value of the signal $x(n)$. Hence we may express the linearly predicted value of $x(n)$ as

$$\hat{x}(n) = -\sum_{k=1}^{m} \alpha_m(k)x(n - k) \qquad (7.2.19)$$

where the $\{-\alpha_m(k)\}$ represent the prediction coefficients. The negative sign in the definition of $\hat{x}(n)$ is used to conform with current practice in the technical literature. The output sequence $y(n)$ may be expressed as

$$y(n) = x(n) - \hat{x}(n) \qquad (7.2.20)$$
$$= x(n) + \sum_{k=1}^{m} \alpha_m(k)x(n - k)$$

Thus the FIR filter output given by (7.2.20) may be interpreted as the error between the true signal value $x(n)$ and the predicted value $\hat{x}(n)$. Two direct-form realizations of the FIR prediction filter are illustrated in Fig. 7.8.

Suppose that we have a filter for which $m = 1$. Clearly, the output of such a filter is

$$y(n) = x(n) + \alpha_1(1)x(n - 1) \qquad (7.2.21)$$

This output can also be obtained from a first-order or single-stage lattice filter, illustrated in Fig. 7.9, by exciting both of the inputs by $x(n)$ and selecting the output from the top branch. Thus the output is exactly (7.2.21) if we select $K_1 = \alpha_1(1)$. The parameter K_1 in the lattice is called a *reflection coefficient* and it is identical to the reflection coefficient introduced in the Schür–Cohn stability test described in Section 3.6.7.

Next, let us consider an FIR filter for which $m = 2$. In this case the output from a direct-form structure is

$$y(n) = x(n) + \alpha_2(1)x(n - 1) + \alpha_2(2)x(n - 2) \qquad (7.2.22)$$

By cascading two lattice stages as shown in Fig. 7.10, it is possible to obtain the same output as (7.2.22). Indeed, the output from the first stage is

$$f_1(n) = x(n) + K_1x(n - 1) \qquad (7.2.23)$$
$$g_1(n) = K_1x(n) + x(n - 1)$$

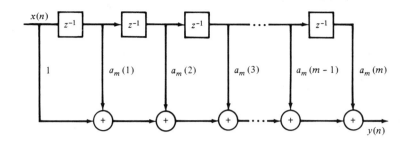

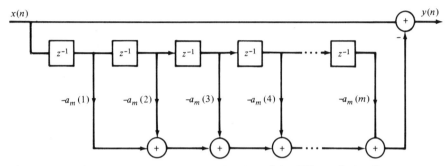

FIGURE 7.8 Direct-form realization of the FIR prediction filter.

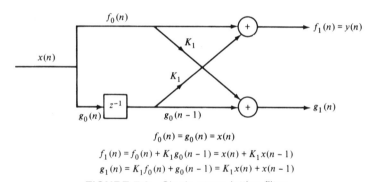

$$f_0(n) = g_0(n) = x(n)$$
$$f_1(n) = f_0(n) + K_1 g_0(n-1) = x(n) + K_1 x(n-1)$$
$$g_1(n) = K_1 f_0(n) + g_0(n-1) = K_1 x(n) + x(n-1)$$

FIGURE 7.9 Single-stage lattice filter.

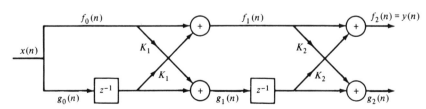

FIGURE 7.10 Two-stage lattice filter.

The output from the second stage is

$$f_2(n) = f_1(n) + K_2 g_1(n - 1)$$
$$g_2(n) = K_2 f_1(n) + g_1(n - 1)$$

(7.2.24)

If we focus our attention on $f_2(n)$ and substitute for $f_1(n)$ and $g_1(n - 1)$ from (7.2.23) into (7.2.24), we obtain

$$f_2(n) = x(n) + K_1 x(n - 1) + K_2[K_1 x(n - 1) + x(n - 2)]$$
$$= x(n) + K_1(1 + K_2)x(n - 1) + K_2 x(n - 2)$$

(7.2.25)

Now (7.2.25) is identical to the output of the direct-form FIR filter as given by (7.2.22) if we equate the coefficients, that is,

$$\alpha_2(2) = K_2 \qquad \alpha_2(1) = K_1(1 + K_2)$$

(7.2.26)

or, equivalently,

$$K_2 = \alpha_2(2) \qquad K_1 = \frac{\alpha_2(1)}{1 + \alpha_2(2)}$$

(7.2.27)

Thus the reflection coefficients K_1 and K_2 of the lattice can be obtained from the coefficients $\{a_m(k)\}$ of the direct-form realization.

By continuing this process, one can easily demonstrate by induction the equivalence between an mth-order direct-form FIR filter and an m-order or m-stage lattice filter. The lattice filter is generally described by the following set of order-recursive equations:

$$f_0(n) = g_0(n) = x(n)$$

(7.2.28)

$$f_m(n) = f_{m-1}(n) + K_m g_{m-1}(n - 1) \qquad m = 1, 2, \ldots, M - 1$$

(7.2.29)

$$g_m(n) = K_m f_{m-1}(n) + g_{m-1}(n - 1) \qquad m = 1, 2, \ldots, M - 1$$

(7.2.30)

Then the output of the $(M - 1)$-stage filter corresponds to the output of an $(M - 1)$-order FIR filter, that is,

$$y(n) = f_{M-1}(n)$$

Figure 7.11 illustrates an $(M - 1)$-stage lattice filter in block diagram form along with a typical stage that shows the computations specified by (7.2.29) and (7.2.30).

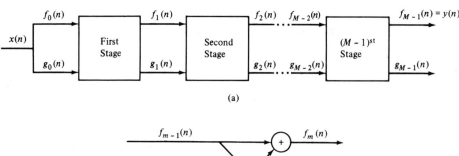

(a)

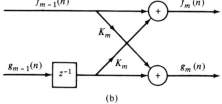

(b)

Typical Stage

FIGURE 7.11 $(M - 1)$-stage lattice filter.

As a consequence of the equivalence between an FIR filter and a lattice filter, the output $f_m(n)$ of an m-stage lattice filter can be expressed as

$$f_m(n) = \sum_{k=0}^{m} \alpha_m(k)x(n - k) \qquad \alpha_m(0) = 1 \qquad (7.2.31)$$

Since (7.2.31) is a convolution sum, it follows that the z-transform relationship is

$$F_m(z) = A_m(z)X(z)$$

or, equivalently,

$$A_m(z) = \frac{F_m(z)}{X(z)} = \frac{F_m(z)}{F_0(z)} \qquad (7.2.32)$$

The other output component from the lattice, namely, $g_m(n)$, may also be expressed in the form of a convolution sum as in (7.2.31), by using another set of coefficients, say $\{\beta_m(k)\}$. That this in fact is the case becomes apparent from observation of (7.2.23) and (7.2.24). From (7.2.23) we note that the filter coefficients for the lattice filter that produces $f_1(n)$ are $\{1, K_1\} = \{1, \alpha_1(1)\}$ while the coefficients for the filter with output $g_1(n)$ are $\{K_1, 1\} = \{\alpha_1(1), 1\}$. We note that these two sets of coefficients are in reverse order. If we consider the two-stage lattice filter, with the output given by (7.2.24), we find that $g_2(n)$ may be expressed in the form

$$
\begin{aligned}
g_2(n) &= K_2 f_1(n) + g_1(n - 1) \\
&= K_2[x(n) + K_1 x(n - 1)] + K_1 x(n - 1) + x(n - 2) \\
&= K_2 x(n) + K_1(1 + K_2)x(n - 1) + x(n - 2) \\
&= \alpha_2(2)x(n) + \alpha_2(1)x(n - 1) + x(n - 2)
\end{aligned}
$$

Consequently, the filter coefficients are $\{\alpha_2(2), \alpha_2(1), 1\}$, whereas the coefficients for the filter that produces the output $f_2(n)$ are $\{1, \alpha_2(1), \alpha_2(2)\}$. Here, again, the two sets of filter coefficients are in reverse order.

From the development above it follows that the output $g_m(n)$ from an m-stage lattice filter may be expressed by the convolution sum of the form

$$g_m(n) = \sum_{k=0}^{m} \beta_m(k)x(n - k) \qquad (7.2.33)$$

where the filter coefficients $\{\beta_m(k)\}$ are associated with a filter that produces $f_m(n) = y(n)$ but operates in reverse order. To elaborate, let us suppose that the data $x(n)$, $x(n - 1), \ldots, x(n - m + 1)$ is used to linearly predict the signal value $x(n - m)$. The predicted value is

$$\hat{x}(n - m) = -\sum_{k=0}^{m-1} \beta_m(k)x(n - k) \qquad (7.2.34)$$

where the coefficients $\beta_m(k)$ in the prediction filter are simply the coefficients $\{\alpha_m(k)\}$ taken in reverse order, that is,

$$\beta_m(k) = \alpha_m(m - k) \qquad k = 0, 1, \ldots, m \qquad (7.2.35)$$

We call the prediction which is performed in (7.2.34) *backward prediction*, that is, the data are run in reverse through a predictor with coefficients $\{-\beta_m(k)\}$. In contrast, the FIR filter with system function $A_m(z)$ performs *forward prediction*.

In the z-transform domain, (7.2.33) becomes

$$G_m(z) = B_m(z)X(z) \qquad (7.2.36)$$

or, equivalently,

$$B_m(z) = \frac{G_m(z)}{X(z)} \tag{7.2.37}$$

where $B_m(z)$ represents the system function of the FIR filter with coefficients $\{\beta_m(k)\}$, that is,

$$B_m(z) = \sum_{k=0}^{m} \beta_m(k)z^{-k} \tag{7.2.38}$$

Since $\beta_m(k) = \alpha_m(m - k)$, (7.2.38) may be expressed as

$$
\begin{aligned}
B_m(z) &= \sum_{k=0}^{m} \alpha_m(m - k)z^{-k} \\
&= \sum_{l=0}^{m} \alpha_m(l)z^{l-m} \\
&= z^{-m} \sum_{l=0}^{m} \alpha_m(l)z^{l} \\
&= z^{-m}A_m(z^{-1})
\end{aligned}
\tag{7.2.39}
$$

The relationship in (7.2.39) implies that the zeros of the FIR filter with system function $B_m(z)$ are simply the reciprocals of the zeros of $A_m(z)$. Hence $B_m(z)$ is called the reciprocal or *reverse* polynomial of $A_m(z)$ and it is associated with the backward prediction of the input data sequence.

Now that we have established these interesting relationships between the direct-form FIR filter and the lattice structure, let us return to the recursive lattice equations in (7.2.28) through (7.2.30) and transfer them to the z-domain. Thus we have

$$
\begin{aligned}
F_0(z) &= G_0(z) = X(z) & & (7.2.40) \\
F_m(z) &= F_{m-1}(z) + K_m z^{-1}G_{m-1}(z) & m = 1, 2, \ldots, M - 1 & (7.2.41) \\
G_m(z) &= K_m F_{m-1}(z) + z^{-1}G_{m-1}(z) & m = 1, 2, \ldots, M - 1 & (7.2.42)
\end{aligned}
$$

If we divide each equation by $X(z)$, we obtain the desired results in the form

$$
\begin{aligned}
A_0(z) &= B_0(z) = 1 & & (7.2.43) \\
A_m(z) &= A_{m-1}(z) + K_m z^{-1}B_{m-1}(z) & m = 1, 2, \ldots, M - 1 & (7.2.44) \\
B_m(z) &= K_m A_{m-1}(z) + z^{-1}B_{m-1}(z) & m = 1, 2, \ldots, M - 1 & (7.2.45)
\end{aligned}
$$

Thus a lattice stage is described in the z-domain by the matrix equation

$$
\begin{bmatrix} A_m(z) \\ B_m(z) \end{bmatrix} = \begin{bmatrix} 1 & K_m \\ K_m & 1 \end{bmatrix} \begin{bmatrix} A_{m-1}(z) \\ z^{-1}B_{m-1}(z) \end{bmatrix} \tag{7.2.46}
$$

Before concluding this discussion, it is desirable to develop the relationships for converting the lattice parameters $\{K_i\}$, that is, the reflection coefficients, to the direct-form filter coefficients $\{\alpha_m(k)\}$, and vice versa.

Conversion of Lattice Coefficients to Direct-Form Filter Coefficients. The direct-form FIR filter coefficients $\{\alpha_m(k)\}$ can be obtained from the lattice coefficients $\{K_i\}$ by using the following relations:

$$
\begin{aligned}
A_0(z) &= B_0(z) = 1 & & (7.2.47) \\
A_m(z) &= A_{m-1}(z) + K_m z^{-1}B_{m-1}(z) & m = 1, 2, \ldots, M - 1 & (7.2.48) \\
B_m(z) &= z^{-m}A_m(z^{-1}) & m = 1, 2, \ldots, M - 1 & (7.2.49)
\end{aligned}
$$

The solution is obtained recursively, beginning with $m = 1$. Thus we obtain a sequence of $(M - 1)$ FIR filters, one for each value of m. The procedure is best illustrated by means of an example.

EXAMPLE 7.2.2 _____

Given a three-stage lattice filter with coefficients $K_1 = \frac{1}{4}$, $K_2 = \frac{1}{2}$, $K_3 = \frac{1}{3}$, determine the FIR filter coefficients for the direct-form structure.

Solution: We solve the problem recursively, beginning with (7.2.45) for $m = 1$. Thus we have

$$A_1(z) = A_0(z) + K_1 z^{-1} B_0(z)$$
$$= 1 + K_1 z^{-1} = 1 + \tfrac{1}{4} z^{-1}$$

Hence the coefficients of an FIR filter corresponding to the single-stage lattice are $\alpha_1(0) = 1$, $\alpha_1(1) = K_1 = \frac{1}{4}$. Since $B_m(z)$ is the reverse polynomial of $A_m(z)$, we have

$$B_1(z) = \tfrac{1}{4} + z^{-1}$$

Next we add the second stage to the lattice. For $m = 2$, (7.2.48) yields

$$A_2(z) = A_1(z) + K_2 z^{-1} B_1(z)$$
$$= 1 + \tfrac{3}{8} z^{-1} + \tfrac{1}{2} z^{-2}$$

Hence the FIR filter parameters corresponding to the two-stage lattice are $\alpha_2(0) = 1$, $\alpha_2(1) = \frac{3}{8}$, $\alpha_2(2) = \frac{1}{2}$. Also,

$$B_2(z) = \tfrac{1}{2} + \tfrac{3}{8} z^{-1} + z^{-2}$$

Finally, the addition of the third stage to the lattice results in the polynomial

$$A_3(z) = A_2(z) + K_3 z^{-1} B_2(z)$$
$$= 1 + \tfrac{13}{24} z^{-1} + \tfrac{5}{8} z^{-2} + \tfrac{1}{3} z^{-3}$$

Consequently, the desired direct-form FIR filter is characterized by the coefficients

$$\alpha_3(0) = 1, \qquad \alpha_3(1) = \tfrac{13}{24} \qquad \alpha_3(2) = \tfrac{5}{8} \qquad \alpha_3(3) = \tfrac{1}{3}$$

As this example illustrates, the lattice structure with parameters $K_1, K_2, \ldots, K_m$, corresponds to a class of m direct-form FIR filters with system functions $A_1(z)$, $A_2(z)$, $\ldots, A_m(z)$. It is interesting to note that a characterization of this class of m FIR filters in direct form requires $m(m + 1)/2$ filter coefficients. In contrast, the lattice-form characterization requires only the m reflection coefficients $\{K_i\}$. The reason that the lattice provides a more compact representation for the class of m FIR filters is simply due to the fact that the addition of stages to the lattice does not alter the parameters of the previous stages. On the other hand, the addition of the mth stage to a lattice with $(m - 1)$ stages results in a FIR filter with system function $A_m(z)$ that has coefficients totally different from the coefficients of the lower-order FIR filter with system function $A_{m-1}(z)$.

A formula for determining the filter coefficients $\{\alpha_m(k)\}$ recursively can be easily derived from polynomial relationships in (7.2.47) through (7.2.49). From the relationship in (7.2.48) we have

$$A_m(z) = A_{m-1}(z) + K_m z^{-1} B_{m-1}(z)$$

$$\sum_{k=0}^{m} \alpha_m(k) z^{-k} = \sum_{k=0}^{m-1} \alpha_{m-1}(k) z^{-k} + K_m \sum_{k=0}^{m-1} \alpha_{m-1}(m - 1 - k) z^{-(k+1)}$$

(7.2.50)

By equating the coefficients of equal powers of z^{-1} and recalling that $\alpha_m(0) = 1$ for $m = 1, 2, \ldots, M - 1$, we obtain the desired recursive equation for the FIR filter coefficients in the form

$$\alpha_m(0) = 1 \tag{7.2.51}$$
$$\alpha_m(m) = K_m \tag{7.2.52}$$
$$\begin{aligned}\alpha_m(k) &= \alpha_{m-1}(k) + K_m\alpha_{m-1}(m - k)\\ &= \alpha_{m-1}(k) + \alpha_m(m)\alpha_{m-1}(m - k) \qquad 1 \le k \le m - 1\\ &\qquad\qquad\qquad\qquad\qquad\qquad m = 1, 2, \ldots, M - 1 \end{aligned} \tag{7.2.53}$$

We note that (7.2.51) through (7.2.53) are simply the Levinson–Durbin recursive equations given in (6A.22) of Appendix 6A. A subroutine that solves (7.2.51) through (7.2.53) recursively is given in Fig. 7.12.

Conversion of Direct-Form FIR Filter Coefficients to Lattice Coefficients. Suppose that we are given the FIR coefficients for the direct-form realization or, equivalently, the polynomial $A_m(z)$ and we wish to determine the corresponding lattice filter parameters $\{K_i\}$. For the m-stage lattice we immediately obtain the parameter $K_m = \alpha_m(m)$. To obtain K_{m-1} we need the polynomials $A_{m-1}(z)$ and, in general, K_m is obtained from the polynomial $A_m(z)$ for $m = M - 1, M - 2, \ldots, 1$. Consequently, we need to compute the polynomials $A_m(z)$ starting from $m = M - 1$ and "stepping down" successively to $m = 1$.

The desired recursive relation for the polynomials is easily determined from (7.2.44) and (7.2.45). We have

$$\begin{aligned}A_m(z) &= A_{m-1}(z) + K_m z^{-1} B_{m-1}(z)\\ &= A_{m-1}(z) + K_m[B_m(z) - K_m A_{m-1}(z)]\end{aligned}$$

```
C
C        CONVERSION FROM LATTICE TO
C        DIRECT STRUCTURE PARAMETERS
C
C        INPUT   : RK(1),RK(2),...,RK(IP)
C        OUTPUT  : A(1),A(2),...A(IP)
C        IP      : FILTER'S ORDER
C
         SUBROUTINE SUP (RK,A,B,IP,IER)
         DIMENSION A(1),B(1),RK(1)
         A(1)=RK(1)
         IF (IP.EQ.1) RETURN
         DO 20 M=2,IP
           IF (ABS(RK(M)).GE.1.0) IER=1
           M1=M-1
           A(M)=RK(M)
           DO 10 J=1,M1
             B(J)=A(M-J)
10         CONTINUE
           DO 20 J=1,M1
             A(J)=A(J)+RK(M)*B(J)
20       CONTINUE
         RETURN
         END
```

FIGURE 7.12 Subroutine for converting lattice coefficients to direct-form FIR filter coefficients.

If we solve for $A_{m-1}(z)$, we obtain

$$A_{m-1}(z) = \frac{A_m(z) - K_m B_m(z)}{1 - K_m^2} \qquad m = M - 1, M - 2, \ldots, 1 \qquad (7.2.54)$$

which is just the step-down recursion used in the Schür–Cohn stability test described in Section 3.6.7. Thus we compute all lower-degree polynomials $A_m(z)$ beginning with $A_{M-1}(z)$ and obtain the desired lattice coefficients from the relation $K_m = \alpha_m(m)$. We observe that the procedure works as long as $|K_m| \neq 1$ for $m = 1, 2, \ldots, M - 1$.

EXAMPLE 7.2.3

Determine the lattice coefficients corresponding to the FIR filter with system function

$$H(z) = A_3(z) = 1 + \tfrac{13}{24} z^{-1} + \tfrac{5}{8} z^{-2} + \tfrac{1}{3} z^{-3}$$

Solution: First we note that $K_3 = \alpha_3(3) = \tfrac{1}{3}$. Furthermore,

$$B_3(z) = \tfrac{1}{3} + \tfrac{5}{8} z^{-1} + \tfrac{13}{24} z^{-2} + z^{-3}$$

The step-down relationship in (7.2.54) with $m = 3$ yields

$$A_2(z) = \frac{A_3(z) - K_3 B_3(z)}{1 - K_3^2}$$
$$= 1 + \tfrac{3}{8} z^{-1} + \tfrac{1}{2} z^{-2}$$

Hence $K_2 = \alpha_2(2) = \tfrac{1}{2}$ and $B_2(z) = \tfrac{1}{2} + \tfrac{3}{8} z^{-1} + z^{-1}$. By repeating the step-down recursion in (7.2.51), we obtain

$$A_1(z) = \frac{A_2(z) - K_2 B_2(z)}{1 - K_2^2}$$
$$= 1 + \tfrac{1}{4} z^{-1}$$

Hence $K_1 = \alpha_1(1) = \tfrac{1}{4}$.

___

From the step-down recursive equation in (7.2.54) it is relatively easy to obtain a formula for recursively computing K_m, beginning with $m = M - 1$ and stepping down to $m = 1$. For $m = M - 1, M - 2, \ldots, 1$ we have

$$K_m = \alpha_m(m) \qquad \alpha_{m-1}(0) = 1 \qquad (7.2.55)$$

$$\alpha_{m-1}(k) = \frac{\alpha_m(k) - K_m B_m(k)}{1 - K_m^2}$$

$$= \frac{\alpha_m(k) - \alpha_m(m)\alpha_m(m - k)}{1 - \alpha_m^2(m)} \qquad 1 \leq k \leq m - 1 \qquad (7.2.56)$$

which is again the recursion we introduced in the Schür–Cohn stability test. Consequently, the subroutine for the Schür–Cohn test, given in Fig. 3.15, implements this recursive equation.

As indicated above, the recursive equation in (7.2.56) breaks down if any lattice parameters $|K_m| = 1$. If this occurs, it is indicative of the fact that the polynomial $A_{M-1}(z)$ has a root on the unit circle. Such a root may be factored out from $A_{M-1}(z)$ and the iterative process in (7.2.56) can be carried out for the reduced-order system.

7.3 Realizations for IIR Systems

In this section we consider different structures for IIR systems described by the difference equation in (7.1.1) or, equivalently, by the system function in (7.1.2). Just as in the case of FIR systems, there are several types of structures or realizations, including direct-form structures, cascade-form structures, lattice structures, and lattice-ladder structures. In addition, IIR systems lend themselves to a parallel-form realization. We begin by describing two direct-form realizations.

7.3.1 Direct-Form Realizations

The rational system function as given by (7.1.2) that characterizes an IIR system may be viewed as two systems in cascade, that is,

$$H(z) = H_1(z)H_2(z) \qquad (7.3.1)$$

where $H_1(z)$ consists of the zeros of $H(z)$ and $H_2(z)$ consists of the poles of $H(z)$, that is,

$$H_1(z) = \sum_{k=0}^{M} b_k z^{-k} \qquad (7.3.2)$$

and

$$H_2(z) = \frac{1}{1 + \sum_{k=1}^{N} a_k z^{-k}} \qquad (7.3.3)$$

In Section 2.4.5 we describe two different direct-form realizations, depending on whether $H_1(z)$ precedes $H_2(z)$, or vice versa. Since $H_1(z)$ is an FIR system, its direct-form realization was illustrated in Fig. 7.1. By attaching the all-pole system in cascade with $H_1(z)$, we obtain the direct form I realization depicted in Fig. 7.13. This realization requires $M + N + 1$ multiplications, $M + N$ additions, and $M + N + 1$ memory locations.

If the all-pole filter $H_2(z)$ is placed before the all-zero filter $H_1(z)$, a more compact structure is obtained as illustrated in Section 2.4.5. Recall that the difference equation for the all-pole filter is

$$w(n) = -\sum_{k=1}^{N} a_k w(n-k) + x(n) \qquad (7.3.4)$$

Since $w(n)$ is the input to the all-zero system, its output is

$$y(n) = \sum_{k=0}^{M} b_k w(n-k) \qquad (7.3.5)$$

We note that both (7.3.4) and (7.3.5) involve delayed versions of the sequence $\{w(n)\}$. Consequently, only a single delay line or a single set of memory locations is required for storing the past values of $\{w(n)\}$. The resulting structure that implements (7.3.4) and (7.3.5) is called a direct form II realization and is depicted in Fig. 7.14. This structure requires $M + N + 1$ multiplications, $M + N$ additions, and the maximum of $\{M, N\}$ memory locations. Since the direct form II realization minimizes the number of memory locations, it is said to be *canonic*. However, we should indicate that other IIR structures also possess this property, so that this terminology is perhaps unjustified.

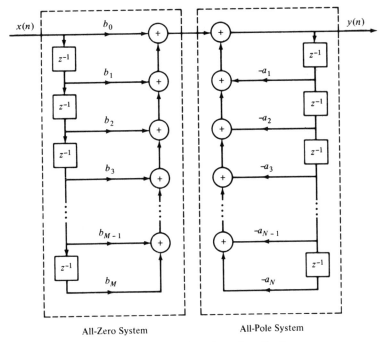

All-Zero System

All-Pole System

FIGURE 7.13 Direct form I realization.

The structures in Figs. 7.13 and 7.14 are both called "direct form" realizations because they are obtained directly from the system function $H(z)$ without any rearrangement of $H(z)$. Unfortunately, both structures are extremely sensitive to parameter quantization, in general, and are not recommended in practical applications. This topic will be discussed in detail in Chapter 10, where we demonstrate that when N is large, a small change in a filter coefficient due to parameter quantization results in a large change in the location of the poles and zeros of the system.

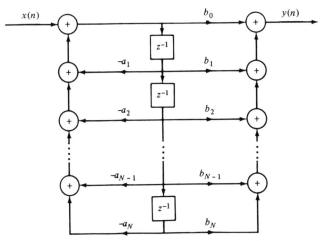

FIGURE 7.14 Direct form II realization ($N = M$).

7.3.2 Signal Flow Graphs and Transposed Structures

A signal flow graph provides an alternative but equivalent graphical representation to a block diagram structure that we have been using to illustrate various system realizations. The basic elements of a flow graph are branches and nodes. A signal flow graph is basically a set of directed branches that connect at nodes. By definition, the signal out of a branch is equal to the branch gain (system function) times the signal into the branch. Furthermore, the signal at a node of a flow graph is equal to the sum of the signals from all branches connecting to the node.

To illustrate some basic notions, let us consider the two-pole and two-zero IIR system depicted in block diagram form in Fig. 7.15a. The system block diagram can be converted to the signal flow graph shown in Fig. 7.15b. We note that the flow graph contains five nodes labeled 1 through 5. Two of the nodes (1, 3) are summing nodes (i.e., they contain adders), while the other three nodes represent branching points. Branch transmittances are indicated for the branches in the flow graph. Note that a delay is indicated by the branch transmittance z^{-1}. When the branch transmittance is unity, it is left unlabeled. The input to the system originates at a *source node* and the output signal is extracted at a *sink node*.

We observe that the signal flow graph contains the same basic information as the block diagram realization of the system. The only apparent difference is that *both* branch points and adders in the block diagram are represented by nodes in the signal flow graph.

The subject of linear signal flow graphs is an important one in the treatment of networks and many interesting results are available. One basic notion involves the transformation of one flow graph into another without changing the basic input–output relationship. Specifically, one technique that is useful in deriving new system structures for FIR and IIR systems stems from the *transposition* or *flow-graph reversal theorem*. This

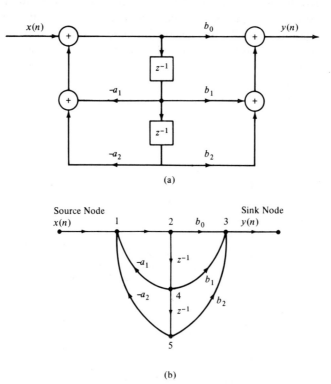

FIGURE 7.15   Second-order filter structure (a) and its signal flow graph (b).

theorem simply states that if we reverse the directions of all branch transmittances and interchange the input and output in the flow graph, the system function remains unchanged. The resulting structure is called a *transposed structure* or a *transposed form*.

For example, the transposition of the signal flow graph in Fig. 7.15b is illustrated in Fig. 7.16a. The corresponding block diagram realization of the transposed form is depicted in Fig. 7.16b. It is interesting to note that the transposition of the original flow graph resulted in branching nodes becoming adder nodes, and vice versa. In Section 7.5 we provide a proof of the transposition theorem by using state-space techniques.

Let us apply the transposition theorem to the direct form II structure. First, we reverse all the signal flow directions in Fig. 7.14. Second, we change nodes into adders and adders into nodes, and finally, we interchange the input and the output. These operations result in the transposed direct form II structure shown in Fig. 7.17. This structure may be redrawn as in Fig. 7.18, which shows the input on the left and the output on the right.

The transposed, direct form II realization we have obtained may be described by the set of difference equations:

$$y(n) = w_1(n - 1) + b_0 x(n) \tag{7.3.6}$$
$$w_k(n) = w_{k+1}(n - 1) - a_k y(n) + b_k x(n) \quad k = 1, 2, \ldots, N - 1 \tag{7.3.7}$$
$$w_N(n) = b_N x(n) - a_N y(n) \tag{7.3.8}$$

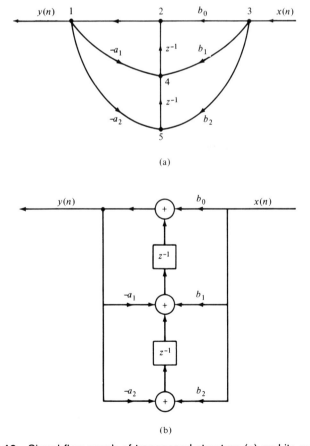

FIGURE 7.16   Signal flow graph of transposed structure (a) and its realization (b)

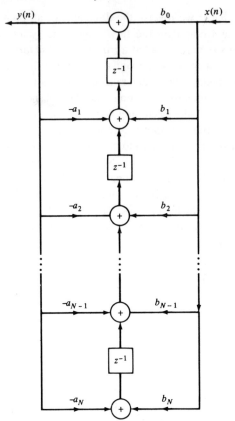

FIGURE 7.17 Transposed direct form II structure.

Without loss of generality, we have assumed that $M = N$ in writing the equations above. It is also clear from observation of Fig. 7.18 that this set of difference equations is equivalent to the single difference equation

$$y(n) = -\sum_{k=1}^{N} a_k y(n-k) + \sum_{k=0}^{M} b_k x(n-k) \qquad (7.3.9)$$

Finally, we observe that the transposed direct form II structure requires the same number of multiplications, additions, and memory locations as the original direct form II structure.

Although our discussion of transposed structures has been concerned with the general form of an IIR system, it is interesting to note that an FIR system, obtained from (7.3.9) by setting the $a_k = 0$, $k = 1, 2, \ldots, N$, also has a transposed direct form as illustrated in Fig. 7.19. This structure is simply obtained from Fig. 7.18 by setting $a_k = 0$, $k = 1, 2, \ldots, N$. This transposed form realization may be described by the set of difference equations

$$w_M(n) = b_M x(n) \qquad (7.3.10)$$
$$w_k(n) = w_{k+1}(n-1) + b_k x(n) \qquad k = M-1, M-2, \ldots, 1 \qquad (7.3.11)$$
$$y(n) = w_1(n-1) + b_0 x(n) \qquad (7.3.12)$$

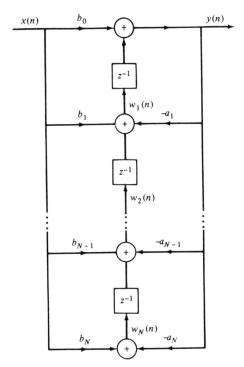

FIGURE 7.18   Transposed direct form II structure.

In summary, Table 7.1 illustrates the direct-form structures and the corresponding difference equations for a basic two-pole and two-zero IIR system with system function

$$H(z) = \frac{b_0 + b_1 z^{-1} + b_2 z^{-2}}{1 + a_1 z^{-1} + a_2 z^{-2}} \qquad (7.3.13)$$

This is the basic building block in the cascade realization of high-order IIR systems, as described in the following section. Of the three direct-form structures given in Table 7.1, the direct form II structures are preferable due to the smaller number of memory locations required in their implementation.

Finally, we note that in the z-domain the set of difference equations describing a linear signal flow graph constitute a linear set of equations. Any rearrangement of such a set of equations is equivalent to a rearrangement of the signal flow graph to obtain a new structure, and vice versa.

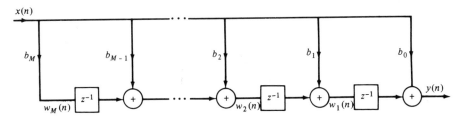

FIGURE 7.19   Transposed FIR structure.

TABLE 7.1 Some Second-Order Modules for Discrete-Time Systems

Structure	Implementation Equations	System Function
Direct Form I	$y(n) = b_0 x(n) + b_1 x(n-1)$ $+ b_2 x(n-2)$ $- a_1 y(n-1) - a_2 y(n-2)$	$H(z) = \dfrac{b_0 + b_1 z^{-1} + b_2 z^{-2}}{1 + a_1 z^{-1} + a_2 z^{-2}}$
Regular Direct Form II	$w(n) = -a_1 w(n-1) - a_2 w(n-2)$ $+ x(n)$ $y(n) = b_0 w(n) + b_1 w(n-1)$ $+ b_2 w(n-2)$	$H(z) = \dfrac{b_0 + b_1 z^{-1} + b_2 z^{-2}}{1 + a_1 z^{-1} + a_2 z^{-2}}$
Transposed Direct Form II	$y(n) = b_0 x(n) + w_1(n-1)$ $w_1(n) = b_1 x(n) - a_1 y(n)$ $+ w_2(n-1)$ $w_2(n) = b_2 x(n) - a_2 y(n)$	$H(z) = \dfrac{b_0 + b_1 z^{-1} + b_2 z^{-2}}{1 + a_1 z^{-1} + a_2 z^{-2}}$

7.3.3 Cascade-Form Realization

Let us consider a high-order IIR system with system function given by (7.1.2). Without loss of generality we assume that $N \geq M$. The system can be factored into a cascade of second-order subsystems, such that $H(z)$ may be expressed as

$$H(z) = G \prod_{k=1}^{K} H_k(z) \tag{7.3.14}$$

where K is the integer part of $(N + 1)/2$, $H_k(z)$ has the general form

$$H_k(z) = \frac{1 + b_{k1}z^{-1} + b_{k2}z^{-2}}{1 + a_{k1}z^{-1} + a_{k2}z^{-2}} \tag{7.3.15}$$

and G is the fixed gain parameter which is easily determined from (7.1.2) to be $G = b_0$. As in the case of FIR systems based on a cascade-form realization, the gain parameter G may be distributed equally among the K filter sections so that $G = G_1 G_2 \ldots, G_K$.

The coefficients $\{a_{ki}\}$ and $\{b_{ki}\}$ in the second-order subsystems are real. This implies that in forming the second-order subsystems or quadratic factors in (7.3.15), we should group together a pair of complex-conjugate poles and we should group together a pair of complex-conjugate zeros. However, the pairing of two complex-conjugate poles with a pair of complex-conjugate zeros or real-valued zeros to form a subsystem of the type given by (7.3.15) may be done arbitrarily. Furthermore, any two real-valued zeros may be paired together to form a quadratic factor and, likewise, any two real-valued poles may be paired together to form a quadratic factor. Consequently, the quadratic factor in the numerator of (7.3.15) may consist of either a pair of real roots or a pair of complex-conjugate roots. The same statement applies to the denominator of (7.3.15).

If $N > M$, some of the second-order subsystems will have numerator coefficients that are zero, that is, either $b_{k2} = 0$ or $b_{k1} = 0$ or both $b_{k2} = b_{k1} = 0$ for some k. Furthermore, if N is odd, one of the subsystems, say $H_K(z)$, must have $a_{k2} = 0$, so that the subsystem is of first order. To preserve the modularity in the implementation of $H(z)$ it is often preferable to use the basic second-order subsystems in the cascade structure and have some zero-valued coefficients in some of the subsystems.

Each of the second-order subsystems with system function of the form (7.3.15) may be realized in either direct form I, or direct form II, or transposed direct form II. Since there are many ways to pair the poles and zeros of $H(z)$ into a cascade of second-order sections, and several ways to order the resulting subsystems, it is possible to obtain a variety of cascade realizations. Although all cascade realizations are equivalent for infinite precision, the various realizations may differ significantly when implemented with finite-precision arithmetic.

The general form of the cascade structure is illustrated in Fig. 7.20. If we use the direct form II structure for each of the subsystems, the computational algorithm for realizing the IIR system with system function $H(z)$ is described by the following set of equations.

$$y_0(n) = x(n) \tag{7.3.16}$$
$$w_k(n) = -a_{k1}w_k(n-1) - a_{k2}w_k(n-2) + y_{k-1}(n) \quad k = 1, 2, \ldots, K \tag{7.3.17}$$
$$y_k(n) = w_k(n) + b_{k1}w_k(n-1) + b_{k2}w_k(n-2) \quad k = 1, 2, \ldots, K \tag{7.3.18}$$
$$y(n) = Gy_K(n) \tag{7.3.19}$$

Thus this set of equations provides a complete description of the cascade structure based on direct form II sections.

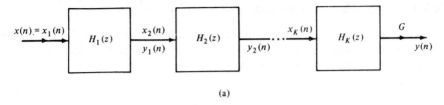

(a)

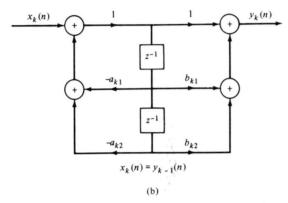

$$x_k(n) = y_{k-1}(n)$$

(b)

FIGURE 7.20   Cascade structure of second-order systems and a realization of each second-order section.

7.3.4 Parallel-Form Realization

A parallel-form realization of an IIR system can be obtained by performing a partial-fraction expansion of $H(z)$. Without loss of generality, we assume again that $N \geq M$ and that the poles are distinct. Then, by performing a partial-fraction expansion of $H(z)$, we obtain the result

$$H(z) = C + \sum_{k=1}^{N} \frac{A_k}{1 - p_k z^{-1}} \qquad (7.3.20)$$

where $\{p_k\}$ are the poles, $\{A_k\}$ are the coefficients (residues) in the partial-fraction expansion, and the constant C is defined as $C = b_N/a_N$. The structure implied by (7.3.20) is shown in Fig. 7.21. It consists of a parallel bank of single-pole filters.

In general, some of the poles of $H(z)$ may be complex valued. In such a case, the corresponding coefficients A_k are also complex valued. To avoid multiplications by complex numbers, we may combine pairs of complex-conjugate poles to form two-pole subsystems. In addition, we may combine, in an arbitrary manner, pairs of real-valued poles to form two-pole subsystems. Each of these subsystems has the form

$$H_k(z) = \frac{b_{k0} + b_{k1}z^{-1}}{1 + a_{k1}z^{-1} + a_{k2}z^{-2}} \qquad (7.3.21)$$

where the coefficients $\{b_{ki}\}$ and $\{a_{ki}\}$ are real-valued system parameters. The overall function can now be expressed as

$$H(z) = C + \sum_{k=1}^{K} H_k(z) \qquad (7.3.22)$$

where K is the integer part of $(N + 1)/2$. When N is odd, one of the $H_k(z)$ is really a single-pole system (i.e., $b_{k1} = a_{k2} = 0$).

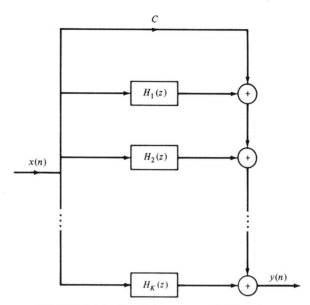

FIGURE 7.21 Parallel structure of IIR system.

The individual second-order sections which are the basic building blocks for $H(z)$ can be implemented in either of the direct forms or in a transposed direct form. The direct form II structure is illustrated in Fig. 7.22. With this structure as a basic building block, the parallel-form realization of the FIR system is described by the following set of equations:

$$w_k(n) = -a_{k1}w_k(n - 1) - a_{k2}w_k(n - 2) + x(n) \qquad k = 1, 2, \dots, K \quad (7.3.23)$$
$$y_k(n) = b_{k0}w_k(n) + b_{k1}w_k(n - 1) \qquad k = 1, 2, \dots, K \quad (7.3.24)$$
$$y(n) = Cx(n) + \sum_{k=1}^{K} y_k(n) \qquad (7.3.25)$$

EXAMPLE 7.3.1 _____

Determine the cascade and parallel realizations for the system described by the system function

$$H(z) = \frac{10(1 - \frac{1}{2}z^{-1})(1 - \frac{2}{3}z^{-1})(1 + 2z^{-1})}{(1 - \frac{3}{4}z^{-1})(1 - \frac{1}{8}z^{-1})[1 - (\frac{1}{2} + j\frac{1}{2})z^{-1}][1 - (\frac{1}{2} - j\frac{1}{2})z^{-1}]}$$

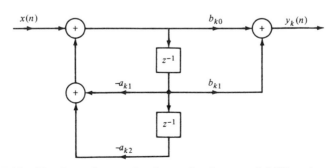

FIGURE 7.22 Structure of second-order section in a parallel IIR system realization.

Solution: The cascade realization is easily obtained from this form. One possible pairing of poles and zeros is

$$H_1(z) = \frac{1 - \frac{2}{3}z^{-1}}{1 - \frac{7}{8}z^{-1} + \frac{3}{32}z^{-2}}$$

$$H_2(z) = \frac{1 + \frac{3}{2}z^{-1} - z^{-2}}{1 - z^{-1} + \frac{1}{2}z^{-2}}$$

and hence

$$H(z) = 10H_1(z)H_2(z)$$

The cascade realization is depicted in Fig. 7.23a.

To obtain the parallel-form realization, $H(z)$ must be expanded in partial fractions. Thus we have

$$H(z) = \frac{A_1}{1 - \frac{3}{4}z^{-1}} + \frac{A_2}{1 - \frac{1}{8}z^{-1}} + \frac{A_3}{1 - (\frac{1}{2} + j\frac{1}{2})z^{-1}} + \frac{A_3^*}{1 - (\frac{1}{2} - j\frac{1}{2})z^{-1}}$$

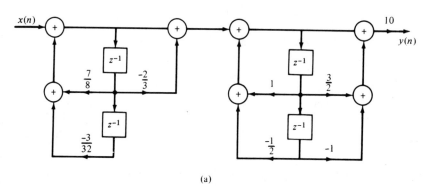

(a)

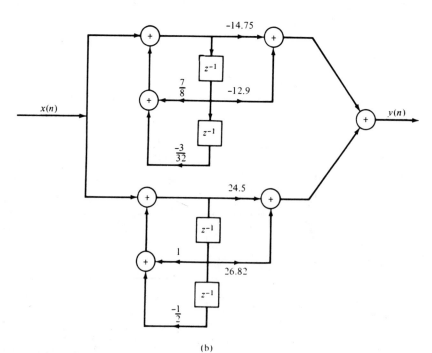

(b)

FIGURE 7.23 Cascade and parallel realizations for the system in Example 7.3.1.

where A_1, A_2, A_3, and A_3^* are to be determined. After some arithmetic we find that

$$A_1 = 2.93 \qquad A_2 = -17.68 \qquad A_3 = 12.25 - j14.57 \qquad A_3^* = 12.25 + j14.57$$

Upon recombining pairs of poles, we obtain

$$H(z) = \frac{-14.75 - 12.90z^{-1}}{1 - \frac{7}{8}z^{-1} + \frac{3}{32}z^{-2}} + \frac{24.50 + 26.82z^{-1}}{1 - z^{-1} + \frac{1}{2}z^{-2}}$$

The parallel-form realization is illustrated in Fig. 7.23b.

7.3.5 Lattice and Lattice-Ladder Structures for IIR Systems

In Section 7.2.4 we developed a lattice filter structure that is equivalent to an FIR system. In this section we extend the development to IIR systems.

Let us begin with an all-pole system with system function

$$H(z) = \frac{1}{1 + \displaystyle\sum_{k=1}^{N} a_N(k)z^{-k}} = \frac{1}{A_N(z)} \tag{7.3.26}$$

The direct form realization of this system is illustrated in Fig. 7.24. The difference equation for this IIR system is

$$y(n) = -\sum_{k=1}^{N} a_N(k)y(n - k) + x(n) \tag{7.3.27}$$

It is interesting to note that if we interchange the roles of input and output [i.e., interchange $x(n)$ with $y(n)$ in (7.3.27)], we obtain

$$x(n) = -\sum_{k=1}^{N} a_N(k)x(n - k) + y(n)$$

or, equivalently,

$$y(n) = x(n) + \sum_{k=1}^{N} a_N(k)x(n - k) \tag{7.3.28}$$

We note that the equation in (7.3.28) describes an FIR system having the system function $H(z) = A_N(z)$, while the system described by the difference equation in (7.3.27) represents an IIR system with system function $H(z) = 1/A_N(z)$. One system can be obtained from the other simply by interchanging the roles of the input and output.

Based on this observation, we shall use the all-zero (FIR) lattice described in Section 7.2.4 to obtain a lattice structure for an all-pole IIR system by interchanging the roles of the input and output. First, we take the all-zero lattice filter illustrated in Fig. 7.11 and redefine the input as

$$x(n) = f_N(n) \tag{7.3.29}$$

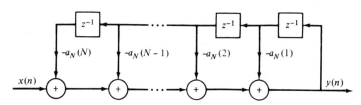

FIGURE 7.24   Direct-form realization of an all-pole system.

and the output as

$$y(n) = f_0(n) \tag{7.3.30}$$

These are exactly the opposite of the definitions for the all-zero lattice filter. These definitions dictate that the quantities $\{f_m(n)\}$ be computed in descending order [i.e., $f_N(n)$, $f_{N-1}(n)$, . . .]. This computation can be accomplished by rearranging the recursive equation in (7.2.29) and thus solving for $f_{m-1}(n)$ in terms of $f_m(n)$, that is,

$$f_{m-1}(n) = f_m(n) - K_m g_{m-1}(n-1) \qquad m = N, N-1, \ldots, 1$$

The equation (7.2.30) for $g_m(n)$ remains unchanged.

The result of these changes is the set of equations

$$\begin{aligned}
f_N(n) &= x(n) & &\text{(7.3.31)}\\
f_{m-1}(n) &= f_m(n) - K_m g_{m-1}(n-1) & m &= N, N-1, \ldots, 1 & &\text{(7.3.32)}\\
g_m(n) &= K_m f_{m-1}(n) + g_{m-1}(n-1) & m &= N, N-1, \ldots, 1 & &\text{(7.3.33)}\\
y(n) &= f_0(n) = g_0(n) & & & &\text{(7.3.34)}
\end{aligned}$$

which correspond to the structure shown in Fig. 7.25.

To demonstrate that the set of equations (7.3.31) through (7.3.34) represent an all-pole IIR system, let us consider the case where $N = 1$. The equations above reduce to

$$\begin{aligned}
x(n) &= f_1(n)\\
f_0(n) &= f_1(n) - K_1 g_0(n-1)\\
g_1(n) &= K_1 f_0(n) + g_0(n-1) \qquad\qquad (7.3.35)\\
y(n) &= f_0(n)\\
&= x(n) - K_1 y(n-1)
\end{aligned}$$

Furthermore, the equation for $g_1(n)$ may be expressed as

$$g_1(n) = K_1 y(n) + y(n-1) \tag{7.3.36}$$

We observe that (7.3.35) represents a first-order all-pole IIR system while (7.3.36) represents a first-order FIR system. The pole is a result of the feedback that was introduced by the solution of the $\{f_m(n)\}$ in descending order. This feedback is depicted in Fig. 7.26a.

Next, let us consider the case $N = 2$, which corresponds to the structure in Fig. 7.26b. The equations corresponding to this structure are

$$\begin{aligned}
f_2(n) &= x(n)\\
f_1(n) &= f_2(n) - K_2 g_1(n-1)\\
g_2(n) &= K_2 f_1(n) + g_1(n-1)\\
f_0(n) &= f_1(n) - K_1 g_0(n-1) \qquad\qquad (7.3.37)\\
g_1(n) &= K_1 f_0(n) + g_0(n-1)\\
y(n) &= f_0(n) = g_0(n)
\end{aligned}$$

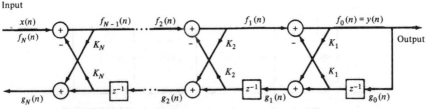

FIGURE 7.25 Lattice structure for an all-pole IIR system.

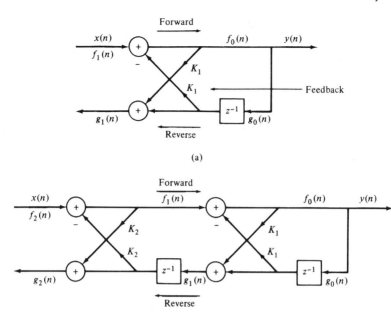

FIGURE 7.26   Single-pole and two-pole lattice system.

After some simple substitutions and manipulations we obtain

$$y(n) = -K_1(1 + K_2)y(n - 1) - K_2 y(n - 2) + x(n) \qquad (7.3.38)$$
$$g_2(n) = K_2 y(n) + K_1(1 + K_2)y(n - 1) + y(n - 2) \qquad (7.3.39)$$

Clearly, the difference equation in (7.3.38) represents a two-pole IIR system and the relation in (7.3.39) is the input–output equation for a two-zero FIR system. Note that the coefficients for the FIR system are identical to those in the IIR system except that they occur in reverse order.

In general, the conclusions above hold for any N. Indeed, with the definition of $A_m(z)$ given in (7.2.32), the system function for the all-pole IIR system is

$$H_a(z) = \frac{Y(z)}{X(z)} = \frac{F_0(z)}{F_m(z)} = \frac{1}{A_m(z)} \qquad (7.3.40)$$

Similarly, the system function of the all-zero (FIR) system is

$$H_b(z) = \frac{G_m(z)}{Y(z)} = \frac{G_m(z)}{G_0(z)} = B_m(z) = z^{-m} A_m(z^{-1}) \qquad (7.3.41)$$

where we have used the previously established relationships in (7.2.38) and (7.2.40). Thus the coefficients in the FIR system $H_b(z)$ are identical to the coefficients in $A_m(z)$, except that they occur in reverse order.

It is interesting to note that the all-pole lattice structure has an all-zero path with input $g_0(n)$ and output $g_N(n)$, which is identical to its counterpart all-zero path in the all-zero lattice structure. The polynomial $B_m(z)$, which represents the system function of the all-zero path that is common to both lattice structures, is usually called the *backward system function*, because it provides a backward path in the all-pole lattice structure.

From the discussion above the reader should observe that the all-zero and all-pole lattice structures are characterized by the same set of lattice parameters, namely, K_1, K_2,

..., K_N. The two lattice structures differ only in the interconnections of their signal flow graphs. Consequently, the algorithms for converting between the system parameters $\{\alpha_m(k)\}$ in the direct form realization of an FIR system and the parameters of its lattice counterpart apply as well to the all-pole structure.

We recall that the roots of the polynomial $A_N(z)$ lie inside the unit circle if and only if the lattice parameters $|K_m| < 1$ for all $m = 1, 2, \ldots, N$. Therefore, the all-pole lattice structure is a stable system if and only if its parameters $|K_m| < 1$ for all m.

In practical applications the all-pole lattice structure has been used to model the human vocal tract and a stratified earth. In such cases the lattice parameters, $\{K_m\}$ have the physical significance of being identical to reflection coefficients in the physical medium. This is the reason that the lattice parameters are often called *reflection coefficients*. In such applications, a stable model of the medium requires that the reflection coefficients, which are obtained by performing measurements on output signals from the medium, be less than unity.

The all-pole lattice provides the basic building block for lattice-type structures that implement IIR systems that contain both poles and zeros. To develop the appropriate structure, let us consider an IIR system with system function

$$H(z) = \frac{\displaystyle\sum_{k=0}^{M} c_M(k)z^{-k}}{1 + \displaystyle\sum_{k=1}^{N} a_N(k)z^{-k}} = \frac{C_M(z)}{A_N(z)} \tag{7.3.42}$$

where the notation for the numerator polynomial has been changed to avoid confusion with our previous development. Without loss of generality, we assume that $N \geq M$.

In the direct form II structure, the system in (7.3.42) is described by the difference equations

$$w(n) = -\sum_{k=1}^{N} a_N(k)w(n - k) + x(n) \tag{7.3.43}$$

$$y(n) = \sum_{k=0}^{M} c_M(k)w(n - k) \tag{7.3.44}$$

Note that (7.3.43) is the input–output of an all-pole IIR system and (7.3.44) is the input–output of an all-zero system. Furthermore, we observe that the output of the all-zero system is simply a linear combination of delayed outputs from the all-pole system. This is easily seen by observing the direct form II structure redrawn as in Fig. 7.27.

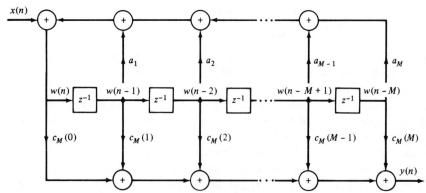

FIGURE 7.27 Direct form II realization of IIR system.

Since zeros will result from forming a linear combination of previous outputs we may carry over this observation to construct pole–zero IIR system using the all-pole lattice structure as the basic building block. We have already observed that $g_m(n)$ is a linear combination of present and past outputs. In fact, the system

$$H_b(z) = \frac{G_m(z)}{Y(z)} = B_m(z)$$

is an all-zero system. Therefore, any linear combination of $\{g_m(n)\}$ is also an all-zero system.

Thus we begin with an all-pole lattice structure with parameters K_m, $1 \leq m \leq N$, and we add a *ladder* part by taking as the output a weighted linear combination of $\{g_m(n)\}$. The result is a pole–zero IIR system which has the *lattice-ladder* structure shown in Fig. 7.28 for $M = N$. Its output is

$$y(n) = \sum_{m=0}^{M} v_m g_m(n) \tag{7.3.45}$$

where $\{v_m\}$ are the parameters that determine the zeros of the system. The system function corresponding to (7.3.45) is

$$H(z) = \frac{Y(z)}{X(z)} \tag{7.3.46}$$

$$= \sum_{m=0}^{M} v_m \frac{G_m(z)}{X(z)}$$

Since $X(z) = F_N(z)$ and $F_0(z) = G_0(z)$, (7.3.46) may be written as

$$H(z) = \sum_{m=0}^{M} v_m \frac{G_m(z)}{G_0(z)} \frac{F_0(z)}{F_N(z)}$$

$$= \sum_{m=0}^{M} v_m \frac{B_m(z)}{A_N(z)} \tag{7.3.47}$$

$$= \frac{\sum_{m=0}^{M} v_m B_m(z)}{A_N(z)}$$

If we compare (7.3.41) with (7.3.47), we conclude that

$$C_M(z) = \sum_{m=0}^{M} v_m B_m(z) \tag{7.3.48}$$

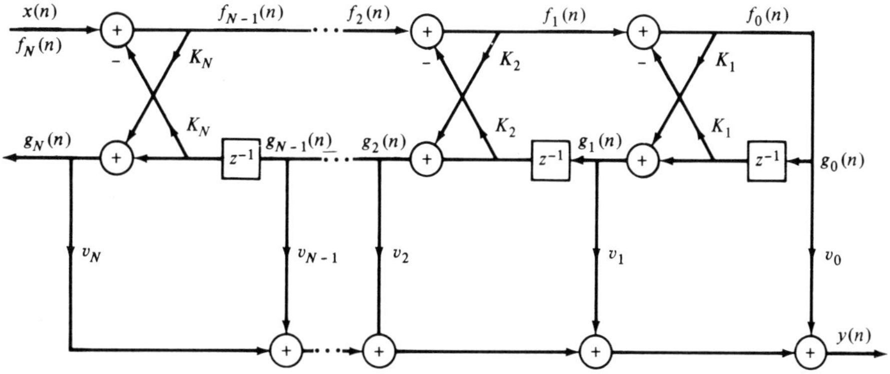

FIGURE 7.28 Lattice-ladder structure for the realization of a pole–zero system.

This is the desired relationship that can be used to determine the weighting coefficients $\{v_m\}$. Thus we have demonstrated that the coefficients of the numerator polynomial $C_M(z)$ determine the ladder parameters $\{v_m\}$, whereas the coefficients in the denominator polynomial $A_N(z)$ determine the lattice parameters $\{K_m\}$.

Given the polynomials $C_M(z)$ and $A_N(z)$, where $N \geq M$, the parameters of the all-pole lattice are determined first, as described previously, by the conversion algorithm given in Section 7.2.4, that converts the direct form coefficients into lattice parameters. By means of the step-down recursive relations given by (7.2.54) we obtain the lattice parameters $\{K_m\}$ and the polynomials $B_m(z)$, $m = 1, 2, \ldots, N$.

The ladder parameters are determined from (7.3.48), which can be expressed as

$$C_m(z) = \sum_{k=0}^{m-1} v_k B_k(z) + v_m B_m(z) \tag{7.3.49}$$

or, equivalently, as

$$C_m(z) = C_{m-1}(z) + v_m B_m(z) \tag{7.3.50}$$

Thus $C_m(z)$ can be computed recursively from the reverse polynomials $B_m(z)$, $m = 1$, $2, \ldots, M$. Since $\beta_m(m) = 1$ for all m, the parameters v_m, $m = 0, 1, \ldots, M$ can be determined by first noting that

$$v_m = c_m(m) \qquad m = 0, 1, \ldots, M \tag{7.3.51}$$

Then, by rewriting (7.3.50) as

$$C_{m-1}(z) = C_m(z) - v_m B_m(z) \tag{7.3.52}$$

and running this recursive relation backward in m (i.e., $m = M, M - 1, \ldots, 2$), we obtain $c_m(m)$ and thus the ladder parameters according to (7.3.51). A subroutine that implements the recursive procedure for converting the parameters of a direct form pole–zero IIR system into the parameters of a lattice-ladder structure is given in Fig. 7.29. For simplicity we have assumed that $N = M$.

The lattice-ladder filter structures that we have presented require the minimum amount of memory but not the minimum number of multiplications. Although lattice structures with only one multiplier per lattice stage exist, the two-multiplier-per-stage lattice that we have described is by far the most widely used in practical applications. In conclusion, the modularity, the built-in stability characteristics embodied in the coefficients $\{K_m\}$, and its robustness to finite-word-length effects make the lattice structure very attractive in many practical applications, including speech processing systems, adaptive filtering, and geophysical signal processing.

7.4 Software Implementation of Discrete-Time Systems

In Section 2.4.7 we described the software implementation of the direct form FIR structure and the direct form I IIR structure. The emphasis in the previous discussion was on understanding the basic structures and their implementation without any concern for the optimality of the implementation.

In this section we present software implementations for the direct form II structure for an IIR system, as well as the more practical cascade and parallel realizations. Finally, we describe the software implementation of all-zero (FIR) and all-pole lattice filters.

```
C
C          CONVERSION FROM POLE - ZERO DIRECT
C          TO LATTICE - LADDER PARAMETERS
C
C          INPUT  : A(1),A(2),...,A(IP)
C                   B(1),B(2),...,B(IP+1)
C          OUTPUT : RK(1),RK(2),...,RK(IP)
C                   RG(1),RG(2),...,RG(IP+1)
C          IP = M = N : FILTER'S ORDER
C
           SUBROUTINE DLAT (A,Q,IP,RK,RG,B,IER)
           DIMENSION A(1),B(1),Q(1),RK(1),RG(1)
           IER=0
           IP2=IP+2
           DO 10 M=1,IP
             RK(M)=A(M)
             RG(M)=Q(M)
        10 CONTINUE
           RG(M+1)=Q(M+1)
           DO 50 L=2,IP
             M=IP2-L
             M1=M+1
             DO 20 J=1,M
               B(J)=RK(M1-J)
        20     CONTINUE
             IF (ABS(RK(M)).GE.1.0) IER=1
             HELP=1.0-RK(M)*RK(M)
             HELP=1.0/HELP
             M1=M-1
             DO 30 J=1,M1
               RK(J)=RK(J)-RK(M)*B(J+1)
               RK(J)=HELP*RK(J)
        30     CONTINUE
             DO 40 J=1,M
               RG(J)=RG(J)-RG(M+1)*B(J)
        40     CONTINUE
        50 CONTINUE
           RETURN
           END
```

FIGURE 7.29 Subroutine for converting the parameters of a direct-form, pole–zero IIR system into a lattice-ladder filter with paramters $\{R_i\}$.

7.4.1 Direct Form II IIR Systems

As shown previously, the direct form II realization is specified by the difference equations

$$w(n) = -\sum_{k=1}^{N} a_k w(n-k) + x(n) \tag{7.4.1}$$

$$y(n) = \sum_{k=0}^{M} b_k w(n-k) \tag{7.4.2}$$

where $x(n)$ is the input sequence, $y(n)$ is the output sequence, and $w(n)$ is an intermediate sequence that is computed, which forms the link that couples the input to the output.

The computation of the sequences $w(n)$ and $y(n)$ as given in (7.4.1) and (7.4.2) requires storage for the variables $w(n-k)$, $k = 0, 1, \ldots,$ max $\{M, N\}$, as well as storage for the system parameters $\{a_k\}$ and $\{b_k\}$. The subroutine IIRDF1 given in Section 2.5 has been modified to incorporate the difference equations in (7.4.1) and (7.4.2). The resulting subroutine, called IIRDF2, is listed in Fig. 7.30.

```
C
C         DIRECT FORM II DIGITAL FILTER
C
C         INPUT  : B(1),B(2),...,B(M+1),A(1),A(2),...A(N),XIN
C         OUTPUT : YOUT
C         NOTE THAT :B(K+1) = b   ,K=0,1,...,M
C
          SUBROUTINE IIRDF2 (XIN,YOUT,M,N,B,A,W,INIT)
          DIMENSION B(1),A(1),W(1)
C         INITIALIZATION
          IF (INIT.NE.0) GOTO 6
          MP1=M+1
          MP2=M+2
          MAX=N
          IF (M.GT.N) MAX=M
          DO 5 K=1,MAX
            W(K)=0.0
        5 CONTINUE
          W(MAX+1)=0.0
          RETURN
        6 CONTINUE
C         COMPUTE CURRENT OUTPUT SAMPLE
          S=XIN
          DO 10 K=1,N
            S=S-A(K)*W(K+1)
       10 CONTINUE
          W(1)=S
          S=0.0
          DO 20 K=1,MP1
            S=S+B(K)*W(K)
       20 CONTINUE
          YOUT=S
C         SHIFT STORED DATA
          DO 30 K=1,MAX
            W(MAX+2-K)=W(MAX+1-K)
       30 CONTINUE
          RETURN
          END
```

FIGURE 7.30 Subroutine for direct form II IIR filter.

IIRDF2 is initialized by calling it once with INIT $= 0$. Thus all the intermediate variables $w(n - k)$, $k = 0, 1, \ldots$, max $\{M, N\}$ are set to zero initially. We observe that IIRDF2 requires less data memory than IIRDF1 but the same number of numerical operations as IIRDF1.

7.4.2 Cascade- and Parallel-Form Realizations

To reduce the sensitivity to quantization of its coefficients, high-order pole–zero systems are usually implemented as either a cascade or a parallel configuration of second-order IIR systems as indicated previously. These realizations were described in detail in Section 7.3.

In the cascade realization, there is an optimum ordering of the second-order modules that results in a low sensitivity to coefficient quantization errors. On the other hand, the parallel realization does not involve ordering of the second-order modules. Furthermore, there is no interaction among the modules, which may be considered an advantage. In addition, the parallel realization easily lends itself to multiprocessing or parallel process-ing implementation. Parallel (pipeline) processing can also be applied to the cascade realization as well, but in this case, the processors must communicate with one another to ensure the proper flow of data from one processor to another.

In this section we limit our attention to the software implementation of the cascade realization based on the use of transposed direct form II modules. The sequential computational algorithm may be described by the following three-step process.

1. Load the new input sample $x(n)$; set $k = 1$.

$$y_0(n) = x(n) \tag{7.4.3}$$

2. Compute the output of the kth module:

$$y_k(n) = b_{k0} y_{k-1}(n) + w_{k1}(n - 1) \tag{7.4.4}$$
$$w_{k1}(n) = b_{k1} y_{k-1}(n) - a_{k1} y_k(n) + w_{k2}(n - 1) \tag{7.4.5}$$
$$w_{k2}(n) = b_{k2} y_{k-1}(n) - a_{k2} y_k(n) \tag{7.4.6}$$

3. If $k \leq K$, increment k by one and jump to step 2 to compute the output of the module $k + 1$. If $k > K$, output the value $y_k(n)$ and jump to step 1, in order to process the next input sample.

Two arrays A(K,2) and B(K,3) may be used to store the coefficients $\{a_{k1}, a_{k2}, b_{k0}, b_{k1}, b_{k2}, k = 1, 2, \ldots, K\}$ and two additional arrays W1(k) and W2(k) may be used to store the intermediate variables $w_{k1}(n)$ and $w_{k2}(n)$ for $k = 1, 2, \ldots, K$. Then the Jth module can be realized with the following FORTRAN statements:

$$\text{YJ} = \text{B(J, 1)} * \text{YJM1} + \text{W1(J)} \tag{7.4.7}$$
$$\text{W1(J)} = \text{B(J, 1)} * \text{YJM1} - \text{A(J, 1)} * \text{YJ} + \text{W2(J)} \tag{7.4.8}$$
$$\text{W2(J)} = \text{B(J, 2)} * \text{YJM1} - \text{A(J, 2)} * \text{YJ} \tag{7.4.9}$$
$$\text{YJM1} = \text{YJ} \tag{7.4.10}$$

Note that we use W1(J) to represent both $w_k(n)$ and $w_{k1}(n - 1)$. This is possible because $w_{k1}(n - 1)$ is used only in (7.4.4), but not in the subsequent computations. The same applies to the array W2(J). The statement in (7.4.10) feeds the output of module J to the input of the next module.

A subroutine CTDFII that implements the cascade realization of an IIR system with transposed direct form II modules is given in Fig. 7.31.

7.4.3 Lattice Realizations

In this section we describe the implementation of the all-zero and all-pole lattice filters in FORTRAN software. Although lattice filters are highly modular, their software implementation is more complex than the direct-form structures. For illustrative purposes, we describe the steps in the implementation of a lattice filter having four stages.

The column on the left side of Table 7.2 lists the equations for the four stages in an all-zero lattice. These equations are obtained from (7.2.28)–(7.2.30). The column on the right side of Table 7.2 lists the FORTRAN code for each stage. The arrays {F(M), G(M), OLDG(M), M = 1, 2, \ldots IP} are used to store the signals $\{f_m(m), g_m(n), g_m(n - 1), m = 1, 2, \ldots, p\}$, where p denotes the number of lattice stages. The lattice parameters $K_m, m = 1, 2, \ldots, p$ are stored in the array {RK(M), M = 1, 2, \ldots IP}. With the aid of the relations given in Table 7.2, we have written the subroutine AZLAT, given in Fig. 7.32, that implements an all-zero lattice with $p = $ IP stages.

The filter is initialized by calling the subroutine once with UNIT $\neq$ 1. In general, the all-zero lattice with p stages requires only p storage locations. This implies that one of the arrays, either G(M) or OLDG(M) can be avoided. However, in an attempt to simplify

```
C
C        CASCADE IIR FILTER USING TRANSPOSED
C        DIRECT FORM II SECOND ORDER MODULES
C
C        SUBROUTINE PARAMETERS DESCRIPTION : SEE TEXT
C
         SUBROUTINE CTDFII (XIN,YOUT,K,B,A,W1,W2,INIT)
         DIMENSION B(K,3),A(K,2),W1(K),W2(K)
         IF (INIT.NE.0) GOTO 6
C
C        INITIALIZATION
C
         DO 5 J=1,K
           B(J,1)=0.0
           B(J,2)=0.0
           B(J,3)=0.0
           A(J,1)=0.0
           A(J,2)=0.0
       5 CONTINUE
         RETURN
       6 CONTINUE
C
C        COMPUTE THE CURRENT OUTPUT SAMPLE
C
         YJM1=XIN
         DO 20 J=1,K
           YJ=B(J,I)*YJM1+W1(J)
           W1(J)=B(J,1)*YJM1-A(J,1)*YJ+W2(J)
           W2(J)=B(J,2)*YJM1-A(J,2)*YJ
           YJM1=YJ
      20 CONTINUE
         YOUT=YJ
         RETURN
         END
```

FIGURE 7.31 Subroutine for cascade-form realization of IIR filter with transposed direct form II second-order sections.

TABLE 7.2 Software for Implementation of an All-Zero Lattice

Stage	Describing Equations	FORTRAN Code
Input	$f_0(n) = g_0(n) = x(n)$	XIN
1	$f_1(n) = f_0(n) + K_1 g_0(n - 1)$	F(I) = XIN + RK(1) * OLDX
	$g_1(n) = K_1 f_0(n) + g_0(n - 1)$	G(I) = RK(1) * XIN + OLDX
		OLDX = XIN
2	$f_2(n) = f_1(n) + K_2 g_1(n - 1)$	F(2) = F(1) + RK(2) * OLDG(1)
	$g_2(n) = K_2 f_1(n) + g_1(n - 1)$	G(2) = RK(2) * F(1) +
		OLDG(1)
		OLDG(1) = G(1)
3	$f_3(n) = f_2(n) + K_3 g_2(n - 1)$	F(3) = F(2) + RK(3) * OLDG(2)
	$g_3(n) = K_3 f_2(n) + g_2(n - 1)$	G(3) = RK(3) * F(2) +
		OLDG(2)
		OLDG(2) = G(2)
4	$f_4(n) = f_3(n) + K_4 g_3(n - 1)$	F(4) = F(3) + RK(4) * OLDG(3)
	$g_4(n) = K_4 f_3(n) + g_3(n - 1)$	G(4) = RK(4) * F(3) +
		OLDG(3)
		OLDG(3) = G(3)
Output	$y(n) = f_4(n)$	YOUT = F(4)

```
C
C       ALL - ZERO   LATTICE FILTER
C
C       INPUT  : IP, RK(1),...,RK(IP),XIN
C       OUTPUT : YOUT
C
        SUBROUTINE AZLAT (IP,RK,XIN,YOUT,INIT)
        DIMENSION RK(IP),F(20),G(20),OLDG(20)
        IF (INIT.EQ.1) GOTO 6
C
C       INITIALIZATION
C
        OLDX=0.0
        DO 5 M=1,IP
          OLDG(M)=0.0
      5 CONTINUE
        RETURN
      6 CONTINUE
C
C       COMPUTE THE CURRENT OUTPUT SAMPLE
C
        F(1)=XIN+RK(1)*OLDX
        G(1)=RK(1)*XIN+OLDX
        OLDX=XIN
        DO 10 M=2,IP
          F(M)=F(M-1)+RK(M)*OLDG(M-1)
          G(M)=RK(M)*F(M-1)+OLDG(M-1)
          OLDG(M-1)=G(M-1)
     10 CONTINUE
        YOUT=F(IP)
        RETURN
        END
```

FIGURE 7.32 Subroutine for implementtion of an all-zero lattice.

the programming, we have used two separate arrays in the implementation of the lattice structure.

Next, we turn our attention to the software implementation of the all-pole lattice described by the basic equations in (7.3.31) through (7.3.34). In Table 7.3 we list the equations for a four-stage filter and the corresponding FORTRAN code. Observe that in this case, we do not need the array OLDG(M). The reason is that we first use the sample $g_m(n-1)$ and then we compute the new value $g_m(n)$. In contrast, in the all-zero lattice,

TABLE 7.3 Software for Implementation of an All-Pole Lattice

Stage	Describing Equations	FORTRAN Code
Input	$f_4(n) = x(n)$	F(4) = XIN
4	$f_3(n) = f_4(n) - K_4 g_3(n-1)$	F(3) = F(4) - RK(4) * G(3)
	$g_4(n) = K_4 f_3(n) + g_3(n-1)$	G(4) = RK(4) * F(3) + G(3)
3	$f_2(n) = f_3(n) - K_3 g_2(n-1)$	F(2) = F(3) - RK(3) * G(2)
	$g_3(n) = K_3 f_2(n) + g_2(n-1)$	G(3) = RK(3) * F(2) + G(2)
2	$f_1(n) = f_2(n) - K_2 g_1(n-1)$	F(1) = F(2) - RK(2) * G(1)
	$g_2(n) = K_2 f_1(n) + g_1(n-1)$	G(2) = RK(2) * F(1) + G(1)
1	$f_0(n) = f_1(n) - K_1 g_0(n-1)$	YOUT = F(1) - RK(1) * YOLD
	$g_1(n) = K_1 f_0(n) + g_0(n-1)$	G(1) = RK(1) * YOUT + YOLD
		YOLD = YOUT
Output	$y(n) = f_0(n) = g_0(n)$	YOUT

```
C
C      ALL - POLE LATTICE FILTER
C
C      INPUT  : IP,RK(1),...,RK(IP),XIN
C      OUTPUT : YOUT
C
       SUBROUTINE APLAT (IP,RK,XIN,YOUT,INIT)
       DIMENSION RK(IP),F(20),G(20)
       IF (INIT.EQ.1) GOTO 6
C
C      INITIALIZATION
C
       OLDY=0.0
       DO 5 M=1,IP
         G(M)=0.0
     5 CONTINUE
       RETURN
     6 CONTINUE
C
C      COMPUTE CURRENT OUTPUT SAMPLE
C
       NM1=IP-1
       DO 10 J=1,NM1
         M=IP+1-J
         F(M-1)=F(M)-RK(M)*G(M-1)
         G(M)=RK(M)*F(M-1)+G(M-1)
    10 CONTINUE
       YOUT=F(1)-RK(1)*OLDY
       G(1)=RK(1)*YOUT+OLDY
       OLDY=YOUT
       RETURN
       END
```

FIGURE 7.33 Subroutine for implementation of an all-pole lattice.

we first compute $g_m(n)$ and then we use $g_m(n - 1)$. The FORTRAN Subroutine APLAT that implements the all-pole lattice is given in Fig. 7.33.

7.5 State-Space System Analysis and Structures

Our treatment of linear time-invariant systems up to this point has been limited to an *input–output* or *external description* of the characteristics of the system. In other words, the system was characterized by mathematical equations that relate the input signal to the output signal. In this section we introduce the basic concepts in the state-space description of linear time-invariant causal systems. Although the *state-space* or *internal description* of the system still involves a relationship between the input and output signals, it also involves an additional set of variables, called *state variables*. Furthermore, the mathematical equations describing the system, its input, and its output are usually divided into two parts:

1. A set of mathematical equations relating the state variables to the input signal.
2. A second set of mathematical equations relating the state variables and the current input to the output signal.

The state variables provide information about all the internal signals in the system. As a result, the state-space description provides a more detailed description of the system as compared to the input–output description. Although our treatment of state-space analysis is confined primarily to single input–single output linear time-invariant causal systems, the state-space techniques may be applied to nonlinear systems, time-variant systems, and multiple input–multiple output systems. In fact, it is in the characterization and analysis of multiple input–multiple output systems that the power and importance of state-space methods are clearly evident.

Both the input–output description and the state-variable description of a system are useful in practice. Which description we use depends on the problem, the available information, and the questions to be answered. In our presentation, the emphasis is on the use of state-space techniques in system analysis and in the development of state-space structures for the realization of discrete-time systems.

7.5.1 The Concept of State

As we have already observed, the determination of the output of a system requires that we know the input signal and the set of initial conditions at the time the input is applied. If a system is not relaxed initially, say at time n_0, then knowledge of the input signal $x(n)$ for $n \geq n_0$ is not sufficient to uniquely determine the output $y(n)$ for $n \geq n_0$. The initial conditions of the system at $n = n_0$ must also be known and taken into account. This set of initial conditions is called the state of the system at $n = n_0$.

DEFINITION: *the* state *of a system at time n_0 is the amount of information that must be provided at time n_0, which, together with the input signal $x(n)$ for $n \geq n_0$, uniquely determine the output of the system for all $n \geq n_0$.*

From this definition we infer that the concept of state leads to a decomposition of a system into two parts, a part that contains memory and a memoryless component. The information stored in the memory components constitutes the set of initial conditions and is called the *state of the system*. Then the current output of the system becomes a function of the current value of the input and the current state. Thus, to determine the output of the system at a given time, we need the current value of the state and the current input. Since the current value of the input is available, we only need to provide a mechanism for updating the state of the system recursively. Consequently, the state of the system at time $n_0 + 1$ should depend on the state of the system at time n_0 and the value of the input signal $x(n)$ at $n = n_0$.

The following example illustrates the approach in formulating a state-space description of a system. Let us consider a linear time-invariant causal system described by the difference equation

$$y(n) = \tfrac{1}{2}y(n - 1) + x(n) + 2x(n - 1) \qquad (7.5.1)$$

Figure 7.34a shows the direct form II realization of this system. We note that the system contains only one delay element, which is, in fact, a storage register (memory location). The output $v(n)$ of the delay denotes the *present* contents of the memory, whereas the input $v(n + 1)$ to the delay element denotes the *next* value to be stored in memory. Clearly, this storage register includes all of the memory required to compute the current output $y(n)$.

To verify this argument, we first obtain the equations for the realization shown in Fig. 7.34a. These are

$$v(n + 1) = \tfrac{1}{2}v(n) + x(n) \qquad (7.5.2)$$
$$y(n) = v(n + 1) + 2v(n) \qquad (7.5.3)$$

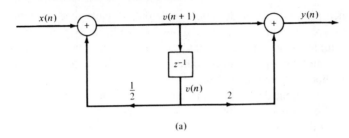

(a)

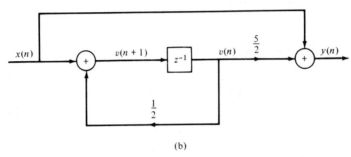

(b)

FIGURE 7.34 Two realizations of the system described by the difference equations in (7.5.1).

Substituting (7.5.2) into (7.5.3) gives

$$y(n) = \tfrac{5}{2}v(n) + x(n)$$

which is a memoryless function of the input $x(n)$ and the contents $v(n)$ of the memory. On the other hand, (7.5.2) provides a mechanism for updating the content of the storage register, using the current content and the current input value.

The pair of equations

$$v(n + 1) = \tfrac{1}{2}v(n) + x(n) \tag{7.5.4}$$
$$y(n) = \tfrac{5}{2}v(n) + x(n) \tag{7.5.5}$$

provide a complete description of the system. Furthermore, the variable $v(n)$, which summarizes all the necessary past information is called the *state variable* and, in fact, is the state of this system. Since there is only a single state variable, the state equation is one-dimensional and its value at any time is simply represented as a point in one-dimensional (the real line) space. We also observe that as indicated previously, equations (7.5.4) and (7.5.5) split the system into two component parts, a dynamic (memory) subsystem and a static (memoryless) subsystem. We say that this set of equations provides a *state-space description* of the system. This description leads to an alternative equivalent realization of the system shown in Fig. 7.34b.

The state-space description can also be applied to causal systems which are time variant and/or nonlinear. To illustrate this point, suppose that we have a system that computes the "running variance" of a signal. The system may be described by the equation

$$y(n) = \frac{1}{n} \sum_{k=0}^{n-1} [x(k) - \mu(n)]^2 \tag{7.5.6}$$

where

$$\mu(n) = \frac{1}{n} \sum_{k=0}^{n-1} x(k) \tag{7.5.7}$$

is the "running mean value." By expanding the square in (7.5.6) and using (7.5.7) it easily follows that

$$y(n) = \frac{1}{n} \sum_{k=0}^{n-1} x^2(k) - \mu^2(n) \qquad (7.5.8)$$

Clearly, the system with input $x(n)$ and output $y(n)$ is both time variant and nonlinear. To obtain a state-space description, we define the following state variables:

$$v_1(n) = \sum_{k=0}^{n-1} x(k) \qquad (7.5.9)^*$$

$$v_2(n) = \sum_{k=0}^{n-1} x^2(k) \qquad (7.5.10)$$

Then, by combining these relations with (7.5.8) and (7.5.7), we obtain

$$y(n) = -\frac{1}{n^2} v_1^2(n) + \frac{1}{n} v_2(n) \qquad (7.5.11)$$

which is the memoryless function expressing the output in terms of the current state variables. The updating of the state variables can be done through the equations

$$v_1(n + 1) = v_1(n) + x(n) \qquad (7.5.12)$$
$$v_2(n + 1) = v_2(n) + x^2(n) \qquad (7.5.13)$$

which easily follow from the definitions of $v_1(n)$ and $v_2(n)$. Clearly, equation (7.5.13) is nonlinear, whereas (7.5.11) is nonlinear and time varying.

In general, the state-space description of a causal system consists of two sets of mathematical equations: (1) a set of equations, called the *state equations*, that relate the state variables at time $n + 1$ to the state variables and the input at time n, and (2) an equation, called the *output equation*, that expresses the output at time n in terms of the state variables and the input at the same time. In particular, if we have a causal system with N state variables $v_1(n)$, $v_2(n)$, . . . , $v_N(n)$, the state-space description can be expressed by the following two sets of equations:

State equations

$$v_i(n + 1) = f_i[v_1(n), v_2(n), \dots , v_N(n), x(n)] \qquad i = 1, 2, \dots , N \quad (7.5.14)$$

Output equation

$$y(n) = g[v_1(n), v_2(n), \dots , v_N(n), x(n)] \qquad (7.5.15)$$

The N state variables $v_i(n)$, $i = 1, 2, \dots , N$, may be viewed as a vector of dimension N, and the value of the vector at any instant may be viewed as a point in the N-dimensional space called the *state space*, spanned by the state variables. The state equations in (7.5.14) specify the sequence of values of the state vector as a function of time. A plot of successive values of the state vector in the N-dimensional state space constitutes a *trajectory* for the vector of state variables. Figure 7.35 illustrates a trajectory for the state of a second-order system in two-dimensional space. In general, the state equations describe the dynamic part of the system, whereas the output equation describes the memoryless (static) part. The number N of the state variables will be referred as the *order* of the system.

Although the state-space description can be easily generalized to systems with multiple inputs and multiple outputs (MIMO), our discussion will be limited to single input–single output (SISO) systems.

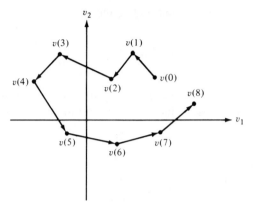

FIGURE 7.35 Trajectory for the state of a second-order system.

7.5.2 State-Space Description of Systems Characterized by Difference Equations

In this section we derive state-space equations for discrete-time systems described by linear constant-coefficient difference equations. For purposes of illustration, we consider a third-order system. However, the results are easily generalized to systems of any arbitrary finite order N.

Let us consider the causal linear time-invariant system characterized by the difference equation

$$y(n) = -\sum_{k=1}^{3} a_k y(n-k) + \sum_{k=0}^{3} b_k x(n-k) \qquad (7.5.16)$$

The direct form II realization for the system is shown in Fig. 7.36.

As state variables, we use the contents of the system memory registers, counting them from the bottom, as shown in Fig. 7.36. We recall that the output of a delay element represents the present value stored in the register and the input represents the next value to be stored in the memory. Consequently, with the aid of Fig. 7.36, we may write

$$
\begin{aligned}
v_1(n+1) &= v_2(n) \\
v_2(n+1) &= v_3(n) \\
v_3(n+1) &= -a_3 v_1(n) - a_2 v_2(n) - a_1 v_3(n) + x(n)
\end{aligned}
\qquad (7.5.17)
$$

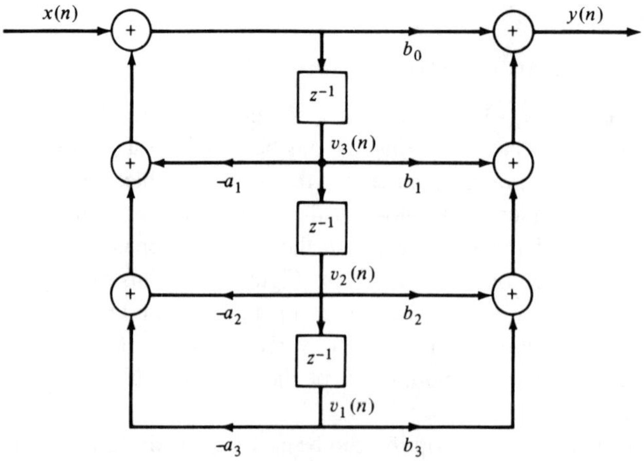

FIGURE 7.36 Direct form II realization of system described by the difference equation in (7.5.16).

It is interesting to note that the state-variable formulation for the third-order system of (7.5.16) involves three first-order difference equations given by (7.5.17). In general, an nth-order system may be described by n first-order difference equations.

The output equation, which expresses $y(n)$ in terms of the state variables and the present input value $x(n)$, can also be obtained by referring to Fig. 7.36. We have

$$y(n) = b_0 v_3(n + 1) + b_3 v_1(n) + b_2 v_2(n) + b_1 v_3(n)$$

We can eliminate $v_3(n + 1)$ by using the last equation in (7.5.17). Thus we obtain the desired output equation

$$y(n) = (b_3 - b_0 a_3)v_1(n) + (b_2 - b_0 a_2)v_2(n) + (b_1 - b_0 a_1)v_3(n) + b_0 x(n) \quad (7.5.18)$$

An alternative state-space description for the system in (7.5.16) can be obtained by starting with the transposed direct form II structure shown in Fig. 7.37. The validity of this structure can be seen if we rewrite (7.5.16) as

$$y(n) = \sum_{k=1}^{3} [b_k x(n - k) - a_k y(n - k)] + b_0 x(n)$$

Due to the linearity and time invariance of the system, instead of first delaying the signals $x(n)$ and $y(n)$ and then computing the terms $b_k x(n - k) - a_k y(n - k)$ as in Fig. 7.36, we first compute the terms $b_k x(n) - a_k y(n)$ and then we delay them.

If we use the state variables indicated in Fig. 7.37, we obtain

$$\begin{aligned} v_1(n + 1) &= b_3 x(n) - a_3 y(n) \\ v_2(n + 1) &= v_1(n) + b_2 x(n) - a_2 y(n) \\ v_3(n + 1) &= v_2(n) + b_1 x(n) - a_1 y(n) \end{aligned} \quad (7.5.19)$$

The output equation is

$$y(n) = b_0 x(n) + v_3(n) \quad (7.5.20)$$

We can use the output equation to eliminate $y(n)$ from (7.5.19). Thus we obtain the following state equations:

$$\begin{aligned} v_1(n + 1) &= -a_3 v_3(n) + (b_3 - b_0 a_3)x(n) \\ v_2(n + 1) &= v_1(n) - a_2 v_3(n) + (b_2 - b_0 a_2)x(n) \\ v_3(n + 1) &= v_2(n) - a_1 v_3(n) + (b_1 - b_0 a_1)x(n) \end{aligned} \quad (7.5.21)$$

From the two examples given above, it is clear that a state-space description of a system can be easily obtained from a block diagram realization of the linear, time-invariant system. As the order of the system increases, the more complex this description becomes. However, by introducing matrix notation, we can express the state-space equa-

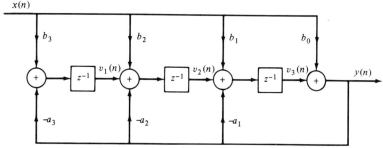

FIGURE 7.37   State-space realization for the system described by (7.5.16).

tions in a more compact form that simplifies their manipulations and allows us to use the powerful matrix algebra tools for state-space analysis.

If we put (7.5.17) and (7.5.18) into matrix form we have

$$
\begin{bmatrix} v_1(n+1) \\ v_2(n+1) \\ v_3(n+1) \end{bmatrix} = \begin{bmatrix} 0 & 1 & 0 \\ 0 & 0 & 1 \\ -a_3 & -a_2 & -a_1 \end{bmatrix} \begin{bmatrix} v_1(n) \\ v_2(n) \\ v_3(n) \end{bmatrix} + \begin{bmatrix} 0 \\ 0 \\ 1 \end{bmatrix} x(n) \qquad (7.5.22)
$$

and

$$
y(n) = [(b_3 - b_0 a_3) \ (b_2 - b_0 a_2) \ (b_1 - b_0 a_1)] \begin{bmatrix} v_1(n) \\ v_2(n) \\ v_3(n) \end{bmatrix} + b_0 x(n) \qquad (7.5.23)
$$

We call these equations the *type 1 state-space realization*.

Similarly, (7.5.20) and (7.5.21) can be expressed in matrix notation. We have

$$
\begin{bmatrix} v_1(n+1) \\ v_2(n+1) \\ v_3(n+1) \end{bmatrix} = \begin{bmatrix} 0 & 0 & -a_3 \\ 1 & 0 & -a_2 \\ 0 & 1 & -a_1 \end{bmatrix} \begin{bmatrix} v_1(n) \\ v_2(n) \\ v_3(n) \end{bmatrix} + \begin{bmatrix} (b_3 - b_0 a_3) \\ (b_2 - b_0 a_2) \\ (b_1 - b_0 a_1) \end{bmatrix} x(n) \qquad (7.5.24)
$$

$$
y(n) = [0 \ 0 \ 1] \begin{bmatrix} v_1(n) \\ v_2(n) \\ v_3(n) \end{bmatrix} + b_0 x(n) \qquad (7.5.25)
$$

We call this description a *type 2 state-space realization*. We note that in both cases the state-space equations have the desired functional form as described by equations (7.5.14) and (7.5.15).

Suppose now that we have a system with N state variables $v_1(n)$, $v_2(n)$, . . . , $v_N(n)$. We define the state $\mathbf{v}(n)$ as an N-dimensional column vector given by

$$
\mathbf{v}(n) = \begin{bmatrix} v_1(n) \\ v_2(n) \\ \vdots \\ v_N(n) \end{bmatrix} \qquad (7.5.26)
$$

Also, let F denote an $N \times N$ matrix, and $\mathbf{q}$ and $\mathbf{g}$ be N-dimensional column vectors, defined by

$$
\mathbf{F} = \begin{bmatrix} f_{11} & f_{12} & \cdots & f_{1N} \\ f_{21} & f_{22} & \cdots & f_{2N} \\ \vdots & \vdots & \ddots & \vdots \\ f_{N1} & f_{N2} & \cdots & f_{NN} \end{bmatrix} \qquad \mathbf{q} = \begin{bmatrix} q_1 \\ q_2 \\ \vdots \\ q_N \end{bmatrix} \qquad \mathbf{g} = \begin{bmatrix} g_1 \\ g_2 \\ \vdots \\ g_N \end{bmatrix} \qquad (7.5.27)
$$

where $\{f_{ij}\}$, $\{q_k\}$, $\{g_k\}$ are all constant coefficients. The transpose of a column vector, say $\mathbf{g}$, is a row vector denoted by $\mathbf{g}'$, given by

$$
\mathbf{g}' = [g_1 \ g_2 \ \cdots \ g_N] \qquad (7.5.28)
$$

By using these quantities, we can easily see that the general state-space equations, which include type 1 and type 2 realizations as special cases, are given by

State Equation

$$\mathbf{v}(n + 1) = \mathbf{F}\mathbf{v}(n) + \mathbf{q}x(n) \tag{7.5.29}$$

Output equation

$$y(n) = \mathbf{g}'\mathbf{v}(n) + dx(n) \tag{7.5.30}$$

where the elements of $\mathbf{F}$, $\mathbf{q}$, $\mathbf{g}$, and d are constants (i.e., they do not change as a function of the time index n).

Any discrete-time system whose input $x(n)$, output $y(n)$, and state $\mathbf{v}(n)$, for all $n \geq n_0$, are related by the state-space equations above will be called *linear* and *time invariant*. If at least one of the quantities in $\mathbf{F}$, $\mathbf{q}$, $\mathbf{g}$, or d, depends on time, the system becomes *time variant*.

We will refer to (7.5.29)–(7.5.30) as the *linear time-invariant state-space model*, which can be represented by the simple vector-matrix block diagram in Fig. 7.38. In this figure the double lines represent vector quantities and the blocks represent the vector or matrix coefficients.

By generalizing the previous examples, it can easily be seen that the Nth-order system described by

$$y(n) = - \sum_{k=1}^{N} a_k y(n - k) + \sum_{k=0}^{N} b_k x(n - k) \tag{7.5.31}$$

can be expressed as a linear time-invariant type 1 state-space realization by setting

$$\mathbf{F} = \begin{bmatrix} 0 & 1 & 0 & \cdot & \cdot & \cdot & 0 \\ 0 & 0 & 1 & 0 & \cdot & \cdot & 0 \\ \vdots & \vdots & & & & & \vdots \\ 0 & 0 & & \cdot & \cdot & 0 & 1 \\ -a_N & -a_{N-1} & \cdot & \cdot & \cdot & -a_2 & -a_1 \end{bmatrix} \qquad \mathbf{q} = \begin{bmatrix} 0 \\ 0 \\ \vdots \\ 0 \\ 1 \end{bmatrix}$$

$$\mathbf{g} = \begin{bmatrix} b_N - b_0 a_N \\ b_{N-1} - b_0 a_{N-1} \\ \vdots \\ b_1 - b_0 a_1 \end{bmatrix} \qquad d = b_0 \tag{7.5.32}$$

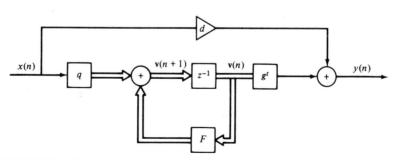

FIGURE 7.38 General state-space description of a linear time-invariant system.

Similarly, the linear time-invariant type 2 state-space realization is

$$
\mathbf{F} = \begin{bmatrix}
0 & 0 & \cdots & & 0 & -a_N \\
1 & 0 & \cdots & & 0 & -a_{N-1} \\
\vdots & \vdots & & & & \vdots \\
0 & 0 & \cdots & 0 & 1 & 0 & -a_2 \\
0 & 0 & \cdots & \cdot & 0 & 1 & -a_1
\end{bmatrix}
\qquad
\mathbf{q} = \begin{bmatrix}
b_N - b_0 a_N \\
b_{N-1} - b_0 a_{N-1} \\
\vdots \\
b_2 - b_0 a_2 \\
b_1 - b_0 a_1
\end{bmatrix}
$$

$$
\mathbf{g} = \begin{bmatrix}
0 \\
0 \\
\vdots \\
0 \\
1
\end{bmatrix}
\qquad d = b_0
\tag{7.5.33}
$$

There are many other ways of selecting state variables and formulating the corresponding state-space structures which are equivalent for the same system. Additional state-space structures will be introduced in Section 7.5.6. The reasons for studying a variety of models, and hence structures, is to find those that are insensitive to finite-word-length arithmetic in actual hardware or software implementations or to reduce the complexity of the implementation.

So far we have described the methods for obtaining state-space models given a system specified by a block diagram realization. We will next illustrate, by means of a simple example, how to move in the other direction, that is, how to obtain a block diagram realization of a system given its state-space description.

EXAMPLE 7.5.1 _____

Determine the block diagram realization of a system described by the following state-space model

$$
\begin{bmatrix} v_1(n+1) \\ v_2(n+1) \end{bmatrix} = \begin{bmatrix} 1.35 & 0.55 \\ -0.45 & 0.35 \end{bmatrix} \begin{bmatrix} v_1(n) \\ v_2(n) \end{bmatrix} + \begin{bmatrix} 0.5 \\ 0.5 \end{bmatrix} x(n)
$$

$$
y(n) = \begin{bmatrix} 3 & 1 \end{bmatrix} \begin{bmatrix} v_1(n) \\ v_2(n) \end{bmatrix} + x(n)
$$

Solution: By writing the equations above explicitly, we have

$$
v_1(n+1) = 1.35 v_1(n) + 0.55 v_2(n) + 0.5 x(n)
$$
$$
v_2(n+1) = -0.45 v_1(n) + 0.35 v_2(n) + 0.5 x(n)
$$
$$
y(n) = 3 v_1(n) + v_2(n) + x(n)
$$

The equations lead to the block diagram realization shown in Fig. 7.39. Clearly, the equations provide a computational algorithm for the implementation of the system since, at any given time, we use the stored state variables $v_1(n)$ and $v_2(n)$ together with the present input to compute the output. Then we update the state variables to be used at the next instant of time.

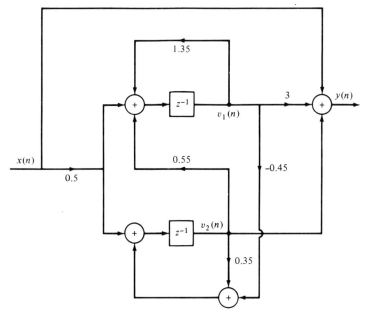

FIGURE 7.39 Realization of system in Example 7.5.1.

Example 7.5.2 _____

Determine the direct form II, transpose direct form II, state-space type 1, and state-space type 2 realizations for the system described by the difference equation

$$y(n) = 3y(n - 1) - 2y(n - 2) + x(n) + x(n - 1)$$

Solution: By comparing this equation to (7.5.31), we obtain the following parameters:

$$N = 2 \qquad a_1 = -3 \qquad a_2 = 2 \qquad b_0 = 1 \qquad b_1 = 1 \qquad b_2 = 0$$

By referring to Figs. 7.36 and 7.37 we easily obtain the realizations shown in Fig. 7.40a and b.

To derive the state-space structures we first note that substitution into (7.5.32) gives, for the type 1 structure,

$$\mathbf{F} = \begin{bmatrix} 0 & 1 \\ -2 & 3 \end{bmatrix} \qquad \mathbf{q} = \begin{bmatrix} 0 \\ 1 \end{bmatrix} \qquad \mathbf{g} = \begin{bmatrix} -2 \\ 4 \end{bmatrix} \qquad d = 1$$

Then substitution into (7.5.29) and (7.5.30) results in

$$v_1(n + 1) = v_2(n)$$
$$v_2(n + 1) = -2v_1(n) + 3v_2(n) + x(n)$$
$$y(n) = -2v_1(n) + 4v_2(n) + x(n)$$

These equations imply the realization shown in Fig. 7.40c.

Similarly, for the type 2 state-space structure, we have

$$\mathbf{F} = \begin{bmatrix} 0 & -2 \\ 1 & 3 \end{bmatrix} \qquad \mathbf{q} = \begin{bmatrix} -2 \\ 4 \end{bmatrix} \qquad \mathbf{g} = \begin{bmatrix} 0 \\ 1 \end{bmatrix} \qquad d = 1$$

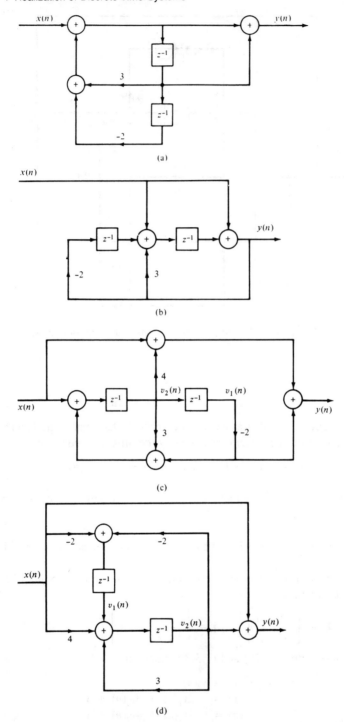

FIGURE 7.40 Direct-form and state-space realizations of system in Example 7.5.2.

or

$$v_1(n + 1) = -2v_2(n) - 2x(n)$$
$$v_2(n + 1) = v_1(n) + 3v_2(n) + 4x(n)$$
$$y(n) = v_2(n) + x(n)$$

which leads to the realization shown in Fig. 7.40d. Note that all these realizations are different.

7.5.3 Solution of the State-Space Equations

There are several methods for solving the state-space equations. Here we discuss a recursive solution which makes use of the fact that the state-space equations are a set of linear first-order difference equations.

To illustrate the basic approach we will first deal with the one-dimensional state-space model.

$$v(n + 1) = fv(n) + qx(n) \tag{7.5.34}$$
$$y(n) = gv(n) + dx(n) \tag{7.5.35}$$

where f, q, g, and d are fixed scalar system parameters. The problem is to determine the output $y(n)$ for $n \geq n_0$, given the input $x(n)$, $n \geq n_0$, and the initial state $v(n_0)$. This can be done first by solving (7.5.34) recursively, as follows:

$$
\begin{aligned}
v(n_0 + 1) &= fv(n_0) + qx(n_0) \\
v(n_0 + 2) &= fv(n_0 + 1) + qx(n_0 + 1) \\
&= f^2 v(n_0) + fqx(n_0) + qx(n_0 + 1) \\
v(n_0 + 3) &= fv(n_0 + 2) + qx(n_0 + 2) \\
&= f^3 v(n_0) + f^2 qx(n_0) + fqx(n_0 + 1) + qx(n_0 + 2) \\
&\ \ \vdots \\
v(n_0 + m) &= f^m v(n_0) + f^{m-1} qx(n_0) + f^{m-2} qx(n_0 + 1) + \cdots \\
&\quad + fqx(n_0 + m - 2) + qx(n_0 + m - 1)
\end{aligned}
$$

If we define $n \equiv n_0 + m$ so that $m = n - n_0$, the equation above becomes

$$
\begin{aligned}
v(n) &= f^{n-n_0} v(n_0) + f^{n-n_0-1} qx(n_0) + f^{n-n_0-2} qx(n_0 + 1) + \cdots \\
&\quad + fqx(n - 2) + qx(n - 1)
\end{aligned}
$$

Thus, for any $n > n_0$, we have

$$v(n) = f^{n-n_0} v(n_0) + \sum_{k=n_0}^{n-1} f^{n-1-k} qx(k) \tag{7.5.36}$$

The output equation is obtained by substituting (7.5.36) into (7.5.35). This substitution yields

$$y(n) = gf^{n-n_0} v(n_0) + \sum_{k=n_0}^{n-1} gf^{n-1-k} qx(k) + dx(n) \tag{7.5.37}$$

which represents the total response of the system.

If we set the initial state $v(n_0) = 0$, we obtain the *zero-state response* of the system as

$$y_{zs}(n) = \sum_{k=n_0}^{n-1} gf^{n-1-k} qx(k) + dx(n) \tag{7.5.38}$$

On the other hand, if we set $x(n) = 0$ for $n \geq n_0$, we obtain the *zero-input response*

$$y_{zi}(n) = gf^{n-n_0} v(n_0) \tag{7.5.39}$$

It follows that the total response given in (7.5.37) is the sum of the responses given by (7.5.38) and (7.5.39), that is,

$$y(n) = y_{zi}(n) + y_{zs}(n) \tag{7.5.40}$$

It also follows from (7.5.38) and (7.5.39) that the system is zero-state linear and zero-input linear. Hence, according to the linearity definition given in Section 2.4.2, the system is linear.

These results can be easily generalized to the N-dimensional state-space model

$$\mathbf{v}(n + 1) = \mathbf{F}\mathbf{v}(n) + \mathbf{q}x(n) \tag{7.5.41}$$

$$y(n) = \mathbf{g}'\mathbf{v}(n) + dx(n) \tag{7.5.42}$$

Indeed, given $\mathbf{v}(n_0)$, we have for $n > n_0$,

$$\mathbf{v}(n_0 + 1) = \mathbf{F}\mathbf{v}(n_0) + \mathbf{q}x(n)$$

$$\mathbf{v}(n_0 + 2) = \mathbf{F}\mathbf{v}(n_0 + 1) + \mathbf{q}x(n_0 + 1)$$

$$= \mathbf{F}^2\mathbf{v}(n_0) + \mathbf{F}\mathbf{q}x(n_0) + \mathbf{q}x(n_0 + 1)$$

where $\mathbf{F}^2$ represents the matrix product $\mathbf{FF}$ and $\mathbf{Fq}$ is the product of the matrix $\mathbf{F}$ and the vector $\mathbf{q}$. If we continue as in the one-dimensional case, we obtain, for $n > n_0$,

$$\mathbf{v}(n) = \mathbf{F}^{n-n_0}\mathbf{v}(n_0) + \sum_{k=n_0}^{n-1} \mathbf{F}^{n-1-k}\mathbf{q}x(k) \tag{7.5.43}$$

The matrix $\mathbf{F}^0$ is defined as the $N \times N$ identity matrix, having unity on the main diagonal and zeros elsewhere. The matrix $\mathbf{F}^{i-j}$ is often denoted as $\mathbf{\Phi}(i - j)$, that is,

$$\mathbf{\Phi}(i - j) = \mathbf{F}^{i-j} \tag{7.5.44}$$

for any positive integers $i \geq j$. This matrix is called the *state transition matrix* of the system.

The output of the system is obtained by substituting (7.5.43) into (2.7.42). The result of this substitution is

$$y(n) = \mathbf{g}'\mathbf{F}^{n-n_0}\mathbf{v}(n_0) + \sum_{k=n_0}^{n-1} \mathbf{g}'\mathbf{F}^{n-1-k}\mathbf{q}x(k) + dx(n)$$

$$= \mathbf{g}'\mathbf{\Phi}(n - n_0)\mathbf{v}(n_0) + \sum_{k=n_0}^{n-1} \mathbf{g}'\mathbf{\Phi}(n - 1 - k)\mathbf{q}x(k) + dx(n) \tag{7.5.45}$$

From this general result, we can determine the output for two special cases. First, the zero-input response of the system is

$$y_{zi}(n) = \mathbf{g}'\mathbf{F}^{n-n_0}\mathbf{v}(n_0) = \mathbf{g}'\mathbf{\Phi}(n - n_0)\mathbf{v}(n_0) \tag{7.5.46}$$

On the other hand, the zero-state response is

$$y_{zs}(n) = \sum_{k=n_0}^{n-1} \mathbf{g}'\mathbf{\Phi}(n - 1 - k)\mathbf{q}x(k) + dx(n) \tag{7.5.47}$$

Clearly, the N-dimensional state-space system is zero-input linear, zero-state linear and since $y(n) = y_{zi}(n) + y_{zs}(n)$, it is linear. Furthermore, since any system described by a linear constant-coefficient difference equation can be put in the state-space form above, it is linear, in agreement with the results obtained in Section 2.4.2.

Let us now work through some examples to gain familiarity with the results above.

EXAMPLE 7.5.3 _____

Compute the zero-state response of the system

$$\mathbf{v}(n + 1) = \begin{bmatrix} 0 & 1 \\ -2 & 3 \end{bmatrix} \mathbf{v}(n) + \begin{bmatrix} 0 \\ 1 \end{bmatrix} x(n)$$

$$y(n) = [-2 \quad 4]\mathbf{v}(n) + x(n)$$

when the input is a unit step.

Solution: The input signal is given by

$$x(n) = u(n) = \begin{cases} 1 & n \geq 0 \\ 0 & n < 0 \end{cases}$$

Since we are interested in obtaining the zero-state response, we set the initial state vector to zero, that is,

$$\mathbf{v}(0) = \begin{bmatrix} 0 \\ 0 \end{bmatrix}$$

Then

$$y(0) = [-2 \quad 4] \begin{bmatrix} 0 \\ 0 \end{bmatrix} + x(0) = 1$$

$$\mathbf{v}(1) = \begin{bmatrix} 0 & 1 \\ -2 & 3 \end{bmatrix} \begin{bmatrix} 0 \\ 0 \end{bmatrix} + \begin{bmatrix} 0 \\ 1 \end{bmatrix} x(0) = \begin{bmatrix} 0 \\ 1 \end{bmatrix}$$

$$y(1) = [-2 \quad 4] \begin{bmatrix} 0 \\ 1 \end{bmatrix} + x(1) = 5$$

$$\mathbf{v}(2) = \begin{bmatrix} 0 & 1 \\ -2 & 3 \end{bmatrix} \begin{bmatrix} 0 \\ 1 \end{bmatrix} + \begin{bmatrix} 0 \\ 1 \end{bmatrix} x(1) = \begin{bmatrix} 1 \\ 4 \end{bmatrix}$$

$$y(2) = [-2 \quad 4] \begin{bmatrix} 1 \\ 4 \end{bmatrix} + x(2) = 15$$

By continuing this iterative procedure, we obtain $y(3) = 47$, $y(4) = 113$, and so on. The results can be checked using the difference equation description of this system given in Example 7.5.2.

EXAMPLE 7.5.4

Compute the response of the FIR system

$$y(n) = x(n) + 2x(n - 1) + x(n - 2)$$

to the arbitrary input signal $x(n)$, $n \geq 0$.

Solution: The state-space description of the system is

$$\mathbf{v}(n + 1) = \begin{bmatrix} 0 & 1 \\ 0 & 0 \end{bmatrix} \mathbf{v}(n) + \begin{bmatrix} 0 \\ 1 \end{bmatrix} x(n)$$

$$y(n) = [1 \quad 2]\mathbf{v}(n) + x(n)$$

where $v_1(n) = x(n - 2)$ and $v_2(n) = x(n - 1)$. From the state equation we have

$$\mathbf{v}(1) = \begin{bmatrix} 0 & 1 \\ 0 & 0 \end{bmatrix} \mathbf{v}(0) + \begin{bmatrix} 0 \\ 1 \end{bmatrix} x(0)$$

$$\mathbf{v}(2) = \begin{bmatrix} 0 & 1 \\ 0 & 0 \end{bmatrix} \mathbf{v}(1) + \begin{bmatrix} 0 \\ 1 \end{bmatrix} x(1)$$

$$= \begin{bmatrix} 0 & 0 \\ 0 & 0 \end{bmatrix} \mathbf{v}(0) + \begin{bmatrix} 1 \\ 0 \end{bmatrix} x(0) + \begin{bmatrix} 0 \\ 1 \end{bmatrix} x(1)$$

We find that $\mathbf{F}^n = 0$ for $n \geq 2$. Hence the initial state $\mathbf{v}(0)$ does not affect the state of the system after two steps. This is obvious because the FIR system has finite memory equal to 2. Consequently, the influence of the initial state on the future states and outputs vanishes after two steps. For $n_0 = 0$, equation (7.5.43) gives

$$y(n) = \mathbf{g}'\mathbf{F}^n\mathbf{v}(0) + \mathbf{g}'\mathbf{F}^{n-1}\mathbf{q}x(0) + \cdots$$
$$+ \mathbf{g}'\mathbf{F}\mathbf{q}x(n - 2) + \mathbf{g}'\mathbf{q}x(n - 1) + dx(n)$$

Since $\mathbf{F}^n = 0$ for $n \geq 2$, the output becomes

$$y(n) = \mathbf{g}'\mathbf{F}\mathbf{q}x(n - 2) + \mathbf{g}'\mathbf{q}x(n - 1) + dx(n) \qquad n \geq 2$$

When we substitute the values for $\mathbf{g}$, $\mathbf{F}$, $\mathbf{q}$, and d the equation above reduces to

$$y(n) = x(n) + 2x(n - 1) + x(n - 2) \qquad n \geq 2$$

which is the input–output description of the FIR system.

As a generalization of the results in the example above, it follows that for an Nth-order FIR system, $\mathbf{F}^n = 0$ for $n \geq N$. Consequently, the initial state of the system affects only the first N outputs.

EXAMPLE 7.5.5 _____

Compute the step response of the system

$$\mathbf{v}(n + 1) = \begin{bmatrix} \frac{1}{2} & 0 \\ 0 & \frac{1}{3} \end{bmatrix} \mathbf{v}(n) + \begin{bmatrix} 1 \\ 1 \end{bmatrix} x(n)$$

$$y(n) = [1 \quad 1]\mathbf{v}(n) + 2x(n)$$

Solution: The two components in the state equation are

$$v_1(n + 1) = \tfrac{1}{2}v_1(n) + x(n)$$
$$v_2(n + 1) = \tfrac{1}{3}v_2(n) + x(n)$$

This is a set of two decoupled equations that can easily be solved recursively, as in the one-dimensional case. Indeed, from (7.5.36) we have, for $n > 0$,

$$v_1(n) = (\tfrac{1}{2})^n v_1(0) + \sum_{k=0}^{n-1} (\tfrac{1}{2})^{n-1-k} x(k)$$

$$v_2(n) = (\tfrac{1}{3})^n v_2(0) + \sum_{k=0}^{n-1} (\tfrac{1}{3})^{n-1-k} x(k)$$

The output of the system is

$$y(n) = v_1(n) + v_2(n) + 2x(n)$$

$$= (\tfrac{1}{2})^n v_1(0) + (\tfrac{1}{3})^n v_2(0) + \sum_{k=0}^{n-1} (\tfrac{1}{2})^{n-1-k} + \sum_{k=0}^{n-1} (\tfrac{1}{3})^{n-1-k} + 2 \qquad n > 0$$

which simplifies to

$$y(n) = (\tfrac{1}{2})^n v_1(0) + (\tfrac{1}{3})^n v_2(0) + 2(1 - 2^{-n}) + \tfrac{3}{2}(1 - 3^{-n}) + 2 \qquad n > 0$$

We note in Example 7.5.5 that we could obtain a closed-form solution for the output of the system because the matrix $\mathbf{F}$ is diagonal. In general, when $\mathbf{F}$ is diagonal an N-dimensional state-space system can be equivalently described by a set of N independent one-dimensional systems. In such a case the state equations become simple first-order difference equations that are easily solved.

7.5.4 Relationships Between Input–Output and State-Space Descriptions

From our previous discussion we have seen that there is no unique choice for the state variables of a causal system. Furthermore, different choices for the state vector lead to different structures for the realization of the same system. Hence, in general, the input–output relationship does not uniquely describe the internal structure of the system.

To illustrate these assertions, let us consider an N-dimensional SISO system with the state-space representation

$$\mathbf{v}(n + 1) = \mathbf{F}\mathbf{v}(n) + \mathbf{q}x(n) \tag{7.5.48}$$
$$y(n) = \mathbf{g}'\mathbf{v}(n) + dx(n) \tag{7.5.49}$$

Let $\mathbf{P}$ be any $N \times N$ matrix whose inverse matrix $\mathbf{P}^{-1}$ exists. We define a new state vector $\hat{\mathbf{v}}(n)$ as

$$\hat{\mathbf{v}}(n) = \mathbf{P}\mathbf{v}(n) \tag{7.5.50}$$

Then

$$\mathbf{v}(n) = \mathbf{P}^{-1}\hat{\mathbf{v}}(n) \tag{7.5.51}$$

If (7.5.48) is premultiplied by $\mathbf{P}$, we obtain

$$\mathbf{P}\mathbf{v}(n + 1) = \mathbf{P}\mathbf{F}\mathbf{v}(n) + \mathbf{P}\mathbf{q}x(n)$$

By using (7.5.51), the state equation above becomes

$$\hat{\mathbf{v}}(n + 1) = (\mathbf{P}\mathbf{F}\mathbf{P}^{-1})\hat{\mathbf{v}}(n) + (\mathbf{P}\mathbf{q})x(n) \tag{7.5.52}$$

Similarly, with the aid of (7.5.51) the output equation (7.5.49) becomes

$$y(n) = (\mathbf{g}'\mathbf{P}^{-1})\hat{\mathbf{v}}(n) + dx(n) \tag{7.5.53}$$

Now, we define a new system parameter matrix $\hat{\mathbf{F}}$ and the vectors $\hat{\mathbf{q}}$ and $\hat{\mathbf{g}}$ as

$$\begin{aligned}
\hat{\mathbf{F}} &= \mathbf{P}\mathbf{F}\mathbf{P}^{-1} \\
\hat{\mathbf{q}} &= \mathbf{P}\mathbf{q} \\
\hat{\mathbf{g}}' &= \mathbf{g}'\mathbf{P}^{-1}
\end{aligned} \tag{7.5.54}$$

With these definitions, the state equations can be expressed in terms of the new system quantities as

$$\hat{\mathbf{v}}(n + 1) = \hat{\mathbf{F}}\hat{\mathbf{v}}(n) + \hat{\mathbf{q}}x(n) \tag{7.5.55}$$
$$y(n) = \hat{\mathbf{g}}'\hat{\mathbf{v}}(n) + dx(n) \tag{7.5.56}$$

If we compare (7.5.48) and (7.5.49) with (7.5.55) and (7.5.56), we observe that by a simple linear transformation of the state variables we have generated a new set of state equations and output equation, in which the input $x(n)$ and the output $y(n)$ are unchanged. Since there is an infinite number of choices of the transformation matrix $\mathbf{P}$, there is an infinite number of state-space equations and structures for a system. Some of these structures are different, while some others are very similar, differing only by scale factors.

Associated with any state-space realization of a system is the concept of a *minimal realization*. A state-space realization is said to be *minimal* if the dimension of the state space (the number of state variables) is the smallest of all possible realizations. Since each state variable represents a quantity that must be stored and updated at every time instant n, it follows that a minimal realization is one that requires the smallest number of delays (storage registers). We recall that the direct form II realization requires the smallest number of storages registers, and consequently, a state-space realization based on the contents of the delay elements results in a minimal realization. Similarly, an FIR system realized as a direct form structure leads to a minimal state-space realization if the values of the storage registers are defined as the state variables. On the other hand, the direct form I realization of an IIR system does not lead to a minimal realization.

Now, let us determine the impulse response of the system from the state-space realization. The impulse response provides one of the links between the input–output and state-space description of systems.

By definition the impulse response $h(n)$ of a system is the zero-state response of the system to the excitation $x(n) = \delta(n)$. Hence it can be obtained from equation (7.5.45) if we set $n_0 = 0$ (the time we apply the input), $v(0) = 0$, and $x(n) = \delta(n)$. Thus the impulse response of the system described by (7.5.48) and (7.5.49) is given by

$$h(n) = \mathbf{g}'\mathbf{F}^{n-1}\mathbf{q}u(n-1) + d\,\delta(n) \tag{7.5.57}$$
$$= \mathbf{g}'\mathbf{\Phi}(n-1)\mathbf{q}u(n-1) + d\,\delta(n)$$

Given a state-space description, it is straightforward to determine the impulse response from (7.5.57). However, the inverse is not easy, because there is an infinite number of state-space realizations for the same input–output description.

EXAMPLE 7.5.6

The Fibonacci sequence is given by

$$\left\{ \begin{matrix} 1,\ 1,\ 2,\ 3,\ 5,\ 8,\ 13,\ \ldots \\ \uparrow \end{matrix} \right\}$$

Determine the seventeenth term without computing the previous terms.

Solution: The Fibonacci sequence can be thought as the impulse response of the system

$$y(n) = y(n-1) + y(n-2) + x(n)$$

Indeed, by setting $y(-1) = y(-2) = 0$, and $x(n) = \delta(n)$, we obtain $h(0) = 1$, $h(1) = 1$, $h(2) = 2$, $h(3) = 3$, $h(4) = 5$, and so on.

The type 1 state-space realization is described by

$$\mathbf{F} = \begin{bmatrix} 0 & 1 \\ 1 & 1 \end{bmatrix} \qquad \mathbf{q} = \begin{bmatrix} 0 \\ 1 \end{bmatrix} \qquad \mathbf{g} = \begin{bmatrix} 1 \\ 1 \end{bmatrix} \qquad d = 1$$

From (7.5.57), we have

$$h(17) = \mathbf{g}'\mathbf{F}^{16}\mathbf{q}$$

By computing $\mathbf{F}^2$, $\mathbf{F}^4$, $\mathbf{F}^8$, and then $\mathbf{F}^{16}$, we obtain

$$\mathbf{F}^{16} = \begin{bmatrix} 610 & 987 \\ 987 & 1597 \end{bmatrix}$$

Then it easily follows that

$$h(17) = 2584$$

The Transpose System. The transpose of a matrix $\mathbf{F}$ is obtained by interchanging its columns and rows, and it is denoted by $\mathbf{F}'$. For example,

$$\mathbf{F} = \begin{bmatrix} f_{11} & f_{12} & \cdots & f_{1N} \\ f_{21} & f_{22} & \cdots & f_{2N} \\ \vdots & \vdots & \cdots & \vdots \\ f_{N1} & f_{N2} & \cdots & f_{NN} \end{bmatrix} \qquad \mathbf{F}' = \begin{bmatrix} f_{11} & f_{21} & \cdots & f_{N1} \\ f_{12} & f_{22} & \cdots & f_{N2} \\ \vdots & \vdots & \cdots & \vdots \\ f_{1N} & f_{2N} & \cdots & f_{NN} \end{bmatrix}$$

Now define the *transpose system* (7.5.48)–(7.5.49) as

$$\mathbf{v}'(n + 1) = \mathbf{F}'\mathbf{v}'(n) + \mathbf{q}x(n) \tag{7.5.58}$$
$$y'(n) = \mathbf{g}'\mathbf{v}'(n) + dx(n) \tag{7.5.59}$$

According to (7.5.57), the impulse response of this system is given as

$$h'(n) = \mathbf{g}'(\mathbf{F}')^{n-1}\mathbf{q}u(n - 1) + d\,\delta(n) \tag{7.5.60}$$

From matrix algebra we know that $(\mathbf{F}')^{n-1} = (\mathbf{F}^{n-1})'$. Hence

$$h'(n) = \mathbf{q}'(\mathbf{F}^{n-1})'\mathbf{g}u(n - 1) + d\,\delta(n)$$

We claim that $h'(n) = h(n)$. Indeed, the term $\mathbf{q}'(\mathbf{F}^{n-1})'\mathbf{g}$ is a scalar. Hence it is equal to its transpose. Consequently,

$$[\mathbf{q}'(\mathbf{F}^{n-1})'\mathbf{g}]' = \mathbf{g}'(\mathbf{F}')^{n-1}\mathbf{q}$$

Since this is true, it follows that (7.5.60) is identical to (7.5.57) and, therefore, $h'(n) = h(n)$. Thus *a SISO system and its transpose have identical impulse responses and hence the same input–output relationship*. To support this claim further, we note that the type 1 and type 2 state-space realizations, described by (7.5.22) through (7.5.25) are transpose structures, which stem from the same input–output relationship (7.5.16).

EXAMPLE 7.5.7 _____

Sketch the block diagram for the transpose of the system given in Example 7.5.1 (see Fig. 7.39).

Solution: By reversing the signal flow direction in all branches and by replacing branch nodes by adders and adders by branch nodes in the block diagram of Fig. 7.39, we obtain the block diagram in Fig. 7.41.
   The transpose system is described by

$$\begin{bmatrix} v'(n + 1) \\ v'_2(n + 1) \end{bmatrix} = \begin{bmatrix} 1.35 & -0.45 \\ 0.55 & 0.35 \end{bmatrix} \begin{bmatrix} v'_1(n) \\ v'_2(n) \end{bmatrix} + \begin{bmatrix} 3 \\ 1 \end{bmatrix} x(n)$$

$$y(n) = [0.5 \quad 0.5] \begin{bmatrix} v'_1(n) \\ v'_2(n) \end{bmatrix} + x(n)$$

which leads directly to the block diagram realization of Fig. 7.42. The reader is invited to show that the realizations in Figs. 7.41 and 7.42 are identical.

_____

   We have introduced the transpose structure because it provides an easy method for generating a new structure. However, sometimes this new structure may either differ trivially or be identical to the original one.

   The Diagonal System. In Example 7.5.5 we easily obtained a closed-form solution of the state-space equations because the system matrix $\mathbf{F}$ was diagonal. Hence, by finding a matrix $\mathbf{P}$ so that $\hat{\mathbf{F}} = \mathbf{PFP}^{-1}$ is diagonal, the solution of the state equations is simplified considerably. The diagonalization of the matrix $\mathbf{F}$ can be accomplished by first determining the eigenvalues and eigenvectors of the matrix.

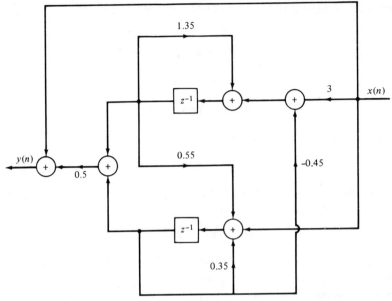

FIGURE 7.41 Transpose of the system in Figure 7.39.

A number λ is an *eigenvalue* of $\mathbf{F}$ and a nonzero vector $\mathbf{u}$ is the associated *eigenvector* if

$$\mathbf{Fu} = \lambda\mathbf{u} \qquad (7.5.61)$$

To determine the eigenvalues of $\mathbf{F}$, we note that

$$(\mathbf{F} - \lambda\mathbf{I})\mathbf{u} = 0 \qquad (7.5.62)$$

This equation has a (nontrivial) nonzero solution $\mathbf{u}$ if the matrix $\mathbf{F} - \lambda\mathbf{I}$ is singular [i.e., if $(\mathbf{F} - \lambda\mathbf{I})$ is noninvertible], which is the case if the determinant of $(\mathbf{F} - \lambda\mathbf{I})$ is zero, that is, if

$$\det (\mathbf{F} - \lambda\mathbf{I}) = 0 \qquad (7.5.63)$$

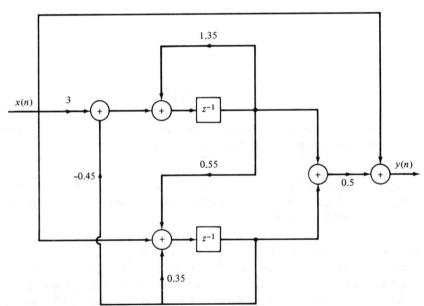

FIGURE 7.42 Realization of system in Example 7.5.7.

This determinant in (7.5.63) yields the *characteristic polynomial* of the matrix $\mathbf{F}$. For an $N \times N$ matrix $\mathbf{F}$, the characteristic polynomial of $\mathbf{F}$ is degree N and hence it has N roots, say λ_i, $i = 1, 2, \ldots, N$. The roots may be distinct or some roots may be repeated. In any case, for each root λ_i, we can determine a vector $\mathbf{u}_i$, called the eigenvector corresponding to the eigenvalue λ_i, from the equation

$$\mathbf{Fu}_i = \lambda_i \mathbf{u}_i$$

These eigenvectors are orthogonal, that is, $\mathbf{u}_i'\mathbf{u}_j = 0$, for $i \neq j$.

If we form a matrix $\mathbf{U}$ whose columns consist of the eigenvectors $\{\mathbf{u}_i\}$, that is,

$$\mathbf{U} = \left\{ \begin{matrix} \uparrow & \uparrow & & \uparrow \\ \mathbf{u}_1 & \mathbf{u}_2 & \cdots & \mathbf{u}_N \\ \downarrow & \downarrow & & \downarrow \end{matrix} \right\}$$

then the matrix $\mathbf{F} = \mathbf{U}^{-1}\mathbf{FU}$ is diagonal. Thus we have solved for the matrix that diagonalizes $\mathbf{F}$.

The following example illustrates the procedure of diagonalizing $\mathbf{F}$.

EXAMPLE 7.5.8

Find an explicit formula for the Fibonacci sequence discussed in Example 7.5.6.

Solution: In Example 7.5.6 we noted that the Fibonacci sequence can be considered as the impulse response of the system

$$\mathbf{v}(n + 1) = \begin{bmatrix} 0 & 1 \\ 1 & 1 \end{bmatrix} \mathbf{v}(n) + \begin{bmatrix} 0 \\ 1 \end{bmatrix} x(n)$$

$$y(n) = [1 \quad 1]\mathbf{v}(n) + x(n)$$

Now we wish to determine an equivalent system

$$\hat{\mathbf{v}}(n + 1) = \hat{\mathbf{F}}\hat{\mathbf{v}}(n) + \hat{\mathbf{q}}x(n)$$

$$y(n) = \mathbf{g}'\hat{\mathbf{v}}(n) + dx(n)$$

such that the matrix $\hat{\mathbf{F}}$ is diagonal. From (7.5.54) we recall that the two systems are equivalent if

$$\hat{\mathbf{F}} = \mathbf{PFP}^{-1} \qquad \hat{\mathbf{q}} = \mathbf{P}\mathbf{q} \qquad \hat{\mathbf{g}}' = \mathbf{g}'\mathbf{P}^{-1}$$

Given $\mathbf{F}$ the problem is to determine a matrix $\mathbf{P}$ such that $\hat{\mathbf{F}} = \mathbf{PFP}^{-1}$ is a diagonal matrix.

First, we compute the determinant in (7.5.63). We have

$$\det{(\mathbf{F} - \lambda\mathbf{I})} = \det \begin{bmatrix} -\lambda & 1 \\ 1 & 1 - \lambda \end{bmatrix} = \lambda^2 - \lambda - 1 = 0$$

or

$$\lambda_1 = \frac{1 + \sqrt{5}}{2} \qquad \lambda_2 = \frac{1 - \sqrt{5}}{2}$$

To find the eigenvector $\mathbf{u}_1$ corresponding to λ_1, we have

$$\begin{bmatrix} 0 & 1 \\ 1 & 1 \end{bmatrix} \mathbf{u}_1 = \lambda_1 \mathbf{u}_1 \qquad \text{or} \qquad \mathbf{u}_1 = \begin{bmatrix} 1 \\ \lambda_1 \end{bmatrix}$$

Similarly, we obtain

$$\mathbf{u}_2 = \begin{bmatrix} 1 \\ \lambda_2 \end{bmatrix}$$

We observe that $\mathbf{u}_1'\mathbf{u}_2 = 1 + \lambda_1\lambda_2 = 0$ (i.e., the eigenvectors are orthogonal). Now matrix $\mathbf{U}$, whose columns are the eigenvectors of $\mathbf{F}$, is

$$\mathbf{U} = \begin{bmatrix} 1 & 1 \\ \lambda_1 & \lambda_2 \end{bmatrix}$$

Then the matrix $\mathbf{U}^{-1}\mathbf{F}\mathbf{U}$ is diagonal. Indeed, it easily follows that

$$\hat{\mathbf{F}} = \mathbf{U}^{-1}\mathbf{F}\mathbf{U} = \begin{bmatrix} \lambda_1 & 0 \\ 0 & \lambda_2 \end{bmatrix}$$

and since the transformation matrix is $\mathbf{P} = \mathbf{U}^{-1}$, we have

$$\mathbf{P} = \frac{1}{\lambda_2 - \lambda_1} \begin{bmatrix} \lambda_2 & -1 \\ -\lambda_1 & 1 \end{bmatrix}$$

Thus the diagonal matrix $\mathbf{F}$ has the form

$$\hat{\mathbf{F}} = \begin{bmatrix} \lambda_1 & 0 \\ 0 & \lambda_2 \end{bmatrix}$$

where the diagonal elements are the eigenvalues of the characteristic polynomial. Furthermore, we obtain

$$\hat{\mathbf{q}} = \mathbf{P}\mathbf{q} = \begin{bmatrix} \dfrac{1}{\sqrt{5}} \\ -\dfrac{1}{\sqrt{5}} \end{bmatrix}$$

and

$$\hat{\mathbf{g}}' = \mathbf{g}'\mathbf{P}^{-1} = \mathbf{g}'\mathbf{U}$$
$$= \begin{bmatrix} \dfrac{3 + \sqrt{5}}{2} & \dfrac{3 - \sqrt{5}}{2} \end{bmatrix}$$

The impulse response of this equivalent diagonal system is

$$h(n) = \hat{\mathbf{g}}'\hat{\mathbf{F}}\hat{\mathbf{q}}u(n - 1) + d\delta(n)$$
$$= \frac{1}{\sqrt{5}} \left[\left(\frac{3 + \sqrt{5}}{2} \right)\left(\frac{1 + \sqrt{5}}{2} \right)^{n-1} \right.$$
$$\left. - \left(\frac{3 - \sqrt{5}}{2} \right)\left(\frac{1 - \sqrt{5}}{2} \right)^{n-1} \right] u(n - 1) + \delta(n)$$

which is the general formula for the Fibonacci sequence.

An alternative expression can be found by noting that the Fibonacci sequence can be considered as the zero-input response of the system described by the difference equation

$$y(n) = y(n - 1) + y(n - 2) + x(n)$$

with initial conditions $y(-1) = 1$, $y(-2) = -1$. From the type 1 state-space realization we note that $v_1(0) = y(-2) = -1$ and $v_2(0) = y(-1) = 1$. Hence

$$\begin{bmatrix} \hat{v}_1(0) \\ \hat{v}_2(0) \end{bmatrix} = \mathbf{P} \begin{bmatrix} v_1(0) \\ v_2(0) \end{bmatrix} = \frac{-1}{5} \begin{bmatrix} \dfrac{-3 + \sqrt{5}}{2} \\ \dfrac{3 + \sqrt{5}}{2} \end{bmatrix}$$

and the zero-input response is

$$y_{zi}(n) = \hat{\mathbf{g}}^t \hat{\mathbf{F}}^n \hat{\mathbf{v}}(0)$$

$$= \frac{1}{\sqrt{5}} \left[\left(\frac{1 + \sqrt{5}}{2} \right)^n - \left(\frac{1 - \sqrt{5}}{2} \right)^n \right] u(n)$$

This is the more familiar form for the Fibonacci sequence, where the first term of the sequence is zero, that is, the sequences is $\{0, 1, 1, 2, 3, 5, 8, \ldots\}$.

The example just studied illustrates the method for diagonalizing the matrix $\mathbf{F}$. The diagonal system yields a set of N decoupled, first-order linear difference equations that are easily solved to yield the state and the output of the system.

It is important to note that the eigenvalues of the matrix $\mathbf{F}$ are identical to the roots of the characteristic polynomial obtained from the homogeneous difference equation that characterizes the system. For example, the system that generates the Fibonacci sequence is characterized by the homogeneous difference equation

$$y(n) - y(n - 1) - y(n - 2) = 0 \qquad (7.5.64)$$

Recall that the solution is obtained by assuming that the homogeneous solution has the form

$$y_h(n) = \lambda^n$$

Substitution of this solution into (7.5.64) yields the characteristic polynomial

$$\lambda^2 - \lambda - 1 = 0$$

But this is exactly the same characteristic polynomial obtained from the determinant of $(\mathbf{F} - \lambda \mathbf{I})$.

Since the state-variable realization of the system is not unique, the matrix $\mathbf{F}$ is also not unique. However, the eigenvalues of the system are unique, that is, they are invariant to any nonsingular linear transformation of $\mathbf{F}$. Consequently, the characteristic polynomial of $\mathbf{F}$ can be determined either from evaluating the determinant of $(\mathbf{F} - \lambda \mathbf{I})$ or from the difference equation characterizing the system.

In conclusion, the state-space description provides an alternative characterization of the system that is equivalent to the input–output description. One advantage of the state-variable formulation is that it provides us with the additional information concerning the internal (state) variables of the system, information, that is not easily obtained from the input–output description. Furthermore, the state-variable formulation of a linear time-invariant system allows us to represent the system by a set of (usually coupled) first-order difference equations. The decoupling of the equations can be achieved by means of a linear transformation that can be obtained by solving for the eigenvalues and eigenvectors of the system. Then the decoupled equations are relatively simple to solve. More important, however, the state-space formulation provides a powerful, yet straightforward method for dealing with systems that have multiple inputs and multiple outputs (MIMO). Although we have not considered such systems in our study, it is in the treatment of MIMO systems where the true power and the beauty of the space-space formulation can be fully appreciated.

7.5.5 State-Space Analysis in the z-Domain

The state-space analysis in the previous sections has been performed in the time domain. However, as we have observed previously, the analysis of linear time-invariant discrete-time systems can also be carried out in the z-transform domain, often with greater ease. In this section we treat the state-space representation of linear time-invariant discrete-time systems in the z-transform domain.

Let us consider the state-space equation

$$\mathbf{v}(n + 1) = \mathbf{F}\mathbf{v}(n) + \mathbf{q}x(n) \tag{7.5.65}$$

which is equivalent to the following set of N first-order difference equations:

$$
\begin{aligned}
v_1(n + 1) &= f_{11}v_1(n) + f_{12}v_2(n) + \cdots + f_{1N}v_N(n) + q_1x(n) \\
v_2(n + 1) &= f_{21}v_1(n) + f_{22}v_2(n) + \cdots + f_{2N}v_N(n) + q_2x(n) \\
&\vdots \\
v_N(n + 1) &= f_{N1}v_1(n) + f_{N2}v_2(n) + \cdots + f_{NN}v_N(n) + q_Nx(n)
\end{aligned}
\tag{7.5.66}
$$

Assuming that the initial state of the system is zero, the z-transform of this set of N equations becomes

$$
\begin{bmatrix}
zV_1(z) \\
zV_2(z) \\
\vdots \\
zV_N(z)
\end{bmatrix}
=
\begin{bmatrix}
f_{11} & f_{12} & \cdots & f_{1N} \\
f_{21} & f_{22} & \cdots & f_{2N} \\
\vdots & & & \\
f_{N1} & f_{N2} & \cdots & f_{NN}
\end{bmatrix}
\begin{bmatrix}
V_1(z) \\
V_2(z) \\
\vdots \\
V_N(z)
\end{bmatrix}
+
\begin{bmatrix}
q_1 \\
q_2 \\
\vdots \\
q_N
\end{bmatrix}
X(z)
\tag{7.5.67}
$$

where $V_i(z)$ is the z-transform of $v_i(n)$, $i = 1, 2, \ldots, N$.

If we define the vector $\mathbf{V}(z)$ as

$$
\mathbf{V}(z) =
\begin{bmatrix}
V_1(z) \\
V_2(z) \\
\vdots \\
V_N(z)
\end{bmatrix}
\tag{7.5.68}
$$

then (7.5.67) can be expressed in matrix form as

$$z\mathbf{V}(z) = \mathbf{F}\mathbf{V}(z) + \mathbf{q}X(z) \tag{7.5.69}$$

The two terms involving $\mathbf{V}(z)$ can be collected together and the resulting equation can be used to solve for $\mathbf{V}(z)$. Thus

$$
\begin{aligned}
(z\mathbf{I} - \mathbf{F})\mathbf{V}(z) &= \mathbf{q}X(z) \\
\mathbf{V}(z) &= (z\mathbf{I} - \mathbf{F})^{-1}\mathbf{q}X(z)
\end{aligned}
\tag{7.5.70}
$$

The inverse z-transform of (7.5.70) yields the solution for the state equations.

Next, we turn our attention to the output equation, which is given as

$$y(n) = \mathbf{g}'\mathbf{v}(n) + dx(n) \tag{7.5.71}$$

or, equivalently, as

$$y(n) = \sum_{i=1}^{N} g_i v_i(n) + dx(n) \tag{7.5.72}$$

The z-transform of (7.5.72) is

$$Y(z) = \sum_{i=1}^{N} g_i V_i(z) + dX(z) \tag{7.5.73}$$

or, equivalently,

$$Y(z) = \mathbf{g}'\mathbf{V}(z) + dX(z) \qquad (7.5.74)$$

By using the solution in (7.5.70) we can eliminate the state vector $\mathbf{V}(z)$ in (7.5.74). Thus we obtain

$$Y(z) = [\mathbf{g}'(z\mathbf{I} - \mathbf{F})^{-1}\mathbf{q} + d]X(z) \qquad (7.5.75)$$

which is the z-transform of the zero-state response of the system. The system function is easily obtained from (7.5.75) as

$$H(z) = \frac{Y(z)}{X(z)} = \mathbf{g}'(z\mathbf{I} - \mathbf{F})^{-1}\mathbf{q} + d \qquad (7.5.76)$$

The state equation given by (7.5.70), the output equation given by (7.5.75) and the system function given by (7.5.76) all have in common the factor $(z\mathbf{I} - \mathbf{F})^{-1}$. This is a fundamental quantity that is related to the z-transform of the state transition matrix of the system. The relationship is easily established by computing the z-transform of the impulse response $h(n)$, which is given by (7.5.57). Thus we have

$$H(z) = \sum_{n=0}^{\infty} h(n)z^{-n}$$

$$= \sum_{n=0}^{\infty} [\mathbf{g}'\mathbf{F}^{n-1}\mathbf{q}u(n-1) + d\delta(n)]z^{-n} \qquad (7.5.77)$$

$$= \mathbf{g}' \left(\sum_{n=1}^{\infty} \mathbf{F}^{n-1}\mathbf{z}^{-n} \right) \mathbf{q} + d$$

The term in brackets in (7.5.77) may be written as

$$\sum_{n=1}^{\infty} \mathbf{F}^{n-1}z^{-n} = z^{-1}(\mathbf{I} + \mathbf{F}z^{-1} + \mathbf{F}^2 z^{-2} + \cdots)$$

$$= z^{-1}(\mathbf{I} - \mathbf{F}z^{-1})^{-1} \qquad (7.5.78)$$

$$= (z\mathbf{I} - \mathbf{F})^{-1}$$

If we substitute the result in (7.5.78) into (7.5.77) we obtain the expression for $H(z)$ as given in (7.5.76).

Since the state transition matrix is given by

$$\mathbf{\Phi}(n) = \mathbf{F}^n$$

the z-transform of $\mathbf{\Phi}(n)$ is

$$\sum_{n=0}^{\infty} \mathbf{F}^n z^{-n} = \mathbf{I} + \mathbf{F}z^{-1} + \mathbf{F}^2 z^{-2} + \mathbf{F}^3 z^{-3} + \cdots$$

$$= (\mathbf{I} - \mathbf{F}z^{-1})^{-1} \qquad (7.5.79)$$

$$= z(z\mathbf{I} - \mathbf{F})^{-1}$$

The relation in (7.5.79) provides a simple method for determining the state transition matrix by means of z-transforms. We recall that

$$(z\mathbf{I} - \mathbf{F})^{-1} = \frac{\text{adj } (z\mathbf{I} - \mathbf{F})}{\text{det } (z\mathbf{I} - \mathbf{F})} \qquad (7.5.80)$$

where adj (**A**) denotes the *adjoint matrix* of **A** and det (**A**) denotes the determinant of the matrix **A**. Substitution of (7.5.80) into (7.5.76) yields the result

$$H(z) = \mathbf{g}' \frac{\text{adj } (z\mathbf{I} - \mathbf{F})}{\det (z\mathbf{I} - \mathbf{F})} \mathbf{q} + d \tag{7.5.81}$$

Consequently, the denominator $D(z)$ of the system function $H(z)$, which contains the poles of the system is simply

$$D(z) = \det (z\mathbf{I} - \mathbf{F}) \tag{7.5.82}$$

But the det $(z\mathbf{I} - \mathbf{F})$ is just the characteristic polynomial of **F**. Its roots, which are the poles of system, are the eigenvalues of the matrix **F**.

EXAMPLE 7.5.9 _____

Determine the system function $H(z)$, the impulse response $h(n)$, and the state transition matrix $\mathbf{\Phi}(n)$ of the system that generates the Fibonacci sequence. This system is described by the state-space equation

$$\mathbf{v}(n + 1) = \begin{bmatrix} 0 & 1 \\ 1 & 1 \end{bmatrix} \mathbf{v}(n) + \begin{bmatrix} 0 \\ 1 \end{bmatrix} x(n)$$
$$y(n) = [1 \quad 1]\mathbf{v}(n) + x(n) \tag{7.5.83}$$

Solution: First, we determine $H(z)$ and $h(n)$ by computing $(z\mathbf{I} - \mathbf{F})^{-1}$. We have

$$(z\mathbf{I} - \mathbf{F})^{-1} = \begin{bmatrix} z & -1 \\ -1 & z - 1 \end{bmatrix}^{-1} = \frac{1}{z^2 - z - 1} \begin{bmatrix} z - 1 & 1 \\ 1 & z \end{bmatrix}$$

Hence

$$H(z) = \frac{1}{z^2 - z - 1} [1 \quad 1] \begin{bmatrix} z - 1 & 1 \\ 1 & z \end{bmatrix} \begin{bmatrix} 0 \\ 1 \end{bmatrix} + 1$$

$$= \frac{z^2}{z^2 - z - 1} = \frac{1}{1 - z^{-1} - z^{-2}}$$

By inverting $H(z)$, we obtain $h(n)$ in the form

$$h(n) = \frac{1}{\sqrt{5}} \left[\left(\frac{1 + \sqrt{5}}{2} \right)^{n+1} - \left(\frac{1 - \sqrt{5}}{2} \right)^{n+1} \right] u(n)$$

We note that the poles of $H(z)$ are $p_1 = (1 + \sqrt{5})/2$ and $p_2 = (1 - \sqrt{5})/2$. Since $|p_1| > 1$, the system that generates the Fibonacci sequence is unstable.

The state transition matrix $\mathbf{\Phi}(n)$ has the z-transform

$$z(z\mathbf{I} - \mathbf{F})^{-1} = \frac{1}{z^2 - z - 1} \begin{bmatrix} z^2 - z & z \\ z & z^2 \end{bmatrix}$$

The four elements of $\mathbf{\Phi}(n)$ are obtained by computing the inverse transform of the four elements of $z(z\mathbf{I} - \mathbf{F})^{-1}$. Thus we obtain

$$\mathbf{\Phi}(n) = \begin{bmatrix} \phi_{11}(n) & \phi_{12}(n) \\ \phi_{21}(n) & \phi_{22}(n) \end{bmatrix}$$

where

$$\phi_{11}(n) = \left[\frac{1 + \sqrt{5}}{2\sqrt{5}} \left(\frac{1 - \sqrt{5}}{2} \right)^n - \frac{1 - \sqrt{5}}{2\sqrt{5}} \left(\frac{1 + \sqrt{5}}{2} \right)^n \right] u(n)$$

$$\phi_{12}(n) = \phi_{21}(n) = \frac{1}{\sqrt{5}} \left[\left(\frac{1 + \sqrt{5}}{2} \right)^n - \left(\frac{1 - \sqrt{5}}{2} \right)^n \right] u(n)$$

$$\phi_{22}(n) = \frac{1}{\sqrt{5}} \left[\left(\frac{1 + \sqrt{5}}{2} \right)^{n+1} - \left(\frac{1 - \sqrt{5}}{2} \right)^{n+1} \right] u(n)$$

We note that the impulse response $h(n)$ may also be computed from (7.5.57) by using the state transition matrix.

The analysis method employed above applies specifically to the computation of the zero-state response of the system. This is consequence of the fact that we have used the two-sided z-transform.

If we wish to determine the total response of the system, beginning at a nonzero state, say $v(n_0)$, we must use the one-sided z-transform. Thus for a given initial state $v(n_0)$ and a given input $x(n)$ for $n \geq n_0$, we may determine the state vector $v(n)$ for $n \geq n_0$ and the output $y(n)$ for $n \geq n_0$, by means of the one-sided z-transform.

In this development we assume that $n_0 = 0$, without loss of generality. Then, given $x(n)$ for $n \geq 0$, and a causal system, described by the state equations in (7.5.65), the one-sided z-transform of the state equations is

$$zV^+(z) - zv(0) = FV^+(z) + qX(z)$$

or, equivalently,

$$V^+(z) = z(zI - F)^{-1}v(0) + (zI - F)^{-1}qX(z) \tag{7.5.84}$$

Note that $X^+(z) = X(z)$, since $x(n)$ is assumed to be causal.

Similarly, the z-transform of the output equation given by (7.5.71) is

$$Y^+(z) = g'V^+(z) + dX(z) \tag{7.5.85}$$

If we substitute for $V^+(z)$ from (7.5.84) into (7.5.85), we obtain the result

$$Y^+(z) = zg'(zI - F)^{-1}v(0) + [g'(zI - F)^{-1}q + d]X(z) \tag{7.5.86}$$

Of the terms on the right-hand side of (7.8.86), the first represents the zero-input response of the system due to the initial conditions, while the second represents the zero-state response of the system that we obtained previously. Consequently, (7.8.86) constitutes the total response of the system, which can be expressed in the time domain by inverting (7.8.86). The result of this inversion yields the form for $y(n)$ given previously by (7.5.45).

EXAMPLE 7.5.10

Determine the response of the Fibonacci system for $n \geq 0$ if its initial state is

$$v(0) = \begin{bmatrix} -1 \\ 1 \end{bmatrix}$$

Solution: The zero-state response of this system was determined in Example 7.5.9. Here we need only determine the zero-input response and add it to the zero-state response.

The one-sided z-transform of the zero-input response is

$$Y_{zi}^+(z) = z\mathbf{g}'(z\mathbf{I} - \mathbf{F})^{-1}\mathbf{v}(0)$$

$$= \frac{z}{z^2 - z - 1} \begin{bmatrix} 1 & 1 \end{bmatrix} \begin{bmatrix} z - 1 & 1 \\ 1 & z \end{bmatrix} \begin{bmatrix} -1 \\ 1 \end{bmatrix}$$

$$= \frac{z}{z^2 - z - 1}$$

The inverse transform of $Y_{zi}^+(z)$ is

$$y_{zi}(n) = \frac{1}{\sqrt{5}} \left[\left(\frac{1 + \sqrt{5}}{2} \right)^n - \left(\frac{1 - \sqrt{5}}{2} \right)^n \right] u(n)$$

7.5.6 Additional State-Space Structures

In Section 7.5.2 we described how state-space equations can be obtained from a given structure and, conversely, how to obtain a realization of the system given the state equations. In this section we revisit the parallel-form and cascade-form realizations described previously and consider these structures in the context of a state-space formulation.

The parallel-form state-space structure is obtained by expanding the system function $H(z)$ into a partial-fraction expansion, developing the state-space formulation for each term in the expansion and the corresponding structure, and finally, connecting all the structures in parallel. We illustrate the procedure under the assumption that the poles are distinct and $N = M$.

The system function $H(z)$ may be expressed as

$$H(z) = C + \sum_{k=1}^{N} \frac{B_k}{z - p_k} \tag{7.5.87}$$

Note that this is a different expansion from that given in (7.3.20). The output of the system is

$$Y(z) = H(z)X(z) = CX(z) + \sum_{k=1}^{N} B_k Y_k(z) \tag{7.5.88}$$

where, by definition,

$$Y_k(z) = \frac{X(z)}{z - p_k} \qquad k = 1, 2, \ldots, N \tag{7.5.89}$$

In the time domain, the equations in (7.5.89) become

$$y_k(n + 1) = p_k y_k(n) + x(n) \qquad k = 1, 2, \ldots, N \tag{7.5.90}$$

We define the state variables as

$$v_k(n) = y_k(n) \qquad k = 1, 2, \ldots, N \tag{7.5.91}$$

Then the difference equations in (7.5.90) become

$$v_k(n + 1) = p_k v_k(n) + x(n) \qquad k = 1, 2, \ldots, N \tag{7.5.92}$$

The state equations in (7.5.92) may be expressed in matrix form as

$$
\mathbf{v}(n+1) = \begin{bmatrix} p_1 & 0 & \cdots & 0 \\ 0 & p_2 & \cdots & 0 \\ & & \ddots & \\ 0 & & & p_N \end{bmatrix} \mathbf{v}(n) + \begin{bmatrix} 1 \\ 1 \\ \vdots \\ 1 \end{bmatrix} x(n) \qquad (7.5.93)
$$

and the output equation is

$$
y(n) = [B_1 \quad B_2 \quad \cdots \quad B_N]\mathbf{v}(n) + Cx(n) \qquad (7.5.94)
$$

The block diagram realization for this parallel-form structure is shown in Fig. 7.43.

The parallel-form realization developed above is called the *normal form* representation, because the matrix $\mathbf{F}$ is diagonal, and hence the state variables are uncoupled. An alternative structure is obtained by pairing complex-conjugate poles and any two-real-valued poles to form second-order sections, which can be realized by using either type 1 or type 2 state-space structures.

The *cascade-form* state-space structure can be obtained by factoring $H(z)$ into a product of first-order and second-order sections, as described in Section 7.2.2, and implementing each section by using either type 1 or type 2 state-space structures.

Let us consider the state-space representation of a single second-order section involving a pair of complex-conjugate poles. The system function is

$$
\begin{aligned}
H_k(z) &= \frac{1 + b_{k1}z^{-1} + b_{k2}z^{-2}}{1 + a_{k1}z^{-1} + a_{k2}z^{-2}} \\[2mm]
&= \frac{z^2 + b_{k1}z + b_{k2}}{z^2 + a_{k1}z + a_{k2}} \qquad (7.5.95) \\[2mm]
&= 1 + \frac{A_k}{z - p_k} + \frac{A_k^*}{z - p_k^*}
\end{aligned}
$$

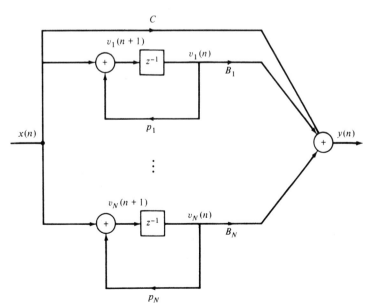

FIGURE 7.43   Parallel-form state-space realization of an IIR system.

The output of this system may be expressed as

$$Y_k(z) = X_k(z) + \frac{A_k X_k(z)}{z - p_k} + \frac{A_k^* X_k(z)}{z - p_k^*} \tag{7.5.96}$$

We define the quantity

$$S_k(z) = \frac{A_k X_k(z)}{z - p_k} \tag{7.5.97}$$

This relationship may be expressed in the time domain as

$$s_k(n + 1) = p_k s_k(n) + A_k x_k(n) \tag{7.5.98}$$

Since $s_k(n)$, p_k, and A_k are complex valued, we define $s_k(n)$ as

$$\begin{aligned} s_k(n) &= v_{k1}(n) + j v_{k2}(n) \\ p_k &= \alpha_{k1} + j\alpha_{k2} \\ A_k &= q_{k1} + j q_{k2} \end{aligned} \tag{7.5.99}$$

Upon substitution of these relations into (7.5.98) and separating its real and imaginary parts, we obtain

$$\begin{aligned} v_{k1}(n + 1) &= \alpha_{k1} v_{k1}(n) - \alpha_{k2} v_{k2}(n) + q_{k1} x_k(n) \\ v_{k2}(n + 1) &= \alpha_{k2} v_{k1}(n) + \alpha_{k1} v_{k2}(n) + q_{k2} x_k(n) \end{aligned} \tag{7.5.100}$$

We choose $v_{k1}(n)$ and $v_{k2}(n)$ as the state variables and thus we obtain the coupled pair of state equations which may be expressed in matrix form as

$$\mathbf{v}_k(n + 1) = \begin{bmatrix} \alpha_{k1} & -\alpha_{k2} \\ \alpha_{k2} & \alpha_{k1} \end{bmatrix} \mathbf{v}_k(n) + \begin{bmatrix} q_{k1} \\ q_{k2} \end{bmatrix} x_k(n) \tag{7.5.101}$$

The output equation may be expressed as

$$y_k(n) = x_k(n) + s_k(n) + s_k^*(n) \tag{7.5.102}$$

Upon substitution for $s_k(n)$ in (7.5.102) we obtain the desired result for the output in the form

$$y_k(n) = [2 \quad 0] \mathbf{v}_k(n) + x_k(n) \tag{7.5.103}$$

A realization for the second-order section is shown in Fig. 7.44. It is simply called the *coupled-form* state-space realization. This structure, which is used as the building block in the implementation of cascade-form realizations for higher-order IIR systems, exhibits low sensitivity to finite-word-length effects. The reader may find it instructive to compare this structure with the coupled-form oscillator described in Section 5.5.6.

7.6 Summary and References

From the treatment in this chapter we have seen that there are various realizations of discrete-time systems. FIR systems may be realized in a direct form, a cascade form, a frequency sampling form, and a lattice form. IIR systems may also be realized in a direct form, a cascade form, a lattice or a lattice-ladder form, and in addition, a parallel form.

For any given system described by a linear constant-coefficient difference equation, these realizations are equivalent in that they represent the same system and produce the same output for any given input, provided that the internal computations are performed with infinite precision. However, the various structures are not equivalent when they are realized with finite-precision arithmetic.

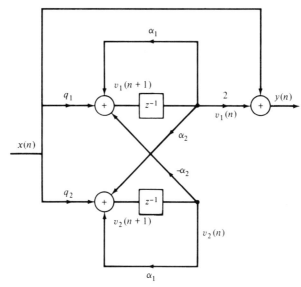

FIGURE 7.44 Coupled-form state-space realization of a two-pole, two-zero IIR system.

Three important factors were presented for choosing among the various FIR and IIR system realizations. These factors are computational complexity, memory requirements, and finite-word-length effects. Depending on either the time-domain or the frequency-domain characteristics of a system, some structures may require less computation and/or less memory than others. Hence our selection must consider these two important factors.

As we shall demonstrate in Chapter 10, finite-word-length effects play an extremely important role in the practical realization of a system. In particular, we demonstrate that direct form realizations based on fixed-point arithmetic should generally be avoided because of their sensitivity to parameter quantization effects. On the other hand, we demonstrate that the cascade-form realization is most robust to parameter quantization effects and hence it is preferable.

The state-space formulation provides an internal description of a system and, as a consequence, we obtained additional system realizations, called *state-space realizations*. These realizations represent additional possible structures that provide good alternative candidate realizations for the system.

Much research has been done over the past two decades on state-space representation and realization of systems. For reference, we cite the books by Chen (1970), DeRusso et al. (1965), Zadeh and Desoer (1963), and Gupta (1966). The use of state-space filter structures in the realization of IIR systems has been proposed by Mullis and Roberts (1976a, b), and further developed by Hwang (1977), Jackson et al. (1979), Jackson (1979), Mills et al. (1981), and Bomar (1985).

In deriving the transposed structures in Section 7.3, we introduced some concepts and operations on signal flow graphs. Signal flow graphs are treated in depth in the books by Mason and Zimmerman (1960) and Chow and Cassignol (1962).

The various FIR and IIR system realizations we have introduced in this chapter have been investigated in the context of finite-word-length implementation by many researchers. In Chapter 10 we provide many references to this literature. In particular, we should mention here the lattice and lattice-ladder structures, which are treated in the book by Markel and Gray (1976), and which find application in speech signal processing. Finally, we should mention another important structure for IIR systems, known as a *wave digital filter*, which has been proposed by Fettweis (1971) and further developed by Sedlmeyer and Fettweis (1973). A treatment of this filter structure can also be found in the book by Antoniou (1979).

PROBLEMS

7.1 Determine a direct form realization for the following linear phase filters.

(a) $h(n) = \left\{ \begin{array}{c} 1, 2, 3, 4, 3, 2, 1 \\ \uparrow \end{array} \right\}$

(b) $h(n) = \left\{ \begin{array}{c} 1, 2, 3, 3, 2, 1 \\ \uparrow \end{array} \right\}$

7.2 Consider an FIR filter with system function

$$H(z) = 1 + 2.88z^{-1} + 3.4048z^{-2} + 1.74z^{-3} + 0.4z^{-4}$$

Sketch the direct form and lattice realizations of the filter and determine in detail the corresponding input–output equations. Is the system minimum phase?

7.3 Determine the system function and the impulse response of the system shown in Fig. P7.3.

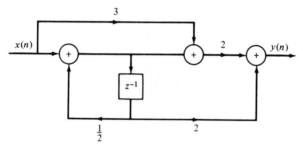

FIGURE P7.3

7.4 Determine the system function and the impulse response of the system shown in Fig. P7.4.

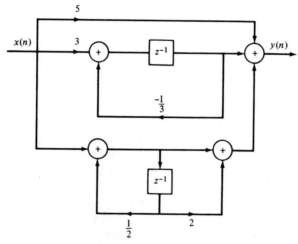

FIGURE P7.4

7.5 A discrete-time system is implemented by the following FORTRAN program.

```
DIMENSION X(100)
NDUR = 100
W1 = 0.0
W2 = 0.0
DO 10 N = 1, NDUR
XIN = X(N)
W = -0.5 * W1 - 0.25 * W2 + XIN
YOUT = 0.5 * W + 2 * W1 + 3 * W2
W2 = W1
W1 = W
10   CONTINUE
END
```

(a) Determine the system function $H(z)$ and the input–output relationship of the system.
(b) Design a realization that requires the minimum amount of memory.
(c) Is the system stable? Why?
(d) The signal

$$x_a(t) = 3 \cos (200\pi t + 60°) \qquad -\infty < t < \infty$$

is sampled with sampling rate $F_s = 1$ kHz and the resulting discrete-time signal $x(n)$ is fed into the system above. Obtain a closed-form expression for the steady-state response $y(n)$ of the system to this input.
(e) Is it theoretically possible to express the output of this system in terms of a convolution summation? Is it possible to do this in practice?

7.6 Determine a_1, a_2, and a_3 in terms of b_1 and b_2 so that the two systems in Fig. P7.6 are equivalent.

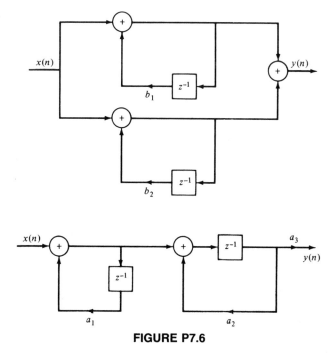

FIGURE P7.6

7.7 Consider the filter shown in Fig. P7.7.

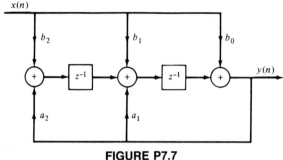

$x(n)$

b_2 b_1 b_0

z^{-1} z^{-1} $y(n)$

a_2 a_1

FIGURE P7.7

(a) Determine its system function.
(b) Sketch the pole–zero plot and check for stability if
   (1) $b_0 = b_2 = 1$, $b_1 = 2$, $a_1 = 1.5$, $a_2 = -0.9$
   (2) $b_0 = b_2 = 1$, $b_1 = 2$, $a_1 = 1$, $a_2 = -2$
(c) Determine the response to $x(n) = \cos(\pi n/3)$ if $b_0 = 1$, $b_1 = b_2 = 0$, $a_1 = 1$, and $a_2 = -0.99$.

7.8 Consider an LTI system, initially at rest, described by the difference equation

$$y(n) = \tfrac{1}{4} y(n - 2) + x(n)$$

(a) Determine the impulse response, $h(n)$, of the system.
(b) What is the response of the system to the input signal

$$x(n) = [(\tfrac{1}{2})^n + (-\tfrac{1}{2})^n]u(n)$$

(c) Determine the direct form II, parallel-form, and cascade-form realizations for this system.
(d) Sketch roughly the magnitude response $|H(\omega)|$ of this system.

7.9 Obtain the direct form I, direct form II, cascade, and parallel structures for the following systems.
(a) $y(n) = \tfrac{3}{4} y(n - 1) - \tfrac{1}{8} y(n - 2) + x(n) + \tfrac{1}{3} x(n - 1)$
(b) $y(n) = -0.1y(n - 1) + 0.72y(n - 2) + 0.7x(n) - 0.252x(n - 2)$
(c) $y(n) = -0.1y(n - 1) + 0.2y(n - 2) + 3x(n) + 3.6x(n - 1) + 0.6x(n - 2)$
(d) $H(z) = \dfrac{2(1 - z^{-1})(1 + \sqrt{2}\, z^{-1} + z^{-2})}{(1 + 0.5z^{-1})(1 - 0.9z^{-1} + 0.81z^{-2})}$
(e) $y(n) = \tfrac{1}{2} y(n - 1) + \tfrac{1}{4} y(n - 2) + x(n) + x(n - 1)$
(f) $y(n) = y(n - 1) - \tfrac{1}{2} y(n - 2) + x(n) - x(n - 1) + x(n - 2)$
Which of the systems above are stable?

7.10 Show that the systems in Fig. P7.10 are equivalent.

7.11 Determine all the FIR filters which are specified by the lattice parameters $K_1 = \frac{1}{2}$, $K_2 = 0.6$, $K_3 = -0.7$, and $K_4 = \frac{1}{3}$.

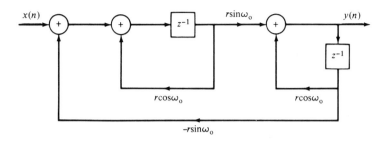

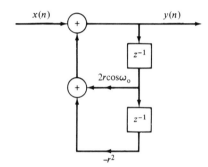

FIGURE P7.10

7.12 Determine the set of difference equations for describing a realization of an IIR system based on the use of the transposed direct form II structure for the second-order subsystems.

7.13 Modify subroutine CTDFII to obtain a subroutine that implements a parallel-form realization based on transposed direct form II second-order modules.

7.14 Write a subroutine that implements a cascade-form realization based on regular direct form II second-order modules.

7.15 Modify the subroutine AZLAT to eliminate the array OLDG(M).

7.16 Determine the parameters $\{K_m\}$ of the lattice filter corresponding to the FIR filter described by the system function

$$H(z) = A_2(z) = 1 + 2z^{-1} + \tfrac{1}{3}z^{-2}$$

7.17 (a) Determine the zeros and sketch the zero pattern for the FIR lattice filter with parameters

$$K_1 = \tfrac{1}{2} \qquad K_2 = -\tfrac{1}{3} \qquad K_3 = 1$$

(b) The same as in part (a) but with $K_3 = -1$.
(c) You should have found that all the zeros lie exactly on the unit circle. Can this result be generalized? How?
(d) Sketch the phase response of the filters in parts (a) and (b). What did you notice? Can this result be generalized? How?

7.18 Determine the coupled-form state-space realization for the digital resonator

$$H(z) = \frac{1}{1 - (2r \cos \omega_0)z^{-1} + r^2 z^{-2}}$$

7.19 (a) Determine the impulse response of an FIR lattice filter with parameters $K_1 = 0.6$, $K_2 = 0.3$, $K_3 = 0.5$, and $K_4 = 0.9$.

(b) Sketch the direct form and lattice all-zero and all-pole filters specified by the K-parameters given in part (a).

7.20 (a) Sketch the lattice realization for the resonator

$$H(z) = \frac{1}{1 - (2r \cos \omega_0)z^{-1} + r^2 z^{-2}}$$

(b) What happens if $r = 1$?

7.21 Sketch the lattice-ladder structure for the system

$$H(z) = \frac{1 - 0.8z^{-1} + 0.15z^{-2}}{1 + 0.1z^{-1} - 0.72z^{-2}}$$

7.22 Determine a state-space model and the corresponding realization for the following FIR system:

$$y(n) = \sum_{k=0}^{M} b_k x(n - k)$$

7.23 Determine the state-space model for the system described by

$$y(n) = y(n - 1) + 0.11y(n - 2) + x(n)$$

and sketch the type 1 and type 2 state-space realizations.

7.24 Determine the type 1 and type 2 state-space realizations for the Fibonacci system and its diagonal form.

7.25 By means of the z-transform, determine the impulse response of the system described by the state-space parameters

$$\mathbf{F} = \begin{bmatrix} 0 & 0.11 \\ 1 & 1 \end{bmatrix} \qquad \mathbf{q} = \begin{bmatrix} 0.11 \\ 1 \end{bmatrix} \qquad \mathbf{g} = \begin{bmatrix} 0 \\ 1 \end{bmatrix} \qquad d = 1$$

7.26 Determine the characteristic polynomial of the coupled-form state-space structure described by (7.5.101) and solve for the roots.

7.27 Determine the transpose structure for the coupled-form state-space structure shown in Fig. 7.44.

7.28 Consider a pole–zero system with system function

$$H(z) = \frac{(1 - 0.5e^{j\pi/4}z^{-1})(1 - 0.5e^{-j\pi/4}z^{-1})}{(1 - 0.8e^{j\pi/3}z^{-1})(1 - 0.8e^{-j\pi/3}z^{-1})}$$

(a) Sketch the regular and transpose direct form II realizations of the system.

(b) Determine and sketch the type 1 and type 2 state-space realizations.

(c) Determine the impulse response of the system by inverting $H(z)$ and by using state-space techniques.

(d) Determine the coupled-form state-space realization.

(e) Repeat parts (a) through (d) for the system obtained by changing the angle of the poles from $\pi/3$ to $\pi/4$.

7.29 (a) Determine a parallel and a cascade realization of the system

$$H(z) = \frac{1 + z^{-1}}{(1 - \frac{1}{2}z^{-1})(1 - 0.8e^{j\pi/4}z^{-1})(1 - 0.8e^{-j\pi/4}z^{-1})}$$

(b) Determine the type 1 and type 2 state-space descriptions of the system in part (a).

7.30 Show how to use a lattice structure to implement the following all-pass filter

$$H(z) = \frac{0.5 + 0.2z^{-1} - 0.6z^{-2} + z^{-3}}{1 - 0.6z^{-1} + 0.2z^{-2} + 0.5z^{-3}}$$

Is the system stable?

7.31 Consider a system described by the following state-space equations:

$$\mathbf{v}(n + 1) = \begin{bmatrix} 0 & 1 \\ -0.81 & 1 \end{bmatrix} \mathbf{v}(n) + \begin{bmatrix} 0 \\ 1 \end{bmatrix} x(n)$$

$$y(n) = [-1.81 \quad 1]\mathbf{v}(n) + x(n)$$

(a) Determine the characteristic polynomial and the eigenvalues of the system
(b) Determine the state transition matrix $\Phi(n)$ for $n \geq 0$.
(c) Determine the system function and the impulse response of the system.
(d) Compute the step response of the system if $\mathbf{v}(0) = [0 \quad 1]'$.
(e) Sketch a state-space realization for the system.

7.32 Repeat Problem 7.31 if the system is described by the state-space equations

$$\mathbf{v}(n + 1) = \begin{bmatrix} 0 & 1 \\ -2 & -3 \end{bmatrix} \mathbf{v}(n) + \begin{bmatrix} 0 \\ 1 \end{bmatrix} x(n)$$

$$y(n) = [1 \quad 0]\, \mathbf{v}(n)$$

7.33 Repeat Problem 7.31 for the system described by the state-space equations

$$\mathbf{v}(n + 1) = \begin{bmatrix} -0.3 & 0.4 \\ 0.4 & -0.3 \end{bmatrix} \mathbf{v}(n) + \begin{bmatrix} 1 \\ 0 \end{bmatrix} x(n)$$

$$y(n) = [1 \quad 1]\mathbf{v}(n) + x(n)$$

7.34 Consider the system

$$y(n) = 0.9y(n - 1) - 0.08y(n - 2) + x(n) + x(n - 1)$$

(a) Determine the type 1 and type 2 state-space realizations of the system.
(b) Determine the parallel and cascade state-space realizations of the system.
(c) Determine the impulse response of the system by at least two different methods.

7.35 Consider the causal system

$$y(n) = \tfrac{3}{4}y(n - 1) - \tfrac{1}{8}y(n - 2) + x(n) + \tfrac{1}{3}x(n - 1)$$

(a) Determine its system function.
(b) Determine the type 1 state-space model.
(c) Determine the state transition matrix $\Phi(n) = \mathbf{F}^n$, for any n, using z-transform techniques.

(d) Determine the system function using the formula

$$H(z) = \mathbf{g}'(z\mathbf{I} - \mathbf{F})^{-1}\mathbf{q} + d$$

Compare the answer with that in part (a).

(e) Compute the characteristic polynomial det $(z\mathbf{I} - \mathbf{F})$ and check if the system is stable.

7.36 Determine the impulse response of the system

$$\mathbf{F} = \begin{bmatrix} 0 & 0.11 \\ 1 & 1 \end{bmatrix} \qquad \mathbf{q} = \begin{bmatrix} 0.11 \\ 1 \end{bmatrix} \qquad \mathbf{g} = \begin{bmatrix} 0 \\ 1 \end{bmatrix} \qquad d = 1$$

using the z-transform approach.

7.37 A discrete-time system is described by the following state-space model:

$$\mathbf{v}(n + 1) = \mathbf{Fv}(n) + \mathbf{q}x(n)$$
$$y(n) = \mathbf{g}'\mathbf{v}(n) + dx(n)$$

where

$$\mathbf{F} = \begin{bmatrix} 0 & 1 \\ -\frac{5}{16} & -1 \end{bmatrix} \qquad \mathbf{q} = \begin{bmatrix} 0 \\ 1 \end{bmatrix} \qquad \mathbf{g} = \begin{bmatrix} \frac{11}{8} \\ 2 \end{bmatrix} \qquad d = 2$$

(a) Sketch the corresponding state-space structure.
(b) Calculate the impulse response for $n = 0, 1, \ldots, 5$ and for $n = 17$ by using the state-space approach
(c) Find the difference equation description of the system.
(d) Repeat part (b) by using the difference equation.
(e) Sketch the direct form II implementation of the system.

7.38 Determine the state-space parameters $\mathbf{F}$, $\mathbf{q}$, $\mathbf{g}$, and d for:
(a) the all-zero lattice structure
(b) the all-pole lattice structure

COMPUTER EXPERIMENTS

7.39 Consider the system specified by the system function

$$H(z) = \frac{B(z)}{A(z)}$$

$$= \left[G_1 \frac{(1 - 0.8e^{j\pi/4})(1 - 0.8e^{-j\pi/4})}{(1 - \frac{1}{2}z^{-1})(1 + \frac{1}{3}z^{-1})} \right]\left[G_2 \frac{(1 + \frac{1}{4}z^{-1})(1 - \frac{1}{5}z^{-1})}{(1 - 0.8e^{j\pi/3})(1 - 0.8e^{-j\pi/3})} \right]$$

(a) Choose G_1 and G_2 so that the gain of each second-order section at $\omega = 0$ is equal to 1.
(b) Sketch the direct form 1, direct form 2, and cascade realizations of the system.
(c) Use the subroutines for the implementation of direct form 1, direct form 2 and cascade structures to compute the first 100 samples of the impulse response and the step response of the system.
(d) Plot the results in part (c) to illustrate the proper functioning of the subroutines.

7.40 Consider the system given in Problem 7.39 with $G_1 = G_2 = 1$.
(a) Determine a lattice realization for the system

$$H(z) = B(z)$$

(b) Determine a lattice realization for the system

$$H(z) = \frac{1}{A(z)}$$

(c) Determine a lattice-ladder realization for the system $H(z) = B(z)/A(z)$.

(d) Write a subroutine for the implementation of the lattice-ladder structure in part (c).

(e) Determine and sketch the first 100 samples of the impulse responses of the systems in parts (a) through (c) by working with the lattice structures.

(f) Compute and sketch the first 100 samples of the convolution of impulse responses in parts (a) and (b). What did you find? Explain your results.

7.41 Consider the system given in Problem 7.39.

(a) Determine the parallel-form structure and write a subroutine for its implementation.

(b) Sketch a parallel structure using second-order coupled-form state-space sections.

(c) Write a subroutine for the implementation of the structure in part (b).

(d) Verify the subroutines in parts (a) and (c) by computing and sketching the impulse response of the system.

7.42 **(a)** Write a FORTRAN subroutine for the implementation of a linear-phase FIR filter with M coefficients. M may be either even or odd.

(b) Test the subroutine by using it to compute the impulse response and the step response of the filters in Problem 7.1.

CHAPTER 8

Design of Digital Filters

With the background that we have developed in the three preceding chapters, we are now in a position to treat the subject of digital filter design that was considered briefly in Chapter 5. We shall describe several methods for designing FIR and IIR digital filters.

The techniques for designing frequency-selective filters treated in this chapter should be distinguished from the methods for designing Wiener filters, inverse filters for de-convolution, and equalizers described in Chapter 6. In the methods treated in Chapter 6, the design criterion was specified in terms of minimizing some performance measure in the time domain (e.g., least-squares error). The frequency-domain characteristics of the desired filter were not specified explicitly. In this chapter we reconsider the least-squares methods and evaluate how well they match a specified frequency response characteristic.

In the design of frequency-selective filters, the desired filter characteristics are speci-fied in the frequency domain in terms of the desired magnitude and phase response of the filter. In the filter design process, we determine the coefficients of a causal FIR or IIR filter that closely approximates the desired frequency response specifications. The issue of which type of filter to design, FIR or IIR, depends on the nature of the problem and on the specifications of the desired frequency response.

In practice, FIR filters are employed in filtering problems where there is a requirement for a linear phase characteristic within the passband of the filter. If there is no requirement for a linear-phase characteristic, either an IIR or an FIR filter may be employed. However, as a general rule, an IIR filter has lower sidelobes in the stopband than an FIR filter having the same number of parameters. For this reason if some phase distortion is either tolerable or unimportant, an IIR filter is preferable, primarily because its implementation involves fewer parameters, less memory requirements, and lower computational complexity.

In conjunction with our discussion of digital filter design, we describe frequency transformations in both the analog and digital domains for transforming a lowpass pro-totype filter into either another lowpass, or a bandpass, or a bandstop or a highpass filter.

We conclude this chapter with a treatment of sampling rate conversion. In particular, we discuss decimation (sampling rate reduction) by an integer factor D, interpolation (sampling rate increase) by a factor U, and sampling rate conversion by a rational factor U/D. The design of filters for sampling rate conversion is also treated.

8.1 Design of FIR Filters

In this section we describe several methods for designing FIR filters. Our treatment is focused on the important class of linear-phase FIR filters, described in Section 5.3.

An FIR filter of length M with input $x(n)$ and output $y(n)$ is described by the difference equation

$$y(n) = b_0 x(n) + b_1 x(n - 1) + \cdots + b_{M-1} x(n - M + 1)$$

$$= \sum_{k=0}^{M-1} b_k x(n - k) \tag{8.1.1}$$

where $\{b_k\}$ is the set of filter coefficients. Alternatively, we may express the output sequence as the convolution of the unit sample response $h(n)$ of the system with the input signal. Thus we have

$$y(n) = \sum_{k=0}^{M-1} h(k) x(n - k) \tag{8.1.2}$$

where the lower and upper limits on the convolution sum reflect the causality and finite-duration characteristics of the filter. Clearly, (8.1.1) and (8.1.2) are identical in form and hence it follows that $b_k = h(k)$, $k = 0, 1, \ldots, M - 1$.

The filter can also be characterized by its system function

$$H(z) = \sum_{k=0}^{M-1} h(k) z^{-k} \tag{8.1.3}$$

which we view as a polynomial of degree $M - 1$ in the variable z^{-1}. The roots of this polynomial constitute the zeros of the filter.

In Chapter 5 we demonstrated that an FIR filter has linear phase if its unit sample response satisfies the condition

$$h(n) = \pm h(M - 1 - n) \qquad n = 0, 1, \ldots, M - 1 \tag{8.1.4}$$

When the symmetry and antisymmetry conditions in (8.1.4) are incorporated into (8.1.3), we have

$$H(z) = h(0) + h(1)z^{-1} + h(2)z^{-2} + \cdots + h(M - 2)z^{-(M-2)} + h(M - 1)z^{-(M-1)}$$

$$= z^{-(M-1)/2} \left\{ h\left(\frac{M-1}{2}\right) + \sum_{n=0}^{(M-3)/2} h(n)[z^{(M-1-2k)/2} \pm z^{-(M-1-2k)/2}] \right\} \quad M \text{ odd}$$

$$= z^{-(M-1)/2} \sum_{n=0}^{(M/2)-1} h(n)[z^{(M-1-2k)/2} \pm z^{-(M-1-2k)/2}] \quad M \text{ even}$$

$$\tag{8.1.5}$$

Now, if we substitute z^{-1} for z in (8.1.3) and multiply both sides of the resulting equation by $z^{-(M-1)}$, we obtain

$$z^{-(M-1)} H(z^{-1}) = \pm H(z) \tag{8.1.6}$$

This result implies that the roots of the polynomial $H(z)$ are identical to the roots of the polynomial $H(z^{-1})$. Consequently, the roots of $H(z)$ must occur in reciprocal pairs. In other words, if z_1 is a root or a zero of $H(z)$, then $1/z_1$ is also a root. Furthermore, if the unit sample response $h(n)$ of the filter is real, complex-valued roots must occur in complex-conjugate pairs. Hence, if z_1 is a complex-valued root, z_1^* is also a root. As a consequence of (8.1.6), $H(z)$ also has zeros at $1/z_1^*$. Figure 8.1 illustrates the symmetry that exists in the location of the zeros of a linear-phase FIR filter.

The frequency response characteristics of linear-phase FIR filters are obtained by evaluating (8.1.5) on the unit circle. This substitution yields the expression for $H(\omega)$, given in Section 5.3.4, in terms of the coefficients $h(n)$. Consequently, the problem of

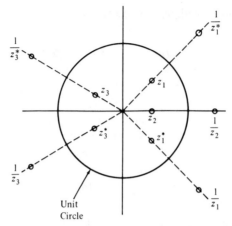

FIGURE 8.1 Symmetry of zero locations for a linear-phase FIR filter.

FIR filter design is simply to determine the M coefficients $h(n)$, $n = 0, 1, \ldots,$ $M - 1$, from a specification of the desired frequency response $H_d(\omega)$ of the FIR filter. The important parameters in the specification of $H_d(\omega)$ were given in Fig. 5.13.

In the following subsections we describe design methods based on specification of $H_d(\omega)$.

8.1.1 Design of Linear-Phase FIR Filters Using Windows

In this method we begin with the desired frequency response specification $H_d(\omega)$ and determine the corresponding unit sample response $h_d(n)$. Indeed, $h_d(n)$ is related to $H_d(\omega)$ by the Fourier transform relation

$$H_d(\omega) = \sum_{n=0}^{\infty} h_d(n)e^{-j\omega n} \tag{8.1.7}$$

where

$$h_d(n) = \frac{1}{2\pi} \int_{-\pi}^{\pi} H_d(\omega)e^{j\omega n} \, d\omega \tag{8.1.8}$$

Thus, given $H_d(\omega)$, we can determine the unit sample response $h_d(n)$ by evaluating the integral in (8.1.8).

In general, the unit sample response $h_d(n)$ obtained from (8.1.8) is infinite in duration and must be truncated at some point, say at $n = M - 1$, to yield an FIR filter of length M. Truncation of $h_d(n)$ to a length $M - 1$ is equivalent to multiplying $h_d(n)$ by a "rectangular window," defined as

$$w(n) = \begin{cases} 1 & n = 0, 1, \ldots, M - 1 \\ 0 & \text{otherwise} \end{cases} \tag{8.1.9}$$

Thus the unit sample response of the FIR filter becomes

$$\begin{aligned} h(n) &= h_d(n)w(n) \\ &= \begin{cases} h_d(n) & n = 0, 1, \ldots, M - 1 \\ 0 & \text{otherwise} \end{cases} \end{aligned} \tag{8.1.10}$$

It is instructive to view the effect of the window function on the desired frequency response $H_d(\omega)$. Recall that multiplication of the window function $w(n)$ with $h_d(n)$ is equivalent to convolution of $H_d(\omega)$ with $W(\omega)$, where $W(\omega)$ is the frequency-domain representation (Fourier transform) of the window function, that is,

$$W(\omega) = \sum_{n=0}^{M-1} w(n)e^{-j\omega n} \qquad (8.1.11)$$

Thus the convolution of $H_d(\omega)$ with $W(\omega)$ yields the frequency response of the (truncated) FIR filter. That is,

$$H(\omega) = \frac{1}{2\pi} \int_{-\pi}^{\pi} H_d(v)W(\omega - v) \, dv \qquad (8.1.12)$$

The Fourier transform of the rectangular window is

$$\begin{aligned} W(\omega) &= \sum_{n=0}^{M-1} e^{-j\omega n} \\ &= \frac{1 - e^{-j\omega M}}{1 - e^{-j\omega}} = e^{-j\omega(M-1)/2} \frac{\sin(\omega M/2)}{\sin(\omega/2)} \end{aligned} \qquad (8.1.13)$$

This window function has a magnitude response

$$|W(\omega)| = \frac{|\sin(\omega M/2)|}{|\sin(\omega/2)|} \qquad \pi \le \omega \le \pi \qquad (8.1.14)$$

and a piecewise linear phase

$$\Theta(\omega) = \begin{cases} -\omega\left(\dfrac{M-1}{2}\right) & \text{when } \sin(\omega M/2) \ge 0 \\[2mm] -\omega\left(\dfrac{M-1}{2}\right) + \pi & \text{when } \sin(\omega M/2) < 0 \end{cases} \qquad (8.1.15)$$

The magnitude of the window function is illustrated in Fig. 8.2 for $M = 31, 61$, and 101. The width of the main lobe [width is measured to the first zero of $W(\omega)$] is $4\pi/M$. Hence as M increases the main lobe becomes narrower. However, the sidelobes of $|W(\omega)|$ are relatively high and remain unaffected by an increase in M. In fact, even though the width of each sidelobe decreases with an increase in M, the height of each sidelobe increases with an increase in M in such a manner that the area under each sidelobe remains invariant to changes in M. This characteristic behavior is not evident from observation of Fig. 8.2, because $W(\omega)$ has been normalized by M, such that the normalized peak values of the sidelobes remain invariant to an increase in M.

The characteristics of the rectangular window play a significant role in the resulting frequency response of the FIR filter obtained by truncating $h_d(n)$ to length M. Specifically, the convolution of $H_d(\omega)$ with $W(\omega)$ has the effect of smoothing $H_d(\omega)$. As M is increased, $W(\omega)$ becomes narrower, and the smoothing provided by $W(\omega)$ is reduced. On the other hand, the large sidelobes of $W(\omega)$ result in some undesirable ringing effects in the FIR filter frequency response $H(\omega)$ and also in relatively larger sidelobes in $H(\omega)$. These undesirable effects are best alleviated by the use of windows that do not contain abrupt discontinuities in their time-domain characteristics, and, correspondingly, low sidelobes in their frequency-domain characteristics.

Table 8.1 lists several window functions that possess desirable frequency response characteristics. Figure 8.3 illustrates the time-domain characteristics of the windows. The frequency response characteristics are illustrated in Figs. 8.4 through 8.10. All of these window functions have significantly lower sidelobes compared with the rectangular window. However, for the same value of M, the width of the main lobe is also wider for

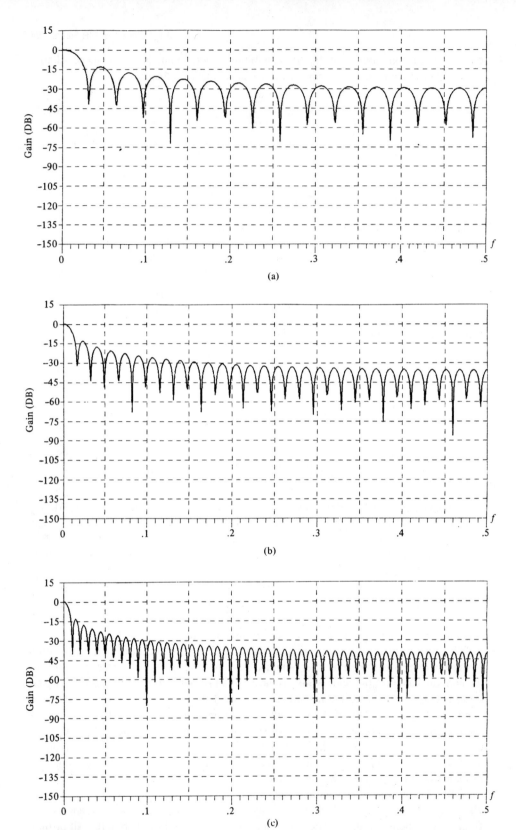

FIGURE 8.2 Frequency response for rectangular window of lengths (a) $M = 31$, (b) $M = 61$, and (c) $M = 101$.

TABLE 8.1 Window Functions for FIR Filter Design

Name of Window	Time-Domain Sequence $h(n), 0 \le n \le M - 1$
Bartlett (triangular)	$1 - \dfrac{2\left(n - \dfrac{M-1}{2}\right)}{M-1}$
Blackman	$0.42 - 0.5 \cos \dfrac{2\pi n}{M-1} + 0.08 \cos \dfrac{4\pi n}{M-1}$
Hamming	$0.54 - 0.46 \cos \dfrac{2\pi n}{M-1}$
Hanning	$\dfrac{1}{2}\left(1 - \cos \dfrac{2\pi n}{M-1}\right)$
Kaiser	$\dfrac{I_0\left[\alpha \sqrt{\left(\dfrac{M-1}{2}\right)^2 - \left(n - \dfrac{M-1}{2}\right)^2}\right]}{I_0\left[\alpha\left(\dfrac{M-1}{2}\right)\right]}$
Lanczos	$\left\{\dfrac{\sin\left[2\pi\left(n - \dfrac{M-1}{2}\right)\Big/(M-1)\right]}{2\pi\left(n - \dfrac{M-1}{2}\right)\Big/\left(\dfrac{M-1}{2}\right)}\right\}^L \quad L > 0$
Tukey	$1, \quad \left\|n - \dfrac{M-1}{2}\right\| \le \alpha \dfrac{M-1}{2} \quad 0 < \alpha < 1$ $\dfrac{1}{2}\left[1 + \cos\left(\dfrac{n - (1+a)(M-1)/2}{(1-\alpha)(M-1)/2}\pi\right)\right]$ $\alpha(M-1)/2 \le \left\|n - \dfrac{M-1}{2}\right\| \le \dfrac{M-1}{2}$

these windows compared to the rectangular window. Consequently, these window functions provide more smoothing through the convolution operation in the frequency domain and, as a result, the transition region in the FIR filter response is wider. To reduce the width of this transition region, we may simply increase the length of the window, which results in a larger filter. Table 8.2 summarizes these important frequency-domain features of the various window functions.

The window technique is best described in terms of a specific example. Suppose that we desire to design a lowpass linear-phase FIR filter having a desired frequency response

$$H_d(\omega) = \begin{cases} 1e^{-j\omega(M-1)/2} & 0 \le |\omega| \le \omega_c \\ 0 & \text{otherwise} \end{cases} \tag{8.1.16}$$

A delay of $(M - 1)/2$ units is incorporated into $H_d(\omega)$ in anticipation of forcing the filter to be of length M. The corresponding unit sample response, obtained by evaluating

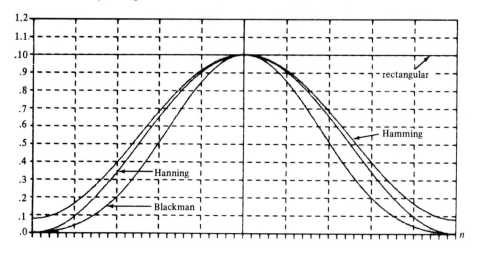

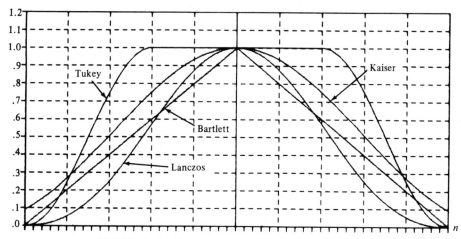

FIGURE 8.3   Shapes of several window functions.

the integral in (8.1.8), is

$$h_d(n) = \frac{1}{2\pi} \int_{-\omega_c}^{\omega_c} e^{j\omega\left(n - \frac{M-1}{2}\right)} d\omega$$

$$= \frac{\sin \omega_c \left(n - \dfrac{M-1}{2}\right)}{\pi \left(n - \dfrac{M-1}{2}\right)} \qquad n \neq \frac{M-1}{2} \qquad (8.1.17)$$

Clearly, $h_d(n)$ is noncausal and infinite in duration.

If we multiply $h_d(n)$ by the rectangular window sequence in (8.1.9), we obtain an FIR filter of length M having the unit sample response

$$h(n) = \frac{\sin \omega_c \left(n - \dfrac{M-1}{2}\right)}{\pi \left(n - \dfrac{M-1}{2}\right)} \qquad \begin{array}{l} 0 \leq n \leq M-1 \\[2mm] n \neq \dfrac{M-1}{2} \end{array} \qquad (8.1.18)$$

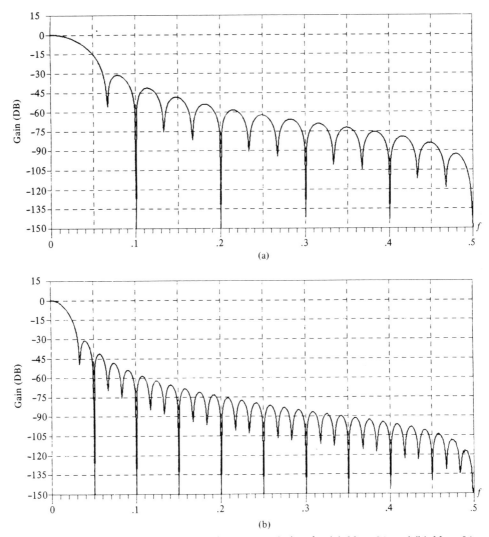

FIGURE 8.4 Frequency responses of Hanning window for (a) $M = 31$ and (b) $M = 61$.

If M is selected to be odd, the value of $h(n)$ at $n = (M - 1)/2$ is

$$h\left(\frac{M - 1}{2}\right) = \frac{\omega_c}{\pi} \tag{8.1.19}$$

The magnitude of the frequency response $H(\omega)$ of this filter is illustrated in Fig. 8.11 for $M = 61$ and $M = 101$. We observe that relatively large oscillations or ripples occur near the band edge of the filter. The oscillations increase in frequency as M increases, but they do not diminish in amplitude. These large oscillations are the direct result of the large sidelobes that exist in the frequency characteristic $W(\omega)$ of the rectangular window. As this window function is convolved with the desired frequency response characteristic $H_d(\omega)$, the oscillations occur as the large constant area sidelobes of $W(\omega)$ move across the discontinuity that exists in $H_d(\omega)$, as illustrated graphically in Fig. 8.12.

Since (8.1.7) is basically a Fourier series representation of $H_d(\omega)$, the multiplication of $h_d(n)$ with a rectangular window is identical to truncating the Fourier series representation of the desired filter characteristic $H_d(\omega)$. The truncation of the Fourier series is

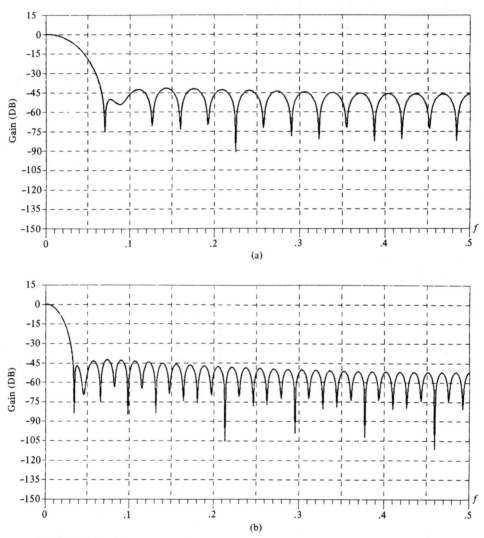

FIGURE 8.5 Frequency responses for Hamming window for (a) $M = 31$ and (b) $M = 61$.

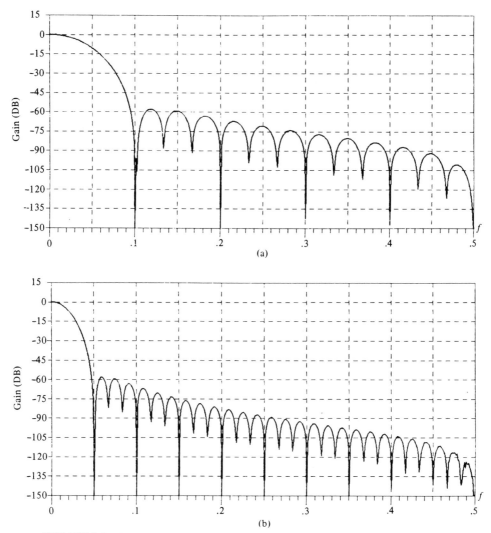

FIGURE 8.6 Frequency responses for Blackman window for (a) $M = 31$ and (b) $M = 61$.

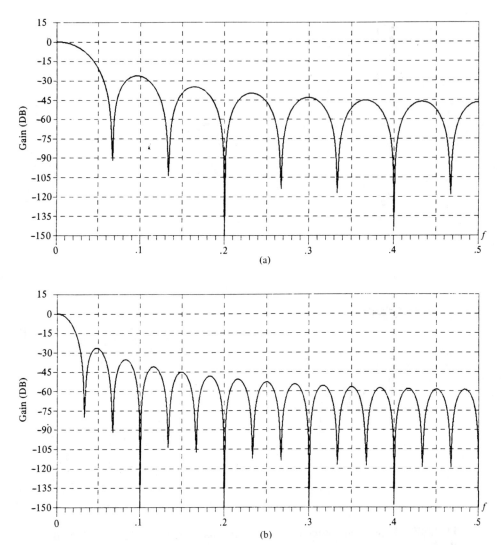

FIGURE 8.7 Frequency responses of Bartlett (triangular) window for (a) $M = 31$ and (b) $M = 61$.

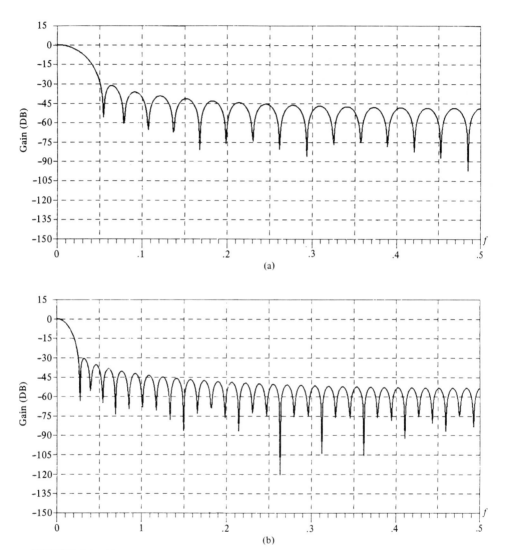

FIGURE 8.8 Frequency responses of Kaiser window for (a) $M = 31$ and (b) $M = 61$ with $\alpha = 4$.

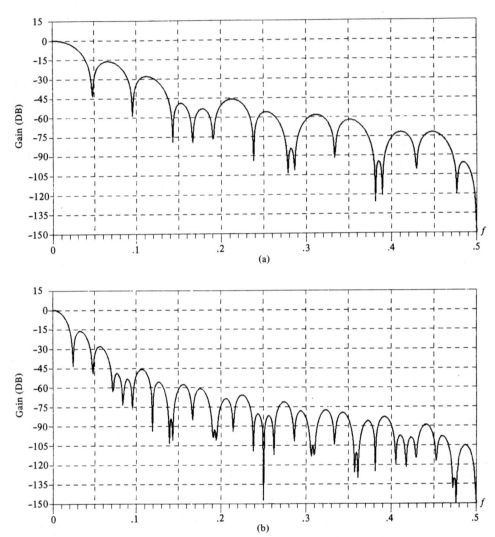

FIGURE 8.9 Frequency responses of Tukey window for (a) *M* = 31 and (b) *M* = 61 with α = 0.4.

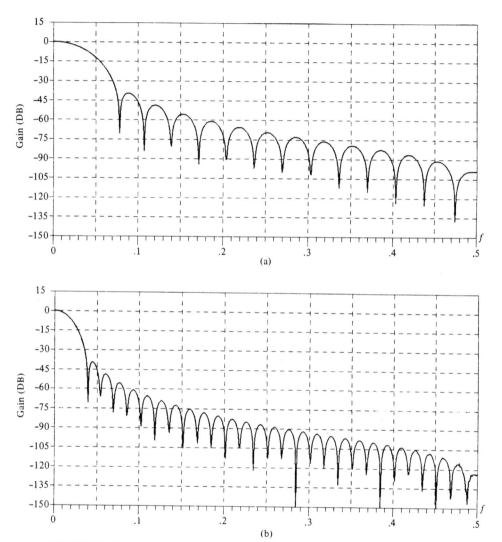

FIGURE 8.10 Frequency responses of Lanczos window for (a) $M = 31$ and (b) $M = 61$, with $L = 2$.

TABLE 8.2 Important Frequency-Domain Characteristics
of Some Window Functions

Type of Window	Approximate Transition Width of Main Lobe	Peak Sidelobe (dB)
Rectangular	$4\pi/M$	-13
Bartlett	$8\pi/M$	-27
Hanning	$8\pi/M$	-32
Hamming	$8\pi/M$	-43
Blackman	$12\pi/M$	-58

known to introduce ripples in the frequency response characteristic $H(\omega)$ due to the nonuniform convergence of the Fourier series at a discontinuity. These ripples or oscillatory behavior near the band edge of the filter is called the *Gibbs phenomenon*.

To alleviate the presence of large oscillations in both the passband and the stopband,

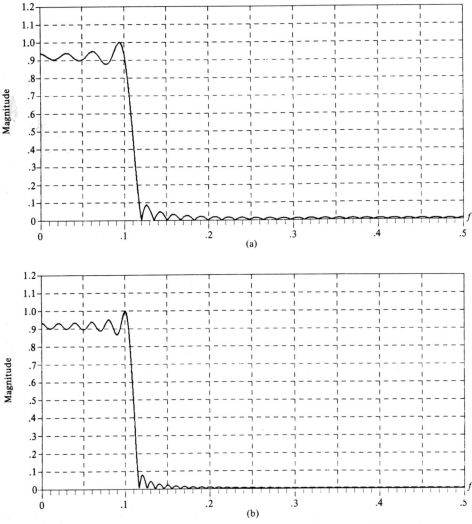

FIGURE 8.11 Lowpass filter designed with a rectangular window (a) $M = 61$ and (b) $M = 101$.

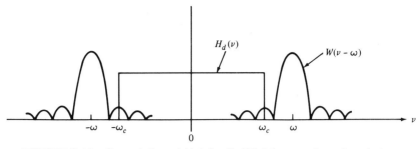

FIGURE 8.12 Convolution of $H_d(\omega)$ with $W(\omega)$ for a rectangular window.

we should use a window function that contains a taper and decays toward zero gradually instead of abruptly, as it occurs in a rectangular window. Figures 8.13 through 8.18, illustrate the frequency response of the resulting filter when some of the window functions listed in Table 8.1 are used to taper $h_d(n)$. As illustrated in these figures, the window functions do indeed eliminate the ringing effects (Gibb's phenomenon) at the band edge and result in lower sidelobes at the expense of an increase in the width of the transition band of the filter.

8.1.2 Design of Linear-Phase FIR Filters by the Frequency-Sampling Method

The frequency sampling method for FIR filter design is basically the technique described in Section 5.3.4. Recall that we specify the desired frequency response at a set of equally spaced frequencies, namely

$$\omega_k = \frac{2\pi}{M}(k + \alpha) \qquad k = 0, 1, \ldots, \frac{M-1}{2}, \quad M \text{ odd}$$

$$k = 0, 1, \ldots, \frac{M}{2} - 1, \quad M \text{ even} \qquad (8.1.20)$$

$$\alpha = 0 \text{ or } \tfrac{1}{2}$$

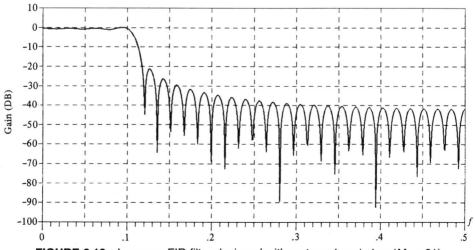

FIGURE 8.13 Lowpass FIR filter designed with rectangular window ($M = 61$).

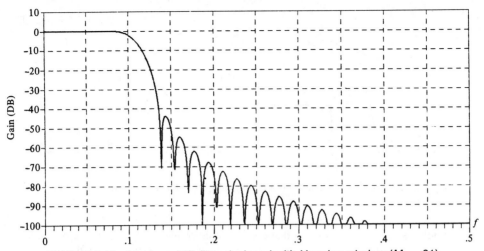

FIGURE 8.14 Lowpass FIR filter designed with Hanning window ($M = 61$).

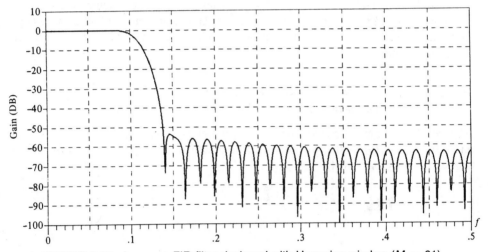

FIGURE 8.15 Lowpass FIR filter designed with Hamming window ($M = 61$).

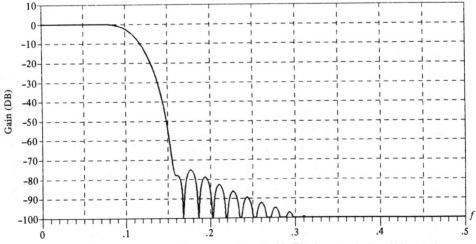

FIGURE 8.16 Lowpass FIR filter designed with Blackman window ($M = 61$).

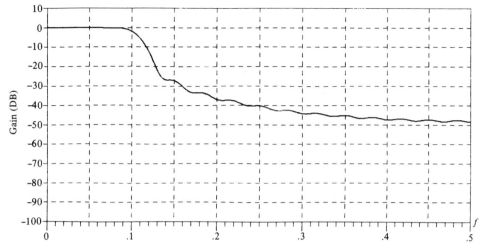

FIGURE 8.17   Lowpass FIR filter designed with Bartlett (triangular) window ($M = 61$).

and solve for the unit sample response $h(n)$ of the FIR filter from these equally spaced frequency specifications. We also recall that it is desirable to optimize the frequency specification in the transition band of the filter. This optimization can be accomplished numerically on a digital computer by means of linear programming techniques.

In this section we consider the frequency sampling method in greater depth and generality. We begin with the frequency response of the FIR filter, which is

$$H(\omega) = \sum_{n=0}^{M-1} h(n)e^{-j\omega n} \tag{8.1.21}$$

Suppose that we specify the frequency response of the filter at the frequencies given by (8.1.20). Then from (8.1.21) we obtain

$$H(k + \alpha) \equiv H\left(\frac{2\pi}{M}(k + \alpha)\right) \tag{8.1.22}$$

$$H(k + \alpha) \equiv \sum_{n=0}^{M-1} h(n)e^{-j2\pi(k+\alpha)n/M} \qquad k = 0, 1, \ldots, M - 1$$

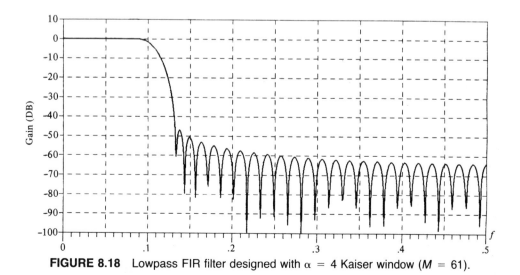

FIGURE 8.18   Lowpass FIR filter designed with $\alpha = 4$ Kaiser window ($M = 61$).

As we shall see below, $H(k + \alpha)$ satisfies a symmetry property that reduces the frequency specifications to $(M + 1)/2$ points if M is odd and to $M/2$ if M is even, as given in (8.1.20). At this point, however, it is desirable to consider the frequency specifications at M points, as indicated in (8.1.22).

It is a simple matter to invert (8.1.22) and express $h(n)$ in terms of $H(k + \alpha)$. If we multiply both sides of (8.1.22) by the exponential $\exp(j2\pi km/M)$, $m = 0, 1, \ldots, M - 1$ and sum over $k = 0, 1, \ldots, M - 1$, the right-hand side of (8.1.22) reduces to $Mh(m) \exp(-j2\pi\alpha m/M)$. Thus we obtain

$$Mh(n)e^{-j2\pi\alpha n/M} = \sum_{k=0}^{M-1} H(k + \alpha)e^{j2\pi kn/M} \tag{8.1.23}$$

or, equivalently,

$$h(n) = \frac{1}{M} \sum_{k=0}^{M-1} H(k + \alpha)e^{j2\pi(k+\alpha)n/M} \qquad n = 0, 1, \ldots, M - 1 \tag{8.1.24}$$

The relationship in (8.1.24) allows us to compute the values of the unit sample response $h(n)$ from the specification of the frequency samples $H(k + \alpha)$, $k = 0, 1, \ldots, M - 1$.

Let us explore the ramifications of the symmetry properties of the sequence $h(n)$. Suppose that we set $\alpha = 0$. Then, since $h(n)$ is real, it is easily established from (8.1.22) that $H(k) = H^*(M - k)$. Furthermore, if $h(n)$ is symmetric [i.e., $h(n) = h(M - 1 - n)$], then it also follows from (8.1.22) that

$$H(k) = \left\{ h\left(\frac{M-1}{2}\right) + 2 \sum_{n=0}^{(M-3)/2} h(n) \cos \frac{2\pi k}{M}\left(\frac{M-1}{2} - n\right) \right\} e^{-j\pi k(M-1)/M}$$

$$M \text{ odd} \tag{8.1.25}$$

$$= \left\{ 2 \sum_{n=0}^{(M/2)-1} h(n) \cos \frac{2\pi k}{M}\left(\frac{M-1}{2} - n\right) \right\} e^{-j\pi k(M-1)/M} \qquad M \text{ even}$$

$$\tag{8.1.26}$$

We observe that the terms in brackets in (8.1.25) and (8.1.26) are simply the results given previously in (5.3.30) and (5.3.31), for M odd and M even, respectively. These terms are simply $H_r(\omega)$ evaluated at the frequencies $\omega_k = 2\pi k/M$. Consequently, we may express $H(k)$ as

$$H(k) = H_r\left(\frac{2\pi k}{M}\right) e^{-j2\pi k(M-1)/2M} \tag{8.1.27}$$

It is more convenient, however, to define the samples $H(k)$ of the frequency response for the FIR filter as

$$H(k) = G(k)e^{j\pi k/M} \tag{8.1.28}$$

where it is easily verified from (8.1.27) that

$$G(k) = (-1)^k H_r\left(\frac{2\pi k}{M}\right) \tag{8.1.29}$$

Since $H_r(\omega)$ is purely real, so is the sequence of frequency samples $G(k)$. Moreover, the condition $H(k) = H^*(M - k)$ leads to the result that

$$G(k) = -G(M - k) \tag{8.1.30}$$

as it is easily verified from (8.1.28). When M is even, this condition requires that

$$G(M/2) = 0$$

In other words, the frequency sample at $\omega = \pi$ must be zero when M is even.

In view of the symmetry in (8.1.30) for the real-valued frequency samples $G(k)$, it is a simple matter to determine an expression for the unit sample response $h(n)$ of the FIR filter in terms of the samples $G(k)$. By beginning with (8.1.24) when $\alpha = 0$, we have

$$
\begin{aligned}
h(n) &= \frac{1}{M} \sum_{k=0}^{M-1} H(k)e^{j2\pi kn/M} \qquad n = 0, 1, \ldots, M-1 \\
&= \frac{1}{M} \sum_{k=0}^{M-1} G(k)e^{j2\pi k \left(n + \frac{1}{2}\right)/M} \\
&= \frac{1}{M} \left\{ G(0) + 2 \sum_{k=1}^{(M-1)/2} G(k) \cos \frac{\pi}{M} k(2n+1) \right\} \qquad M \text{ odd}
\end{aligned}
\tag{8.1.31}
$$

and

$$
h(n) = \frac{1}{M} \left\{ G(0) + 2 \sum_{k=1}^{(M/2)-1} G(k) \cos \frac{\pi k}{M} (2n+1) \right\} \qquad M \text{ even} \tag{8.1.32}
$$

Thus we can compute $h(n)$ directly from the specified frequency samples $G(k)$ or, equivalently, $H_r(2\pi k/M)$. This approach eliminates the need for a matrix inversion, as done in Section 5.3.4, for determining the unit sample response of the FIR filter from its frequency-domain specifications.

Next, we consider the case where $\alpha = 1/2$. Since $h(n)$ is real, it is easily established from (8.1.22) that $H(k + \alpha) = H^*(M - 1 - k + \alpha)$ or, equivalently,

$$H(k + \tfrac{1}{2}) = H^*(M - k - \tfrac{1}{2})$$

When we incorporate the symmetry condition $h(n) = h(M - 1 - n)$ into (8.1.22) we obtain

$$H(k + \alpha) = H_r(\omega_k)e^{-j\pi(k+\alpha)(M-1)/M} \tag{8.1.33}$$

where $\omega_k = 2\pi(k + \alpha)/M$, $k = 0, 1, \ldots, M - 1$, and $H_r(\omega_k)$ is given by (5.3.30) for M odd and (5.3.31) for M even, respectively. Note that (8.1.33) has the form of (8.1.27) with $(k + \alpha)$ substituted for k.

Again, it is convenient to express $H(k + \alpha)$ in the form

$$H(k + \alpha) = G(k + \alpha)e^{-j\pi/2}e^{j\pi(k+\frac{1}{2})/M} \tag{8.1.34}$$

where

$$G(k + \alpha) = (-1)^k H_r(\omega_k) \qquad \omega_k = 2\pi(k + \alpha)/M \qquad k = 0, 1, 2, \ldots, M - 1 \tag{8.1.35}$$

Furthermore, it is easily verified from (8.1.34) that the condition

$$H(k + \tfrac{1}{2}) = H^*(M - k - \tfrac{1}{2})$$

implies that

$$G(k + \tfrac{1}{2}) = G(M - k - \tfrac{1}{2})$$

With the aid of this symmetry condition, the unit sample response $h(n)$ may be expressed in terms of $G(k + \frac{1}{2})$. From (8.1.24) we have

$$h(n) = \frac{1}{M} \sum_{k=0}^{M-1} H(k + \frac{1}{2}) e^{j2\pi n(k+1/2)/M}$$

$$= \frac{1}{M} \sum_{k=0}^{M-1} [-jG(k + \frac{1}{2}) e^{j\pi(k+1/2)/M}] e^{j2\pi n(k+1/2)/M}$$

$$= \frac{1}{M} \sum_{k=0}^{M-1} -jG(k + \frac{1}{2}) e^{j2\pi(k+1/2)(n+1/2)/M}$$

(8.1.36)

$$h(n) = \frac{2}{M} \sum_{k=0}^{(M-1)/2} G(k + \frac{1}{2}) \sin \frac{2\pi}{M} (k + \frac{1}{2})(n + \frac{1}{2}) \qquad M \text{ odd}$$

and

$$h(n) = \frac{2}{M} \sum_{k=0}^{(M/2)-1} G(k + \frac{1}{2}) \sin \frac{2\pi}{M} (k + \frac{1}{2})(n + \frac{1}{2}) \qquad M \text{ even} \qquad (8.1.37)$$

We have now considered the two cases, $\alpha = 0$ and $\alpha = \frac{1}{2}$ when the unit sample response $h(n)$ is symmetric. Let us now consider the antisymmetric unit sample response [i.e., the case where $h(n) = -h(M - 1 - n)$]. For $\alpha = 0$ or $\alpha = \frac{1}{2}$, we have

$$H(k + \alpha) = H_r[2\pi(k + \alpha)/M] e^{j\pi/2} e^{-j\pi(k+\alpha)(M-1)/M} \qquad (8.1.38)$$

where $H_r(\omega)$ is given by (5.3.35) for M odd and (5.3.36) for M even.

Let us define $G(k + \alpha)$ as

$$G(k + \alpha) = (-1)^k H_r[2\pi(k + \alpha)/M] \qquad (8.1.39)$$

Then

$$H(k + \alpha) = G(k + \alpha) e^{j\pi[(1/2)-\alpha]} e^{j\pi(k+\alpha)/M} \qquad (8.1.40)$$

The symmetry conditions $H(k) = H^*(M - k)$ for $\alpha = 0$ and $H(k + \frac{1}{2}) = H^*(M - k - \frac{1}{2})$ for $\alpha = \frac{1}{2}$ still hold for the antisymmetric $h(n)$, since these conditions are based on the assumption that $h(n)$ is real. Consequently, $H(k) = H^*(M - k)$ implies that $G(k) = G(M - k)$ and $H(k + \frac{1}{2}) = H^*(M - k - \frac{1}{2})$ implies that $G(k + \frac{1}{2}) = -G(M - k - \frac{1}{2})$ as can be easily verified from (8.1.40). In the latter case, for M odd and $k = (M - 1)/2$, we have $G(M/2) = -G(M/2)$, which implies that $G(M/2)$ must equal zero.

The expressions for the unit sample response $h(n)$ as a function of the frequency specifications are easily obtained by substituting (8.1.40) into (8.1.24) and making use of the symmetry conditions for the $G(k)$. Thus for $\alpha = 0$ we obtain

$$h(n) = \frac{1}{M} \sum_{k=0}^{M-1} G(k) e^{j\pi/2} e^{j2\pi k(n+1/2)/M}$$

(8.1.41)

$$= -\frac{2}{M} \sum_{k=1}^{(M-1)/2} G(k) \sin \frac{2\pi}{M} k(n + \frac{1}{2}) \qquad M \text{ odd}$$

and

$$h(n) = \frac{1}{M} \left\{ (-1)^{n+1} G(M/2) - 2 \sum_{k=1}^{(M/2)-1} G(k) \sin \frac{2\pi}{M} k(n + \frac{1}{2}) \right\} \qquad M \text{ even}$$

(8.1.42)

Recall that $G(0) = 0$ for an antisymmetric FIR filter. Consequently, this term does not appear in (8.1.41) and (8.1.42).

When $\alpha = \frac{1}{2}$, the expressions for the antisymmetric unit sample response $h(n)$ in terms of the frequency samples $G(k + \frac{1}{2})$ are

$$h(n) = \frac{1}{M} \sum_{k=0}^{M-2} G(k + \tfrac{1}{2}) e^{j2\pi(k + 1/2)(n + 1/2)/M}$$

(8.1.43)

$$= \frac{2}{M} \sum_{k=0}^{(M-3)/2} G(k + \tfrac{1}{2}) \cos \frac{2\pi}{M} (k + \tfrac{1}{2})(n + \tfrac{1}{2}) \qquad M \text{ odd}$$

and

$$h(n) = \frac{2}{M} \sum_{k=0}^{(M/2)-1} G(k + \tfrac{1}{2}) \cos \frac{2\pi}{M} (k + \tfrac{1}{2})(n + \tfrac{1}{2}) \qquad M \text{ even} \qquad (8.1.44)$$

Table 8.3 summarizes all the important relationships between $h(n)$ and the frequency samples that we have derived in this section.

EXAMPLE 8.1.1

Determine the coefficients of a linear-phase FIR filter of length $M = 15$ which has a symmetric unit sample response and a frequency response that satisfies the conditions

$$H_r \left(\frac{2\pi k}{15} \right) = \begin{cases} 1 & k = 0, 1, 2, 3 \\ 0.4 & k = 4 \\ 0 & k = 5, 6, 7 \end{cases}$$

Solution: Since $h(n)$ is symmetric and the frequencies are selected to correspond to the case $\alpha = 0$, we use the formula in (8.1.31) to evaluate $h(n)$. In this case

$$G(k) = (-1)^k H_r \left(\frac{2\pi k}{15} \right) \qquad k = 0, 1, \ldots, 7$$

The result of this computation is

$$\begin{aligned}
h(0) &= h(14) = -0.014112893 \\
h(1) &= h(13) = -0.001945309 \\
h(2) &= h(12) = 0.04000004 \\
h(3) &= h(11) = 0.01223454 \\
h(4) &= h(10) = -0.09138802 \\
h(5) &= h(9) = -0.01808986 \\
h(6) &= h(8) = 0.3133176 \\
h(7) &= = 0.52
\end{aligned}$$

The frequency response characteristic of this filter is shown in Fig. 8.19. We should emphase that $H_r(\omega)$ is exactly equal to the values given by the specifications above at $\omega_k = 2\pi k/15$.

EXAMPLE 8.1.2

Determine the coefficients of a linear-phase FIR filter of length $M = 32$ which has a symmetric unit sample response and a frequency response that satisfies the condition

$$H_r \left(\frac{2\pi(k + \alpha)}{32} \right) = \begin{cases} 1 & k = 0, 1, 2, 3, 4, 5 \\ T_1 & k = 6 \\ 0 & k = 7, 8, \ldots, 15 \end{cases}$$

TABLE 8.3 Unit Sample Response: $h(n) = \pm h(M - 1 - n)$

Symmetric

$\alpha = 0$

$$H(k) = G(k)e^{j\pi k/M}, \ k = 0, 1, \ldots, M - 1$$

$$G(k) = (-1)^k H_r\left(\frac{2\pi k}{M}\right), \ G(k) = -G(M - k)$$

$$h(n) = \frac{1}{M}\left\{G(0) + 2\sum_{k=1}^{U} G(k) \cos\frac{2\pi k}{M}\left(n + \tfrac{1}{2}\right)\right\}$$

$$U = \begin{cases} \dfrac{M - 1}{2} & M \text{ odd} \\[2mm] \dfrac{M}{2} - 1 & M \text{ even} \end{cases}$$

$\alpha = \tfrac{1}{2}$

$$H(k + \tfrac{1}{2}) = G(k + \tfrac{1}{2})e^{-j\pi/2}e^{j\pi(2k+1)/2M}$$

$$G(k + \tfrac{1}{2}) = (-1)^k H_r\left[\frac{2\pi}{M}(k + \tfrac{1}{2})\right]$$

$$G(k + \tfrac{1}{2}) = G(M - k - \tfrac{1}{2})$$

$$h(n) = \frac{2}{M}\sum_{k=0}^{U} G(k + \tfrac{1}{2}) \sin\frac{2\pi}{M}(k + \tfrac{1}{2})(n + \tfrac{1}{2})$$

Antisymmetric

$\alpha = 0$

$$H(k) = G(k)e^{j\pi/2}e^{j\pi k/M}, \ k = 0, 1, \ldots, M - 1$$

$$G(k) = (-1)^k H_r\left(\frac{2\pi k}{M}\right), \ G(k) = G(M - k)$$

$$h(n) = -\frac{2}{M}\sum_{k=1}^{(M-1)/2} G(k) \sin\frac{2\pi k}{M}(n + \tfrac{1}{2}), \ M \text{ odd}$$

$$h(n) = \frac{1}{M}\left\{(-1)^{n+1}G(M/2) - 2\sum_{k=1}^{(M/2)-1} G(k) \sin\frac{2\pi}{M}k\,(n + \tfrac{1}{2})\right\}, \ M \text{ even}$$

$\alpha = \tfrac{1}{2}$

$$H(k + \tfrac{1}{2}) = G(k + \tfrac{1}{2})e^{j\pi(2k+1)/2M}$$

$$G(k + \tfrac{1}{2}) = (-1)^k H_r\left[\frac{2\pi}{M}(k + \tfrac{1}{2})\right]$$

$$G(k + \tfrac{1}{2}) = -G(M - k - \tfrac{1}{2}); \ G(M/2) = 0 \text{ for } M \text{ odd}$$

$$h(n) = \frac{2}{M}\sum_{k=0}^{V} G(k + \tfrac{1}{2}) \cos\frac{2\pi}{M}(k + \tfrac{1}{2})(n + \tfrac{1}{2})$$

$$V = \begin{cases} \dfrac{M - 3}{2} & M \text{ odd} \\[2mm] \dfrac{M}{2} - 1 & M \text{ even} \end{cases}$$

where $T_1 = 0.3789795$ for $\alpha = 0$ and $T_1 = 0.3570496$ for $\alpha = \tfrac{1}{2}$. These values of T_1 were obtained from the tables of optimum transition parameters given in the paper by Rabiner et al. [1970], which are reproduced in Appendix 5A.

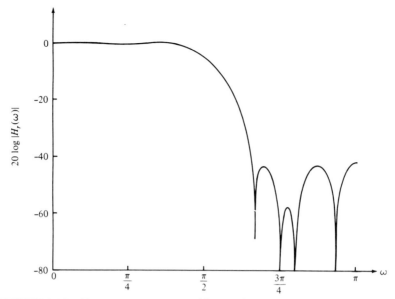

FIGURE 8.19 Frequency response of linear-phase FIR filter in Example 8.1.1.

Solution: The appropriate equations for this computation are (8.1.32) for $\alpha = 0$ and (8.1.37) for $\alpha = \frac{1}{2}$. These computations yield the unit sample response shown in Table 8.4 for $\alpha = 0$ and Table 8.5 for $\alpha = \frac{1}{2}$. The corresponding frequency response characteristics are illustrated in Figs. 8.20 and 8.21, respectively.

Although the frequency sampling method provides us with another means for designing FIR digital filters, its major advantage lies in an efficient alternative realization of the filter, called the frequency sampling realization, which was derived in Section 7.2.3. Recall that this alternative realization is obtained by substituting (8.1.24) for $h(n)$ in the expression for the system function $H(z)$ of the filter and rearranging the terms. Thus we

TABLE 8.4

$M = 32$
$\text{ALPHA} = 0.$
$\text{T1} = 0.3789795E+00$

h(0)	=	$-0.7141978E-02$
h(1)	=	$-0.3070801E-02$
h(2)	=	$0.5891327E-02$
h(3)	=	$0.1349923E-01$
h(4)	=	$0.8087033E-02$
h(5)	=	$-0.1107258E-01$
h(6)	=	$-0.2420687E-01$
h(7)	=	$-0.9446550E-02$
h(8)	=	$0.2544464E-01$
h(9)	=	$0.3985050E-01$
h(10)	=	$0.2753036E-02$
h(11)	=	$-0.5913959E-01$
h(12)	=	$-0.6841660E-01$
h(13)	=	$0.3175741E-01$
h(14)	=	$0.2080981E+00$
h(15)	=	$0.3471138E+00$

TABLE 8.5

M	=	32
ALPHA	=	0.5
T1	=	0.3570496E+00

h(0)	=	-0.4089120E-02
h(1)	=	-0.9973779E-02
H(2)	=	-0.7379891E-02
h(3)	=	0.5949799E-02
h(4)	=	0.1727056E-01
h(5)	=	0.7878412E-02
h(6)	=	-0.1798590E-01
h(7)	=	-0.2670584E-01
h(8)	=	0.3778549E-02
h(9)	=	0.4191022E-01
h(10)	=	0.2839344E-01
h(11)	=	-0.4163144E-01
h(12)	=	-0.8254962E-01
h(13)	=	0.2802212E-02
h(14)	=	0.2013655E+00
h(15)	=	0.3717532E+00

express the system function in the form

$$H(z) = \frac{1 - z^{-M}e^{j2\pi\alpha}}{M} \sum_{k=0}^{M-1} \frac{H(k + \alpha)}{1 - e^{j2\pi(k+\alpha)/M}z^{-1}} \quad (8.1.45)$$

and interpreted (8.1.45) as a cascade of a comb filter having the system function

$$H_1(z) = \frac{1}{M}[1 - z^{-M}e^{j2\pi\alpha}] \quad (8.1.46)$$

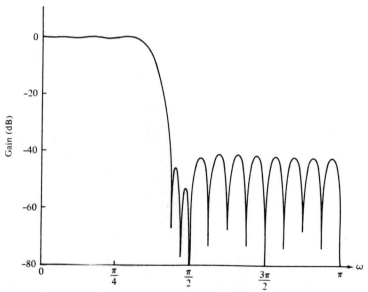

FIGURE 8.20 Frequency response of linear-phase FIR filter in Example 8.1.2 ($M = 32$ and $\alpha = 0$).

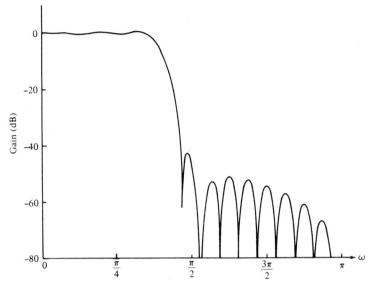

FIGURE 8.21 Frequency response of linear-phase FIR filter in Example 8.1.2 ($M = 32$ and $\alpha = \frac{1}{2}$).

and a parallel bank of single-pole filters having the system function

$$H_2(z) = \sum_{k=0}^{M-1} \frac{H(k + \alpha)}{1 - e^{j2\pi(k+\alpha)/M}z^{-1}} \qquad (8.1.47)$$

The zeros of the comb filter $H_1(z)$ coincide and thus cancel the poles of the parallel bank of filters $H_2(z)$, so that in effect the system function of the cascade connection $H_1(z)H_2(z)$ contains no poles, but it is in fact FIR.

Alternative forms to (8.1.45) can be obtained as derived in Section 7.2.3 by combining a pair of complex-conjugate poles or two real-valued poles to form two-pole filters with real-valued coefficients. Thus for $\alpha = 0$ we obtained the system functions for $H_2(z)$ in the forms

$$H_2(z) = \frac{H(0)}{1 - z^{-1}} + \sum_{k=1}^{(M-1)/2} \frac{A(k) + B(k)z^{-1}}{1 - 2\cos(2\pi k/M)z^{-1} + z^{-2}} \qquad M \text{ odd} \qquad (8.1.48)$$

$$H_2(z) = \frac{H(0)}{1 - z^{-1}} + \frac{H(M/2)}{1 + z^{-1}} + \sum_{k=1}^{(M/2)-1} \frac{A(k) + B(k)z^{-1}}{1 - 2\cos(2\pi k/M)z^{-1} + z^{-2}} \qquad M \text{ even}$$
$$(8.1.49)$$

where

$$A(k) = H(k) + H(M - k)$$
$$B(k) = H(k)e^{-j2\pi k/M} + H(M - k)e^{j2\pi k/M}$$

Similar expressions can be obtained for $\alpha = \frac{1}{2}$.

The frequency response of the FIR filter can be obtained by evaluating the system function $H(z)$ on the unit circle. Thus from (8.1.45) we obtain

$$H(\omega) = \left\{ \frac{\sin\left(\dfrac{\omega M}{2} - \pi\alpha\right)}{M} \sum_{k=0}^{M-1} \frac{H(k + \alpha)e^{-j\pi(k+\alpha)/M}e^{j\pi\alpha}}{\sin\left(\dfrac{\omega}{2} - \dfrac{\pi(k + \alpha)}{M}\right)} \right\} e^{-j\omega(M - 1)/2} \qquad (8.1.50)$$

However, when $h(n)$ is symmetric,

$$H(k + \alpha) = G(k + \alpha)e^{j\pi(k+\alpha)/M}e^{-j\pi\alpha} \qquad (8.1.51)$$

and when $h(n)$ is antisymmetric,

$$H(k + \alpha) = G(k + \alpha)e^{j\pi(k+\alpha)/M}e^{-j\pi(\alpha-1/2)} \qquad (8.1.52)$$

When these relationships are substituted into (8.1.50), the frequency response of the linear-phase FIR filter becomes a function of the real-valued parameters. From (8.1.51) and (8.1.50) we obtain, for the symmetric FIR filter,

$$H(\omega) = \left\{ \frac{\sin\left(\frac{\omega M}{2} - \pi\alpha\right)}{M} \sum_{k=0}^{M-1} \frac{G(k + \alpha)}{\sin\left[\frac{\omega}{2} - \frac{\pi}{M}(k + \alpha)\right]} \right\} e^{-j\omega(M-1)/2} \qquad (8.1.53)$$

where

$$G(k + \alpha) = \begin{cases} -G(M - k) & \alpha = 0 \\ G(M - k - \frac{1}{2}) & \alpha = \frac{1}{2} \end{cases} \qquad (8.1.54)$$

Similarly, for the antisymmetric linear-phase FIR filter we obtain, from (8.1.52) and (8.1.50),

$$H(\omega) = \left\{ \frac{\sin\left(\frac{\omega M}{2} - \pi\alpha\right)}{M} \sum_{k=0}^{M-1} \frac{G(k + \alpha)}{\sin\left[\frac{\omega}{2} - \frac{\pi}{M}(k + \alpha)\right]} \right\} e^{-j\omega(M-1)/2}e^{j\pi/2} \qquad (8.1.55)$$

where

$$G(k + \alpha) = \begin{cases} G(M - k) & \alpha = 0 \\ -G(M - k - \frac{1}{2}) & \alpha = \frac{1}{2} \end{cases} \qquad (8.1.56)$$

With these expressions for the frequency response $H(\omega)$ given in terms of the desired frequency samples $\{G(k + \alpha)\}$, we can easily explain the method for selecting the parameters $\{G(k + \alpha)\}$ in the transition band which result in minimizing the peak sidelobe in the stopband. In brief, the values of $G(k + \alpha)$ in the passband are set to $(-1)^k$ and in the stopband are set to zero. For any choice of $G(k + \alpha)$ in the transition band the value of $H(\omega)$ is computed at a dense set of frequencies (e.g., at $\omega_n = 2\pi n/K$, $n = 0$, $1, \dots, K - 1$, where, for example, $K = 10M$). The value of the maximum sidelobe is determined, and the values of the parameters $\{G(k + \alpha)\}$ in the transition band are changed in a direction of steepest descent, which in effect reduces the maximum sidelobe. The computation of $H(\omega)$ is now repeated with the new choice of $\{G(k + \alpha)\}$. The maximum sidelobe of $H(\omega)$ is again determined and the values of the parameters $\{G(k + \alpha)\}$ in the transition band are adjusted in a direction of steepest descent that reduces the sidelobe. This interactive process is performed until it converges to the optimum choice of the parameters $\{G(k + \alpha)\}$ in the transition band.

There is a potential problem in the frequency-sampling realization of the FIR linear-phase filter. As indicated in (8.1.45), the frequency sampling realization of the FIR filter introduces poles and zeros at equally spaced points on the unit circle. In the ideal situation, the zeros cancel the poles and, consequently, the actual zeros of $H(z)$ are determined by the selection of the frequency samples $\{H(k + \alpha)\}$. In a practical implementation of the frequency-sampling realization, however, quantization effects preclude a perfect cancellation of the poles and zeros. In fact, the location of poles on the unit circle provide no damping of the round-off noise that is introduced in the computations. As a result, such noise tends to increase with time and, ultimately, may destroy the normal operation of the filter.

To mitigate this problem, we can move both the poles and zeros from the unit circle to a circle just inside the unit circle, say at radius $r = 1 - \epsilon$, where ϵ is a very small number. Thus the system function of the linear-phase FIR filter becomes

$$H(z) = \frac{1 - r^M z^{-M} e^{j2\pi\alpha}}{M} \sum_{k=0}^{M-1} \frac{H(k + \alpha)}{1 - re^{j2\omega\pi(k+\alpha)/M} z^{-1}} \qquad (8.1.57)$$

The corresponding two-pole filter realization given in (8.1.48) can be modified accordingly. The damping provided by selecting $r < 1$ ensures that roundoff noise will be bounded and thus instability is avoided.

8.1.3 Design of Equiripple Linear-Phase FIR Filters

The window method and the frequency-sampling method are relatively simple techniques for designing linear-phase FIR filters. However, they also possess some minor disadvantages, described in Section 8.1.6, which may render them undesirable for some applications. A major problem is the lack of precise control of the critical frequencies such as ω_p and ω_s.

The filter design method described in this section is formulated as a Chebyshev approximation problem. It is viewed as an optimum design criterion in the sense that the weighted approximation error between the desired frequency response and the actual frequency response is spread evenly across the passband and evenly across the stopband of the filter and the maximum error is minimized. The resulting filter designs have ripples in both the passband and the stopband.

To describe the design procedure, let us consider the design of a lowpass filter with passband edge frequency ω_p and stopband edge frequency ω_s. From the general specifications given in Fig. 5.13, in the passband, the filter frequency response satisfies the condition

$$1 - \delta_1 \leq H_r(\omega) \leq 1 + \delta_1 \qquad |\omega| \leq \omega_p \qquad (8.1.58)$$

Similarly, in the stopband, the filter frequency response is specified to fall between the limits $\pm \delta_2$, that is,

$$-\delta_2 \leq H_r(\omega) \leq \delta_2 \qquad |\omega| > \omega_s \qquad (8.1.59)$$

Thus δ_1 represents the ripple in the passband and δ_2 represents the attenuation or ripple in the stopband. The remaining filter parameter is M, the filter length or the number of filter coefficients.

As we recall, there are four different cases that result in a linear phase FIR filter. These are summarized below.

CASE 1: SYMMETRIC UNIT SAMPLE RESPONSE $h(n) = h(M - 1 - n)$ AND M ODD. In this case, the real-valued frequency response characteristic $H_r(\omega)$, derived in Section 5.3.4, is

$$H_r(\omega) = h\left(\frac{M-1}{2}\right) + 2 \sum_{n=0}^{(M-3)/2} h(n) \cos \omega \left(\frac{M-1}{2} - n\right) \qquad (8.1.60)$$

If we let $k = (M - 1)/2 - n$ and define a new set of filter parameters $\{a(k)\}$ as

$$a(k) = \begin{cases} h\left(\dfrac{M-1}{2}\right) & k = 0 \\[2mm] 2h\left(\dfrac{M-1}{2} - k\right) & k = 1, 2, \ldots, \dfrac{M-1}{2} \end{cases} \tag{8.1.61}$$

then (8.1.60) reduces to the compact form

$$H_r(\omega) = \sum_{k=0}^{(M-1)/2} a(k) \cos \omega k \tag{8.1.62}$$

CASE 2: SYMMETRIC UNIT SAMPLE RESPONSE $h(n) = h(M - 1 - n)$ AND M even. In this case, $H_r(\omega)$ is expressed as

$$H_r(\omega) = 2 \sum_{n=0}^{(M/2)-1} h(n) \cos \omega \left(\frac{M-1}{2} - n\right) \tag{8.1.63}$$

Again, we change the summation index from n to $k = M/2 - n$ and define a new set of filter parameters $\{b(k)\}$ as

$$b(k) = 2h\left(\frac{M}{2} - k\right) \qquad k = 1, 2, \ldots, M/2 \tag{8.1.64}$$

With these substitutions (8.1.63) becomes

$$H_r(\omega) = \sum_{k=1}^{M/2} b(k) \cos \omega \left(k - \frac{1}{2}\right) \tag{8.1.65}$$

In carrying out the optimization, it is convenient to rearrange (8.1.65) further into the form

$$H_r(\omega) = \cos \frac{\omega}{2} \sum_{k=0}^{(M/2)-1} \tilde{b}(k) \cos \omega k \tag{8.1.66}$$

where the coefficients $\{\tilde{b}(k)\}$ are linearly related to the coefficients $\{b(k)\}$. In fact, it can be shown that the relationship is

$$\tilde{b}(0) = \tfrac{1}{2}b(1)$$
$$\tilde{b}(k) = 2b(k) - \tilde{b}(k - 1) \qquad k = 1, 2, 3, \ldots, \frac{M}{2} - 2 \tag{8.1.67}$$
$$\tilde{b}\left(\frac{M}{2} - 1\right) = 2b\left(\frac{M}{2}\right)$$

CASE 3: ANTISYMMETRIC UNIT SAMPLE RESPONSE $h(n) = -h(M - 1 - n)$ AND M odd. The real-valued frequency response characteristic $H_r(\omega)$ for this case is

$$H_r(\omega) = 2 \sum_{n=0}^{(M-3)/2} h(n) \sin \omega \left(\frac{M-1}{2} - n\right) \tag{8.1.68}$$

If we change the summation in (8.1.68) from n to $k = (M - 1)/2 - n$ and define a new set of filter parameters $\{c(k)\}$ as

$$c(k) = 2h\left(\frac{M-1}{2} - k\right) \qquad k = 1, 2, \ldots, (M - 1)/2 \tag{8.1.69}$$

then (8.1.68) becomes

$$H_r(\omega) = \sum_{k=1}^{(M-1)/2} c(k) \sin \omega k \qquad (8.1.70)$$

As in the previous case, it is convenient to rearrange (8.1.70) into the form

$$H_r(\omega) = \sin \omega \sum_{k=0}^{(M-3)/2} \tilde{c}(k) \cos \omega k \qquad (8.1.71)$$

where the coefficients $\{\tilde{c}(k)\}$ are linearly related to the parameters $\{c(k)\}$. This desired relationship can be derived from (8.1.70) and (8.1.71) and is simply given as

$$\tilde{c}\left(\frac{M-3}{2}\right) = c\left(\frac{M-1}{2}\right)$$

$$\tilde{c}\left(\frac{M-5}{2}\right) = 2c\left(\frac{M-3}{2}\right)$$

$$\vdots \qquad \qquad \vdots \qquad \qquad (8.1.72)$$

$$\tilde{c}(k-1) - \tilde{c}(k+1) = 2c(k) \qquad 2 \le k \le \frac{M-5}{2}$$

$$\tilde{c}(0) + \tfrac{1}{2}\tilde{c}(2) = c(1)$$

CASE 4: ANTISYMMETRIC UNIT SAMPLE RESPONSE $h(n) = h(M - 1 - n)$ AND M even. In this case, the real-valued frequency response characteristic $H_r(\omega)$ is

$$H_r(\omega) = 2 \sum_{n=0}^{(M/2)-1} h(n) \sin \omega \left(\frac{M-1}{2} - n\right) \qquad (8.1.73)$$

A change in the summation index from n to $k = M/2 - n$ combined with a definition of a new set of filter coefficients $\{d(k)\}$, related to $\{h(n)\}$ according to

$$d(k) = 2h\left(\frac{M}{2} - k\right) \qquad k = 1, 2, \ldots, \frac{M}{2} \qquad (8.1.74)$$

results in the expression

$$H_r(\omega) = \sum_{k=1}^{M/2} d(k) \sin \omega \left(k - \frac{1}{2}\right) \qquad (8.1.75)$$

As in the previous two cases, we find it convenient to rearrange (8.1.75) into the form

$$H_r(\omega) = \sin \frac{\omega}{2} \sum_{k=0}^{(M/2)-1} \tilde{d}(k) \cos \omega k \qquad (8.1.76)$$

where the new filter parameters $\{\tilde{d}(k)\}$ are related to $\{d(k)\}$ as follows:

$$\tilde{d}\left(\frac{M}{2} - 1\right) = 2d\left(\frac{M}{2}\right)$$

$$\tilde{d}(k-1) - \tilde{d}(k) = 2d(k) \qquad 2 \le k \le \frac{M}{2} - 1 \qquad (8.1.77)$$

$$\tilde{d}(0) - \tfrac{1}{2}\tilde{d}(1) = d(1)$$

The expressions for $H_r(\omega)$ in these four cases are summarized in Table 8.6. We note that the rearrangements that we made in cases 2, 3, and 4 have allowed us to express

TABLE 8.6 Real-Valued Frequency Response Functions
for Linear-Phase FIR Filters

Filter Type	$Q(\omega)$	$P(\omega)$
$h(n) = h(M - 1 - n)$ M odd (case 1)	1	$\displaystyle\sum_{k=0}^{(M-1)/2} a(k) \cos \omega k$
$h(n) = h(M - 1 - n)$ M even (case 2)	$\cos \dfrac{\omega}{2}$	$\displaystyle\sum_{k=0}^{(M/2)-1} \tilde{b}(k) \cos \omega k$
$h(n) = -h(M - 1 - n)$ M odd (case 3)	$\sin \omega$	$\displaystyle\sum_{k=0}^{(M-3)/2} \tilde{c}(k) \cos \omega k$
$h(n) = -h(M - 1 - n)$ M even (case 4)	$\sin \dfrac{\omega}{2}$	$\displaystyle\sum_{k=0}^{(M/2)-1} \tilde{d}(k) \cos \omega k$

$H_r(\omega)$ as

$$H_r(\omega) = Q(\omega)P(\omega) \tag{8.1.78}$$

where

$$Q(\omega) = \begin{cases} 1 & \text{case 1} \\ \cos \dfrac{\omega}{2} & \text{case 2} \\ \sin \omega & \text{case 3} \\ \sin \dfrac{\omega}{2} & \text{case 4} \end{cases} \tag{8.1.79}$$

and $P(\omega)$ has the common form

$$P(\omega) = \sum_{k=0}^{L} \alpha(k) \cos \omega k \tag{8.1.80}$$

with $\{\alpha(k)\}$ representing the parameters of the filter, which are linearly related to the unit sample response $h(n)$ of the FIR filter, and the upper limit L in the sum is $L = (M - 1)/2$ for case 1, $L = (M - 3)/2$ for case 3, and $L = M/2 - 1$ for case 2 and case 4.

In addition to the common framework given above for the representation of $H_r(\omega)$, we also define the real-valued desired frequency response $H_{dr}(\omega)$ and the weighting function $W(\omega)$ on the approximation error. The real-valued desired frequency response $H_{dr}(\omega)$ is simply defined to be unity in the passband and zero in the stopband. For example, Fig. 8.22 illustrates several different types of characteristics for $H_{dr}(\omega)$. The weighting function on the approximation error allows us to choose the relative size of the errors in the different frequency bands (i.e., in the passband and in the stopband). In particular, it is convenient to normalize $W(\omega)$ to unity in the stopband and set $W(\omega) = \delta_2/\delta_1$ in the passband, that is,

$$W(\omega) = \begin{cases} \delta_2/\delta_1 & \omega \text{ in the passband} \\ 1 & \omega \text{ in the stopband} \end{cases} \tag{8.1.81}$$

Then we simply select $W(\omega)$ in the passband to reflect our emphasis on the relative size of the ripple in the stopband to the ripple in the passband.

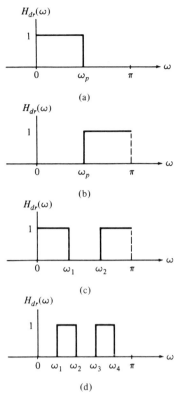

FIGURE 8.22 Desired frequency response characteristics for different types of filters.

With the specification of $H_{dr}(\omega)$ and $W(\omega)$, we can now define the weighted approximation error as

$$
\begin{aligned}
E(\omega) &= W(\omega)[H_{dr}(\omega) - H_r(\omega)] \\
&= W(\omega)[H_{dr}(\omega) - Q(\omega)P(\omega)] \qquad (8.1.82) \\
&= W(\omega)Q(\omega)\left[\frac{H_{dr}(\omega)}{Q(\omega)} - P(\omega)\right]
\end{aligned}
$$

For mathematical convenience, we define a modified weighting function $\hat{W}(\omega)$ and a modified desired frequency response $\hat{H}_{dr}(\omega)$ as

$$
\begin{aligned}
\hat{W}(\omega) &= W(\omega)Q(\omega) \\
\hat{H}_{dr}(\omega) &= \frac{H_{dr}(\omega)}{Q(\omega)} \qquad (8.1.83)
\end{aligned}
$$

Then the weighted approximation error may be expressed as

$$
E(\omega) = \hat{W}(\omega)[\hat{H}_{dr}(\omega) - P(\omega)] \qquad (8.1.84)
$$

for all four different types of linear-phase FIR filters.

Given the error function $E(\omega)$, the Chebyshev approximation problem is basically to determine the filter parameters $\{\alpha(k)\}$ that minimize the maximum absolute value of $E(\omega)$ over the frequency bands in which the approximation is to be performed. In mathematical terms we seek the solution to the problem

$$
\min_{\text{over } \{\alpha(k)\}} \left[\max_{\omega \in S} |E(\omega)| \right] = \min_{\text{over } \{a(k)\}} \left[\max_{\omega \in S} \left|\hat{W}(\omega)[\hat{H}_{dr}(\omega) - \sum_{k=0}^{L} \alpha(k) \cos \omega k]\right| \right] \qquad (8.1.85)
$$

where S represents the set (disjoint union) of frequency bands over which the optimization is to be performed. Basically, the set S consists of the passbands and stopbands of the desired filter.

The solution to this problem is due to Parks and McClellan (1972a), who applied a theorem in the theory of Chebyshev approximation. It is called the *alternation theorem*, which we state without proof.

ALTERNATION THEOREM: *Let S be a compact subset of the interval $[0, \pi)$. A necessary and sufficient condition for*

$$P(\omega) = \sum_{k=0}^{L} \alpha(k) \cos \omega k$$

to be the unique, best weighted Chebyshev approximation to $\hat{H}_{dr}(\omega)$ in S is that the error function $E(\omega)$ exhibit at least $L + 2$ extremal frequencies in S. That is, there must exist at least $L + 2$ frequencies $\{\omega_i\}$ in S such that $\omega_1 < \omega_2 < \cdots < \omega_{L+2}$, $E(\omega_i) = -E(\omega_{i+1})$, and

$$|E(\omega_i)| = \max_{\omega \in S} |E(\omega)| \qquad i = 1, 2, \ldots, L + 2$$

We note that the error function $E(\omega)$ alternates in sign between two successive extremal frequencies. Hence the theorem is called the alternation theorem.

To elaborate on the alternation theorem, let us consider the design of a lowpass filter with passband $0 \leq \omega \leq \omega_p$ and stopband $\omega_s \leq \omega \leq \pi$. Since the desired frequency response $H_{dr}(\omega)$ and the weighting function $W(\omega)$ are piecewise constant, we have

$$\frac{dE(\omega)}{d\omega} = \frac{d}{d\omega} \{W(\omega)[H_{dr}(\omega) - H_r(\omega)]\}$$

$$= -\frac{dH_r(\omega)}{d\omega} = 0$$

Consequently, the frequencies $\{\omega_i\}$ corresponding to the peaks of $E(\omega)$ also correspond to peaks at which $H_r(\omega)$ meets the error tolerance. Since $H_r(\omega)$ is a trigonometric polynomial of degree L, for example, for case 1,

$$H_r(\omega) = \sum_{k=0}^{L} \alpha(k) \cos \omega k$$

$$= \sum_{k=0}^{L} \alpha(k) \left[\sum_{n=0}^{k} \beta_{nk}(\cos \omega)^n \right] \qquad (8.1.86)$$

$$= \sum_{k=0}^{L} \alpha'(k)(\cos \omega)^k$$

it follows that $H_r(\omega)$ can have at most $L - 1$ local maxima and minima in the open interval $0 < \omega < \pi$. In addition, $\omega = 0$ and $\omega = \pi$ are usually extrema of $H_r(\omega)$ and, also, of $E(\omega)$. Therefore, $H_r(\omega)$ has at most $L + 1$ extremal frequencies. Furthermore, the band-edge frequencies ω_p and ω_s are also extrema of $E(\omega)$, since $|E(\omega)|$ is maximum at $\omega = \omega_p$ and $\omega = \omega_s$. As a consequence, there are at most $L + 3$ extremal frequencies in $E(\omega)$ for the unique, best approximation of the ideal lowpass filter. On the other hand, the alternation theorem states that there are at least $L + 2$ extremal frequencies in $E(\omega)$. Thus the error function for the lowpass filter design will have either $L + 3$ or $L + 2$ extrema. In general, filter designs that contain more than $L + 2$ alternations or ripples are called *extra ripple filters*. When the filter design contains the maximum number of alternations, it is called a *maximal ripple filter*.

The alternation theorem guarantees a unique solution for the Chebyshev optimization problem in (8.1.85). At the desired extremal frequencies $\{\omega_n\}$, we have the set of equations

$$\hat{W}(\omega_n)[\hat{H}_{dr}(\omega_n) - P(\omega_n)] = (-1)^n\delta \qquad n = 0, 1, \ldots, L + 1 \qquad (8.1.87)$$

where δ represents the maximum value of the error function $E(\omega)$. In fact, if we select $W(\omega)$ as indicated by (8.1.81), it follows that $\delta = \delta_2$.

The set of linear equations in (8.1.87) can be rearranged as

$$P(\omega_n) + \frac{(-1)^n\delta}{\hat{W}(\omega_n)} = \hat{H}_{dr}(\omega_n) \qquad n = 0, 1, \ldots, L + 1$$

or, equivalently, in the form

$$\sum_{k=0}^{L} \alpha(k) \cos \omega_n k + \frac{(-1)^n\delta}{\hat{W}(\omega_n)} = \hat{H}_{dr}(\omega_n) \qquad n = 0, 1, \ldots, L + 1 \qquad (8.1.88)$$

If we treat the $\{a(k)\}$ and δ as the parameters to be determined, (8.1.88) may be expressed in matrix form as

$$\begin{bmatrix} 1 & \cos \omega_0 & \cos 2\omega_0 & \cdots & \cos L\omega_0 & \frac{1}{\hat{W}(\omega_0)} \\ 1 & \cos \omega_1 & \cos 2\omega_1 & \cdots & \cos L\omega_1 & \frac{-1}{\hat{W}(\omega_1)} \\ & \vdots & & & & \vdots \\ 1 & \cos \omega_{L+1} & \cos 2\omega_{L+1} & \cdots & \cos L\omega_{L+1} & \frac{(-1)^{L+1}}{\hat{W}(\omega_{L+1})} \end{bmatrix} \begin{bmatrix} \alpha(0) \\ \alpha(1) \\ \vdots \\ \alpha(L) \\ \delta \end{bmatrix} = \begin{bmatrix} \hat{H}_{dr}(\omega_0) \\ \hat{H}_{dr}(\omega_1) \\ \vdots \\ \hat{H}_{dr}(\omega_{L+1}) \end{bmatrix}$$

$$(8.1.89)$$

Initially, we know neither the set of extremal frequencies $\{\omega_n\}$ nor the parameters $\{\alpha(k)\}$ and δ. To solve for the parameters, we use an iterative algorithm, called the *Remez exchange algorithm* [see Rabiner et al. (1975)], in which we begin by guessing at the set of extremal frequencies, determine $P(\omega)$ and δ, and then compute the error function $E(\omega)$. From $E(\omega)$ we determine another set of $L + 2$ extremal frequencies and repeat the iterative process until it converges to the optimal set of extremal frequencies. Although the matrix equation in (8.1.89) can be used in the iterative procedure, matrix inversion is time consuming and inefficient.

A more efficient procedure, suggested in the paper by Rabiner et al. (1975), is to compute δ analytically, according to the formula

$$\delta = \frac{\gamma_0 \hat{H}_{dr}(\omega_0) + \gamma_1 \hat{H}_{dr}(\omega_1) + \cdots + \gamma_{L+1} \hat{H}_{dr}(\omega_{L+1})}{\dfrac{\gamma_0}{\hat{W}(\omega_0)} - \dfrac{\gamma_1}{\hat{W}(\omega_1)} + \cdots + \dfrac{(-1)^{L+1}\gamma_{L+1}}{\hat{W}(\omega_{L+1})}} \qquad (8.1.90)$$

where

$$\gamma_k = \prod_{n=0}^{L+1} \frac{1}{\cos \omega_k - \cos \omega_n} \qquad (8.1.91)$$

The expression for δ in (8.1.90) follows immediately from the matrix equation in (8.1.89). Thus with an initial guess at the $L + 2$ extremal frequencies, we compute δ.

Now since $P(\omega)$ is a trigometric polynomial of the form

$$P(\omega) = \sum_{k=0}^{L} \alpha(k)x^k \qquad x = \cos \omega$$

and since we know that the polynomial at the points $x_n \equiv \cos \omega_n$, $n = 0, 1, \ldots,$ $L + 1$, has the corresponding values

$$P(\omega_n) = \hat{H}_{dr}(\omega_n) - \frac{(-1)^n \delta}{\hat{W}(\omega_n)} \qquad n = 0, 1, \ldots, L + 1 \qquad (8.1.92)$$

we may use the Lagrange interpolation formula for $P(\omega)$. Thus $P(\omega)$ may be expressed as [see Hamming (1962)]

$$P(\omega) = \frac{\displaystyle\sum_{k=0}^{L} P(\omega_k)[\beta_k/(x - x_k)]}{\displaystyle\sum_{k=0}^{L} [\beta_k/(x - x_k)]} \qquad (8.1.93)$$

where $P(\omega_n)$ is given by (8.1.92), $x = \cos \omega$, $x_k = \cos \omega_k$, and

$$\beta_k = \prod_{\substack{n=0 \\ n \neq k}}^{L} \frac{1}{x_k - x_n} \qquad (8.1.94)$$

Having the solution for $P(\omega)$, we can now compute the error function $E(\omega)$ from

$$E(\omega) = \hat{W}(\omega)[\hat{H}_{dr}(\omega) - P(\omega)] \qquad (8.1.95)$$

on a dense set of frequency points. Usually, a number of points equal to $16M$, where M is the length of the filter, suffices. If $|E(\omega)| \geq \delta$ for some frequencies on the dense set, then a new set of frequencies corresponding to the $L + 2$ largest peaks of $|E(\omega)|$ are selected and the computational procedure beginning with (8.1.90) is repeated. Since the new set of $L + 2$ extremal frequencies are selected to correspond to the peaks of the error function $|E(\omega)|$, the algorithm forces δ to increase in each iteration until it converges to the upper bound and hence to the optimum solution for the Chebyshev approximation problem. In other words, when $|E(\omega)| \leq \delta$ for all frequencies on the dense set, the optimal solution has been found in terms of the polynomial $H(\omega)$.

A flowchart of the algorithm is shown in Fig. 8.23 and is due to Remez (1957).

Once the optimal solution has been obtained in terms of $P(\omega)$, the unit sample response $h(n)$ can be computed directly, without having to compute the parameters $\{\alpha(k)\}$. In effect, we have determined

$$H_r(\omega) = Q(\omega)P(\omega)$$

which can be evaluated at $\omega = 2\pi k/M$, $k = 0, 1, \ldots, (M - 1)/2$ for M odd or $M/2$ for M even. Then $h(n)$ can be determined from either (8.1.31) or (8.1.32) or (8.1.41) or (8.1.42), depending on the type of filter being designed.

A computer program written by Parks and McClellan (1972b) is available for designing linear phase FIR filters based on the Chebyshev approximation criterion and implemented with the Remez exchange algorithm. This program may be used to design lowpass, highpass, or bandpass filters, differentiators, and Hilbert transformers. The latter two types of filters are described in the following sections.

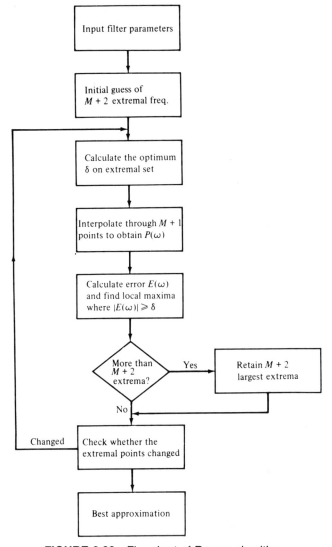

FIGURE 8.23   Flowchart of Remez algorithm.

The program requires a number of input parameters which determine the filter characteristics. In particular, the following parameters must be specified:

LINE 1

NFILT: The filter length, denoted above as M.

JTYPE: Type of filter:

JTYPE = 1 results in a multiple passband/stopband filter.

JTYPE = 2 results in a differentiator.

JTYPE = 3 results in a Hilbert transformer.

NBANDS: The number of frequency bands from 2 (for a lowpass filter) to a maximum of 10 (for a multiple-band filter).

LGRID: The grid density for interpolating the error function $E(\omega)$. The default value is 16 if left unspecified.

LINE 2

EDGE: The frequency bands specified by lower and upper cutoff frequencies, up to a maximum of 10 bands (an array of size 20, maximum). The frequencies are given in terms of the variable $f = \omega/2\pi$, where $f = 0.5$ corresponds to the folding frequency.

LINE 3

FX: An array of maximum size 10 that specifies the desired frequency response $H_{dr}(\omega)$ in each band.

LINE 4

WTX: An array of maximum size 10 that specifies the weight function in each band.

The following examples demonstrate the use of this program to design a lowpass and a bandpass filter.

EXAMPLE 8.1.3 _____

Design a lowpass filter of length $M = 61$ with a passband edge frequency $f_p = 0.1$ and a stopband edge frequency $f_s = 0.15$.

Solution: The lowpass filter is a two-band filter with passband edge frequencies $(0, 0.1)$ and stopband edge frequencies $(0.15, 0.5)$. The desired response is $(1, 0)$ and the weight function is arbitrarily selected as $(1, 1)$.

$$61, 1, 2$$
$$0.0, 0.1, 0.15, 0.5$$
$$1.0, 0.0$$
$$1.0, 1.0$$

The result of this design is illustrated in Table 8.7, which gives the filter coefficients. The frequency response is shown in Fig. 8.24 the resulting filter has a stopband attenuation of -56 dB and a passband ripple of 0.0135 dB.

If we increase the length of the filter to $M = 101$, while maintaining all the other parameters given above the same, the resulting filter that we obtain has the frequency response characteristic shown in Fig. 8.25. Now, the stopband attenuation is -85 dB and the passband ripple is reduced to 0.00046 dB.

We should indicate that it is possible to increase the attenuation in the stopband by keeping the filter length fixed, say at $M = 61$, and decreasing the weighting function $W(\omega) = \delta_2/\delta_1$ in the passband. With $M = 61$ and a weighting function $(0.1, 1)$, we obtain a filter that has the frequency response characteristic shown in Fig 8.26. This filter has a stopband attenuation of -65 dB and a passband rippling of 0.049 dB.

EXAMPLE 8.1.4 _____

Design a bandpass filter of length $M = 32$ with passband edge frequencies $f_{p1} = 0.2$ and $f_{p2} = 0.35$ and stopband edge frequencies of $f_{s1} = 0.1$ and $f_{s2} = 0.425$.

Solution: This passband filter is a three-band filter with a stopband range of $(0, 0.1)$, a passband range of $(0.2, 0.35)$ and a second stopband range of $(0.425, 0.5)$. The weighting function is selected as $(10.0, 1.0, 10.0)$, or as $(1.0, 0.1, 1.0)$, and the desired response

TABLE 8.7 Parameters for Lowpass Filter Design in Example 8.1.3

FINITE IMPULSE RESPONSE (FIR)
LINEAR PHASE DIGITAL FILTER DESIGN
REMEZ EXCHANGE ALGORITHM

FILTER LENGTH = 61

***** IMPULSE RESPONSE *****

H(1)	=	-0.12109351E-02	=	H(61)
H(2)	=	-0.67270687E-03	=	H(60)
H(3)	=	0.98090240E-04	=	H(59)
H(4)	=	0.13536664E-02	=	H(58)
H(5)	=	0.22969784E-02	=	H(57)
H(6)	=	0.19963495E-02	=	H(56)
H(7)	=	0.97026095E-04	=	H(55)
H(8)	=	-0.26466695E-02	=	H(54)
H(9)	=	-0.45133103E-02	=	H(53)
H(10)	=	-0.37704944E-02	=	H(52)
H(11)	=	0.13079655E-04	=	H(51)
H(12)	=	0.51791356E-02	=	H(50)
H(13)	=	0.84883478E-02	=	H(49)
H(14)	=	0.69532110E-02	=	H(48)
H(15)	=	0.71037059E-04	=	H(47)
H(16)	=	-0.90407897E-02	=	H(46)
H(17)	=	-0.14723047E-01	=	H(45)
H(18)	=	-0.11958945E-01	=	H(44)
H(19)	=	-0.29799214E-04	=	H(43)
H(20)	=	0.15713422E-01	=	H(42)
H(21)	=	0.25657151E-01	=	H(41)
H(22)	=	0.21057373E-01	=	H(40)
H(23)	=	0.68637768E-04	=	H(39)
H(24)	=	-0.28902054E-01	=	H(38)
H(25)	=	-0.49118541E-01	=	H(37)
H(26)	=	-0.42713970E-01	=	H(36)
H(27)	=	-0.50114304E-04	=	H(35)
H(28)	=	0.73574215E-01	=	H(34)
H(29)	=	0.15782040E+00	=	H(33)
H(30)	=	0.22465512E+00	=	H(32)
H(31)	=	0.25007001E+00	=	H(31)

	BAND 1	BAND 2
LOWER BAND EDGE	0.0000000	0.1500000
UPPER BAND EDGE	0.1000000	0.5000000
DESIRED VALUE	1.0000000	0.0000000
WEIGHTING	1.0000000	1.0000000
DEVIATION	0.0015537	0.0015537
DEVIATION IN DB	0.0134854	-56.1724014

EXTREMAL FREQUENCIES--MAXIMA OF THE ERROR CURVE

0.0000000	0.0252016	0.0423387	0.0584677	0.0735887
0.0866935	0.0957661	0.1000000	0.1500000	0.1540323
0.1631048	0.1762097	0.1903225	0.2054435	0.2215725
0.2377015	0.2538306	0.2699596	0.2860886	0.3022176
0.3183466	0.3354837	0.3516127	0.3677417	0.3848788
0.4010078	0.4171368	0.4342739	0.4504029	0.4665320
0.4836690	0.5000000			

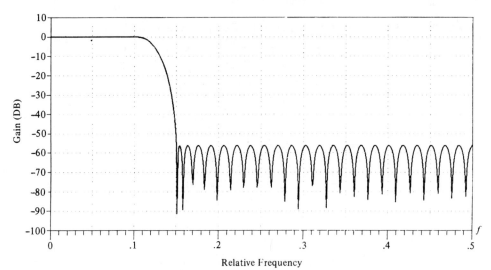

FIGURE 8.24 Frequency response of $M = 61$ FIR filter in Example 8.1.3.

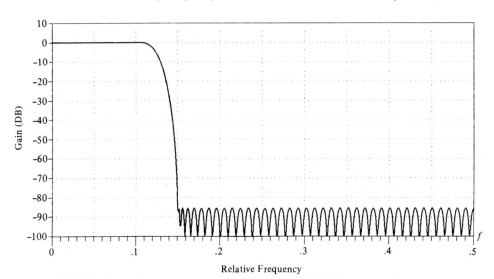

FIGURE 8.25 Frequency response of $M = 101$ FIR filter in Example 8.1.3.

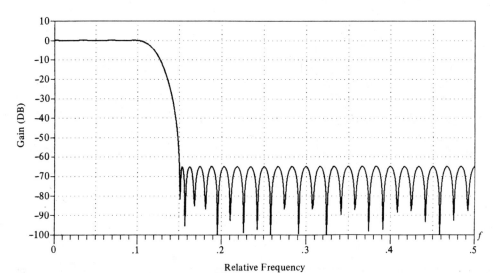

FIGURE 8.26 Frequency response of $M = 61$ FIR filter in Example 8.1.3.

in the three bands is (0.0, 1.0, 0.0). Thus the input parameters to the program are

$$32, 1, 3$$
$$0.0, 0.1, 0.2, 0.35, 0.425, 0.5$$
$$0.0, 1.0, 0.0$$
$$10.0, 1.0, 10.0$$

The results of this design are shown in Table 8.8, which gives the filter coefficients. We note that the ripple in the stopbands δ_2 is 10 times smaller than the ripple in the passband due to the fact that errors in the stopband were given a weight of 10 compared to the passband weight of unity. The frequency response of the bandpass filter is illustrated in Fig. 8.27.

The examples above serve to illustrate the relative ease with which optimal lowpass, highpass, bandstop, bandpass, and more general multiband linear-phase FIR filters can be designed based on the Chebyshev approximation criterion implemented by means of the Remez exchange algorithm. In the next two sections we consider the design of differentiators and Hilbert transformers.

TABLE 8.8 Parameters for Bandpass Filter in Example 8.1.4

```
                FINITE IMPULSE RESPONSE (FIR)
              LINEAR PHASE DIGITAL FILTER DESIGN
                 REMEZ EXCHANGE ALGORITHM

                     BANDPASS FILTER

                  FILTER LENGTH = 32

             ***** IMPULSE RESPONSE *****
        H( 1) =  -0.57534026E-02 = H( 32)
        H( 2) =   0.99026691E-03 = H( 31)
        H( 3) =   0.75733471E-02 = H( 30)
        H( 4) =  -0.65141204E-02 = H( 29)
        H( 5) =   0.13960509E-01 = H( 28)
        H( 6) =   0.22951644E-02 = H( 27)
        H( 7) =  -0.19994041E-01 = H( 26)
        H( 8) =   0.71369656E-02 = H( 25)
        H( 9) =  -0.39657373E-01 = H( 24)
        H(10) =   0.11260066E-01 = H( 23)
        H(11) =   0.66233635E-01 = H( 22)
        H(12) =  -0.10497202E-01 = H( 21)
        H(13) =   0.85136160E-01 = H( 20)
        H(14) =  -0.12024988E+00 = H( 19)
        H(15) =  -0.29678580E+00 = H( 18)
        H(16) =   0.30410913E+00 = H( 17)
```

	BAND 1	BAND 2	BAND 3
LOWER BAND EDGE	0.0000000	0.2000000	0.4250000
UPPER BAND EDGE	0.1000000	0.3500000	0.5000000
DESIRED VALUE	0.0000000	1.0000000	0.0000000
WEIGHTING	10.0000000	1.0000000	10.0000000
DEVIATION	0.0015131	0.0151312	0.0015131
DEVIATION IN DB	-56.4025536	0.1304428	-56.4025536

```
EXTREMAL FREQUENCIES--MAXIMA OF THE ERROR CURVE
0.0000000    0.0273438    0.0527344    0.0761719    0.0937500
0.1000000    0.2000000    0.2195313    0.2527344    0.2839844
0.3132813    0.3386719    0.3500000    0.4250000    0.4328125
0.4503906    0.4796875
```

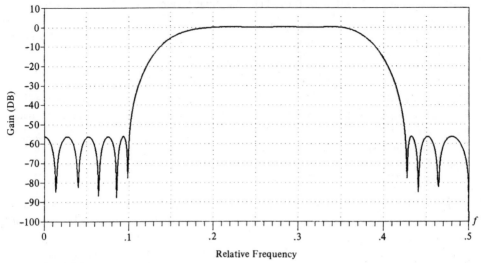

FIGURE 8.27   Frequency response of $M = 32$ FIR filter in Example 8.1.4.

8.1.4 Design of FIR Differentiators

Differentiators are used in many analog and digital systems to take the derivative of a signal. An ideal differentiator has a frequency response that is linearly proportional to frequency. Similarly, an ideal digital differentiator is defined as one that has the frequency response

$$H_d(\omega) = j\omega \qquad 0 \le \omega \le \pi \qquad (8.1.96)$$

The unit sample response corresponding to $H_d(\omega)$ is

$$h_d(n) = \frac{1}{2\pi} \int_{-\pi}^{\pi} H_d(\omega) e^{j\omega n} \, d\omega$$

$$= \frac{1}{2\pi} \int_{-\pi}^{\pi} j\omega e^{j\omega n} \, d\omega \qquad (8.1.97)$$

$$= \frac{\cos \pi n}{n} \qquad -\infty < n < \infty, \, n \ne 0$$

We observe that the ideal differentiator has an antisymmetric unit sample response [i.e., $h_d(n) = -h_d(-n)$]. Hence, $h_d(n) = 0$.

In this section we consider the design of linear-phase FIR differentiators based on the Chebyshev approximation criterion and the frequency sampling method. In view of the fact that the ideal differentiator has an antisymmetric unit sample response, we shall confine our attention to FIR designs in which

$$h(n) = -h(M - 1 - n) \qquad (8.1.98)$$

Hence we consider the filter types classified as case 3 and case 4, in the preceding section.

We recall that in case 3, where M is odd, the real-valued frequency response of the FIR filter $H_r(\omega)$ has the characteristic that $H_r(0) = 0$. A zero response at zero frequency is just the condition that the differentiator should satisfy, and we see from Table 8.6 that both filter types satisfy this condition. However, if a full-band differentiator is desired, this is impossible to achieve with an FIR filter having an odd number of coefficients, since $H_r(\pi) = 0$ for M odd. In practice, however, full-band differentiators are rarely required.

In most cases of practical interest, the desired frequency response characteristic need only be linear over the limited frequency range $0 \le \omega \le 2\pi f_p$, where f_p is called the bandwidth of the differentiator. In the frequency range $2\pi f_p < \omega \le \pi$, the desired response may be either left unconstrained or constrained to be zero.

In the design of FIR differentiators based on the Chebyshev approximation criterion, the weighting function $W(\omega)$ is specified in the program as

$$W(\omega) = \frac{1}{\omega} \qquad 0 \le \omega \le 2\pi f_p \tag{8.1.99}$$

in order that the relative ripple in the passband be a constant. Thus the absolute error between the desired response ω and the approximation $H_r(\omega)$ increases as ω varies from 0 to $2\pi f_p$. However, the weighting function in (8.1.99) ensures that the relative error

$$\delta = \max_{0 \le \omega \le 2\pi f_p} \{W(\omega)[\omega - H_r(\omega)]\}$$

$$= \max_{0 \le \omega \le 2\pi F_p} \left[1 - \frac{H_r(\omega)}{\omega} \right] \tag{8.1.100}$$

is fixed within the passband of the differentiator.

EXAMPLE 8.1.5 ───

Use the Remez algorithm to design a linear-phase FIR differentiator of length $M = 60$. The passband edge frequency is 0.1 and the stopband edge frequency is 0.15.

Solution: The input parameters to the program are

$$\begin{array}{cccc} 60, & 2, & 2 \\ 0.0, & 0.1, & 0.15, & 0.5 \\ 1.0, & 0.0 \\ 1.0, & 1.0 \end{array}$$

The results of this design are shown in Table 8.9, which includes the filter coefficients. The frequency response characteristic is illustrated in Fig. 8.28. Also shown in the same figure is the approximation error over the passband $0 \le f \le 0.1$ of the filter.

───

The important parameters in a differentiator are its length M, its bandwidth {band-edge frequency} f_p, and the peak relative error δ of the approximation. The interrelationship among these three parameters can be easily displayed parametrically. In particular, the value of $20 \log_{10} \delta$ versus f_p with M as a parameter is shown in Fig. 8.29 for M even and in Fig. 8.30 for M odd. These results, due to Rabiner and Schafter (1974a), are useful in the selection of the filter length, given specifications on the inband ripple and the cutoff frequency f_p.

A comparison of the graphs in Figs. 8.29 and 8.30 reveals that even length differentiators result in a significantly smaller approximation error δ than comparable odd-length differentiators. Designs based on M odd are particularly poor if the bandwidth exceeds $f_p = 0.45$. The problem is basically the zero in the frequency response at $\omega = \pi$ ($f = 1/2$). When $f_p < 0.45$, good designs are obtained for M odd, but comparable-length differentiators with M even are always better in the sense that the approximation error is smaller.

In view of the obvious advantage of even-length over odd-length differentiators, a conclusion might be that even-length differentiators are always preferable in practical systems. This is certainly true for many applications. However, we should indicate that

TABLE 8.9 Parameters for FIR Differentiator in Example 8.1.5

FINITE IMPULSE RESPONSE (FIR)
LINEAR–PHASE DIGITAL FILTER DESIGN
REMEZ EXCHANGE ALGORITHM

DIFFERENTIATOR

FILTER LENGTH = 60

***** IMPULSE RESPONSE *****

```
H( 1) =  -0.12478075E-02 = -H( 60)
H( 2) =  -0.15713560E-02 = -H( 59)
H( 3) =   0.36846737E-02 = -H( 58)
H( 4) =   0.19298020E-02 = -H( 57)
H( 5) =   0.14264141E-02 = -H( 56)
H( 6) =  -0.17615277E-02 = -H( 55)
H( 7) =  -0.43110573E-02 = -H( 54)
H( 8) =  -0.46953405E-02 = -H( 53)
H( 9) =  -0.14105244E-02 = -H( 52)
H(10) =   0.41694222E-02 = -H( 51)
H(11) =   0.85736215E-02 = -H( 50)
H(12) =   0.79813031E-02 = -H( 49)
H(13) =   0.11833385E-02 = -H( 48)
H(14) =  -0.87396065E-02 = -H( 47)
H(15) =  -0.15401847E-01 = -H( 46)
H(16) =  -0.12878445E-01 = -H( 45)
H(17) =  -0.18826872E-03 = -H( 44)
H(18) =   0.16620506E-01 = -H( 43)
H(19) =   0.26741523E-01 = -H( 42)
H(20) =   0.20892018E-01 = -H( 41)
H(21) =  -0.18584095E-02 = -H( 40)
H(22) =  -0.31109909E-01 = -H( 39)
H(23) =  -0.48822176E-01 = -H( 38)
H(24) =  -0.38673453E-01 = -H( 37)
H(25) =   0.36760122E-02 = -H( 36)
H(26) =   0.65462478E-01 = -H( 35)
H(27) =   0.12066317E+00 = -H( 34)
H(28) =   0.14182134E+00 = -H( 33)
H(29) =   0.11403757E+00 = -H( 32)
H(30) =   0.43620080E-01 = -H( 31)
```

	BAND 1	BAND 2
LOWER BAND EDGE	0.0000000	0.1500000
UPPER BAND EDGE	0.1000000	0.5000000
DESIRED SLOPE	10.0000000	0.0000000
WEIGHTING	1.0000000	1.0000000
DEVIATION	0.0073580	0.0073580

EXTREMAL FREQUENCIES—MAXIMA OF THE ERROR CURVE

0.0010417	0.0156250	0.0312500	0.0468750	0.0614583
0.0750000	0.0875000	0.0968750	0.1000000	0.1500000
0.1552083	0.1666667	0.1822916	0.1979166	0.2156249
0.2322916	0.2489582	0.2666668	0.2843754	0.3020839
0.3187508	0.3364594	0.3541680	0.3718765	0.3906268
0.4083354	0.4260439	0.4447942	0.4625027	0.4812530
0.5000000				

the signal delay introduced by any linear-phase FIR differentiator is $(M - 1)/2$, which is not an integer when M is even. In many practical applications, this is unimportant. In some applications where it is desirable to have an integer-valued delay in the signal at the output of the differentiator, we must select M to be odd.

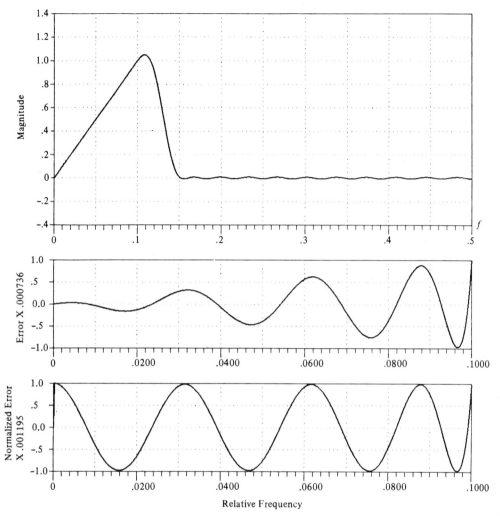

FIGURE 8.28 Frequency response and approximation error for $M = 60$ FIR differentiator of Example 8.1.6.

The numerical results given above are based on designs resulting from the Chebyshev approximation criterion. We wish to indicate it is also possible and relatively easy to design linear phase FIR differentiators based on the frequency sampling method. For example, Fig. 8.31 illustrates the frequency response characteristics of a wideband ($f_p = 0.5$) differentiator, of length $M = 30$. The graph of the absolute value of the approximation error as a function of frequency is also shown in this figure.

Narrowband differentiators can also be designed by means of the frequency sampling method. For example, Fig. 8.32 illustrates a design for $M = 100$, $f_p = 0.1$, and a single transition point selected (but not optimized) as 0.50. The absolute error is also plotted in this figure. If we neglect the error in the transition band, we note that the peak error in the passband is about -28 dB. This error is sufficiently low for many practical applications. However, the Remez algorithm yields a better design (i.e., smaller approximation error).

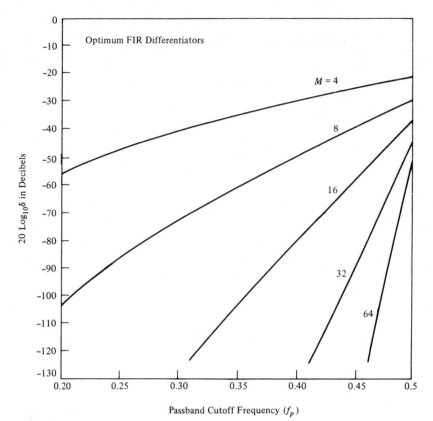

FIGURE 8.29 Curves of 20 $\log_{10} \delta$ versus f_p for M = 4, 8, 16, 32, and 64. [From paper by Rabiner and Schafer (1974a). Reprinted with permission of AT&T.]

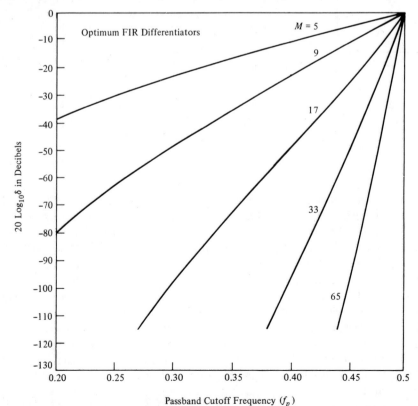

FIGURE 8.30 Curves of 20 $\log_{10} \delta$ versus F_p for N = 5, 9, 17, 33, and 65. [From paper by Rabiner and Schafer (1974a). Reprinted with permission of AT&T.]

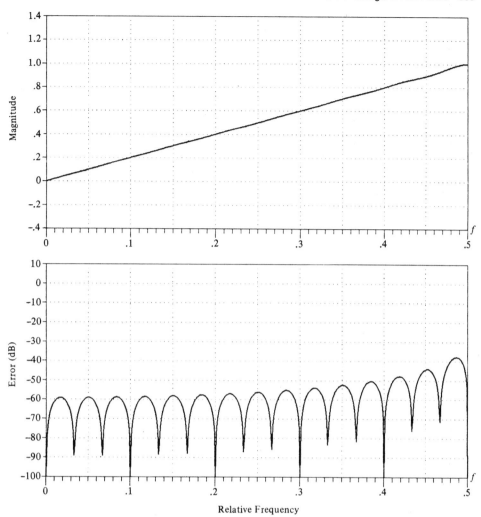

FIGURE 8.31 Frequency response and approximation error for $M = 30$ FIR differentiator designed by frequency sampling method.

8.1.5 Design of Hilbert Transformers

An ideal Hilbert transformer, as we recall from our discussion in Section 6.1, is an all-
pass filter that imparts a 90° phase shift on the signal at its input. Hence the frequency
response of the ideal Hilbert transformer is specified as

$$H_d(\omega) = \begin{cases} -j & 0 < \omega \le \pi \\ j & -\pi < \omega < 0 \end{cases} \qquad (8.1.101)$$

Hilbert transformers are frequently used in communication systems and signal processing,
as, for example, in the generation of single-sideband modulated signals, radar signal
processing, and speech signal processing.

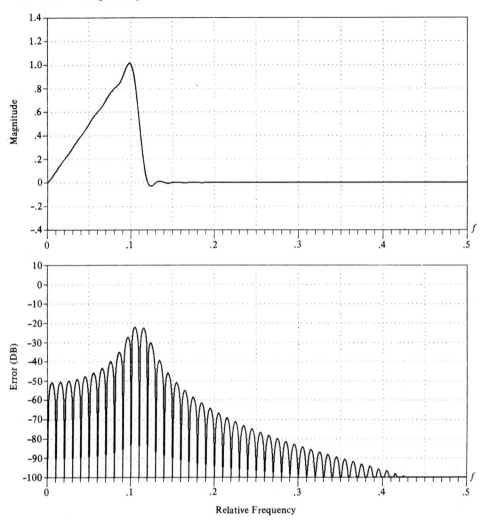

FIGURE 8.32 Frequency response and approximation error for $M = 100$ FIR differentiator designed by frequency sampling method.

The unit sample response of an ideal Hilbert transformer is

$$h_d(n) = \frac{1}{2\pi} \int_{-\pi}^{\pi} H_d(\omega) e^{j\omega n}\, d\omega$$

$$= \frac{1}{2\pi} \left(\int_{-\pi}^{0} je^{j\omega n}\, d\omega - \int_{0}^{\pi} je^{j\omega n}\, d\omega \right) \qquad (8.1.102)$$

$$= \begin{cases} \dfrac{2}{\pi} \dfrac{\sin^2{(\pi n/2)}}{n} & n \neq 0 \\[2ex] 0 & n = 0 \end{cases}$$

As expected, $h_d(n)$ is infinite in duration and noncausal. We note that $h_d(n)$ is antisymmetric [i.e., $h_d(n) = -h_d(-n)$]. In view of this characteristic, we focus our attention on the design of linear-phase FIR Hilbert transformers with antisymmetric unit sample response [i.e., $h(n) = -h(M - 1 - n)$]. We also observe that our choice of an antisymmetric unit sample response is consistent with having a purely imaginary frequency response characteristic $H_d(\omega)$.

We recall once again that when $h(n)$ is antisymmetric, the real-valued frequency response characteristic $H_r(\omega)$ is zero at $\omega = 0$ for both M odd and even and at $\omega = \pi$ when M is odd. Clearly, then, it is impossible to design an all-pass digital Hilbert transformer. Fortunately, in practical signal processing applications, an all-pass Hilbert transformer is unnecessary. Its bandwidth need only cover the bandwidth of the desired signal to be phase shifted. Consequently, we specify the desired real-valued frequency response of a Hilbert transform filter as

$$H_{dr}(\omega) = 1 \qquad 2\pi f_l \le \omega \le 2\pi f_u \qquad (8.1.103)$$

where f_l and f_u are the lower and upper cutoff frequencies, respectively.

It is interesting to note that the ideal Hilbert transformer with unit sample response $h_d(n)$ as given in (8.1.102) is zero for n even. This property is retained by the FIR Hilbert transformer under some symmetry conditions. In particular, let us consider the case 3 filter type for which

$$H_r(\omega) = \sum_{k=1}^{(M-1)/2} c(k) \sin \omega k \qquad (8.1.104)$$

and suppose that $f_l = 0.5 - f_u$. This ensures a symmetric passband about the midpoint frequency $f = 0.25$. If we have this symmetry in the frequency response, $H_r(\omega) = H_r(\pi - \omega)$ and hence (8.1.104) yields

$$\sum_{k=1}^{(M-1)/2} c(k) \sin \omega k = \sum_{k=1}^{(M-1)/2} c(k) \sin k(\pi - \omega)$$

$$= \sum_{k=1}^{(M-1)/2} c(k) \sin \omega k \cos \pi k$$

$$= \sum_{k=1}^{(M-1)/2} c(k)(-1)^{k+1} \sin \omega k$$

or equivalently,

$$\sum_{k=1}^{(M-1)/2} [1 - (-1)^{k+1}] c(k) \sin \omega k = 0 \qquad (8.1.105)$$

Clearly, $c(k)$ must be equal to zero for $k = 0, 2, 4, \ldots$.

Now, the relationship between $\{c(k)\}$ and the unit sample response $\{h(n)\}$ is, from (8.1.69),

$$c(k) = 2h \left(\frac{M-1}{2} - k \right)$$

or, equivalently,

$$h \left(\frac{M-1}{2} - k \right) = \frac{1}{2} c(k) \qquad (8.1.106)$$

If $c(k)$ is zero for $k = 0, 2, 4, \ldots$, then (8.1.106) yields

$$h(k) = \begin{cases} 0 & k = 0, 2, 4, \ldots, \text{ for } \dfrac{M-1}{2} \text{ even} \\[2ex] 0 & k = 1, 3, 5, \ldots, \text{ for } \dfrac{M-1}{2} \text{ odd} \end{cases} \qquad (8.1.107)$$

Unfortunately, (8.1.107) holds only for M odd. It does not hold for M even. This means that for comparable values of M, the case M odd is preferable since the computational complexity (number of multiplications and additions per output point) is roughly one half of that for M even.

When the design of the Hilbert transformer is performed by the Chebyshev approximation criterion using the Remez algorithm, we select the filter coefficients to minimize the peak approximation error

$$\delta = \max_{2\pi f_l \leq \omega \leq 2\pi f_u} [H_{dr}(\omega) - H_r(\omega)]$$

$$= \max_{2\pi f_l \leq \omega \leq 2\pi f_u} [1 - H_r(\omega)]$$

(8.1.108)

Thus the weighting function is set to unity and the optimization is performed over the single frequency band (i.e., the passband of the filter).

EXAMPLE 8.1.6

Design a Hilbert transformer with parameters $M = 31$, $f_l = 0.05$, and $f_u = 0.45$.

Solution: We observe that the frequency response is symmetric, since $f_u = 0.5 - f_l$. The parameters for executing the Remez algorithm are

$$31, \quad 3, \quad 1$$
$$0.05, \quad 0.45$$
$$1.0$$
$$1.0$$

The result of this design is the unit sample response coefficients and the peak approximation error $\delta = 0.0026803$ or -51.4 dB given in Table 8.10. We observe that, indeed, every other value of $h(n)$ is essentially zero (these values are of the order of 10^{-7}). The frequency response of the Hilbert transformer is shown in Fig. 8.33.

Rabiner and Schafer (1974b) have investigated the characteristics of Hilbert transformer designs for both M odd and M even. If the filter design is restricted to a symmetric frequency response, then there are basically three parameters of interest, M, δ, and f_l. Figure 8.34 is a plot of $20 \log_{10} \delta$ versus f_l with M as a parameter. We observe that for comparable values of M, there is no performance advantage of using M odd over M even, and vice versa. However, the computational complexity in implementing a filter for M odd is less by a factor of 2 over M even as indicated above. Therefore, M odd is preferable in practice.

For design purposes, the graphs in Fig. 8.34 suggest that, as a rule of thumb,

$$Mf_l \approx -0.61 \log_{10} \delta$$

(8.1.109)

Hence this formula can be used to estimate the size of one of the three basic filter parameters when the other two parameters are specified.

In concluding this section, we wish to indicate that Hilbert transformers can also be designed by the window method and the frequency sampling method. For example, Fig. 8.35 illustrates the frequency response of an $M = 31$ Hilbert transformer designed by means of the frequency sampling method. The corresponding values of the unit sample response are given in Table 8.11. A comparison of these filter parameters with those given in Table 8.10 indicates some small differences. In particular it appears that the Chebyshev approximation criterion gives significantly smaller values for the filter coefficients that should be zero. In general, the Chebyshev approximation criterion results in better filter designs.

TABLE 8.10 Parameters for FIR Hilbert Transform Filter in Example 8.1.6

```
                      FINITE IMPULSE RESPONSE (FIR)
                    LINEAR PHASE DIGITAL FILTER DESIGN
                        REMEZ EXCHANGE ALGORITHM

                          HILBERT TRANSFORMER

                        FILTER LENGTH = 31

                  ***** IMPULSE RESPONSE *****
             H( 1) =     0.41957516E-02 =  -H( 31)
             H( 2) =     0.64310257E-07 =  -H( 30)
             H( 3) =     0.92822444E-02 =  -H( 29)
             H( 4) =     0.52693927E-07 =  -H( 28)
             H( 5) =     0.18835988E-01 =  -H( 27)
             H( 6) =     0.82308283E-07 =  -H( 26)
             H( 7) =     0.34401190E-01 =  -H( 25)
             H( 8) =     0.93328794E-07 =  -H( 24)
             H( 9) =     0.59551738E-01 =  -H( 23)
             H(10) =     0.50821171E-07 =  -H( 22)
             H(11) =     0.10303782E+00 =  -H( 21)
             H(12) =     0.17612138E-07 =  -H( 20)
             H(13) =     0.19683167E+00 =  -H( 19)
             H(14) =    -0.23977606E-07 =  -H( 18)
             H(15) =     0.63135374E+00 =  -H( 17)
             H(16) =     0.0
```

	BAND 1
LOWER BAND EDGE	0.0500000
UPPER BAND EDGE	0.4500000
DESIRED VALUE	1.0000000
WEIGHTING	1.0000000
DEVIATION	0.0026803

```
EXTREMAL FREQUENCIES—MAXIMA OF THE ERROR CURVE
0.0500000      0.0562500      0.0750000      0.1000000      0.1291666
0.1583333      0.1874999      0.2187499      0.2499998      0.2812498
0.3124998      0.3416664      0.3708331      0.3999997      0.4249997
0.4437497
```

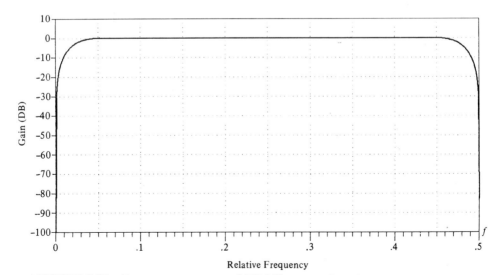

FIGURE 8.33 Frequency response of FIR Hilbert transform filter in Example 8.1.6.

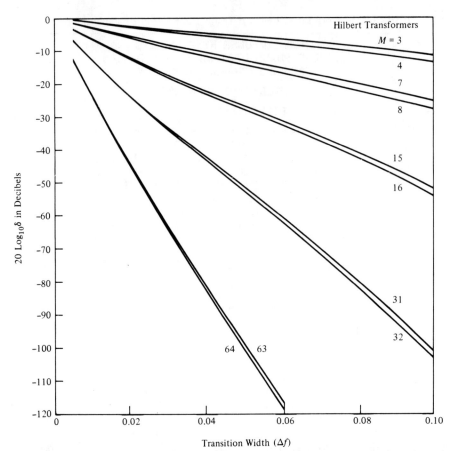

Transition Width (Δf)

FIGURE 8.34 Curves of 20 $\log_{10} \delta$ versus Δf for M = 3, 4, 7, 8, 15, 16, 31, 32, 63, 64. [From paper by Rabiner and Schafer (1974b). Reprinted with permission of AT&T.]

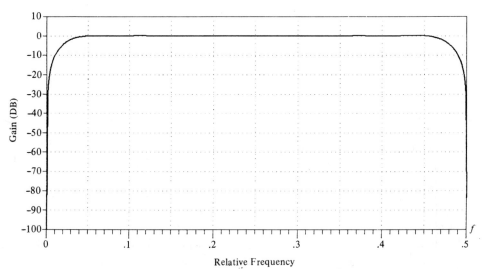

Relative Frequency

FIGURE 8.35 Frequency response of M = 31 FIR Hilbert transform filter designed by the frequency sampling method.

TABLE 8.11 Parameters a for $M = 31$ Hilbert Transform Filter Designed
by the Frequency-Sampling Method

```
LINEAR PHASE FIR HILBERT TRANSFORM
FREQUENCY SAMPLING METHOD

FILTER LENGTH  =31
LOWER CUTOFF FREQUENCY (RELATIVE)  = 0.5000000E-01
UPPER CUTOFF FREQUENCY (RELATIVE)  = 0.4500000E+00

IMPULSE RESPONSE:
                    H( 0)  =  -0.1342662E-03
                    H( 1)  =   0.2133148E-02
                    H( 2)  =   0.4848863E-02
                    H( 3)  =   0.2286159E-02
                    H( 4)  =   0.1423532E-01
                    H( 5)  =   0.1517075E-02
                    H( 6)  =   0.3001805E-01
                    H( 7)  =   0.5263533E-03
                    H( 8)  =   0.5574721E-01
                    H( 9)  =  -0.2281570E-03
                    H(10)  =   0.1001032E+00
                    H(11)  =  -0.5338326E-03
                    H(12)  =   0.1949848E+00
                    H(13)  =  -0.3994641E-03
                    H(14)  =   0.6307253E+00
                    H(15)  =  -0.9335956E-06
                    H(16)  =  -0.6307245E+00
                    H(17)  =   0.3996222E-03
                    H(18)  =  -0.1949853E+00
                    H(19)  =   0.5341307E-03
                    H(20)  =  -0.1001035E+00
                    H(21)  =   0.2285338E-03
                    H(22)  =  -0.5574735E-01
                    H(23)  =  -0.5263340E-03
                    H(24)  =  -0.3001794E-01
                    H(25)  =  -0.1517240E-02
                    H(26)  =  -0.1423557E-01
                    H(27)  =  -0.2285915E-02
                    H(28)  =  -0.4848215E-02
                    H(29)  =  -0.2133800E-02
                    H(30)  =   0.1344162E-03
```

8.1.6 Comparison of Design Methods for Linear-Phase FIR Filters

Historically, the design method based on the use of windows to truncate the impulse response $h_d(n)$ and to obtain the desired spectral shaping was the first method proposed for designing linear-phase FIR filters. The frequency-sampling method and the Chebyshev approximation method were developed in the 1970s and have since become very popular in the design of practical linear-phase FIR filters.

The major disadvantage of the window design method is the lack of precise control of the critical frequencies, such as ω_p and ω_s, for example, in the design of a lowpass FIR filter. The values of ω_p and ω_s, in general, depend on the type of window and the filter length M.

The frequency sampling method provides an improvement over the window design method, since $H_r(\omega)$ is specified at the frequencies $\omega_k = 2\pi k/M$, and the transition band is a multiple of $2\pi/M$. This filter design method is particularly attractive when the FIR filter is realized either in the frequency domain by means of the DFT or in any of the frequency sampling realizations given by (8.1.45) or (8.1.48) or (8.1.49). The attractive feature of these realizations is that $H_r(\omega_k)$ is either zero or unity at all frequencies, except in the transition band.

The Chebyshev approximation method provides total control of the filter specifications, and, as a consequence, it is usually preferable over the other two methods. For a lowpass filter, the specifications are given in terms of the parameters ω_p, ω_s, δ_1, δ_2, and M. We may specify the parameters ω_p, ω_s, M and δ, and optimize the filters relative to δ_2. By spreading the approximation error over the passband and the stopband of the filter, this method results in an optimal filter design, in the sense that for a given set of specifications as indicated above, the maximum sidelobe level is minimized.

The Chebyshev design procedure based on the Remez exchange algorithm requires that we specify the length of the filter, the critical frequencies ω_p and ω_s, and the ratio δ_2/δ_1. However, it is more natural in filter design to specify ω_p, ω_s, δ_1, and δ_2 and to determine the filter length that satisfies the specifications. Although there is no simple formula to determine the filter length from these specifications, a number of approximations have been proposed for estimating M from ω_p, ω_s, δ_1, and δ_2. A particularly simple formula attributed to Kaiser for approximating M is

$$\hat{M} = \frac{-20 \log_{10} (\sqrt{\delta_1 \delta_2}) - 13}{14.6 \, \Delta f} + 1 \qquad (8.1.110)$$

where Δf is the transition band, defined as $\Delta f = (\omega_s - \omega_p)/2\pi$. This formula has been given in the paper by Rabiner et al. (1975). A more accurate formula proposed by Herrmann et al. (1973) is

$$\hat{M} = \frac{D_\infty(\delta_1, \delta_2) - f(\delta_1, \delta_2)(\Delta f)^2}{\Delta f} + 1 \qquad (8.1.111)$$

where, by definition,

$$D_\infty(\delta_1, \delta_2) = [0.005309(\log_{10} \delta_1)^2 + 0.07114(\log_{10} \delta_1) - 0.4761](\log_{10} \delta_2)$$
$$- [0.00266(\log_{10} \delta_1)^2 + 0.5941 \log_{10} \delta_1 + 0.4278] \qquad (8.1.112)$$
$$f(\delta_1, \delta_2) = 11.012 + 0.51244(\log_{10} \delta_1 - \log_{10} \delta_2) \qquad (8.1.113)$$

These formulas are extremely useful in obtaining a good estimate of the filter length required to achieve the given specifications Δf, δ_1, and δ_2. The estimate is used to carry out the design and if the resulting δ exceeds the specified δ_2, the length may be increased until we obtain a sidelobe level that meets the specifications.

8.2 Design of IIR Filters from Analog Filters

Just as in the design of FIR filters, there are several methods that can be used to design digital filters having an infinite-duration unit sample response. The techniques to be described in this section are all based on taking an analog filter and converting it to a digital filter. Analog filter design is a mature and well-developed field, so it is not surprising that we begin the design of a digital filter in the analog domain and then convert the design into the digital domain.

An analog filter may be described by its system function.

$$H_a(s) = \frac{B(s)}{A(s)} = \frac{\sum\limits_{k=0}^{M} \beta_k s^k}{\sum\limits_{k=0}^{N} \alpha_k s^k} \tag{8.2.1}$$

where $\{\alpha_k\}$ and $\{\beta_k\}$ are the filter coefficients, or by its impulse response, which is related to $H_a(s)$ by the Laplace transform

$$H_a(s) = \int_{-\infty}^{\infty} h(t)e^{-st} \, dt \tag{8.2.2}$$

Alternatively, the analog filter having the rational system function $H(s)$ given in (8.2.1) can be described by the linear constant-coefficient differential equation

$$\sum_{k=0}^{N} \alpha_k \frac{d^k y(t)}{dt^k} = \sum_{k=0}^{M} \beta_k \frac{d^k x(t)}{dt^k} \tag{8.2.3}$$

where $x(t)$ denotes the input signal and $y(t)$ denotes the output of the filter.

Each of these three equivalent characterizations of an analog filter leads to alternative methods for converting the filter into the digital domain, as will be described below. We recall that an analog linear time-invariant system with system function $H(s)$ is stable if all its poles lie in the left-half of the s-plane. Consequently, if the conversion technique is to be effective, it should possess the following desirable properties:

1. The $j\Omega$ axis in the s-plane should map into the unit circle in the z-plane. Thus there will be a direct relationship between the two frequency variables in the two domains.
2. The left-half plane (LHP) of the s-plane should map into the inside of the unit circle in the z-plane. Thus a stable analog filter will be converted to a stable digital filter.

We mentioned in the preceding section that physically realizable and stable IIR filters cannot have linear phase. Recall that a linear-phase filter must have a system function that satisfies the condition

$$H(z) = \pm z^{-N} H(z^{-1}) \tag{8.2.4}$$

where z^{-N} represents a delay of N units of time. But if this were the case, the filter would have a mirror-image pole outside the unit circle for every pole inside the unit circle. Hence the filter would be unstable. Consequently, a causal and stable IIR filter cannot have linear phase.

If the restriction on physical realizability is removed, it is possible to obtain a linear phase IIR filter, at least in principle. This approach involves performing a time reversal of the input signal $x(n)$, passing $x(-n)$ through a digital filter $H(z)$, time-reversing the output of $H(z)$, and finally, passing the result through $H(z)$ again. This signal processing is computationally cumbersome and appears to offer no advantages over linear-phase FIR filters. Consequently, when an application requires a linear-phase filter, it should be an FIR filter.

In the design of IIR filters, we shall specify the desired filter characteristics for the magnitude response only. This does not mean that we consider the phase response unimportant. Since the magnitude and phase characteristics are related, as indicated in Section 5.3.2, we specify the desired magnitude characteristics and accept the phase response that is obtained from the design methodology.

8.2.1 IIR Filter Design by Approximation of Derivatives

One of the simplest methods for converting an analog filter into a digital filter is to approximate the differential equation in (8.2.3) by an equivalent difference equation. This approach is often used to solve a linear constant-coefficient differential equation numerically on a digital computer.

For the derivative $dy(t)/dt$ at time $t = nT$ we substitute the *backward difference* $[y(nT) - y(nT - 1)]/T$. Thus

$$\left. \frac{dy(t)}{dt} \right|_{t=nT} = \frac{y(nT) - y(nT - T)}{T}$$

$$= \frac{y(n) - y(n - 1)}{T} \tag{8.2.5}$$

where T represents the sampling interval and $y(n) \equiv y(nT)$. The analog differentiator with output $dy(t)/dt$ has the system function $H(s) = s$, while the digital system that produces the output $[y(n) - y(n - 1)]/T$ has the system function $H(z) = (1 - z^{-1})/T$. Consequently, as shown in Fig. 8.36, the frequency-domain equivalent for the relationship in (8.2.5) is

$$s = \frac{1 - z^{-1}}{T} \tag{8.2.6}$$

The second derivative $d^2y(t)/dt^2$ is replaced by the second difference, which is derived as follows:

$$\left. \frac{d^2y(t)}{dt^2} \right|_{t=nT} = \frac{d}{dt} \left[\frac{dy(t)}{dt} \right]_{t=nT}$$

$$= \frac{[y(nT) - y(nT - T)]/T - [y(nT - T) - y(nT - 2T)]/T}{T} \tag{8.2.7}$$

$$= \frac{y(n) - 2y(n - 1) + y(n - 2)}{T^2}$$

In the frequency domain, (8.2.7) is equivalent to

$$s^2 = \frac{1 - 2z^{-1} + z^{-2}}{T^2} = \left(\frac{1 - z^{-1}}{T} \right)^2 \tag{8.2.8}$$

It easily follows from the above that the substitution for the kth derivative of $y(t)$ results in the equivalent frequency-domain relationship

$$s^k = \left(\frac{1 - z^{-1}}{T} \right)^k \tag{8.2.9}$$

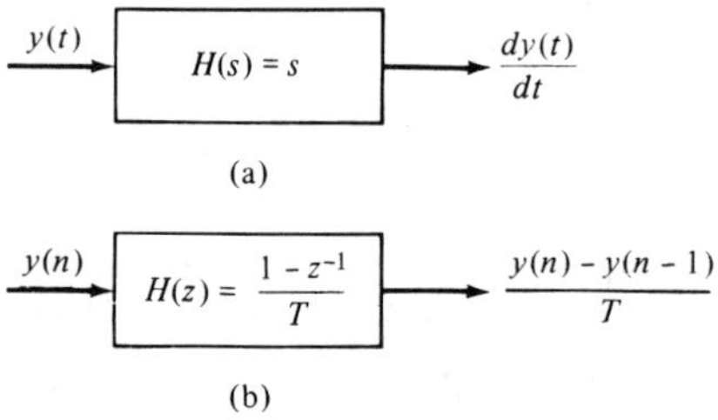

(a)

(b)

FIGURE 8.36 Substitution of the backward difference for the derivative implies the mapping $s = (1 - z^{-1})/T$.

Consequently, the system function for the digital IIR filter obtained as a result of the approximation of the derivatives by finite differences is

$$H(z) = H_a(s)\big|_{s=(1-z^{-1})/T} \qquad (8.2.10)$$

where $H_a(s)$ is the system function of the analog filter characterized by the differential equation given in (8.2.3).

Let us investigate the implications of the mapping from the s-plane to the z-plane as given by (8.2.6) or, equivalently,

$$z = \frac{1}{1 - sT} \qquad (8.2.11)$$

If we substitute $s = j\Omega$ in (8.2.11), we find that

$$z = \frac{1}{1 - j\Omega T} \qquad (8.2.12)$$
$$= \frac{1}{1 + \Omega^2 T^2} + j\frac{\Omega T}{1 + \Omega^2 T^2}$$

As Ω varies from $-\infty$ to ∞, the corresponding locus of points in the z-plane is a circle of radius $\frac{1}{2}$ and with center at $z = \frac{1}{2}$, as illustrated in Fig. 8.37.

It is easily demonstrated that the mapping in (8.2.11) takes points in the LHP of s into corresponding points inside this circle in the z-plane and points in the RHP plane in s are mapped into points outside this circle. Consequently, this mapping has the desirable property that a stable analog filter is transformed into a stable digital filter. However, the possible location of the poles of the digital filter are confined to relatively small frequencies and, as a consequence, the mapping is restricted to design of lowpass filters and bandpass filters having relatively small resonant frequencies. It is not possible, for example, to transform a highpass analog filter into a corresponding highpass digital filter.

The relationship in (8.2.11) for mapping the s-plane into the z-plane resulted from using the *backward difference* in (8.2.5) to substitute for the derivative. Instead of the backward difference, suppose we use the *forward difference* in the substitution for the derivative, that is,

$$\frac{dy(t)}{dt} = \frac{y(nT + T) - y(nT)}{T} \qquad (8.2.13)$$
$$= \frac{y(n + 1) - y(n)}{T}$$

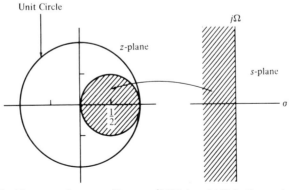

FIGURE 8.37   The mapping $s = (1 - z^{-1})/T$ takes LHP in the s-plane into points inside the circle of radius $\frac{1}{2}$ and center $z = \frac{1}{2}$ in the z-plane.

This approximation is equivalent to the mapping

$$s = \frac{z - 1}{T} \qquad (8.2.14)$$

or, equivalently,

$$z = 1 + sT \qquad (8.2.15)$$

The characteristics of this mapping are worse than the mapping obtained with the backward difference. For example, when $s = j\Omega$, the corresponding points in the z-plane lie on a vertical line with coordinates $(z_r, z_i) = (1, \Omega T)$, as illustrated in Fig. 8.38. More generally, any point $s = \sigma + j\Omega$ maps into

$$z_r + jz_i = 1 + \sigma T + j\Omega T$$
$$= \sqrt{(1 + \sigma T)^2 + \Omega^2 T^2} \; e^{j\tan^{-1}\Omega T/(1+\sigma T)} \qquad (8.2.16)$$

Hence stable analog filters do not always map into stable digital filters when the forward difference is used to substitute for the derivative. Consequently, this is not generally a good method for converting an analog filter into an equivalent digital filter.

In an attempt to overcome the limitations in the two mappings given above, more complex substitutions for the derivatives have been proposed. In particular, an Lth-order difference of the form

$$\left.\frac{dy(t)}{dt}\right|_{t=nT} = \frac{1}{T}\sum_{k=1}^{L}\alpha_k \frac{y(nT + kT) - y(nT - kT)}{T} \qquad (8.2.17)$$

has been proposed, where $\{\alpha_k\}$ are a set of parameters that can be selected to optimize the approximation. The resulting mapping between the s-plane and the z-plane is now

$$s = \frac{1}{T}\sum_{k=1}^{L}\alpha_k(z^k - z^{-k}) \qquad (8.2.18)$$

When $z = e^{j\omega}$, we have

$$s = j\frac{2}{T}\sum_{k=1}^{L}\alpha_k \sin \omega k \qquad (8.2.19)$$

which is purely imaginery. Thus

$$\Omega = \frac{2}{T}\sum_{k=1}^{L}\alpha_k \sin \omega k \qquad (8.2.20)$$

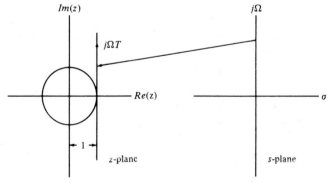

FIGURE 8.38 The mapping $s = (z - 1)/T$ takes the $j\Omega$-axis in the s-plane into the line $1 + j\Omega T$ in the z-plane.

is the resulting mapping between the two frequency variables. By proper choice of the coefficients $\{\alpha_k\}$ it is possible to map the $j\Omega$-axis into the unit circle. Furthermore, points in the LHP in s can be mapped into points inside the unit circle in z.

Despite achieving the two desirable characteristics with the mapping of (8.2.18), the problem of selecting the set of coefficients $\{\alpha_k\}$ remains. In general, this is a difficult problem. Since simpler techniques exist for converting analog filters into IIR digital filters, we shall not emphasize the use of the Lth-order difference as a substitute for the derivative.

EXAMPLE 8.2.1

Use the backward difference for the derivative to convert the analog lowpass filter with system function

$$H_a(s) = \frac{1}{s + 1}$$

into a digital IIR filter.

Solution: The mapping from s to z is

$$s = \frac{1 - z^{-1}}{T}$$

Hence

$$H(z) = \frac{1}{[(1 - z^{-1})/T] + 1}$$
$$= \frac{zT/(1 + T)}{z - [1/(1 + T)]}$$

The digital filter has a pole at $1/(1 + T)$. In order to obtain a resonance at low frequencies, we must select T as small as possible, so as to place the pole close to the unit circle. For example, we may select $T = 0.1$, so that

$$H(z) = \frac{0.09z}{z - 0.909}$$
$$= \frac{0.09}{1 - 0.909z^{-1}}$$

Since the pole of the system is at $z = 0.909$, the frequency response $H(\omega)$ has a peak at $\omega = 0$. Figure 8.39 illustrates the frequency responses of the analog and digital filters.

EXAMPLE 8.2.2

Convert the analog bandpass filter with system function

$$H_a(s) = \frac{1}{(s + 0.1)^2 + 9}$$

into a digital IIR filter by use of the backward difference for the derivative.

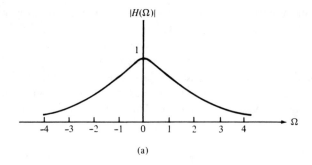

(a)

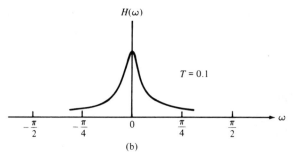

(b)

FIGURE 8.39 Frequency responses of analog and digital filters in Example 8.2.1.

Solution: Substitution for s from (8.2.6) into $H(s)$ yields

$$H(z) = \frac{1}{\left(\dfrac{1 - z^{-1}}{T} + 0.1\right)^2 + 9}$$

$$= \frac{T^2/(1 + 0.2T + 9.01T^2)}{1 - \dfrac{2(1 + 0.1T)}{1 + 0.2T + 9.01T^2}z^{-1} + \dfrac{1}{1 + 0.2T + 9.01T^2}z^{-2}}$$

The system function $H(z)$ has the form of a resonator provided that T is selected small enough (e.g., $T \le 0.1$), in order for the poles to be near the unit circle. Note that the condition $a_1^2 < 4a_2$ is satisfied, so that the poles are complex valued.

For example, if $T = 0.1$, the poles are located at

$$p_{1,2} = 0.91 \pm j0.27$$
$$= 0.949e^{\pm j16.5°}$$

We note that the range of resonant frequencies is limited to low frequencies, due to the characteristics of the mapping. The reader is encouraged to plot the frequency response $H(\omega)$ of the digital filter for different values of T and compare the results with the frequency response of the analog filter.

EXAMPLE 8.2.3 _____

Convert the analog bandpass filter in Example 8.2.2 into a digital IIR filter by use of the mapping

$$s = \frac{1}{T}(z - z^{-1})$$

Solution: By substituting for s in $H(s)$, we obtain

$$H(z) = \cfrac{1}{\left(\cfrac{z - z^{-1}}{T} + 0.1\right)^2 + 9}$$

$$= \frac{z^2 T^2}{z^4 + 0.2Tz^3 + (2 + 9.01T^2)z^2 - 0.2Tz + 1}$$

We observe that this mapping has introduced two additional poles in the conversion from $H_a(s)$ to $H(z)$. As a consequence, the digital filter is significantly more complex than the analog filter. This is a major drawback to the mapping given above.

8.2.2 IIR Filter Design by Impulse Invariance

In the impulse invariance method, our objective is to design an IIR filter having a unit sample response $h(n)$ that is the sampled version of the impulse response of the analog filter. That is,

$$h(n) \equiv h(nT) \qquad n = 0, 1, 2, \ldots \qquad (8.2.21)$$

where T is the sampling interval.

To examine the implications of (8.2.21), we refer back to Section 4.5. Recall that when a continuous time signal $x_a(t)$ with spectrum $X_a(F)$ is sampled at a rate $F_s = 1/T$ samples per second, the spectrum of the sampled signal is the periodic repetition of the scaled spectrum $F_s X_a(F)$ with period F_s. Specifically, the relationship is

$$X(f) = F_s \sum_{k=-\infty}^{\infty} X_a[(f - k)F_s] \qquad (8.2.22)$$

where $f = F/F_s$ is the normalized frequency. Aliasing occurs if the sampling rate F_s is less than twice the highest frequency contained in $X_a(F)$.

Expressed in the context of sampling the impulse response of an analog filter with frequency response $H_a(F)$, the digital filter with unit sample response $h(n) \equiv h_a(nT)$ has the frequency response

$$H(f) = F_s \sum_{k=-\infty}^{\infty} H_a[(f - k)F_s] \qquad (8.2.23)$$

or, equivalently,

$$H(\omega) = F_s \sum_{k=-\infty}^{\infty} H_a[(\omega - 2\pi k)F_s] \qquad (8.2.24)$$

or

$$H(\Omega T) = \frac{1}{T} \sum_{k=-\infty}^{\infty} H_a\left(\Omega - \frac{2\pi k}{T}\right) \qquad (8.2.25)$$

Figure 8.40 depicts the frequency response of a lowpass analog filter and the frequency response of the corresponding digital filter.

It is clear that the digital filter with frequency response $H(\omega)$ will possess the frequency response characteristics of the corresponding analog filter if the sampling interval T is selected sufficiently small to avoid completely or minimize the effects of aliasing. It is also clear that the impulse invariance method is inappropriate for designing highpass filters due to spectrum aliasing that results from the sampling process.

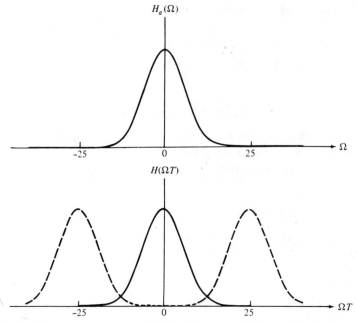

FIGURE 8.40   Frequency response $H_a(\Omega)$ of the analog filter and frequency response of the corresponding digital filter with aliasing.

To investigate the mapping between the z-plane and the s-plane implied by the sampling process, we rely on a generalization of (8.2.25) which relates z-transform of $h(n)$ to the Laplace transform of $h_a(t)$. This relationship is

$$H(z)|_{z=e^{sT}} = \frac{1}{T} \sum_{k=-\infty}^{\infty} H_a\left(s - j\frac{2\pi k}{T}\right) \qquad (8.2.26)$$

where

$$H(z) = \sum_{n=0}^{\infty} h(n)z^{-n} \qquad (8.2.27)$$

$$H(z)|_{z=e^{sT}} = \sum_{n=0}^{\infty} h(n)e^{-sTn}$$

Note that when $s = j\Omega$, (8.2.26) reduces to (8.2.25), where the factor of j in $H_a(\Omega)$ is suppressed in our notation.

The general characteristic of the mapping

$$z = e^{sT} \qquad (8.2.28)$$

can be obtained by substituting $s = \sigma + j\Omega$ and expressing the complex variable z in polar form as $z = re^{j\omega}$. With these substitutions (8.2.28) becomes

$$re^{j\omega} = e^{\sigma T}e^{j\Omega T}$$

Clearly, we must have

$$r = e^{\sigma T} \qquad (8.2.29)$$
$$\omega = \Omega T$$

Consequently, $\sigma < 0$ implies that $0 < r < 1$ and $\sigma > 0$ implies that $r > 1$. When $\sigma = 0$, we have $r = 1$. Therefore, the LHP in s is mapped inside the unit circle in z and the

RHP in s is mapped into points that fall outside the unit circle in z. This is one of the desirable properties of a good mapping.

Also, the $j\Omega$-axis is mapped into the unit circle in z as indicated above. However, the mapping of the $j\Omega$-axis into the unit circle is not one-to-one. Since ω is unique over the range $(-\pi, \pi)$, the mapping $\omega = \Omega T$ implies that the interval $-\pi/T \le \Omega \le \pi/T$ maps into the corresponding values of $-\pi \le \omega \le \pi$. Furthermore, the frequency interval $\pi/T \le \Omega \le 3\pi/T$ also maps into the interval $-\pi \le \omega \le \pi$ and, in general, so does the interval $(2k - 1)\pi/T \le \Omega \le (2k + 1)\pi/T$, when k is an integer. Thus the mapping from the analog frequency Ω to the frequency variable ω in the digital domain is many-to-one, which simply reflects the effects of aliasing due to sampling. Figure 8.41 illustrates the mapping from the s-plane to the z-plane.

To explore further the effect of the impulse invariance design method on the characteristics of the resulting filter, let us express the system function of the analog filter in partial-fraction form. On the assumption that the poles of the analog filter are distinct, we may write

$$H_a(s) = \sum_{k=1}^{N} \frac{c_k}{s - p_k} \qquad (8.2.30)$$

where $\{p_k\}$ are the poles of the analog filter and $\{c_k\}$ are the coefficients in the partial-fraction expansion. Consequently,

$$h_a(t) = \sum_{k=1}^{N} c_k e^{p_k t} \qquad t \ge 0 \qquad (8.2.31)$$

If we sample $h_a(t)$ periodically at $t = nT$, we have

$$\begin{aligned} h(n) &= h_a(nT) \\ &= \sum_{k=1}^{N} c_k e^{p_k T n} \end{aligned} \qquad (8.2.32)$$

Now, with the substitution of (8.2.32), the system function of the resulting digital IIR filter becomes

$$\begin{aligned} H(z) &= \sum_{n=0}^{\infty} h(n) z^{-n} \\ &= \sum_{n=0}^{\infty} \left(\sum_{k=1}^{N} c_k e^{p_k T n} \right) z^{-n} \qquad (8.2.33) \\ &= \sum_{k=1}^{N} c_k \sum_{n=0}^{\infty} (e^{p_k T} z^{-1})^n \end{aligned}$$

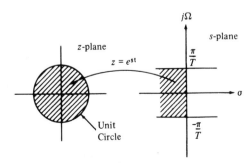

FIGURE 8.41 The mapping of $z = e^{sT}$ maps strips of width $2\pi/T$ (for $\sigma < 0$) in the s-plane into points in the unit circle in the z-plane.

The inner sum in (8.2.33) converges because $p_k < 0$ and yields

$$\sum_{n=0}^{\infty} (e^{p_k T} z^{-1})^n = \frac{1}{1 - e^{p_k T} z^{-1}} \tag{8.2.34}$$

Therefore, the system function of the digital filter is

$$H(z) = \sum_{k=1}^{N} \frac{c_k}{1 - e^{p_k T} z^{-1}} \tag{8.2.35}$$

We observe that the digital filter has poles at

$$z_k = e^{p_k T} \qquad k = 1, 2, \ldots, N \tag{8.2.36}$$

With this form for the system function $H(z)$, the IIR filter is easily realized as a parallel bank of single-pole filters. If some of the poles are complex valued, they may be paired together and combined to form two-pole filter sections. In addition, two factors containing real-valued poles may be combined to form two-pole filters. Consequently, the resulting filter may be realized as a parallel bank of two-pole filters.

Although the development that resulted in $H(z)$ given by (8.2.35) was based on a filter having distinct poles, it can be generalized to include multiple-order poles. For brevity, however, we shall not attempt to generalize (8.2.35).

EXAMPLE 8.2.4 _____

Convert the analog filter with system function

$$H_a(s) = \frac{s + 0.1}{(s + 0.1)^2 + 9}$$

into a digital IIR filter by means of the impulse invariance method.

Solution: We note that the analog filter has a zero at $s = -0.1$ and a pair of complex-conjugate poles at

$$p_k = -0.1 \pm j3$$

as illustrated in Fig. 8.42.

We do not have to determine the impulse response $h_a(t)$ in order to design the digital IIR filter based on the method of impulse invariance. Instead, we directly determine $H(z)$, as given by (8.2.35), from the partial-fraction expansion of $H_a(s)$. Thus we have

$$H(s) = \frac{\frac{1}{2}}{s + 0.1 - j3} + \frac{\frac{1}{2}}{s + 0.1 + j3}$$

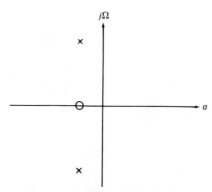

FIGURE 8.42   Pole–zero locations for analog filter in Example 8.2.4.

Then

$$H(z) = \frac{\frac{1}{2}}{1 - e^{0.1T}e^{j3T}z^{-1}} + \frac{\frac{1}{2}}{1 - e^{0.1T}e^{-j3T}z^{-1}}$$

Since the two poles are complex, conjugates we can combine them to form a single two-pole filter with system function

$$H(z) = \frac{1 - e^{0.1T}\cos 3Tz^{-1}}{1 - 2e^{0.1T}\cos 3Tz^{-1} + e^{0.2T}z^{-1}}$$

The magnitude of the frequency response characteristic of this filter is plotted in Fig. 8.43 for $T = 0.1$ and $T = 0.5$. For purpose of comparison, we have also plotted the magnitude of the frequency response of the analog filter in Fig. 8.44, We note that aliasing is significantly more prevalent when $T = 0.5$ than when $T = 0.1$. Also, note the shift of the resonant frequency as T changes.

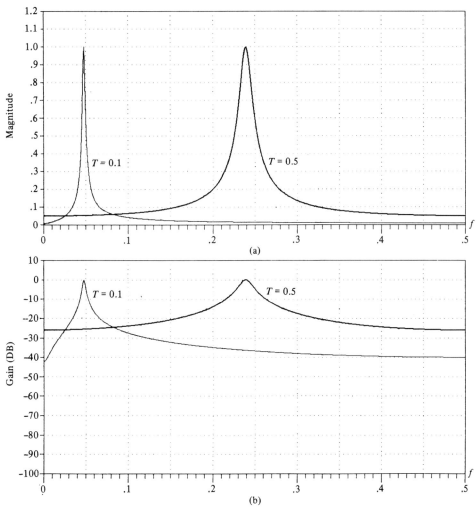

FIGURE 8.43   Frequency response of digital filter in Example 8.2.4.

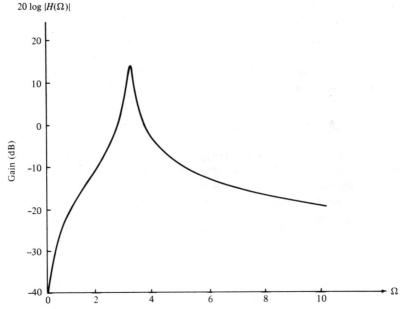

FIGURE 8.44 Frequency response of analog filter in Example 8.2.4.

The preceding example illustrates the importance of selecting a small value for T to minimize the effect of aliasing. Due to the presence of aliasing, the impulse invariance method is appropriate for the design of lowpass and bandpass filters.

8.2.3 IIR Filter Design by the Bilinear Transformation

The IIR filter design techniques described in the preceding two sections have a severe limitation in that they are appropriate only for lowpass filters and a limited class of bandpass filters. This limitation is a result of the mapping that converts points in the s-plane to corresponding points in the z-plane.

In this section we describe a mapping from the s-plane to the z-plane, called the bilinear transformation, that overcomes the limitation of the other two design methods described previously. The bilinear transformation is a conformal mapping that transforms the $j\Omega$-axis into the unit circle in the z-plane only once, thus avoiding aliasing of frequency components. Furthermore, all points in the LHP of s are mapped inside the unit circle in the z-plane and all points in the RHP of s are mapped into corresponding points outside the unit circle in the z-plane.

The bilinear transformation can be linked to the trapezoidal formula for numerical integration. For example, let us consider an analog linear filter with system function

$$H(s) = \frac{b}{s + a} \tag{8.2.37}$$

This system is also characterized by the differential equation

$$\frac{dy(t)}{dt} + ay(t) = bx(t) \tag{8.2.38}$$

Instead of substituting a finite difference for the derivative, suppose that we integrate the derivative and approximate the integral by the trapezoidal formula. Thus

$$y(t) = \int_{t_0}^{t} y'(\tau) \, d\tau + y(t_0) \tag{8.2.39}$$

where $y'(t)$ denotes the derivative of $y(t)$. The approximation of the integral in (8.2.39) by the trapezoidal formula at $t = nT$ and $t_0 = nT - T$ yields

$$y(nT) = \frac{T}{2} [y'(nT) + y'(nT - T)] + y(nT - T) \tag{8.2.40}$$

Now the differential equation in (8.2.38) evaluated at $t = nT$ yields

$$y'(nT) = -ay(nT) + bx(nT) \tag{8.2.41}$$

We use (8.2.41) to substitute for the derivative in (8.2.40) and thus obtain a difference equation for the equivalent discrete-time system. With $y(n) \equiv y(nT)$ and $x(n) \equiv x(nT)$, we obtain the result

$$\left(1 + \frac{aT}{2}\right) y(n) - \left(1 - \frac{aT}{2}\right) y(n - 1) = \frac{bT}{2} [x(n) + x(n - 1)] \tag{8.2.42}$$

The z-transform of this difference equation is

$$\left(1 + \frac{aT}{2}\right) Y(z) - \left(1 - \frac{aT}{2}\right) z^{-1} Y(z) = \frac{bT}{2} (1 + z^{-1}) X(z)$$

Consequently, the system function of the equivalent digital filter is

$$H(z) = \frac{Y(z)}{X(z)} = \frac{(bT/2)(1 + z^{-1})}{1 + aT/2 - (1 - aT/2)z^{-1}}$$

or, equivalently,

$$H(z) = \frac{b}{\dfrac{2}{T}\left(\dfrac{1 - z^{-1}}{1 + z^{-1}}\right) + a} \tag{8.2.43}$$

Clearly, the mapping from the s-plane to the z-plane is

$$s = \frac{2}{T}\left(\frac{1 - z^{-1}}{1 + z^{-1}}\right) \tag{8.2.44}$$

This is called the *bilinear transformation*.

Although our derivation of the bilinear transformation was performed for a first-order differential equation, it holds, in general, for an Nth-order differential equation.

To investigate the characteristics of the bilinear transformation, let

$$z = re^{j\omega}$$
$$s = \sigma + j\Omega$$

Then (8.2.44) can be expressed as

$$s = \frac{2}{T} \frac{z - 1}{z + 1}$$
$$= \frac{2}{T} \frac{re^{j\omega} - 1}{re^{j\omega} + 1}$$
$$= \frac{2}{T} \left(\frac{r^2 - 1}{1 + r^2 + 2r \cos \omega} + j \frac{2r \sin \omega}{1 + r^2 + 2r \cos \omega}\right)$$

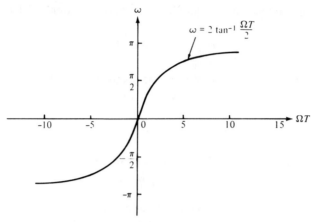

FIGURE 8.45 Mapping between the frequency variables ω and Ω resulting from the bilinear transformation.

Consequently,

$$\sigma = \frac{2}{T} \frac{r^2 - 1}{1 + r^2 + 2r \cos \omega} \tag{8.2.45}$$

$$\Omega = \frac{2}{T} \frac{2r \sin \omega}{1 + r^2 + 2r \cos \omega} \tag{8.2.46}$$

First, we note that if $r < 1$, then $\sigma < 0$, and if $r > 1$, then $\sigma > 0$. Consequently, the LHP in s maps into the inside of the unit circle in the z-plane and the RHP in s maps into the outside of the unit circle. When $r = 1$, then $\sigma = 0$ and

$$\begin{aligned}
\Omega &= \frac{2}{T} \frac{\sin \omega}{1 + \cos \omega} \\
&= \frac{2}{T} \tan \frac{\omega}{2}
\end{aligned} \tag{8.2.47}$$

or, equivalently,

$$\omega = 2 \tan^{-1} \frac{\Omega T}{2} \tag{8.2.48}$$

The relationship in (8.2.48) between the frequency variables in the two domains is illustrated in Fig. 8.45. We observe that the entire range in Ω is mapped only once into the range $-\pi \le \omega \le \pi$. However, the mapping is highly nonlinear. We observe a frequency compression or *frequency warping,* as it is usually called, due to the nonlinearity of the arctangent function.

It is also interesting to note that the bilinear transformation maps the point $s = \infty$ into the point $z = -1$. Consequently, the single-pole lowpass filter in (8.2.37), which has a zero at $s = \infty$, results in a digital filter that has a zero at $z = -1$.

EXAMPLE 8.2.5 _____

Convert the analog filter with system function

$$H_a(s) = \frac{s + 0.1}{(s + 0.1)^2 + 16}$$

into a digital IIR filter by means of the bilinear transformation. The digital filter is to have a resonant frequency of $\omega_r = \pi/2$.

Solution: First, we note that the analog filter has a resonant frequency $\Omega_r = 4$. This frequency is to be mapped into $\omega_r = \pi/2$ by selecting the value of the parameter T. From the relationship in (8.2.47), we must select $T = \frac{1}{2}$ in order to have $\omega_r = \pi/2$. Thus the desired mapping is

$$s = 4 \left(\frac{1 - z^{-1}}{1 + z^{-1}} \right)$$

The resulting digital filter has the system function

$$H(z) = \frac{0.128 + 0.006z^{-1} - 0.122z^{-1}}{1 + 0.0006z^{-1} + 0.975z^{-2}}$$

We note that the coefficient of the z^{-1} term in the denominator of $H(z)$ is extremely small and may be approximated by zero. Thus we have the system function

$$H(z) = \frac{0.128 + 0.006z^{-1} - 0.122z^{-2}}{1 + 0.975z^{-2}}$$

This filter has poles at

$$p_{1,2} = 0.987e^{\pm j\pi/2}$$

and zeros at

$$z_{1,2} = -1, 0.95$$

Therefore, we have succeeded in designing a two-pole filter that resonates at $\omega = \pi/2$.

In this example the parameter T was selected to map the resonant frequency of the analog filter into the desired resonant frequency of the digital filter. Usually, the design of the digital filter begins with specifications in the digital domain, which involve the frequency variable ω. These specifications in frequency are converted to the analog domain by means of the relation in (8.2.47). The analog filter is then designed that meets these specifications and converted to a digital filter by means of the bilinear transformation in (8.2.44). In this procedure, the parameter T is transparent and may be set to any arbitrary value (e.g., $T = 1$). The following example illustrates this point.

EXAMPLE 8.2.6 ───

Design a single-pole lowpass digital filter with a 3-dB bandwidth of 0.2π, by use of the bilinear transformation applied to the analog filter

$$H(s) = \frac{\Omega_c}{s + \Omega_c}$$

where Ω_c is the 3-dB bandwidth of the analog filter.

Solution: The digital filter is specified to have its -3-dB gain at $\omega_c = 0.2\pi$. In the frequency domain of the analog filter $\omega_c = 0.2\pi$ corresponds to

$$\Omega_c = \frac{2}{T} \tan 0.1\pi$$

$$= \frac{0.65}{T}$$

Thus the analog filter has the system function

$$H(s) = \frac{0.65/T}{s + 0.65/T}$$

This represents our filter design in the analog domain.

Now, we apply the bilinear transformation given by (8.2.44) to convert the analog filter into the desired digital filter. Thus we obtain

$$H(z) = \frac{0.245(1 + z^{-1})}{1 - 0.509z^{-1}}$$

where the parameter T has been divided out.

The frequency response of the digital filter is

$$H(\omega) = \frac{0.245(1 + e^{-j\omega})}{1 - 0.509e^{-j\omega}}$$

At $\omega = 0$, $H(0) = 1$, and at $\omega = 0.2\pi$, we have $|H(0.2\pi)| = 0.707$, which is the desired response.

8.2.4 The Matched-z Transformation

Another method for converting an analog filter into an equivalent digital filter is to map the poles and zeros of $H(s)$ directly into poles and zeros in the z-plane. Suppose that the system function of the analog filter is expressed in the factored form

$$H(s) = \frac{\displaystyle\prod_{k=1}^{M} (s - z_k)}{\displaystyle\prod_{k=1}^{N} (s - p_k)} \qquad (8.2.49)$$

where $\{z_k\}$ are the zeros and $\{p_k\}$ are the poles of the filter. Then the system function for the digital filter is

$$H(z) = \frac{\displaystyle\prod_{k=1}^{M} (1 - e^{z_k T}z^{-1})}{\displaystyle\prod_{k=1}^{N} (1 - e^{p_k T}z^{-1})} \qquad (8.2.50)$$

where T is the sampling interval. Thus each factor of the form $(s - a)$ in $H(s)$ is mapped into the factor $(1 - e^{aT}z^{-1})$. This mapping is called the *matched-z transformation*.

We observe that the poles obtained from the matched-z transformation are identical to the poles obtained with the impulse invariance method. However, the two techniques result in different zero positions.

To preserve the frequency response characteristic of the analog filter, the sampling interval in the matched-z transformation must be selected properly to yield the pole and zero locations at the equivalent position in the z-plane. Thus aliasing must be avoided by selecting T sufficiently small.

8.2.5 Characteristics of Commonly Used Analog Filters

As we have seen from our discussion above, IIR digital filters can easily be obtained by beginning with an analog filter and using a mapping to transform the s-plane into the

z-plane. Thus the design of a digital filter reduces to designing an appropriate analog filter and then performing the conversion from $H(s)$ to $H(z)$, in such a way so as to preserve as much as possible the desired characteristics of the analog filter.

Analog filter design is a well-developed field and many books have been written on the subject. In this section we briefly describe the important characteristics of commonly used analog filters and introduce the relevant filter parameters. Our discussion is limited to lowpass filters. Subsequently, we describe several frequency transformations that convert a lowpass prototype filter into either a bandpass or a highpass or a band-elimination filter.

Butterworth Filters. Lowpass Butterworth filters are all-pole filters characterized by the magnitude-squared frequency response

$$|H(\Omega)|^2 = \frac{1}{1 + (\Omega/\Omega_c)^{2N}} \tag{8.2.51}$$

where N is the order of the filter and Ω_c is its -3-dB frequency (usually called the cutoff frequency). Since $H(s)H(-s)$ evaluated at $s = j\Omega$ is simply equal to $|H(\Omega)|^2$, it follows

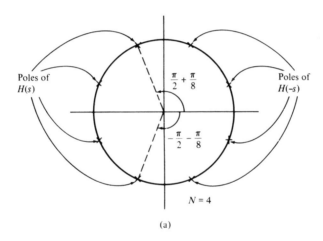

(a)

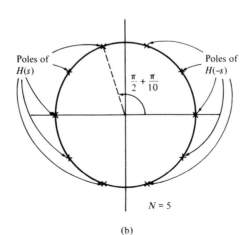

(b)

FIGURE 8.46 Pole positions for Butterworth filters.

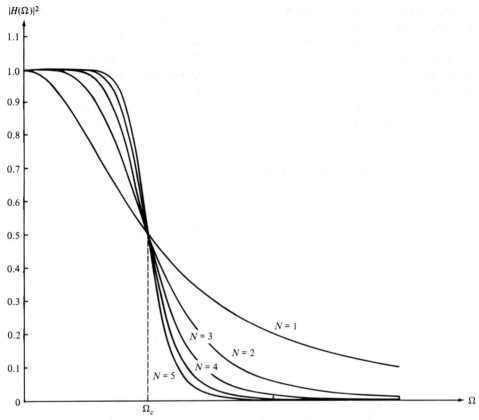

FIGURE 8.47 Frequency responses of Butterworth filters.

that

$$H(s)H(-s) = \frac{1}{1 + (-s^2/\Omega_c^2)^N} \tag{8.2.52}$$

The poles of $H(s)H(-s)$ occur on a circle of radius Ω_c at equally spaced points. From (8.2.52) we find that

$$\frac{-s^2}{\Omega_c^2} = (-1)^{1/N} = e^{j(2k+1)\pi/N} \qquad k = 0, 1, \ldots, N-1$$

and hence

$$s_k = \Omega_c e^{j\pi/2} e^{j(2k+1)\pi/2N} \qquad k = 0, 1, \ldots, N-1 \tag{8.2.53}$$

For example, Fig. 8.46 illustrates the pole positions for an $N = 4$ and $N = 5$ Butterworth filters.

The frequency response characteristics of the class of Butterworth filters are shown in Fig. 8.47 for several values of N. We note that $|H(\Omega)|^2$ is monotonic in both the passband and stopband. The order of the filter rquired to meet an attenuation δ_2 at a specified frequency Ω_s is easily determined from (8.2.51). Thus at $\Omega = \Omega_s$ we have

$$\frac{1}{1 + (\Omega_s/\Omega_c)^{2N}} = \delta_2^2$$

and hence

$$N = \frac{\log_{10}\left[(1/\delta_2^2) - 1\right]}{2\log_{10}(\Omega_s/\Omega_c)} \tag{8.2.54}$$

Thus the Butterworth filter is completely characterized by the parameters N, δ_2 and the ratio Ω_s/Ω_c.

EXAMPLE 8.2.7 _____

Determine the order and the poles of a lowpass Butterworth filter that has a -3-dB bandwidth of 500 Hz and an attenuation of 40 dB at 1000 Hz.

Solution: The critical frequencies are the -3-dB frequency Ω_c and the stopband frequency Ω_s, which are

$$\Omega_c = 1000\pi$$
$$\Omega_s = 2000\pi$$

For an attenuation of 40 dB, $\delta_2 = 0.01$. Hence from (8.2.54) we obtain

$$N = \frac{\log_{10}(10^4 - 1)}{2\log_{10} 2}$$
$$= 6.64$$

To meet the desired specifications, we select $N = 7$. The pole positions are

$$s_k = 1000\pi e^{j[\pi/2 + (2k+1)\pi/14]} \qquad k = 0, 1, 2, \ldots, 6$$

Chebyshev Filters. There are two types of Chebyshev filters. Type I Chebyshev filters are all-pole filters that exhibit equiripple behavior in the passband and a monotonic characteristic in the stopband. On the other hand, the family of type II Chebyshev filters contain both poles and zeros and exhibits a monotonic behavior in the passband and an equiripple behavior in the stopband. The zeros of this class of filters lie on the imaginary axis in the s-plane.

The magnitude squared of the frequency response characteristic of a type I Chebyshev filter is given as

$$|H(\Omega)|^2 = \frac{1}{1 + \epsilon^2 T_N^2(\Omega/\Omega_c)} \tag{8.2.55}$$

where ϵ is a parameter of the filter that is related to the ripple in the passband and $T_N(x)$ is the Nth-order Chebyshev polynomial defined as

$$T_N(x) = \begin{cases} \cos(N\cos^{-1}x) & |x| \le 1 \\ \cosh(N\cosh x) & |x| > 1 \end{cases} \tag{8.2.56}$$

The Chebyshev polynomials can be generated by the recursive equation

$$T_{N+1}(x) = 2xT_N(x) - T_{N-1}(x) \qquad N = 1, 2, \ldots \tag{8.2.57}$$

where $T_0(x) = 1$ and $T_1(x) = x$. From (8.2.57) we obtain $T_2(x) = 2x^2 - 1$, $T_3(x) = 4x^3 - 3x$, and so on.

Some of the properties of these polynomials are as follows:

1. $|T_N(x)| \le 1$ for all $|x| \le 1$.
2. $T_N(1) = 1$ for all N.
3. All the roots of the polynomial $T_N(x)$ occur in the interval $-1 \le x \le 1$.

The filter parameter ϵ is related to the ripple in the passband, as illustrated in Fig. 8.48, for N odd and N even. For N odd, $T_N(0) = 0$ and hence $|H(0)|^2 = 1$. On the other hand, for N even $T_N(0) = 1$ and hence $|H(0)|^2 = 1/(1 + \epsilon^2)$. At the band edge frequency $\Omega = \Omega_c$, we have $T_N(1) = 1$, so that

$$\frac{1}{\sqrt{1 + \epsilon^2}} = 1 - \delta_1$$

or, equivalently,

$$\epsilon^2 = \frac{1}{(1 - \delta_1)^2} - 1 \qquad (8.2.58)$$

where δ_1 is the value of the passband ripple.

The poles of a type I Chebyshev filter lie on an ellipse in the s-plane with major axis

$$r_1 = \Omega_c \frac{\beta^2 + 1}{2\beta} \qquad (8.2.59)$$

and minor axis

$$r_2 = \Omega_c \frac{\beta^2 - 1}{2\beta} \qquad (8.2.60)$$

where β is related to ϵ according to the equation

$$\beta = \left[\frac{\sqrt{1 + \epsilon^2} + 1}{\epsilon} \right]^{1/N} \qquad (8.2.61)$$

The pole locations are most easily determined for a filter of order N by first locating the poles for an equivalent Nth-order Butterworth filter that lie on circles of radius r_1 or radius r_2, as illustrated in Fig. 8.49. If we denote the angular positions of the poles of the Butterworth filter as

$$\phi_k = \frac{\pi}{2} + \frac{(2k + 1)\pi}{2N} \qquad k = 0, 1, 2, \ldots, N - 1 \qquad (8.2.62)$$

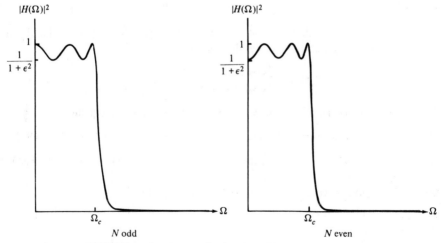

N odd   N even

FIGURE 8.48 Type I Chebyshev filter characteristic.

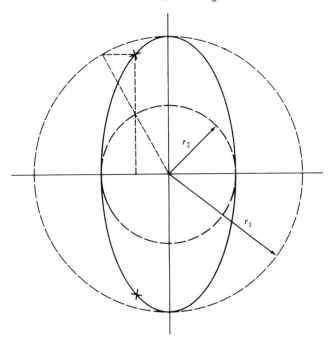

FIGURE 8.49 Determination of the pole locations for a Chebyshev filter.

then the positions of the poles for the Chebyshev filter lie on the ellipse at the coordinates (x_k, y_k), $k = 0, 1, \ldots, N - 1$, where

$$x_k = r_2 \cos \phi_k \qquad k = 0, 1, \ldots, N - 1$$
$$y_k = r_1 \sin \phi_k \qquad k = 0, 1, \ldots, N - 1$$
(8.2.63)

A type II Chebyshev filter contains zeros as well as poles. The magnitude squared of its frequency response is given as

$$|H(\Omega)|^2 = \frac{1}{1 + \epsilon^2 [T_N^2(\Omega_s/\Omega_c)/T_N^2(\Omega_s/\Omega)]}$$
(8.2.64)

where $T_N(x)$ is, again, the Nth-order Chebyshev polynomial and Ω_s is the stopband frequency as illustrated in Fig. 8.50. The zeros are located on the imaginary axis at the points

$$s_k = j \frac{\Omega_s}{\sin \phi_k} \qquad k = 0, 1, \ldots, N - 1$$
(8.2.65)

The poles are located at the points (v_k, w_k), where

$$v_k = \frac{\Omega_s x_k}{\sqrt{x_k^2 + y_k^2}} \qquad k = 0, 1, \ldots, N - 1$$
(8.2.66)

$$w_k = \frac{\Omega_s y_k}{\sqrt{x_k^2 + y_k^2}} \qquad k = 0, 1, \ldots, N - 1$$
(8.2.67)

where $\{x_k\}$ and $\{y_k\}$ are defined in (8.2.63) with β now related to the ripple in the stopband through the equation

$$\beta = \left[\frac{1 + \sqrt{1 - \delta_2^2}}{\delta_2} \right]^{1/N}$$
(8.2.68)

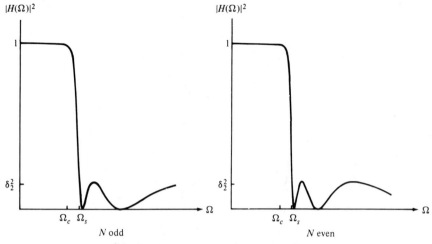

FIGURE 8.50 Type II Chebyshev filters.

From the description above, we observe that the Chebyshev filters are characterized by the parameters N, ϵ, δ_2 and the ratio Ω_s/Ω_c. For a given set of specifications on ϵ, δ_2, and Ω_s/Ω_c we can determine the order of the filter from the equation

$$N = \frac{\log_{10} [(\sqrt{1 - \delta_2^2} + \sqrt{1 - \delta_2^2(1 + \epsilon^2)})/\epsilon\delta_2]}{\log_{10} [(\Omega_s/\Omega_c) + \sqrt{(\Omega_s/\Omega_c)^2 - 1}]} \qquad (8.2.69)$$

EXAMPLE 8.2.8

Determine the order and the poles of a type I lowpass Chebyshev filter that has a 1-dB ripple in the passband, a cutoff frequency $\Omega_c = 1000\pi$, a stopband frequency of 2000π, and an attenuation of 40 dB or more for $\Omega \geq \Omega_s$.

Solution: First, we determine the order of the filter. We have

$$10 \log_{10} (1 + \epsilon^2) = 1$$
$$1 + \epsilon^2 = 1.259$$
$$\epsilon^2 = 0.259$$
$$\epsilon = 0.5088$$

Also,

$$20 \log_{10} \delta_2 = -40$$
$$\delta_2 = 0.01$$

Hence from (8.2.69) we obtain

$$N = \frac{\log_{10} 196.54}{\log_{10} (2 + \sqrt{3})}$$
$$= 4.0$$

Thus a type I Chebyshev filter having four poles meets the specifications.

The pole positions are determined from the relations in (8.2.59) through (8.2.63). First, we compute β, r_1, and r_2. Hence

$$\beta = 1.429$$
$$r_1 = 1.06\Omega_c$$
$$r_2 = 0.365\Omega_c$$

The angles $\{\phi_k\}$ are

$$\phi_k = \frac{\pi}{2} + \frac{(2k + 1)\pi}{8} \qquad k = 0, 1, 2, 3$$

Therefore, the poles are located at

$$x_1 + jy_1 = -0.1397\Omega_c \pm j0.979\Omega_c$$
$$x_2 + jy_2 = -0.337\Omega_c \pm j0.4056\Omega_c$$

The filter specifications in the example above were identical to the specifications given in Example 8.2.7, which involved the design of a Butterworth filter. In that case the number of poles required to meet the specifications was seven. On the other hand, the Chebyshev filter required only four. This result is typical of such comparisons. In general, the Chebyshev filter meets the specifications with a fewer number of poles than the corresponding Butterworth filter. Alternatively, if we compare a Butterworth filter to a Chebyshev filter having the same number of poles and the same passband and stopband specifications, the Chebyshev filter will have a smaller transition bandwidth. For a tabulation of the characteristics of Chebyshev filters and their pole–zero locations, the interested reader is referred to the handbook of Zverev (1967).

Elliptic Filters. Elliptic (or Cauer) filters exhibit equiripple behavior in both the passband and the stopband, as illustrated in Fig. 8.51 for N odd and N even. This class of filters contain both poles and zeros and are characterized by the magnitude-squared frequency response

$$|H(\Omega)|^2 = \frac{1}{1 + \epsilon^2 U_N(\Omega/\Omega_c)} \qquad (8.2.70)$$

where $U_N(x)$ is the Jacobian elliptic function of order N, which has been tabulated by Zverev (1967), and ϵ is a parameter related to the passband ripple. The zeros lie on the $j\Omega$-axis.

We recall from our discussion of FIR filters that the most efficient designs occur when we spread the approximation error equally over the passband and equally over the stopband. Elliptic filters accomplish this objective and, as a consequence, are most efficient from the viewpoint of yielding the smallest-order filter for a given set of specifications.

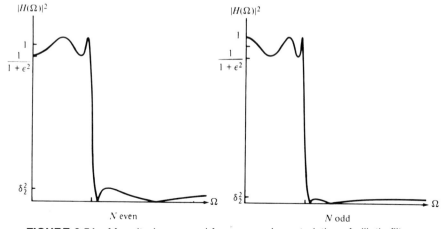

FIGURE 8.51 Magnitude-squared frequency characteristics of elliptic filters.

Equivalently, we may say that for a given order and a given set of specifications, an elliptic filter has the smallest transition bandwidth.

The filter order required to achieve a given set of specifications in passband ripple δ_1, stopband ripple δ_2, and transition ratio Ω_c/Ω_s is given as

$$N = \frac{K(\Omega_c/\Omega_s)K(\sqrt{1 - \delta_2^2(1 + \epsilon^2)}/\sqrt{1 - \delta_2^2})}{K(\epsilon\,\delta_2/\sqrt{1 - \delta_2^2})K(\sqrt{1 - (\Omega_c/\Omega_s)^2})} \tag{8.2.71}$$

where $K(x)$ is the complete elliptic integral of the first kind, defined as

$$K(x) = \int_0^{\pi/2} \frac{d\theta}{\sqrt{1 - x^2 \sin^2 \theta}} \tag{8.2.72}$$

Values of this integral have been tabulated in a number of texts, [e.g., the books by Jahnke and Emde (1945) and Dwight (1957)]. The passband ripple is $10 \log_{10}(1 + \epsilon^2)$.

We shall not attempt to describe elliptic functions in any detail because such a discussion would take us too far afield. Suffice it to say that computer programs are available for designing elliptic filters from the frequency specifications indicated above.

In view of the optimality of elliptic filters the reader may question the reason for considering the class of Butterworth or the class of Chebyshev filters in practical applications. One important reason that these others types of filters might be preferable in some applications is that they possess better phase response characteristics. The phase response of elliptic filters is more nonlinear in the passband than a comparable Butterworth filter or a Chebyshev filter, especially near the band edge.

Bessel Filters. Bessel filters are a class of all-pole filters that are characterized by the system function

$$H(s) = \frac{1}{B_N(s)} \tag{8.2.73}$$

where $B_N(s)$ is the Nth-order Bessel polynomial. These polynomials may be expressed in the form

$$B_N(s) = \sum_{k=0}^{N} a_k s^k \tag{8.2.74}$$

where the coefficients $\{a_k\}$ are given as

$$a_k = \frac{(2N - k)!}{2^{N-k} k!(N-k)!} \qquad k = 0, 1, \ldots, N \tag{8.2.75}$$

Alternatively, the Bessel polynomials may be generated recursively from the relation

$$B_N(s) = (2N - 1)B_{N-1}(s) + s^2 B_{N-2}(s) \tag{8.2.76}$$

with $B_0(s) = 1$ and $B_1(s) = s + 1$ as initial conditions.

An important characteristic of Bessel filters is the linear phase response over the passband of the filter. For example, Fig. 8.52 shows a comparison of the magnitude and phase responses of a Bessel filter and Butterworth of order $N = 4$. We note that the Bessel filter has a larger transition bandwidth, but its phase is linear within the passband. However, we should emphasize that the linear-phase chacteristics of the analog filter are destroyed in the process of converting the filter into the digital domain by means of the transformations decribed previously.

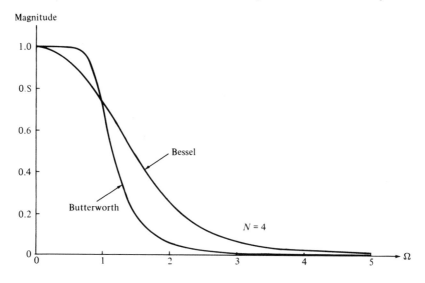

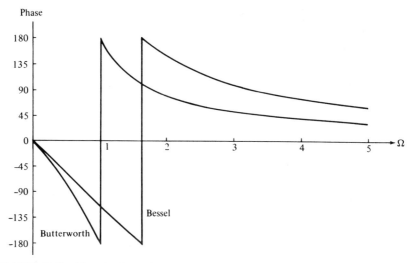

FIGURE 8.52 Magnitude and phase responses of Bessel and Butterworth filters of order $N = 4$.

8.2.6 Some Examples of Digital Filter Designs Based on the Bilinear Transformation

In this section we present several examples of digital filter designs obtained from analog filters by applying the bilinear transformation to convert $H(s)$ to $H(z)$. These filters designs were performed with the aid of one of several software packages now available for use on a personal computer.

A lowpass filter was designed to meet specifications of a maximum ripple of $\frac{1}{2}$ dB in the passband, 60-dB attenuation in the stopband, a passband edge frequency of $\omega_p = 0.25\pi$, and a stopband edge frequency of $\omega_s = 0.30\pi$.

A Butterworth filter of order $N = 37$ is required to satisfy the specifications. Its frequency response characteristics are illustrated in Fig. 8.53. If a Chebyshev filter is

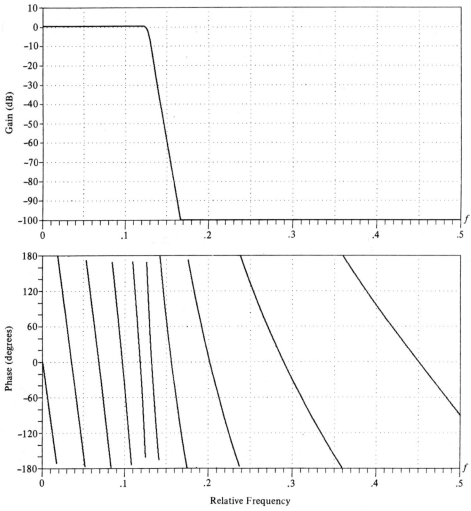

FIGURE 8.53 Frequency response characteristics of a 37-order Butterworth filter.

used, a filter of order $N = 13$ satisfies the specifications. The frequency response characteristics for a type I and type II Chebyshev filters are shown in Figs. 8.54 and 8.55, respectively. The type I filter has a passband ripple of 0.31 dB. Finally, an elliptic filter of order $N = 7$ was designed which also satisfied the specifications. For illustrative purposes, we show in Table 8.12 the numerical values for the filter parameters and the resulting frequency specifications are shown in Fig. 8.56. The following notation is used for the parameters in the function $H(z)$:

$$H(z) = \prod_{i=1}^{K} \frac{b(i, 0) + b(i, 1)z^{-1} + b(i, 2)z^{-2}}{1 + a(i, 1)z^{-1} + a(i, 2)z^{-2}} \qquad (8.2.77)$$

For illustrative purposes, we have also performed filter designs of bandpass, bandstop, and highpass filters by applying the bilinear transformation to the corresponding analog filter. Although we have described only lowpass analog filters in the preceding section, it is a simple matter to convert a lowpass analog filter into a bandpass, or a bandstop, or a highpass analog filter by a frequency transformation, as will be described in Section 8.3, which follows. The bilinear transformation is then applied to convert the analog

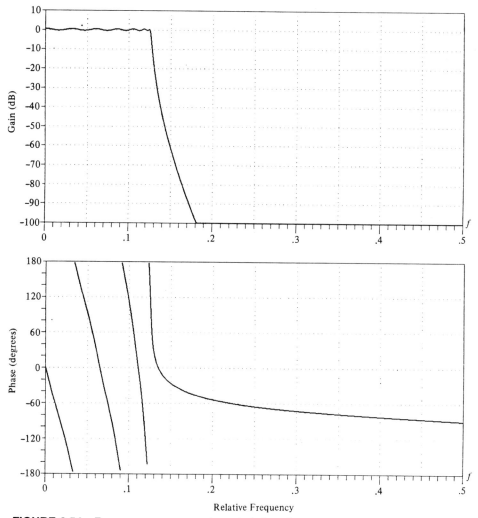

FIGURE 8.54 Frequency response characteristics of a 13-order type I Chebyshev filter.

filter into an equivalent digital filter. As in the case of the lowpass filters described above, the entire design was carried out on a personal computer.

The specifications for the bandpass filter are a passband in the frequency range $0.25\pi \leq \omega_p \leq 0.35\pi$, stopband edge frequencies at $\omega_s = 0.20\pi$ and $\omega_s = 0.40\pi$, 60-dB attenuation in the stopband and a maximum of 0.5-dB ripple in the passband.

A Butterworth filter of order $N = 24$ satisfies these frequency specifications, and its frequency response is shown in Fig. 8.57. The corresponding Chebyshev type I and type II filters of order $N = 14$ have the frequency response characteristics shown in Figs. 8.58 and 8.59, respectively. Finally, an elliptic filter of order $N = 10$ satisfies these specifications. Its frequency response characteristics are illustrated in Fig. 8.60.

The bandstop filter is designed to have -60-dB attenuation in the frequency range $0.25\pi \leq \omega_s \leq 0.35\pi$, passbands in the range $0 \leq \omega_p \leq 0.2\pi$ and $0.4\pi \leq \omega_p \leq \pi$, and a maximum ripple of 0.5 dB in the two passbands. The frequency response characteristics for an elliptic filter design are illustrated in Fig. 8.61.

Finally, the highpass filter was designed to have a passband $0.6\pi \leq \omega_p \leq \pi$, a stopband $0 \leq \omega_s \leq 0.55\pi$, an attenuation of at least 60 dB and a maximum passband

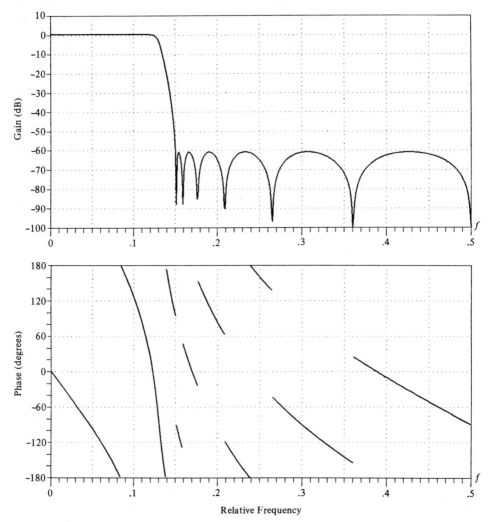

FIGURE 8.55 Frequency response characteristics of a 13-order type II Chebyshev filter.

TABLE 8.12 Filter Coefficients for a 7-Order Elliptic Filter

```
                    INFINITE IMPULSE RESPONSE (IIR)
                        ELLIPTIC LOWPASS FILTER
                       UNQUANTIZED COEFFICIENTS

        FILTER ORDER = 7
        SAMPLING FREQUENCY = 2.000 KILOHERTZ
```

I.	A(I, 1)	A(I, 2)	B(I, 0)	B(I, 1)	B(I, 2)
1	− .790103	.000000	.104948	.104948	.000000
2	−1.517223	.714088	.102450	− .007817	.102232
3	−1.421773	.861895	.420100	− .399842	.419864
4	−1.387447	.962252	.714929	− .826743	.714841

```
            *** CHARACTERISTICS OF DESIGNED FILTER ***
```

	BAND 1	BAND 2
LOWER BAND EDGE	.00000	.30000
UPPER BAND EDGE	.25000	1.00000
NOMINAL GAIN	1.00000	.00000
NOMINAL RIPPLE	.05600	.00100
MAXIMUM RIPPLE	.04910	.00071
RIPPLE IN DB	.41634	−63.00399

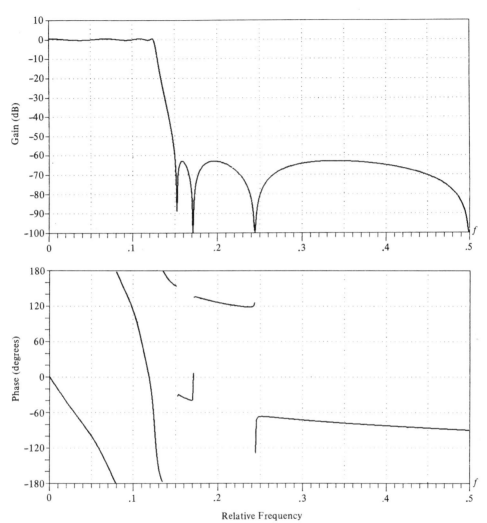

FIGURE 8.56 Frequency response characteristics of a 7-order elliptic filter.

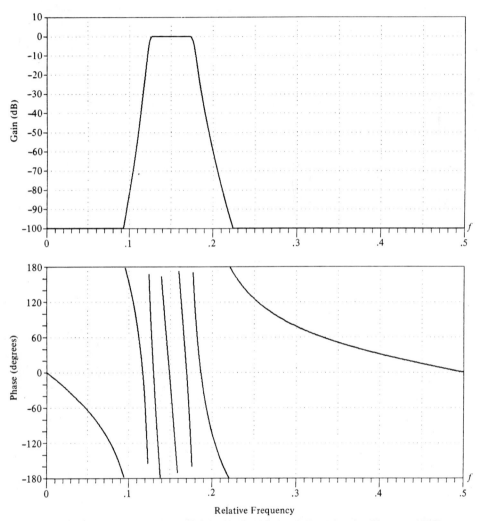

FIGURE 8.57 Frequency response characteristics for a 24-order Butterworth filter.

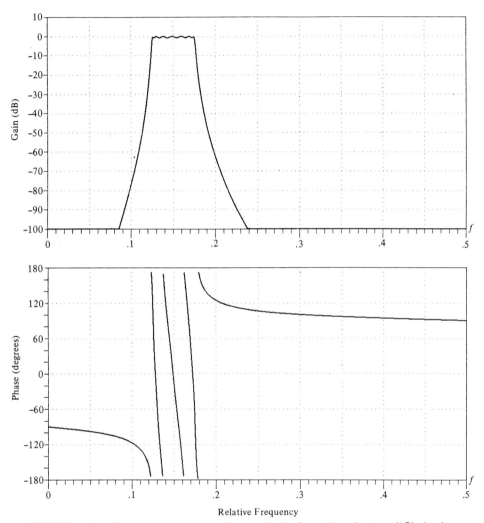

FIGURE 8.58 Frequency response characteristics for a 14-order type I Chebyshev filter.

627

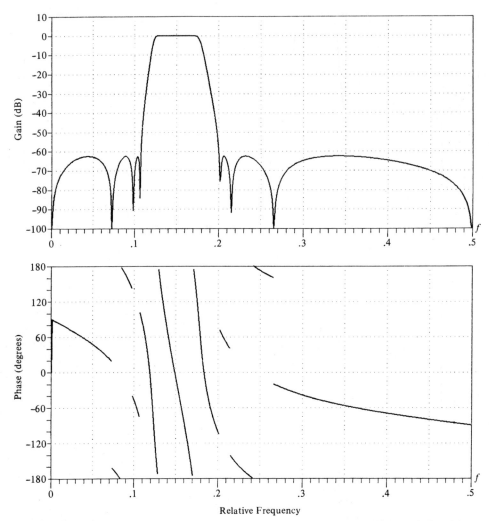

FIGURE 8.59 Frequency response characteristics for a 14-order type II Chebyshev filter.

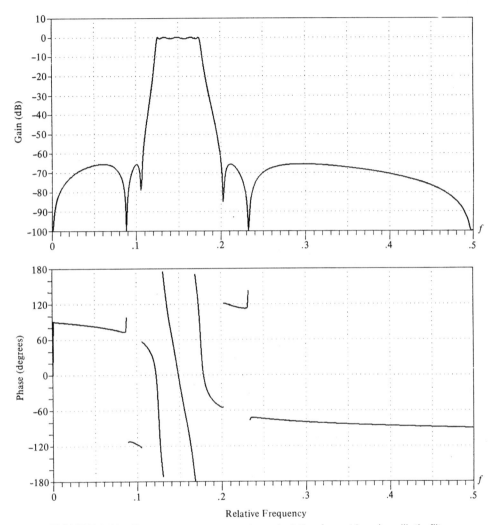

FIGURE 8.60 Frequency response characteristics for a 10-order elliptic filter.

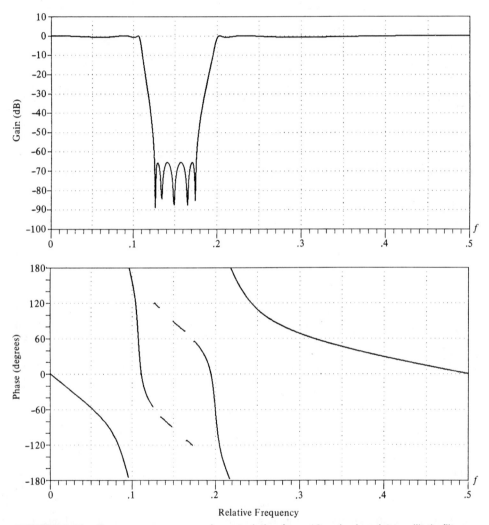

FIGURE 8.61 Frequency response characteristics for a 10-order bandstop elliptic filter.

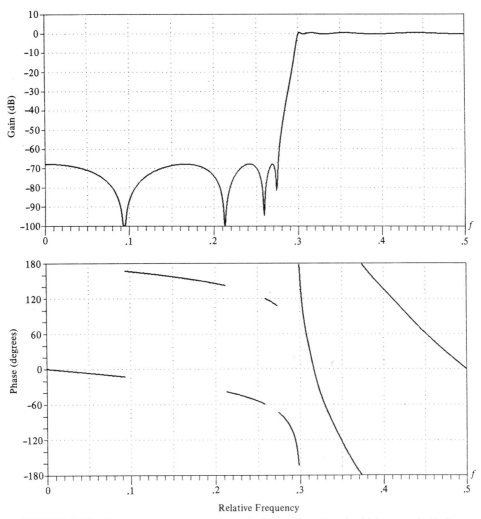

FIGURE 8.62 Frequency response characteristics of an 8-order highpass elliptic filter.

ripple of 0.5 dB. The frequency repsonse characteristics of the elliptic filter resulting from the design specifications are illustrated in Fig. 8.62.

8.3 Frequency Transformations

The treatment in the preceding section was focused primarily on the design of lowpass IIR filters. If we wish to design a highpass or a bandpass or a bandstop filter, it is a simple matter to take a lowpass prototype filter (Butterworth, Chebyshev, elliptic, Bessel) and perform a frequency transformation.

One possibility is to perform the frequency transformation in the analog domain and then to convert the analog filter into a corresponding digital filter by a mapping of the s-plane into the z-plane. An alternative approach is first to convert the analog lowpass filter into a lowpass digital filter and then to transform the lowpass digital filter into the desired digital filter by a digital transformation. In general, these two approaches yield different results, except for the bilinear transformation, in which case the filter designs are identical. These two approaches are described below.

8.3.1 Frequency Transformations in the Analog Domain

First, we consider frequency transformations in the analog domain. Suppose that we have a lowpass filter with cutoff frequency Ω_c and we wish to convert it to another lowpass filter with cutoff frequency Ω_c'. The transformation that accomplishes this is

$$s \longrightarrow \frac{\Omega_c}{\Omega_c'} s \qquad \text{(lowpass to lowpass)} \qquad (8.3.1)$$

Thus we obtain a lowpass filter with system function $H_l(s) = H_p[(\Omega_c/\Omega_c')s]$, where $H_p(s)$ is the system function of the prototype filter with cutoff frequency Ω_c.

If we wish to convert a lowpass filter into a highpass filter with cutoff frequency Ω_c', the desired transformation is

$$s \longrightarrow \frac{\Omega_c\Omega_c'}{s} \qquad \text{(lowpass to highpass)} \qquad (8.3.2)$$

The system function of the highpass filter is $H_h(s) = H_p(\Omega_c\Omega_c'/s)$.

The transformation for converting a lowpass analog filter with cutoff frequency Ω_c into a bandpass filter, having a lower cutoff frequency Ω_l and an upper cutoff frequency Ω_u, may be accomplished by first converting the lowpass filter into another lowpass filter having a cutoff frequency $\Omega_c' = 1$ and then performing the transformation

$$s \longrightarrow \frac{s^2 + \Omega_l\Omega_u}{s(\Omega_u - \Omega_l)} \qquad \text{(lowpass to bandpass)} \qquad (8.3.3)$$

Equivalently, we may accomplish the same result in a single step by means of the transformation

$$s \longrightarrow \Omega_c \frac{s^2 + \Omega_l\Omega_u}{s(\Omega_u - \Omega_l)} \qquad \text{(lowpass to bandpass)} \qquad (8.3.4)$$

where

$$\Omega_l = \text{lower cutoff frequency}$$
$$\Omega_u = \text{upper cutoff frequency}$$

Thus we obtain

$$H_b(s) = H_p\left(\Omega_c \frac{s^2 + \Omega_l\Omega_u}{s(\Omega_u - \Omega_u)}\right)$$

Finally, if we wish to convert a lowpass analog filter with cutoff frequency Ω_c into a bandstop filter, the transformation is simply the inverse of (8.3.3) with the additional factor Ω_c which serves to normalize for the cutoff frequency of the lowpass filter. Thus the transformation is

$$s \longrightarrow \Omega_c \frac{s(\Omega_u - \Omega_l)}{s^2 + \Omega_u\Omega_l} \qquad \text{(lowpass to bandstop)} \qquad (8.3.5)$$

which leads to

$$H_{bs}(s) = H_p\left(\Omega_c \frac{s(\Omega_u - \Omega_l)}{s^2 + \Omega_u\Omega_l}\right)$$

The mappings in (8.3.1), (8.3.2), (8.3.3), and (8.3.5) are summarized in Table 8.13. The mappings in (8.3.4) and (8.3.5) are nonlinear and may appear to distort the frequency response characteristics of the lowpass filter. However, the effects of the nonlinearity on

TABLE 8.13 Frequency Transformations for Analog Filters (Prototype Lowpass Filter Has Cutoff Frequency Ω_c)

Type of Transformation	Transformation	Cutoff Frequencies of New Filter
Lowpass	$s \longrightarrow \dfrac{\Omega_c}{\Omega_c'} s$	Ω_c'
Highpass	$s \longrightarrow \dfrac{\Omega_c \Omega_c'}{s}$	Ω_c'
Bandpass	$s \longrightarrow \Omega_c \dfrac{s^2 + \Omega_l \Omega_u}{s(\Omega_u - \Omega_l)}$	Ω_l, Ω_u
Bandstop	$s \longrightarrow \Omega_c \dfrac{s(\Omega_u - \Omega_c)}{s^2 + \Omega_u \Omega_l}$	Ω_l, Ω_u

the frequency response are minor, affecting primarily the frequency scale but preserving the amplitude response characteristics of the filter. Thus an equiripple lowpass filter is transformed into an equiripple bandpass or bandstop or highpass filter.

EXAMPLE 8.3.1

Transform the single-pole lowpass Butterworth filter with system function

$$H(s) = \frac{\Omega_c}{s + \Omega_c}$$

into a bandpass filter with upper and lower cutoff frequencies Ω_u and Ω_l, respectively.

Solution: The desired transformation is given by (8.3.4). Thus we have

$$H(s) = \frac{1}{\dfrac{s^2 + \Omega_l \Omega_u}{s(\Omega_u - \Omega_l)} + 1}$$

$$= \frac{(\Omega_u - \Omega_l)s}{s^2 + (\Omega_u - \Omega_l)s + \Omega_l \Omega_u}$$

The resulting filter has a zero at $s = 0$ and poles at

$$s = \frac{-(\Omega_u - \Omega_l) \pm \sqrt{\Omega_u^2 + \Omega_l^2 - 6\Omega_u \Omega_l}}{2}$$

8.3.2 Frequency Transformations in the Digital Domain

As in the analog domain, frequency transformations can be performed on a digital lowpass filter to convert it to either a bandpass, bandstop, or highpass filter. The transformation involves replacing the variable z^{-1} by a rational function $g(z^{-1})$, which must satisfy the following properties.

1. The mapping $z^{-1} \longrightarrow g(z^{-1})$ must map points inside the unit circle in the z-plane into itself.
2. The unit circle must also be mapped into itself.

Condition (2) implies that for $r = 1$,

$$e^{-j\omega} = g(e^{-j\omega}) \equiv g(\omega)$$
$$= |g(\omega)|e^{j\arg[g(\omega)]}$$

It is clear that we must have $|g(\omega)| = 1$ for all ω. That is, the mapping must be all-pass. Hence it is of the form

$$g(z^{-1}) = \pm \prod_{k=1}^{n} \frac{z^{-1} - \alpha_k}{1 - \alpha_k z^{-1}} \qquad (8.3.6)$$

where $|\alpha_k| < 1$ to ensure that a stable filter is transformed into another stable filter (i.e., to satisfy condition 1).

From the general form in (8.3.6) we obtain the desired set of digital transformations for converting a prototype digital lowpass filter into either a bandpass, a bandstop, a highpass, or another lowpass digital filter. These transformations are tabulated in Table 8.14.

TABLE 8.14 Frequency Transformation for Digital Filters
(Prototype Lowpass Filter Has Cutoff Frequency ω_c)

Type of Transformation	Transformation	Parameters
Lowpass	$z^{-1} \longrightarrow \dfrac{z^{-1} - a}{1 - az^{-1}}$	$\omega_c' = $ cutoff frequency of new filter $a = \dfrac{\sin[(\omega_c - \omega_c')/2]}{\sin[(\omega_c + \omega_c')/2]}$
Highpass	$z^{-1} \longrightarrow -\dfrac{z^{-1} + a}{1 + az^{-1}}$	$\omega_c' = $ cutoff frequency new filter $a = -\dfrac{\cos[(\omega_c - \omega_c')/2]}{\cos[(\omega_c + \omega_c')/2]}$
Bandpass	$z^{-1} \longrightarrow -\dfrac{z^{-2} - a_1 z^{-1} + a_2}{a_2 z^{-2} - a_1 z^{-1} + 1}$	$\omega_l = $ lower cutoff frequency $\omega_u = $ upper cutoff frequency $a_1 = -2\alpha K/(K + 1)$ $a_2 = (K - 1)/(K + 1)$ $\alpha = \dfrac{\cos[(\omega_u + \omega_l)/2]}{\cos[(\omega_u - \omega_l)/2]}$ $K = \cot \dfrac{\omega_u - \omega_l}{2} \tan \dfrac{\omega_c}{2}$
Bandstop	$z^{-1} \longrightarrow \dfrac{z^{-2} - a_1 z^{-1} + a_2}{a_2 z^{-1} - a_1 z^{-1} + 1}$	$\omega_l = $ lower cutoff frequency $\omega_u = $ upper cutoff frequency $a_1 = -2\alpha/(K + 1)$ $a_2 = (1 - K)/(1 + K)$ $\alpha = \dfrac{\cos[(\omega_u + \omega_l)/2]}{\cos[(\omega_u - \omega_l)/2]}$ $K = \tan \dfrac{\omega_u - \omega_l}{2} \tan \dfrac{\omega_c}{2}$

EXAMPLE 8.3.2

Convert the single-pole lowpass Butterworth filter with system function

$$H(z) = \frac{0.245(1 + z^{-1})}{1 - 0.509z^{-1}}$$

into a bandpass filter with upper and lower cutoff frequencies ω_u and ω_l, respectively. The lowpass filter has 3-dB bandwidth $\omega_c = 0.2\pi$ (see Example 8.2.6).

Solution: The desired transformation is

$$z^{-1} \longrightarrow -\frac{z^{-2} - a_1 z^{-1} + a_2}{a_2 z^{-2} - a_1 z^{-1} + 1}$$

where a_1 and a_2 are defined in Table 8.14. Substitution into $H(z)$ yields

$$H(z) = \frac{0.245 \left[1 - \dfrac{z^{-2} - a_1 z^{-1} + a_2}{a_2 z^{-2} - a_1 z^{-1} + 1}\right]}{1 + 0.509 \left(\dfrac{z^{-2} - a_1 z^{-1} + a_2}{a_2 z^{-2} - a_1 z^{-1} + 1}\right)}$$

$$= \frac{0.245(1 - a_2)(1 - z^{-2})}{(1 + 0.509a_2) - 1.509a_1 z^{-1} + (a_2 + 0.509)z^{-2}}$$

Note that the resulting filter has zeros at $z = \pm 1$ and a pair of poles which depend on the choice of ω_u and ω_l.

For example, suppose that $\omega_u = 3\pi/5$ and $\omega_l = 2\pi/5$. Since $\omega_c = 0.2\,\pi$, we find that $k = 1$, $a_2 = 0$, and $a_1 = 0$. Then

$$H(z) = \frac{0.245(1 - z^{-2})}{1 + 0.509z^{-2}}$$

This filter has poles at $z = \pm j0.713$ and hence resonates at $\omega = \pi/2$.

Since a frequency transformation may be performed either in the analog domain or in the digital domain, the filter designer has a choice as to which approach to take. However, some caution must be exercised depending on the types of filters being designed. In particular, we know that the impulse invariance method and the mapping of derivatives are inappropriate to use in designing highpass and many bandpass filters, due to the aliasing problem. Consequently, one would not employ an analog frequency transformation followed by conversion of the result into the digital domain by use of these two mappings. Instead, it is much better to perform the mapping from an analog lowpass filter into a digital lowpass filter by either of these mappings and then to perform the frequency transformation in the digital domain. Thus the problem of aliasing is avoided.

In the case of the bilinear transformation, where aliasing is not a problem, it does not matter whether the frequency transformation is performed in the analog domain or in the digital domain. In fact, in this case only, the two approaches result in identical digital filters.

8.4 Direct Design Techniques for Digital IIR Filters

The design techniques for IIR filters described in Section 8.2 involved the conversion of an analog filter into a digital filter by some mapping from the s-plane to the z-plane. As an alternative, one may design digital IIR filters directly in the z-domain without reference to the analog domain.

Below we describe several methods for designing digital IIR filters directly. In the first two techniques, the Padé approximation method and least-squares design methods, the specifications are given in the time domain and the design is carried out in the time domain. The least-squares design methods are related to the system identification methods based on least squares, discussed in Section 6.3. In this section we evaluate the least-squares design methods from the viewpoint of how well the design matches the desired frequency response characteristics. The final section describes a technique in which the design is carried out in the frequency domain.

8.4.1 Padé Approximation Method

designed has the system function

$$H(z) = \frac{\sum\limits_{k=0}^{M} b_k z^{-k}}{1 + \sum\limits_{k=1}^{N} a_k z^{-k}}$$

$$= \sum_{k=0}^{\infty} h(k)z^{-k}$$

(8.4.1)

where $h(k)$ is its unit sample response. The filter has $L = M + N + 1$ parameters, namely, the coefficients $\{a_k\}$ and $\{b_k\}$, which can be selected to minimize some error criterion.

The least-squares error criterion is often used in optimization problems of this type. Suppose that we minimize the sum of the squared errors

$$\mathcal{E} = \sum_{n=0}^{U} [h_d(n) - h(n)]^2$$

(8.4.2)

with respect to the filter parameters $\{a_k\}$ and $\{b_k\}$, where U is some preselected upper limit in the summation.

In general, $h(n)$ is a nonlinear function of the filter parameters and hence the minimization of $\mathcal{E}$ involves the solution of a set of nonlinear equations. However, if we select the upper limit as $U = L - 1$, it is possible to match $h(n)$ perfectly to the desired response $h_d(n)$ for $0 \le n \le M + N$. This can be achived in the following manner.

The difference equation for the desired filter is

$$y(n) = -a_1 y(n-1) - a_2 y(n-2) - \cdots - a_N y(n-N) \\ + b_0 x(n) + b_1 x(n-1) + \cdots + b_M x(n-M)$$

(8.4.3)

Suppose that the input to the filter is a unit sample [i.e., $x(n) = \delta(n)$]. Then the response of the filter is $y(n) = h(n)$ and hence (8.4.3) becomes

$$h(n) = -a_1 h(n-1) - a_2 h(n-2) - \cdots - a_N h(n-N) \\ + b_0\,\delta(n) + b_1\,\delta(n-1) + \cdots + b_M\,\delta(n-M)$$

(8.4.4)

Since $\delta(n-k) = 0$ except for $n = k$, (8.4.4) reduces to

$$h(n) = -a_1 h(n-1) - a_2 h(n-2) - \cdots - a_N h(n-N) + b_n \qquad 0 \le n \le M$$

(8.4.5)

For $n > M$, (8.4.4) becomes

$$h(n) = -a_1 h(n-1) - a_2 h(n-2) - \cdots - a_N h(n-N) \qquad (8.4.6)$$

The set of linear equations in (8.4.5) and (8.4.6) can be used to solve for the filter parameters $\{a_k\}$ and $\{b_k\}$. We set $h(n) = h_d(n)$ for $0 \le n \le M + N$, and use the linear equations in (8.4.6) to solve for the filter parameters $\{a_k\}$. Then we use values for the $\{a_k\}$ in (8.4.5) and solve for the parameters $\{b_k\}$. Thus we obtain a perfect match between $h(n)$ and the desired response $h_d(n)$ for the first L values of the impulse response. This design technique is usually called the *Padé approximation procedure*.

The degree to which this design technique produces acceptable filter designs depends in part on the number of filter coefficients selected. Since the design method matches $h_d(n)$ only up to the number of filter parameters, the more complex the filter, the better the approximation to $h_d(n)$ for $0 \le n \le M + N$. However, this is also the major limitation with the Padé approximation method, namely, the resulting filter must contain a large number of poles and zeros. For this reason, the Padé approximation method has found limited use in filter designs for practical applications.

EXAMPLE 8.4.1

Suppose that the desired unit sample response is

$$h_d(n) = 2(\tfrac{1}{2})^n u(n)$$

Determine the parameters of the filter with system function

$$H(z) = \frac{b_0 + b_1 z^{-1}}{1 + a_1 z^{-1}}$$

using the Padé approximation technique.

Solution: In this simple example, $H(z)$ can provide a perfect match to $H_d(z)$, by selecting $b_0 = 2$, $b_1 = 0$, and $a_1 = -\tfrac{1}{2}$. Let us apply the Padé approximation to see if we indeed obtain the same result.

With $\delta(n)$ as the input to $H(z)$, we obtain the output

$$h(n) = -a_1 h(n-1) + b_0 \delta(n) + b_1 \delta(n-1)$$

For $n > 1$, we have

$$h(n) = -a_1 h(n-1)$$

or, equivalently,

$$h_d(n) = -a_1 h_d(n-1)$$

With the substitution for $h_d(n)$, we obtain $a_1 = -\tfrac{1}{2}$. To solve for b_0 and b_1, we use the form (8.4.5) with $h(n) = h_d(n)$. Thus

$$h_d(n) = \tfrac{1}{2} h_d(n-1) + b_0 \delta(n) + b_1 \delta(n-1)$$

For $n = 0$ this equation yields $b_0 = 2$. For $n = 1$ we obtain the result $b_1 = 0$. Thus $H(z) = H_d(z)$.

This example illustrates that the Padé approximation results in a perfect match to $H_d(z)$ when the desired system function is rational and we have prior knowledge of the number of poles and zeros in the system. In general, however, this will not be the case in practice, since $h_d(n)$ will be determined from some desired frequency response specifications

$H_d(\omega)$. In such a case the Padé approximation may not result in a good filter design. To illustrate a potential problem and suggest a solution, let us consider the following examples.

EXAMPLE 8.4.2 _____

A fourth-order Butterworth filter has the system function

$$H_d(z) = \frac{4.8334 \times 10^{-3} (z + 1)^4}{(z^2 - 1.3205z + 0.6326)(z^2 - 1.0482z + 0.2959)}$$

The unit sample response corresponding to $H_d(z)$ is illustrated in Fig. 8.63. Use the Padé approximation method to approximate $H_d(z)$.

Solution: We observe that the desired filter has $M = 4$ zeros and $N = 4$ poles. It is instructive to determine the coefficients in the Padé approximation when the number of zeros and/or poles are not identical to the desired number of filter parameters.
  In Fig. 8.64 we plot the frequency response of the filter obtained by the Padé approximation method. We have considered four cases: $M = 3, N = 5$; $M = 3, N = 4$; $M = 4, N = 4$; $M = 4, N = 5$. We observe that when $M = 3$, the resulting frequency response is a relatively poor approximation to the desired response. However, an increase in the number of poles from $N = 4$ to $N = 5$ appears to compensate in part for the lack of the one zero. When M is increased from three to four, we obtain a perfect match with the desired Butterworth filter not only for $N = 4$ but for $N = 5$, and, in fact, for larger values of N.

_____

EXAMPLE 8.4.3 _____

A three-pole and three-zero type II lowpass Chebyshev digital filter has the system function

$$H_d(z) = \frac{0.3060(1 + z^{-1})(0.2652 - 0.09z^{-1} + 0.2652z^{-2})}{(1 - 0.3880z^{-1})(1 - 1.1318z^{-1} + 0.5387z^{-2})}$$

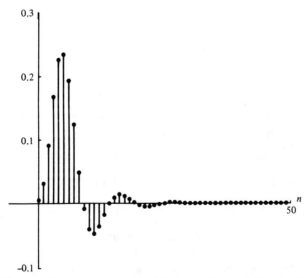

FIGURE 8.63 Impulse response $h_d(n)$ of digital Butterworth filter in Example 8.4.2.

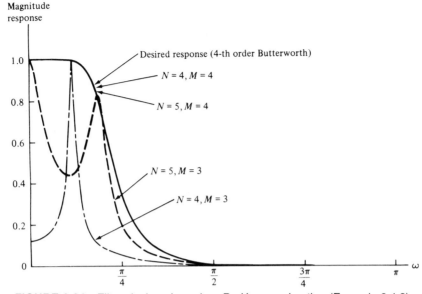

FIGURE 8.64   Filter designs based on Padé approximation (Example 8.4.2).

Its unit sample response is illustrated in Fig. 8.65. Use the Padé approximation method to approximate $H_d(z)$.

Solution: By following the same procedure as in Example 8.4.2, we determined the Padé approximation of $H_d(z)$ based on the selection of $M = 2$, $N = 3$; $M = 2$, $N = 4$; $M = 3$, $N = 3$; $M = 3$, $N = 4$. The frequency responses of the resulting designs are illustrated in Fig. 8.66.

As in Example 8.4.2, we note that when we underestimate the number of zeros we obtain a relatively poor design, as evidenced by the two cases in which $M = 2$. On the other hand, if $M = 3$, we obtain a perfect match for $N = 3$ and $N = 4$.

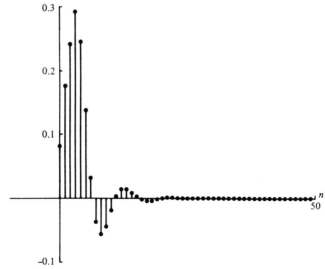

FIGURE 8.65   Impulse response $h_d(n)$ of type II Chebyshev digital filter given in Example 8.4.3.

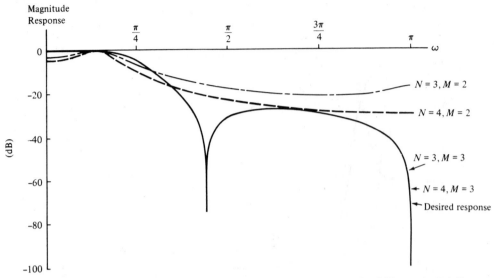

FIGURE 8.66 Filter designs based on Padé approximation method (Example 8.4.3).

The two examples given above suggest that an effective approach in using the Padé approximation is to try different values of M and N until the frequency responses of the resulting filters have converged to the desired frequency response within some small, acceptable approximation error. However, in practice, this approach appears to be cumbersome.

8.4.2 Least-Squares Design Method

In Section 6.3 we introduced the least-squares method in the context of system identification of an unknown system. In this section we employ the same criterion to the direct design of digital filters. The results obtained in Sections 6.3.5 and 6.3.6 are directly applicable to the filter design problem.

Again, let us assume that $h_d(n)$ is specified for $n \geq 0$. We begin with the simple case in which the digital filter contains only poles, that is,

$$H(z)) = \frac{b_0}{1 + \displaystyle\sum_{k=1}^{N} a_k z^{-k}} \qquad (8.4.7)$$

Now, consider the cascade connection of the desired filter $H_d(z)$ with the reciprocal, all-zero filter $1/H(z)$, as illustrated in Fig. 8.67, which is identical in form to Fig. 6.40. Consequently, this is exactly the same least-squares problem formulation as for the inverse filter identification performed in Section 6.3.5. Hence, the parameter b_0 is $b_0 = h_d(0)$ and the $\{a_k\}$ are obtained by solving the set of linear equations

$$\sum_{l=1}^{N} a_l r_{dd}(k - l) = -r_{dd}(k) \qquad k = 1, 2, \ldots, N \qquad (8.4.8)$$

where $\{r_{dd}(h)\}$ is the autocorrelation sequence of $\{h_d(n)\}$, defined as

$$r_{dd}(k) = \sum_{n=0}^{\infty} h_d(n)h_d(n + k) \qquad (8.4.9)$$

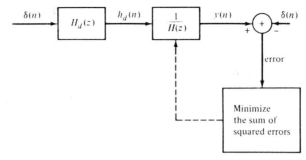

FIGURE 8.67 Least-squares inverse filter design method.

The set of linear equations in (8.4.8) can also be expressed in matrix form as

$$\mathbf{Ra} = \mathbf{r} \tag{8.4.10}$$

where $\mathbf{R}$ is an $N \times N$ correlation matrix with elements $\{r_{dd}(n)\}$, $\mathbf{a}$ is the $N \times 1$ vector of filter coefficients, and $\mathbf{r}$ is an $N \times 1$ correlation vector with elements $(-r_{dd}(n))$.

The set of linear equations in (8.4.10) can easily be solved on a digital computer to yield the coefficients of the all-pole filter. Thus the solution may be expressed as

$$\mathbf{a} = \mathbf{R}^{-1}\mathbf{r} \tag{8.4.11}$$

This design method is called the *least-squares inverse filter design method*.

In a particular design problem, the desired impulse response $h_d(n)$ is specified for a finite set of points, say $0 \le n \le L$, where $L >> N$. In such a case, the correlation sequence $r_{dd}(k)$ can be computed from the finite sequence $h_d(n)$ as

$$\hat{r}_{dd}(k - l) = \sum_{n=0}^{L-|k-l|} h_d(n)h_d(n + k - l) \qquad 0 \le k - l \le N \tag{8.4.12}$$

and these values can be used to form $\mathbf{R}$ and $\mathbf{r}$. Then the set of linear equations in (8.4.10) become

$$\hat{\mathbf{R}}\mathbf{a} = \hat{\mathbf{r}} \tag{8.4.13}$$

which yield the solution

$$\mathbf{a} = \hat{\mathbf{R}}^{-1}\hat{\mathbf{r}} \tag{8.4.14}$$

An alternative method for solving the all-pole approximation problem is based on the concept of *least-squares linear prediction,* described in Section 6.5.7 in the context of deconvolution. As illustrated in Fig. 8.68, the output of the all-pole filter to the impulse $\delta(n)$ is

$$y(n) = -\sum_{k=1}^{N} a_k h(n - k) + b_0\,\delta(n)$$

or, equivalently,

$$h(n) = -\sum_{k=1}^{N} a_k h(n - k) + b_0\,\delta(n) \qquad n = 0, 1, \ldots \tag{8.4.15}$$

The desired response is $h_d(n)$. Since $h(0) = b_0$, we may set $b_0 = h_d(0)$. For $n \ge 1$, (8.4.15) becomes

$$h(n) = -\sum_{k=1}^{N} a_k h(n - k) \tag{8.4.16}$$

with initial conditions $h(n) = 0$, $n < 0$.

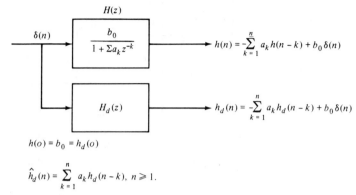

$$h(o) = b_0 = h_d(o)$$

$$\hat{h}_d(n) = \sum_{k=1}^{n} a_k h_d(n-k), \quad n \geq 1.$$

FIGURE 8.68 Least-squares filter design based on linear prediction.

Now, if $H_d(z)$ is itself an all-pole filter, then

$$h_d(n) = -\sum_{k=1}^{N} a_k h_d(n-k) \qquad n \geq 1 \tag{8.4.17}$$

In general, however, this is not the case. Indeed, the linear combination on the right-hand side of (8.4.17) may be considered as an estimate of $h_d(n)$. That is,

$$\hat{h}_d(n) = -\sum_{k=1}^{N} a_k h_d(n-k) \qquad n \geq 1 \tag{8.4.18}$$

We call $\hat{h}_d(n)$ the *linear prediction value* of $h_d(n)$. Then the sum of squares of the prediction error between $h_d(n)$ and $\hat{h}_d(n)$ is

$$\begin{aligned}
\mathcal{E} &= \sum_{n=1}^{\infty} [h_d(n) - \hat{h}_d(n)]^2 \\
&= \sum_{n=1}^{\infty} \left[h_d(n) + \sum_{k=1}^{N} a_k h_d(n-k) \right]^2
\end{aligned} \tag{8.4.19}$$

Since this is exactly the same error function that resulted in the set of linear equations given by (8.4.8), the *least-squares linear prediction method* leads to the same result as the least-squares inverse filter design method.

The least-squares linear prediction method is helpful in extending this method to a pole–zero approximation for $H_d(z)$. If the filter $H(z)$ that approximates $H_d(z)$ has both poles and zeros, its response to the unit impulse $\delta(n)$ is

$$h(n) = -\sum_{k=1}^{M} a_k h(n-k) + \sum_{k=0}^{M} b_k \delta(n-k) \qquad n \geq 0 \tag{8.4.20}$$

or, equivalently,

$$h(n) = -\sum_{k=1}^{N} a_k h(n-k) + b_n \qquad 0 \leq n \leq M \tag{8.4.21}$$

For $n > M$, (8.4.20) reduces to

$$h(n) = -\sum_{k=1}^{N} a_k h(n-k) \qquad n > M \tag{8.4.22}$$

Clearly, if $H_d(z)$ is a pole–zero filter, its response to $\delta(n)$ would satisfy the same equations

(8.4.20) through (8.4.22). In general, however, it will not. Based on (8.4.22), we define the linear prediction value of $h_d(n)$ as

$$\hat{h}_d(n) = -\sum_{k=1}^{M} a_k h_d(n - k) \qquad n > M \qquad (8.4.23)$$

Then, just as in the all-pole filter, the sum of squares of the prediction error is

$$\begin{aligned} \mathcal{E}_1 &= \sum_{n=M+1}^{\infty} [h_d(n) - \hat{h}_d(n)]^2 \\ &= \sum_{n=M+1}^{\infty} \left[h_d(n) + \sum_{k=1}^{N} a_k h_d(n - k) \right]^2 \end{aligned} \qquad (8.4.24)$$

The minimization of $\mathcal{E}_1$ with respect to the pole parameters $\{a_k\}$ leads to the set of linear equations

$$\sum_{l=1}^{N} a_l r_{dd}(k, l) = -r_{dd}(k, 0) \qquad k = 1, 2, \ldots, N \qquad (8.4.25)$$

where $r(k, l)$ is now defined as

$$r_{dd}(k, l) = \sum_{n=M+1}^{\infty} h_d(n - k) h_d(n - l) \qquad (8.4.26)$$

Thus these linear equations yield the filter parameters $\{a_k\}$. Note that these equations reduce to the all-pole filter approximation when M is set to zero.

The parameters $\{b_k\}$ that determine the zeros of the filter can be obtained simply from (8.4.21), where $h(n) = h_d(n)$ by substitution of the values $\{\hat{a}_k\}$ obtained by solving (8.4.25). Thus

$$b_n = h_d(n) + \sum_{k=1}^{N} \hat{a}_k h_d(n - k) \qquad 0 \leq n \leq M \qquad (8.4.27)$$

Therefore, the parameters $\{\hat{a}_k\}$ that determine the poles are obtained by the method of least squares while the parameters $\{b_k\}$ that determine the zeros are obtained as in the Padé approximation method. The foregoing approach for determining the poles and zeros of $H(z)$ is sometimes called *Prony's method*.

The least-squares method provides good estimates for the pole parameters $\{a_k\}$. However, Prony's method may not be as effective in estimating the parameters $\{b_k\}$, primarily because the computation in (8.4.27) is not based on the least-squares method.

An alternative method in which both sets of parameters $\{a_k\}$ and $\{b_k\}$ are determined by application of the least-squares method has been proposed by Shanks (1967). In Shanks' method, the parameters $\{a_k\}$ are computed on the basis of the least-squares criterion, according to (8.4.25), as indicated above. This yields the estimates $\{\hat{a}_k\}$, which allow us to synthesize the all-pole filter.

$$H_1(z) = \frac{1}{1 + \sum_{k=1}^{N} \hat{a}_k z^{-k}} \qquad (8.4.28)$$

The response of this filter to the impulse $\delta(n)$ is

$$v(n) = -\sum_{k=1}^{N} \hat{a}_k v(n - k) + \delta(n) \qquad n \geq 0 \qquad (8.4.29)$$

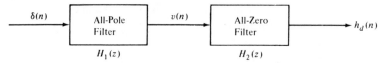

FIGURE 8.69 Least-squares method for determining the poles and zeros of a filter.

If the sequence $\{v(n)\}$ is used to excite an all-zero filter with system function

$$H_2(z) = \sum_{k=0}^{M} b_k z^{-k} \tag{8.4.30}$$

as illustrated in Fig. 8.69, its response is

$$\hat{h}_d(n) = \sum_{k=0}^{M} b_k v(n - k) \tag{8.4.31}$$

Now we can define an error sequence $e(n)$ as

$$e(n) = h_d(n) - \hat{h}_d(n) \tag{8.4.32}$$

$$= h_d(n) - \sum_{k=0}^{M} b_k v(n - k)$$

and, consequently, the parameters $\{b_k\}$ can also be determined by means of the least-squares criterion, namely, from the minimization of

$$\mathcal{E}_2 = \sum_{n=0}^{\infty} \left[h_d(n) - \sum_{k=0}^{M} b_k v(n - k) \right]^2 \tag{8.4.33}$$

Thus we obtain a set of linear equations for the parameters $\{b_k\}$, in the form

$$\sum_{k=0}^{M} b_k r_{vv}(k, l) = r_{vh}(l) \qquad l = 0, 1, \ldots, M \tag{8.4.34}$$

where, by definition,

$$r_{vv}(k, l) = \sum_{n=0}^{\infty} v(n - k)v(n - l) \tag{8.4.35}$$

$$r_{hv}(k) = \sum_{n=0}^{\infty} h_d(n)v(n - k) \tag{8.4.36}$$

EXAMPLE 8.4.4 _____

Approximate the fourth-order Butterworth filter given in Example 8.4.2 by means of the least-squares inverse design method.

Solution: From the desired impulse response $h_d(n)$, which is illustrated in Fig. 8.63, we computed the autocorrelation sequence $r_{dd}(k, l) = r_{dd}(k - l)$ and solved to set of linear equations in (8.4.8) to obtain the filter coefficients. The results of this computation are given in Table 8.15 for $N = 3, 4, 5, 10,$ and 15. In Table 8.16 we list the poles of the filter designs for $N = 3, 4,$ and 5 along with the actual poles of the fourth-order Butterworth filter. We note that the poles obtained from the designs are far from the actual poles of the desired filter.

The frequency responses of the filter designs are plotted in Fig. 8.70. We note that when N is small, the approximation to the desired filter is poor. As N is increased to

TABLE 8.15 Estimates of Filter Coefficients $\{a_k\}$ in Least-Squares Inverse Filter Design Method

$N = 3$		$N = 15$	
$a_1 =$	$0.254295E+01$	$a_1 =$	2.993620
$a_2 =$	$-0.241800E+01$	$a_2 =$	-1.143053
$a_3 =$	$0.853829E+00$	$a_3 =$	-12.132861
$N = 4$		$a_4 =$	39.663433
$a_1 =$	$0.319047E+01$	$a_5 =$	-75.749001
$a_2 =$	$-0.425176E+01$	$a_6 =$	109.247757
$a_3 =$	$0.278234E+01$	$a_7 =$	-129.513794
$a_4 =$	$0.758375E+00$	$a_8 =$	131.026794
$N = 5$		$a_9 =$	-114.905266
$a_1 =$	$0.368733E+01$	$a_{10} =$	87.449211
$a_2 =$	$-0.607422E+01$	$a_{11} =$	-57.031906
$a_3 =$	$0.556726E+01$	$a_{12} =$	30.915134
$a_4 =$	$-0.284813E+01$	$a_{13} =$	-13.124536
$a_5 =$	$0.654996E+00$	$a_{14} =$	3.879295
		$a_{15} =$	-0.597313

$N = 10$	
$a_1 =$	5.008451
$a_2 =$	-12.660761
$a_3 =$	21.557365
$a_4 =$	-27.804110
$a_5 =$	28.683949
$a_6 =$	-24.058558
$a_7 =$	16.156847
$a_8 =$	-8.247148
$a_9 =$	2.854789
$a_{10} =$	-0.502956

TABLE 8.16 Estimates of Pole Positions in Least-Squares Inverse Filter Design Method (Example 8.4.4)

Number of Poles	Pole Positions
$N = 3$	0.9305
	$0.8062 \pm j0.5172$
$N = 4$	$0.8918 \pm j0.2601$
	$0.7037 \pm j0.6194$
$N = 5$	0.914
	$0.8321 \pm j0.4307$
	$0.5544 \pm j0.7134$
$N = 4$ Butterworth filter	$0.6603 \pm j0.4435$
	$0.5241 \pm j0.1457$

$N = 10$ and $N = 15$, the approximation improves significally. However, even for $N = 15$, there are large ripples in the passband of the filter response. It is apparent that this method, which is based on an all-pole approximation, does not provide good approximations to filters that contain zeros.

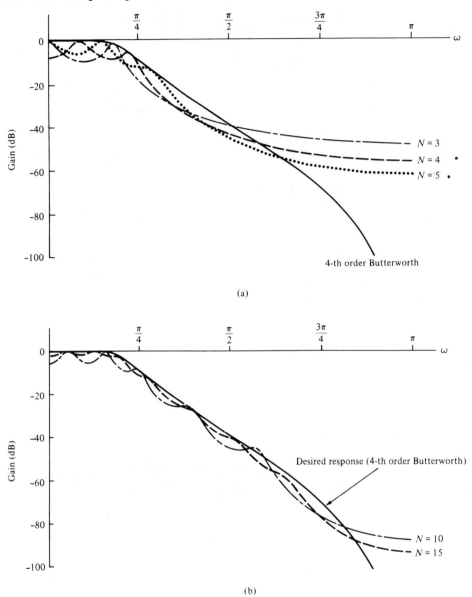

FIGURE 8.70 Magnitude responses for filter designs based on the least-squares inverse filter method.

EXAMPLE 8.4.5

Approximate the type II Chebyshev lowpass filter given in Examples 8.4.3 by means of the three least-squares methods described above.

Solution: The results of the filter designs obtained by means of the least-squares inverse method, Prony's method and Shanks' method are illustrated in Fig. 8.71. The filter parameters obtained from these design methods are listed in Table 8.17.

The frequency response characteristics in Fig. 8.71 illustrate that the least-squares inverse (all-pole) design method yields poor designs when the filter contains zeros. On

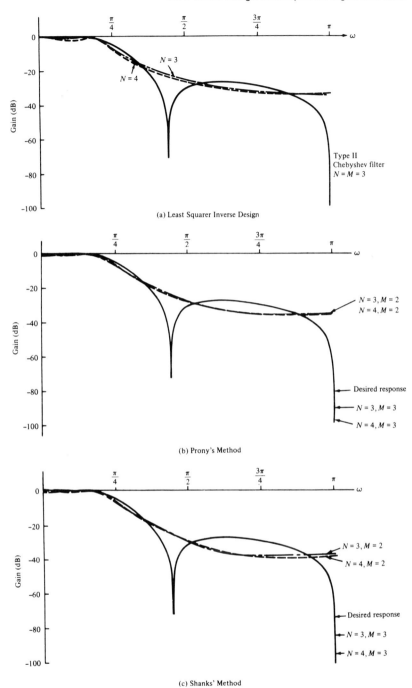

FIGURE 8.71   Filter designs based on least-squares methods (Example 8.4.5):
(a) least-squares inverse design; (b) Prony's method; (c) Shank's method.

the other hand, both Prony's method and Shanks' method yield very good designs when the number of poles and zeros equals or exceeds the number of poles and zeros in the actual filter. Thus the inclusion of zeros in the approximation has a significant effect in the resulting filter design.

TABLE 8.17 Pole–Zero Locations for Filter Designs in Example 8.4.5

Chebyshev Filter:

Zeros: -1, $0.1738311 \pm j0.9847755$
Poles: 0.3880, $0.5659 \pm j0.467394$

Filter Order	Poles in Least Squares Inverse
$N = 3$	0.8522
	$0.6544 \pm j0.6224$
$N = 4$	$0.7959 \pm j0.3248$
	$0.4726 \pm j0.7142$

Filter Order	Prony's Method		Shanks' Method	
	Poles	Zeros	Poles	Zeros
$N = 3$	0.5332		0.5348	
$M = 2$	$0.6659 \pm j0.4322$	$-0.1497 \pm j0.4925$	$0.6646 \pm j0.4306$	$-0.2437 \pm j0.5918$
$N = 4$	0.7092		0.7116	
	-0.2919		-0.2921	
$M = 2$	$0.6793 \pm j0.4863$	$-0.1982 \pm j0.37$	$0.6783 \pm j0.4855$	$0.306 \pm j0.4482$
$N = 3$	0.3881	-1	0.3881	-1
$M = 3$	$0.5659 \pm j0.4671$	$0.1736 \pm j0.9847$	$0.5659 \pm j0.4671$	$0.1738 \pm j0.9848$
$N = 4$	-0.00014	-1	-0.00014	-1
	0.388		0.388	
$M = 3$	$0.5661 \pm j0.4672$	$0.1738 \pm j0.9848$	$0.566 \pm j0.4671$	$0.1738 \pm j0.9848$

8.4.3 Design of IIR Filters in the Frequency Domain

The two IIR filter design methods described above were carried out in the time domain. There are also direct design techniques for IIR filters that can be performed in the frequency domain. In this section we describe a filter parameter optimization technique carried out in the frequency domain that is representative of frequency-domain design methods.

The design is most easily carried out with the system function for the IIR filter expressed in the cascade form as

$$H(z) = G \prod_{k=1}^{K} \frac{1 + \beta_{k1}z^{-1} + \beta_{k2}z^{-2}}{1 + \alpha_{k1}z^{-1} + \alpha_{k2}z^{-2}} \tag{8.4.37}$$

where the filter gain G and the filter coefficients $\{\alpha_{k1}\}$, $\{\alpha_{k2}\}$, $\{\beta_{k1}\}$, $\{\beta_{k2}\}$ are to be determined. The frequency response of the filter may be expressed as

$$H(\omega) = GA(\omega)e^{j\Theta(\omega)} \tag{8.4.38}$$

where

$$A(\omega) = \prod_{k=1}^{K} \left| \frac{1 + \beta_{k1}z^{-1} + \beta_{k2}z^{-2}}{1 + \alpha_{k1}z^{-1} + \alpha_{k2}z^{-2}} \right|_{z = e^{j\omega}} \tag{8.4.39}$$

and $\Theta(\omega)$ is the phase response.

Instead of dealing with the phase of the filter, it is more convenient to deal with the envelope delay as a function of frequency, which is defined as

$$\tau(\omega) = -\frac{d\Theta(\omega)}{d\omega} \tag{8.4.40}$$

or, equivalently,

$$\tau(\omega) = \tau(z)|_{z=e^{j\omega}} \tag{8.4.41}$$

$$= -\left[\frac{d\Theta(z)}{dz}\right]_{z=e^{j\omega}} \frac{dz}{d\omega}$$

It can be shown that $\tau(z)$ can be expressed as

$$\tau(z) = \mathrm{Re}\left\{\sum_{k=1}^{K}\left[\frac{\beta_{k1}z + 2\beta_{k2}}{z^2 + \beta_{k1}z + \beta_{k2}} - \frac{\alpha_{k1}z + 2\alpha_{k2}}{z^2 + \alpha_{k1}z + \alpha_{k2}}\right]\right\} \tag{8.4.42}$$

where Re (u) denotes the real part of the complex-valued quantity u.

Now suppose that the desired magnitude and delay characteristics $A(\omega)$ and $\tau(\omega)$ are specified at arbitrarily chosen discrete frequencies $\omega_1, \omega_2, \ldots, \omega_L$ in the range $0 \le |\omega| \le \pi$. Then the error in magnitude at the frequency ω_k is $GA(\omega_k) - A_d(\omega_k)$ where $A_d(\omega_k)$ is the desired magnitude response at ω_k. Similarly, the error in delay at ω_k may be defined as $\tau(\omega_k) - \tau_d(\omega_k)$, where $\tau_d(\omega_k)$ is the desired delay response. However, the choice of $\tau_d(\omega_k)$ is complicated by the difficulty in assigning a nominal delay to the filter. Hence we are led to define the error in delay as $\tau(\omega_k) - \tau(\omega_0) - \tau_d(\omega_k)$, where $\tau(\omega_0)$ is the filter delay at some nominal center frequency in the passband of the filter and $\tau_d(\omega_k)$ is the desired delay response of the filter relative to $\tau(\omega_0)$. By defining the error in delay in this manner, we are willing to accept a filter having whatever nominal delay $\tau(\omega_0)$ results from the optimization procedure.

As a performance index for determining the filter parameters, one may choose any arbitrary function of the errors in magnitude and delay. To be specific, let us select the total weighted least-squares error over all frequencies $\omega_1, \omega_2, \ldots, \omega_L$, that is,

$$\mathcal{E}(\mathbf{p}, G) = (1 - \lambda) \sum_{n=1}^{L} w_n[GA(\omega_n) - A_d(\omega_n)]^2 \tag{8.4.43}$$

$$+ \lambda \sum_{N=1}^{L} v_n[\tau(\omega_n) - \tau(\omega_0) - \tau_d(\omega_n)]^2$$

where $\mathbf{p}$ denotes the $4K$-dimensional vector of filter coefficients $\{\alpha_{k1}\}$, $\{\alpha_{k2}\}$, $\{\beta_{k1}\}$, and $\{\beta_{k2}\}$, and λ, $\{w_n\}$, and $\{v_n\}$ are weighting factors selected by the designer. Thus the emphasis in the errors affecting the design may be placed entirely on the magnitude ($\lambda = 0$), or in the delay ($\lambda = 1$) or, perhaps, equally weighted between magnitude and delay ($\lambda = 1/2$). Similarly, the weighting factors in frequency $\{w_n\}$ and $\{v_n\}$ determine the relative emphasis on the errors as a function of frequency.

The squared-error function $\mathcal{E}(\mathbf{p}, G)$ is a nonlinear function of $(4K + 1)$ parameters. The gain G that minimizes $\mathcal{E}$ is easily determined and given by the relation

$$\hat{G} = \frac{\displaystyle\sum_{n=1}^{L} w_n A(\omega_n) A_d(\omega_n)}{\displaystyle\sum_{n=1}^{L} w_n A^2(\omega_n)} \tag{8.4.44}$$

The optimum gain G may be substituted in (8.43) to yield

$$\mathcal{E}(\mathbf{p}, \hat{G}) = (1 - \lambda) \sum_{n=1}^{L} w_n [\hat{G} A(\omega_n) - A_d(\omega_n)]^2$$

$$+ \lambda \sum_{n=1}^{L} v_n [\tau(\omega_n) - \tau(\omega_0) - \tau_d(\omega_n)]^2$$

(8.4.45)

Due to the nonlinear nature of $\mathcal{E}(\mathbf{p}, \hat{G})$, its minimization over the remaining $4K$ parameters is performed by an iterative numerical optimization method such as the Fletcher and Powell method (1963). One begins the iterative process by assuming an initial set of parameter values, say $\mathbf{p}^{(0)}$. With the initial values substituted in (8.4.45), we obtain the least-squares error $\mathcal{E}(\mathbf{p}^{(0)}, \hat{G})$. If we also evaluate the partial derivatives $\partial \mathcal{E} / \partial \alpha_{k1}$, $\partial \mathcal{E} / \partial \alpha_{k2}$, $\partial \mathcal{E} / \partial \beta_{k1}$, and $\partial \mathcal{E} / \partial \beta_{k2}$ at the initial value $\mathbf{p}^{(0)}$, we can use this first derivative information to change the initial values of the parameters in a direction that leads toward the minimum of the function $\mathcal{E}(\mathbf{p}, \hat{G})$ and thus to a new set of parameters $\mathbf{p}^{(1)}$.

Repetition of the above steps results in an iterative algorithm which is described mathematically by the recursive equation

$$\mathbf{p}^{(m+1)} = \mathbf{p}^{(m)} - \Delta^{(m)} \mathbf{Q}^{(m)} \mathbf{g}^{(m)} \qquad m = 0, 1, 2, \dots$$

where $\Delta^{(m)}$ is a scalar representing the step size of the iteration, $\mathbf{Q}^{(m)}$ is a $(4K \times 4K)$ matrix, which is an estimate of the Hessian, and $\mathbf{g}^{(m)}$ is a $(4K \times 1)$ vector consisting of the four K-dimensional vectors of gradient components of $\mathcal{E}$ (i.e., $\partial \mathcal{E} / \partial \alpha_{k1}$, $\partial \mathcal{E} / \partial \alpha_{k2}$, $\partial \mathcal{E} / \partial \beta_{k1}$, $\partial \mathcal{E} / \partial \beta_{k2}$), evaluated at $\alpha_{k1} = \alpha_{k1}^{(m)}$, $\alpha_{k2} = \alpha_{k2}^{(m)}$, $\beta_{k1} = \beta_{k1}^{(m)}$, and $\beta_{k2} = \beta_{k2}^{(m)}$. The iterative process described above is terminated when the gradient components are nearly zero and the value of the function $\mathcal{E}(p, \hat{G})$ does not change appreciably from one iteration to another.

The stability constraint is easily incorporated into the computer program through the parameter vector p. When $|\alpha_{k2}| > 1$ for any $k = 1, \dots, K$, the parameter α_{k2} is forced back inside the unit circle and the iterative process continues. A similar process can be used to force zeros inside the unit circle if a minimum-phase filter is desired.

The major difficulty with any iterative procedure that searches for the parameter values that minimize a nonlinear function is that the process may converge to a local minimum instead of a global minimum. Our only recourse around this problem is to start the iterative process with different starting values for the parameters and observe the end result.

EXAMPLE 8.4.6

Let us design a lowpass filter using the Fletcher–Powell optimization procedure described above. The filter is to have a bandwidth of 0.3π and a rejection band commencing at 0.45π. The delay distortion may be ignored by selecting the weighting factor $\lambda = 0$.

Solution: We have selected a two-stage $(K = 2)$ or four-pole and four-zero filter which we believe is adequate to meet the transition band and rejection requirements. The magnitude response is specified at 19 equally spaced frequencies, which is considered a sufficiently dense set of points to realize a good design. Finally, a set of uniform weights is selected.

The filter designed has the response shown in Fig. 8.72. It has a remarkable resemblance to the response of the elliptic lowpass filter shown in Fig. 8.73, which was

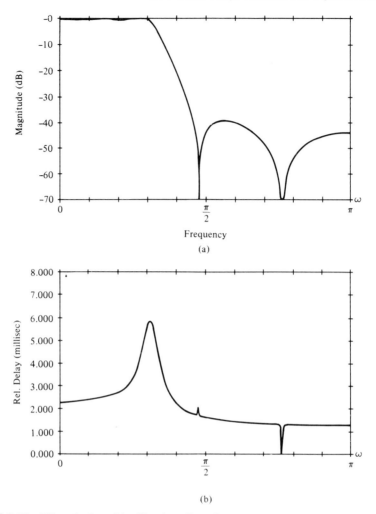

FIGURE 8.72 Filter designed by Fletcher–Powell optimization method (Example 8.4.6).

designed to have the same passband ripple and transition region as the computer-generated filter. A small but noticeable difference between the elliptic filter and the computer-generated filter is the somewhat flatter delay response of the latter relative to the former.

EXAMPLE 8.4.7 _____

Design an IIR filter with magnitude characteristics

$$A_d(\omega) = \begin{cases} \sin \omega & 0 \le |\omega| \le \dfrac{\pi}{2} \\ 0 & \dfrac{\pi}{2} < |\omega| < \pi \end{cases}$$

and a constant envelope delay in the passband.

Solution: The desired filter is called a modified duobinary filter and finds application in high-speed modems for digital communications. The frequency response was specified

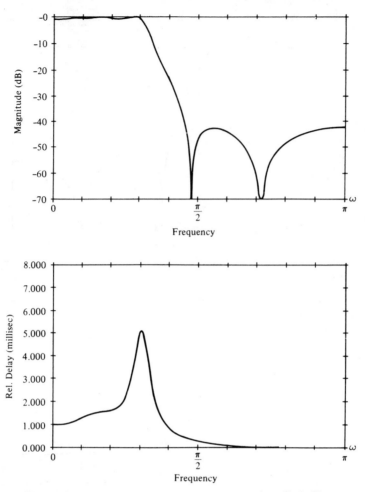

FIGURE 8.73 Amplitude and delay response for elliptic filter.

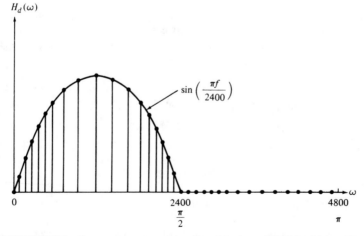

FIGURE 8.74 Frequency response of an ideal modified duobinary filter.

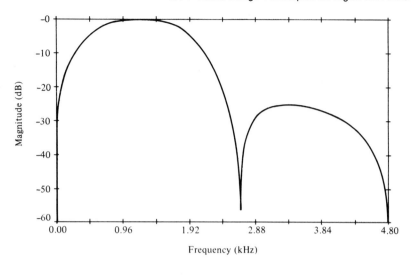

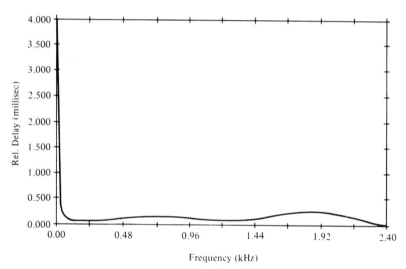

FIGURE 8.75   Frequency response of filter in Example 8.4.7. Designed by the Fletcher–Powell optimization method.

at the frequencies illustrated in Fig. 8.74. The envelope delay was left unspecified in the stopband and selected to be flat in the passband. Equal weighting coefficients $\{w_n\}$ and $\{v_n\}$ were selected. A weighting factor of $\lambda = 1/2$ was selected.

A two-stage (four-pole, four-zero) filter was designed to meet the foregoing specifications. The result of the design is illustrated in Fig. 8.75. We note that the magnitude characteristic is reasonably well matched to $\sin \omega$ in the passband, but the stopband attenuation peaks at about -25 dB, which is rather large. The envelope delay characteristic is relatively flat in the passband.

A four-stage (eight-pole, eight-zero) filter having the same frequency response specifications was also designed. This design produced better results, especially in the stopband where the attenuation peaked at -36 dB. The envelope delay was also considerably flatter.

8.5 Decimation and Interpolation

In many practical applications of digital signal processing, one is faced with the problem of changing the sampling rate of a signal, either increasing it or decreasing it by some amount. For example, in telecomunication systems that transmit and receive different types of signals (e.g., teletype, facsimile, speech, video, etc.), there is a requirement to process the various signals at different rates that are commensurate with the corresponding bandwidths of the signals. The process of converting a signal from a given rate to a different rate is called *sampling rate conversion*. In turn, systems that employ multiple sampling rates in the processing of signals are called *multirate digital signal processing systems*.

Sampling rate conversion of a digital signal can be accomplished in one of two general methods. One method is to pass the digital signal through a D/A converter, filter it if necessary, and then to resample the resulting analog signal at the desired rate (i.e., to pass the analog signal through an A/D converter). The second method is to perform the sampling rate conversion entirely in the digital domain.

One advantage of the first method is that the new sampling rate can be selected arbitrarily and need not have any special relationship to the old sampling rate. A major disadvantage, however, is the signal distortion introduced by the D/A in the signal reconstruction and by the quantization effects in the A/D conversion. On the other hand, the only apparent problem in performing the sampling rate conversion in the digital domain is that the ratio of new to old sampling rates is constrained to be rational. However, this constraint does not pose a limitation in most practical applications. As a consequence, sampling rate conversion is usually performed in the digital domain. Our discussion below is limited to this case.

The process of sampling-rate conversion in the digital domain can be viewed as a linear filtering operation, as illustrated in Fig. 8.76. The input signal $x(n)$ is characterized by the sampling rate $F_x = 1/T_x$ and the output signal $y(m)$ is characterized by the sampling rate $F_y = 1/T_y$, where T_x and T_y are the corresponding sampling intervals. The ratio F_y/F_x is constrained to be rational, that is,

$$\frac{F_y}{F_x} = \frac{U}{D} \qquad (8.5.1)$$

where D and U are integers. As we shall observe from our discussion below, the linear filter is characterized by a time-variant impulse response, denoted as $h(n, m)$. Hence the input $x(n)$ and the output $y(m)$ are related by the convolution summation for time-variant systems.

Before considering the general case of sampling rate conversion by the factor U/D, we shall consider two special cases. One is the case of sampling rate reduction by a factor D and the second is the case of a sampling rate increase by a factor U. The process of reducing the sampling rate by an integer factor D (down-sampling by D) is called *decimation*. The process of increasing the sampling rate by an integer factor U (up-sampling by U) is called *interpolation*.

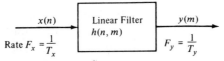

FIGURE 8.76   Sampling rate conversion viewed as a linear filtering operation.

8.5.1 Decimation by a Factor D

Let us assume that the signal $x(n)$ with spectrum $X(\omega)$ is to be down-sampled by a factor D. The spectrum $X(\omega)$ is assumed to be nonzero in the frequency interval $0 \le |\omega| \le \pi$ or, equivalently, $|F| \le F_x/2$. We know that if we reduce the sampling rate simply by selecting every Dth value of $x(n)$, the resulting signal will be an aliased version of $x(n)$, with a folding frequency of $F_x/2D$. To avoid aliasing, we must first reduce the bandwidth of $x(n)$ to $F_{max} = F_x/2D$ or equivalently, to $\omega_{max} = \pi/D$. Then, we may down-sample by D and thus avoid aliasing.

The decimation process is illustrated in Fig. 8.77. The input sequence $x(n)$ is passed through a lowpass filter characterized by the impulse response $h(n)$ and a frequency response $H(\omega)$, which ideally satisfies the condition

$$H(\omega) = \begin{cases} 1 & |\omega| \le \pi/D \\ 0 & \text{otherwise} \end{cases} \tag{8.5.2}$$

Thus the filter eliminates the spectrum of $X(\omega)$ in the range $\pi/D < \omega < \pi$. Of course, the implication is that only the frequency components of $x(n)$ in the range $|\omega| \le \pi/D$ are of interest in further processing of the signal.

The output of the filter is a sequence $v(n)$ given as

$$v(n) = \sum_{k=0}^{\infty} h(k)x(n-k) \tag{8.5.3}$$

which is then down-sampled by the factor D to produce $y(m)$. Thus

$$y(m) = v(mD) \tag{8.5.4}$$

$$= \sum_{k=0}^{\infty} h(k)x(mD-k)$$

Although the filtering operation on $x(n)$ is linear and time invariant, the down-sampling operation in combination with the filtering results in a time-variant system. This is easily verified. Given the fact that $x(n)$ produces $y(m)$, we note that $x(n-n_0)$ does not imply $y(n-n_0)$ unless n_0 is a multiple of D. Consequently, the overall linear operation (linear filtering followed by down-sampling) on $x(n)$ is not time invariant.

The frequency-domain characteristics of the output sequence $y(m)$ may be obtained by relating the spectrum of $y(m)$ to the spectrum of the input sequence $x(n)$. First, it is convenient to define a sequence $v(n)$ as

$$v'(n) = \begin{cases} v(n) & n = 0, \pm D, +2D, \ldots \\ 0 & \text{otherwise} \end{cases} \tag{8.5.5}$$

Clearly, $v'(n)$ may be viewed as a sequence obtained by multiplying $v(n)$ with a periodic train of impulses $p(n)$, with period D, as illustrated in Fig. 8.78. The discrete Fourier series representation of $p(n)$ is

$$p(n) = \frac{1}{D} \sum_{k=0}^{D-1} e^{j2\pi kn/D} \tag{8.5.6}$$

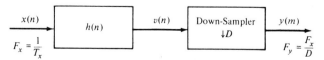

FIGURE 8.77 Decimation by a factor D.

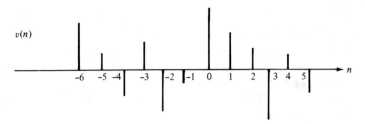

$v(n)$

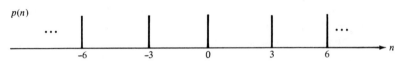

$p(n)$

FIGURE 8.78 Multiplication of $v(n)$ with a periodic impulse train $p(n)$ with period $D = 3$.

Hence

$$v'(n) = v(n)p(n) \tag{8.5.7}$$

and

$$y(m) = v'(mD) = v(mD)p(mD) = v(mD) \tag{8.5.8}$$

Now the z-transform of the output sequence $y(m)$ is

$$Y(z) = \sum_{m=-\infty}^{\infty} y(m)z^{-m}$$

$$= \sum_{m=-\infty}^{\infty} v'(mD)z^{-m}$$

$$Y(z) = \sum_{m=-\infty}^{\infty} v'(m)z^{-m/D} \tag{8.5.9}$$

where the last step follows from the fact that $v'(m) = 0$ except at multiples of D. By making use of the relations in (8.5.6) and (8.5.7) in (8.5.9), we obtain

$$Y(z) = \sum_{m=-\infty}^{\infty} v(m) \left[\frac{1}{D} \sum_{k=0}^{D-1} e^{j2\pi mk/D} \right] z^{-m/D}$$

$$= \frac{1}{D} \sum_{k=0}^{D-1} \sum_{m=-\infty}^{\infty} v(m)(e^{-j2\pi k/D}z^{1/D})^{-m}$$

$$= \frac{1}{D} \sum_{k=0}^{D-1} V(e^{-j2\pi k/D}z^{1/D})$$

$$= \frac{1}{D} \sum_{k=0}^{D-1} H(e^{-j2\pi k/D}z^{1/D})X(e^{-j2\pi k/D}z^{1/D}) \tag{8.5.10}$$

where the last step follows from the fact that $V(z) = H(z)X(z)$.

By evaluating $Y(z)$ on the unit circle, we obtain the spectrum of the output signal $y(m)$. Since the rate of $y(m)$ is $F_y = 1/T_y$, the frequency variable, which we denote as ω_y, is in radians relative to the sampling rate F_y, that is,

$$\omega_y = 2\pi F/F_y = 2\pi FT_y \tag{8.5.11}$$

Since the sampling rates are related by the expression

$$F_y = F_x/D \tag{8.5.12}$$

it follows that the frequency variables ω_y and

$$\omega_x = 2\pi F/F_x = 2\pi F T_x \tag{8.5.13}$$

are related by

$$\omega_y = D\omega_x \tag{8.5.14}$$

Thus the frequency range $0 \le |\omega_x| \le \pi/D$ is stretched into the corresponding frequency range $0 \le |\omega_y| \le \pi$ by the down-sampling process as expected.

We conclude that the spectrum $Y(\omega_y)$, which is obtained by evaluating (8.5.10) on the unit circle, may be expressed as

$$Y(\omega_y) = \frac{1}{D} \sum_{k=0}^{D-1} H\left(\frac{\omega_y - 2\pi k}{D}\right) X\left(\frac{\omega_y - 2\pi k}{D}\right) \tag{8.5.15}$$

With a properly designed filter $H(\omega)$ the aliasing is eliminated and, consequently, all but the first term in (8.5.15) vanish. Hence

$$Y(\omega_y) = \frac{1}{D} H\left(\frac{\omega_y}{D}\right) X\left(\frac{\omega_y}{D}\right) \tag{8.5.16}$$

$$= \frac{1}{D} X\left(\frac{\omega_y}{D}\right)$$

for $0 \le |\omega_y| \le \pi$. The spectra for the sequences $x(n)$, $v(n)$, and $y(m)$ are illustrated in Fig. 8.79.

8.5.2 Interpolation by a Factor U

An increase in the sampling rate by a factor of U can be accomplished by interpolating $U - 1$ new samples between successive values of the signal. The interpolation process may be accomplished in a variety of ways. We shall describe a process that preserves the spectral shape of the signal sequence $x(n)$.

Let $v(m)$ denote a sequence with a rate $F_y = UF_x$, which is obtained from $x(n)$ by adding $U - 1$ zeros between successive values of $x(n)$. Thus

$$v(m) = \begin{cases} x(m/U) & m = 0, \pm U, \pm 2U, \ldots \\ 0 & \text{otherwise} \end{cases} \tag{8.5.17}$$

and its sampling rate is identical to the rate of $y(m)$. This sequence has a z-transform

$$V(z) = \sum_{m=-\infty}^{\infty} v(m)z^{-m}$$

$$= \sum_{m=-\infty}^{\infty} x(m)z^{-mU} \tag{8.5.18}$$

$$= X(z^U)$$

The corresponding spectrum of $v(m)$ is obtained by evaluating (8.5.18) on the unit circle. Thus

$$V(\omega_y) = X(\omega_y U) \tag{8.5.19}$$

where ω_y denotes the frequency variable relative to the new sampling rate F_y (i.e., $\omega_y = 2\pi F/F_y$). Now the relationship between sampling rates is $F_y = UF_x$ and hence

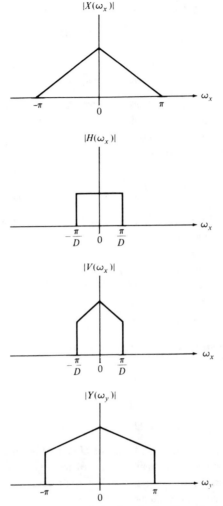

FIGURE 8.79 Spectra of signals in the decimation of $x(n)$ by a factor D.

the frequency variables ω_x and ω_y are related according to the formula

$$\omega_y = \frac{\omega_x}{U} \tag{8.5.20}$$

The spectra $X(\omega_x)$ and $V(\omega_y)$ are illustrated in Fig. 8.80. We observe that the sampling rate increase obtained by the addition of $U - 1$ zero samples between successive values of $x(n)$ results in a signal whose spectrum $V(\omega_y)$ is a U-fold periodic repetition of the input signal spectrum $X(\omega_x)$.

Since only the frequency components of $x(n)$ in the range $0 \le \omega_y \le \pi/U$ are unique, the images of $X(\omega)$ above $\omega_y = \pi/U$ should be rejected by passing the sequence $v(m)$ through a lowpass filter with frequency response $H(\omega_y)$, which ideally has the characteristic

$$H(\omega_y) = \begin{cases} C & 0 \le |\omega_y| \le \pi/U \\ 0 & \text{otherwise} \end{cases} \tag{8.5.21}$$

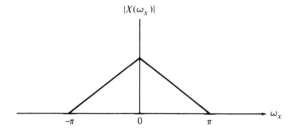

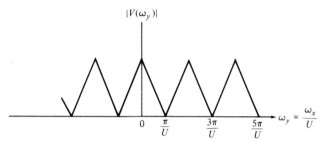

FIGURE 8.80 Spectra of $x(n)$ and $v(n)$ where $V(\omega_y) = X(\omega_y U)$.

where C is a scale factor that is required in order to normalize the output sequence $y(m)$ properly. Consequently, the output spectrum is

$$Y(\omega_y) = \begin{cases} CX(\omega_y U) & 0 \le |\omega_y| \le \pi/U \\ 0 & \text{otherwise} \end{cases} \qquad (8.5.22)$$

The scale factor C is selected so that the output $y(m) = x(m/U)$ for $m = 0, \pm U, \pm 2U, \dots$. For mathematical convenience, we select the point $m = 0$. Thus

$$\begin{aligned} y(0) &= \frac{1}{2\pi} \int_{-\pi}^{\pi} Y(\omega_y) \, d\omega_y \\ &= \frac{C}{2\pi} \int_{-\pi/U}^{\pi/U} X(\omega_y U) \, d\omega_y \end{aligned} \qquad (8.5.23)$$

Since $\omega_y = \omega_x/U$, (8.5.23) may be expressed as

$$\begin{aligned} y(0) &= \frac{C}{U} \frac{1}{2\pi} \int_{-\pi}^{\pi} X(\omega_x) \, d\omega_x \\ &= \frac{C}{U} x(0) \end{aligned} \qquad (8.5.24)$$

Therefore, $C = U$ is the desired normalization factor.

Finally, we indicate that the output sequence $y(m)$ can be expressed as a convolution of the sequence $v(n)$ with the unit sample response $h(n)$ of the lowpass filter. Thus

$$y(m) = \sum_{k=-\infty}^{\infty} h(m - k)v(k) \qquad (8.5.25)$$

Since $v(k) = 0$ except at multiples of U, where $v(kU) = x(k)$, (8.5.25) becomes

$$y(m) = \sum_{k=-\infty}^{\infty} h(m - kU)x(k) \qquad (8.5.26)$$

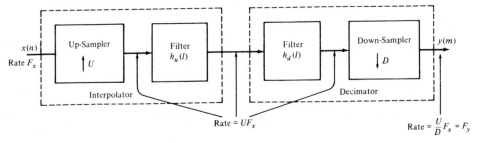

FIGURE 8.81   Method for sampling rate conversion by a factor U/D.

8.5.3 Sampling Rate Conversion by a Rational Factor U/D

Having discussed the special cases of decimation (down-sampling by a factor D) and interpolation (up-sampling by a factor U), we now consider the general case of sampling-rate conversion by a rational factor U/D. Basically, we can achieve this sampling-rate conversion by first performing interpolation by the factor U and then decimating the output of the interpolator by the factor D. In other words, a sampling-rate conversion by the rational factor U/D is accomplished by cascading an interpolator with a decimator, as illustrated in Fig. 8.81.

We emphasize the importance of performing the interpolation first and the decimation second, in order to preserve the desired spectral characteristics of $x(n)$. Furthermore, with the cascade configuration illustrated in Fig. 8.81, the two filters with impulse response $\{h_u(l)\}$ and $\{h_d(l)\}$ are operated at the same rate, namely, UF_x and hence can be combined into a single lowpass filter with impulse response $h(l)$ as illustrated in Fig. 8.82. The frequency response $H(\omega_v)$ of the combined filter must incorporate the filtering operations for both interpolation and decimation and hence it should ideally possess the frequency response characteristic

$$H(\omega_v) = \begin{cases} U & 0 \le |\omega_v| \le \min\left(\dfrac{\pi}{D}, \dfrac{\pi}{U}\right) \\ 0 & \text{otherwise} \end{cases} \tag{8.5.27}$$

where $\omega_v = 2\pi F/F_v = 2\pi F/UF_x = \omega_x/U$.

In the time domain, the output of the up-sampler is the sequence

$$v(l) = \begin{cases} x(l/U) & l = 0, \pm U, \pm 2U, \dots \\ 0 & \text{otherwise} \end{cases} \tag{8.5.28}$$

and the output of the linear, time-invariant filter is

$$\begin{aligned} w(l) &= \sum_{k=-\infty}^{\infty} h(l - k)v(k) \\ &= \sum_{k=-\infty}^{\infty} h(l - kU)x(k) \end{aligned} \tag{8.5.29}$$

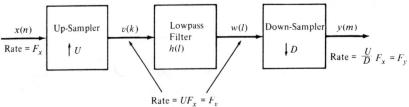

FIGURE 8.82   Method for sampling rate conversion by a factor U/D.

Finally, the output of the sampling-rate converter is the sequence $\{y(m)\}$, which is obtained by down-sampling the sequence $\{w(l)\}$ by a factor of D. Thus

$$y(m) = w(mD) \tag{8.5.30}$$

$$= \sum_{k=-\infty}^{\infty} h(mD - kU)x(k)$$

It is illuminating to express (8.5.30) in a different form by making a change in variable. Let

$$k = \left\lfloor \frac{mD}{U} \right\rfloor - n \tag{8.5.31}$$

where the notation $\lfloor r \rfloor$ denotes the largest integer contained in r. With this change in variable, (8.5.30) becomes

$$y(m) = \sum_{n=-\infty}^{\infty} h\left(mD - \left\lfloor \frac{mD}{U} \right\rfloor U + nU\right) x\left(\left\lfloor \frac{mD}{U} \right\rfloor - n\right) \tag{8.5.32}$$

We note that

$$mD - \left\lfloor \frac{mD}{U} \right\rfloor U = mD \quad \text{modulo } U$$

$$= (mD)_U$$

Consequently, (8.5.32) may be expressed as

$$y(m) = \sum_{n=-\infty}^{\infty} h(nU + (mD)_U)x\left(\left\lfloor \frac{mD}{U} \right\rfloor - n\right) \tag{8.5.33}$$

It is apparent from this form that the output $y(m)$ is obtained by passing the input sequence $x(n)$ through a time-variant filter with impulse response

$$g(n, m) = h(nU + (mD)_U) \qquad -\infty < m, \quad n < \infty \tag{8.5.34}$$

where $h(k)$ is the impulse response of the time-invariant lowpass filter which is operating at the sampling rate UF_x. We further observe that for any integer k,

$$g(n, m + kU) = h(nU + (mD + kDU)_U)$$
$$= h(nU + (mD)_U) \tag{8.5.35}$$
$$= g(n, m)$$

Hence $g(n, m)$ is periodic in the variable m with period U.

The frequency-domain relationships may be obtained by combining the results of the interpolation and decimation processes. Thus the spectrum at the output of the linear filter with impulse response $h(l)$ is

$$W(\omega_v) = H(\omega_v)X(\omega_v U)$$

$$= \begin{cases} UX(\omega_v U) & 0 \le |\omega_v| \le \min\left(\dfrac{\pi}{D}, \dfrac{\pi}{U}\right) \\[2mm] 0 & \text{otherwise} \end{cases} \tag{8.5.36}$$

The spectrum of the output sequence $y(m)$, which is obtained by decimating the sequence $w(n)$ by a factor of D is

$$Y(\omega_y) = \frac{1}{D} \sum_{k=0}^{D-1} W\left(\frac{\omega_y - 2\pi k}{D}\right) \tag{8.5.37}$$

where $\omega_y = D\omega_v$. Since the linear filter prevents aliasing as implied by (8.5.36), the spectrum of the output sequence given by (8.5.37) reduces to

$$Y(\omega_y) = \begin{cases} \dfrac{U}{D} X\left(\dfrac{\omega_y}{D}\right) & 0 \le |\omega_y| \le \min\left(\pi, \dfrac{\pi D}{U}\right) \\ 0 & \text{otherwise} \end{cases} \qquad (8.5.38)$$

8.5.4 Filter Design and Implementation for Sampling-Rate Conversion

As indicated in the discussion above, sampling-rate conversion by a factor U/D can be achieved by first increasing the sampling rate by U, which is achieved by inserting $U - 1$ zeros between successive values of the input signal $x(n)$, followed by linear filtering of the resulting sequence to eliminate the unwanted images of $X(\omega)$, and, finally, down-sampling the filtered signal by the factor D. In this section we consider the design and implementation of the linear filter.

In principle, the simplest realization of the filter is the direct form FIR structure with system function

$$H(z) = \sum_{k=0}^{M-1} h(k) z^{-k} \qquad (8.5.39)$$

where $\{h(k)\}$ is the unit sample response of the FIR filter. The lowpass filter can be designed to have linear phase and a specified passband ripple and stopband attenuation. Any of the FIR filter design techniques described earlier may be used to carry out this design. Thus we will have the filter parameters $\{h(k)\}$, which allow us to implement the FIR filter directly as shown in Fig. 8.83.

Although the direct form FIR filter realization illustrated in Fig. 8.83 is simple, it is also very inefficient. The inefficiency results from the fact that the up-sampling process introduces $U - 1$ zeros between successive points of the input signal. If U is large, most of the signal components in the FIR filter are zero. Consequently, most of the multipli-

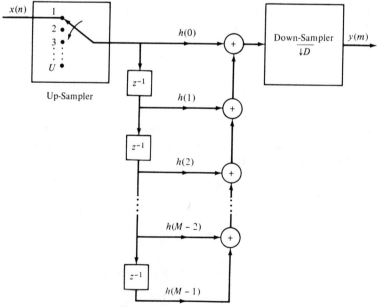

FIGURE 8.83   Direct-form realization of FIR filter in sampling rate conversion by factor U/D.

cations and additions result in zeros. Furthermore, the down-sampling process at the output of the filter implies that only one out of every D output samples is required at the output of the filter. Consequently, only one out of every D possible values at the output of the filter should be computed.

To develop a more efficient filter structure, let us begin with a decimator that reduces the sampling rate by an integer factor D. From our previous discussion, the decimator is obtained by passing the input sequence $x(n)$ through an FIR filter and then down-sampling the filter output by a factor D, as illustrated in Fig. 8.84a. In this configuration, the filter is operating at the high sampling rate F_x, while only one out of every D output samples is actually needed. The logical solution to this inefficiency problem is to embed the down-sampling operation within the filter, as illustrated in the filter realization given in Fig. 8.84b. In this filter structure, all the multiplications and additions are performed at the lower sampling rate F_x/D. Thus we have achieved the desired efficiency. Additional

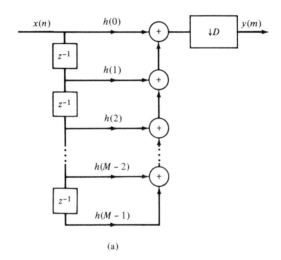

(a)

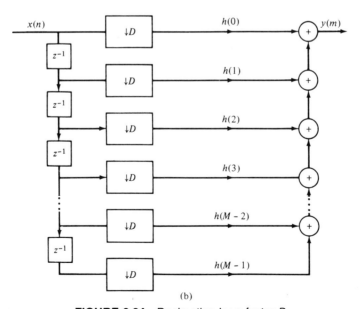

(b)

FIGURE 8.84   Decimation by a factor D.

reduction in computation can be achieved by exploiting the symmetry characteristics of $\{h(k)\}$. Figure 8.85 illustrates an efficient realization of the decimator in which the FIR filter has linear phase and hence $\{h(k)\}$ is symmetric.

Let us consider next the efficient implementation of an interpolator, which is realized by first inserting $U - 1$ zeros between samples of $x(n)$ and then filtering the resulting sequence. The direct form realization is illustrated in Fig. 8.86. The major problem with this structure is that the filter computations are performed at the high sampling rate UF_x. The desired simplification is achieved by first using the transposed form of the FIR filter, as illustrated in Fig. 8.87a and then embedding the up-sampler within the filter, as shown in Fig. 8.87b. Thus all the filter multiplications are performed at the low rate F_x, while the up-sampling process introduces $U - 1$ zeros in each of the filter branches of the

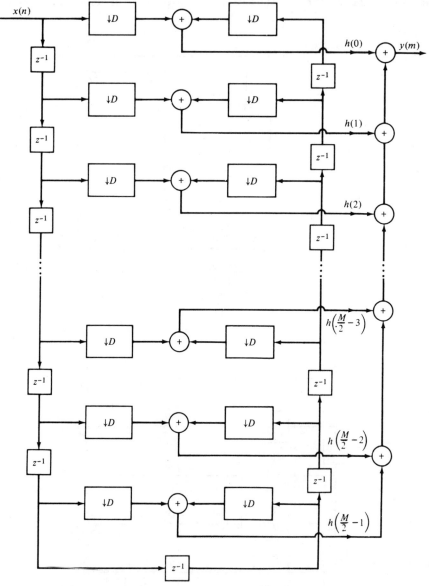

FIGURE 8.85   Efficient realization of a decimator that exploits the symmetry in the FIR filter.

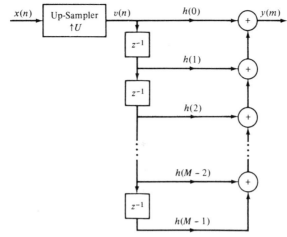

FIGURE 8.86 Direct-form realization of FIR filter in interpolation by a factor U.

structure shown in Fig. 8.87b. The reader may easily verify that the two filter structures in Fig. 8.87 are equivalent.

It is interesting to note that the structure of the interpolator shown in Fig. 8.87b can be obtained by transposing the structure of the decimator shown in Fig. 8.84b. We observe that the transpose of a decimator is an interpolator, and vice versa. These relationships are illustrated in Fig. 8.88, where (b) is obtained by transposing (a) and (d) is obtained by transposing (c). Consequently, a decimator is the dual of an interpolator, and vice versa. From these relationships it follows that there is an interpolator whose structure is the dual of the decimator shown in Fig. 8.85, which exploits the symmetry in $h(n)$.

The computational efficiency of the filter structure above can also be achieved by reducing the large FIR filter of length M into a set of smaller filters of length $K = M/U$, where M is selected to be a multiple of U. To demonstrate this point, let us consider the interpolator given in Fig. 8.83. Since the up-sampling process inserts $U - 1$ zeros between successive values of $x(n)$, only K out of the M input values stored in the FIR filter at any one time are nonzero. At one time instant, these nonzero values coincide and are multipled by the filter coefficients $h(0)$, $h(U)$, $h(2U)$, . . . , $h(M - U)$. In the following time instant, the nonzero values of the input sequence coincide and are multiplied by the filter coefficients $h(1)$, $h(U + 1)$, $h(2U + 1)$, . . . , $h(M - U + 1)$, etc. This observation leads us to define a set of smaller filters, called *polyphase filters,* with unit sample responses

$$p_k(n) = h(k + nU) \quad \begin{aligned} k &= 0, 1, \ldots, U - 1 \\ n &= 0, 1, \ldots, K - 1 \end{aligned} \tag{8.5.40}$$

where $K = M/U$ is an integer.

From the discussion above it follows that the set of U polyphase filters can be arranged as a parallel realization and the output of each filter can be selected by a commutator as illustrated in Fig. 8.89. The rotation of the commutator is in the counterclockwise direction beginning with the point at $m = 0$. Thus the polyphase filters perform computations at the low sampling rate F_x and the rate conversion results from the fact that U output samples are generated, one from each of the filters, for each input sample.

The decomposition of $\{h(k)\}$ into the set of U filters with impulse response $p_k(n)$, $k = 0, 1, \ldots, U - 1$, is consistent with our previous observation that the input signal

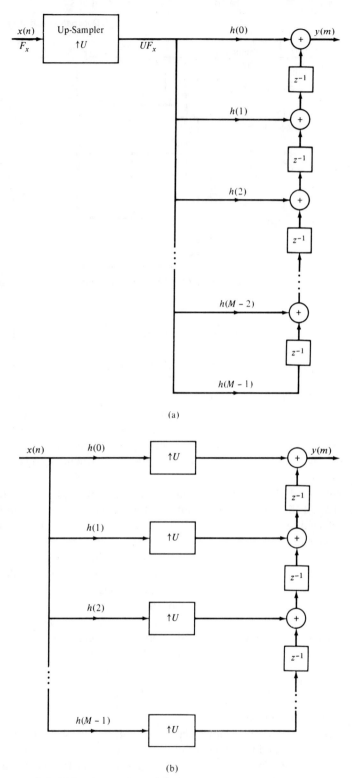

(a)

(b)

FIGURE 8.87 Efficient realization of an interpolator.

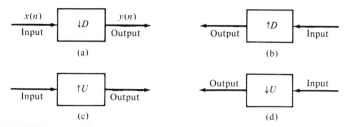

FIGURE 8.88 Duality relationships obtained through transposition.

is being filtered by a periodically time-variant linear filter with impulse response

$$g(n, m) = h(nU + (mD)_U) \qquad (8.5.41)$$

where $D = 1$ in the case of the interpolator. We noted previously that $g(n, m)$ varies periodically with period U. Consequently, a different set of coefficients are used to generate the set of U output samples $y(m)$, $m = 0, 1, \ldots, U - 1$.

Additional insight can be gained about the characteristics of the set of polyphase filters by noting that $p_k(n)$ is obtained from $h(n)$ by decimation with a factor U. Consequently, if the original filter frequency response $H(\omega)$ is flat over the range $0 \leq |\omega| \leq \pi/U$, then each of the polyphase filters will possess a relatively flat response over the range $0 \leq |\omega| \leq \pi$, that is, the polyphase filters are basically all-pass filters and differ primarily in their phase characteristics. This explains the reason for the term "polyphase" in describing these filters.

By transposing the interpolator structure in Fig. 8.89 we obtain a commutator structure for a decimator that is based on the parallel bank of polyphase filters, as illustrated in Fig. 8.90. The unit sample responses of the polyphase filters are now defined as

$$p_k(n) = h(k + nD) \qquad \begin{matrix} k = 0, 1, \ldots, D - 1 \\ n = 0, 1, \ldots, K - 1 \end{matrix} \qquad (8.5.42)$$

where $K = M/D$ is an integer when M is selected to be a multiple of D. The commutator rotates in a counterclockwise direction starting with the filter $p_0(n)$ at $m = 0$.

Although the two commutator structures for the interpolator and the decimator described above rotate in a counterclockwise direction, it is also possible to derive an equivalent pair of commutator structures having a clockwise rotation. In this alternative

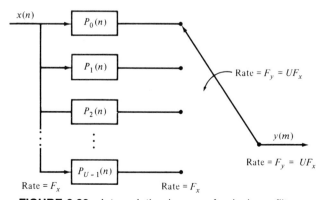

FIGURE 8.89 Interpolation by use of polyphase filters.

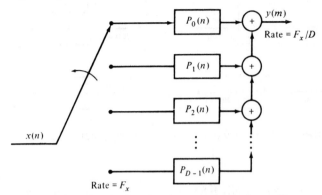

FIGURE 8.90 Decimation by use of polyphase filters.

formulation, the sets of polyphase filters are defined to have impulse responses

$$p_k(n) = h(nU - k) \qquad k = 0, 1, \ldots, U - 1 \qquad (8.5.43)$$
$$p_k(n) = h(nD - k) \qquad k = 0, 1, \ldots, D - 1 \qquad (8.5.44)$$

for the interpolator and decimator, respectively.

Having described the filter implementation for a decimator and an interpolator, let us now consider the general problem of sampling rate conversion by the factor U/D.

In the general case of sampling rate conversion by a factor U/D, the filtering can be accomplished by means of the linear time-variant filter described by the response function

$$g(n, m) = h(nU + (mD)_U) \qquad (8.5.45)$$

where $h(n)$ is the impulse response of the low-pass FIR filter, which, ideally, has the frequency response specified by (8.5.27). For convenience we select the length of the FIR filter $\{h(n)\}$ to a multiple of U (i.e., $M = KU$). As a consequence, the set of coefficients $\{g(n, m)\}$ for each $m = 0, 1, 2, \ldots, U - 1$, contains K elements. Since $g(n, m)$ is also periodic with period U, as previously demonstrated in (8.5.35), it follows that the output $y(m)$ may be expressed as

$$y(m) = \sum_{n=0}^{K-1} g\left(n, m - \left\lfloor \frac{m}{U} \right\rfloor U\right) x\left(\left\lfloor \frac{mD}{U} \right\rfloor - n\right) \qquad (8.5.46)$$

Conceptually, we can think of performing the computations specified by (8.5.46) by processing blocks of data of length K by a set of K filter coefficients

$$g\left(n, m - \left\lfloor \frac{m}{U} \right\rfloor U\right) \qquad n = 0, 1, \ldots, K - 1$$

There are U such sets of coefficients, one set for each block of U output points of $y(m)$. For each block of U output points, there is a corresponding block of D input points of $x(n)$ that enter in the computation.

The block processing algorithm for computing (8.5.46) can be visualized as illustrated in Fig. 8.91. A block of D input samples is buffered and shifted into a second buffer, of length K, one sample at a time. The shifting from the input buffer to the second buffer occurs at a rate of one sample each time the quantity $\lfloor mD/U \rfloor$ increases by one. For each output sample $y(l)$, the samples from the second buffer are multiplied by the corresponding set of filter coefficients $g(n, l)$ for $n = 0, 1, \ldots, K - 1$, and the K products are accumulated to give $y(l)$ for $l = 0, 1, \ldots, U - 1$. Thus U outputs result from this computation, which is then repeated for a new set of D input samples, and so on.

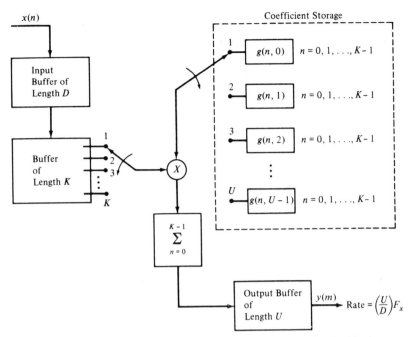

FIGURE 8.91 Efficient implementation of sampling-rate conversion by block processing.

An alternative method for computing the output of the sample rate converter specified by (8.5.46) is by means of an FIR filter structure with periodically varying filter coefficients. Such a structure is illustrated in Fig. 8.92. The input samples $x(n)$ are passed into a shift register that operates at the sampling rate F_x and is of length $K = M/U$, where M is the length of the time-invariant FIR filter, specified by the frequency response given by (8.5.27). Each stage of the register is connected to a hold-and-sample device which serves to couple the input sample rate F_x to the output sample rate $F_y = (U/D)F_x$.

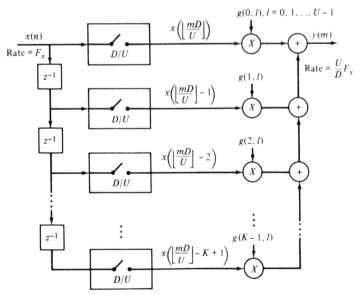

FIGURE 8.92 Efficient realization of sampling-rate conversion by a factor U/D.

The sample at the input to each hold-and-sample device is held until the next input sample arrives and then discarded. The output samples of the hold-and-sample device are taken at times mD/U, $m = 0, 1, 2, \ldots$ When both the input and output sampling times coincide (i.e., when mD/U is an integer), the input to the hold-and-sample is changed first and then the output samples the new input. The K outputs from the K hold-and-sample devices are multiplied by the periodically time-varying coefficients

$$g\left(n, m - \left(\frac{m}{U}\right)U\right) \qquad \text{for } n = 0, 1, \ldots, K - 1$$

and the resulting products are summed to yield $y(m)$. The computations at the output of the hold-and-sample devices are repeated at the output sampling rate of $F_y = (U/D)F_x$.

Let us now demonstrate the filter design procedure, first in the design of a decimator, second in the design of an interpolator, and finally, in the design of a rational sample-rate converter.

EXAMPLE 8.5.1

Design a decimator that down-samples an input signal $x(n)$ by a factor $D = 2$. Use the Remez algorithm to determine the coefficients of the FIR filter that has a 0.1-dB ripple

TABLE 8.18 Coefficients of Linear-Phase FIR Filter in Example 8.5.1

```
                    FINITE IMPULSE RESPONSE (FIR)
                  LINEAR-PHASE DIGITAL FILTER DESIGN
                      REMEZ EXCHANGE ALGORITHM

                     FILTER LENGTH = 30

              ***** IMPULSE RESPONSE *****
       H( 1) =    0.60256165E-02 = H( 30)
       H( 2) =   -0.12817143E-01 = H( 29)
       H( 3) =   -0.28582066E-02 = H( 28)
       H( 4) =    0.13663346E-01 = H( 27)
       H( 5) =   -0.46688961E-02 = H( 26)
       H( 6) =   -0.19704415E-01 = H( 25)
       H( 7) =    0.15984623E-01 = H( 24)
       H( 8) =    0.21384886E-01 = H( 23)
       H( 9) =   -0.34979440E-01 = H( 22)
       H(10) =   -0.15615522E-01 = H( 21)
       H(11) =    0.64006113E-01 = H( 20)
       H(12) =   -0.73451772E-02 = H( 19)
       H(13) =   -0.11873185E+00 = H( 18)
       H(14) =    0.98047845E-01 = H( 17)
       H(15) =    0.49225068E+00 = H( 16)
```

	BAND 1	BAND 2
LOWER BAND EDGE	0.0000000	0.3100000
UPPER BAND EDGE	0.2500000	0.5000000
DESIRED VALUE	1.0000000	0.0000000
WEIGHTING	2.0000000	1.0000000
DEVIATION	0.0107151	0.0214302
DEVIATION IN DB	0.0925753	-33.3794746

```
EXTREMAL FREQUENCIES—MAXIMA OF THE ERROR CURVE
0.0000000    0.0416667    0.0791667    0.1166666    0.1520833
0.1854166    0.2145832    0.2395832    0.2500000    0.3100000
0.3225000    0.3495833    0.3808333    0.4141666    0.4474999
0.4829165
```

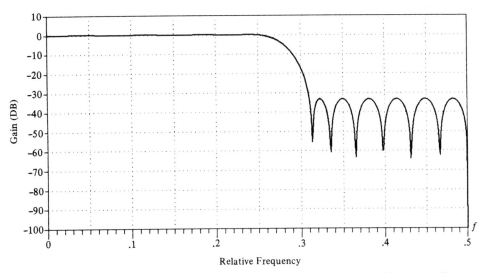

FIGURE 8.93 Magnitude response of linear-phase FIR filter of length $M = 30$ in Example 8.5.1.

in the passband and is down by at least 30 dB in the stopband. Also determine the polyphase filter structure in a decimator realization that employs polyphase filters.

Solution: A filter of length $M = 30$ achieves the design specifications given above. The impulse response of the FIR filter is given in Table 8.18 and the frequency response is illustrated in Fig. 8.93. Note that the cutoff frequency is $\omega_c = \pi/2$.

The polyphase filters obtained from $h(n)$ have impulse responses

$$p_k(n) = h(2n + k) \qquad k = 0, 1, \quad n = 0, 1, \ldots, 14$$

Note that $p_0(n) = h(2n)$ and $p_1(n) = h(2n + 1)$. Hence one filter consists of the even-numbered samples of $h(n)$ and the other filter consists of the odd-numbered samples of $h(n)$.

EXAMPLE 8.5.2 ——————————————————————————————

Design an interpolator that increases the input sampling rate by a factor of $U = 5$. Use the Remez algorithm to determine the coefficients of the FIR filter that has a 0.1-dB ripple in the passband and is down by at least 30 dB in the stopband. Also, determine the polyphase filter structure in an interpolator realization based on polyphase filters.

Solution: A filter of length $M = 30$ achieves the design specifications given above. The frequency response of the FIR filter is illustrated in Fig. 8.94 and its coefficients are given in Table 9.19. The cutoff frequency is $\omega_c = \pi/5$.

The polyphase filters obtained from $h(n)$ have impulse responses

$$p_k(n) = h(5n + k) \qquad k = 0, 1, 2, 3, 4$$

Consequently, each filter has length 6.

EXAMPLE 8.5.3 ——————————————————————————————

Design a sample-rate converter that increases the sampling rate by a factor 2.5. Use the Remez algorithm to determine the coefficients of the FIR filter that has 0.1-dB ripple in

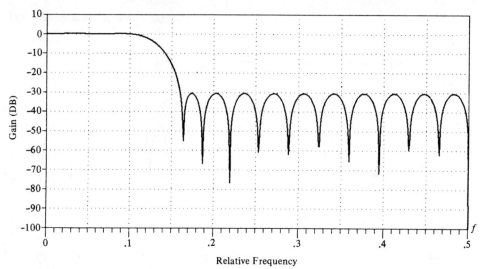

FIGURE 8.94 Magnitude response of linear-phase FIR filter of length $M = 30$ in Example 8.5.2.

TABLE 8.19 Coefficients of Linear-Phase FIR Filter in Example 8.5.2

FINITE IMPULSE RESPONSE (FIR)
LINEAR–PHASE DIGITAL FILTER DESIGN
REMEZ EXCHANGE ALGORITHM

FILTER LENGTH = 30

***** IMPULSE RESPONSE *****

H(1)	=	0.63987216E−02	=	H(30)
H(2)	=	−0.14761304E−01	=	H(29)
H(3)	=	−0.10886577E−02	=	H(28)
H(4)	=	−0.28714957E−02	=	H(27)
H(5)	=	0.10486430E−01	=	H(26)
H(6)	=	0.21477142E−01	=	H(25)
H(7)	=	0.19479362E−01	=	H(24)
H(8)	=	−0.31067431E−03	=	H(23)
H(9)	=	−0.30053033E−01	=	H(22)
H(10)	=	−0.49877029E−01	=	H(21)
H(11)	=	−0.37371285E−01	=	H(20)
H(12)	=	0.18482896E−01	=	H(19)
H(13)	=	0.10747141E+00	=	H(18)
H(14)	=	0.19951098E+00	=	H(17)
H(15)	=	0.25794828E+00	=	H(16)

	BAND 1	BAND 2
LOWER BAND EDGE	0.0000000	0.1600000
UPPER BAND EDGE	0.1000000	0.5000000
DESIRED VALUE	1.0000000	0.0000000
WEIGHTING	3.0000000	1.0000000
DEVIATION	0.0097524	0.0292572
DEVIATION IN DB	0.0842978	−30.6753349

EXTREMAL FREQUENCIES—MAXIMA OF THE ERROR CURVE

0.0000000	0.0333333	0.0645834	0.0895833	0.1000000
0.1600000	0.1745833	0.2016666	0.2370833	0.2704166
0.3058332	0.3412498	0.3766665	0.4120831	0.4474997
0.4829164				

the passband and is down by at least 30 dB in the stopband. Specify the sets of time-varying coefficients $g(n, m) - (m/U)$ used in the realization of the sampling-rate converter according to the structure in Fig. 8.92.

Solution: The FIR filter that meets the specifications of this problem is exactly the same as the filter designed in Example 8.5.2. Its bandwidth is $\pi/5$.

The coefficients of the filter are given by (8.5.34) as

$$g(n, m) = h(nU + (mD)_U)$$

$$= h\left(nU + mD - \left|\frac{mD}{U}\right| U\right)$$

By substituting $U = 5$ and $D = 2$, we obtain

$$g(n, m) = h\left(5n + 2m - 5\left|\frac{2m}{5}\right|\right)$$

By evaluating $g(n, m)$ for $n = 0, 1, \ldots, 5$ and $m = 0, 1, \ldots, 4$ we obtain the following coefficients for the time-variant filter:

$$
\begin{aligned}
g(0, m) &= \{h(0) & h(2) & h(4) & h(1) & h(3)\} \\
g(1, m) &= \{h(5) & h(7) & h(9) & h(6) & h(8)\} \\
g(2, m) &= \{h(10) & h(12) & h(14) & h(11) & h(13)\} \\
g(3, m) &= \{h(15) & h(17) & h(19) & h(16) & h(18)\} \\
g(4, m) &= \{h(20) & h(22) & h(24) & h(21) & h(23)\} \\
g(5, m) &= \{h(25) & h(27) & h(29) & h(26) & h(28)\}
\end{aligned}
$$

8.6 Summary and References

We have described in some detail the most important techniques for designing FIR and IIR digital filters based on either frequency-domain specifications expressed in terms of a desired frequency response $H_d(\omega)$ or in terms of the desired impulse response $h_d(n)$.

As a general rule, FIR filters are used in applications where there is a need for a linear-phase filter. This requirement occurs in many applications, especially in telecommunications, where there is a requirement to separate (demultiplex) signals such as data, that have been frequency-division multiplexed, without distorting these signals in the process of demultiplexing. Of the several methods described for designing FIR filters, the frequency sampling design method and the optimum Chebyshev approximation method yield the best designs.

IIR filters are generally used in applications where some phase distortion is tolerable. Of the class of IIR filters, elliptic filters are the most efficient to implement in the sense that for a given set of specifications, an elliptic filter will have a lower order or fewer coefficients than any other IIR filter type. When compared with FIR filters, elliptic filters are also considerably more efficient. In view of this, one might consider the use of an elliptic filter to obtain the desired frequency selectivity, followed by an all-pass phase equalizer that compensates for the phase distortion in the elliptic filter. However, attempts to accomplish this have resulted in filters with a number of coefficients in the cascade combination that equaled or exceeded the number of coefficients in an equivalent linear-phase FIR filter. Consequently, no reduction in complexity is achievable in using phase-equalized elliptic filters.

In addition to the filter design methods based on the transformation of analog filters into the digital domain, we also presented several methods in which the design is done directly in the discrete-time domain. The least-squares method previously described in Chapter 6 is particularly appropriate for designing IIR filters. The least-squares method has already been considered in Chapter 6 for the design of FIR Wiener filters.

Finally, we treated the important topic of sampling-rate conversion, which arises in many applications of digital signal processing where multiple-rate processing is employed within a system. Our treatment focused on the design of decimators and interpolators for either decreasing or increasing the sampling rate, respectively, by integer multiples.

Such a rich literature now exists on the design of digital filters that it is not possible to cite all the important references. We shall cite only a few. Some of the early work on digital filter design was done by Kaiser (1963, 1966), Steiglitz (1965), Golden and Kaiser (1964), Rader and Gold (1967a), Shanks (1967), Helms (1968), Gibbs (1969, 1970), and Gold and Rader (1969).

The design of analog filters is treated in the classic books by Storer (1957), Guillemin (1957), Weinberg (1962), and Daniels (1974).

The frequency sampling method for filter design was first proposed by Gold and Jordan (1968, 1969), and optimized by Rabiner et al. (1970). Additional results were published by Herrmann (1970), Herrmann and Schuessler (1970a), and Hofstetter et al. (1971). The Chebyshev (minimax) approximation method for designing linear-phase FIR filters was proposed by Parks and McClellan (1972a,b) and discussed further by Rabiner et al. (1975). The design of elliptic digital filters is treated in the book by Gold and Rader (1969) and in the paper by Gray and Merkel (1976). The latter includes a computer program for designing digital elliptic filters.

The use of frequency transformations in the digital domain was proposed by Constantinides (1967, 1968, 1970). These transformations are appropriate only for IIR filters. The reader should note that when these transformations are applied to a lowpass FIR filter the resulting filter is IIR.

Direct design techniques for digital filters have been considered in a number of papers, including Shanks (1967), Burrus and Parks (1970), Steiglitz (1970), Deczky (1972), Brophy and Salazar (1973), and Bandler and Bardakjian (1973).

Sampling-rate conversion is also a topic that has received considerable attention in the literature. For reference we cite the papers by Schafer and Rabiner (1973) and Crochiere and Rabiner (1975, 1976, 1981, 1983). Our treatment in Section 8.5 follows the approach given in the paper by Crochiere and Rabiner (1981).

PROBLEMS

8.1 Design an FIR linear phase, digital filter approximating the ideal frequency response

$$H_d(\omega) = \begin{cases} 1 & \text{for } |\omega| \le \dfrac{\pi}{6} \\ 0 & \text{for } \dfrac{\pi}{6} < |\omega| \le \pi \end{cases}$$

(a) Determine the coefficients of a 25-tap filter based on the window method with a rectangular window.
(b) Determine and plot the magnitude and phase response of the filter.
(c) Repeat parts (a) and (b) using the Hamming window.
(d) Repeat parts (a) and (b) using a Bartlett window.

8.2 Repeat Problem 8.1 for a bandstop filter having the ideal response

$$H_d(\omega) = \begin{cases} 1 & \text{for } |\omega| \le \dfrac{\pi}{6} \\ 0 & \text{for } \dfrac{\pi}{6} < |\omega| < \dfrac{\pi}{3} \\ 1 & \text{for } \dfrac{\pi}{3} \le |\omega| \le \pi \end{cases}$$

8.3 Redesign the filter of Problem 8.1 using the Hanning and Blackman windows.

8.4 Redesign the filter of Problem 8.2 using the Hanning and Blackman windows.

8.5 Use the window method with a Hamming window to design a 21-tap differentiator as shown in Fig. P8.5. Compute and plot the magnitude and phase response of the resulting filter.

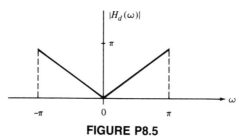

FIGURE P8.5

8.6 Use the matched-z transformation to convert the analog filter with system function

$$H(s) = \frac{s + 0.1}{(s + 0.1)^2 + 9}$$

into a digital IIR filter. Select $T = 0.1$ and compare the location of the zeros in $H(z)$ with the locations of the zeros obtained by applying the impulse invariance method in the conversion of $H(s)$.

8.7 Convert the analog bandpass filter designed in Example 8.3.1 into a digital filter by means of the bilinear transformation. Thus derive the digital filter characteristic obtained in Example 8.3.2 by the alternative approach and verify that the bilinear transformation applied to the analog filter results in the same digital bandpass filter.

8.8 An ideal analog integrator is described by the system function $H_a(s) = 1/s$. A digital integrator with system function $H(z)$ can obtained by use of the bilinear transformation. That is,

$$H(z) = \frac{T}{2} \frac{1 - z^{-1}}{1 + z^{-1}} \equiv H_a(s)\big|_{s = (2/T)(1 - z^{-1})/(1 + z^{-1})}$$

(a) Write the difference equation for the digital integrator relating the input $x(n)$ to the output $y(n)$.
(b) Roughly sketch the magnitude $|H_a(j\Omega)|$ and phase $\theta(\Omega)$ of the analog integrator.
(c) It is easily verified that the frequency response of the digital integrator is

$$H(\omega) = -j \frac{T}{2} \frac{\cos(\omega/2)}{\sin(\omega/2)} = -j \frac{T}{2} \cot \frac{\omega}{2}$$

Roughly sketch $|H(\omega)|$ and $\theta(\omega)$.
(d) Compare the magnitude and phase characteristics obtained in parts (b) and (c). How well does the digital integrator match the magnitude and phase characteristics of the analog integrator?

(e) The digital integrator has a pole at $z = 1$. If you implement this filter on a digital computer, what restrictions might you place on the input signal sequence $x(n)$ to avoid computational difficulties?

8.9 A z-plane pole–zero plot for a certain digital filter is shown in Fig. P8.9. The filter has unity gain at dc.

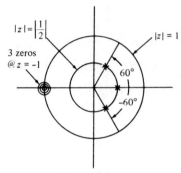

FIGURE P8.9

(a) Determine the system function in the form

$$H(z) = A \left[\frac{(1 + a_1 z^{-1})(1 + b_1 z^{-1} + b_2 z^{-2})}{(1 + c_1 z^{-1})(1 + d_1 z^{-1} + d_2 z^{-2})} \right]$$

giving numerical values for the parameters A, a_1, b_1, b_2, c_1, d_1, and d_2.

(b) Draw block diagrams showing numerical values for path gains in the following forms:
   (1) Direct form II (canonic form)
   (2) Cascade form (make each section canonic, with real coefficients)

8.10 Consider the pole–zero plot shown in Fig. P8.10.

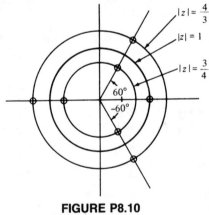

FIGURE P8.10

(a) Does it represent an FIR filter?
(b) Is it a linear-phase system?
(c) Give a direct form realization that exploits all symmetries to minimize the number of multiplications. Show all path gains.

8.11 A digital low-pass filter is required to meet the following specifications:
   Passband ripple: ≤ 1 dB
   Passband edge: 4 kHz
   Stopband attenuation: ≥ 40 dB

Stopband edge: 6 kHz
Sample rate: 24 kHz

The filter is to be designed by performing a bilinear transformation on an analog system function. Determine what order Butterworth, Chebyshev, and elliptic analog designs must be used to meet the specifications in the digital implementation.

8.12 An IIR digital low-pass filter is required to meet the following specifications:
    Passband ripple (or peak-to-peak ripple): ≤ 0.5 dB
    Passband edge: 1.2 kHz
    Stopband attenuation: ≥ 40 dB
    Stopband edge: 2.0 kHz
    Sample rate: 8.0 kHz

Use the design formulas in the book to determine the required filter order for
(a) A digital Butterworth filter
(b) A digital Chebyshev filter
(c) A digital elliptic filter

8.13 Determine the system function $H(z)$ of the lowest-order Chebyshev digital filter that meets the following specifications:
(a) 1-dB ripple in the passband $0 \leq |\omega| \leq 0.3\pi$.
(b) At least 60 dB attentuation in the stopband $0.35\pi \leq |\omega| \leq \pi$. Use the bilinear transformation.

8.14 Determine the system function $H(z)$ of the lowest-order Chebyshev digital filter that meets the following specifications:
(a) $\frac{1}{2}$-dB ripple in the passband $0 \leq |\omega| \leq 0.24\pi$.
(b) At least 50-dB attenuation in the stopband $0.35\pi \leq |\omega| \leq \pi$. Use the bilinear transformation.

8.15 An analog signal $x(t)$ consists of the sum of two components $x_1(t)$ and $x_2(t)$. The spectral characteristics of $x(t)$ are shown in the sketch in Fig. P8.15. The signal $x(t)$ is bandlimited to 40 kHz and it is sampled at a rate of 100 kHz to yield the sequence $x(n)$.

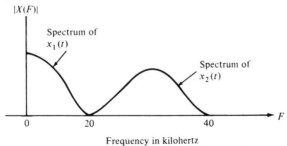

Frequency in kilohertz

FIGURE P8.15

It is desired to suppress the signal $x_2(t)$ by passing the sequence $x(n)$ through a digital lowpass filter. The allowable amplitude distortion on $|X_1(f)|$ is $\pm 2\%$ ($\delta_1 = 0.02$) over the range $0 \leq |F| \leq 15$ kHz. Above 20 kHz, the filter must have an attenuation of at least 40 dB ($\delta_2 = 0.01$).
(a) Use the Remez exchange algorithm to design the *minimum*-order linear-phase FIR filter that meets the specifications above. From the plot of the magnitude characteristic of the filter frequency response, give the actual specifications achieved by the filter.

(b) Compare the order M obtained in part (a) with the approximate formulas given in equations (8.1.110) and (8.1.111).

(c) For the order M obtained in part (a), design an FIR digital lowpass filter using the window technique and the Hamming window. Compare the frequency response characteristics of this design with those obtained in part (a).

(d) Design the *minimum*-order elliptic filter that meets the amplitude specifications given above. Compare the frequency response characteristics of the elliptic filter with that of the FIR filter in part (a).

(e) Compare the complexity of implementing the FIR filter in part (a) versus the elliptic filter obtained in part (d). Assume that the FIR filter is implemented in the direct form and the elliptic filter is implemented as a cascade of two-pole filters. Use storage requirements and the number of multiplications per output point in the comparison of complexity.

8.16 The impulse response of an analog filter is shown in Fig. P8.16.

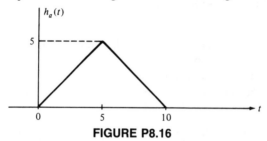

FIGURE P8.16

(a) Let $h(n) = h_a(nT)$, where $T = 1$, be the impulse response of a discrete-time filter. Determine the system function $H(z)$ and the frequency response $H(\omega)$ for this FIR filter.

(b) Sketch (roughly) $|H(\omega)|$ and compare this frequency response characteristic with $|H_a(j\Omega)|$.

(c) The FIR filter with unit sample response $h(n)$ given above is to be approximated by a second-order IIR filter of the form

$$G(z) = \frac{b_0 z^{-1}}{1 - a_1 z^{-1} - a_2 z^{-2}}$$

Use the least-squares inverse design procedure to determine the values of the coefficients b_0, a_1, and a_2.

8.17 In this problem you will be comparing some of the characteristics of analog and digital implementations of the single-pole low-pass analog system

$$H_a(s) = \frac{\alpha}{s + \alpha} \Leftrightarrow h_a(t) = e^{-\alpha t}$$

(a) What is the gain at dc? At what radian frequency is the analog frequency response 3 dB down from its dc value? At what frequency is the analog frequency reponse zero? At what time has the analog impulse response decayed to $1/e$ of its initial value?

(b) Give the digital system function $H(z)$ for the impulse-invariant design for this filter. What is the gain at dc? Give an expression for the 3-dB radian frequency. At what (real-valued) frequency is the response zero? How many samples are there in the unit sample time-domain response before it has decayed to $1/e$ of its initial value?

(c) "Prewarp" the parameter α and perform the bilinear transformation to obtain the digital system function $H(z)$ from the analog design. What is the gain at dc? At what (real-valued) frequency is the response zero? Give an expression for the 3-dB radian frequency. How many samples in the unit sample time-domain response before it has decayed to $1/e$ of its initial value?

8.18 We wish to design a FIR bandpass filter having a duration $M = 201$. $H_d(\omega)$ represents the ideal characteristic of the noncausal bandpass filter as shown in Fig. P8.18.

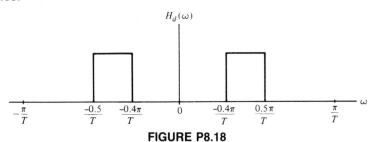

$H_d(\omega)$

$-\dfrac{\pi}{T}$ $\dfrac{-0.5}{T}$ $\dfrac{-0.4\pi}{T}$ 0 $\dfrac{-0.4\pi}{T}$ $\dfrac{0.5\pi}{T}$ $\dfrac{\pi}{T}$ ω

FIGURE P8.18

(a) Determine the unit sample (impulse) response $h_d(n)$ corresponding to $H_d(\omega)$.
(b) Explain how you would use the Hamming window

$$w(n) = 0.54 \text{ to } 0.46 \cos\left(\frac{2\pi}{N-1}\right) n \qquad -\frac{M-1}{2} \leq n \leq \frac{M-1}{2}$$

to design a FIR bandpass filter having an impulse response $h(n)$ for $0 \leq n \leq 200$.
(c) Suppose that you were to design the FIR filter with $M = 201$ by using the frequency sampling technique in which the DFT coefficients $H(k)$ are specified instead of $h(n)$. Give the values of $H(k)$ for $0 \leq k \leq 200$ corresponding to $H_d(e^{j\omega})$ and indicate how the frequency response of the actual filter will differ from the ideal. Would the actual filter represent a good design? Explain your answer.

8.19 We wish to design a digital bandpass filter from a second-order analog lowpass Butterworth filter prototype using the bilinear transformation. The specifications on the digital filter are shown in Fig. P8.19(a). The cutoff frequencies (measured at the half power points) for the digital filter should lie at $\omega = 5\pi/12$ and $\omega = 7\pi/12$.

The analog protoype is given by

$$H(s) = \frac{1}{s^2 + \sqrt{2}\, s + 1}$$

with the half-power point at $\Omega = 1$.
(a) Determine the system function for the digital bandpass filter.
(b) Using the same specs on the digital filter as in part (a), determine which of the analog bandpass prototype filters shown in Fig. P.8.19(b) could be transformed directly using the bilinear transformation to give the proper digital filter. Only the plot of the magnitude squared of the frequency is given.

8.20 Figure P8.20 shows a digital filter designed using the frequency sampling method.
(a) Sketch a z-plane pole–zero plot for this filter.
(b) Is the filter lowpass, highpass, or bandpass?
(c) Determine the magnitude response $|H(\omega)|$ at the frequencies $\omega_k = \pi k/6$ for $k = 0, 1, 2, 3, 4, 5, 6$.

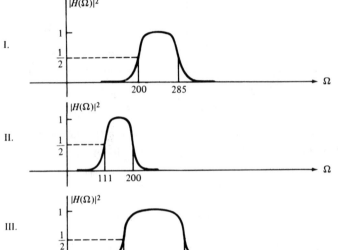

(a)

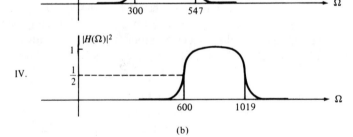

(b)

FIGURE P8.19

(d) Use the results of part (c) to sketch the magnitude response for $0 \le \omega \le \pi$ and confirm your answer to part (b).

8.21 An analog signal $x_a(t)$ is bandlimited to the range $900 \le F \le 1100$ Hz. It is used as an input to the system shown in Fig. P8.21. In this system, $H(\omega)$ is an ideal lowpass filter with cutoff frequency $F_c = 125$ Hz.
(a) Determine and sketch the spectra for the signals $x(n)$, $w(n)$, $v(n)$, and $y(n)$.
(b) Show that it is possible to obtain $y(n)$ by sampling $x_a(t)$ with period $T = 4$ milliseconds.

8.22 Consider the signal $x(n) = a^n u(n)$, $|a| < 1$.
(a) Determine the spectrum $X(\omega)$.
(b) The signal $x(n)$ is applied to a decimator which reduces the rate by a factor of 2. Determine the output spectrum.
(c) Show that the spectrum is simply the Fourier transform of $x(2n)$.

8.23 The sequence $x(n)$ is obtained by sampling an analog signal with a period T. From this signal a new signal is derived having the sampling period $T/2$ by use of a

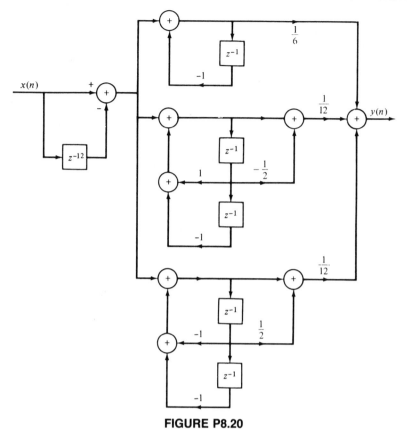

FIGURE P8.20

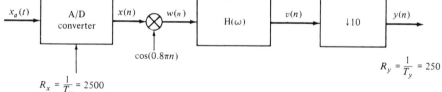

FIGURE P8.21

linear interpolation method described by the equation

$$y(n) = \begin{cases} x(n/2) & n \text{ even} \\ \dfrac{1}{2}\left[x\left(\dfrac{n-1}{2}\right) + x\left(\dfrac{n+1}{2}\right) \right] & n \text{ odd} \end{cases}$$

(a) Show that this linear interpolation scheme can be realized by basic digital signal processing elements.

(b) Determine the spectrum of $y(n)$ when the spectrum of $x(n)$ is

$$X(\omega) = \begin{cases} 1 & 0 \le |\omega| \le 0.2\pi \\ 0 & \text{otherwise} \end{cases}$$

(c) Determine the spectrum of $y(n)$ when the spectrum of $x(n)$ is

$$X(\omega) = \begin{cases} 1 & 0.7\pi \le |\omega| \le 0.9\pi \\ 0 & \text{otherwise} \end{cases}$$

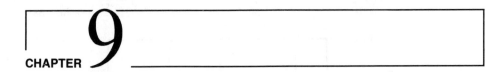

CHAPTER 9

The Discrete Fourier Transform: Its Properties and Computation

The Discrete Fourier Transform (DFT) plays an important role in many applications of digital signal processing, including linear filtering, correlation analysis, and spectrum analysis. A major reason for its importance is the existence of efficient algorithms for computing the DFT.

The main topic of this chapter is the description of computationally efficient algorithms for evaluating the DFT. Two different approaches are described. One is a divide-and-conquer approach in which a DFT of size N, where N is a composite number, is reduced to the computation of smaller DFTs from which the larger DFT is computed. In particular, we present important computational algorithms, called fast Fourier transform (FFT) algorithms, for computing the DFT when the size N is a power of 2 and when it is a power of 4.

The second approach is based on the formulation of the DFT as a linear filtering operation on the data. This approach leads to two algorithms, the Goertzel algorithm and the chirp-z transform algorithm for computing the DFT via linear filtering of the data sequence.

We begin the chapter with a discussion of the relationship between the DFT and other transforms. Section 9.2 is concerned with the derivation of the most important properties of the DFT. The FFT algorithms are presented in Section 9.3. The application of the FFT algorithms to linear filtering and correlation is discussed in Section 9.4. Finally, Section 9.5 treats the computation of the DFT as a linear filtering operation.

9.1 The DFT and Its Relationship to Other Transforms

In Section 4.5.4 we introduced the DFT as a set of N samples $\{X(k)\}$ of the Fourier transform $X(\omega)$ for a finite-duration sequence $\{x(n)\}$ of length $L \leq N$. The sampling of $X(\omega)$ occurs at the N equally spaced frequencies $\omega_k = 2\pi k/N$, $k = 0, 1, 2, \ldots, N - 1$. We demonstrated that the N samples $\{X(k)\}$ uniquely represent the sequence $\{x(n)\}$ in the frequency domain. The use of the DFT to perform linear filtering with FIR filters was described in Section 5.4.

In this section we explore the relationship of the DFT to other frequency-domain representations of a signal. Since the DFT is often used as an analysis tool for both analog and discrete-time signals, it is important for us to understand its relationship to the various frequency-domain representations of signals. These relationships are described below.

For reference, we recall from Section 4.5.4 that the N-point DFT of a finite-duration sequence $x(n)$ of length $L \leq N$ is defined as

$$X(k) = \sum_{n=0}^{N-1} x(n)e^{-j2\pi kn/N} \qquad k = 0, 1, \ldots, N - 1 \qquad (9.1.1)$$

and the IDFT is

$$x(n) = \frac{1}{N}\sum_{k=0}^{N-1} X(k)e^{j2\pi kn/N} \qquad n = 0, 1, \ldots, N - 1 \qquad (9.1.2)$$

9.1.1 Interpretations of the DFT

We have interpreted the DFT as the sampled version of the Fourier transform $X(\omega)$ for a finite-duration sequence $\{x(n)\}$, where the sampling is done at equally spaced frequencies $\omega_k = 2\pi k/N$, $k = 0, 1, \ldots, N - 1$. Thus the DFT is viewed as a frequency-domain representation of the sequence $x(n)$. This is certainly an important interpretation which is useful in linear FIR filtering via the DFT. However, there are other interpretations which prove useful in digital signal processing applications. These interpretations are described below.

Relationship to the Fourier Series Coefficients of a Periodic Sequence. A periodic sequence $x_p(n)$ with fundamental period N can be represented as a Fourier series of the form

$$x_p(n) = \sum_{k=0}^{N-1} c_k e^{j2\pi kn/N} \qquad -\infty < n < \infty \qquad (9.1.3)$$

where the Fourier series coefficients are given by the expression

$$c_k = \frac{1}{N}\sum_{n=0}^{N-1} x_p(n)e^{-j2\pi kn/N} \qquad k = 0, 1, \ldots, N - 1 \qquad (9.1.4)$$

If we compare (9.1.1) and (9.1.2) with (9.1.3) and (9.1.4), we observe that the formula for the Fourier series coefficients has the form of a DFT. In fact, if we define a sequence $x(n)$ which is identical to $x_p(n)$ over a single period, the DFT of this sequence is simply

$$X(k) = Nc_k \qquad (9.1.5)$$

Furthermore, (9.1.3) has the form of an IDFT. Thus the DFT provides an important link between the frequency-domain characterization of periodic sequences and aperiodic, finite-duration sequences.

The relationships above suggest that the DFT be viewed as the discrete frequency spectrum of a periodic sequence $x_p(n)$. In such an interpretation, a finite-duration sequence $x(n)$ of length N is viewed as a single period of a periodic sequence $x_p(n)$, which is formed by periodically repeating $x(n)$ every N samples, that is,

$$x_p(n) = \sum_{l=-\infty}^{\infty} x(n - lN) \qquad (9.1.6)$$

Then the discrete frequency spectrum of $x_p(n)$ is

$$X(k) = \sum_{n=0}^{N-1} x_p(n)e^{-j2\pi nk/N} = Nc_k \qquad k = 0, 1, \ldots, N - 1 \qquad (9.1.7)$$

and the IDFT becomes

$$x_p(n) = \frac{1}{N} \sum_{k=0}^{N-1} X(k) e^{j2\pi nk/N} \qquad -\infty < n < \infty \qquad (9.1.8)$$

Relationship to the Spectrum of an Infinite-Duration Signal. Suppose that $x(n)$ is an aperiodic finite-energy sequence with Fourier transform

$$X(\omega) = \sum_{n=-\infty}^{\infty} x(n) e^{-j\omega n} \qquad (9.1.9)$$

If $X(\omega)$ is sampled at N equally spaced frequencies, $\omega_k = 2\pi k/N$, $k = 0, 1, 2, \ldots ,$ $N - 1$, then

$$X(k) \equiv X(\omega)|_{\omega = 2\pi k/N}$$

$$= \sum_{n=-\infty}^{\infty} x(n) e^{-j2\pi kn/N} \qquad k = 0, 1, \ldots , N - 1$$

The spectral components $\{X(k)\}$ correspond to the spectrum of a periodic sequence of period N, given by

$$x_p(n) = \sum_{l=-\infty}^{\infty} x(n - lN) \qquad (9.1.10)$$

This result was derived in Section 4.5.3, where we treated the frequency sampling of discrete-time signals. The finite-duration sequence

$$\hat{x}(n) = \begin{cases} x_p(n) & 0 \le n \le N - 1 \\ 0 & \text{otherwise} \end{cases} \qquad (9.1.11)$$

bears no resemblance to the original sequence $x(n)$, unless $x(n)$ is of finite duration and length $L \le N$, in which case

$$x(n) = \hat{x}(n) \qquad 0 \le n \le N - 1 \qquad (9.1.12)$$

Only in this case will the IDFT of $\{X(k)\}$ yield the original sequence $x(n)$.

In general, $x(n)$ may be viewed as an aliased version of the sequence $x(n)$ when the length of the latter exceeds N. The following example illustrates this important point.

EXAMPLE 9.1.1 _____

The spectrum of the six-point sequence

$$x(n) = \left\{ \begin{array}{c} 1, 2, 3, 4, 5, 6 \\ \uparrow \end{array} \right\}$$

is sampled at the frequencies $\omega_k = 2\pi k/4$, $k = 0, 1, 2, 3$. Determine the sequence $\hat{x}(n)$ which has a discrete spectrum $\{X(k)\}$, identical to the four-point sampled spectrum of $x(n)$.

Solution: Without having to compute the spectrum of $x(n)$ or its sampled version for $N = 4$, and its IDFT, we can immediately determine $\hat{x}(n)$ by forming $x_p(n)$ according to (9.1.10) and then using (9.1.11). The result is an aliased version of $x(n)$ in which $x(4)$ is aliased into $x(0)$ and $x(5)$ is aliased into $x(1)$. Thus we obtain the sequence

$$\hat{x}(n) = \left\{ 6, 8, 3, 4 \atop \uparrow \right\}$$

Note that $x(0) \neq \hat{x}(0)$, $x(1) \neq \hat{x}(1)$, $x(2) = \hat{x}(2)$, and $x(3) = \hat{x}(3)$.

9.1.2 Relationship of the DFT to the z-Transform

Let us consider a sequence $x(n)$ having the z-transform

$$X(z) = \sum_{n=-\infty}^{\infty} x(n)z^{-n} \tag{9.1.13}$$

with a ROC that includes the unit circle. If $X(z)$ is sampled at the N equally spaced points on the unit circle $z_k = e^{j2\pi k/N}$, $k = 0, 1, 2, \ldots, N - 1$, we obtain

$$X(k) \equiv X(z)\big|_{z=e^{j2\pi nk/N}} \qquad k = 0, 1, \ldots, N - 1 \tag{9.1.14}$$

$$= \sum_{n=-\infty}^{\infty} x(n)e^{-j2\pi nk/N}$$

The expression in (9.1.14) is identical to the Fourier transform $X(\omega)$ evaluated at the N equally spaced frequencies $\omega_k = 2\pi k/N$, $k = 0, 1, \ldots, N - 1$, which is the topic treated in Section 9.1.1.

If the sequence $x(n)$ has a finite duration of length N, the z-transform $X(z)$ can be expressed as a function of the DFT. That is,

$$X(z) = \sum_{n=0}^{N-1} x(n)z^{-n}$$

$$X(z) = \sum_{n=0}^{N-1} \left[\frac{1}{N} \sum_{k=0}^{N-1} X(k)e^{j2\pi kn/N} \right] z^{-n}$$

$$X(z) = \frac{1}{N} \sum_{k=0}^{N-1} X(k) \sum_{n=0}^{N-1} (e^{j2\pi k/N}z^{-1})^n$$

$$X(z) = \frac{1 - z^{-N}}{N} \sum_{k=0}^{N-1} \frac{X(k)}{1 - e^{j2\pi k/N}z^{-1}} \tag{9.1.15}$$

The expression in (9.1.15) is identical to the frequency-sampling form for the system function of an FIR filter, given in Section 7.2.3. When evaluated on the unit circle, (9.1.15) yields the Fourier transform of the finite-duration sequence in terms of its DFT, in the form

$$X(\omega) = \frac{1 - e^{-j\omega N}}{N} \sum_{k=0}^{N-1} \frac{X(k)}{1 - e^{-j(\omega - 2\pi k/N)}} \tag{9.1.16}$$

This expression for the Fourier transform is a polynomial (Lagrange) interpolation formula for $X(\omega)$ expressed in terms of the values $\{X(k)\}$ of the polynomial at a set of equally spaced discrete frequencies $\omega_k = 2\pi k/N$, $k = 0, 1, \ldots, N - 1$.

9.1.3 Relationship of the DFT to the Fourier Transform of a Continuous-Time Signal

Suppose that $x_a(t)$ is a finite-energy continuous-time signal with Fourier transform $X_a(F)$. From Section 4.5.1 we recall that when $x_a(t)$ is sampled periodically at a rate F_s samples per second, the corresponding discrete-time signal $x(n) = x_a(nT)$ has a Fourier transform $X(f)$ which is related to the spectrum $X_a(F)$ by the formula

$$X\left(\frac{F}{F_s}\right) = F_s \sum_{m=-\infty}^{\infty} X_a(F - mF_s) \qquad (9.1.17)$$

or, equivalently,

$$X(f) = F_s \sum_{m=-\infty}^{\infty} X_a[(f - m)F_s] \qquad (9.1.18)$$

$$X(\omega) = F_s \sum_{m=-\infty}^{\infty} X_a[(\omega - 2\pi m)F_s] \qquad (9.1.19)$$

where f and ω are the normalized frequencies, defined as $f = F/F_s$ and $\omega = 2\pi f = 2\pi F/F_s$, respectively.

The spectral aliasing resulting from the time-domain sampling process is clearly evident in the expressions (9.1.17) through (9.1.19). Such effects can be minimized by prefiltering the analog signal prior to sampling and by sampling at a sufficiently high rate.

Now suppose that we sample the spectrum $X(\omega)$ of the discrete-time signal $x(n)$ at the N equally spaced frequencies $\omega_k = 2\pi k/N$, $k = 0, 1, \ldots, N - 1$. Then

$$X(k) \equiv X(\omega)|_{\omega = 2\pi k/N}$$

$$= F_s \sum_{m=-\infty}^{\infty} X_a\left[\left(\frac{2\pi k}{N} - 2\pi m\right)F_s\right] \qquad k = 0, 1, \ldots, N - 1 \quad (9.1.20)$$

$$= F_s \sum_{m=-\infty}^{\infty} X_a\left[\left(\frac{kF_s}{N} - mF_s\right)\right] \qquad k = 0, 1, \ldots, N - 1 \quad (9.1.21)$$

We may view the samples $\{X(k)\}$ of the spectrum $X(\omega)$ as the DFT of a periodic sequence $x_p(n)$ which is related to the sampled analog signal $x_a(nT)$ by the formula

$$x_p(n) = \sum_{l=-\infty}^{\infty} x(n - lN) = \sum_{l=-\infty}^{\infty} x_a(nT - lNT) \qquad (9.1.22)$$

From the relationships above we observe that sampling an analog signal $x_a(t)$ with a sampling period $T = 1/F_s$ or rate F_s, and the corresponding spectrum with the sampling period F_s/N results in the relationship

$$\sum_{l=-\infty}^{\infty} x_a(nT - lNT) \underset{N}{\overset{\text{DFT}}{\longleftrightarrow}} F_s \sum_{m=-\infty}^{\infty} X_a\left[\left(\frac{kF_s}{N} - mF_s\right)\right] \qquad (9.1.23)$$

which defines an N-point DFT pair.

The N-point DFT pair in (9.1.23) indicates the time-domain aliasing and the frequency-domain aliasing that are inherent in the two sampling processes described above. The relationship in (9.1.23) illustrates the potential pitfalls in the computation of the spectrum of an analog signal by use of the DFT. The reader should refer back to Section 4.5, especially to Examples 4.5.2 and 4.5.3, which illustrate the time-domain and frequency-domain aliasing effects and their dependence on the choice of F_s and N.

Before concluding this section, let us consider the practical problem in which we are given an analog signal $x_a(t)$, which is to be sampled in time, and the spectrum of the signal is to be computed on the basis of L samples of $x_a(t)$. This situation differs from our treatment of frequency-domain sampling given above in that the DFT of the finite-duration sequence is different from the sampled version of the spectrum $X(\omega)$. In fact, the discrete-time signal obtained by sampling $x_a(t)$ periodically at a rate $F_s = 1/T$ and truncating the sequence after L samples may be expressed as

$$\tilde{x}(n) = x(n)w(n) \tag{9.1.24}$$

where $x(n) \equiv x_a(nT)$ and $w(n)$ is a rectangular window function defined as

$$w(n) = \begin{cases} 1 & 0 \le n \le L - 1 \\ 0 & \text{otherwise} \end{cases} \tag{9.1.25}$$

Consequently, the frequency-domain relationship between $\tilde{x}(n)$ and $x(n)$ is the convolution of $X(\omega)$ and the spectrum $W(\omega)$ of the window function, that is,

$$\tilde{X}(\omega) = X(\omega) * W(\omega)$$

$$= \frac{1}{2\pi} \int_{-\pi}^{\pi} X(\sigma)W(\omega - \sigma)\, d\sigma \tag{9.1.26}$$

The DFT of the truncated sequence $\tilde{x}(n)$ is the sampled version of the spectrum $\tilde{X}(\omega)$. Thus we have

$$\tilde{X}(k) \equiv \tilde{X}(\omega)\big|_{\omega = 2\pi k/N}$$

$$= \frac{1}{2\pi} \int_{-\pi}^{\pi} X(\sigma)W\left(\frac{2\pi k}{N} - \sigma\right) d\sigma \qquad k = 0, 1, \ldots, N - 1 \tag{9.1.27}$$

The effect of the window function is basically the same as that encountered previously in the design of FIR filters using the window method. If the spectrum of the window is relatively narrow in width compared to the spectrum $X(\omega)$ of the signal, the window function has only a small (smoothing) effect on the spectrum $X(\omega)$. On the other hand, if the window function has a wide spectrum compared to the width of $X(\omega)$, as would be the case when the number of samples of $x_a(t)$ is small, the window spectrum masks the signal spectrum and, consequently, the DFT of the data reflects the spectral characteristics of the window function. Of course, this situation should be avoided.

Although windowing of the data is undesirable, it cannot be avoided, since the data record must be truncated at some finite length L. However, just as in the cases of FIR filter design, windows can be selected that have a smooth time-domain cutoff instead of the abrupt one encountered in the rectangular window. Although such window functions tend to increase the smoothing effects on the signal spectrum, they offer the advantages of small sidelobes.

The use of window functions in spectrum estimation is considered in more detail in Chapter 11. At this point we shall demonstrate, via an example, the spectral characteristics of a truncated sequence.

EXAMPLE 9.1.2 _____

The exponential signal

$$x_a(t) = \begin{cases} e^{-t} & t \ge 0 \\ 0 & t < 0 \end{cases}$$

is sampled at the rate $F_s = 20$ samples per second and a block of 100 samples is used to estimate its spectrum. Determine the spectral characteristics of the signal $x_a(t)$ by computing the DFT of the finite-duration sequence. Compare the spectrum of the truncated discrete-time signal to the spectrum of the analog signal.

Solution: The spectrum of the analog signal is

$$X_a(F) = \frac{1}{1 + j2\pi F}$$

The exponential analog signal sampled at the rate of 20 samples per second yields the sequence

$$
\begin{aligned}
x(n) &= e^{-nT} = e^{-n/20} && n \geq 0 \\
&= (e^{-1/20})^n = (0.95)^n && n \geq 0
\end{aligned}
$$

Now, let

$$
\tilde{x}(n) = \begin{cases} (0.95)^n & 0 \leq n \leq 99 \\ 0 & \text{otherwise} \end{cases}
$$

The N-point DFT of the $L = 100$ point sequence is

$$\tilde{X}(k) = \sum_{k=0}^{99} \tilde{x}(n)e^{-j2\pi k/N} \qquad k = 0, 1, \ldots, N - 1$$

To obtain sufficient detail in the spectrum we choose $N = 200$. This is equivalent to padding the sequence $x(n)$ with 100 zeros.

The graph of the analog signal $x_a(t)$ and its magnitude spectrum $|X_a(F)|$ are illustrated in Fig. 9.1a and b, respectively. The truncated sequence $x(n)$ and its $N = 200$ point DFT (magnitude) are illustrated in Fig. 9.1c and d, respectively. In this case the DFT $\{X(k)\}$ bears a close resemblance to the spectrum of the analog signal. The effect of the window function is relatively small.

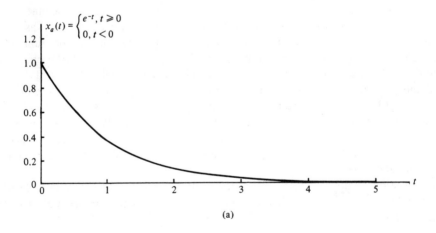

$$x_a(t) = \begin{cases} e^{-t}, & t \geq 0 \\ 0, & t < 0 \end{cases}$$

(a)

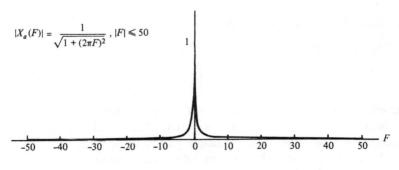

$$|X_a(F)| = \frac{1}{\sqrt{1 + (2\pi F)^2}}, \quad |F| \leq 50$$

(b)

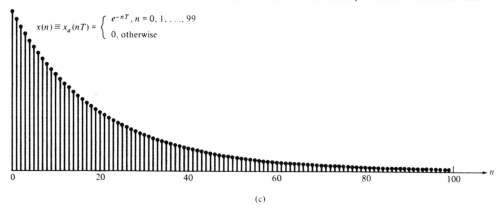

$$x(n) \equiv x_a(nT) = \begin{cases} e^{-nT}, n = 0, 1, \dots, 99 \\ 0, \text{otherwise} \end{cases}$$

(c)

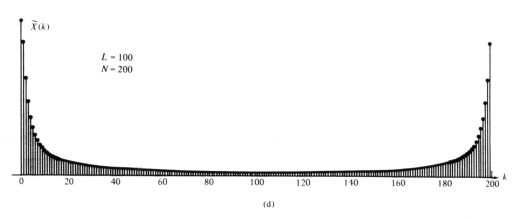

$\tilde{X}(k)$

$L = 100$
$N = 200$

(d)

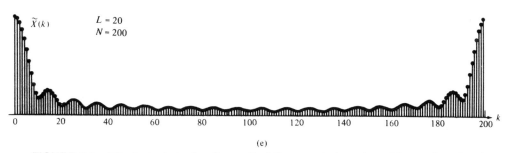

$\tilde{X}(k)$ $L = 20$
$N = 200$

(e)

FIGURE 9.1 Effect of windowing (truncating) the sampled version of the analog signal in Example 9.1.2.

On the other hand, suppose that a window function of length $L = 20$ is selected. Then the truncated sequence $x(n)$ is now given as

$$\tilde{x}(n) = \begin{cases} (0.95)^n & 0 \le n \le 19 \\ 0 & \text{otherwise} \end{cases}$$

Its $N = 200$ point DFT is illustrated in Fig. 9.1e. Now the effect of the wider spectral window function is clearly evident. First, the main peak is very wide as a result of the wide spectral window. Second, the sinusoidal envelope variations in the spectrum away from the main peak are due to the large sidelobes of the rectangular window spectrum. Consequently, the DFT is no longer a good approximation of the analog signal spectrum.

9.2 Properties of the DFT

In this section we present the important properties of the DFT. In view of the relationships established in the preceding section between the DFT and Fourier series, Fourier transforms and z-transforms of discrete-time signals, we expect the properties of the DFT to resemble the properties of these other transforms and series. However, some important differences exist, one of which is the circular convolution property derived in Section 5.4. A good understanding of these properties is extremely helpful in the application of the DFT to practical problems.

The notation used below to denote the N-point DFT pair $x(n)$ and $X(k)$ is

$$x(n) \xleftrightarrow[N]{\text{DFT}} X(k)$$

Periodicity. If $x(n)$ and $X(k)$ are an N-point DFT pair, then

$$
\begin{aligned}
x(n + N) &= x(n) & \text{for all } n & \qquad (9.2.1) \\
X(k + N) &= X(k) & \text{for all } k & \qquad (9.2.2)
\end{aligned}
$$

These periodicities in $x(n)$ and $X(k)$ follow immediately from formulas (9.1.1) and (9.1.2) for the DFT and IDFT, respectively.

We illustrated previously the periodicity property in the sequence $x(n)$ for a given DFT. However, we had not previously viewed the DFT $X(k)$ as a periodic sequence. In some applications it is advantageous to do this.

Linearity. If

$$x_1(n) \xleftrightarrow[N]{\text{DFT}} X_1(k)$$

and

$$x_2(n) \xleftrightarrow[N]{\text{DFT}} X_2(k)$$

then for any real-valued or complex-valued constants a_1 and a_2,

$$a_1 x_1(n) + a_2 x_2(n) \xleftrightarrow[N]{\text{DFT}} a_1 X_1(k) + a_2 X_2(k) \qquad (9.2.3)$$

This property follows immediately from the definition of the DFT given by (9.1.1).

Circular Shift and Circular Symmetries of a Sequence. As we have seen, the N-point DFT of a finite duration sequence, $x(n)$ of length $L \leq N$ is equivalent to the N-point DFT of a periodic sequence $x_p(n)$, of period N, which is obtained by periodically extending $x(n)$, that is,

$$x_p(n) = \sum_{l=-\infty}^{\infty} x(n - lN) \qquad (9.2.4)$$

Now suppose that we shift the periodic sequence $x_p(n)$ by k units to the right. Thus we obtain another periodic sequence

$$x'_p(n) = x_p(n - k) = \sum_{l=-\infty}^{\infty} x(n - k - lN) \qquad (9.2.5)$$

The finite-duration sequence

$$x'(n) = \begin{cases} x'_p(n) & 0 \leq n \leq N - 1 \\ 0 & \text{otherwise} \end{cases} \qquad (9.2.6)$$

is related to the original sequence $x(n)$ by a circular shift. This relationship is illustrated in Fig. 9.2 for $N = 4$.

In general, the circular shift of the sequence may be represented as the index modulo N. Thus we may write

$$x'(n) = x(n - k, (\text{mod } N)) \qquad (9.2.7)$$

For example, if $k = 2$ and $N = 4$, we have

$$x'(n) = x(n - 2, (\text{mod } 4))$$

which implies that

$$\begin{aligned}
x'(0) &= x(-2, (\text{mod } 4)) = x(2) \\
x'(1) &= x(-1, (\text{mod } 4)) = x(3) \\
x'(2) &= x(0, (\text{mod } 4)) = x(0) \\
x'(3) &= x(1, (\text{mod } 4)) = x(1)
\end{aligned}$$

Hence $x'(n)$ is simply $x(n)$ shifted circularly by two units in time, where the counterclockwise direction has been arbitrarily selected as the positive direction. Thus we conclude that a circular shift of an N-point sequence is equivalent to a linear shift of its periodic extension, and vice versa.

The inherent periodicity resulting from the arrangement of the N-point sequence on the circumference of a circle dictates a different definition of even and odd symmetry and time reversal of a sequence.

An N-point sequence is called *even* if it is symmetric about the point zero on the circle. This implies that

$$x(N - n) = x(n) \qquad 0 \leq n \leq N - 1 \qquad (9.2.8)$$

An N-point sequence is called *odd* if it is antisymmetric about the point zero on the circle. This implies that

$$x(N - n) = -x(n) \qquad 0 \leq n \leq N - 1 \qquad (9.2.9)$$

The time reversal of an N-point sequence is attained by reversing its samples about the point zero on the circle. Thus the sequence $x(-n, (\text{mod } N))$ is simply given as

$$x(-n, (\text{mod } N)) = x(N - n) \qquad 0 \leq n \leq N - 1 \qquad (9.2.10)$$

This time reversal is equivalent to plotting $x(n)$ in a clockwise direction on a circle.

Time Reversal of a Sequence. If

$$x(n) \xleftrightarrow[N]{\text{DFT}} X(k)$$

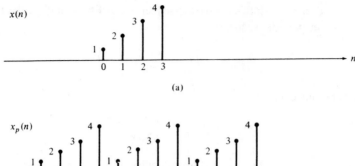

(a)

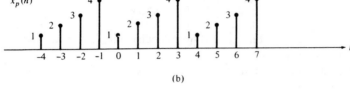

(b)

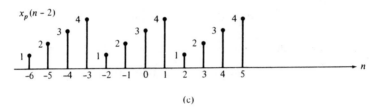

(c)

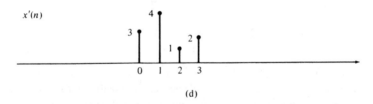

(d)

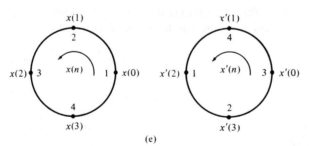

(e)

FIGURE 9.2 Circular shift of a sequence.

then

$$x(-n, \,(\text{mod } N)) = x(N - n) \xleftarrow[N]{\text{DFT}} X(-k, \,(\text{mod } N)) = X(N - k) \quad (9.2.11)$$

Hence reversing the N-point sequence in time is equivalent to reversing the DFT values. Time reversal of a sequence $x(n)$ is illustrated in Fig. 9.3.

PROOF. From the definition of the DFT in (9.1.1) we have

$$\text{DFT } \{x(N - n)\} = \sum_{n=0}^{N-1} x(N - n)e^{-j2\pi kn/N}$$

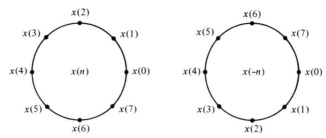

FIGURE 9.3   Time reversal of a sequence.

If we change the index from n to $m = N - n$, then

$$\text{DFT } \{x(N - n)\} = \sum_{m=0}^{N-1} x(m)e^{-j2\pi k(N-m)/N}$$

$$= \sum_{m=0}^{N-1} x(m)e^{j2\pi km/N}$$

$$= \sum_{m=0}^{N-1} x(m)e^{-j2\pi m(N-k)/N} = X(N - k)$$

We note that $X(N - k) = X(-k, (\text{mod } N)), 0 \leq k \leq N - 1$.

Circular Time Shift of a Sequence. If

$$x(n) \xleftarrow[N]{\text{DFT}} X(k)$$

then

$$x(n - l, (\text{mod } N)) \xleftarrow[N]{\text{DFT}} X(k)e^{-j2\pi kl/N} \qquad (9.2.12)$$

PROOF. From the definition of the DFT we have

$$\text{DFT}\{x(n - l, (\text{mod } N)) = \sum_{n=0}^{N-1} x(n - l, (\text{mod } N))e^{-j2\pi kn/N}$$

$$= \sum_{n=0}^{l-1} x(n - l, (\text{mod } N))e^{-j2\pi kn/N}$$

$$+ \sum_{n=l}^{N-1} x(n - l)e^{-j\pi kn/N}$$

But $x(n - l, (\text{mod } N)) = x(N - l + n)$. Consequently,

$$\sum_{n=0}^{l-1} x(n - l, (\text{mod } N))e^{-j2\pi kn/N} = \sum_{n=0}^{l-1} x(N - l + n)e^{-j2\pi kn/N}$$

$$= \sum_{m=N-l}^{N-1} x(m)e^{-j2\pi k(m+l)/N}$$

Furthermore,

$$\sum_{n=l}^{N-1} x(n - l)e^{-j2\pi kn/N} = \sum_{m=0}^{N-1-l} x(m)e^{-j2\pi k(m+l)/N}$$

Therefore,

$$\text{DFT}\{x(n - l, (\text{mod } N)\} = \sum_{m=0}^{N-1} x(m)e^{-j2\pi k(m+l)/N}$$

$$= X(k)e^{-j2\pi kl/N}$$

Circular Frequency Shift. If

$$x(n) \xleftarrow[N]{\text{DFT}} X(k)$$

then

$$x(n)e^{j2\pi ln/N} \xleftarrow[N]{\text{DFT}} X(k - l, (\text{mod } N)) \qquad (9.2.13)$$

Hence the multiplication of the sequence $x(n)$ with the complex exponential sequence $e^{j2\pi kn/N}$ is equivalent to the circular shift of the DFT by l units in frequency. This is the dual to the circular time-shifting property and its proof is similar to the latter.

Complex-Conjugate Properties. If

$$x(n) \xleftarrow[N]{\text{DFT}} X(k)$$

then

$$x^*(n) \xleftarrow[N]{\text{DFT}} X^*(-k, (\text{mod } N)) = X^*(N - k) \qquad (9.2.14)$$

The proof of this property is left as an exercise for the reader. The IDFT of $X^*(k)$ is

$$\frac{1}{N} \sum_{k=0}^{N-1} X^*(k)e^{j2\pi kn/N} = \left[\frac{1}{N} \sum_{k=0}^{N-1} X(k)e^{j2\pi k(N-n)/N} \right]^*$$

Therefore,

$$x^*(-n, (\text{mod } N)) = x^*(N - n) \xleftarrow[N]{\text{DFT}} X^*(k) \qquad (9.2.15)$$

Circular Convolution. If

$$x_1(n) \xleftarrow[N]{\text{DFT}} X_1(k)$$

and

$$x_2(n) \xleftarrow[N]{\text{DFT}} X_2(k)$$

then

$$x_1(n) \,\text{Ⓝ}\, x_2(n) \xleftarrow[N]{\text{DFT}} X_1(k)X_2(k) \qquad (9.2.16)$$

where $x_1(n)\,\text{Ⓝ}\,x_2(n)$ denotes the circular convolution of the sequence $x_1(n)$ and $x_2(n)$, defined as

$$x_3(n) = \sum_{m=0}^{N-1} x_1(m)x_2(n - m, (\text{mod } N)) \tag{9.2.17}$$

$$= \sum_{m=0}^{N-1} x_2(m)x_1(n - m, (\text{mod } N))$$

The proof of this property was given in Section 5.4.1.

Circular Correlation. In general, for complex-valued sequences $x(n)$ and $y(n)$, if

$$x(n) \xleftrightarrow[N]{\text{DFT}} X(k)$$

and

$$y(n) \xleftrightarrow[N]{\text{DFT}} Y(k)$$

then

$$\tilde{r}_{xy}(l) \xleftrightarrow[N]{\text{DFT}} \tilde{R}_{xy}(k) = X(k)Y^*(k) \tag{9.2.18}$$

where $\tilde{r}_{xy}(l)$ is the (unnormalized) circular crosscorrelation sequence, defined as

$$\tilde{r}_{xy}(l) = \sum_{n=0}^{N-1} x(n)y^*(n - l, (\text{mod } N))$$

PROOF. We may write $\tilde{r}_{xy}(l)$ as the circular convolution of $x(n)$ with $y^*(-n)$, that is,

$$\tilde{r}_{xy}(l) = x(l) \ \text{(N)} \ y^*(-l)$$

Then with the aid of the properties in (9.2.15) and (9.2.16), the N-point DFT of $\tilde{r}_{xy}(l)$ is

$$\tilde{R}_{xy}(k) = X(k)Y^*(k)$$

In the special case where $y(n) = x(n)$, we have the corresponding expression for the circular autocorrelation of $x(n)$, which is

$$\tilde{r}_{xx}(l) \xleftrightarrow[N]{\text{DFT}} \tilde{R}_{xx}(k) = |X(k)|^2 \tag{9.2.19}$$

Multiplication of Two Sequences. If

$$x_1(n) \xleftrightarrow[N]{\text{DFT}} X_1(k)$$

and

$$x_2(n) \xleftrightarrow[N]{\text{DFT}} X_2(k)$$

then

$$x_1(n)x_2(n) \xleftrightarrow[N]{\text{DFT}} \frac{1}{N} X_1(k) \ \text{(N)} \ X_2(k) \tag{9.2.20}$$

This property is the dual of (9.2.16). Its proof follows simply by interchanging the roles of time and frequency in the expression for the circular convolution of two sequences.

Parseval's Theorem. For complex-valued sequences $x(n)$ and $y(n)$, in general, if

$$x_1(n) \xleftarrow[N]{\text{DFT}} X(k)$$

and

$$y(n) \xleftarrow[N]{\text{DFT}} Y(k)$$

then

$$\sum_{n=0}^{N-1} x(n)y^*(n) = \frac{1}{N} \sum_{k=0}^{N-1} X(k)Y^*(k) \tag{9.2.21}$$

PROOF. The property follows immediately from the circular correlation property in (9.2.18). We have

$$\sum_{n=0}^{N-1} x(n)y^*(n) = \tilde{r}_{xy}(0)$$

and

$$\tilde{r}_{xy}(l) = \frac{1}{N} \sum_{k=0}^{N-1} \tilde{R}_{xy}(k)e^{j2\pi kl/N}$$

$$= \frac{1}{N} \sum_{k=0}^{N-1} X(k)Y^*(k)e^{j2\pi kl/N}$$

Hence (9.2.21) follows by evaluating the IDFT at $l = 0$.

The expression in (9.2.21) is the general form of Parseval's theorem. In the special case where $y(n) = x(n)$, (9.2.21) reduces to

$$\sum_{n=0}^{N-1} |x(n)|^2 = \frac{1}{N} \sum_{k=0}^{N-1} |X(k)|^2 \tag{9.2.22}$$

which is an expression that relates the energy in the finite-duration sequence $x(n)$ to the power in the frequency components $\{X(k)\}$.

Symmetry Properties of the DFT. The symmetry properties for the DFT can be obtained by applying the methodology previously used in Section 4.3 for the Fourier transform. Let us assume that the N-point sequence $x(n)$ and its DFT are both complex valued. Then the sequences can be expressed as

$$x(n) = x_R(n) + jx_I(n) \qquad 0 \le n \le N - 1 \tag{9.2.23}$$
$$X(k) = X_R(k) + jX_I(k) \qquad 0 \le k \le N - 1 \tag{9.2.24}$$

By substituting (9.2.23) into the expression for the DFT given by (9.1.11), we obtain

$$X_R(k) = \sum_{n=0}^{N-1} \left[x_R(n) \cos \frac{2\pi kn}{N} + x_I(n) \sin \frac{2\pi kn}{N} \right] \tag{9.2.25}$$

$$X_I(k) = - \sum_{n=0}^{N-1} \left[x_R(n) \sin \frac{2\pi kn}{N} - x_I(n) \cos \frac{2\pi kn}{N} \right] \tag{9.2.26}$$

Similarly, by substituting (9.2.24) into the expression for the IDFT given by (9.1.2), we obtain

$$x_R(n) = \frac{1}{N} \sum_{k=0}^{N-1} \left[X_R(k) \cos \frac{2\pi kn}{N} - X_I(k) \sin \frac{2\pi kn}{N} \right] \tag{9.2.27}$$

$$x_I(n) = \frac{1}{N} \sum_{k=0}^{N-1} \left[X_R(k) \sin \frac{2\pi kn}{N} + X_I(k) \cos \frac{2\pi kn}{N} \right] \tag{9.2.28}$$

REAL-VALUED SEQUENCES. If the sequence $x(n)$ is real, it follows directly from (9.1.1) that

$$X(N - k) = X^*(k) = X(-k) \tag{9.2.29}$$

Consequently, $|X(N - k)| = |X(k)|$ and $\angle X(N - k) = -\angle X(k)$. Furthermore, $x_I(n) = 0$ and hence $x(n)$ can be determined from (9.2.27), which is another form for the IDFT.

REAL AND EVEN SEQUENCES. If $x(n)$ is real and even, that is,

$$x(n) = x(N - n) \qquad 0 \le n \le N - 1$$

then (9.2.26) yields $X_I(k) = 0$. Hence the DFT reduces to

$$X(k) = \sum_{n=0}^{N-1} x(n) \cos \frac{2\pi kn}{N} \qquad 0 \le k \le N - 1 \tag{9.2.30}$$

which is itself real-valued and even. Furthermore, since $X_I(k) = 0$, the IDFT reduces to

$$x(n) = \frac{1}{N} \sum_{k=0}^{N-1} X(k) \cos \frac{2\pi kn}{N} \qquad 0 \le n \le N - 1 \tag{9.2.31}$$

REAL AND ODD SEQUENCES. If $x(n)$ is real and odd, that is,

$$x(n) = -x(N - n) \qquad 0 \le n \le N - 1$$

then (9.2.25) yields $X_R(k) = 0$. Hence

$$X(k) = -j \sum_{n=0}^{N-1} x(n) \sin \frac{2\pi kn}{N} \qquad 0 \le k \le N - 1 \tag{9.2.32}$$

which is purely imaginery and odd. Since $X_R(k) = 0$, the IDFT reduces to

$$x(n) = j \frac{1}{N} \sum_{k=0}^{N-1} X(k) \sin \frac{2\pi kn}{N} \qquad 0 \le n \le N - 1 \tag{9.2.33}$$

PURELY IMAGINARY SEQUENCES. In this case $x(n) = jx_I(n)$. Consequently, (9.2.25) and (9.2.26) reduce to

$$X_R(k) = \sum_{n=0}^{N-1} x_I(n) \sin \frac{2\pi kn}{N} \tag{9.2.34}$$

$$X_I(k) = \sum_{n=0}^{N-1} x_I(n) \cos \frac{2\pi kn}{N} \tag{9.2.35}$$

We observe that $X_R(k)$ is odd and $X_I(k)$ is even.

If $x_I(n)$ is odd, then $X_I(k) = 0$ and hence $X(k)$ is purely real. On the other hand, if $x_I(n)$ is even, then $X_R(k) = 0$ and hence $X(k)$ is purely imaginary. These symmetry properties are summarized in Table 9.1.

TABLE 9.1   Symmetry Properties of the DFT

$x(n)$	$X(k)$
Real	Real part is even
	Imaginary part is odd
Imaginary	Real part is odd
	Imaginary part is even
Real and even	Real and even
Real and odd	Imaginary and odd
Imaginary and even	Imaginary and even
Imaginary and odd	Real and odd

The properties of the DFT presented in this section are summarized in Table 9.2.

TABLE 9.2   Properties of the DFT

Property	Time Domain	Frequency Domain
Notation	$x(n), y(n)$	$X(k), Y(k)$
Periodicity	$x(n) = x(n + N)$	$X(k) = X(k + N)$
Linearity	$a_1 x_1(n) + a_2 x_2(n)$	$a_1 X_1(k) + a_2 X_2(k)$
Time reversal	$x(N - n)$	$X(N - k)$
Circular time shift	$x(n - l, (\text{mod } N))$	$X(k)e^{-j2\pi kl/N}$
Circular frequency shift	$x(n)e^{j2\pi ln/N}$	$X(k - l, (\text{mod } N))$
Complex conjugate	$x^*(n)$	$X^*(N - k)$
Circular convolution	$x(n) \,\text{Ⓝ}\, x_2(n)$	$X_1(k)X_2(k)$
Circular correlation	$x(n) \,\text{Ⓝ}\, y^*(-n)$	$X(k)Y^*(k)$
Multiplication of two sequences	$x_1(n)x_2(n)$	$\dfrac{1}{N} X_1(k) \,\text{Ⓝ}\, X_2(k)$
Parseval's theorem	$\displaystyle\sum_{n=0}^{N-1} x(n)y^*(n)$	$\dfrac{1}{N}\displaystyle\sum_{k=0}^{N-1} X(k)Y^*(k)$

9.3 Efficient Computation of the DFT: FFT Algorithms

In this section we present several methods for computing the DFT efficiently. In view of the importance of the DFT in various digital signal processing applications, such as linear filtering, correlation analysis, and spectrum analysis, its efficient computation is a topic that has received considerable attention by many mathematicians, engineers, and applied scientists.

Basically, the computational problem for the DFT is to compute the sequence $\{X(k)\}$ of N complex-valued numbers given another sequence of data $\{x(n)\}$ of length N, according to the formula

$$X(k) = \sum_{n=0}^{N-1} x(n)e^{-j2\pi nk/N} \qquad 0 \le k \le N - 1 \qquad (9.3.1)$$

In general, the data sequence $x(n)$ is also assumed to be complex valued.

To simplify the notation, it is desirable to define the complex-valued phase factor W_N, which is an Nth root of unity, as

$$W_N = e^{-j2\pi/N} \qquad (9.3.2)$$

Then (9.3.1) becomes

$$X(k) = \sum_{n=0}^{N-1} x(n)W_N^{kn} \qquad 0 \le k \le N - 1 \qquad (9.3.3)$$

Similarly, the IDFT becomes

$$x(n) = \frac{1}{N} \sum_{k=0}^{N-1} x(k)W_N^{-nk} \qquad 0 \le n \le N - 1 \qquad (9.3.4)$$

Since the DFT and IDFT involve basically the same type of computations, our discussion of efficient computational algorithms for the DFT applies as well to the efficient computation of the IDFT.

We observe that for each value of k, direct computation of $X(k)$ involves N complex multiplications ($4N$ real multiplications) and $N - 1$ complex additions ($4N - 2$ real additions). Consequently, to compute all N values of the DFT requires N^2 complex multiplications and $N^2 - N$ complex additions.

Direct computation of the DFT is basically inefficient primarily because it does not exploit the symmetry and periodicity properties of the phase factor W_N. In particular, these two properties are:

$$\text{Symmetry property: } W_N^{k+N/2} = -W_N^k \qquad (9.3.5)$$
$$\text{Periodicity property: } W_N^{k+N} = W_N^k \qquad (9.3.6)$$

The computationally efficient algorithms described in this section, known collectively as fast Fourier transform (FFT) algorithms, exploit these two basic properties of the phase factor.

9.3.1 Direct Computation of the DFT

For a complex-valued sequence $x(n)$ of N points the DFT may be expressed as

$$X_R(k) = \sum_{n=0}^{N-1} \left[x_R(n) \cos \frac{2\pi kn}{N} + x_I(n) \sin \frac{2\pi kn}{N} \right] \qquad (9.3.7)$$

$$X_I(k) = -\sum_{n=0}^{N-1} \left[x_R(n) \sin \frac{2\pi kn}{N} - x_I(n) \cos \frac{2\pi kn}{N} \right] \qquad (9.3.8)$$

The direct computation of (9.3.7) and (9.3.8) requires:

1. $2N^2$ evaluations of trigonometric functions.
2. $4N^2$ real multiplications.
3. $4N(N - 1)$ real additions.
4. A number of indexing and addressing operations.

These operations are typical of DFT computational algorithms. The operations in items 2 and 3 result in the DFT values $X_R(k)$ and $X_I(k)$. The indexing and addressing operations are necessary to fetch the data $x(n)$, $0 \le n \le N - 1$ and the phase factors and to store the results. The variety of DFT algorithms optimize each of these computational processes in a different way.

A simple FORTRAN subroutine for the direct computation of the DFT according to

```
C
C   DFT SUBROUTINE
C     ISEL = 0 : DFT
C     ISEL = 1 : INVERSE DFT
C
      SUBROUTINE DFT(N,XR,XI,XFR,XFI,ISEL)
      DIMENSION XR(N),XI(N),XFR(N),XFI(N)
      WN = 6.2831853 / FLOAT(N)
      IF (ISEL.EQ.1) WN = -WN
      DO 20 K = 1, N
        XFR(K) = 0.
        XFI(K) = 0.
        KM1 = K-1
        DO 10 I = 1, N
          IM1 = I-1
          ARG=WN*KM1*IM1
          C = COS(ARG)
          S = SIN(ARG)
          XFR(K) = XFR(K) + XR(I)*C + XI(I)*S
          XFI(K) = XFI(K) - XR(I)*S + XI(I)*C
10      CONTINUE
        IF (ISEL - 1) 20,30,20
30      XFR(K) = XFR(K)/FLOAT(N)
        XFI(K) = XFI(K)/FLOAT(N)
20    CONTINUE
      RETURN
      END
```

FIGURE 9.4   Subroutine for the direct computation of the DFT and IDFT.

(9.3.7) and (9.3.8) is given in Fig. 9.4. In this subroutine the phase factors are computed as needed and the size N of the DFT is arbitrary.

In applications where the N-point DFT of several sequences is to be computed, it is inefficient to recompute the phase factors each time. In such a case it is more efficient to compute the phase factors once and to store them in a table in memory. In general, it is not necessary to store all phase factors in all four quadrants. However, a simpler program can be obtained by storing one full cycle of the cosine and sine functions. The subroutine in Fig. 9.5 accomplishes this computation.

A DFT subroutine that uses the table of phase factors created by subroutine WTABLE is given in Fig. 9.6. The table look-up approach is probably the best way to deal with the phase factors in real-time applications of the DFT.

```
C
C SINE AND COSINE TABLE INITIALIZATION
C
      SUBROUTINE WTABLE(N,WR,WI)
      DIMENSION WR(N),WI(N)
      WN = 6.2831853 / FLOAT(N)
      DO 10 K = 1, N
        PHI = WN*(K-1)
        WR(K) = COS(PHI)
        WI(K) = SIN(PHI)
10    CONTINUE
      RETURN
      END
```

FIGURE 9.5   Subroutine for generating a sine and cosine table for the computation of the DFT.

```
C
C DFT SUBROUTINE WITH TABLE LOOK-UP
C
      SUBROUTINE DFT2(N,XR,XI,XFR,XFI,WR,WI)
      DIMENSION XR(N),XI(N),XFR(N),XFI(N),WR(N),WI(N)
      DO 20 K = 1, N
        XFR(K) = XR(1)
        XFI(K) = XI(1)
        L = 1
        DO 10 I = 2, N
          L = L + K - 1
          IF (L.GT.N) L = L-N
          C = WR(L)
          S = WI(L)
          XFR(K) = XFR(K) + XR(I)*C + XI(I)*S
          XFI(K) = XFI(K) - XR(I)*S + XI(I)*C

10        CONTINUE
20 CONTINUE
      RETURN
      END
```

FIGURE 9.6   Subroutine for computing the DFT with a table look-up.

9.3.2 Divide-and-Conquer Approach to the Computation of the DFT

The development of computationally efficient algorithms for the DFT is made possible if we adopt a divide-and-conquer approach. This approach is based on the decomposition of an N-point DFT into successively smaller DFTs. This basic approach leads to a family of computationally efficient algorithms known collectively as FFT algorithms.

To illustrate the basic notions let us consider the computation of an N-point DFT, where N can be factored as a product of two integers, that is,

$$N = LM \qquad (9.3.9)$$

The assumption that N is not a prime number is not restrictive, since we can pad any sequence with zeros to ensure a factorization of the form (9.3.9).

Now the sequence $x(n)$, $0 \leq n \leq N - 1$, can be stored in either a one-dimensional array indexed by n or as a two-dimensional array indexed by l and m, where $0 \leq l \leq L - 1$ and $0 \leq m \leq M - 1$ as illustrated in Fig. 9.7. Note that l is the row index and m is the column index. Thus the sequence $x(n)$ can be stored in a rectangular array in a variety of ways each of which depends on the mapping of index n to the indexes (l, m).

For example, suppose that we select the mapping

$$n = Ml + m \qquad (9.3.10)$$

This leads to an arrangement in which the first row consists of the first M elements of $x(n)$, the second row consists of the next M elements of $x(n)$, and so on, as illustrated in Fig. 9.8a. On the other hand, the mapping

$$n = l + mL \qquad (9.3.11)$$

stores the first L elements of $x(n)$ in the first column, the next L elements in the second column, and so on, as illustrated in Fig. 9.8b.

A similar arrangement can be used to store the computed DFT values. In particular, the mapping is from the index k to a pair of indices (p, q), where $0 \leq p \leq L - 1$ and $0 \leq q \leq M - 1$. If we select the mapping

$$k = Mp + q \qquad (9.3.12)$$

$n \longrightarrow$	0	1		$\cdots$	$N-1$
	$x(0)$	$x(1)$	$x(2)$	$\cdots$	$x(N-1)$

(a)

row index l / m column index

	0	1		$M-1$
0	$x(0,0)$	$x(0,1)$	$\cdots$	
1	$x(1,0)$	$x(1,1)$	$\cdots$	
2	$x(2,0)$	$x(2,1)$	$\cdots$	
$\vdots$	$\vdots$	$\vdots$	$\cdots$	$\vdots$
$L-1$			$\cdots$	

(b)

FIGURE 9.7 Two-dimensional data array for storing the sequence $x(n)$, $0 \leq n \leq N-1$.

Row wise: $n = Ml + m$

l / m	0	1	2		$M-1$
0	$x(0)$	$x(1)$	$x(2)$	$\cdots$	$x(M-1)$
1	$x(M)$	$x(M+1)$	$x(M+2)$	$\cdots$	$x(2M-1)$
2	$x(2M)$	$x(2M+1)$	$x(2M+2)$	$\cdots$	$x(3M-1)$
$\vdots$	$\vdots$	$\vdots$	$\vdots$	$\cdots$	$\vdots$
$L-1$	$x((L-1)M)$	$x((L-1)M+1)$	$x((L-1)M+2)$	$\cdots$	$x(LM-1)$

(a)

Column wise: $n = l + mL$

l / m	0	1	2		$M-1$
0	$x(0)$	$x(L)$	$x(2L)$	$\cdots$	$x((M-1)L)$
1	$x(1)$	$x(L+1)$	$x(2L+1)$	$\cdots$	$x((M-1)L+1)$
2	$x(2)$	$x(L+2)$	$x(2L+2)$	$\cdots$	$x((M-1)L+2)$
$\vdots$	$\vdots$	$\vdots$	$\vdots$	$\cdots$	$\vdots$
$L-1$	$x(L-1)$	$x(2L-1)$	$x(3L-1)$	$\cdots$	$x(LM-1)$

(b)

FIGURE 9.8 Two arrangements for the data arrays.

the DFT is stored on a row-wise basis, where the first row contains the first M elements of the DFT $X(k)$, the second row contains the next set of M elements, and so on. On the other hand, the mapping

$$k = qL + p \qquad (9.3.13)$$

results in a column-wise storage of $X(k)$, where the first L elements are stored in the first column, the second set of L elements are stored in the second column, and so on.

Now suppose that $x(n)$ is mapped into the rectangular array $x(l, m)$ and $X(k)$ is mapped into a corresponding rectangular array $X(p, q)$. Then the DFT can be expressed as a double sum over the elements of the rectangular array multiplied by the corresponding phase factors. To be specific, let us adopt a column-wise mapping for $x(n)$ given by (9.3.11) and the row-wise mapping for the DFT given by (9.3.12). Then

$$X(p, q) = \sum_{m=0}^{M-1} \sum_{l=0}^{L-1} x(l, m) W_N^{(Mp+q)(mL+l)} \qquad (9.3.14)$$

But

$$W_N^{(Mp+q)(mL+l)} = W_N^{MLmp} W_N^{mLq} W_N^{Mpl} W_N^{lq}$$

However, $W_N^{Nmp} = 1$, $W_N^{mqL} = W_{N/L}^{mq} = W_M^{mq}$, and $W_N^{Mpl} = W_{N/M}^{pl} = W_L^{pl}$.

With these simplifications, (9.3.14) may be expressed as

$$X(p, q) = \sum_{l=0}^{L-1} \left\{ W_N^{lq} \left[\sum_{m=0}^{M-1} x(l, m) W_M^{mq} \right] \right\} W_L^{lp} \qquad (9.3.15)$$

The expression in (9.3.15) involves the computation of DFTs of length M and length L. To elaborate, let us subdivide the computation into three steps:

1. First, we compute the M-point DFTs

$$F(l, q) \equiv \sum_{m=0}^{M-1} x(l, m) W_M^{mq} \qquad 0 \le q \le M - 1 \qquad (9.3.16)$$

for each of the rows $l = 0, 1, \ldots, L - 1$.

2. Second, we compute a new rectangular array $G(l, q)$ defined as

$$G(l, q) = W_N^{lq} F(l, q) \qquad \begin{matrix} 0 \le l \le L - 1 \\ 0 \le q \le M - 1 \end{matrix} \qquad (9.3.17)$$

3. Finally, we compute the L-point DFTs

$$X(p, q) = \sum_{l=0}^{L-1} G(l, q) W_L^{lp} \qquad (9.3.18)$$

for each column $q = 0, 1, \ldots, M - 1$, of the array $G(l, q)$.

On the surface it may appear that the computational procedure outlined above is more complex than the direct computation of the DFT. However, let us evaluate the computational complexity of (9.3.15). The first step involves the computation of L DFTs, each of M points. Hence this step requires LM^2 complex multiplications and $LM(M - 1)$ complex additions. The second step requires LM complex multiplications. Finally, the third step in the computation requires ML^2 complex multiplications and $ML(L - 1)$ complex additions. Therefore, the computational complexity is

$$\begin{array}{ll} \text{Complex multiplications:} & N(M + L + 1) \\ \text{Complex additions:} & N(M + L - 2) \end{array}$$

where $N = ML$. Thus the number of multiplications has been reduced from N^2 to $N(M + L + 1)$ and the number of additions has been reduced from $N(N - 1)$ to $N(M + L - 2)$.

For example, suppose that $N = 1000$ and we select $L = 2$ and $M = 500$. Then, instead of having to perform 10^6 complex multiplications via direct computation of the DFT, the approach above leads to 503,000 complex multiplications. This represents a reduction by approximately a factor of 2. The number of additions is also reduced by about a factor of 2.

When N is a highly composite number, that is, N can be factored into a product of prime numbers of the form

$$N = r_1 r_2 \cdots r_\nu$$

then the decomposition above can be repeated $(\nu - 1)$ more times. This procedure results in smaller DFTs, which, in turn, leads to a more efficient computational algorithm.

In effect, the first segmentation of the sequence $x(n)$ into a rectangular array of M columns with L elements in each column resulted in DFTs of sizes L and M. Further decomposition of the data in effect involves the segmentation of each row (or column) into smaller rectangular arrays which result in smaller DFTs. This procedure terminates when N is factored into its prime factors.

EXAMPLE 9.3.1

To illustrate the computational procedure above, let us consider the computation of an $N = 15$ point DFT. Since $N = 5 \times 3 = 15$, we select $L = 5$ and $M = 3$. In other words, we store the 15-point sequence $x(n)$ column-wise as follows:

Row 1: $x(0, 0) = x(0)$ $x(0, 1) = x(5)$ $x(0, 2) = x(10)$
Row 2: $x(1, 0) = x(1)$ $x(1, 1) = x(6)$ $x(1, 2) = x(11)$
Row 3: $x(2, 0) = x(2)$ $x(2, 1) = x(7)$ $x(2, 2) = x(12)$
Row 4: $x(3, 0) = x(3)$ $x(3, 1) = x(8)$ $x(3, 2) = x(13)$
Row 5: $x(4, 0) = x(4)$ $x(4, 1) = x(9)$ $x(4, 2) = x(14)$

Now, we compute the three-point DFTs for each of the five rows. This leads to the following 5×3 array:

$$
\begin{array}{ccc}
F(0, 0) & F(0, 1) & F(0, 2) \\
F(1, 0) & F(1, 1) & F(1, 2) \\
F(2, 0) & F(2, 1) & F(2, 2) \\
F(3, 0) & F(3, 1) & F(3, 2) \\
F(4, 0) & F(4, 1) & F(4, 2)
\end{array}
$$

The next step is to multiply each of the terms $F(l, q)$ by the phase factors $W_N^{lq} = W_{15}^{lq}$, $0 \le l \le 4$ and $0 \le q \le 2$. This computation results in the 5×3 array:

Column 1	Column 2	Column 3
$G(0, 0)$	$G(0, 1)$	$G(0, 2)$
$G(1, 0)$	$G(1, 1)$	$G(1, 2)$
$G(2, 0)$	$G(2, 1)$	$G(2, 2)$
$G(3, 0)$	$G(3, 1)$	$G(3, 2)$
$G(4, 0)$	$G(4, 1)$	$G(4, 2)$

The final step is to compute the five-point DFTs for each of the three columns. This computation yields the desired values of the DFT in the form

$$
\begin{array}{lll}
X(0, 0) = X(0) & X(0, 1) = X(1) & X(0, 2) = X(2) \\
X(1, 0) = X(3) & X(1, 1) = X(4) & X(1, 2) = X(5) \\
X(2, 0) = X(6) & X(2, 1) = X(7) & X(2, 2) = X(8) \\
X(3, 0) = X(9) & X(3, 1) = X(10) & X(3, 2) = X(11) \\
X(4, 0) = X(12) & X(4, 1) = X(13) & X(4, 2) = X(14)
\end{array}
$$

Figure 9.9 illustrates the steps in the computation.

It is interesting to view the segmented data sequence and the resulting DFT in terms of one-dimensional arrays. When the input sequence $x(n)$ and the output DFT $X(k)$ in the two-dimensional arrays are read across from row 1 through row 5, we obtain the following sequences:

INPUT ARRAY

$$x(0) \; x(5) \; x(10) \; x(1) \; x(6) \; x(11) \; x(2) \; x(7) \; x(12) \; x(3) \; x(8) \; x(13) \; x(14) \; x(9) \; x(14)$$

OUTPUT ARRAY

$$X(0) \; X(1) \; X(2) \; X(3) \; X(4) \; X(5) \; X(6) \; X(7) \; X(8) \; X(9) \; X(10) \; X(11) \; X(12) \; X(13) \; X(14)$$

We observe that the input data sequence is shuffled from the normal order in the computation of the DFT. On the other hand, the output sequence occurs in normal order. In this case the rearrangement of the input data array is due to the segmentation of the one-dimensional array into a rectangular array and the order in which the DFTs are computed. This shuffling of either the input data sequence or the output DFT sequence is a characteristic of most FFT algorithms.

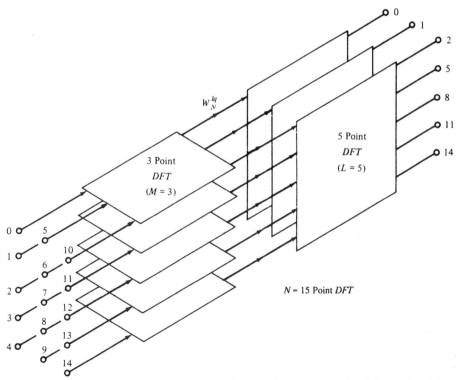

FIGURE 9.9 Computation of $N = 15$-point DFT by means of 3-point and 5-point DFTs.

It is possible to rearrange the double summation in (9.3.15) and, as a consequence, another computational algorithm can be obtained for the DFT. In particular, by interchanging the order of the summations, (9.3.15) can be expressed as

$$X(p, q) = \sum_{m=0}^{M-1} \left\{ \sum_{l=0}^{L-1} [x(l, m)W_N^{lq}]W_L^{lp} \right\} W_M^{mq} \qquad (9.3.19)$$

Now the computation of the DFT involves the following steps.

1. First, we compute a new array

$$g(l, m) = x(l, m)W_N^{lm} \qquad (9.3.20)$$

where q has been replaced by m in the factor W_N^{lq}, since both indices cover the same range.

2. Second, we compute the L-point DFTs of the data $g(l, m)$, that is,

$$F(p, m) = \sum_{l=0}^{L-1} g(l, m)W_L^{lp}$$

for each column $m = 0, 1, \ldots, M - 1$.

3. Finally, we compute the M-point DFTs

$$X(p, q) = \sum_{m=0}^{M-1} F(p, m)W_M^{mq}$$

for each row, $l = 0, 1, \ldots, L - 1$.

The computational complexity of this algorithm is the same as the previous one, since both algorithms involve similar computational steps carried out in different order.

To summarize, the two algorithms that we have introduced involve the following computations:

ALGORITHM 1
1. Store the signal column-wise.
2. Compute the M-point DFT of each row.
3. Multiply the resulting array by the phase factors W_N^{lq}.
4. Compute the L-point DFT of each column
5. Read the resulting array row-wise.

ALGORITHM 2
1. Store the signal column-wise.
2. Multiply the signal $x(l, m)$ by the phase factors W_N^{lq}.
3. Compute the L-point DFT at each column.
4. Compute the M-point DFT of each row.
5. Read the resulting array row-wise.

Of course, the steps above represent the first step in the factorization of N. They are repeated by further factorization of L and M.

Two additional algorithms with a similar computational structure can be obtained if the input signal is stored row-wise and the resulting transformation is column-wise. In this case we select as

$$\begin{aligned} n &= Ml + m \\ k &= qL + p \end{aligned} \qquad (9.3.21)$$

This choice of indices leads to the formula for the DFT in the form

$$X(p, q) = \sum_{m=0}^{M-1} \sum_{l=0}^{L-1} x(l, m) W_N^{pm} W_L^{pl} W_M^{qm} \tag{9.3.22}$$

The two additional algorithms result from the two different ways in which the summations can be arranged. These algorithms are:

ALGORITHM 3
1. Store the signal row-wise.
2. Compute the L-point DFT at each column.
3. Multiply the resulting array by the factors W_N^{pm}.
4. Compute the M-point DFT of each row.
5. Read the resulting array column-wise.

ALGORITHM 4
1. Store the signal row-wise.
2. Multiply the signal $x(l, m)$ by the factors W_N^{pm}.
3. Compute the M-point DFT of each row.
4. Compute the L-point DFT of each column.
5. Read the resulting array column-wise.

The four algorithms given above have the same complexity. However, they differ in the arrangement of the computations. In the following sections we exploit the divide-and-conquer approach to derive fast algorithms when the size of the DFT is restricted to be a power of 2 or a power of 4.

9.3.3 Radix-2 FFT Algorithms

In the preceding section we described four algorithms for efficient computation of the DFT based on the divide-and-conquer approach. Such an approach is applicable when the number N of data points is not a prime. In particular, the approach is very efficient when N is highly composite, that is, when N can be factored as $N = r_1 r_2 r_3 \cdots r_\nu$, where the $\{r_j\}$ are prime.

Of particular importance as the case in which $r_1 = r_2 = \cdots = r_\nu \equiv r$, so that $N = r^\nu$. In such a case the DFTs are of size r, so that the computation of the N-point DFT has a regular pattern. The number r is called the *radix* of the FFT algorithm.

In this section we describe radix-2 algorithms, which are by far the most widely used FFT algorithms. Radix-4 algorithms are described in the following section.

Let us consider the computation of the $N = 2^\nu$ point DFT by the divide-and-conquer approach specified by (9.3.16) through (9.3.18). We select $M = N/2$ and $L = 2$. This selection results in a split of the N-point data sequence into two $N/2$-point data sequences $f_1(n)$ and $f_2(n)$, corresponding to the even-numbered and odd-numbered samples of $x(n)$, respectively, that is,

$$f_1(n) = x(2n) \tag{9.3.23}$$

$$f_2(n) = x(2n + 1) \qquad n = 0, 1, \ldots, \frac{N}{2} - 1$$

Thus $f_1(n)$ and $f_2(n)$ are obtained by decimating $x(n)$ by a factor of 2, and hence the resulting FFT algorithm is called a *decimation-in-time algorithm*.

Now the N-point DFT can be expressed in terms of the DFTs of the decimated sequences as follows:

$$X(k) = \sum_{n=0}^{N-1} x(n)W_N^{kn} \qquad k = 0, 1, \ldots, N - 1$$

$$= \sum_{n \text{ even}} x(n)W_N^{kn} + \sum_{n \text{ odd}} x(n)W_N^{kn}$$

$$= \sum_{m=0}^{(N/2)-1} x(2m)W_N^{2mk} + \sum_{m=0}^{(N/2)-1} x(2m + 1)W_N^{k(2m+1)} \qquad (9.3.24)$$

But $W_N^2 = W_{N/2}$. With this substitution, (9.3.24) can be expressed as

$$X(k) = \sum_{m=0}^{(N/2)-1} f_1(m)W_{N/2}^{km} + W_N^k \sum_{m=0}^{(N/2)-1} f_2(m)W_{N/2}^{km}$$

$$= F_1(k) + W_N^k F_2(k) \qquad k = 0, 1, \ldots, N - 1 \qquad (9.3.25)$$

where $F_1(k)$ and $F_2(k)$ are the $N/2$-point DFTs of the sequences $f_1(m)$ and $f_2(m)$, respectively.

Since $F_1(k)$ and $F_2(k)$ are periodic, with period $N/2$, we have $F_1(k + N/2) = F_1(k)$ and $F_2(k + N/2) = F_2(k)$. In addition, the factor $W_N^{k+N/2} = -W_N^k$. Hence (9.3.25) may be expressed as

$$X(k) = F_1(k) + W_N^k F_2(k) \qquad k = 0, 1, \ldots, \frac{N}{2} - 1 \qquad (9.3.26)$$

$$X\left(k + \frac{N}{2}\right) = F_1(k) - W_N^k F_2(k) \qquad k = 0, 1, \ldots, \frac{N}{2} - 1 \qquad (9.3.27)$$

We observe that the direct computation of $F_1(k)$ requires $(N/2)^2$ complex multiplications. The same applies to the computation of $F_2(k)$. Furthermore, there are $N/2$ additional complex multiplications required to compute $W_N^k F_2(k)$. Hence the computation of $X(k)$ requires $2(N/2)^2 + N/2 = N^2/2 + N/2$ complex multiplications. This first step results in a reduction of the number of multiplications from N^2 to $N^2/2 + N/2$, which is about a factor of 2 for N large.

To be consistent with our previous notation, we may define

$$G_1(k) = F_1(k) \qquad k = 0, 1, \ldots, \frac{N}{2} - 1$$

$$G_2(k) = W_N^k F_2(k) \qquad k = 0, 1, \ldots, \frac{N}{2} - 1$$

Then the DFT $X(k)$ may be expressed as

$$X(k) = G_1(k) + G_2(k) \qquad k = 0, 1, \ldots, \frac{N}{2} - 1$$
$$X\left(k + \frac{N}{2}\right) = G_1(k) - G_2(k) \qquad k = 0, 1, \ldots, \frac{N}{2} - 1 \qquad (9.3.28)$$

This computation is illustrated in Fig. 9.10.

Having performed the decimation-in-time once, we can repeat the process for each of sequences $f_1(n)$ and $f_2(n)$. Thus $f_1(n)$ would result in the two $N/4$-point sequences

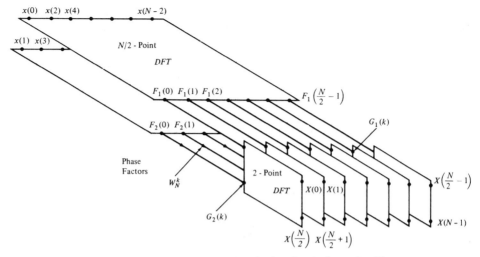

FIGURE 9.10   First step in the decimation-in-time algorithm.

$$v_{11}(n) = f_1(2n) \qquad n = 0, 1, \ldots, \frac{N}{4} - 1$$

$$v_{12}(n) = f_1(2n + 1) \qquad n = 0, 1, \ldots, \frac{N}{4} - 1 \qquad (9.3.29)$$

and $f_2(n)$ would yield

$$v_{21}(n) = f_2(2n) \qquad n = 0, 1, \ldots, \frac{N}{4} - 1$$

$$v_{22}(n) = f_2(2n + 1) \qquad n = 0, 1, \ldots, \frac{N}{4} - 1 \qquad (9.3.30)$$

By computing $N/4$-point DFTs, we would obtain the $N/2$-point DFTs $F_1(k)$ and $F_2(k)$ from the relations

$$F_1(k) = V_{11}(k) + W_{N/2}^k V_{12}(k) \qquad k = 0, 1, \ldots, \frac{N}{4} - 1$$

$$F_1\left(k + \frac{N}{4}\right) = V_{11}(k) - W_{N/2}^k V_{12}(k) \qquad k = 0, 1, \ldots, \frac{N}{4} - 1 \qquad (9.3.31)$$

$$F_2(k) = V_{21}(k) + W_{N/2}^k V_{22}(k) \qquad k = 0, 1, \ldots, \frac{N}{4} - 1$$

$$F_2\left(k + \frac{N}{4}\right) = V_{21}(k) - W_{N/2}^k V_{22}(k) \qquad k = 0, \ldots, \frac{N}{4} - 1 \qquad (9.3.32)$$

where the $\{V_{ij}(k)\}$ are the $N/4$-point DFTs of the sequences $\{v_{ij}(n)\}$.

We observe that the computation of $\{V_{ij}(k)\}$ requires $4(N/4)^2$ multiplications and hence the computation of $F_1(k)$ and $F_2(k)$ can be accomplished with $N^2/4 + N/2$ complex multiplication. An additional $N/2$ complex multiplications are required to compute $X(k)$ from $F_1(k)$ and $F_2(k)$. Consequently, the total number of multiplications is reduced approximately by a factor of 2 again to $N^2/4 + N$.

TABLE 9.3 Comparison of Computational Complexity for the Direct Computation of the DFT versus the FFT Algorithm

Number of Points N	Complex Multiplications in Direct Computation N^2	Complex Multiplications in FFT Algorithm $(N/2) \log_2 N$	Speed Improvement Factor
4	16	4	4.0
8	64	12	5.3
16	256	32	8.0
32	1,024	80	12.8
64	4,096	192	21.3
128	16,384	448	36.6
256	65,536	1,024	64.0
512	262,144	2,304	113.8
1,024	1,048,576	5,120	204.8

The decimation of the data sequence can be repeated again and again until the resulting sequences are reduced to one-point sequences. For $N = 2^{\nu}$, this decimation can be performed $\nu = \log_2 N$ times. Thus the total number of complex multiplications is reduced to $(N/2) \log_2 N$. The number of complex additions is $N \log_2 N$. Table 9.3 presents a comparison of the number of complex multiplications in the FFT and in the direct computation of the DFT.

For illustrative purposes, Fig. 9.11 depicts the computation of an $N = 8$ point DFT. We observe that the computation is performed in three stages, beginning with the computations of four two-point DFTs, then two four-point DFTs, and finallly, one eight-point DFT. The combination of the smaller DFTs to form the larger DFT is illustrated in Fig. 9.12 for $N = 8$.

Observe that the basic computation performed at every stage, as illustrated in Fig. 9.12, is to take two complex numbers, say the pair (a, b), multiply b by W_N^r, and then to add and subtract the product from a to form two new complex numbers (A, B). This basic computation, which is shown in Fig. 9.13, is called a *butterfly* because the flow graph resembles a butterfly.

In general, each butterfly involves one complex multiplication and two complex additions. For $N = 2^{\nu}$, there are $N/2$ butterflies per stage of the computation process and $\log_2 N$ stages. Therefore, the total number of complex multiplications is $(N/2) \log_2 N$ and complex additions is $N \log_2 N$, as indicated previously.

Once a butterfly operation is performed on a pair of complex numbers (a, b) to produce

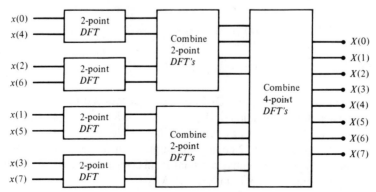

FIGURE 9.11 Three stages in the computation of an $N = 8$-point DFT.

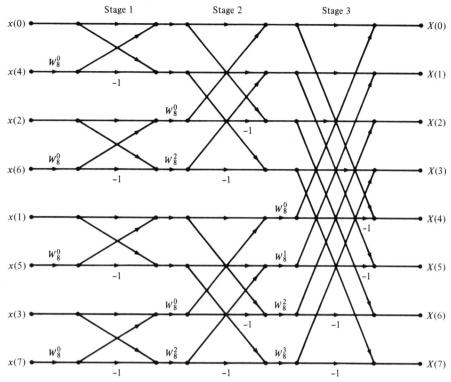

FIGURE 9.12   Eight-point decimation-in-time FFT algorithm.

(A, B), there is no need to save the input pair (a, b). Hence we may store the result (A, B) in the same locations as (a, b). Consequently, we require a fixed amount of storage, namely, $2N$ storage registers, in order to store the results (N complex numbers) of the computations at each stage. Since the same $2N$ storage locations are used throughout the computation of the N-point DFT, we say that *the computations are done in place*.

A second important observation is concerned with the order of the input data sequence after it is decimated $(v - 1)$ times. For example, if we consider the case where $N = 8$, we know that the first decimation yields the sequence $x(0)$, $x(2)$, $x(4)$, $x(6)$, $x(1)$, $x(3)$, $x(5)$, $x(7)$, and the second decimation results in the sequence $x(0)$, $x(4)$, $x(2)$, $x(6)$, $x(1)$, $x(5)$, $x(3)$, $x(7)$. This *shuffling* of the input data sequence has a well-defined order as can be ascertained from observing Fig. 9.14, which illustrates the decimation of the eight-point sequence. By expressing the index n, in the sequence $x(n)$, in binary form we note that the order of the decimated data sequence is easily obtained by reading the binary representation of the index n in reverse order. Thus the data point $x(3) \equiv x(011)$ is placed in position $m = 110$ or $m = 6$ in the decimated array. Thus we say that the data $x(n)$ after decimation is stored in bit-reversed order.

With the input data sequence stored in bit-reversed order and the butterfly computations performed in place, the resulting DFT sequence $X(k)$ is obtained in natural order

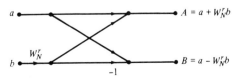

FIGURE 9.13   Basic butterfly computation in the decimation-in-time FFT algorithm.

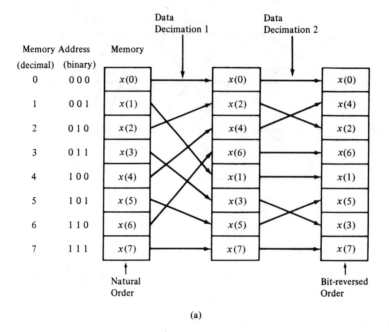

(a)

$$(n_2 n_1 n_0) \to (n_0 n_2 n_1) \to (n_0 n_1 n_2)$$

(0 0 0)	→	(0 0 0)	→	(0 0 0)
(0 0 1)	→	(1 0 0)	→	(1 0 0)
(0 1 0)	→	(0 0 1)	→	(0 1 0)
(0 1 1)	→	(1 0 1)	→	(1 1 0)
(1 0 0)	→	(0 1 0)	→	(0 0 1)
(1 0 1)	→	(1 1 0)	→	(1 0 1)
(1 1 0)	→	(0 1 1)	→	(0 1 1)
(1 1 1)	→	(1 1 1)	→	(1 1 1)

(b)

FIGURE 9.14   Shuffling of the data and bit reversal.

(i.e., $k = 0, 1, \ldots, N - 1$). On the other hand, we should indicate that it is possible to arrange the FFT algorithm such that the input is left in natural order and the resulting output DFT will occur in bit-reversed order. Furthermore, we may impose the restriction that both the input data $x(n)$ and the output DFT $X(k)$ be in natural order, and derive an FFT algorithm in which the computations are not done in place. Hence such an algorithm requires additional storage.

Another important radix-2 FFT algorithm, called the decimation-in-frequency algorithm, is obtained by using the divide-and-conquer approach described in Section 9.3.2 with the choice of $M = 2$ and $L = N/2$. This choice of parameters implies a columnwise storage of the input data sequence. To derive the algorithm, we begin by splitting the DFT formula into two summations, one of which involves the sum over the first $N/2$ data points and the second sum involves the last $N/2$ data points. Thus we obtain

$$X(k) = \sum_{n=0}^{(N/2)-1} x(n)W_N^{kn} + \sum_{n=N/2}^{N-1} x(n)W_N^{kn}$$

$$= \sum_{n=0}^{(N/2)-1} x(n)W_N^{kn} + W_N^{Nk/2} \sum_{n=0}^{(N/2)-1} x\left(n + \frac{N}{2}\right) W_N^{kn} \qquad (9.3.33)$$

Since $W_N^{kN/2} = (-1)^k$, the expression (9.3.33) can be rewritten as

$$X(k) = \sum_{n=0}^{(N/2)-1} \left[x(n) + (-1)^k x\left(n + \frac{N}{2}\right) \right] W_N^{kn} \qquad (9.3.34)$$

Now, let us split (decimate) $X(k)$ into the even- and odd-numbered samples. Thus we obtain

$$X(2k) = \sum_{n=0}^{(N/2)-1} \left[x(n) + x\left(n + \frac{N}{2}\right) \right] W_{N/2}^{kn} \qquad k = 0, 1, \ldots, \frac{N}{2} - 1$$

$$(9.3.35)$$

and

$$X(2k+1) = \sum_{n=0}^{(N/2)-1} \left\{ \left[x(n) - x\left(n + \frac{N}{2}\right) \right] W_N^n \right\} W_{N/2}^{kn} \qquad k = 0, 1, \ldots, \frac{N}{2} - 1$$

$$(9.3.36)$$

where we have used the fact that $W_N^2 = W_{N/2}$.

If we define the $N/2$-point sequences $g_1(n)$ and $g_2(n)$ as

$$g_1(n) = x(n) + x\left(n + \frac{N}{2}\right)$$

$$g_2(n) = \left[x(n) - x\left(n + \frac{N}{2}\right) \right] W_N^n \qquad n = 0, 1, 2, \ldots, \frac{N}{2} - 1 \qquad (9.3.37)$$

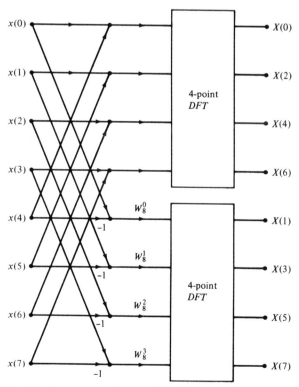

FIGURE 9.15   First stage of the decimation-in-frequency FFT algorithm.

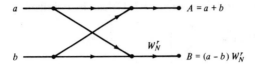

FIGURE 9.16   Basic butterfly computation in the decimation-in-frequency FFT algorithm.

then

$$X(2k) = \sum_{n=0}^{(N/2)-1} g_1(n)W_{N/2}^{kn}$$

$$(9.3.38)$$

$$X(2k+1) = \sum_{n=0}^{(N/2)-1} g_2(n)W_{N/2}^{kn}$$

The computation of the sequences $g_1(n)$ and $g_2(n)$ according to (9.3.37) and the subsequent use of these sequences to compute the $N/2$-point DFTs are depicted in Fig. 9.15. We observe that the basic computation in this figure involves the butterfly operation illustrated in Fig. 9.16.

The computational procedure above can be repeated through decimation of the $N/2$-point DFTs $X(2k)$ and $X(2k+1)$. The entire process involves $\nu = \log_2 N$ stages of decimation, where each stage involves $N/2$ butterflies of the type shown in Fig. 9.16. Consequently, the computation of the N-point DFT via the decimation-in-frequency FFT algorithm requires $(N/2) \log_2 N$ complex multiplications and $N \log_2 N$ complex additions, just as in the decimation-in-time algorithm. For illustrative purposes, the eight-point decimation-in-frequency algorithm is given in Fig. 9.17.

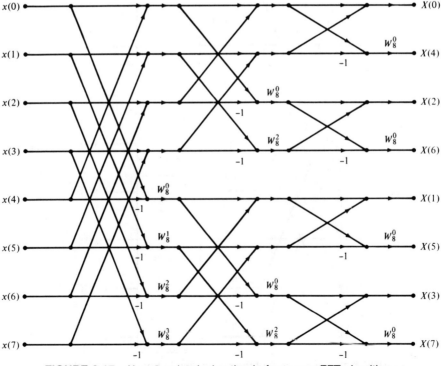

FIGURE 9.17 $N = 8$-point decimation-in-frequency FFT algorithm.

We observe from Fig. 9.17 that the input data $x(n)$ occurs in natural order, but the output DFT occurs in bit-reversed order. We also note that the computations are performed in place. However, it is possible to reconfigure the decimation-in-frequency algorithm so that the input sequence occurs in bit-reversed order while the output DFT occurs in normal order. Furthermore, if we abandon the requirement that the computations be done in place, it is also possible to have both the input data and the output DFT in normal order.

9.3.4 Radix-4 FFT Algorithms

When the number of data points N in the DFT is a power of 4 (i.e., $N = 4^\nu$), we can, of course, always use a radix-2 algorithm for the computation. However, for this case, it is more efficient computationally to employ a radix-4 FFT algorithm.

Let us begin by describing a radix-4 decimation-in-time FFT algorithm, which is obtained by selecting $L = 4$ and $M = N/4$ in the divide-and-conquer approach described in Section 9.3.2. For this choice of L and M, we have $l, p = 0, 1, 2, 3; m, q = 0, 1, \ldots, N/4 - 1; n = 4m + l;$ and $k = (N/4)p + q$. Thus we split or decimate the N-point input sequence into four subsequences, $x(4n), x(4n + 1), x(4n + 2), x(4n + 3), n = 0, 1, \ldots, N/4 - 1$.

By applying (9.3.15) we obtain

$$X(p, q) = \sum_{l=0}^{3} [W_N^{lq} F(l, q)] W_4^{lp} \qquad p = 0, 1, 2, 3 \qquad (9.3.39)$$

where $F(l, q)$ is given by (9.3.16), that is,

$$F(l, q) = \sum_{m=0}^{(N/4)-1} x(l, m) W_{N/4}^{mq} \qquad \begin{array}{l} l = 0, 1, 2, 3, \\[4pt] q = 0, 1, 2, \ldots, \dfrac{N}{4} - 1 \end{array} \qquad (9.3.40)$$

and

$$x(l, m) = x(4m + l) \qquad (9.3.41)$$

$$X(p, q) = X\left(\frac{N}{4}p + q\right) \qquad (9.3.42)$$

Thus the four $N/4$-point DFTs obtained from (9.3.40) are combined according to (9.3.39) to yield the N-point DFT. The expression in (9.3.39) for combining the $N/4$-point DFTs defines a radix-4 decimation-in-time butterfly, which can be expressed in matrix form as

$$\begin{bmatrix} X(0, q) \\ X(1, q) \\ X(2, q) \\ X(3, q) \end{bmatrix} = \begin{bmatrix} 1 & 1 & 1 & 1 \\ 1 & -j & -1 & j \\ 1 & -1 & 1 & -1 \\ 1 & j & -1 & -j \end{bmatrix} \begin{bmatrix} W_N^0 F(0, q) \\ W_N^q F(1, q) \\ W_N^{2q} F(2, q) \\ W_N^{3q} F(3, q) \end{bmatrix} \qquad (9.3.43)$$

The radix-4 butterfly is depicted in Fig. 9.18a and in a more compact form in Fig. 9.18b. Note that each butterfly involves three complex multiplications, since $W_N^0 = 1$, and 12 complex additions.

The decimation-in-time procedure described above can be repeated recursively ν times. Hence the resulting FFT algorithm consists of ν stages, where each stage contains $N/4$ butterflies. Consequently, the computational burden, for the algorithm is $3\nu N/4 = (3N/8) \log_2 N$ complex multiplications and $(3N/2) \log_2 N$ complex additions. We note that the number of multiplications is reduced by 25%, but the number of additions has increased by 50% from $N \log_2 N$ to $(3N/2) \log_2 N$.

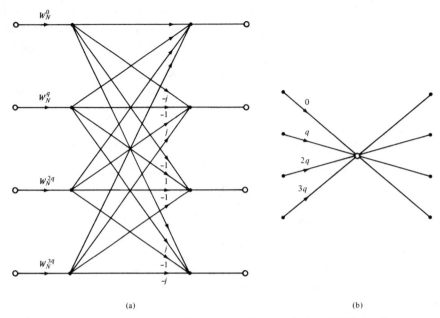

(a)   (b)

FIGURE 9.18   Basic butterfly computation in a radix-4 FFT algorithm.

It is interesting to note, however, that by performing the additions in two steps, it is possible to reduce the number of additions per butterfly from 12 to 8. This can be accomplished by expressing the matrix of the linear transformation in (9.3.43) as a product of two matrices as follows:

$$
\begin{bmatrix} X(0, q) \\ X(1, q) \\ X(2, q) \\ X(3, q) \end{bmatrix} = \begin{bmatrix} 1 & 0 & 1 & 0 \\ 0 & 1 & 0 & -j \\ 1 & 0 & -1 & 0 \\ 0 & 1 & 0 & j \end{bmatrix} \begin{bmatrix} 1 & 0 & 1 & 0 \\ 1 & 0 & -1 & 0 \\ 0 & 1 & 0 & 1 \\ 0 & 1 & 0 & -1 \end{bmatrix} \begin{bmatrix} W_N^0 F(0, q) \\ W_N^q F(1, q) \\ W_N^{2q} F(2, q) \\ W_N^{3q} F(3, q) \end{bmatrix} \tag{9.3.44}
$$

Now each matrix multiplication involves four additions, for a total of eight additions. Thus the total number of complex additions is reduced to $N \log_2 N$, which is identical to the radix-2 FFT algorithm. The computational savings results from the 25% reduction in the number of complex multiplications.

An illustration of a radix-4 decimation-in-time FFT algorithm is shown in Fig. 9.19 for $N = 16$. Note that in this algorithm the input sequence is in normal order while the output DFT is shuffled. In the radix-4 FFT algorithm, where the decimation is by a factor of 4, the order of the decimated sequence can be determined by reversing the order of the number that represents the index n in a quaternary number system (i.e., the number system based on the digits 0, 1, 2, 3).

A radix-4 decimation-in-frequency FFT algorithm can be obtained by selecting $L = N/4$, $M = 4$, l; $p = 0, 1, \ldots, N/4 - 1$; $m, q = 0, 1, 2, 3$; $n = (N/4)m + l$ and $k = 4p + q$. With this choice of parameters, the general equation given by (9.3.15) may be expressed as

$$
X(p, q) = \sum_{l=0}^{(N/4)-1} G(l, q) W_{N/4}^{lp} \tag{9.3.45}
$$

where

$$
G(l, q) = W_N^{lq} F(l, q) \qquad q = 0, 1, 2, 3 \tag{9.3.46}
$$

$$
l = 0, 1, \ldots, \frac{N}{4} - 1
$$

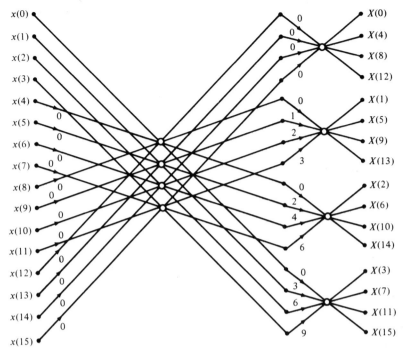

FIGURE 9.19   Sixteen-point radix-4 decimation-in-time algorithm with input in normal order and output in digit-reversed order.

and

$$F(l, q) = \sum_{m=0}^{3} x(l, m)W_4^{mq} \qquad q = 0, 1, 2, 3 \qquad (9.3.47)$$

$$l = 0, 1, 2, 3, \ldots, \frac{N}{4} - 1$$

We note that $X(p, q) = X(4p + q)$, $q = 0, 1, 2, 3$. Consequently, the N-point DFT is decimated into four $N/4$-point DFTs and hence we have a decimation-in-frequency FFT algorithm. The computations in (9.3.46) and (9.3.47) define the basic radix-4 butterfly for the decimation-in-frequency algorithm. Note that the multiplications by the factors W_N^{lq} occur after the combination of the data points $x(l, m)$, just as in the case of the radix-2, decimation-in-frequency algorithm.

A 16-point, radix-4 decimation-in-frequency FFT algorithm is shown in Fig. 9.20. Its input is in normal order and its output is in digit-reversed order. It has exactly the same computational complexity as the decimation-in-time radix-4 FFT algorithm.

9.3.5 Implementation of FFT Algorithms

Now that we have described the basic radix-2 and radix-4 FFT algorithms, let us consider some of the implementation issues. Our remarks apply directly to radix-2 algorithms, although similar comments may be made about radix-4 and higher-radix algorithms.

Basically, the radix-2 FFT algorithm consists of taking two data points at a time from memory, performing the butterfly computations and returning the resulting numbers to memory. This procedure is repeated many times ($(N \log_2 N)/2$ times), in the computation of an N-point DFT.

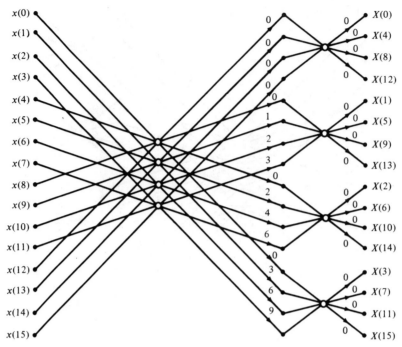

FIGURE 9.20 Sixteen-point, radix-4 decimation-in-frequency algorithm with input in normal order and output in digit-reversed order.

The butterfly computations require the phase factors $\{W_N^k\}$ at various stages, in either natural or bit-reversed order. In an efficient implementation of the algorithm, the phase factors are computed once and stored in a table, either in normal order or in bit-reversed order, depending on the specific implementation of the algorithm.

Memory requirements is another factor that must be considered. If the computations are performed in place, the number of memory locations is $2N$, since the numbers are complex. However, we may instead double the memory, thus simplifying the indexing and control operations in the FFT algorithms. In this case we simply alternate in the use of the two sets of memory locations from one stage of the FFT algorithm to the other. Doubling of the memory also allows us to have both the input sequence and the output sequence in normal order.

There are a number of other implementation issues regarding indexing, bit reversal, and the degree of parallelism in the computations. To a large extent, these issues are a function of the specific algorithm and the type of implementation, namely, a hardware or software implementation. In implementations based on a fixed-point arithmetic, or floating-point arithmetic on small machines, there is also the issue of round-off errors in the computation. This topic is considered in the following chapter.

Although the FFT algorithms described previously were presented in the context of computing the DFT efficiently, they may also be used to compute the IDFT, which is

$$x(n) = \frac{1}{N} \sum_{k=0}^{N-1} X(k) W_N^{-nk} \qquad (9.3.48)$$

The only difference between the two transforms is the normalization factor $1/N$ and the sign of the phase factor W_N. Consequently, an FFT algorithm for computing the DFT may be converted to an FFT algorithm for computing the IDFT by changing the sign on all the phase factors and dividing each output of the algorithm by N.

In fact, if we take the decimation-in-time algorithm that we described in Section 9.3.3, reverse the direction of the flow graph, change the sign on the phase factors, interchange the output and input, and finally, divide the output by N, we obtain a decimation-in-frequency FFT algorithm for computing the IDFT. On the other hand, if we begin with the decimation-in-frequency FFT algorithm described in Section 9.3.3 and repeat the changes described above, we obtain a decimation-in-time FFT algorithm for computing the IDFT. Thus it is a simple matter to devise FFT algorithms for computing the IDFT.

A radix-2 decimation-in-frequency FFT algorithm for computing the DFT or IDFT for complex signals is given in Fig. 9.21. The algorithm uses the subroutine given in Fig. 9.22 for bit reversal and a table look-up technique to obtain the phase factors.

Finally, we note that the emphasis in our discussion of FFT algorithms was on radix-2 and radix-4 algorithms. These are by far the most widely used in practice. When the number of data points is not a power of 2 or 4, it is a simple matter to pad the sequence $x(n)$ with zeros such that $N = 2^v$ or $N = 4^v$.

```
C
C RADIX-2, DIF FFT SUBROUTINE WITH TABLE LOOK-UP
C
      SUBROUTINE FFT(N,XR,XI,WR,WI)
      DIMENSION XR(N),XI(N),WR(N),WI(N)
      DO 11 I = 1, 15
        M = I
        N2 = 2**I
        IF (N.EQ.N2) GO TO 21
   11 CONTINUE
      WRITE (*,99)
   99 FORMAT(24H N IS NOT A POWER OF TWO)
      STOP
   21 N2 = N
      DO 10 I = 1, M
        N1 = N2
        N2 = N2/2
        I1 = 1
        I2 = N/N1
        DO 20 J = 1, N2
          C = WR(I1)
          S = WI(I1)
          I1 = I1 + I2
          DO 30 K = J, N, N1
            L = K + N2
            TEMPR = XR(K) - XR(L)
            XR(K) = XR(K) + XR(L)
            TEMPI = XI(K) - XI(L)
            XI(K) = XI(K) + XI(L)
            XR(L) = C*TEMPR + S*TEMPI
            XI(L) = C*TEMPI - S*TEMPR
   30     CONTINUE
   20   CONTINUE
   10 CONTINUE
      CALL BITRV(N,XR,XI)
      RETURN
      END
```

FIGURE 9.21 Subroutine for radix-2 decimation-in-frequency FFT algorithm.

```
C
C BIT-REVERSAL SUBROUTINE
C
      SUBROUTINE BITRV(N,XR,XI)
      DIMENSION XR(N),XI(N)
      J = 1
      N1 = N-1
      DO 40 I = 1, N1
        IF (I.GE.J) GO TO 10
        TEMP = XR(J)
        XR(J) = XR(I)
        XR(I) = TEMP
        TEMP = XI(J)
        XI(J) = XI(I)
        XI(I) = TEMP
   10   K = N/2
   20   IF (K.GE.J) GO TO 30
        J = J-K
        K = K/2
        GO TO 20
   30   J = J + K
   40 CONTINUE
      RETURN
      END
```

FIGURE 9.22   Subroutine for bit reversal.

9.4 Applications of FFT Algorithms

The FFT algorithms described in the preceding section find application in a variety of areas, including linear filtering, correlation, and spectrum analysis. Basically, the FFT algorithm is used as an efficient means to compute the DFT and the IDFT.

   In this section we consider the use of the FFT algorithm in linear filtering and in the computation of the crosscorrelation of two sequences. The use of the FFT in spectrum analysis is considered in Chapter 11. In addition, we illustrate how to enhance the efficiency of the FFT algorithm by forming complex-valued sequences from real-valued sequences prior to the computation of the DFT.

9.4.1 Efficient Computation of the DFT of Two Real Sequences

The FFT algorithm is designed to perform complex multiplications and additions, even though the input data may be real valued. The basic reason for this situation is that the phase factors are complex and hence after the first stage of the algorithm all variables are basically complex valued.

   In view of the fact that the algorithm can handle complex-valued input sequences, we can exploit this capability in the computation of the DFT of two real-valued sequences.

   Suppose that $x_1(n)$ and $x_2(n)$ are two real-valued sequences of length N, and let $x(n)$ be a complex-valued sequence defined as

$$x(n) = x_1(n) + jx_2(n) \qquad 0 \le n \le N - 1 \tag{9.4.1}$$

The DFT operation is linear and hence the DFT of $x(n)$ may be expressed as

$$X(k) = X_1(k) + jX_2(k) \tag{9.4.2}$$

The sequences $x_1(n)$ and $x_2(n)$ may be expressed in terms of $x(n)$ as follows:

$$x_1(n) = \frac{x(n) + x^*(n)}{2} \tag{9.4.3}$$

$$x_2(n) = \frac{x(n) - x^*(n)}{2j} \tag{9.4.4}$$

Hence the DFTs of $x_1(n)$ and $x_2(n)$ are

$$X_1(k) = \frac{1}{2}\{\text{DFT}[x(n)] + \text{DFT}[x^*(n)]\} \tag{9.4.5}$$

$$X_2(k) = \frac{1}{2j}\{\text{DFT}[x(n)] - \text{DFT}[x^*(n)]\} \tag{9.4.6}$$

Recall that the DFT of $x^*(n)$ is $X^*(N - k)$, as given in (9.2.14). Therefore,

$$X_1(k) = \frac{1}{2}[X(k) + X^*(N - k)] \tag{9.4.7}$$

$$X_2(k) = \frac{1}{j2}[X(k) - X^*(N - k)] \tag{9.4.8}$$

Thus, by performing a single DFT on the complex-valued sequence $x(n)$, we have obtained the DFT of the two real sequences with only a small amount of additional computation that is involved in computing $X_1(k)$ and $X_2(k)$ from $X(k)$ by use of (9.4.7) and (9.4.8).

9.4.2 Efficient Computation of the DFT of a 2N-Point Real Sequence

Suppose that $g(n)$ is a real-valued sequence of $2N$ points. We demonstrate below how to obtain the $2N$-point DFT of $g(n)$ from computation of one N-point DFT involving complex-valued data. First, we define

$$\begin{aligned} x_1(n) &= g(2n) \\ x_2(n) &= g(2n + 1) \end{aligned} \tag{9.4.9}$$

Thus we have subdivided the $2N$-point real sequence into two N-point real sequences. Now we can apply the method described in the preceding section.

Let $x(n)$ be the N-point complex-valued sequence

$$x(n) = x_1(n) + jx_2(n) \tag{9.4.10}$$

From the results of the preceding section, we have

$$X_1(k) = \frac{1}{2}[X(k) + X^*(N - k)]$$
$$\tag{9.4.11}$$
$$X_2(k) = \frac{1}{2j}[X(k) - X^*(N - k)]$$

Finally, we must express the $2N$-point DFT in terms of the two N-point DFTs, $X_1(k)$, and $X_2(k)$. To accomplish this we proceed as in the decimation-in-time FFT algorithm, namely,

$$\begin{aligned} G(k) &= \sum_{n=0}^{N-1} g(2n)W_{2N}^{2nk} + \sum_{n=0}^{N-1} g(2n + 1)W_{2N}^{(2n+1)k} \\ &= \sum_{n=0}^{N-1} x_1(n)W_N^{nk} + W_{2N}^{k}\sum_{n=0}^{N-1} x_2(n)W_N^{nk} \end{aligned}$$

Consequently,

$$G(k) = X_1(k) + W_{2N}^k X_2(k) \qquad k = 0, 1, \ldots, N - 1 \qquad (9.4.12)$$
$$G(k + N) = X_1(k) - W_{2N}^k X_2(k) \qquad k = 0, 1, \ldots, N - 1$$

Thus we have computed the DFT of a $2N$-point real sequence from one N-point DFT and some additional computation as indicated by (9.4.11) and (9.4.12).

9.4.3 Use of the FFT Algorithm in Linear Filtering and Correlation

An important application of the FFT algorithm is in FIR linear filtering of long data sequences. In Section 5.4.3 we described two methods, the overlap-add and the overlap-save methods for filtering a long data sequence with an FIR filter, based on the use of the DFT. In this section we consider the use of these two methods in conjunction with the FFT algorithm for computing the DFT and the IDFT.

Let $h(n)$, $0 \leq n \leq M - 1$ be the unit sample response of the FIR filter and let $x(n)$ denote the input data sequence. The block size of the FFT algorithm is N, where $N = L + M - 1$ and L is the number of new data samples being processed by the filter. We assume that for any given value of M, the number L of data samples is selected so that N is a power of 2. For purposes of this discussion, we consider only radix-2 FFT algorithms.

The N-point DFT of $h(n)$, which is padded by $L - 1$ zeros, is denoted as $H(k)$. This computation is performed once via the FFT and the resulting N complex numbers are stored. To be specific we assume that the decimation-in-frequency FFT algorithm is used to compute $H(k)$. This yields $H(k)$ in bit-reversed order, which is the way it is stored in memory.

In the overlap-save method, the first $M - 1$ data points of each data block are the last $M - 1$ data points of the previous data block. Each data block contains L new data points, such that $N = L + M - 1$. The N-point DFT of each data block is performed by the FFT algorithm. If the decimation-in-frequency algorithm is employed, the input data block requires no shuffling and the values of the DFT occur in bit-reversed order. Since this is exactly the order of $H(k)$, we may multiply the DFT of the data, say $X_m(k)$, with $H(k)$ and thus the result

$$Y_m(k) = H(k)X_m(k)$$

is also a bit-reversed order.

The inverse DFT (IDFT) may be computed by use of an FFT algorithm that takes the input in bit-reversed order and produces an output in normal order. Thus there is no need to shuffle any block of data either in computing the DFT or the IDFT.

If the overlap-add method is used to perform the linear filtering, the computational method for use of the FFT algorithm is basically the same. The only difference is that the N-point data blocks consist of L new data points and $M - 1$ additional zeros. After the IDFT is computed for each data block, the N-point filtered blocks are overlapped as indicated in Section 5.4.3 and the $M - 1$ overlapping data points between successive output records are added together.

Let us assess the computational complexity of the FFT method for linear filtering. For this purpose, the one-time computation of $H(k)$ is insignificant and can be ignored. Each FFT requires $(N/2) \log_2 N$ complex multiplications and $N \log_2 N$ additions. Since the FFT is performed twice, once for the DFT and once for the IDFT, the computational burden is $N \log_2 N$ complex multiplications and $2N \log_2 N$ additions. There are also N complex multiplications and $N - 1$ additions required to compute $Y_m(k)$. Therefore, we have $(N \log_2 2N)/L$ complex multiplications per output data point and approximately $(2N \log_2 2N)/L$ additions per output data point. The overlap-add method requires an incremental increase of $(M - 1)/L$ in the number of additions.

By way of comparison, a direct form realization of the FIR filter involves M real multiplications per output point if the filter is not linear phase and $M/2$ if it is linear phase. Also, the number of additions is $M - 1$ per output point.

It is interesting to compare the efficiency of the FFT algorithm with the direct form realization of the FIR filter. Let us focus on the number of multiplications, which are more time consuming than additions. Suppose that $M = 128 = 2^7$ and $N = 2^\nu$. Then the number of complex multiplications per output point for an FFT size of $N = 2^\nu$ is

$$c(\nu) = \frac{N \log_2 2N}{L} = \frac{2^\nu(\nu + 1)}{N - M + 1}$$

$$\approx \frac{2^\nu(\nu + 1)}{2^\nu - 2^7}$$

The values of $c(\nu)$ for different values of ν are given in Table 9.4. We observe that there is an optimum value of ν which minimizes $c(\nu)$. For the FIR filter of size $M = 128$, the optimum occurs at $\nu = 10$.

We should emphasize that $c(\nu)$ represents the number of complex multiplications for the FFT-based method. The number of real multiplications is four times this number. However, even if the FIR filter has linear phase, the number of computations per output point is still less with the FFT-based method. Furthermore, the efficiency of the FFT method can be improved by computing the DFT of two successive data blocks simultaneously according to the method described in Section 9.4.1. Consequently, the FFT-based method is indeed superior from a computational point of view when the filter length is relatively large.

The computation of the cross correlation between two sequences by means of the FFT algorithm is similar to the linear FIR filtering problem described above. In practical applications involving crosscorrelation, at least one of the sequences has finite duration and is akin to the impulse response of the FIR filter. The second sequence may be a long sequence which contains the desired sequence corrupted by additive noise. Hence the second sequence is akin to the input to the FIR filter. By folding the first sequence and computing its DFT, we have reduced the crosscorrelation to an equivalent convolution problem (i.e., a linear FIR filtering problem). Therefore, the methodology described above for linear FIR filtering by use of the FFT applies directly.

TABLE 9.4 Computational Complexity
for the FFT Method

Size of FFT $\nu = \log_2 N$	$c(\nu)$ Number of Complex Multiplication per Output Point
9	13.3
10	12.6
11	12.8
12	13.4
14	15.1

9.5 A Linear Filtering Approach to the Computation of the DFT

The FFT algorithm takes N points of input data and produces an output sequence of N points corresponding to the DFT of the input data. As we have shown, the radix-2 FFT algorithm performs the computation of the DFT in $(N/2) \log_2 N$ multiplications and $N \log_2 N$ additions for an N-point sequence.

There are some applications where only a selected number of values of the DFT are desired, but the entire DFT is not required. In such a case, the FFT algorithm may no longer be more efficient than a direct computation of the desired values of the DFT. In fact, when the desired number of values of the DFT is less than $\log_2 N$, a direct computation of the desired values is more efficient.

The direct computation of the DFT may be formulated as a linear filtering operation on the input data sequence. As we demonstrate below, the linear filter takes the form of a parallel bank of resonators where each resonator selects one of the frequencies $\omega_k = 2\pi k/N$, $k = 0, 1, \ldots, N - 1$ corresponding to the N frequencies in the DFT.

There are other applications in which we require the evaluation of the z-transform of a finite-duration sequence at points other than the unit circle. If the set of desired points in the z-plane possesses some regularity, it is possible to also express the computation of the z-transform as a linear filtering operation. In this connection we introduce another algorithm, called the chirp-z transform algorithm, which is suitable for evaluating the z-transform of a set of data on a variety of contours in the z-plane. This algorithm is also formulated as a linear filtering of a set of input data. As a consequence, the FFT algorithm may be used to compute the chirp-z transform and thus to evaluate the z-transform at various contours in the z-plane, including the unit circle.

9.5.1 The Goertzel Algorithm

The Goertzel algorithm exploits the periodicity of the phase factors $\{W_N^k\}$ and allows us to express the computation of the DFT as a linear filtering operation. Since $W_N^{-kN} = 1$, we can multiply the DFT by this factor. Thus

$$X(k) = W_N^{-kN} \sum_{m=0}^{N-1} x(m) W_N^{km} = \sum_{m=0}^{N-1} x(m) W_N^{-k(N-m)} \tag{9.5.1}$$

We note that (9.5.1) is in the form of a convolution. Indeed, if we define the sequence $y_k(n)$ as

$$y_k(n) = \sum_{m=0}^{N-1} x(m) W_N^{-k(n-m)} \tag{9.5.2}$$

then it is clear that $y_k(n)$ is the convolution of the finite-duration input sequence $x(n)$ of length N with a filter that has an impulse response

$$h_k(n) = W_N^{-kn} u(n) \tag{9.5.3}$$

The output of this filter at $n = N$ yields the value of the DFT at the frequency $\omega_k = 2\pi k/N$. That is,

$$X(k) = y_k(n)|_{n=N} \tag{9.5.4}$$

as can be verified by comparing (9.5.1) with (9.5.2).

The filter with impulse response $h_k(n)$ has the system function

$$H_k(z) = \frac{1}{1 - W_N^{-k}z^{-1}} \tag{9.5.5}$$

This filter has a pole on the unit circle at the frequency $\omega_k = 2\pi k/N$. Thus the entire DFT can be computed by passing the block of input data into a parallel bank of N single-pole filters (resonators), where each filter has a pole at the corresponding frequency of the DFT.

Instead of performing the computation of the DFT as in (9.5.2), via convolution, we can use the difference equation corresponding to the filter given by (9.5.5) to compute $y_k(n)$ recursively. Thus we have

$$y_k(n) = W_N^{-k}y_k(n - 1) + x(n) \qquad y_k(-1) = 0 \tag{9.5.6}$$

The desired output is $X(k) = y_k(N)$, for $k = 0, 1, \ldots, N - 1$. To perform this computation, we can compute once and store the phase factor W_N^{-k}.

The complex multiplications and additions inherent in (9.3.6) can be avoided by combining the pairs of resonators possessing complex-conjugate poles. This leads to two-pole filters with system functions of the form

$$H_k(z) = \frac{1 - W_N^k z^{-1}}{1 - 2 \cos (2\pi k/N)z^{-1} + z^{-2}} \tag{9.5.7}$$

The direct form II realization of the system illustrated in Fig. 9.23 is described by the difference equation

$$v_k(n) = 2 \cos \frac{2\pi k}{N} v_k(n - 1) - v_k(n - 2) + x(n) \tag{9.5.8}$$

$$y_k(n) = v_k(n) - W_N^k v_k(n - 1) \tag{9.5.9}$$

with initial conditions $v_k(-1) = v_k(-2) = 0$.

The recursive relation in (9.5.8) is iterated for $n = 0, 1, \ldots, N$, but the equation in (9.5.9) is computed only once at time $n = N$. Each iteration requires one real multiplication and two additions. Consequently, for a real input sequence $x(n)$, this algorithm requires $N + 1$ real multiplications to yield not only $X(k)$ but also, due to symmetry, the value of $X(N - k)$. A subroutine for computing the DFT via (9.5.8) and (9.5.9) is given in Fig. 9.24.

The Goertzel algorithm is particularly attractive when the DFT is to be computed at a relatively small number M of values, where $M \leq \log_2 N$. Otherwise, the FFT algorithm is a more efficient method.

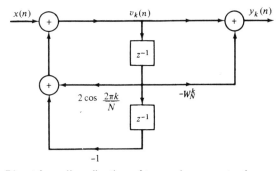

FIGURE 9.23   Direct form II realization of two-pole resonator for computing the DFT.

Substitution of (9.5.15) into (9.5.13) yields

$$X(z_k) = V^{-k^2/2} \sum_{n=0}^{N-1} [x(n)(r_0 e^{j\theta_0})^{-n} V^{-n^2/2}] V^{(k-n)^2/2} \qquad (9.5.16)$$

Let us define a new sequence $g(n)$ as

$$g(n) = x(n)(r_0 e^{j\theta_0})^{-n} V^{-n^2/2} \qquad (9.5.17)$$

Then (9.5.16) can be expressed as

$$X(z_k) = V^{-k^2/2} \sum_{n=0}^{N-1} g(n) V^{(k-n)^2/2} \qquad (9.5.18)$$

The summation in (9.5.18) may be interpreted as the convolution of the sequence $g(n)$ with the impulse response $h(n)$ of a filter, where

$$h(n) = V^{n^2/2} \qquad (9.5.19)$$

Consequently, (9.5.18) may be expressed as

$$\begin{aligned} X(z_k) &= V^{-k^2/2} y(k) \\ &= \frac{y(k)}{h(k)} \qquad k = 0, 1, \ldots, L - 1 \end{aligned} \qquad (9.5.20)$$

where $y(k)$ is the output of the filter

$$y(k) = \sum_{n=0}^{N-1} g(n)h(k - n) \qquad k = 0, 1, \ldots, L - 1 \qquad (9.5.21)$$

We observe that both $h(n)$ and $g(n)$ are complex-valued sequences.

The sequence $h(n)$ with $R_0 = 1$ has the form of a complex exponential with argument $\omega n = n^2 \phi_0/2 = (n\phi_0/2)n$. The quantity $n\phi_0/2$ represents the frequency of the complex exponential signal, which we note is increasing linearly with time. Such signals are used in radar systems and are called *chirp signals*. Hence the z-transform evaluated as in (9.5.18) is called the *chirp-z transform*.

The linear convolution in (9.5.21) is most efficiently done by use of the FFT algorithm. The sequence $g(n)$ is of length N. However, $h(n)$ has infinite duration. Fortunately, only a portion $h(n)$ is required to compute the L values of $X(z)$.

Since we will compute the convolution in (9.5.1) via the FFT, let us consider the circular convolution of the N-point sequence $g(n)$ with an M-point section of $h(n)$, where $M > N$. In such a case, we know that the first $N - 1$ points contain aliasing and the remaining $M - N + 1$ points are identical to the result that would be obtained from linear convolution of $h(n)$ with $g(n)$. In view of this, we should select a DFT of size

$$M = L + N - 1$$

which would yield L valid points and $N - 1$ points corrupted by aliasing.

The section of $h(n)$ that is needed for this computation corresponds to the values of $h(n)$ for $-(N - 1) \le n \le (L - 1)$, which is of length $M = L + N - 1$, as observed from (9.5.21). Let us define the sequence $h_1(n)$ of length M as

$$h_1(n) = h(n - N + 1) \qquad n = 0, 1, \ldots, M - 1 \qquad (9.5.22)$$

and compute its M-point DFT via the FFT algorithm to obtain $H_1(k)$. From $x(n)$ we compute $g(n)$ as specified by (9.5.17), pad $g(n)$ with $L - 1$ zeros and compute its M-point DFT to yield $G(k)$. The IDFT of the product $Y_1(k) = G(k)H_1(k)$ yields the M-point sequence $y_1(n)$, $n = 0, 1, \ldots, M - 1$. The first $N - 1$ points of $y_1(n)$ are corrupted by aliasing and are discarded. The desired values are $y_1(n)$ for $N - 1 \leq n \leq M - 1$, which correspond to the range $0 \leq n \leq L - 1$ in (9.5.21), that is,

$$y(n) = y_1(n + N - 1) \qquad n = 0, 1, \ldots, L - 1 \qquad (9.5.23)$$

Alternatively, we may define a sequence $h_2(n)$ as

$$h_2(n) = \begin{cases} h(n) & 0 \leq n \leq L - 1 \\ h(n - N - L + 1) & L \leq n \leq M - 1 \end{cases} \qquad (9.5.24)$$

The M-point DFT of $h_2(n)$ yields $H_2(k)$, which when multiplied by $G(k)$ yields $Y_2(k) = G(k)H_2(k)$. The IDFT of $Y_2(k)$ yields the sequence $y_2(n)$ for $0 \leq n \leq M - 1$. Now the desired values of $y_2(n)$ are in the range $0 \leq n \leq L - 1$, that is,

$$y(n) = y_2(n) \qquad n = 0, 1, \ldots, L - 1 \qquad (9.5.25)$$

Finally, the complex values $X(z_k)$ are computed by dividing $y(k)$ by $h(k)$, $k = 0, 1, \ldots, L - 1$, as specified by (9.5.20).

In general, the computational complexity of the chirp-z transform algorithm described above is of the order of $M \log_2 M$ complex multiplications, where $M = N + L - 1$. This number should be compared with the number $N \cdot L$, if the computations are performed by direct evaluation of the z-transform. Clearly, if L is small, direct computation is more efficient. However, if L is large, then the chirp-z transform algorithm is more efficient.

The chirp-z transform method has been implemented in hardware to compute the DFT

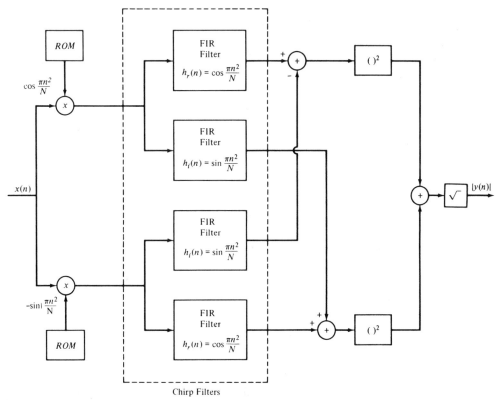

FIGURE 9.26 Block diagram illustrating the implementation of the chirp-z transform for computing the DFT (magnitude only).

of signals. For the computation of the DFT, we select $r_0 = R_0 = 1$, $\theta_0 = 0$, $\phi_0 = 2\pi/N$, and $L = N$. In this case

$$V^{-n^2/2} = e^{-j\pi n^2/N}$$
$$= \cos\frac{\pi n^2}{N} - j\sin\frac{\pi n^2}{N} \qquad (9.5.26)$$

The chirp filter with impulse response

$$h(n) = V^{n^2/2}$$
$$= \cos\frac{\pi n^2}{N} + j\sin\frac{\pi n^2}{N} \qquad (9.5.27)$$
$$= h_r(n) + jh_i(n)$$

has been implemented as a pair of FIR filters with coefficients $h_r(n)$ and $h_i(n)$, respectively. Both *surface acoustic wave* (SAW) devices and *charge coupled devices* (CCD) have been used for the FIR filters. The cosine and sine sequences given in (9.5.26) that are needed for the premultiplications and postmultiplications are usually stored in a read-only memory (ROM). Furthermore, we note that if only the magnitude of the DFT is desired, the postmultiplications are unnecessary. In this case,

$$|X(z_k)| = |y(k)| \qquad k = 0, 1, \ldots, N - 1 \qquad (9.5.28)$$

as illustrated in Figure 9.26. Thus the linear FIR filtering approach using the chirp-z transform has been used to compute the DFT.

9.6 Summary and References

The focus of this chapter was on the efficient computation of the DFT. We demonstrated that by taking advantage of the symmetry and periodicity properties of the exponential factors W_N^{kn}, we can reduce the number of complex multiplications needed to compute the DFT from N^2 to $N \log_2 N$ when N is a power of 2. As we indicated, any sequence can be augmented with zeros, such that $N = 2^\nu$.

For decades, FFT-type algorithms were of interest to mathematicians who were con-cerned with computing values of Fourier series by hand. However, it was not until Cooley and Tukey (1965) published their well-known paper that the impact and significance of the efficient computation of the DFT was recognized. Since then the Cooley–Tukey FFT algorithm and its various forms, for example, the algorithms of Singleton (1967, 1969), have had a tremendous influence on the use of the DFT in convolution, correlation, and spectrum analysis. For a historical perspective on the FFT algorithm, the reader is referred to the paper by Cooley et al. (1967).

Over the years, a number of tutorial papers have been published on FFT algorithms. We cite the early papers by Brigham and Morrow (1967), Cochran et al. (1967), Bergland (1969), and Cooley et al. (1969).

The recognition that the DFT can be arranged and computed as a linear convolution is also highly significant. Goertzel (1968) indicated that the DFT can be computed recursively, although the computational savings of this approach is rather modest, as we have observed. More significant is the work of Bluestein (1970), who demonstrated that the computation of the DFT can be formulated as a chirp linear filtering operation. This work led to the development of the chirp-z transform algorithm by Rabiner et al. (1969).

In addition to the FFT algorithms described in this chapter, there are other efficient algorithms for computing the DFT, some of which further reduce the number of multiplications, but usually require more additions. Of particular importance is an algorithm due to Rader and Brenner (1976), the class of prime factor algorithms, such as the Good algorithm (1971), and the Winograd algorithm (1976, 1978). For a description of these and related algorithms, the reader may refer to the text by Blahut (1985).

PROBLEMS

9.1 Show that each of the numbers

$$e^{j(2\pi/N)k} \qquad 0 \le k \le N - 1$$

corresponds to an Nth root of unity. Plot these numbers as phasors in the complex plane and illustrate, by means of this figure, the orthogonality property

$$\sum_{n=0}^{N-1} e^{j(2\pi/N)kn} e^{-j(2\pi/N)ln} = \begin{cases} N & \text{if } k = l \\ 0 & \text{if } k \ne l \end{cases}$$

9.2 (a) Show that the phase factors can be computed recursively by

$$W_N^{ql} = W_N^q W_N^{q(l-1)}$$

(b) Perform this computation once using single-precision floating-point arithmetic and once using only four significant digits. Note the deterioration due to the accumulation of round-off errors in the later case.

(c) Show how the results in part (b) can be improved by resetting the result to the correct value $-j$, each time $ql = N/4$.

9.3 The first five points of the eight-point DFT of a real-valued sequence are $\{0.25, 0.125 - j0.3018, 0, 0.125 - j0.0518, 0\}$. Determine the remaining three points.

9.4 Compute the eight-point circular convolution for the following sequences.

(a) $x_1(n) = \{1, 1, 1, 1, 0, 0, 0, 0\}$

$$x_2(n) = \sin\frac{3\pi}{8} n, \ 0 \le n \le 7$$

(b) $x_1(n) = (\frac{1}{4})^n, \ 0 \le n \le 7$

$$x_2(n) = \cos\frac{3\pi}{8} n, \ 0 \le n \le 7$$

(c) Compute the DFT of the two circular convolution sequences using the DFTs of $x_1(n)$ and $x_2(n)$.

9.5 Let $X(k), 0 \le k \le N - 1$ be the N-point DFT of the sequence $x(n), 0 \le n \le N - 1$. We define

$$\hat{X}(k) = \begin{cases} X(k) & 0 \le k \le k_c, N - k_c \le k \le N - 1 \\ 0 & k_c < k < N - k_c \end{cases}$$

and we compute the inverse N-point DFT of $\hat{X}(k), 0 \le k \le N - 1$. What is the effect of this process on the sequence $x(n)$? Explain.

9.6 For the sequences

$$x_1(n) = \cos\frac{2\pi}{N} n \qquad x_2(n) = \sin\frac{2\pi}{N} n \qquad 0 \le n \le N - 1$$

determine the N-point:

(a) Circular convolution $x_1(n) \; \textcircled{N} \; x_2(n)$

(b) Circular correlation of $x_1(n)$ and $x_2(n)$

(c) Circular autocorrelation of $x_1(n)$

(d) Circular autocorrelation of $x_2(n)$

9.7 Compute the quantity

$$\sum_{n=0}^{N-1} x_1(n) x_2^*(n)$$

for the following pairs of sequences.

(a) $x_1(n) = x_2(n) = \cos \dfrac{2\pi}{N} n, \quad 0 \le n \le N - 1$

(b) $x_1(n) = \cos \dfrac{2\pi}{N} n, \quad x_2(n) = \sin \dfrac{2\pi}{N} n, \quad 0 \le n \le N - 1$

(c) $x_1(n) = \delta(n) + \delta(n - 8), \quad x_2(n) = u(n) - u(n - N)$

9.8 Let $x(n)$ be a real-valued N-point $(N = 2^\nu)$ sequence. Develop a method to compute an N-point DFT $X'(k)$, which contains only the odd harmonics [i.e., $X'(k) = 0$ if k is even] by using only a real $N/2$-point DFT.

9.9 Compute the N-point DFT of the Blackman window

$$w(n) = 0.42 - 0.5 \cos \dfrac{2\pi n}{N - 1} + 0.08 \cos \dfrac{4\pi n}{N - 1} \qquad 0 \le n \le N - 1$$

9.10 If $X(k)$ is the DFT of the sequence $x(n)$, determine the N-point DFTs of the sequences

$$x_c(n) = x(n) \cos \dfrac{2\pi kn}{N} \qquad 0 \le n \le N - 1$$

and

$$x_s(n) = x(n) \sin \dfrac{2\pi kn}{N} \qquad 0 \le n \le N - 1$$

in terms of $X(k)$.

9.11 Determine the circular convolution of the sequences

$$x_1(n) = \left\{ \begin{matrix} 1, 2, 3, 1 \\ \uparrow \end{matrix} \right\}$$

$$x_2(n) = \left\{ \begin{matrix} 4, 3, 2, 2 \\ \uparrow \end{matrix} \right\}$$

using the time-domain formula in (9.2.17).

9.12 Use the four-point DFT and IDFT to determine the sequence

$$x_3(n) = x_1(n) \; \textcircled{N} \; x_2(n)$$

where $x_1(n)$ and $x_2(n)$ are the sequence given in Problem 9.11.

9.13 Compute the energy of the N-point sequence

$$x(n) = \cos \dfrac{2\pi kn}{N} \qquad 0 \le n \le N - 1$$

9.14 Given the eight-point DFT of the sequence

$$x(n) = \begin{cases} 1 & 0 \le n \le 3 \\ 0 & 4 \le n \le 7 \end{cases}$$

compute the DFT of the sequences:

(a) $x_1(n) = \begin{cases} 1 & n = 0 \\ 0 & 1 \le n \le 4 \\ 1 & 5 \le n \le 7 \end{cases}$

(b) $x_2(n) = \begin{cases} 0 & 0 \le n \le 1 \\ 1 & 2 \le n \le 5 \\ 0 & 6 \le n \le 7 \end{cases}$

9.15 A designer has available a number of eight-point FFT chips. Show explicitly how he should interconnect three such chips in order to compute a 24-point DFT.

9.16 The impulse response of an LTI system is given by $h(n) = \delta(n) - \frac{1}{4}\delta(n - k_0)$. To determine the impulse response $g(n)$ of the inverse system, an engineer computes the N-point DFT $H(k)$, $N = 4k_0$, of $h(n)$ and then defines $g(n)$ as the inverse DFT of $G(k) = 1/H(k)$, $k = 0, 1, 2, \ldots, N - 1$. Determine $g(n)$ and the convolution $h(n) * g(n)$, and comment on whether the system with impulse response $g(n)$ is the inverse of the system with impulse response $h(n)$.

9.17 Determine the eight-point DFT of the signal

$$x(n) = \{1, 1, 1, 1, 1, 1, 0, 0\}$$

and sketch its magnitude and phase.

9.18 Derive the radix-2 decimation-in-time FFT algorithm given by (9.3.26) and (9.3.27) as a special case of the more general algorithmic procedure given by (9.3.16) through (9.3.18).

9.19 Compute the eight-point DFT of the sequence

$$x(n) = \begin{cases} 1 & 0 \le n \le 7 \\ 0 & \text{otherwise} \end{cases}$$

by using the decimation-in-frequency FFT algorithm described in the text.

9.20 Derive the signal flow graph for the $N = 16$ point, radix-4 decimation-in-time FFT algorithm in which the input sequence is in normal order and the computations are done in place.

9.21 Derive the signal flow graph for the $N = 16$ point, radix-4 decimation-in-frequency FFT algorithm in which the input sequence is in digit-reversed order and the output DFT is in normal order.

9.22 Modify the FFT subroutine given in Fig. 9.21 to compute the IDFT.

9.23 Compute the eight-point DFT of the sequence

$$x(n) = \{\tfrac{1}{2}, \tfrac{1}{2}, \tfrac{1}{2}, \tfrac{1}{2}, 0, 0, 0, 0\}$$

using the in-place radix-2 decimation-in-time and radix-2 decimation-in-frequency algorithms. Follow exactly the corresponding signal flow graphs and keep track of all the intermediate quantities by putting them on the diagrams.

9.24 Compute the 16-point DFT of the sequence

$$x(n) = \cos \frac{\pi}{2} n \qquad 0 \le n \le 15$$

using the radix-4 decimation-in-time algorithm.

9.25 Explain how the DFT can be used to compute N equispaced samples of the z-transform, of an N-point sequence, on a circle of radius r.

9.26 A real-valued N-point sequence $x(n)$ is called DFT bandlimited if its DFT $X(k) = 0$ for $k_0 \le k \le N - k_0$. We insert $(L - 1)N$ zeros in the middle of $X(k)$ to obtain the following LN-point DFT

$$X'(k) = \begin{cases} X(k) & 0 \le k \le k_0 - 1 \\ 0 & k_0 \le k \le LN - k_0 \\ X(k + N - LN) & LN - k_0 + 1 \le k \le LN - 1 \end{cases}$$

Show that

$$Lx'(Ln) = x(n) \qquad 0 \le n \le N - 1$$

where

$$x'(n) \xleftarrow[\quad LN \quad]{\text{DFT}} X'(k)$$

Explain the meaning of this type of processing by working out an example with $N = 4$, $L = 1$, and $X(k) = \{1, 0, 0, 1\}$.

9.27 Let $X(k)$ be the N-point DFT of the sequence $x(n)$, $0 \le n \le N - 1$. What is the N-point DFT of the sequence $s(n) = X(n)$, $0 \le n \le N - 1$?

9.28 A linear time-invariant system with frequency response $H(\omega)$ is excited with the periodic input

$$x(n) = \sum_{k=-\infty}^{\infty} \delta(n - kN)$$

Suppose that we compute the N-point DFT $Y(k)$ of the samples $y(n)$, $0 \le n \le N - 1$ of the output sequence. How is $Y(k)$ related to $H(\omega)$?

9.29 Let $X(k)$ be the N-point DFT of the sequence $x(n)$, $0 \le n \le N - 1$. We define a $2N$-point sequence $y(n)$ as

$$y(n) = \begin{cases} x\left(\dfrac{n}{2}\right) & n \text{ even} \\ 0 & n \text{ odd} \end{cases}$$

Express the $2N$-point DFT of $y(n)$ in terms of $X(k)$.

9.30 **(a)** Determine the z-transform $W(z)$ of the Hanning window $w(n)$.
(b) Determine a formula to compute the N-point DFT $X_w(k)$ of the signal $x_w(n) = w(n)x(n)$, $0 \le n \le N - 1$, from the N-point DFT $X(k)$ of the signal $x(n)$.

9.31 Create a DFT coefficient table that uses only $N/4$ memory locations to store the first quadrant of the sine sequence (assume N even).

9.32 Show how the DFT of four N-point real even or real odd sequences can be computed from one N-point DFT based on complex-valued sequences.

9.33 Determine the computational burden of the algorithm given by (9.4.12) and compare it with the computational burden required in the $2N$-point DFT of $g(n)$. Assume that the FFT algorithm is a radix-2 algorithm.

9.34 Determine the system function $H(z)$ and the difference equation for the system that uses the Goertzel algorithm to compute the DFT value $X(N - k)$.

9.35 An $N = 51$ point FIR filter is characterized by the following DFT coefficients:

$$H(0) = 2$$
$$H(1) = e^{-j(2\pi/51)(25)} = e^{-j(50\pi/51)}$$
$$H(50) = e^{j(50\pi/51)}$$

(a) Sketch a block diagram of the frequency sampling realization of the FIR filter using single-pole filters for the building blocks. *Do not* include any single-pole filters having zero gain.
(b) Determine the impulse response $h(n)$ of the filter.
(c) Determine the phase characteristic $\theta(\omega)$ of the filter.
(d) Suppose that the filter is to be used for filtering an infinite-length data sequence $x(n)$. Three possible realizations of the filter are (1) direct-form realization, (2) frequency-sampling realization, and (3) DFT realization using the FFT algorithm. Determine *precisely* the number of *real multiplications* per filtered output point for these three realizations. For the FFT algorithm, assume that 1024-point FFTs are computed and the overlap-add method is used for filtering.

9.36 (a) Suppose that $x(n)$ is a finite-duration sequence of $N = 1024$ points. It is desired to evaluate the z-transform $X(z)$ of the sequence at the points

$$z_k = e^{j(2\pi/1024)k} \qquad k = 0, 100, 200, \dots, 1000$$

by using the most efficient method or algorithm possible. Describe an algorithm for performing this computation efficiently. Explain how you arrived at your answer by giving the various options or algorithms that can be used.
(b) Repeat part (a) if $X(z)$ is to be evaluated at

$$z_k = 2(0.9)^k e^{j[(2\pi/5000)k + \pi/2]} \qquad k = 0, 1, 2, \dots, 999$$

9.37 Recall that the Fourier transform of $x(t) = e^{j\Omega_0 t}$ is $X(j\Omega) = 2\pi \delta(\Omega - \Omega_0)$ and the Fourier transform of

$$p(t) = \begin{cases} 1 & 0 \le t \le T_0 \\ 0 & \text{otherwise} \end{cases}$$

is

$$P(j\Omega) = T_0 \frac{\sin \Omega T_0/2}{\Omega T_0/2} e^{-j\Omega T_0/2}$$

(a) Determine the Fourier transform $Y(j\Omega)$ of

$$y(t) = p(t)e^{j\Omega_0 t}$$

and roughly sketch $|Y(j\Omega)|$ versus Ω.
(b) Now consider the exponential sequence

$$x(n) = e^{j\omega_0 n}$$

where ω_0 is some arbitrary frequency in the range $0 < \omega_0 < \pi$ radians. Give the most general condition that ω_0 must satisfy in order for $x(n)$ to be periodic with period P (P is a positive integer).
(c) Let $y(n)$ be the finite-duration sequence

$$y(n) = x(n)w_N(n) = e^{j\omega_0 n}w_N(n)$$

where $w_N(n)$ is a finite-duration rectangular sequence of length N and where $x(n)$ is *not necessarily periodic*. Determine $Y(\omega)$ and roughly sketch $|Y(\omega)|$ for $0 \le \omega \le 2\pi$. What effect does N have in $|Y(\omega)|$? Briefly comment on the similarities and differences between $|Y(\omega)|$ and $|Y(j\Omega)|$.

(d) Suppose that

$$x(n) = e^{j(2\pi/P)n} \qquad P \text{ a positive integer}$$

and

$$y(n) = w_N(n)x(n)$$

where $N = lP$, l a positive integer. Determine and *sketch* the N-point DFT of $y(n)$. Relate your answer to the characteristics of $|Y(\omega)|$.

(e) Is the frequency sampling for the DFT in part (d) adequate for obtaining a rough approximation of $|Y(\omega)|$ directly from the magnitude of the DFT sequence $|Y(k)|$? If not, explain briefly how the sampling can be increased so that it will be possible to obtain a rough sketch of $|Y(\omega)|$ from an appropriate sequence $|Y(k)|$.

COMPUTER EXPERIMENT

9.38 *Computation of the DFT* Use the FFT subroutine to compute the following DFTs and plot the magnitudes $|X(k)|$ of the DFTs.

(a) The 64-point DFT of the sequence

$$x(n) = \begin{cases} 1 & n = 0, 1, \dots, 15 \quad (N_1 = 16) \\ 0 & \text{otherwise} \end{cases}$$

(b) The 64-point DFT of the sequence

$$x(n) = \begin{cases} 1 & n = 0, 1, \dots, 7 \quad (N_1 = 8) \\ 0 & \text{otherwise} \end{cases}$$

(c) The 128-point DFT of the sequence in part (a).

(d) The 64-point DFT of the sequence

$$x(n) = \begin{cases} 10e^{j(\pi/8)n} & n = 0, 1, \dots, 63 \quad (N_1 = 64) \\ 0 & \text{otherwise} \end{cases}$$

Answer the following questions.

(1) What is the frequency interval between succesive samples for the plots in parts (a), (b), (c), and (d)?

(2) What is the value of the spectrum at zero frequency (dc value) obtained from the plots in parts (a), (b), (c), (d)?
From the formula

$$X(k) = \sum_{n=0}^{N-1} x(n)e^{-j(2\pi/N)nk}$$

compute the theoretical values for the dc value and check these with the computer results.

(3) In plots (a), (b), and (c) what is the *frequency interval* between successive nulls in the spectrum? What is the relationship between N_1 of the sequence $x(n)$ and the frequency interval between successive nulls?

(4) Explain the difference between the plots obtained from parts (a) and (c).

9.39 *Identification of pole positions in a system* Consider the system described by the difference equation

$$y(n) = -r^2 y(n - 2) + x(n)$$

(a) Let $r = 0.9$ and $x(n) = \delta(n)$. Generate the output sequence $y(n)$ for $0 \le n \le 127$. Compute the $N = 128$ point DFT $\{Y(k)\}$ and plot $\{|Y(k)|\}$.

(b) Compute the $N = 128$ point DFT of the sequence

$$w(n) = (0.92)^{-n} y(n)$$

where $y(n)$ is the sequence generated in part (a). Plot the DFT values $|W(k)|$. What can you conclude from the plots in parts (a) and (b)?

(c) Let $r = 0.5$ and repeat part (a).

(d) Repeat part (b) for the sequence

$$w(n) = (0.55)^{-n} y(n)$$

where $y(n)$ is the sequence generated in part (c). What can you conclude from the plots in parts (c) and (d)?

(e) Now let the sequence generated in part (c) be corrupted by a sequence of "measurement" noise which is Gaussian with zero mean and variance $\sigma^2 = 0.1$. Repeat parts (c) and (d) for the noise-corrupted signal.

9.40 *Zoom-frequency analysis* Consider the system in Fig. P9.40a.

(a) Sketch the spectrum of the signal $y(n) = y_R(n) + jy_I(n)$ if the input signal $x(n)$ has the spectrum shown in Fig. P9.40b.

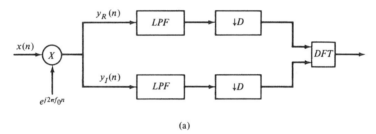

(a)

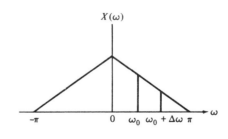

(b)

FIGURE P9.40

(b) Suppose that we are interested in the analysis of the frequencies in the band $f_0 \le f \le f_0 + \Delta f$, where $f_0 = \pi/6$ and $\Delta f = \pi/3$. Determine the cutoff of a lowpass filter and the decimation factor D required to retain the information contained in this band of frequencies.

(c) Assume that

$$x(n) = \sum_{k=0}^{p-1} \left(1 - \frac{k}{2p}\right) \cos 2\pi f_k n$$

where $p = 40$ and $f_k = k/p$, $k = 0, 1, \ldots, p - 1$. Compute and plot the 1024-point DFT of $x(n)$.

(d) Repeat part (b) for the signal $x(n)$ given in part (c) by using an appropriately designed lowpass linear phase FIR filter to determine the signal $s(n) = s_R(n) + js_I(n)$.

(e) Compute the 1024-point DFT of $s(n)$ and investigate if you have obtained the expected results.

CHAPTER 10

Quantization Effects in Digital Signal Processing

Up to this point we have considered the implementation of digital signal processing algorithms without being concerned about finite-word-length effects which are inherent in any digital realization, whether it be in hardware or in software. In fact, we have analyzed systems that are modeled as linear when, in fact, digital realizations of such systems are inherently nonlinear.

In this chapter we treat the various forms of quantization effects that arise in digital signal processing. Although we describe floating-point arithmetic operations briefly, our major concern is with fixed-point realizations of digital filters and in the computation of the DFT via fixed-point arithmetic.

In the case of recursive digital filters we demonstrate that the nonlinear characteristics resulting from quantization in multipliers may cause oscillatory behavior in the output of the filters, even in the absence of an input signal. Furthermore, overflows can occur in adders, which also can lead to oscillations in the output. In addition, we consider the effects of quantization errors in A/D conversion and in the coefficients of digital filters. A statistical model is adopted for quantization errors in multipliers and A/D converters which leads to relatively simple and useful results on the effects of such errors in the performance of digital filters and in the computation of the DFT. As an aid to the reader, a brief review of random processes and statistical concepts is provided in Appendix 10A.

10.1 Binary Fixed-Point and Floating-Point Representation of Numbers

In digital processing of analog signals, the samples of the analog signal are represented in digital form. Basically, the A/D conversion process involves sampling of the analog signal and mapping the samples into a sequence of binary digits that represent the quantized amplitude of the signal.

Within the binary number system, there are several methods by which a sample of an analog signal can be represented in binary form. The general class of binary representations may be subdivided into fixed-point and floating-point representations. Within the subclass of fixed-point representations, there are several binary fixed-point representations (e.g., the natural binary, sign and magnitude, offset binary, one's complement, and two's complement). These binary representations are described briefly below.

10.1.1 Binary Fixed-Point Representation

The best known binary representation or *binary code* is *natural binary*. In the natural binary code, a code word of b bits, such as 10011, corresponds to the decimal number

$$A = 1 \cdot 2^0 + 1 \cdot 2^1 + 0 \cdot 2^2 + 0 \cdot 2^3 + 1 \cdot 2^4$$
$$= 1 + 2 + 16 = 19$$

On the other hand, the binary code word 0.10011 represents a fraction corresponding to the decimal number.

$$B = 1 \cdot 2^{-1} + 0 \cdot 2^{-2} + 0 \cdot 2^{-3} + 1 \cdot 2^{-4} + 1 \cdot 2^{-5}$$
$$= \tfrac{1}{2} + \tfrac{1}{16} + \tfrac{1}{32} = \tfrac{19}{32}$$

Note that a shift of the binary point to the left by n bit positions corresponds to dividing the number by 2^n. A shift of the binary point to the right by n bit positions corresponds to multiplying the number by 2^n.

Whether the binary code word represents a fraction, an integer or a mixed number, the leftmost bit is called the most significant bit (MSB) and the rightmost bit is called the least significant (LSB) bit in the code word. In the representation of a fraction, the MSB has a weight of $2^{-1} = 1/2$ and the LSB has weight $2^{-b} = 1/2^b$, where b is the number of bits in the fractional binary representation. Note that the weight 2^{-b} assigned to the LSB is the *resolution* inherent in the representation of a fraction by b bits. Furthermore, the binary code word $0.11 \cdots 1$ corresponds to the decimal number $1 - 2^{-b}$.

In the conversion of bipolar analog signals, an additional bit is required to carry the sign information. The result is a *bipolar code*. There are four commonly used methods to represent bipolar numbers: sign-magnitude, offset binary, two's-complement, and one's-complement representations. Table 10.1 illustrates these four representations with four bits.

Sign-magnitude is the simplest and the most straightforward method for representing signed numbers in digital form. A zero in the leftmost bit position represents a positive

TABLE 10.1 Bipolar Codes

Number	Sign-Magnitude	Offset Binary	Two's Complement	One's Complement
7	0111	1111	0111	0111
6	0110	1110	0110	0110
5	0101	1101	0101	0101
4	0100	1100	0100	0100
3	0011	1011	0011	0011
2	0010	1010	0010	0010
1	0001	1001	0001	0001
0	0000	1000	0000	0000
0	1000	1000	0000	1111
-1	1001	0111	1111	1110
-2	1010	0110	1110	1101
-3	1011	0101	1101	1100
-4	1100	0100	1100	1011
-5	1101	0011	1011	1010
-6	1110	0010	1010	1001
-7	1111	0001	1001	1000

number and a one in that position represents a negative number. The remaining b bits represent the magnitude. We should indicate that the number zero has two representations: $00 \cdots 0$ or $100 \cdots 0$.

The *offset binary* representation is similar to the natural binary representation. In this representation, the most negative number is represented by the all-zero code word of $b + 1$ bits and the largest positive number is represented by the code word consisting of $b + 1$ ones. Thus we may view the offset binary representation as identical to the natural binary representation which would be obtained by adding a positive bias to the bipolar signal such that the resulting signal becomes unipolar with smallest amplitude of zero. Note that there is only one code word corresponding to the value zero. Consequently, this code avoids the ambiguity inherent in the sign-magnitude representation. On the other hand, there is a major drawback to the offset binary representation, namely, an error in reading the leftmost bit as a one instead of a zero, or vice versa, results in a large-amplitude error. For example, from Table 10.1 we note that 1001 and 0001 represent the amplitudes of $+1$ and -7, respectively. Consequently, an error in the leftmost bit corresponds to an amplitude error of $+8$, which is one-half of full scale (i.e., a large number).

Two's-complement representation is identical to the sign-magnitude representation for positive numbers. Hence all positive numbers are represented with a zero in the sign bit position. A negative number is represented by forming the two's complement of the corresponding positive number. In other words, the negative number is obtained by subtracting the positive number from 2.0. More simply, the two's complement is formed by complementing the number (change all zeros to ones and all ones to zeros) and adding one LSB. For example, the number $-\frac{3}{8}$ is obtained simply by complementing 0011 ($\frac{3}{8}$) to obtain 1100 and then adding 0001. This yields 1101, which represents $-\frac{3}{8}$ in two's complement.

If we compare two's complement with offset binary, we find that they differ only in the most significant bit (i.e., the leftmost bit). Hence it is a simple matter to change from two's complement to offset binary and vice versa.

One's-complement representation is identical to two's complement and sign magnitude for positive numbers, but differs in the way in which negative numbers are formed. In one's complement, a negative number is obtained by complementing its corresponding positive-number representation. For example, the representation of $-\frac{3}{8}$ is 1100, which is the one's complement of 0011 ($\frac{3}{8}$). Note that a zero is now represented as $00 \cdots 0$ or $11 \cdots 1$, which is an undesired ambiguity.

It is interesting to note the effect of a carry in the MSB with two's-complement and one's-complement arithmetic. For example, $\frac{4}{8} - \frac{3}{8} = \frac{1}{8}$. In two's complement we have

$$0100 \oplus 1101 = 0001$$

where $\oplus$ indicates modulo-2 addition. Note that the carry bit, if present in the MSB, is dropped. On the other hand, in one's-complement arithmetic, the carry in the MSB, if present, is carried around to the LSB. Thus the computation $\frac{4}{8} - \frac{3}{8} = \frac{1}{8}$ becomes

$$0100 \oplus 1100 = 0000 \oplus 0001 = 0001$$

Although a variety of additional fixed-point representations are possible, the ones described above are most often used in practice.

In addition or subtraction of two fixed-point numbers, each of b bits long (with an additional bit for the sign), the result is a b-bit number. If the result of the addition exceeds the largest number that can be represented with b bits, an overflow occurs. The only way to avoid this problem is to increase the number of bits in the accumulator and thus increase the dynamic range that can be accommodated.

The multiplication of two fixed-point numbers each of b bits in length results in a product of $2b$ bits in length. In fixed-point arithmetic, the product is either truncated or rounded back to b bits. As a result we have a truncation or round-off error in the b least significant bits. The characterization of such errors is treated below.

10.1.2 Binary Floating-Point Representation

A fixed-point representation of numbers allows us to cover a range of numbers, say, $x_{max} - x_{min}$ with a resolution

$$\Delta = \frac{x_{max} - x_{min}}{m - 1}$$

where $m = 2^b$ is the number of levels and b is the number of bits. A basic characteristic of the fixed-point representation is that the resolution is fixed. Furthermore, Δ increases in direct proportion to an increase in the dynamic range.

A floating-point representation may be employed as a means for covering a larger dynamic range. The binary floating-point representation commonly used in practice consists of a mantissa M, which is the fractional part of the number and falls in the range $\frac{1}{2} \le M < 1$, multiplied by the exponential factor 2^p, where the exponent p is either a positive or negative integer. Hence a number N is represented as

$$N = M \cdot 2^p$$

The mantissa requires a sign bit for representing positive and negative numbers, and the exponent requires an additional sign bit. Since the mantissa is a signed fraction, we may use any of the four fixed-point representations described above.

For example, the number $N_1 = 5$ is represented by the following mantissa and exponent:

$$M_1 = 0.101000$$
$$p_1 = 011$$

while the number $N_2 = \frac{3}{8}$ is represented by the following mantissa and exponent:

$$M_2 = 0.110000$$
$$p_2 = 101$$

where the leftmost bit in the exponent represents the sign bit.

If the two numbers are to be multiplied, the mantissas are multiplied and the exponents are added. Thus the product of the two numbers given above is

$$N_1 N_2 = M_1 M_2 \cdot 2^{p_1 + p_2}$$
$$= (0.011110) \cdot 2^{010}$$
$$= (0.111100) \cdot 2^{001}$$

On the other hand, the addition of the two floating-point numbers requires that the exponents be equal. This can be accomplished by shifting the mantissa of the smaller number to the right and compensating by increasing the corresponding exponent. Thus the number N_2 may be expressed as

$$M_2 = 0.000011$$
$$p_2 = 011$$

With $p_2 = p_1$, we may add the two numbers N_1 and N_2. The result is

$$N_1 + N_2 = (0.101011) \cdot 2^{011}$$

It should be observed that the shifting operation required to equalize the exponent of N_2 with that for N_1 results in loss of precision, in general. In the example above, the six-bit mantissa was sufficiently long to accommodate a shift of four bits to the right for M_2 without dropping any of the ones. However, a shift of five bits would have caused the loss of a single bit and a shift of six bits to the right would have resulted in a mantissa of $M_2 = 0.000000$, unless we round upward after shifting so that $M_2 = 0.000001$.

Overflow occurs in the multiplication of two floating-point numbers when the sum of the exponents exceeds the dynamic range of the fixed-point representation of the exponent.

In comparing a fixed-point representation with a floating-point representation, with the same number of total bits, it is apparent that the floating-point representation allows us to cover a larger dynamic range by varying the resolution across the range. The resolution decreases with an increase in the size of successive numbers. In other words, the distance between two successive floating-point numbers increases as the numbers increase in size. It is this variable resolution that results in a larger dynamic range. Alternatively, if we wish to cover the same dynamic with both fixed-point and floating-point representations, the floating-point representation provides finer resolution for small numbers but coarser resolution for the larger numbers. In contrast, the fixed-point representation provides a uniform resolution throughout the range of numbers.

For example, if we have a computer with a word size of 32 bits, it is possible to represent 2^{32} numbers. If we wish to represent the positive integers beginning with zero, the largest possible integer that can be accommodated is

$$2^{32} - 1 = 4{,}294{,}967{,}295$$

The distance between successive numbers (the resolution) is 1. Alternatively, we may designate the leftmost bit as the sign bit and use the remaining 31 bits for the magnitude. In such a case a fixed-point representation allows us to cover the range

$$-(2^{31} - 1) = -2{,}147{,}483{,}647 \quad \text{to} \quad (2^{31} - 1) = 2{,}147{,}483{,}647$$

again with a resolution of 1.

On the other hand, suppose that we increase the resolution by allocating 10 bits for a fractional part, 21 bits for the integer part, and 1 bit for the sign. Then this representation allows us to cover the dynamic range

$$-(2^{31} - 1) \cdot 2^{-10} = -(2^{21} - 2^{-10}) \quad \text{to} \quad (2^{31} - 1) \cdot 2^{-10} = 2^{21} - 2^{-10}$$

or, equivalently,

$$-2{,}097{,}151.999 \quad \text{to} \quad 2{,}097{,}151.999$$

In this case, the resolution is 2^{-10}. Thus the dynamic range has been decreased by a factor of approximately 1000 (actually 2^{10}), while the resolution has been increased by the same factor.

For comparison, suppose that the 32-bit word is used to represent floating-point numbers. In particular, let the mantissa be represented by 23 bits plus a sign bit and let the exponent be represented by 7 bits plus a sign bit. Now the smallest number in magnitude will have the representation.

sign	23 bits	sign	7 bits	
0.	$100 \cdots 0$	1	$1111111 = \frac{1}{2} \times 2^{-127} \approx 0.3 \times 10^{-38}$	

At the other extreme, the largest number that can be represented with this floating-point representation is

sign	23 bits	sign	7 bits	
0	111 $\cdots$ 1	0	1111111	$= (1 - 2^{-23}) \times 2^{127} \approx 1.7 \times 10^{38}$

Thus we have achieved a dynamic range of approximately 10^{76}, but with variable resolution. In particular, we have fine resolution for small numbers and coarse resolution for the larger numbers.

10.1.3 Errors Resulting from Rounding and Truncation

In performing computations such as multiplications with either fixed-point or floating-point arithmetic, we are usually faced with the problem of quantizing a number via truncation or rounding, from a level of some precision to a level of lower precision. The effect of rounding and truncation is to introduce an error whose value depends on the number of bits in the original number to the number of bits after quantization. The characteristics of the errors introduced through either truncation or rounding depend on the particular form of number representation.

To be specific, let us consider a fixed-point representation in which a number x is quantized from b_u bits to b bits. Thus the number

$$\overbrace{x = 0.1011 \cdots 01}^{b_u}$$

consisting of b_u bits prior to quantization is represented as

$$\overbrace{x = 0.101 \cdots 1}^{b}$$

after quantization, where $b < b_u$. For example, if x represents the sample of an analog signal, then b_u may be taken as infinite. In any case if the quantizer truncates the value of x, the truncation error is defined as

$$E_t = Q_t(x) - x \tag{10.1.1}$$

First, we consider the range of values of the error for sign-magnitude and two's-complement representation. In both of these representations the positive numbers have identical representations. For positive numbers, truncation results in a number that is smaller than the unquantized number. Consequently, the truncation error resulting from a reduction of the number of significant bits from b_u to b is

$$-(2^{-b} - 2^{-b_u}) \le E_t \le 0 \tag{10.1.2}$$

where the largest error arises from discarding $b_u - b$ bits, all of which are ones.

In the case of negative fixed-point numbers based on the sign-magnitude representation, the truncation error is positive, since truncation basically reduces the magnitude of the numbers. Consequently, for negative numbers, we have

$$0 \le E_t \le (2^{-b} - 2^{-b_u}) \tag{10.1.3}$$

In the two's-complement representation, the negative of a number is obtained by subtracting the corresponding positive number from 2. As a consequence, the effect of truncation on a negative number is to increase the magnitude of the negative number. Consequently, $x > Q_t(x)$ and hence

$$-(2^{-b} - 2^{-b_u}) \le E_t \le 0 \tag{10.1.4}$$

Hence we conclude that *the truncation error for the sign-magnitude representation is symmetric about zero and falls in the range*

$$-(2^{-b} - 2^{-b_u}) \le E_t \le (2^{-b} - 2^{-b_u}) \tag{10.1.5}$$

On the other hand, *for two's-complement representation, the truncation error is always negative and falls in the range*

$$-(2^{-b} - 2^{-b_u}) \le E_t \le 0 \tag{10.1.6}$$

Next, let us consider the quantization errors due to rounding of a number. A number x represented by b_u bits before quantization and b bits after quantization incurs a quantization error

$$E_r = Q_r(x) - x \tag{10.1.7}$$

Basically, rounding involves only the magnitude of the number and, consequently, the round-off error is independent of the type of fixed-point representation. The maximum error that can be introduced through rounding is $(2^{-b} - 2^{-b_u})/2$ and this may be either positive or negative, depending on the value of x. Therefore, *the round-off error is symmetric about zero and falls in the range*

$$-\tfrac{1}{2}(2^{-b} 2^{-b_u}) \le E_r \le \tfrac{1}{2}(2^{-b} - 2^{-b_u}) \tag{10.1.8}$$

These relationships are summarized in Fig. 10.1 when x is a continuous signal amplitude $(b_u = \infty)$.

In a floating-point representation, the mantissa is either rounded or truncated. Due to the nonuniform resolution, the corresponding error in a floating-point representation is proportional to the number being quantized. An appropriate representation for the quantized value is

$$Q(x) = x + ex \tag{10.1.9}$$

where e is called the relative error. Now

$$Q(x) - x = ex \tag{10.1.10}$$

In the case of truncation based on two's-complement representation of the mantissa, we have

$$-2^p 2^{-b} < e_t x < 0 \tag{10.1.11}$$

for positive numbers. Since $2^{p-1} \le x < 2^p$, it follows that

$$-2^{-b+1} < e_t \le 0 \qquad x > 0 \tag{10.1.12}$$

On the other hand, for a negative number in two's-complement representation the error is

$$0 \le e_t x < 2^p 2^{-b}$$

and hence

$$0 \le e_t < 2^{-b+1} \qquad x < 0 \tag{10.1.13}$$

In the case where the mantissa is rounded, the resulting error is symmetric relative to zero and has a maximum value of $\pm 2^{-b}/2$. Consequently, the round-off error becomes

$$-2^p \cdot 2^{-b}/2 < e_r x \le 2^p \cdot 2^{-b}/2 \tag{10.1.14}$$

Again, since x falls in the range $2^{p-1} \le x < 2^p$, we divide through by 2^{p-1} so that

$$-2^{-b} < e_r \le 2^{-b} \tag{10.1.15}$$

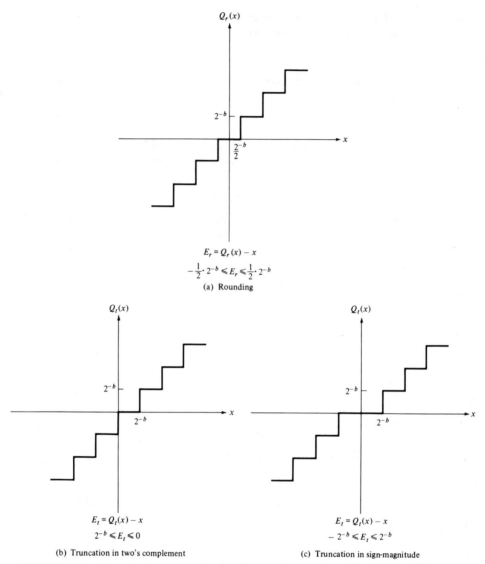

FIGURE 10.1   Quantization errors in rounding and truncation: (a) rounding; (b) truncation in two's complement; (c) truncation in sign-magnitude.

In arithmetic computations involving quantization via truncation and rounding, it is
convenient to adopt a statistical approach to the characterization of such errors. The
quantizer may be modeled as introducing an additive noise to the unquantized value x.
Thus we may write

$$Q(x) = x + \epsilon$$

where $\epsilon = E_r$ for rounding and $\epsilon = E_t$ for truncation. This model is illustrated in Fig.
10.2.

Since x may be any number that falls within any of the levels of the quantizer, the
quantization error is usually modeled as a random variable that falls within the limits
specified above. This random variable is assumed to be uniformly distributed within the
ranges specified for the fixed-point representations given above. Furthermore, in practice,
$b_u \gg b$, so that we may neglect the factor of 2^{-b_u} in the formulas given below. Under

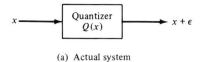

(a) Actual system

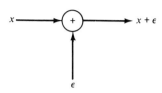

(b) Model for quantization

FIGURE 10.2 Additive noise model for the nonlinear quantization process: (a) actual system; (b) model for quantization.

these conditions, the probability density functions for the round-off and truncation errors in the two fixed-point representations are illustrated in Fig. 10.3. We note that in the case of truncation of the two's-complement representation of the number, the average value of the error has a bias of $2^{-b}/2$, whereas in all other cases illustrated above, the error has an average value of zero.

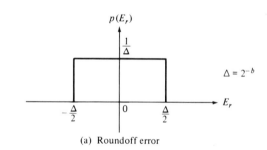

(a) Roundoff error

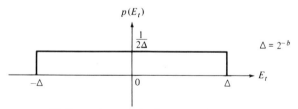

(b) Truncation error for sign-magnitude

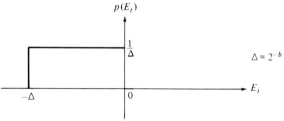

(c) Truncation error for two's complement

FIGURE 10.3 Statistical characterization of quantization errors: (a) round-off error; (b) truncation error for sign-magnitude; (c) truncation error for two's complement.

We shall use this statistical characterization of the quantization errors in our treatment of such errors in digital filtering and in the computation of the DFT for fixed-point implementation.

10.2 Quantization Effects in A/D Conversion of Signals

In Chapter 1 we stated that the basic functional operations performed by an A/D converter are (1) to sample the analog signal periodically at a sufficiently high rate to avoid aliasing, and (2) to quantize the amplitude of the sampled signal into a set of discrete amplitude levels. Thus an analog signal $x_a(t)$ sampled at a rate $F_s = 1/T$ results in the sequence $x(n) = x_a(nT)$, which when quantized in amplitude yields the sequence

$$x_q(n) \equiv Q[x(n)] \tag{10.2.1}$$

The input–output characteristic of a uniform quantizer is illustrated in Fig. 10.4. We observe that $Q[x(n)]$ is a many-to-one mapping of $x(n)$ into $x_q(n)$ and hence the quantization operation is both nonlinear and noninvertible.

The quantizer shown in Fig. 10.4 rounds the sampled signal to the nearest quantized output level. Thus if Δ is the distance between two successive output levels, $x(n)$ is quantized to the nearest output level from the set $l\,\Delta, l = 0, \pm 1, \pm 2, \ldots$, of possible output levels. In other words, the quantizer rounds the signal amplitude to the nearest quantization level.

The difference between the quantized signal amplitude $x_q(n)$ and the actual signal amplitude $x(n)$ is the error incurred in the quantization process. It may be represented as the sequence

$$e(n) = x_q(n) - x(n) \tag{10.2.2}$$

Suppose that we have a signal whose amplitude is normalized to the range $-\frac{1}{2} \le x(n) \le \frac{1}{2}$. If we employ a quantizer of b bits, the quantization step size or the resolution Δ is

$$\Delta = \frac{1}{2^b} = 2^{-b} \tag{10.2.3}$$

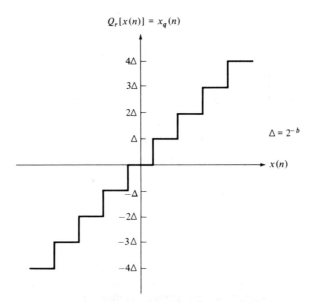

FIGURE 10.4 Rounding of signal samples.

The quantization error $e(n)$ for any given value of n is in the range $(-\Delta/2, \Delta/2)$, that is,

$$-\frac{\Delta}{2} < e(n) \leq \frac{\Delta}{2} \qquad (10.2.4)$$

or, equivalently,

$$-\tfrac{1}{2}2^{-b} < e(n) < \tfrac{1}{2}2^{-b} \qquad (10.2.5)$$

We may obtain a simple expression for the effect of the quantization error due to rounding by treating the error $e(n)$ as a random sequence which satisfies the following properties:

1. The error $e(n)$ is uniformly distributed over the range given by either (10.2.4) or (10.2.5).
2. The error sequence $\{e(n)\}$ is a stationary white noise sequence. In other words, the error $e(n)$ and the error $e(m)$ for $m \neq n$ are uncorrelated.
3. The error sequence $\{e(n)\}$ is uncorrelated with the signal sequence $x(n)$.

The assumptions given above do not hold, in general. However, they do hold when the quantization step size is small and the signal sequence $x(n)$ traverses several quantization levels between two successive samples.

Under the assumptions given above, we model the quantization error sequence $e(n)$ as an additive noise corrupting the desired signal $x(n)$, that is,

$$x_q(n) = x(n) + e(n) \qquad (10.2.6)$$

as illustrated in Fig. 10.5. The effect of this additive noise $e(n)$ on the desired signal can be quantified by evaluating the signal-to-noise (power) ratio (SNR), which may be expressed on a logarithmic scale (in decibels or dB) as

$$SNR = 10 \log_{10} \frac{P_x}{P_n} \qquad (10.2.7)$$

where P_x is the signal power and P_n is the power of the quantization noise.

If the quantization error is uniformly distributed in the range $(-\Delta/2, \Delta/2)$ as shown in Fig. 10.6, the mean value of the error is zero and the variance (the quantization noise

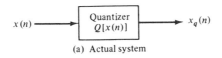

$x(n) \longrightarrow \boxed{\begin{array}{c} \text{Quantizer} \\ Q[x(n)] \end{array}} \longrightarrow x_q(n)$

(a) Actual system

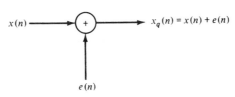

$x(n) \longrightarrow \oplus \longrightarrow x_q(n) = x(n) + e(n)$

$e(n)$

(b) Mathematical model

FIGURE 10.5 Additive noise model for the quantization errors in A/D conversion: (a) actual system; (b) mathematical model.

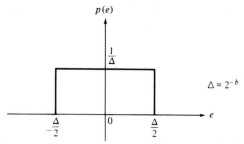

FIGURE 10.6 Probability density function for the quantization roundoff error in A/D conversion.

power) is

$$P_n = \sigma_e^2 = \int_{-\Delta/2}^{\Delta/2} e^2 p(e)\, de = \frac{1}{\Delta} \int_{-\Delta/2}^{\Delta/2} e^2\, de$$

$$P_n = \frac{\Delta^2}{12} = \frac{2^{-2b}}{12} \tag{10.2.8}$$

Consequently, the SNR is

$$\text{SNR} = 10 \log_{10} \frac{P_x}{P_n} = 10 \log_{10} P_x + 10 \log_{10} (12 \times 2^{2b})$$

$$\text{SNR} = 10 \log_{10} P_x + 10.8 + 6b \tag{10.2.9}$$

This expression for the SNR indicates that each bit used in the A/D converter, or quantizer, increases the signal-to-quantization noise ratio by 6 dB. Alternatively, we may compute the logarithm of P_n, that is,

$$10 \log_{10} P_n = 10 \log_{10} \frac{2^{-2b}}{12}$$

$$= -6b - 10.8 \tag{10.2.10}$$

From this expression we conclude that each additional bit used in the quantization process reduces the quantization noise power by 6 dB. For example, if we set the quantization noise power level at -70 dB below the signal power level, we must use a 10-bit quantizer (10-bit A/D converter).

10.3 Quantization of Filter Coefficients

In the realization of FIR and IIR filters in hardware or in software on a general-purpose computer, the accuracy with which filter coefficients can be specified is limited by the word length of the computer or the length of the register that is provided to store the coefficients. Since the coefficients used in implementing a given filter are not exact, the poles and zeros of the system function will in general be different from the desired poles and zeros. Consequently, we obtain a filter having a frequency response that is different from the frequency response of the filter with unquantized coefficients.

Below we demonstrate that the sensitivity of the filter frequency response characteristics to quantization of the filter coefficients is minimized by realizing the filter having a large number of poles and zeros as an interconnection of second-order filter sections. This leads us to the parallel-form and cascade-form realizations in which the basic building blocks are second-order filter sections, as described in Chapter 7.

10.3.1 Analysis of Sensitivity to Quantization of Filter Coefficients

To illustrate the effect of quantization of the filter coefficients in a direct-form realization of an IIR filter, let us consider a general IIR filter with system function

$$H(z) = \frac{\displaystyle\sum_{k=0}^{M} b_k z^{-k}}{1 + \displaystyle\sum_{k=1}^{N} a_k z^{-k}} \tag{10.3.1}$$

The direct-form realization of the IIR filter with quantized coefficients has the system function

$$\overline{H}(z) = \frac{\displaystyle\sum_{k=0}^{M} \overline{b}_k z^{-k}}{1 + \displaystyle\sum_{k=1}^{N} \overline{a}_k z^{-k}} \tag{10.3.2}$$

where the quantized coefficients $\{\overline{b}_k\}$ and $\{\overline{a}_k\}$ may be related to the unquantized coefficients $\{b_k\}$ and $\{a_k\}$ by the relations

$$\begin{aligned}
\overline{a}_k &= a_k + \Delta a_k & k &= 1, 2, \ldots, N \\
\overline{b}_k &= b_k + \Delta b_k & k &= 0, 1, \ldots, M
\end{aligned} \tag{10.3.3}$$

and $\{\Delta a_k\}$ and $\{\Delta b_k\}$ represent the quantization errors.

The denominator of $H(z)$ may be expressed in the form

$$D(z) = 1 + \sum_{k=0}^{N} a_k z^{-k} = \prod_{k=1}^{N} (1 - p_k z^{-1}) \tag{10.3.4}$$

where $\{p_k\}$ are the poles of $H(z)$. Similarly, we may express the denominator of $\overline{H}(z)$ as

$$\overline{D}(z) = \prod_{k=1}^{N} (1 - \overline{p}_k z^{-1}) \tag{10.3.5}$$

where $\overline{p}_k = p_k + \Delta p_k$, $k = 1, 2, \ldots, N$, and Δp_k is the error or perturbation resulting from the quantization of the filter coefficients.

We shall now relate the perturbation Δp_k to the quantization errors in the $\{a_k\}$. The perturbation error Δp_i can be expressed as

$$\Delta p_i = \sum_{k=1}^{N} \frac{\partial p_i}{\partial a_k} \Delta a_k \tag{10.3.6}$$

where $\partial p_i / \partial a_k$, the partial derivative of p_i with respect to a_k, represents the incremental change in the pole p_i due to a change in the coefficient a_k. Thus the total error Δp_i is expressed as a sum of the incremental errors due to changes in each of the coefficients $\{a_k\}$.

The partial derivatives $\partial p_i / \partial a_k$, $k = 1, 2, \ldots, N$, can be obtained by differentiating $D(z)$ with respect to each of the $\{a_k\}$. First we have

$$\left(\frac{\partial D(z)}{\partial a_k} \right)_{z=p_i} = \left(\frac{\partial D(z)}{\partial z} \right)_{z=p_i} \left(\frac{\partial p_i}{\partial a_k} \right) \tag{10.3.7}$$

Then

$$\frac{\partial p_i}{\partial a_k} = \frac{(\partial D(z)/\partial a_k)_{z=p_i}}{(\partial D(z)/\partial z)_{z=p_i}} \tag{10.3.8}$$

The numerator of (10.3.8) is

$$\left(\frac{\partial D(z)}{\partial a_k}\right)_{z=p_i} = -z^{-k}\big|_{z=p_i} = -p_i^{-k} \tag{10.3.9}$$

The denominator of (10.3.8) is

$$\left(\frac{\partial D(z)}{\partial z}\right)_{z=p_i} = \left\{\frac{\partial}{\partial z}\left[\prod_{l=1}^{N}(1 - p_l z^{-1})\right]\right\}_{z=p_i}$$

$$= \left\{\sum_{k=1}^{N}\frac{p_k}{z^2}\prod_{\substack{l=1 \\ l\neq k}}^{N}(1 - p_l z^{-1})\right\}_{z=p_i}$$

$$= \frac{1}{p_i^N}\prod_{\substack{l=1 \\ l\neq i}}^{N}(p_i - p_l) \tag{10.3.10}$$

Therefore, (10.3.8) may be expressed as

$$\frac{\partial p_i}{\partial a_k} = \frac{-p_i^{N-k}}{\displaystyle\prod_{\substack{l=1 \\ l\neq i}}^{N}(p_i - p_l)} \tag{10.3.11}$$

Substitution of the result in (10.3.11) into (10.3.6) yields the total perturbation error Δp_i in the form

$$\Delta p_i = -\sum_{k=1}^{N}\frac{p_i^{N-k}}{\displaystyle\prod_{\substack{l=1 \\ l\neq i}}^{N}(p_i - p_l)}\Delta a_k \tag{10.3.12}$$

This expression provides a measure of the sensitivity of the ith pole to changes in the coefficients $\{a_k\}$. An analogous result can be obtained for the sensitivity of the zeros to errors in the parameters $\{b_k\}$.

The terms $(p_i - p_l)$ in the denominator of (10.3.12) represent vectors in the z-plane from the poles $\{p_l\}$ to the pole p_i. If the poles are tightly clustered as they will be in a narrowband filter, as illustrated in Fig. 10.7, the lengths $|p_i - p_l|$ will be small for the poles in the vicinity of p_i. These small lengths will contribute to large errors and hence a large perturbation error Δp_i.

The error Δp_i can be minimized by maximizing the lengths $|p_i - p_l|$. This can be accomplished by realizing the high-order filter with either single-pole or double-pole filter sections. In general, however, single-pole (and single-zero) filters have complex-valued poles and require complex-valued arithmetic operations for their realization. This problem can be avoided by combining complex-valued poles (and zeros) to form second-order filter sections. Since the complex-valued poles are usually sufficiently far apart, the perturbation errors $\{\Delta p_i\}$ are minimized. As a consequence, the resulting filter with quantized coefficients approximate more closely the frequency response characteristics of the filter with unquantized coefficients.

It is interesting to note that even in the case of a two-pole filter section, the structure used to realize the filter section plays an important role in the errors caused by coefficient quantization. To be specific, let us consider a two-pole filter with system function

$$H(z) = \frac{1}{1 - (2r\cos\theta)z^{-1} + r^2 z^{-2}} \tag{10.3.13}$$

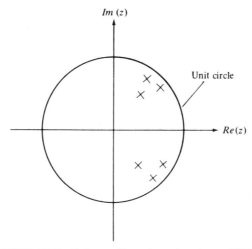

FIGURE 10.7   Pole positions for a bandpass IIR filter.

This filter has poles at $z = re^{\pm j\theta}$. When realized as shown in Fig. 10.8, it has two coefficients, $a_1 = 2r \cos \theta$ and $a_2 = -r^2$. With infinite precision it is possible to achieve an infinite number of pole positions. Clearly, with finite precision (i.e., quantized coefficients a_1 and a_2), the possible pole positions are also finite. In fact, when b bits are used to represent the magnitudes of a_1 and a_2, there are *at most* $(2^b - 1)^2$ possible positions for the poles in each quadrant, excluding the case $a_1 = 0$ and $a_2 = 0$.

For example, suppose that $b = 4$. Then there are 15 possible nonzero values for a_1. There are also 15 possible values for r^2. We illustrate these possible values in Fig. 10.9, for the first quadrant of the z-plane only. There are 169 possible pole positions in this case. The nonuniformity in their positions is due to the fact that we are quantizing r^2, whereas the pole positions lie on a circular arc of radius r. Of particular significance is the sparse set of poles for values of θ near zero and, due to symmetry, near $\theta = \pi$. This situation would be highly unfavorable for lowpass filters and highpass filters which normally have poles clustered near $\theta = 0$ and $\theta = \pi$, respectively.

An alternative realization of the two-pole filter is the coupled-form realization illustrated in Fig. 10.10. The two coupled equations are

$$y_1(n) = x(n) + r \cos \theta \, y_1(n - 1) - r \sin \theta \, y(n - 1) \qquad (10.3.14)$$
$$y(n) = r \sin \theta \, y_1(n - 1) + r \cos \theta \, y(n - 1)$$

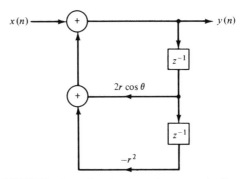

FIGURE 10.8   Realization of a two-pole filter.

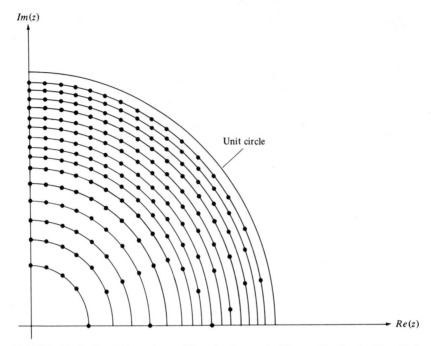

FIGURE 10.9   Possible pole positions for two-pole filter realization in Fig. 10.8.

By transforming these two equations into the z-domain, it is a simple matter to show that

$$\frac{Y(z)}{X(z)} = H(z) = \frac{(r \sin \theta)z^{-1}}{1 - (2r \cos \theta)z^{-1} + r^2 z^{-2}} \qquad (10.3.15)$$

In the coupled form we observe that there are also two coefficients, $\alpha_1 = r \sin \theta$ and $\alpha_2 = r \cos \theta$. Since they are both linear in r, the possible pole positions are now equally spaced points on a rectangular grid, as shown in Fig. 10.11. As a consequence, the pole positions are now uniformly distributed inside the unit circle, which is a more desirable situation than the previous realization, especially for lowpass filters. (There are 198 possible pole positions in this case.) However, the price that we pay for this uniform

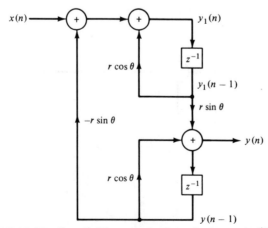

FIGURE 10.10   Coupled-form realization of a two-pole IIR filter.

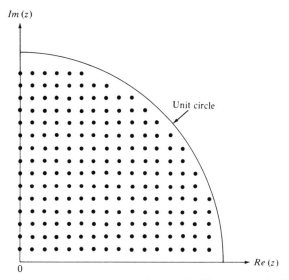

FIGURE 10.11   Possible pole positions for the coupled-form two-pole filter in Fig. 10.10.

distribution of pole positions is an increase in computations. The coupled-form realization requires four multiplications per output point, whereas the realization in Fig. 10.8 requires only two multiplications per output point.

Since there are various ways in which one can realize a second-order filter section, there are obviously many possibilities for different pole locations with quantized coefficients. Ideally, we should select a structure that provides us with a dense set of points in the regions where the poles lie. Unfortunately, however, there is no simple and systematic method for determining the filter realization that yields this desired result.

Given that a higher-order IIR filter should be implemented as a combination of second-order sections, we still must decide whether to employ a parallel configuration or a cascade configuration. In other words, we must decide between the realization

$$H(z) = \prod_{k=1}^{K} \frac{b_{k0} + b_{k1}z^{-1} + b_{k2}z^{-2}}{1 + a_{k1}z^{-1} + a_{k2}z^{-2}} \tag{10.3.16}$$

and the realization

$$H(z) = \sum_{k=1}^{K} \frac{c_{k0} + c_{k1}z^{-1}}{1 + a_{k1}z^{-1} + a_{k2}z^{-2}} \tag{10.3.17}$$

If the IIR filter has zeros on the unit circle, as is the case in general with elliptic and Chebyshev type II filters, each second-order section in the cascade configuration of (10.3.16) contains a pair of complex-conjugate zeros. The coefficients $\{b_k\}$ directly determine the location of these zeros. If the $\{b_k\}$ are quantized, the sensitivity of the system response to the quantization errors is easily and directly controlled by allocating a sufficiently large number of bits to the representation of the $\{b_{ki}\}$. In fact, we can easily evaluate the perturbation effect resulting from quantizing the coefficients $\{b_{ki}\}$ to some specified precision. Thus we have direct control of both the poles and the zeros that result from the quantization process.

On the other hand, the parallel realization of $H(z)$ provides direct control of the poles of the system only. The numerator coefficients $\{c_{k0}\}$ and $\{c_{k1}\}$ do not specify the location of the zeros directly. In fact, the $\{c_{k0}\}$ and $\{c_{k1}\}$ are obtained by performing a partial-fraction expansion of $H(z)$. Hence they do not directly influence the location of the zeros, but only indirectly through a combination of all the factors of $H(z)$. As a consequence, it is more difficult to determine the effect of quantization errors in the coefficients $\{c_{ki}\}$ on the location of the zeros of the system.

It is apparent that quantization of the parameters $\{c_{ki}\}$ is likely to produce a significant perturbation of the zero positions and usually, it will be sufficiently large in fixed-point implementations to move the zeros off the unit circle. This is a highly undesirable situation, which can be easily remedied by use of a floating-point representation. In any case the cascade form is more robust in the presence of coefficient quantization and should be the preferred choice in practical applications, especially where a fixed-point representation is employed.

EXAMPLE 10.3.1 _____

Determine the effect of parameter quantization on the frequency response of the 7-order elliptic filter given in Table 8.12 when it is realized as a cascade of second-order sections.

Solution: The coefficients for the elliptic filter given in Table 8.12 are specified for the cascade form to six significant digits. We quantized these coefficients to four and then three significant digits (by rounding) and plotted the magnitude (in decibels) and the phase of the frequency response. The results are shown in Fig. 10.12 along the frequency

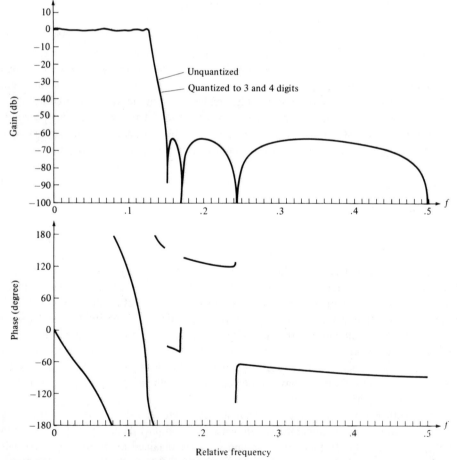

FIGURE 10.12 Effect of coefficient quantization on the magnitude and phase response of an $N = 7$ elliptic filter realized in cascade form.

response of the filter with unquantized (six significant digits) coefficients. We observe that there is an insignificant degradation due to coefficient quantization for the cascade realization.

EXAMPLE 10.3.2

Repeat the computation of the frequency response for the elliptic filter considered in Example 10.3.1 when it is realized in the parallel form with second-order sections.

Solution: The system function for the 7-order elliptic filter given in Table 8.12 is

$$H(z) = \frac{0.2781304 + 0.0054373108z^{-1}}{1 - 0.790103z^{-1}}$$

$$+ \frac{-0.3867805 + 0.3322229z^{-1}}{1 - 1.517223z^{-1} + 0.714088z^{-2}}$$

$$+ \frac{0.1277036 - 0.1558696z^{-1}}{1 - 1.421773z^{-1} + 0.861895z^{-1}}$$

$$+ \frac{-0.015824186 + 0.38377356z^{-1}}{1 - 1.387447z^{-1} + 0.962242z^{-1}}$$

The frequency response of this filter with coefficients quantized to four digits is shown in Fig. 10.13a. When this result is compared with the frequency response in Fig. 10.12, we observe that the zeros in the parallel realization have been perturbed sufficiently so that the nulls in the magnitude response are now at -80, -85 and -92 dB. The phase response has also been perturbed by a small amount.

When the coefficients are quantized to three significant digits the frequency response characteristic deteriorates significantly, in both magnitude and phase, as illustrated in Fig. 10.13b. It is apparent from the magnitude response that the zeros are no longer on the unit circle as a result of the quantization of the coefficients. This result clearly illustrates the sensitivity of the zeros to quantization of the coefficients in the parallel form.

When compared with the results of Example 10.3.1, it is also apparent that the cascade form is definitely more robust to parameter quantization than the parallel form.

10.3.2 Quantization of Coefficients in FIR Filters

As indicated in the preceding section, the sensitivity analysis performed on the poles of a system also applies directly to the zeros of the IIR filters. Consequently, an expression analogous to (10.3.12) can be obtained for the zeros of an FIR filter. In effect, we should generally realize FIR filters with a large number of zeros as a cascade of second-order and first-order filter sections in order to minimize the sensitivity to coefficient quantization.

Of particular interest in practice is the realization of linear phase FIR filters. The direct-form realizations shown in Figs. 7.1 and 7.2 maintain the linear-phase property even when the coefficients are quantized. This follows easily from the observation that the system function of a linear-phase FIR filter satisfies the property

$$H(z) = \pm z^{-(M-1)}H(z^{-1})$$

independent of whether the coefficients are quantized or unquantized. Consequently, coefficient quantization does not affect the phase characteristic of the FIR filter, but

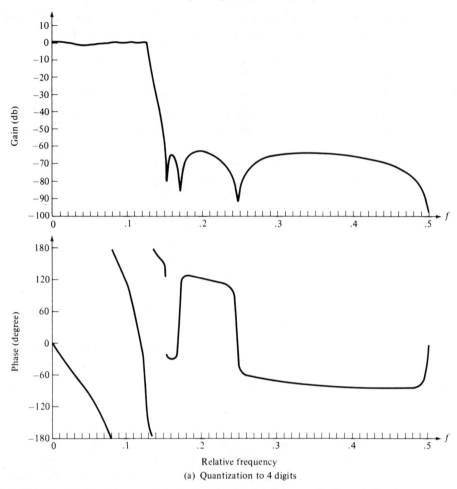

FIGURE 10.13 Effect of coefficient quantization on the magnitude and phase response of an $N = 7$ elliptic filter realized in parallel form: (a) quantization to four digits; (b) quantization to three digits.

affects only the magnitude. As a result, coefficient quantization effects are not as severe on a linear-phase FIR filter, since the only effect is in the magnitude.

EXAMPLE 10.3.3 ──────────────────────────────

Determine the effect of parameter quantization on the frequency response of an $M = 32$ linear-phase FIR bandpass filter. The filter is realized in the direct form.

Solution: The frequency response of a linear-phase FIR bandpass filter with unquantized coefficients is illustrated in Fig. 10.14a. When the coefficients were quantized to four significant digits, the effect on the frequency response was insignificant. However, when the coefficients are quantized to three significant digits, the sidelobes increased by several decibels, as illustrated in Fig. 10.14b. This result indicates that we should use a minimum of 10 bits to represent the coefficients of this FIR filter and, preferably, 12 to 14 bits, if possible.

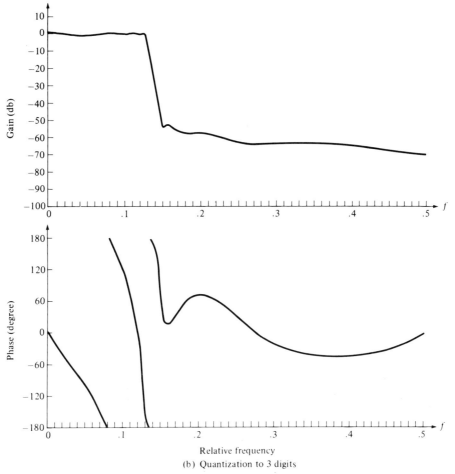

FIGURE 10.13   (continued)

In a cascade realization of the form

$$H(z) = G \prod_{k=1}^{K} H_k(z) \tag{10.3.18}$$

where the second-order sections are given as

$$H_k(z) = 1 + b_{k1}z^{-1} + b_{k2}z^{-2} \tag{10.3.19}$$

the coefficients of complex-valued zeros are expressed as $b_{k1} = -2r_k \cos \theta_k$ and $b_{k2} = r_k^2$. Quantization of b_{k1} and b_{k2} result in zero locations as shown in Fig. 10.9, except that the grid extends to points outside the unit circle.

A problem may arise, in this case, in maintaining the linear-phase property, because the quantized pair of zeros at $z = (1/r_k)e^{\pm j\theta_k}$ may not be the mirror image of the quantized zeros at $z = r_k e^{\pm j\theta_k}$. This problem can be avoided by rearranging the factors corresponding to the mirror-image zero. That is, we may write the mirror-image factor

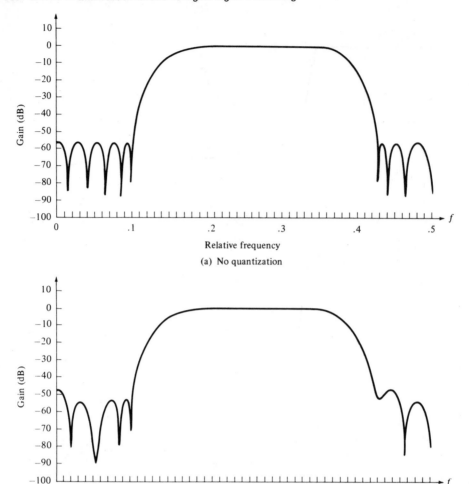

FIGURE 10.14   Effect of coefficient quantization on the magnitude of an $M = 32$ linear-phase FIR filter realized in direct form: (a) no quantization; (b) quantization to three digits.

as

$$\left(1 - \frac{2}{r_k}\cos\theta_k z^{-1} + \frac{1}{r_k^2}z^{-2}\right) = \frac{1}{r_k^2}(r_k^2 - 2r_k\cos\theta_k z^{-1} + z^{-2}) \quad (10.3.20)$$

The factors $\{1/r_k^2\}$ may be combined with the overall gain factor G, or they can be distributed in each of the second-order filters. The factor in (10.3.20) contains exactly the same parameters as the factor $(1 - 2r_k\cos\theta_k z^{-1} + r_k^2 z^{-2})$, and consequently, the zeros now occur in mirror-image pairs even when the parameters are quantized.

In this brief treatment we have given the reader an introduction to the problems of coefficient quantization in IIR and FIR filters. We have demonstated that a high-order filter should be reduced to a cascade (for FIR or IIR filters) or a parallel (for IIR filters) realization in order to minimize the effects of quantization errors in the coefficients. This is especially important in fixed-point realizations in which the coefficients are represented by a relatively small number of bits.

10.4 Quantization Effects in Digital Filters

In Section 10.1 we characterized the errors that occur in arithmetic operations performed in a digital filter. The presence of one or more quantizers in the realization of a digital filter results in a nonlinear device with characteristics that may be significantly different from the ideal linear filter. For example, a recursive digital filter may exhibit undesirable oscillations in its output, as shown below, even in the absence of an input signal.

As a result of the finite-precision arithmetic operations performed in the digital filter, some registers may overflow if the input signal level becomes large. Overflow represents another form of undesirable nonlinear distortion on the desired signal at the output of the filter. Consequently, special care must be exercised to scale the input signal properly so as either to prevent overflow completely or, at least, to minimize its rate of occurrence.

The nonlinear effects due to finite-precision arithmetic make it extremely difficult to analyze precisely the performance of a digital filter. To perform an analysis of quantization effects, we adopt a statistical characterization of quantization errors which, in effect, results in a linear model for the filter. Thus we are able to quantify the effects of quantization errors in the implementation of digital filters. Our treatment is limited to fixed-point realizations where quantization effects are very important.

10.4.1 Limit-Cycle Oscillations in Recursive Systems

In the realization of a digital filter either in digital hardware or in software on a digital computer, the quantization inherent in the finite-precision arithmetic operations render the system nonlinear. In recursive systems, the nonlinearities due to the finite-precision arithmetic operations often cause periodic oscillations to occur in the output, even when the input sequence is zero or some nonzero constant value. Such oscillations in recursive systems are called *limit cycles* and are directly attributable to round-off errors in multiplication and overflow in addition.

To illustrate the characteristics of a limit-cycle oscillation, let us consider a single-pole system described by the linear difference equation

$$y(n) = ay(n - 1) + x(n) \qquad (10.4.1)$$

where the pole is at $z = a$. The ideal system is realized as shown in Fig. 10.15. On the other hand, the actual system, which is described by the nonlinear difference equation

$$v(n) = Q[av(n - 1)] + x(n) \qquad (10.4.2)$$

is realized as shown in Fig. 10.16.

Suppose that the actual system in Fig. 10.16 is implemented with fixed-point arithmetic based on four bits for the magnitude plus a sign bit. The quantization that takes place after multiplication is assumed to round the product upward.

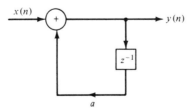

FIGURE 10.15   Ideal single-pole recursive system.

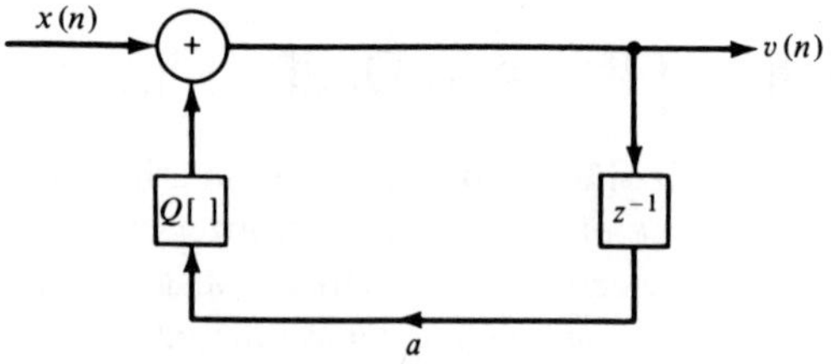

FIGURE 10.16   Actual nonlinear system.

In Table 10.2 we list the response of the actual system for four different locations of the pole $z = a$ and an input $x(n) = \beta\,\delta(n)$, where $\beta = 15/16$, which has the binary representation 0.1111. Ideally, the response of the system should decay toward zero exponentially [i.e., $y(n) = a^n \to 0$ as $n \to \infty$]. In the actual system, however, the response $v(n)$ reaches a steady-state periodic output sequence having a period that depends on the value of the pole. When the pole is positive, the oscillations occur with a period $N_p = 1$, so that the output reaches a constant value of $\frac{1}{16}$ for $a = \frac{1}{2}$ and $\frac{1}{8}$ for $a = \frac{3}{4}$. On the other hand, when the pole is negative, the output sequence oscillates between positive and negative values ($\pm\frac{1}{16}$ for $a = -\frac{1}{2}$ and $\pm\frac{1}{8}$ for $a = -\frac{3}{4}$). Hence the period is $N_p = 2$.

These limit cycles occur as a result of the quantization effects in multiplications. When the input sequence $x(n)$ to the filter becomes zero, the output of the filter enters into the limit cycle after a number of iterations. The output remains in the limit cycle until another input of sufficient size is applied that drives the system out of the limit cycle. Similarly, zero-input limit cycles occur from nonzero initial conditions with the input $x(n) = 0$. The amplitudes of the output during a limit cycle are confined to a range of values that is called the *dead band* of the filter.

It is interesting to note that when the response of the single-pole filter is in the limit cycle, the actual nonlinear system operates as an equivalent linear system with a pole at $z = 1$ when the pole is positive and $z = -1$ when the pole is negative. That is,

$$Q_r[av(n - 1)] = \begin{cases} v(n - 1) & a > 0 \\ -v(n - 1) & a < 0 \end{cases} \qquad (10.4.3)$$

Since the quantized product $av(n - 1)$ is by rounding, it follows that the quantization error is bounded as

$$|Q_r[av(n - 1)] - av(n - 1)| \le \tfrac{1}{2} \cdot 2^{-b} \qquad (10.4.4)$$

TABLE 10.2   Limit Cycles for Lowpass Single-Pole Filter

n	$a = 0.1000$ $= \frac{1}{2}$		$a = 1.1000$ $= -\frac{1}{2}$		$a = 0.1100$ $= \frac{3}{4}$		$a = 1.1100$ $= -\frac{3}{4}$	
0	0.1111	$(\frac{15}{16})$	0.1111	$(\frac{15}{16})$	0.1011	$(\frac{11}{16})$	0.1011	$(\frac{11}{16})$
1	0.0111	$(\frac{7}{16})$	1.0111	$(-\frac{7}{16})$	0.1000	$(\frac{8}{16})$	1.1000	$(-\frac{8}{16})$
2	0.0011	$(\frac{3}{16})$	0.0011	$(\frac{3}{16})$	0.0110	$(\frac{6}{16})$	0.0110	$(\frac{6}{16})$
3	0.0001	$(\frac{1}{16})$	1.0001	$(-\frac{1}{16})$	0.0101	$(\frac{5}{16})$	1.0101	$(-\frac{5}{16})$
4	0.0001	$(\frac{1}{16})$	0.0001	$(\frac{1}{16})$	0.0100	$(\frac{4}{16})$	0.0100	$(\frac{4}{16})$
5	0.0001	$(\frac{1}{16})$	1.0001	$(-\frac{1}{16})$	0.0011	$(\frac{3}{16})$	1.0011	$(-\frac{3}{16})$
6	0.0001	$(\frac{1}{16})$	0.0001	$(\frac{1}{16})$	0.0010	$(\frac{2}{16})$	0.0010	$(\frac{2}{16})$
7	0.0001	$(\frac{1}{16})$	1.0001	$(-\frac{1}{16})$	0.0010	$(\frac{2}{16})$	1.0010	$(-\frac{2}{16})$
8	0.0001	$(\frac{1}{16})$	0.0001	$(\frac{1}{16})$	0.0010	$(\frac{2}{16})$	0.0010	$(\frac{2}{16})$

where b is the number of bits (exclusive of sign) used in the representation of the pole a and $v(n)$. Consequently, (10.4.4) and (10.4.3) leads to

$$|v(n - 1)| - |av(n - 1)| \le \tfrac{1}{2} \cdot 2^{-b}$$

and hence

$$|v(n - 1)| \le \frac{\tfrac{1}{2} \cdot 2^{-b}}{1 - |a|} \tag{10.4.5}$$

The expression in (10.4.5) defines the dead band for a single-pole filter. For example, when $b = 4$ and $|a| = \tfrac{1}{2}$ we have a dead band with a range of amplitudes $(-\tfrac{1}{16}, \tfrac{1}{16})$. When $b = 4$ and $|a| = \tfrac{3}{4}$, the dead band increases to $(-\tfrac{1}{8}, \tfrac{1}{8})$.

The limit-cycle behavior in a two-pole filter is much more complex and a larger variety of oscillations can occur. In this case the ideal two-pole system is described by the linear difference equation,

$$y(n) = a_1 y(n - 1) + a_2 y(n - 2) + x(n) \tag{10.4.6}$$

whereas the actual system is described by the nonlinear difference equation

$$v(n) = Q_r[a_1 v(n - 1)] + Q_r[a_2 v(n - 2)] + x(n) \tag{10.4.7}$$

When the filter coefficients satisfy the condition $a_1^2 < -4a_2$, the poles of the system occur at

$$z = re^{\pm j\theta}$$

where $a_2 = -r^2$ and $a_1 = 2r \cos \theta$. As in the case of the single-pole filter, when the system is in a zero-input or zero-state limit cycle,

$$Q_r[a_2 v(n - 2)] = -v(n - 2) \tag{10.4.8}$$

In other words, the system behaves as an oscillator with complex-conjugate poles on the unit circle (i.e., $a_2 = -r^2 = -1$). Rounding the product $av(n - 2)$ implies that

$$|Q_r[a_2 v(n - 2)] - a_2 v(n - 2)| \le \tfrac{1}{2} \cdot 2^{-b} \tag{10.4.9}$$

Upon substitution of (10.4.8) into (10.4.9), we obtain the result

$$|v(n - 2)| - |a_2 v(n - 2)| \le \tfrac{1}{2} \cdot 2^{-b}$$

or equivalently,

$$|v(n - 2)| \le \frac{\tfrac{1}{2} \cdot 2^{-b}}{1 - |a_2|} \tag{10.4.10}$$

The expression in (10.4.10) defines the dead band of the two-pole filter with complex-congugate poles. We observe that the dead-band limits depend only on $|a_2|$. The parameter $a_1 = 2r \cos \theta$ determines the frequency of oscillation.

Another possible limit-cycle mode with zero input, which occurs as a result of rounding the multiplications, corresponds to an equivalent second-order system with poles at $z = \pm 1$. In this case it was shown by Jackson (1969) that the two-pole filter exhibits oscillations with an amplitude that falls in the dead band bounded by $2^{-b}/(1 - |a_1| - a_2)$.

It is interesting to note that the limit cycles described above result from rounding the product of the filter coefficients with the previous outputs $v(n - 1)$ and $v(n - 2)$. Instead of rounding, we may choose to truncate the products to b bits. With truncation, we can eliminate many, although not all, of the limit cycles as shown by Claasen et al. (1973).

However, recall that truncation results in a biased error unless the sign-magnitude representation is used, in which case the truncation error is symmetric about zero. In general, this bias is undesirable in digital filter implementation.

In a parallel realization of a high-order IIR system, each second-order filter section exhibits its own limit-cycle behavior, with no interaction among the second-order filter sections. Consequently, the output is the sum of the zero-input limit cycles from the individual sections. In the case of a cascade realization for a high-order IIR system, the limit cycles are much more difficult to analyze. In particular, when the first filter section exhibits a zero-input limit cycle, the output limit cycle is filtered by the succeeding sections. If the frequency of the limit cycle falls near a resonance frequency in a succeeding filter section, the amplitude of the sequence will be enhanced by the resonance characteristic. In general, we must be careful to avoid such situations.

In addition to limit cycles caused by rounding the result of multiplications, there are limit cycles caused by overflows in addition. An overflow in addition of two or more binary numbers occurs when the sum exceeds the word size available in the digital implementation of the system. For example, let us consider the second-order filter section illustrated in Fig. 10.17, in which the addition is performed in two's-complement arithmetic. Thus we may write the output $y(n)$ as

$$y(n) = g[a_1 y(n - 1) + a_2 y(n - 2) + x(n)] \qquad (10.4.11)$$

where the function $g[\cdot]$ represents the two's-complement addition. It is easily verified that the function $g(v)$ versus v is described by the graph in Fig. 10.18.

Recall that the range of values of the parameters (a_1, a_2) for a stable filter was given by the stability triangle in Fig. 3.16. However, these conditions are no longer sufficient to prevent overflow oscillation with two's-complement arithmetic. In fact, it can easily be shown that a necessary and sufficient condition for ensuring that no zero-input overflow limit cycles occur is

$$|a_1| + |a_2| < 1 \qquad (10.4.12)$$

which is extremely restrictive and hence an unreasonable constraint to impose on any second-order section.

An effective remedy for curing the problem of overflow oscillations is to modify the adder characteristic, as illustrated in Fig. 10.19, so that it performs saturation arithmetic. Thus when an overflow (or underflow) is sensed, the output of the adder will be the full-scale value of ± 1. The distortion caused by this nonlinearity in the adder is usually small provided that saturation occurs infrequently. The use of such a nonlinearity does not preclude the need for scaling of the signals and the system parameters, as described in the following section.

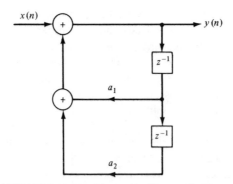

FIGURE 10.17 Two-pole filter realization.

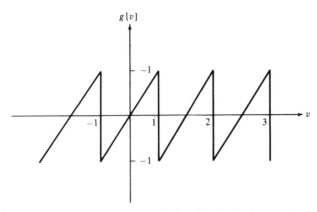

FIGURE 10.18 Characteristic functional relationship for two's complement addition of two or more numbers.

10.4.2 Scaling to Prevent Overflow

Saturation arithmetic as described above eliminates limit cycles due to overflow, on the one hand, but on the other hand, it causes undesirable signal distortion due to the nonlinearity of the clipper. In order to limit the amount of nonlinear distortion, it is important to scale the input signal and the unit sample response between the input and any internal summing node in the system such that overflow becomes a rare event.

For fixed-point arithmetic, let us first consider the extreme condition that no overflow should occur at any node of the system. Let $y_k(n)$ denote the response of the system at the kth node when the input sequence is $x(n)$ and the unit sample response between the node and the input is $h_k(n)$. Then

$$|y_k(n)| = \left| \sum_{m=-\infty}^{\infty} h_k(m)x(n-m) \right| \leq \sum_{m=-\infty}^{\infty} |h_k(m)| \, |x(n-m)|$$

Suppose that $x(n)$ is upper bounded by A_x. Then

$$|y_k(n)| \leq A_x \sum_{m=-\infty}^{\infty} |h_k(m)| \qquad \text{for all } n \qquad (10.4.13)$$

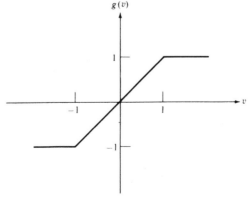

FIGURE 10.19 Characteristic functional relationship for addition with clipping at ± 1.

Now, if the dynamic range of the computer is limited to $(-1, 1)$, the condition

$$|y_k(n)| < 1$$

can be satisfied by requiring that the input $x(n)$ be scaled such that

$$A_x < \frac{1}{\displaystyle\sum_{m=-\infty}^{\infty} |h_k(m)|} \qquad (10.4.14)$$

for all possible nodes in the system. The condition in (10.4.14) is both necessary and sufficient to prevent overflow.

The condition in (10.4.14) is overly conservative, however, to the point where the input signal may be scaled too much. In such a case, much of the precision used to represent $x(n)$ is lost. This is especially true for narrowband sequences, such as sinusoids, where the scaling implied by (10.4.14) is extremely severe. For narrowband signals we may use the frequency response characteristics of the system in determining the appropriate scaling. Since $|H(\omega)|$ represents the gain of the system at frequency ω, a less severe and practically adequate scaling is to require that

$$A_x < \frac{1}{\displaystyle\max_{0 \le \omega \le \pi} |H_k(\omega)|} \qquad (10.4.15)$$

where $H_k(\omega)$ is the Fourier transform of $\{h_k(n)\}$.

In the case of an FIR filter, the condition in (10.4.14) reduces to

$$A_x < \frac{1}{\displaystyle\sum_{m=0}^{M-1} |h_k(m)|} \qquad (10.4.16)$$

which is now a sum over the M nonzero terms of the filter unit sample response.

In the following section we observe the ramifications of this scaling on the output signal-to-noise (power) ratio (SNR) from a first-order and a second-order filter section.

10.4.3 Statistical Characterization of Quantization Effects in Fixed-Point Realizations of Digital Filters

It is apparent from our treatment above that an analysis of quantization errors in digital filtering, based on deterministic models of quantization effects, is not a very fruitful approach. The basic problem is that the nonlinear effects in quantizing the products of two numbers and in clipping the sum of two numbers to prevent overflow are not easily modeled in large systems that contain many multipliers and many summing nodes.

To obtain more general results on quantization effects in digital filters, we shall model the quantization errors in multiplication as an additive noise sequence $e(n)$, just as we did in characterizing the quantization errors in A/D conversion of an analog signal. For addition, we consider the effect of scaling the input signal to prevent overflow.

Let us begin our treatment with the characterization of the round-off noise in a single-pole filter which is implemented in fixed-point arithmetic and is described by the nonlinear difference equation

$$v(n) = Q_r[av(n-1)] + x(n) \qquad (10.4.17)$$

The effect of rounding the product $av(n-1)$ is modeled as a noise sequence $e(n)$ added to the actual product $av(n - 1)$, that is,

$$Q_r[av(n - 1)] = av(n - 1) + e(n) \qquad (10.4.18)$$

With this model for the quantization error, the system under consideration is described by the *linear difference equation*

$$v(n) = av(n - 1) + x(n) + e(n) \qquad (10.4.19)$$

The corresponding system is illustrated in block diagram form in Fig. 10.20.

It is apparent from (10.4.19) that the output sequence $v(n)$ of the filter may be separated into two components. One is the response of the system to the input sequence $x(n)$. The second is the response of the system to the additive quantization noise $e(n)$. In fact, we may express the output sequence $v(n)$ as a sum of these two components, that is,

$$v(n) = y(n) + q(n) \qquad (10.4.20)$$

where $y(n)$ represents the response of the system to $x(n)$ and $q(n)$ represents the response of the system to the quantization error $e(n)$. Upon substitution from (10.4.20) for $v(n)$ into (10.4.19), we obtain

$$y(n) + q(n) = ay(n - 1) + aq(n - 1) + x(n) + e(n) \qquad (10.4.21)$$

To simplify the analysis, we make the following assumptions about the error sequence $e(n)$.

1. For any n, the error sequence $\{e(n)\}$ is uniformly distributed over the range $(-\frac{1}{2} \cdot 2^{-b}, \frac{1}{2} \cdot 2^{-b})$. This implies that the mean value of $e(n)$ is zero and its variance

$$\sigma_e^2 = \frac{2^{-2b}}{12} \qquad (10.4.22)$$

2. The error $\{e(n)\}$ is a stationary white noise sequence. In other words, the error $e(n)$ and the error $e(m)$ are uncorrelated for $n \neq m$.
3. The error sequence $\{e(n)\}$ is uncorrelated with the signal sequence $\{x(n)\}$.

The last assumption allows us to separate the difference equation in (10.4.21) into two uncoupled difference equations, namely,

$$y(n) = ay(n - 1) + x(n) \qquad (10.4.23)$$
$$q(n) = aq(n - 1) + e(n) \qquad (10.4.24)$$

The difference equation in (10.4.23) represents the input–output relation for the desired system and the difference equation in (10.4.24) represents the relation for the quantization error at the output of the system.

To complete the analyis, we will make use of two important relationships developed

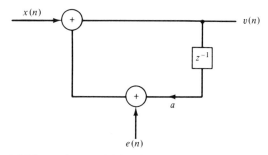

FIGURE 10.20 Additive noise model for the quantization error in a single-pole filter.

in Appendix 10A. The first is the relationship for the mean value of the output $q(n)$ of a linear shift-invariant filter with impulse response $h(n)$ when excited by a random sequence $e(n)$ having a mean value m_e. The result is

$$m_q = m_e \sum_{n=-\infty}^{\infty} h(n) \tag{10.4.25}$$

or, equivalently,

$$m_q = m_e H(0) \tag{10.4.26}$$

where $H(0)$ is the value of the frequency response $H(\omega)$ of the filter evaluated at $\omega = 0$.

The second important relationship is the expression for the autocorrelation sequence of the output $q(n)$ of the filter with impulse response $h(n)$ when the input random sequence $e(n)$ has an autocorrelation $\gamma_{ee}(n)$. This result is

$$\gamma_{qq}(n) = \sum_{k=-\infty}^{\infty} \sum_{l=-\infty}^{\infty} h(k)h(l)\gamma_{ee}(k - l + n) \tag{10.4.27}$$

In the important special case where the random sequence is white (spectrally flat), the autocorrelation $\gamma_{ee}(n)$ is a unit sample sequence scaled by the variance σ_e^2, that is,

$$\gamma_{ee}(n) = \sigma_e^2 \, \delta(n) \tag{10.4.28}$$

Upon substituting (10.4.28) into (10.4.27), we obtain the desired result for the autocorrelation sequence at the output of a filter excited by white noise, namely,

$$\gamma_{qq}(n) = \sigma_e^2 \sum_{k=-\infty}^{\infty} h(k)h(k + n) \tag{10.4.29}$$

The variance σ_q^2 of the output noise is simply obtained by evaluating $\gamma_{qq}(n)$ at $n = 0$. Thus

$$\sigma_q^2 = \sigma_e^2 \sum_{k=-\infty}^{\infty} h^2(k) \tag{10.4.30}$$

and with the aid of Parseval's theorem, we have the alternative expression

$$\sigma_q^2 = \frac{\sigma_e^2}{2\pi} \int_{-\pi}^{\pi} |H(\omega)|^2 \, d\omega \tag{10.4.31}$$

In the case of the single-pole filter under consideration, the unit sample response is

$$h(n) = a^n u(n) \tag{10.4.32}$$

Since the quantization error due to rounding has zero mean, the mean value of the error at the output of the filter is $m_q = 0$. The variance of the error at the output of the filter is

$$\sigma_q^2 = \sigma_e^2 \sum_{k=0}^{\infty} a^{2k}$$

$$= \frac{\sigma_e^2}{1 - a^2} \tag{10.4.33}$$

We observe that the noise power σ_q^2 at the output of the filter is enhanced relative to the input noise power σ_e^2 by the factor $1/(1 - a^2)$. This factor increases as the pole is moved closer to the unit circle.

To obtain a clearer picture of the effect of the quantization error, we should also consider the effect of scaling the input. Let us assume that the input sequence $\{x(n)\}$ is a white noise sequence (wideband signal) whose amplitude has been scaled according to (10.4.14) to prevent overflows in addition. Then

$$A_x < 1 - |a|$$

If we assume that $x(n)$ is uniformly distributed in the range $(-A_x, A_x)$, then, according to (10.4.27) and (10.4.30), the signal power at the output of the filter is

$$\sigma_y^2 = \sigma_x^2 \sum_{k=0}^{\infty} a^{2k}$$

$$= \frac{\sigma_x^2}{1 - a^2} \qquad (10.4.34)$$

where $\sigma_x^2 = (1 - |a|)^2/3$ is the variance of the input signal. The ratio of the signal power σ_y^2 to the quantization error power σ_q^2, which is called the signal-to-noise ratio (SNR), is simply

$$\frac{\sigma_y^2}{\sigma_q^2} = \frac{\sigma_x^2}{\sigma_e^2} \qquad (10.4.35)$$

$$= (1 - |a|)^2 \cdot 2^{2(b+1)}$$

This expression for the output SNR clearly illustrates the severe penalty paid as a consequence of the scaling of the input, expecially when the pole is near the unit circle. By comparison, if the input is not scaled and the adder has a sufficient number of bits to avoid overflow, the signal amplitude may be confined to the range $(-1, 1)$. In this case $\sigma_x^2 = \frac{1}{3}$, which is independent of the pole position. Then

$$\frac{\sigma_y^2}{\sigma_q^2} = 2^{2(b+1)} \qquad (10.4.36)$$

The difference between the SNRs in (10.4.36) and (10.4.35) clearly demonstrates the need to use more bits in addition than in multiplication. The number of additional bits depends on the position of the pole and should be increased as the pole is moved closer to the unit circle.

Next, let us consider a two-pole filter which, with infinite precision, is described by the linear difference equation

$$y(n) = a_1 y(n - 1) + a_2 y(n - 2) + x(n) \qquad (10.4.37)$$

where $a_1 = 2r \cos \theta$ and $a_2 = -r^2$. When the two products are rounded, we have a system which is described by the nonlinear difference equation

$$v(n) = Q_r[a_1 v(n - 1)] + Q_r[a_2 v(n - 2)] + x(n) \qquad (10.4.38)$$

This system is illustrated in block diagram form in Fig. 10.21.

Now there are two multiplications, and hence two quantization errors are produced for each output. Consequently, we should introduce two noise sequences $e_1(n)$ and $e_2(n)$, which correspond to the quantizer outputs

$$Q_r[a_1 v(n - 1)] = a_1 v(n - 1) + e_1(n) \qquad (10.4.39)$$
$$Q_r[a_2 v(n - 2)] = a_2 v(n - 2) + e_2(n)$$

A block diagram for the corresponding model is shown in Fig. 10.22. Note that the error sequences $e_1(n)$ and $e_2(n)$ can be moved directly to the input of the filter.

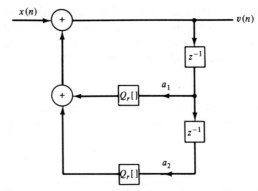

FIGURE 10.21 Two-pole digital filter with rounding quantizers.

As in the case of the first-order filter, the output of the second-order filter can be separated into two components, the desired signal component and the quantization error component. The former is described by the difference equation

$$y(n) = a_1 y(n - 1) + a_2 y(n - 2) + x(n) \qquad (10.4.40)$$

while the latter satisfies the difference equation

$$q(n) = a_1 q(n - 1) + a_2 q(n - 2) + e_1(n) + e_2(n) \qquad (10.4.41)$$

It is reasonable to assume that the two sequences $e_1(n)$ and $e_2(n)$ are uncorrelated.

Now the second-order filter has a unit sample response

$$h(n) = \frac{r^n}{\sin \theta} \sin (n + 1)\theta \, u(n) \qquad (10.4.42)$$

Hence

$$\sum_{n=0}^{\infty} h^2(n) = \frac{1 + r^2}{1 - r^2} \frac{1}{r^4 + 1 - 2r^2 \cos 2\theta} \qquad (10.4.43)$$

By applying (10.4.30) we obtain the variance of the quantization errors at the output of the filter in the form

$$\sigma_q^2 = \sigma_e^2 \left(\frac{1 + r^2}{1 - r^2} \frac{1}{r^4 + 1 - 2r^2 \cos 2\theta} \right) \qquad (10.4.44)$$

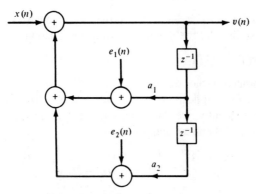

FIGURE 10.22 Additive noise model for the quantization errors in a two-pole filter realization.

In the case of the signal component, if we scale the input as in (10.4.16) to avoid overflow, the power in the output signal is

$$\sigma_y^2 = \sigma_x^2 \sum_{n=0}^{\infty} h^2(n) \tag{10.4.45}$$

where the power in the input signal $x(n)$ is given by the variance

$$\sigma_x^2 = \frac{1}{3\left[\sum_{n=0}^{\infty} |h(n)|\right]^2} \tag{10.4.46}$$

Consequently, the SNR at the output of the two-pole filter is

$$\frac{\sigma_y^2}{\sigma_q^2} = \frac{\sigma_x^2}{\sigma_e^2} = \frac{2^{2(b+1)}}{\left[\sum_{n=0}^{\infty} |h(n)|\right]^2} \tag{10.4.47}$$

Although it is difficult to determine the exact value of the denominator term in (10.4.47), it is easy to obtain an upper and a lower bound. In particular, $|h(n)|$ is upper bounded as

$$|h(n)| \leq \frac{1}{\sin \theta} r^n \qquad n \geq 0 \tag{10.4.48}$$

so that

$$\sum_{n=0}^{\infty} |h(n)| \leq \frac{1}{\sin \theta} \sum_{n=0}^{\infty} r^n = \frac{1}{(1 - r) \sin \theta} \tag{10.4.49}$$

The lower bound may be obtained by noting that

$$|H(\omega)| = \left| \sum_{n=0}^{\infty} h(n) e^{-j\omega n} \right| \leq \sum_{n=0}^{\infty} |h(n)|$$

But

$$H(\omega) = \frac{1}{(1 - re^{j\theta}e^{-j\omega})(1 - re^{-j\theta}e^{-j\omega})}$$

At $\omega = \theta$, which is the resonant frequency of the filter, we obtain the largest value of $|H(\omega)|$. Hence

$$\sum_{n=0}^{\infty} |h(n)| \geq |H(\theta)| = \frac{1}{(1 - r)\sqrt{1 + r^2 - 2r \cos 2\theta}} \tag{10.4.50}$$

Therefore, the SNR is bounded from above and below according to the relation

$$2^{2(b+1)}(1 - r)^2 \sin^2\theta \leq \frac{\sigma_y^2}{\sigma_q^2} \leq 2^{2(b+1)}(1 - r)^2(1 + r^2 - 2r \cos 2\theta) \tag{10.4.51}$$

For example, when $\theta = \pi/2$, the expression in (10.4.51) reduces to

$$2^{2(b+1)}(1 - r)^2 \leq \frac{\sigma_y^2}{\sigma_q^2} \leq 2^{2(b+1)}(1 - r)^2(1 + r)^2 \tag{10.4.52}$$

The dominant term in this bound is $(1 - r)^2$, which acts to reduce the SNR dramatically as the poles move toward the unit circle. Hence the effect of scaling in the second-order filter is more severe than in the single-pole filter. Note that if $d = 1 - r$ is the distance of the pole from the unit circle, the SNR in (10.4.52) is reduced by d^2, whereas in the single-pole filter the reduction is proportional to d. These results serve to reinforce the earlier statement regarding the use of more bits in addition than multiplication as a mechanism for avoiding the severe penalty due to scaling.

The analysis of the quantization effects in a second-order filter can be applied directly to higher-order filters based on a parallel realization. In this case each second-order filter section is independent of all the other sections, and hence the total quantization noise power at the output of the parallel bank is simply the linear sum of the quantization noise powers of each of the individual sections. On the other hand, the cascade realization is more difficult to analyze. For the cascade interconnection, the noise generated in any second-order filter section is filtered by the succeeding sections. As a consequence, there is the issue of how to pair together real-valued poles to form second-order sections and how to arrange the resulting second-order filters to minimize the total noise power at the output of the high-order filter. This general topic was investigated by Jackson (1970a, b), who showed that poles close to the unit circle should be paired with nearby zeros so as to reduce the gain of each second-order section. In ordering the second-order sections in cascade, a reasonable strategy is to place the sections in the order of decreasing maximum frequency gain. In this case the noise power generated in the early high-gain section will not be boosted significantly by the latter sections.

The following example illustrates the point that proper ordering of sections in a cascade realization is important in controlling the round-off noise at the output of the overall filter.

EXAMPLE 10.4.1 _____

Determine the variance of the round-off noise at the output of the two cascade realizations of the filter with system function

$$H(z) = H_1(z)H_2(z)$$

where

$$H_1(z) = \frac{1}{1 - \frac{1}{2}z^{-1}}$$

$$H_2(z) = \frac{1}{1 - \frac{1}{4}z^{-1}}$$

Solution: Let $h(n)$, $h_1(n)$, and $h_2(n)$ represent the unit sample responses corresponding to the system functions $H(z)$, $H_1(z)$, and $H_2(z)$, respectively. It follows that

$$h_1(n) = (\tfrac{1}{2})^n u(n) \qquad h_2(n) = (\tfrac{1}{4})^n u(n)$$
$$h(n) = [2(\tfrac{1}{2})^n - (\tfrac{1}{4})^n]u(n)$$

The two cascade realizations are shown in Fig. 10.23.

In the first cascade realization, the variance of the output is

$$\sigma_{q1}^2 = \sigma_e^2 \left[\sum_{n=0}^{\infty} h^2(n) + \sum_{n=0}^{\infty} h_2^2(n) \right]$$

In the second cascade realization, the variance of the output noise is

$$\sigma_{q2}^2 = \sigma_e^2 \left[\sum_{n=0}^{\infty} h^2(n) + \sum_{n=0}^{\infty} h_1^2(n) \right]$$

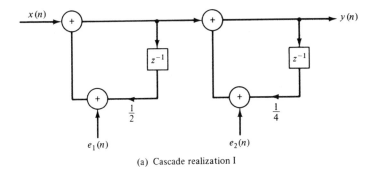

(a) Cascade realization I

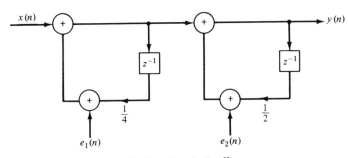

(b) Cascade realization II

FIGURE 10.23   Two cascade realizations in Example 10.4.1: (a) cascade realization I; (b) cascade realization II.

Now

$$\sum_{n=0}^{\infty} h_1^2(n) = \frac{1}{1 - \frac{1}{4}} = \frac{4}{3}$$

$$\sum_{n=0}^{\infty} h_2^2(n) = \frac{1}{1 - \frac{1}{16}} = \frac{16}{15}$$

$$\sum_{m=0}^{\infty} h^2(n) = \frac{4}{1 - \frac{1}{4}} - \frac{4}{1 - \frac{1}{8}} + \frac{1}{1 - \frac{1}{16}} = 1.83$$

Therefore,

$$\sigma_{q1}^2 = 2.90\sigma_e^2$$
$$\sigma_{q2}^2 = 3.16\sigma_e^2$$

and the ratio of noise variances is

$$\frac{\sigma_{q2}^2}{\sigma_{q1}^2} = 1.09$$

Consequently, the noise power in the second cascade realization is 9% larger than the first realization.

10.5 Quantization Effects in the Computation of the DFT

As we have observed in our previous discussions, the DFT plays an important role in many digital signal processing applications, including FIR filtering, in the computation of the correlation between signals, and in spectral analysis. For this reason it is important for us to know the effect of quantization errors in its computation. In particular, we shall consider the effect of round-off errors due to the multiplications performed in the DFT with fixed-point arithmetic.

The model that we shall adopt for characterizing round-off errors in multiplication is the additive white noise model that we used in the statistical analysis of round-off errors in IIR and FIR filters. Although the statistical analysis is performed for rounding, the analysis can be easily modified to apply to truncation in two's-complement arithmetic.

Of particular interest is the analysis of round-off errors in the computation of the DFT via the FFT algorithm. However, we shall first establish a benchmark by determining the roundoff errors in the direct computation of the DFT.

10.5.1 Quantization Errors in the Direct Computation of the DFT

Given a finite-duration sequence $\{x(n)\}$, $0 \leq n \leq N - 1$, the DFT of $\{x(n)\}$ is defined as

$$X(k) = \sum_{n=0}^{N-1} x(n)W_N^{kn} \qquad k = 0, 1, \ldots, N - 1 \qquad (10.5.1)$$

where $W_N = e^{-j2\pi/N}$. We assume that $\{x(n)\}$ is a complex-valued sequence, in general. Consequently, the computation of the product $x(n)W_N^{kn}$ requires four real multiplications. Each multiplication is rounded from $2b$ bits to b bits, and hence there are four quantization errors for each complex-valued multiplication.

In the direct computation of the DFT, there are N complex-valued multiplications for each point in the DFT. Therefore, the total number of real multiplications in the computation of a single point in the DFT is $4N$. Consequently, there are $4N$ quantization errors.

Let us evaluate the variance of the quantization errors in a fixed-point computation of the DFT. First, we make the following assumptions about the statistical properties of the quantization errors.

1. The quantization errors due to rounding are uniformly distributed random variables in the range $(-\Delta/2, \Delta/2)$ where $\Delta = 2^{-b}$.
2. The $4N$ quantization errors are mutually uncorrelated.
3. The $4N$ quantization errors are uncorrelated with the sequence $\{x(n)\}$.

Since each of the quantization errors has a variance

$$\sigma_e^2 = \frac{\Delta^2}{12} = \frac{2^{-2b}}{12} \qquad (10.5.2)$$

the variance of the quantization errors from the $4N$ multiplication is

$$\sigma_q^2 = 4N\sigma_e^2$$
$$= \frac{N}{3} \cdot 2^{-2b} \qquad (10.5.3)$$

Hence the variance of the quantization error is proportional to the size of DFT. Note that when N is a power of 2 (i.e., $N = 2^v$), the variance may be expressed as

$$\sigma_q^2 = \frac{2^{-2(b-v/2)}}{3} \tag{10.5.4}$$

This expression implies that every fourfold increase in the size N of the DFT requires an additional bit in computational precision to offset the additional quantization errors.

To prevent overflow, the input sequence to the DFT requires scaling. Clearly, an upper bound on $|X(k)|$ is

$$|X(k)| \le \sum_{n=0}^{N-1} |x(n)| \tag{10.5.5}$$

If the dynamic range in addition is $(-1, 1)$, then $|X(k)| < 1$ requires that

$$\sum_{n=0}^{N-1} |x(n)| < 1 \tag{10.5.6}$$

If $|x(n)|$ is initially scaled such that $|x(n)| < 1$ for all n, then each point in the sequence can be divided by N to ensure that (10.5.6) is satisfied.

The scaling implied by (10.5.6) is extremely severe. For example, suppose that the signal sequence $\{x(n)\}$ is white and, after scaling, each value $|x(n)|$ of the sequence is uniformly distributed in the range $(-1/N, 1/N)$. Then the variance of the signal sequence is

$$\sigma_x^2 = \frac{(2/N)^2}{12} = \frac{1}{3N^2} \tag{10.5.7}$$

and the variance of the output DFT coefficients $|X(k)|$ is

$$\sigma_X^2 = N\sigma_x^2 \tag{10.5.8}$$

$$= \frac{1}{3N}$$

Thus the signal-to-noise power ratio is

$$\frac{\sigma_X^2}{\sigma_q^2} = \frac{2^{2b}}{N^2} \tag{10.5.9}$$

We observe that the scaling is responsible for reducing the SNR by N and the combination of scaling and quantization errors result in a total reduction that is proportional to N^2. Hence scaling the input sequence $\{x(n)\}$ to satisfy (10.5.6) imposes a severe penalty on the signal-to-noise ratio in the DFT.

EXAMPLE 10.5.1

Use (10.5.9) to determine the number of bits required to compute the DFT of a 1024-point sequence with a SNR of 30 dB.

Solution: The size of the sequence is $N = 2^{10}$. Hence the SNR is

$$10 \log_{10} \frac{\sigma_X^2}{\sigma_q^2} = 10 \log_{10} 2^{2b-20}$$

For an SNR of 30 dB, we have

$$3(2b - 20) = 30$$
$$b = 15 \text{ bits}$$

Note that the 15 bits is the precision for both multiplication and addition.

Instead of scaling the input sequence $\{x(n)\}$, suppose we simply require that $|x(n)| < 1$. Then we must provide a sufficiently large dynamic range for addition such that $|X(k)| < N$. In such a case the variance of the sequence $\{|x(n)|\}$ is $\sigma_x^2 = \frac{1}{3}$, and hence the variance of $|X(k)|$ is

$$\sigma_X^2 = N\sigma_x^2 = \frac{N}{3} \tag{10.5.10}$$

Consequently, the SNR is

$$\frac{\sigma_X^2}{\sigma_q^2} = 2^{2b} \tag{10.5.11}$$

If we repeat the computation in Example 10.5.1, we find that the number of bits required to achieve a SNR of 30 dB is $b = 5$ bits. However, we need an additional 10 bits for the accumulator (the adder) to accommodate the increase in the dynamic range for addition. Although we did not achieve any reduction in the dynamic range for addition, we have managed to reduce the precision in multiplication from 15 bits to 5 bits, which is highly significant.

10.5.2 Quantization Errors in FFT Algorithms

As we have shown, the FFT algorithms require significantly fewer multiplications than the direct computation of the DFT. In view of this we might conclude that the computation of the DFT via an FFT algorithm will result in smaller quantization errors. Unfortunately, that is not the case, as we demonstrate below.

Let us consider the use of fixed-point arithmetic in the computation of a radix-2 FFT algorithm. To be specific, we select the radix-2, decimation-in-time algorithm illustrated in Fig. 10.24 for the case $N = 8$. The results on quantization errors that we obtain for this radix-2 FFT algorithm are typical of the results obtained with other radix-2 and higher radix algorithms.

We observe that each butterfly computation involves one complex-valued multiplication or, equivalently, four real multiplications. We ignore the fact that some butterflies contain a trivial multiplication by ± 1. If we consider the butterflies that affect the computation of any one value of the DFT, we find that, in general, there are $N/2$ in the first stage of the FFT, $N/4$ in the second stage, $N/8$ in the third state, and so on, until the last stage, where there is only one. Consequently, the number of butterflies per output point is

$$2^{\nu-1} + 2^{\nu-2} + \cdots + 2 + 1 = 2^{\nu-1}[1 + (\tfrac{1}{2}) + \cdots + (\tfrac{1}{2})^{\nu-1}] \tag{10.5.12}$$
$$= 2^{\nu}[1 - (\tfrac{1}{2})^{\nu}] = N - 1$$

For example, the butterflies that affect the computation of $X(3)$ in the eight-point FFT algorithm of Fig. 10.24 are illustrated in Fig. 10.25.

The quantization errors introduced in each butterfly propagate to the output. Note that the quantization errors introduced in the first stage propagate through $(\nu - 1)$ stages, those introduced in the second stage propagate through $(\nu - 2)$ stages, and so on. As these quantization errors propagate through a number of subsequent stages, they are phase shifted (phase rotated) by the phase factors W_N^{kn}. These phase rotations do not

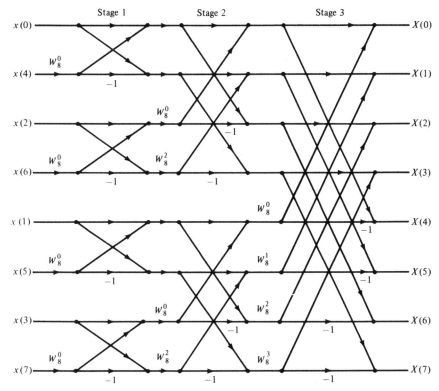

Stage 1   Stage 2   Stage 3

FIGURE 10.24   Decimation-in-time FFT algorithm.

change the statistical properties of the quantization errors and, in particular, the variance of each quantization error remains invariant.

If we assume that the quantization errors in each butterfly are uncorrelated with the errors in other butterflies, then there are $4(N - 1)$ errors that affect the output of each point of the FFT. Consequently, the variance of the total quantization error at the output is

$$\sigma_q^2 = 4(N - 1)\frac{\Delta^2}{12} \approx \frac{N \Delta^2}{3} \tag{10.5.13}$$

where $\Delta = 2^{-b}$. Hence

$$\sigma_q^2 = \frac{N}{3} \cdot 2^{-2b} \tag{10.5.14}$$

This is exactly the same result that we obtained for the direct computation of the DFT.

The result in (10.5.14) should not be surprising. In fact, the FFT algorithm does not reduce the number of multiplications required to compute a single point of the DFT. It does, however, exploit the periodicities in W_N^{kn} and thus reduces the number of multiplications in the computation of the entire block of N points in the DFT.

As in the case of the direct computation of the DFT, we must scale the input sequence to prevent overflow. Recall that if $|x(n)| < 1/N$, $0 \le n \le N - 1$, then $|X(k)| < 1$ for $0 \le k \le N - 1$. Thus overflow is avoided. With this scaling, the relations in (10.5.7), (10.5.8), and (10.5.9) obtained previously for the direct computation of the DFT apply to the FFT algorithm as well. Consequently, the same SNR is obtained for the FFT.

Since the FFT algorithm consists of a sequence of stages, where each stage contains butterflies that involve pairs of points, it is possible to devise a different scaling strategy that is not as severe as dividing each input point by N. This alternative scaling strategy is motivated by the observation that the intermediate values $|X_n(k)|$ in the $n = 1, 2, \ldots$,

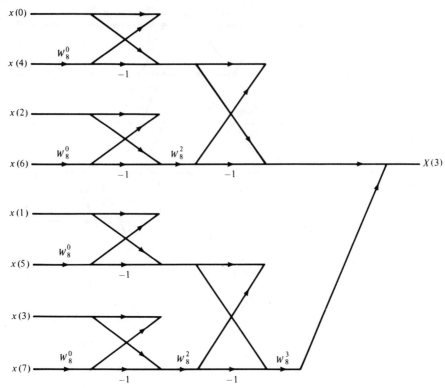

FIGURE 10.25 Butterflies that affect the computation of $X(3)$.

v stages of the FFT algorithm satisfy the conditions (see Problem 10.9)

$$\max[|X_{n+1}(k)|, |X_{n+1}(l)|] \geq \max[|X_n(k)|, |X_n(l)|] \qquad (10.5.15)$$
$$\max[|X_{n+1}(k)|, |X_{n+1}(l)|] \leq 2 \max[|X_n(k)|, |X_n(l)|]$$

In view of these relations, we can distribute the total scaling of $1/N$ into each of the stages of the FFT algorithm. In particular, if $|x(n)| < 1$, we apply a scale factor of $\frac{1}{2}$ in the first stage so that $|x(n)| < \frac{1}{2}$. Then the output of each subsequent stage in the FFT algorithm is scaled by $\frac{1}{2}$, so that after v stages we have achieved an overall scale factor of $(\frac{1}{2})^v = 1/N$. Thus overflow in the computation of the DFT is avoided.

This scaling procedure does not affect the signal level at the output of the FFT algorithm, but it significantly reduces the variance of the quantization errors at the output. Specifically, each factor of $\frac{1}{2}$ reduces the variance of a quantization error term by a factor of $\frac{1}{4}$. Thus the $4(N/2)$ quantization errors introduced in the first stage are reduced in variance by $(\frac{1}{4})^{v-1}$, the $4(N/4)$ quantization errors introduced in the second stage are reduced in variance by $(\frac{1}{4})^{v-2}$, and so on. Consequently, the total variance of the quantization errors at the output of the FFT algorithm is

$$\sigma_q^2 = \frac{\Delta^2}{12} \left\{ 4\left(\frac{N}{2}\right)\left(\frac{1}{4}\right)^{v-1} + 4\left(\frac{N}{4}\right)\left(\frac{1}{4}\right)^{v-2} + 4\left(\frac{N}{8}\right)\left(\frac{1}{4}\right)^{v-3} + \cdots + 4 \right\}$$

$$= \frac{\Delta^2}{3} \left\{ \left(\frac{1}{2}\right)^{v-1} + \left(\frac{1}{2}\right)^{v-2} + \cdots + \frac{1}{2} + 1 \right\}$$

$$= \frac{2\Delta^2}{3} \left[1 - \left(\frac{1}{2}\right)^v \right] \approx \frac{2}{3} \cdot 2^{-2b}$$

$$(10.5.16)$$

where the factor $(\frac{1}{2})^v$ is negligible.

Now we observe that (10.5.16) is no longer proportional to N. On the other hand, the signal has the variance $\sigma_X^2 = 1/3N$, as given in (10.5.8). Hence the SNR is

$$\frac{\sigma_X^2}{\sigma_q^2} = \frac{1}{2N} \cdot 2^{2b}$$
$$= 2^{2b - v - 1} \tag{10.5.17}$$

Thus, by distributing the scaling of $1/N$ uniformly throughout the FFT algorithm, we have achieved an SNR that is inversely proportional to N instead of N^2.

EXAMPLE 10.5.2 _____

Determine the number of bits required to compute an FFT of 1024 points with an SNR of 30 dB when the scaling is distributed as described above.

Solution: The size of the FFT is $N = 2^{10}$. Hence the SNR according to (10.5.17) is

$$10 \log_{10} 2^{2b - v - 1} = 30$$
$$3(2b - 11) = 30$$
$$b = \frac{21}{2} \quad (11 \text{ bits})$$

This may be compared with the 15 bits required if all the scaling is performed in the first stage of the FFT algorithm.

10.6 Summary and References

Finite-word-length effects are an important factor in the implementation of digital signal processing systems. In this chapter we described the effects of a finite word length in digital filtering and in the computation of the DFT. In particular, we considered the following problems dealing with finite-word length effects:

1. Quantization errors in A/D conversion
2. Parameter quantization in digital filters
3. Round-off noise in multiplication
4. Overflow in addition
5. Limit cycles

Of these five types of quantization effects, the first one takes place outside the filter, prior to the internal computations. The remaining four effects are internal to the filter and influence the method by which the system will be implemented. In particular, we demonstrated that high-order systems, especially IIR systems, should be realized by using second-order sections as building blocks. We advocated the use of the direct form II realization, either the conventional or the transposed form.

Effects of round-off errors in fixed-point implementations of FIR and IIR filter structures have been investigated by many researchers. We cite the papers by Gold and Rader (1966), Rader and Gold (1967b), Jackson (1970a,b), Liu (1971), Chan and Rabiner (1973a,b,c), and Oppenheim and Weinstein (1972).

As an alternative to the use of direct form II second-order filters as building blocks for high-order filters, we may use second-order state-variable forms. Such state-variable forms can be optimized with respect to the state transition matrix to minimize round-off errors. The optimization leads to minimum-round-off-noise second-order state-variable filters that are highly robust for implementing both narrowband and wideband filters.

For a treatment of minimum-round-off-noise second-order state-space realizations, the reader may refer to the papers of Mullis and Roberts (1976a,b), Hwang (1977), Jackson et al. (1979), Mills et al. (1981), Bomar (1985), and the book by Roberts and Mullis (1987).

Limit-cycle oscillations occur in IIR filters as a result of quantization effects in fixed-point multiplication and rounding. Investigation of limit cycles in digital filtering and their characteristic behavior is treated in the papers by Parker and Hess (1971), Brubaker and Gowdy (1972), Sandberg and Kaiser (1972), and Jackson (1969, 1979). The latter paper deals with limit cycles in state-space structures. Methods have also been devised to eliminate limit cycles caused by round-off errors. For example, the papers by Barnes and Fam (1977), Fam and Barnes (1979), Chang (1981), Butterweck et al. (1984), and Auer (1987) discuss this problem. Overflow oscillations have been treated in the paper by Ebert et al. (1969).

Effects of round-off errors in the computation of the DFT are treated in detail in the paper by Oppenheim and Weinstein (1972). Our treatment parallels the approach used in this paper.

The effects of parameter quantization has been treated in a number of papers. We cite for reference the work of Rader and Gold (1967b), Knowles and Olcayto (1968), Avenhaus and Schuessler (1970), Herrmann and Schuessler (1970b), Chan and Rabiner (1973c), and Jackson (1976).

Finally, we mention that the lattice and lattice-ladder filter structures, which were introduced in Chapter 7, are known to be robust in fixed-point implementations. For a treatment of these types of filters, the reader is referred to the papers of Gray and Markel (1973), Makhoul (1978), and Morf et al. (1977) and to the book by Markel and Gray (1976).

APPENDIX 10A: RANDOM SIGNALS, CORRELATION FUNCTIONS, AND POWER SPECTRA

In this appendix we provide a brief review of the characterization of random signals in terms of statistical averages expressed in both the time domain and the frequency domain. The reader is assumed to have a background in probability theory and random processes, at the level given in the books of Helstrom (1984) and Peebles (1987).

Random Processes

Many physical phenomena encountered in nature are best characterized in statistical terms. For example, meteorological phenomena such as air temperature and air pressure fluctuate randomly as a function of time. Thermal noise voltages generated in the resistors of electronic devices, such as a radio or television receiver, are also randomly fluctuating phenomena. These are just a few examples of random signals. Such signals are usually modeled as infinite-duration infinite-energy signals.

Suppose that we take the set of waveforms corresponding to the air temperature in different cities around the world. For each city there is a corresponding waveform that is a function of time, as illustrated in Fig. 10A.1. The set of all possible waveforms is called an *ensemble* of time functions or, equivalently, a *random process*. The waveform for the temperature in any particular city is a *single realization* or a *sample function* of the random process.

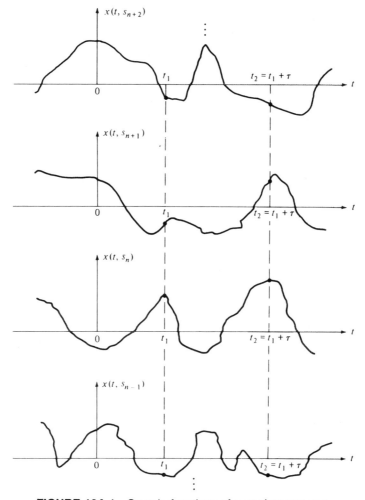

FIGURE 10A.1   Sample functions of a random process.

Similarly, the thermal noise voltage generated in a resistor is a single realization or a sample function of the random process that consists of all noise voltage waveforms generated by the set of all resistors.

The set (ensemble) of all possible noise waveforms of a random process is denoted as $X(t, S)$, where t represents the time index and S represents the set (sample space) of all possible sample functions. A single waveform in the ensemble is denoted by $x(t, s)$. Usually, we drop the variable s (or S) for notational convenience, so that the random process is denoted as $X(t)$ and a single realization is denoted as $x(t)$.

Having defined a random process $X(t)$ as an ensemble of sample functions, let us consider the values of the process for any set of time instants $t_1 < t_2 < \ldots < t_n$, where n is any positive integer. In general, the samples $X_{t_i} \equiv x(t_i)$, $i = 1, 2, \ldots, n$ are n random variables characterized statistically by their joint probability density function (PDF) denoted as $p(X_{t_1}, X_{t_2}, \ldots, X_{t_n})$ for any n.

Stationary Random Processes

Suppose that we have n samples of the random process $X(t)$ at $t = t_i$, $i = 1, 2, \ldots$, n, and another set of n samples displaced in time from the first set by an amount τ. Thus the second set of samples are $X_{t_i+\tau} \equiv X(t_i + \tau)$, $i = 1, 2, \ldots, n$, as shown in Fig. 10A.1. This second set of n random variables are characterized by the joint probability density function $p(X_{t_1+\tau}, \ldots, X_{t_n+\tau})$. The joint PDFs of the two sets of random variables may or may not be identical. When they are identical, that is when

$$P(X_{t_1}, X_{t_2}, \ldots, X_{t_n}) = p(X_{t_1+\tau}, X_{t_2+\tau}, \ldots, X_{t_n+\tau}) \tag{10A.1}$$

for all τ and all n, the random process is said to be *stationary in the strict sense*. In other words, the statistical properties of a stationary random process are invariant to a translation of the time axis. On the other hand, when the joint PDFs are different, the random process is nonstationary.

Statistical (Ensemble) Averages

Let us consider a random process $X(t)$ sampled at time instant $t = t_i$. Thus $X(t_i)$ is a random variable with PDF $p(X_{t_i})$. The *l*th *moment* of the random variable is defined as the *expected value* of $X^l(t_i)$, that is,

$$E(X_{t_i}^l) = \int_{-\infty}^{\infty} X_{t_i}^l p(X_{t_i}) \, dX_{t_i} \tag{10A.2}$$

In general, the value of the *l*th moment will depend on the time instant t_i, if the PDF of X_{t_i} depends on t_i. When the process is stationary, however, $p(X_{t_i+\tau}) = p(X_{t_i})$ for all τ. Hence the PDF is independent of time and, consequently, the *l*th moment is independent of time (a constant).

Next, let us consider the two random variables $X_{t_i} = X(t_i)$, $i = 1, 2$, corresponding to samples of $X(t)$ taken at $t = t_1$ and $t = t_2$. The statistical (ensemble) correlation between X_{t_1} and X_{t_2} is measured by the joint moment

$$E(X_{t_1}X_{t_2}) = \int_{-\infty}^{\infty} \int_{-\infty}^{\infty} X_{t_1}X_{t_2}p(X_{t_1}X_{t_2}) \, dX_1 \, dX_2 \tag{10A.3}$$

Since the joint moment depends on the time instants t_1 and t_2, it is denoted as $\gamma_{xx}(t_1, t_2)$ and is called the *autocorrelation function* of the random process. When the process $X(t)$ is stationary, the joint PDF of the pair (X_{t_1}, X_{t_2}) is identical to the joint PDF of the pair $(X_{t_1+\tau}, X_{t_1+\tau})$ for any arbitrary τ. This implies that the autocorrelation function of $X(t)$ depends on the time difference $t_1 + \tau - t_1 = \tau$. Hence for a stationary real-valued random process the autocorrelation function is

$$\gamma_{xx}(\tau) = E[X_{t_1+\tau}X_{t_1}] \tag{10A.4}$$

On the other hand,

$$\gamma_{xx}(-\tau) = E(X_{t_1-\tau}X_{t_1}) = E(X_{t_i'}X_{t_i'+\tau}) = \gamma_{xx}(\tau) \tag{10A.5}$$

Therefore, $\gamma_{xx}(\tau)$ is an even function. We also note that $\gamma_{xx}(0) = E(X_{t_1}^2)$ is the *average power* of the random process.

There exist nonstationary processes with the property that the mean value of the process is a constant and the autocorrelation function satisfies the property $\gamma_{xx}(t_1, t_2) = \gamma_{xx}(t_1 - t_2)$. Such a process is called *wide-sense stationary*. Clearly, wide-sense stationarity is a less stringent condition than strict-sense stationarity. In our treatment we shall require only that the processes be wide-sense stationary.

Related to the autocorrelation function is the *autocovariance function,* which is defined as

$$c_{xx}(t_1, t_2) = E\{[X_{t_1} - m(t_1)][X_{t_2} - m(t_2)]\} \tag{10A.6}$$
$$= \gamma_{xx}(t_1, t_2) - m(t_1)m(t_2)$$

where $m(t_1) = E(X_{t_1})$ and $m(t_2) = E(X_{t_2})$ are the mean values of X_{t_1} and X_{t_2}, respectively. When the process is stationary,

$$c_{xx}(t_1, t_2) = c_{xx}(t_1 - t_2) = c_{xx}(\tau) = \gamma_{xx}(\tau) - m_x^2 \tag{10A.7}$$

where $\tau = t_1 - t_2$. Furthermore, the variance of the process is $\sigma_x^2 = c_{xx}(0) = \gamma_{xx}(0) - m_x^2$.

Statistical Averages for Joint Random Processes

Let $X(t)$ and $Y(t)$ be two random processes and let $X_{t_i} \equiv X(t_i)$, $i = 1, 2, \ldots, n$ and $Y_{t_j} \equiv Y(t'_j)$, $j = 1, 2, \ldots, m$ represent the random variables at times $t_1 < t_2 < \cdots < t_n$ and $t'_1 < t'_2 < \cdots < t'_m$, respectively. The two sets of random variables are characterized statistically by the joint PDF

$$p(X_{t_1}, X_{t_2}, \ldots, X_{t_n}, Y_{t'_1}, Y_{t'_2}, \ldots, Y_{t'_m})$$

for any set of time instants $\{t_i\}$ and $\{t'_i\}$ and for any positive integer values of m and n.

The *crosscorrelation function* of $X(t)$ and $Y(t)$, denoted as $\gamma_{xy}(t_1, t_2)$ is defined by the joint moment

$$\gamma_{xy}(t_1, t_2) \equiv E(X_{t_1}Y_{t_2}) = \int_{-\infty}^{\infty} \int_{-\infty}^{\infty} X_{t_1}Y_{t_2}p(X_{t_1}, Y_{t_2}) \, dX_{t_1} \, dY_{t_2} \tag{10A.8}$$

and the crosscovariance is

$$c_{xy}(t_1, t_2) = \gamma_{xy}(t_1, t_2) - m_x(t_1)m_y(t_2) \tag{10A.9}$$

When the random processes are jointly and individually stationary, we have $\gamma_{xy}(t_1, t_2) = \gamma_{xy}(t_1 - t_2)$ and $c_{xy}(t_1, t_2) = c_{xy}(t_1 - t_2)$. In this case

$$\gamma_{xy}(-\tau) = E(X_{t_1}Y_{t_1+\tau}) = E(X_{t_i-\tau}Y_{t_i}) = \gamma_{yx}(\tau) \tag{10A.10}$$

The random processes $X(t)$ and $Y(t)$ are said to be statistically independent if and only if

$$p(X_{t_1}, X_{t_2}, \ldots, X_{t_n}, Y_{t'_1}, Y_{t'_2}, \ldots, Y_{t'_m}) = p(X_{t_1}, \ldots, X_{t_n})p(Y_{t'_1}, \ldots, Y_{t'_m})$$

for all choices of t_i, t'_i and for all positive integers n and m. The processes are said to be *uncorrelated* if

$$\gamma_{xy}(t_1, t_2) = E(X_{t_1})E(Y_{t_2}) \tag{10A.11}$$

so that $c_{xy}(t_1, t_2) = 0$.

A complex-valued random process $Z(t)$ is defined as

$$Z(t) = X(t) + jY(t) \tag{10A.12}$$

where $X(t)$ and $Y(t)$ are random processes. The joint PDF of the complex-valued random variables $Z_{t_i} \equiv Z(t_i)$, $i = 1, 2, \ldots$, is given by the joint PDF of the components (X_{t_i}, Y_{t_i}), $i = 1, 2, \ldots, n$. Thus the PDF that characterizes Z_{t_i}, $i = 1, 2, \ldots, n$ is

$$p(X_{t_1}, X_{t_2}, \ldots, X_{t_n}, Y_{t_1}, Y_{t_2}, \ldots, Y_{t_n})$$

A complex-valued random process $Z(t)$ is encountered in the representation of the in-phase and quadrature components of the lowpass equivalent of a narrowband random signal or noise. An important characteristic of such a process is its autocorrelation function, which is defined as

$$
\begin{aligned}
\gamma_{zz}(t_1, t_2) &= E(Z_{t_1} Z_{t_2}^*) \\
&= E[(X_{t_1} + jY_{t_1})(X_{t_2} - jY_{t_2})] \\
&= \gamma_{xx}(t_1, t_2) + \gamma_{yy}(t_1, t_2) + j[\gamma_{yx}(t_1, t_2) - \gamma_{xy}(t_1, t_2)]
\end{aligned}
\tag{10A.13}
$$

When the random processes $X(t)$ and $Y(t)$ are jointly and individually stationary, the autocorrelation function of $Z(t)$ becomes

$$
\gamma_{zz}(t_1, t_2) = \gamma_{zz}(t_1 - t_2) = \gamma_{zz}(\tau)
$$

where $\tau = t_1 - t_2$. The complex conjugate of (10A.13) is

$$
\gamma_{zz}^*(\tau) = E(Z_{t_1}^* Z_{t_1 - \tau}) = \gamma_{zz}(-\tau)
\tag{10A.14}
$$

Now, suppose that $Z(t) = X(t) + jY(t)$ and $W(t) = U(t) + jV(t)$ are two complex-valued random processes. Their crosscorrelation function is defined as

$$
\begin{aligned}
\gamma_{zw}(t_1, t_2) &= E(Z_{t_1} W_{t_2}^*) \\
&= E[(X_{t_1} + jY_{t_1})(U_{t_2} - jV_{t_2})] \\
&= \gamma_{xu}(t_1, t_2) + \gamma_{yv}(t_1, t_2) + j[\gamma_{yu}(t_1, t_2) - \gamma_{xv}(t_1, t_2)]
\end{aligned}
\tag{10A.15}
$$

When $X(t)$, $Y(t)$, $U(t)$, and $V(t)$ are pairwise stationary, the crosscorrelation functions in (10A.15) become functions of the time difference $\tau = t_1 - t_2$. In addition, we have

$$
\gamma_{zw}^*(\tau) = E(Z_{t_1}^* W_{t_1 - \tau}) = E(Z_{t_1 + \tau}^* W_{t_1}) = \gamma_{wz}(-\tau)
\tag{10A.16}
$$

Power Density Spectrum

A stationary random process is an infinite-energy signal and hence its Fourier transform does not exist. The spectral characteristic of a random process is obtained according to the Wiener–Khinchine theorem, by computing the Fourier transform of the autocorrelation function. That is, the distribution of power with frequency is given by the function

$$
\Gamma_{xx}(F) = \int_{-\infty}^{\infty} \gamma_{xx}(\tau) e^{-j2\pi F \tau} \, d\tau
\tag{10A.17}
$$

The inverse Fourier transform is given as

$$
\gamma_{xx}(\tau) = \int_{-\infty}^{\infty} \Gamma_{xx}(F) e^{j2\pi F \tau} \, dF
\tag{10A.18}
$$

We observe that

$$
\begin{aligned}
\gamma_{xx}(0) &= \int_{-\infty}^{\infty} \Gamma_{xx}(F) \, dF \\
&= E(X_t^2) \geq 0
\end{aligned}
\tag{10A.19}
$$

Since $E(X_t^2) = \gamma_{xx}(0)$ represents the average power of the random process, which is in the area under $\Gamma_{xx}(F)$, it follows that $\Gamma_{xx}(F)$ is the distribution of power as a function of frequency. For this reason, $\Gamma_{xx}(F)$ is called the *power density spectrum* of the random process.

If the random process is real, $\gamma_{xx}(\tau)$ is real and even and hence $\Gamma_{xx}(F)$ is real and even. If the random process is complex valued, $\gamma_{xx}(\tau) = \gamma_{xx}^*(-\tau)$ and, hence

$$\Gamma_{xx}^*(F) = \int_{-\infty}^{\infty} \gamma_{xx}^*(\tau)e^{j2\pi F\tau} \, d\tau = \int_{-\infty}^{\infty} \gamma_{xx}^*(-\tau)e^{-j2\pi F\tau} \, d\tau$$

$$= \int_{-\infty}^{\infty} \gamma_{xx}(\tau)e^{-j2\pi F\tau} \, d\tau = \Gamma_{xx}(F)$$

Therefore, $\Gamma_{xx}(F)$ is always real.

The definition of the power density spectrum can be extended to two jointly stationary random processes $X(t)$ and $Y(t)$, which have a crosscorrelation function $\gamma_{xy}(\tau)$. The Fourier transform of $\gamma_{xy}(\tau)$ is

$$\Gamma_{xy}(F) = \int_{-\infty}^{\infty} \gamma_{xy}(\tau)e^{-j2\pi F\tau} \, d\tau \tag{10A.20}$$

which is called the *cross-power density spectrum*. It is easily shown that $\Gamma_{xy}^*(F) = \Gamma_{yx}(-F)$. For real random processes, the condition is $\Gamma_{yx}(F) = \Gamma_{xy}(-F)$.

Discrete-Time Random Signals

The characterization of continuous-time random signals given above can be easily carried over to discrete-time signals. Such signals are usually obtained by uniformly sampling a continuous-time random process.

A discrete-time random process $X(n)$ consists of an ensemble of sample sequences $x(n)$. The statistical properties of $X(n)$ are similar to the characterization of $X(t)$, with the restriction that n is now an integer (time) variable. To be specific, we state the form for the important moments that we will use in this text.

The lth moment of $X(n)$ is defined as

$$E(X_n^l) = \int_{-\infty}^{\infty} X_n^l p(X_n) \, dX_n \tag{10A.21}$$

and the autocorrelation sequence is

$$\gamma_{xx}(n, k) = E(X_n X_k) = \int_{-\infty}^{\infty} \int_{-\infty}^{\infty} X_n X_k p(X_n, X_k) \, dX_n \, dX_k \tag{10A.22}$$

Similarly, the autocovariance is

$$c_{xx}(n, k) = \gamma_{xx}(n, k) - E(X_n)E(X_k) \tag{10A.23}$$

For a stationary process, we have the special forms ($m = n - k$)

$$\gamma_{xx}(n - k) = \gamma_{xx}(m) \tag{10A.24}$$
$$c_{xx}(n - k) = c_{xx}(m) = \gamma_{xx}(m) - m_x^2$$

where $m_x = E(X_n)$ is the mean of the random process. The variance is defined as $\sigma^2 = c_{xx}(0) = \gamma_{xx}(0) - m_x^2$.

For a complex-valued stationary process $Z(n) = X(n) + jY(n)$, we have

$$\gamma_{zz}(m) = \gamma_{xx}(m) + \gamma_{yy}(m) + j[\gamma_{yx}(m) - \gamma_{xy}(m)] \tag{10A.25}$$

and the crosscorrelation sequence of two complex-valued stationary sequences is

$$\gamma_{zw}(m) = \gamma_{xu}(m) + \gamma_{yv}(m) + j[\gamma_{yu}(m) - \gamma_{xv}(m)] \tag{10A.26}$$

As in the case of a continuous-time random process, a discrete-time random process has infinite energy but a finite average power, which is given as

$$E(X_n^2) = \gamma_{xx}(0) \tag{10A.27}$$

By use of the Wiener–Khinchine theorem, we obtain the power density spectrum of the discrete-time random process by computing the Fourier transform of the autocorrelation sequence $\gamma_{xx}(m)$, that is,

$$\Gamma_{xx}(f) = \sum_{m=-\infty}^{\infty} \gamma_{xx}(m)e^{-j2\pi fm} \tag{10A.28}$$

The inverse transform relationship is

$$\gamma_{xx}(m) = \int_{-1/2}^{1/2} \Gamma_{xx}(f)e^{j2\pi fm}\,df \tag{10A.29}$$

We observe that the average power is

$$\gamma_{xx}(0) = \int_{-1/2}^{1/2} \Gamma_{xx}(f)\,df \tag{10A.30}$$

so that $\Gamma_{xx}(f)$ is the distribution of power as a function of frequency, that is, $\Gamma_{xx}(f)$ is the power density spectrum of the random process $X(n)$. The properties we have stated for $\Gamma_{xx}(F)$ also hold for $\Gamma_{xx}(f)$.

Time Averages for a Discrete-Time Random Process

Although we have characterized a random process in terms of statistical averages, such as the mean and the autocorrelation sequence, in practice we usually have available a single realization of the random process. Let us consider the problem of obtaining the averages of the random process from a single realization. To accomplish this, the random process must be *ergodic*.

By definition, a random process $X(n)$ is ergodic if, with probability 1, all the statistical averages can be determined from a single sample function of the process. In effect, the random process is ergodic if time averages obtained from a single realization are equal to the statistical (ensemble) averages. Under this condition we can attempt to estimate the ensemble averages by use of time averages from a single realization.

To illustrate this point, let us consider the estimation of the mean and the autocorrelation of the random process from a single realization $x(n)$. Since we are interested only in these two moments, we will define engodicity with respect to these parameters. For additional details on the requirements for mean ergodicity and autocorrelation ergodicity which are given below, the reader is referred to the book of Papoulis (1984).

Mean-Ergodic Process

Given a stationary random process $X(n)$ with mean

$$m_x = E(X_n)$$

let us form the *time average*

$$\hat{m}_x = \frac{1}{2N+1} \sum_{n=-N}^{N} x(n) \tag{10A.31}$$

In general, we view $\hat{m}_x$ in (10A.31) as an estimate of the statistical mean whose value will vary with the different realizations of the random process. Hence $\hat{m}_x$ is a random variable with a PDF $p(\hat{m}_x)$. Let us compute the expected value of $\hat{m}_x$ over all possible realizations of $X(n)$. Since the summation and the expectation are linear operations we may interchange them, so that

$$E(\hat{m}_x) = \frac{1}{2N + 1} \sum_{n=-N}^{N} E[x(n)] = \frac{1}{2N + 1} \sum_{n=-N}^{N} m_x = m_x \qquad (10A.32)$$

Since the mean value of the estimate is equal to the statistical mean, we say that the estimate $\hat{m}_x$ is *unbiased*.

Next, we compute the variance of $\hat{m}_x$. We have

$$\text{var } (\hat{m}_x) = E(|\hat{m}_x|^2) - |m_x|^2$$

But

$$E(|\hat{m}_x|^2) = \frac{1}{(2N + 1)^2} \sum_{n=-N}^{N} \sum_{k=-N}^{N} E[x^*(n)x(k)]$$

$$= \frac{1}{(2N + 1)^2} \sum_{n=-N}^{N} \sum_{k=-N}^{N} \gamma_{xx}(k - n)$$

$$= \frac{1}{2N + 1} \sum_{m=-2N}^{2N} \left(1 - \frac{|m|}{2N}\right) \gamma_{xx}(m)$$

Therefore,

$$\text{var } (\hat{m}_x) = \frac{1}{2N + 1} \sum_{m=-2N}^{2N} \left(1 - \frac{|m|}{2N}\right) \gamma_{xx} - |m_x|^2$$

$$= \frac{1}{2N + 1} \sum_{m=-2N}^{2N} \left(1 - \frac{|m|}{2N}\right) c_{xx}(m) \qquad (10A.33)$$

If $\text{var } (m_x) \to 0$ as $N \to \infty$, the estimate converges with probability 1 to the statistical mean m_x. Therefore, the process $X(n)$ is mean ergodic if

$$\lim_{N \to \infty} \frac{1}{2N + 1} \sum_{m=-2N}^{2N} \left(1 - \frac{|m|}{2N}\right) c_{xx}(m) = 0 \qquad (10A.34)$$

Under this condition, the estimate $\hat{m}_x$ in the limit as $N \to \infty$ becomes equal to the statistical mean, that is,

$$m_x = \lim_{N \to \infty} \frac{1}{2N + 1} \sum_{n=-N}^{N} x(n) \qquad (10A.35)$$

Thus the time-averaged mean, in the limit as $N \to \infty$, is equal to the ensemble mean.

A sufficient condition for (10A.34) to hold is if

$$\sum_{m=-\infty}^{\infty} |c_{xx}(m)| < \infty \qquad (10A.36)$$

which implies that $c_{xx}(m) \to 0$ as $m \to \infty$. This condition holds for most zero-mean processes encountered in the physical world.

Correlation-Ergodic Processes

Now, let us consider the estimate of the autocorrelation $\gamma_{xx}(m)$ from a single realization of the process. Following our previous notation, we denote the estimate (for a complex-valued signal, in general) as

$$r_{xx}(m) = \frac{1}{2N + 1} \sum_{n=-N}^{N} x^*(n)x(n + m) \tag{10A.37}$$

Again, we regard $r_{xx}(m)$ as a random variable for any given lag m, since it is a function of the particular realization. The expected value (mean value over all realizations) is

$$E[r_{xx}(m)] = \frac{1}{2N + 1} \sum_{n=-N}^{N} E[x^*(n)x(n + m)] \tag{10A.38}$$

$$= \frac{1}{2N + 1} \sum_{n=-N}^{N} \gamma_{xx}(m) = \gamma_{xx}(m)$$

Therefore, the expected value of time-average autocorrelation is equal to the statistical average. Hence we have an unbiased estimate of $\gamma_{xx}(m)$.

To determine the variance of the estimate $r_{xx}(m)$, we compute the expected value of $|r_{xx}(m)|^2$ and subtract the square of the mean value. Thus

$$\text{var } [r_{xx}(m)] = E[|r_{xx}(m)|^2] - |\gamma_{xx}(m)|^2 \tag{10A.39}$$

But

$$E[|r_{xx}(m)|] = \frac{1}{(2N + 1)^2} \sum_{n=-N}^{N} \sum_{k=-N}^{N} E[x^*(n)x(n + m)x(k)x^*(k + m)] \tag{10A.40}$$

The expected value of the term $x^*(n)x(n + m)x(k)x^*(k + m)$ is just the autocorrelation sequence of a random process defined as

$$v_m(n) = x^*(n)x(n + m)$$

Hence (10A.40) may be expressed as

$$E[|r_{xx}(m)|^2] = \frac{1}{(2N + 1)^2} \sum_{n=-N}^{N} \sum_{k=-N}^{N} \gamma_{vv}^{(m)}(n - k) \tag{10A.41}$$

$$= \frac{1}{2N + 1} \sum_{n=-2N}^{2N} \left(1 - \frac{|n|}{2N}\right) \gamma_{vv}^{(m)}(n)$$

and the variance is

$$\text{var } [r_{xx}(m)] = \frac{1}{2N + 1} \sum_{n=-2N}^{2N} \left(1 - \frac{|n|}{2N}\right) \gamma_{vv}^{(m)}(n) - |\gamma_{xx}(m)|^2 \tag{10A.42}$$

If var $[r_{xx}(m)] \to 0$ as $N \to \infty$, the estimate $r_{xx}(m)$ converges with probability 1 to the statistical autocorrelation $\gamma_{xx}(m)$. Under these conditions, the process is *correlation ergodic* and the time-average correlation is identical to the statistical average, that is,

$$\lim_{N \to \infty} \frac{1}{2N + 1} \sum_{n=-\infty}^{\infty} x^*(n)x(n + m) = \gamma_{xx}(m) \tag{10A.43}$$

In our treatment of random signals in this and the next two chapters, we assume that the random processes are mean ergodic and correlation ergordic, so that we can deal with time averages of the mean and autocorrelation obtained from a single realization of the process.

Response of Linear Systems to Random Input Signals

Let us consider a discrete-time linear time-invariant system with unit sample response $h(n)$ and frequency response $H(f)$. For this development we assume that $h(n)$ is real. Let

$x(n)$ be a sample function of a stationary random process $X(n)$ that excites the system and let $y(n)$ denote the response of the system to $x(n)$.

From the convolution summation that relates the output to the input we have

$$y(n) = \sum_{k=-\infty}^{\infty} h(k)x(n-k) \tag{10A.44}$$

Since $x(n)$ is a random input signal, the output is also a random sequence. In other words, for each sample sequence $x(n)$ of the process $X(n)$ there is a corresponding sample sequence $y(n)$ of the output random process $Y(n)$. We wish to relate the statistical characteristics of the output random process $Y(n)$ to the statistical characterization of the input process and the characteristics of the system.

The expected value of the output $y(n)$ is

$$m_y \equiv E[y(n)] = E\left[\sum_{k=-\infty}^{\infty} h(k)x(n-k)\right]$$

$$= \sum_{k=-\infty}^{\infty} h(k)E[x(n-k)]$$

$$m_y = m_x \sum_{k=-\infty}^{\infty} h(k) \tag{10A.45}$$

From the Fourier transform relationship

$$H(f) = \sum_{k=-\infty}^{\infty} h(k)e^{-2\pi fk} \tag{10A.46}$$

we have

$$H(0) = \sum_{k=-\infty}^{\infty} h(k) \tag{10A.47}$$

which is the dc gain of the system. The relationship in (10A.47) allows us to express the mean value in (10A.45) as

$$m_y = m_x H(0) \tag{10A.48}$$

The autocorrelation sequence for the output random process is

$$\gamma_{yy}(m) = E[y^*(n)y(n+m)]$$

$$= E\left[\sum_{k=-\infty}^{\infty} h(k)x^*(n-k) \sum_{j=-\infty}^{\infty} h(j)x(n+m-j)\right]$$

$$= \sum_{k=-\infty}^{\infty} \sum_{j=-\infty}^{\infty} h(k)h(j)E[x^*(n-k)x(n+m-j)]$$

$$= \sum_{k=-\infty}^{\infty} \sum_{j=-\infty}^{\infty} h(k)h(j)\gamma_{xx}(k-j+m) \tag{10A.49}$$

This is the general form for the autocorrelation of the output in terms of the autocorrelation of the input and the impulse response of the system.

A special form of (10A.49) is obtained when the input random process is white, that is, when $m_x = 0$ and

$$\gamma_{xx}(m) = \sigma_x^2 \delta(m) \tag{10A.50}$$

where $\sigma_x^2 \equiv \gamma_{xx}(0)$ is the input signal power. Then (10A.49) reduces to

$$\gamma_{yy}(m) = \sigma_x^2 \sum_{k=-\infty}^{\infty} h(k)h(k+m) \tag{10A.51}$$

Under this condition the output process has the average power

$$\gamma_{yy}(0) = \sigma_x^2 \sum_{n=-\infty}^{\infty} h^2(n) = \sigma_x^2 \int_{-1/2}^{1/2} |H(f)|^2 \, df \tag{10A.52}$$

where we have applied Parseval's theorem.

The relationship in (10A.49) can be transformed into the frequency domain by determining the power density spectrum of $\gamma_{yy}(m)$. We have

$$\begin{aligned}
\Gamma_{yy}(f) &= \sum_{m=-\infty}^{\infty} \gamma_{yy}(m) e^{-j2\pi fm} \\
&= \sum_{m=-\infty}^{\infty} \left[\sum_{k=-\infty}^{\infty} \sum_{l=-\infty}^{\infty} h(k)h(l)\gamma_{xx}(h-l+m) \right] e^{-j2\pi fm} \\
&= \sum_{k=-\infty}^{\infty} \sum_{l=-\infty}^{\infty} h(k)h(l) \left[\sum_{m=-\infty}^{\infty} \gamma_{xx}(k-l+m)e^{-j2\pi fm} \right] \tag{10A.53} \\
&= \Gamma_{xx}(f) \left[\sum_{k=-\infty}^{\infty} h(k)e^{j2\pi fk} \right] \left[\sum_{l=-\infty}^{\infty} h(l)e^{-j2\pi fl} \right] \\
&= |H(f)|^2 \Gamma_{xx}(f)
\end{aligned}$$

This is the desired relationship for the power density spectrum of the output process in terms of the power density spectrum of the input process and the frequency response of the system.

The equivalent expression for continuous-time systems with random inputs is

$$\Gamma_{yy}(F) = |H(F)|^2 \Gamma_{xx}(F) \tag{10A.54}$$

where the power density spectra $\Gamma_{yy}(F)$ and $\Gamma_{xx}(F)$ are the Fourier transforms of the autocorrelation functions $\gamma_{yy}(\tau)$ and $\gamma_{xx}(\tau)$, respectively, and $H(F)$ is the frequency response of the system, which is related to the impulse response by the Fourier transform, that is,

$$H(F) = \int_{-\infty}^{\infty} h(t)e^{-j2\pi Ft} \, dt \tag{10A.55}$$

As a final exercise, we determine the crosscorrelation of the output $y(n)$ with the input signal $x(n)$. If we multiply both sides of (10A.44) by $x^*(n-m)$ and take the expected value, we obtain

$$\begin{aligned}
E[y(n)x^*(n-m)] &= E\left[\sum_{k=-\infty}^{\infty} h(k)x^*(n-m)x(n-k) \right] \\
\gamma_{yx}(m) &= \sum_{k=-\infty}^{\infty} h(k)E[x^*(n-m)x(n-k)] \\
&= \sum_{k=-\infty}^{\infty} h(k)\gamma_{xx}(m-k) \tag{10A.56}
\end{aligned}$$

Since (10A.56) has the form of a convolution, the frequency-domain equivalent expression is

$$\Gamma_{yx}(f) = H(f)\Gamma_{xx}(f) \tag{10A.57}$$

In the special case where $x(n)$ is white noise, (10A.57) reduces to

$$\Gamma_{yx}(f) = \sigma_x^2 H(f) \tag{10A.58}$$

where σ_x^2 is the input noise power. This result forms the basis for the system identification procedure discussed in Chapter 6. Simply, it means that an unknown system with frequency response $H(f)$ can be identified by exciting the input with white noise, cross-correlating the input sequence with the output sequence to obtain $\gamma_{yx}(m)$, and finally, computing the Fourier transform of $\gamma_{yx}(m)$. The result of these computations is proportional to $H(f)$. In Chapter 6 we used the method of least squares to obtain basically the same result from time averages.

PROBLEMS

10.1 Let x be a number such that $|x| < 1$. The $(b + 1)$-bit binary representation of the number is $x = (s.xxx \ldots x)_2$ where $s = 0$ or 1 is the sign bit and the remaining b bits correspond to the magnitude of the number.

(a) Show that the one's complement of this number is given by

$$x_1 = \begin{cases} |x| & x > 0 \\ 2 - |x| - 2^{-b} & x \le 0 \end{cases}$$

(b) Show that the two's complement of the number is given by

$$x_2 = \begin{cases} |x| & x \ge 0 \\ 2 - |x| & x < 0 \end{cases}$$

10.2 Let x_1 and x_2 be $(b + 1)$-bit binary numbers with magnitude less than 1. To compute the sum of x_1 and x_2 using two's-complement representation we treat them as $(b + 1)$-bit unsigned numbers, we perform addition modulo-2 and ignore any carry after the sign bit.

(a) Show that if the sum of two numbers with the same sign has the opposite sign, this corresponds to overflow.

(b) Show that when we compute the sum of several numbers using two's-complement representation, the result will be correct, even if there are overflows, if the correct sum is less than 1 in magnitude. Illustrate this argument by constructing a simple example with three numbers.

10.3 Consider the system described by the difference equation

$$y(n) = ay(n - 1) - ax(n) + x(n - 1)$$

(a) Show that it is all-pass.
(b) Obtain the direct form II realization of the system
(c) If you quantize the coefficients of the system in part (b), is it still all-pass?
(d) Obtain a realization by rewriting the difference equation as

$$y(n) = a[y(n - 1) - x(n)] + x(n - 1)$$

(e) If you quantize the coefficients of the system in part (d), is it still all-pass?

10.4 Consider the system

$$y(n) = \tfrac{1}{2}y(n - 1) + x(n)$$

(a) Compute its response to the input $x(n) = (\tfrac{1}{4})^n u(n)$ assuming infinite-precision arithmetic.

(b) Compute the response of the system $y(n)$, $0 \leq n \leq 5$ to the same input, assuming finite-precision sign-and-magnitude fractional arithmetic with five bits (i.e., the sign bit plus four fractional bits). The quantization is performed by truncation.

(c) Compare the results obtained in parts (a) and (b).

10.5 The input to the system

$$y(n) = 0.999y(n - 1) + x(n)$$

is quantized to $b = 8$ bits. What is the power produced by the quantization noise at the output of the filter?

10.6 Consider the system

$$y(n) = 0.875y(n - 1) - 0.125y(n - 2) + x(n)$$

(a) Compute its poles and design the cascade realization of the system.

(b) Quantize the coefficients of the system using truncation, maintaining a sign bit plus three other bits. Determine the poles of the resulting system.

(c) Repeat part (b) for the same precision using rounding.

(d) Compare the poles obtained in parts (b) and (c) with those in part (a). Which realization is better? Sketch the frequency responses of the systems in parts (a), (b), and (c).

10.7 Consider the system

$$H(z) = \frac{1 - \frac{1}{2}z^{-1}}{(1 - \frac{1}{4}z^{-1})(1 + \frac{1}{4}z^{-1})}$$

(a) Draw all possible realizations of the system.

(b) Suppose now that we implement the filter with fixed-point sign-and-magnitude fractional arithmetic using $(b + 1)$ bits (one bit is used for the sign). Each resulting product is rounded into b bits. Determine the variance of the round-off noise created by the multipliers at the output of each one of the realizations in part (a).

10.8 Repeat the analysis carried out in Section 10.5.2 for the decimation-in-frequency radix-2 FFT algorithm.

10.9 The basic butterfly in the radix-2 decimation-in-time FFT algorithm is

$$X_{n+1}(k) = X_n(k) + W_N^m X_n(l)$$
$$X_{n+1}(l) = X_n(k) - W_N^m X_n(l)$$

(a) If we require that $|X_n(k)| < \frac{1}{2}$ and $|X_n(l)| < \frac{1}{2}$, show that

$$|\text{Re } [X_{n+1}(k)]| < 1 \qquad |\text{Re } [X_{n+1}(l)]| < 1$$
$$|\text{Im } [X_{n+1}(k)]| < 1 \qquad |\text{Im } [X_{n+1}(l)]| < 1$$

Thus overflow does not occur.

(b) Prove that

$$\max \left[|X_{n+1}(k)|, |X_{n+1}(l)| \right] \geq \max \left[|X_n(k)|, |X_n(l)| \right]$$
$$\max \left[|X_{n+1}(k)|, |X_{n+1}(l)| \right] \leq 2 \max \left[|X_n(k)|, |X_n(l)| \right]$$

10.10 The first-order filter shown in Fig. P10.10 is implemented in four-bit (including sign) fixed-point two's-complement fractional arithmetic. Products are rounded to four-bit representation. Using the input $x(n) = 0.10 \, \delta(n)$, determine:

(a) The first five outputs if $\alpha = 0.5$. Does the filter go into a limit cycle?

(b) The first five outputs if $\alpha = 0.75$. Does the filter go into a limit cycle?

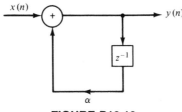

x(n) ——— + ——— y(n)

z^{-1}

α

FIGURE P10.10

10.11 The digital system shown in Fig. P10.11 uses a six-bit (including sign) fixed-point two's-complement A/D converter with rounding, and the filter $H(z)$ is implemented using eight-bit (including sign) fixed-point two's-complement fractional arithmetic with rounding. The input $x(t)$ is a zero-mean uniformly distributed random process having autocorrelation $\gamma_{xx}(\tau) = 3 \, \delta(\tau)$. Assume that the A/D converter can handle input values up to ± 1.0 without overflow.

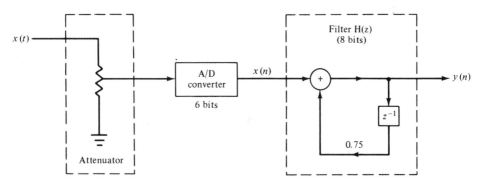

FIGURE P10.11

(a) What value of attenuation should be applied prior to the A/D converter to assure that it does not overflow?

(b) With the attenuation above, what is the signal-to-quantization noise ratio (SNR) at the A/D converter output?

(c) The six-bit A/D samples can be left-justified, right-justified, or centered in the eight-bit word used as the input to the digital filter. What is the correct strategy to use for maximum SNR at the filter output without overflow?

(d) What is the SNR at the output of the filter due to all quantization noise sources?

10.12 Shown in Fig. P10.12 is the coupled-form implementation of a two-pole filter with poles at $x = re^{\pm j\theta}$. There are four real multiplications per output point. Let $e_i(n)$, $i = 1, 2, 3, 4$ represent the round-off noise in a fixed-point implementation of the filter. Assume that the noise sources are zero-mean mutually uncorrelated stationary white noise sequences. For each n the probability density function $p(e)$ is uniform in the range $-\Delta/2 \le e \le \Delta/2$, where $\Delta = 2^{-b}$.

(a) Write the two coupled difference equations for $y(n)$ and $v(n)$, including the noise sources and the input sequence $x(n)$.

(b) From these two difference equations, show that the filter system functions $H_1(z)$ and $H_2(z)$ between the input noise terms $e_1(n) + e_2(n)$ and $e_3(n) + e_4(n)$ and the output $y(n)$ are:

$e_4(n)$ and the output $y(n)$ are:

$$H_1(z) = \frac{r \sin \theta z^{-1}}{1 - 2r \cos \theta z^{-1} + r^2 z^{-2}}$$

$$H_2(z) = \frac{1 - r \cos \theta z^{-1}}{1 - 2r \cos \theta z^{-1} + r^2 z^{-2}}$$

We know that

$$H(z) = \frac{1}{1 - 2r \cos \theta z^{-1} + r^2 z^{-2}} \Rightarrow h(n) = \frac{1}{\sin \theta} r^n \sin (n + 1)\theta u(n)$$

Determine $h_1(n)$ and $h_2(n)$.

(c) Determine a closed-form expression for the variance of the total noise from $e_i(n)$, $i = 1, 2, 3, 4$ at the output of the filter.

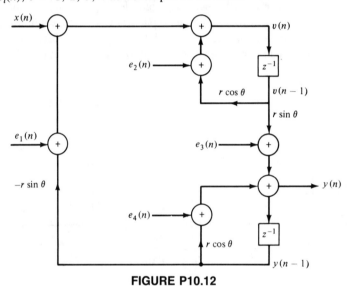

FIGURE P10.12

COMPUTER EXPERIMENT

10.13 Design a linear-phase FIR filter with a transition band from 0.2 to 0.3 rad, pass-band ripple 0.5 dB, and stop-band attenuation -30 dB, using the Remez exchange algorithm.

(a) Compute and sketch its magnitude and phase response.

(b) Quantize the filter coefficients using sign-and-magnitude representation with rounding at 16, 12, 8, 6, and 4 bits (including the sign bit), and sketch the magnitude and phase responses of the corresponding filters.

(c) Comment on the results obtained in parts (a) and (b).

CHAPTER 11

Power Spectrum Estimation

Spectral analysis for deterministic signals was introduced in Chapter 4 as a means of characterizing signals in the frequency domain. We recall that the basic mathematical tools for performing spectral analysis are the Fourier series, which is particularly suitable for characterizing periodic signals, and the Fourier transform, which was introduced to characterize finite-energy signals in the frequency domain.

In this chapter we are concerned with the estimation of the spectral characteristics of signals that are characterized as random processes. Many of the phenomena that occur in nature are best characterized statistically in terms of averages. For example, meteorological phenomena such as the fluctuations in air temperature and pressure are best characterized statistically as random processes. Thermal noise voltages generated in resistors and electronic devices are additional examples of physical signals that are well modeled as random processes.

Due to the random fluctuations in such signals, we cannot simply apply directly the Fourier analysis methods developed in Chapter 4. Instead, we must adopt a statistical viewpoint, which deals with the average characteristics of random signals, as described in Appendix 10A. In particular, the autocorrelation function of a random process is the appropriate statistical average that we will be concerned with for characterizing random signals in the time domain, and the Fourier transform of the autocorrelation function, which yields the power density spectrum, provides the transformation from the time domain to the frequency domain.

Power spectrum estimation methods have a relatively long history. For a historical perspective, the reader is referred to the paper by Robinson (1982) and the book by Marple (1987). Our treatment of this subject covers the classical power spectrum estimation methods based on the periodogram originally introduced by Schuster (1898), and the modern model-based or parametric methods that originated with the work of Yule (1927). These methods were subsequently developed and applied by Walker (1931), Bartlett (1948), Parzen (1957), Blackman and Tukey (1958), Burg (1967), and others. We also describe the method of Capon (1969) and the harmonic decomposition method of Pisarenko (1973).

11.1 Estimation of Spectra from Finite-Duration Observations of Signals

The basic problem that we consider in this chapter is the estimation of the power density spectrum of a signal from observation of the signal over a finite time interval. As we will see, the finite record length of the data sequence is a major limitation on the quality of the power spectrum estimate. When dealing with signals that are statistically stationary, the longer the data record, the better the estimate that can be extracted from the data. On the other hand, if the signal statistics are nonstationary, we cannot select an arbitrarily long data record to estimate the spectrum. In such a case, the length of the data record that we select is determined by the rapidity of the time variations in the signal statistics. Ultimately, our goal is to select as short a data record as possible that will allow us to resolve spectral characteristics of different signal components, contained in the data record, that have closely spaced spectra.

One of the problems that we encounter with classical power spectrum estimation methods based on a finite-length data record is distortion of the spectrum that we are attempting to estimate. This problem occurs in both the computation of the spectrum for a deterministic signal and the estimation of the power spectrum of a random signal. Since it is easier to observe the effect of the finite length of the data record on a deterministic signal, we treat this case first. Thereafter, we consider only random signals and the estimation of their power spectra.

11.1.1 Computation of the Energy Density Spectrum

Let us consider the computation of the spectrum of a deterministic signal from a finite sequence of data. The sequence $x(n)$ is usually the result of sampling a continuous-time signal $x_a(t)$ at some uniform sampling rate F_s. Our objective is to obtain an estimate of the true spectrum from a finite-duration sequence $x(n)$.

Recall that if $x(t)$ is a finite-energy signal, that is,

$$E = \int_{-\infty}^{\infty} |x_a(t)|^2 \, dt < \infty$$

then its Fourier transform exists and is given as

$$X_a(F) = \int_{-\infty}^{\infty} x_a(t) e^{-j2\pi Ft} \, dt$$

From Parseval's theorem we have

$$E = \int_{-\infty}^{\infty} |x_a(t)|^2 \, dt = \int_{-\infty}^{\infty} |X_a(F)|^2 \, dF \qquad (11.1.1)$$

The quantity $|X_a(F)|^2$ represents the distribution of signal energy as a function of frequency, and hence it is called the energy density spectrum of the signal, that is,

$$S_{xx}(F) = |X_a(F)|^2 \qquad (11.1.2)$$

as described in Chapter 4. Thus the total energy in the signal is simply the integral of $S_{xx}(F)$ over all F [i.e., the total area under $S_{xx}(F)$].

It is also interesting to note that $S_{xx}(F)$ may be viewed as the Fourier transform of another function, $R_{xx}(\tau)$, called the *autocorrelation function* of the finite-energy signal $x_a(t)$, defined as

$$R_{xx}(\tau) = \int_{-\infty}^{\infty} x_a^*(t) x_a(t + \tau) \, dt \qquad (11.1.3)$$

Indeed, it easily follows that

$$\int_{-\infty}^{\infty} R_{xx}(\tau) e^{-j2\pi F\tau} \, d\tau = S_{xx}(F) = |X_a(F)|^2 \qquad (11.1.4)$$

so that $R_{xx}(\tau)$ and $S_{xx}(F)$ are a Fourier transform pair.

Now suppose that we compute the energy density spectrum of the signal $x_a(t)$ from its samples taken at the rate F_s samples per second. To ensure that there is no spectral aliasing resulting from the sampling process, the signal is assumed to be prefiltered, so that, for practical purposes, its bandwidth is limited to B hertz. Then the sampling frequency F_s is selected such that $F_s > 2B$.

The sampled version of $x_a(t)$ is a sequence $x(n)$, $-\infty < n < \infty$, which has a Fourier transform (voltage spectrum)

$$X(\omega) = \sum_{n=-\infty}^{\infty} x(n)e^{-j\omega n}$$

or, equivalently,

$$X(f) = \sum_{n=-\infty}^{\infty} x(n)e^{-j2\pi fn} \qquad (11.1.5)$$

Recall that $X(f)$ may be expressed in terms of the voltage spectrum of the analog signal $x_a(t)$ as

$$X\left(\frac{F}{F_s}\right) = F_s \sum_{k=-\infty}^{\infty} X_a(F - kF_s) \qquad (11.1.6)$$

where $f = F/F_s$ is the normalized frequency variable.

In the absence of aliasing, within the fundamental range $|F| \le F_s/2$, we have

$$X\left(\frac{F}{F_s}\right) = F_s X_a(F) \qquad |F| \le F_s/2 \qquad (11.1.7)$$

Hence the voltage spectrum of the sampled signal is identical to the voltage spectrum of the analog signal. As a consequence, the energy density spectrum of the sampled signal is

$$S_{xx}\left(\frac{F}{F_s}\right) = \left|X\left(\frac{F}{F_s}\right)\right|^2 = F_s^2 \, |X_a(F)|^2 \qquad (11.1.8)$$

We may proceed further by noting that the autocorrelation of the sampled signal, which is defined as

$$r_{xx}(k) = \sum_{n=-\infty}^{\infty} x^*(n)x(n + k) \qquad (11.1.9)$$

has the Fourier transform (Wiener–Khintchine theorem)

$$S_{xx}(f) = \sum_{k=-\infty}^{\infty} r_{xx}(k)e^{-j2\pi kf} \qquad (11.1.10)$$

Hence the energy density spectrum may be obtained by Fourier transforming the auto-correlation of the sequence $\{x(n)\}$.

The relations above lead us to distinguish between two distinct methods for computing the energy density spectrum of a signal $x_a(t)$ from its samples $x(n)$. One is the *direct method*, which involves computing the Fourier transform of $\{x(n)\}$, and then

$$S_{xx}(f) = |X(f)|^2$$

$$= \left|\sum_{n=-\infty}^{\infty} x(n)e^{-j2\pi fn}\right|^2 \qquad (11.1.11)$$

The second approach is called the *indirect method* because it requires two steps. First, the autocorrelation $r_{xx}(k)$ is computed from $x(n)$ and then the Fourier transform of the autocorrelation is computed as in (11.1.10) to obtain the energy density spectrum.

In practice, however, only the finite-duration sequence $x(n)$, $0 \le n \le N - 1$, is available for computing the spectrum of the signal. In effect, limiting the duration of the sequence $x(n)$ to N points is equivalent to multiplying $x(n)$ by a rectangular window. Thus we have

$$\tilde{x}(n) = x(n)w(n) = \begin{cases} x(n) & 0 \le n \le N - 1 \\ 0 & \text{otherwise} \end{cases} \tag{11.1.12}$$

From our discussion of FIR filter design based on the use of windows to limit the duration of the impulse response, we recall that multiplication of two sequences is equivalent to convolution of their voltage spectra. Consequently, the frequency-domain relation corresponding to (11.1.12) is

$$\tilde{X}(f) = X(f) * W(f)$$

$$= \int_{-1/2}^{1/2} X(\alpha)W(f - \alpha)\, d\alpha \tag{11.1.13}$$

Convolution of the window function $W(f)$ with $X(f)$ smooths the spectrum $X(f)$, provided that the spectrum $W(f)$ is relatively narrow compared to $X(f)$. But this condition implies that the window $w(n)$ be sufficiently long (i.e., N must be sufficiently large) such that $W(f)$ is narrow compared to $X(f)$. Even if $W(f)$ is narrow compared to $X(f)$, the convolution of $X(f)$ with the sidelobes of $W(f)$ results in sidelobe energy in $\tilde{X}(f)$, in frequency bands where the true signal spectrum $X(f) = 0$. This sidelobe energy is called *leakage*. The following example illustrates the leakage problem.

EXAMPLE 11.1.1

A signal with (voltage) spectrum

$$X(f) = \begin{cases} 1 & |f| \le 0.1 \\ 0 & \text{otherwise} \end{cases}$$

is convolved with the rectangular window of length $N = 61$. Determine the spectrum of $\tilde{X}(f)$ given by (11.1.13).

Solution: The spectral characteristic $W(f)$ for the length $N = 61$ rectangular window is illustrated in Fig. 8.2(b). Note that the width of the main lobe of the window function is $\Delta\omega = 4\pi/61$ or $\Delta f = 2/61$, which is narrow compared to $X(f)$.

The convolution of $X(f)$ with $W(f)$ is illustrated in Fig. 11.1. We note that energy has leaked into the frequency band $0.1 < |f| \le 0.5$, where $X(f) = 0$. A part of this is due to the width of the main lobe in $W(f)$, which causes a broading or smearing of $X(f)$ outside the range $|f| \le 0.1$. However, the sidelobe energy in $X(f)$ is due to the presence of the sidelobes of $W(f)$, which are convolved with $X(f)$. The smearing of $X(f)$ for $|f| > 0.1$ and the sidelobes in the range $0.1 \le |f| \le 0.5$ constitute the leakage.

Just as in the case of FIR filter design, we can reduce sidelobe leakage by selecting windows that have low sidelobes. This implies that the windows have a smooth time-domain cutoff instead of the abrupt cutoff in the rectangular window. Although such window functions reduce sidelobe leakage, they result in an increase in smoothing or broadening of the spectral characteristic $X(f)$. For example, the use of a Blackman window of length $N = 61$ in Example 11.1.1 results in the spectral characteristic $X(f)$

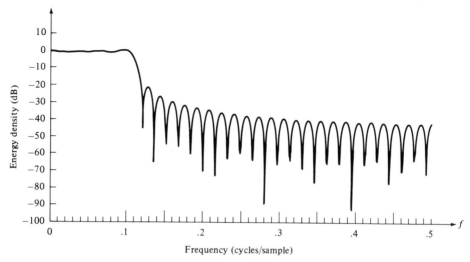

FIGURE 11.1 Spectrum obtained by convolving an $M = 61$ rectangular window with the ideal lowpass spectrum in Example 11.1.1.

shown in Fig. 11.2. The sidelobe leakage has certainly been reduced, but the spectral width has been increased by about 50%.

The broadening of the spectrum being estimated due to windowing is particularly a problem when we wish to resolve signals with closely spaced frequency components. For example, the signal with spectral characteristic $X(f) = X_1(f) + X_2(f)$, as shown in Fig. 11.3, cannot be resolved as two separate signals unless the width of the window function is significantly narrower than the frequency separation Δf. Thus we observe that using smooth time-domain windows reduces leakage at the expense of a decrease in frequency resolution.

It is clear from the discussion above that the energy density spectrum of the windowed sequence $\{x(n)\}$ is an approximation of the desired spectrum of the sequence $\{x(n)\}$. The

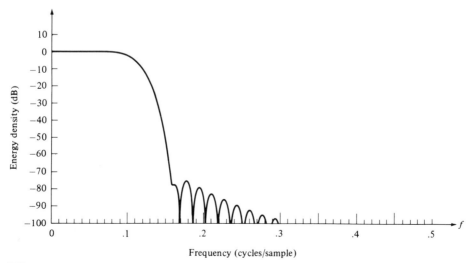

FIGURE 11.2 Spectrum obtained by convolving an $M = 61$ Blackman window with the ideal lowpass spectrum in Example 11.1.1.

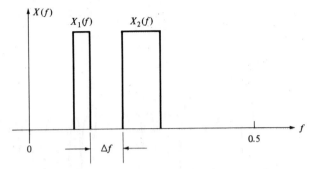

FIGURE 11.3   Two narrowband signal spectra.

spectral density obtained from $\{\tilde{x}(n)\}$ is

$$S_{\tilde{x}\tilde{x}}(f) = |\tilde{X}(f)|^2 = \left| \sum_{n=0}^{N-1} \tilde{x}(n)e^{-j2\pi fn} \right|^2 \qquad (11.1.14)$$

The spectrum given by (11.1.14) can be computed numerically at a set of N frequency points by means of the DFT. Thus

$$\tilde{X}(k) = \sum_{n=0}^{N-1} \tilde{x}(n)e^{-j2\pi kn/N} \qquad (11.1.15)$$

Then

$$|\tilde{X}(k)|^2 = S_{\tilde{x}\tilde{x}}(f)|_{f=k/N} = S_{\tilde{x}\tilde{x}}\left(\frac{k}{N}\right) \qquad (11.1.16)$$

and hence

$$S_{\tilde{x}\tilde{x}}\left(\frac{k}{N}\right) = \left| \sum_{n=0}^{N-1} \tilde{x}(n)e^{-j2\pi kn/N} \right|^2 \qquad (11.1.17)$$

which is a diostorted version of the true spectrum $S_{xx}(k/N)$.

11.1.2 Estimation of the Autocorrelation and Power Spectrum of Random Signals: The Periodogram

The finite-energy signals considered in the preceding section possess a Fourier transform and were characterized in the spectral domain by their energy density spectrum. On the other hand, the important class of signals characterized as stationary random processes do not have finite energy and hence do not possess a Fourier transform. Such signals have finite average power and hence are characterized by a *power density spectrum*. If $x(t)$ is a stationary random process, its autocorrelation function is

$$\gamma_{xx}(\tau) = E[x^*(t)x(t+\tau)] \qquad (11.1.18)$$

where $E[\cdot]$ denotes the statistical average. Then via the Wiener–Khintchine theorem, the power density spectrum of the stationary random process is the Fourier transform of the autocorrelation function, that is,

$$\Gamma_{xx}(F) = \int_{-\infty}^{\infty} \gamma_{xx}(\tau)e^{-j2\pi F\tau}\, dt \qquad (11.1.19)$$

In practice, we deal with a single realization of the random process from which we estimate the power spectrum of the process. We do not know the true autocorrelation function $\gamma_{xx}(\tau)$, and as a consequence, we cannot compute the Fourier transform in (11.1.19) to obtain $\Gamma_{xx}(F)$. On the other hand, from a single realization of the random process we can compute the time-average autocorrelation function

$$R_{xx}(\tau) = \frac{1}{2T_0} \int_{-T_0}^{T_0} x^*(t)x(t + \tau)\, dt \tag{11.1.20}$$

where $2T_0$ is the observation interval. If the stationary random process is *ergodic* in the first and second moments (mean and autocorrelation function), then

$$\gamma_{xx}(\tau) = \lim_{T_0 \to \infty} R_{xx}(\tau) \tag{11.1.21}$$

$$= \lim_{T_0 \to \infty} \frac{1}{2T_0} \int_{-T_0}^{T_0} x^*(t)x(t + \tau)\, dt$$

This relation justifies the use of the time-average autocorrelation $R_{xx}(\tau)$ as an estimate of the statistical autocorrelation function $\gamma_{xx}(\tau)$. Furthermore, the Fourier transform of $R_{xx}(\tau)$ provides an estimate $P_{xx}(F)$ of the power density spectrum, that is,

$$P_{xx}(F) = \int_{-T_0}^{T_0} R_{xx}(\tau)e^{-j2\pi F\tau}\, d\tau$$

$$= \frac{1}{2T_0} \int_{-T_0}^{T_0} \left[\int_{-T_0}^{T_0} x^*(t)x(t + \tau)\, dt \right] e^{-j2\pi F\tau}\, d\tau$$

$$= \frac{1}{2T_0} \left| \int_{-T_0}^{T_0} x(t)e^{-j2\pi Ft}\, dt \right|^2 \tag{11.1.22}$$

The actual power density spectrum is the expected value of $P_{xx}(F)$ in the limit as $T_0 \to \infty$, that is,

$$\Gamma_{xx}(F) = \lim_{T_0 \to \infty} E[P_{xx}(F)]$$

$$= \lim_{T_0 \to \infty} E\left[\frac{1}{2T_0} \left| \int_{-T_0}^{T_0} x(t)e^{-j2\pi Ft}\, dt \right|^2 \right] \tag{11.1.23}$$

From (11.1.20) and (11.1.22) we again note the two possible approaches to computing $P_{xx}(F)$, the direct method as given by (11.1.22) or the indirect method, in which we obtain $R_{xx}(\tau)$ first and then compute the Fourier transform.

We shall consider the estimation of the power density spectrum from samples of a single realization of the random process. In particular, we assume that $x_a(t)$ is sampled at a rate $F_s > 2B$, where B is the highest frequency contained in the power density spectrum of the random process. Thus we obtain a finite-duration sequence $x(n)$, $0 \le n \le N - 1$, by sampling $x_a(t)$. From these samples we may compute the time-average autocorrelation sequence

$$r'_{xx}(m) = \frac{1}{N - |m|} \sum_{n=0}^{N-|m|-1} x^*(n)x(n + m) \qquad |m| = 0, 1, \ldots, N - 1 \tag{11.1.24}$$

and then compute the Fourier transform

$$P'_{xx}(f) = \sum_{m=-N+1}^{N-1} r'_{xx}(m)e^{-j2\pi fm} \tag{11.1.25}$$

The normalization factor $N - |m|$ in (11.1.24) results in an estimate with mean value

$$E[r'_{xx}(m)] = \frac{1}{N - |m|} \sum_{n=0}^{N-|m|-1} E[x^*(n)x(n + m)] \tag{11.1.26}$$
$$= \gamma_{xx}(m)$$

where $\gamma_{xx}(m)$ is the true (statistical) autocorrelation sequence of $x(n)$. Hence $r'_{xx}(m)$ is an unbiased estimate of the autocorrelation function $\gamma_{xx}(m)$. The variance of the estimate $r'_{xx}(m)$ is approximately

$$\text{var } [r'_{xx}(m)] \approx \frac{N}{[N - |m|]^2} \sum_{n=-\infty}^{\infty} [|\gamma_{xx}(n)|^2 + \gamma_{xx}^*(n - m)\gamma_{xx}(n + m)] \tag{11.1.27}$$

which is a result given by Jenkins and Watts (1968). Clearly,

$$\lim_{N \to \infty} \text{var } [r'_{xx}(m)] = 0 \tag{11.1.28}$$

provided that

$$\sum_{n=-\infty}^{\infty} |\gamma_{xx}(n)|^2 < \infty$$

Since $E[r'_{xx}(m)] = \gamma_{xx}(m)$ and the variance of the estimate converges to zero as $N \to \infty$, the estimate $r'_{xx}(m)$ is said to be *consistent*.

For large values of the lag parameter m, the estimate $r'_{xx}(m)$ given by (11.1.24) has a large variance, especially as m approaches N. This is due to the fact that fewer data points enter into the estimate for large lags. As an alternative to (11.1.24) we may use the estimate

$$r_{xx}(m) = \frac{1}{N} \sum_{n=0}^{N-|m|-1} x^*(n)x(n + m) \tag{11.1.29}$$

which has a bias of $|m|\gamma_{xx}(m)/N$, since its mean value is

$$E[r_{xx}(m)] = \frac{1}{N} \sum_{n=0}^{N-|m|-1} E[x^*(n)x(n + m)]$$

$$= \frac{N - |m|}{N} \gamma_{xx}(m) = \left(1 - \frac{|m|}{N}\right) \gamma_{xx}(m) \tag{11.1.30}$$

However, this estimate has a smaller variance, given approximately as

$$\text{var } [r_{xx}(m)] \approx \frac{1}{N} \sum_{n=-\infty}^{\infty} [|\gamma_{xx}(n)|^2 + \gamma_{xx}^*(n - m)\gamma_{xx}(n + m)] \tag{11.1.31}$$

We observe that $r_{xx}(m)$ is *asymptotically unbiased*, that is,

$$\lim_{N \to \infty} E[r_{xx}(m)] = \gamma_{xx}(m) \tag{11.1.32}$$

and its variance converges to zero as $N \to \infty$. Therefore, the estimate $r_{xx}(m)$ is also a *consistent estimate* of $\gamma_{xx}(m)$.

We shall use the estimate $r_{xx}(m)$ given by (11.1.29) in our treatment of power spectrum estimation. The corresponding estimate of the power density spectrum is

$$P_{xx}(f) = \sum_{m=-(N-1)}^{N-1} r_{xx}(m)e^{-j2\pi fm} \tag{11.1.33}$$

If we substitute for $r_{xx}(m)$ from (11.1.29) into (11.1.33), the estimate $P_{xx}(f)$ may also be expressed as

$$P_{xx}(f) = \frac{1}{N} \left| \sum_{n=0}^{N-1} x(n)e^{-j2\pi fn} \right|^2 = \frac{1}{N} |X(f)|^2 \qquad (11.1.34)$$

where $X(f)$ is the Fourier transform of the sample sequence $x(n)$. This well-known form of the power density spectrum estimate is called the *periodogram*. It was originally introduced by Schuster (1898) to detect and measure "hidden periodicities" in data.

From (11.1.33), the average value of the periodogram estimate $P_{xx}(f)$ is

$$E[P_{xx}(f)] = E \left[\sum_{m=-(N-1)}^{N-1} r_{xx}(m)e^{-j2\pi fm} \right] = \sum_{m=-(N-1)}^{N-1} E[r_{xx}(m)]e^{-j2\pi fm}$$

$$E[P_{xx}(f)] = \sum_{m=-(N-1)}^{N-1} \left(1 - \frac{|m|}{N} \right) \gamma_{xx}(m)e^{-j2\pi fm} \qquad (11.1.35)$$

The interpretation that we give to (11.1.35) is that the mean of the estimated spectrum is the Fourier transform of the windowed autocorrelation function

$$\tilde{\gamma}_{xx}(m) = \left(1 - \frac{|m|}{N} \right) \gamma_{xx}(m) \qquad (11.1.36)$$

where the window function is the (triangular) Bartlett window. Hence the mean of the estimated spectrum is

$$E[P_{xx}(f)] = \sum_{m=-\infty}^{\infty} \tilde{\gamma}_{xx}(m)e^{-j2\pi fm}$$

$$= \int_{-1/2}^{1/2} \Gamma_{xx}(\alpha)W_B(f - \alpha)\, d\alpha \qquad (11.1.37)$$

where $W_B(f)$ is the spectral characteristic of the Bartlett window. The relation (11.1.37) illustrates that the mean of the estimated spectrum is the convolution of the true power density spectrum $\Gamma_{xx}(f)$ with the Fourier transform $W_B(f)$ of the Bartlett window. Consequently, the mean of the estimated spectrum is a smoothed version of the true spectrum and suffers from the same spectral leakage problems which are due to the finite number of data points.

We observe that the estimated spectrum is asymptotically unbiased, that is,

$$\lim_{N\to\infty} E \left[\sum_{m=-(N-1)}^{N-1} r_{xx}(m)e^{-j2\pi fm} \right] = \sum_{m=-\infty}^{\infty} \gamma_{xx}(m)e^{-j2\pi fm} = \Gamma_{xx}(f)$$

However, in general, the variance of the estimate $P_{xx}(f)$ does not decay to zero as $N \to \infty$. For example, when the data sequence is a Gaussian random process, the variance is easily shown to be (see Problem 11.4)

$$\text{var}\,[P_{xx}(f)] = \Gamma_{xx}^2(f) \left[1 + \left(\frac{\sin 2\pi fN}{N \sin 2\pi f} \right)^2 \right] \qquad (11.1.38)$$

which, in the limit as $N \to \infty$, becomes

$$\lim_{N\to\infty} \text{var}\,[P_{xx}(f)] = \Gamma_{xx}^2(f) \qquad (11.1.39)$$

Hence we conclude that the *periodogram is not a consistent estimate of the true power density spectrum* (i.e., it does not converge to the true power density spectrum).

In summary, the estimated autocorrelation $r_{xx}(m)$ is a consistent estimate of the true autocorrelation function $\gamma_{xx}(m)$. However, its Fourier transform $P_{xx}(f)$, the periodogram, is not a consistent estimate of the true power density spectrum. We observed that $P_{xx}(f)$ is an asymptotically unbiased estimate of $\Gamma_{xx}(f)$, but for a finite-duration sequence, the mean value of $P_{xx}(f)$ contains a bias, which from (11.1.37) is evident as a distortion of the true power density spectrum. Thus the estimated spectrum suffers from the smoothing effects and the leakage embodied in the Bartlett window. The smoothing and leakage ultimately limit our ability to resolve closely spaced spectra.

The problems of leakage and frequency resolution that we have described above, as well as the problem that the periodogram is not a consistent estimate of the power spectrum, provide the motivation for the power spectrum estimation methods described in Sections 11.2, 11.3, and 11.4. The methods described in Section 11.2 are classical nonparametric methods, which make no assumptions about the data sequence. The emphasis of the classical methods is on obtaining a consistent estimate of the power spectrum through some averaging or smoothing operations performed directly on the periodogram or on the autocorrelation. As we will see, the effect of these operations is to reduce the frequency resolution further, while the variance of the estimate is decreased.

The spectrum estimation methods described in Section 11.3 are based on some model of how the data were generated. In general, the model-based methods that have been developed over the past two decades provide significantly higher resolution than do the classical methods.

Two additional methods are described in Section 11.4. One of these methods is based on a model for the observed process, while the other is not.

11.1.3 The Use of the DFT in Power Spectrum Estimation

As given by (11.1.14) and (11.1.34), the estimated energy density spectrum $S_{xx}(f)$ and the periodogram $P_{xx}(f)$, respectively, can be computed by use of the DFT, which in turn is efficiently computed by the FFT algorithm. If we have N data points, we compute as a minimum the N-point DFT. For example, the computation yields samples of the periodogram

$$P_{xx}\left(\frac{k}{N}\right) = \frac{1}{N}\left|\sum_{n=0}^{N-1} x(n)e^{-j2\pi nk/N}\right|^2 \qquad k = 0, 1, \ldots, N-1 \qquad (11.1.40)$$

at the frequencies $f_k = k/N$.

In practice, however, such a sparse sampling of the spectrum does not provide a very good representation or a good picture of the continuous spectrum estimate $P_{xx}(f)$. This is easily remedied by evaluating $P_{xx}(f)$ at additional frequencies. Equivalently, we may effectively increase the length of the sequence by means of zero padding and then evaluate $P_{xx}(f)$ at a more dense set of frequencies. Thus if we increase the data sequence length to L points by means of zero padding and evaluate the L-point DFT, we have

$$P_{xx}\left(\frac{k}{L}\right) = \frac{1}{N}\left|\sum_{n=0}^{N-1} x(n)e^{-j2\pi nk/L}\right|^2 \qquad k = 0, 1, \ldots, L-1 \qquad (11.1.41)$$

We emphasize that zero padding and evaluating the DFT at $L > N$ points does not improve the frequency resolution in the spectral estimate. It simply provides us with a method for interpolating the values of the measured spectrum at more frequencies. The frequencey resolution in the spectral estimate $P_{xx}(f)$ is determined by the length N of the data record.

EXAMPLE 11.1.2 _____

A sequence of $N = 16$ samples is obtained by sampling an analog signal consisting of two frequency components. The resulting discrete-time sequence is

$$x(n) = \sin 2\pi(0.135)n + \cos 2\pi(0.135 + \Delta f)n \qquad n = 0, 1, \ldots, 15$$

where Δf is the frequency separation. Evaluate the power spectrum $P(f) = (1/N)|X(f)|^2$ at the frequencies $f_k = k/L$, $k = 0, 1, \ldots, L - 1$, for $L = 16, 32, 64,$ and 128 for values of $\Delta f = 0.06$ and $\Delta f = 0.01$.

Solution: By zero padding we increase the data sequence to obtain the power spectrum estimate $P_{xx}(k/L)$. The results for $\Delta f = 0.06$ are plotted in Fig. 11.4. Note that zero padding does not change the resolution, but it does have the effect of interpolating the spectrum $P_{xx}(f)$. In this case the frequency separation Δf is sufficiently large so that the two frequency components are resolvable.

The spectral estimates for $\Delta f = 0.01$ are shown in Fig. 11.5. In this case the two spectral components are not resolvable. Again, the effect of zero padding is to provide

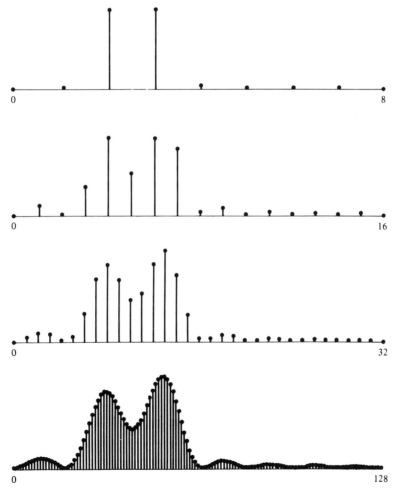

FIGURE 11.4   Spectra of two sinusoids with frequency separation $\Delta f = 0.06$.

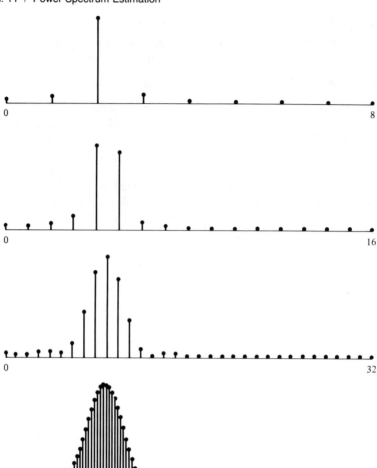

FIGURE 11.5 Spectra of two sinusoids with frequency separation $\Delta f = 0.01$.

more interpolation, thus giving us a better picture of the estimated spectrum. It does not improve the frequency resolution.

When only a few points of the periodogram are needed, the Goertzel algorithm described in Chapter 9 may provide a more efficient computation. Since the Goertzel algorithm has been interpreted as a linear filtering approach to computing the DFT, it is clear that the periodogram estimate can be obtained by passing the signal through a bank of parallel tuned filters and squaring their outputs (see Problem 11.5).

11.2 Nonparametric Methods for Power Spectrum Estimation

The power spectrum estimation methods described in this section are the classical methods developed by Bartlett (1948), Blackman and Tukey (1958), and Welch (1967). These methods make no assumption about how the data were generated and hence are called *nonparametric*.

Since the estimates are based entirely on a finite record of data, the frequency reso-lution of these methods is, at best, equal to the spectral width of the rectangular window of length N, which is approximately $1/N$ at the -3-dB points. We shall be more precise in specifying the frequency resolution of the specific methods. All the estimation tech-niques described in this section decrease the frequency resolution in order to reduce the variance in the spectral estimate.

First, we will describe the estimates and derive the mean and variance of each. A comparison of the three methods is given in Section 11.2.4. Although the spectral esti-mates are expressed as a function of the continuous frequency variable f, in practice, the estimates are computed at discrete frequencies via the FFT algorithm. The FFT-based computational requirements are considered in Section 11.2.5.

11.2.1 The Bartlett Method: Averaging Periodograms

Bartlett's method for reducing the variance in the periodogram involves three steps. First, the N-point sequence is subdivided into K nonoverlapping segments, where each segment has length M. This results in the K data segments

$$x_i(n) = x(n + iM) \qquad i = 0, 1, \ldots, K - 1 \tag{11.2.1}$$
$$n = 0, 1, \ldots, M - 1$$

For each segment, we compute the periodogram

$$P_{xx}^{(i)}(f) = \frac{1}{M} \left| \sum_{n=0}^{M-1} x_i(n) e^{-j2\pi fn} \right|^2 \qquad i = 0, 1, \ldots, K - 1 \tag{11.2.2}$$

Finally, we average the periodograms for the K segments to obtain the Bartlett power spectrum estimate

$$P_{xx}^B(f) = \frac{1}{K} \sum_{i=0}^{K-1} P_{xx}^{(i)}(f) \tag{11.2.3}$$

The statistical properties of this estimate are easily obtained. First, the mean value is

$$E[P_{xx}^B(f)] = \frac{1}{K} \sum_{i=0}^{K-1} E[P_{xx}^{(i)}(f)]$$
$$= E[P_{xx}^{(i)}(f)] \tag{11.2.4}$$

From (11.1.35) and (11.1.37) we have the expected value for the single periodogram as

$$E[P_{xx}^{(i)}(f)] = \sum_{m=-(M-1)}^{M-1} \left(1 - \frac{|m|}{M}\right) \gamma_{xx}(m) e^{-j2\pi fm}$$
$$= \frac{1}{M} \int_{-1/2}^{1/2} \Gamma_{xx}(\alpha) \left(\frac{\sin \pi(f - \alpha)M}{\sin \pi(f - \alpha)}\right)^2 d\alpha \tag{11.2.5}$$

where

$$W_B(f) = \frac{1}{M} \left(\frac{\sin \pi fM}{\sin \pi f}\right)^2 \tag{11.2.6}$$

is the frequency characteristics of the Bartlett window

$$w_B(n) = \begin{cases} 1 - \dfrac{|m|}{M} & |m| \le M - 1 \\ 0 & \text{otherwise} \end{cases} \tag{11.2.7}$$

From (11.2.5) we observe that the true spectrum is now convolved the frequency characteristic $W_B(f)$ of the Bartlett window. The effect of reducing the length of the data from N points to $M = N/K$ results in a window whose spectral width has been increased by a factor of K. Consequently, the frequency resolution has been reduced by a factor K.

In return for this reduction in resolution, we have reduced the variance. The variance of the Bartlett estimate is

$$\text{var } [P_{xx}^B(f)] = \frac{1}{K^2} \sum_{i=0}^{K-1} \text{var } [P_{xx}^{(i)}(f)]$$

$$= \frac{1}{K} \text{var } [P_{xx}^{(i)}(f)]$$

(11.2.8)

If we make use of (11.1.38) in (11.2.8), we obtain

$$\text{var } [P_{xx}^B(f)] = \frac{1}{K} \Gamma_{xx}^2(f) \left[1 + \left(\frac{\sin 2\pi fM}{M \sin 2\pi f} \right)^2 \right]$$

(11.2.9)

Therefore, the variance of the Bartlett power spectrum estimate has been reduced by the factor K.

11.2.2 The Welch Method: Averaging Modified Periodograms

Welch (1967) made two basic modifications to the Bartlett method. First, he allowed the data segments to overlap. Thus the data segments may be represented as

$$x_i(n) = x(n + iD) \qquad n = 0, 1, \ldots, M - 1$$
$$i = 0, 1, \ldots, L - 1$$

(11.2.10)

where iD is the starting point for the ith sequence. Observe that if $D = M$, the segments do not overlap and the number L of data segments is identical to the number K in the Bartlett method. However, if $D = M/2$, there is 50% overlap between successive data segments and $L = 2K$ segments are obtained. Alternatively, we can form K data segments each of length $2M$.

The second modification made by Welch to the Bartlett method is to window the data segments prior to computing the periodogram. The result is a "modified" periodogram

$$\tilde{P}_{xx}^{(i)}(f) = \frac{1}{MU} \left| \sum_{n=0}^{M-1} x_i(n) w(n) e^{-j2\pi fn} \right|^2 \qquad i = 0, 1, \ldots, L - 1 \quad (11.2.11)$$

where U is a normalization factor for the power in the window function and is selected as

$$U = \frac{1}{M} \sum_{n=0}^{M-1} w^2(n)$$

(11.2.12)

The Welch power spectrum estimate is the average of these modified periodograms, that is,

$$P_{xx}^W(f) = \frac{1}{L} \sum_{i=0}^{L-1} \tilde{P}_{xx}^{(i)}(f)$$

(11.2.13)

The mean value of the Welch estimate is

$$E[P_{xx}^W(f)] = \frac{1}{L} \sum_{i=0}^{L-1} E[\tilde{P}_{xx}^{(i)}(f)]$$
$$= E[\tilde{P}_{xx}^{(i)}(f)]$$

(11.2.14)

But the expected value of the modified periodogram is

$$E[\tilde{P}_{xx}^{(i)}(f)] = \frac{1}{MU} \sum_{n=0}^{M-1} \sum_{m=0}^{M-1} w(n)w(m)E[x_i(n)x_i^*(m)]e^{-j2\pi f(n-m)}$$

$$= \frac{1}{MU} \sum_{n=0}^{M-1} \sum_{m=0}^{M-1} w(n)w(m)\gamma_{xx}(n-m)e^{-j2\pi f(n-m)} \qquad (11.2.15)$$

Since

$$\gamma_{xx}(n) = \int_{-1/2}^{1/2} \Gamma_{xx}(\alpha)e^{j2\pi\alpha n}\, d\alpha \qquad (11.2.16)$$

substitution for $\gamma_{xx}(n)$ from (11.2.16) into (11.2.15) yields

$$E[\tilde{P}_{xx}^{(i)}(f)] = \frac{1}{MU} \int_{-1/2}^{1/2} \Gamma_{xx}(\alpha) \left[\sum_{n=0}^{M-1} \sum_{m=0}^{M-1} w(n)w(m)e^{-j2\pi(n-m)(f-\alpha)} \right] d\alpha$$

$$\qquad (11.2.17)$$

$$= \int_{-1/2}^{1/2} \Gamma_{xx}(\alpha)W(f-\alpha)\, d\alpha$$

where, by definition,

$$W(f) = \frac{1}{MU} \left| \sum_{n=0}^{M-1} w(n)e^{-j2\pi fn} \right|^2 \qquad (11.2.18)$$

The normalization factor U ensures that

$$\int_{-1/2}^{1/2} W(f)\, df = 1 \qquad (11.2.19)$$

The variance of the Welch estimate is

$$\text{var}\,[P_{xx}^W(f)] = \frac{1}{L^2} \sum_{i=0}^{L-1} \sum_{j=0}^{L-1} E[\tilde{P}_{xx}^{(i)}(f)\tilde{P}_{xx}^{(j)}(f)] - \{E[P_{xx}^W(f)]\}^2 \qquad (11.2.20)$$

In the case of no overlap between successive data segments ($L = K$) Welch has shown that

$$\text{var}\,[P_{xx}^W(f)] = \frac{1}{L} \text{var}\,[\tilde{P}_{xx}^{(i)}(f)]$$

$$\qquad (11.2.21)$$

$$\approx \frac{1}{L} \Gamma_{xx}^2(f)$$

In the case of 50% overlap between successive data segments ($L = 2K$), the variance of the Welch power spectrum estimate with the Bartlett (triangular) window, also derived in the paper by Welch, is

$$\text{var}\,[P_{xx}^W(f)] \approx \frac{9}{8L} \Gamma_{xx}^2(f) \qquad (11.2.22)$$

Although we considered only the triangular window in the computation of the variance, other window functions may be used. In general, they will yield a different variance. In addition, one may also overlap the data segments by either more or less than the 50% considered in this section, in an attempt to improve the relevant characteristics of the estimate.

11.2.3 The Blackman and Tukey Method: Smoothing the Periodogram

Blackman and Tukey (1958) proposed and analyzed the method in which the sample autocorrelation sequence is windowed first and then Fourier transformed to yield the estimate of the power spectrum. The rationale for windowing the estimated autocorrelation sequence $r_{xx}(m)$ is that, for large lags, the estimates are less reliable because a smaller number $(N - m)$ of data points enter into the estimate. For values of m approaching N, the variance of these estimates is very high, and hence these estimates should be given a smaller weight in the formation of the estimated power spectrum. Thus the Blackman–Tukey estimate is

$$P_{xx}^{BT}(f) = \sum_{m=-(M-1)}^{M-1} r_{xx}(m)w(m)e^{-j2\pi fm} \qquad (11.2.23)$$

where the window function $w(n)$ has length $2M - 1$ and is zero for $|m| \geq M$. With this definition for $w(n)$, the limits on the sum in (11.2.23) may be extended to $(-\infty, \infty)$. Hence the frequency-domain equivalent expression for (11.2.23) is the convolution integral

$$P_{xx}^{BT}(f) = \int_{-1/2}^{1/2} P_{xx}(\alpha)W(f - \alpha)\, d\alpha \qquad (11.2.24)$$

where $P_{xx}(f)$ is the periodogram. It is clear from (11.2.24) that the effect of windowing the autocorrelation is to smooth the periodogram estimate, thus decreasing the variance in the estimate at the expense of reducing the resolution.

The window sequence $w(n)$ should be symmetric (even) about $m = 0$ to ensure that the estimate of the power spectrum is real. Furthermore, it is desirable to select the window spectrum to be nonnegative, that is,

$$W(f) \geq 0 \qquad |f| \leq 1/2 \qquad (11.2.25)$$

This condition will ensure that $P_{xx}^{BT}(f) \geq 0$ for $|f| \leq 1/2$, which is a desirable property for any power spectrum estimate. We should indicate, however, that some of the window functions we have introduced do not satisfy this condition. For example, in spite of their low sidelobe levels, the Hamming and Hann (or Hanning) windows do not satisfy the property in (11.2.25) and, consequently, may result in negative spectrum estimates in some parts of the frequency range.

The expected value of the Blackman–Tukey power spectrum estimate is

$$E[P_{xx}^{BT}(f)] = \int_{-1/2}^{1/2} E[P_{xx}(\alpha)]W(f - \alpha)\, d\alpha \qquad (11.2.26)$$

where from (11.1.37) we have

$$E[P_{xx}(\alpha)] = \int_{-1/2}^{1/2} \Gamma_{xx}(\theta)W_B(\alpha - \theta)\, d\theta \qquad (11.2.27)$$

and $W_B(f)$ is the Fourier transform of the Bartlett window. Substitution of (11.2.27) into (11.2.26) yields the double convolution integral

$$E[P_{xx}^{BT}(f)] = \int_{-1/2}^{1/2}\int_{-1/2}^{1/2} \Gamma_{xx}(\theta)W_B(\alpha - \theta)W(f - \alpha)\, d\alpha\, d\theta \qquad (11.2.28)$$

Equivalently, by working in the time domain, the expected value of the Blackman–Tukey power spectrum estimate is

$$E[P_{xx}^{BT}(f)] = \sum_{m=-(M-1)}^{M-1} E[r_{xx}(m)]w(m)e^{-j2\pi fm}$$

(11.2.29)

$$= \sum_{m=-(M-1)}^{M-1} \gamma_{xx}(m)w_B(m)w(m)e^{-j2\pi fm}$$

where the Bartlett window is

$$w_B(m) = \begin{cases} 1 - \dfrac{|m|}{N} & |m| < N \\ 0 & \text{otherwise} \end{cases}$$

(11.2.30)

Clearly, we should select the window length for $w(n)$ such that $M \ll N$, that is, $w(n)$ should be narrower than $w_B(m)$ in order to provide additional smoothing of the periodogram. Under this condition, (11.2.28) becomes

$$E[P_{xx}^{BT}(f)] \approx \int_{-1/2}^{1/2} \Gamma_{xx}(\theta)W(f - \theta)\, d\theta$$

(11.2.31)

since

$$\int_{-1/2}^{1/2} W_B(\alpha - \theta)W(f - \alpha)\, d\alpha = \int_{-1/2}^{1/2} W_B(\alpha)W(f - \theta - \alpha)\, d\alpha \approx W(f - \theta)$$

(11.2.32)

The variance of the Blackman–Tukey power spectrum estimate is

$$\text{var}\,[P_{xx}^{BT}(f)] = E\{[P_{xx}^{BT}(f)^2]\} - \{E[P_{xx}^{BT}(f)]\}^2$$

(11.2.33)

where the mean may be approximated as in (11.2.31). The second moment in (11.2.33) is

$$E\{[P_{xx}^{BT}(f)]^2\} = \int_{-1/2}^{1/2}\int_{-1/2}^{1/2} E[P_{xx}(\alpha)P_{xx}(\theta)]W(f - \alpha)W(f - \theta)\, d\alpha\, d\theta$$

(11.2.34)

On the assumption that the random process is Gaussian (see Problem 11.5), we find that

$$E[P_{xx}(\alpha)P_{xx}(\theta)] = \Gamma_{xx}(\alpha)\Gamma_{xx}(\theta)\left\{1 + \left[\frac{\sin \pi(\theta + \alpha)N}{N \sin \pi(\theta + \alpha)}\right]^2 + \left[\frac{\sin \pi(\theta - \alpha)N}{N \sin \pi(\theta - \alpha)}\right]^2\right\}$$

(11.2.35)

Substitution of (11.2.35) into (11.2.34) yields

$$E\{[P_{xx}^{BT}(f)]^2\} = \left[\int_{-1/2}^{1/2} \Gamma_{xx}(\theta)W(f - \theta)\, d\theta\right]^2$$

$$+ \int_{-1/2}^{1/2}\int_{-1/2}^{1/2} \Gamma_{xx}(\alpha)\Gamma_{xx}(\theta)W(f - \alpha)W(f - \theta)$$

$$\times \left\{\left[\frac{\sin \pi(\theta + \alpha)N}{N \sin \pi(\theta + \alpha)}\right]^2 + \left[\frac{\sin \pi(\theta - \alpha)N}{N \sin \pi(\theta - \alpha)}\right]^2\right\}\, d\alpha\, d\theta$$

(11.2.36)

The first term in (11.2.36) is simply the square of the mean of $P_{xx}^{BT}(f)$, which is to be subtracted out according to (11.2.33). This leaves the second term in (11.2.36), which constitutes the variance. For the case in which $N \gg M$, the functions $\sin \pi(\theta + \alpha)N/N$

$\sin \pi(\theta + \alpha)$ and $\sin \pi(\theta - \alpha)N/N \sin \pi(\theta - \alpha)$ are relatively narrow compared to $W(f)$ in the vicinity of $\theta = -\alpha$ and $\theta = \alpha$, respectively. Therefore,

$$\int_{-1/2}^{1/2} \Gamma_{xx}(\theta)W(f - \theta) \left\{ \left[\frac{\sin \pi(\theta + \alpha)N}{N \sin \pi(\theta + \alpha)} \right]^2 + \left[\frac{\sin \pi(\theta - \alpha)N}{N \sin \pi(\theta - \alpha)} \right]^2 \right\} d\theta$$

$$\approx \frac{\Gamma_{xx}(-\alpha)W(f + \alpha) + \Gamma_{xx}(\alpha)W(f - \alpha)}{N} \qquad (11.2.37)$$

With this approximation, the variance of $P_{xx}^{BT}(f)$ becomes

$$\text{var} \left[P_{xx}^{BT}(f)\right] \approx \frac{1}{N} \int_{-1/2}^{1/2} \Gamma_{xx}(\alpha)W(f - \alpha)\left[\Gamma_{xx}(-\alpha)W(f + \alpha) + \Gamma_{xx}(\alpha)W(f - \alpha)\right] d\alpha$$

$$\approx \frac{1}{N} \int_{-1/2}^{1/2} \Gamma_{xx}^2(\alpha)W^2(f - \alpha) \, d\alpha \qquad (11.2.38)$$

where in the last step we made the approximation

$$\int_{-1/2}^{1/2} \Gamma_{xx}(\alpha)\Gamma_{xx}(-\alpha)W(f - \alpha)W(f + \alpha) \, d\alpha \approx 0 \qquad (11.2.39)$$

We shall make one additional approximation in (11.2.38). When $W(f)$ is narrow compared to the true power spectrum $\Gamma_{xx}(f)$, (11.2.38) is further approximated as

$$\text{var} \left[P_{xx}^{BT}(f)\right] \approx \Gamma_{xx}^2(f) \left[\frac{1}{N} \int_{-1/2}^{1/2} W^2(\theta) \, d\theta \right]$$

$$\approx \Gamma_{xx}^2(f) \left[\frac{1}{N} \sum_{m=-(M-1)}^{M-1} w^2(m) \right] \qquad (11.2.40)$$

11.2.4 Performance Characteristics of Nonparametric Power Spectrum Estimators

In this section we compare the *quality* of the Bartlett, Welch, and Blackman and Tukey power spectrum estimates. As a measure of quality, we use the ratio of the square of the mean of the power spectrum estimate to its variance, that is,

$$Q_A = \frac{\{E[P_{xx}^A(f)]\}^2}{\text{var} \, [P_{xx}^A(f)]} \qquad (11.2.41)$$

where $A = B$, W, or BT for the three power spectrum estimates. The reciprocal of this quantity, called the *variability*, may also be used as a measure of performance.

For reference, the periodogram has a mean and variance

$$E[P_{xx}(f)] = \int_{-1/2}^{1/2} \Gamma_{xx}(\theta)W_B(f - \theta) \, d\theta \qquad (11.2.42)$$

$$\text{var} \, [P_{xx}(f)] = \Gamma_{xx}^2(f) \left[1 + \left(\frac{\sin 2\pi fN}{N \sin 2\pi f} \right)^2 \right] \qquad (11.2.43)$$

where

$$W_B(f) = \frac{1}{N} \left(\frac{\sin \pi fN}{\sin \pi f} \right)^2 \qquad (11.2.44)$$

For large N (i.e., $N \rightarrow \infty$),

$$E[P_{xx}(f)] \rightarrow \Gamma_{xx}(f) \int_{-1/2}^{1/2} W_B(\theta) \, d\theta = w_B(0)\Gamma_{xx}(f) = \Gamma_{xx}(f) \tag{11.2.45}$$

$$\text{var } [P_{xx}(f)] \rightarrow \Gamma_{xx}^2(f)$$

Hence, as indicated previously, the periodogram is an asymptotically unbiased estimate of the power spectrum, but it is not consistent because its variance does not approach zero as N increases toward infinity.

Asymptotically, the periodogram is characterized by the quality factor

$$Q_P = \frac{\Gamma_{xx}^2(f)}{\Gamma_{xx}^2(f)} = 1 \tag{11.2.46}$$

The fact that Q_P is fixed and independent of the data length N is another indication of the poor quality of this estimate.

Bartlett Power Spectrum Estimate. The mean and variance of the Bartlett power spectrum estimate are

$$E[P_{xx}^B(f)] = \int_{-1/2}^{1/2} \Gamma_{xx}(\theta) W_B(f - \theta) \, d\theta \tag{11.2.47}$$

$$\text{var } [P_{xx}^B(f)] = \frac{1}{K} \Gamma_{xx}^2(f) \left[1 + \left(\frac{\sin 2\pi f M}{M \sin 2\pi f} \right)^2 \right] \tag{11.2.48}$$

and

$$W_B(f) = \frac{1}{M} \left(\frac{\sin \pi f M}{\sin \pi f} \right)^2 \tag{11.2.49}$$

As $N \rightarrow \infty$ and $M \rightarrow \infty$, while $K = N/M$ remains fixed, we find that

$$E[P_{xx}^B(f)] \rightarrow \Gamma_{xx}(f) \int_{1/2}^{1/2} W_B(f) \, df = \Gamma_{xx}(f)w_B(0) = \Gamma_{xx}(f) \tag{11.2.50}$$

$$\text{var } [P_{xx}^B(f)] \rightarrow \frac{1}{K} \Gamma_{xx}^2(f)$$

We observe that the Bartlett power spectrum estimate is asymptotically unbiased and if K is allowed to increase with an increase in N, the estimate is also consistent. Hence, asymptotically, this estimate is characterized by the quality factor

$$Q_B = K = \frac{N}{M} \tag{11.2.51}$$

The frequency resolution of the Bartlett estimate, measured by taking the 3-dB width of the main lobe of the rectangular window, is

$$\Delta f = \frac{0.9}{M} \tag{11.2.52}$$

Hence $M = 0.9/\Delta f$ and, therefore, the quality factor becomes

$$Q_B = \frac{N}{0.9/\Delta f} = 1.1N \, \Delta f \tag{11.2.53}$$

Welch Power Spectrum Estimate. The mean and variance of the Welch power spectrum estimate are

$$E[P_{xx}^W(f)] = \int_{-1/2}^{1/2} \Gamma_{xx}(\theta)W(f - \theta) \, d\theta \tag{11.2.54}$$

where

$$W(f) = \frac{1}{MU} \left| \sum_{n=0}^{M-1} w(n)e^{-j2\pi fn} \right|^2 \tag{11.2.55}$$

and

$$\text{var } [P_{xx}^W(f)] = \begin{cases} \dfrac{1}{L} \Gamma_{xx}^2(f) & \text{for no overlap} \\[3mm] \dfrac{9}{8L} \Gamma_{xx}^2(f) & \begin{array}{l} \text{for 50\% overlap and} \\ \text{triangular window} \end{array} \end{cases} \tag{11.2.56}$$

As $N \to \infty$ and $M \to \infty$, the mean converges to

$$E[P_{xx}^W(f)] \to \Gamma_{xx}(f) \tag{11.2.57}$$

and the variance converges to zero, so that the estimate is consistent.
    Under the two conditions given by (11.2.56) the quality factor is

$$Q_W = \begin{cases} L = \dfrac{N}{M} & \text{for no overlap} \\[3mm] \dfrac{8L}{9} = \dfrac{16N}{9M} & \begin{array}{l} \text{for 50\% overlap and} \\ \text{triangular window} \end{array} \end{cases} \tag{11.2.58}$$

On the other hand, the spectral width of the triangular window at the 3-dB points is

$$\Delta f = \frac{1.28}{M} \tag{11.2.59}$$

Consequently, the quality factor expressed in terms of Δf and N is

$$Q_W = \begin{cases} 0.78N \, \Delta f & \text{for no overlap} \\[3mm] 1.39N \, \Delta f & \begin{array}{l} \text{for 50\% overlap and} \\ \text{triangular window} \end{array} \end{cases} \tag{11.2.60}$$

Blackman–Tukey Power Spectrum Estimate. The mean and variance of this estimate are approximated as

$$E[P_{xx}^{BT}(f)] \approx \int_{-1/2}^{1/2} \Gamma_{xx}(\theta)W(f - \theta) \, d\theta \tag{11.2.61}$$

$$\text{var } [P_{xx}^{BT}(f)] \approx \Gamma_{xx}^2(f) \left[\frac{1}{N} \sum_{m=-(M-1)}^{M-1} w^2(m) \right]$$

where $w(m)$ is the window sequence which is used to taper the estimated autocorrelation sequence. For the rectangular and Bartlett (triangular) windows we have

$$\frac{1}{N} \sum_{n=-(M-1)}^{M-1} w^2(n) = \begin{cases} 2M/N & \text{rectangular window} \\[3mm] 2M/3N & \text{triangular window} \end{cases} \tag{11.2.62}$$

TABLE 11.1 Quality of Power
Spectrum Estimates

Estimate	Quality Factor
Bartlett	$1.11N \, \Delta f$
Welch	$1.39N \, \Delta f$
(50% overlap)	
Blackman–Tukey	$2.34N \, \Delta f$

It is clear from (11.2.61) that the mean value of the estimate is asymptotically un-biased. Its quality factor for the triangular window is

$$Q_{BT} = 1.5 \frac{N}{M} \tag{11.2.63}$$

Since the window length is $2M - 1$, the frequency resolution measured at the 3-dB points is

$$\Delta f = \frac{1.28}{2M} = \frac{0.64}{M} \tag{11.2.64}$$

and hence

$$Q_{BT} = \frac{1.5}{0.64} N\Delta f = 2.34N\Delta f \tag{11.2.65}$$

These results are summarized in Table 11.1. It is apparent from the results we have obtained that the Welch and Blackman–Tukey power spectrum estimates are somewhat better than the Bartlett estimate. However, the differences in performance are relatively small. The main point is that the quality factor increases with an increase in the length N of the data. This characteristic behavior is not shared by the periodogram estimate. Furthermore, the quality factor depends on the product of the data length N and the frequency resolution Δf. For a desired level of quality, Δf can be decreased (frequency resolution increased) by increasing the length N of the data, and vice versa.

11.2.5 Computational Requirements of Nonparametric Power Spectrum Estimates

The other important aspect of the nonparametric power spectrum estimates is their com-putational requirements. For this comparison we assume the estimates are based on a fixed amount of data N and a specified resolution Δf. The radix-2 FFT algorithm is assumed in all the computations. We shall count only the number of complex multipli-cations required to compute the power spectrum estimate.

Bartlett Power Spectrum Estimate

$$\text{FFT length} = M = 0.9/\Delta f$$

$$\text{Number of FFTs} = \frac{N}{M} = 1.11N \, \Delta f$$

$$\text{Number of computations} = \frac{N}{M} \left(\frac{M}{2} \log_2 M \right) = \frac{N}{2} \log_2 \frac{0.9}{\Delta f}$$

Welch Power Spectrum Estimate (50% Overlap)

$$\text{FFT length} = M = 1.28/\Delta f$$

$$\text{Number of FFTs} = \frac{2N}{M} = 1.56N\,\Delta f$$

$$\text{Number of computations} = \frac{2N}{M}\left(\frac{M}{2}\log_2 M\right) = N\log_2 \frac{1.28}{\Delta f}$$

In addition to the $2N/M$ FFTs there are additional multiplications required for windowing the data. Each data record requires M multiplications. Therefore, the total number of computations is

$$\text{Total computations} = 2N + N\log_2 \frac{1.28}{\Delta f} = N\log_2 \frac{5.12}{\Delta f}$$

Blackman–Tukey Power Spectrum Estimate. In the Blackman–Tukey method, the autocorrelation $r_{xx}(m)$ can be computed efficiently via the FFT algorithm. However, if the number of data points is large, it may not be possible to compute one N-point DFT. For example, we may have $N = 10^5$ data points but only the capacity to perform 1024-point DFTs. Since the autocorrelation sequence is windowed to $2M - 1$ points where $M \ll N$, it is possible to compute the desired $2M - 1$ points of $r_{xx}(m)$ by segmenting the data into $K = N/2M$ records and computing $2M$-point DFTs and one $2M$-point IDFT via the FFT algorithm. Rader (1970) has described a method for performing this computation (see Problem 11.7).

If we base the computational complexity of the Blackman–Tukey method on this approach, we obtain the following computational requirements.

$$\text{FFT length} = 2M = 1.28/\Delta f$$

$$\text{Number of FFTs} = 2K + 1 = 2\left(\frac{N}{2M}\right) + 1 \approx \frac{N}{M}$$

$$\text{Number of computations} = \frac{N}{M}(M\log_2 2M) = N\log_2 \frac{1.28}{\Delta f}$$

We may neglect the additional M multiplications required to window the autocorrelation sequence $r_{xx}(m)$, since this is a relatively small number. Finally, there is the additional computation required to Fourier transform the windowed autocorrelation sequence. The FFT algorithm may be used for this computation with some zero padding for purposes of interpolating the spectral estimate. As a result of these additional computations, the number of computations given above is increased by a small amount.

From these results we conclude that the Welch method requires a little more computational power than do the other two methods. The Bartlett method apparently requires the smallest number of computations. However, the differences in the computational requirements of the three methods are relatively small.

11.3 Parametric Methods for Power Spectrum Estimation

The nonparametric power spectrum estimation methods described in the preceding section are relatively simple, well understood, and easy to compute via the FFT algorithm. However, these methods require the availability of long data records in order to yield the necessary frequency resolution that is required in many applications. Furthermore, these methods suffer from spectral leakage effects due to windowing that are inherent in finite-length data records. Often, the spectral leakage masks weak signals that are present in the data.

From one point of view, the basic limitation of the nonparametric methods is the inherent assumption that the autocorrelation estimate $r_{xx}(m)$ is zero for $m \geq N$, as implied by (11.1.33). This assumption severely limits the frequency resolution and the quality of the power spectrum estimate that is achieved. From another viewpoint, the inherent assumption in the periodogram estimate is that the data are periodic with period N. Neither one of these inherent assumptions is realistic.

In this section we describe power spectrum estimation methods that do not require such assumptions. In fact, these methods *extrapolate* the values of the autocorrelation for lags $m \geq N$. Extrapolation is possible if we have some a priori information on how the data were generated. In such a case a model for the signal generation may be constructed with a number of parameters that can be estimated from the observed data. From the model and the estimated parameters we can compute the power density spectrum implied by the model.

In effect, the modeling approach eliminates the need for window functions and the assumption that the autocorrelation sequence is zero for $|m| \geq N$. As a consequence, *parametric* (model-based) power spectrum estimation methods provide better frequency resolution than do the FFT-based, nonparametric methods described in the preceding section and avoid the problem of leakage. This is especially true in applications where short data records are available due to time-variant or transient phenomena.

The parametric methods considered in this section are based on modeling the data sequence $x(n)$ as the output of a linear system characterized by a rational system function of the form

$$H(z) = \frac{B(z)}{A(z)}$$

$$= \frac{\displaystyle\sum_{k=0}^{q} b_k z^{-k}}{1 + \displaystyle\sum_{k=1}^{p} a_k z^{-k}} \tag{11.3.1}$$

The corresponding difference equation is

$$x(n) = -\sum_{k=1}^{p} a_k x(n-k) + \sum_{k=0}^{q} b_k w(n-k) \tag{11.3.2}$$

where $w(n)$ is the input sequence to the system and the observed data, $x(n)$, represents the output sequence.

In power spectrum estimation, the input sequence is not observable. However, if the observed data are characterized as a stationary random process, then the input sequence is also assumed to be a stationary random process. In such a case the power density spectrum of the data is

$$\Gamma_{xx}(f) = |H(f)|^2 \Gamma_{ww}(f)$$

where $\Gamma_{ww}(f)$ is the power density spectrum of the input sequence and $H(f)$ is the frequency response of the model.

Since our objective is to estimate the power density spectrum $\Gamma_{xx}(f)$, it is convenient to assume that the input sequence $w(n)$ is a zero-mean white noise sequence with auto-correlation

$$\gamma_{ww}(m) = \sigma_w^2 \, \delta(m)$$

where σ_w^2 is the variance (i.e., $\sigma_w^2 = E[|w(n)|^2]$). Then the power density spectrum of the observed data is simply

$$\Gamma_{xx}(f) = \sigma_w^2 |H(f)|^2 = \sigma_w^2 \frac{|B(f)|^2}{|A(f)|^2} \tag{11.3.3}$$

A discrete-time random sequence can be represented by (11.3.2) if its power density spectrum satisfies the Paley–Wiener condition [see Papoulis (1984)]

$$\int_{-1/2}^{1/2} \log \Gamma_{xx}(f) \, df < \infty$$

In the model-based approach, the spectrum estimation procedure consists of two steps. Given the data sequence $x(n)$, $0 \le n \le N - 1$, we estimate the parameters $\{a_k\}$ and $\{b_k\}$ of the model. Then from these estimates, we compute the power spectrum estimate according to (11.3.3).

The random process $x(n)$ generated by the pole–zero model in (11.3.1) or (11.3.2) is called an *autoregressive–moving average* (ARMA) *process* of order (p, q) and it is usually denoted as ARMA (p, q). If $q = 0$ and $b_0 = 1$, the resulting system model has a system function $H(z) = 1/A(z)$ and its output $x(n)$ is called an *autoregressive* (AR) *process* of order p. This is denoted as AR(p). The third possible model is obtained by setting $A(z) = 1$, so that $H(z) = B(z)$. Its output $x(n)$ is called a *moving average* (MA) *process* of order q and denoted as MA(q).

Of these three linear models the AR model is by far the most widely used. The reasons are twofold. First, the AR model is suitable for representing spectra with narrow peaks (resonances). Second, the AR model results in very simple linear equations for the AR parameters. On the other hand, the MA model, as a general rule, requires many more coefficients to represent a narrow spectrum. Consequently, it is rarely used alone as a model for spectrum estimation. By combining poles and zeros, the ARMA model provides a more efficient representation from the viewpoint of the number of model parameters to represent the spectrum of a random process.

The decomposition theorem due to Wold (1938) asserts that any ARMA or MA process may be represented uniquely by an AR model of possibly infinite order, and any ARMA or AR process may be represented by a MA model of possibly infinite order. In view of this theorem, the issue of model selection reduces to selecting the model that requires the smallest number of parameters which are also easy to compute. Usually, the choice in practice is the AR model. The ARMA model is used to a lesser extent.

Before describing methods for estimating the parameters in an AR(p), MA(q), and ARMA(p, q) models, it is useful to establish the basic relationships between the model parameters and the autocorrelation sequence $\gamma_{xx}(m)$. In addition, we relate the AR model parameters to the coefficients in a linear predictor for the process $x(n)$.

11.3.1 Relationships Between the Autocorrelation and the Model Parameters

There are several relationships that exist between the autocorrelation sequence $\gamma_{xx}(m)$ and the system model parameters. These relationships are established below.

First, the z-transform of $\gamma_{xx}(m)$ is defined as

$$\Gamma_{xx}(z) = \sum_{m=-\infty}^{\infty} \gamma_{xx}(m)z^{-m} \qquad (11.3.4)$$

where the region of convergence of $\Gamma_{xx}(z)$ is the annulus $r_1 < |z| < r_2$, which includes the unit circle. Since

$$\Gamma_{xx}(z) = \sigma_w^2 H(z)H(z^{-1}) \qquad (11.3.5)$$

it follows that

$$\gamma_{xx}(m) = \sigma_w^2 \sum_{n=0}^{\infty} h^*(n)h(n+m) \qquad (11.3.6)$$

Second, in Section 5.1.4 we demonstrated that the relationship

$$\Gamma_{xx}(z) = \sigma_w^2 H(z)H(z^{-1}) = \sigma_w^2 \frac{B(z)B(z^{-1})}{A(z)A(z^{-1})} \equiv \sigma_w^2 \frac{D(z)}{C(z)} \qquad (11.3.7)$$

implies that

$$c_m = \sum_{k=0}^{p-|m|} a_k a_{k+m} \qquad |m| \le p$$
$$d_m = \sum_{k=0}^{q-|m|} b_k b_{k+m} \qquad |m| \le q \qquad (11.3.8)$$

where

$$C(z) = \sum_{m=-p}^{p} c_m z^{-m}$$
$$D(z) = \sum_{m=-q}^{q} d_m z^{-m} \qquad (11.3.9)$$

An explicit relationship between the autocorrelation sequence $\gamma_{xx}(m)$ and the parameters $\{a_k\}$, $\{b_k\}$ can be obtained from (11.3.2). If we multiply (11.3.2) by $x^*(n-m)$ and take the expected value of both sides of the equation, we obtain

$$E[x(n)x^*(n-m)] = -\sum_{k=1}^{p} a_k E[x(n-k)x^*(n-m)] + \sum_{k=0}^{q} b_k E[w(n-k)x^*(n-m)]$$

Hence

$$\gamma_{xx}(m) = -\sum_{k=1}^{p} a_k \gamma_{xx}(m-k) + \sum_{k=0}^{q} b_k \gamma_{wx}(m-k) \qquad (11.3.10)$$

But the crosscorrelation $\gamma_{wx}(m)$ may be expressed as

$$\gamma_{wx}(m) = E[x^*(n)w(n+m)]$$
$$= E\left[\sum_{k=0}^{\infty} h(k)w^*(n-k)w(n+m) \right]$$
$$= \sum_{k=0}^{\infty} h(k)E[w^*(n-k)w(n+m)]$$
$$= \sigma_w^2 h(-m)$$

where, in the last step, we have used the fact that the sequence $w(n)$ is white [i.e., $E[w^*(n)w(n+m)] = \sigma_w^2 \delta(m)$]. Therefore,

$$\gamma_{wx}(m) = \begin{cases} 0 & m > 0 \\ \sigma_w^2 h(-m) & m \le 0 \end{cases} \qquad (11.3.11)$$

With the result in (11.3.11) substituted into (11.3.10), we obtain the desired relationship

$$\gamma_{xx}(m) = \begin{cases} -\displaystyle\sum_{k=1}^{p} a_k \gamma_{xx}(m - k) & m > q \\[2mm] -\displaystyle\sum_{k=1}^{p} a_k \gamma_{xx}(m - k) + \sigma_w^2 \displaystyle\sum_{k=0}^{q-m} h(k) b_{k+m} & 0 \le m \le q \\[2mm] \gamma_{xx}^*(-m) & m < 0 \end{cases} \qquad (11.3.12)$$

The relationships in (11.3.12) provide a formula for determining the model parameters $\{a_k\}$ by restricting our attention to the case $m > q$. Thus the set of linear equations

$$\begin{bmatrix} \gamma_{xx}(q) & \gamma_{xx}(q - 1) & \cdots & \gamma_{xx}(q - p + 1) \\ \gamma_{xx}(q + 1) & \gamma_{xx}(q) & \cdots & \gamma_{xx}(q + p + 2) \\ \vdots & \vdots & & \\ \gamma_{xx}(q + p - 1) & \gamma_{xx}(q + p - 2) & \cdots & \gamma_{xx}(q) \end{bmatrix} \begin{bmatrix} a_1 \\ a_2 \\ \vdots \\ a_p \end{bmatrix} =$$

$$-\begin{bmatrix} \gamma_{xx}(q + 1) \\ \gamma_{xx}(q + 2) \\ \vdots \\ \gamma_{xx}(q + p) \end{bmatrix} \qquad (11.3.13)$$

may be used to solve for the model parameters $\{a_k\}$ by using estimates of the autocorrelation sequence in place of $\gamma_{xx}(m)$ for $m \ge q$. This problem is discussed in Section 11.3.9.

Another interpretation of the relationship in (11.3.13) is that the values of the autocorrelation $\gamma_{xx}(m)$ for $m > q$ are uniquely determined from the pole parameters $\{a_k\}$ and the values of $\gamma_{xx}(m)$ for $0 \le m \le p$. Consequently, the linear system model automatically extends the values of the autocorrelation sequence $\gamma_{xx}(m)$ for $m > p$.

If the pole parameters $\{a_k\}$ are obtained from (11.3.13), the result does not help us in determining the MA parameters $\{b_k\}$, because the equation

$$\sigma_w^2 \sum_{k=0}^{q-m} h(k) b_{k+m} = \gamma_{xx}(m) + \sum_{k=1}^{p} a_k \gamma_{xx}(m - k) \qquad 0 \le m \le q \quad (11.3.14)$$

depends on the impulse response $h(n)$. Although the impulse response can be expressed in terms of the parameters $\{b_k\}$ by long division of $B(z)$ with the known $A(z)$, this approach results in a set of nonlinear equations for the MA parameters.

If we adopt an AR(p) model for the observed data, the relationship between the AR parameters and the autocorrelation sequence is obtained by setting $q = 0$ in (11.3.12). Thus we obtain

$$\gamma_{xx}(m) = \begin{cases} -\displaystyle\sum_{k=1}^{p} a_k \gamma_{xx}(m - k) & m > 0 \\[2mm] -\displaystyle\sum_{k=1}^{p} a_k \gamma_{xx}(m - k) + \sigma_w^2 & m = 0 \\[2mm] \gamma_{xx}^*(-m) & m < 0 \end{cases} \qquad (11.3.15)$$

In this case, the AR parameters $\{a_k\}$ are obtained from the solution of the Yule–Walker or normal equations

$$
\begin{bmatrix}
\gamma_{xx}(0) & \gamma_{xx}(-1) & \cdots & \gamma_{xx}(-p+1) \\
\gamma_{xx}(1) & \gamma_{xx}(0) & \cdots & \gamma_{xx}(-p+2) \\
\vdots & \vdots & & \vdots \\
\gamma_{xx}(p-1) & \gamma_{xx}(p-2) & \cdots & \gamma(0)
\end{bmatrix}
\begin{bmatrix}
a_1 \\ a_2 \\ \vdots \\ a_p
\end{bmatrix}
= -
\begin{bmatrix}
\gamma_{xx}(1) \\ \gamma_{xx}(2) \\ \vdots \\ \gamma_{xx}(p)
\end{bmatrix}
$$

$$(11.3.16)$$

and the variance σ_w^2 can be obtained from the equation

$$
\sigma_w^2 = \gamma_{xx}(0) + \sum_{k=1}^{p} a_k \gamma_{xx}(-k) \tag{11.3.17}
$$

The equations in (11.3.16) and (11.3.17) are usually combined into a single matrix equation of the form

$$
\begin{bmatrix}
\gamma_{xx}(0) & \gamma_{xx}(-1) & \cdots & \gamma_{xx}(-p) \\
\gamma_{xx}(1) & \gamma_{xx}(0) & \cdots & \gamma_{xx}(-p+1) \\
\vdots & \vdots & & \vdots \\
\gamma_{xx}(p) & \gamma_{xx}(p-1) & \cdots & \gamma_{xx}(0)
\end{bmatrix}
\begin{bmatrix}
1 \\ a_1 \\ \vdots \\ a_p
\end{bmatrix}
=
\begin{bmatrix}
\sigma_w^2 \\ 0 \\ \vdots \\ 0
\end{bmatrix}
\tag{11.3.18}
$$

Since the correlation matrix in (11.3.16), or in (11.3.18), is Toeplitz, it can be efficiently inverted by use of the Levinson–Durbin algorithm, described in Appendix 6A.

Thus all the system parameters in the $AR(p)$ model are easily determined from knowledge of the autocorrelation sequence $\gamma_{xx}(m)$ for $0 \le m \le p$. Furthermore, (11.3.15) may be used to extend the autocorrelation sequence for $m > p$, once the $\{a_k\}$ are determined.

Finally, for completeness, we indicate that in a $MA(q)$ model for the observed data, the autocorrelation sequence $\gamma_{xx}(m)$ is related to the MA parameters $\{b_k\}$ by the equation

$$
\gamma_{xx}(m) =
\begin{cases}
\sigma_w^2 \sum\limits_{k=0}^{q} b_k b_{k+m} & 0 \le m \le q \\
0 & m > q \\
\gamma_{xx}^{*}(-m) & m < 0
\end{cases}
\tag{11.3.19}
$$

This result is easily established from (11.3.12) by setting $a_k = 0$ for $k = 1, 2, \ldots, p$ and substituting $\{b_k\}$ for $h(k)$. Furthermore, from (11.3.8) and (11.3.9) it is clear that $\gamma_{xx}(m) = \sigma_w^2 d_m$, and therefore the spectrum of the MA process is simply

$$
\Gamma_{xx}^{MA}(f) = \sigma_w^2 \sum_{m=-q}^{q} d_m e^{-j2\pi fm}
$$

Since $\gamma_{xx}(m) = 0$ for $m > q$, this spectrum has the same form as the (estimated) periodogram spectrum.

11.3.2 Relationship of AR Process to Linear Prediction

The parameters in an $AR(p)$ process are intimately related to a predictor of order p for the same process. The relationship is established in this section.

Let $x(n)$ be a sample sequence of a stationary random process and let us consider a linear predictor of order m for the data point $x(n)$. Thus we have

$$\hat{x}(n) = -\sum_{k=1}^{m} a_m(k)x(n - k) \tag{11.3.20}$$

where $a_m(k)$, $k = 1, 2, \ldots, m$ are the prediction coefficients for the mth-order predictor. We call $\hat{x}(n)$ in (11.3.20) the *forward predictor*, since it predicts forward in time. The forward prediction error is defined as (see Section 7.2.4)

$$f_m(n) = x(n) - \hat{x}(n) \tag{11.3.21}$$

$$= x(n) + \sum_{k=1}^{m} a_m(k)x(n - k)$$

and its mean-square value is

$$\mathcal{E}_m^f = E[|f_m(n)|^2]$$

$$= E\left[\left|x(n) + \sum_{k=1}^{m} a_m(k)x(n - k)\right|^2\right]$$

$$= \gamma_{xx}(0) + 2\,\text{Re}\left[\sum_{l=1}^{m} a_m^*(l)\gamma_{xx}(l)\right] + \sum_{k=1}^{m}\sum_{l=1}^{m} a_m^*(l)a_m(k)\gamma_{xx}(l - k) \tag{11.3.22}$$

$\mathcal{E}_m^f$ is a quadratic function of the predictor coefficients and its minimization leads to the set of linear equations (see Problem 11.12)

$$\gamma_{xx}(l) = -\sum_{k=1}^{m} a_m(k)\gamma_{xx}(l - k) \qquad l = 1, 2, \ldots, m \tag{11.3.23}$$

The minimum mean-square prediction error is

$$\min [\mathcal{E}_m^f] \equiv E_m^f = \gamma_{xx}(0) + \sum_{k=1}^{m} a_m(k)\gamma_{xx}(-k) \tag{11.3.24}$$

If we compare (11.3.23) and (11.3.24) to the equations in (11.3.16) and (11.3.17), we observe that the prediction coefficients in a linear predictor of order p are identical to the parameters of an AR(p) model. Furthermore, the minimum mean-square prediction error $E_p^f = \sigma_w^2$.

The Levinson–Durbin algorithm derived in Appendix 6A may be used to solve (11.3.23) for the linear predictor coefficients. This algorithm is initialized by solving for the coefficient in the first-order ($m = 1$) predictor

$$a_1(1) = \frac{-\gamma_{xx}(1)}{\gamma_{xx}(0)} \tag{11.3.25}$$

$$E_1^f = (1 - |a_1(1)|^2)\gamma_{xx}(0)$$

and the prediction coefficients for the higher-order predictors are given recursively as

$$a_m(k) = a_{m-1}(k) + a_m(m)a_{m-1}^*(m - k) \tag{11.3.26}$$

where

$$a_m(m) = -\frac{\gamma_{xx}(m) + \sum_{k=1}^{m-1} a_{m-1}(k)\gamma_{xx}(m-k)}{E^f_{m-1}} \tag{11.3.27}$$

$$E^f_m = (1 - |a_m(m)|^2)E^f_{m-1}$$

$$= \gamma_{xx}(0) \prod_{k=1}^{m} [1 - |a_k(k)|^2] \tag{11.3.28}$$

Note that the recursion in (11.3.26) involves the complex conjugate of $a_{m-1}(m-1)$, which is appropriate for the case in which the autocorrelation $\gamma_{xx}(m)$ is complex valued. In addition, we recall from Section 7.2.4 that $a_m(m) = K_m$, where K_m is the mth reflection coefficient in the equivalent lattice realization of the predictor.

If we substitute the recursion for $a_m(k)$ given in (11.3.26) into (11.3.21), we obtain an order-recursive equation for the forward prediction error, as

$$f_m(n) = x(n) + \sum_{k=1}^{m-1} a_{m-1}(k)x(n-k) + a_m(m)$$

$$\times \left[x(n-m) + \sum_{k=1}^{m-1} a^*_{m-1}(m-k)x(n-k) \right]$$

$$f_m(n) = f_{m-1}(n) + K_m g_{m-1}(n-1) \tag{11.3.29}$$

where $g_m(n)$ is the error in an mth-order *backward predictor*, defined as

$$g_m(n) = x(n-m) + \sum_{k=1}^{m} a^*_m(k)x(n-m+k) \tag{11.3.30}$$

It is easy to show (see Problem 11.14) that the mean-square value of the backward prediction error,

$$\mathscr{E}^b_m = E[|g_m(n)|^2] \tag{11.3.31}$$

when minimized with respect to the prediction coefficients yields the same set of equations as (11.3.23), and

$$\min [\mathscr{E}^b_m] \equiv E^b_m = E^f_m \tag{11.3.32}$$

Furthermore, if we substitute the recursion in (11.3.26) into (11.3.30), we obtain the order-recursive equation for the backward error as

$$g_m(n) = g_{m-1}(n-1) + K^*_m f_{m-1}(n) \tag{11.3.33}$$

The pair of order-recursive relations

$$f_m(n) = f_{m-1}(n) + K_m g_{m-1}(n-1) \tag{11.3.34}$$
$$g_m(n) = g_{m-1}(n-1) + K^*_m f_{m-1}(n)$$

define a lattice filter, as illustrated in Fig. 7.11, which is a direct consequence of the Levinson–Durbin recursion. The minimization of $\mathscr{E}^f_m = E[|f_m(n)|^2]$ and $\mathscr{E}^b_m = E[|g_m(n)|^2]$ with respect to the reflection coefficients $\{K_m\}$ yields the results

$$K_m = \frac{-E[f_{m-1}(n)g^*_{m-1}(n-1)]}{E[|g_{m-1}(n-1)|^2]} \tag{11.3.35}$$

and

$$K_m^* = \frac{-E[f_{m-1}^*(n)g_{m-1}(n-1)]}{E[|f_{m-1}(n)|^2]} \tag{11.3.36}$$

Since the denominators terms in (11.3.35) and (11.3.36) are equal, as indicated by (11.3.32), it follows that the reflection coefficients may also be expressed as

$$K_m = \frac{-E[f_{m-1}(n)g_{m-1}^*(n-1)]}{\sqrt{E[|g_{m-1}(n-1)|^2]E[|f_{m-1}(n)|^2]}} \tag{11.3.37}$$

From this form it is apparent that the reflection coefficients in the lattice filter are the negative of the (normalized) correlation coefficients between the forward and backward errors in the lattice. It is also apparent from (11.3.37) that $|K_m| \leq 1$. Since $|K_m| \leq 1$, it follows that the minimum mean-square value of the prediction error, which is given recursively as

$$E_m^f = (1 - |K_m|^2)E_{m-1}^f$$

or, equivalently, as

$$\sigma_{wm}^2 = (1 - |K_m|^2)\sigma_{wm-1}^2 \tag{11.3.38}$$

is a monotonically decreasing sequence and can be used in the AR model for determining the order (number of poles p) that provides a good fit to the data. Clearly, when σ_{wm}^2 does not change significantly with an increase in the order of the model, we can terminate the recursive algorithm.

In summary, we have shown that the linear prediction of the process $x(n)$ from either past or future samples yields predictor coefficients that are identical to the parameters in the AR model for representing the process $x(n)$. Hence the Levinson–Durbin algorithm that provides the prediction coefficients in the linear predictor also yields the AR model parameters of all orders from $m = 1$ to $m = p$. This property, coupled with (11.3.38), helps us to adaptively determine the model order that provides a good fit to the data. On the other hand, if the process $x(n)$ is AR(p), we will find that the model parameters $a_m(m) = 0$ for $m > p$, and $\sigma_{wp}^2 = \sigma_w^2$ for all $m > p$.

The parameters $K_m = a_m(m)$, $m = 1, 2, \ldots, p$, which are a by-product of the Levinson–Durbin algorithm, are the reflection coefficients of the lattice filter that is equivalent to the linear FIR predictor. These coefficients satisfy the condition $|K_m| \leq 1$, which implies that the roots of

$$A_p(z) = 1 + \sum_{k=1}^{p} a_p(k)z^{-k} \tag{11.3.39}$$

lie inside the unit circle [see Papoulis (1984)]. Therefore, the AR filter model is stable.

The filter with the system function $A(z)$ given by (11.3.39) is called the *forward prediction error filter*. Its input is $x(n)$ and its output is the forward prediction error $f_p(n)$. If $x(n)$ is actually an AR process, then $A(z)$ is the inverse filter or the noise whitening filter to $x(n)$. That is, $f_p(n)$ is a white noise sequence. Similarly, the *backward prediction error filter* with system function

$$B_p(z) = \sum_{k=0}^{p} \beta_p(k)z^{-k} \tag{11.3.40}$$

where $\beta_p(k) = a_p^*(p - k)$ is also a noise whitening filter when the process $x(n)$ is an AR(p) process.

With the background established above, we will now describe the power spectrum estimation methods for AR(p) and ARMA(p, q) models.

11.3.3 The Yule–Walker Method for the AR Model Parameters

In the Yule–Walker method we simply estimate the autocorrelation from the data and use the estimates in (11.3.16) to solve for the AR model parameters. In this method it is desirable to use the biased form of the autocorrelation estimate,

$$r_{xx}(m) = \frac{1}{N} \sum_{n=0}^{N-|m|} x^*(n)x(n + m) \tag{11.3.41}$$

to ensure that the autocorrelation matrix is positive semidefinite. The result will be a stable AR model. Although stability is not a critical issue in power spectrum estimation, it is conjectured that a stable AR model best represents the data.

The Levinson–Durbin algorithm given by (11.3.25) through (11.3.28) with $r_{xx}(m)$ substituted for $\gamma_{xx}(m)$ yields the AR parameters. The corresponding power spectrum estimate is

$$P_{xx}^{YW}(f) = \frac{\hat{\sigma}_{wp}^2}{\left| 1 + \sum_{k=1}^{p} \hat{a}_p(k)e^{-j2\pi fk} \right|^2} \tag{11.3.42}$$

where $\hat{a}_p(k)$ are estimates of the AR parameters obtained from the Levinson–Durbin recursions and

$$\hat{\sigma}_{wp}^2 = \hat{E}_p^f = r_{xx}(0) \prod_{k=1}^{p} [1 - |\hat{a}_k(k)|^2] \tag{11.3.43}$$

is the estimated minimum mean-square value for the pth-order predictor. An example illustrating the frequency resolution capabilities of this estimator is given in Section 11.3.10.

In estimating the power spectrum of sinusoidal signals via AR models, Lacoss (1971) showed that spectral peaks in an AR spectrum estimate are proportional to the square of the power of the sinusoidal signal. On the other hand, the area under the peak in the power density spectrum is linearly proportional to the power of the sinusoid. This characteristic behavior holds for all AR model-based estimation methods.

11.3.4 The Burg Method for the AR Model Parameters

The method devised by Burg (1968) for estimating the AR parameters may be viewed as an order-recursive least-squares lattice method based on the minimization of the forward and backward errors in linear predictors, with the constraint that the AR parameters satisfy the Levinson–Durbin recursion.

To derive the estimator, suppose that we are given the data $x(n)$, $n = 0, 1, \ldots$, $N - 1$, and let us consider the forward and backward linear prediction estimates of order m, which are given as

$$\hat{x}(n) = -\sum_{k=1}^{m} a_m(k)x(n - k)$$

$$\hat{x}(n - m) = -\sum_{k=1}^{m} a_m^*(k)x(n + k - m) \tag{11.3.44}$$

and the corresponding forward and backward errors $f_m(n)$ and $g_m(n)$ given by (11.3.21) and (11.3.30), respectively, where $a_m(k)$, $0 \le k \le m - 1$, $m = 1, 2, \ldots, p$ are the prediction coefficients. The total squared error is

$$\mathscr{E}_m = \sum_{n=m}^{N-1} [|f_m(n)|^2 + |g_m(n)|^2] \qquad (11.3.45)$$

This error is to be minimized by selecting the prediction coefficients, subject to the constraint that they satisfy the Levinson–Durbin recursion given by

$$a_m(k) = a_{m-1}(k) + K_m a_{m-1}^*(m - k) \qquad \begin{array}{l} 1 \le k \le m - 1 \\ 1 \le m \le p \end{array} \qquad (11.3.46)$$

where $K_m = a_m(m)$ is the mth reflection coefficient in the lattice filter realization of the predictor. Recall that when (11.3.46) is substituted into the expressions for $f_m(n)$ and $g_m(n)$, the result is the pair of order-recursive equations for the forward and backward prediction errors given by (11.3.34).

Now, if we substitute from (11.3.34) into (11.3.45) and perform the minimization of $\mathscr{E}_m$ with respect to the complex-valued reflection coefficient K_m, we obtain the result

$$\hat{K}_m = \frac{-\displaystyle\sum_{n=m}^{N-1} f_{m-1}(n) g_{m-1}^*(n - 1)}{\dfrac{1}{2} \displaystyle\sum_{n=m}^{N-1} [|f_{m-1}(n)|^2 + |g_{m-1}(n - 1)|^2]} \qquad m = 1, 2, \ldots, p \qquad (11.3.47)$$

The term in the numerator of (11.3.47) is an estimate of the crosscorrelation between the forward and backward prediction errors. With the normalization factors in the denominator of (11.3.47), it is apparent that $|K_m| < 1$, so that the all-pole model obtained from the data is stable. The reader should note the similarity of (11.3.47) with the statistical counterparts given by (11.3.35) through (11.3.37).

We note that the denominator in (11.3.47) is simply the least-squares estimate of the forward and backward errors, E_{m-1}^f and E_{m-1}^b, respectively. Hence (11.3.47) may be expressed as

$$\hat{K}_m = \frac{-\displaystyle\sum_{n=m}^{N-1} f_{m-1}(n) g_{m-1}^*(n - 1)}{\frac{1}{2}[\hat{E}_{m-1}^f + \hat{E}_{m-1}^b]} \qquad m = 1, 2, \ldots, p \qquad (11.3.48)$$

where $\hat{E}_{m-1}^f + \hat{E}_{m-1}^b$ is an estimate of the total squared error E_m. We leave as an exercise for the reader to verify that the denominator term in (11.3.48) can be computed in an order-recursive fashion according to the relation

$$\hat{E}_m = (1 - |\hat{K}_{m-1}|^2)\hat{E}_{m-1} - |f_{m-1}(N)|^2 - |g_{m-1}(N - m)|^2 \qquad (11.3.49)$$

where $\hat{E}_m \equiv \hat{E}_m^f + \hat{E}_m^b$ is the total least-squares error. This result is due to Andersen (1978).

To summarize, the Burg algorithm computes the reflection coefficients in the equivalent lattice structure as specified by (11.3.48) and (11.3.49), and the Levinson–Durbin algorithm is used to obtain the AR model parameters. From the estimates of the AR parameters, we form the power spectrum estimate

$$P_{xx}^{BU}(f) = \frac{\hat{E}_p}{\left| 1 + \displaystyle\sum_{k=1}^{p} \hat{a}_p(k) e^{-j2\pi fk} \right|^2} \qquad (11.3.50)$$

The major advantages of the Burg method for estimating the parameters of the AR model are (1) it results in high frequency resolution, (2) it yields a stable AR model, and (3) it is computationally efficient.

The Burg method is known to have several disadvantages, however. First, it exhibits spectral line splitting at high signal-to-noise ratios. [see the paper by Fougere et al. (1976)]. By line splitting, we mean that the spectrum of $x(n)$ may have a single sharp peak, but the Burg method may result in two or more closely spaced peaks. For high-order models, the method also introduces spurious peaks. Furthermore, for sinusoidal signals in noise, the Burg method exhibits a sensitivity to the initial phase of a sinusoid, especially in short data records. This sensitivity is manifest as a frequency shift from the true frequency, resulting in a frequency bias that is phase dependent. For more details on some of these limitations the reader is referred to the papers of Chen and Stegen (1974), Ulrych and Clayton (1976), Fougere et al. (1976), Kay and Marple (1979), Swingler (1979a, 1980), Herring (1980), and Thorvaldsen (1981).

Several modifications have been proposed to overcome some of the more important limitations of the Burg method: namely, the line splitting, spurious peaks, and frequency bias. Basically, the modifications involve the introduction of a weighting (window) sequence on the squared forward and backward errors. That is, the least-squares optimization is performed on the weighted squared errors

$$\mathcal{E}_m^{WB} = \sum_{n=m}^{N-1} w_m(n)[|f_m(n)|^2 + |g_m(n)|^2] \tag{11.3.51}$$

which, when minimized, results in the reflection coefficient estimates

$$\hat{K}_m = \frac{\displaystyle\sum_{n=m}^{N-1} w_{m-1}(n)f_{m-1}(n)g_{m-1}^*(n-1)}{\dfrac{1}{2}\displaystyle\sum_{n=m}^{N-1} w_{m-1}(n)[|f_{m-1}(n)|^2 + |g_{m-1}(n-1)|^2]} \tag{11.3.52}$$

In particular, we mention the use of a Hamming window used by Swingler (1979b), a quadratic or parabolic window used by Kaveh and Lippert (1983), the energy weighting method used by Nikias and Scott (1982), and the data-adaptive energy weighting used by Helme and Nikias (1985).

These windowing and energy weighting methods have proved effective in reducing the occurrence of line splitting and spurious peaks, and are also effective in reducing frequency bias.

The Burg method for power spectrum estimation is usually associated with *maximum entropy spectrum estimation*, which is a criterion used by Burg (1967, 1975) as a basis for AR modeling in parametric spectrum estimation. The problem considered by Burg was how best to extrapolate from the given values of the autocorrelation sequence $\gamma_{xx}(m)$, $0 \le m \le p$, the values for $m > p$, such that the entire autocorrelation sequence is positive semidefinite. Since an infinite number of extrapolations are possible, Burg postulated that the extrapolation be made on the basis of maximizing uncertainty (entropy) or randomness, in the sense that the spectrum $\Gamma_{xx}(f)$ of the process is the flattest of all spectra which have the given autocorrelation values $\gamma_{xx}(m)$, $0 \le m \le p$. In particular the entropy per sample is proportional to the integral [see Burg (1975)]

$$\int_{-1/2}^{1/2} \ln \Gamma_{xx}(f) \, df \tag{11.3.53}$$

Burg found that the maximum of this integral subject to the $(p + 1)$ constraints

$$\int_{-1/2}^{1/2} \Gamma_{xx}(f)e^{j2\pi fm}\, df = \gamma_{xx}(m) \qquad 0 \le m \le p \qquad (11.3.54)$$

is the AR(p) process for which the given autocorrelation sequence $\gamma_{xx}(m)$, $0 \le m \le p$ is related to the AR parameters by the equation (11.3.15). This solution provides an additional justification for the use of the AR model in power spectrum estimation.

In view of Burg's basic work in maximum entropy spectral estimation, the Burg power spectrum estimation procedure is often called the *maximum entropy method* (MEM). We should emphasize, however, that the maximum entropy spectrum is identical to the AR-model spectrum only when the exact autocorrelation $\gamma_{xx}(m)$ is known. When only an estimate of $\gamma_{xx}(m)$ is available for $0 \le m \le p$, the AR-model estimates of Yule–Walker and Burg are not maximum entropy spectral estimates. The general formulation for the maximum entropy spectrum based on estimates of the autocorrelation sequence results in a set of nonlinear equations. Solutions for the maximum entropy spectrum with measurement errors in the correlation sequence have been obtained by Newman (1981) and Schott and McClellan (1984).

11.3.5 Unconstrained Least-Squares Method for the AR Model Parameters

As described in the preceding section, the Burg method for determining the parameters of the AR model is basically a least-squares lattice algorithm with the added constraint that the predictor coefficients satisfy the Levinson recursion. As a result of this constraint, an increase in the order of the AR model requires only a single parameter optimization at each stage. In contrast to this approach, we may use an unconstrained least-squares algorithm to determine the AR parameters.

To elaborate, we form the forward and backward linear prediction estimate and their corresponding forward and backward errors as indicated in (11.3.44). Then we minimize the sum of squares of both errors, that is,

$$
\begin{aligned}
\mathcal{E}_p &= \sum_{n=p}^{N-1} [|f_p(n)|^2 + |g_p(n)|^2] \\
&= \sum_{n=p}^{N-1} \left[\left| x(n) + \sum_{k=1}^{p} a_p(k)x(n-k) \right|^2 + \left| x(n-p) + \sum_{k=1}^{p} a_p^*(k)x(n+k-p) \right|^2 \right]
\end{aligned}
$$
$$(11.3.55)$$

which is the same performance index as in the Burg method. However, we will not impose the Levinson–Durbin recursion in (11.3.55) for the AR parameters. The unconstrained minimization of $\mathcal{E}_p$ with respect to the prediction coefficients yields the set of linear equations

$$\sum_{k=1}^{p} a_p(k)r_{xx}(l, k) = -r_{xx}(l, 0) \qquad l = 1, 2, \ldots, p \qquad (11.3.56)$$

where, by definition, the autocorrelation $r_{xx}(l, k)$ is

$$r_{xx}(l, k) = \sum_{n=p}^{N-1} [x(n-k)x^*(n-l) + x(n-p+l)x^*(n-p+k)] \qquad (11.3.57)$$

The resulting residual least-squares error is

$$\mathscr{E}_p^{LS} = r_{xx}(0, 0) + \sum_{k=1}^{p} \hat{a}_p(k)r_{xx}(0, k) \tag{11.3.58}$$

Hence the unconstrained least-squares power spectrum estimate is

$$P_{xx}^{LS}(f) = \frac{E_p^{LS}}{\left| 1 + \sum_{k=1}^{p} \hat{a}_p(k)e^{-j2\pi fk} \right|^2} \tag{11.3.59}$$

The correlation matrix in (11.3.57), with elements $r_{xx}(l, k)$, is not Toeplitz, so that the Levinson–Durbin algorithm cannot be applied. However, the correlation matrix has sufficient structure to make it possible to devise computationally efficient algorithms with computational complexity proportional to p^2. Marple (1980) devised such an algorithm, which has a lattice structure and employs Levinson–Durbin-type order recursions and additional time recursions. This algorithm is similar in form to the adaptive lattice algorithms described in Chapter 12 and will not be discussed further in this section. The interested reader is referred to the paper by Marple (1980).

The form of the unconstrained least-squares method described above has also been called the *unwindowed data* least-squares method. It has been proposed for spectrum estimation in several papers, including the papers by Burg (1967), Nuttall (1976), and Ulrych and Clayton (1976). Its performance characteristics have been found to be superior to the Burg method, in the sense that the unconstrained least-squares method does not exhibit the same sensitivity to such problems as line splitting, frequency bias, and spurious peaks. In view of the computational efficiency of Marple's algorithm, which is comparable to the efficiency of the Levinson–Durbin algorithm, the unconstrained least-squares method is very attractive. With this method there is no guarantee that the estimated AR parameters yield a stable AR model. However, in spectrum estimation, this is not considered to be a problem.

11.3.6 Sequential Estimation Methods for the AR Model Parameters

The three power spectrum estimation methods described in the preceding sections for the AR model may be classified as block processing methods. These methods obtain estimates of the AR parameters from a block of data, say $x(n)$, $n = 0, 1, \ldots, N - 1$. The AR parameters based on the block of N data points is then used to obtain the power spectrum estimate.

In situations where data are available on a continuous basis, we can still segment the data into blocks of N points and perform spectrum estimation on a block-by-block basis. This is often done in practice, for both real-time and non-real-time applications. However, in such applications, there is an alternative approach based on sequential (in time) estimation of the AR model parameters as each new data point becomes available. By introducing a weighting function into past data samples, it is possible to deemphasize the effect of older data samples as new data are received.

Sequential estimation methods for AR models have been developed over the past 20 years as a result of adaptive FIR filtering applications. Adaptive FIR filters and adaptive filtering algorithms are intimately related to linear prediction and linear estimation methods. In view of the relationships we have already established between the coefficients in a linear prediction filter and the parameters of the AR model, it should not be surprising that the adaptive filtering algorithms are also applicable directly to power spectrum estimation based on the AR model.

Chapter 12 is devoted entirely to the derivation of sequential adaptive filtering algorithms. Both the LMS algorithm and the recursive least-squares algorithms presented in Chapter 12 are applicable sequential algorithms for estimating the AR model parameters. In particular, the sequential lattice methods, described in Chapter 12, directly and optimally estimate the prediction coeffficients and the reflection coefficients in the lattice realization of the forward and backward linear predictors. The recursive equations for the prediction coefficients relate directly to the AR model parameters. In addition to the order-recursive nature of these equations, as implied by the lattice structure, we also obtain time-recursive equations for the reflection coefficients in the lattice and for the forward and backward prediction coefficients.

The sequential recursive least-squares algorithms described in Chapter 12 are equivalent to the unconstrained least-squares block processing method described in the preceding section. Hence the power spectrum estimates obtained by the sequential recursive least-squares method retain the desirable properties of the block processing algorithm described in Section 11.3.5. Since the AR parameters are being estimated continuously in a sequential estimation algorithm, power spectrum estimates may be obtained as often as desired, from once per sample to once every N samples. By properly weighting past data samples, the sequential estimation methods are particularly suitable for estimating and tracking time-variant power spectra resulting from nonstationary signal statistics.

The computational complexity of the sequential estimation methods described in Chapter 12 is proportional to p, the order of the AR process. As a consequence, the sequential estimation algorithms are computationally efficient and, from this viewpoint, may offer some advantage over the block processing methods.

Many references to sequential estimation methods in adaptive filtering are given in the following chapter. The papers by Griffiths (1975), Friedlander (1982b) and Kalouptsidis and Theodoridis (1987) are particularly relevant to the spectrum estimation problem.

11.3.7 Selection of AR Model Order

One of the most important aspects of the use of the AR model is the selection of the order p. As a general rule, if we select a model with too low an order, we obtain a highly smoothed spectrum. On the other hand, if p is selected too high, we run the risk of introducing spurious low-level peaks in the spectrum. We mentioned previously that one indication of the performance of the AR model is the mean-square value of the residual error, which, in general, is different for each of the estimators described above. The characteristic of this residual error is that it decreases as the order of the AR model is increased. We can monitor the rate of decrease and decide to terminate the process when the rate of decrease becomes relatively slow. It is apparent, however, that this approach may be imprecise and ill-defined, and other methods should be investigated.

Much work has been done by various researchers on this problem and many experimental results have been given in the literature [e.g., the papers by Gersch and Sharpe (1973), Ulrych and Bishop (1975), Tong (1975, 1977), Jones (1976), Nuttall (1976), Berryman (1978), Kaveh and Bruzzone (1979), and Kashyap (1980)].

Two of the better known criterion for selecting the model order have been proposed by Akaike (1969, 1974). With the first, called the *final prediction error (FPE) criterion*, the order is selected to minimize the performance index

$$FPE(p) = \hat{\sigma}_{wp}^2 \left(\frac{N + p + 1}{N - p - 1} \right) \qquad (11.3.60)$$

where $\hat{\sigma}_{wp}^2$ is the estimated variance of the linear prediction error. This performance index is based on minimizing the mean-square error for a one-step predictor.

The second criterion proposed by Akaike (1974), called the *Akaike information criterion* (AIC), is based on selecting the order that minimizes

$$\text{AIC}(p) = \ln \hat{\sigma}_{wp}^2 + 2p/N \tag{11.3.61}$$

Note that the term $\hat{\sigma}_{wp}^2$ decreases and hence $\ln \hat{\sigma}_{wp}^2$ also decreases, as the order of the AR model is increased. However, $2p/N$ increases with an increase in p. Hence a minimum value is obtained for some p.

An alternative information criterion, proposed by Rissanen (1983), is based on selecting the order that *minimizes the description length* (MDL), where MDL is defined as

$$\text{MDL}(p) = N \ln \hat{\sigma}_{wp}^2 + p \ln N \tag{11.3.62}$$

A fourth criterion has been porposed by Parzen (1974). This is called the *criterion autoregressive transfer* (CAT) function and is defined as

$$\text{CAT}(p) = \left(\frac{1}{N} \sum_{k=1}^{p} \frac{1}{\bar{\sigma}_{wk}^2} \right) - \frac{1}{\bar{\sigma}_{wp}^2} \tag{11.3.63}$$

where

$$\bar{\sigma}_{wk}^2 = \frac{N}{N-k} \hat{\sigma}_{wk}^2 \tag{11.3.64}$$

The order p is selected to minimize $\text{CAT}(p)$.

In applying the criteria given above, the mean should be removed from the data. Since $\hat{\sigma}_{wk}^2$ depends on the type of spectrum estimate we obtain, the model order is also a function of the criterion.

The experimental results given in the references cited above indicate that the model-order selection criteria do not yield definitive results. For example, Ulrych and Bishop (1975), Jones (1976), and Berryman (1978), found that the FPE(p) criterion tends to underestimate the model order. Kashyap (1980) showed that the AIC criterion is statistically inconsistent as $N \to \infty$. On the other hand, the MDL information criterion proposed by Rissanen is statistically consistent. Other experimental results indicate that for small data lengths, the order of the AR model should be selected to be in the range $N/3$ to $N/2$ for good results. It is apparent that in the absence of any prior information regarding the physical process that resulted in the data, one should try different model orders and different criteria and, ultimately, interpret the different results.

11.3.8 MA Model for Power Spectrum Estimation

As shown in Section 11.3.1, the parameters in a MA(q) model are related to the statistical autocorrelation $\gamma_{xx}(m)$ by (11.3.19). According to (11.3.7) and (11.3.8),

$$B(z)B(z^{-1}) = D(z) = \sum_{m=-q}^{q} d_m z^{-m} \tag{11.3.65}$$

where the coefficients $\{d_m\}$ are related to the MA parameters by the equation (11.3.8). Clearly, then,

$$\gamma_{xx}(m) = \begin{cases} \sigma_w^2 d_m & |m| \le q \\ 0 & |m| > q \end{cases} \tag{11.3.66}$$

and the power spectrum for the MA(q) process is

$$\Gamma_{xx}^{MA}(f) = \sum_{m=-q}^{q} \gamma_{xx}(m)e^{-j2\pi fm} \tag{11.3.67}$$

It is apparent from these expressions that we do not have to solve for the MA parameters $\{b_k\}$ to estimate the power spectrum. The estimates of the autocorrelation $\gamma_{xx}(m)$ for $|m| \leq q$ suffice. From such estimates we compute the estimated MA power spectrum, given as

$$P_{xx}^{MA}(f) = \sum_{m=-q}^{q} r_{xx}(m)e^{-j2\pi fm} \tag{11.3.68}$$

which is identical to the classical (nonparametric) power spectrum estimate described in Section 11.1.

Since $\gamma_{xx}(m) = 0$ for $|m| > q$, the order of the MA model may be determined empirically by noting if the values of the (unbiased) autocorrelation estimates are nearly zero for large lags. If this is not the case, within the limitations imposed by the data length N, the MA model will result in poor frequency resolution and should be abandoned in favor of either the AR model or ARMA model.

11.3.9 ARMA Model for Power Spectrum Estimation

The Burg algorithm, and its variations, and the least-squares method described in the previous sections provide reliable high-resolution spectrum estimates based on the AR model. An ARMA model provides us with an opportunity to improve on the AR spectrum estimate, perhaps, by using fewer model parameters.

The ARMA model is particularly appropriate when our data have been corrupted by noise. For example, suppose that the data $x(n)$ are generated by an AR system, where the system output is corrupted by additive white noise. The z-transform of the autocorrelation of the resultant signal may be expressed as

$$\begin{aligned}
\Gamma_{xx}(z) &= \frac{\sigma_w^2}{A(z)A(z^{-1})} + \sigma_n^2 \\
&= \frac{\sigma_w^2 + \sigma_n^2 A(z)A(z^{-1})}{A(z)A(z^{-1})}
\end{aligned} \tag{11.3.69}$$

where σ_n^2 is the variance of the additive noise. Therefore, the process $x(n)$ is ARMA(p, p), where p is the order of the autocorrelation process. This relationship provides some motivation for investigating ARMA models for power spectrum estimation.

As we have demonstrated in Section 11.3.1, the parameters of the ARMA model are related to the autocorrelation by the equation in (11.3.12). For lags $|m| > q$, the equation involves only the AR parameters $\{a_k\}$. With estimates substituted in place of $\gamma_{xx}(m)$, we can solve the p equations in (11.3.13) to obtain $\hat{a}_k$. For high-order models, however, this approach is likely to yield poor estimates of the AR parameters due to the poor estimates of the autocorrelation for large lags. Consequently, this approach is not recommended.

Instead, a more reliable method is to construct an overdetermined set of linear equations for $m > q$, and to use the method of least squares on the set of overdetermined equations, as proposed by Cadzow (1979). To elaborate, we may write (as in linear prediction)

$$\hat{r}_{xx}(m) = -\sum_{k=1}^{p} a_k r_{xx}(m-k) \qquad m = q+1, q+2, \ldots, M < N \quad (11.3.70)$$

where $r_{xx}(m)$ is the estimated autocorrelation sequence, for which we may use either the biased or the unbiased form. Then, we select the parameters $\{a_k\}$ that minimize the squared error

$$\mathscr{E} = \sum_{n=q+1}^{M} |e(n)|^2 \tag{11.3.71}$$

$$= \sum_{n=q+1}^{M} \left| r_{xx}(m) + \sum_{k=1}^{p} a_k r_{xx}(m-k) \right|^2$$

This is the familiar least-squares problem that we have considered previously in other contexts. The minimization of $\mathscr{E}$ yields a set of linear equations for the AR parameters $\{a_k\}$. This procedure is called the *least-squares modified Yule–Walker method*. A weighting factor may also be applied to the autocorrelation sequence to deemphasize the less reliable estimates for large lags.

Once the parameters for the AR part of the model have been estimated as indicated above, we have the system

$$\hat{A}(z) = 1 + \sum_{k=1}^{p} \hat{a}_k z^{-k} \tag{11.3.72}$$

The sequence $x(n)$ may now be filtered by the FIR filter $\hat{A}(z)$ to yield the sequence

$$v(n) = x(n) + \sum_{k=1}^{p} \hat{a}_k x(n-k) \qquad n = 0, 1, \ldots, N-1 \tag{11.3.73}$$

The cascade of the ARMA(p, q) model with $\hat{A}(z)$ is approximately the MA(q) process generated by the model $B(z)$. Hence we may apply the MA estimate given in the preceding section to obtain the MA spectrum. To be specific, the filtered sequence $v(n)$ for $p \leq n \leq N-1$ is used to form the estimated correlation sequences $r_{vv}(m)$, from which we obtain the MA spectrum

$$P_{vv}^{MA}(f) = \sum_{m=-q}^{q} r_{vv}(m) e^{-j2\pi fm} \tag{11.3.74}$$

First, we observe that the parameters $\{b_k\}$ are not required to determine the power spectrum. Second, we observe that $r_{vv}(m)$ is an estimate of the autocorrelation for the MA model given by (11.3.19). In forming the estimate $r_{vv}(m)$, weighting (e.g., with the Bartlett window) may be used to deemphasize correlation estimates for large lags. In addition, the data may be filtered by a backward filter, thus creating another sequence, say $v^b(n)$, so that both $v(n)$ and $v^b(n)$ can be used in forming the estimate of the autocorrelation $r_{vv}(m)$, as proposed by Kay (1980). Finally, the estimated ARMA power spectrum is

$$\hat{P}_{xx}^{ARMA}(f) = \frac{P_{vv}^{MA}(f)}{\left| 1 + \sum_{k=1}^{p} \hat{a}_k e^{-j2\pi fk} \right|^2} \tag{11.3.75}$$

The problem of order selection for the ARMA(p, q) model has been investigated by Chow (1972) and Bruzzone and Kaveh (1980). For this purpose the minimum of the

AIC index

$$\text{AIC}(p, q) = \ln \hat{\sigma}^2_{wpq} + \frac{2(p + q)}{N} \tag{11.3.76}$$

may be used, where $\hat{\sigma}^2_{wpq}$ is an estimate of the variance of the input error. An additional test on the adequacy of a particular ARMA(p, q) model is to filter the data through the model and test for whiteness of the output data. This would require that the parameters of the MA model be computed from the estimated autocorrelation, using spectral factorization to determine $B(z)$ from $D(z) = B(z)B(z^{-1})$.

Another approach to order determination in ARMA modeling has been proposed by Cadzow (1982). In this paper he demonstrates that the use of a singular value (eigenvalue) decomposition of an extended autocorrelation matrix provides a good estimate for the order of the ARMA model.

For additional reading on ARMA power spectrum estimation, the reader is referred to the papers by Graupe et al. (1975), Cadzow (1981, 1982), Kay (1980), and Friedlander (1982b).

11.3.10 Some Experimental Results

In this section we present some experimental results on the performance of AR and ARMA power spectrum estimates which were obtained with artificially generated data. Our objective is to compare the spectral estimation methods on the basis of their frequency resolution, bias, and their robustness in the presence of additive noise.

The data consist of either one or two sinusoids and additive Gaussian noise. The two

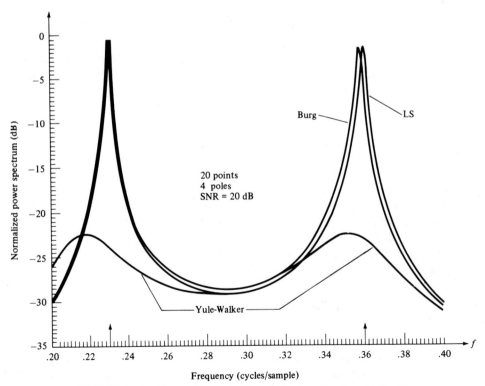

FIGURE 11.6   Comparison of AR spectrum estimation methods.

sinousoids are spaced Δf apart. Clearly, the underlying process is ARMA(4, 4). The results that are shown employ an AR(p) model for these data. For high signal-to-noise ratios (SNRs) we expect the AR(4) to be adequate. However, for low SNRs, a higher-order AR model is needed to approximate the ARMA(4, 4) process. The results given below are consistent with this statement. The SNR is defined as $10 \log_{10} A^2/2\sigma^2$, where σ^2 is variance of the additive noise and A is the amplitude of the sinusoid.

In Fig. 11.6 we illustrate the results for $N = 20$ data points based on an AR(4) model with a SNR = 20 dB and $\Delta f = 0.13$. Note that the Yule–Walker method gives an extremely smooth (broad) spectral estimate with small peaks. If Δf is decreased to $\Delta f = 0.07$, the Yule–Walker method no longer resolves the peaks as illustrated in Fig. 11.7. Some bias is also evident in the Burg method. Of course, by increasing the number of data points the Yule–Walker method will eventually resolve the peaks. However, the Burg and least-squares methods are clearly superior for short data records.

The effect of additive noise on the estimate is illustrated in Fig. 11.8 for the least-squares method. The effect of filter order on the Burg and least-squares methods is illustrated in Figs. 11.9 and 11.10, respectively. Both methods exhibit spurious peaks as the order is increased.

The effect of initial phase is illustrated in Figs. 11.11 and 11.12, for the Burg and least-squares methods. It is clear that the least-squares method exhibits less sensitivity to initial phase than the Burg algorithm.

An example of line splitting for the Burg method is shown in Fig. 11.13 with $p = 12$. It does not occur for the AR(8) model. The least-squares method did not exhibit line splitting under the same conditions. On the other hand, the line splitting on the Burg method disappeared with an increase in the number of data points N.

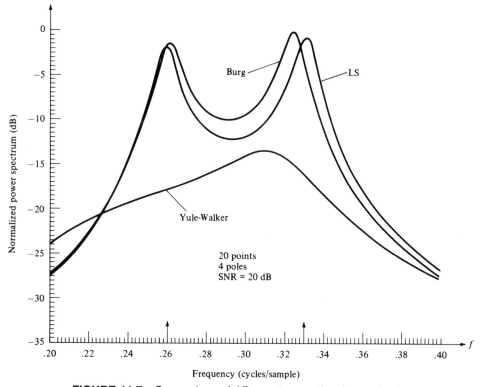

FIGURE 11.7   Comparison of AR spectrum estimation methods.

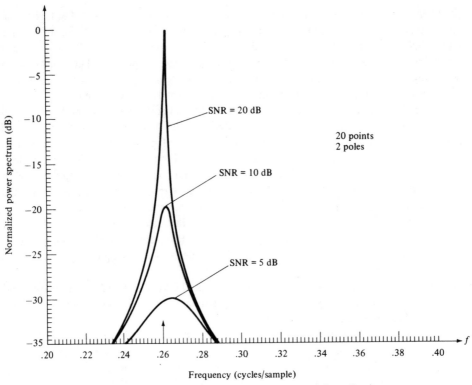

FIGURE 11.8 Effect of additive noise on LS method.

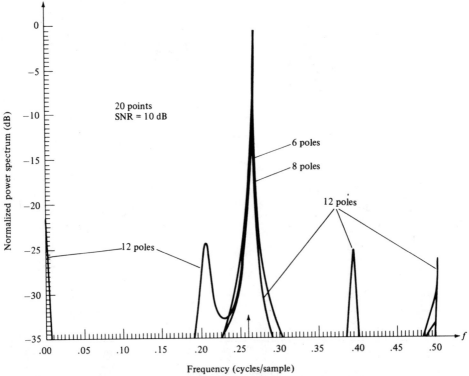

FIGURE 11.9 Effect of filter order on Burg method.

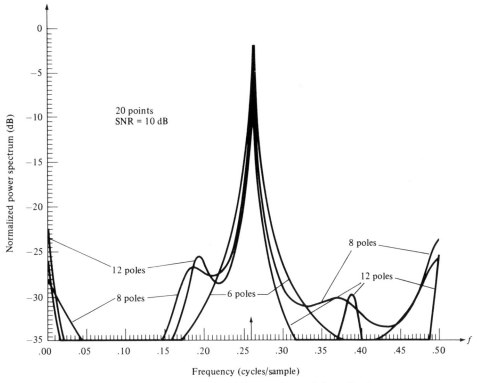

FIGURE 11.10   Effect of filter order on LS method.

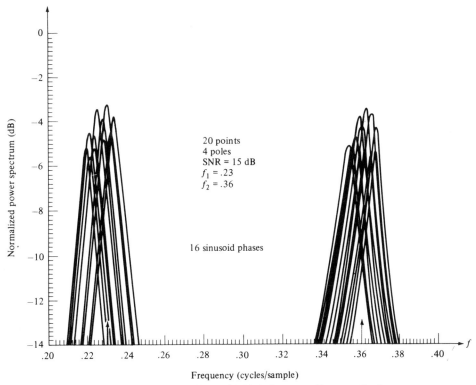

FIGURE 11.11   Effect of initial phase on Burg method.

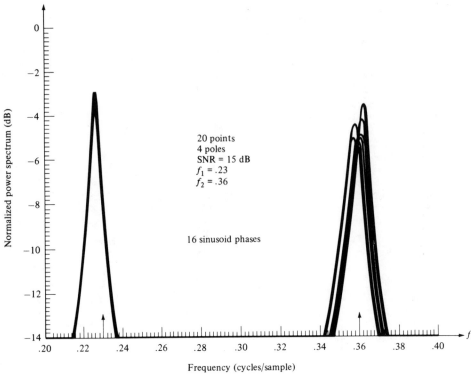

FIGURE 11.12   Effect of initial phase on LS method.

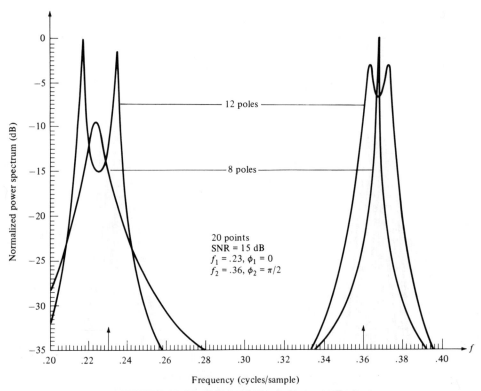

FIGURE 11.13   Line splitting in Burg method.

838

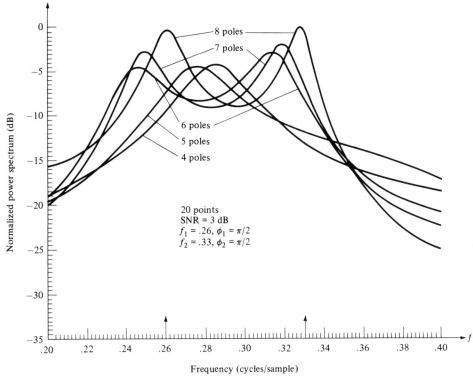

FIGURE 11.14 Frequency resolution of Burg method with $N = 20$ points.

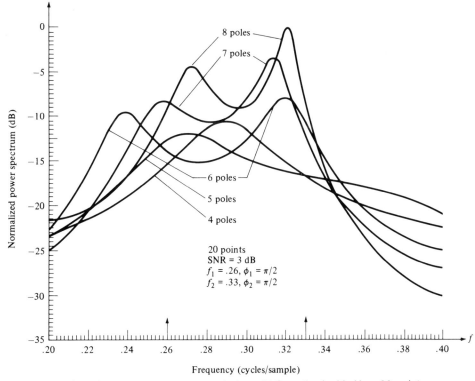

FIGURE 11.15 Frequency resolution of LS method with $N = 20$ points.

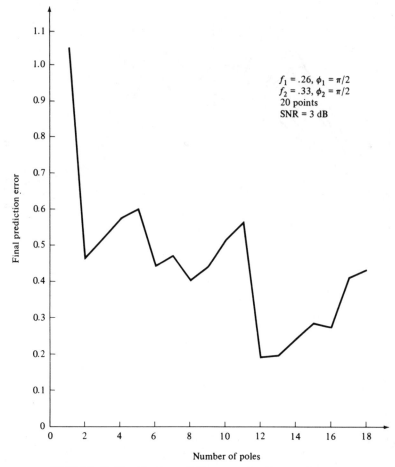

FIGURE 11.16 Final prediction error for Burg estimate.

Figures 11.14 and 11.15 illustrate the resolution properties of the Burg and least-squares methods for $\Delta f = 0.07$ and $N = 20$ points at low SNR (3 dB). Since the additive noise process is ARMA, a higher-order AR model is required to provide a good approximation at low SNR. Hence the frequency resolution improves as the order is increased.

The FPE for the Burg method is illustrated in Fig. 11.16 for an SNR = 3 dB. For this SNR the optimum value is $p = 12$ according to the FPE criterion.

The Burg and least-squares methods were also tested with data from a narrowband process, which was obtained by exciting a four-pole (two pairs of complex-conjugate poles) narrowband filter and selecting a portion of the output sequence for the data record. Figure 11.17 illustrates the superposition of 20 data records of 20 points each. We observe a relatively small variability. In contrast, the Burg method exhibited a much larger variability, approximately a factor of 2 compared to the least-squares method. The results shown in Figs. 11.6 through 11.17 are taken from Poole (1981).

Finally, we show in Fig. 11.18 the ARMA(10, 10) spectral estimates obtained by Kay (1980) for two sinusoids in noise using the least-squares ARMA method described

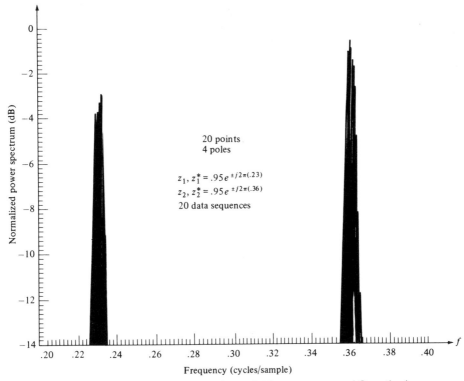

FIGURE 11.17 Effect of starting point in sequence on LS method.

in Section 11.3.9, as an illustration of the quality of power spectrum estimation obtained with the ARMA model.

11.4 Other Spectrum Estimation Methods

The nonparametric and parametric spectrum estimation methods described in Sections 11.2 and 11.3 are widely used in practice. However, there are two additional methods that have received considerable attention in the literature and have been used in practical applications. One is the method proposed by Capon (1969), termed the *maximum likelihood method*. The second method is based on a harmonic decomposition method originally proposed by Pisarenko (1973) for estimating sinusoidal signals in additive noise. These two methods are described briefly in this section.

11.4.1 The Capon Method

The spectral estimator proposed by Capon (1969) was intended for use in large seismic arrays for frequency–wave number estimation. It was later adapted to single-time-series spectrum estimation by Lacoss (1971), who demonstrated that the method provides a minimum variance unbiased estimate of the spectral components in the signal.

Following the development of Lacoss, let us consider an FIR filter with coefficients a_k, $0 \le k \le p$, to be determined. Unlike the linear prediction problem, we do not

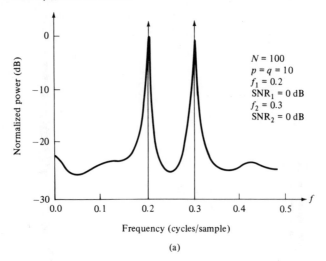

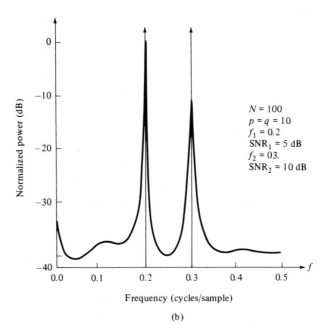

FIGURE 11.18 ARMA (10, 10) power spectrum estimates from paper by Kay (1980). Reprinted with permission from the IEEE.

constrain a_0 to be unity. Then, if the observed data $x(n)$, $0 \le n \le N - 1$, are passed through the filter, the response is

$$y(n) = \sum_{k=0}^{p} a_k x(n - k) \equiv \mathbf{X}^t(n)\mathbf{a} \tag{11.4.1}$$

$\mathbf{X}^t(n) = [x(n) \quad x(n - 1) \quad \cdots \quad x(n - p)]$ is the data vector and $\mathbf{a}$ is the filter coefficient vector. If we assume that $E[x(n)] = 0$, the variance of the output sequence is

$$\sigma_y^2 = E[|y(n)|^2] = F[\mathbf{a}^{*t}\mathbf{X}^*(n)\mathbf{X}^t(n)\mathbf{a}] \tag{11.4.2}$$
$$= \mathbf{a}^{*t}\mathbf{\Gamma}_{xx}\mathbf{a}$$

where $\mathbf{\Gamma}_{xx}$ is the autocorrelation matrix of the sequence $x(n)$, with elements $\gamma_{xx}(m)$.

The filter coefficients are selected so that at the frequency f_l, the frequency response of the FIR filter is normalized to unity, that is,

$$\sum_{k=0}^{p} a_k e^{-j2\pi k f_l} = 1$$

This constraint may also be written in matrix form as

$$\mathbf{E}^{*t}(f_l)\mathbf{a} = 1 \qquad (11.4.3)$$

where

$$\mathbf{E}^t(f_l) = [1 \quad e^{j2\pi f_l} \quad \cdots \quad e^{j2\pi p f_l}]$$

By minimizing the variance σ_y^2 subject to the constraint (11.4.3), we obtain an FIR filter that passes the frequency component f_l undistorted, while components distant from f_l are severely attenuated. The result of this minimization is shown by Lacoss to lead to the coefficient vector

$$\hat{\mathbf{a}} = \boldsymbol{\Gamma}_{xx}^{-1}\mathbf{E}^*(f_l)/\mathbf{E}^t(f_l)\boldsymbol{\Gamma}_{xx}^{-1}\mathbf{E}^*(f_l) \qquad (11.4.4)$$

If $\hat{\mathbf{a}}$ is substituted into (11.4.2), we obtain the minimum variance

$$\sigma_{\min}^2 = \frac{1}{\mathbf{E}^t(f_l)\boldsymbol{\Gamma}_{xx}^{-1}\mathbf{E}^*(f_l)} \qquad (11.4.5)$$

The expression in (11.4.5) is the minimum variance power spectrum estimate at the frequency f_l. By changing f_l over the range $0 \leq f_l \leq 0.5$ we can obtain the power spectrum estimate. It should be noted that although $\mathbf{E}(f)$ changes with the choice of frequency, $\boldsymbol{\Gamma}_{xx}^{-1}$ is computed only once. As demonstrated by Lacoss (1971), the computation of the quadratic form $\mathbf{E}^t(f)\boldsymbol{\Gamma}_{xx}^{-1}\mathbf{E}^*(f)$ can be done with a single DFT.

With an estimate $\mathbf{R}_{xx}$ of the autocorrelation matrix substituted in place of $\boldsymbol{\Gamma}_{xx}$, we obtain the minimum variance power spectrum estimate of Capon as

$$P_{xx}^{\text{MV}}(f) = \frac{1}{\mathbf{E}^t(f)\mathbf{R}_{xx}^{-1}\mathbf{E}^*(f)} \qquad (11.4.6)$$

It has been shown by Lacoss (1971) that this power spectrum estimator yields estimates of the spectral peaks which are proportional to the power at that frequency. In constrast, the AR methods described in Section 11.3 result in estimates of the spectral peaks which are proportional to the square of the power at that frequency.

The minimum variance method as described above is basically a filter bank implementation for the spectrum estimator. It differs basically from the filter bank interpretation of the periodogram in that the filter coefficients in the Capon method are optimized.

Experiments on the performance of this method compared with the performance of the Burg method have been done by Lacoss (1971) and others. In general, the minimum variance estimate in (11.4.6) outperforms the nonparametric spectral estimators in frequency resolution, but it does not provide the high frequency resolution obtained with the AR methods of Burg and the unconstrained least squares. Extensive comparisons between the Burg method and the minimum variance method have been made in the paper by Lacoss. Furthermore, Burg (1972) demonstrated that for a known correlation sequence, the minimum variance spectrum is related to the AR model spectrum through the equation

$$\frac{1}{\Gamma_{xx}^{MV}(f)} = \frac{1}{p}\sum_{k=0}^{p}\frac{1}{\Gamma_{xx}^{AR}(f, k)} \qquad (11.4.7)$$

where $\Gamma_{xx}^{AR}(f, k)$ is the AR power spectrum obtained with an AR(k) model. Thus the reciprocal of the minimum variance estimate is equal to the average of the reciprocals of all spectra obtained with AR(k) models for $1 \leq k \leq p$. Since low-order AR models, in general, will not provide good resolution, the averaging operation in (11.4.7) reduces the frequency resolution in the spectral estimate. Hence we conclude that the AR power spectrum estimate of order p is superior to the minimum variance estimate of order $p + 1$.

The relationship given by (11.4.7) represents a frequency-domain relationship between the Capon minimum variance estimate and the Burg AR estimate. A time-domain relationship between these two estimates also can be established as shown by Musicus (1985). This has led to a computationally efficient algorithm for the minimum variance estimate.

Additional references to the method of Capon and comparisons with other estimators can be found in the literature. We cite the papers of Capon and Goodman (1971), Marzetta (1983), Marzetta and Lang (1983, 1984), Capon (1983), and McDonough (1983).

11.4.2 The Pisarenko Harmonic Decomposition Method

In Section 11.3.9 we demonstrated that an AR(p) process corrupted by additive (white) noise is equivalent to an ARMA(p, p) process. In this section we consider the special case in which the signal components are sinusoids corrupted by additive white noise. The technique for estimating the spectrum of the noise-corrupted signal was proposed by Pisarenko (1973).

From our previous discussion on the generation of sinusoids in Chapter 5, we recall that a real sinusoidal signal can be generated via the difference equation.

$$x(n) = -a_1 x(n - 1) - a_2 x(n - 2) \tag{11.4.8}$$

where $a_1 = 2 \cos 2\pi f_k$ and $a_2 = 1$, and, initially, $x(-1) = -1$, $x(-2) = 0$. This system has a pair of complex-conjugate poles (at $f = f_k$ and $f = -f_k$) and thus generates the sinusoid $x(n) = \cos 2\pi f_k n$, for $n \geq 0$.

In general, a signal consisting of p sinusoidal components satisfies the difference equation

$$x(n) = -\sum_{m=1}^{2p} a_m x(n - m) \tag{11.4.9}$$

and corresponds to the system with system function

$$H(z) = \frac{1}{1 + \displaystyle\sum_{m=1}^{2p} a_m z^{-m}} \tag{11.4.10}$$

The polynomial

$$A(z) = 1 + \sum_{m=1}^{2p} a_m z^{-m} \tag{11.4.11}$$

has $2p$ roots on the unit circle which correspond to the frequencies of the sinusoids.

Now, suppose that the sinusoids are corrupted by a white noise sequence $w(n)$ with $E[|w(n)|^2] = \sigma_w^2$. Then we observe that

$$y(n) = x(n) + w(n) \tag{11.4.12}$$

If we substitute $x(n) = y(n) - w(n)$ in (11.4.9), we obtain

$$y(n) - w(n) = -\sum_{m=1}^{2p} [y(n - m) - w(n - m)]a_m$$

or, equivalently,

$$\sum_{m=0}^{2p} a_m y(n - m) = \sum_{m=0}^{2p} a_m w(n - m) \qquad (11.4.13)$$

where, by definition, $a_0 = 1$.

We observe that (11.4.13) is the difference equation for an ARMA(p, p) process in which both the AR and MA parameters are identical. This symmetry is a characteristic of the sinusoidal signals in white noise. The difference equation in (11.4.13) may be expressed in matrix form as

$$\mathbf{Y}'\mathbf{a} = \mathbf{W}'\mathbf{a} \qquad (11.4.14)$$

where $\mathbf{Y}' = [y(n) \quad y(n - 1) \quad \cdots \quad y(n - 2p)]$ is the observed data vector of dimension $(2p + 1)$, $\mathbf{W}' = [w(n) \quad w(n - 1) \quad \cdots \quad w(n - 2p)]$ is the noise vector, and $\mathbf{a} = [1 \quad a_1 \quad \cdots \quad a_{2p}]$ is the coefficient vector.

If we premultiply (11.4.14) by $\mathbf{Y}$ and take the expected value, we obtain

$$E(\mathbf{YY}')\mathbf{a} = E(\mathbf{YW}')\mathbf{a} = E[(\mathbf{X} + \mathbf{W})\mathbf{W}']\mathbf{a}$$
$$\mathbf{\Gamma}_{yy}\mathbf{a} = \sigma_w^2 \mathbf{a} \qquad (11.4.15)$$

where we have used the assumption that the sequence $w(n)$ is zero mean and white, and $\mathbf{X}$ is a deterministic signal.

The equation in (11.4.15) is in the form of an eigenequation, that is,

$$(\mathbf{\Gamma}_{yy} - \sigma_w^2 \mathbf{I})\mathbf{a} = \mathbf{O} \qquad (11.4.16)$$

where σ_w^2 is an eigenvalue of the autocorrelation matrix $\mathbf{\Gamma}_{yy}$. Then the parameter vector $\mathbf{a}$ is an eigenvector associated with the eigenvalue σ_w^2. The eigenequation in (11.4.16) forms the basis for the Pisarenko harmonic decomposition method.

For p sinusoids in additive white noise the autocorrelation values are

$$\gamma_{yy}(0) = \sigma_w^2 + \sum_{i=1}^{p} P_i \qquad (11.4.17)$$

$$\gamma_{yy}(k) = \sum_{i=1}^{p} P_i \cos 2\pi f_i k \qquad k \neq 0$$

where $P_i = A_i^2/2$ is the average power in the ith sinusoid and A_i is the corresponding amplitude. Hence we may write

$$\begin{bmatrix} \cos 2\pi f_1 & \cos 2\pi f_2 & \cdots & \cos 2\pi f_p \\ \cos 4\pi f_1 & \cos 4\pi f_2 & \cdots & \cos 4\pi f_p \\ \vdots & \vdots & & \vdots \\ \cos 2\pi p f_1 & \cos 2\pi p f_2 & \cdots & \cos 2\pi p f_p \end{bmatrix} \begin{bmatrix} P_1 \\ P_2 \\ \vdots \\ P_p \end{bmatrix} = \begin{bmatrix} \gamma_{yy}(1) \\ \gamma_{yy}(2) \\ \vdots \\ \gamma_{yy}(p) \end{bmatrix} \qquad (11.4.18)$$

If we know the frequencies f_i, $1 \leq i \leq p$, we may use this equation to determine the powers of the sinusoids. In place of $\gamma_{xx}(m)$, we use the estimates $r_{xx}(m)$. Once the powers are known, the noise variance can be obtained from (11.4.17) as

$$\sigma_w^2 = r_{yy}(0) - \sum_{i=1}^{p} P_i \qquad (11.4.19)$$

The problem that remains is to determine the p frequencies f_i, $1 \leq i \leq p$, which, in turn, require knowledge of the eigenvector $\mathbf{a}$ corresponding to the eigenvalue σ_w^2. Pisarenko (1973) observed [see also Papoulis (1984) and Grenander and Szegö (1958)] that for an ARMA process consisting of p sinusoids in additive white noise, the variance σ_w^2 corresponds to the minimum eigenvalue of Γ_{yy}, when the dimension of the autocorrelation matrix equals or exceeds $(2p + 1) \times (2p + 1)$. The desired ARMA coefficient vector corresponds to the eigenvector associated with the minimum eigenvalue. Therefore, the frequencies f_i, $1 \le i \le p$ are obtained from the roots of the polynomial in (11.4.11), where the coefficients are the elements of the eigenvector $\mathbf{a}$ corresponding to the minimum eigenvalue σ_w^2.

EXAMPLE 11.4.1 _____

Suppose that we are given the autocorrelation values $\gamma_{yy}(0) = 3$, $\gamma_{yy}(1) = 1$, and $\gamma_{yy}(2) = 0$ for a process consisting of a single sinusoid in additive white noise. Determine the frequency, its power, and the variance of the additive noise.

Solution: The correlation matrix is

$$\Gamma_{yy} = \begin{bmatrix} 3 & 1 & 0 \\ 1 & 3 & 1 \\ 0 & 1 & 3 \end{bmatrix}$$

The minimum eigenvalue is the smallest root of the characteristic polynomial

$$g(\lambda) = \begin{bmatrix} 3 - \lambda & 1 & 0 \\ 1 & 3 - \lambda & 1 \\ 0 & 1 & 3 - \lambda \end{bmatrix} = (3 - \lambda)(\lambda^2 - 6\lambda + 7) = 0$$

Therefore, the eigenvalues are $\lambda_1 = 3$, $\lambda_2 = 3 + \sqrt{2}$, $\lambda_3 = 3 - \sqrt{2}$.
The variance of the noise is

$$\sigma_w^2 = \lambda_{\min} = 3 - \sqrt{2}$$

The corresponding eigenvector is the vector that satisfies (11.4.16), that is,

$$\begin{bmatrix} \sqrt{2} & 1 & 0 \\ 1 & \sqrt{2} & 1 \\ 0 & 1 & \sqrt{2} \end{bmatrix} \begin{bmatrix} 1 \\ a_1 \\ a_2 \end{bmatrix} = \begin{bmatrix} 0 \\ 0 \\ 0 \end{bmatrix}$$

The solution is $a_1 = -\sqrt{2}$ and $a_2 = 1$.
The next step is to use the value a_1 and a_2 to determine the roots of the polynomial in (11.4.11). We have

$$z^2 - \sqrt{2}\,z + 1 = 0$$

Thus

$$z_1, z_2 = \frac{1}{\sqrt{2}} \pm j\frac{1}{\sqrt{2}}$$

Note that $|z_1| = |z_2| = 1$, so that the roots are on the unit circle. The corresponding frequency is obtained from

$$z_i = e^{j2\pi f_i} = \frac{1}{\sqrt{2}} + j\frac{1}{\sqrt{2}}$$

which yields $f_1 = \frac{1}{8}$. Finally, the power of the sinusoid is

$$P_1 \cos 2\pi f_1 = \gamma_{yy}(1) = 1$$
$$P_1 = \sqrt{2}$$

and its amplitude is $A = \sqrt{2P_1} = \sqrt{2\sqrt{2}}$.

As a check on our computations, we have

$$\sigma_w^2 = \gamma_{yy}(0) - P_1$$
$$= 3 - \sqrt{2}$$

which agrees with $\lambda_{\min}$.

In this example, we were given the values of the autocorrelation $\gamma_{yy}(m)$. In practice, we use the estimates $r_{yy}(m)$. In addition, the number of sinusoids is not known. In such a case we begin with $p = 1$ and compute the minimum eigenvalue of the estimated correlation matrix $R_{yy}(m)$. Then p is increased in order successively, and the minimum eigenvalue is computed for each value of p. This process is terminated when the value of the minimum eigenvalue changes very little from the previous iteration. Once σ_w^2 is obtained, we solve for **a**, then for the roots of (11.4.11), then for the frequencies, and finally, for the powers of the sinusoids.

The Pisarenko method has its limitations. First, it is suitable for estimating the power spectrum for sinusoids in noise. It is not very effective in measuring broadband spectra. Second, if the order is selected too high, spurious frequency components may be introduced in the estimate. On the other hand, if p is low, the estimated spectral components are biased (i.e., they appear at the wrong frequencies). Finally, if the noise is colored, the technique must be modified as indicated in the paper by Satorius and Alexander (1978). Unfortunately, it is not always known a priori if the additive noise is colored or white.

In addition to Pisarenko's method, other high-resolution methods have been proposed for estimating the spectrum of sinusoids in noise. The interested reader is referred to the paper by Tufts and Kumaresan (1982) for a modification of the forward-backward linear prediction approach used in the Burg algorithm, and to the papers by Schmidt (1981, 1986) for a description of the *multiple signal classification* (MUSIC) algorithm.

11.5 Summary and References

Power spectrum estimation is one of the most important areas of research and applications in digital signal processing. In this chapter we have described the most important power spectrum estimation techniques and algorithms that have been developed over the past century, beginning with the nonparametric or classical methods based on the periodogram and concluding with the more modern parametric methods based on AR, MA, and ARMA linear models. Our treatment was limited in scope to single-time-series spectrum estimation methods, based on second moments (autocorrelation) of the statistical data.

The parametric and nonparametric methods that we have described have been extended to multichannel and multidimensional spectrum estimation. The tutorial paper by Mc-Clellan (1982) treats the multidimensional spectrum estimation problem, while the paper by Johnson (1982) treats the multichannel spectrum estimation problem. Additional spectrum estimation methods have been developed for use with higher-order cumulants that involve the bispectrum and the trispectrum. A tutorial paper on these topics has been published by Nikias and Raghuveer (1987).

As evidenced from our previous discussion, power spectrum estimation is an area that has attracted many researchers and, as a result, thousands of papers have been published in the technical literature on this subject. Much of this work has been concerned with new algorithms and techniques, and modifications of existing techniques. Other work has been concerned with obtaining an understanding of the capabilities and limitations of the various power spectrum methods. In this context the statistical properties and limitations of the classical nonparametric methods have been thoroughly analyzed and are well understood. The parametric methods have also been investigated by many researchers, but the analysis of their performance is difficult and, consequently, fewer results are available. Some of the papers that have addressed the problem of performance characteristics of parametric methods are those of Kromer (1969), Lacoss (1971), Berk (1974), Baggeroer (1976), Sakai (1979), Swingler (1980), and Lang and McClellan (1980).

In addition to the references already given in this chapter on the various methods for spectrum estimation and their performance, we should include for reference some of the tutorial and survey papers. In particular, we cite the tutorial paper by Kay and Marple (1981), which includes about 280 references, the paper by Brillinger (1974), and the Special Issue on Spectral Estimation of the *IEEE Proceedings*, September 1982. Another indication of the widespread interest in the subject of spectrum estimation and analysis is the recent publication of texts by Gardner (1987), Kay (1987), and Marple (1987), and the IEEE books edited by Childers (1978) and Kesler (1986).

Many computer programs as well as software packages that implement the various spectrum estimation methods described in this chapter are available. One software package is available through the IEEE (*Programs for Digital Signal Processing*, IEEE Press, 1979); others are available commercially.

PROBLEMS

11.1 **(a)** By expanding (11.1.23), taking the expected value and finally, taking the limit as $T_0 \to \infty$, show that the right-hand side converges to $\Gamma_{xx}(F)$.

(b) Prove that

$$\sum_{m=-N}^{N} r_{xx}(m)e^{-j2\pi fm} = \frac{1}{N}\left|\sum_{n=0}^{N-1} x(n)e^{-j2\pi fn}\right|^2$$

11.2 For zero mean, jointly Gaussian random variables, X_1, X_2, X_3, X_4, it is well known [see Papoulis (1984)] that

$$E(X_1 X_2 X_3 X_4) = E(X_1 X_2)E(X_3 X_4) + E(X_1 X_3)E(X_2 X_4) + E(X_1 X_4)E(X_2 X_3)$$

Use this result to derive the mean-square value of $r_{xx}(m)$, given by (11.1.24) and the variance, which is

$$\text{var}\,[r_{xx}(m)] = E[|r_{xx}(m)|^2] - |E[r_{xx}(m)]|^2$$

11.3 By use of the expression for the fourth joint moment for Gaussian random variables, show that

(a) $E[P_{xx}(f_1)P_{xx}(f_2)] = \sigma_x^4 \left\{ 1 + \left[\dfrac{\sin \pi(f_1 + f_2)N}{N \sin \pi(f_1 + f_2)} \right]^2 \right.$

$\left. + \left[\dfrac{\sin \pi(f_1 - f_2)N}{N \sin \pi(f_1 - f_2)} \right]^2 \right\}$

(b) $\text{cov}\,[P_{xx}(f_1)P_{xx}(f_2)] = \sigma_x^4 \left\{ \left[\dfrac{\sin \pi(f_1 + f_2)N}{N \sin \pi(f_1 + f_2)} \right]^2 + \left[\dfrac{\sin \pi(f_1 - f_2)N}{N \sin \pi(f_1 - f_2)} \right]^2 \right\}$

(c) $\text{var}\,[P_{xx}(f)] = \sigma_x^4 \left\{ 1 + \left(\dfrac{\sin 2\pi fN}{N \sin 2\pi f} \right)^2 \right\}$

under the condition that the sequence $x(n)$ is a zero-mean white Gaussian noise sequence with variance σ_x^2.

11.4 Generalize the results in Problem 11.3 to a zero-mean Gaussian noise process with power density spectrum $\Gamma_{xx}(f)$. Then derive the variance of the periodogram $P_{xx}(f)$, as given by (11.1.38). (*Hint*: Assume that the colored Gaussian noise process is the output of a linear system excited by white Gaussian noise. Then use the appropriate relations given in Appendix 10A.)

11.5 Show that the periodogram values at frequencies $f_k = k/L$, $k = 0, 1, \ldots,$ $L - 1$, given by (11.1.41) can be computed by passing the sequence through a bank of L IIR filters, where each filter has an impulse response

$$h_k(n) = e^{-j2\pi nk/N}u(n)$$

and then computing the magnitude-squared value of the filter outputs at $n = N$. Note that each filter has a pole on the unit circle at the frequency f_k.

11.6 Prove that the normalization factor given by (11.2.12) ensures that (11.2.19) is satisfied.

11.7 Let us consider the use of the DFT (computed via the FFT algorithm) to compute the autocorrelation of the complex-valued sequence $x(n)$, that is,

$$r_{xx}(m) = \frac{1}{N}\sum_{n=0}^{N-|m|-1} x^*(n)x(n + m)$$

Suppose the size M of the FFT is much smaller than that of the data length N. Specifically, assume that $N = KM$.
(a) Determine the steps needed to section $x(n)$ and compute $r_{xx}(m)$ for $-(M/2) + 1 \leq m \leq (M/2) - 1$, by using $4K$ M-point DFTs and one M-point IDFT
(b) Now consider the following three sequences $x_1(n)$, $x_2(n)$, and $x_3(n)$, each of duration M. Let the sequences $x_1(n)$ and $x_2(n)$ have arbitrary values in the range $0 \leq n \leq (M/2) - 1$, but are zero for $(M/2) \leq n \leq M - 1$. The sequence $x_3(n)$ is defined as

$$x_3(n) = \begin{cases} x_1(n) & 0 \leq \dfrac{M}{2} - 1 \\[2mm] x_2\left(n - \dfrac{M}{2}\right) & \dfrac{M}{2} \leq n \leq M - 1 \end{cases}$$

Determine a simple relationship among the M-point DFTs $X_1(k)$, $X_2(k)$, and $X_3(k)$.
(c) By using the result in part (b), show how the computation of the DFTs in part (a) can be reduced in number from $4K$ to $2K$.

11.8 The Bartlett method is used to estimate the power spectrum of a signal $x(n)$. We know that the power spectrum consists of a single peak with a 3-dB bandwidth of 0.01 cycle per sample, but we do not know the location of the peak.
(a) Assuming that N is large, determine the value of $M = N/K$ so that the spectral window is narrower than the peak.
(b) Explain why it is not advantageous to increase M beyond the value obtained in part (a).

11.9 Suppose we have $N = 1000$ samples from a sample sequence of a random process.
(a) Determine the frequency resolution of the Bartlett, Welch (50% overlap), and Blackman–Tukey methods for a quality factor $Q = 10$.

(b) Determine the record lengths (M) for the Bartlett, Welch (50% overlap, and Blackman–Tukey methods.

11.10 Consider the problem of continuously estimating the power spectrum from a sequence $x(n)$ based on averaging periodograms with exponential weighting into the past. Thus with $P_{xx}^{(0)}(f) = 0$, we have

$$P_{xx}^{(m)}(f) = wP_{xx}^{(m-1)}(f) + \frac{1-w}{M}\left|\sum_{n=0}^{M-1} x_m(n)e^{-j2\pi fn}\right|^2$$

where successive periodograms are assumed to be uncorrelated and w is the (exponential) weighting factor.

(a) Determine the mean and variance of $P_{xx}^{(m)}(f)$ for a Gaussian random process.

(b) Repeat the analysis of part (a) for the case in which the modified periodogram defined by Welch is used in the averaging, with no overlap.

11.11 The periodogram in the Bartlett method may be expressed as

$$P_{xx}^{(i)}(f) = \sum_{m=-(M-1)}^{M-1}\left(1 - \frac{|m|}{M}\right) r_{xx}^{(i)}(m)e^{-j2\pi fm}$$

where $r_{xx}^{(i)}(m)$ is the estimated autocorrelation sequence obtained from the ith block of data. Show that $P_{xx}^{(i)}(f)$ may be expressed as

$$P_{xx}^{(i)}(f) = E^{*t}(f)R_{xx}^{(i)}E(f)$$

where

$$E(f) = [1 \quad e^{j2\pi f} \quad e^{j4\pi f} \quad \cdots \quad e^{j2\pi(M-1)f}]^t$$

and therefore,

$$P_{xx}^B(f) = \frac{1}{K}\sum_{k=1}^{K} E^{*t}(f)R_{xx}^{(k)}E(f)$$

11.12 Derive the set of linear equations in (11.3.23) by substituting $a_m(k) = a_{mr}(k) + ja_{mi}(k)$ in (11.3.22) and differentiating the resulting equation with respect to $a_{mr}(k)$ and $a_{mi}(k)$, that is,

$$\frac{\partial \mathcal{E}_m^f}{\partial a_{mr}(k)} = 0 \qquad \frac{\partial \mathcal{E}_m^f}{\partial a_{mi}(k)} = 0$$

11.13 The set of linear equations in (11.3.23) may also be derived by invoking the orthogonality principle in mean-square estimation [see Papoulis (1984)], which asserts that the mean-square error $\mathcal{E}_m^f$ is minimized by making the error $f_m(n)$ orthogonal to each of the data samples $x(n-l)$, $l = 1, 2, \ldots, m$ that are used to form the estimate. That is,

$$E[f_m(n)x^*(n-l)] = 0 \qquad l = 1, 2, \ldots, m$$

(a) Demonstrate that (11.3.23) is easily derived by invoking the orthogonality principle.

(b) Derive the expression for the minimum mean-square error in (11.3.24) by substituting the optimum prediction coefficients in (11.3.22) and combining terms.

(c) Derive (11.3.24) by noting that

$$E_m^f = E[f_m(n)x^*(n)]$$

11.14 Minimize (11.3.31) with respect to the prediction coefficients using the form of $g_m(n)$ given by (11.3.30). Also, prove (11.3.32).

11.15 Derive the recursive order-update equation given in (11.3.49).

11.16 Determine the mean and the autocorrelation of the sequence $x(n)$, which is the output of a ARMA (1, 1) process described by the difference equation

$$x(n) = \tfrac{1}{2}x(n - 1) + w(n) - w(n - 1)$$

where $w(n)$ is a white noise process with variance σ_w^2.

11.17 Determine the mean and the autocorrelation of the sequence $x(n)$ generated by the MA(2) process described by the difference equation

$$x(n) = w(n) - 2w(n - 1) + w(n - 2)$$

where $w(n)$ is a white noise process with variance σ_w^2.

11.18 An MA(2) process has the autocorrelation sequence

$$\gamma_{xx}(m) = \begin{cases} 6\sigma_w^2 & m = 0 \\ -4\sigma_w^2 & m = \pm 1 \\ -2\sigma_w^2 & m = \pm 2 \\ 0 & \text{otherwise} \end{cases}$$

(a) Determine the coefficients of the MA(2) process that have the foregoing autocorrelation.

(b) Is the solution unique? If not, give all the possible solutions.

11.19 An MA(2) process has the autocorrelation sequence

$$\gamma_{xx}(m) = \begin{cases} \sigma_w^2 & m = 0 \\ -\dfrac{35}{62}\sigma_w^2 & m = \pm 1 \\ \dfrac{6}{62}\sigma_w^2 & m = \pm 2 \end{cases}$$

(a) Determine the coefficients of the minimum-phase system for the MA(2) process.

(b) Determine the coefficients of the maximum-phase system for the MA(2) process.

(c) Determine the coefficients of the mixed-phase system for the MA(2) process.

11.20 Consider the linear system described by the difference equation

$$y(n) = 0.8y(n - 1) + x(n) + x(n - 1)$$

where $x(n)$ is a wide-sense stationary random process with zero mean and auto-correlation

$$\gamma_{xx}(m) = (\tfrac{1}{2})^{|m|}$$

(a) Determine the power density spectrum of the output $y(n)$.

(b) Determine the autocorrelation $\gamma_{yy}(m)$ of the output.

(c) Determine the variance σ_y^2 of the output.

11.21 From (11.3.15) and (11.3.18) we note that an AR(p) stationary random process satisfies the equation

$$\gamma_{xx}(m) + \sum_{k=1}^{p} a_p(k)\gamma_{xx}(m-k) = \begin{cases} \sigma_w^2 & m = 0 \\ 0 & 1 \le m \le p \end{cases}$$

where $a_p(k)$ are the prediction coefficients of the linear predictor of order p and σ_w^2 is the minimum mean-square prediction error. If the $(p+1) \times (p+1)$ autocorrelation matrix Γ_{xx} in (11.3.18) is positive definite, prove that:

(a) The reflection coefficients $|K_m| < 1$ for $1 \le m \le p$.

(b) The polynomial

$$A_p(z) = 1 + \sum_{k=1}^{p} a_p(k)z^{-k}$$

has all its roots inside the unit circle (i.e., it is minimum phase).

11.22 Consider the AR(3) process generated by the equation

$$x(n) = \tfrac{14}{24}x(n-1) + \tfrac{9}{24}x(n-2) - \tfrac{1}{24}x(n-3) + w(n)$$

where $w(n)$ is a stationary white noise process with variance σ_w^2.

(a) Determine the coefficients of the optimum $p = 3$ linear predictor.

(b) Determine the autocorrelation sequence $\gamma_{xx}(m)$, $0 \le m \le 5$.

(c) Determine the reflection coefficients corresponding to the $p = 3$ linear predictor.

11.23 An AR(2) process is described by the difference equation

$$x(n) = 0.81x(n-2) + w(n)$$

where $w(n)$ is a white noise process with variance σ_w^2.

(a) Determine the parameters of the MA(2), MA(4), and MA(8) models which provide a minimum mean-square error fit to the data $x(n)$.

(b) Plot the true spectrum and those of the MA(q), $q = 2, 4, 8$ spectra and compare the results. Comment on how well the MA(q) models approximate the AR(2) process.

11.24 An MA(2) process is described by the difference equation

$$x(n) = w(n) + 0.81w(n-2)$$

where $w(n)$ is a white noise process with variance σ_w^2.

(a) Determine the parameters of the AR(2), AR(4), and AR(8) models that provide a minimum mean-square error fit to the data $x(n)$.

(b) Plot the true spectra m and those of the AR(p), $p = 2, 4, 8$, and compare the results. Comment on how well the AR(p) models approximate the MA(2) process.

11.25 The z-transform of the autocorrelation $\gamma_{xx}(m)$ of an ARMA(1, 1) process is

$$\Gamma_{xx}(z) = \sigma_w^2 H(z)H(z^{-1})$$

$$\Gamma_{xx}(z) = \frac{4\sigma_w^2}{9}\frac{5 - 2z - 2z^{-1}}{10 - 3z^{-1} - 3z}$$

(a) Determine the minimum-phase system function $H(z)$.

(b) Determine the system function $H(z)$ for a mixed-phase stable system.

11.26 Consider a FIR filter with coefficient vector

$$[1 \quad -2r \cos \theta \quad r^2]$$

(a) Determine the reflection coefficients for the corresponding FIR lattice filter.

(b) Determine the values of the reflection coefficients in the limit as $r \to 1$.

11.27 An AR(3) process is characterized by the prediction coefficients

$$a_3(1) = -1.25 \qquad a_3(2) = 1.3125 \qquad a_3(3) = -1$$

(a) Determine $\gamma_{xx}(m)$ for $0 \le m \le 3$.

(b) Determine the reflection coefficients.

(c) Determine the mean-square prediction error.

11.28 The autocorrelation sequence for a random process is

$$\gamma_{xx}(m) = \begin{cases} 1 & m = 0 \\ -0.5 & m = \pm 1 \\ 0.625 & m = \pm 2 \\ -0.6875 & m = \pm 3 \\ 0 & \text{otherwise} \end{cases}$$

Determine the system functions $A_m(z)$ for the prediction-error filters for $m = 1$, 2, 3, the reflection coefficients $\{K_m\}$, and the corresponding mean-square prediction errors.

11.29 (a) Determine the power spectra for the random processes generated by the following difference equations.

(1) $x(n) = -0.81x(n - 2) + w(n) - w(n - 1)$

(2) $x(n) = w(n) - w(n - 2)$

(3) $x(n) = -0.81x(n - 2) + w(n)$

where $w(n)$ is a white noise process with variance σ_w^2.

(b) Sketch the spectra for the processes given in part (a).

(c) Determine the autocorrelation $\gamma_{xx}(m)$ for the processes in (2) and (3).

11.30 The autocorrelation sequence for an AR process $x(n)$ is

$$\gamma_{xx}(m) = (\tfrac{1}{4})^{|m|}$$

(a) Determine the difference equation for $x(n)$.

(b) Is your answer unique? If not, give any other possible solutions.

11.31 Repeat Problem 11.28 for an AR process with autocorrelation

$$\gamma_{xx}(m) = a^{|m|} \cos \frac{\pi m}{2}$$

where $0 < a < 1$.

11.32 The Bartlett method is used to estimate the power spectrum of a signal from a sequence $x(n)$ consisting of $N = 2400$ samples.

(a) Determine the smallest length M of each segment in the Bartlett method that yields a frequency resolution of $\Delta f = 0.01$.

(b) Repeat part (a) for $\Delta f = 0.02$.

(c) Determine the quality factors Q_B for parts (a) and (b).

11.33 Prove that a FIR filter with system function

$$A_p(z) = 1 + \sum_{k=1}^{p} a_p(k)z^{-k}$$

and reflection coefficients $|K_k| < 1$ for $1 \le k \le p - 1$ and $|K_p| > 1$ is maximum phase [all the roots of $A_p(z)$ lie outside the unit circle].

11.34 A random process $x(n)$ is characterized by the power density spectrum

$$\Gamma_{xx}(f) = \sigma_w^2 \frac{|e^{j2\pi f} - 0.9|^2}{|e^{j2\pi f} - j0.9|^2 |e^{j2\pi f} + j0.9|^2}$$

where σ_w^2 is a constant (scale factor).

(a) If we view $\Gamma_{xx}(f)$ as the power spectrum at the output of a linear pole–zero system $H(z)$ driven by white noise, determine $H(z)$.

(b) Determine the system function of a stable system that produces a white noise output when excited by $x(n)$. This is called a *noise-whitening filter*.

11.35 The N-point DFT of a random sequence $x(n)$ is

$$X(k) = \sum_{n=0}^{N-1} x(n) e^{-j2\pi nk/N}$$

Assume that $E[x(n)] = 0$ and $E[x(n)x(n + m)] = \sigma_x^2 \delta(m)$ [i.e., $x(n)$ is a white noise process].

(a) Determine the variance of $X(k)$.

(b) Determine the autocorrelation of $X(k)$.

11.36 A useful relationship is obtained by representing an ARMA(p, q) process as a cascade of a MA(q) followed by an AR(p) model. The input–output equation for the MA(q) model is

$$v(n) = \sum_{k=0}^{q} b_k w(n - k)$$

where $w(n)$ is a white noise process. The input–output equation for the AR(p) model is

$$x(n) + \sum_{k=1}^{p} a_k x(n - k) = v(n)$$

By computing the autocorrelation of $v(n)$, show that

$$\gamma_{vv}(m) = \sigma_w^2 \sum_{k=0}^{q-m} b_k^* b_{k+m} = \sigma_w^2 d_m$$

and hence the power spectrum for the MA(q) process is

$$\Gamma_{vv}(f) = \sigma_w^2 \sum_{m=-q}^{q} d_m e^{-j2\pi fm}$$

11.37 Determine the autocorrelation $\gamma_{xx}(m)$ of the random sequence

$$x(n) = A \cos (\omega_1 n + \phi)$$

where the amplitude A and the frequency ω_1 are (known) constants and ϕ is a uniformly distributed random phase over the interval $(0, 2\pi)$.

11.38 Suppose that the AR(2) process in Problem 11.23 is corrupted by an additive white noise process $v(n)$ with variance σ_v^2. Thus we have

$$y(n) = x(n) + v(n)$$

(a) Determine the difference equation for $y(n)$ and thus demonstrate that $y(n)$ is an ARMA(2, 2) process. Determine the coefficients of the ARMA process.

(b) Generalize the result in part (a) to an AR(p) process

$$x(n) = -\sum_{k=1}^{p} a_k(xn - k) + w(n)$$

and

$$y(n) = x(n) + v(n)$$

11.39 (a) Determine the autocorrelation of the random sequence

$$x(n) = \sum_{k=1}^{K} A_k \cos(\omega_k n + \phi_k) + w(n)$$

where $\{A_k\}$ are constant amplitudes, $\{\omega_k\}$ are constant frequencies, and $\{\phi_k\}$ are mutually statistically independent and uniformly distributed random phases. The noise sequence $w(n)$ is white with variance σ_w^2.
(b) Determine the power density spectrum of $x(n)$.

11.40 The harmonic decomposition problem considered by Pisarenko may be expressed as the solution to the equation [see (11.4.15)]

$$\mathbf{a}^{*\prime}\Gamma_{yy}\mathbf{a} = \sigma_w^2\mathbf{a}^{*\prime}\mathbf{a}$$

The solution for $\mathbf{a}$ may be obtained by minimizing the quadratic form $\mathbf{a}^{*\prime}\Gamma_{yy}\mathbf{a}$ subject to the constraint that $\mathbf{a}^{*\prime}\mathbf{a} = 1$. The constraint can be incorporated into the performance index by means of a Lagrange multiplier. Thus the performance index becomes

$$\mathscr{E} = \mathbf{a}^{*\prime}\Gamma_{yy}\mathbf{a} + \lambda(1 - \mathbf{a}^{*\prime}\mathbf{a})$$

By minimizing $\mathscr{E}$ with respect to $\mathbf{a}$ show that this formulation is equivalent to the Pisarenko eigenvalue problem given in (11.4.16) with the Lagrange multiplier playing the role of the eigenvalue. Thus show that the minimum of $\mathscr{E}$ is the minimum eigenvalue σ_w^2.

11.41 The autocorrelation of a sequence consisting of a sinusoid with random phase in noise is

$$\gamma_{xx}(m) = P \cos 2\pi f_1 m + \sigma_w^2 \delta(m)$$

where f_1 is the frequency of the sinusoidal, P its power, and σ_w^2 the variance of the noise. Suppose that we attempt to fit an AR(2) model to the data.
(a) Determine the optimum coefficients of the AR(2) model as a function of σ_w^2 and f_1.
(b) Determine the reflection coefficients K_1 and K_2 corresponding to the AR(2) model parameters.
(c) Determine the limiting values of the AR(2) parameters and (K_1, K_2) as $\sigma_w^2 \to 0$.

11.42 This problem involves the use of crosscorrelation to detect a signal in noise and estimate the time delay in the signal. A signal $x(n)$ consists of a pulsed sinusoid corrupted by a stationary zero-mean white noise sequence. That is,

$$x(n) = y(n - n_0) + w(n) \qquad 0 \le n \le N - 1$$

where $w(n)$ is the noise with variance σ_w^2 and the signal is

$$y(n) = A \cos \omega_0 n \qquad 0 \le n \le M - 1$$
$$= 0 \qquad\qquad \text{otherwise}$$

The frequency ω_0 is known but the delay n_0, which is a positive integer, is unknown, and is to be determined by crosscorrelating $x(n)$ with $y(n)$. Assume that $N > M + n_0$. Let

$$r_{xy}(m) = \sum_{n=0}^{N-1} y(n - m)x(n)$$

denote the crosscorrelation sequence between $x(n)$ and $y(n)$. In the absence of noise this function exhibits a peak at delay $m = n_0$. Thus n_0 is determined with no error. The presence of noise can lead to errors in determining the unknown delay.

(a) For $m = n_0$, determine $E[r_{xy}(n_0)]$. Also, determine the variance, var $[r_{xy}(n_0)]$, due to the presence of the noise. In both calculations, assume that the double-frequency term averages to zero. That is, $M \gg 2\pi/\omega_0$.

(b) Determine the signal-to-noise ratio, defined as

$$\text{SNR} = \frac{\{E[r_{xy}(n_0)]\}^2}{\text{var}\,[r_{xy}(n_0)]}$$

(c) What is the effect of the pulse duration M on the SNR?

COMPUTER EXPERIMENTS

11.43 Generate 100 samples of a zero-mean white noise sequence $w(n)$ with variance $\sigma_w^2 = \frac{1}{12}$, by using a uniform random number generator.

(a) Compute the autocorrelation of $w(n)$ for $0 \le m \le 15$.

(b) Compute the periodogram estimate $P_{xx}(f)$ and plot it.

(c) Generate 10 different realizations of $w(n)$ and compute the corresponding sample autocorrelation sequences $r_k(m)$, $1 \le k \le 10$ and $0 \le m \le 15$.

(d) Compute and plot the average periodogram for part (c):

$$r_{av}(m) = \frac{1}{10} \sum_{k=1}^{10} r_k(m)$$

(e) Comment on the results in parts (a) through (d).

11.44 A random signal is generated by passing zero-mean white Gaussian noise with unit variance through a filter with system function

$$H(z) = \frac{1}{(1 + az^{-1} + 0.99z^{-2})(1 - az^{-1} + 0.98z^{-2})}$$

(a) Sketch a typical plot of the theoretical power spectrum $\Gamma_{xx}(f)$ for a small value of the parameter a (i.e., $0 < a < 0.1$). Pay careful attention to the value of the two spectral peaks and the value of $P_{xx}(\omega)$ for $\omega = \pi/2$.

(b) Let $a = 0.1$. Determine the section length M required to resolve the spectral peaks of $\Gamma_{xx}(f)$ when using Bartlett's method.

(c) Consider the Blackman–Tukey method of smoothing the periodogram. How many lags of the correlation estimate must be used to obtain resolution comparable to that of the Bartlett estimate considered in part (b)? How many data must be used if the variance of the estimate is to be comparable to that of a four-section Bartlett estimate?

(d) For $a = 0.05$, fit an AR(4) model to 100 samples of the data based on the Yule–Walker method and plot the power spectrum. Avoid transient effects by discarding the first 200 samples of the data.

(e) Repeat part (d) with the Burg method.

(f) Repeat part (d) and (e) for 50 data samples and comment on similarities and differences in the results.

CHAPTER 12

Adaptive Filters

In Chapter 8 we described in detail various methods for designing FIR and IIR digital filters to satisfy some desired specifications. Our goal was to determine the coefficients of the digital filter that met the desired specifications. The resulting filter is realized by one of the several structures described in Chapter 7.

In contrast to the filter design techniques considered in Chapter 8, there are many digital signal processing applications in which the filter coefficients cannot be specified a priori. We have already encountered such applications in Chapter 6 where we described problems dealing with equalization of the distortion introduced by a communication channel and the removal of reverberation effects through deconvolution in seismic signal processing. In these cases the coefficients of the channel equalization filter and the deconvolution filter depend on the characteristics of the medium and cannot be specified a priori. Instead, they are determined from measurements performed by transmitting signals through the physical media.

For example, let us consider a high-speed modem that is designed to transmit data over telephone channels. Such a modem employs a channel equalizer to compensate for the channel distortion. The modem must effectively transmit data through communication channels that have different frequency response characteristics and hence result in different distortion effects. The only way in which this is possible is if the channel equalizer has *adjustable coefficients* that can be optimized to minimize some measure of the distortion, on the basis of measurements performed on the characteristics of the channel. Such a filter with adjustable parameters is called an *adaptive filter*, in this case an *adaptive equalizer*.

Adaptive filters have received considerable attention by many researchers over the past 15 to 20 years. As a result, many computationally efficient algorithms for adaptive filtering have been developed during this period. In this chapter we describe two basic algorithms, the least-mean-square (LMS) algorithm, which is based on a gradient optimization for determining the coefficients, and the class of recursive least-squares algorithms, which include both direct-form FIR and lattice realizations. Before we describe the algorithms, we shall present several practical applications in which adaptive filters have been successfully used in the estimation of signals corrupted by noise and other interference.

12.1 Applications of Adaptive Filters

Adaptive filters have been widely used in communication systems, control systems and various other systems in which the statistical characteristics of the signals to be filtered are either unknown a priori or, in some cases, slowly time-variant (nonstationary signals).

Numerous applications of adaptive filters have been described in the literature. Some of the more noteworthy applications include (1) adaptive antenna systems in which adaptive filters are used for beam steering and for providing nulls in the beam pattern to remove undesired interference [for a reference, see the paper by Widrow et al. (1967)]; (2) digital communication receivers in which adaptive filters are used to provide equalization of intersymbol interference and for channel identification [for references, see papers by Lucky (1965), Proakis and Miller (1969), Gersho (1969), George et al. (1971), Proakis (1970, 1975), Magee and Proakis (1973), Picinbono (1978), and Nichols et al. (1977)]; (3) adaptive noise canceling techniques in which an adaptive filter is used to estimate and eliminate a noise component in some desired signal [for references, see papers by Widrow et al. (1975), Hsu and Giordano (1978), and Ketchum and Proakis (1982)]; (4) system modeling, in which an adaptive filter is used as a model to estimate the characteristics of an unknown system. These are just a few of the best known examples on the use of adaptive filters.

Although both IIR and FIR filters have been considered for adaptive filtering, the FIR filter is by far the most practical and widely used. The reason for this preference is quite simple. The FIR filter has only adjustable zeros and hence it is free of stability problems associated with adaptive IIR filters that have adjustable poles as well as zeros. We should not conclude, however, that adaptive FIR filters are always stable. On the contrary, the stability of the filter depends critically on the algorithm for adjusting its coefficients, as will be demonstrated in Sections 12.2 and 12.3.

Of the various FIR filter structures that we have presented in Chapter 7, the direct form and the lattice form are the ones often used in adaptive filtering applications. The direct-form FIR filter structure with adjustable coefficients $h(0), h(1), \ldots, h(M-1)$ is illustrated in Fig. 12.1. On the other hand, the adjustable parameters in an FIR lattice structure are the reflection coefficients $\{K_n\}$. The algorithms given in Figs. 7.12 and 3.15 may be used to relate the parameters $\{K_n\}$ to $\{h(n)\}$, and vice versa.

An important consideration in the use of an adaptive filter is the criterion for optimizing the adjustable filter parameters. The criterion must not only provide a meaningful measure of filter performance, but it must also result in a practically realizable algorithm.

For example, a desirable performance index in a digital communication system is the average probability of error. Consequently, in implementing an adaptive equalizer, we might consider the selection of the equalizer coefficients to minimize the average probability of error as the basis for our optimization criterion. Unfortunately, however, the performance index (average probability of error) for this criterion is a highly nonlinear

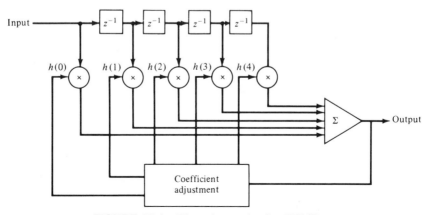

FIGURE 12.1   Direct-form adaptive FIR filter.

function of the filter coefficients. Although a number of algorithms are known for finding a minimum or a maximum of a nonlinear function of several variables, such algorithms are unsuitable for adaptive filtering primarily because the signal statistical characteristics are unknown and, possibly, time variant.

In some cases a performance index that is a nonlinear function of the filter parameters possesses many relative minima (or maxima), so that one is not certain whether the adaptive filter has converged to the optimum solution or to one of the relative minima (or maxima). For these reasons, some desirable performance indices, such as the average probability of error in a digital communication system, must be rejected on the grounds that they are impractical to implement.

One criterion that provides a good measure of performance in adaptive filtering applications is the least-squares criterion, and its counterpart in a statistical formulation of the problem, namely, the mean-square-error (MSE) criterion. The least-squares (and MSE) criterion results in a quadratic performance index as a function of the filter coefficients and hence it possesses a single minimum. The resulting algorithms for adjusting the coefficients of the filter are relatively easy to implement as we shall demonstrate in Sections 12.2 and 12.3.

Below we describe several applications of adaptive filters which serve as a motivation for the mathematical development of algorithms derived in Sections 12.2 and 12.3. We find it convenient to use the direct-form FIR structure in these examples. Although we will not develop the recursive algorithms for automatically adjusting the filter coefficients in this section, it is instructive to formulate the optimization of the filter coefficients as a least-squares optimization problem. This development will serve to establish a common framework for the algorithms derived in the next two sections.

System Identification or System Modeling. The formulation of the problem has already been given in Section 6.3.5 and illustrated in Fig. 6.38. We have an unknown system, called a *plant*, that we wish to identify. The system is modeled by an FIR filter with adjustable coefficients. Both the plant and model are excited by an input sequence $x(n)$. If $y(n)$ denotes the output of the plant and $\hat{y}(n)$ denotes the output of the model, that is,

$$\hat{y}(n) = \sum_{k=0}^{M-1} h(k)x(n-k) \qquad (12.1.1)$$

we may form the error sequence

$$e(n) = y(n) - \hat{y}(n) \qquad (12.1.2)$$

and select the coefficients $\{h(k)\}$ to minimize

$$\mathscr{E}_M = \sum_{n=0}^{\infty} \left[y(n) - \sum_{k=0}^{M-1} h(k)x(n-k) \right]^2 \qquad (12.1.3)$$

The least-squares criterion leads to the set of linear equations derived in Section 6.3.5 for determining the filter coefficients:

$$\sum_{k=0}^{M-1} h(k)r_{xx}(l-k) = r_{yx}(l) \qquad l = 0, 1, \ldots, M-1 \qquad (12.1.4)$$

In (12.1.4), $r_{xx}(l)$ is the autocorrelation of the sequence $x(n)$ and $r_{yx}(l)$ is the crosscorrelation of the system output with the input sequence.

By solving (12.1.4) we obtain the filter coefficients for the model. Since the filter parameters are obtained directly from measurement data at the input and output of the

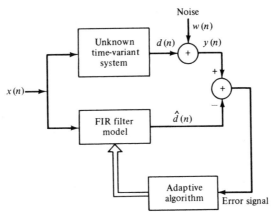

FIGURE 12.2 Application of adaptive filtering to system identification.

system, without prior knowledge of the plant, we call the FIR filter model an adaptive filter.

If our only objective is to identify the system by use of the FIR model, the solution of (12.1.4) would suffice. In control systems applications, however, the system being modeled may be time variant, changing slowly with time, and our purpose for having a model is ultimately to use it for designing a controller that controls the plant. Furthermore, measurement noise is usually present at the output of the plant. This noise introduces uncertainty in the measurements and corrupts the estimates of the filter coefficients in the model. Such a scenario is illustrated in Fig. 12.2. Now, the adaptive filter must identify and track the time-variant characteristics of the plant in the presence of measurement noise at the output of the plant. The algorithms described in Sections 12.2 and 12.3 are applicable to this system identification problem.

Adaptive Channel Equalization. Figure 12.3 shows a block diagram of a digital communication system in which an adaptive equalizer is used to compensate for the distortion caused by the transmission medium (channel). The digital sequence of information symbols $a(n)$ is fed to the transmit filter whose output is

$$s(t) = \sum_{k=0}^{\infty} a(k)p(t - kT_s) \qquad (12.1.5)$$

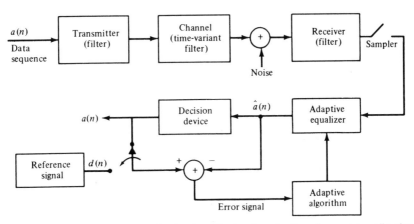

FIGURE 12.3 Application of adaptive filtering to adaptive channel equalization.

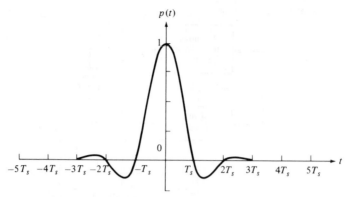

FIGURE 12.4 Pulse shape for digital transmission of symbols at a rate of $1/T_s$ symbols per second.

where $p(t)$ is the impulse response of the filter at the transmitter and T_s is the time interval between information symbols (i.e., $1/T_s$ is the symbol rate). For purposes of this discussion we may assume that $a(n)$ is a multilevel sequence which takes on values from the set $\pm 1, \pm 3, \pm 5, \ldots, \pm(L - 1)$, where L is the number of possible symbol values.

Typically, the pulse $p(t)$ is designed to have the characteristics illustrated in Fig. 12.4. Note that $p(t)$ has amplitude $p(0) = 1$ at $t = 0$ and $p(nT_s) = 0$ at $t = nT_s$, $n = \pm 1$, $\pm 2, \ldots$ As a consequence, a series of successive pulses transmitted sequentially every T_s seconds do not interfere with one another when sampled at the time instants $t = nT_s$. Thus $a(n) = s(nT_s)$.

The channel, which is usually well modeled as a linear filter, distorts the pulse and thus causes intersymbol interference. For example, in telephone channels, filters are used throughout the system to separate signals in different frequency ranges. These filters cause phase and amplitude distortion. Figure 12.5 illustrates the effect of channel distortion on the pulse $p(t)$ as it might appear at the output of a telephone channel. Now

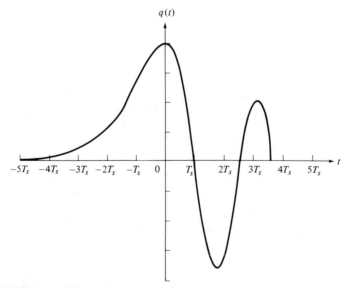

FIGURE 12.5 Effect of channel distortion on the signal pulse in Fig. 12.4.

we observe that the samples taken every T_s seconds are corrupted by interference from several adjacent symbols. The distorted signal is also corrupted by additive noise, which is usually wideband.

At the receiving end of the communication system the signal is first passed through a filter which is designed primarily to eliminate the noise outside of the frequency band occupied by the signal. We may assume that this filter is a linear-phase FIR filter that limits the bandwidth of the noise but causes negligible additional distortion on the channel-corrupted signal.

Samples of the received signal at the output of this filter reflect the presence of intersymbol interference and additive noise. If we ignore the possible time variations in the channel for the moment, we may express the sampled output at the receiver as

$$x(nT_s) = \sum_{k=0}^{\infty} a(k)q(nT_s - kT_s) + w(nT_s)$$

$$= a(n)q(0) + \sum_{\substack{k=0 \\ k \neq n}}^{\infty} a(k)q(nT_s - kT_s) + w(nT_s)$$

$$(12.1.6)$$

where $w(t)$ represents the additive noise and $q(t)$ represents the distorted pulse at the output of the receiver filter.

To simplify our discussion, we assume that the sample $q(0)$ is normalized to unity by means of an automatic gain control (AGC) contained in the receiver. Then the sampled signal given in (12.1.6) may be expressed as

$$x(n) = a(n) + \sum_{\substack{k=0 \\ k \neq n}}^{\infty} a(k)q(n - k) + w(n)$$

$$(12.1.7)$$

where $x(n) \equiv x(nT_s)$, $q(n) \equiv q(nT_s)$, and $w(n) \equiv w(nT_s)$. The term $a(n)$ in (12.1.7) is the desired symbol at the nth sampling instant. The second term

$$\sum_{\substack{k=0 \\ k \neq n}}^{\infty} a(k)q(n - k)$$

constitutes the intersymbol interference due to the channel distortion, and $w(n)$ represents the additive noise in the system.

In general, the channel distortion effects embodied through the sampled values $q(n)$ are unknown at the receiver. Furthermore, the channel may vary slowly with time so that the intersymbol interference effects are time variant. The purpose of the adaptive equalizer is to compensate the signal for the channel distortion so that the resulting signal can be detected reliably.

Let us assume that the equalizer is an FIR filter with M adjustable coefficients $\{h(n)\}$. Its output may be expressed as

$$\hat{a}(n - D) = \sum_{k=0}^{M-1} h(k)x(n - k)$$

$$(12.1.8)$$

where D is some nominal delay in processing the signal through the filter and $\hat{a}(n)$ represents an estimate of the nth information symbol. Initially, the equalizer is trained by transmitting a known data sequence $d(n)$. Then the equalizer output is compared with $d(n)$ and an error is generated which is used to optimize the filter coefficients.

If we adopt the least-squares error criterion again, we select the coefficients $\{h(k)\}$ to minimize the quantity

$$\mathscr{E}_M = \sum_{n=0}^{\infty} [d(n) - \hat{a}(n)]^2$$

$$= \sum_{n=0}^{\infty} \left[d(n) - \sum_{k=0}^{\infty} h(k)x(n + D - k) \right]^2$$

(12.1.9)

The results of the optimization is a set of linear equations of the form

$$\sum_{k=0}^{M-1} h(k)r_{xx}(l - k) = r_{dx}(l + D) \qquad l = 0, 1, 2, \ldots, M - 1 \quad (12.1.10)$$

where $r_{xx}(l)$ is the autocorrelation of the sequence $x(n)$ and $r_{dx}(l)$ is the crosscorrelation between the desired sequence $d(n)$ and the received sequence $x(n)$.

Although the solution of (12.1.10) is obtained recursively in practice, as will be demonstrated in the following two sections, in principle, we observe that these equations result in values of the coefficients for the initial adjustment of the equalizer. After the short training period, which usually lasts less than 1 second for most channels, the transmitter begins to transmit the information sequence $a(n)$. To track the possible time variations in the channel, the equalizer coefficients must continue to be adjusted in an adaptive manner while receiving data. This is usually accomplished, as illustrated in Fig. 12.3, by treating the decisions at the output of the decision device as correct, and using the decisions in place of the reference $d(n)$ to generate the error signal. This approach works quite well when decision errors occur infrequently (e.g., less than one decision error per 100 symbols). The occasional decision errors cause only small misadjustments in the equalizer coefficients. In Sections 12.2 and 12.3 we describe the adaptive algorithms for recursively adjusting the equalizer coefficients.

Echo Cancellation in Data Transmission Over Telephone Channels. In the transmission of data over telephone channels, modems (modulator/demodulator) are used to provide an interface between the digital data sequence and the analog channel. Recall that a QAM modem was previously described in Section 6.1. Shown in Fig. 12.6 is a block diagram of a communication system in which two terminals, labeled A and B, transmit data by using modems A and B to interface to a telephone channel. As shown, a digital sequence $a(n)$ is transmitted from terminal A to terminal B while a digital sequence $b(n)$ is transmitted from terminal B to A. This simultaneous transmission in both directions is called *full-duplex transmission*.

As described above, the two transmitted signals may be represented as

$$s_A(t) = \sum_{k=0}^{\infty} a(k)p(t - kT_s)$$

(12.1.11)

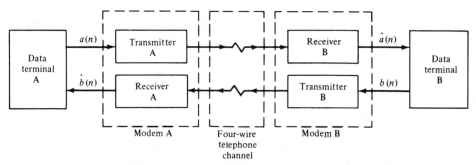

FIGURE 12.6 Full-duplex data transmission over telephone channels.

and

$$s_B(t) = \sum_{k=0}^{\infty} b(k)p(t - kT_s) \qquad (12.1.12)$$

where $p(t)$ is a pulse as shown in Fig. 12.4.

When a subscriber leases a private line from a telephone company for the purpose of transmitting data between terminals A and B, the telephone line provided is a four-wire line, which is equivalent to having two dedicated telephone (two-wire) channels, one (pair of wires) for transmitting data in one direction and one (pair of wires) for receiving data from the other direction. In such a case the two transmission paths are isolated, and consequently, there is no "crosstalk" or mutual interference between the two signal paths. Channel distortion is compensated by use of an adaptive equalizer, as described above, at the receiver of each modem.

The major problem with the system shown in Fig. 12.6 is the cost of leasing a four-wire telephone channel. If the volume of traffic is high and the telephone channel is used either continuously or frequently, as in banking transactions systems or airline reservation systems, the system pictured in Fig. 12.6 is cost-effective. Otherwise, it is not.

An alternative solution for low-volume, infrequent transmission of data is to use the dial-up switched telephone network. In this case, the local communication link between the subscriber and the local central telephone office is a two-wire line, called the *local loop*. At the central office, the subscriber two-wire line is connected to the main four-wire telephone channels that interconnect different central offices, called *trunk lines*, by a device called a *hybrid*. By using transformer coupling, the hybrid is tuned to provide isolation between the transmit and receiver channels in full-duplex operation. However, due to impedance mismatch between the hybrid and the telephone channel, the level of isolation is often insufficient, and consequently, some of the signal on the transmit side leaks back and corrupts the signal on the receiver side, causing an "echo" that is often heard in voice communications over telephone channels.

To mitigate the echoes in voice transmissions, the telephone companies employ a device called an *echo suppressor*. In data transmission, the solution is to use an *echo canceller* within each modem. The echo cancellers are implemented as adaptive FIR filters with automatically adjustable coefficients.

With the use of hybrids to couple a two-wire channel to a four-wire channel and echo cancellers at each modem to estimate and subtract out the echoes, the data communication system for the dial-up switched network takes the form shown in Fig. 12.7. A hybrid is needed at each modem to isolate the transmitter from the receiver and to couple to the two-wire local loop. Hybrid A is physically located at the central office of subscriber A, while hybrid B is located at the central office to which subscriber B is connected. The two central offices are connected by a four-wire line, one pair used for transmission from A to B and the other pair is used for transmission in the reverse direction, from B to A. An echo at terminal A due to the hybrid A is called a *near-end-echo*, while an echo at terminal A due to the hybrid B is termed a *far-end echo*. Both types of echoes are usually present in data transmission and must be removed by the echo canceller.

Suppose that we neglect the channel distortion for purposes of this discussion, and let us deal with echoes only. The signal received at modem A may be expressed as

$$s_{RA}(t) = A_1 s_B(t) + A_2 s_A(t - d_1) + A_3 s_A(t - d_2) \qquad (12.1.13)$$

where $s_B(t)$ is the desired signal to be demodulated at modem A, $s_A(t - d_1)$ is the near-end echo due to hybrid A, $s_A(t - d_2)$ is the far-end echo due to hybrid B, and A_i, $i = 1, 2, 3$ are the corresponding amplitudes of the three signal components and (d_1, d_2)

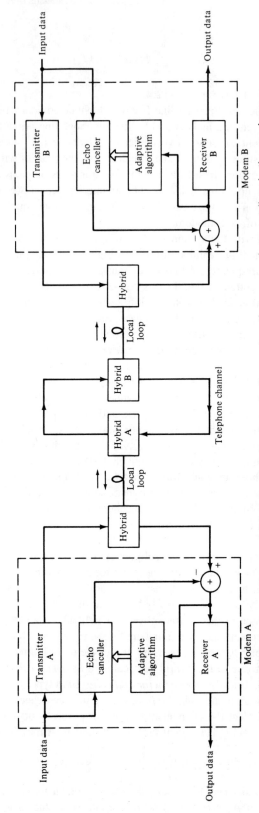

FIGURE 12.7 Block diagram model of a digital communication system that uses echo cancellers in the modems.

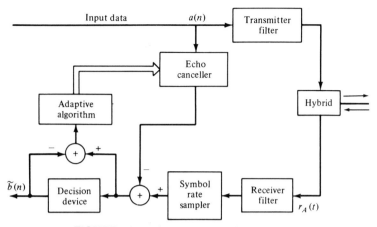

FIGURE 12.8 Symbol-rate echo canceller.

are the delays associated with the echo components. A further disturbance that corrupts the received signal is additive noise, so that the received signal at modem A is

$$r_A(t) = s_{RA}(t) + w(t) \tag{12.1.14}$$

where $w(t)$ represents the additive noise process.

The adaptive echo canceller attempts to estimate adaptively the two echo components. If its coefficients are $h(n)$, $n = 0, 1, \ldots, M - 1$, its output is

$$\hat{s}_A(n) = \sum_{k=0}^{M-1} h(k)a(n - k) \tag{12.1.15}$$

which is an estimate of the echo signal components. This estimate is subtracted from the sampled received signal and the resulting error signal can be minimized in the least-squares sense to adjust optimally the coefficients of the echo canceller. There are several possible configurations for placement of the echo canceller in the modem and for forming the corresponding error signal. Figure 12.8 illustrates one configuration in which the canceller output is subtracted from the sampled output of the receiver filter with input $r_A(t)$. Figure 12.9 illustrates a second configuration, in which the echo canceller is generating samples at the Nyquist rate instead of the symbol rate. In this case the error

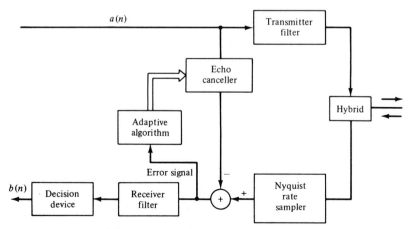

FIGURE 12.9 Nyquist rate echo canceller.

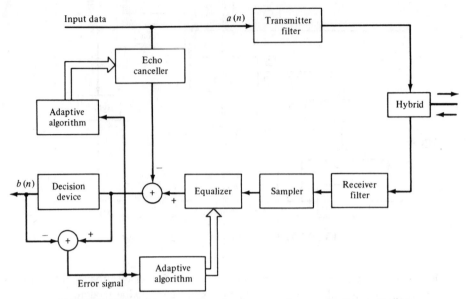

FIGURE 12.10 Modem with adaptive equalizer and echo canceller.

signal used to adjust the coefficients is simply the difference between $r_A(n)$, the sampled received signal, and the canceller output. Finally, Fig. 12.10 illustrates the canceller operating in combination with an adaptive equalizer.

Application of the least-squares criterion in any of the configurations shown in Figs. 12.8, 12.9, or 12.10 leads to a set of linear equations for the coefficients of the echo canceller. The reader is encouraged to derive these equations corresponding to the configurations above.

Suppression of Narrowband Interference in a Wideband Signal. This problem arises often in practice, especially in signal detection and in digital communications. Let us assume that we have a signal sequence $v(n)$ which consists of a desired wideband signal sequence $w(n)$ corrupted by an additive narrowband interference sequence $x(n)$. The two sequences are uncorrelated. These sequences result from sampling an analog signal $v(t)$ at the Nyquist rate (or faster) of the wideband signal $w(t)$. Figure 12.11 illustrates the spectral characteristics of $w(n)$ and $x(n)$. Usually, the interference $|X(f)|$ is much larger than $|W(f)|$ within the narrow frequency band that it occupies.

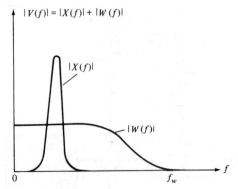

FIGURE 12.11 Strong narrowband interference $X(f)$ in a wideband signal $W(f)$.

In digital communications and in signal detection problems that fit the model above, the desired signal sequence $w(n)$ is often a *spread-spectrum signal* while the narrowband interference represents a signal from another user of the frequency band or some intentional interference from a jammer who is trying to disrupt the communications or detection system.

Our objective from a filtering viewpoint is to employ a filter that suppresses the narrowband interference. In effect, such a filter will have a notch in the frequency band occupied by $|X(f)|$. In practice, the band occupied by $|X(f)|$ is unknown. Moreover, if the interference is nonstationary, its frequency band occupancy may vary with time. Hence an adaptive filter is desired.

From another viewpoint, the narrowband characteristics of the interference allow us to estimate $x(n)$ from past samples of the sequence $v(n)$ and to subtract the estimate from $v(n)$. Since the bandwidth of $x(n)$ is narrow compared to the bandwidth of the sequence $w(n)$, the samples $x(n)$ are highly correlated due to the high sampling rate. On the other hand, the samples $w(n)$ are not highly correlated, since the samples are taken at the Nyquist rate of $w(n)$. By exploiting the high correlation between $x(n)$ and past samples of the sequence $v(n)$, it is possible to obtain an estimate of $x(n)$, which can be subtracted from $v(n)$.

The general configuration is illustrated in Fig. 12.12. The signal $v(n)$ is delayed by D samples, where D is selected sufficiently large so that the wideband signal components $w(n)$ and $w(n - D)$ contained in $v(n)$ and $v(n - D)$, respectively, are uncorrelated.

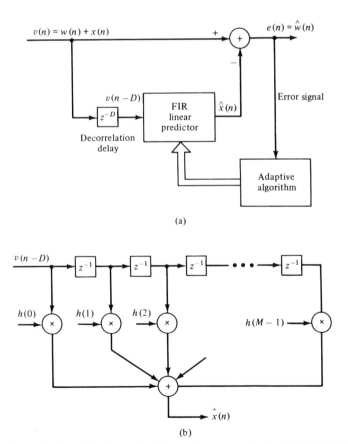

(a)

(b)

FIGURE 12.12 Adaptive filter for estimating and suppressing a narrowband interference in a wideband signal.

Usually, a choice of $D = 1$ or 2 is adequate. The delayed signal sequence $v(n - D)$ is passed through an FIR filter, which is best characterized as a linear predictor of the value $x(n)$ based on M samples $v(n - D - k)$, $k = 0, 1, \ldots, M - 1$. The output of the linear predictor is

$$\hat{x}(n) = \sum_{k=0}^{M-1} h(k)v(n - D - k) \tag{12.1.16}$$

This predicted value of $x(n)$ is subtracted from $v(n)$ to yield an estimate of $w(n)$, as illustrated in Fig. 12.12. Clearly, the quality of the estimate $x(n)$ determines how well the narrowband interference is suppressed. It is also apparent that the delay D must be kept as small as possible in order to obtain a good estimate of $x(n)$, but must be sufficiently large so that $w(n)$ and $w(n - D)$ are uncorrelated.

Let us define the error sequence

$$\begin{aligned}
e(n) &= v(n) - \hat{x}(n) \\
&= v(n) - \sum_{k=0}^{M-1} h(k)v(n - D - k)
\end{aligned} \tag{12.1.17}$$

If we apply the least-squares criterion to select the optimal prediction coefficients, we obtain the set of linear equations

$$\sum_{k=0}^{M-1} h(k)r_{vv}(l - k) = r_{vv}(l + D) \qquad l = 0, 1, \ldots, M - 1 \tag{12.1.18}$$

where $r_{vv}(l)$ is the autocorrelation sequence of $v(n)$. Note, however, that the right-hand side of (12.1.18) may be expressed as

$$\begin{aligned}
r_{vv}(l + D) &= \sum_{n=0}^{\infty} v(n)v(n - l - D) \\
&= \sum_{n=0}^{\infty} [w(n) + x(n)][w(n - l - D) + x(n - l - D)] \\
&= r_{ww}(l + D) + r_{xx}(l + D) + r_{wx}(l + D) + r_{xw}(l + D) \tag{12.1.19}
\end{aligned}$$

The correlations in (12.1.19) are sample correlation sequences. The expected value of $r_{ww}(l + D)$ is

$$E[r_{ww}(l + D)] = 0 \qquad l = 0, 1, \ldots, M - 1 \tag{12.1.20}$$

because $w(n)$ is wideband and D is large enough so that $w(n)$ and $w(n - D)$ are uncorrelated. Also,

$$E[r_{xw}(l + D)] = E[r_{wx}(l + D)] = 0 \tag{12.1.21}$$

by assumption. Finally,

$$E[r_{xx}(l + D)] = \gamma_{xx}(l + D) \tag{12.1.22}$$

Therefore, the expected value of $r_{vv}(l + D)$ is simply the statistical autocorrelation of the narrowband signal $x(n)$. Furthermore, if the wideband signal is weak relative to the interference, the autocorrelation $r_{vv}(l)$ on the left-hand side of (12.1.18) is approximately $r_{xx}(l)$. The major influence of $w(n)$ is to the diagonal elements of $r_{vv}(l)$. Consequently, the values of the filter coefficients determined from the linear equations in (12.1.18) are a function of the statistical characteristics of the interference $x(n)$.

We may view the overall filter structure in Fig. 12.12 as an equivalent FIR filter with coefficients

$$h'(k) = \begin{cases} 1 & k = 0 \\ -h(k - D) & k = D, D + 1, \ldots, D + M \\ 0 & \text{otherwise} \end{cases} \qquad (12.1.23)$$

and a frequency response

$$H(\omega) = \sum_{k=0}^{D+M} h'(k)e^{-j\omega k} \qquad (12.1.24)$$

This overall filter acts as a notch filter for the interference. For example, Fig. 12.13 illustrates the magnitude of the frequency response of an adaptive filter with $M = 15$ coefficients, which attempts to suppress a narrowband interference that occupies 20% of the frequency band of a desired spread-spectrum signal sequence. The data were generated pseudorandomly by adding a narrowband interference consisting of 100 randomly phased equal-amplitude sinusoids to a pseudonoise spread-spectrum signal. The coefficients of the filter were obtained by solving the equations in (12.1.18), with $D = 1$, where the correlation $r_{vv}(l)$ was obtained from the data. We observe that the overall interference suppression filter has the characteristics of a notch filter. The depth of the notch depends on the relative power of the interference to the wideband signal. The stronger the interference, the deeper the notch.

The algorithms presented in Sections 12.2 and 12.3 are appropriate for estimating the predictor coefficients continuously, in order to track a nonstationary narrowband interference signal.

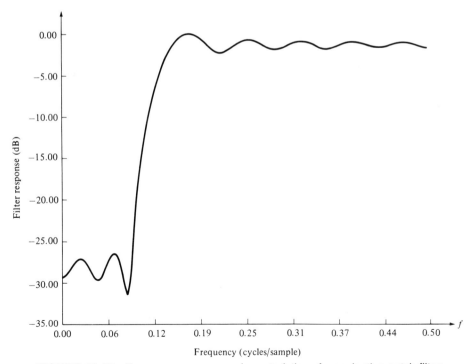

FIGURE 12.13   Frequency response characteristics of an adaptive notch filter.

Adaptive Line Enhancer. In the preceding example, the adaptive linear predictor was used to estimate the narrowband interference for the purpose of suppressing the interference from the input sequence $v(n)$. An adaptive line enhancer (ALE) has the same configuration as the interference suppression filter in Fig. 12.12, except that the objective is different.

In the adaptive line enhancer $x(n)$ is the desired signal and $w(n)$ represents a wideband noise component that masks $x(n)$. The desired signal $x(n)$ is either a spectral line or a relatively narrowband signal. The linear predictor shown in Fig. 12.12b operates in exactly the same fashion as in Fig. 12.12a, and provides an estimate of the narrowband signal $x(n)$. It is apparent that the ALE (i.e., the FIR prediction filter) is a self-tuning filter that has a peak in its frequency response at the frequency of the sinusoid or, equivalently, in the frequency band of the narrowband signal $x(n)$. By having a narrow bandwidth, the noise $w(n)$ outside the band is suppressed and thus the spectral line is enhanced in amplitude relative to the noise power in $w(n)$. This explains why the FIR predictor is called an ALE. Its coefficients are determined by the solution of (12.1.18).

Adaptive Noise Canceling. Echo cancellation, the suppression of narrowband interference in a wideband signal, and the ALE are related to another form of adaptive filtering called *adaptive noise canceling*. A model for the adaptive noise canceller is illustrated in Fig. 12.14.

The primary input signal consists of a desired signal sequence $x(n)$ corrupted by an additive noise sequence $w_1(n)$ and an additive interference (noise) $w_2(n)$. The additive interference (noise) is also observable after it has been filtered by some unknown linear system that yields $v_2(n)$ and is further corrupted by an additive noise sequence $w_3(n)$. Thus we have available a secondary signal sequence, which may be expressed as $v(n) = v_2(n) + w_3(n)$. The sequences $w_1(n)$, $w_2(n)$, and $w_3(n)$ are assumed to be mutually uncorrelated and zero mean.

As shown in Fig. 12.14, an adaptive FIR filter is used to estimate the interference sequence $w_2(n)$ from the secondary signal $v(n)$ and subtract the estimate $\hat{w}_2(n)$ from the primary signal. The output sequence, which represents an estimate of the desired signal $x(n)$, is the error signal

$$e(n) = y(n) - \hat{w}_2(n) \tag{12.1.25}$$

$$= y(n) - \sum_{k=0}^{M-1} h(k)v(n-k)$$

This error sequence is used to adjust adaptively the coefficients of the FIR filter.

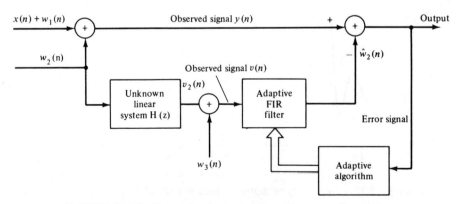

FIGURE 12.14 Example of an adaptive noise-canceling system.

If the least-squares criterion is used to determine the filter coefficients, the result of the optimization is the set of linear equations

$$\sum_{k=0}^{M-1} h(k)r_{vv}(l - k) = r_{yv}(l) \qquad l = 0, 1, \ldots, M - 1 \qquad (12.1.26)$$

where $r_{vv}(l)$ is the sample autocorrelation of the sequence $v(n)$ and $r_{yv}(l)$ is the sample crosscorrelation of the sequences $y(n)$ and $v(n)$. Clearly, the noise canceling problem is similar to the last three adaptive filtering applications described above.

Linear Predictive Coding of Speech Signals. Various methods have been developed over the past four decades for digital encoding of speech signals. In the telephone system, for example, two commonly used methods for speech encoding are pulse code modulation (PCM) and differential PCM (DPCM). These are examples of *waveform coding* methods. Other waveform coding methods have also been developed, such as delta modulation (DM) and adaptive DPCM. Adaptive DPCM has been selected for use in new telephone network installations.

Since the digital speech signal is ultimately transmitted from the source to a destination, a primary objective in devising speech encoders is to minimize the number of bits required to represent the speech signal, while maintaining speech intelligibility. This objective has led to the development of class of low-bit-rate (2400 bits per second and below) speech encoding methods, which are based on constructing a model of the speech source and transmitting the model parameters. Adaptive filtering finds application in these model-based speech coding systems. We shall describe one very effective method called *linear predictive coding* (LPC).

In LPC the vocal tract is modeled as a linear all-pole filter having the system function

$$H(z) = \frac{G}{1 + \displaystyle\sum_{k=1}^{p} a_k z^{-k}} \qquad (12.1.27)$$

where p is the number of poles, G is the filter gain, and $\{a_k\}$ are the parameters that determine the poles. There are two mutually exclusive excitation functions to model voiced and unvoiced speech sounds. On a short-time basis, voiced speech is periodic with a fundamental frequency F_0, or a pitch period $1/F_0$, which depends on the speaker. Thus voiced speech is generated by exciting the all-pole filter model by a periodic impulse train with a period equal to the desired pitch period. Unvoiced speech sounds are generated by exciting the all-pole filter model by the output of a random-noise generator. This model is shown in Fig. 12.15.

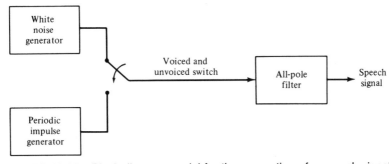

FIGURE 12.15 Block diagram model for the generation of a speech signal.

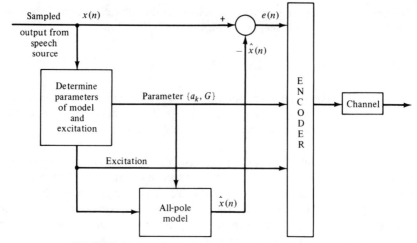

FIGURE 12.16 Source encoder for a speech signal.

Given a short-time segment of a speech signal, the speech encoder at the transmitter must determine the proper excitation function, the pitch period for voiced speech, the gain parameter G, and the coefficients $\{a_k\}$. A block diagram that illustrates the source encoding system is given in Fig. 12.16. The parameters of the model are determined adaptively from the data. Then the speech samples are synthesized by using the model, and an error signal sequence is generated as shown in Fig. 12.16 by taking the difference between the actual and the synthesized sequence. The error signal and the model parameters are encoded into a binary sequence and transmitted to the destination. At the receiver, the speech signal is synthesized from the model and the error signal.

The parameters of the all-pole filter model are easily determined from the speech samples by means of linear prediction. To be specific, consider the system shown in Fig. 12.17 and assume that we have N signal samples. The output of the FIR filter is

$$\hat{x}(n) = -\sum_{k=1}^{p} a_k x(n-k) \tag{12.1.28}$$

and the corresponding error between the observed sample $x(n)$ and the estimate $\hat{x}(n)$ is

$$e(n) = x(n) + \sum_{k=1}^{p} a_k x(n-k) \tag{12.1.29}$$

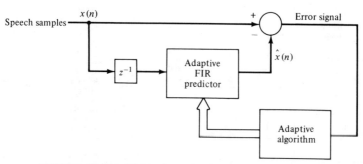

FIGURE 12.17 Estimation of pole parameters in LPC.

By applying the least-squares criterion, we can determine the model parameters $\{a_k\}$. The result of this optimization is a set of linear equations

$$\sum_{k=1}^{p} a_k r_{xx}(l - k) = -r_{xx}(l) \qquad l = 1, 2, \ldots, p \qquad (12.1.30)$$

where $r_{xx}(l)$ is the autocorrelation of the sequence $x(n)$. The gain parameter for the filter can be obtained by noting that its input–output equation is

$$x(n) = -\sum_{k=1}^{p} a_k x(n - k) + Gv(n) \qquad (12.1.31)$$

where $v(n)$ is the input sequence. Clearly,

$$Gv(n) = x(n) + \sum_{k=1}^{p} a_k x(n - k)$$

$$= e(n)$$

Then

$$G^2 \sum_{n=0}^{N-1} v^2(n) = \sum_{n=0}^{N-1} e^2(n) \qquad (12.1.32)$$

If the input excitation is normalized to unit energy by design, then

$$G^2 = \sum_{n=0}^{N-1} e^2(n) \qquad (12.1.33)$$

$$= r_{xx}(0) + \sum_{k=1}^{p} a_k r_{xx}(k)$$

Thus G^2 is set equal to the residual energy resulting from the least-squares optimization.

In this development we have described the use of linear prediction to determine adaptively the pole parameters and the gain of an all-pole filter model for speech generation. In practice, due to the nonstationary character of speech signals, this model is applied to short-time segments (10 to 20 milliseconds) of a speech signal. Usually, a new set of parameters is determined for each short time segment. However, it is often advantageous to use the model parameters measured from previous segments to smooth out sharp discontinuities that usually exist in estimates of model parameters obtained from segment to segment. Although our discussion was totally in terms of the FIR filter structure, we should mention that speech analysis is usually performed by using the FIR lattice structure and the reflection coefficients $\{K_i\}$. Since the dynamic range of the $\{K_i\}$ is significantly smaller than that of the $\{a_k\}$, the reflection coefficients require fewer bits to represent them. Hence the $\{K_i\}$ are transmitted over the channel. Consequently, it is natural to synthesize the speech at the destination using the lattice structure.

In our treatment of LPC for speech coding, we have not considered algorithms for the estimation of the excitation and the pitch period. A discussion of appropriate algorithms for these parameters of the model would take us too far afield and hence are omitted. The interested reader is referred to the book by Rabiner and Schafer (1978) for a detailed treatment of speech analysis and synthesis methods.

Adaptive Arrays. In the previous examples we considered adaptive filtering performed on a single data sequence. However, adaptive filtering has also been widely applied to multiple data sequences which result from antenna arrays and seismometer arrays, where the sensors (antennas or seismometers) are arranged in some spatial configuration. Each element of the array of sensors provides a signal sequence. By properly combining the signals from the various sensors it is possible to change the directivity pattern of the array.

For example, consider the linear antenna array consisting of five elements, as shown in Fig 12.18a. If the signals are simply linearly summed, we obtain the sequence

$$x(n) = \sum_{k=1}^{5} x_k(n) \qquad (12.1.34)$$

which results in the antenna directivity pattern shown in Fig. 12.18a. Now suppose that an interference signal is received from a direction corresponding to one of the sidelobes

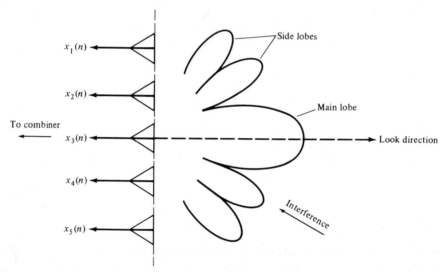

(a) Linear antenna array with antenna pattern

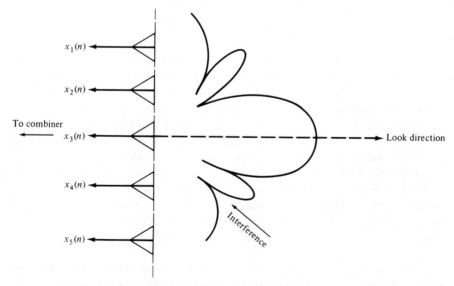

(b) Linear antenna array with a null placed in the direction of the interference

FIGURE 12.18 Linear antenna array: (a) linear antenna array with antenna pattern; (b) linear antenna array with a null placed in the direction of the interference.

in the array. By properly weighting the sequences $\{x_k(n)\}$ prior to combining, it is possible to alter the sidelobe pattern such that the array contains a null in the direction of the interference as shown in Fig. 12.18b. Thus we obtain

$$x(n) = \sum_{k=1}^{5} h_k x_k(n) \qquad (12.1.35)$$

where $\{h_k\}$ are the weights.

We may also change or steer the direction of the main antenna lobe by simply introducing delays in the output of the sensor signals prior to combining. Hence from N sensors we have a combined signal of the form

$$x(n) = \sum_{k=1}^{N} h_k x_k(n - n_k) \qquad (12.1.36)$$

where $\{h_k\}$ are the weights and n_k corresponds to an n_k-sample delay in the signal $x_k(n)$. The choice of weights may be used to place nulls in specific directions.

More generally, we may simply filter each sequence prior to combining. In such a case the output sequence has the general form

$$y(n) = \sum_{k=1}^{N} y_k(n)$$

$$= \sum_{k=1}^{N} \sum_{l=0}^{M-1} h_k(l) x_k(n - n_k - l) \qquad (12.1.37)$$

where $h_k(l)$ is the impulse response of the filter for processing the kth sensor output and $\{n_k\}$ are the delays which steer the beam pattern.

The LMS algorithm described in Section 12.2.2 is frequently used in adaptively selecting the weights $\{h_k\}$ or the impulse responses $\{h_k(l)\}$. The more powerful recursive least-squares algorithms can also be applied to the multisensor (multichannel) data problem.

In our treatment we deal with single-channel (sensor) signals. However, it is relatively easy to modify the algorithms given in Sections 12.2 and 12.3, so they apply to multichannel signals.

12.2 Adaptive Direct-Form FIR Filters

From the examples of the preceding section we have observed that there is a common framework in all the adaptive filtering applications. The least-squares criterion that we have adopted leads to a set of linear equations for the filter coefficients, which may be expressed as

$$\sum_{k=0}^{M-1} h(k) r_{xx}(l - k) = r_{dx}(l + D) \qquad l = 0, 1, 2, \ldots, M - 1 \qquad (12.2.1)$$

where $r_{xx}(l)$ is the autocorrelation of the sequence $x(n)$ and $r_{dx}(l)$ is the crosscorrelation of the sequences $d(n)$ and $x(n)$. The delay parameter D is zero in some cases (e.g., Examples 1, 6, and 7) and nonzero in others (e.g., Examples 2, 3, 4, and 5).

We observe that the autocorrelation $r_{xx}(l)$ and the crosscorrelation $r_{dx}(l)$ are obtained from the data and hence represent estimates of the true (statistical) autocorrelation and crosscorrelation sequences. As a result, the coefficients $\{h(k)\}$ obtained from (12.2.1) are estimates of the true coefficients. The quality of the estimates depend on the length of the data record that is available for estimating $r_{xx}(l)$ and $r_{dx}(l)$. This is one problem that must be considered in the implementation of an adaptive filter.

A second problem that must be considered is that the underlying random process $x(n)$ is usually nonstationary. For example, in channel equalization, the frequency response characteristics of the channel may vary with time. As a consequence, the statistical autocorrelation and crosscorrelation sequences and hence their estimates vary with time. This implies that the coefficients of the adaptive filter must change with time to reflect the time-variant statistical characteristics of the signal into the filter. This also implies that the quality of the estimates cannot be made arbitrarily high simply by increasing the number of signal samples used in the estimation of the autocorrelation and crosscorrelation sequences.

There are several ways by which the coefficients of the adaptive filter can be varied with time to track the time-variant statistical characteristics of the signal. The most popular method is to adapt the filter recursively on a sample-by-sample basis, as each new signal sample is received. A second approach is to estimate $r_{xx}(l)$ and $r_{dx}(l)$ on a block-by-block basis, with no attempt to maintain continuity in the values of the filter coefficients from one block of data to another. In such a scheme the block size must be relatively small, encompassing a time interval that is short compared to the time interval over which the statistical characteristics of the data change significantly. In addition to this block processing method, other block processing schemes can be devised which incorporate some continuity in the filter coefficients from block to block.

In our treatment of adaptive filtering algorithms, we consider only time-recursive algorithms which update the filter coefficients on a sample-by-sample basis. In particular, we consider two types of algorithms, the LMS algorithm, which is based on a gradient-type search for tracking the time-variant signal characteristics, and the class of recursive least-squares algorithms, which are significantly more complex than the LMS algorithm, but which provide faster tracking of time-variant signal statistics.

12.2.1 The Minimum Mean-Square-Error (MMSE) Criterion

The LMS algorithm which is described in the following subsection is most easily obtained by formulating the optimization of the FIR filter coefficients as an estimation problem based on the minimization of the mean-square error (MSE). Let us assume that we have available the (possibly complex-valued) data sequence $\{x(n)\}$, which are samples from a stationary random process with autocorrelation sequence

$$\gamma_{xx}(m) = E[x(n)x^*(n - m)] \qquad (12.2.2)$$

From these samples we form an estimate of the desired sequence $\{d(n)\}$ by passing the observed data $x(n)$ through an FIR filter with coefficients $h(n)$, $0 \le n \le M - 1$. The filter output may be expressed as

$$\hat{d}(n) = \sum_{k=0}^{M-1} h(k)x(n - k) \qquad (12.2.3)$$

where $\hat{d}(n)$ represents an estimate of $d(n)$. The estimation error is defined as

$$e(n) = d(n) - \hat{d}(n)$$
$$= d(n) - \sum_{k=0}^{M-1} h(k)x(n - k) \qquad (12.2.4)$$

The mean-square error as a function of the filter coefficients is

$$J(\mathbf{h}_M) = E\{|e(n)|^2\}$$

$$= E\left\{\left|d(n) - \sum_{k=0}^{M-1} h(k)x(n-k)\right|^2\right\}$$

$$= E\left\{|d(n)|^2 - 2\operatorname{Re}\left[\sum_{l=0}^{M-1} h^*(l)\,d(n)x^*(n-l)\right]\right.$$

$$\left. + \sum_{k=0}^{M-1}\sum_{l=0}^{M-1} h^*(l)h(k)x^*(n-l)x(n-k)\right\}$$

$$= \sigma_d^2 - 2\operatorname{Re}\left[\sum_{l=0}^{M-1} h^*(l)\gamma_{dx}(l)\right]$$

$$+ \sum_{k=0}^{M-1}\sum_{l=0}^{M-1} h^*(l)h(k)\gamma_{xx}(l-k) \qquad (12.2.5)$$

where, by definition, $\sigma_d^2 = E[|d(n)|^2]$, $\gamma_{dx}(l) = E[d(n)x^*(n-l)]$, and $\mathbf{h}_M$ denotes the vector of coefficients. The complex conjugate of $\mathbf{h}_M$ is denoted as $\mathbf{h}_M^*$ and the transpose as $\mathbf{h}_M'$.

We observe that the MSE is a quadratic function of the filter coefficients. Consequently, the minimization of $J(\mathbf{h}_M)$ with respect to the coefficients leads to the set of M linear equations, which is the discrete-time equivalent of the Wiener–Hopf equation [see Paopulis (1984)].

$$\sum_{k=0}^{M-1} h(k)\gamma_{xx}(l-k) = \gamma_{dx}(l) \qquad l = 0, 1, \ldots, M-1 \qquad (12.2.6)$$

The filter with coefficients obtained from (12.2.6) is called the *Wiener filter*.

If we compare (12.2.6) with (12.2.1), it is apparent that these equations are similar in form. In (12.2.1) we use estimates of the autocorrelation and crosscorrelation to determine the filter coefficients, whereas in (12.2.6) the statistical autocorrelation and crosscorrelation are employed. Hence (12.2.6) yields the optimum (Wiener) filter coefficients in the MSE sense, whereas (12.2.1) yields estimates of the optimum coefficients.

The equations in (12.2.6) may be expressed in matrix form as

$$\mathbf{\Gamma}_M \mathbf{h}_M = \mathbf{\gamma}_d \qquad (12.2.7)$$

where $\mathbf{\Gamma}_M$ is an $M \times M$ (Hermitian) Toeplitz matrix with elements $\Gamma_{lk} = \gamma_{xx}(l-k)$ and $\mathbf{\gamma}_d$ is an $M \times 1$ crosscorrelation vector with elements $\gamma_{dx}(l)$, $l = 0, 1, \ldots,$ $M-1$. The solution for the optimum filter coefficients is

$$\mathbf{h}_{\mathrm{opt}} = \mathbf{\Gamma}_M^{-1}\mathbf{\gamma}_d \qquad (12.2.8)$$

and the resulting minimum MSE achieved with the optimum coefficients given by (12.2.8) is

$$J_{\min} \equiv J(\mathbf{h}_{\mathrm{opt}}) = \sigma_d^2 - \sum_{k=0}^{M-1} h_{\mathrm{opt}}(k)\gamma_{dx}^*(k)$$

$$= \sigma_d^2 - \mathbf{\gamma}_d^{*\prime}\mathbf{\Gamma}_M^{-1}\mathbf{\gamma}_d \qquad (12.2.9)$$

For future reference, we wish to note that the set of linear equations in (12.2.6) can also be obtained by invoking the *orthogonality principle* in mean-square estimation [see

Papoulis (1984)]. According to the orthogonality principle, the mean-square estimation error in minimized when the error $e(n)$ is orthogonal, in the statistical sense, to the data in the estimate $\hat{d}(n)$, that is,

$$E[e(n)x^*(n - l)] = 0 \qquad l = 0, 1, \ldots, M - 1 \qquad (12.2.10)$$

If we substitute for $e(n)$ from (12.2.4) into (12.2.10), we obtain

$$E[d(n)x^*(n - l) - \sum_{k=0}^{M-1} h(k)x(n - k)x^*(n - l)] = 0$$

But the expected value of the terms in brackets is simply the set of M linear equations

$$\sum_{k=0}^{M-1} h(k)\gamma_{xx}(l - k) = \gamma_{dx}(l) \qquad l = 0, 1, \ldots, M - 1 \qquad (12.2.11)$$

which were obtained from the minimization of (12.2.5).

Since $\hat{d}(n)$ is orthogonal to $e(n)$, the residual (minimum) mean-square error is

$$J_{min} = E[e(n)\, d^*(n)]$$
$$= E[|d(n)|^2] - \sum_{k=0}^{M-1} h_{opt}(k)\gamma_{dx}^*(k) \qquad (12.2.12)$$

which is the result given in (12.2.9).

The optimum filter coefficients given by (12.2.8) can be obtained by inverting the correlation matrix Γ_M. In the following section we consider the use of a gradient method for solving for $\mathbf{h}_{opt}$, iteratively. This development leads to the LMS algorithm.

12.2.2 The Widrow LMS Algorithm

There are various numerical methods that can be used to solve the set of linear equations given by (12.2.6) or (12.2.7) for the optimum FIR filter coefficients. Below we consider iterative methods that have been devised for finding the minimum of a function of several variables. In our problem, the performance index is the MSE given by (12.2.5), which is a quadratic function of the filter coefficients. Hence this function has a unique minimum which we shall determine by an iterative search.

For the moment, let us assume that the autocorrelation matrix Γ_M and the crosscorrelation vector γ_d are known. Hence $J(\mathbf{h}_M)$ is a known function of the coefficients $h(n)$, $0 \leq n \leq M - 1$. Algorithms for iteratively computing the filter coefficients and thus searching for the minimum of $J(\mathbf{h})$ have the form

$$\mathbf{h}_M(n + 1) = \mathbf{h}_M(n) + \tfrac{1}{2}\Delta(n)\mathbf{S}(n) \qquad n = 0, 1, \ldots \qquad (12.2.13)$$

where $\mathbf{h}_M(n)$ is the vector of filter coefficients at the nth iteration, $\Delta(n)$ is a step size at the nth iteration, and $\mathbf{S}(n)$ is a direction vector for the nth iteration. The initial vector $\mathbf{h}_M(0)$ is chosen arbitrarily. In this treatment we exclude methods that require the computations of Γ_M^{-1}, such as Newton's method. We only consider search methods based on the use of gradient vectors.

The simplest method for finding the minimum of $J(\mathbf{h}_M)$ recursively is based on a steepest-descent search [for reference, see Murray (1972)]. In the method of steepest descent, the direction vector $\mathbf{S}(n) = -\mathbf{g}(n)$, where $\mathbf{g}(n)$ is the gradient vector at the nth iteration, defined as

$$\mathbf{g}(n) = \frac{dJ(\mathbf{h}_M(n))}{d\mathbf{h}_M(n)}$$
$$= 2[\Gamma_M \mathbf{h}_M(n) - \gamma_d] \qquad n = 0, 1, 2, \ldots \qquad (12.2.14)$$

Hence we compute the gradient vector at each iteration and change the values of $\mathbf{h}_M(n)$ in a direction opposite the gradient. Thus the recursive algorithm based on the method of steepest descent is

$$\mathbf{h}_M(n + 1) = \mathbf{h}_M(n) - \tfrac{1}{2}\Delta(n)\mathbf{g}(n) \qquad (12.2.15)$$

or, equivalently,

$$\mathbf{h}_M(n + 1) = [\mathbf{I} - \Delta(n)\Gamma_M]\mathbf{h}_M(n) + \Delta(n)\gamma_d \qquad (12.2.16)$$

We state without proof that the algorithm leads to the convergence of $\mathbf{h}_M(n)$ to $\mathbf{h}_{\text{opt}}$ in the limit as $n \to \infty$, provided that the sequence of step sizes $\Delta(n)$ is absolutely summable, with $\Delta(n) \to 0$ as $n \to \infty$. It follows that as $n \to \infty$, $\mathbf{g}(n) \to \mathbf{0}$.

Other candidate algorithms that provide faster convergence are the conjugate-gradient algorithm and the Fletcher–Powell algorithm. In the conjugate-gradient algorithm, the direction vectors are given as

$$\mathbf{S}(n) = \beta(n - 1)\mathbf{S}(n - 1) - \mathbf{g}(n) \qquad (12.2.17)$$

where $\beta(n)$ is a scalar function of the gradient vectors [for a reference, see Beckman (1960)]. In the Fletcher–Powell algorithm, the direction vectors are given as

$$\mathbf{S}(n) = -\mathbf{H}(n)\mathbf{g}(n) \qquad (12.2.18)$$

where $H(n)$ is an $M \times M$ positive definite matrix, computed iteratively, that converges to the inverse of Γ_M [for a reference, see Fletcher and Powell (1963)]. Clearly, the three algorithms differ in the manner in which the direction vectors are computed.

The three algorithms described above are appropriate when Γ_M and γ_d are known. However, this is not the case in adaptive filtering applications as we have previously indicated. In the absence of knowledge of Γ_M and γ_d we may substitute estimates $\hat{\mathbf{S}}(n)$ of the direction vectors in place of the actual vectors $\mathbf{S}(n)$. Below we consider this approach for the steepest-descent algorithm.

First, we note that the gradient vector given by (12.2.14) may also be expressed in terms of the orthogonality conditions given by (12.2.10). In fact, the conditions in (12.2.10) are equivalent to the expression

$$E[e(n)\mathbf{X}_M^*(n)|] = \gamma_d - \Gamma_M\mathbf{h}_M(n) \qquad (12.2.19)$$

where $\mathbf{X}_M(n)$ is the vector with elements $x(n - l)$, $l = 0, 1, \ldots, M - 1$. Therefore, the gradient vector is simply

$$\mathbf{g}(n) = -2E[e(n)\mathbf{X}_M^*(n)] \qquad (12.2.20)$$

Clearly, the gradient vector $\mathbf{g}(n) = \mathbf{0}$ when the error is orthogonal to the data in the estimate $\hat{d}(n)$.

An unbiased estimate of the gradient vector at the nth iteration is simply obtained from (12.2.20) as

$$\hat{\mathbf{g}}(n) = -2e(n)\mathbf{X}_M^*(n) \qquad (12.2.21)$$

where $e(n) = d(n) - \hat{d}(n)$ and $\mathbf{X}_M(n)$ is the set of M signal samples in the filter at the nth iteration. Thus with $\hat{\mathbf{g}}(n)$ substituted for $\mathbf{g}(n)$ we have the algorithm

$$\mathbf{h}_M(n + 1) = \mathbf{h}_M(n) + \Delta(n)e(n)\mathbf{X}_M^*(n) \qquad (12.2.22)$$

This is called a *stochastic-gradient-descent algorithm*. As given by (12.2.22) it has a variable step size.

It has become common practice in adaptive filtering to use a fixed-step-size algorithm for two reasons. One is that a fixed-step-size algorithm is easily implemented in either hardware or software. Second, a fixed step size is appropriate for tracking time-variant signal statistics, whereas if $\Delta(n) \to 0$ as $n \to \infty$, adaptation to signal variations cannot occur. For these reasons, (12.2.22) is modified to the algorithm

$$\mathbf{h}_M(n + 1) = \mathbf{h}_M(n) + \Delta e(n)\mathbf{X}_M^*(n) \qquad (12.2.23)$$

where Δ is now the fixed step size. This algorithm was first proposed by Widrow and Hoff (1960) and is now widely known as the *LMS (least-mean-square) algorithm*. Clearly, it is a stochastic-gradient algorithm.

The LMS algorithm is a relatively simple algorithm to implement. For this reason it has been widely used in many adaptive filtering applications. Its properties and limitations have also been thoroughly investigated. In the following section we provide a brief treatment of its important properties concerning convergence, stability, and the noise resulting from the use of estimates of the gradient vectors. Later, we shall compare its properties with the more complex recursive least-squares algorithm.

Several variations of the basic LMS algorithm have been proposed in the literature and implemented in some adaptive filtering applications. One variation is obtained if we average the gradient vectors over several iterations prior to making adjustments of the filter coefficients. For example, the average over K gradient vectors is

$$\bar{\mathbf{g}}(nK) = -\frac{2}{K} \sum_{k=0}^{K-1} e(nK + k)\mathbf{X}_M^*(nK + k) \qquad (12.2.24)$$

and the corresponding recursive equation for updating the filter coefficients once every K iterations is

$$\mathbf{h}_M((n + 1)K) = \mathbf{h}_M(nK) - \tfrac{1}{2}\Delta\bar{\mathbf{g}}(nK) \qquad (12.2.25)$$

In effect, the averaging operation performed in (12.2.24) reduces the noise in the estimate of the gradient vector as shown by Gardner (1984).

An alternative approach is to filter the gradient vectors by a lowpass filter and use the output of the filter as an estimate of the gradient vector. For example, a simple lowpass filter for the gradients yields as an output

$$\hat{\mathbf{S}}(n) = \beta\hat{\mathbf{S}}(n - 1) - \hat{\mathbf{g}}(n) \qquad \hat{\mathbf{S}}(0) = -\hat{\mathbf{g}}(0) \qquad (12.2.26)$$

where the choice of $0 \le \beta < 1$ determines the bandwidth of the lowpass filter. When β is close to unity, the filter bandwidth is small and the effective averaging is performed over many gradient vectors. On the other hand, when β is small, the lowpass filter has a large bandwidth, and hence it provides little averaging of the gradient vectors. With the filtered gradient vectors given by (12.2.26) in place of $\hat{\mathbf{g}}(n)$ we obtain the filtered version of the LMS algorithm given by

$$\mathbf{h}_M(n + 1) = \mathbf{h}_M(n) + \tfrac{1}{2}\Delta\hat{\mathbf{S}}(n) \qquad (12.2.27)$$

An analysis of the filtered-gradient LMS algorithm is given in the paper by Proakis (1974).

12.2.3 Properties of the LMS Algorithm

In this section we consider the basic properties of the LMS algorithm given by (12.2.23). In particular, we focus on its convergence properties, its stability, and the excess noise generated as a result of using noisy gradient vectors in place of the actual gradient vectors. The use of noisy estimates of the gradient vectors implies that the filter coefficients will fluctuate randomly and hence an analysis of the characteristics of the algorithm should be performed in statistical terms.

The convergence and stability of the LMS algorithm may be investigated by determining how the mean value of $\mathbf{h}_M(n)$ converges to the optimum coefficients $\mathbf{h}_{\text{opt}}$. If we take the expected value of (12.2.23), we obtain

$$
\begin{aligned}
\bar{\mathbf{h}}_M(n + 1) &= \bar{\mathbf{h}}_M(n) + \Delta E[e(n)\mathbf{X}_M^*(n)] \\
&= \bar{\mathbf{h}}_M(n) + \Delta[\boldsymbol{\gamma}_d - \boldsymbol{\Gamma}_M\bar{\mathbf{h}}_M(n)] \\
&= (\mathbf{I} - \Delta\boldsymbol{\Gamma}_M)\bar{\mathbf{h}}_M(n) + \Delta\boldsymbol{\gamma}_d
\end{aligned}
\tag{12.2.28}
$$

where $\bar{\mathbf{h}}_M(n) = E[\mathbf{h}_M(n)]$ and $\mathbf{I}$ is the identity matrix.

The recursive relation in (12.2.28) may be represented as a closed-loop control system, as shown in Fig. 12.19. The convergence rate and the stability of this closed-loop system are governed by our choice of the step size parameter Δ. To determine the convergence behavior, it is convenient to decouple the M simultaneous difference equations given in (12.2.28), by performing a linear transformation of the mean coefficient vector $\bar{\mathbf{h}}_M(n)$. The appropriate transformation is obtained by noting that the autocorrelation matrix $\boldsymbol{\Gamma}_M$ is Hermitian and hence it can be represented as [see Gantmacher (1960)]

$$
\boldsymbol{\Gamma}_M = \mathbf{U\Lambda U}^{*t}
\tag{12.2.29}
$$

where $\mathbf{U}$ is the normalized model matrix of $\boldsymbol{\Gamma}_M$ and $\boldsymbol{\Lambda}$ is a diagonal matrix with diagonal elements λ_k, $0 \le k \le M - 1$ equal to the eigenvalues of $\boldsymbol{\Gamma}_M$.

When (12.2.29) is substituted into (12.2.28), the latter may be expressed as

$$
\bar{\mathbf{h}}_M^0(n + 1) = (\mathbf{I} - \Delta\boldsymbol{\Lambda})\bar{\mathbf{h}}_M^0(n) + \Delta\boldsymbol{\gamma}_d^0
\tag{12.2.30}
$$

where the transformed (orthogonalized) vectors are $\bar{\mathbf{h}}_M^0(n) = \mathbf{U}^*\bar{\mathbf{h}}_M(n)$ and $\boldsymbol{\gamma}_d^0 = \mathbf{U}^{*t}\boldsymbol{\gamma}_d$. The set of M first-order difference equations in (12.2.30) is now decoupled. Their convergence and their stability are determined from the homogeneous equation

$$
\bar{\mathbf{h}}_M^0(n + 1) = (\mathbf{I} - \Delta\boldsymbol{\Lambda})\bar{\mathbf{h}}_M^0(n)
\tag{12.2.31}
$$

If we focus our attention on the solution of kth equation in (12.2.31), we observe that

$$
\bar{h}^0(k, n) = C(1 - \Delta\lambda_k)^n u(n) \qquad k = 0, 1, 2, \ldots, M - 1
\tag{12.2.32}
$$

where C is some arbitrary constant. Clearly, $\bar{h}^0(k, n)$ converges to zero exponentially provided that

$$
|1 - \Delta\lambda_k| < 1
$$

or, equivalently,

$$
0 < \Delta < \frac{2}{\lambda_k} \qquad k = 0, 1, \ldots, M - 1
\tag{12.2.33}
$$

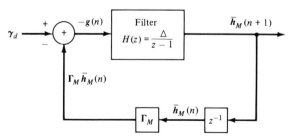

FIGURE 12.19 Closed-loop control system representation of recursive equation (12.2.28).

The condition given by (12.2.33) for convergence of the homogeneous difference equation for the kth normalized filter coefficient (kth mode of the closed-loop system) must be satisfied for all $k = 0, 1, \ldots, M - 1$. Therefore, the range of values of Δ that ensures the convergence of the mean of the coefficient vector in the LMS algorithm is

$$0 < \Delta < \frac{2}{\lambda_{\max}} \qquad (12.2.34)$$

where $\lambda_{\max}$ is the largest eigenvalue of Γ_M.

Since Γ_M is an autocorrelation matrix, its eigenvalues are nonnegative. Hence an upper bound on $\lambda_{\max}$ is

$$\lambda_{\max} < \sum_{k=0}^{M-1} \lambda_k = \text{trace } \Gamma_M = M\gamma_{xx}(0) \qquad (12.2.35)$$

where $\gamma_{xx}(0)$ is the input signal power that is easily estimated from the received signal. Therefore, an upper bound on the step size Δ is $2/M\gamma_{xx}(0)$.

From (12.2.32) we observe that rapid convergence of the LMS algorithm occurs when $|1 - \Delta\lambda_k|$ is small, that is, when the poles of the closed-loop system in Fig. 12.19 are far from the unit circle. However, we cannot achieve this desirable condition and still satisfy the upper bound in (12.2.33) when there is a large difference between the largest and smallest eigenvalues of Γ_M. In other words, even if we select Δ to be near the upper bound $2/\lambda_{\max}$, the convergence rate of the LMS algorithm will be determined by the decay of the mode corresponding to the smallest eigenvalue $\lambda_{\min}$. For this mode, with $\Delta = 2/\lambda_{\max}$ substituted in (12.2.32) we have

$$h_M^0(k, n) = C\left(1 - 2\frac{\lambda_{\min}}{\lambda_{\max}}\right)^n u(n)$$

Consequently, the ratio $\lambda_{\max}/\lambda_{\min}$ ultimately determines the convergence rate. If $\lambda_{\max}/\lambda_{\min}$ is small, the convergence will be rapid. On the other hand, if $\lambda_{\max}/\lambda_{\min}$ is large, the convergence rate of the algorithm is slow.

The other important characteristic of the LMS algorithm is the noise resulting from the use of estimates of the gradient vectors. The noise in the gradient vector estimates causes random fluctuations in the coefficients about their optimal values and thus leads to an increase in the minimum MSE at the output of the adaptive filter. Hence the total MSE is $J_{\min} + J_\Delta$, where J_Δ is called the *excess mean-square error*.

For any given set of filter coefficients $\mathbf{h}_M(n)$, the total MSE at the output of the adaptive filter may be expressed as

$$J(\mathbf{h}_M(n)) = J_{\min} + (\mathbf{h}_M(n) - \mathbf{h}_{\text{opt}})^t \Gamma_M (\mathbf{h}_M(n) - \mathbf{h}_{\text{opt}})^* \qquad (12.2.36)$$

where $\mathbf{h}_{\text{opt}}$ represents the optimum filter coefficients defined by (12.2.8). A plot of $J(\mathbf{h}_M(n))$ as a function of the iteration n is called a *learning curve*. If we substitute (12.2.29) for Γ_M and perform the linear orthogonal transformation used previously, we obtain

$$J(\mathbf{h}_M(n)) = J_{\min} + \sum_{k=0}^{M-1} \lambda_k |h^0(k, n) - h_{\text{opt}}^0(k)|^2 \qquad (12.2.37)$$

where the term $[h^0(k, n) - h_{\text{opt}}^0(k)]$ represents the error in the kth filter coefficient (in the orthogonal coordinate system). The excess mean-square error is defined as the expected value of the second term in (12.2.37), that is,

$$J_\Delta = \sum_{k=0}^{M-1} \lambda_k E[|h^0(k, n) - h^0_{\text{opt}}(k)|^2] \tag{12.2.38}$$

To derive an expression for the excess mean square error J_Δ, we assume that the mean values of the filter coefficients $\mathbf{h}_M(n)$ have converged to their optimum values $\mathbf{h}_{\text{opt}}$. Then the term $\Delta e(n)\mathbf{X}_M^*(n)$ in the LMS algorithm given by (12.2.23) is a zero-mean noise vector. Its covariance is

$$\text{cov} [\Delta e(n)\mathbf{X}_M^*(n)] = \Delta^2 E[|e(n)|^2 \mathbf{X}_M^*(n)\mathbf{X}_M^t(n)] \tag{12.2.39}$$

To a first approximation, we assume that $|e(n)|^2$ is uncorrelated with the signal vector. Although this assumption is not strictly true, it simplifies the derivation and yields useful results. [The reader may refer to the papers by Mazo (1979), Jones et al. (1982), and Gardner (1984) for further discussion on this assumption.] Then

$$\begin{aligned} \text{cov} [\Delta e(n)\mathbf{X}_M^*(n)] &= \Delta^2 E[|e(n)|^2]E[\mathbf{X}_M^*(n)\mathbf{X}_M^t(n)] \\ &= \Delta^2 J_{\text{min}}\Gamma_M \end{aligned} \tag{12.2.40}$$

For the orthogonalized coefficient vector $\mathbf{h}_M^0(n)$, with additive noise, we have the equation

$$\mathbf{h}_M^0(n + 1) = (\mathbf{I} - \Delta\Lambda)\mathbf{h}_M^0(n) + \Delta\gamma_d^0 + \mathbf{w}^0(n) \tag{12.2.41}$$

where $\mathbf{w}^0(n)$ is the additive noise vector, which is related to the noise vector $\Delta e(n)\mathbf{X}_M^*(n)$ through the transformation

$$\begin{aligned} \mathbf{w}^0(n) &= \mathbf{U}^{*t}[\Delta e(n)\mathbf{X}_M^*(n)] \\ &= \Delta e(n)\mathbf{U}^{*t}\mathbf{X}_M^*(n) \end{aligned} \tag{12.2.42}$$

It is easily seen that the covariance matrix of the noise vector is

$$\begin{aligned} \text{cov} [\mathbf{w}^0(n)] &= \Delta^2 J_{\text{min}}\mathbf{U}^{*t}\Gamma_M\mathbf{U} \\ &= \Delta^2 J_{\text{min}}\Lambda \end{aligned} \tag{12.2.43}$$

Therefore, the M components of $\mathbf{w}^0(n)$ are uncorrelated and each component has the variance $\sigma_k^2 = \Delta^2 J_{\text{min}}\lambda_k$, $k = 0, 1, \ldots, M - 1$.

Since the noise components of $\mathbf{w}^0(n)$ are uncorrelated, we may consider the M uncoupled difference equations in (12.2.41) separately. Each first-order difference equation represents a filter with impulse response $(1 - \Delta\lambda_k)^n$. When such a filter is excited with a noise sequence $w_k^0(n)$, the variance of the noise at the output of the filter is

$$E|h^0(k, n) - h^0_{\text{opt}}(k)|^2 = \sum_{n=0}^{\infty} \sum_{m=0}^{\infty} (1 - \Delta\lambda_k)^n(1 - \Delta\lambda_k)^m E[w_k^0(n)w_k^{0*}(m)] \tag{12.2.44}$$

We make the simplifying assumption that the noise sequence $w_k^0(n)$ is white. Then (12.2.44) reduces to

$$E|h^0(k, n) - h^0_{\text{opt}}(k)|^2 = \frac{\sigma_k^2}{1 - (1 - \Delta\lambda_k)^2} = \frac{\Delta^2 J_{\text{min}}\lambda_k}{1 - (1 - \Delta\lambda_k)^2} \tag{12.2.45}$$

If we substitute the result of (12.2.45) into (12.2.38) we obtain the expression for the excess mean-square error as

$$J_\Delta = \Delta^2 J_{\text{min}} \sum_{k=0}^{M-1} \frac{\lambda_k^2}{1 - (1 - \Delta\lambda_k)^2} \tag{12.2.46}$$

This expression can be simplified if we assume that Δ is selected such that $\Delta\lambda_k \ll 1$

for all k. Then

$$J_\Delta \approx \Delta^2 J_{min} \sum_{k=0}^{M-1} \frac{\lambda_k^2}{2\Delta\lambda_k}$$

$$\approx \tfrac{1}{2}\Delta J_{min} \sum_{k=0}^{M-1} \lambda_k$$

$$\approx \frac{\Delta M J_{min}\gamma_{xx}(0)}{2} \qquad\qquad (12.2.47)$$

where $\gamma_{xx}(0)$ is the power of the input signal.

The expression for J_Δ indicates that the excess mean-square error is proportional to the step-size parameter Δ. Hence our choice of Δ must be based on a compromise between fast convergence and a small excess mean-square error. In practice, it is desirable to have $J_\Delta < J_{min}$. Hence

$$\frac{J_\Delta}{J_{min}} \approx \frac{\Delta M\gamma_{xx}(0)}{2} < 1$$

or, equivalently,

$$\Delta < \frac{2}{M\gamma_{xx}(0)} \qquad\qquad (12.2.48)$$

But this is just the upper bound for Δ that we had obtained previously based on the upper bound for λ_{max}, given by (12.2.35). In steady-state operation, Δ should satisfy the upper bound in (12.2.48); otherwise, the excess mean-square error causes significant degradation in the performance of the adaptive filter.

The analysis given above on the excess mean-square error is based on the assumption that the mean value of the filter coefficients have converged to the optimum solution $\mathbf{h}_{opt}$. Under this condition the step size Δ should satisfy the bound in (12.2.48). On the other hand, we have determined that convergence of the mean coefficient vector requires that $\Delta < 2/\lambda_{max}$. While a choice of Δ near the upper bound $2/\lambda_{max}$ may lead to initial convergence of the deterministic (known) gradient algorithm, such a large value of Δ will usually result in instability of the stochastic LMS gradient algorithm.

The initial convergence or transient behavior of the LMS algorithm has been investigated by several researchers. Their results clearly indicate that the step size must be reduced in direct proportion to the length of the adaptive filter, as in (12.2.48). The upper bound given in (12.2.48) is necessary to ensure the initial convergence of the stochastic-gradient LMS algorithm. In practice, a choice of $\Delta \le 1/M\gamma_{xx}(0)$ is usually made. The papers by Gitlin and Weinstein (1979) and Ungerboeck (1972) contain an analysis of the transient behavior and the convergence properties of the LMS algorithm.

In a digital implementation of the LMS algorithm, the choice of the step-size parameter becomes even more critical. In an attempt to reduce the excess mean-square error, it is possible to reduce the step-size parameter to the point where the total output mean-square error actually increases. This condition occurs when the estimated gradient components $e(n)x^*(n - l), l = 0, 1, M - 1$ after multiplication by the small step-size parameter Δ are smaller than one-half of the least significant bit in the fixed-point representation of the filter coefficients. In such a case, adaptation ceases. Consequently, it is important for the step size to be large enough to bring the filter coefficients in the vicinity of $\mathbf{h}_{opt}$. If it is desired to decrease the step size significantly, it is necessary to increase the precision in the filter coefficients. Typically, 20 to 24 bits of precision may be used for the filter coefficients, with the 12 most significant bits used for arithmetic operations in the filtering of the data. The 8 to 12 least significant bits are required to provide the necessary precision for the adaptation process. Thus the scaled, estimated gradient components $\Delta e(n)x^*(n - l)$ usually affect only the least significant bits. In effect, the added precision also allows for the noise to be averaged out, since many incremental changes in the least significant bits are required before any change occurs in the upper more significant bits used in arithmetic operations for filtering of the data. For an analysis of round-off errors in a digital implementation of the LMS algorithm, the reader is referred to the papers by Gitlin and Weinstein (1979), Gitlin et al. (1982), and Caraiscos and Liu (1984).

As a final point, we should indicate that the LMS algorithm is appropriate for tracking slowly time-variant signal statistics. In such a case, the minimum MSE and the optimum coefficient vector will be time variant. In other words, $J_{min}(n)$ is a function of time and the M-dimensional error surface is moving with the time index n. The LMS algorithm attempts to follow the moving minimum $J_{min}(n)$ in the M-dimensional space, but it is always lagging behind due to its use of (estimated) gradient vectors. As a consequence, the LMS algorithm incurs another form of error, called the *lag error*, whose mean-square value decreases with an increase in the step size Δ. The total MSE error can now be expressed as

$$J_{total} = J_{min}(n) + J_\Delta + J_l \tag{12.2.49}$$

where J_l denotes the mean-square error due to the lag.

In any given nonstationary adaptive filtering problem, if we plot the J_Δ and J_l as a function of Δ, we expect these errors to behave as illustrated in Fig. 12.20. We observe that J_Δ increases with an increase in Δ while J_l decreases with an increase in Δ. The total error will exhibit a minimum, which will determine the optimum choice of the step-size parameter.

When the statistical time variations of the signals occur rapidly, the lag error will dominate the performance of the adaptive filter. In such a case, $J_l \gg J_{min} + J_\Delta$, even when the largest possible value of Δ is used. When this condition occurs, the LMS algorithm is inappropriate for the application and one must rely on the more complex recursive least-squares algorithms described in Sections 12.2.4 and 12.3, to obtain faster convergence and tracking.

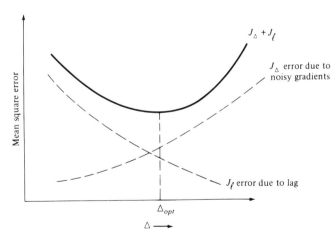

FIGURE 12.20 Excess mean-square error J_Δ and lag error J_l as a function of the step size Δ.

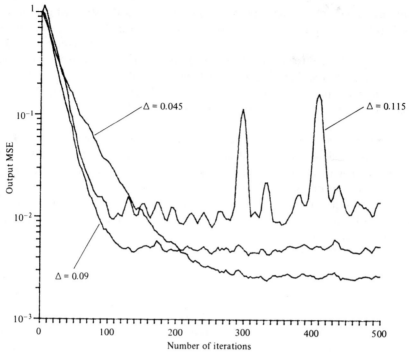

FIGURE 12.21 Learning curves for the LMS algorithm applied to an adaptive equalizer of length $M = 11$ and a channel with eigenvalue spread $\lambda_{max}/\lambda_{min} = 11$.

EXAMPLE 12.2.1 _____

Learning curves for the LMS algorithm when used to adaptively equalize a communication channel are illustrated in Fig. 12.21. The FIR equalizer was realized in direct form and had a length $M = 11$. The autocorrelation matrix Γ_M has an eigenvalue spread of $\lambda_{max}/\lambda_{min} = 11$. These three learning curves have been obtained with step sizes $\Delta = 0.045, 0.09$, and 0.115, by averaging the (estimated) MSE in 200 simulation runs. The input signal power was normalized to unity. Hence the upper bound in (12.2.48) is equal to 0.18. By selecting $\Delta = 0.09$ (one-half of the upper bound) we obtain a fast-decaying learning course, as shown in Fig. 12.21. If we divide Δ by 2 to 0.045, the convergence rate is reduced but the excess mean-square error is' also reduced, so the algorithm performs better in the time-invariant signal environment. Finally, we note that a choice of $\Delta = 0.115$ causes large undesirable fluctuations in the output MSE of the algorithm. Note that $\Delta = 0.115$ is significantly lower than the upper bound given in (12.2.48).

12.2.4 Recursive Least-Squares Algorithms for Direct-Form FIR Filters

The major advantage of the LMS algorithm lies in its computational simplicity. However, the price paid for this simplicity is slow convergence, especially when the eigenvalues of the autocorrelation matrix Γ_M have a large spread (i.e., $\lambda_{max}/\lambda_{min} \gg 1$). From another point of view, the LMS algorithm has only a single adjustable parameter for controlling the convergance rate, namely, the step-size parameter Δ. Since Δ is limited for purposes of stability to be less than the upperbound in (12.2.48), the modes corresponding to the smaller eigenvalues converge very slowly.

To obtain faster convergence, it is necessary to devise more complex algorithms which involve additional parameters. In particular, if the correlation matrix Γ_M has unequal eigenvalues $\lambda_0, \lambda_1, \ldots, \lambda_{M-1}$, we should use an algorithm that contains M parameters, one for each of the eigenvalues. The derivation of optimum algorithms to achieve rapid convergence is the topic of this section.

In deriving faster converging adaptive filtering algorithms, we shall adopt the least-squares criterion instead of the statistical approach based on the MSE criterion. Thus we deal directly with the data sequence $x(n)$ and obtain estimates of correlations from the data.

It is convenient to express the least-squares algorithms in matrix form, to simplfy the notation. Since the algorithms will be recursive in time, it is also necessary to introduce a time index in the filter-coefficient vector and in the error sequence. Hence we define the filter-coefficient vector at time n as

$$\mathbf{h}_M(n) = \begin{bmatrix} h(0, n) \\ h(1, n) \\ h(2, n) \\ \vdots \\ h(M-1, n) \end{bmatrix} \tag{12.2.50}$$

where the subscript M denotes the length of the filter. Similarly, the input signal vector to the filter at time n is denoted as

$$\mathbf{X}_M(n) = \begin{bmatrix} x(n) \\ x(n-1) \\ x(n-2) \\ \vdots \\ x(n-M+1) \end{bmatrix} \tag{12.2.51}$$

We assume that $x(n) = 0$ for $n < 0$. This is usually called *prewindowing* of the input data. The transpose of $\mathbf{X}_M(n)$ is denoted as $\mathbf{X}_M^t(n)$ and its complex conjugate is denoted as $\mathbf{X}_M^*(n)$.

The recursive least-squares problem may now be formulated as follows. Suppose that we have observed the vectors $\mathbf{X}_M(l)$, $l = 0, 1, \ldots, n$ and we wish to determine the filter-coefficient vector $\mathbf{h}_M(n)$ that minimizes the weighted sum of magnitude-squared errors

$$\mathscr{E}_M = \sum_{l=0}^{n} w^{n-l} |e_M(l, n)|^2 \tag{12.2.52}$$

where the error is defined as the difference between the desired sequence $d(l)$ and the estimate $\hat{d}(l, n)$, that is,

$$\begin{aligned} e_M(l, n) &= d(l) - \hat{d}(l, n) \\ &= d(l) - \mathbf{h}_M^t(n)\mathbf{X}_M(l) \end{aligned} \tag{12.2.53}$$

and w is a weighting factor in the range $0 < w \leq 1$.

The purpose of the factor w is to weight the most recent data points more heavily and thus allow the filter coefficients to adapt to time-varying statistical characteristics of the data. This is accomplished by using the exponential weighting factor with the past data. Alternatively, we may use a finite-duration sliding window with uniform weighting over the window length. We find the exponential weighting factor more convenient, both mathematically and practically. For comparison, an exponentially weighted window sequence has an effective memory of

$$\overline{N} = \frac{\displaystyle\sum_{n=0}^{\infty} n w^n}{\displaystyle\sum_{n=0}^{\infty} w^n} = \frac{w}{1 - w} \tag{12.2.54}$$

and hence should be appproximately equivalent to a sliding window of length $\overline{N}$.

The minimization of $\mathscr{E}_M$ with respect to the filter coefficient vector $\mathbf{h}_M(n)$ yields the set of linear equations

$$\mathbf{R}_M(n)\mathbf{h}_M(n) = \mathbf{D}_M(n) \tag{12.2.55}$$

where $\mathbf{R}_M(n)$ is the signal (estimated) correlation matrix defined as

$$\mathbf{R}_M(n) = \sum_{l=0}^{n} w^{n-l}\mathbf{X}_M^*(l)\mathbf{X}_M^t(l) \tag{12.2.56}$$

and $\mathbf{D}_M(n)$ is the (estimated) crosscorrelation vector

$$\mathbf{D}_M(n) = \sum_{l=0}^{n} w^{n-l}\mathbf{X}_M^*(l) \, d(l) \tag{12.2.57}$$

The solution of (12.2.55) is

$$\mathbf{h}_M(n) = \mathbf{R}_M^{-1}(n) \, \mathbf{D}_M(n) \tag{12.2.58}$$

Clearly, the matrix $\mathbf{R}_M(n)$ is akin to the statistical autocorrelation matrix $\mathbf{\Gamma}_M$ while the vector $\mathbf{D}_M(n)$ is akin to the crosscorrelation vector $\mathbf{\gamma}_d$, defined previously. We emphasize, however, that $\mathbf{R}_M(n)$ is not a Toeplitz matrix, whereas $\mathbf{\Gamma}_M$ is. We should also mention that for small values of n, $\mathbf{R}_M(n)$ may be ill conditioned, so that its inverse is not computable. In such a case it is customary to add initially the matrix $\delta\mathbf{I}_M$ to $\mathbf{R}_M(n)$, where $\mathbf{I}_M$ is an identity matrix and δ is a small positive constant. With exponential weighting into the past, the effect of adding $\delta\mathbf{I}_M$ dissipates with time.

Now suppose that we have the solution of (12.2.58) at time $n - 1$ [i.e., we have $\mathbf{h}_M(n - 1)$], and we wish to compute $\mathbf{h}_M(n)$. It is inefficient and hence impractical to solve the set of M linear equations for each new signal component. Instead, we may compute the matrix and vectors recursively. First, $\mathbf{R}_M(n)$ may be computed recursively as

$$\mathbf{R}_M(n) = w\mathbf{R}_M(n - 1) + \mathbf{X}_M^*(n)\mathbf{X}_M^t(n) \tag{12.2.59}$$

We call (12.2.59) the *time-update equation* for $\mathbf{R}_M(n)$.

Since, the inverse of $\mathbf{R}_M(n)$ is needed, we use the matrix inversion lemma [see Householder (1964)]

$$\mathbf{R}_M^{-1}(n) = \frac{1}{w}\left[\mathbf{R}_M^{-1}(n - 1) - \frac{\mathbf{R}_M^{-1}(n - 1)\mathbf{X}_M^*(n)\mathbf{X}_M^t(n)\mathbf{R}_M^{-1}(n - 1)}{w + \mathbf{X}_M^t(n)\mathbf{R}_M^{-1}(n - 1)\mathbf{X}_M^*(n)}\right] \tag{12.2.60}$$

Thus $\mathbf{R}_M^{-1}(n)$ may be computed recursively.

For convenience, we define $\mathbf{P}_M(n) = \mathbf{R}_M^{-1}(n)$. It is also convenient to define an M-dimensional vector $\mathbf{K}_M(n)$, sometimes called the *Kalman gain vector*, as

$$\mathbf{K}_M(n) = \frac{1}{w + \mu_M(n)} \, \mathbf{P}_M(n - 1)\mathbf{X}_M^*(n) \tag{12.2.61}$$

where $\mu_M(n)$ is a scalar defined as

$$\mu_M(n) = \mathbf{X}_M^t(n)\mathbf{P}_M(n - 1)\mathbf{X}_M^*(n) \tag{12.2.62}$$

With these definitions, (12.2.60) becomes

$$\mathbf{P}_M(n) = \frac{1}{w}[\mathbf{P}_M(n - 1) - \mathbf{K}_M(n)\mathbf{X}_M^t(n)\mathbf{P}_M(n - 1)] \tag{12.2.63}$$

Let us postmultiply (12.2.63) by $\mathbf{X}_M^*(n)$. Then

$$\mathbf{P}_M(n)\mathbf{X}_M^*(n) = \frac{1}{w}[\mathbf{P}_M(n - 1)\mathbf{X}_M^*(n) - \mathbf{K}_M(n)\mathbf{X}_M^t(n)\mathbf{P}_M(n - 1)\mathbf{X}_M^*(n)]$$

$$= \frac{1}{w}\{[w + \mu_M(n)]\mathbf{K}_M(n) - \mathbf{K}_M(n)\mu_M(n)\} = \mathbf{K}_M(n) \tag{12.2.64}$$

Therefore, the Kalman gain vector may also be defined as $\mathbf{P}_M(n)\mathbf{X}_M^*(n)$.

Now we may use the matrix inversion lemma to derive an equation for computing the filter coefficients recursively. Since

$$\mathbf{h}_M(n) = \mathbf{P}_M(n)\mathbf{D}_M(n) \tag{12.2.65}$$

and

$$\mathbf{D}_M(n) = w\mathbf{D}_M(n - 1) + d(n)\mathbf{X}_M^*(n) \tag{12.2.66}$$

we have, upon substitution of (12.2.63) and (12.2.66) into (12.2.58),

$$\mathbf{h}_M(n) = \frac{1}{w}[\mathbf{P}_M(n - 1) - \mathbf{K}_M(n)\mathbf{X}_M^t(n)\mathbf{P}_M(n - 1)]$$

$$\times [w\mathbf{D}_M(n - 1) + d(n)\mathbf{X}_M^*(n)]$$

$$= \mathbf{P}_M(n - 1)\mathbf{D}_M(n - 1) + \frac{1}{w}d(n)\mathbf{P}_M(n - 1)\mathbf{X}_M^*(n)$$

$$- \mathbf{K}_M(n)\mathbf{X}_M^t(n)\mathbf{P}_M(n - 1)\mathbf{D}_M(n - 1)$$

$$- \frac{1}{w}d(n)\mathbf{K}_M(n)\mathbf{X}_M^t(n)\mathbf{P}_M(n - 1)\mathbf{X}_M^*(n)$$

$$= \mathbf{h}_M(n - 1) + \mathbf{K}_M(n)[d(n) - \mathbf{X}_M^t(n)\mathbf{h}_M(n - 1)] \tag{12.2.67}$$

We observe that $\mathbf{X}_M^t(n)\mathbf{h}_M(n - 1)$ is the output of the adaptive filter at time n based on use of the filter coefficients at time $n - 1$. Since

$$\mathbf{X}_M^t(n)\mathbf{h}_M(n - 1) = \hat{d}(n, n - 1) \equiv \hat{d}(n) \tag{12.2.68}$$

and

$$e_M(n, n - 1) = d(n) - \hat{d}(n, n - 1) \equiv e_M(n) \tag{12.2.69}$$

it follows that the time-update equation for $\mathbf{h}_M(n)$ may be expressed as

$$\mathbf{h}_M(n) = \mathbf{h}_M(n - 1) + \mathbf{K}_M(n)e_M(n) \tag{12.2.70}$$

or, equivalently,

$$\mathbf{h}_M(n) = \mathbf{h}_M(n - 1) + \mathbf{P}_M(n)\mathbf{X}_M^*(n)e_M(n) \tag{12.2.71}$$

To summarize, suppose that we have the optimum filter coefficients $\mathbf{h}_M(n - 1)$, the matrix $\mathbf{P}_M(n - 1)$, and the vector $\mathbf{X}_M(n - 1)$. When the new signal component $x(n)$ is obtained, we form the vector $\mathbf{X}_M(n)$ by dropping the term $x(n - M)$ from $\mathbf{X}_M(n - 1)$

and adding the term $x(n)$ as the first element. Then the recursive computation for the filter coefficients proceeds as follows:

1. Compute the filter output:

$$\hat{d}(n) = \mathbf{X}_M^t(n)\mathbf{h}_M(n - 1) \tag{12.2.72}$$

2. Compute the error:

$$e_M(n) = d(n) - \hat{d}(n) \tag{12.2.73}$$

3. Compute the Kalman gain vector:

$$\mathbf{K}_M(n) = \frac{\mathbf{P}_M(n-1)\mathbf{X}_M^*(n)}{w + \mathbf{X}_M^t(n)\mathbf{P}_M(n - 1)\mathbf{X}_M^*(n)} \tag{12.2.74}$$

4. Update the inverse of the correlation matrix:

$$\mathbf{P}_M(n) = \frac{1}{w}[\mathbf{P}_M(n - 1) - \mathbf{K}_M(n)\mathbf{X}_M^t(n)\mathbf{P}_M(n - 1)] \tag{12.2.75}$$

5. Update the coefficient vector of the filter:

$$\mathbf{h}_M(n) = \mathbf{h}_M(n - 1) + \mathbf{K}_M(n)e_M(n) \tag{12.2.76}$$

The recursive algorithm specified by (12.2.72) through (12.2.76) is called the direct-form *recursive least-squares (RLS) algorithm*. It is initialized by setting $\mathbf{h}_M(-1) = \mathbf{0}$ and $\mathbf{P}_M(-1) = \delta\mathbf{I}_M$, where δ is a small positive number.

The residual mean-square error resulting from the optimization above is

$$\mathcal{E}_{M\ \min} = \sum_{l=0}^{n} w^{n-l}|d(l)|^2 - \mathbf{h}_M^t(n)\mathbf{D}_M^*(n) \tag{12.2.77}$$

From (12.2.76) we observe that the filter coefficients vary with time by an amount equal to the error $e_M(n)$ multiplied by the Kalman gain vector $\mathbf{K}_M(n)$. Since $\mathbf{K}_M(n)$ is an M-dimensional vector, each filter coefficient is controlled by one of the elements of $\mathbf{K}_M(n)$. Consequently, rapid convergence is obtained. In contrast, the time-update equation for the coefficients of the filter adjusted by use of the LMS algorithm is

$$\mathbf{h}_M(n) = \mathbf{h}_M(n - 1) + \Delta\mathbf{X}_M^*(n)e_M(n) \tag{12.2.78}$$

which has only the single parameter Δ for controlling the adjustment rate of the coefficients.

The Cholesky Factorization and Square-Root Algorithms. We shall observe in the next section that the RLS algorithm given above is very susceptible to round-off noise in implementation of the algorithm with finite-precision arithmetic. The major problem with round-off errors occurs in the updating of $\mathbf{P}_M(n)$. To remedy this problem, algorithms have been developed which avoid the computation of $\mathbf{P}_M(n)$ according to (12.2.75). To describe this class of algorithms, we first consider a factorization of $\mathbf{R}_M(n)$ as

$$\mathbf{R}_M(n) = \mathbf{L}_M(n)\mathbf{\Delta}_M(n)\mathbf{L}_M^{*t}(n) \tag{12.2.79}$$

where $\mathbf{L}_M(n)$ is a lower-triangular matrix with elements $\{l_{ik}\}$, $\mathbf{\Delta}_M(n)$ is a diagonal matrix with diagonal elements $\{\delta_k\}$ and $\mathbf{L}_M^{*t}(n)$ is an upper-triangular matrix. The diagonal elements of $\mathbf{L}_M(n)$ are set to unity (i.e., $l_{ii} = 1$).

Let us use this factorization to solve the set of linear equations $\mathbf{R}_M\mathbf{h}_M = \mathbf{D}_M$, where we have suppressed the dependence of the time index n, in order to simplify the notation.

Given $\mathbf{R}_M(n)$, the factorization in (12.2.79), is equivalent to the set of linear equations

$$r_{ij} = \sum_{k=1}^{j} l_{ik} \, \delta_k \, l_{jk}^* \qquad 1 \le j \le i-1, \, i \ge 2 \qquad (12.2.80)$$

where $\{r_{ij}\}$ are the elements of $\mathbf{R}_M$. Consequently, the elements l_{ik} and δ_k are determined from (12.2.80) according to the equations

$$\delta_1 = r_{11}$$

$$l_{ij} \, \delta_j = r_{ij} - \sum_{k=1}^{j-1} l_{ik} \, \delta_k \, l_{jk}^* \qquad \begin{array}{c} 1 \le j \le i-1 \\ 2 \le i \le M \end{array} \qquad (12.2.81)$$

$$\delta_i = r_{ii} - \sum_{k=1}^{j-1} |l_{ik}|^2 \, \delta_k \qquad 2 \le i \le M$$

Once the elements of $\mathbf{L}_M$ and $\mathbf{\Delta}_M$ are determined, the solution of the linear equations for the coefficients are easily obtained in the following manner. First we have

$$\mathbf{L}_M \mathbf{\Delta}_M \mathbf{L}_M^{*t} \mathbf{h}_M = \mathbf{D}_M \qquad (12.2.82)$$

Let us define the vector

$$\mathbf{Y}_M = \mathbf{\Delta}_M \mathbf{L}_M^{*t} \mathbf{h}_M \qquad (12.2.83)$$

Then

$$\mathbf{L}_M \mathbf{Y}_M = \mathbf{D}_M \qquad (12.2.84)$$

Since $\mathbf{L}_M$ is a lower-triangular matrix, (12.2.84) immediately yields the solution

$$y_1 = d_1 \qquad (12.2.85)$$

$$y_i = d_i - \sum_{j=1}^{i-1} l_{ij} y_j \qquad 2 \le i \le M$$

Having obtained $\mathbf{Y}_M$, the last step is to compute $\mathbf{h}_M$. That is,

$$\mathbf{\Delta}_M \mathbf{L}_M^{*t} \mathbf{h}_M = \mathbf{Y}_M$$

or, equivalently,

$$\mathbf{L}_M^{*t} \mathbf{h}_M = \mathbf{\Delta}_M^{-1} \mathbf{Y}_M \qquad (12.2.86)$$

We begin with

$$h_M = \frac{y_M}{\delta_M} \qquad (12.2.87)$$

The remaining $M-1$ coefficients of $\mathbf{h}_M$ are obtained recursively as follows:

$$h_i = \frac{y_i}{\delta_i} - \sum_{j=i+1}^{M} l_{ji}^* h_j \qquad 1 \le i \le M-1 \qquad (12.2.88)$$

The computation of the FIR filter coefficients $\mathbf{h}_M(n)$ based on the Cholesky factorization as described above is not at all computationally efficient. The number of multiplications and divisions required to compute the elements of $\mathbf{L}_M$ and $\mathbf{\Delta}_M$ is proportional to M^3. The number of multiplications and divisions required to compute $\mathbf{h}_M(n)$ once $\mathbf{L}_M$ and $\mathbf{\Delta}_M$ are determined is proportional to M^2. Hence the computational burden is proportional to M^3.

Obviously, the time updating of $\mathbf{R}_M(n)$ followed by the factorization given by

(12.2.79) and (12.2.81) is an inefficient method and should be avoided. This is indeed possible as demonstrated in the book by Bierman (1977) and the papers by Carlson and Culmone (1979) and Hsu (1982). The key is to update the matrices $\mathbf{L}_M$ and $\mathbf{\Delta}_M$ directly, without computing $\mathbf{R}_M(n)$ and the factorization from $\mathbf{R}_M(n)$, based on the equation

$$\mathbf{L}_M(n)\mathbf{\Delta}_M(n)\mathbf{L}_M^{*t}(n) = \mathbf{L}_M(n-1)\mathbf{\Delta}_M(n-1)\mathbf{L}_M^{*t}(n-1) + \mathbf{X}_M^t(n)\mathbf{X}_M^*(n) \quad (12.2.89)$$

The algorithms for this updating given in the papers by Carlson and Culmone (1979) and Hsu (1982) have a computational complexity proportional to M^2. Once $\mathbf{L}_M(n)$ and $\mathbf{\Delta}_m(n)$ have been updated, the equations (12.2.85) through (12.2.88) are used to compute the filter coefficients. The computational complexity of the resulting algorithm is $2M^2 + 6M$ multiplications and divisions and $2M^2 - M$ additions and subtractions.

The type of factorization described in (12.2.79) for $\mathbf{R}_M(n)$ can also be applied to $\mathbf{P}_M(n)$ = $\mathbf{R}_M^{-1}(n)$. This leads to another computationally efficient algorithm for determining $\mathbf{h}_M(n)$ by directly updating the matrices $\mathbf{L}_M(n)$ and $\mathbf{\Delta}_M(n)$ in the factorization of $\mathbf{P}_M(n)$. The interested reader is referred to the book by Bierman (1977) for this algorithm.

The factorization of $\mathbf{R}_M(n)$ or $\mathbf{P}_M(n)$ as given by (12.2.79) may also be expressed as

$$\mathbf{R}_M(n) = (\mathbf{L}_M(n)\mathbf{\Delta}_M^{1/2}(n))(\mathbf{\Delta}_M^{1/2}(n)\mathbf{L}_M^{*t}(n)) \quad (12.2.90)$$

This factorization is generally called a *square-root factorization* and the resulting RLS algorithms that stem from such a factorization are called *RLS square-root algorithms*.

Fast RLS Algorithms. The RLS direct-form algorithm and the square-root algorithms have a computational complexity proportional to M^2, as indicated above. On the other

---

TABLE 12.1   Fast RLS Algorithm: Version A

$$f_{M-1}(n) = x(n) + \mathbf{a}_{M-1}^t(n-1)\mathbf{X}_{M-1}(n-1)$$

$$g_{M-1}(n) = x(n-M+1) + \mathbf{b}_{M-1}^t(n-1)\mathbf{X}_{M-1}(n)$$

$$\mathbf{a}_{M-1}(n) = \mathbf{a}_{M-1}(n-1) - \mathbf{K}_{M-1}(n-1)f_{M-1}(n)$$

$$f_{M-1}(n, n) = x(n) + \mathbf{a}_{M-1}^t(n)\mathbf{X}_{M-1}(n-1)$$

$$E_{M-1}^f(n) = wE_{M-1}^f(n-1) + f_{M-1}(n)f_{M-1}^*(n, n)$$

$$\begin{bmatrix} \mathbf{C}_{M-1}(n) \\ c_{MM}(n) \end{bmatrix} \equiv \mathbf{K}_M(n) = \begin{bmatrix} 0 \\ \mathbf{K}_{M-1}(n-1) \end{bmatrix} + \frac{f_{M-1}^*(n,n)}{E_{M-1}^f(n)} \begin{bmatrix} 1 \\ \mathbf{a}_{M-1}(n) \end{bmatrix}$$

$$\mathbf{K}_{M-1}(n) = \frac{\mathbf{C}_{M-1}(n) - c_{MM}(n)\mathbf{b}_{M-1}(n-1)}{1 - c_{MM}(n)g_{M-1}(n)}$$

$$\mathbf{b}_{M-1}(n) = \mathbf{b}_{M-1}(n-1) - \mathbf{K}_{M-1}(n)g_{M-1}(n)$$

$$\hat{d}(n) = \mathbf{h}_M^t(n-1)\mathbf{X}_M(n)$$

$$e_M(n) = d(n) - \hat{d}(n)$$

$$\mathbf{h}_M(n) = \mathbf{h}_M(n-1) + \mathbf{K}_M(n)e_M(n)$$

Initialization

$$\mathbf{a}_{M-1}(-1) = \mathbf{b}_{M-1}(-1) = 0$$

$$\mathbf{K}_{M-1}(-1) = 0$$

$$\mathbf{h}_{M-1}(-1) = 0$$

$$E_{M-1}^f(-1) = \epsilon, \, \epsilon > 0$$

hand, the RLS lattice algorithms derived in Section 12.3 have a computational complexity proportional to M. Basically, the lattice algorithms avoid the matrix multiplications involved in computing the Kalman gain vector $\mathbf{K}_M(n)$.

By using the forward and backward prediction formulas derived in Section 12.3 for the RLS lattice, it is possible to obtain time-update equations for the Kalman gain vector that completely avoid matrix multiplications. The resulting algorithms have a complexity that is proportional to M (multiplications and divisions) and hence they are called *fast RLS algorithms* for direct-form FIR filters.

There are several versions of fast algorithms which differ in minor ways. Two versions are given in Tables 12.1 and 12.2 for complex-valued signals. The variables used in the fast algorithms listed in Tables 12.1 and 12.2 are defined in Section 12.3. The computational complexity for the algorithm in Table 12.1 is $10M - 5$ (complex) multiplications and divisions, whereas the one in Table 12.2 has a complexity of $9M + 3$ multiplications and divisions. Further reduction of computational complexity to $7M$ is possible. For example, Carayannis et al. (1983) describe a fast RLS algorithm, termed the *FAEST algorithm*, with a computational complexity of the order of $7M$. This algorithm is given in Section 12.3. Other versions of these algorithms with a complexity of $7M$ have been proposed, but many of these algorithms are extremely sensitive to round-off noise and, exhibit instability problems [for references, see the papers by Falconer and Ljung (1978), Carayannis et al. (1983, 1986) and Cioffi and Kailath (1984)].

TABLE 12.2 Fast RLS Algorithm: Version B

$$f_{M-1}(n) = x(n) + \mathbf{a}_{M-1}^t(n-1)\mathbf{X}_{M-1}(n-1)$$

$$g_{M-1}(n) = x(n-M+1) + \mathbf{b}_{M-1}^t(n-1)\mathbf{X}_{M-1}(n)$$

$$\mathbf{a}_{M-1}(n) = \mathbf{a}_{M-1}(n-1) - \mathbf{K}_{M-1}(n-1)f_{M-1}(n)$$

$$f_{M-1}(n,n) = \alpha_{M-1}(n-1)f_{M-1}(n)$$

$$E_{M-1}^f(n) = wE_{M-1}^f(n-1) + \alpha_{M-1}(n-1)\,|f_{M-1}(n)|^2$$

$$\begin{bmatrix} \mathbf{C}_{M-1}(n) \\ c_{MM}(n) \end{bmatrix} \equiv \mathbf{K}_M(n) = \begin{bmatrix} 0 \\ \mathbf{K}_{M-1}(n-1) \end{bmatrix} + \frac{f_{M-1}^*(n,n)}{E_{M-1}^f(n)} \begin{bmatrix} 1 \\ \mathbf{a}_{M-1}(n) \end{bmatrix}$$

$$\mathbf{K}_{M-1}(n) = \frac{\mathbf{C}_{M-1}(n) - c_{MM}(n)\mathbf{b}_{M-1}(n-1)}{1 - c_{MM}(n)g_{M-1}(n)}$$

$$\mathbf{b}_{M-1}(n) = \mathbf{b}_{M-1}(n-1) - \mathbf{K}_{M-1}(n)g_{M-1}(n)$$

$$\alpha_{M-1}(n) = \alpha_{M-1}(n-1)\left[\dfrac{1 - \dfrac{f_{M-1}(n)f_{M-1}^*(n,n)}{E_{M-1}^f(n)}}{1 - c_{MM}(n)g_{M-1}(n)}\right]$$

$$\hat{d}(n) = \mathbf{h}_M^t(n-1)\mathbf{X}_M(n)$$

$$e_M(n) = d(n) - \hat{d}(n)$$

$$\mathbf{h}_M(n) = \mathbf{h}_M(n-1) + \mathbf{K}_M(n)e_M(n)$$

Initialization

$$\mathbf{a}_{M-1}(-1) = \mathbf{b}_{M-1}(-1) = \mathbf{0}$$

$$\mathbf{K}_{M-1}(-1) = \mathbf{0}, \quad \mathbf{h}_{M-1}(-1) = \mathbf{0}$$

$$E_{M-1}^f(-1) = \epsilon > 0$$

12.2.5 Properties of Direct-Form RLS Algorithms

A major advantage of the direct-form RLS algorithms over the LMS algorithm is their faster convergence rate. This characteristic behavior is illustrated in Fig. 12.22, which shows the convergence rate of the LMS algorithm and the direct-form RLS algorithm for an adaptive FIR channel equalizer of length $M = 11$. The statistical autocorrelation matrix Γ_M for the received signal has an eigenvalue ratio of $\lambda_{max}/\lambda_{min} = 11$. All the equalizer coefficients were initially set to zero. The step size for the LMS algorithm was selected as $\Delta = 0.02$, which represents a good compromise between convergence rate and excess mean-square error.

The superiority of the RLS algorithm in achieving faster convergence is clearly evident. The algorithm converges in less than 70 iterations (70 signal samples) while the LMS algorithm has not converged in over 600 iterations. This rapid rate of convergence of the RLS algorithm is extremely important in applications where the signal statistics vary rapidly with time. For example, the time variations of the characteristics of an ionospheric high-frequency (HF) radio channel are too rapid to be adaptively followed by the LMS algorithm. However, the RLS algorithm adapts sufficiently fast to track such rapid variations [see the paper by Hsu (1982)].

In spite of their superior tracking performance the RLS algorithms for FIR adaptive filtering described in the preceding section have two important disadvantages. One is their computational complexity. The square-root algorithms have a complexity proportional to M^2. The fast RLS algorithms have a computational complexity proportional to M, but the proportionality factor is four to five times that of the LMS algorithm.

The second disadvantage of the algorithms is their sensitivity to round-off errors that accumulate as a result of the recursive computations. In some case, the round-off errors cause these algorithms to become unstable.

The numerical properties of the RLS algorithms have been investigated by several researchers, including Ling and Proakis (1984a, b), Ljung and Ljung (1985), and Cioffi (1987). For illustrative purposes, Table 12.3 includes simulation results on the steady-state (time-averaged) square error for the RLS square-root algorithm, the fast RLS al-

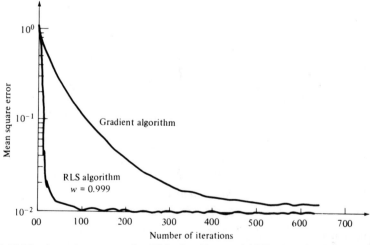

FIGURE 12.22 Learning curves for RLS algorithm and LMS algorithm for adaptive equalizer of length $M = 11$. The eigenvalue spread of the channel is $\lambda_{max}/\lambda_{min} = 11$. The step size for the LMS algorithm is $\Delta = 0.02$. (From *Digital Communication* by John G. Proakis, © 1983 by McGraw-Hill Book Company. Reprinted with permission of the publisher.)

TABLE 12.3 Numerical Accuracy of FIR Adaptive Filtering Algorithms
(Least-Squares Error $\times\ 10^{-3}$)

Number of Bits (Including Sign)	Algorithm		
	RLS Square Root	Fast RLS	LMS
16	2.17	2.17	2.30
13	2.33	2.21	2.30
11	6.14	3.34	19.0
10	17.6	a	77.2
9	75.3	a	311.0
8	a	a	1170.0

[a]Algorithm does not converge to optimum coefficients.

gorithm in Table 12.2, and the LMS algorithm for different word lengths. The simulation was performed with a linear adaptive equalizer having $M = 11$ coefficients. The channel had an eigenvalue ratio of $\lambda_{max}/\lambda_{min} = 11$. The exponential weighting factor used in the RLS algorithms was $w = 0.975$ and the step size for the LMS algorithm was $\Delta = 0.025$. The additive noise has a variance of 0.001. The output MSE with infinite precision is 2.1×10^{-3}. We should indicate that the direct-form RLS algorithm becomes unstable and hence does not work properly with 16-bit fixed-point arithmetic. For this algorithm, we found experimentally that approximately 24 bits of precision are needed for the algorithm to work properly. On the other hand, the square-root algorithm works down to about 9 bits, but the degradation in performance below 11 bits is significant. The fast RLS algorithm works well down to 11 bits for short durations of the order of 500 iterations. For a much larger number of iterations, the algorithm becomes unstable due to the accumulation of round-off errors. In such a case several methods have been proposed to restart the algorithm in order to prevent overflow in the coefficients. The interested reader may refer to the papers of Eleftheriou and Falconer (1987), Cioffi and Kailath (1984), and Hsu (1982). We also observe from the results of Table 12.3 that the LMS algorithm is quite robust to round-off noise. It deteriorates as expected with a decrease in the precision of the filter coefficients, but we did not observe any catastrophic failure (instability) with 8 or 9 bits of precision. However, the degradation in performance below 12 bits is significant.

12.3 Adaptive Lattice-Ladder Filters

In our treatment of filter structures given in Chapter 7, we demonstrated that an FIR filter may also be realized as a lattice structure in which the lattice parameters, called the reflection coefficients, are related to the filter coefficients in the direct-form FIR structure. Formulas were derived for converting the filter coefficients into the reflection coefficients, and vice versa.

In this section we derive adaptive filtering algorithms in which the filter structure is a lattice or a lattice-ladder. Adaptive lattice-ladder filter algorithms based on the method of least squares are derived which have several desirable properties, including computational efficiency and robustness to round-off errors.

From the development of the RLS lattice-ladder algorithms, we obtain the fast RLS algorithms which were described in Section 12.2. We also develop a gradient-type lattice-ladder algorithm which possesses a number of desirable properties that will be described in this section.

12.3.1 Recursive Least-Squares Lattice-Ladder Algorithms

We have already shown in Section 7.2.4 the relationship between the lattice filter structure and a linear predictor and have derived the equations that relate the predictor coefficients to the reflection coefficients of the lattice, and vice versa. We have also established the relationship between the Levinson–Durbin recursions for the linear predictor coefficients and the reflection coefficients in the lattice filter. From these developments we would expect to obtain the recursive least-squares lattice filter by formulating the least-squares estimation problem in terms of linear prediction. This is the approach that we take below.

The recursive least-squares algorithms for the direct-form FIR structures described in Section 12.2.4 are recursive in time only. The length of the filter is fixed. A change (increase or decrease) in the filter length results in a new set of filter coefficients that is totally different from the previous set.

In contrast, the lattice filter is order recursive. As a consequence, the number of sections that it contains can be easily increased or decreased without affecting the reflection coefficients of the remaining sections. This and several other advantages described in this and subsequent sections make the lattice filter very attractive for adaptive filtering applications.

To begin, suppose that we observe the signal $x(n - l)$, $l = 1, 2, \ldots, n$ and let us consider the linear prediction of $x(n)$. Let $f_m(l, n)$ denote the forward prediction error for an mth-order predictor, defined as

$$f_m(l, n) = x(l) + \mathbf{a}_m^t(n)\mathbf{X}_m(l - 1) \tag{12.3.1}$$

where the vector $\mathbf{a}_m(n)$ consists of the forward prediction coefficients, that is,

$$\mathbf{a}_m^t(n) = [a_m(1, n) \quad a_m(2, n) \quad \cdots \quad a_m(m, n)] \tag{12.3.2}$$

and the data vector $\mathbf{X}_m(l - 1)$ is

$$\mathbf{X}_m^t(l - 1) = [x(l - 1) \quad x(l - 2) \quad \cdots \quad x(l - m)] \tag{12.3.3}$$

The predictor coefficients $\mathbf{a}_m(n)$ are selected to minimize the time-average weighted squared error

$$\mathscr{E}_m^f(n) = \sum_{l=0}^{n} w^{n-l} |f_m(l, n)|^2 \tag{12.3.4}$$

The minimization of $\mathscr{E}_m^f(n)$ with respect to $\mathbf{a}_m(n)$ leads to the following set of linear equations:

$$\mathbf{R}_m(n - 1)\mathbf{a}_m(n) = -\mathbf{Q}_m(n) \tag{12.3.5}$$

where $\mathbf{R}_m(n)$ is defined by (12.2.56) and $\mathbf{Q}_m(n)$ is defined as

$$\mathbf{Q}_m(n) = \sum_{l=0}^{n} w^{n-l}x(l)\mathbf{X}_m^*(l - 1) \tag{12.3.6}$$

The solution of (12.3.5) is

$$\mathbf{a}_m(n) = -\mathbf{R}_m^{-1}(n - 1)\mathbf{Q}_m(n) \tag{12.3.7}$$

The minimum value of $\mathscr{E}_m^f(n)$, obtained with the linear predictor specified by (12.3.7), is denoted as $E_m^f(n)$ and is given by

$$E_m^f(n) - \sum_{l=0}^{n} w^{n-l}x^*(l)[x(l) + \mathbf{a}_m^t(n)\mathbf{X}_m(l - 1)]$$

$$= q(n) + \mathbf{a}_m^t(n)\mathbf{Q}_m^*(n) \tag{12.3.8}$$

where $q(n)$ is defined as

$$q(n) = \sum_{l=0}^{n} w^{n-l} |x(l)|^2 \tag{12.3.9}$$

The linear equations in (12.3.5) and the equation for $E_m^f(n)$ in (12.3.8) can be combined in a single matrix equation of the form

$$\begin{bmatrix} q(n) & \mathbf{Q}_m^{*t}(n) \\ \mathbf{Q}_m(n) & \mathbf{R}_m(n-1) \end{bmatrix} \begin{bmatrix} 1 \\ \mathbf{a}_m(n) \end{bmatrix} = \begin{bmatrix} E_m^f(n) \\ \mathbf{O}_m \end{bmatrix} \tag{12.3.10}$$

where $\mathbf{O}_m$ is the m-dimensional null vector. It is interesting to note that

$$
\begin{aligned}
\mathbf{R}_{m+1}(n) &= \sum_{l=0}^{n} w^{n-l} \mathbf{X}_{m+1}^{*}(l) \mathbf{X}_{m+1}^{t}(l) \\
&= \sum_{l=0}^{n} w^{n-l} \begin{bmatrix} x^*(l) \\ \mathbf{X}_m^*(l-1) \end{bmatrix} [x(l) \mathbf{X}_m^t(l-1)] \\
&= \begin{bmatrix} q(n) & \mathbf{Q}_m^*(n) \\ \mathbf{Q}_m(n) & \mathbf{R}_m(n-1) \end{bmatrix}
\end{aligned}
\tag{12.3.11}
$$

which is the matrix in (12.3.10).

In a completely parallel development to (12.3.1) through (12.3.11), we minimize the backward time-average weighted squared error for an mth-order backward predictor defined as

$$\mathcal{E}_m^b(n) = \sum_{l=0}^{n} w^{n-l} |g_m(l, n)|^2 \tag{12.3.12}$$

where the backward error is defined as

$$g_m(l, n) = x(l - m) + \mathbf{b}_m^t(n) \mathbf{X}_m(l) \tag{12.3.13}$$

and $\mathbf{b}_m^t(n) = [b_m(1, n) \; b_m(2, n) \; \cdots \; b_m(m, n)]$ is the vector of coefficients for the backward predictor. The minimization of $\mathcal{E}_m^b(n)$ leads to the equation

$$\mathbf{R}_m(n) \mathbf{b}_m(n) = -\mathbf{V}_m(n) \tag{12.3.14}$$

and hence to the solution

$$\mathbf{b}_m(n) = -\mathbf{R}_m^{-1}(n) \mathbf{V}_m(n) \tag{12.3.15}$$

where

$$\mathbf{V}_m(n) = \sum_{l=0}^{n} w^{n-l} x(l - m) \mathbf{X}_m^*(l) \tag{12.3.16}$$

The minimum value of $\mathcal{E}_m^b(n)$, denoted as $E_m^b(n)$, is

$$
\begin{aligned}
E_m^b(n) &= \sum_{l=0}^{n} w^{n-l} [x(l - m) + \mathbf{b}_m^t(n) \mathbf{X}_m(n)] x^*(l - m) \\
&= v(n) + \mathbf{b}_m^t(n) \mathbf{V}_m^*(n)
\end{aligned}
\tag{12.3.17}
$$

where the scalar quantity $v(n)$ is defined as

$$v(n) = \sum_{l=0}^{n} w^{n-l} |x(l - m)|^2 \tag{12.3.18}$$

If we combine (12.3.14) and (12.3.17) into a single equation, we obtain

$$\begin{bmatrix} \mathbf{R}_m(n) & \mathbf{V}_m(n) \\ \mathbf{V}_m^{*\prime}(n) & v(n) \end{bmatrix} \begin{bmatrix} \mathbf{b}_m(n) \\ 1 \end{bmatrix} = \begin{bmatrix} \mathbf{0}_m \\ E_m^b(n) \end{bmatrix} \tag{12.3.19}$$

We also note that the (estimated) autocorrelation matrix $\mathbf{R}_{m+1}(n)$ can be expressed as

$$\mathbf{R}_{m+1}(n) = \sum_{l=0}^{n} w^{n-l} \begin{bmatrix} \mathbf{X}_m^*(l) \\ x^*(l-m) \end{bmatrix} [\mathbf{X}_m^t(l) \quad x(l-m)] \tag{12.3.20}$$

$$= \begin{bmatrix} \mathbf{R}_m(n) & \mathbf{V}_m(n) \\ \mathbf{V}_m^{*\prime}(n) & v(n) \end{bmatrix}$$

Thus we have obtained the equations for the forward and backward least-squares predictors of order m.

Next, we derive the order-update equations for these predictors, which will lead us to the lattice filter structure. In deriving the order-update equations for $\mathbf{a}_m(n)$ and $\mathbf{b}_m(n)$, we will make use of two matrix inversion identities for a matrix of the form

$$\mathbf{A} = \begin{bmatrix} \mathbf{A}_{11} & \mathbf{A}_{12} \\ \mathbf{A}_{21} & \mathbf{A}_{22} \end{bmatrix} \tag{12.3.21}$$

where $\mathbf{A}$, $\mathbf{A}_{11}$, and $\mathbf{A}_{22}$ are square matrices. The inverse of $\mathbf{A}$ is expressible in two different forms, namely,

$$\mathbf{A}^{-1} = \begin{bmatrix} \mathbf{A}_{11}^{-1} + \mathbf{A}_{11}^{-1}\mathbf{A}_{12}\tilde{\mathbf{A}}_{22}^{-1}\mathbf{A}_{21}\mathbf{A}_{11}^{-1} & -\mathbf{A}_{11}^{-1}\mathbf{A}_{12}\tilde{\mathbf{A}}_{22}^{-1} \\ -\tilde{\mathbf{A}}_{22}^{-1}\mathbf{A}_{21}\mathbf{A}_{11}^{-1} & \tilde{\mathbf{A}}_{22}^{-1} \end{bmatrix} \tag{12.3.22}$$

and

$$\mathbf{A}^{-1} = \begin{bmatrix} \tilde{\mathbf{A}}_{11}^{-1} & -\tilde{\mathbf{A}}_{11}^{-1}\mathbf{A}_{12}\mathbf{A}_{22}^{-1} \\ -\mathbf{A}_{22}^{-1}\mathbf{A}_{21}\tilde{\mathbf{A}}_{11}^{-1} & \mathbf{A}_{22}^{-1}\mathbf{A}_{21}\tilde{\mathbf{A}}_{11}^{-1}\mathbf{A}_{12}\mathbf{A}_{22}^{-1} + \mathbf{A}_{22}^{-1} \end{bmatrix} \tag{12.3.23}$$

where $\tilde{\mathbf{A}}_{11}$ and $\tilde{\mathbf{A}}_{12}$ are defined as

$$\tilde{\mathbf{A}}_{11} = \mathbf{A}_{11} - \mathbf{A}_{12}\mathbf{A}_{22}^{-1}\mathbf{A}_{21} \tag{12.3.24}$$
$$\tilde{\mathbf{A}}_{22} = \mathbf{A}_{22} - \mathbf{A}_{21}\mathbf{A}_{11}^{-1}\mathbf{A}_{12}$$

provided that $\mathbf{A}_{11}$, $\mathbf{A}_{22}$, $\tilde{\mathbf{A}}_{11}$, and $\tilde{\mathbf{A}}_{22}$ are nonsingular.

Order-Update Recursions. Now let us use the formula in (12.3.22) to obtain the inverse of $\mathbf{R}_{m+1}(n)$ by using the form in (12.3.20). First we have

$$\tilde{\mathbf{A}}_{22} = v(n) - \mathbf{V}_m^{*\prime}(n)\mathbf{R}_m^{-1}(n)\mathbf{V}_m(n)$$
$$= v(n) + \mathbf{b}_m^t(n)\mathbf{V}_m^*(n) = E_m^b(n) \tag{12.3.25}$$

and

$$\mathbf{A}_{11}^{-1}\mathbf{A}_{12} = \mathbf{R}_m^{-1}(n)\mathbf{V}_m(n) = -\mathbf{b}_m(n) \tag{12.3.26}$$

Hence

$$\mathbf{R}_{m+1}^{-1}(n) \equiv \mathbf{P}_{m+1}(n) = \begin{bmatrix} \mathbf{P}_m(n) + \dfrac{\mathbf{b}_m(n)\mathbf{b}_m^{*\prime}(n)}{E_m^b(n)} & \dfrac{\mathbf{b}_m(n)}{E_m^b(n)} \\[3mm] \dfrac{\mathbf{b}_m^{*\prime}(n)}{E_m^b(n)} & \dfrac{1}{E_m^b(n)} \end{bmatrix}$$

or, equivalently,

$$P_{m+1}(n) = \begin{bmatrix} P_m(n) & 0 \\ 0 & 0 \end{bmatrix} + \frac{1}{E_m^b(n)} \begin{bmatrix} b_m(n) \\ 1 \end{bmatrix} [b_m^{*t}(n) \quad 1] \qquad (12.3.27)$$

By substituting $n-1$ for n in (12.3.27) and postmultiplying the result by $Q_m(n)$, we obtain the order update for $a_m(n)$. Thus

$$a_{m+1}(n) = -P_{m+1}(n-1)Q_{m+1}(n)$$

$$= \begin{bmatrix} P_m(n-1) & 0 \\ 0 & 0 \end{bmatrix} \begin{bmatrix} -Q_m(n) \\ \cdots \end{bmatrix}$$

$$- \frac{1}{E_m^b(n-1)} \begin{bmatrix} b_m(n-1) \\ 1 \end{bmatrix} [b_m^{*t}(n-1) \quad 1]Q_m(n)$$

$$= \begin{bmatrix} a_m(n) \\ 0 \end{bmatrix} - \frac{k_{m+1}(n)}{E_m^b(n-1)} \begin{bmatrix} b_m(n-1) \\ 1 \end{bmatrix} \qquad (12.3.28)$$

where the scalar quantity $k_{m+1}(n)$ is defined as

$$k_{m+1}(n) = [b_m^{*t}(n-1) \quad 1]Q_{m+1}(n) \qquad (12.3.29)$$

The reader should observe that (12.3.28) is a Levinson-type recursion for the predictor coefficients.

To obtain the corresponding order update for $b_m(n)$, we use the matrix inversion formula in (12.3.23) for the inverse of $R_{m+1}(n)$, along with the form in (12.3.11). In this case we have

$$\tilde{A}_{11} = q(n) - Q_m^{*t}(n)R_m^{-1}(n-1)Q_m(n)$$
$$= q(n) + a_m^t(n)Q_m^*(n) = E_m^f(n) \qquad (12.3.30)$$

and

$$A_{22}^{-1}A_{21} = R_m^{-1}(n-1)Q_m(n) = -a_m(n) \qquad (12.3.31)$$

Hence

$$P_{m+1}(n) = \begin{bmatrix} \dfrac{1}{E_m^f(n)} & \dfrac{a_m^{*t}(n)}{E_m^f(n)} \\[2ex] \dfrac{a_m(n)}{E_m^f(n)} & P_m(n-1) + \dfrac{a_m(n)a_m^{*t}(n)}{E_m^f(n)} \end{bmatrix}$$

or, equivalently,

$$P_{m+1}(n) = \begin{bmatrix} 0 & 0 \\ 0 & P_m(n-1) \end{bmatrix} + \frac{1}{E_m^f(n)} \begin{bmatrix} 1 \\ a_m(n) \end{bmatrix} [1 \quad a_m^{*t}(n)] \qquad (12.3.32)$$

Now, if we postmultiply (12.3.32) by $-V_{m+1}(n)$, we obtain

$$b_{m+1}(n) = \begin{bmatrix} 0 & 0 \\ 0 & P_m(n-1) \end{bmatrix} \begin{bmatrix} \cdots \\ -V_m(n-1) \end{bmatrix}$$

$$- \frac{1}{E_m^f(n)} \begin{bmatrix} 1 \\ a_m(n) \end{bmatrix} [1 \quad a_m^{*t}(n)]V_{m+1}(n)$$

$$= \begin{bmatrix} 0 \\ b_m(n-1) \end{bmatrix} - \frac{k_{m+1}^*(n)}{E_m^f(n)} \begin{bmatrix} 1 \\ a_m(n) \end{bmatrix} \qquad (12.3.33)$$

where

$$[1 \quad \mathbf{a}_m^{*t}(n)]\mathbf{V}_{m+1}(n) = [\mathbf{b}_m^t(n-1) \quad 1]\mathbf{Q}_{m+1}^*(n) = k_{m+1}^*(n) \qquad (12.3.34)$$

The proof of (12.3.34) and its relation to (12.3.29) is left as an exercise for the reader. Thus (12.3.28) and (12.3.33) specify the order update equations for $\mathbf{a}_m(n)$ and $\mathbf{b}_m(n)$, respectively.

The order-update equations for $E_m^f(n)$ and $E_m^b(n)$ may now be obtained. From the definition of $E_m^f(n)$ given by (12.3.8) we have

$$E_{m+1}^f(n) = q(n) + \mathbf{a}_{m+1}^t(n)\mathbf{Q}_{m+1}^*(n) \qquad (12.3.35)$$

By substituting from (12.3.28) for $\mathbf{a}_{m+1}(n)$ into (12.3.35), we obtain

$$E_{m+1}^f(n) = q(n) + \left\{ [\mathbf{a}_m^t(n) \quad 0] \begin{bmatrix} Q_m^*(n) \\ \cdots \end{bmatrix} \right.$$

$$\left. - \frac{k_{m+1}(n)}{E_m^b(n-1)} [\mathbf{b}_m^t(n-1) \quad 1]\mathbf{Q}_{m+1}^*(n) \right\}$$

$$= E_m^f(n) - \frac{|k_{m+1}(n)|^2}{E_m^b(n-1)} \qquad (12.3.36)$$

Similarly, by using (12.3.17) and (12.3.33), we obtain the order update for $E_{m+1}^b(n)$ in the form

$$E_{m+1}^b(n) = E_m^b(n-1) - \frac{|k_{m+1}(n)|^2}{E_m^f(n)} \qquad (12.3.37)$$

The lattice filter is specified by two coupled equations involving the forward and backward errors $f_m(n, n-1)$ and $g_m(n, n-1)$, respectively. From the definition of the forward error in (12.3.1) we have

$$f_{m+1}(n, n-1) = x(n) + \mathbf{a}_{m+1}^t(n-1)\mathbf{X}_{m+1}(n-1) \qquad (12.3.38)$$

Substituting for $\mathbf{a}_{m+1}^t(n-1)$ from (12.3.28) into (12.3.38) yields

$$f_{m+1}(n, n-1) = x(n) + [\mathbf{a}_m^t(n-1) \quad 0] \begin{bmatrix} \mathbf{X}_m(n-1) \\ \cdots \end{bmatrix}$$

$$- \frac{k_{m+1}(n-1)}{E_m^b(n-2)} [\mathbf{b}_m^t(n-2) \quad 1]\mathbf{X}_{m+1}(n-1)$$

$$= f_m(n, n-1) - \frac{k_{m+1}(n-1)}{E_m^b(n-2)}$$

$$\times [x(n-m-1) + \mathbf{b}_m^t(n-2)\mathbf{X}_m(n-1)]$$

$$= f_m(n, n-1) - \frac{k_{m+1}(n-1)}{E_m^b(n-2)} g_m(n-1, n-2) \qquad (12.3.39)$$

To simplify the notation, we define

$$f_m(n) \equiv f_m(n, n-1) \qquad (12.3.40)$$
$$g_m(n) \equiv g_m(n, n-1)$$

Then (12.3.39) may be expressed as

$$f_{m+1}(n) = f_m(n) - \frac{k_{m+1}(n-1)}{E_m^b(n-2)} g_m(n-1) \qquad (12.3.41)$$

Similarly, beginning with the definition of the backward error given by (12.3.13), we have

$$g_{m+1}(n, n-1) = x(n-m-1) + \mathbf{b}_{m+1}^t(n-1)\mathbf{X}_{m+1}(n) \qquad (12.3.42)$$

Substituting for $\mathbf{b}_{m+1}(n-1)$ from (12.3.33) and simplifying the result, we obtain

$$g_{m+1}(n, n-1) = g_m(n-1, n-2) - \frac{k_{m+1}^*(n-1)}{E_m^f(n-1)} f_m(n, n-1) \qquad (12.2.43)$$

or equivalently,

$$g_{m+1}(n) = g_m(n-1) - \frac{k_{m+1}^*(n-1)}{E_m^f(n-1)} f_m(n) \qquad (12.3.44)$$

The two recursive equations in (12.3.41) and (12.3.44) specify the lattice filter illustrated in Fig. 12.23, where, for notational convenience, we have defined the *reflection coefficients* for the lattice as

$$\mathcal{H}_m^f(n) = \frac{-k_m(n)}{E_{m-1}^b(n-1)} \qquad (12.3.45)$$

$$\mathcal{H}_m^b(n) = \frac{-k_m^*(n)}{E_{m-1}^f(n)}$$

The initial conditions on the order updates are

$$f_0(n) = g_0(n) = x(n)$$

$$E_0^f(n) = E_0^b(n) = \sum_{l=0}^{n} w^{n-l}|x(l)|^2$$

$$= wE_0^f(n-1) + |x(n)|^2 \qquad (12.3.46)$$

We note that (12.3.46) is also a time-update equation for $E_0^f(n)$ and $E_0^b(n)$.

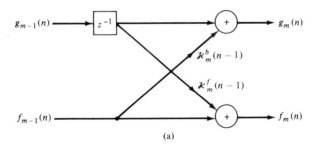

(a)

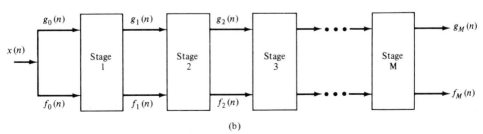

(b)

FIGURE 12.23 Least-squares lattice filter.

Time-Update Recursions. Our goal is to determine a time-update equation for $k_m(n)$ which is necessary if the lattice filter is to be adaptive. This derivation will require time-update equations for the prediction coefficients.

We begin with the form

$$k_{m+1}(n) = V_{m+1}^{*t}(n) \begin{bmatrix} 1 \\ \mathbf{a}_m(n) \end{bmatrix} \tag{12.3.47}$$

The time-update equation for $V_{m+1}(n)$ is

$$V_{m+1}(n) = wV_{m+1}(n-1) + x(n-m-1)X_{m+1}^*(n) \tag{12.3.48}$$

The time update for the prediction coefficients is determined as follows. From (12.3.6), (12.3.7), and (12.2.63), we have

$$\mathbf{a}_m(n) = -\mathbf{P}_m(n-1)\mathbf{Q}_m(n)$$

$$= -\frac{1}{w}[\mathbf{P}_m(n-2) - \mathbf{K}_m(n-1)X_m^t(n-1)\mathbf{P}_m(n-2)]$$

$$\times [w\mathbf{Q}_m(n-1) + x(n)X_m^*(n-1)]$$

$$= \mathbf{a}_m(n-1) - \mathbf{K}_m(n-1)[x(n) + \mathbf{a}_m^t(n-1)X_m(n-1)] \tag{12.3.49}$$

where $\mathbf{K}_m(n-1)$ is the Kalman gain vector at iteration $n-1$. But from (12.3.38) we have

$$x(n) + \mathbf{a}_m^t(n-1)X_m(n-1) = f_m(n, n-1) \equiv f_m(n)$$

Therefore, the time update for $\mathbf{a}_m(n)$ is

$$\mathbf{a}_m(n) = \mathbf{a}_m(n-1) - \mathbf{K}_m(n-1)f_m(n) \tag{12.3.50}$$

In a parallel development to the above, using (12.3.15), (12.3.16), and (12.2.63), we obtain the time-update equations for the coefficients of the backward predictor, in the form

$$\mathbf{b}_m(n) = \mathbf{b}_m(n-1) - \mathbf{K}_m(n)g_m(n) \tag{12.3.51}$$

Now, from (12.3.48) and (12.3.50), the time-update equation for $k_{m+1}(n)$ is

$$k_{m+1}(n) = [wV_{m+1}^{*t}(n-1) + x^*(n-m-1)X_{m+1}^t(n)]$$

$$\times \left\{ \begin{bmatrix} 1 \\ \mathbf{a}_m(n-1) \end{bmatrix} - \begin{bmatrix} 0 \\ \mathbf{K}_m(n-1)f_m(n) \end{bmatrix} \right\}$$

$$= wk_{m+1}(n-1) - wV_{m+1}^{*t}(n-1)\begin{bmatrix} 0 \\ \mathbf{K}_m(n-1) \end{bmatrix}f_m(n) \tag{12.3.52}$$

$$+ x^*(n-m-1)X_{m+1}^t(n)\begin{bmatrix} 1 \\ \mathbf{a}_m(n-1) \end{bmatrix}$$

$$- x^*(n-m-1)X_{m+1}^t(n)\begin{bmatrix} 0 \\ \mathbf{K}_m(n+1) \end{bmatrix}f_m(n)$$

But

$$X_{m+1}^t(n)\begin{bmatrix} 1 \\ \mathbf{a}_m(n-1) \end{bmatrix} = [x(n) \ X_m^t(n-1)]\begin{bmatrix} 1 \\ \mathbf{a}_m(n-1) \end{bmatrix} = f_m(n) \tag{12.3.53}$$

and

$$\mathbf{V}_{m+1}^{*t}(n-1) \begin{bmatrix} 0 \\ \mathbf{K}_m(n-1) \end{bmatrix} = \mathbf{V}_m^{*t}(n-2)\mathbf{K}_m(n-1)$$

$$= \frac{\mathbf{V}_m^{*t}(n-2)\mathbf{P}_m(n-2)\mathbf{X}_m^*(n-1)}{w + \mu_m(n-1)}$$

$$= \frac{-\mathbf{b}_m^{*t}(n-2)\mathbf{X}_m^*(n-1)}{w + \mu_m(n-1)}$$

$$= -\frac{g_m^*(n-1) - x^*(n-m-1)}{w + \mu_m(n-1)} \quad (12.3.54)$$

where $\mu_m(n)$ was defined in (12.2.62). Finally,

$$\mathbf{X}_{m+1}^t(n) \begin{bmatrix} 0 \\ \mathbf{K}_m(n-1) \end{bmatrix} = \frac{\mathbf{X}_m^t(n-1)\mathbf{P}_m(n-2)\mathbf{X}_m^*(n-1)}{w + \mu_m(n)} = \frac{\mu_m(n-1)}{w + \mu_m(n-1)} \quad (12.3.55)$$

Substituting the results of (12.3.53), (12.3.54), and (12.3.55) into (12.3.52), we obtain the desired time-update equation in the form

$$k_{m+1}(n) = wk_{m+1}(n-1) + \frac{w}{w + \mu_m(n-1)} f_m(n)g_m^*(n-1) \quad (12.3.56)$$

It is convenient to define a new variable

$$\alpha_m(n) = \frac{w}{w + \mu_m(n)} \quad (12.3.57)$$

Clearly, $\alpha_m(n)$ is real valued and has a range $0 < \alpha_m(n) < 1$. Then the time update (12.3.56) becomes

$$k_{m+1}(n) = wk_{m+1}(n-1) + \alpha_m(n-1)f_m(n)g_m^*(n-1) \quad (12.3.58)$$

Order Update for $\alpha_m(n)$. Although $\alpha_m(n)$ can be computed directly for each value of m and for each n, it is more efficient to use an order-update equation which is determined as follows. First, from the definition of $\mathbf{K}_m(n)$ given in (12.2.61), it is easily seen that

$$\alpha_m(n) = 1 - \mathbf{X}_m^t(n)\mathbf{K}_m(n) \quad (12.3.59)$$

To obtain an order-update equation for $\alpha_m(n)$, we need an order-update equation for the Kalman gain vector $\mathbf{K}_m(n)$. But $\mathbf{K}_{m+1}(n)$ may be expressed as

$$\mathbf{K}_{m+1}(n) = \mathbf{P}_{m+1}(n)\mathbf{X}_{m+1}^*(n)$$

$$= \left\{ \begin{bmatrix} \mathbf{P}_m(n) & 0 \\ 0 & 0 \end{bmatrix} + \frac{1}{E_m^b(n)} \begin{bmatrix} \mathbf{b}_m(n) \\ 1 \end{bmatrix} [\mathbf{b}_m^{*t}(n) \quad 1] \right\} \begin{bmatrix} \mathbf{X}_m^*(n) \\ x^*(n-m) \end{bmatrix}$$

$$= \begin{bmatrix} \mathbf{K}_m(n) \\ 0 \end{bmatrix} + \frac{g_m^*(n,n)}{E_m^b(n)} \begin{bmatrix} \mathbf{b}_m(n) \\ 1 \end{bmatrix} \quad (12.3.60)$$

But the term $g_m(n,n)$ may also be expressed as

$$\begin{aligned} g_m(n,n) &= x(n-m) + \mathbf{b}_m^t(n)\mathbf{X}_m(n) \\ &= x(n-m) + [\mathbf{b}_m^t(n-1) - \mathbf{K}_{m,}^t(n)g_m(n)]\mathbf{X}_m(n) \\ &= x(n-m) + \mathbf{b}_m^t(n-1)\mathbf{X}_m(n) - g_m(n)\mathbf{K}_m^t(n)\mathbf{X}_m(n) \\ &= g_m(n)[1 - \mathbf{K}_m^t(n)\mathbf{X}_m(n)] \\ &= \alpha_m(n)g_m(n) \quad (12.3.61) \end{aligned}$$

Hence the order-update equation for $\mathbf{K}_m(n)$ in (12.3.60) may also be written

$$\mathbf{K}_{m+1}(n) = \begin{bmatrix} \mathbf{K}_m(n) \\ 0 \end{bmatrix} + \frac{\alpha_m(n)g_m^*(n)}{E_m^b(n)} \begin{bmatrix} \mathbf{b}_m(n) \\ 1 \end{bmatrix} \qquad (12.3.62)$$

By using (12.3.62) and the relation in (12.3.59), we obtain the order-update equation for $\alpha_m(n)$ as follows:

$$\begin{aligned} \alpha_{m+1}(n) &= 1 - \mathbf{X}_{m+1}^t(n)\mathbf{K}_{m+1}(n) \\ &= 1 - [\mathbf{X}_m^t(n) \quad x(n-m)]\left\{ \begin{bmatrix} \mathbf{K}_m(n) \\ 0 \end{bmatrix} + \frac{\alpha_m(n)g_m^*(n)}{E_m^b(n)} \begin{bmatrix} \mathbf{b}_m(n) \\ 1 \end{bmatrix} \right\} \\ &= \alpha_m(n) - \frac{\alpha_m(n)g_m^*(n)}{E_m^b(n)} [\mathbf{X}_m^t(n) \quad x(n-m)] \begin{bmatrix} \mathbf{b}_m(n) \\ 1 \end{bmatrix} \\ &= \alpha_m(n) - \frac{\alpha_m(n)g_m^*(n)}{E_m^b(n)} g_m(n, n) \\ &= \alpha_m(n) - \frac{\alpha_m^2(n)|g_m(n)|^2}{E_m^b(n)} \qquad (12.3.63) \end{aligned}$$

Thus we have obtained both the order-update and time-update equations for the basic least-squares lattice shown in Fig. 12.23. The basic equations are (12.3.41) and (12.3.44) for the forward and backward errors, usually called the *residuals*, (12.3.36) and (12.3.37) for the corresponding least-squares errors, the time-update equation (12.3.58) for $k_m(n)$ and the order-update equation (12.3.63) for the parameter $\alpha_m(n)$. Initially, we have

$$\begin{aligned} E_m^f(-1) &= E_m^b(-1) = E_m^b(-2) = \epsilon > 0 \\ f_m(-1) &= g_m(-1) = k_m(-1) = 0 \\ \alpha_m(-1) &= 1 \qquad \alpha_{-1}(n) = \alpha_{-1}(n-1) = 1 \end{aligned} \qquad (12.3.64)$$

Joint Process Estimation. The last step in the derivation is to obtain the least-squares estimate of the desired signal $d(n)$ from the lattice. Suppose that the adaptive filter has $m + 1$ coefficients which are determined to minimize the average weighted squared error

$$\mathscr{E}_{m+1} = \sum_{l=0}^{n} w^{n-l}|e_{m+1}(l, n)|^2 \qquad (12.3.65)$$

where

$$e_{m+1}(l, n) = d(l) - \mathbf{h}_{m+1}^t(n)\mathbf{X}_{m+1}(l) \qquad (12.3.66)$$

The linear estimate

$$\hat{d}(l, n) = \mathbf{h}_{m+1}^t(n)\mathbf{X}_{m+1}(l) \qquad (12.3.67)$$

which will be obtained from the lattice by using the residuals $\{g_m(n)\}$ is called the *joint process estimate*.

From the results of Section 12.2.4 we have already established that the coefficients of the adaptive filter that minimize (12.3.65) are given by the equation

$$\mathbf{h}_{m+1}(n) = \mathbf{P}_{m+1}(n)\mathbf{D}_{m+1}(n) \qquad (12.3.68)$$

We have also established that $\mathbf{h}_m(n)$ satisfies the time-update equation given in (12.2.76).

Now, let us obtain an order-update equation for $\mathbf{h}_m(n)$. From (12.3.68) and (12.3.27), we have

$$\mathbf{h}_{m+1}(n) = \begin{bmatrix} \mathbf{P}_m(n) & 0 \\ 0 & 0 \end{bmatrix} \begin{bmatrix} \mathbf{D}_m(n) \\ \cdots \end{bmatrix} + \frac{1}{E_m^b(n)} \begin{bmatrix} \mathbf{b}_m(n) \\ 1 \end{bmatrix} [\mathbf{b}_m^{*t}(n) \quad 1]\mathbf{D}_{m+1}(n)$$

$$(12.3.69)$$

We define a complex-valued scalar quantity $\delta_m(n)$ as

$$\delta_m(n) = [\mathbf{b}_m^{*t}(n) \quad 1]\mathbf{D}_{m+1}(n) \tag{12.3.70}$$

Then (12.3.69) may be expressed as

$$\mathbf{h}_{m+1}(n) = \begin{bmatrix} \mathbf{h}_m(n) \\ 0 \end{bmatrix} + \frac{\delta_m(n)}{E_m^b(n)} \begin{bmatrix} \mathbf{b}_m(n) \\ 1 \end{bmatrix} \tag{12.3.71}$$

The scalar $\delta_m(n)$ satisfies a time-update equation which is obtained from the time-update equation for $\mathbf{b}_m(n)$ and $\mathbf{D}_m(n)$, given by (12.3.51) and (12.2.66), respectively. Thus

$$\delta_m(n) = [\mathbf{b}_m^{*t}(n-1) - \mathbf{K}_m^{*t}(n)g_m^*(n) \quad 1][w\mathbf{D}_{m+1}(n-1) + d(n)\mathbf{X}_{m+1}^*(n)]$$
$$= w\,\delta_m(n-1) + [\mathbf{b}_m^{*t}(n-1) \quad 1]\mathbf{X}_{m+1}^*(n)\,d(n)$$
$$- wg_m^*(n)[\mathbf{K}_m^{*t}(n) \quad 0]\mathbf{D}_{m+1}(n-1) - g_m^*(n)\,d(n)[\mathbf{K}_m^{*t}(n) \quad 0]\mathbf{X}_{m+1}^*(n) \tag{12.3.72}$$

But

$$[\mathbf{b}_m^{*t}(n-1) \quad 1]\mathbf{X}_{m+1}^*(n) = x^*(n-m) + \mathbf{b}_m^{*t}(n-1)\mathbf{X}_m^*(n) = g_m^*(n) \tag{12.3.73}$$

Also,

$$[\mathbf{K}_m^{*t}(n) \quad 0]\mathbf{D}_{m+1}(n-1) = \frac{1}{w + \mu_m(n)}[\mathbf{X}_m^t(n)\mathbf{P}_m(n-1) \quad 0]\begin{bmatrix} \mathbf{D}_m(n-1) \\ \cdots \end{bmatrix}$$
$$= \frac{1}{w + \mu_m(n)}\mathbf{X}_m^t(n)\mathbf{h}_m(n-1) \tag{12.3.74}$$

The last term in (12.3.72) may be expressed as

$$[\mathbf{K}_m^{*t}(n) \quad 0]\begin{bmatrix} \mathbf{X}_m^*(n) \\ \cdots \end{bmatrix} = \frac{1}{w + \mu_m(n)}\mathbf{X}_m^t(n)\mathbf{P}_m(n-1)\mathbf{X}_m^*(n) \tag{12.3.75}$$
$$= \frac{\mu_m(n)}{w + \mu_m(n)}$$

Upon substituting the results in (12.3.73) through (12.3.75) into (12.3.72), we obtain the desired time-update equation for $\delta_m(n)$ as

$$\delta_m(n) = w\,\delta_m(n-1) + \alpha_m(n)g_m^*(n)e_m(n) \tag{12.3.76}$$

Order-update equations for $\alpha_m(n)$ and $g_m(n)$ have already been derived. With $e_0(n) = d(n)$, the order-update equation for $e_m(n)$ is obtained as follows:

$$e_m(n) \equiv e_m(n, n-1) = d(n) - \mathbf{h}_m^t(n-1)\mathbf{X}_m(n)$$
$$= d(n) - [\mathbf{h}_{m-1}^t(n-1) \quad 0]\begin{bmatrix} \mathbf{X}_{m-1}(n) \\ \cdots \end{bmatrix}$$
$$- \frac{\delta_{m-1}(n-1)}{E_{m-1}^b(n-1)}[\mathbf{b}_{m-1}^t(n) \quad 1]\mathbf{X}_m(n)$$
$$= e_{m-1}(n) - \frac{\delta_{m-1}(n-1)g_{m-1}(n)}{E_{m-1}^b(n-1)} \tag{12.3.77}$$

Finally, the output estimate $\hat{d}(n)$ of the least-squares lattice is

$$\hat{d}(n) = \mathbf{h}_{m+1}^t(n-1)\mathbf{X}_{m+1}(n) \tag{12.3.78}$$

But $\mathbf{h}'_{m+1}(n-1)$ is not computed explicitly. By repeated use of the order-update equation for $\mathbf{h}_{m+1}(n)$ given by (12.3.71) in (12.3.78), we obtain the desired expression for $\hat{d}(n)$ in the form

$$\hat{d}(n) = \sum_{k=0}^{M-1} \frac{\delta_k(n-1)}{E_k^b(n-1)} g_k(n) \tag{12.3.79}$$

In other words, the output estimate $\hat{d}(n)$ is a linear weighted sum of the backward residuals $\{g_k(n)\}$.

The adaptive least-squares lattice/joint-process (ladder) estimator is illustrated in Fig. 12.24. This lattice-ladder structure is mathematically equivalent to the RLS direct-form FIR filter. The recursive equations are summarized in Table 12.4. This is called the *a priori* form of the *RLS lattice-ladder* algorithm in order to distinguish it from another form of the algorithm, called the *a posteriori* form, in which the coefficient vector $\mathbf{h}_M(n)$ is used in place of $\mathbf{h}_m(n-1)$ to compute the estimate $\hat{d}(n)$. In some adaptive filtering problems, such as channel equalization, the a posteriori form cannot be used, because $\mathbf{h}_M(n)$ cannot be computed prior to the computation of $\hat{d}(n)$.

A number of modifications can be made to the "conventional" RLS lattice-ladder algorithm given in Table 12.4. Below, we describe some of these modifications.

Modified RLS Lattice Algorithms. The recursive equations in the RLS lattice algorithm given in Table 12.4 are by no means unique. Modifications can be made to some of the equations without affecting the optimality of the algorithm. However, some modifications result in algorithms that are more robust numerically when fixed-point arithmetic is used in implementation of the algorithms. We shall give a number of basic relationships that are easily established from the foregoing developments.

First, we have a relationship between the a priori error residuals and the a posteriori residuals:

A PRIORI ERRORS

$$\begin{aligned} f(n, n-1) &\equiv f_m(n) = x(n) + \mathbf{a}'_m(n-1)\mathbf{X}_m(n-1) \\ g(n, n-1) &\equiv g_m(n) = x(n-m) + \mathbf{b}'_m(n-1)\mathbf{X}_m(n) \end{aligned} \tag{12.3.80}$$

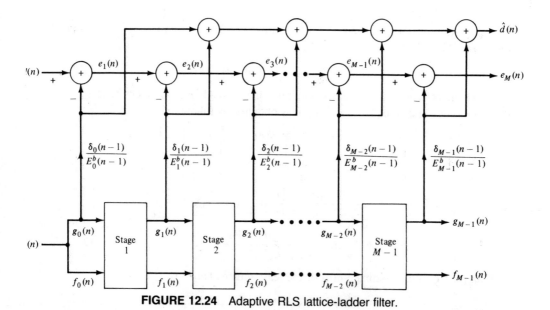

FIGURE 12.24 Adaptive RLS lattice-ladder filter.

TABLE 12.4 A Priori Form of the RLS Lattice-Ladder Algorithm

Lattice Predictor: Begin with $n = 1$ and compute the order updates for $m = 0, 1, \ldots, M - 2$

$$k_{m+1}(n - 1) = w k_{m+1}(n - 2) + \alpha_m(n - 2) f_m(n - 1) g_m^*(n - 2)$$

$$\mathcal{K}_{m+1}^b(n - 1) = -\frac{k_{m+1}(n - 1)}{E_m^b(n - 2)}$$

$$\mathcal{K}_{m+1}^b(n - 1) = -\frac{k_{m+1}^*(n - 1)}{E_m^f(n - 1)}$$

$$f_{m+1}(n) = f_m(n) + \mathcal{K}_{m+1}^f(n - 1) g_m(n - 1)$$

$$g_{m+1}(n) = g_m(n - 1) + \mathcal{K}_{m+1}^b(n - 1) f_m(n)$$

$$E_{m+1}^f(n - 1) = E_m^f(n - 1) - \frac{|k_{m+1}(n - 1)|^2}{E_m^b(n - 2)}$$

$$E_{m+1}^b(n - 1) = E_m^b(n - 2) - \frac{|k_{m+1}(n - 1)|^2}{E_m^f(n - 1)}$$

$$\alpha_{m+1}(n - 1) = \alpha_m(n - 1) - \frac{\alpha_m^2(n - 1) |g_m(n - 1)|^2}{E_m^b(n - 1)}$$

Ladder Filter: Begin with $n = 1$ and compute the order updates for $m = 0, 1, \ldots, M - 1$

$$\delta_m(n - 1) = w \, \delta_m(n - 2) + \alpha_m(n - 1) g_m^*(n - 1) e_m(n - 1)$$

$$\xi_m(n - 1) = -\frac{\delta_m(n - 1)}{E_m^b(n - 1)}$$

$$e_{m+1}(n) = e_m(n) + \xi_m(n - 1) g_m(n)$$

Initialization

$$\alpha_0(n - 1) = 1, \quad e_0(n) = d(n), \quad f_0(n) = g_0(n) = x(n)$$

$$E_0^f(n) = E_0^b(n) = w E_0^f(n - 1) + |x(n)|^2$$

$$\alpha_m(-1) = 1, \quad k_m(-1) = 0$$

$$E_m^b(-1) = E_m^f(0) = \epsilon > 0; \quad \delta_m(-1) = 0$$

A POSTERIORI ERRORS

$$f_m(n, n) = x(n) + \mathbf{a}_m^t(n) \mathbf{X}_m(n - 1)$$
$$g_m(n, n) = x(n - m) + \mathbf{b}_m^t(n - 1) \mathbf{X}_m(n) \tag{12.3.81}$$

The basic relations between (12.3.80) and (12.3.81) are

$$f_m(n, n) = \alpha_m(n - 1) f_m(n)$$
$$g_m(n, n) = \alpha_m(n) g_m(n) \tag{12.3.82}$$

These relations follow easily by using (12.3.50) and (12.3.51) in (12.3.81).

Second, we may obtain time-update equations for the least-squares forward and backward errors. For example, from (12.3.8) and (12.3.50) we obtain

$$E_m^f(n) = q(n) + \mathbf{a}_m^t(n) \mathbf{Q}_m^*(n)$$
$$= q(n) + [\mathbf{a}_m^t(n - 1) - \mathbf{K}_m^t(n - 1) f_m(n)]$$
$$\times [w \mathbf{Q}_m^*(n - 1) + x^*(n) \mathbf{X}_m(n - 1)]$$
$$= w E_m^f(n - 1) + \alpha_m(n - 1) |f_m(n)|^2 \tag{12.3.83}$$

Similarly, from (12.3.17) and (12.3.51) we obtain

$$E_m^b(n) = wE_m^b(n-1) + \alpha_m(n)|g_m(n)|^2 \qquad (12.3.84)$$

Usually, (12.3.83) and (1.3.84) are used in place of (6) and (7) in Table 12.4.

Third, we obtain a time-update equation for the Kalman gain vector, which is not explicitly used in the lattice algorithm, but which is used in the fast algorithms. For this derivation we also use the time-update equations for the forward and backward prediction coefficients given by (12.3.50) and (12.3.51). Thus we have

$$
\begin{aligned}
\mathbf{K}_m(n) &= \mathbf{P}_m(n)\mathbf{X}_m^*(n) \\[2mm]
&= \begin{bmatrix} 0 & 0 \\ 0 & \mathbf{P}_{m-1}(n-1) \end{bmatrix} \begin{bmatrix} x^*(n) \\ \mathbf{X}_{m-1}^*(n-1) \end{bmatrix} \\[2mm]
&\quad + \frac{1}{E_{m-1}^f(n)} \begin{bmatrix} 1 \\ \mathbf{a}_{m-1}(n) \end{bmatrix} [1 \quad \mathbf{a}_{m-1}^{*t}(n)] \begin{bmatrix} x^*(n) \\ \mathbf{X}_{m-1}^*(n-1) \end{bmatrix} \\[2mm]
&= \begin{bmatrix} 0 \\ \mathbf{K}_{m-1}(n-1) \end{bmatrix} + \frac{f_{m-1}^*(n,\,n)}{E_{m-1}^f(n)} \begin{bmatrix} 1 \\ \mathbf{a}_{m-1}(n) \end{bmatrix} \equiv \begin{bmatrix} \mathbf{C}_{m-1}(n) \\ c_{mm}(n) \end{bmatrix} \qquad (12.3.85)
\end{aligned}
$$

where, by definition, $\mathbf{C}_{m-1}(n)$ consists of the first $(m-1)$ elements of $\mathbf{K}_m(n)$ and $c_{mm}(n)$ is the last element. From (12.3.60) we also have the order-update equation for $\mathbf{K}_m(n)$ as

$$\mathbf{K}_m(n) = \begin{bmatrix} \mathbf{K}_{m-1}(n) \\ 0 \end{bmatrix} + \frac{g_{m-1}^*(n,\,n)}{E_{m-1}^b(n)} \begin{bmatrix} \mathbf{b}_{m-1}(n) \\ 1 \end{bmatrix} \qquad (12.3.86)$$

By equating (12.3.85) to (12.3.86) we obtain the result

$$c_{mm}(n) = \frac{g^*(n,\,n)}{E_{m-1}^b(n)} \qquad (12.3.87)$$

and

$$\mathbf{K}_{m-1}(n) + c_{mm}(n)\mathbf{b}_{m-1}(n) = \mathbf{C}_{m-1}(n) \qquad (12.3.88)$$

By substituting from (12.3.51) into (12.3.88) for $\mathbf{b}_{m-1}(n)$ we obtain the time-update equation for the Kalman-gain vector in (12.3.85) as

$$\mathbf{K}_{m-1}(n) = \frac{\mathbf{C}_{m-1}(n) + c_{mm}(n)\mathbf{b}_{m-1}(n-1)}{1 - c_{mm}(n)g_{m-1}(n)} \qquad (12.3.89)$$

There is also a time-update equation for the scalar $\alpha_m(n)$. From (12.3.63) we have

$$
\begin{aligned}
\alpha_m(n) &= \alpha_{m-1}(n) - \frac{\alpha_{m-1}^2(n)|g_{m-1}(n)|^2}{E_{m-1}^b(n)} \\[2mm]
&= \alpha_{m-1}(n)[1 - c_{mm}(n)g_{m-1}(n)] \qquad (12.3.90)
\end{aligned}
$$

A second relation is obtained by using (12.3.85) to eliminate $\alpha_{m-1}(n)$ in the expression for $\alpha_m(n)$. Then

$$
\begin{aligned}
\alpha_m(n) &= 1 - \mathbf{X}_m^t(n)\mathbf{K}_m(n) \\[2mm]
&= \alpha_{m-1}(n-1)\left[1 - \frac{f_{m-1}^*(n,\,n)f_{m-1}(n)}{E_{m-1}^f(n)}\right] \qquad (12.3.91)
\end{aligned}
$$

By equating (12.3.90) to (12.3.91) we obtain the desired time-update equation for $\alpha_m(n)$

as

$$\alpha_{m-1}(n) = \alpha_{m-1}(n-1)\left[\frac{1 - f_{m-1}^*(n, n)f_{m-1}(n)/E_{m-1}^f(n)}{1 - c_{mm}(n)g_{m-1}(n)}\right] \quad (12.3.92)$$

Finally, we wish to distinguish between two different methods for updating the reflection coefficients in the lattice filter and the ladder part, the *conventional (indirect) method* and the *direct method*. In the conventional (indirect) method,

$$\mathcal{K}_{m+1}^f(n) = -\frac{k_{m+1}(n)}{E_m^b(n-1)} \quad (12.3.93)$$

$$\mathcal{K}_{m+1}^b(n) = -\frac{k_{m+1}^*(n)}{E_m^f(n)} \quad (12.3.94)$$

$$\xi_m(n) = -\frac{\delta_m(n)}{E_m^b(n)} \quad (12.3.95)$$

where $k_{m+1}(n)$ is time updated from (12.3.58), $\delta_m(n)$ is updated according to (12.3.76), and $E_m^f(n)$ and $E_m^b(n)$ are updated according to (12.3.83) and (12.3.84). By substituting for $k_{m+1}(n)$ from (12.3.58) into (12.3.93) and using (12.3.84) and equation (8) in Table 12.4, we obtain

$$\mathcal{K}_{m+1}^f(n) = -\frac{k_{m+1}(n-1)}{E_m^b(n-2)}\frac{wE_m^b(n-2)}{E_m^b(n-1)}$$

$$-\frac{\alpha_m(n-1)f_m(n)g_m^*(n-1)}{E_m^b(n-1)}$$

$$= \mathcal{K}_{m+1}^f(n-1)\left[1 - \frac{\alpha_m(n-1)|g_m(n-1)|^2}{E_m^b(n-1)}\right]$$

$$-\frac{\alpha_m(n-1)f_m(n)g_m^*(n-1)}{E_m^b(n-1)}$$

$$= \mathcal{K}_{m+1}^f(n-1) - \frac{\alpha_m(n-1)f_{m+1}(n)g_m^*(n-1)}{E_m^b(n-1)} \quad (12.3.96)$$

which is a formula for directly updating the reflection coefficient in the lattice. Similarly, by substituting (12.3.58) into (12.3.94) and using (12.3.83) and equation (8) in Table 12.4, we obtain

$$\mathcal{K}_{m+1}^b(n) = \mathcal{K}_{m+1}^b(n-1) - \frac{\alpha_m(n-1)f_m^*(n)g_{m+1}(n)}{E_m^f(n)} \quad (12.3.97)$$

Finally, the ladder gain can also be updated directly according to the relation

$$\xi_m(n) = \xi_m(n-1) - \frac{\alpha_m(n)g_m^*(n)e_{m+1}(n)}{E_m^b(n)} \quad (12.3.98)$$

The RLS lattice-ladder algorithm that uses the direct update relations in (12.3.96)–(12.3.98) and (12.3.83)–(12.3.84) is listed in Table 12.5.

An important characteristic of the algorithm in Table 12.5 is that the forward and backward residuals are fed back to time-update the reflection coefficients in the lattice stage and $e_{m+1}(n)$ is fed back to update the ladder gain $\xi_m(n)$. For this reason, this RLS lattice-ladder algorithm has been called the *error-feedback form*. A similar form can be

TABLE 12.5   Direct Update (Error-Feedback) Form of the A Priori RLS Lattice-Ladder Algorithm

---

Lattice Predictor: Begin with $n = 1$ and compute the order updates for $m = 0, 1, \ldots, M - 2$

$$\mathcal{H}^f_{m+1}(n-1) = \mathcal{H}^f_{m+1}(n-2) - \frac{\alpha_m(n-2)f_{m+1}(n-1)g_m^*(n-2)}{E_m^b(n-2)}$$

$$\mathcal{H}^b_{m+1}(n-1) = \mathcal{H}^b_{m+1}(n-2) - \frac{\alpha_m(n-2)f_m^*(n-1)g_{m+1}(n-1)}{E_m^f(n-1)}$$

$$f_{m+1}(n) = f_m(n) + \mathcal{H}^f_{m+1}(n-1)g_m(n-1)$$

$$g_{m+1}(n) = g_m(n-1) + \mathcal{H}^b_{m+1}(n-1)f_m(n)$$

$$E^f_{m+1}(n-1) = wE^f_{m+1}(n-2) + \alpha_{m+1}(n-2)|f_{m+1}(n-1)|^2$$

$$\alpha_{m+1}(n-1) = \alpha_m(n-1) - \frac{\alpha_m^2(n-1)|g_m(n-1)|^2}{E_m^b(n-1)}$$

$$E^b_{m+1}(n-1) = wE^b_{m+1}(n-2) + \alpha_{m+1}(n-1)|g_m(n-1)|^2$$

Ladder Filter: Begin with $n = 1$ and compute the order updates $m = 0, 1, \ldots, M - 1$

$$\xi_m(n-1) = \xi_m(n-2) - \frac{\alpha_m(n-1)g_m^*(n-1)e_{m+1}(n-1)}{E_m^b(n-1)}$$

$$e_{m+1}(n) = e_m(n) + \xi_m(n-1)g_m(n)$$

Initialization

$$\alpha_0(n-1) = 1, \quad e_0(n) = d(n), \quad f_0(n) = g_0(n) = x(n)$$

$$E_0^f(n) = E_0^b(n) = wE_0^f(n-1) + |x(n)|^2$$

$$\alpha_m(-1) = 1, \quad \mathcal{H}^f_m(-1) = \mathcal{H}^b_m(-1) = 0$$

$$E_m^b(-1) = E_m^f(0) = \epsilon > 0$$

---

obtained for the a posteriori RLS lattice-ladder algorithm. For more details on the error-feedback form of the RLS lattice-ladder algorithm, the interested reader is referred to the paper by Ling et al. (1986).

Fast RLS Algorithms. The two versions of the fast RLS algorithms given in the preceding section follow directly from the relationships we have obtained in this section. In particular, we fix the size of the lattice and the associated forward and backward predictors at $M - 1$ stages. Thus we obtain the first seven recursive equations in the two versions of the algorithm. The remaining problem is to determine the time-update equation for the Kalman gain vector, which was determined above in (12.3.85) through (12.3.89). In version B of the algorithm, given in Table 12.2, we used the scalar $\alpha_m(n)$ to reduce the computations. Version A of the algorithm, given in Table 12.1, avoids the use of this parameter. These algorithms have a computational complexity of $10M - 5$ and $9M + 3$. Since these algorithms provide a direct updating of the Kalman gain vector, they have also been called *fast Kalman algorithms* (for references, see Falconer and Ljung, 1978, and Proakis, 1983).

Further reduction of computational complexity to the order of $7M$ is possible by directly updating the following *alternative gain vector* (see Carayannis et al., 1983):

$$\tilde{\mathbf{K}}_M(n) = \frac{1}{w}\mathbf{P}_M(n-1)\mathbf{X}_M^*(n)$$

TABLE 12.6 Fast A Posteriori Error Sequential Technique (FAEST) Algorithm

$$f_{M-1}(n) = x(n) + \mathbf{a}'_{M-1}(n-1)\mathbf{X}_{M-1}(n-1)$$

$$\hat{f}_{M-1}(n, n) = \frac{f_{M-1}(n)}{\tilde{\alpha}_{M-1}(n-1)}$$

$$\mathbf{a}_{M-1}(n) = \mathbf{a}_{M-1}(n-1) - \tilde{\mathbf{K}}_{M-1}(n-1)\hat{f}_{M-1}(n, n)$$

$$E^f_{M-1}(n) = wE^f_{M-1}(n-1) + f_{M-1}(n)f^*_{M-1}(n, n)$$

$$\tilde{\mathbf{K}}_M(n) = \begin{bmatrix} 0 \\ \tilde{\mathbf{K}}_{M-1}(n-1) \end{bmatrix} + \frac{f^*_{M-1}(n)}{wE^f_{M-1}(n)} \begin{bmatrix} 1 \\ \mathbf{a}_{M-1}(n) \end{bmatrix} \equiv \begin{bmatrix} \tilde{\mathbf{C}}_{m-1}(n) \\ \tilde{c}_{MM}(n) \end{bmatrix}$$

$$g_{M-1}(n) = -wE^b_{M-1}(n-1)\tilde{c}_{MM}(n)$$

$$\tilde{\mathbf{K}}_{M-1}(n) = \tilde{\mathbf{C}}_{M-1}(n) + \mathbf{b}_{M-1}(n-1)\tilde{c}_{MM}(n)$$

$$\tilde{\alpha}_M(n) = \tilde{\alpha}_{M-1}(n-1) + \frac{f^2_{M-1}(n)}{wE^f_{M-1}(n)}$$

$$\tilde{\alpha}_{M-1}(n) = \tilde{\alpha}_M(n) + g_{M-1}(n)\tilde{c}_{MM}(n)$$

$$g_{M-1}(n, n) = \frac{g_{M-1}(n)}{\tilde{\alpha}_{M-1}(n)}$$

$$E^b_{M-1}(n) = wE^b_{M-1}(n-1) + g_{M-1}(n)g^*_{M-1}(n, n)$$

$$\mathbf{b}_{M-1}(n) = \mathbf{b}_{M-1}(n-1) + \tilde{\mathbf{K}}_{M-1}(n)g_{M-1}(n, n)$$

$$e_M(n) = d(n) - \mathbf{h}'_M(n-1)\mathbf{X}_M(n)$$

$$e_M(n, n) = \frac{e_M(n)}{\tilde{\alpha}_M(n)}$$

$$\mathbf{h}_M(n) = \mathbf{h}_M(n-1) + \tilde{\mathbf{K}}_M(n)e_M(n, n)$$

Initialization: Set all vectors to zero

$$E^f_{M-1}(-1) = E^b_{M-1}(-1) = \epsilon > 0$$

$$\tilde{\alpha}_{M-1}(-1) = 1$$

Several fast algorithms using this gain have been proposed with a complexity of the order of $7M$. In general, many of these algorithms are sensitive to round-off noise and exhibit instability problems (for references, see the papers by Falconer and Ljung, 1978; Carayannis et al., 1983, 1986; and Cioffi and Kailath, 1984). Table 12.6 lists the FAEST (fast a posteriori error sequential technique) algorithm with a computational complexity $7M + 9$ (for a derivation, see Carayannis et al., 1983, 1986, and Problem 12.8).

12.3.2 Gradient Lattice-Ladder Algorithm

The RLS lattice-ladder algorithms described in the preceding section are significantly more complicated than the LMS algorithm. However, they do result in superior performance, as we shall observe in the following section.

In an attempt to simplify the computational aspects of this class of algorithms but yet retain many of their optimal properties, we shall consider a lattice-ladder filter structure in which the number of filter parameters is significantly reduced. In particular, the lattice-ladder filter structure is illustrated in Fig. 12.25. Each stage of the lattice is characterized by the output–input relations

$$\begin{align} f_m(n) &= f_{m-1}(n) + k_m(n)g_{m-1}(n-1) \\ g_m(n) &= g_{m-1}(n-1) + k^*_m(n)f_{m-1}(n) \end{align} \quad (12.3.99)$$

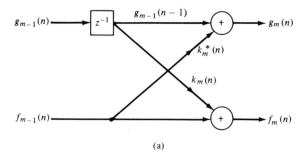

(a)

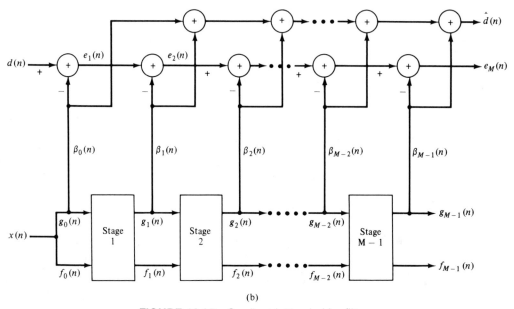

(b)

FIGURE 12.25   Gradient lattice-ladder filter.

where $k_m(n)$ is the reflection coefficient in the mth stage of the lattice, and $f_m(n)$ and $g_m(n)$ are the forward and backward residuals, respectively.

This form of the lattice filter is identical to that obtained from the Levinson–Durbin algorithm, except that now $k_m(n)$ is allowed to vary with time so that the lattice filter adapts to the time variations in the signal statistics. In comparison with the RLS lattice filter, the lattice described by (12.3.99) is more restrictive in that the forward and backward predictors have identical (complex-conjugate) coefficients.

The lattice filter parameters $\{k_m(n)\}$ may be optimized according to a MSE criterion or by employing the method of least squares. Suppose that we adopt the MSE criterion and select the parameters to minimize the sum of the mean-square forward and backward errors, that is,

$$\begin{aligned}\mathcal{E}_M &= E[|f_m(n)|^2 + |g_m(n)|^2] \\ &= E\{|f_{m-1}(n) + k_m g_{m-1}(n-1)|^2 + |g_{m-1}(n-1) + k_m^* f_{m-1}(n)|^2\}\end{aligned} \tag{12.3.100}$$

where we have dropped the time dependence on the parameters $\{k_m\}$ for this optimization, since the statistics now are assumed to be time invariant. The minimization of $\mathcal{E}_m$ with respect to the filter parameters results in the solution

$$k_m = \frac{-2E[f_{m-1}(n)g^*_{m-1}(n-1)]}{[E|f_{m-1}(n)|^2 + E|g_{m-1}(n-1)|^2]} \qquad (12.3.101)$$

Note that k_m has the form of a normalized correlation coefficient.

When the statistical properties of the signal are unknown, we adopt the least-squares criterion for determining $\{k_m(n)\}$. The performance index to be minimized is

$$\mathscr{E}^{LS}_m = \sum_{l=0}^{n} w^{n-l}[|f_m(n)|^2 + |g_m(n)|^2]$$

$$= \sum_{l=0}^{n} w^{n-l}\{|f_{m-1}(n) + k_m(n)g_{m-1}(n-1)|^2 \qquad (12.3.102)$$

$$+ |g_{m-1}(n-1) + k^*_m(n)f_{m-1}(n)|^2\}$$

Minimization of $\mathscr{E}^{LS}_m$ with respect to $k_m(n)$ yields the solution

$$k_m(n) = \frac{-2\sum_{l=0}^{n} w^{n-l}f_{m-1}(n)g^*_{m-1}(n-1)}{\sum_{l=0}^{n} w^{n-l}[|f_{m-1}(n)|^2 + |g_{m-1}(n-1)|^2]} \qquad (12.3.103)$$

Clearly, (12.3.103) is the appropriate expression for estimating $k_m(n)$ in an adaptive filtering application.

In a recursive implementation of the computation in (12.3.103), the numerator and denominator terms may be updated in time as follows:

$$u_m(n) = wu_m(n-1) + 2f_{m-1}(n)g^*_{m-1}(n-1) \qquad (12.3.104)$$
$$v_m(n) = wv_m(n-1) + |f_{m-1}(n)|^2 + |g_{m-1}(n-1)|^2$$

Then

$$k_m(n) = \frac{-u_m(n)}{v_m(n)} \qquad (12.3.105)$$

Equivalently, $k_m(n)$ may be updated recursively in time according to the relation

$$k_m(n) = k_m(n-1) + \frac{f_{m-1}(n-1)g^*_m(n-1) + g^*_{m-1}(n-2)f_m(n-1)}{v_m(n-1)} \qquad (12.3.106)$$

Following the form of the solution obtained in the RLS lattice, we form the output of the lattice in Fig. 12.25 as a linear combination of the backward residuals. Thus

$$\hat{d}(n) = \sum_{k=0}^{M-1} \beta_k(n)g_k(n) \qquad (12.3.107)$$

where $\{\beta_k(n)\}$ are the weighting coefficients of the ladder part. The optimum values of the weighting coefficients are obtained by minimizing the MSE between the desired signal $d(n)$ and the estimate. Let $e_{m+1}(n)$ denote the error between $d(n)$ and the estimate at the output of the m-stage lattice. Then, with $e_0(n) = d(n)$, we have

$$e_{m+1}(n) = d(n) - \sum_{k=0}^{m} \beta_k(n)g_k(n)$$

$$= d(n) - \sum_{k=0}^{m-1} \beta_k(n)g_k(n) - \beta_m(n)g_m(n)$$

$$= e_m(n) - g_m(n)\beta_m(n) \qquad (12.3.108)$$

The error in (12.3.108) may also be expressed in matrix form as

$$e_{m+1}(n) = d(n) - \boldsymbol{\beta}_{m+1}^t(n)\mathbf{G}_{m+1}(n) \tag{12.3.109}$$

where $\boldsymbol{\beta}_{m+1}(n)$ is the vector of ladder weights and $\mathbf{G}_{m+1}(n)$ is the vector of backward residuals.

If we assume for the moment that the signal statistics are stationary, we may drop the time dependence on the coefficient vector, and select $\boldsymbol{\beta}_M$ to satisfy the orthogonality condition

$$E[e_M(n)g_k^*(n)] = 0 \qquad k = 0, 1, \ldots, M - 1 \tag{12.3.110}$$

If we substitute from (12.3.109) into (12.3.110) and perform the expectation operation, we obtain

$$E[d(n)\mathbf{G}_M^*(n)] - E[\mathbf{G}_M^*(n)\mathbf{G}_M^t(n)]\boldsymbol{\beta}_M = 0$$

or, equivalently,

$$\boldsymbol{\beta}_M = \{E[\mathbf{G}_M^*(n)\mathbf{G}_M^t(n)]\}^{-1}E[d(n)\mathbf{G}_M^*(n)] \tag{12.3.111}$$

An important property of the backward residuals in a lattice filter described by (12.3.99) is that they are orthogonal (see Makhoul, 1978), that is,

$$E[g_k(n)g_j^*(n)] = \begin{cases} \mathcal{E}_k^b & k = j \\ 0 & \text{otherwise} \end{cases} \tag{12.3.112}$$

Consequently, the matrix $E[\mathbf{G}_M^*(n)\mathbf{G}_M^t(n)]$ is diagonal and hence the optimum ladder gains are given as

$$\beta_m = \frac{1}{\mathcal{E}_m^b} E[d(n)g_m^*(n)] \tag{12.3.113}$$

There remains the problem of adjusting the ladder gains $\{\beta_m(n)\}$ adaptively. Since the desired $\{\beta_m\}$ minimize the MSE between $d(n)$ and $\hat{d}(n)$, the error will be orthogonal to the backward residuals $g_n(n)$ in $\hat{d}(n)$. This suggests a gradient algorithm of the form

$$\beta_m(n + 1) = \beta_m(n) + \frac{e_m(n)g_m^*(n)}{\hat{\mathcal{E}}_m^b(n)} \tag{12.3.114}$$

where $\hat{\mathcal{E}}_m^b(n)$ is an estimate of $\mathcal{E}_m^b$, which may be computed recursively as

$$\hat{\mathcal{E}}_m^b(n) = w\hat{\mathcal{E}}_m^b(n - 1) + |g_m(n)|^2 \tag{12.3.115}$$

However, the computation in (12.3.115) can be avoided. Since the forward and backward residuals have identical mean-square values, the variable $v_m(n)$ in (12.3.104), which represents the combined residual noise power in $f_m(n)$ and $g_m(n)$, is an estimate of $2\mathcal{E}_m^b$. Hence (12.3.114) is replaced by the recursive equation

$$\beta_m(n + 1) = \beta_m(n) + \frac{2e_m(n)g_m^*(n)}{v_m(n)} \tag{12.3.116}$$

In summary, the adaptive lattice-ladder algorithm is listed in Table 12.7. Since the algorithm in (12.3.116) for updating the ladder gains is a gradient algorithm, this filter is called a gradient lattice-ladder filter. The factor $2/v_m(n)$ plays the role of the step-size parameter.

This algorithm was originally proposed by Griffiths (1977) and considered for noise-canceling applications by Griffiths (1978) and for adaptive equalization by Satorius and Pack (1981) and Proakis (1983).

TABLE 12.7   Gradient Lattice-Ladder Algorithm

$$v_m(n) = wv_m(n - 1) + |f_{m-1}(n)|^2 + |g_{m-1}(n - 1)|^2$$

$$k_m(n) = k_m(n - 1) + \frac{f_{m-1}(n - 1)g_m^*(n - 1) + g_{m-1}^*(n - 2)f_m(n - 1)}{v_m(n - 1)}$$

$$f_m(n) = f_{m-1}(n) + k_m(n)g_{m-1}(n - 1)$$

$$g_m(n) = g_{m-1}(n) + k_m^*(n)f_{m-1}(n)$$

$$e_{m+1}(n) = e_m(n) - g_m(n)\beta_m(n)$$

$$\hat{d}(n) = \beta_M'(n)G_M(n)$$

$$\beta_m(n + 1) = \beta_m(n) + \frac{2e_m(n)g_m^*(n)}{v_m(n)}$$

Initialization

$$f_0(n) = g_0(n) = x(n), \quad f_m(-1) = g_m(-1) = g_m(-2) = 0$$

$$e_0(n) = d(n), \quad e_m(0) = 0, \quad m > 1$$

$$v_m(-1) = \epsilon > 0$$

$$\beta_m(0) = 0$$

$$k_m(-1) = 0$$

12.3.3 Properties of Lattice-Ladder Algorithms

The lattice algorithms that we have derived in the two preceding sections have a number of desirable properties. In this section we consider the properties of these algorithms and compare them with the corresponding properties of the LMS algorithm and the RLS direct-form FIR filtering algorithms.

Convergence Rate. The RLS lattice-ladder algorithms basically have the same convergence rate as the RLS direct-form FIR filter structures. This characteristic behavior is not surprising since both filter structures are optimum in the least-squares sense. Although the gradient lattice algorithm retains some of the optimal characteristics of the RLS lattice, nevertheless, the former is not optimum in the least-squares sense and hence its convergence rate is slower.

For comparison purposes, Figs. 12.26 and 12.27 illustrate the learning curves for an adaptive equalizer of length $M = 11$, implemented as a RLS lattice-ladder filter, as a gradient lattice-ladder filter, and a direct-form FIR filter using the LMS algorithm, for two channel autocorrelation matrices that have eigenvalue ratios of $\lambda_{max}/\lambda_{min} = 11$ and $\lambda_{max}/\lambda_{min} = 21$. From these learning curves, we observe that the gradient lattice algorithm takes about twice as many iterations to converge as the optimum RLS lattice algorithm. Furthermore, the gradient lattice algorithm provides significantly faster convergence than the LMS algorithm. For both lattice structures, the convergence rate does not depend on the eigenvalue spread of the correlation matrix.

Computational Requirements. The RLS lattice algorithms described in the preceding section have a computational complexity that is proportional to M. In contrast, the computational complexity of the RLS square-root algorithms have a complexity proportional to M^2. On the other hand, the direct-form fast algorithms, which are a derivative of the lattice algorithm have a complexity proportional to M, and they are a little more efficient than the lattice-ladder algorithms.

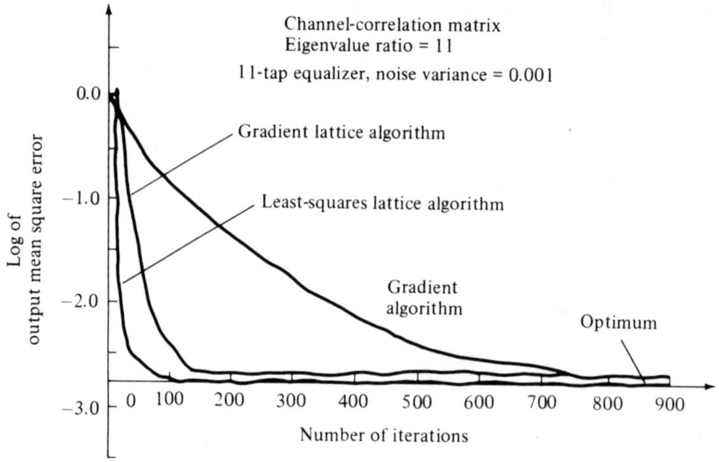

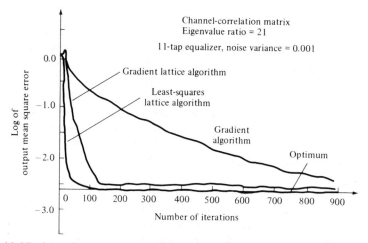

FIGURE 12.26 Learning curves for RLS lattice, gradient lattice, and LMS algorithm for adaptive equalizer of length $M = 11$. (From *Digital Communications* by John G. Proakis, © 1983 by McGraw-Hill Book Company. Reprinted with permission of the publisher.)

In Fig. 12.28 we illustrate the computational complexity (number of complex multiplications and divisions) of the various adaptive filtering algorithms we have described. Clearly, the LMS algorithm requires the smallest number of computations. The fast RLS algorithm in Table 12.2 is the most efficient of the RLS algorithms shown, closely followed by the gradient lattice algorithm, then the RLS lattice algorithms, and finally, the square-root algorithms. Note that for small values of M, there is little difference in complexity among the rapidly convergent algorithm.

Numerical Properties. In addition to providing fast convergence, the RLS and gradient lattice algorithms are numerically robust. First, these lattice algorithms are numerically stable. The term "numerically stable" means that the output estimation error from the computational procedure is bounded when a bounded error signal is introduced at the input. Second, the numerical accuracy of the optimum solution is also relatively good compared to the LMS and the RLS direct-form FIR algorithms.

FIGURE 12.27 Learning curves for RLS lattice, gradient lattice and LMS algorithms for adaptive equalizer of length $M = 11$. (From *Digital Communications* by John G. Proakis, © 1983 by McGraw-Hill Book Company. Reprinted with permission of the publisher.)

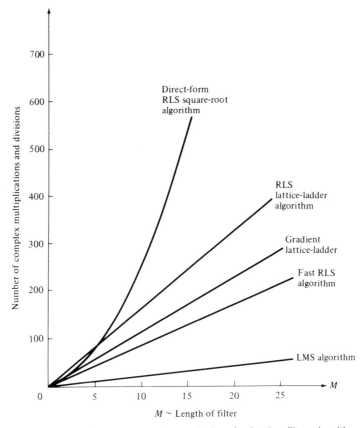

FIGURE 12.28 Computational complexity of adaptive filter algorithms.

For purposes of comparison, we illustrate in Table 12.8 the steady-state average squared error or (estimated) minimum MSE obtained through computer simulation from the two RLS lattice algorithms and the direct-form FIR filter algorithms described in Section 12.2. The striking result in Table 12.8 is the superior performance obtained with the RLS lattice-ladder algorithm, in which the reflection coefficients and the ladder gain are updated directly according to (12.3.96) through (12.3.98). This is the error-feedback form of the RLS lattice algorithm. It is clear that the direct updating of these coefficients is significantly more robust to round-off errors than all the other adaptive algorithms,

TABLE 12.8 Numerical Accuracy, in Terms of Output MSE for Channel with $\lambda_{max}/\lambda_{min} = 11$ and $w = 0.975$, MSE $\times 10^{-3}$

	Algorithm				
Number Bits (Including Sign)	**RLS Square Root**	**Fast RLS**	**Conventional RLS Lattice**	**Error Feedback RLS Lattice**	**LMS**
16	2.17	2.17	2.16	2.16	2.30
13	2.33	2.21	3.09	2.22	2.30
11	6.14	3.34	25.2	3.09	19.0
9	75.3	[a]	365	31.6	311

[a]Algorithm did not converge.

TABLE 12.9 Numerical Accuracy, in Terms of Output MSE, of A Priori LS Lattice Algorithm with Different Values of the Weighting Factor w, MSE $\times 10^{-3}$

Number of Bits (With Sign)	Algorithm					
	$w = 0.99$		$w = 0.975$		$w = 0.95$	
	Conventional	Error Feedback	Conventional	Error Feedback	Conventional	Error Feedback
16	2.14	2.08	2.18	2.16	2.66	2.62
13	7.08	2.11	3.09	2.22	3.65	2.66
11	39.8	3.88	25.2	3.09	15.7	2.78
9	750	44.1	365	31.6	120	15.2

including the LMS algorithm. It is also apparent that the two-step process used in the conventional RLS lattice algorithm to estimate the reflection coefficients is not as accurate. Furthermore, the estimation errors that are generated in the coefficients at each stage propragate from stage to stage, causing additional errors.

The effect of changing the weighting factor w is illustrated in the numerical results given in Table 12.9. In this table we give the minimum (estimated) MSE obtained with the conventional and error-feedback forms of the RLS lattice algorithm. We observe that the output MSE decreases with an increase in the weighting factor when the precision is high (13 bits and 16 bits). This reflects the improvement in performance obtained by increasing the observation interval. As the number of bits of precision is decreased, we observe that the weighting factor should also be decreased in order to maintain good performance. In effect, with low precision, the effect of a longer averaging time results in a larger round-off noise. Of course, these results were obtained with time-invariant signal statistics. If the signal statistics are time variant, the rate of the time variations will also influence our choice of w.

In the gradient lattice algorithm, the reflection coefficients and the ladder gains are also updated directly. Consequently, the numerical accuracy of the gradient lattice algorithm is comparable to that obtained with the direct update form of the RLS lattice.

Analytical and simulation results on numerical stability and numerical accuracy in fixed-point implementation of these algorithms can be found in the papers by Ling and Proakis (1984a, b), Ling et al. (1985, 1986), Ljung and Ljung (1985), and Gardner (1984).

Implementation Considerations. As we have observed, the lattice filter structure is highly modular and allows for the computations to be pipelined. Because of the high degree of modularity, the RLS and gradient lattice algorithms are particularly suitable for implementation in VLSI. As a result of this advantage in implementation and the desirable properties of stability, excellent numerical accuracy, and fast convergence, we anticipate that in the near future, more and more adaptive filters will be implemented as lattice-ladder structures.

12.4 Summary and References

We have presented adaptive algorithms for direct-form FIR and lattice filter structures. The algorithms for the direct-form FIR filter consisted of the simple LMS algorithm due to Widrow and Hoff (1960) and the direct-form, time-recursive least-squares (RLS) algorithms, including the conventional RLS form given by (12.2.72) through (12.2.76), the square-root RLS forms described by Bierman (1977), Carlson and Culmone (1979), and Hsu (1982), and the RLS fast algorithms, one form of which was described by Falconer and Ljung (1978), and other forms later derived by Carayannis et al. (1983), Proakis (1983), and Cioffi and Kailath (1984).

Of these algorithms, the LMS algorithm is the simplest. It is used in many applications where its slow speed of convergence is adequate. Of the direct-form RLS algorithms, the square-root algorithms have been used in applications where fast convergence is required. The algorithms have good numerical properties. The family of fast algorithms is very attractive from the viewpoint of computational efficiency, but they are extremely sensitive to round-off noise. Methods for reinitializing these algorithms have been proposed by Hsu (1982), Cioffi and Kailath (1984), Lin (1984), and Eleftheriou and Falconer (1987).

The adaptive lattice-ladder filter algorithms derived in this chapter included the optimum RLS lattice-ladder algorithm, both the conventional form and the error-feedback form, and the gradient lattice-ladder algorithm. Only the *a priori* form of the lattice-ladder algorithms was derived, which is the form used most often in applications. In addition, there is an *a posteriori* form of the RLS lattice-ladder algorithm, both conventional and error-feedback types, as described by Ling et al. (1986). The error-feedback form of the RLS lattice-ladder algorithms and the gradient lattice-ladder algorithm have excellent numerical properties and are particularly suitable for implementation in fixed-point arithmetic and in VLSI.

In the direct-form and lattice RLS algorithms we used exponential weighting into the past in order to reduce the effective memory in the adaptation process. As an alternative to exponential weighting we may employ finite-length uniform weighting into the past. This approach leads to the class of finite-memory RLS direct-form and lattice structures described in the papers of Cioffi and Kalaith (1985) and Manolakis et al. (1987).

Although only single-channel adaptive filtering algorithms have been presented, the approach we have used can be followed to easily derive multi-channel versions of these algorithms. For reference, the interested reader is referred to the papers by Morf et al. (1977), Lee et al. (1981), Ling and Proakis (1984c), and Ling et al. (1986) for generalization to the multi-channel case. A comprehensive treatment of adaptive filtering algorithms is given in the text by Haykin (1986).

In addition to the various algorithms that we have presented in this chapter, there is currently considerable research into efficient implementation of these algorithms using systolic arrays and other parallel architectures. For reference, the reader is referred to the publications by Kung (1982) and Kung et al. (1985).

PROBLEMS

12.1 Use the least-squares criterion to determine the equations for the parameters of the FIR filter model in Fig. 12.2 when the plant output is corrupted by additive noise $w(n)$.

12.2 Determine the equations for the coefficients of an adaptive echo canceller based on the least-squares criterion. Use the configuration in Fig. 12.8 and assume the presence of a near-end echo only.

12.3 If the sequences $w_1(n)$, $w_2(n)$, and $w_3(n)$ in the adaptive noise-cancelling system shown in Fig. 12.14 are mutually uncorrelated, determine the expected value of the estimated correlation sequences $r_{vv}(k)$ and $r_{yv}(k)$ contained in (12.1.26).

12.4 Prove the result in (12.3.34)

12.5 Derive the equation for the direct update of the ladder gain given by (12.3.98).

12.6 Derive the direct update equation for the reflection coefficients in a gradient lattice algorithm given in (12.3.106).

12.7 Derive the time-update relations given in (12.3.104).

12.8 Derive the FAEST algorithm given in Table 12.6 by using the alternative Kalman gain vector

$$\tilde{\mathbf{K}}_M(n) = \frac{1}{w} \mathbf{P}_M(n - 1)\mathbf{X}_M^*(n)$$

instead of the Kalman gain vector $\mathbf{K}_M(n)$.

12.9 Prove the orthogonality property of the backward residual errors given by (12.3.112).

12.10 The "tap-leakage LMS algorithm" proposed in the paper by Gitlin et al. (1982) may be expressed as

$$\mathbf{h}_M(n + 1) = w\mathbf{h}_M(n) + \Delta e(n)\mathbf{X}_M^*(n)$$

where $0 < w < 1$, Δ is the step size, and $\mathbf{X}_M(n)$ is the data vector at time n. Determine the condition for the convergence of the mean value of $\mathbf{h}_M(n)$.

12.11 By using the alternative Kalman gain vector given in Problem 12.8, modify the a priori fast least-squares algorithms given in Table 12.1 and 12.2 and thus reduce the number of computations.

12.12 Consider the random process

$$x(n) = gv(n) + w(n) \qquad n = 0, 1, \ldots, M - 1$$

where $v(n)$ is a known sequence, g is a random variable with $E(g) = 0$, and $E(g^2) = G$. The process $w(n)$ is a white noise sequence with

$$\gamma_{ww}(m) = \sigma_w^2 \, \delta(m)$$

Determine the coefficients of the linear estimator for g, that is,

$$\hat{g} = \sum_{n=0}^{M-1} h(n)x(n)$$

which minimize the mean-square error

$$\mathscr{E} = E[(g - \hat{g})^2]$$

12.13 Recall that an FIR filter can be realized in the frequency-sampling form with system function

$$H(z) = \frac{1 - z^{-M}}{M} \sum_{k=0}^{M-1} \frac{H_k}{1 - e^{j2\pi k/M}z^{-1}}$$

$$= H_1(z)H_2(z)$$

where $H_1(z)$ is the comb filter and $H_2(z)$ is the parallel bank of resonators.
(a) Suppose that this structure is implemented as an adaptive filter using the LMS algorithm to adjust the filter (DFT) parameters $\{H_k\}$. Give the time-update equation for these parameters. Sketch the adaptive filter structure.

(b) Suppose that this structure is used as an adaptive channel equalizer in which the desired signal is

$$d(n) = \sum_{k=0}^{M-1} A_k \cos \omega_k n \qquad \omega_k = \frac{2\pi k}{M}$$

With this form for the desired signal what advantages are there in the (LMS) adaptive algorithm for the DFT coefficients $\{H_k\}$ over the direct-form structure with coefficients $\{h(n)\}$? [*Hint*: Refer to paper by Proakis (1970).]

12.14 Consider the performance index

$$J = h^2 + 40h + 28$$

Suppose that we search for the minimum of J by using the steepest descent algorithm

$$h(n + 1) = h(n) - \tfrac{1}{2}\Delta g(n)$$

where $g(n)$ is the gradient.
(a) Determine the range of values of Δ that provides an overdamped system for the adjustment process.
(b) Plot the expression for J as a function of n for a value of Δ in this range.

12.15 Consider the noise-canceling adaptive filter shown in Fig. 12.14. Assume that the additive noise processes are white and mutually uncorrelated with equal variances σ_w^2. Suppose that the linear system has a known system function

$$H(z) = \frac{1}{1 - \tfrac{1}{2}z^{-1}}$$

Determine the optimum weights of a three-tap noise canceller that minimizes the MSE.

12.16 Determine the coefficients a_1 and a_2 for the linear predictor shown in Fig. P12.16, given that the autocorrelation $\gamma_{xx}(m)$ of the input signal is

$$\gamma_{xx}(m) = a^{|m|} \qquad 0 < a < 1$$

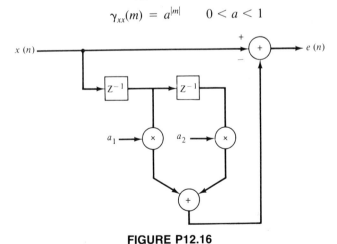

FIGURE P12.16

12.17 Determine the lattice filter and its optimum reflection coefficients corresponding to the linear predictor in Problem 12.16.

12.18 A communication channel is characterized by the system function

$$C(z) = 1 - 2z^{-1} + 2z^{-2}$$

Note that the channel is nonminimum phase.

(a) Design an FIR equalizer (Wiener filter) of length $M = 3$ and $M = 5$ based on the mean-square error criterion. Assume that the desired response is z^{-D}, where the tolerable delay $D = 5$ for $M = 3$ and $D = 7$ for $M = 5$.

(b) Determine the resulting minimum MSE for these two cases.

(c) Compute the equalized channel responses $C(z)H(z)$. Comment on how well the equalizer is performing.

12.19 Consider the adaptive FIR filter shown in Fig. P12.19. The system $C(z)$ is characterized by the system function

$$C(z) = \frac{1}{1 - 0.9z^{-1}}$$

Determine the optimum coefficients of the adaptive FIR filter $B(z) = b_0 + b_1 z^{-1}$ which minimize the mean-square error. The additive noise is white with variance $\sigma_w^2 = 0.1$.

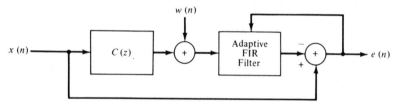

FIGURE P12.19

12.20 Consider the system illustrated in Fig. P12.20. The signal $x(n)$ is causal and the desired signal sequence $d(n)$ is also causal. The objective is to design an FIR filter with system function

$$A(z) = \sum_{k=0}^{p-1} a(k)z^{-k}$$

such that the output $e(n)$ is the best (minimum) mean-square-error approximation to $d(n)$.

(a) Derive the Yule–Walker equations for the optimum filter coefficients $\{a(k)\}$.

(b) Suppose that $x(n) = (\frac{1}{2})^n u(n)$ and $d(n) = \delta(n) - \frac{1}{4}\delta(n - 1)$. Determine the parameters of $A(z)$ for $p = 4$.

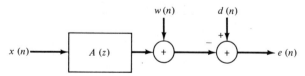

FIGURE P12.20

12.21 An $N \times N$ correlation matrix Γ has eigenvalues $\lambda_1 > \lambda_2 > \cdots > \lambda_N > 0$ and associated eigenvectors $v_1, v_2, \ldots, v_N$. Such a matrix can be represented as

$$\Gamma = \sum_{i=1}^{N} \lambda_i v_i v_i^{*t}$$

(a) If $\Gamma = \Gamma^{1/2}\Gamma^{1/2}$, where $\Gamma^{1/2}$ is the square root of Γ, show that $\Gamma^{1/2}$ can be represented as

$$\Gamma^{1/2} = \sum_{i=1}^{N} \lambda_i^{1/2} \mathbf{v}_i \mathbf{v}_i^{*t}$$

(b) Using this representation, determine a procedure for computing $\Gamma^{1/2}$.

12.22 Prove the following properties of least-squares lattice filters [for a reference, see the paper by Markhoul (1978)]. The signal $x(n)$ is real.

(a) $E[f_m(n)x(n-i)] = 0, \quad 1 \le i \le m$

(b) $E[g_m(n)x(n-i)] = 0, \quad 0 \le i \le m-1$

(c) $E[f_m(n)x(n)] = E[g_m(n)x(n-m)] = E_m$

(d) $E[f_i(n)f_j(n)] = E_{\max(i,j)}$

(e) $E[g_i(n)g_j(n)] = \begin{cases} E_i & i = j \\ 0 & i \ne j \end{cases}$

(f) $E[f_i(n)f_j(n-t)] = 0$ for $\begin{cases} 1 \le t \le i-j & i > j \\ -1 \ge t \ge i-j & i < j \end{cases}$

(g) $E[g_i(n)g_j(n-t)] = 0$ for $\begin{cases} 0 \le t \le i-j-1 & i > j \\ 0 \ge t \ge i-j+1 & i < j \end{cases}$

(h) $E[f_i(n+i)f_j(n+j)] = \begin{cases} E_i & i = j \\ 0 & i \ne j \end{cases}$

(i) $E[g_i(n+i)g_j(n+j)] = E_{\max(i,j)}$

(j) $E[f_i(n)g_j(n)] = \begin{cases} K_jE_i & i \ge j \\ 0 & i < j \end{cases} \quad i, j \ge 0, \quad K_0 = 1$

(k) $E[f_i(n)g_i(n-1)] = -K_{i+1}E_i$

(l) $E[g_i(n-1)x(n)] = E[f_i(n+1)x(n-i)] = -K_{i+1}E_i$

(m) $E[f_i(n)g_j(n-1)] = \begin{cases} 0 & i > j \\ -K_{j+1}E_j & i \le j \end{cases}$

COMPUTER EXPERIMENTS

12.23 Consider the adaptive predictor shown in Fig. P12.16.

(a) Determine the quadratic performance index and the optimum parameters for the signal

$$x(n) = \sin \frac{n\pi}{4} + w(n)$$

where $w(n)$ is white noise with variance $\sigma_w^2 = 0.1$.

(b) Generate a sequence of 1000 samples of $x(n)$ and use the LMS algorithm to adaptively obtain the predictor coefficients. Compare the experimental results with the theoretical values obtained in part (a). Use a step size of $\Delta \le \frac{1}{10}\Delta_{\max}$.

(c) Repeat the experiment in part (b) for $N = 10$ trials with different noise sequences and compute the average values of the predictor coefficients. Comment on how these results compare with the theoretical values in part (a).

12.24 An autoregressive process is described by the difference equation

$$x(n) = 1.26x(n - 1) - 0.81x(n - 2) + w(n)$$

(a) Generate a sequence of $N = 1000$ samples of $x(n)$, where $w(n)$ is a white noise sequence with variance $\sigma_w^2 = 0.1$. Use the LMS algorithm to determine the parameters of a second-order ($p = 2$) linear predictor. Begin with $a_1(0) = a_2(0) = 0$. Plot the coefficients $a_1(n)$ and $a_2(n)$ as a function of the iteration number.

(b) Repeat part (a) for 10 trials using different noise sequences and superimpose the 10 plots of $a_1(n)$ and $a_2(n)$.

(c) Plot the learning curve for the average (over the 10 trials) mean-square error for the data in part (b).

12.25 A random process $x(n)$ is given as

$$x(n) = s(n) + w(n)$$
$$= \sin(\omega_0 n + \phi) + w(n) \qquad \omega_0 = \pi/4, \quad \phi = 0$$

where $w(n)$ is an additive white noise sequence with variance $\sigma_w^2 = 0.1$.

(a) Generate $N = 1000$ samples of $x(n)$ and simulate an adaptive line enhancer (ALE) of length $L = 4$. Use the LMS algorithm to adapt the ALE.

(b) Plot the output of the ALE.

(c) Compute the autocorrelation $\gamma_{xx}(m)$ of the sequence $x(n)$.

(d) Determine the theoretical values of the ALE coefficients and compare them with the experimental values.

(e) Compute the frequency response of the linear predictor (ALE).

(f) Compute the frequency response of the prediction-error filter.

(g) Compute the experimental values of the autocorrelation $r_{ee}(m)$ of the output error sequence for $0 \leq m \leq 10$.

(h) Repeat the experiment for 10 trials using different noise sequences and superimpose the frequency response plots on the same graph.

(i) Comment on the result in parts (a) through (h).

References and Bibliography

AKAIKE, H. (1969). "Power Spectrum Estimation Through Autoregression Model Fitting," *Ann. Inst. Stat. Math.*, vol. 21, pp. 407–419.

AKAIKE, H. (1974). "A New Look at the Statistical Model Identification," *IEEE Trans. Automatic Control*, vol. AC-19, pp. 716–723, December.

ANDERSEN, N. O. (1978). "Comments on the Performance of Maximum Entropy Algorithm," *Proc. IEEE*, vol. 66, pp. 1581–1582, November.

ANTONIOU, A. (1979). *Digital Filters: Analysis and Design*, McGraw-Hill, New York.

AUER, E. (1987). "A Digital Filter Structure Free of Limit Cycles," *Proc. 1987 ICASSP*, pp. 21.11.1–21.11.4, Dallas, Tex., April.

AVENHAUS, E., and SCHUESSLER, H. W. (1970). "On the Approximation Problem in the Design of Digital Filters with Limited Wordlength," *Arch. Elek. Ubertragung*, vol. 24, pp. 571–572.

BAGGEROER, A. B. (1976). "Confidence Intervals for Regression (MEM) Spectral Estimates," *IEEE Trans. Information Theory*, vol. IT-22, pp. 534–545, September.

BANDLER, J. W., and BARDAKJIAN, B. J. (1973). "Least pth Optimization of Recursive Digital Filters," *IEEE Trans. Audio and Electroacoustics*, vol. AU-21, pp. 460–470, October.

BARNES, C. W., and FAM, A. T. (1977). "Minimum Norm Recursive Digital Filters That Are Free of Overflow Limit Cycles," *IEEE Trans. Circuits and Systems*, vol. CAS-24, pp. 569–574, October.

BARTLETT, M. S. (1948). "Smoothing Periodograms from Time Series with Continuous Spectra," *Nature* (London), vol. 161, pp. 686–687, May.

BARTLETT, M. S. (1961). *Stochastic Processes*, Cambridge University Press, Cambridge.

BECKMAN, F. S. (1960). "The Solution of Linear Equations by the Conjugate Gradient Method," in *Mathematical Methods for Digital Computers*, A. Ralston and H. S. Wilf, Eds., Wiley, New York.

BERGLAND, G. D. (1969). "A Guided Tour of the Fast Fourier Transform," *IEEE Spectrum*, vol. 6, pp. 41–52, July.

BERK, K. N. (1974). "Consistent Autoregressive Spectral Estimates," *Ann. Stat.*, vol. 2, pp. 489–502.

BERNHARDT, P. A., ANTONIADIS, D. A., and DA ROSA, A. V. (1976). "Lunar Perturbations in Columnar Electron Content and Their Interpretation in Terms of Dynamo Electrostatic Fields," *J. Geophys. Res.*, vol. 81, pp. 5957–5963, December.

BERRYMAN, J. G. (1978). "Choice of Operator Length for Maximum Entropy Spectral Analysis," *Geophysics*, vol. 43, pp. 1384–1391, December.

BIERMAN, G. J. (1977). *Factorization Methods for Discrete Sequential Estimation*, Academic, New York.

BLACK, H. S. (1953). *Modulation Theory*, D. Van Nostrand, Princeton, N.J.

BLACKMAN, R. B., and TUKEY, J. W. (1958). *The Measurement of Power Spectra*, Dover, New York.

BLAHUT, R. E. (1985). *Fast Algorithms for Digital Signal Processing*, Addison-Wesley, Reading, Mass.

BLUESTEIN, L. I. (1970). "A Linear Filtering Approach to the Computation of the Discrete Fourier Transform," *IEEE Trans. Audio and Electroacoustics*, vol. AU-18, pp. 451–455, December.

BOMAR, B. W. (1985). "New Second-Order State-Space Structures for Realizing Low Roundoff Noise Digital Filters," *IEEE Trans. Acoustics, Speech, and Signal Processing*, vol. ASSP-33, pp. 106–110, February.

BRACEWELL, R. N. (1978). *The Fourier Transform and Its Applications*, 2nd ed., McGraw-Hill, New York.

BRIGHAM, E. O., and MORROW, R. E. (1967). "The Fast Fourier Transform," *IEEE Spectrum*, vol. 4, pp. 63–70, December.

BRILLINGER, D. R. (1974). "Fourier Analysis of Stationary Processes," *Proc. IEEE*, vol. 62, pp. 1628–1643, December.

BROPHY, F., and SALAZAR, A. C. (1973). "Considerations of the Padé Approximant Technique in the Synthesis of Recursive Digital Filters," *IEEE Trans. Audio and Electroacoustics*, vol. AU-21, pp. 500–505, December.

BROWN, J. L., JR. (1980). "First-Order Sampling of Bandpass Signals—A New Approach," *IEEE Trans. Information Theory*, vol. IT-26, pp. 613–615, September.

BRUBAKER, T. A., and GOWDY, J. N. (1972). "Limit Cycles in Digital Filters," *IEEE Trans. Automatic Control*, vol. AC-17, pp. 675–677, October.

BRUZZONE, S. P., and KAVEH, M. (1980). "On Some Suboptimum ARMA Spectral Estimators," *IEEE Trans. Acoustics, Speech, and Signal Processing*, vol. ASSP-28, pp. 753–755, December.

BURG, J. P. (1967). "Maximum Entropy Spectral Analysis," *Proc. 37th Meeting of the Society of Exploration Geophysicists*, Oklahoma City, Okla., October. Reprinted in *Modern Spectrum Analysis*, D. G. Childers, Ed., IEEE Press, New York.

BURG, J. P. (1968). "A New Analysis Technique for Time Series Data," NATO Advanced Study Institute on Signal Processing with Emphasis on Underwater Acoustics, August 12–23. Reprinted in *Modern Spectrum Analysis*, D. G. Childers, Ed., IEEE Press, New York.

BURG, J. P. (1972). "The Relationship Between Maximum Entropy and Maximum Likelihood Spectra," *Geophysics*, vol. 37, pp. 375–376, April.

BURG, J. P. (1975). "Maximum Entropy Spectral Analysis," Ph.D. dissertation, Department of Geophysics, Stanford University, Stanford, Calif., May.

BURRUS, C. S., and PARKS, T. W. (1970). "Time-Domain Design of Recursive Digital Filters," *IEEE Trans. Audio and Electroacoustics*, vol. 18, pp. 137–141, June.

BUTTERWECK, H. J., VAN MEER, A. C. P., and VERKROOST, G. (1984). "New Second-Order Digital Filter Sections Without Limit Cycles," *IEEE Trans. Circuits and Systems*, vol. CAS-31, pp. 141–146, February.

CADZOW, J. A. (1979). "ARMA Spectral Estimation: An Efficient Closed-form Procedure," *Proc. RADC Spectrum Estimation Workshop*, pp. 81–97, Rome, N.Y., October.

CADZOW, J. A. (1981). "Autoregressive–Moving Average Spectral Estimation: A Model Equation Error Procedure," *IEEE Trans. Geoscience Remote Sensing*, vol. GE-19, pp. 24–28, January.

CADZOW, J. A. (1982). "Spectral Estimation: An Overdetermined Rational Model Equation Approach," *Proc. IEEE*, vol. 70, pp. 907–938, September.

CAPON, J. (1969). "High-Resolution Frequency–Wavenumber Spectrum Analysis," *Proc. IEEE*, vol. 57, pp. 1408–1418, August.

CAPON, J. (1983). "Maximum-Likelihood Spectral Estimation," in *Nonlinear Methods of Spectral Analysis*, 2nd ed., S. Haykin, Ed., Springer-Verlag, New York.

CAPON, J., and GOODMAN, N. R. (1971). "Probability Distribution for Estimators of the Frequency–Wavenumber Spectrum, *Proc. IEEE*, vol. 58, pp. 1785–1786, October.

CARAISCOS, C., and LIU, B. (1984). "A Roundoff Error Analysis of the LMS Adaptive Algorithm," *IEEE Trans. Acoustics, Speech, and Signal Processing*, vol. ASSP-32, pp. 34–41, January.

CARAYANNIS, G., MANOLAKIS, D. G., and KALOUPTSIDIS, N. (1983). "A Fast Sequential Algorithm for Least-Squares Filtering and Prediction," *IEEE Trans. Acoustics, Speech, and Signal Processing*, vol. ASSP-31, pp. 1394–1402, December.

CARAYANNIS, G., MANOLAKIS, D. G., and KALOUPTSIDIS, N. (1986). "A Unified View of Parametric Processing Algorithms for Prewindowed Signals," *Signal Processing*, vol. 10, pp. 335–368, June.

CARLSON, A. B. (1975). *Communication Systems*, 2nd ed., McGraw-Hill, New York.

CARLSON, N. A., and CULMONE, A. F. (1979). "Efficient Algorithms for On-Board Array Processing," *Record 1979 International Conference on Communications*, pp. 58.1.1–58.1.5, Boston, June 10–14.

CHAN, D. S. K., and RABINER, L. R. (1973a). "Theory of Roundoff Noise in Cascade Realizations of Finite Impulse Response Digital Filters," *Bell Syst. Tech. J.*, vol. 52, pp. 329–345, March.

CHAN, D. S. K., and RABINER, L. R. (1973b). "An Algorithm for Minimizing Roundoff Noise in Cascade Realizations of Finite Impulse Response Digital Filters," *Bell Syst. Tech. J.*, vol. 52, pp. 347–385, March.

CHAN, D. S. K., and RABINER, L. R. (1973c). "Analysis of Quantization Errors in the Direct Form for Finite Impulse Response Digital Filters," *IEEE Trans. Audio and Electroacoustics*, vol. AU-21, pp. 354–366, August.

CHANG, T. (1981). "Suppression of Limit Cycles in Digital Filters Designed with One Magnitude-Truncation Quantizer," *IEEE Trans. Circuits and Systems*, vol. CAS-28, pp. 107–111, February.

CHEN, C. T. (1970). *Introduction to Linear System Theory*, Holt, Rinehart and Winston, New York.

CHEN, W. Y., and STEGEN, G. R. (1974). "Experiments with Maximum Entropy Power Spectra of Sinusoids," *J. Geophys. Res.*, vol. 79, pp. 3019–3022, July.

CHILDERS, D. G., ed. (1978). *Modern Spectrum Analysis*, IEEE Press, New York.

CHOW, J. C. (1972). "On Estimating the Orders of an Autoregressive-Moving Average Process with Uncertain Observations," *IEEE Trans. Automatic Control*, vol. AC-17, pp. 707–709, October.

CHOW, Y., and CASSIGNOL, E. (1962). *Linear Signal Flow Graphs and Applications*, Wiley, New York.

CIOFFI, J. M. (1987). "Limited Precision Effects in Adaptive Filtering," *IEEE Trans. Circuits and Systems*, vol. CAS-34, pp. 821–833, July.

CIOFFI, J. M., and KAILATH, T. (1984). "Fast Recursive-Least-Squares Transversal Filters for Adaptive Filtering," *IEEE Trans. Acoustics, Speech, and Signal Processing*, vol. ASSP-32, pp. 304–337, April.

CIOFFI, J. M., and KAILATH, T. (1985). "Windowed Fast Transversal Filters Adaptive Algorithms with Normalization," *IEEE Trans. Acoustics, Speech, and Signal Processing*, vol. ASSP-33, pp. 607–625, June.

CLAASEN, T. A. C. M., MECKLENBRAUKER, W. F. G., and PEEK, J. B. H. (1973). "Second-Order Digital Filter with Only One Magnitude-Truncation Quantizer and Having Practically No Limit Cycles," *Electron. Lett.*, vol. 9, November.

COCHRAN, W. T., COOLEY, J. W., FAVIN, D. L., HELMS, H. D., KAENEL, R. A., LANG, W. W., MALING, G. C., NELSON, D. E., RADER, C. E., and WELCH, P. D. (1967). "What Is the Fast Fourier Transform," *IEEE Trans. Audio and Electroacoustics*, vol. AU-15, pp. 45–55, June.

CONSTANTINIDES, A. G. (1967). "Frequency Transformations for Digital Filters," *Electron. Lett.*, vol. 3, pp. 487–489, November.

CONSTANTINIDES, A. G. (1968). "Frequency Transformations for Digital Filters," *Electron. Lett.*, vol. 4, pp. 115–116, April.

CONSTANTINIDES, A. G. (1970). "Spectral Transformations for Digital Filters," *Proc. IEE*, vol. 117, pp. 1585–1590. August.

COOLEY, J. W., and TUKEY, J. W. (1965). "An Algorithm for the Machine Computation of Complex Fourier Series," *Math. Comp.*, vol. 19, pp. 297–301, April.

COOLEY, J. W., LEWIS, P., and WELCH, P. D. (1967). "Historical Notes on the Fast Fourier Transform," *IEEE Trans. Audio and Electroacoustics*, vol. AU-15, pp. 76–79, June.

COOLEY, J. W., LEWIS, P., and WELCH, P. D. (1969). "The Fast Fourier Transform and Its Applications," *IEEE Trans. Education*, vol. E-12 pp. 27–34, March.

CROCHIERE, R. E., and RABINER, L. R. (1975). "Optimum FIR Digital Filter Implementations for Decimation, Interpolation, and Narrowband Filtering," *IEEE Trans. Acoustics, Speech, and Signal Processing*, vol. ASSPP-23, pp. 444–456, October.

CROCHIERE, R. E., and RABINER, L. R. (1976). "Further Considerations in the Design of Decimators and Interpolators," *IEEE Trans. Acoustics, Speech, and Signal Processing*, vol. ASSP-24, pp. 296–311, August.

CROCHIERE, R. E., and RABINER, L. R. (1981). "Interpolation and Decimation of Digital Signals— A Tutorial Review," *Proc. IEEE*, vol. 69, pp. 300–331, March.

CROCHIERE, R. E., and RABINER, L. R. (1983). *Multirate Digital Signal Processing*, Prentice-Hall, Englewood Cliffs, N.J.

DANIELS, R. W. (1974). *Approximation Methods for the Design of Passive, Active and Digital Filters*, McGraw-Hill, New York.

DAVENPORT, W. B., JR. (1970). *Probability and Random Processes: An Introduction for Applied Scientists and Engineers*, McGraw-Hill, New York.

DAVIS, H. F. (1963). *Fourier Series and Orthogonal Functions*, Allyn and Bacon, Boston.

DECZKY, A. G. (1972). "Synthesis of Recursive Digital Filters Using the Minimum p-Error Criterion," *IEEE Trans. Audio and Electroacoustics*, vol. AU-20, pp. 257–263, October.

DE RUSSO, P. M., ROY, R. J., and CLOSE, C. M. (1965). *State Variables for Engineers*, Wiley, New York.

DURBIN, J. (1959). "Efficient Estimation of Parameters in Moving-Average Models," *Biometrika*, vol. 46, pp. 306–316.

DWIGHT, H. B. (1957). *Tables of Integrals and Other Mathematical Data*, 3rd ed., Macmillan, New York.

DYM, H., and McKEAN, H. P. (1972). *Fourier Series and Integrals*, Academic, New York.

EBERT, P. M., MAZO, J. E., and TAYLOR, M. G. (1969). "Overflow Oscillations in Digital Filters," *Bell Syst. Tech. J.*, vol. 48, pp. 2999–3020, November.

ELEFTHERIOU, E., and FALCONER, D. D. (1987). "Adaptive Equalization Techniques for HF Channels," *IEEE J. Selected Areas in Communications*, vol. SAC-5, pp. 238–247. February.

FAM, A. T., and BARNES, C. W. (1979). "Non-minimal Realizations of Fixed-Point Digital Filters That Are Free of All Finite Wordlength Limit Cycles," *IEEE Trans. Acoustics, Speech, and Signal Processing*, vol. ASSP-27, pp. 149–153, April.

FALCONER, D. D., and LJUNG, L. (1978). "Application of Fast Kalman Estimation to Adaptive Equalization," *IEEE Trans. Communications*, vol. COM-26, pp. 1439–1446, October.

FETTWEIS, A. (1971). "Some Principles of Designing Digital Filters Imitating Classical Filter Structures," *IEEE Trans. Circuit Theory*, vol. CT-18, pp. 314–316, March.

FLETCHER, R., and POWELL, M. J. D. (1963). "A Rapidly Convergent Descent Method for Minimization," *Comput. J.*, vol. 6, pp. 163–168.

FRIEDLANDER, B. (1982a). "Lattice Filters for Adaptive Processing," *Proc. IEEE*, vol. 70, pp. 829–867, August.

FRIEDLANDER, B. (1982b). "Lattice Methods for Spectral Estimation," *Proc. IEEE*, vol. 70, pp. 990–1017, September.

FOUGERE, P. F., ZAWALICK, E. J., and RADOSKI, H. R. (1976). "Spontaneous Line Splitting in Maximum Entropy Power Spectrum Analysis," *Phys. Earth Planet. Inter.*, vol. 12, 201–207, August.

GANTMACHER, F. R. (1960). *The Theory of Matrices*, vol. I., Chelsea, New York.

GARDNER, W. A. (1984). "Learning Characteristics of Stochastic-Gradient-Descent Algorithms: A General Study, Analysis and Critique," *Signal Processing*, vol. 6, pp. 113–133, April.

GARDNER, W. A. (1987). *Statistical Spectral Analysis: A Nonprobabilistic Theory*, Prentice-Hall, Englewood Cliffs, N.J.

GEORGE, D. A., BOWEN, R. R., and STOREY, J. R. (1971). "An Adaptive Decision-Feedback Equalizer," *IEEE Trans. Communication Technology*, vol. COM-19, pp. 281–293, June.

GERSCH, W., and SHARPE, D. R. (1973). "Estimation of Power Spectra with Finite-Order Autoregressive Models," *IEEE Trans. Automatic Control*, vol. AC-18, pp. 367–369, August.

GERSHO, A. (1969). "Adaptive Equalization of Highly Dispersive Channels for Data Transmission," *Bell Syst. Tech. J.*, vol. 48, pp. 55–70, January.

GIBBS, A. J. (1969). "An Introduction to Digital Filters," *Aust. Tellcommun. Res.*, vol. 3, pp. 3–14, November.

GIBBS, A. J. (1970). "The Design of Digital Filters," *Austr. Telecommun. Res.*, vol. 4, pp. 29–34, May.

GITLIN, R. D., and WEINSTEIN, S. B. (1979). "On the Required Tap-Weight Precision for Digitally Implemented Mean-Squared Equalizers," *Bell Syst. Tech. J.*, vol. 58, pp. 301–321, February.

GITLIN, R. D., MEADORS, H. C., and WEINSTEIN, S. B. (1982). "The Tap-Leakage Algorithm: An Algorithm for the Stable Operation of a Digitally Implemented Fractionally Spaced, Adaptive Equalizer," *Bell Syst. Tech. J.*, vol. 61, pp. 1817–1839, October.

GOERTZEL, G. (1968). "An Algorithm for the Evaluation of Finite Trigonometric Series," *Am. Math. Monthly*, vol. 65, pp. 34–35, January.

GOLD, B., and JORDAN, K. L., JR. (1968). "A Note on Digital Filter Synthesis," *Proc. IEEE*, vol. 56, pp. 1717–1718, October.

GOLD, B., and JORDAN, K. L., JR. (1969). "A Direct Search Procedure for Designing Finite Duration Impulse Response Filters," *IEEE Trans. Audio and Electroacoustics*, vol. AU-17, pp. 33–36, March.

GOLD, B., and RADER, C. M. (1966). "Effects of Quantization Noise in Digital Filters," *Proc. AFIPS 1966 Spring Joint Computer Conference*, vol. 28, pp. 213–219.

GOLD, B., and RADER, C. M. (1969). *Digital Processing of Signals*, McGraw-Hill, New York.

GOLDEN, R. M., and KAISER, J. F. (1964). "Design of Wideband Sampled Data Filters," *Bell Syst. Tech. J.*, vol. 43, pp. 1533–1546, July.

GORSKI-POPIEL, J., ED. (1975). *Frequency Synthesis: Techniques and Applications*, IEEE Press, New York.

GOOD, I. J. (1971). "The Relationship Between Two Fast Fourier Transforms," *IEEE Trans. Computers*, vol. C-20, pp. 310–317.

GRAUPE, D., KRAUSE, D. J., and MOORE, J. B. (1975). "Identification of Autoregressive-Moving Average Parameters of Time Series," *IEEE Trans. Automatic Control*, vol. AC-20, pp. 104–107, February.

GRENANDER, O., and SZEGÖ, G. (1958). *Toeplitz Forms and Their Applications*, University of California Press, Berkeley, Calif.

GRIFFITHS, L. J. (1975). "Rapid Measurements of Digital Instantaneous Frequency," *IEEE Trans. Acoustics, Speech, and Signal Processing*, vol. ASSP-23, pp. 207–222, April.

GRIFFITHS, L. J. (1977). "A Continuously Adaptive Filter Implemented as a Lattice Structure," *Proc. ICASSP-77*, pp. 683–686, Hartford, Conn., May.

GRIFFITHS, L. J. (1978). "An Adaptive Lattice Structure for Noise Cancelling Applications," *Proc. ICASSP-78*, pp. 87–90, Tulsa, Okla., April.

GUILLEMIN, E. A. (1957). *Synthesis of Passive Networks*, Wiley, New York.

GUPTA, S. C. (1966). *Transform and State Variable Methods in Linear Systems*, Wiley, New York.

HAMMING, R. W. (1962). *Numerical Methods for Scientists and Engineers*, McGraw-Hill, New York.

HAYKIN, S. (1986). *Adaptive Filter Theory*, Prentice-Hall, Englewood Cliffs, N.J.

HELME, B., and NIKIAS, C. L. (1985). "Improved Spectrum Performance via a Data-Adaptive Weighted Burg Technique," *IEEE Trans. Acoustics, Speech, and Signal Processing*, vol. ASSP-33, pp. 903–910, August.

HELMS, H. D. (1967). "Fast Fourier Transforms Method of Computing Difference Equations and Simulating Filters," *IEEE Trans. Audio and Electroacoustics*, vol. AU-15, pp. 85–90, June.

HELMS, H. D. (1968). "Nonrecursive Digital Filters: Design Methods for Achieving Specifications on Frequency Response," *IEEE Trans. Audio and Electroacoustics*, vol. AU-16, pp. 336–342, September.

HELSTROM, C. W. (1984). *Probability and Stochastic Processes for Engineers*, Macmillan, New York.

HERMANN, O. (1970). "Design of Nonrecursive Digital Filters with Linear Phase," *Electron. Lett.*, vol. 6, pp. 328–329, November.

HERMANN, O., and SCHUESSLER, H. W. (1970a). "Design of Nonrecursive Digital Filters with Minimum Phase," *Electron. Lett.*, vol. 6, pp. 329–330, November.

HERMANN, O., and SCHUESSLER, H. W. (1970b). "On the Accuracy Problem in the Design of Nonrecursive Digital Filters," *Arch. Elek. Ubertragung*, vol. 24, pp. 525–526.

HERRING, R. W. (1980). "The Cause of Line Splitting in Burg Maximum-Entropy Spectral Analysis," *IEEE Trans. Acoustics, Speech, and Signal Processing*, vol. ASSP-28, pp. 692–701, December.

HILDEBRAND, F. B. (1952). *Methods of Applied Mathematics*, Prentice-Hall, Englewood Cliffs, N.J.

HOFSTETTER, E., OPPENHEIM, A. V., and SIEGEL, J. (1971). "A New Technique for the Design of Nonrecursive Digital Filters," *Proc. 5th Annual Princeton Conference on Information Sciences and Systems*, pp. 64–72.

HOUSEHOLDER, A. S. (1964). *The Theory of Matrices in Numerical Analysis*, Blaisdell, Waltham, Mass.

HSU, F. M. (1982). "Square-Root Kalman Filtering for High-Speed Data Received over Fading Dispersive HF Channels," *IEEE Trans. Information Theory*, vol. IT-28, pp. 753–763, September.

HSU, F. M., and GIORDANO, A. A. (1978). "Digital Whitening Techniques for Improving Spread Spectrum Communications Performance in the Presence of Narrowband Jamming and Interference," *IEEE Trans. Communications*, vol. COM-26, pp. 209–216, February.

HWANG, S. Y. (1977). "Minimum Uncorrelated Unit Noise in State Space Digital Filtering," *IEEE Trans. Acoustics, Speech, and Signal Processing*, vol. ASSP-25, pp. 273–281, August.

JACKSON, L. B. (1969). "An Analysis of Limit Cycles Due to Multiplication Rounding in Recursive Digital (Sub) Filters," *Proc. 7th Annual Allerton Conference on Circuit and System Theory*, pp. 69–78.

JACKSON, L. B. (1970a). "On the Interaction of Roundoff Noise and Dynamic Range in Digital Filters," *Bell Syst. Tech. J.*, vol. 49, pp. 159–184, February.

JACKSON, L. B. (1970b). "Roundoff Noise Analysis for Fixed-Point Digital Filters Realized in Cascade or Parallel Form," *IEEE Trans. Audio and Electroacoustics*, vol. AU-18, pp. 107–122, June.

JACKSON, L. B. (1976). "Roundoff Noise Bounds Derived from Coefficient Sensitivities in Digital Filters," *IEEE Trans. Circuits and Systems*, vol. CAS-23, pp. 481–485, August.

JACKSON, L. B. (1979). "Limit Cycles on State-Space Structures for Digital Filters," *IEEE Trans. Circuits and Systems*, vol. CAS-26, pp. 67–68, January.

JACKSON, L. B., LINDGREN, A. G., and KIM, Y. (1979). "Optimal Synthesis of Second-Order State-Space Structures for Digital Filters," *IEEE Trans. Circuits and Systems*, vol. CAS-26, pp. 149–153, March.

JAHNKE, E., and EMDE, F. (1945). *Tables of Functions*, 4th ed., Dover, New York.

JENKINS, G. M., and WATTS, D. G. (1968). *Spectral Analysis and Its Applications*, Holden-Day, San Francisco.

JOHNSON, D. H. (1982). "The Application of Spectral Estimation Methods to Bearing Estimation Problems," *Proc. IEEE*, vol. 70, pp. 1018–1028, September.

JONES, R. H. (1976). "Autoregression Order Selection," *Geophysics*, vol. 41, pp. 771–773, August.

JONES, S. K., CAVIN, R. K., and REED, W. M. (1982). "Analysis of Error-Gradient Adaptive Linear Equalizers for a Class of Stationary-Dependent Processes," *IEEE Trans. Information Theory*, vol. IT-28, pp. 318–329, March.

JURY, E. I. (1964). *Theory and Applications of the z-Transform Method*, Wiley, New York.

KAISER, J. F. (1963). "Design Methods for Sampled Data Filters," *Proc. First Allerton Conference on Circuit System Theory*, pp. 221–236, November.

KAISER, J. F. (1966). "Digital Filters," in *System Analysis by Digital Computer*, F. F. Kuo and J. F. Kaiser, Eds., Wiley, New York.

KALOUPTSIDIS, N., and THEODORIDIS, S. (1987). "Fast Adaptive Least-Squares Algorithms for Power Spectral Estimation," *IEEE Trans. Acoustics, Speech, and Signal Processing*, vol. ASSP-35, pp. 661–670, May.

KASHYAP, R. L. (1980). "Inconsistency of the AIC Rule for Estimating the Order of Autoregressive Models," *IEEE Trans. Automatic Control*, vol. AC-25, pp. 996–998, October.

KAVEH, M., and BRUZZONE, S. P. (1979). "Order Determination for Autoregressive Spectral Estimation," *Record of the 1979 RADC Spectral Estimation Workshop*, pp. 139–145, Griffin Air Force Base, Rome, N.Y.

KAVEH, M., and LIPPERT, G. A. (1983). "An Optimum Tapered Burg Algorithm for Linear Prediction and Spectral Analysis," *IEEE Trans. Acoustics, Speech, and Signal Processing*, vol. ASSP-31, pp. 438–444, April.

KAY, S. M. (1980). "A New ARMA Spectral Estimator," *IEEE Trans. Acoustics, Speech, and Signal Processing*, vol. ASSP-28, pp. 585–588, October.

KAY, S. M. (1987). *Modern Spectral Estimation*, Prentice-Hall, Englewood Cliffs, N. J.

KAY, S. M., and MARPLE, S. L., JR. (1981). "Spectrum Analysis: A Modern Perspective," *Proc. IEEE*, vol. 69, pp. 1380–1419, November.

KAY, S. M., and MARPLE, S. L., JR. (1979). "Sources of and Remedies for Spectral Line Splitting in Autoregressive Spectrum Analysis," *Proc. 1979 ICASSP*, pp. 151–154.

KESLER, S. B., ED. (1986). *Modern Spectrum Analysis II*, IEEE Press, New York.

KETCHUM, J. W., and PROAKIS, J. G. (1982). "Adaptive Algorithms for Estimating and Suppressing Narrow-Band Interference in PN Spread-Spectrum Systems," *IEEE Trans. Communications*, vol. COM-30, pp. 913–923, May.

KNOWLES, J. B., and OLCAYTO, E. M. (1968). "Coefficient Accuracy and Digital Filter Response," *IEEE Trans. Circuit Theory*, vol. CT-15, pp. 31–41, March.

KROMER, R. E. (1969). "Asymptotic Properties of the Autoregressive Spectral Estimator," Ph.D. dissertation, Department of Statistics, Stanford University, Stanford, Calif.

KUNG, H. T. (1982). "Why Systolic Architectures?" *IEEE Computer*, vol. 15, pp. 37–46.

KUNG, S. Y., WHITEHOUSE, H. J., and KAILATH, T., EDS. (1985). *VLSI and Modern Signal Processing*, Prentice-Hall, Englewood Cliffs, N.J.

LACOSS, R. T. (1971). "Data Adaptive Spectral Analysis Methods," *Geophysics*, vol. 36, pp. 661–675, August.

LANG, S. W., and MCCLELLAN, J. H. (1980). "Frequency Estimation with Maximum Entropy Spectral Estimators," *IEEE Trans. Acoustics, Speech, and Signal Processing*, vol. ASSP-28, pp. 716–724, December.

LEE, D. T., MORF, M., and FRIEDLANDER, B. (1981). "Recursive Least-Squares Ladder Estimation Algorithms," *IEEE Trans. Acoustics, Speech, and Signal Processing*, vol. ASSP-29, pp. 627–641, June.

LEVINSON, N. (1947). "The Wiener RMS Error Criterion in Filter Design and Prediction," *J. Math. Phys.*, vol. 25, pp. 261–278.

LEVY, H., and LESSMAN, F. (1961). *Finite Difference Equations*, Macmillan, New York.

LIN, D. W. (1984). "On Digital Implementation of the Fast Kalman Algorithm," *IEEE Trans. Acoustics, Speech, and Signal Processing*, vol. ASSP-32, pp. 998–1005, October.

LING, F., and PROAKIS, J. G. (1984a). "Numerical Accuracy and Stability: Two Problems of Adaptive Estimation Algorithms Caused by Round-Off Error," *Proc. ICASSP-84*, pp. 30.3.1–30.3.4, San Diego, Calif., March.

LING, F., and PROAKIS, J. G. (1984b). "Nonstationary Learning Characteristics of Least-Squares Adaptive Estimation Algorithms," *Proc. ICASSP-84*, pp. 3.7.1–3.7.4, San Diego, Calif., March.

LING, F., and PROAKIS, J. G. (1984c). "A Generalized Multichannel Least-Squares Lattice Algorithm with Sequential Processing Stages," *IEEE Trans. Acoustics, Speech, and Signal Processing*, vol. ASSP-32, pp. 381–389, April.

LING, F., MANOLAKIS, D., and PROAKIS, J. G. (1985). "New Forms of LS Lattice Algorithms and an Analysis of Their Round-Off Error Characteristics," *Proc. ICASSP-85*, pp. 1739–1742, Tampa, Fla., April.

LING, F., MANOLAKIS, D., and PROAKIS, J. G. (1986). "Numerically Robust Least-Squares Lattice-Ladder Algorithms with Direct Updating of the Reflection Coefficients," *IEEE Trans. Acoustics, Speech, and Signal Processing*, vol. ASSP-34, pp. 837–845, August.

LIU, B. (1971). "Effect of Finite Word Length on the Accuracy of Digital Filters—A Review," *IEEE Trans. Circuit Theory*, vol. CT-18, pp. 670–677, November.

LJUNG, S., and LJUNG, L. (1985). "Error Propagation Properties of Recursive Least-Squares Adaptation Algorithms," *Automatica*, vol. 21, pp. 157–167.

LUCKY, R. W. (1965). Automatic Equalization for Digital Communications," *Bell Syst. Tech. J.*, vol. 44, pp. 547–588, April.

MAGEE, F. R., and PROAKIS, J. G. (1973). "Adaptive Maximum-Likelihood Sequence Estimation for Digital Signaling in the Presence of Intersymbol Interference," *IEEE Trans. Information Theory*, vol. IT-19, pp. 120–124, January.

MAKHOUL, J. (1975). "Linear Prediction: A Tutorial Review," *Proc. IEEE*, vol. 63, pp. 561–580, April.

MAKHOUL, J. (1978). "A Class of All-Zero Lattice Digital Filters: Properties and Applications," *IEEE Trans. Acoustics, Speech, and Signal Processing*, vol. ASSP-26, pp. 304–314, August.

MANOLAKIS, D., LING, F., and PROAKIS, J. G. (1987). "Efficient Time-Recursive Least-Squares Algorithms for Finite-Memory Adaptive Filtering," *IEEE Trans. Circuits and Systems*, vol. CAS-34, pp. 400–408, April.

MARKEL, J. D., and GRAY, A. H., JR. (1976). *Linear Prediction of Speech*, Springer-Verlag, New York.

MARPLE, S. L., JR. (1980). "A New Autoregressive Spectrum Analysis Algorithm," *IEEE Trans. Acoustics, Speech, and Signal Processing*, vol. ASSP-28, pp. 441–454, August.

MARPLE, S. L., JR. (1987). *Digital Spectral Analysis with Applications*, Prentice-Hall, Englewood Cliffs, N.J.

MARZETTA, T. L. (1983). "A New Interpretation for Capon's Maximum Likelihood Method of Frequency–Wavenumber Spectral Estimation," *IEEE Trans. Acoustics, Speech, and Signal Processing*, vol. ASSP-31, pp. 445–449, April.

MARZETTA, T. L., and LANG, S. W. (1983). "New Interpretations for the MLM and DASE Spectral Estimators," *Proc. 1983 ICASSP*, pp. 844–846, Boston, April.

MARZETTA, T. L., and LANG, S. W. (1984). "Power Spectral Density Bounds," *IEEE Trans. Information Theory*, vol. IT-30, pp. 117–122, January.

MASON, S. J., and ZIMMERMAN, H. J. (1960). *Electronic Circuits, Signals and Systems*, Wiley, New York.

MAZO, J. E. (1979). "On the Independence Theory of Equalizer Convergence," *Bell Syst. Tech. J.*, vol. 58, pp. 963–993, May.

McCLELLAN, J. H. (1982). "Multidimensional Spectral Estimation," *Proc. IEEE*, vol. 70, pp. 1029–1039, September.

McDONOUGH, R. N. (1983). "Application of the Maximum-Likelihood Method and the Maximum Entropy Method to Array Processing," in *Nonlinear Methods of Spectral Analysis*, 2nd ed., S. Haykin, Ed., Springer-Verlag, New York.

McGILLEM, C. D., and COOPER, G. R. (1984). *Continuous and Discrete Signal and System Analysis*, 2nd ed., Holt, Rinehart and Winston, Inc., New York.

MILLS, W. L., MULLIS, C. T., and ROBERTS, R. A. (1981). "Low Roundoff Noise and Normal Realizations of Fixed-Point IIR Digital Filters," *IEEE Trans. Acoustics, Speech, and Signal Processing*, vol. ASSP-29, pp. 893–903, August.

MORF, M., and LEE, D. T. (1979). "Recursive Least-Squares Ladder Forms for Fast Parameter Tracking," *Proc. 1979 IEEE Conference on Decision and Control*, San Diego, Calif., pp. 1362–1367, January.

MORF, M., VIEIRA, A., and LEE, D. T. (1977). "Ladder Forms for Identification and Speech Processing," *Proc. 1977 IEEE Conference Decision and Control*, pp. 1074–1078, New Orleans, La., December.

MULLIS, C. T., and ROBERTS, R. A. (1976a). "Synthesis of Minimum Roundoff Noise Fixed-Point Digital Filters," *IEEE Trans. Circuits and Systems*, vol. CAS-23, pp. 551–561, September.

MULLIS, C. T., and ROBERTS, R. A. (1976b). "Roundoff Noise in Digital Filters: Frequency Transformations and Invariants," *IEEE Trans. Acoustics, Speech and Signal Processing*, vol. ASSP-24, pp. 538–549, December.

MURRAY, W., ED. (1972). *Numerical Methods for Unconstrained Minimization*, Academic, New York.

MUSICUS, B. (1985). "Fast MLM Power Spectrum Estimation from Uniformly Spaced Correlations," *IEEE Trans. Acoustics, Speech, and Signal Proc.*, vol. ASSP-33, pp. 1333–1335, October.

NEWMAN, W. I. (1981). "Extension to the Maximum Entropy Method III," *Proc. 1st ASSP Workshop on Spectral Estimation*, pp. 1.7.1–1.7.6, Hamilton, Ontario, Canada, August.

NICHOLS, H. E., GIORDANO, A. A., and PROAKIS, J. G. (1977). "MLD and MSE Algorithms for Adaptive Detection of Digital Signals in the Presence of Interchannel Interference," *IEEE Trans. Information Theory,* vol. IT-23, pp. 563–575, September.

NIKIAS, C. L., and RAGHUVEER, M. R. (1987). "Bispectrum Estimation: A Digital Signal Processing Framework," *Proc. IEEE,* vol. 75, pp. 869–891, July.

NIKIAS, C. L., and SCOTT, P. D. (1982). "Energy-Weighted Linear Predictive Spectral Estimation: A New Method Combining Robustness and High Resolution," *IEEE Trans. Acoustics, Speech, and Signal Processing,* vol. ASSP-30, pp. 287–292, April.

NUTTALL, A. H. (1976). "Spectral Analysis of a Univariate Process with Bad Data Points, via Maximum Entropy and Linear Predictive Techniques," *NUSC Technical Report TR-5303,* New London, Conn., March.

NYQUIST, H. (1928). "Certain Topics in Telegraph Transmission Theory," *Trans. AIEE,* vol. 47, pp. 617–644, April.

OPPENHEIM, A. V. (1978). *Applications of Digital Signal Processing,* Prentice-Hall, Englewood Cliffs, N.J.

OPPENHEIM, A. V., and WEINSTEIN, C. W. (1972). "Effects of Finite Register Length in Digital Filters and the Fast Fourier Transform," *Proc. IEEE,* vol. 60, pp. 957–976, August.

OPPENHEIM, A. V., and WILLSKY, A. S. (1983). *Signals and Systems,* Prentice-Hall, Englewood Cliffs, N.J.

PAPOULIS, A. (1962). *The Fourier Integral and Its Applications,* McGraw-Hill, New York.

PAPOULIS, A. (1984). *Probability, Random Variables, and Stochastic Processes,* 2nd ed., McGraw-Hill, New York.

PARKER, S. R., and HESS, S. F. (1971). "Limit-Cycle Oscillations in Digital Filters," *IEEE Trans. Circuit Theory,* vol. CT-18, pp. 687–696, November.

PARKS, T. W., and MCCLELLAN, J. H. (1972a). "Chebyshev-Approximation for Nonrecursive Digital Filters with Linear Phase," *IEEE Trans. Circuit Theory,* vol. CT-19, pp. 189–194, March.

PARKS, T. W., and MCCLELLAN, J. H. (1972b). "A Program for the Design of Linear Phase Finite Impulse Response Digital Filters," *IEEE Trans. Audio and Electroacoustics,* vol. AU-20, pp. 195–199, August.

PARZEN, E. (1957). "On Consistent Estimates of the Spectrum of a Stationary Time Series," *Am. Math. Stat.,* vol. 28, pp. 329–348.

PARZEN, E. (1974). "Some Recent Advances in Time Series Modeling," *IEEE Trans. Automatic Control,* vol. AC-19, pp. 723–730, December.

PEACOCK, K. L., and TREITEL, S. (1969). "Predictive Deconvolution—Theory and Practice," *Geophysics,* vol. 34, pp. 155–169.

PEEBLES, P. Z., JR. (1987). *Probability, Random Variables, and Random Signal Principles,* 2nd ed., McGraw-Hill, New York.

PICINBONO, B. (1978). "Adaptive Signal Processing for Detection and Communication," in *Communication Systems and Random Process Theory,* J. K. Skwirzynski, Ed., Sijthoff en Noordhoff, Alphen aan den Rijn, The Netherlands.

PISARENKO, V. F. (1973). "The Retrieval of Harmonics from a Covariance Function," *Geophys. J. R. Astron. Soc.,* vol. 33, pp. 347–366.

POOLE, M. A. (1981). *Autoregressive Methods of Spectral Analysis,* EE degree thesis, Department of Electrical and Computer Engineering, Northeastern University, Boston, May.

PROAKIS, J. G. (1970). "Adaptive Digital Filters for Equalization of Telephone Channels," *IEEE Trans. Audio and Electroacoustics,* vol. AU-18, pp. 195–200, June.

PROAKIS, J. G. (1974). "Channel Identification for High Speed Digital Communications," *IEEE Trans. Automatic Control,* vol. AC-19, pp. 916–922, December.

PROAKIS, J. G. (1975). "Advances in Equalization for Intersymbol Interference," in *Advances in Communication Systems,* vol. 4, A. J. Viterbi, Ed., Academic, New York.

PROAKIS, J. G. (1983). *Digital Communications,* McGraw-Hill, New York.

PROAKIS, J. G. and MILLER, J. H. (1969). "Adaptive Receiver for Digital Signaling Through Channels with Intersymbol Interference," *IEEE Trans. Information Theory,* vol. IT-15, pp. 484–497, July.

RABINER, L. R., and SCHAFER, R. W. (1974a). "On the Behavior of Minimax Relative Error FIR Digital Differentiators," *Bell Syst. Tech. J.*, vol. 53, pp. 333–362, February.

RABINER, L. R., and SCHAFER, R. W. (1974b). "On the Behavior of Minimax FIR Digital Hilbert Transformers," *Bell Syst. Tech. J.*, vol. 53, pp. 363–394, February.

RABINER, L. R., and SCHAFER, R. W. (1978). *Digital Processing of Speech Signals*, Prentice-Hall, Englewood Cliffs, N.J.

RABINER, L. R., SCHAFER, R. W., and RADER, C. M. (1969). "The Chirp z-Transform Algorithm and Its Applications," *Bell Syst. Tech. J.*, vol. 48, pp. 1249–1292, May-June.

RABINER, L. R., GOLD, B., and McGONEGAL, C. A. (1970). "An Approach to the Approximation Problem for Nonrecursive Digital filters," *IEEE Trans. Audio and Electroacoustics*, vol. AU-18, pp. 83–106, June.

RABINER, L. R., McCLELLAN, J. H., and PARKS, T. W. (1975). "FIR Digital Filter Design Techniques Using Weighted Chebyshev Approximation," *Proc. IEEE*, vol. 63, pp. 595–610, April.

RADER, C. M. (1970). An Improved Algorithm for High-Speed Auto-correlation with Applications to Spectral Estimation," *IEEE Trans. Audio and Electroacoustics*, vol. AU-18, pp. 439–441, December.

RADER, C. M., and BRENNER, N. M. (1976). "A New Principle for Fast Fourier Transformation," *IEEE Trans. Acoustics, Speech, and Signal Processing*, vol. ASSP-24, pp. 264–266, June.

RADER, C. M., and GOLD, B. (1967a). "Digital Filter Design Techniques in the Frequency Domain," *Proc. IEEE*, vol. 55, pp. 149–171, February.

RADER, C. M., and GOLD, B. (1967b). "Effects of Parameter Quantization on the Poles of a Digital Filter," *Proc. IEEE*, vol. 55, pp. 688–689, May.

REMEZ, E. YA. (1957). *General Computational Methods of Chebyshev Approximation*, Atomic Energy Translation 4491, Kiev, USSR.

RISSANEN, J. (1983). "A Universal Prior for the Integers and Estimation by Minimum Description Length," *Ann. Stat.*, vol. 11, pp. 417–431.

ROBERTS, R. A., and MULLIS, C. T. (1987). *Digital Signal Processing*, Addison-Wesley, Reading, Mass.

ROBINSON, E. A. (1962). *Random Wavelets and Cybernetic Systems*, Charles Griffin, London.

ROBINSON, E. A. (1982). "A Historical Perspective of Spectrum Estimation," *Proc. IEEE*, vol. 70, pp. 885–907, September.

ROBINSON, E. A., and TREITEL, S. (1978). "Digital Signal Processing in Geophysics," in *Applications of Digital Signal Processing*, A. V. Oppenheim, Ed., Prentice-Hall, Englewood Cliffs, N.J.

SAKAI, H. (1979). "Statistical Properties of AR Spectral Analysis," *IEEE Trans. Acoustics, Speech, and Signal Processing*, vol. ASSP-27, pp. 402–409, August.

SANDBERG, I. W., and KAISER, J. F. (1972). "A Bound on Limit Cycles in Fixed-Point Implementations of Digital Filters," *IEEE Trans. Audio and Electroacoustics*, vol. AU-20, pp. 110–112, June.

SATORIUS, E. H., and ALEXANDER J. T. (1978). "High Resolution Spectral Analysis of Sinusoids in Correlated Noise," *Proc. 1978 ICASSP*, pp. 349–351, Tulsa, Okla., April 10–12.

SATORIUS, E. H., and PACK, J. D. (1981). "Application of Least-Squares Lattice Algorithms to Adaptive Equalization, *IEEE Trans. Communications*, vol. COM-29, pp. 136–142, February.

SCHAFER, R. W., and RABINER, L. R. (1973). "A Digital Signal Processing Approach to Interpolation," *Proc. IEEE*, vol. 61, pp. 692–702, June.

SCHMIDT, R. D. (1981). "A Signal Subspace Approach to Multiple Emitter Location and Spectral Estimation," Ph.D. dissertation, Department of Electrical Engineering, Stanford University, Stanford, Calif., November.

SCHMIDT, R. D. (1986). "Multiple Emitter Location and Signal Parameter Estimation," *IEEE Trans. Antennas and Propagation*, vol. AP 34, pp. 276–280, March.

SCHOTT, J. P., and McCLELLAN, J. H. (1984). "Maximum Entropy Power Spectrum Estimation with Uncertainty in Correlation Measurements," *IEEE Trans. Acoustics, Speech, and Signal Processing*, vol. ASSP-32, pp. 410–418, April.

SCHUSTER, SIR ARTHUR. (1898). "On the Investigation of Hidden Periodicities with Application to a Supposed Twenty-Six-Day Period of Meteorological Phenomena," *Terr. Mag.*, vol. 3, pp. 13–41, March.

SCHWARTZ, M. (1980). *Information Transmission, Modulation and Noise*, 3rd ed., McGraw-Hill, New York.

SEDLMEYER, A., and FETTWEIS, A. (1973). "Digital Filters with True Ladder Configuration," *Int. J. Circuit Theory Appl.*, vol. 1, pp. 5–10, March.

SHANKS, J. L. (1967). "Recursion Filters for Digital Processing," *Geophysics*, vol. 32, pp. 33–51, February.

SHANNON, C. E. (1949). "Communication in the Presence of Noise," *Proc. IRE*, pp. 10–21, January.

SIEBERT, W. M. (1986). *Circuits, Signals and Systems*, McGraw-Hill, New York.

SINGLETON, R. C. (1967). "A Method for Computing the Fast Fourier Transform with Auxiliary Memory and Limited High Speed Storage," *IEEE Trans. Audio and Electroacoustics*, vol. AU-15, pp. 91–98, June.

SINGLETON, R. C. (1969). "An Algorithm for Computing the Mixed Radix Fast Fourier Transform," *IEEE Trans. Audio and Electroacoustics*, vol. AU-17, pp. 93–103, June.

STEIGLITZ, K. (1965). "The Equivalence of Digital and Analog Signal Processing," *Inf. Control*, vol. 8, pp. 455–467, October.

STEIGLITZ, K. (1970). "Computer-Aided Design of Recursive Digital Filters," *IEEE Trans. Audio and Electroacoustics*, vol. AU-18, pp. 123–129, June.

STOCKHAM, T. G. (1966). "High Speed Convolution and Correlation," *1966 Spring Joint Computer Conference, AFIPS Proc.*, vol. 28, pp. 229–233.

STORER, J. E. (1957). *Passive Network Synthesis*, McGraw-Hill, New York.

STREMLER, F. G. (1982). *Introduction to Communication Systems*, 2nd ed., Addison-Wesley, Reading, Mass.

SWINGLER, D. N. (1979a). "A Comparison Between Burg's Maximum Entropy Method and a Nonrecursive Technique for the Spectral Analysis of Deterministic Signals," *J. Geophys. Res.*, vol. 84, pp. 679–685, February.

SWINGLER, D. N. (1979b). "A Modified Burg Algorithm for Maximum Entropy Spectral Analysis," *Proc. IEEE*, vol. 67, pp. 1368–1369, September.

SWINGLER, D. N. (1980). "Frequency Errors in MEM Processing," *IEEE Trans. Acoustics, Speech, and Signal Processing*, vol. ASSP-28, pp. 257–259, April.

TAUB, H., and SCHILLING, D. L. (1986). *Principles of Communication Systems*, 2nd ed., McGraw-Hill, New York.

THORVALDSEN, T. (1981). "A Comparison of the Least-Squares Method and the Burg Method for Autoregressive Spectral Analysis," *IEEE Trans. Antennas and Propagation*, vol. AP-29, pp. 675–679, July.

TONG, H. (1975). "Autoregressive Model Fitting with Noisy Data by Akaike's Information Criterion," *IEEE Trans. Information Theory*, vol. IT-21, pp. 476–480, July.

TONG, H. (1977). "More on Autoregressive Model Fitting with Noisy Data by Akaike's Information Criterion," *IEEE Trans. Information Theory*, vol. IT-23, pp. 409–410, May.

TUFTS, D. W., and KUMARESAN, R. (1982). "Estimation of Frequencies of Multiple Sinusoids: Making Linear Prediction Perform Like Maximum Likelihood," *Proc. IEEE*, vol. 70, pp. 975–989, September.

ULRYCH, T. J., and BISHOP, T. N. (1975). "Maximum Entropy Spectral Analysis and Autoregressive Decomposition," *Rev. Geophys. Space Phys.*, vol. 13, pp. 183–200, February.

ULRYCH, T. J., and CLAYTON, R. W. (1976). "Time Series Modeling and Maximum Entropy," *Phys. Earth Planet. Inter.*, vol. 12, pp. 188–200, August.

UNGERBOECK, G. (1972). "Theory on the Speed of Convergence in Adaptive Equalizers for Digital Communication," *IBM J. Res. Devel.*, vol. 16, pp. 546–555, November.

WALKER, G. (1931). "On Periodicity in Series of Related Terms," *Proc. R. Soc.*, Ser. A, vol. 313, pp. 518–532.

WEINBERG, L. (1962). *Network Analysis and Synthesis*, McGraw-Hill, New York.

WELCH, P. D. (1967). "The Use of Fast Fourier Transform for the Estimation of Power Spectra: A Method Based on Time Averaging over Short Modified Periodograms," *IEEE Trans. Audio and Electroacoustics*, vol. AU-15, pp. 70–73, June.

WIDROW, B. (1970). "Adaptive Filters," in *Aspects of Network and System Theory*, R. E. Kalman and N. DeClaris, Eds., Holt, Rinehart and Winston, New York.

WIDROW, B., and HOFF, M. E., JR. (1960). "Adaptive Switching Circuits," *IRE WESCON Conv. Rec.*, pt. 4, pp. 96–104.

WIDROW, B., MANTEY, P., and GRIFFITHS, L. J. (1967). "Adaptive Antenna Systems, *Proc. IEEE*, vol. 55, pp. 2143–2159, December.

WIDROW, B., ET AL. (1975). "Adaptive Noise Cancelling: Principles and Applications," *Proc. IEEE*, vol. 63, pp. 1692–1716, December.

WIDROW, B., McCOOL, J. M., LARIMORE, M. G., and JOHNSON, C. R., JR. (1976). "Stationary and Nostationary Learning Characteristics of the LMS Adaptive Filters," *Proc. IEEE*, vol. 64, pp. 1151–1162, August.

WIENER, N. (1949). *Extrapolation, Interpolation and Smoothing of Stationary Time Series with Engineering Applications*, Wiley, New York.

WIENER, N., and PALEY, R. E. A. C. (1934). *Fourier Transforms in the Complex Domain*, American Mathematical Society, Providence, R.I.

WINOGRAD, S. (1976). "On Computing the Discrete Fourier Transform," *Proc. Nat. Acad. Sci.*, vol. 73, pp. 105–106.

WINOGRAD, S. (1978). "On Computing the Discrete Fourier Transform," *Math. Comp.*, vol. 32, pp. 177–199.

WOLD, H. (1938). *A Study in the Analysis of Stationary Time Series*, reprinted by Almquist and Wichsells Forlag, Stockholm, Sweden, 1954.

WOOD, L. C., and TREITEL, S. (1975). "Seismic Signal Processing," *Proc. IEEE*, vol. 63, pp. 649–661, April.

YULE, G. U. (1927). "On a Method of Investigating Periodicities in Disturbed Series with Special References to Wolfer's Sunspot Numbers," *Philos. Trans. R. Soc. London*, ser. A, vol. 226, pp. 267–298, July.

ZADEH, L. A., and DESOER, C. A. (1963). *Linear System Theory: The State-Space Approach*, McGraw-Hill, New York.

ZVEREV, A. I. (1967). *Handbook of Filter Synthesis*, Wiley, New York.

Index